Red Cell Membrane Transport in Health and Disease

Springer
Berlin
Heidelberg
New York
Honkong
London
Mailand
Paris
Tokio

I. Bernhardt · J. C. Ellory (Eds.)

Red Cell Membrane Transport in Health and Disease

With 103 figures and 31 tables.

Springer

Professor Dr. Ingolf Bernhardt
Universität des Saarlandes
Naturwissenschaftlich-Technische Fakultät III
Arbeitsgruppe Biophysik
P.O.Box 151150
66041 Saarbrücken
Germany

Professor Dr. J. Clive Ellory
University of Oxford
Laboratory of Physiology
Parks Road
Oxford OX1 3PT
United Kingdom

ISBN 3-540-44227-8 Springer Verlag Berlin Heidelberg New York

Cataloging-in-Publication Data applied for
A catalog record for this book is available from the Library of Congress.
Bibliographic information published by Die Deutsche Bibliothek
Die Deutsche Bibliothek lists this publication in the Deutsche Nationalbibliografie;
detailed bibliographic data is available in the Internet at http://dnb.ddb.de

Springer-Verlag is a company in the specialist publishing group BertelsmannSpringer
Springer-Verlag Berlin Heidelberg 2003
Printed in Germany
http://www.springer.de

Einbandgestaltung: Design & Production, Heidelberg

57/3020 UW – Gedruckt auf säurefreiem Papier – 5 4 3 2 1 0

Preface

The red cell has been a focus for scientific and medical investigation since the earliest times. A higher erythrocyte sedimentation rate was associated with diseases (usually pyrexias) before the thermometer was invented. Furthermore, ever since the early observers Swammerdam and Leeuvenhoek saw discrete corpuscles in samples of blood using the first microscopes, there has been a significant scientific interest in the structure and function of red blood cells. The later discovery that red cells were not spherical, but biconcave discs introduced a scientific puzzle which is still not completely resolved today, and identified the need for a detailed knowledge of the plasma membrane composition and structure, and its interaction with the cytoskeleton. Important concepts like the lipid bilayer, together with its more recent refinement as asymmetric in phospholipid composition led to the identification of translocases involved in actively maintaining its composition. Understanding the mechanics of red cell deformation as these biconcave discs traverse capillaries was advanced by the pioneering work of Rand and Burton in the Sixties, and progressed by Evans, Skalak and others. Based on the bilayer-couple hypothesis, the shape changes that are possible for a human red cell from echinocyte to stomatocyte were described by Sheetz and Singer in the Seventies in terms of alterations in the individual halves of the bilayer. Certain clinical conditions are associated with obvious changes in red cell morphology. Although alterations in membrane lipid or protein can be responsible, mutant haemoglobins (Hb SS), or changes in other cytoplasmic constituents can also be responsible for dramatic alterations in red cell shape, size and lifetime. Another area of importance which has attracted much recent attention is the intraerythrocytic life of malaria, and the way *Plasmodium falciparum* affects red cell contents, function and membrane transport.

In fact red cell contents, and in particular the intracellular concentrations of cations and anions represents an enduring topic of scientific interest. Over a hundred years ago, Abderhalden made meticulous chemical analyses of red cell ionic composition, and established two important concepts. He identified the fundamental principle of a non-equilibrium distribution of ions across the membrane, i.e. a low intracellular sodium and high intracellular potassium concentration. He also identified differences in internal cation levels between species' red cells, a finding taken further in classical work on HK and LK sheep red cells by Kerr, then Evans and later Tosteson and Hoffman and in the work on carnivore red cells leading to the seminal work of Parker with dog red cells.

The obvious conclusion from the finding of non-equilibrium ion distribution across the red cell membrane was that the cell membrane was impermeable (e.g. Gürber 1895). However, van Slyke et al. in the Twenties showed that the red cell membrane has a very high chloride permeability, whilst amongst others Dean et al. and Eisenmann in the early Forties demonstrated a significant sodium and potassium permeability in the red cells of various species. With this information, and the consequent interest in membrane properties and intracellular composition, it was realised that the red cell represents a convenient model for studying membrane transport processes, since it was assumed to be an easily accessible homogeneous population of isolated cells. The further discovery of the resealed red cell ghost allowed the incorporation of selected molecules, and the alteration of ion concentrations. This allowed red cells to be used to investigate the stoichiometry and energetics of the Na^+/K^+ pump and for radioisotope flux studies to characterise a variety of transporters including glucose and amino acid carriers as well as KCC and NKCC, the Ca^{2+}-activated K^+ channel and band 3-mediated anion transport.

The enormous impact of molecular biology on cell physiology was initially less applicable for studying mammalian red cells which lack intracellular organelles, including a nucleus. However, using progenitor cells, or reticulocytes has allowed a molecular biological approach to red cells, and further, the value of the transgenic mouse in answering haematological questions will continue to grow. In this context, the variety of diseases causing anaemias, resulting from unstable haemoglobins, or metabolic alterations involving particular enzymes and pathways continue to attract considerable research effort and attention.

The rationale for the present volume evolved partially from the meetings of the respective Red Cell Clubs in Europe and the USA. Many of the doyens of the red cell world who have contributed to this volume represent a lifetime of knowledge and authority in the field. Additionally there are contributions from the younger generation of scientists who are mapping significant areas of knowledge and contributing new techniques and developments. As always it is difficult to be comprehensive in coverage with such a volume, but we are pleased with the breadth of contributions, and feel that significant ground has been covered. In terms of editorial input, we have exercised a light touch, which means there are differences in style and density between the various contributions. There are also some inconsistencies between American and Oxford English, and some elements of non-native English style may still exist. The editors have tried to produce a book which is consistent within its scientific content. One of us (IB), in addition, has laboured long and hard, and wrestled with the publishing template to achieve homogeneity as far as possible. In this task he has been assisted by Erwin Weiss in Saarbrücken, and by Maureen Paler in Oxford.

The aim of the present authors is to produce a red cell book which will inform and provide cover of the major biophysical and cell physiological aspects of red cell function. We must thank all the contributors for their input, and also Dr. Rogaschweski (Humboldt University Berlin) for the photograph on the cover of the book. Springer-Verlag have been active in helping to edit and process the text, and we thank Mrs. Cuneus for her assistance.

Finally we hope that everyone will find something interesting or surprising in one or other chapters of this book. We hope it may encourage scientists to continue to work on red cells, and solve some of the mysteries still present. In his paper "Questions for red blood cell physiologists to ponder for the millennium" (Blood cell, molecules and diseases (2001) 27:57-61), Joseph Hoffman posed many fascinating and important questions still to be resolved in the field of red cell physiology. Although it must be admitted that the present volume may not answer them directly, it is by stimulating interest in the subject of red blood cells that we will produce the research and hopefully the answers to his millennium questions.

Saarbrücken and Oxford, March 2003 Ingolf Bernhardt and Clive Ellory

Contents

Contributing Authors

Arese, P., Prof. Dr.
Department of Genetics, Biology and Biochemistry, University of Torino, Medical School,
Via Santena 5 bis, 10126 Torino,
Italy

Baldwin, S. A., Prof. Dr.
School of Biochemistry and Molecular Biology, University of Leeds,
Leeds, LS2 9JT,
UK

Bennekou, P., Prof. Dr.
LCMF, The August Krogh Institute, University of Copenhagen,
Universitetsparken 13, 2100 Copenhagen Ø,
Denmark

Bernhardt, I., Prof. Dr.
Arbeitsgruppe Biophysik, Naturwissenschaftlich-Technische Fakultät III,
Universität des Saarlandes,
Gebäude 6, Postfach 151150, 66041 Saarbrücken,
Germany

Bookchin, R. M., Prof. Dr.
Department of Medicine, Albert Einstein College of Medicine,
Bronx, New York 10461,
USA

Borgese, F., Dr.
Laboratoire de physiologie des membranes cellulaires, Université de Nice-Sophia Antipolis, UMR 6078 CNRS, Bâtiment Jean Maetz,
284 chemin du Lazaret, 06230 Villefranche sur mer,
France

Brovelli, A., Prof. Dr.
Dipartimento di Biochimica "A. Castellani", Università di Pavia,
via Bassi 21, 27100 Pavia,
Italy

Browning, J., Dr.
Laboratory of Physiology, Oxford University,
Parks Road, Oxford, OX1 3PT,
UK

Brugnara, C., Prof. Dr.
Department of Laboratory Medicine, Children's Hospital Boston, Harvard Medical School,
300 Longwood Avenue, Boston, MA 02115,
USA

Brunini, T. M. C., Dr.
Departamento de Farmacologia e Psicobiologia, Instituto de Biologia,
Av. 28 de Setembro 87, CEP 20551-030, Rio de Janeiro,
Brazil

Cass, C. E., Prof. Dr.
Membrane Protein Research Group, Department of Oncology,
University of Alberta, and Cross Cancer Institute,
Edmonton, Alberta T6G 2H7,
Canada

Christophersen, P., Dr.
NeuroSearch A/S, Ballerup,
Pederstrupvej 93, 2750 Ballerup,
Denmark

Deuticke, B., Prof. Dr.
Institut für Physiologie, Universitätsklinikum der RWTH,
Pauwelsstrasse 30, 52057 Aachen,
Germany

Ellory, J. C., Prof. Dr.
Laboratory of Physiology, Oxford University,
Parks Road, Oxford OX1 3PT,
UK

Fischer, T. M., Dr.
Institut für Physiologie, Universitätsklinikum der RWTH,
Pauwelsstrasse 30, 52057 Aachen,
Germany

Flatman, P. W., Dr.
Membrane Biology Group, Division of Biomedical and Clinical Laboratory Sciences, The University of Edinburgh, Hugh Robson Building,
George Square, Edinburgh EH8 9XD, Scotland,
UK

Gibson, J. S., Dr.
Department of Clinical Veterinary Medicine, University of Cambridge,
Madingley Road, Cambridge CB3 0ES,
UK

Grosveld, F. G., Prof. Dr.
MGC Department of Cell Biology and Genetics, Faculty of Medicine, Erasmus University Medical Center Rotterdam,
PO Box 1738, 3000 DR, Rotterdam,
The Netherlands

Guizouarn, H., Dr.
Laboratoire de physiologie des membranes cellulaires, Université de Nice - Sophia Antipolis, UMR 6078 CNRS, Bâtiment Jean Maetz,
284 chemin du Lazaret, 06230 Villefranche sur mer,
France

Haest, C. W. M., Dr.
Institut für Physiologie, Universitätsklinikum der RWTH,
Pauwelsstrasse 30, 52057 Aachen,
Germany

Halestrap, A. P., Prof. Dr.
Department of Biochemistry, School of Medical Sciences,
University of Bristol,
Bristol BS8 1TD,
UK

Heberle, J., Dr.
Forschungszentrum Jülich, IBI-2: Strukturbiologie,
52425 Jülich,
Germany

Horn, N. M., Dr.
School of Biological Sciences, University of Southampton,
Bassett Crescent East, Southampton SO16 7PX,
UK

Jarolim, P., Dr.
Department of Pathology, Brigham and Women's Hospital, Harvard Medical School,
75 Francis Street, Boston MA 02115,
USA

Kirk, K., Prof. Dr.
School of Biochemistry and Molecular Biology, Faculty of Science, Australian National University,
Canberra A.C.T. Australia 0200,
Australia

Knauf, P. A., Prof. Dr.
Department of Biochemistry and Biophysics, University of Rochester Medical Center,
Rochester, NY 14642,
USA

Lauf, P. K., Prof. Dr.
Department of Physiology and Biophysics, Wright State University School of Medicine,
Dayton, OH 45435,
USA

Lepke, S.
Max Planck Institut für Biophysik,
Heinrich-Hoffmann-Str. 7, 60528 Frankfurt am Main,
Germany

Lew, V. L., Dr.
Physiological Laboratory, University of Cambridge,
Cambridge CB2 3EG,
UK

Lytle, C., Prof. Dr.
Division of Biomedical Sciences, University of California, Riverside,
Riverside, California, 92521,
USA

Mendes Ribeiro, A. C., Dr.
Departamento de Farmacologia e Psicobiologia, Instituto de Biologia,
Av. 28 de Setembro 87, CEP 20551-030, Rio de Janeiro,
Brazil

Minetti, G., Dr.
Dipartimento di Biochimica "A. Castellani", Università di Pavia,
via Bassi 21, 27100 Pavia,
Italy

Motais, R., Prof. Dr.
Laboratoire de physiologie des membranes cellulaires, Université de Nice - Sophia Antipolis, UMR 6078 CNRS, Bâtiment Jean Maetz,
284 chemin du Lazaret, 06230 Villefranche sur mer,
France

Naftalin, R. J., Prof. Dr.
Physiology Division, King's College London, Guy's Campus,
London SE1 1UL,
UK

Nikinmaa, M., Prof. Dr.
Department of Biology, University of Turku,
20014 Turku,
Finland

Oakley, F., Dr.
School of Biological Sciences, University of Southampton,
Bassett Crescent East, Southampton SO16 7PX,
UK

Orlov, S. N., Prof. Dr.
Laboratory of Pathophysiology of Ion Transport Disorders, Research Centre,
University of Montreal Hospital, CHUM,
3850 rue St.-Urbain, Montreal, PQ, H2W 1T8,
Canada

Pal, P., M. Sc.
Department of Biochemistry and Biophysics, University of Rochester Medical Center,
Rochester, NY 14642,
USA

Passow, H., Prof. Dr.
Max Planck Institut für Biophysik,
Heinrich-Hoffmann-Str. 7, 60528 Frankfurt am Main, and
Institut für Biochemie und Biophysik, Friedrich Schiller Universität Jena,
Philosophenweg 12, 07743 Jena,
Germany

Patrinos, G. P., Dr.
MGC Department of Cell Biology and Genetics, Faculty of Medicine, Erasmus University Medical Center Rotterdam,
PO Box 1738, 3000 DR, Rotterdam,
The Netherlands

Sachs, J. R., Prof. Dr.
Department of Medicine, State University of New York at Stony Brook,
Stony Brook, NY, 11794-8151,
USA

Saliba, K. J., Dr.
School of Biochemistry and Molecular Biology, Faculty of Science, Australian National University,
Canberra A.C.T. Australia 0200,
Australia

Schwarzer, E., Dr.
Department of Genetics, Biology and Biochemistry, University of Torino, Medical School, Via Santena 5 bis, 10126 Torino,
Italy

Stewart, G. W., Prof. Dr.
Department of Medicine, Rayne Institute, University College London,
University Street, London WC1E 6JJ,
UK

Swietach, P.
Laboratory of Physiology, Oxford University,
Parks Road, Oxford OX1 3PT,
UK

Thomas, A. L., Dr.
School of Biological Sciences, University of Southampton,
Bassett Crescent East, Southampton SO16 7PX,
UK

Tiffert, T., Dr.
Physiological Laboratory, University of Cambridge,
Cambridge CB2 3EG,
UK

Weiss, E., Dipl.-Biophys.
Arbeitsgruppe Biophysik, Naturwissenschaftlich-Technische Fakultät III,
Universität des Saarlandes,
Gebäude 6, Postfach 151150, 66041 Saarbrücken,
Germany

Wilkins, R., Dr.
Laboratory of Physiology, Oxford University,
Parks Road, Oxford, OX1 3PT,
UK

Yao, S. Y. M., Dr.
Membrane Protein Research Group, Department of Physiology,
University of Alberta,
Edmonton, Alberta T6G 2H7,
Canada

Young, J. D., Prof. Dr.
Membrane Protein Research Group, Department of Physiology,
University of Alberta,
Edmonton, Alberta T6G 2H7,
Canada

1 Distribution and Movement of Membrane Lipids

Cees W. M. Haest

Institut für Physiologie, Universitätsklinikum der RWTH, Pauwelsstrasse 30,
52057 Aachen, Germany

1.1 Introduction

1.1.1 Lipid Composition, Structure and Motions

The human red blood cell membrane consists of lipids (41%), proteins (52%), and carbohydrates (7%) (Dodge et al. 1963). Membrane lipids (5.2 mg/ml packed cells or $5.2\cdot10^{-13}$ g/cell) can be divided into three classes: neutral lipids (25.2%), phospholipids (62.7%) and glycosphingolipids (about 12%). Neutral lipids of human erythrocytes represent cholesterol almost exclusively (Nelson 1972). The ratio of cholesterol to phospholipid is about 0.8 (Broekhuyse 1974). Phospholipids (4.3 $\mu mol/ml_{cells}$) consist of sphingomyelin (SM, 26%) and glycerophospho-lipids. Glycerophospholipids can be divided into 3 main fractions – phosphatidylcholine (PC, 30%), phosphatidylethanolamine (PE, 27%), and phosphatidylserine (PS, 13%) as well as several minor fractions – phosphatidic acid, lyso-PC, phosphatidylinositol (PI), and PI mono and diphosphates (Nelson 1972; Broekhuyse 1974). The fatty acid pattern differs among the phospholipid classes and according to the position of the glycerol backbone. PE and PS contain higher fractions of polyunsaturated fatty acids than PC (Gercken et al. 1972). Saturated fatty acids with 16 to 24 carbon atoms are preferentially located at the 1 position of glycerol whereas unsaturated fatty acids with up to 6 double bonds are at the 2 position (Van Deenen 1969). In case of PE, species with vinyl ether bonds at the 1 position of glycerol (plasmalogens) make up the majority (Nelson 1972). SM mainly contains long-chain saturated or monounsaturated fatty acids with up to 24 C atoms (Gercken et al. 1972). In summary therefore this gives a total of 150 to 200 different types of lipid molecules in the red blood cell membrane (Van Deenen 1969).

Erythrocytes of various species differ in their fatty acid and phospholipid compositions. On the one hand, erythrocytes from rat and mouse have a high content of PC (42–45%) and a low content of SM (12%) (Nelson 1972; Renooij et al. 1976; Rawyler et al. 1985). On the other hand, erythrocytes from ruminants such as sheep and ox have a low content of PC (0–4%) and a high content of SM (47–50%) (Nelson 1972; Florin-Christensen et al. 2001). A higher content of PC correlates with a higher content of polyunsaturated fatty acids (Nelson 1972). The low content of PC in ruminant erythrocytes results from an endogenous phospholipase A_2, which is present at the outside of the membrane and cleaves PC (Zwaal et al. 1974; Florin-Christensen et al. 2001). The products lyso-PC and fatty acid then leave the cells into plasma. The low PC content has been suggested to be required in order to prevent binding of antiphospholipid antibodies which are present in ruminant serum probably as a consequence of exposure to phosphonolipids produced by rumen ciliates (Florin-Christensen et al. 2001).

The lipid composition of the red blood cell membrane is rather stable and only alters with diet to a limited extent (Farquar and Ahrens 1963; Rao et al. 1979). This is due to the lack of *de novo* synthesis of phospholipids in the mature erythrocyte. The limited alterations of the fatty acid composition by diet which do occur result from the exchange of phospholipids, primarily PC, between plasma lipoproteins and the cell membrane, as well as the exchange of fatty acids by deacylation of diacylphospholipids and subsequent reacylation of lysophospholipids at the inner membrane surface (Shohet 1972; Renooij et al. 1974).

The membrane lipids are arranged in a bilayer with their polar residues oriented towards the bilayer-water interface and the apolar hydrocarbon tails oriented at the bilayer centre. Water is essentially absent from the apolar core of the membrane and primarily restricted to the polar head group region of the phospholipids (Yeagle 1987). Motions of lipid molecules in the membrane bilayer to be considered are rapid rotational motion of lipid molecules along their long axis (10^8 s^{-1}, Yeagle 1987), exchange of lipids due to lateral diffusion in the plane of the bilayer (about 10^7 s^{-1}, Peters 1988; Marsh and Horvath 1998), tumbling of lipids about their mean axis of orientation (wobbling) relative to the normal of the membrane plane (Yeagle 1987), and a very slow transverse movement of lipids from one lipid leaflet to the other (see Sect. 1.2.1). Membrane proteins restrict lipid lateral motions. The lipid exchange rate with intrinsic membrane proteins lies between 10^7 and 10^6 s^{-1} (Marsh and Horvath 1998). The rotations of the polar head group and of segments of the fatty acyl chains around the carbon-carbon bonds (gauche-trans isomerisation) are faster (~10^9 s^{-1}, Yeagle 1987; Marsh and Horvath 1998) than rotation of the whole phospholipid molecule. Motional freedom of the fatty acyl chains of the phospholipids is highest at the terminal methyl groups of the fatty acids, comparable to liquid oil, and lowest at the ester bonds where order and rigidity are highest. Unsaturation of the fatty acyl chains decreases their packing density, increases hydration of the chains, and increases chain rigidity due to hindered rotations around the double bonds (Yeagle 1987). Cholesterol affects both packing and the rigidity which depends on the structure of fatty acyl chain. It raises both the hydrophobic membrane barrier for polar compounds and the rigidity barrier for apolar compounds (Subczynski et al. 1994).

The bilayer acts not only as a solvent for lipid molecules but also for amphiphiles such as drugs with apolar domains, with an affinity for the hydrophobic bilayer interior, and polar domains, which enforce orientation of the agents at the bilayer surface (Deuticke et al. 1990). In general, the high content of albumin in blood plasma will prevent accumulation of considerable amounts of such amphiphiles in the membrane bilayer of circulating red blood cells due to binding of amphiphiles to albumin.

This chapter will focus on the distribution and movement of lipids between the two membrane bilayer leaflets. Lipids are defined as biomolecules, which are not soluble in water but highly soluble in organic solvents such as chloroform.

1.1.2 Measurements of Distribution and Movement of Lipids between the Membrane Bilayer Leaflets

The method first used to study distribution of phospholipids between the outer and inner membrane leaflet of the erythrocyte was based on the accessibility of membrane aminophospholipids for reaction with amino reagents known to have only slow access to the inner membrane surface. These reagents reacted with aminophospholipids in intact cells to a small extent whilst the reaction was much faster when the reagents had access to the inner leaflet of ghost membranes (Bretcher 1972; Gordesky and Marinetti 1973). Disadvantages of the method are that the reagents available are not fully impermeable, and that the reaction is rather slow at physiological pH. Chemical modification may perturb the membrane, and the procedure is limited to the aminophospholipids, PE and PS. Since the reaction takes significant time even at elevated temperature, this method is only suitable for measurements of the stationary distributions and not useful for measurements of the transbilayer movements of phospholipids. This also applies to measurements of the accessibility of membrane phospholipids to the cleavage by phospholipases or the exchange of outer leaflet phospholipids using transfer proteins. The advantages of the use of phospholipases are their lack of penetration into the cell and the vulnerability of all membrane phospholipids contained in the outer membrane leaflet using different phospholipases with various substrate specificities (Op den Kamp 1979). The presence of PS in the outer membrane leaflet can be detected by the enhancement of procoagulant activity of the cells (Bevers et al. 1998) or the enhanced binding of fluorescein isothiocyanate (FITC)-conjugated annexin V to the cells (Tait and Gibson 1994).

Modern techniques to measure transbilayer movement of lipids and their stationary distributions between the two membrane leaflets use lipid probes with a fluorescent (7-nitrobenz-2-oxa-1,3-diazol-4-yl, NBD) or a paramagnetic (doxyl) reporter group attached to a short chain fatty acid. Trace amounts of these probes are inserted into the outer membrane leaflet taking advantage of their limited water solubility. Subsequently, the time-dependent disappearance of the probes from the outer leaflet is followed by measuring either the decrease of their extractability from the membrane by albumin (Mohandas et al. 1982) or the decrease of fractional chemical quenching of the reporter group in the outer leaflet (Seigneuret and Devaux 1984; Pomorski et al. 1994). Measurements are limited to phospholipid probes with at least one short chain fatty acid in order to enable their rapid insertion into the membrane (within several minutes at low temperature) and their fast extraction by albumin. Such probes, in particular those with small reporter groups, seem to be good models of endogenous phospholipids (Devaux 1991).

A simple alternative qualitative method uses the time-dependent secondary shape changes of erythrocytes following primary echinocyte formation induced by insertion of phospholipid probes into the outer membrane leaflet. This insertion selectively expands the outer leaflet without initially affecting the inner leaflet (Ferrell and Huestis 1985; see also Chap. 2).

Distribution of glycolipids has been detected by measuring their accessibility to catalysed oxidation using galactose oxidase followed by reduction with tritiated borohydride (Gahmberg and Hakomori 1973).

Information on the distribution of cholesterol in the membrane and its flip-flop rate has been derived from measurements of the time-dependent equilibration of radioactively labelled cholesterol between plasma or liposomes and the erythrocyte membrane (Kirby and Green 1977; Lange et al. 1977). In another approach, the accessibility of membrane cholesterol towards oxidation by cholesterol oxidase was measured (Brasaemle et al. 1988; Lange 1992).

1.1.3 Steady State Distribution of Lipids between the Membrane Bilayer Leaflets

The asymmetric stationary distribution of glycerophospholipids and SM between the two membrane leaflets of human erythrocytes was established using a combination of phospholipase A_2 and sphingomyelinase (Verkleij et al. 1973). The aminophospholipids PS and PE are, respectively, solely and predominantly located in the inner membrane leaflet while the choline phospholipids PC and SM are predominantly located in the outer membrane leaflet. PI and its phosphorylated derivatives are predominantly, though probably not exclusively, located in the inner leaflet (Gascard et al. 1991). For other erythrocyte species (Renooij et al. 1976; Rawyler et al. 1985; Van der Schaft et al. 1987b; Florin-Christensen et al. 2001), similar asymmetric distributions of membrane phospholipids have been found (Fig. 1.1).

The development of asymmetry of membrane phospholipids in the course of bone marrow stem cell differentiation has been investigated. A predominant location of SM in the outer membrane leaflet and a predominant but not exclusive location of PS in the inner leaflet is already present at the early stage of erythropoiesis. Asymmetry of PE and PC is present in reticulocytes but absent in normoblast-like cells and is probably established during enucleation (Rawyler et al. 1985; Nijhof et al. 1986; Van der Schaft et al. 1987a).

Glycolipids are localised in the outer membrane leaflet (Gahmberg and Hakomori 1973). The distribution of cholesterol between the two membrane leaflets is still a matter of debate. Both symmetric distribution between the two membrane leaflets and asymmetric distribution of cholesterol in favour of the inner membrane leaflet have been claimed (Zachowski 1993). Difficulties in establishing the distribution are due to the transbilayer movement of cholesterol, which is probably rapid, and the pitfalls in the methods available for the selective determination of cholesterol in the outer leaflet (see Sect. 1.2.2).

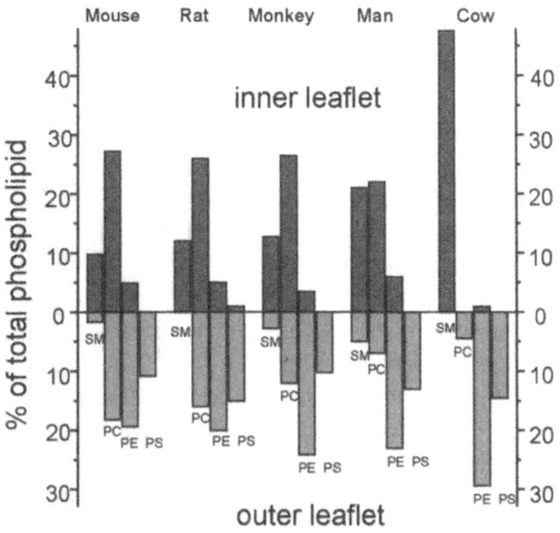

Fig. 1.1. Distributions of the major phospholipids between the outer and inner membrane leaflets of erythrocytes from man (Verkleij et al. 1973), rat (Renooij et al. 1976), mouse (Rawyler et al. 1985), monkey (Van der Schaft et al. 1987b), and cow (Florin-Christensen et al. 2001). PS data for rat and cow include PI

1.1.4 Lipid Domains

Although lateral diffusion of lipids in the membrane bilayer is very fast (see Sect. 1.1.1) evidence has been obtained that the lateral distribution of lipids in both lipid monolayers is not homogeneous and that domains with distinct lipid composition exist in erythrocyte membranes (Kavecansky et al. 1994; Welti and Glaser 1994; Moore et al. 1999; Zuvic-Boterac et al. 1999). Such domains may result from lipid-lipid or lipid-protein interactions. Interactions of lipids with the anion exchanger (AE1) were proposed to play a role in domain formation (Welti and Glaser 1994). The presence of lipid domains has been suggested to depend on the echinocytic shape of erythrocytes (Moore et al. 1999). On the other hand, these lipid domains have been suggested to be produced by the experimental procedures applied and a new model for lateral lipid organisation was proposed. According to this so-called superlattice model different phospholipid classes are not distributed randomly but tend to adopt a regular distribution in the bilayer (Somerharju et al. 1999).

From recent findings on erythrocytes (Salzer and Prohaska 2001) and analogous to membranes of other cell types (Galbiati et al. 2001), it is likely that microdomains rich in the intrinsic membrane proteins somatin and flotillin, as well as in sphingomyelin and proteins anchored to glycosylphosphatidylinositols exist in the erythrocyte membrane. Such so-called lipid rafts are sites for signal transduction and vesicular trafficking (Galbiati et al. 2001).

1.2 Non-Mediated Passive Transbilayer Movements of Lipids

1.2.1 Phospholipids

Since the classic experiments of Kornberg and McConnell (1971) using spin-labelled PC, the transbilayer movement of phospholipids, termed flip-flop, is known to be a very slow process in pure lipid bilayer vesicles with a half time of several hours. This is due to the apolar core of the bilayer, which acts as a barrier for the transbilayer movement of the polar head groups of the lipids (Homan and Pownall 1988). In addition, the high order of the acyl chains at the ester bonds of the phospholipids acts as a rigidity barrier for the movement of the apolar acyl chains of phospholipids during flip-flop (Subczynski et al. 1994). The concerted movement of both polar and apolar residues of the phospholipid across the membrane bilayer thus requires the surmounting of hydrophobic and rigidity barriers. It has been proposed to result from the transient formation of hydrophobic rather than hydrophilic pores in the bilayer (Deuticke et al. 1992).

For the erythrocyte membrane, translocation rates were found to depend on the fatty acid composition of the phospholipid molecule. Half times of translocations at 37 °C varied between 3 h for 1-palmitoyl-2-linoleoyl PC and 26 h for 1,2-dipalmitoyl PC (Middelkoop et al. 1986). A decrease of the chain length of saturated phospholipids reduced the half time (Fujii et al. 1985), in case of 1,2-dioctanoyl PC the half time is less than 10 h. Lysophospholipids move about two-fold faster than diacylphospholipids (Fujii et al. 1985). The half time for palmitoyl lyso-PC is about 10 h (Haest et al. 1986; Vondenhof 1993). Acylation of lyso-PC with an acetyl group slightly increases the flip rate (half time: about 9 h, Vondenhof 1993). Elongation of the acyl chain to butyroyl and hexanoyl strongly increases the flip rate (half time: about 5 h) and a further elongation to octanoyl again decreases the flip rate (half time: about 8 h) (Vondenhof 1993). Platelet-activating factor, 1-*O*-octadecyl-2-acetyl-*sn*-glycero-3-phosphocholine, has a lower flip rate (half time: 17 h; Schneider et al. 1986a) than 1-palmitoyl-2-acetyl PC. In spite of the same polar phosphocholine head groups, SM has an at least ten-fold slower flip rate than PC (Haest et al. 1997) which has to be related to the presence of the amide bond in SM instead of the ester bond in PC. The structure of the polar head group of diacyl glycerophospholipids or lysophospholipids has only small effects on the passive translocation rates (Haest et al. 1986, 1997). The observed faster movement of lyso-PS than that of lyso-PC (Bergmann et al. 1984a) is likely to be due to the contribution of the active flippase-mediated process to translocation (see Sect. 1.5.1). The translocation rate of zwitterionic palmitoylcarnitine with a half time of 2.6 h is considerably faster than that of palmitoyl lysophatidylcholine. The inter-individual variability and enhancement of translocation by acidic pH are peculiarities of this translocation process (Classen et al. 1989).

The activation energies for these passive translocation processes are high: 60–130 kJ/mol (Bergmann et al. 1984a; Haest et al. 1986, 1997; Zachowski et al. 1986; Classen et al. 1989). Cholesterol depletion of the membrane enhances translocation while cholesterol enrichment decreases the translocation rate (Mohandas et al. 1982; Bergmann et al. 1984a). Comparison of translocation rates for various erythrocyte species demonstrates that a low SM content and a correspondingly high content of polyunsaturated fatty acid enhances translocation (Bergmann et al. 1984a).

From the half times of phospholipid flip-flop, it is clear that mechanisms are required to maintain the asymmetric distribution of phospholipids during the lifetime of erythrocytes of 120 days. This is provided by an ATP-consuming translocation system moving aminophospholipids that escaped the inner membrane leaflet back to this leaflet (see Sect. 1.5.1). However, for PC and SM the compensating mechanisms are unknown. There is evidence that newly synthesized PC is rapidly translocated from the inner to the outer membrane leaflet (see Sect. 1.4). An active translocation system for the outward movement of phospholipid probes has been claimed to be involved in the outward translocation of inner leaflet PC (see Sect. 1.5.2).

Mild reversible oxidation of membrane SH groups, aggressive irreversible oxidative or photodynamic treatment of cells (Mohandas et al. 1982; Bergmann et al. 1984b; Jain 1985; Deuticke et al. 1987; Deuticke et al. 1989), electroporation of cells (Haest et al. 1997), and the insertion of amphiphiles, antibiotics or toxins into the membrane (Schneider et al. 1986b; Classen et al. 1988; Rosso et al. 1988; Basse et al. 1992; Schwichtenhövel et al. 1992; Pantaler et al. 2000) enhances the passive non-mediated flip-flop. These enhancements produce a variable decrease of phospholipid asymmetry (Haest et al. 1978; Dressler et al. 1983; Schneider et al. 1986b; Classen et al. 1988; Deuticke et al. 1989) which depends on the extent of flip acceleration and the simultaneous inactivation of the aminophospholipid flippase by the modifications (Franck et al. 1989). Since modifications do not lower the high activation energy for the flip the rate-limiting step in this process is most probably the penetration of the polar head group through the hydrophobic core of the lipid bilayer. Therefore, modifications have been postulated to increase the transient formation of hydrophobic defects in the bilayer (Deuticke et al. 1992).

1.2.2 Cholesterol and Other Neutral Lipids

From the time-dependent equilibration of radioactively-labelled cholesterol between plasma or liposomes and the erythrocyte membrane, an upper boundary of flip-flop rate with a half time of less than 50 min was set (Kirby and Green 1977; Lange et al. 1977). Using cholesterol oxidase to measure the accessibility of membrane cholesterol, very fast flip rates were obtained. All of the cholesterol was oxidised in a first order fashion with a half time of 10 s at 37 °C (Lange et al. 1981). Although oxidation of cholesterol did not result in significant haemolysis it is probable that formation of the oxidation product cholestenone perturbs the membrane and thereby increases the translocation rate of the remaining sterol. This artefact is avoided by a further application of the cholesterol oxidase method in which cells were loaded with labelled cholesterol for a short time and then incubated at 37 °C to measure the time-dependent disappearance of labelled cholesterol from the outer membrane leaflet. In this case, cells were treated with cholesterol oxidase for a short time and the decrease of specific radioactivity of cholestenone measured. The half time of translocation was estimated between 50 and 130 min (Brasaemle 1988). Drawbacks of this method have been discussed in detail (Lange 1992).

A very fast translocation of cholesterol is suggested by the kinetics of the release of sterols from erythrocytes. The half times of translocation for several sex hormones across the erythrocyte membrane ranged from 4 to 150 ms (Koefoed and Brahm 1994). Moreover, quantitative data on the flip-flop rates in pure lipid bilayers of uncharged aliphatic alcohols, diglycerides, non-ionic detergents and fatty acids or bile acids in the undissociated form strongly support the idea of a very fast translocation of cholesterol across the bilayer. For diacylglycerol and the non-ionic detergent octaethylene glycol dodecyl monoether, halftimes for transbilayer movement are less than 15 s and 0.35 s (Ganong and Bell 1984; Le Maire et al. 1987) and for bile acids and fatty acids, less than 1 s and 10 s (Cabral et al. 1987; Kleinfeld 2000). Even the flip rate of phosphatidic acid in its neutral undissociated form is very rapid, having a half time of about 25 s (Eastman et al. 1991).

1.2.3 Fatty Acids: Mediated or Non-Mediated Movement?

In blood plasma, unesterified fatty acids are present in millimolar concentrations but the major fraction of fatty acid is bound to albumin which enables efficient transport. The concentration of unbound fatty acids is estimated at about 5–50 nM for fatty acid to albumin ratios of 0.1–1.0 (Abumrad et al. 1999). The concentration of fatty acid in the red cell membrane will be considerably higher due to the high membrane/water partition coefficient of long-chain fatty acids (about 10^5) even at physiological pH (Kleinfeld et al. 1997). It is, however, still well below the concentration, which would perturb membranes (Pantaler et al. 2000). Transport of fatty acids across cell membranes *in vivo* involves the release of fatty acids from albumin into solution, their binding to the membrane, their movement across the membrane bilayer and their dissociation from the membrane at the inner surface. The unbound fatty acid determines cellular uptake. At low fatty acid to albumin ratios, dissociation of fatty acid from albumin could become rate-limiting for transport while at high fatty acid to albumin ratios translocation across the membrane is rate-limiting. In pure lipid vesicles approaching cell size, translocation of long-chain fatty acids across the bilayer takes 1–10 s and decreases with an increase of fatty acid length (Kleinfeld 2000). In the erythrocyte membrane, translocation rates are comparable to those in lipid vesicles (Bröring et al. 1989; Kleinfeld et al. 1998), increase with an increase of the number of double bonds of the fatty acid, and are about twofold lower in sheep erythrocytes than in human erythrocytes (Bojesen and Bojesen 1999).

Long-chain fatty acids provide a major energy source for many tissues such as myocytes and adipocytes. In these cells, uptake has two components: a non-mediated process via the lipid phase which may dominate at high fatty acid to protein ratios and a protein-mediated process which becomes important at low fatty acid to albumin ratios. The mediated process exhibits saturation, competition between different fatty acids for uptake and inhibition by protein modification with

4,4'-diisothiocyanatostilbene-2,2'-disulphonic acid (DIDS) or the presence of phloretin and sulfosuccinimidyl derivatives of fatty acids (Abumrad et al. 1998). Facilitated uptake could be due to either enhanced binding of fatty acid to a high affinity receptor or mediation of translocation across the hydrophobic membrane barrier (Abumrad et al. 1998).

In the human erythrocyte membrane, DIDS and phloretin which block facilitated fatty acid transport in other cells do not affect transport of long-chain fatty acids (Bröring et al. 1989; Kleinfeld et al. 1998). Furthermore, transport is fast enough to support the turnover rate of fatty acids of erythrocyte membrane lipids (Kleinfeld et al. 1998). Therefore, fatty acid transport is essentially non-mediated in erythrocytes.

For review see Abumrad et al. (1998, 1999) and Kleinfeld (2000).

1.3 Mediated Passive Transbilayer Movements of Lipids

1.3.1 Ca^{2+}-Activated Bidirectional Movement of Lipids via the Phospholipid Scramblase

The asymmetric distribution of phospholipids in erythrocytes and other cells is rapidly lost with an increase of the intracellular Ca^{2+} concentration (Bevers et al. 1999). This loss is the result of activation of a translocation system for endogenous phospholipids, which strongly enhances bidirectional transbilayer movement of all membrane phospholipid species, and of lipid probes, which are inserted into the membrane (Bevers et al. 1999). This so-called scrambling takes place during cell activation and apoptosis and results in the exposure of the inner membrane leaflet phospholipid, PS, to the outer leaflet, which is an important signal for removal of cells by phagocytes (Savill and Fadok 2000). Further signals for phagocytosis are discussed by Brovelli and Minetti (Chap. 29). Scrambling activity can thus be measured by the increase of PS in the outer membrane leaflet using the phospholipase technique (see Sect. 1.1.2), the increase of procoagulant activity (Bevers et al. 1998), and the enhanced binding of FITC-conjugated annexin V to the cells (Tait and Gibson 1994). A more quantitative kinetic analysis of scrambling is obtained by following the time-dependent Ca^{2+}-activated translocation of phospholipid probes between the two membrane leaflets using the albumin extraction technique (see Sect. 1.1.2).

A membrane protein that mediates Ca^{2+}-dependent transbilayer movement of phospholipid probes was isolated and purified from erythrocytes (Basse et al. 1996). The concentration of free intracellular Ca^{2+} for half maximal activation of scrambling is 25–100 μM (Stout et al. 1998; Woon et al. 1999). Other divalent cations also activate phospholipid translocation in the order $Ca^{2+} > Mn^{2+} > Zn^{2+} > Sr^{2+} >> Ba^{2+} > Mg^{2+}$ (Stout et al. 1998). Metal ion binding to the protein results in a conformational change of the protein and self-association to an oligomer. This oligomerization is assumed to be related to the activation of lipid translocation (Stout et al. 1998). The cDNA for the membrane protein with characteristics of a scramblase has been cloned (Zhou et al. 1997). This so-called human phospholipid scramblase (HuPLSCR) type 1 is a 35 kDa protein of 318 residues. In human erythrocytes, about 1000 copies of PLSCR1 are present. PLSCR has a single transmembrane domain (Lys^{288}–Glu^{306}), a potential Ca^{2+}-binding EF hand motif (Asp^{273}–Asp^{285}) adjacent to this membrane domain, a proline-rich N-terminal segment with potential interaction sites for WW or SH3 domains of other proteins (Zhou et al. 1997), a potential protein kinase C phosphorylation site (Thr161) involved in regulation (Frasch et al. 2000), and constitutively tyrosine phosphorylation sites at Tyr^{69}/Tyr^{74} (Sun et al. 2001). Acylation of one or more cysteine residues of PLSCR is required for its function, possibly by promoting the correct conformation for the Ca^{2+}-binding site or accelerating the transbilayer movement of lipids (Zhao et al. 1998). Several other cysteine residues are connected allosterically to the translocation machinery, since the alkylating reagent *N*-ethylmaleimide (NEM) stimulates (Kamp et al. 2001) and pyridyldithioethylamine inhibits scrambling (Bevers et al. 1999). On the other hand, disulphide reducing reagents strongly suppress scrambling and this suppression is reversed by the SH-oxidizing agent diamide. This demonstrates that a disulfide bond within the monomeric PLSCR protein or between two monomers is essential for its activity (Kamp et al. 2001). Besides disulphide-reducing reagents, scrambling in human erythrocytes is also strongly inhibited by a covalent modification of cells with the essentially impermeable amino reagents DIDS, an inhibitor of anion transport, and 2,4,6-trinitrobenzenesulphonate. This inhibition is probably due to reaction of the reagents with an exofacial lysine residue of PLSCR (Kamp et al. 2001). Scrambling is also affected by non-covalent drugs. Clotrimazole, an inhibitor of the Ca^{2+}-activated K^+ channel, and ethanimidothioic acid suppress while oligomycin stimulates scrambling (Bevers et al. 1999; Kamp et al. 2001). Three additional cDNAs encoding proteins with high homology to HuPLSCR1 and the expression of these PLSCRs in different human tissues and blood cells have been reported (Zhou et al. 1997; Wiedmer et al. 2000).

Other mechanisms have been claimed to be involved or to modulate Ca^{2+}-activated flip-flop. Polyamines or analogous motifs on proteins have been suggested to induce phospholipid scrambling by formation of domains of PI-4,5-bisphosphate (Bucki et al. 2000). Furthermore, the ATP-binding cassette (ABC) transporter type A1 has been suggested to promote Ca^{2+}-activated externalization of PS in erythrocytes (Hamon et al. 2000) but this could not be confirmed (Kamp et al. 2001). In fact, ABCA1 is likely to be absent from erythrocytes, since these cells lack the apolipoprotein-mediated lipid release from their cell membranes, a process involving ABCA1 activity (Czarnecka and Yokoyama 1996; Oram 2000).

The physiological relevance of phospholipid scrambling in erythrocytes has been discussed. Exposure of PS to the outer membrane leaflet probably contributes to the removal of senescent erythrocytes from circulation by phagocytosis (Connor and Schroit 1994; Bratosin et al. 1998). Human erythrocytes, which are devoid of intracellular organelles, have Ca^{2+} channels in their membranes that can be opened by lysophosphatidic acid produced in sufficient amounts during platelet activation (Yang et al. 2000). However, the resulting increase of intracellular Ca^{2+} and extent of PS scrambling are unknown while the flip of phospholipid probes in the presence of lysophosphatidic acid is not enhanced (unpublished data; Philip Low, personal communication). Under pathophysiological conditions, intracellular

Ca^{2+} in sickle and thalassaemic erythrocytes is known to increase and probably results in activation of the PLSCR as demonstrated by increased exposure of PS in the outer membrane leaflet (Zwaal and Schroit 1997). This will increase the procoagulant activity of erythrocytes (Bevers et al. 1999) and their susceptibility to phagocytosis (Savill and Fadok 2000).

For review see Sims and Wiedmer (2001).

1.3.2 Movement of Anionic Lipids via the Anion Exchanger AE1

Besides the physiological role of AE1 (band 3) in exchanging chloride against bicarbonate, AE1 transports various monovalent and divalent inorganic and organic anions as well as sulphonated, phosphonated and carboxylated derivatives of aliphatic, aromatic or heterocyclic compounds (Cabantchik and Greger 1992; see also Chap. 11). The latter anionic amphiphiles distribute into the membrane bilayer to a significant extent. In this case, the approach of the anionic group to the transport site from the aqueous phase (flux mode) cannot be taken for granted in the light of the low concentrations of these amphiphiles in the aqueous phase. An approach of the anion to its transport site from the lipid phase (flip mode) has to be considered. This flip mode of operation of AE1 seems very likely for the mediated transport of anionic long-chain amphiphiles with very high membrane-water partition coefficients (> 10^5), such as, alkyl-naphthyl sulphonates, the alkyl sulphate SDS, naphthyl-alkyl phosphates, and anionic phosholipid probes such as lysophosphatidylmethanol, phosphatidic acid with one short-chain fluorescent fatty acid or its esters bearing one or two negative charges (Ortwein et al. 1994; Vondenhof et al. 1994; Serra et al. 1996; Kleinhorst et al. 1998; Voswinkel et al. 2001). The major criterion for transport of anionic amphiphiles via AE1 is full suppression of translocation by a pretreatment of cells with DIDS at stoichiometric numbers of DIDS and AE1 molecules. The activation energy for the flip of anionic long-chain amphiphiles via AE1 is high (80–110 kJ/mol) and similar to those found for the flux of various hydrophilic anions (Ortwein et al. 1994; Vondenhof et al. 1994; Kleinhorst et al. 1998).

The characteristics of translocation of long-chain anions by the flip mode of operation of AE1 differ from those of transport of hydrophilic anions by the flux mode of operation of AE1. Peculiarities comprise weak effects of certain potent inhibitors of exchange of inorganic anions and the enhancement of translocation by a pretreatment of cells with papain or the presence of butanol, instead of inhibition described for transport of inorganic anions (Ortwein et al. 1994; Vondenhof et al. 1994; Kleinhorst et al. 1998). These opposite responses suggested a causal relationship to the different modes of transport (flip *vs.* flux). However, this idea was precluded by measurements of transport of compounds intermediate in structure between hydrophilic inorganic or small organic anions and long-chain amphiphilic anions (Voswinkel et al. 2001). Since transport of various small anionic amphiphiles, which move by flux, is inhibited by papain in some cases but stimulated in others, the opposite responses cannot be due to different pathways of substrate access to the transport site (Voswinkel et al. 2001). Alternatively, different subsites may be involved in interactions of the huge variety of anionic compounds accepted by the transporter. A subsite represents a particular assembly of amino acids on AE1 involved in the formation of the activated state for a particular anion. According to the so-called composite transport site model (Krupka 1989), specificity and rate of transport are determined by the tightness of complex formation between the anion and the carrier in the activated state. For this reason, larger substrates are transported slowly because complexes are less tight. Conformational changes of the transport domain of AE1 by proteolytic cleavage with papain, which produces two large and two small fragments, were proposed to rearrange the subsites for inorganic anions or small anionic amphiphiles in a way which impedes the formation of the activated complex and to rearrange the subsites involved in transport of anionic long-chain amphiphiles in a way which facilitates the formation of the activated complex (Voswinkel et al. 2001). The model also explains the differing changes in sensitivity of transport of the various anions towards inhibitors supposed to act by allosteric mechanisms.

In the flip mode of transport, the amphiphilic anions have to gain access to the transport site from the lipid bilayer. This has structural implications since an opening in the structure of AE1 within the plane of the membrane lipid bilayer is required.

Whether the flip mode of operation of AE1 has physiological relevance remains to be seen.

1.4 Mediated Outward Movement of Newly Synthesized Phosphatidylcholine

Limited fatty acid turnover in the human erythrocyte membrane is due to the incorporation of fatty acids into membrane phospholipids by acylation of lysophospholipids, a process taking place at the inner membrane surface (Renooij et al. 1974). Fatty acids bound to serum albumin will distribute into the outer membrane leaflet and translocate rapidly to the inner leaflet (see Sect. 1.2.3). At the inner membrane surface, fatty acids become transformed to acyl-CoA by an ATP-consuming enzymatic process, and are then transferred to lysophospholipids by an acyltransferase. The newly synthesized PC is rapidly (half time 20 min) translocated to the outer leaflet as detectable by its accessibility towards cleavage by phospholipase A_2 (Andrick et al. 1991). Interestingly, no evidence for an outward movement of PE was obtained which indicates selectivity of the mediated translocation process (Haest and Andrick 1989). Transport is inhibited by a short pretreatment (5 min, 37 °C) of cells with the arginine reagent phenylglyoxal (Andrick et al. 1991), which demonstrates the involvement of proteins. Since the activation energy for the translocation process is low (30 kJ/mol) MRP1-mediated outward translocation, which would have a high activation energy (see Sect. 1.5.2), is unlikely to be involved. Moreover, vanadate did not affect translocation, which is evidence against an active process (Andrick et al. 1991).

1.5 Active Transbilayer Movements of Lipids

1.5.1 Inward Movements of Phosphatidylserine and Phosphatidylethanolamine via the Aminophospholipid Translocase (APLT)

In 1984, Seigneuret and Devaux described an active transport system for PS and PE, that translocates these aminophospholipids from the outer to the inner membrane leaflet. PS is translocated about 10-fold faster than PE (half times: about 5 and 60 min at 37 °C, Zachowski and Devaux 1990). The presence of a negatively charged carboxylate residue on the polar head group of the aminophospholipid thus enhances translocation. Lyso-PS is a poor substrate. The translocation system is highly specific for the aminophospholipids. The presence of an amino group is required for transport to occur. Methylation of PE reduces the translocation. The phosphate and amino residues of the phospholipid head group have to be separated by two methylene groups (Zachowski 1993; Dolis et al. 1997). The transport system exhibits stereo-selectivity for the glycerol backbone (Martin and Pagano 1987) but not for the serine residue (Hall and Huestis 1993). In contrast to the specificity for the polar head group, the translocation system has a relatively low selectivity for the fatty acid moiety (Daleke and Huestis 1985). Saturated and unsaturated species, plasmalogens as well as species containing short spin-labelled (Seigneuret and Devaux 1984) or fluorescent fatty acids (Connor and Schroit 1988) are accepted. A pretreatment of cells with SH reagents such as NEM and pyridyldithioethylamine (Connor and Schroit 1988) and an increase of intracellular calcium (Zachowski 1993) suppresses translocation. The process is Mg^{2+}-dependent and inhibited by P-type ATPase inhibitors such as vanadate, fluoride (Zachowski 1993), suramin, eosin Y, and the antibiotic elaiophylin (Beleznay et al. 1997). One phospholipid molecule is translocated per ATP hydrolysed. The K_m for ATP measured for both PS translocation and Mg^{2+} ATPase activity is 0.2 mM (Beleznay et al. 1997). The activation energy for translocation is surprisingly low (20–30 kJ/mol) (Zachowski et al. 1986; Kamp and Haest 1998).

A PS-stimulated, Mg^{2+}-dependent ATPase with a molecular mass of 80–115 kDa, which translocated PS but not PC, was isolated from erythrocytes, purified, and reconstituted in lipid vesicles (Dolis et al. 1997; Daleke and Lyles 2000). The gene that codes for the Mg^{2+} ATPase II was cloned from bovine chromaffin granules (Tang et al. 1996). The so-called aminophospholipid translocase (APLT) or flippase has a molecular size of 130 kDa with 10 membrane-spanning domains and large cytoplasmic domains reminiscent of the canonical structure of P-type ion transporters. Three consensus sequences diagnostic for P-type transporters are present. These are a sequence for coupling of ATP hydrolysis to transport, another sequence containing the aspartate phosphorylated in the enzyme intermediate and a sequence that may participate in ATP binding. Recently, PS was found to be essential for dephosphorylation of the phosphoenzyme (Ding et al. 2000). APLT belongs to an ancient subfamily of P-type ATPases (Tang et al. 1996; Halleck et al. 1999). Structural differences between APLT and P-type ATPases involved in cation transport are in the membrane domains. Polar and charged amino acids that are important for cation transport are replaced by bulky hydrophobic amino acids in APLT. The ATPase II gene encodes a transporter possibly restricted to secretory vesicles. The gene for the plasma membrane APLT activity remains to be identified.

The physiological role of APLT is to keep PS in the outer leaflet of cell membranes at a very low level by its active translocation to the inner leaflet in order to prevent the recognition and phagocytosis of cells by macrophages (Savill and Fadok 2000), the activation of blood coagulation by cell surfaces (Zwaal and Schroit 1997), and blood cell adherence to vascular endothelium (Manodori et al. 2000). On the other hand, these processes are triggered by the exposure of PS following activation of PLSCR and concomitant inhibition of APLT by an increase of the intracellular Ca^{2+} concentration.

For review see Dolis et al. (1997) and Daleke and Lyles (2000).

1.5.2 Outward Movements of Lipids via the Multidrug Resistance Protein MRP1 (ABCC1)

Phospholipid probes have been shown to be translocated from the inner to the outer membrane leaflet by a Mg^{2+} and ATP-dependent process. Translocation is suppressed by vanadate, fluoride, chemical modification of SH groups or histidines and a high number of non-covalent inhibitors (Bitbol and Devaux 1988; Connor et al. 1992; Serra et al. 1996; Kamp and Haest 1998). Translocation does not exhibit major dependence on the structure of the polar head group of the phospholipid (Serra et al. 1996). In contrast to passive non-mediated translocation, the ATP-dependent outward translocations of PC and SM have similar rates (Haest, unpublished data). The activation energy of this process is high (104 kJ/mol) (Serra et al. 1996). From studies with selective inhibitors of transport via MRP (Kamp and Haest 1998) and using Mrp1 knock out mice (Dekkers et al. 1998) it became evident that MRP1 is involved in this process.

MRP1 is known as the high affinity transporter for conjugates of lipophilic substances with glutathione, glucuronide, or sulphate (König et al. 1999). MRP1 can also transport oxidised glutathione (GSSG) and unconjugated compounds. The lipid mediator of inflamatory response leukotriene C_4, which is a glutathione *S*-conjugate, is the substrate with highest affinity for MRP1 (König et al. 1999). Unconjugated substrates require glutathione (GSH) whereas conjugated substrates do not depend on GSH for transport (Borst et al. 1999). GSH binds to MRP and stimulates its ATPase activity (Hooijberg et al. 2000). ATP hydrolysis is required for transport (Hipfner et al. 1999).

Today, seven MRPs are known (Borst et al. 1999; Hipfner et al. 1999). Location of MRP1 in the body is ubiquitous. MRP1 has the typical structure of ATP binding casette (ABC) transporters with a tandemly arranged pair of membrane-spanning domains each containing 6 membrane-spanning helices (but see Hipfner et al. 1999) and one nucleotide-binding motif (Walker A and Walker B) responsible for ATPase function. Furthermore, it has a conserved motif between the Walker motifs, which is characteristic for ABC proteins. MRP 1, 2, 3 and 6 have an additional N-terminal extension of 5 membrane-spanning domains (Borst et al. 1999; Hipfner et al. 1999).

As for the case of physiological substrates, translocation of NBD-labelled phospholipid probes requires, besides ATP, GSH (but see Dekkers et al. 1998). The K_m for the dependence of transport of NBD-labelled phospholipids on GSH is about 7 mM (Arno Sohnius, personal communication). *S*-methylglutathione also supports translocation (K_m about 5 mM; Haest, unpublished results), which is in line with findings for other substrates (Qian et al. 2001). Elongation of the hydrocarbon group from methyl to hexyl or esterification of the C-terminal glycine of GSH does not support phospholipid translocation (Haest, unpublished results). The K_m for ATP of NBD-phospholipid translocation is about 100 μM (Arno Sohnius, personal communication), which fits into the broad range found for several other translocation substrates (Hipfner et al. 1999, and references therein). While it seems likely that translocation is limited to phospholipids with an unphysiological hydrophobic group, recent experiments suggest that endogenous phospholipids are also accepted by MRP1 since inhibition of the transporter resulted in a slow increase of the PC content in the outer membrane leaflet of erythrocytes (Dekkers et al. 2000).

In view of the high membrane-water partition coefficients of phospholipid probes, their binding to the transport site from the aqueous phase seems unlikely. The transporter thus acts as a floppase, which means that it can bind lipid substrates from the inner membrane lipid leaflet and move them to the outer leaflet.

The physiological role of MRP1 is the protection of cells against toxic compounds, the export of apolar compounds after their conjugation to GSH, and the export of oxidised glutathione (GSSG), which is produced in the defence against oxidative stress (Hipfner et al. 1999). A possible role of MRP1 in the outward translocation of native phospholipids deserves confirmation.

For review see Borst et al. (1999) and Hipfner et al. (1999).

1.6 Disturbance of Distribution and Transbilayer Movements of Membrane Phospholipids in Pathology

In several cases of erythrocyte disorders, passive and/or active phospholipid translocation processes have been found to be altered. While the active translocation rate of aminophospholipids via APLT and the passive translocation of PC in the membrane of patients homozygous for sickle cell anaemia (see also Chap. 23) are normal under aerobic conditions, reversible sickling by deoxygenation of cells increases the rate of passive translocation and decreases the rate of active translocation. These changes are reversed upon reoxygenation (Lubin et al. 1981; Blumenfeld et al. 1991). In irreversibly sickled cells, active translocation of aminophospholipids is decreased even under aerobic conditions (Blumenfeld et al. 1991). This results in a decrease of the asymmetric distribution of PS and its accumulation in the outer membrane leaflet of microvesicles released from sickled cells while the PS level in the outer membrane leaflet of the remnant cells remains low (Bevers et al. 1998). A detailed analysis of sickle cells showed that PS exposure is limited to a subpopulation of the cells, varies widely among sickle cell patients, and takes place at several stages in the life of the sickle cell (De Jong et al.

2001). In thalassaemic patients, the passive transbilayer mobility of phospholipids is enhanced while the active APLT-mediated process is not altered. This enhanced passive transbilayer movement is probably responsible for the observed variable accumulation of PS in the outer leaflet of these cells (Müller et al. 1993; Kuypers et al. 1998). In patients with sickle cell anaemia and thalassaemia, exposure of PS to the outer membrane leaflet enhances adherence of cells to the endothelium (Cines et al. 1998), promotes phagocytosis of cells (Fadok et al. 1998), and stimulates thrombotic events (Zwaal and Schroit 1997).

Rh_{null} erythrocytes lacking the Rhesus blood group antigens, McLeod erythrocytes lacking three minor sialoglycoproteins, and erythrocytes in hereditary pyropoikilocytosis, which have a defective membrane skeleton assembly exhibit an enhanced rate of passive translocation of PC. In Rh_{null} erythrocytes, the asymmetric distribution of PE is decreased (Kuypers et al. 1984) but APLT activity is normal (Geldwerth et al. 1997). In McLeod and pyropoikilocytic cells, membrane phospholipid distributions are normal (Franck et al. 1985; Kuypers et al. 1985). Infection of erythrocytes with *Plasmodium falciparum* induces a rapid translocation of phospholipid probes proposed to be due to membrane-insertion of new translocation mechanisms (Haldar et al. 1989). Exposure of PS to the outer membrane leaflet of the infected cells is controversial and at best concerns a small fraction of cells (Moll et al. 1990; Sherman et al. 1997). Increased exposure of PS has also been reported in diabetes mellitus (Wali et al. 1988) and chronic renal failure (Bonomini et al. 1999). Ca^{2+}-activated scrambling of phospholipids is defective in Scott syndrome but the mutation is still unknown (Bevers et al. 1998; Sims and Wiedmer 2001).

In vivo ageing of erythrocytes does not alter passive mobility but decreases activity of the APLT-mediated translocation and results in minor changes of distribution of aminophospholipids, i.e., exposure of PS to the outer membrane leaflet (Herrmann and Devaux 1990; Connor et al. 1994). This signal contributes to phagocytosis of senescent cells (Connor and Schroit 1994; Bratosin et al. 1998).

For review see Bevers et al. (1998).

1.7 Summary

A summary of the known pathways for the transbilayer movement of lipids in the human erythrocyte membrane is given in Fig. 1.2. The most important process is the active inward translocation of PS and PE via the aminophospholipid flippase (APLT). The function of APLT is to maintain PS in the inner leaflet and thereby prevent procoagulant activity of PS, binding of cells by macrophages and adherence to the endothelium. The other active transporter, MRP1, accepts a very large group of diverse substrates at the inner membrane surface and moves them outward. Physiological lipid substrates are conjugated lipids such as leukotriene C_4. Its possible function as a low affinity transporter for endogenous PC still deserves confirmation. In addition to these active processes, several passive translocation systems for lipids exist. A passive bidirectional translocation system for all membrane phospholipids becomes activated upon an increase of the intracellular Ca^{2+}

concentration and results in the appearance of PS in the outer membrane leaflet. This promotes procoagulant activity and phagocytosis of erythrocytes. Another translocation system of physiological interest concerns the protein-mediated outward movement (passive or active?) of newly synthesised PC. A passive carrier-mediated transport of anionic lipids via the anion exchanger AE1 (band 3), which accepts anionic long-chain and phospholipid probes, is unlikely to have physiological significance. Finally, a slow passive non-mediated flip-flop of lipids takes place at fluctuating hydrophobic pores in the membrane lipid domain.

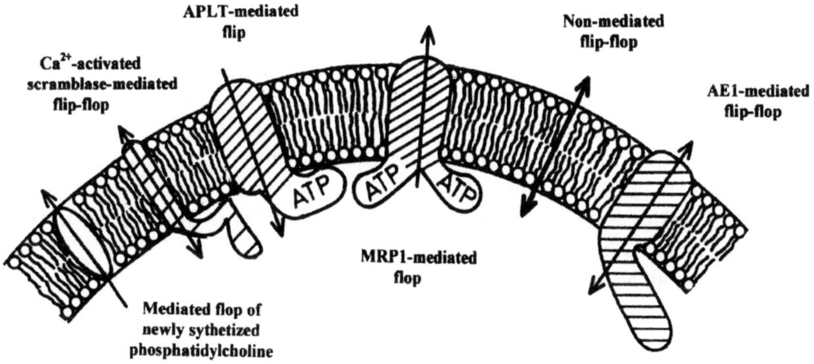

Fig. 1.2. Pathways for the transbilayer movement of lipids

Acknowledgement: I am grateful to Prof. B. Deuticke for his comments and criticism. I thank Mr. F.J. Kaiser for preparation of the figures.

References

Abumrad N, Harmon C, Ibrahimi A (1998) Membrane transport of long-chain fatty acids: evidence for a facilitated process. J Lipid Res 39:2309-2318

Abumrad N, Coburn C, Ibrahimi A (1999) Membrane proteins implicated in long-chain fatty acid uptake by mammalian cells: CD36, FATP and FABPm. Biochim Biophys Acta 1441:4-13

Andrick C, Bröring K, Deuticke B, Haest CWM (1991) Fast translocation of phosphatidylcholine to the outer membrane leaflet after its synthesis at the inner membrane surface in human erythrocytes. Biochim Biophys Acta 1064:235-241

Basse F, Sainte-Marie J, Maurin L, Bienvenue A (1992) Effect of benzylalcohol on phospholipid flip-flop in the human erythrocyte membrane. Eur J Biochem 205:155-162

Basse F, Stout JG, Sims PJ, Wiedmer TW (1996) Isolation of an erythrocyte membrane protein that mediates Ca^{2+}-dependent transbilayer movement of phospholipid. J Biol Chem 271:17205-17210

Beleznay Z, Zachowski A, Devaux PF, Ott P (1997) Characterization of the correlation between ATP-dependent aminophospholipid translocation and Mg^{2+}-ATPase activity in red blood cell membranes. Eur J Biochem 243:58-65

Bergmann WL, Dressler V, Haest CWM, Deuticke B (1984a) Reorientation rates and asymmetry of distribution of lysophospholipids between the inner and outer leaflet of the erythrocyte membrane. Biochim Biophys Acta 772:328-336

Bergmann WL, Dressler V, Haest CWM, Deuticke B (1984b) Cross-linking of SH-groups in the erythrocyte membrane enhances transbilayer reorientation of phospholipids. Biochim Biophys Acta 769:390-398

Bevers EM, Comfurius P, Dekkers DWC, Harmsma M, Zwaal RFA (1998) Transmembrane phospholipid distribution in blood cells: control mechanisms and pathological significance. J Biol Chem 379:973-986

Bevers EM, Comfurius P, Dekkers DWC, Zwaal RFA (1999) Lipid translocation across the plasma membrane of mammalian cells. Biochim Biophys Acta 1439:317-330

Bitbol M, Devaux PF (1988) Measurement of outward translocation of phospholipids across human erythrocyte membrane. Proc Natl Acad Sci 85:6783-6787

Blumenfeld N, Zachowski A, Galacteros F, Beuzard Y, Devaux PF (1991) Transmembrane mobility of phospholipids in sickle erythrocytes: Effect of deoxygenation on diffusion and asymmetry. Blood 77:849-854

Bojesen IN and Bojesen E (1999) Sheep erythrocyte membrane binding and transfer of long-chain fatty acids. J Membrane Biol 171:141-149

Bonomini M, Sirolli V, Settefrati N, Dottori S, Di Liberato L, Arduini A (1999) Increased erythrocyte phosphatidylserine exposure in chronic renal failure. J Am Soc Nephrol 10:1982-1990

Borst P, Evers R, Kool M, Wijnholds J (1999) The multidrug resistance protein family. Biochim Biophys Acta 1461:347-357

Brasaemle DL, Robertson AD, Attie AD (1988) Transbilayer movement of cholesterol in the human erythrocyte membrane. J Lipid Res 29:481-489

Bratosin D, Matzurier J, Tissier JP, Estaquier J, Huart JJ, Ameisen JC, Aminoff D, Montreuil J (1998) Cellular and molecular mechanisms of senescent erythrocyte phagocytosis by macrophages. A review. Biochimie 80:173-195

Bretcher MS (1972) Asymmetrical lipid bilayer structure for biological membranes. Nature New Biology 236:11-12

Broekhuyse RM (1974) Improved lipid extraction of erythrocytes. Clin Chim Acta 51:341-343

Bröring K, Haest CWM, Deuticke B (1989) Translocation of oleic acid across the erythrocyte membrane. Evidence for a fast process. Biochim Biophys Acta 986:321-331

Bucki R, Giraud F, Sulpice JC (2000) Phosphatidylinositol 4,5-bisphosphate domain inducers promote phospholipid transverse redistribution in biological membranes. Biochemistry 39:5838-5844

Cabantchik ZI, Greger R (1992) Chemical probes for anion transporters of mammalian cell membranes. Am J Physiol 262:C803-C827

Cabral DJ, Small DM, Lilly HS, Hamilton JA (1987) Transbilayer movement of bile acids in model membranes. Biochemistry 26:1801-1804

Cines DB, Pollack ES, Buck CA, Loscalzo J, Zimmerman GA, McEver RP, Pober JS, Wick TM, Konkle BA, Schwartz BS, Barnathan ES, McCrae KR, Hug BA, Schmidt AM, Stern DM (1998) Endothelial cells in physiology and in the pathology of vascular disorders. Blood 91:3527-3561

Classen J, Haest CWM, Tournois H, Deuticke B (1987) Gramicidin-induced enhancement of transbilayer reorientation of lipids in the erythrocyte membrane. Biochemistry 26:6604-6612

Classen J, Deuticke B, Haest CWM (1989) Nonmediated flip-flop of phospholipid analogues in the erythrocyte membrane as probed by palmitoylcarnitine: Basic properties and influence of membrane modification. J Membrane Biol 111:169-178

Connor J, Schroit AJ (1988) Transbilayer movement of phosphatidylserine in erythrocytes: Inhibition of transport and preferential labeling of a 31 000-Dalton protein by sulfhydryl reactive reagents. Biochemistry 27:848-851

Connor J, Pak CH, Zwaal RFA, Schroit AJ (1992) Bidirectional transbilayer movement of phospholipid analogs in human red cells. J Biol Chem 267:19412-19417

Connor J, Pak CC, Schroit AJ (1994) Exposure of phosphatidylserine in the outer leaflet of human red blood cells. Relationship to cell density, cell age, and clearance by mononuclear cells. J Biol Chem 269:2399-2404

Czarnecka H, Yokoyama S (1996) Regulation of cellular efflux by lecithin:cholesterol acyltransferase reaction through nonspecific lipid exchange. J Biol Chem 271:2023-2028

Daleke DL, Huestis WH (1985) Incorporation and translocation of aminophospholipids in human erythrocytes. Biochemistry 24:5406-5416

Daleke DL, Lyles JV (2000) Identification and purification of aminophospholipid flippases. Biochim Biophys Acta 1486:108-127

De Jong K, Larkin SK, Styles LA, Bookchin RM, Kuypers FA (2001) Characterization of the phosphatidylserine-exposing subpopulation of sickle cells. Blood 98:860-867

Dekkers DWC, Comfurius P, Schroit AJ, Bevers EM, Zwaal RFA (1998) Transbilayer movement of NBD-labeled phospholipids in red blood cell membranes: outward-directed transport by the multidrug resistance protein 1 (MRP1). Biochemistry 37:14833-14837

Dekkers DWC, Comfurius P, Van Gool RGJ, Bevers EM, Zwaal RFA (2000) Multidrug resistance protein 1 regulates lipid asymmetry in erythrocyte membranes. Biochem J 350:531-535

Deuticke B, Heller KB, Haest CWM (1987) Progressive oxidative membrane damage in erythrocytes after pulse treatment with *t*-butylhydroperoxide. Biochim Biophys Acta 899:113-124

Deuticke B, Henseleit U, Haest CWM, Heller KB, Dubbelman TMAR (1989) Enhancement of transbilayer mobility of a membrane lipid probe accompanies formation of membrane leaks during photodynamic treatment of erythrocytes. Biochim Biophys Acta 982:53-61

Deuticke B, Grebe R, Haest CWM (1990) Action of drugs on the erythrocyte membrane. In: Harris HR (ed) Blood cell biochemistry. Plenum Press, New York, pp 475-529

Deuticke B, Klonk S, Haest CWM (1992) The red cell membrane in health and disease. Erythrocyte membrane barrier function: Role of bilayer polarity and skeletal proteins. Biochem Soc Trans 20:769-773

Devaux PF (1991) Static and dynamic lipid asymmetry in cell membranes. Biochemistry 30:1163-1173

Ding J, Wu Z, Crider BP, Ma Y, Li X, Slaughter C, Gong L, Xie XS (2000) Identification and functional expression of four isoforms of ATPase II, the putative aminophospholipid translocase. J Biol Chem 275:23378-23386

Dodge JT, Mitchell C, Hanahan D (1963) The preparation and chemical characteristics of hemoglobin-free ghosts of human erythrocytes. Arch Biochem Biophys 100:119-130

Dolis D, Moreau C, Zachowski A, Devaux PF (1997) Aminophospholipid translocase and proteins involved in transmembrane traffic. Biophys Chem 68:221-231

Dressler V, Schwister K, Haest CWM, Deuticke B (1983) Dielectric breakdown of the erythrocyte membrane enhances transbilayer mobility of phospholipids. Biochim Biophys Acta 732:304-307

Eastman SJ, Hope MJ, Cullis PR (1991) Transbilayer transport of phosphatidic acid in response to transmembrane pH gradients. Biochemistry 30:1740-1745

Fadok VA, Bratton DL, Frasch SC, Warner ML, Henson PM (1998) The role of phosphatidylserine in recognition of apoptotic cells by phagocytes. Cell Death Differ 5:551-562

Farquahar JW, Ahrens EH Jr (1963) Effects of dietary fats on human erythrocyte fatty acid patterns. J Clin Invest 42:675-685

Ferrell JE, Lee KJ, Huestis WH (1985) Membrane bilayer balance and erythrocyte shape: A quantitative assessment. Biochemistry 24:2849-2857

Florin-Christensen J, Suarez CE, Florin-Christensen M, Wainszelbaum M, Brown WC, McElwain TF, Palmer GH (2001) A unique phospholipid organization in bovine erythrocyte membranes. Proc Natl Acad Sci USA 98:7736-7741

Franck PFH, Op den Kamp JAF, Lubin B, Berendsen W, Joosten P, Briet E, Van Deenen LLM, Roelofsen B (1985) Abnormal transbilayer mobility of phosphatidylcholine in hereditary pyropoikilocytosis reflects the increased heat sensitivity of membrane skeleton. Biochim Biophys Acta 815:259-267

Frasch SC, Henson PM, Kailey JM, Richter DA, Janes MS, Fadok VA, Bratton DL (2000) Regulation of phospholipid scramblase activity during apoptosis and cell activation by protein kinase CΔ. J Biol Chem 275:23065-23073

Fujii T, Tamura A, Yamane T (1985) Trans-bilayer movement of added phosphatidylcholine and lysophosphatidylcholine species with various acyl chain lengths in plasma membrane of intact human erythrocytes. J Biochem 98:1221-1227

Gahmberg CG, Hakomori S (1973) External labelling of cell surface galactose and galactoseamine in glycolipid and glycoprotein of human erythrocytes. J Biol Chem 248:4311-4317

Galbiati F, Razani B, Lisanti MP (2001) Emerging themes in lipid rafts and caveolae. Cell 106:403-411

Ganong BR, Bell RM (1984) Transmembrane movement of phosphatidylglycerol and diacylglycerol sulfhydryl analogues. Biochemistry 23:4977-4983

Gascard P, Tran D, Sauvage M, Sulpice JC, Fukami K, Takenawa T, Claret M, Giraud F (1991) Asymmetric distribution of phosphoinositides and phophatidic acid in the human erythrocyte membrane. Biochim Biophys Acta 1069:27-36

Geldwerth D, Cherif-Zahar B, Helley D, Gane P, Freyssinet JM, Colin Y, Devaux PF, Cartron JP (1997) Phosphatidylserine exposure and aminophospholipid translocase activity in Rh-deficient erythrocytes. Mol Membr Biol 14:125-132

Gercken G, Tiling T, Brockmann U, Schröter W (1972) Fatty acid composition of phospholipids in erythrocytes of adults, normal newborn infants and neonates with Rh erythroblastosis. Pediat Res 6:487-494

Gordesky SE, Marinetti GV (1973) The asymmetric arrangement of phospholipids in the human erythrocyte membrane. Biochem Biophys Res Commun 50:1027-1031

Haest CWM, Andrick C (1989) Newly synthetized phosphatidylcholine is very rapidly translocated from the inner to the outer membrane leaflet in human erythrocytes. Studia Biophysica 134:77-82

Haest CWM, Plasa G, Kamp D, Deuticke B (1978) Spectrin as a stabilizer of the phospholipid asymmetry in the human erythrocyte membrane. Biochim Biophys Acta 509:21-32

Haest CWM, Kunze I, Classen J, Schneider E, Deuticke B (1986) Flip-flop rates and origin of phospholipid asymmetry in the native and modified red cell membrane can be studied by phospholipid analogues. In: Klein R, Schmitz B (eds) Topics in lipid research. The Royal Society of Chemistry, Burlington House, London, pp 235-242

Haest CWM, Kamp D, Deuticke B (1997) Transbilayer reorientation of phospholipid probes in the human erythrocyte membrane. Lessons from studies on electroporated and resealed cells. Biochim Biophys Acta 1325:17-33

Haldar K, De Amorim AF, Cross GA (1989) Transport of fluorescent phospholipid analogues from the erythrocyte membrane to the parasite in *Plasmodium falciparum*-infected cells. J Cell Biol 108:2183-2192

Hall MP, Huestis WH (1994) Phosphatidylserine headgroup diastereomers translocate equivalently across human erythrocyte membranes. Biochim Biophys Acta 1190:243-247

Halleck MS, Lawler JF Jr, Blackshaw S, Gao L, Nagarajan P, Hacker C, Scott P, Newman JT, Nakanishi Y, Ando H, Weinstock D, Williamson P, Schlegel R (1999) Differential expression of putative transbilayer amphipath transporters. Physiol Genomics 1:139-150

Hamon Y, Broccardo C, Chambenoit O, Luciani MF, Toti F, Chaslin S, Freyssinet JM, Devaux PF, McNeish J, Marguet D, Chimini G (2000) ABC1 promotes engulfment of apoptotic cells and transbilayer redistribution of phosphatidylserine. Nature Cell Biol 2:399-406

Herrmann A, Devaux PF (1990) Alteration of the aminophospholipid translocase activity during *in vivo* and artificial aging of human erythrocytes. Biochim Biophys Acta 1027:41-46

Hipfner DR, Deeley RG, Cole SPC (1999) Structural, mechanistic and clinical aspects of MRP1. Biochim Biophys Acta 1461:359-376

Homan R, Pownall HJ (1988) Transbilayer diffusion of phospholipids: dependence on head group structure and acyl chain length. Biochim Biophys Acta 938:155-166

Hooijberg JH, Pinedo HM, Vrasdonk C, Priebe W, Lankelma J, Broxterman HJ (2000) The effect of glutathione on the ATPase activity of MRP1 in its natural membranes. FEBS Letters 469:47-51

Jain SK (1985) *In vivo* externalization of phosphatidylserine and phosphatidylethanolamine in the membrane bilayer and hypercoagulability by the lipid peroxidation of erythrocytes in rats. J Clin Invest 76:281-286

Kamp D, Haest CWM (1998) Evidence for a role of the multidrug resistance protein (MRP) in the outward translocation of NBD-phospholipids in the erythrocyte membrane. Biochim Biophys Acta 1372:91-101

Kamp D, Sieberg T, Haest CWM (2001) Inhibition and stimulation of phospholipid scrambling activity. Consequences for lipid asymmetry, echinocytosis and microvesiculation of erythrocytes. Biochemistry 40:9438-9446

Kavecansky J, Joiner CH, Schroeder F (1994) Erythrocyte membrane lateral sterol domains: a dehydroergosterol fluorescence polarization study. Biochemistry 33:2880-2890

Kirby CJ, Green C (1977) Transmembrane migration ('Flip-Flop') of cholesterol in erythrocyte membranes. Biochem J 168:575-577

Kleinfeld AM (2000) Lipid phase fatty acid flip-flop, is it fast enough for cellular transport? J Membrane Biol 175:79-86

Kleinfeld AM, Chu P, Romero C (1997) Transport of long-chain native fatty acids across lipid bilayer membranes indicates that transbilayer flip-flop is rate limiting. Biochemistry 36:14146-14158

Kleinfeld AM, Storms S, Watts M (1998) Transport of long-chain fatty acids across human erythrocyte ghost membranes. Biochemistry 37:8011-8019

Kleinhorst A, Oslender A, Haest CWM, Deuticke B (1998) Band 3-mediated flip-flop and phosphatase-catalyzed cleavage of a long-chain alkyl phosphate anion in the human erythrocyte membrane. J Membrane Biol 165:111-124

König J, Nies AT, Cui Y, Leier I, Keppler D (1999) Conjugate export pumps of the multidrug resistance protein (MRP) family: localization, substrate specificity, and MRP2-mediated drug resistance. Biochim Biophys Acta 1461:377-394

Koefoed P, Brahm J (1994) The permeability of the human red cell membrane to steroid sex hormones. Biochim Biophys Acta 1195:55-62

Kornberg RD, McConnell HM (1971) Inside-outside transitions of phospholipids in vesicle membranes. Biochemistry 14:2809-2816

Krupka RM (1989) Role of substrate binding forces in exchange-only transport systems: II. Implications for the mechanism of the anion exchanger of red cells. J Membrane Biol 109:159-171

Kuypers FA, Van Linde-Sibenius Trip M, Roelofsen B, Tanner MJA, Anstee DJ, Op den Kamp JAF (1984) Rh_{null} human erythrocytes have an abnormal membrane phospholipid organization. Biochem J 221:931-934

Kuypers FA, Van Linde-Sibenius-Trip M, Roelofsen B, Op den Kamp JAF, Tanner MJA, Anstee DJ (1985) The phospholipid organisation in the membranes of McLeod and Leach phenotype erythrocytes. FEBS Letters 184:20-24

Kuypers FA, Yuan J, Lewis RA, Snyder LM, Kiefer CR, Bunyaratvej A, Fucharoen S, Ma L, Styles L, De Jong K, Schrier SL (1998) Membrane phospholipid asymmetry in human thalassemia. Blood 91:3044-3051

Lange Y (1992) Tracking cell cholesterol with cholesterol oxidase. J Lipid Res 33:315-321

Lange Y, Cohen CM, Poznansky MJ (1977) Transmembrane movement of cholesterol in human erythrocytes. Proc Natl Acad Sci 74:1538-1542

Lange Y, Dolde J, Steck TL (1981) The rate of transmembrane movement of cholesterol in the human erythrocyte membrane. J Biol Chem 256:5321-5323

Le Maire M, Moller JV, Champeil P (1987) Binding of a nonionic detergent to membranes: Flip-flop rate and location on the bilayer. Biochemistry 26:4803-4810

Lubin B, Chiu D, Bastacky J, Roelofsen B, Van Deenen LLM (1981) Abnormalities in membrane phospholipid organization in sickled erythrocytes. J Clin Invest 67:1643-1649

Manodori AB, Barabino GA, Lubin BH, Kuypers FA (2000) Adherence of phosphatidylserine-exposing erythrocytes to endothelial matrix thrombospondin. Blood 95:1293-1300

Marsh D, Horvath LI (1998) Structure, dynamics and composition of the lipid-protein interface. Perspectives from spin-labelling. Biochim Biophys Acta 1376:267-296

Martin OC, Pagano RE (1987) Transbilayer movement of fluorescent analogs of phosphatdiylserine and phosphatidylethanolamine at the plasma membrane of cultured cells. J Biol Chem 262:5890-5898

Middelkoop E, Lubin BH, Op den Kamp JAF, Roelofsen B (1986) Flip-flop rates of individual molecular species of phosphatidylcholine in the human red cell membrane. Biochim Biophys Acta 855:421-424

Mohandas N, Wyatt J, Mel SF, Rossi ME, Shohet SB (1982) Lipid translocation across the human erythrocyte membrane. Regulatory factors. J Biol Chem 257:6537-6543

Moll GN, Vial HJ, Ancelin ML, Roelofsen B, Comfurius P, Slotboom AJ, Zwaal RF, Op den Kamp JA, Van Deenen LL (1990) Phospholipid asymmetry in the plasma membrane of malaria infected cells. Biochem Cell Biol 68:579-585

Moore DJ, Giosioso S, Sills RH, Mendelsohn R (1999) Some relationships between membrane phospholipid domains, conformational order, and cell shape in intact human erythrocytes. Biochim Biophys Acta 1415:342-348

Müller P, Zachowski A, Beuzard Y, Devaux PF (1993) Transmembrane mobility and distribution of phospholipids in the membrane of mouse ß-thalassemic red blood cells. Biochim Biophys Acta 1151:7-12

Nelson GJ (1972) Lipid composition and metabolism of erythrocytes. In: Nelson GJ (ed) Blood lipids and lipoproteins: Quantitation, composition, and metabolism. Wiley-Interscience, New York, pp 317-386

Nijhof W, Van der Schaft PH, Wierenga PK, Roelofsen B, Op den Kamp JAF, Van Deenen LLM (1986) The transbilayer distribution of phosphatidylethanolamine in erythroid plasma membranes during erythropoiesis. Biochim Biophys Acta 862:273-277

Op den Kamp JAF (1979) Lipid asymmetry in membranes. Annu Rev Biochem 48:47-71

Oram JF (2000) Tangier disease and ABCA1. Biochim Biophys Acta 1529:321-330

Ortwein R, Oslender-Kohnen A, Deuticke B (1994) Band 3, the anion exchanger of the erythrocyte membrane, is also a flippase. Biochim Biophys Acta 1191:317-323

Pantaler E, Kamp D, Haest CWM (2000) Acceleration of flip-flop in the erythrocyte membrane by detergents differing in polar head group and alkyl chain length. Biochim Biophys Acta 1509:397-408

Peters R (1988) Lateral mobility of proteins and lipids in the red cell membrane and the activation of adenylate cyclase by ß-adrenergic receptors. FEBS Letters 234:1-7

Pomorski T, Herrmann A, Zachowski A, Devaux PF, Müller P (1994) Rapid determination of the transbilayer distribution of NBD-phospholipids in erythrocyte membranes by dithionite. Mol Membr Biol 11:39-44

Qian YM, Song WC, Cui H, Cole SP, Deeley RG (2001) Glutathione stimulates sulfated estrogen transport by multidrug resistance protein 1. J Biol Chem 276:6404-6411

Rao GA, Siler K, Larkin EC (1979) Diet-induced alterations in the discoid shape and phospholipid compositions of rat erythrocytes. Lipids 14:30-38

Rawyler A, Van der Schaft PH, Roelofsen B, Op den Kamp JAF (1985) Phospholipid localization in the plasma membrane of Friend erythroleukemic cells and mouse erythrocytes. Biochemistry 24:1777-1783

Renooij W, Van Golde LMG, Zwaal RFA, Roelofsen B, Van Deenen LLM (1974) Preferential incorporation of fatty acids at the inside of human erythrocyte membranes. Biochim Biophys Acta 363:287-292

Renooij W, Van Golde LMG, Zwaal RFA, Van Deenen LLM (1976) Topological asymmetry of phospholipid metabolism in rat erythrocyte membranes. Eur J Biochem 61:53-58

Rosso J, Zachowski A, Devaux PF (1988) Influence of chlorpromazine on the transverse mobility of phospholipids in the human erythrocyte membrane: relation to shape changes. Biochim Biophys Acta 942:271-279

Salzer U, Prohaska R (2001) Stomatin, flotillin-1, and flotillin-2 are major integral proteins of erythrocyte lipid rafts. Blood 97:1141-1143

Savill J, Fadok V (2000) Corpse clearance defines the meaning of cell death. Nature 407:784-788

Schneider E, Haest CWM, Deuticke B (1986a) Transbilayer reorientation of platelet-activating factor in the erythrocyte membrane. FEBS Letters 198:311-314

Schneider E, Haest CWM, Plasa G, Deuticke B (1986b) Bacterial toxins, amphotericin B and local anesthetics enhance transbilayer mobility of phospholipids in erythrocyte membranes. Consequences for phospholipid asymmetry. Biochim Biophys Acta 855:325-336

Schroeder F, Nemecz G, Wood WG, Joiner C, Morrot G, Ayaut-Jarrier M, Devaux PF (1991) Transmembrane distribution of sterol in human erythrocyte. Biochim Biophys Acta 1066:183-192

Schwichtenhövel C, Deuticke B, Haest CWM (1992) Alcohols produce reversible and irreversible acceleration of phospholipid flip-flop in the human erythrocyte membrane. Biochim Biophys Acta 1111:35-44

Seigneuret M, Devaux PF (1984) ATP-dependent asymmetric distribution of spin-labeled phospholipids in the erythrocyte membrane: Relation to shape changes. Proc Natl Acad Sci USA 81:3751-3755

Serra MV, Kamp D, Haest CWM (1996) Pathways for flip-flop of mono- and di-anionic phospholipids in the erythrocyte membrane. Biochim Biophys Acta 1282:263-273

Sherman IW, Prudhomme J, Tait JT (1997) Altered membrane phosphlolipid asymmetry in *Plasmodium falciparum*-infected erythrocytes. Parasitology Today 13:242-243

Shohet SB (1972) Hemolysis and changes in erythrocyte membrane lipids. New Engl J Med 286:577-583

Sims PJ, Wiedmer T (2001) Unraveling the mysteries of phospholipid scrambling. Thromb Haemost 86:266-275

Somerharju P, Virtanen JA, Cheng KH (1999) Lateral organisation of membrane lipids. The superlattice model. Biochim Biophys Acta 1440:32-48

Stout JG, Zhou Q, Wiedmer T, Sims PJ (1998) Change in conformation of plasma membrane phospholipid scramblase induced by occupancy of its Ca^{2+} binding site. Biochemistry 37:14860-14866

Subsczynski WK, Wisniewska A, Yin JJ, Hyde JS, Kusumi A (1994) Hydrophobic barriers of lipid bilayer membranes formed by reduction of water penetration by alkyl chain unsaturation and cholesterol. Biochemistry 33:7670-7681

Sun J, Zhao J, Schwartz MA, Wang JYJ, Wiedmer T (2001) c-Abl tyrosine kinase binds and phosphorylates phospholipid scramblase 1. J Biol Chem 276:28984-28990

Tait JF, Gibson D (1994) Measurement of membrane phospholipid asymmetry in normal and sickle-cell erythrocytes by means of annexin V binding. J Lab Clin Med 123:741-748

Tang X, Halleck MS, Schlegel RA, Williamson P (1996) A subfamily of P-type ATPases with aminophospholipid transporting activity. Science 272:1495-1497

Van Deenen LLM (1969) Membrane lipids and lipophilic proteins. In: Tosteson DC (ed) The molecular basis of membrane function. Printice-Hall, Englewood Cliffs, pp 47-78

Van der Schaft PH, Roelofsen B, Op den Kamp JAF, Van Deenen LLM (1987a) Phospholipid asymmetry during erythropoiesis. A study on Friend erythroleukemic cells and mouse reticulocytes. Biochim Biophys Acta 900:103-115

Van der Schaft PH, Beaumelle B, Vial H, Roelofsen B, Op den Kamp JAF, Van Deenen LLM (1987b) Phospholipid organization in monkey erythrocytes upon *Plasmodium knowlesi* infection. Biochim Biophys Acta 901:1-14

Verkleij AJ, Zwaal RFA, Roelofsen B, Comfurius P, Kastelijn D, Van Deenen LLM (1973) The asymmetric distribution of phospholipids in the human red cell membrane. A

combined study using phospholipases and freeze-etch electron microscopy. Biochim Biophys Acta 323:178-193
Vondenhof A (1993) Mechanismen der Phospholipid-Translokation in der Erythrozytenmembran. Ph.D. thesis, RWTH Aachen
Vondenhof A, Oslender A, Deuticke B, Haest CWM (1994) Band 3, an accidental flippase for anionic phospholipids. Biochemistry 33:4517-4520
Voswinkel St, Haest CWM, Deuticke B (2001) Complex effects of papain on function and inhibitor sensitivity of the red cell anion exchanger AE1 suggest the presence of different transport subsites. J Membrane Biol 179:205-221
Wali RK, Jaffe S, Kumar D, Kalra VK (1988) Alterations in organization of phospholipids in erythrocytes as factor in adherence to endothelial cells in diabetic mellitus. Diabetes 37:104-111
Welti R, Glaser M (1994) Lipid domains in model and biological membranes. Chem Phys Lipids 73:121-137
Wiedmer T, Zhou Q, Kwoh DY, Sims P (2000) Identification of three new members of the phospholipid scramblase gene family. Biochim Biophys Acta 1467:244-253
Woon LA, Holland JW, Kable EP, Roufagalis BD (1999) Ca^{2+} sensitivity of phospholipid scrambling in human red cell ghosts. Cell Calcium 25:313-320
Yang L, Andrews DA, Low PS (2000) Lysophosphatidic acid opens Ca^{++} channels in human erythrocytes. Blood 95:2420-2425
Yeagle P (1987) The membranes of cells. Academic Press, Orlando
Zachowski A (1993) Phospholipids in animal eukaryotic membranes: transverse asymmetry and movement. Biochem J 294:1-14
Zachowski A, Devaux PF (1990) Transmembrane movements of lipids. Experientia 46:644-656
Zachowski A, Favre E, Cribier S, Herve P, Devaux PF (1986) Outside-inside translocation of aminophospholipids in the human erythrocyte membrane is mediated by a specific enzyme. Biochemistry 25:2585-2590
Zhou Q, Zhao J, Stout JG, Luhm RA, Wiedmer T, Sims PJ (1997) Molecular cloning of human plasma membrane phospholipid scramblase. J Biol Chem 272:18240-18244
Zhao J, Zhou Q, Wiedmer T, Sims PJ (1998) Palmitoylation of phospholipid scramblase is required for normal function in promoting Ca^{2+}-activated transbilayer movement of membrane phospholipids. Biochemistry 37:6361-6366
Zuvic-Butorac M, Müller P, Pomorski T, Libera J, Herrmann A, Schara M (1999) Lipid domains in the exoplasmic and cytoplasmic leaflet of the human erythrocyte membrane: a spin label approach. Eur Biophys J 28:302-311
Zwaal RFA, Flückinger R, Moser S, Zahler P (1974) Lecithinase activities at the external surface of ruminant erythrocyte membranes. Biochim Biophys Acta 373:416-424
Zwaal RFA, Schroit AJ (1997) Pathophysiological implications of membrane phospholipid asymmetry in blood cells. Blood 89:1121-1132

2 Membrane Lipids and Proteins as a Basis of Red Cell Shape and its Alterations

Bernhard Deuticke

Institut für Physiologie, Universitätsklinikum der RWTH, Pauwelsstrasse 30, 52057 Aachen, Germany

2.1 Introduction

The peculiar, biconcave-discoidal, shape of the mammalian erythrocyte has intrigued scientists from many disciplines between theoretical physics and clinical medicine ever since its first observation following the invention of sufficiently magnifying microscopes.

While the basic prerequisite for all non-spherical cell shapes is an excess in surface area produced by active removal of ions and water from the cytoplasmic compartment, it has become evident during the last century that the particular shape assumed by the mature mammalian red cell is based on static and dynamic properties of the plasma membrane arising from the chemical composition, supramolecular organisation and physical properties of its major components, lipids and proteins.

Contemporary concepts of the origin of red cell shape are mainly based on contributions from four sources:

- analysis of the mechanical properties of the cell and its membrane,
- analysis of composition and structure of the membrane,
- experimental modification of the shape of the resting cell,
- study of altered shapes in pathological states.

The concepts derived from the analysis and theory of membrane mechanics are dealt with in Chap. 3 of this book by Fischer. The following sections will therefore put major emphasis on biochemical and cell physiological aspects of cell shape and its alterations, but also consider, though briefly, pathological shapes.

In order to keep the number of references within reasonable limits, reviews and even textbooks had to be cited in numerous instances. For earlier reviews of the subject, the reader might consult Steck (1989), Bull and Brailsford (1989) and Nakao (1990). Excellent collections of the morphology of normal, experimentally modified and pathological erythrocytes have been compiled by Bessis (1972, 1973, 1974, 1977).

2.2 Early History

Following the development of light microscopy in the second half of the 17th century, Malpighi first observed, in lung capillaries, the flow of "corpuscles", which were characterised by van Leeuwenhoek in 1688 as "sometimes hard, sometimes pliable", i.e. capable of shape changes. The further development of our knowledge of the membrane properties and the shape of the resting red cell has recently been summarised by Wong (1994) and Kleinzeller (1996).

Aspects of red cell shape that puzzle investigators up to the present day became first evident, when Hamburger reported in 1895 (as quoted by Ponder (1948)), that the discocytic erythrocyte reversibly assumes a spherical shape when brought into contact with glass, e.g. on a microscopic slide. The diverse ideas concerning the mechanistic basis of this "glass effect" have been compiled by Ponder in his monograph of 1948 on "Hemolysis and related phenomena". This classical review of red cell biology also describes Ponder's pioneering studies on the "disc-sphere transformation" of red cells by "lysins", which mark the onset of experimental approaches to understand the basis of shape changes.

Only much later attempts could be undertaken to understand the origin of alterations of the shape of the resting cell in haematological diseases. Likewise, the analysis and causal explanation of the shape changes of the red cell in flow could only be addressed after the advent of suitable microrheological techniques of experimentation and of appropriate theoretical models (Evans and Skalak 1979). Since these dynamic aspects of red cell shape will not be treated in the following, it should at least be pointed out here, that the red cell in its natural habitat, the circulation, spends a considerable part of its life-time in a state of deformation by flow, dominated by the mechanical properties of the cell.

2.3 Experimental Changes of Resting Shape

2.3.1 Outline of Observations

The basic differentiation of experimental shape changes given by Ponder (1948) is still a valuable criterion. Ponder distinguished two types of shape changes:

- those accompanied, or caused, by changes of cell volume, and
- those occurring without volume changes.

The first type arises from the fact that the native biconcave-discoid erythrocyte has a volume smaller than a sphere of the same surface area. Consequently, increases of cell volume will affect shape by decreasing the extent of biconcavity. Corresponding shape changes will result from losses of surface area at constant volume, e.g. under pathological conditions. Both processes ultimately lead to the formation of "spherocytes". Moderate decreases of cell volume at constant surface area will primarily only increase biconcavity. Pronounced shrinking of the cell, however, leads to a wrinkling of the membrane.

The second type defined by Ponder, which will be the major subject of this contribution, consists of shape changes *without* changes of cell volume. This type, observed under very diverse experimental conditions, comprises only two basic morphologies, originally termed "crenated" and "cup-shaped", but nowadays almost generally called *echinocytic* and *stomatocytic* following proposals of the French haematologist M. Bessis (1972, 1973) (Fig. 2.1).

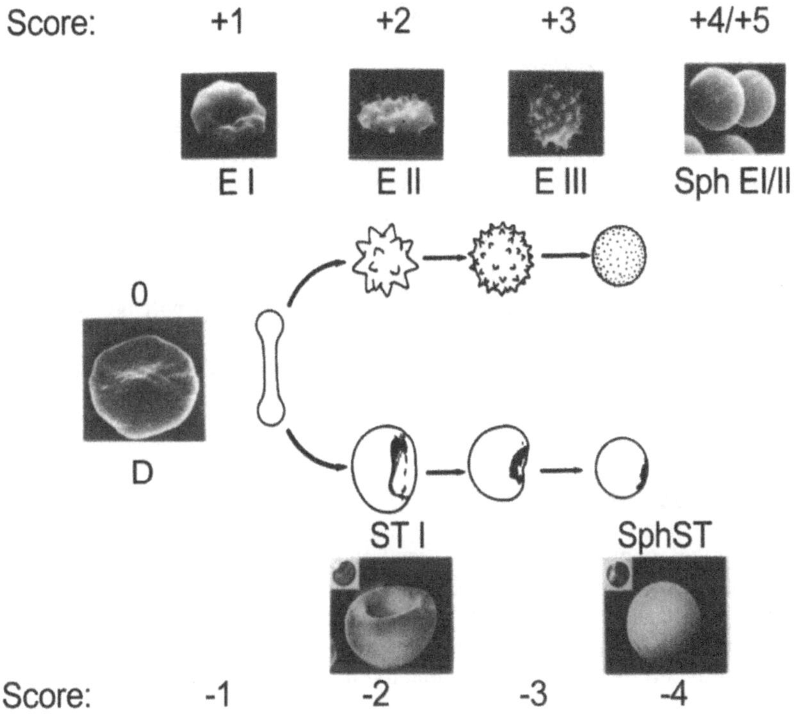

Fig. 2.1. Classification and scoring of the echinocytic (*E*) and stomatocytic (*ST*) shape transformation of discocytic (*D*) human erythrocytes. Scheme modified after Bessis (1973) and Daleke and Huestis (1989)

Both types may vary in intensity in a cell population inspected under the microscope in a given experimental set-up, but also as a function of the set-up. Echino- or stomatocytosis can be quantified subjectively by defining shape classes, characterised by scores (Bessis 1972). From these scores a "mean morphological index" can than be calculated (e.g. Ferrell and Huestis 1984; Schwarz et al. 1999b). More objective techniques to quantify shapes are also available (Grebe and Schmid-Schönbein 1985). As will become evident in greater detail below, this second type of shape changes is essentially due to membrane modifications. Such modifications will sometimes also lead to changes of cell volume, giving rise to ambiguities concerning the attribution of shape-transformed cells to one of Ponder's two types.

Shape changes observed in haematological diseases have to be considered as additional types, although in some cases they share features with the above defined types.

2.3.2 Outline of the Possible Origins of Shape Changes

Cytoplasmic Parameters

Since mammalian erythrocytes lack a true cytoskeleton, the amount of intracellular fluid, which determines cell volume, is the only cytoplasmic parameter affecting the shape of the normal cell directly. Other cytoplasmic constituents, like ATP, cations, or H^+ play of course also a role, but do so indirectly by controlling cell volume or modifying the membrane. Lew and Bookchin (1986) may be consulted for a deeper insight into the parameters, which influence the cellular content of water and their complex interrelationships. This subject will not be treated in detail here, since the effects of volume on the resting shape are of minor importance in comparison to its effects on the shape changes during flow through the microvasculature (Murdock et al. 2000) and to the relevance of membrane properties for changes of the resting shape.

Membrane Parameters

The chemical composition of the erythrocyte membrane as well as the supramolecular organisation and mechanical properties of its components are the major determinants of the shape of the resting cell. Composition and supramolecular organisation of the red cell membrane, as far as they concern the present discussion, are therefore shown schematically in Fig. 2.2, and briefly described in the following.

Membrane Lipids. Red cell membrane lipids (phospholipids, glycolipids and cholesterol) are arranged as a bilayer as in almost all biomembranes (see Chap. 1 for details). When classified according to their polar headgroups and their apolar long-chain tails, more than 400 molecular species can be identified, differing in structure, volume, geometry and charge distribution as well as lateral, transversal and rotational mobility (Cevc and Marsh 1987; Nagle and Tristram-Nagle 2000; Chap. 1). Moreover, these lipid patterns also vary between different mammalian species (see Deuticke 1977).

Most important in our context, membrane lipids are distributed unevenly between the two leaflets of the bilayer, constituting a transverse membrane asymmetry. Simplifying, the glycolipids and the choline phospholipids, i.e. phosphatidylcholine (lecithin, PC) and sphingomyelin (SM) are oriented towards the outer surface of the membrane, while the aminophospholipids, i.e. phosphatidylserine (PS), phosphatidylethanolamine (PE) and the phosphoinositolphospholipids (PI), are oriented towards the cytoplasmic surface. A more precise distribution pattern is given in Fig. 2.3. The transbilayer distribution of cholesterol (Chol) is not yet unambiguously clear (see Chap. 1).

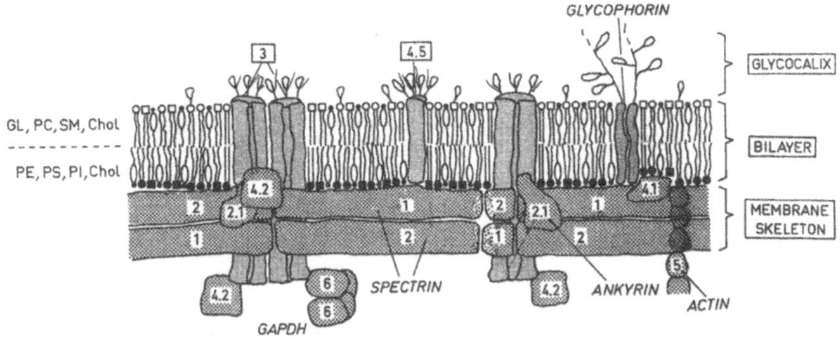

Fig. 2.2. Schematised structure of the human erythrocyte membrane. The cartoon emphasises the asymmetric transbilayer distribution of the phospholipids and comprises only the quantitatively dominating integral and peripheral proteins. The relative thicknesses of the constitutive layers are not given correctly scaled (but see [1]). *GAPDH*: Glyceraldehyde phosphate dehydrogenase

[1] In contrast to the fairly well established thickness of the glycocalyx (average about 6 nm (Linss et al. 1991), single projections possibly up to 15 nm (Viitala and Järnefelt 1985), and the well-known thickness of the lipid bilayer (around 4 nm (Nagle and Tristram-Nagle 2000)), it is a matter of considerable and continuous uncertainty, how far the sub-bilayer, membrane skeletal, lamina reaches into the cytoplasmic space. An approx. thickness between 20 and 30 nm is suggested not only by the molecular geometry and orientation of the major constituents of this lamina such as spectrin (McGough and Josephs 1990; Yan et al. 1993), actin (Picart and Discher 1999), the pole proteins, band 3 (Zhang et al. 2000) and glycophorin C (Lux and Palek 1995), and the anchor proteins ankyrin and protein 4.1 (Lux and Palek 1995), but also by the arrangement of the network in the plane of the membrane, established by ultramicroscopic technologies applied to isolated membranes after spreading, fixation and dehydration (see e.g. Terada et al. 1996; Takeuchi et al. 1998). Model calculations based on statistical mechanics resulted in a mean thickness of the network proper of about 15 nm (Boal 1994), which have to be added to the cytoplasmic extensions of the poles and the anchor proteins of about 10 nm. In contrast, total thicknesses between 50 and 100 nm were obtained by Bull et al. (1986) and Heinrich et al. (2001) using biomechanical approaches involving the application of (variable) compression of the membrane under fairly native conditions. An extended discussion of these rather divergent results is beyond the scope of this chapter. In any case, the reader should be aware that the sub-bilayer lamina considerably exceeds the other layers of the membrane in thickness!

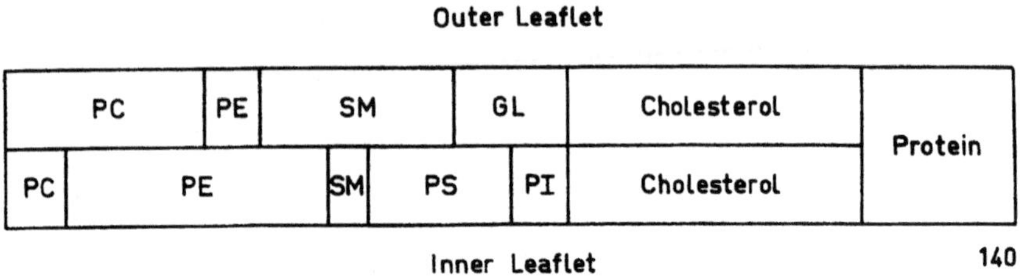

Fig. 2.3. Fractional contribution of lipids and proteins to the surface area of the human erythrocyte membrane and transbilayer distribution of phospholipids. Scheme based on data from Broekhuyse (1974) and Verkleij et al. (1973). For further details see Chap. 1

The origin of lipid asymmetry is not fully understood. Present concepts assign a major significance to an active, outward directed translocation of the aminophospholipids, mediated by an ATP dependent aminophospholipid translocase, "flippase" for short, combined with a very low passive transbilayer mobility of all phospholipids. This concept, however, cannot account for all aspects of asymmetry, in particular the distribution of the choline phospholipids. A stabilising role of lipid-protein interactions cannot yet be excluded.

The lipid bilayer contributes decisively to the spontaneous curvature of the membrane. As discussed in detail by Fischer (Chap. 3, Sect. 3.4.4) this curvature has two components:

First, the *double-layer-based curvature*, arising from differences in area between the two leaflets of the bilayer. Changes in area of individual leaflets can be induced by addition or removal of components, by their redistribution between the two leaflets, and by chemical or metabolic modification of their components, going along with changes of their molecular geometry or distribution pattern.

Second, the *single-layer-derived curvature*, based on the sum of the curvatures of the two leaflets, which, in turn, are functions of the sum of the molecular shapes of the individual phospholipid species forming each leaflet, and their "bending stiffnesses". Molecular shapes may be described as "cylinders", "truncated cones" or "inverted truncated cones" (for details see Chap. 3).

Membrane proteins. Membrane proteins are conceptually divided into two classes, depending on their relation to the lipid bilayer. First, *integral*, membrane-spanning proteins, dominated quantitatively by band 3 protein[2], now identified as the anion exchanger AE 1, but also comprising the large number of other membrane transporters and channels described in this book as well as the functionally ill-defined glycophorins (GP's, in particular GP A and C) and related glycoproteins (see Lux and Palek (1995) for details). Integral proteins may affect membrane curvature by virtue of their mere molecular geometry, like the lipids. More-

[2] Based on the pioneering work of Fairbanks et al. (1971), the quantitatively dominant erythrocyte membrane proteins are still named according to their sequence of running distances in SDS polyacrylamide gel electrophoresis. For a recent survey see Lux and Palek (1995).

over, carbohydrate moieties coupled exofacially to integral proteins form an additional leaflet, usually termed "glycocalyx" in analogy to other types of cells where this "fuzzy coat" is much more developed (Viitala and Järnefeldt 1985; Linss et al. 1991). Band 3 has a large cytoplasmic domain (Zhang et al. 2000) linked to its membrane-spanning domain, which serves as a "pole" of attachment for the sub-bilayer lamina of proteins acting as membrane skeleton. Glycophorin C has a functionally corresponding cytoplasmic extension.

The membrane skeleton is part of the second class of proteins, termed *"peripheral"*. The skeleton, described in more detail in Sect. 2.6.1, plays an important role as a stabiliser of bilayer integrity and thereby as a basic prerequisite of native cell shape. A direct effect of this further layer on the spontaneous curvature of the membrane, comparable to the role of the two lipid layers, is usually denied by the argument that the skeletal protein network is highly compressible (has little resistance to biaxial strain), in contrast to the lipid layers (Lenormand et al. 2001). As long as the structural details of the interaction of the skeletal proteins, particularly spectrin, with the inner leaflet of the bilayer are still unsettled (Haest 1982; O'Toole 2000; Bennett and Baines 2001) one should, however, not dismiss skeletal proteins from the list of membrane components affecting membrane curvature (see also Sect 2.6.6). In any case, a relevant influence of the skeleton on shape changes derives from "restoring forces" it exerts against the shear deformations arising from changes of membrane curvature originating in the lipid bilayer (Evans 1974; Waugh 1996; see also Chap. 3). Moreover, alterations of the network, by chemical modification or due to genetic defects, interfere with or allow for shape changes, as discussed in more detail below.

Summing up this brief survey of red cell membrane structure it may be stated that the chemical composition and the physical properties of at least two, but possibly even four coupled laminae determine the resting shape(s) of the mammalian erythrocyte. The following sections of this chapter present in more detail the available information on which this statement is based.

2.4 A Role for the Lipid Bilayer: the Bilayer Couple Concept of Shape Changes

2.4.1 Underlying Observations and Definition

In his monograph, Ponder (1948) summarised his observations that numerous agents which lyse erythrocytes primarily induce a sequence of shape transformations of the discocytic cell, beginning with a crenated disc and proceeding via a crenated sphere to a smooth sphere before ending with the formation of a "ghost" following the release of haemoglobin. He also reported that the initial steps of the process of crenation (formation of echinocytes) are reversible. Interestingly, the alternative type of shape change, formation of cup-shaped cells (stomatocytes) by membrane active agents, is not mentioned by Ponder. Stomatocyte formation was only analyzed in detail, classified in its steps of progress (Fig. 2.1), and appreciated in its conceptual relevance some 30 years ago by Deuticke (1968) and Bessis

(1973, 1977). It became also clear in these investigations that cell lysis is not an obligatory consequence of such shape transformations. Deuticke (1968) could further demonstrate that formation of echinocytes and of stomatocytes are antagonistic processes in the sense that one type of transformation is attenuated or even abolished by agents producing the other type of transformation. Cells remain discocytic in the presence of a combination of appropriate concentrations of an echino- and a stomatocytogenic agent. Moreover, all shape transforming agents proved to be amphiphilic, i.e. to have a hydrophobic and a hydrophilic domain. Most important, echinocytogenic agents were shown to be mostly anionic, stomatocytogenic agents mostly cationic.

A mechanism causing these phenomena was proposed by Sheetz and Singer (1974) after two important additional findings had become known:

- Shape transforming compounds are membrane-active agents, which insert into and expand the lipid domain of biomembranes (Seeman 1972).
- The lipid domain of the erythrocyte membrane is transversely asymmetric (see Fig. 2.3 and Chap. 1).

Sheetz and Singer (1974) argued that anionic agents produce the spicules in echinocytes due to their asymmetric insertion into and subsequent expansion of the outer leaflet of the lipid bilayer, being withheld from the negatively charged inner leaflet by electrostatic repulsion. Since the two leaflets cannot separate from each other due to their coupling by hydrophobic interactions, expansion of the outer leaflet relative to the inner one will change the double-layer-based curvature of the membrane to more positive values, i.e. bend the membrane outward (Fig. 2.4), giving rise to the formation of spicules. Stomatocytogenic effects of cationic amphiphiles such as local anaesthetics were interpreted in corresponding terms as the consequence of a preferential accumulation of these agents in the negatively charged inner leaflet. The process was termed "bilayer couple mechanism".

This concept has proven extremely useful for the further experimental analysis and model building in red cell shape research. In a general form it postulates that all changes in the difference, or the ratio, of the areas of its two coupled leaflets will change the spontaneous curvature of the bilayer and thereby transform the discocyte either in the echinocytic or the stomatocytic direction.

The original observations have meanwhile been confirmed and extended to many agents known as drugs or detergents. In line with the bilayer couple mechanism it could be shown that

- amphiphilic compounds with identical hydrophobic domains will produce echinocytes when negatively charged, but stomatocytes when positively charged (Mohandas and Feo 1975),
- agents inserting spontaneously into the outer leaflet but unable to cross the bilayer to the inner leaflet will produce echinocytes even if uncharged, zwitterionic or cationic (Sheetz and Singer 1976; Isomaa et al. 1987; Schwarz et al. 1999b),
- cationic amphiphiles inserting rapidly into the outer leaflet but permeating only slowly to the inner leaflet will initially produce echinocytes, which then slowly return to the discocytic shape and finally become stomatocytes in dependence

of the state of distribution of the amphiphile between the two leaflets (Sheetz and Singer 1976; Isomaa et al. 1987),

- quantitative estimates of the steady-state distribution of echinocytogenic agents between the outer and the inner leaflet of the bilayer agree with the predictions of the hypothesis. The distribution ratios (inner leaflet/outer leaflet) for the echinocytogenic anions pyrenebutyrate (Matayoshi 1980) and dodecylsulphate (Poser and Deuticke 1999) were determined to be 0.29/0.71 and 0.18/0.82, respectively.

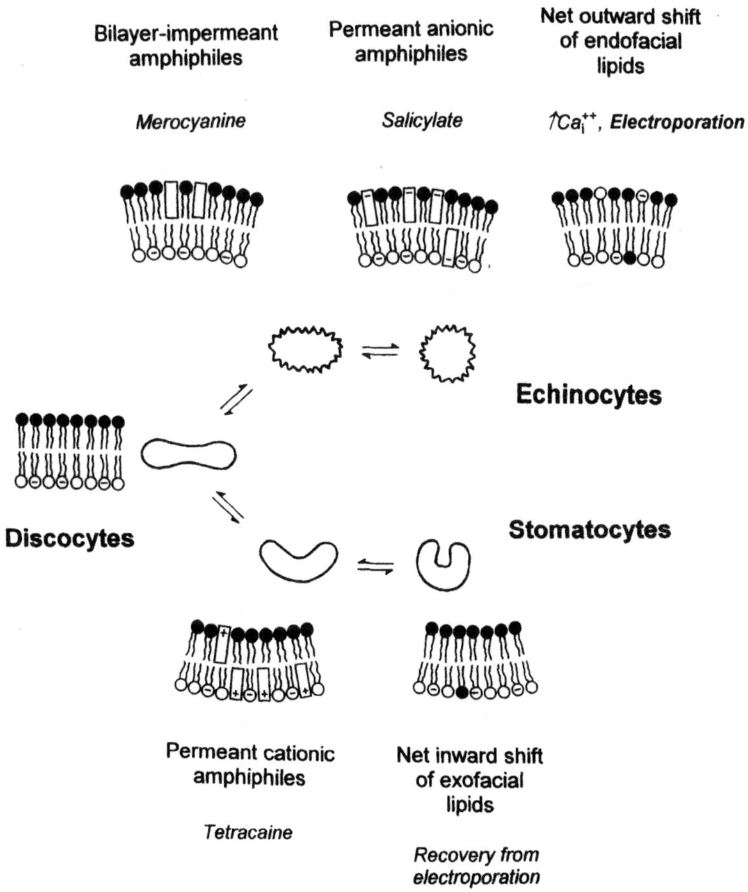

Fig. 2.4. Examples of shape transformations according to the bilayer couple mechanism, i.e. based primarily on changes of the double-layer-based curvature of the membrane. Phospholipids with "open" heads: PC, SM; with "closed" heads: PE, PS$^-$, PI. Rectangles: inserted amphiphiles.

Other quantitative details, however, of shape transformation by amphiphilic agents necessitate additional assumptions. Thus, it has been shown by Mohandas and Feo (1975) that quite different membrane concentrations of amphiphiles seem to be required for the formation of echinocytes and stomatocytes of the same score. About 15 times fewer copies per cell of a cationic phenothiazine were needed to induce stomatocytes III than of the corresponding anion to induce echinocytes III. Similar observations were made for the shape-transforming phospholipids to be discussed in Sect 2.4.2 (Daleke and Huestis 1989). Numerous explanations for this peculiarity may be considered. These include, e.g., erroneous assumptions underlying the assignment of the scores. It is by no means self-evident that stomato- and echinocytes of the same, rather arbitrary, absolute score value should be produced by the same absolute difference in area of the two leaflets. Alternatively, true differences in the "gain in area per inserted molecule" between the two leaflets might be involved. Moreover, the steady-state transbilayer distribution ratios of anionic and cationic amphiphiles may not be inversely equal. Those of cationic species are unknown. A solution of the problem is presently not at hand.

2.4.2 Shape Changes Following Alterations of the Transbilayer Balance of Membrane Lipids

The bilayer couple mechanism is not only supported by data obtained with non-biological agents. Its validity could also be demonstrated for the physiological lipid constituents of the red cell membrane, phospholipids and cholesterol.

Phospholipids

Pattern and transbilayer distribution of phospholipids can be altered by various procedures, which in many cases cause shape changes predictable by the bilayer couple concept. Additional phospholipids have been inserted into the membrane using the long-chain mono-acyl phospholipid lysolecithin (Mohandas et al. 1978) as well as various short-chain diacylphospholipids (Daleke and Huestis 1989). As expected, insertion of these analogs of the endogenous phospholipids into the outer leaflet converts native discocytes into echinocytes varying in degree (score) with the concentration applied. In the case of PC analogs these echinocytes either maintain this shape (diacyl-PC) or return to the discocyte shape (lyso-PC), in line with the incapability (diacyl-PC) or capability (lyso-PC) of these analogs to "flip" to the inner leaflet of the bilayer (see also Chap. 1). Insertion of short-chain diacyl PS also first induces echinocytes. These, however, turn subsequently into stomatocytes due to the active transbilayer flip of the PS analog depending on the availability of ATP and the activity of the above-mentioned aminophospholipid translocase (for details concerning this transporter see Chap. 1).

Since the molecular dimensions of PC analogs can be approximated from monolayer studies (for references see Kuypers et al. (1984); Ferrell et al. (1985); Nagle and Tristram-Nagle (2000)), and numbers of inserted copies of PC present in the membrane of echinocytes of a defined score can be measured with sufficient accuracy, attempts have been made to quantify the area differences between inner and outer leaflet required for the formation of echinocytes of a certain score (Ferrell et al. 1985).

These studies have shown that very small differences in area, probably below 1%, suffice to induce considerable shape changes. This, in turn, supports a significant role of the transbilayer distribution of endogenous phospholipids for the shape of the resting erythrocyte. Further evidence for this concept could be provided by the experimental removal of endogenous phospholipids from the outer leaflet of the bilayer, induced by first cleaving exofacial PC (and some PE) *in situ* by phospholipase A_2 and subsequently removing the products, fatty acid and lysophospholipid, from the membrane with serum albumin (Haest et al. 1980a). As predictable, this procedure induced stomatocytes, due to the resulting excess of material in the inner leaflet.

Cholesterol

Since cholesterol can be incorporated into or removed from the erythrocyte membrane experimentally (Grunze et al. 1980; Lange and Slayton 1982; Chabanel et al. 1983), its shape-affecting potency has also been tested. Although the reported results may not be fully unambiguous, they seem to indicate that removal of cholesterol induces stomatocytes while additional cholesterol produces echinocytes (Lange and Slayton 1982; Chabanel et al. 1983).

2.4.3 Shape Changes Following *in situ* Modification of Phospholipid Patterns

As mentioned above and discussed by Fischer (Chap. 3), the shape-determining spontaneous curvature of the red cell membrane should also be sensitive to the molecular geometry of the components present in the bilayer. The components can have the shape of a cylinder, a truncated cone, or an inverted truncated cone. Summed up over the about $2.5 \cdot 10^8$ copies of lipid per leaflet, these shapes will induce a positive (outward), negative (inward) or zero (flat) curvature of the leaflet. This is the basis of the "*single-layer based curvature*" as defined by Fischer (see Chap. 3).

The concept of a single-layer-based curvature is presently supported by only one type of experiment. Kuypers et al. (1984) have shown that replacement, with the aid of a phospholipid exchange protein, of endogenous phospholipids in the outer leaflet by more cylindrical or more inverted-cone-shaped phospholipid species produces echinocytes in the first, stomatocytes in the second case, in line with theoretical predictions considering the shape of the components of the outer leaflet (see also Chap. 3). Since single-layer-based curvature seems to provide a lower driving force for shape changes than double-layer-based curvature (Waugh 1996), its overall contribution to membrane curvature can be neglected in most cases (but see Chap. 3).

2.4.4 Shape Changes Arising from Experimentally Induced Transbilayer Redistribution of Endogenous Phospholipids

The evidence given in the preceding section for the bilayer couple concept was based on the insertion of exogenous or the removal of endogenous membrane components, leading to changes in area requirement or the spontaneous curvature of the two leaflets. Evidence is also available, however, based on the mere transbilayer redistribution of endogenous components, which goes along with shape changes. This evidence will be presented in the following.

Transbilayer Redistribution of Phospholipids Induced by Elevated Cytoplasmic Calcium

White (1974) was probably the first to report that loading of erythrocytes with calcium (Ca^{2+}), using the ionophore A 23187, will induce echinocytes (Fig. 2.4). Morphologically, the spicules of the shape-transformed cells differ to some extent from echinocytes produced by insertion of amphiphiles. Moreover, Ca^{2+} loading leads finally to shedding of microvesicles from the tip of the spicules (Allan and Thomas 1981), an event not regularly observed in echinocytes produced by amphiphiles. Nevertheless, there is good evidence that the bilayer couple mechanism can account for the gross shape changes induced by Ca^{2+}. Elevated cytoplasmic Ca^{2+} increases the transbilayer mobility of membrane phospholipids (Williamson et al. 1992; Henseleit et al. 1990) by activating a catalytic membrane protein, termed "scramblase" (Bassè et al. 1996). The increased transbilayer mobility promotes a (partial) decay of phospholipid asymmetry, i.e. a compositional scrambling of the two leaflets (see e.g., Henseleit et al. 1990). Details of the scrambling process are reviewed in Chap. 1.

The excess in area of the outer leaflet required for Ca^{2+}-induced echinocyte formation arises from the fact that the mobility of the three glycerophospholipids (PC, PE, PS) is much higher in the Ca^{2+}-exposed membrane than that of SM. Since SM is located in the outer leaflet (Haest et al. 1997; Schwarz et al. 1999a), the content of phospholipids in the outer leaflet will be augmented at the expense of the inner leaflet. This shift of the mass balance within the bilayer is supported by an inhibitory effect of Ca^{2+} on the aminophospholipid translocase, which would normally mediate, as a flippase, an active reaccumulation of PS and PE in the inner leaflet. Removal of Ca^{2+} from the cytoplasm deactivates the scramblase, reactivates the flippase, and thus allows a partial restoration of the asymmetric distribution of PS (and PE) sufficient to normalise the shape of the cells, provided the necessary supply of ATP is available (Lin et al. 1994).

This reversible echinocyte formation can be repeated numerous times (Schwarz 2000), a finding which probably excludes an alternative interpretation of the effect of Ca^{2+}. It has been speculated (Ferrell and Huestis 1984) that Ca^{2+} produces echinocytes via the enzymatic dephosphorylation of phosphoinositolphospholipids located in the inner leaflet of the bilayer, leading to the formation of phosphatidic acid and highly mobile diacylglycerol and to a consecutive increase in membrane curvature. Since, however, this enzymatic dephosphorylation is irreversible (Lin et al. 1994), it cannot be responsible for a reversible echinocyte formation. Nevertheless, although Ca^{2+}-induced echinocytosis is thus clearly a consequence of a scrambling of phospholipids, it is not yet clear to what extent the subsequent microvesiculation results from the mere loss of phospholipid asymmetry. Recent data suggest that additional processes may be involved (Bucki et al. 1998; Smith et al. 2001), but scrambling is certainly required for microvesiculation, since inhibition of scramblase (by dithioerythritol) prevents Ca^{2+}-induced echinocytogenesis as well as microvesiculation (Kamp et al. 2001).

Transbilayer Redistribution of Phospholipids Following Exposure to Strong Electric Field Pulses

Electric field pulses of μs duration and field strengths in the range of 3–10 kV/cm produce in the erythrocyte membrane long-lasting defects with the properties of aqueous pores of definable radii (Zimmermann et al. 1986; Deuticke and Schwister 1989), which are stable at 0 °C but reseal slowly at physiological temperature. This defect formation also goes along with a reversible increase of the transbilayer mobility of membrane phospholipids (Haest et al. 1997).

Moreover, field-pulse-exposed red cells become echinocytes in a biphasic process (Henszen et al. 1997). An initial phase in the ms range (Baumann 2001) is followed by a subsequent, much slower ($t_{½}$ up to many minutes) process (Schwarz et al. 1999b). Field-pulse-induced echinocytes are essentially time-stable at 0 °C. At higher temperatures they return slowly to the discocytic shape followed by spontaneous transformation into stomatocytes (Fig. 2.4). As in the case of the Ca^{2+} effect, this recovery depends on the availability of ATP and a functioning aminophospholipid translocase (for details see Schwarz et al. (1999b) and Chap. 1). It could be demonstrated (Haest et al. 1997; Schwarz et al. 1999a) that the initial echinocyte formation is due to an "electroscrambling" of phospholipids analogous to Ca^{2+}-induced scrambling but not involving scramblase activation. While Ca^{2+}-induced scrambling is, however, a quantitatively limited process that does not lead to a complete loss of phospholipid asymmetry (Henseleit et al. 1990), present evidence indicates that the less specific electric pulses induce – under appropriate conditions – an essentially complete scrambling of phospholipids (Schwarz et al. 1999b). This has provided a promising new tool to test the range of applicability of the bilayer couple concept for understanding the mechanism behind a number of causally unresolved shape changes.

Shape Changes Induced by Isotropic Stress

Artmann et al. (1997) have recently shown that transitory aspiration of red cells into a narrow pipette transforms discocytes into echinocytes, which subsequently return to the discocytic shape in a process sensitive to the same inhibitory measures also inhibiting the reversal of echinocytes produced by internal Ca^{2+} or electric field pulses. A "mechanoscrambling" of phospholipids may be responsible for this shape transformation, but this interpretation requires further confirmation.

2.5 Shape Changes of Unresolved Origin

2.5.1 Depletion of ATP or Magnesium and Exposure to Vanadate

Although known as a phenomenon for more than 40 years (Nakao et al. 1960; see also Nakao (1990)), it is still not clear why metabolic starvation as well as ATP depletion leads to the formation of echinocytes. The slow progress ($t_{½}$ ~ 10–15 h) of this shape change, lagging behind the disappearance of ATP, but also its reversibility (Feo and Mohandas 1977) have suggested to investigators the involvement of a dephosphorylation process, either of a protein, e.g. spectrin (Birchmeier and Singer 1977), or of lipids, particularly phosphoinositolphospholipids (Ferrell and Huestis 1984). Evidence in favour of these explanations remained, however, ambiguous (Anderson and Tyler 1980; Patel and Fairbanks 1986; Lin et al. 1994). Following the discovery of phospholipid asymmetry and the role of an ATP-dependent transport in its maintenance, interpretations based on the bilayer couple concept, i.e. an echinocytogenic loss of phospholipid asymmetry were considered (Lin et al. 1994) but the issue is unresolved, since available data on phospholipid distribution in ATP depleted cells (Haest et al. 1980b; Henseleit et al. 1990) are inconclusive.

Like ATP depletion, removal of Mg^{2+} from the cytoplasm, brought about by exposure of the cells to the ionophore A 23187 and EDTA (see Chap. 16) produces echinocytes (Truong et al. 1993; Schwarz et al. 1999b). Whether this phenomenon is due to the same processes also involved in the echinocytogenic effect of metabolic depletion is not yet clear.

The same is true for shape effects of vanadate, which not only acts echinocytogenic during prolonged contact with erythrocytes (Backman 1986), but also inhibits shape normalization and subsequent stomatocyte formation of echinocytes produced by elevated cytoplasmic Ca^{2+} or electric field exposure (Lin et al. 1994; Henszen et al. 1997; Schwarz et al. 1999b). Vanadate is well-known to inhibit various P-type ATPases including the aminophospholipid translocase (see Simons (1979) and Chap. 1). Moreover, it induces dephosphorylation and subsequent degradation of phosphoinositolphospholipids, located predominantly in the inner leaflet of the lipid bilayer (Backman 1986). This has given rise to the proposal that the state of phosphorylation of the inositolphospholipids may control red cell shape by affecting the spontaneous curvature of the membrane. Since, however, vanadate may interfere with the transbilayer distribution of phospholipids by inhibiting the active inward movement of aminophospholipids, its echinocytogenic effect might also arise from a loss of aminophospholipid asymmetry. In this case a phospholipid-symmetrised membrane should not respond to vanadate.

2.5.2 Ligands of Band 3 Protein

The bilayer domain of the red cell membrane also contains the integral proteins, in particular band 3, the anion exchanger. It has been speculated that shape changes could originate from this component (Wong 1994). More specifically, it was suggested (Blank et al. 1994; Gimsa and Ried 1995) that echinocytogenic effects of stilbene disulphonates (e.g. DIDS and DNDS) and pyrimido-pyrimidines (such as dipyridamole) result from their exofacial interaction with band 3. Both classes of compounds are well-known inhibitors of anion transport via band 3 (Deuticke et al. 1990). Schwarz et al. (1999b) have presented further experimental data supporting the concept of shape changes induced via this protein.

Gimsa and Ried (1995) interpreted the echinocytogenic effect of the two classes of agents qualitatively in terms of a ligand-induced "recruitment" of band 3 protein to an outward-oriented conformation, similar to that established by transport studies (see Chap. 11), and a subsequent expansion of the outer leaflet. Recent, more quantitative, model calculations (Deuticke, unpublished) have shown that in order to produce the observed shape changes a ligand-induced change of the conformation of band 3 would only have to go along with a 5–10% increase of the cross-sectional area of the membrane domain of each band 3 monomer in the outer leaflet, possibly at the expense of a corresponding decrease in the inner leaflet. Since band 3 seems to have a considerable internal flexibility as indicated, e.g., by its high volume of activation (Canfield and Macey 1972), such an internal shift of molecular volume between the two leaflets would not seem unrealistic.

2.5.3 Effects of Glass Contact and of Serum Albumin

As mentioned above, red cells suspended in saline media immediately become spherocytic (more precisely, spheroechinocytic (Bessis 1973)) when brought into close contact with glass surfaces (see Ponder (1948) and Trotter (1956) for the pertinent early references). According to more recent data (Eriksson 1990) other artificial surfaces have the same effect. There is presently no generally accepted explanation for this phenomenon.

It has also long been known (see Ponder 1948), that low concentrations of serum albumin (< 0.2 g/dl) suppress this echinocytogenic "glass effect", possibly due to a "surface-coating" activity of this protein. However, serum albumin at higher concentrations has a stomatocytogenic effect of its own (Mehta 1983), which is mechanistically unresolved.

2.5.4 Phospholipid Symmetrization Following Field Pulse Exposure: a Tool for Testing the Requirement of Bilayer Asymmetry for Shape Changes of Unresolved Origin

The obvious question, whether the causally unresolved types of shape transformation presented above might result from changes in the transbilayer asymmetry of phospholipids, has recently become accessible to experimental investigation. As shown by Schwarz et al. (1999a) and Schwarz (2000), red cells subjected to 20 sequential field pulses at 7 kV/cm lose their capacity to maintain or establish phospholipid asymmetry during subsequent prolonged (4–20 h) incubation at 37 °C as a consequence of the very dramatic increase of phospholipid transbilayer mobility (Schwarz et al. 1999b). Phospholipid "symmetrization" was proven by measurements of the transbilayer distribution ratio of suitable phospholipid probes and the loss of the permanent echinocytogenic effect of phospholipid probes restricted in native cells to the outer leaflet of the bilayer. Symmetrization of the endogenous phospholipids could not be established directly, since the selective enzymatic cleavage of the phospholipids in the outer leaflet required for this purpose was precluded by the field-induced high transbilayer mobility of phospholipids.

Phospholipid-symmetrised cells are discocytic and morphologically indistinguishable from native discocytes[3]. This similarity in shapes allows the interesting conclusion that the lipid-asymmetric native bilayer probably has a spontaneous curvature very similar to that of the symmetrised bilayer.

In agreement with the bilayer couple concept, phospholipid-symmetrised cells no longer respond to experimental procedures normally inducing shape changes based on that mechanism, in particular the insertion of permeable echinocytogenic or stomatocytogenic agents (Schwarz et al. 1999b). Such agents will distribute symmetrically in a bilayer in which the anionic phospholipids are no longer asymmetric. Moreover, echinocyte formation following exposure to electric field pulses and to an increase of cytoplasmic Ca^{2+}, both based on the partial symmetrization of phospholipids and a consecutive increase in area of the outer leaflet (see above), is no longer possible (Schwarz et al. 1999b).

In an attempt to elucidate the origin of causally unresolved shape changes, phospholipid-symmetrised cells were subjected to treatments producing such shape changes (Schwarz et al. 1999b). When symmetrised cells are depleted – *after symmetrization* – of ATP or Mg^{2+}, or exposed to vanadate, they do not become echinocytes. This finding adds new evidence to the concept, that in the native, phospholipid-asymmetric cell these procedures lead to echinocyte formation as a consequence of an outward reorientation of lipid constituents of the inner leaflet, most likely PS or PE, following the blockade of the energy-dependent processes normally keeping these lipids in the inner leaflet. Since only about 1% of the endofacial lipids has to reorient outward to produce echinocytes (Ferrell et al. 1985), which is below reliable detectability by available analytical techniques, even the unaltered low passive transbilayer mobility of phospholipids in energy-depleted cells might suffice to produce the required shift of material. In the fully symmetrised membrane such an echinocytogenic shift could not occur. Additional, more direct, evidence will certainly be needed before echinocyte formation due to metabolic perturbations can be considered as causally resolved. This is even more necessary, since, surprisingly, echinocyte formation induced by ligands of band 3 protein is also strongly suppressed in phospholipid-symmetrised cells (Schwarz et al. 1999b). Even more surprising, the glass effect also disappears. In contrast, the stomatocytogenic effect of albumin can still be elicited (Schwarz et al. 1999b).

At present, no convincing explanation is at hand why phospholipid symmetrization should interfere with the shape effects of ligands of band 3 or the glass effect. It might be speculated that a field-pulse-induced rigidification of the membrane suppresses the shape transformations (see below). This can be excluded on the ba-

[3] This is only true if cell volume is kept at its physiological value by addition, to the suspension medium, of colloid-osmotic stabilizers such as dextrans (Henszen et al. 1997; Schwarz 2000).

sis of the observation (Schwarz et al. 1999b), that impermeant echinocytogenic agents, such as dodecylmaltoside are still effective in the symmetrised cells. As a serious alternative, yet undefined field-pulse-induced alterations of the membrane laminae adjacent to the bilayer on both sides might be proposed. This leads us to the question of the contribution of these layers to red cell shape and shape transformations.

2.6 From the Bilayer to a Quadrilaminar Membrane

2.6.1 Constituents of the Exo- and the Endofacial Membrane Surface Coat

Although for the purpose of the present discussion a very simplistic model of the surface coats flanking the lipid bilayer will be sufficient (Fig. 2.5), it seems appropriate to describe these two additional laminae in some detail.

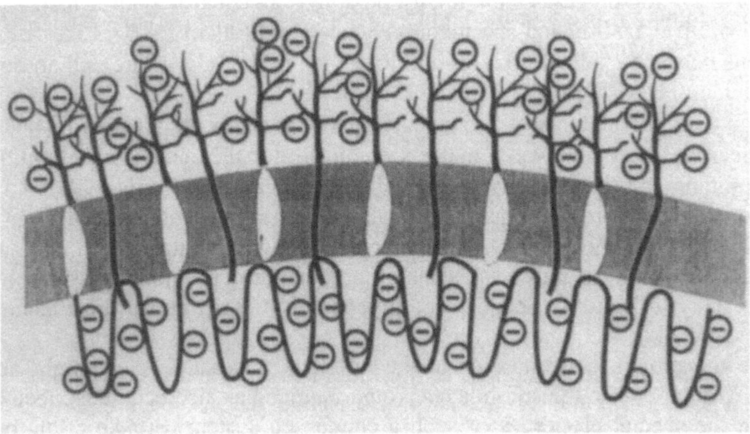

Fig. 2.5. Cartoon visualising the red cell membrane as a quadrilaminar structure. The exofacial lamina (glycocalyx) and the endofacial lamina (skeletal proteins) contain net negative charges in high density

The red cells' glycocalyx on the *exofacial* surface is dominated by the carbohydrate domains of glycolipids and integral glycoproteins. These oligosaccharides contain, besides simple neutral hexoses and pentoses and *N*-acetylhexosamines, fully ionised sialic acids, in particular *N*-acetylneuraminic acid (King 1994; Lux and Palek 1995). Branched and extending into the extracellular space to a varying extent, the glycocalyx forms a fuzzy coat of an average thickness of about 6 nm (Linss et al. 1991), stabilised and structured by ionic and weaker interactions among sugar moieties and polypeptide chains and with the extracellular medium (Viitala and Järnefelt 1985). Hydration may play a particularly important role besides coulombic forces. Density and steric arrangement of the anionic charges in this surface coat determine the negative surface potential of the cell (Donath and Voigt 1986) and also seem to affect the shape of the red cell (see below).

The *endofacial* sub-bilayer lamina of the membrane is composed of the cytoplasmic domains of integral proteins and the skeletal proteins, dominated by spectrin and actin. Spectrin forms antiparallel double-stranded rod-like flexible heterodimers about 100 nm in extended length. By head-to-head interactions most of these heterodimers associate to tetramers and hexamers. With its tail each dimer is linked to a short filament of F-actin. Since each filament binds, on average, six spectrin tails, a hexagonal network is formed, consisting of equilateral triangles. This network is stabilised by additional proteins such as protein 4.1 and 4.2, tropomyosin, adducin and myosin, and coupled to proteins of the bilayer domain (band 3, glycophorin C) by the anchor proteins ankyrin and protein 4.1 (Bennett 1990; Mohandas and Chasis 1993; Zhang et al. 2000; Bennett and Baines 2001; Chang and Low 2001; see also Lux and Palek 1995). In electron micrographs, the skeleton appears as a plane hexagonal network (Shen 1989; Terada et al. 1996). The end-to-end distance of the spectrin heterotetramers in this network, however, is considerably lower than their extended length (Waugh 1982). The nature of the underlying folding process of the highly flexible segmented polymer chain is still a matter of investigation and speculation (Bennett and Baines 2001). Likewise, interactions between skeletal proteins and the lipids in the bilayer are still an unsettled problem (Haest 1982; Bennett and Baines 2001).

In any case, this extensible protein skeleton is not only responsible for the red cells' shear elasticity, but also a major prerequisite for the mechanical stability of the membrane and the integrity of the cell (Mohandas and Chasis 1993). To account for this function, two types of interaction within the skeleton have been defined (Tse and Lux 1999). *Horizontal* interactions involving the spectrin tetramer, actin, and protein 4.1 stabilise the skeleton proper. *Vertical* interactions involving spectrin, ankyrin, band 4.1, band 3 and glycophorin C provide for the attachment of the meshwork to the lipid bilayer.

Since the protein concentration at the interface is certainly very high (about 30 g/dl, Waugh 1982), the sub-bilayer protein lamina has also been described as a gel with the special properties of such a condensed system (Stokke et al. 1986; Elgsaeter et al. 1986). It carries excess negative charges at a high density like the exofacial glycocalyx (Viitala and Järnefelt 1985; Heinrich et al. 1981). This accounts for a high sensitivity of this innermost layer to ionic strength and pH (see below).

Glycocalyx and membrane skeleton may be considered as laminae coupled to, but differing chemically and physically from, the two laminae of the bilayer. This justifies the concept of a "quadrilaminar" structure of the membrane (see Grebe in Deuticke et al. 1990; Elgsaeter and Mikkelsen 1991). There is even evidence indicating an interaction between these distant laminae (see Sect. 2.6.6). The following sections will deal with the observations suggesting a role of these two additional layers in altered cell shapes.

2.6.2 Influence of the Exofacial Lamina: Shape Effects of Enzyme Treatment

The high density of net negative fixed charges in the exofacial lamina would be expected to induce or increase – by electrostatic repulsion – a positive (echinocytogenic) curvature of the membrane. If this is true, a decrease of charge density should be stomatocytogenic. This has been demonstrated by enzymatic removal of the charge-carrying residues using neuraminidase or proteases such as trypsin etc., which makes cells more stomatocytic (Grebe et al. 1988; Schwarz et al. 1999b) and suppresses echinocytogenesis of various origins (Schmid-Schönbein et al. 1983). This stomatocytogenetic effect must originate from a negative curvature of the three innermost laminae of the quadrilaminar membrane. Surprisingly, the stomatocytogenic effect of trypsin was considerably increased after field-pulse-induced phospholipid symmetrization (Schwarz et al. 1999b), although the symmetrised bilayer is no longer under the influence of the potentially stomatocytogenic effect of an excess of negative surface charges in the inner lipid leaflet. A similar increased stomatocyte formation was observed in trypsin-treated red cells from spectrin-deficient mice (Schmid-Schönbein et al. 1983). One might therefore hypothesise that extensive exposure to electric field pulses does not only symmetrise the bilayer, but also somehow interferes with the yet undefined interaction between skeleton and bilayer.

2.6.3 Influence of the Endofacial Lamina: the Role of the Membrane Skeleton in Shape Transformations

The contribution of the membrane skeleton to the native discocytic shape and its causal role in experimental shape changes belong to the somewhat controversial subjects of red cell physiology. This is partly due to the fact that many experiments dealing with this problem will provide different results depending on whether they are carried out on intact cells on the one hand, or red cell ghosts or isolated skeletons on the other hand. Isolation of the skeleton can be achieved by removing lipids and – most – integral proteins from the skeleton by the nonionic detergent, Triton X 100 (Sheetz 1983). Model calculations based on a gel-like nature of the sub-bilayer lamina have led to the conclusion (Elgsaeter et al. 1986) that shape changes may result from modifications of this lamina. In view of the mechanical characteristics of the skeleton, in particular its low resistance to biaxial expansion, shear-induced extension and bending deformations (see Chap. 3), shape changes driven directly by this flexible polymer network are, however, hard to envisage.

The shape changes induced by depletion and repletion of ATP (see Sect. 2.5.1) and the presence, in the skeleton, of proteins also involved in contractile processes in muscle might suggest an "active", ATP-dependent function of the skeletal proteins (Shen 1989; Bennett 1990). Experimental data in favour of this concept were, however, mostly obtained on ghosts or isolated skeletal shells. Both react, indeed, to exposure to ATP by assuming stomatocyte-like contracted shapes (Birchmeier et al. 1979; Jinbu et al. 1984; Nakao 1990). On the other hand, in ghosts at least, in which phospholipid asymmetry is partly lost during preparation (Haest et al. 1980b; Connor et al. 1990; Schrier et al. 1992), ATP-induced formation of stomatocytic shapes can be attributed to a normalization of the asymmetric transbilayer distribution of aminophospholipids, comparable to the situation in whole cells recovering from field-pulse-induced echinocytosis. In line with this assumption, ghosts derived from field-pulse-symmetrised cells lose the capacity to become stomatocytic in the presence of ATP (Schwarz 2000), probably due to the drastically enhanced transbilayer mobility of phospholipids. These observations, like those reported in Sect. 2.5.4, support the concept that shape changes in energy-depleted erythrocytes are related to changes in the transbilayer distribution of phospholipids.

Nevertheless, additional shape effects of ATP, somehow mediated by skeletal or anchor proteins (Jinbu et al. 1983), but independent of the established phosphorylation of spectrin (Patel and Fairbanks 1981) cannot be excluded. Whether an interaction of ATP with skeletal elements also triggers the final formation of endocytotic vesicles in ATP-loaded stomatocytic ghosts (Hardy and Schrier 1978; Birchmeier et al. 1979) is not yet clear. Indirect evidence compatible with an involvement of ATP in a particular type of rapid "shape changes", mediated by a skeletal element, the F-actin ATPase, has recently been reported by Tuvia et al. (1998). They observed ATP-dependent fluctuations of the red cell membrane.

Shape studies on ghosts and isolated skeletons are complicated by the fact that the shape of these structures is strongly influenced by the composition, pH and ionic strength of their (internal) environment (Johnson and Robinson 1976; Johnson et al. 1980; Raval et al. 1989). Ionic strengths > 0.02 M and acid pH produce crenated ghosts, that can be converted into cup forms not only by ATP but also by replacement of chloride by zwitterionic buffers, e.g. HEPES, in the incubation medium. Since ghosts are prepared by procedures likely not to leave the membrane ultrastructure unperturbed, it is not readily clear whether this type of shape changes will also occur in whole cells. Their inner ionic milieu cannot be greatly modified without transient permeabilisation and thus perturbation of the membrane. An exception is manipulation of the cytoplasmic pH. Changes of this parameter have therefore attracted growing interest over the past decades.

2.6.4 Shape Effects of Cytoplasmic pH

Rand and Burton (1965) were probably the first to report that discocytes turn into stomatocytes when incubated in media of acid pH < 6.5, and into echinocytes in media of alkaline pH > 8.5. In a thorough investigation of this phenomenon, Gedde and Huestis (1997) and Gedde et al. (1997) could demonstrate that changes of the intracellular pH but not of other parameters co-varying with pH (e.g. transmembrane potential or cellular water and chloride content) can account for these shape transformations. The direction of these shape changes is opposite to that expected on the basis of the shape changes observed in isolated membranes (i.e. unresealed ghosts), which evaginate at low pH and invaginate at high pH. This direction of shape changes agrees with a contraction at low pH and expansion at high pH of the isolated membrane skeleton (Johnson et al. 1980). The pH-induced shape changes of the whole cell are thus not simply explicable by skeletal properties, although alterations within the skeleton or its association with the bilayer domain (Lelkes and Fodor 1991; Low et al. 1991) provide a reasonable explanation for the shedding of vesicles from whole cells observed at both extremes of pH (Leonards and Ohki 1983; Lelkes and Fodor 1991; Gros et al. 1996; Bobrowska-Hägerstrand et al. 1997, 1998).

A first clue to the mechanistic basis of the "paradoxical" direction of the pH dependency of the shape of the whole cell has recently been given by Gedde et al. (1999). They reported a pH-dependent hydrophobic insertion of a membrane-attached cytoplasmic protein, possibly glyceraldehyde-3-phosphate dehydrogenase, into the bilayer. This was proposed to result in a progressive stomatocytogenic expansion of the inner leaflet with decreasing pH, compensating or overcompensating the above-described skeletal reactions to pH. In an analogous concept, pH-dependent changes of the intrabilayer conformation or orientation of band 3 have been proposed by Gimsa (1998) to cause the pH dependency of red cell shape.

It is this type of mechanism (variable extent of insertion of protein into the bilayer), which might also quite generally account for any shape changes resulting from physico-chemical alterations of skeletal and other membrane proteins. Alternative models invoking contraction or expansion of the skeletal lamina as causes for changes of the curvature of the bilayer are not easily reconcilable with the isotropic modulus of the skeletal network, which is very low in relation to the moduli governing bilayer bending (see Chap. 3).

Nonetheless, mechanisms additional to those discussed here may contribute to the shape changes induced by changes of internal pH. This view is also supported by the observation (Schwarz et al. 1999b; Schwarz 2000) that phospholipid-symmetrising electric-field-pulse treatment, while not affecting the low pH stomatocytosis, interferes in a complex way, suppressing and enhancing, with the high pH echinocytosis.

2.6.5 Influence of Ion Gradients and Transmembrane Potential

Besides intracellular pH, changes of the extracellular anion milieu affect red cell shape. In particular, replacement of chloride, the dominant permeant extracellular anion, by impermeant, buffering or non-buffering, anions or by sucrose is strongly stomatocytogenic (Deuticke 1968; Rasia and Bollini 1998). Depending on the extracellular buffer capacity such replacements will have either no influence on the cytoplasmic pH or shift it to more alkaline, i.e. more echinocytogenic, values resulting from an increased (inside positive) membrane potential and an enforced uptake of external hydroxyl anions in exchange for internal chloride (Macey et al. 1978; Eidmann 1997; see also Chap. 4). Changes of cytoplasmic pH cannot therefore account for this stomatocyte formation. In extended studies, Glaser and associates (for references see Glaser 1998) have developed the concept that altered transmembrane potentials *per se* may cause shape changes such as those produced by replacement of chloride and in cells hyperpolarised by an ionophore-mediated cation conductance. Other investigators (Bifano et al. 1984; Gedde and Huestis 1997) have refuted this concept on the basis of their data and interpretations.

2.6.6 Limitation of Shape Changes by the Membrane Skeleton

As stated above, changes of cell shape likely to involve the membrane skeleton are probably not caused by its mechanical properties. However, these properties, in particular the shear elasticity of the skeleton, *counteract* shape transformations based on changes of the spontaneous curvature of the bilayer according to the bilayer couple mechanism. This becomes readily evident when the elastic modulus of the skeleton is increased experimentally. Chemical crosslinking as well as side chain modifications of spectrin by oxidizing and alkylating SH-reagents (Fischer et al. 1978; Haest et al. 1980a) but also by other group-specific reactions (Thelen and Deuticke 1988), while not *per se* altering cell shape, make the cell resistant, due to a "rigidification" of the membrane skeleton, to all shape changing procedures (Haest et al. 1980a). Only very few sites in spectrin need to be modified to greatly increase the shear resistance of the membrane. Decreases of the shear resistance of the skeleton, on the other hand, e.g. by urea, sensitise the cells to shape changing procedures (Khodadad et al. 1996).

Interestingly, red cells become also resistant to shape changing procedures following the exofacial binding, to transmembrane glycoproteins, of certain lectins, such as wheat germ agglutinin (WGA) or concanavalin A (Con A) (Lovrien and Anderson 1980; Anderson and Lovrien 1981; Lin and Huestis 1995). WGA interacts with specific oligosaccharide residues of glycophorin C (and A), while Con A binds to oligosaccharides on band 3. Other lectins or IgG antibodies against exofacial epitopes on glycophorin do not exhibit this inhibitory effect. Rheological measurements on WGA-treated cells revealed the expected increase in shear resistance. Nevertheless, the effect of WGA seems not to be due to a "simple" rigidification, since it was restricted to the suppression of echinocytogenic effects, while stomatocytogenic effects could still be elicited (Chasis and Schrier 1989). The inhibitory effect of Con A even seems not to be associated at all with a measurable increase of shear resistance (Chasis et al. 1985). The widely accepted explanation for these effects of lectins proposes that by interacting with specific sites on the exofacial domain of transmembrane proteins they induce conformational changes that are transmitted to the cytoplasmic face of the membrane. The nature of the consecutive alterations in the sub-bilayer lamina is still largely enigmatic. Interestingly, antibodies against skeletal and anchor proteins and even the cytoplasmic domain of band 3 *per se* also seem to affect shape and shape changes in a complex way (Carter and Fairbanks 1984; Pestonjamasp and Mehta 1995). In terms of the bilayer concept one would tend to interpret such effects as the result of altered protein-bilayer interactions such as depth of insertion or intrabilayer geometry of integral or membrane-attached proteins, following the binding of antibodies or lectin-induced changes of conformation.

2.6.7 Membrane "Stabilization" by the Skeletal Network and its Alterations

Besides its contributions to the various resting shapes of the red cell considered above, the skeletal lamina is involved in the maintenance of cell shape in a very fundamental way by providing for membrane stability. It prevents spontaneous shedding of vesicles and safeguards against fragmentation. Since such vesiculation processes and cell fragmentation are shape changes only in a rather loose sense, it will have to suffice to summarise conditions inducing vesiculation and to outline present hypotheses concerning the molecular basis of membrane stability.

Vesicles are shed *in vitro* from red cells upon metabolic depletion, storage, heating to > 49 °C, oxidative membrane damage (Wagner et al. 1986), exposure to extremes of pH (see Sect. 2.6.4), Ca^{2+} loading (see Sect. 2.4.4) and – to the extent of complete fragmentation of the cells – by exposure to the protein denaturant, urea (Fuhrmann 1968). Likewise, formation of spheroechino- or -stomatocytes by amphiphilic agents will often go along with the occurrence of exo- or endovesicles (Hägerstrand and Isomaa 1992). *In vivo*, vesicles are released from aged cells (Dumasvala and Greenwalt 1984; Waugh et al. 1992), and from cells in numerous haematological disorders resulting from defective membrane proteins (see Sect. 2.7).

It is generally agreed that vesiculation is initiated by a dissociation of the skeleton from the bilayer. Vesicles are therefore in most cases free of skeletal proteins and contain integral proteins, e.g. band 3 or glycophorin, in inverse relation to the intensity of the attachment of these integral proteins to the skeleton (Knowles et al. 1997). It is not clear yet, how a detachment of the skeleton destabilises the bilayer. On the one hand, a direct interaction between spectrin and proteins or lipids of the bilayer might prevent "budding" of the bilayer and subsequent vesiculation. Alternatively, it has been proposed, on the basis of studies on band 3-knock-out mice (Peters et al. 1996) and findings in haematological disorders (see below and Lux and Palek 1995), that the bilayer owes its integrity to the interaction of the abundant (tetrameric) band 3 protein with its surrounding lipids (about 500 per tetramer, corresponding to about 9 concentric layers). Since the skeleton contributes to a stable lateral distribution of the band 3 tetramers (cf. Tomishige et al. 1998), disintegration of the skeleton by experimental procedures or in pathological states may lead to aggregation of band 3 tetramers, create protein-depleted lipid domains and thereby destabilise the bilayer. In summary, membrane stability may thus be based on proteins in different layers of the membrane.

2.7 Pathological Alterations of Red Cell Shape *in vivo*

This brief survey of shape transformations cannot provide a thorough introduction to red cell membrane pathology. Only some aspects of this subject will be covered here. For a more comprehensive treatment, including the medical aspects and a compilation of the pertinent literature, the reader should consult textbooks of haematology, e.g. Handin et al. 1995. The morphology of red cells in many haematological disorders has been excellently documented by Bessis (1973, 1974).

Major pathological changes of red cell shape may arise from or go along with

- changes of cell volume,
- alterations of haemoglobin,
- alterations of membrane lipids or (skeletal) proteins.

2.7.1 Stomatocytosis Going Along with Alterations of Cell Volume

Stomatocytic red cells are observed in a number of genetic haematological disorders going along with changes in cell volume. An increased content of water ("overhydration" in clinical terminology) due to an increased leak to and resulting uptake of sodium accompanies the change of shape in "Hereditary stomatocytosis" and may even play a causal role, since experimental blockade of the leak also reverses the shape change (Mentzer et al. 1976). On the other hand, stomatocytes have also been observed in haematological disorders not going along with cell swelling or even accompanied by cell shrinking due to an increased leak to and loss of potassium. For the details and references see Stewart et al. (1993), Lux and Palek (1995) and Chap. 21.

2.7.2 Shape Changes Due to Altered Properties of Haemoglobin

For reasons of completeness it should be mentioned that red cell shape might also be affected by the physical state of haemoglobin (Hb). Mutation of a single amino acid in the beta chain of globin (beta 6 glu $\rightarrow$ val) produces Hb S, which undergoes polymerisation and forms solid fibers upon deoxygenation and thereby imposes shape changes on the cell. They appear mainly sickle-like, which has given the disorder its name, but other bizarre shapes may also be adopted. Flow-induced deformations of the cells during deoxygenation and polymerisation of Hb S in the microvasculature probably contribute to this heterogeneity. These shape changes are primarily reversible upon reoxygenation and depolymerisation of Hb S. They will become irreversible and thus be evident even after reoxygenation following secondary, persistent alterations of the membrane, which affect the lipid bilayer and the proteins. Shrinking of the cells due to loss of ions and water induces further deformation of the cells and makes them extremely rigid. For all details and references see Chap. 23 and Platt (1995).

2.7.3 Lipid-Based Alterations of Cell Shape

Disturbances of lipid metabolism, e.g. in hepato-biliary diseases, but also due to the inborn lack of serum beta-lipoproteins or of the enzyme, lecithin cholesterol acyl transferase (LCAT), go along with two types of altered shapes (for details and references see Lux and Palek (1995); Tse and Lux (1999)):

- *acanthocytes* (spur cells), looking somewhat like echinocytes but having fewer spicules of variable size, which project irregularly from the surface, and
- *target cells* (from their appearance in dried smears), which look bowl-shaped in wet smears.

Both are primarily due to the acquisition, from the serum, of additional phospholipids (PC or SM) and/or cholesterol. The resulting increase of surface area, sometimes combined with a loss of volume, produces the target cell, e.g. in obstructive liver disease. Acanthocytes either arise from a secondary (asymmetric?) removal of material from the lipid-enriched membrane during passage of the cells through the spleen (in liver diseases), or (presumably) from an asymmetric transbilayer distribution of the inserted lipids (in abetalipoproteinemia).

Acanthocytic shapes can be normalised by stomatocytogenic agents, indicating that the shape alteration is related to the bilayer couple mechanism. On the other hand, acanthocytes have been observed independent of lipid alterations, e.g. in cells lacking certain blood group antigens, and in neurological diseases, reminding us that the mechanisms controlling red cell shape are still far from being fully understood.

2.7.4 Protein-Based Alterations of Red Cell Shape

As outlined in Sect. 2.6.1, integrity and stability of the red cell membrane require an adequate horizontal and vertical organisation of the skeletal network including its regular connection with the bilayer domain. Deficiencies and disturbances in this support give rise to haematological disorders (see Lux and Palek 1995; Tse and Lux 1999).

Defects in the vertical interactions lead to disconnection of the bilayer from the skeleton and subsequent loss of surface area, removed as vesicles. The prototype of this disturbance is "Hereditary spherocytosis". Cells of afflicted people are spherical or slightly stomatocytic due to the loss of membrane during their passage through the spleen, a complex process termed "conditioning". The underlying destabilisation arises from deficiencies, due to mutations, of either spectrin, ankyrin (or both), band 4.2 (pallidin), or band 3. In view of this diversity of causes, both of the above-mentioned mechanisms of membrane stabilisation (see Sect. 2.6.7) may be impaired, i.e. the direct effect of the skeleton on the bilayer, or the skeleton-controlled effect of band 3 on the surrounding lipids.

Defective horizontal interactions in the skeletal network either arise from deficiencies, which prevent the oligomerisation of spectrin into the hetero-dimers and -tetramers forming the skeletal network, or from a deficiency of protein 4.1, the stabiliser of the spectrin-actin interaction. Both defects disrupt the membrane skeleton. For reasons yet unclear this leads to the irreversible formation of elongated elliptic cells, which have a decreased deformability but normal stability (Mohandas and Chasis 1993). If such cells exhibit also a quantitative lack of spectrin, a vertical will add to the horizontal instability, the cells will lose membrane material and adopt irregular but more spherical shapes. *In vitro* they will fragment, upon heating, at lower temperatures than normal cells, which has given the disorder its name "Pyropoikilocytosis".

Elliptic, or oval shapes also characterize red cells in a hereditary disorder widespread in the Pacific region, termed "Southeast Asian ovalocytosis" (SAO). Membranes are extremely rigid in these cells and thus resistant to shape-changing procedures in spite of no detectable abnormality of skeletal proteins. The molecular basis of this disorder is a deletion of 9 amino acids (residues 400–408) in band 3. These residues are located at the boundary between the cytoplasmic and the intramembrane domain of the protein. In addition to several other functional consequences, this defect is thought to restrict the angular mobility of the cytoplasmic domain and to affect its attachment to the skeletal network. Why the cells are elliptic under these conditions is as uncertain in this case as in the other cases of elliptocytosis.

2.8 Outlook

Hopefully, the preceding pages have shown that the joint efforts of many scientific disciplines over the past three decades have widened our understanding of the molecular basis of red cell shape and its experimental and pathological alterations. Cellular biochemistry and membrane biomechanics (to be covered in Chap. 3), complementing each other, have provided rather consistent concepts for the contributions of the lipid bilayer domain. In somewhat disappointing contrast, the role of membrane proteins is less well characterised and the subject of controversial models. While there is general agreement that peripheral and integral proteins influence red cell shape, it still remains to be established, how and to what extent they exert their influence. Probable manifestations of this influence comprise the dependence of shape on the metabolic state of the cell and its ionic conditions, but particularly on possible genetic disorders going along with pathological shape changes of the erythrocyte. We may expect that the recent and future achievements of molecular genetics, but also more refined techniques of biophysical and biomechanical investigation will close parts of this gap in the near future, and help us to gain better insight into the complex interplay between lipids and proteins underlying the shape of the red blood cell.

Ironically, the oldest known experimental shape transformation, the "glass effect" may, nevertheless, remain as far from a causal understanding as a century ago!

Acknowledgement: The author is indebted to Drs. C.W.M. Haest and T.M. Fischer for helpful suggestions and inspiring discussions.

References

Agre P, Parker JC, (eds) (1989) Red blood cell membranes. Marcel Dekker, New York

Allan D, Thomas P (1981) Ca^{2+}-induced biochemical changes in human erythrocytes and their relation to microvesiculation. Biochem J 198:433-440

Anderson RA, Lovrien RE (1981) Erythrocyte membrane sidedness in lectin control of the Ca^{2+}-A23187- mediated diskocyte echinocyte conversion. Nature 292:158-161
Artmann GM, Sung KL, Horn T, Whittemore D, Norwich G, Chien S (1997) Micropipette aspiration of human erythrocytes induces echinocytes via membrane phospholipid translocation. Biophys J 72:1434-1441
Backman L (1986) Shape control in the human red blood cell. J Cell Sci 80:281-298
Bassé F, Stout JG, Sims PJ, Wiedmer T (1996) Isolation of an erythrocyte membrane protein that mediates Ca^{2+}-dependent transbilayer movement of phospholipid. J Biol Chem 271:17205-17210
Baumann M (2001) Early stage shape change of human erythrocytes after application of electric field pulses. Mol Membr Biol 18:153-160
Bennett V (1990) Spectrin-based membrane skeleton: A multipotential adaptor between plasma membrane and cytoplasm. Physiol Rev 70:1029-1065
Bennett V, Baines AJ (2001) Spectrin and ankyrin-based pathways: metazoan inventions for integrating cells into tissues. Physiol Rev 81:1353-1392
Bessis M (1972) Red cell shapes. An illustrated classification and its rationale. Nouv Rev fr Hémat 12:721-746
Bessis M (1973) Living blood cells and their ultrastructure. Springer-Verlag, Berlin
Bessis M (1974) Corpuscles. Atlas of Red Blood Cell Shapes. Springer-Verlag, Berlin
Bessis M (1977) La forme et la déformabilité des érythrocytes normaux et dans certaines anémies hémolytiques congénitales. Nouv Rev fr Hémat 18:75-94
Bifano EM, Novak TS, Freedman JC (1984) Relationship between the shape and the membrane potential of human red blood cells. J Membrane Biol 82:1-13
Birchmeier W, Singer SJ (1977) On the mechanism of ATP-induced shape changes in human erythrocyte membranes. II. The role of ATP. J Cell Biol 73:647-659
Birchmeier W, Lanz JH, Winterhalter KH, Conrad MJ (1979) ATP-induced endocytosis in human erythrocyte ghosts. Characterization of the process and isolation of the endocytosed vesicles. J Biol Chem 254:9298-9304
Blank ME, Hoefner DM, Diedrich DF (1994) Morphology and volume alterations of human erythrocytes caused by the anion transporter inhibitors, DIDS and p-azidobenzylphlorizin. Biochim Biophys Acta 1192:223-233
Boal DH (1994) Computer simulation of a model network for the erythrocyte cytoskeleton. Biophys J 67:521-529
Bobrowska-Hägerstrand M, Iglic A, Hägerstrand H (1997) Erythrocytes vesiculate at high pH. Cellular and Molecular Biology Letters 2:9-13
Bobrowska-Hägerstrand M, Hägerstrand H, Iglic A (1998) Membrane skeleton and red blood cell vesiculation at low pH. Biochim Biophys Acta 1371:123-128
Broekhuyse RM (1974) Improved lipid extraction of erythrocytes. Clin Chim Acta 51:341-343
Bucki R, Bachelot-Loza C, Zachowski A, Giraud F, Sulpice JC (1998) Calcium induces phospholipid redistribution and microvesicle release in human erythrocyte membranes by independent pathways. Biochemistry 37:15383-15391
Bull BS, Brailsford D (1989) Red blood cell shape. In: Agre P, Parker JC (eds) Red blood cell membranes. Marcel Dekker, New York, pp 401-421
Bull BS, Weinstein RS, Korpman RA (1986) On the thickness of the red cell membrane skeleton: Quantitative electron microscopy of maximally narrowed isthmus regions of intact cells. Blood Cells 12:25-42

Canfield VA, Macey RI (1984) Anion exchange in human erythrocytes has a large activation volume. Biochim Biophys Acta 778:379-384

Carter DP, Fairbanks G (1984) Inhibition of erythrocyte membrane shape change by band 3 cytoplasmic fragment. J Cell Biochem 24:385-393

Cevc G, Marsh D (1987) Phospholipid bilayers: Physical principles and models. John Wiley & Sons, USA

Chabanel A, Flamm M, Sung KL, Lee MM, Schachter D, Chien S (1983) Influence of cholesterol content on red cell membrane viscoelasticity and fluidity. Biophys J 44:171-176

Chang SH, Low PS (2001) Regulation of the glycophorin C-protein 4.1 membrane-to-skeleton bridge and evaluation of its contribution to erythrocyte membrane stability. J Biol Chem 276:22223-22230

Chasis JA, Schrier SL (1989) Membrane deformability and the capacity for shape change in the erythrocyte. Blood 74:2562-2568

Chasis JA, Mohandas N, Shohet SB (1985) Erythrocyte membrane rigidity induced by glycophorin A-ligand interaction. Evidence for a ligand-induced association between glycophorin A and skeletal proteins. J Clin Invest 75:1919-1926

Connor J, Gillum K, Schroit AJ (1990) Maintenance of lipid asymmetry in red blood cells and ghosts: effect of divalent cations and serum albumin on the transbilayer distribution of phosphatidylserine. Biochim Biophys Acta 1025:82-86

Daleke DL, Huestis WH (1989) Erythrocyte morphology reflects the transbilayer distribution of incorporated phospholipids. J Cell Biol 108:1375-1385

Deuticke B (1968) Transformation and restoration of biconcave shape of human erythrocytes induced by amphiphilic agents and changes of ionic environment. Biochim Biophys Acta 163:494-500

Deuticke B (1977) Properties and structural basis of simple diffusion pathways in the erythrocyte membrane. Rev Physiol Biochem Pharmacol 78:1-97

Deuticke B, Schwister K (1989) Leaks induced by electrical breakdown in the erythrocyte membrane. In: Neumann E, Sowers A, Jordan CA (eds) Electroporation and Electrofusion in Cell Biology. Plenum Press, New York, pp 127-148

Deuticke B, Grebe R, Haest CWM (1990) Action of drugs on the erythrocyte membrane. In: Harris JR (ed) Blood Cell Biochemistry, vol 1: Erythroid cells. Plenum Press, New York, pp 475-529

Donath E, Voigt A (1986) Electrophoretic mobility of human erythrocytes. On the applicability of the charged layer model. Biophys J 49:493-499

Dumaswala VJ, Greenwalt TJ (1984) Human erythrocytes shed exocytic vesicles *in vivo*. Transfusion 24:490-492

Eidmann K (1997) Membranpotentialabhängigkeit der Translokation eines anionischen Lysophospholipids in der Erythrozytenmembran. MD Thesis RWTH Aachen, Shaker-Verlag

Elgsaeter A, Mikkelsen A (1991) Shapes and shape changes *in vitro* in normal red blood cells. Biochim Biophys Acta 1071:273-290

Elgsaeter A, Stokke BT, Mikkelsen A, Branton D (1986) The molecular basis of erythrocyte shape. Science 234:1217-1223

Eriksson LEG (1990) On the shape of human red blood cells interacting with flat artificial surfaces - the 'glass effect'. Biochim Biophys Acta 1036:193-201

Evans EA, Skalak R (1979) Mechanics and thermodynamics of biomembranes. CRC Crit Rev Bioeng 3:181-184

Fairbanks G, Steck TL, Wallach DF (1971) Electrophoretic analysis of the major polypeptides of the human erythrocyte membrane. Biochemistry 10:2606-2617

Feo C, Mohandas N (1977) Clarification of role of ATP in red-cell morphology and function. Nature 265:166-168

Ferrell JE Jr, Huestis WH (1984) Phosphoinositide metabolism and the morphology of human erythrocytes. J Cell Biol 98:1992-1998

Ferrell JE Jr, Lee KJ, Huestis WH (1985) Membrane bilayer balance and erythrocyte shape: a quantitative assessment. Biochemistry 24:2849-2857

Fischer TM, Haest CWM, Stöhr M, Kamp D, Deuticke B (1978) Selective alteration of red cell deformability by SH-reagents. Evidence for an involvement of spectrin in membrane shear elasticity. Biochim Biophys Acta 510:270-282

Fuhrmann GF (1968) Kationentransport, Haemolyse und Fragmentieren von menschlichen Erythrozyten in Harnstofflösungen. Blut 16:321-327

Gedde MM, Huestis WH (1997) Membrane potential and human erythrocyte shape. Biophys J 72:1220-1233

Gedde MM, Davis DK, Huestis WH (1997) Cytoplasmic pH and human erythrocyte shape. Biophys J 72:1234-1246

Gedde MM, Yang E, Huestis WH (1999) Resolution of the paradox of red cell shape changes in low and high pH. Biochim Biophys Acta 1417:246-253

Gimsa J (1998) A possible molecular mechanism governing human erythrocyte shape. Biophys J 75:568-569

Gimsa J, Ried C (1995) Do band 3 protein conformational changes mediate shape changes of human erythrocytes? Mol Membr Biol 12:247-254

Glaser R (1998) Does the transmembrane potential ($\Delta\Psi$) or the intracellular pH (pHi) control the shape of human erythrocytes? Biophys J 75:569-570

Grebe R, Schmid-Schönbein H (1985) Tangent counting for objective assessment of erythrocyte shape changes. Biorheology 22:455-469

Grebe R, Wolff H, Schmid-Schönbein H (1988) Influence of red cell surface charge on red cell membrane curvature. Pflügers Arch 413:77-83

Gros M, Vrhovec S, Brumen M, Svetina S, Zeks B (1996) Low pH induced shape changes and vesiculation of human erythrocytes. Gen Physiol Biophys 15:145-163

Grunze M, Forst B, Deuticke B (1980) Dual effect of membrane cholesterol on simple and mediated transport processes in human erythrocytes. Biochim Biophys Acta 600:860-869

Haest CWM (1982) Interactions between membrane skeleton proteins and the intrinsic domain of the erythrocyte membrane. Biochim Biophys Acta 694:331-352

Haest CWM, Kamp D, Deuticke B (1997) Transbilayer reorientation of phospholipid probes in the human erythrocyte membrane. Lessons from studies on electroporated and resealed cells. Biochim Biophys Acta 1325:17-33

Haest CWM, Fischer TM, Plasa G, Deuticke B (1980a) Stabilization of erythrocyte shape by a chemical increase of membrane shear stiffness. Blood Cells 6:539-553

Haest CWM, Plasa G, Kamp D, Deuticke B (1980b) Protein-lipid interaction in the erythrocyte membrane: relevance for structural properties. In: Lassen U, Ussing HH, Wieth JO (eds) Membrane transport in erythrocytes. Alfred Benzon Symposium 14, Munksgaard, Kopenhagen, pp 108-119

Hardy B, Schrier SL (1978) The role of spectrin in erythrocyte ghost endocytosis. Biochem Biophys Res Commun 81:1153-1161

Hägerstrand H, Isomaa B (1992) Morphological characterization of exovesicles and endovesicles released from human erythrocytes following treatment with amphiphiles. Biochim Biophys Acta 1109:117-126

Heinrich R, Gaestel M, Glaser R (1981) The electric potential across the erythrocyte membrane: A mathematical model. Acta biol med germ 40:765-770

Heinrich V, Ritchie K, Mohandas N, Evans E (2001) Elastic thickness compressibility of the red cell membrane. Biophys J 81:1452-1463

Henseleit U, Plasa G, Haest CWM (1990) Effects of divalent cations on lipid flip-flop in the human erythrocyte membrane. Biochim Biophys Acta 1029:127-135

Henszen MMM, Weske M, Schwarz St, Haest CWM, Deuticke B (1997) Electric field pulses induce reversible shape transformation of human erythrocytes. Mol Membr Biol 14:195-204

Isomaa B, Hägerstrand H, Paatero G (1987) Shape transformations induced by amphiphiles in erythrocytes. Biochim Biophys Acta 899:93-103

Jinbu Y, Nakao M, Otsuka M, Sato S (1983) Two steps in ATP-dependent shape changes of human erythrocyte ghosts. Biochem Biophys Res Commun 112:384-390

Jinbu Y, Sato S, Nakao M (1984) Reversible shape change of Triton-treated erythrocyte ghosts induced by Ca^{2+} and Mg-ATP. Nature 307:376-378

Johnson RM, Robinson J (1976) Morphological changes in asymmetric erythrocyte membranes induced by electrolytes. Biochem Biophys Res Commun 70:925-931

Johnson RM, Taylor G, Meyer DB (1980) Shape and volume changes in erythrocyte ghosts and spectrin-actin networks. J Cell Biol 86:371-376

Kamp D, Sieberg T, Haest CWM (2001) Inhibition and stimulation of phospholipid scrambling activity. Consequences for lipid asymmetry, echinocytosis, and microvesiculation of erythrocytes. Biochemistry 40:9438-9446

Khodadad JK, Waugh RE, Podolski JL, Josephs R, Steck TL (1996) Remodeling the shape of the skeleton in the intact red cell. Biophys J 70:1036-1044

King MJ (1994) Blood group antigens on human erythrocytes - Distribution, structure and possible functions. Biochim Biophys Acta 1197:15-44

Kleinzeller A (1996) William Hewson's studies of red blood corpuscles and the evolving concept of a cell membrane. Am J Physiol 271:C1-C8

Knowles DW, Tilley L, Mohandas N, Chasis JA (1997) Erythrocyte membrane vesiculation: model for the molecular mechanism of protein sorting. Proc Natl Acad Sci USA 94:12969-12974

Kuypers FA, Roelofsen B, Berendsen W, Op den Kamp JA, Van Deenen LL (1984) Shape changes in human erythrocytes induced by replacement of the native phosphatidylcholine with species containing various fatty acids. J Cell Biol 99:2260-2267

Lange Y, Slayton JM (1982) Interaction of cholesterol and lysophosphatidylcholine in determining red cell shape. J Lipid Res 23:1121-1127

Lelkes G, Fodor I (1991) Formation of large, membrane skeleton-free erythrocyte vesicles as a function of the intracellular pH and temperature. Biochim Biophys Acta 1065:135-144

Lenormand G, Henon S, Richert A, Simeon J, Gallet F (2001) Direct measurement of the area expansion and shear moduli of the human red blood cell membrane skeleton. Biophys J 81:43-56

Leonards KS, Ohki S (1983) Isolation and characterization of large (0.5-1.0 micron) cytoskeleton-free vesicles from human and rabbit erythrocytes. Biochim Biophys Acta 728:383-393

Lew VL, Bookchin RM (1986) Volume, pH, and ion-content regulation in human red cells: analysis of transient behavior with an integrated model. J Membrane Biol 92:57-74

Lin S, Huestis WH (1995) Wheat germ agglutinin stabilization of erythrocyte shape: role of bilayer balance and the membrane skeleton. Biochim Biophys Acta 1233:47-56

Lin S, Yang E, Huestis WH (1994) Relationship of phospholipid distribution to shape change in Ca^{2+}-crenated and recovered human erythrocytes. Biochemistry 33:7337-7344

Linss W, Pilgrim C, Feuerstein H (1991) How thick is the glycocalyx of human erythrocytes? Acta Histochem 91:101-104

Lovrien RE, Anderson RA (1980) Stoichiometry of wheat germ agglutinin as a morphology controlling agent and as a morphology protective agent for the human erythrocyte. J Cell Biol 85:534-548

Low PS, Willardson BM, Mohandas N, Rossi M, Shohet S (1991) Contribution of the band-3-ankyrin interaction to erythrocyte membrane mechanical stability. Blood 77:1581-1586

Lux SE, Palek J (1995) Disorders of the red cell membrane. In: Handin RI, Lux SE, Stossel TP (eds) Blood. Principles and practice of hematology. J B Lippincott Company, Philadelphia, pp 1701-1818

Macey RI, Adorante JS, Orme FW (1978) Erythrocyte membrane potentials determined by hydrogen ion distribution. Biochim Biophys Acta 512:284-295

Matayoshi ED (1980) Distribution of shape-changing compounds across the red cell membrane. Biochemistry 19:3414-3422

McGough AM, Josephs R (1990) On the structure of erythrocyte spectrin in partially expanded membrane skeletons. Proc Natl Acad Sci USA 87:5208-5212

Mehta NG (1983) Role of membrane integral proteins in the modulation of red cell shape by albumin, dinitrophenol and the glass effect. Biochim Biophys Acta 762:9-18

Mentzer WC, Lubin BH, Emmons S (1976) Correction of the permeability defect in hereditary stomatocytosis by dimethyladipimidate. N Engl J Med 294:1200-1204

Mohandas N, Chasis JA (1993) Red blood cell deformability, membrane material properties and shape: regulation by transmembrane, skeletal and cytosolic proteins and lipids. Semin Hematol 30:171-192

Mohandas N, Feo C (1975) A quantitative study of the red cell shape changes produced by anionic and cationic derivatives of phenothiazine. Blood Cells 11:375-384

Mohandas N, Greenquist AC, Shohet SB (1978) Bilayer balance and regulation of red cell shape changes. J Supramol Struct 9:453-458

Murdock RC, Reynolds C, Sarelius IH, Waugh RE (2000) Adaptation and survival of surface-deprived red blood cells in mice. Am J Physiol 279:C970-C980

Nagle JF, Tristram-Nagle S (2000) Structure of lipid bilayers. Biochim Biophys Acta 1469:159-195

Nakao M (1990) Function and structure of red blood cell cytoskeleton. In: Harris JR (ed) Blood cell biochemistry. vol 1: Erythroid cells. Plenum Press, New York, pp 195-225

Nakao M, Makao T, Yamazoe S (1960) Adenosine triphosphate and maintenance of shape of the human red cells. Nature 187:945-946

O'Toole PJ, Morrison IE, Cherry RJ (2000) Investigations of spectrin-lipid interactions using fluoresceinphosphatidylethanolamine as a membrane probe. Biochim Biophys Acta 1466:39-46

Patel VP, Fairbanks G (1981) Spectrin phosphorylation and shape change of human erythrocyte ghosts. J Cell Biol 88:430-440

Patel VP, Fairbanks G (1986) Relationship of major phosphorylation reactions and Mg-ATPase activities to ATP-dependent shape change of human erythrocyte membranes. J Biol Chem 261:3170-3177

Pestonjamasp KN, Mehta NG (1995) Effect of antibodies to membrane skeletal proteins on the shape of erythrocytes and their ability to respond to shape-modulating agents. Important role of 4.1 protein in the determination/maintenance of the discoid shape of erythrocytes. Exp Cell Res 219:74-81

Peters LL, Shivdasani RA, Liu SC, Hanspal M, John KM, Gonzalez JM, Brugnara C, Gwynn B, Mohandas N, Alper SL, Orkin SH, Lux SE (1996) Anion exchanger 1 (band 3) is required to prevent erythrocyte membrane surface loss but not to form the membrane skeleton. Cell 86:917-927

Picart C, Discher DE (1999) Actin protofilament orientation at the erythrocyte membrane. Biophys J 77:865-878

Platt OS (1995) The sickle syndromes. In: Handin RI, Lux SE, Stossel TP (eds) Blood. Principles and practice of hematology. J B Lippincott Company, Philadelphia, pp 1645-1700

Ponder E (1948) Hemolysis and related phenomena. Grune & Stratton, New York

Poser B, Deuticke B (1999) Transbilayer flip-flop and steady state distribution of dodecylsulfate (SDS) in the erythrocyte membrane: flip *vs.* flux. Biol Chem 380:S56-S56

Rand RP, Burton AC, Canham P (1965) Reversible changes in shape of red cells in electrical fields. Nature 205:977-978

Rasia M, Bollini A (1998) Red blood cell shape as a function of medium's ionic strength and pH. Biochim Biophys Acta 1372:198-204

Raval PJ, Carter DP, Fairbanks G (1989) Relationship of hemolysis buffer structure, pH and ionic strength to spontaneous contour smoothing of isolated erythrocyte membranes. Biochim Biophys Acta 983:230-240

Schmid-Schönbein H, Grebe R, Heidtmann H (1983) A new membrane concept for viscous RBC deformation in shear: spectrin oligomer complexes as a Bingham-fluid in shear and a dense periodic colloidal system in bending. Ann New York Acad Sci 416:225-254

Schrier SL, Zachowski A, Herve P, Kader JC, Devaux PF (1992) Transmembrane redistribution of phospholipids of the human red cell membrane during hypotonic hemolysis. Biochim Biophys Acta 1105:170-176

Schwarz S, Deuticke B, Haest CWM (1999a) Passive transmembrane redistributions of phospholipids as a determinant of erythrocyte shape change. Studies on electroporated cells. Mol Membr Biol 16:247-255

Schwarz S, Haest CWM, Deuticke B (1999b) Extensive electroporation abolishes experimentally induced shape transformations of erythrocytes: a consequence of phospholipid symmetrization? Biochim Biophys Acta 1421:361-379

Schwarz S (2000) Zum Einfluss von elektrischen Feldpulsen auf Dynamik und transversale Orientierung der Phospholipide in der Erythrocytenmembran. MD thesis RWTH Aachen

Seeman P (1972) The membrane actions of anesthetics and tranquilizers. Pharmacol Rev 24:583-655

Sheetz MP (1983) Membrane skeletal dynamics: role in modulation of red cell deformability, mobility of transmembrane proteins, and shape. Semin Hematol 20:175-188

Sheetz MP, Sawyer D (1978) Triton shells of intact erythrocytes. J Supramol Struct 8:399-412

Sheetz MP, Singer SJ (1974) Biological membranes as bilayer couples. A molecular mechanism of drug-erythrocyte interactions. Proc Natl Acad Sci USA 71:4457-4461

Sheetz MP, Singer SJ (1976) Equilibrium and kinetic effects of drugs on the shapes of human erythrocytes. J Cell Biol 70:247-251

Shen BW (1989) Ultrastructure and function of membrane skeleton. In: Agre P, Parker JC (eds) Red blood cell membranes. Marcel Dekker, New York, pp 261-297

Simons TJ (1979) Vanadate - a new tool for biologists. Nature 281:337-338

Smith SK, Farnbach AR, Harrist FM, Hawes AC, Jackson LR, Judd AM, Vest RS, Sanchez S, Bell JD (2001) Mechanisms by which intracellular calcium induces susceptibility to secretory phospholipase A2 in human erythrocytes. J Biol Chem 276:22732-22741

Steck TL (1989) Red cell shape. In: Stein W, Brouner F (eds) Cell shape: Determinants, regulation and regulatory role. Academic Press, New York, pp 205-246

Stewart GW, Argent AC, Dash BC (1993) Stomatin: a putative cation transport regulator in the red cell membrane. Biochim Biophys Acta 1225:15-25

Stokke BT, Mikkelsen A, Elgsaeter A (1986) Spectrin, human erythrocyte shapes, and mechanochemical properties. Biophys J 49:319-327

Takeuchi M, Miyamoto H, Sako Y, Komizu H, Kusumi A (1998) Structure of the erythrocyte membrane skeleton as observed by atomic force microscopy. Biophys J 74:2171-2183

Terada N, Fujii Y, Ohno S (1996) Three-dimensional ultrastructure of *in situ* membrane skeletons in human erythrocytes by quick-freezing and deep-etching method. Histol Histopathol 11:787-800

Thelen B, Deuticke B (1988) Chemo-mechanical leak formation in human erythrocytes upon exposure to a water-soluble carbodiimide followed by very mild shear stress. I. Basic characteristics of the process. Biochim Biophys Acta 944:285-296

Tomishige M, Sako Y, Kusumi A (1998) Regulation mechanism of the lateral diffusion of band 3 in erythrocyte membranes by the membrane skeleton. J Cell Biol 142:989-1000

Trotter WD (1956) The slide-coverslip disc-sphere transformation in mammalian erythrocytes. Brit J Haemat 2:65-74

Truong HTN, Daleke DL, Huestis WH (1993) Human erythrocyte shape regulation: interaction of metabolic and redox status. Biochim Biophys Acta 1150:51-56

Tse WT, Lux SE (1999) Red blood cell membrane disorders. Brit J Haematol 104:2-13

Tuvia S, Levin S, Bitler A, Korenstein R (1998) Mechanical fluctuations of the membrane-skeleton are dependent on F-actin ATPase in human erythrocytes. J Cell Biol 141:1551-1561

Verkleij AJ, Zwaal RF, Roelofsen B, Comfurius P, Kastelijn D, Van Deenen LL (1973) The asymmetric distribution of phospholipids in the human red cell membrane. A combined study using phospholipases and freeze-etch electron microscopy. Biochim Biophys Acta 323:178-193

Viitala J, Järnefelt J (1985) The red cell surface revisited. Trends Biochem Sci 10:392-395

Wagner GM, Chiu DT, Yee MC, Lubin BH (1986) Red cell vesiculation – a common membrane physiologic event. J Lab Clin Med 108:315-324

Waugh RE (1982) Temperature dependence of the yield shear resultant and the plastic viscosity coefficient of erythrocyte membrane. Implications about molecular events during membrane failure. Biophys J 39:273-278

Waugh RE (1996) Elastic energy of curvature-driven bump formation on red blood cell membrane. Biophys J 70:1027-1035

Waugh RE, Narla N, Jackson CN, Mueller TJ, Suzuki T, Dale GL (1992) Rheologic properties of senescent erythrocytes. Loss of surface area and volume with red blood cell age. Blood 79:1351-1358

White JG (1974) Effects of an ionophore, A 23187, on the surface morphology of normal erythrocytes. Am J Pathol 77:507-518

Williamson P, Kulick A, Zachowski A, Schlegel RA, Devaux PF (1992) Ca^{2+} induces transbilayer redistribution of all major phospholipids in human erythrocytes. Biochemistry 31:6355-6360

Wong P (1994) Mechanism of control of erythrocyte shape: a possible relationship to band 3. J Theor Biol 171:197-205

Yan Y, Winograd E, Viel A, Cronin T, Harrison SC, Branton D (1993) Crystal structure of the repetitive segments of spectrin. Science 262:2027-2030

Zhang D, Kiyatkin A, Bolin JT, Low PS (2000) Crystallographic structure and functional interpretation of the cytoplasmic domain of erythrocyte membrane band 3. Blood 96:2925-2933

Zimmermann U (1986) Electrical breakdown, electropermeabilization and electrofusion. Rev Physiol Biochem Pharmacol 105:175-256

3 Human Red Cell Shape and the Mechanical Characteristics of the Membrane

Thomas M. Fischer

Institut für Physiologie, Universitätsklinikum der RWTH, Pauwelsstrasse 30, 52057 Aachen, Germany

3.1 Introduction

The main task of the red cell is to transport oxygen and CO_2. It fulfils this task as a passive transport vehicle in the circulating blood. When the forces exerted by the suspending fluid on the red cell membrane are large the red cell shows liquid behaviour; i.e. it behaves similarly to a liquid drop in that its interior participates in flow (Schmid-Schönbein and Wells 1969; Fischer et al. 1978; Fischer 1978). This behaviour rests on the fact that the cytoplasm of the red cell is liquid and contains no cell skeleton. When the forces are small the red cell behaves as a solid body. It may rotate in flow but its shape is the same as in the absence of external forces. The reason for this solid behaviour is the shear stiffness of the membrane (Sect. 3.5.3) and the fact that its relaxed shape (Sect. 3.6.3) is not spherical.

In this chapter we consider the shape of the red cell in the absence of external forces, e.g. shear flow or pipet suction. Such shapes are called resting shapes. In particular we consider changes of the resting shape, which are induced *in vitro*. The transients of shape are not considered.

The human red cell in blood plasma has the shape of a biconcave disk. This shape can be converted *in vitro* to so-called echinocytes or stomatocytes by a plethora of exogenous manipulations (Chap. 2). It is intriguing that the sequence of shapes observed upon increasing the intensity of a manipulation is with one exception (Sect. 3.9.2) practically independent of the type of manipulation. This observation has two consequences. First, the sequence of shapes can be quantified by a single variable. A score (I to V) for echinocytes as well as for stomatocytes was introduced by Bessis (1972); see also Fig. 2.1 (Chap. 2). Second, a description of the origin of these shape changes should be based on the change of a single parameter. Such a description will be given in this chapter.

The liquid cytoplasm does not influence the resting shape. Accordingly the shape of the red cell will be described via the mechanical properties of its membrane. For a more general readership, physical principles are presented without quantitative detail. For a formula-based description the reader is referred to the literature (Evans and Skalak 1979; Fischer 1992a,b, 1993; Mohandas and Evans 1994; Waugh and Hochmuth 1995).

Finally, this chapter does not claim to give complete reference to the literature relevant to this subject.

3.2 Continuum Mechanics *vs.* Membrane Composition

A continuum mechanical approach has been adopted in the present work. More specifically, the red cell membrane will be described as a continuum in two dimensions. To this end we subdivide the surface of the membrane conceptually into small patches, elements, which contain just sufficient molecules so that variations due to motions on a molecular scale are averaged out. We describe the behaviour of the molecules within such patches by only a few parameters to reduce the complexity.

In the direction perpendicular to the membrane surface, a continuum mechanical description is not possible because the membrane is only a few molecules thick. These molecules are classified into three distinct layers:

(a) the outer lipid (phospholipid and cholesterol) monolayer with its integral proteins and the attached glycocalyx,
(b) the inner lipid monolayer with its integral proteins,
(c) the peripheral proteins on the cytoplasmic side of the membrane, in particular the spectrin-actin network. This network is also known as the membrane skeleton (Gilligan and Bennet 1993).

This definition of layers differs to some extent from what is used in a more biochemical approach to the red cell shape (Chap. 2). For the present mechanical approach the consideration of three layers is appropriate. These layers are tightly coupled in the transmembrane axis, but may slide relative to each other in the plane of the membrane. In the following (a) and (b) are referred to as the outer and inner monolayer, respectively, (c) is referred to as the skeleton, (a + b) are referred to as the bilayer, and (a + b + c) are referred to as the membrane.

In principle the mechanical properties of these layers could be described without reference to their chemical composition. To ease understanding, reference is made to their structure and molecular constituents. The chemical composition of the layers and the properties of their constituents are, however, not described in detail. These are treated in Chap. 2 and in other reviews, e.g. Haest (1982) or Bennet (1993).

Most shape changes of red cells can be explained by a membrane with uniform mechanical properties, i.e. these properties are the same everywhere on the membrane surface (Sects. 3.6, 3.7, and 3.8). Cases in which this description is not sufficient are briefly discussed in Sect. 3.9.

3.3 Modes of Membrane Deformation

3.3.1 General Remarks

Let us consider a change of the red cell from some shape A (e.g. echinocyte II) to some shape B (e.g. echinocyte III). Such a change in shape of the whole cell is accompanied by a change in shape of most if not all surface elements of the red cell membrane. As an example of such a surface element we choose a patch of membrane, which is circular in shape A and has a diameter of about a quarter of a micron. The change in shape of this patch when the red cell goes from shape A to shape B will be called a *membrane deformation*. Any deformation can be considered as being composed of three kinds of "pure" deformations. These pure deformations are considered in the next sections. They apply to the membrane as well as to just one or two of its constituent layers.

3.3.2 Isotropic or Biaxial Deformation

If the patch diameter is changed but it remains circular, the deformation is *isotropic*. What alters is the area of the patch. Such changes only occur following a change in the pressure difference between cytoplasm and suspending medium. In the shape changes considered in this chapter this pressure difference remains constant at essentially zero. Therefore the surface area of the membrane does not change either. However, a change in the surface area of the individual layers may occur due to bending of the membrane (Sect. 3.5.4).

3.3.3 Shear or Uniaxial Deformation

If the area of the patch remains constant but the aspect ratio changes, i.e. the circle turns into an ellipse, it is called a shear deformation or briefly *shear*. Concomitant to a shape change of the red cell most, if not all, membrane patches are sheared.

3.3.4 Bending Deformation

Before considering bending, it is necessary to define membrane curvature. In a sphere the curvature is defined as the reciprocal of the radius (1/r). The curvature has the same value everywhere on the surface of the sphere and in all directions when the location on the sphere is kept constant.

For the red cell, the curvature at most locations on the membrane depends on direction. For example, at the rim of the biconcave disk the curvature in the meridional direction is much larger than along a latitude circle.

In further analyses we define a *mean curvature* of a patch where the mean of curvatures is taken over all directions. We will refer to this mean curvature simply by *curvature*. In a biconcave red cell, the curvature varies from a positive value at the rim to a negative value in the dimple. In between the curvature passes through zero.

A change in curvature is called a bending deformation or in short *bending*. In the shape changes considered in this chapter, large bending deformations occur. The minimal radii of curvature that are observed in red cell shape changes are below light microscopic resolution.

3.4 Resting Shape of Membrane Patches

3.4.1 General Remarks

The concept of the resting shape of membrane patches can be explained by a thin rod, which in the absence of external forces is straight. Imagine this rod bent into a circle, with the ends joined. If a small piece of the rod is excised and let free, the piece will relax to the original straight configuration. This configuration is defined as the resting shape of the piece. When the piece is incorporated in the circle, bending moments at both ends are responsible for the fact that the piece is bent.

An experimental determination of the resting shape of membrane patches would require the excision of a patch and observation of the shape to which it relaxes. This is, however, not possible because cutting out a piece of membrane would produce tremendous edge effects which would dominate the shape of the patch. So, presently, the changes in shape a membrane patch would undergo in the absence of edge effects after releasing the forces or bending moments at its borders will be considered. The three types of deformations as well as the three layers constituting the membrane are discussed separately.

3.4.2 Surface Area

Since there is no pressure difference across the red cell membrane, the area of the patches would not change upon excision. If the monolayers were allowed to assume their respective resting area separately, the two areas would change inversely. Either the outer monolayer would expand and the inner shrink by about the same amount, or the outer would shrink and the inner would expand. Of course, no change of area is also possible. The resting area of each monolayer is determined mainly by the sum of the cross sectional areas of all molecules constituting the monolayer.

Whether the skeleton would shrink or expand upon excision of the patch is not known. There is some evidence that it would expand (Discher et al. 1998). Due to its coupling to the bilayer and its much lower isotropic stiffness (Sect. 3.5.2) the skeleton is compliant with the surface area of the bilayer.

3.4.3 Aspect Ratio

Upon excision, a patch would change its aspect ratio and relax from circular to elliptic. Such a relaxation requires that the membrane behaves as a two dimensional solid. This solid behaviour is caused by the skeleton. In a patch that has assumed its resting shape the texture of the skeleton is isotropic, i.e. the spectrin tetramers are equally stretched in all directions. The monolayers behave like two-dimensional liquids (Sect. 3.5.3), complying with the aspect ratio prescribed by the skeleton.

3.4.4 Spontaneous Curvature

If the two monolayers are kept flush at the border of the patch after excision from the closed membrane, the patch would relax to the so-called *spontaneous curvature* of the membrane. To distinguish the spontaneous curvature from the curvature before excision the latter is sometimes referred to as the *actual curvature*.

The spontaneous curvature of the membrane can be subdivided conceptually into two contributions as detailed in the following. If we permitted a separation of the monolayers, each monolayer would assume its own spontaneous curvature, the value of which would depend on the packing behaviour of the lipids and integral proteins constituting the respective monolayer. In a graphical view each monolayer can be considered to be composed of closely fitting cylindrical or truncated conical bodies (Israelachvilli et al. 1980). In isolation these bodies are usually drawn rotationally symmetric although such bodies would not fill the space. The nomenclature is such, that in "truncated cones" the cross sectional area at the headgroup region is larger than at the hydrocarbon end, whereas in "inverted truncated cones" the opposite is the case (see Appendix, Fig. 3.2). If the outer monolayer is constituted of truncated cones its spontaneous curvature will be positive. If the inner monolayer is constituted of truncated cones its spontaneous curvature will be negative.

If the two monolayers are put together without keeping them flush at the border of the patch the first of the above mentioned two contributions to the spontaneous curvature of the membrane would be seen. This contribution is called the *single-layer based spontaneous curvature* of the bilayer. Mathematically its value is obtained by averaging the spontaneous curvatures of the two monolayers. For a symmetrical bilayer the single-layer based spontaneous curvature of the bilayer is zero. If the monolayers are different in composition this value is in general non zero and the values of the bending stiffness (Sect. 3.5.4) of the two monolayers serve as weight factors in calculating the average.

If the two monolayers were kept flush and their spontaneous curvatures are neglected, the bilayer would assume the second contribution to the spontaneous curvature of the membrane. This contribution is called the *double-layer based spontaneous curvature*. It is a function of the resting areas of the two monolayers (Sect. 3.4.2) and the distance of the neutral planes of the two monolayers. If the resting area of the outer monolayer is larger than that of the inner one, the double-layer based spontaneous curvature is positive.

The contribution of the skeleton to the bending stiffness of the membrane can be neglected (Sect. 3.5.4). For this reason the skeleton does not exert a direct influence on the single-layer or on the double-layer based spontaneous curvature of the membrane. It could, however, exert an indirect influence by an interaction with the molecules of the inner monolayer. This influence could either change the resting area or the spontaneous curvature of the inner monolayer. These are probably the only ways in which a modification of the skeleton could influence the spontaneous curvature of the membrane. Similarly modifications of the glycocalyx may influence indirectly the outer monolayer. In this chapter the glycocalyx and the outer monolayer are lumped together conceptually.

The spontaneous curvature of the bilayer and thus of the membrane is given by the average of its single-layer based and double-layer based spontaneous curvatures. Again the values for the respective bending stiffnesses have to be accounted for.

3.5 Membrane Stiffness

3.5.1 General Remarks

To hold the shape of a patch of membrane or of one or two of its constituent layers in a shape different from the respective resting shape requires either forces or bending moments acting at the border of the patch. The ratio between (i) these forces or bending moments and (ii) the deformation relative to the resting shape is called *stiffness* (synonymous with rigidity or modulus). The stiffer the membrane or layer, the smaller is the deformation for the same force or bending moment. In the following sections we treat the three kinds of deformation introduced in Sect. 3.3 separately and show which of the three layers of the red cell membrane are involved in the resistance against the respective deformation.

3.5.2 Isotropic Stiffness

Isotropic deformation is caused by forces acting in the plane of the membrane. These forces are distributed over the border of the membrane patch. The quantity "force per length element along the border" is called tension and has the dimension N/m. An isotropic deformation results if the tension is isotropic, i.e. if it acts equally in all directions. Its three-dimensional analog is pressure. The respective stiffness is called isotropic modulus. Its value for the red cell membrane is 0.5 N/m (Waugh and Hochmuth 1995). This means that an isotropic tension of 0.005 N/m increases the surface area of the membrane by 1% with respect to the resting area. The bilayer dominates the isotropic stiffness of the membrane.

Two opposing effects are mainly responsible for the isotropic stiffness of the bilayer. An increase in surface area is resisted because hydrophobic moieties of membrane lipids would then become exposed to the aqueous medium surrounding the membrane. A decrease in surface area is resisted by repulsive forces between adjacent molecules. Both monolayers contribute to approximately the same extent.

The isotropic modulus of the skeleton has been shown to be 5 orders of magnitude smaller than the isotropic modulus of the bilayer (Lenormand et al. 2001). Therefore the skeleton does not contribute appreciably to the isotropic stiffness of the membrane.

3.5.3 Shear Stiffness

A shear deformation is also caused by tensions, which are, however, not distributed isotropically along the border of the membrane patch. Rather, they act as to extend the patch in one direction and to compress it at right angles to this direction.

Shear deformations are only resisted by the membrane skeleton. In schematic cartoons of the red cell membrane the skeleton is often depicted as a plane structure comparable to a mesh. Actually, the spectrin tetramers extend into the cytoplasm. This is possible because the end-to-end distance of the spectrin tetramers in the resting shape of a patch is much smaller than their extended length. Mathematical modelling has suggested that the resistance against shear is caused by a restriction of thermal fluctuations of the spectrin tetramers concomitant with a shear deformation (Boal 1994). Experimental measurements of the shear modulus gave values between 2 and 9 μN/m (Waugh and Hochmuth 1995; Lenormand et al. 2001). As will be shown in Sect. 3.7.3 it is an important determinant in red cell shape changes.

Since the lipids are laterally mobile each monolayer and thus the bilayer behave as liquids in two dimensions. Therefore, the bilayer does not contribute to the shear stiffness of the membrane. One could of course ask whether a resistance to shear arises from the fixation of the skeleton to integral proteins. However, when the membrane is sheared the lipids of the bilayer will rearrange laterally relative to the integral proteins to which the skeleton is fixed and thus allow the bilayer to comply to shear deformations (Fischer 1992b). This rearrangement of lipids leads to a viscous resistance during transients between equilibrium shapes or during a continuous deformation but does not play a role when resting shapes of red cells are considered.

3.5.4 Bending Stiffness

In order to understand the causes of the bending stiffness of the membrane consider a simple analogy. Imagine a bar supported at both ends and loaded by a weight in the middle. Due to the weight the bar will bend. In a cross section of the bar, the material will be compressed in the upper part while in the lower part it will be extended. There is one plane (the neutral plane) in between which is neither extended nor compressed. The bending stiffness of the bar results from the resistance of the material against compression and extension. Next consider a circular plate supported at the circumference and loaded in the centre. It will deform into a shape similar to the segment of a sphere. The resistance to bending results from the resistance of the upper layers of the plate against a decrease in area and of the lower layers against an increase in area. Again a neutral plane will exist.

Turning to the membrane, as in Sect. 3.4.4 there will be two conceptual contributions. The first derives from the two monolayers separately. Their bending stiffness can be considered by analogy to that of the plate. If the membrane curvature is increased, the headgroup region of the outer monolayer gets expanded while the extreme tail region will be compressed. In the inner monolayer the headgroups are compressed and the tails are expanded. The values for the bending stiffness of the two monolayers add up to the *single-layer based stiffness* of the bilayer. Its experimental value is $2 \cdot 10^{-19}$ Nm (Waugh and Hochmuth 1995).

The second contribution arises from the cooperation of the two monolayers. When the curvature of the bilayer is increased the whole outer monolayer is isotropically expanded and the whole inner one isotropically compressed. We will refer to this type of resistance against bending as *double-layer based stiffness*[1]. Its value is about 3 times larger than the value of the single-layer based stiffness of the bilayer (Raphael and Waugh 1996). The reason for the preponderance of double-layer based stiffness is that the distance of the layers from the neutral plane enters the bending stiffness quadratically. The factor of 3 has important consequences, which will be discussed in the Appendix.

The contribution of the skeleton to the bending stiffness of the red cell membrane can be neglected. Experimental evidence is the similarity of the bending stiffness of the red cell membrane and the bilayer of lipid vesicles. Theoretically, a single-layer based stiffness is not expected since the skeleton, when considered as a layer, contains a large proportion of water. Upon bending some of this water would flow from compressed regions to expanded regions thus relaxing any resistance to bending. As to the cooperative bending, due to its 5 orders of magnitude lower isotropic stiffness compared to that of the monolayers the contribution of the skeleton can be neglected.

The total bending stiffness is the sum of single-layer based and double-layer based stiffness.

[1] For resting shapes and also for slowly changing shapes single-layer based and double-layer based bending stiffness have also been called local and global (or non local) bending stiffness, respectively. This designation originates from the respective type of mathematical description of the two contributions to the bending stiffness of a bilayer (Božič et al. 1992). The single-layer based stiffness has also been called intrinsic bending stiffness (Raphael and Waugh 1996).

3.6 Uniformity of the Membrane

3.6.1 General Remarks

In the previous sections we considered the properties of membrane patches. We now turn to the whole membrane. It is called *uniform* when all its patches have the *same* properties.

3.6.2 Surface Area

Since each monolayer behaves as a liquid in two dimensions (Sect. 3.5.3) any lateral difference in isotropic tension and therefore in area/lipid would equilibrate by a lateral flow of lipids. For this reason the isotropic tension and, as a consequence, the isotropic deformation is laterally uniform in each monolayer. This holds true for all resting shapes.

3.6.3 Aspect Ratio

Of all possible shapes the red cell membrane can assume, one is unique where the aspect ratio of all membrane patches would not change upon excision. This means the skeleton is not sheared in this particular shape of the red cell identified as the *relaxed shape of the skeleton*. In the relaxed shape the structure of the skeleton is isotropic, i.e. the distance of network nodes does, at least locally, not depend on direction in the plane of the membrane. In order to distinguish the shape of the red cell from the relaxed shape of the skeleton the first is sometimes referred to as the *actual shape*.

3.6.4 Single-Layer Based Spontaneous Curvature

The single-layer based spontaneous curvature will be uniform laterally if the lipid and protein species constituting a monolayer are well mixed. It might be tempting to postulate that, e.g. in the outer monolayer, truncated conical lipids accumulate in regions of positive curvature because the local curvature fits the lipid shape better than smaller curvatures elsewhere on the membrane. However, it can be shown that in the absence of additional mechanisms, thermal energy is sufficient to randomise the lateral distribution of lipids (Song and Waugh 1993). The spontaneous curvature of a well-mixed monolayer is uniform and therefore the single-layer based spontaneous curvature of the membrane is uniform as well. Mechanisms able to overcome the randomising influence of thermal energy and to induce a lateral segregation may be e.g. attractive interactions of one or several molecular species in a monolayer. We will come back to this point in Sect. 3.9.2.

3.6.5 Double-Layer Based Spontaneous Curvature

As detailed in Sect. 3.6.2 the isotropic deformation is uniform laterally in each monolayer. For this reason the double-layer based spontaneous curvature is not uniform on the surface. However, its difference from the actual curvature will be uniform laterally. This sounds complicated and is mentioned here just for completeness. It is, however, of little relevance for the further treatment since it is the *change* (Sect. 3.7.3) in double-layer based spontaneous curvature that matters in shape changes of red cells.

3.7 Resting Shapes of Red Cells

3.7.1 General Remarks

Since its cytoplasm is liquid, the shape of the red cell is completely determined by its volume, surface area, and the mechanical characteristics of the membrane. The volume enclosed by the membrane depends on the amount and type of solutes in the cytoplasm and their concentration in the suspending medium. Volume changes may occur concomitant to shape changes but do not interfere with the formation of echinocytes or stomatocytes, unless the volume increase is so large that the pressure difference between cytoplasm and medium rises and the membrane gets under isotropic tension. In this case the shape would become spherical and the red cell would haemolyse when the isotropic deformation exceeds a few percent. Such extreme cases are not covered in the present chapter. The mechanical characteristics of the membrane, which determine red cell shape, will be discussed in Sect. 3.7.3.

3.7.2 The Biconcave Discocyte

The biconcave disk is a special resting shape. It is located between the echinocytes on one side of a scale and the stomatocytes on the other side. It has been shown by mathematical modelling that a biconcave shape can be generated assuming a membrane with zero shear stiffness, a finite bending rigidity, and zero spontaneous curvature (Canham 1970). The values for volume and surface area were assumed as in a typical red cell. Since the curvature almost everywhere on the membrane of a biconcave disk is different from zero, the actual curvature of most membrane patches deviates from their spontaneous curvature. This is analogous to the rod bent into a circle (Sect. 3.4.1) and is due to the fact that the membrane forms a closed surface.

To reconcile this mathematical model with the observed finite shear stiffness of the red cell membrane one has to assume that the relaxed shape of the skeleton is identical to the biconcave shape of the cell. This in turn requires that during maturation of the red cell from the stage of the early reticulocyte (which has a quite wrinkled shape) to the late reticulocyte (which is biconcave) a remodelling of the skeleton takes place in the sense that it adapts its relaxed shape to the shape which the bilayer assumes due to its resistance to bending.

Human red cells exhibit a variation in their discoidal shape intraindividually as well as interindividually. These different shapes can be modelled by adjusting surface area and volume and by applying either one or two modifications of the above model. First, the spontaneous curvature can be given values different from zero (Deuling and Helfrich 1976). Second, a relaxed shape different from the actual shape can be assumed (Fischer et al. 1981).

3.7.3 Echinocytes and Stomatocytes

For the following discussion we introduce a new quantity, the *average curvature*. It is the actual curvature averaged over the whole surface area of the membrane.

Many different manipulations *in vitro* have been shown to transform discocytes into either echinocytes or stomatocytes (Chap. 2). It is intuitively appreciated and has been shown in calculations (Beck 1978; Ferrell et al. 1985) that the average curvature is larger in echinocytes than in discocytes and smaller in stomatocytes. Starting from this observed change in average curvature several similar hypotheses were put forward in order to explain the mechanism of the shape change (Wilbrandt 1953; Evans 1974; Sheetz and Singer 1974). The difference between these hypotheses depends mainly on the respective type of manipulation, which was considered. Based on the description given in this chapter these hypotheses can be put on a common basis as follows: (i) an exogenous manipulation causes an increase or a decrease in spontaneous curvature of the membrane relative to the value prevailing in the discocyte, (ii) as a consequence a new resting shape is assumed which is echinocytic if the spontaneous curvature is increased and stomatocytic if it is decreased.

Theoretical confirmation of this hypothesis by a mathematical model has been accomplished only for moderate stomatocytes (Evans and Buxbaum 1981) but neither for more pronounced stomatocytes nor for any echinocytic shape. However, compensation experiments (Deuticke 1968) make it very likely that echinocytes are induced by an increase in spontaneous curvature.

The change in spontaneous curvature induced by the exogenous manipulation does not show up to the same extent in a change of the average curvature. This is because the new red cell shape (echinocyte or stomatocyte) is the result of a mechanical equilibrium between shear tensions and bending moments in the membrane, as discussed in the following.

In the new red cell shape the skeleton is deformed relative to its relaxed shape. Therefore, shear tensions in the skeleton pull the shape back to its relaxed shape, which is biconcave. These shear tensions are counteracted in mechanical equilibrium by bending moments. The existence of these bending moments indicates that the change in the average curvature is smaller than the change in spontaneous curvature. And indeed, although not appreciated in this way by the authors, the two quantities were found to be related by a factor of 2 in an experimental study (Ferrell et al. 1985).

Direct evidence for the existence of a mechanical equilibrium between shear and bending elasticity was presented by first increasing the shear stiffness of the membrane by chemical modification of the skeleton and then changing the spontaneous curvature (Haest et al. 1980b). The spontaneous curvature was assumed to be changed to the same extent in control cells with native shear stiffness as in the pre-treated cells. As expected, the changed shape was closer to the biconcave disk in the cells with shear-stiffened membranes than in controls.

It can be shown mathematically that it is the *ratio* of shear stiffness to bending stiffness, which determines the extent of shape change of the red cell after a given change in spontaneous curvature. It happens that the value of this ratio is such as to allow for the impressive changes of red cell shape. If this ratio was much larger the red cell would never deviate very much from the biconcave shape. If the ratio was much smaller the red cell would behave similarly to artificial giant lipid vesicles which have no shear elasticity at all. Such lipid vesicles form spherical daughter vesicles upon a change in spontaneous curvature (Käs and Sackmann 1991; Farge and Devaux 1992).

3.7.4 Conclusions

As detailed above, the variation of only a *single* parameter, the spontaneous curvature of the membrane, can explain most experimental shape changes of red cells. This is in keeping with the finding that the sequence of shapes observed upon increasing the intensity of a manipulation is independent of the type of the manipulation. Deviations from this sequence are expected when a second parameter besides the spontaneous curvature is varied. An example is the above mentioned increase in shear stiffness and subsequent change in spontaneous curvature. Although it was not interpreted by the authors in this way, the sequence of shapes after this double manipulation is somewhat different from that observed after solely changing the spontaneous curvature of the membrane (Haest et al. 1980b).

It is also physiologically relevant that the ratio of shear to bending stiffness in native cells has a value that allows for the observed changes in shape. If the ratio was much larger the red cell would deform less well while flowing and thus impede the flow of blood (Haest et al. 1980a; Johnson et al. 1980). If the ratio was much lower the membrane would be in danger of shedding vesicles (Waugh and Agre 1998). This in turn would increase the ratio of volume to surface area and impair the ability of red cells to flow (Fischer 1989) even more than an increase in the ratio of shear to bending stiffness.

3.8 Relative Importance of Single-Layer and Double-Layer Based Spontaneous Curvature

Conceptually, there are three ways to change the spontaneous curvature of the membrane: (i) by incorporating additional material asymmetrically into the two monolayers or extracting native material asymmetrically from the two monolayers, which is just a negative incorporation, (ii) by inducing transverse redistribution of lipids from one monolayer to the other, or (iii) by exchanging native molecules of the outer monolayer against exogenous molecules in a one-to-one fashion. A modification of lipids, e.g. by enzyme treatment, or of integral proteins (Schwarz et al. 1999), is equivalent to type (iii) in that it also leaves the number of molecules constant. In real experiments type (i), (ii), or (iii) can be induced by external manipulations. However, in many cases type (ii) occurs secondarily after a primary manipulation.

The spontaneous curvature of the membrane has two contributions, i.e. the single-layer based and the double-layer based spontaneous curvature (Sects. 3.4.4, 3.6.4, and 3.6.5). Accordingly, changes in spontaneous curvature can be caused by changes in either of these contributions or in both. An asymmetric change in the area of the neutral planes of the two monolayers results in a change in the double-layer based spontaneous curvature. This mechanism has been termed the bilayer couple mechanism (Sheetz and Singer 1974). A disproportionate change in the areas of the neutral plane of a monolayer and the plane at the free end of its hydrocarbon chains causes a change in the spontaneous curvature of that monolayer. Unless the changes in the two monolayers compensate each other, the single-layer based spontaneous curvature of the membrane is altered.

So far, there seems to be only a single published study, which demonstrates the effect of a change in the single-layer based spontaneous curvature on red cell shape (Kuypers et al. 1984). In all other studies (Chap. 2) the bilayer couple mechanism, i.e. a change in the double-layer based spontaneous curvature, was used to explain the experimental findings. It therefore appears that in the majority of cases the bilayer couple mechanism predicts at least the correct sign (either more echinocytic or more stomatocytic) of the shape change. In order to verify this conjecture and to arrive at quantitative relations a geometrical model was explored (Fischer, unpublished) which is presented in the Appendix.

3.9 Non-Uniform Mechanical Properties of the Membrane

3.9.1 Spherical Echinocytes and Stomatocytes

The most severe echinocytes and stomatocytes are called spheroechinocytes or spherostomatocytes, respectively. The shape of spheroechinocytes can be described as spheres the surface of which is covered with very fine spikes. In order to comply with this shape the skeleton would have to be deformed extremely. Under such deformations the density of the network becomes non-uniform (Discher and Mohandas 1996) a subject omitted in this chapter for reasons of simplicity. As long as the behaviour of the skeleton under such extreme shape changes is not known, a description of spheroechinocytes in terms of continuum mechanics is not possible. The same applies to spherostomatocytes.

3.9.2 Effects of Raising the Cytoplasmic Concentration of Calcium

As mentioned in the Introduction, the sequence of echinocytic shapes upon increasing the extent of modification is the same for most manipulations. The same holds true for stomatocytes. This was the basis for a common mechanical description of most of the published shape changes. An exception to this general scheme is the shape change upon raising the cytoplasmic concentration of calcium. In contrast to most manipulations, the ensuing shape change starts by formation of tiny and pointed spikes (White 1976). In a later stage, the spikes become larger and the overall cell shape comes closer to that of an ordinary echinocyte. Nevertheless, a difference in the shape of the spikes remains. A likely explanation for the occurrence of the tiny spikes is that the single-layer based spontaneous curvature of the membrane has become non-uniform, more specifically that in these regions the single-layer based spontaneous curvature is much more positive than in laterally adjacent regions.

The origin of the envisaged non-uniformity can only be a matter of speculation. It could e.g. be caused by a clustering of inverted truncated conical lipids in the inner monolayer due to an interaction of calcium with negatively charged lipid headgroups.

As discussed in Sect. 3.7.4, the deviation of shape changes from the usual sequence requires that the value of more than a single variable has been changed. The envisaged lateral inhomogeneity of the spontaneous curvature requires a change of at least two additional variables. The first would be the single-layer based spontaneous curvature of the membrane at the location of the spikes. The second would be the ratio of membrane areas with high and low single-layer based spontaneous curvature of the membrane.

Appendix

A Model of the Relative Contributions of Single-Layer and Double-Layer Based Spontaneous Curvature to the Spontaneous Curvature of the Membrane

General Remarks

In the following, surface areas of monolayers or cross sectional areas of molecules are meant to be areas in the neutral plane (Sect. 3.5.4). The shape of amphiphilic molecules either added to or removed from a monolayer is characterized by a *shape coefficient*. This is the ratio of two volumes (Fig. 3.1): (i) the partial volume of these molecules on the hydrocarbon side of the neutral plane and (ii) the volume of a cylinder with the same cross sectional area as these molecules, extending from the neutral plane to the free end of the hydrocarbon chains of the monolayer. The shape coefficient of truncated cones is less than unity, that of inverted truncated cones is greater than unity (Fig. 3.2). A cylindrical molecule has a value of one. This value applies to a phosphatidylcholine with two saturated hydrocarbon chains, which are as long as the average chain length in the surrounding monolayer. If the two saturated hydrocarbon chains were shorter, the shape coefficient would decrease (Fig. 3.1). Such a molecule may act as a truncated conical molecule in the red cell membrane although in a monolayer constituted exclusively by such molecules it may act as a cylindrical molecule.

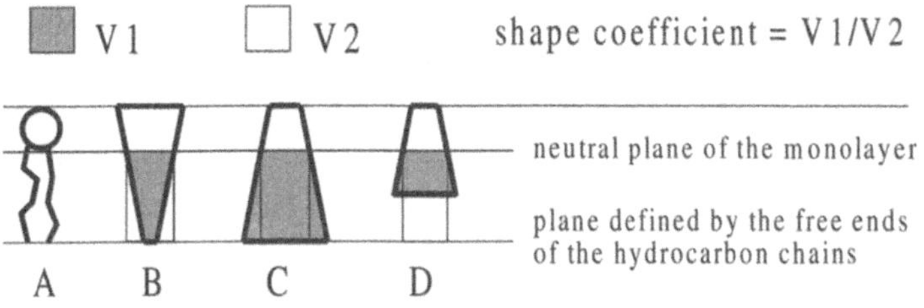

Fig. 3.1. Definition of the shape coefficient of molecules embedded in a monolayer. For explanation see text. *A*: schematic phospholipid to define the orientation of the monolayer. *B*, *C*, *D*: cross sections of truncated cones, inverted truncated cones, and cylinders. The thick line indicates the molecular shape. *B*: a truncated conical molecule. *C*: an inverted truncated conical molecule. *D*: a molecule that is inverted truncated conical in a pure monolayer but in a thicker monolayer and at low concentrations acts as a truncated conical molecule

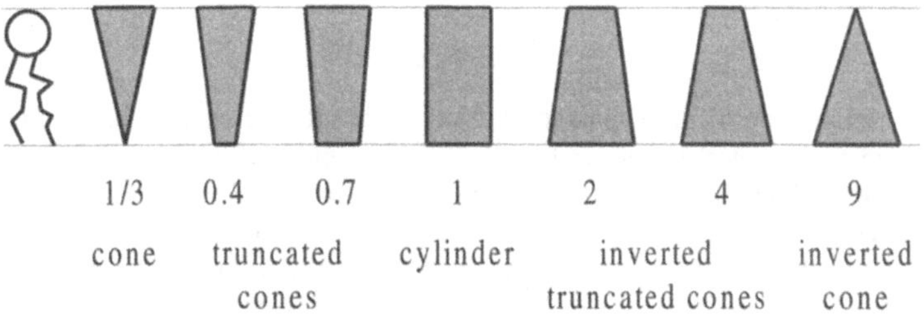

Fig. 3.2. Cross sections of rotationally symmetric bodies, as models of lipid shape, and the respective shape coefficients. For details see text. The schematic phospholipid defines the orientation of the monolayer

In order to relate shape coefficients to geometrical shapes the position of the neutral plane has to be known. Measurements are available only for pure phospholipid systems (Rand and Fuller 1994). In these systems the distance between the neutral plane and the hydrocarbon end of the monolayer was found to be 2/3 of the monolayer thickness. If this number is adopted a shape coefficient of 1/3 corresponds to a cone (Fig. 3.2). An inverted cone has a shape coefficient of 9. In Figs. 3.3, 3.4, and 3.5 a logarithmic scale was chosen for plotting the shape index because on a linear scale the inverted truncated cones would have been overemphasized.

Starting from a membrane in which all spontaneous curvatures are zero the change in these curvatures following a manipulation was calculated. The values for the bending stiffness of the two monolayers were assumed to be equal. The double-layer based stiffness was taken to be 3 times larger than the single-layer based stiffness (Sect. 3.5.4). The results of the calculations are independent of the position of the neutral plane because all spontaneous curvatures were normalized by multiplication with the distance between the neutral plane of a monolayer and the plane defined by the free ends of the hydrocarbon chains (Fig. 3.1). The three experimental strategies introduced above are treated separately in the next three sections. They are assumed to occur in pure form, i.e. not to be accompanied or followed by a secondary redistribution of material.

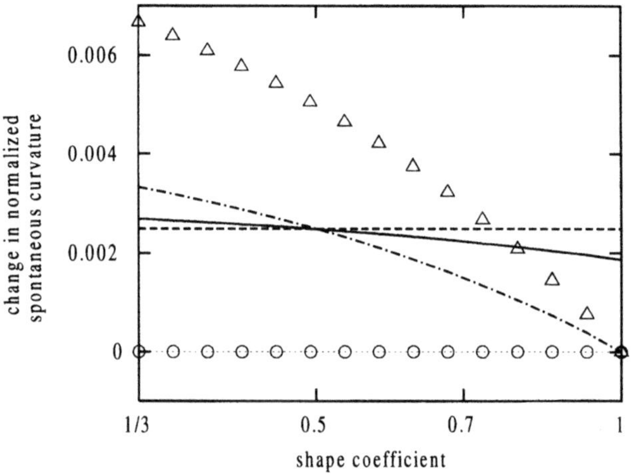

Fig. 3.3. Change in normalized spontaneous curvatures after an increase in the area of the neutral plane of the outer monolayer by 1%, *vs.* the shape coefficient of the incorporated molecules (logarithmic scale). The change in double-layer based spontaneous curvature of the membrane (dashed line) is independent of the shape coefficient because this curvature is determined by the increase in the area of the neutral plane. The change in spontaneous curvature of the outer monolayer (triangles) decreases with decreasing conicity (increasing shape coefficient) and passes through zero at the cylindrical shape. The spontaneous curvature of the inner monolayer (circles) does not change because nothing was incorporated there. The change in single-layer based spontaneous curvature (dash-dotted line) is the mean of the change in the spontaneous curvature of both monolayers because the values of their single-layer stiffness were assumed to be equal. The change in spontaneous curvature of the membrane (solid line) is closer to the change in double-layer based spontaneous curvature than to the change in single-layer based spontaneous curvature because of the ratio of 3 between the values for the two bending stiffnesses

Incorporation of Exogenous Molecules

To model the first type of experiment the area of the outer monolayer was increased by 1%. Fig. 3.3 shows how the various contributions to the change in spontaneous curvature of the membrane depend on the shape coefficient of the inserted molecules (see the legend for a detailed description). The change in the spontaneous curvature of the membrane is positive over the plotted range of shape indices. This means that in this range only echinocytes would occur.

The salient result is that the change in double-layer based spontaneous curvature differs by less than 10% from the change in spontaneous curvature of the membrane. If no secondary redistribution of material occurs the error made by interpreting an experimental shape change in terms of a change in double-layer spontaneous curvature, i.e. in terms of the bilayer couple mechanism alone, is therefore smaller than 10%. There are two causes for this finding. First, the double-layer based bending stiffness is three times larger than the single-layer based bending stiffness of the bilayer. This gives the change in double-layer based spontaneous curvature a weight 3 times higher than that of the single-layer based spontaneous curvature in the determination of the spontaneous curvature of the membrane. Second, for membrane active substances to incorporate from the suspending medium into the outer monolayer clearly their shape coefficient must be smaller than unity. The same limit applies for an extraction of lipids from the outer monolayer, e.g. by albumin. The abscissa in Fig. 3.3 is therefore not continued beyond unity. If incorporation is achieved via lipid vesicles it is possible to insert molecules with shape coefficients larger than unity. However, such molecules redistribute readily to the inner monolayer. The results of Fig. 3.3 would therefore not apply.

Redistribution of Endogenous or Inserted Molecules between the Two Monolayers

In principle, redistribution is equivalent to a negative incorporation in one monolayer and a positive incorporation to the same extent in the other monolayer. The result of a redistribution of 1% of the area of the neutral plane from the inner monolayer to the outer one is shown in Fig. 3.4. The abscissa is extended to a shape coefficient of 9 because lipids with small hydrophilic portions readily redistribute and are likely to have inverted truncated conical shapes.

According to the bilayer couple mechanism, one would expect the score (Introduction) of echinocytes not to depend on the shape coefficient of the redistributing molecules. This expectation corresponds to the behaviour of the change in double-layer based spontaneous curvature, i.e. a horizontal line. The change in spontaneous curvature of the membrane (Fig. 3.4, solid line) indicates the true shape change. For shape coefficients < 2.5 the model predicts echinocytes. At a shape coefficient of about 2.5 the change in spontaneous curvature of the membrane is zero. This means the effects of the change in double-layer and single-layer based spontaneous curvature compensate each other. Therefore no shape change is predicted despite a redistribution of molecules from the inner to the outer monolayer. For shape coefficients > 2.5 stomatocytes are predicted.

The occurrence of stomatocytes is contrary to the expectation based on the bilayer couple mechanism alone. To document such "opposite" shape changes would require the redistribution to be restricted to inverted truncated conical molecules. Two scenarios may be envisaged. First, redistribution is initiated by increases of the transbilayer mobility of endogenous phospholipids via Ca^{2+}-induced facilitation by a scramblase (Chap. 1) or following exposure of red cells to electric

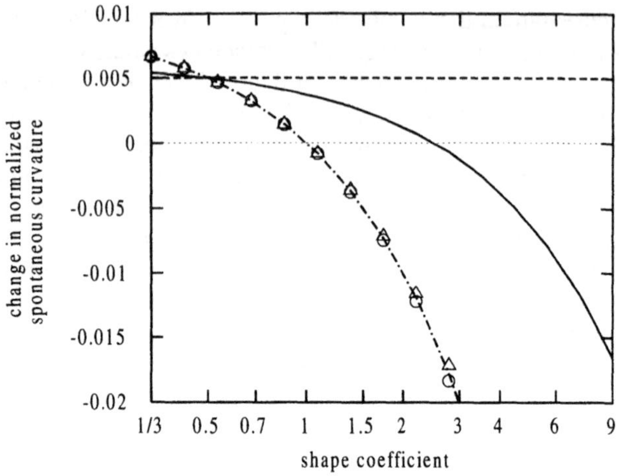

Fig. 3.4. Change in normalized spontaneous curvatures after a redistribution of material corresponding to 1% of the area of the neutral plane from the inner monolayer to the outer monolayer, *vs.* the shape coefficient of the redistributed molecules (logarithmic scale). The change in double-layer based spontaneous curvature of the membrane (dashed line) is independent of the shape coefficient because it is determined by the change in the area of the neutral planes of both monolayers. The changes in the spontaneous curvatures of the outer monolayer (triangles) and the inner monolayer (circles) are essentially the same because of the opposite orientation of lipids in the two monolayers. (The minute difference between the two monolayer curves relates to the fact, that although the molecular volumes subtracted and added are the same, the product of the respective volume ratios is not equal to unity.) The change in single-layer based spontaneous curvature of the membrane (dash-dotted line) lies in between because it is the average of the two monolayer changes. The change in spontaneous curvature of the membrane (solid line) is closer to the double-layer based spontaneous curvature of the membrane because the double-layer based bending stiffness is 3 times larger than the single-layer based stiffness

field pulses (Chap. 2). These procedures, however, are not selective for inverted truncated cones. Second, by enzymatic removal of phospholipid headgroups inverted truncated conical molecules are formed which then redistribute. However, this redistribution is so fast that its shape effect cannot be separated from the shape effect of the primary modification, i.e. the enzyme treatment. It is therefore unlikely to find conditions in which the change in single-layer based spontaneous curvature outweighs the effect of double-layer based spontaneous curvature. The results of Fig. 3.4 show, however, that the change in single-layer based spontaneous curvature must not be neglected when inverted truncated conical lipids redistribute.

Exchange of Native Against Exogenous Molecules in a One-To-One Fashion

An exchange of native against exogenous molecules in a one-to-one fashion is equivalent to a sequential negative and positive incorporation in the outer monolayer. To reduce the complexity of the presentation the shape coefficient of the native molecules that are replaced was not varied. A value above unity is suggested by the experiments of Kuypers et al. (1984). In the calculations it is assumed that molecules with a shape coefficient of 2 and which comprise 1% of the area of the outer monolayer are replaced. Since the exchange is done in a one-to-one fashion, the area of the inserted molecules may be different from the replaced area. Therefore the area ratio of the exchanged molecules appears as a second independent variable besides their shape coefficient (Fig. 3.5)

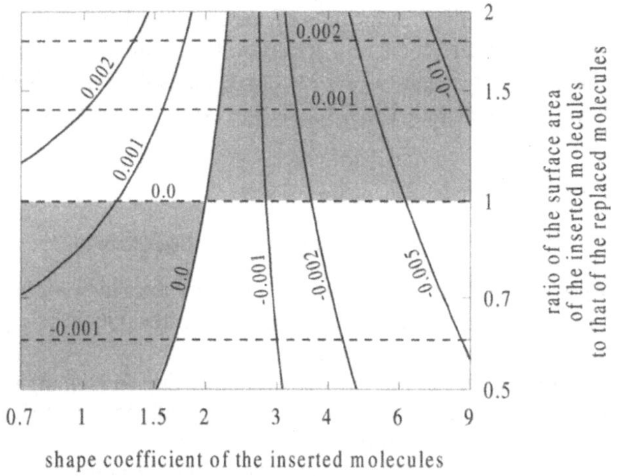

Fig. 3.5. Contour plots of the change in normalized spontaneous curvature after a replacement of native lipids having a shape coefficient of 2 and constituting 1% of the area of the neutral plane of the outer leaflet in a one-to-one fashion, *vs.* (i) the shape coefficient of the inserted molecules (logarithmic scale) and (ii) the ratio of the surface area of the exchanged molecules (logarithmic scale). The change in double-layer based spontaneous curvature of the membrane (dashed line) is independent of the shape coefficient because it is determined by the change in the area of the neutral plane of the outer monolayer. The change in spontaneous curvature of the membrane is shown by solid lines

The changes in normalized spontaneous curvatures are shown as contour plots (Fig. 3.5). For clarity only two spontaneous curvatures are shown: the double-layer based spontaneous curvature (dashed lines) and the spontaneous curvature of the membrane (solid lines). Above the level zero of the double-layer based spontaneous curvature (dashed lines) one would expect echinocytes on the basis of the bilayer couple mechanism alone. Below the level zero one would expect stomatocytes. The change in spontaneous curvature of the membrane (Fig. 3.5 solid lines) indicates the real shape change. Left of the line indicating level zero the shape change would be echinocytic, to the right it would be stomatocytic. The shading indicates regions in which the bilayer couple mechanism alone predicts the opposite shape change from that predicted by the change in spontaneous curvature of the membrane, i.e. what would be observed in an experiment. Therefore, provided that no redistribution or spontaneous incorporation occurs in parallel, a one-to-one exchange bears the chance to demonstrate an effect of molecular shape on red cell shape change, which cannot be attributed alternatively to an effect of molecular area (bilayer couple mechanism).

As mentioned in the Appendix above, lipids with a shape coefficient clearly smaller than unity would spontaneously incorporate into the membrane. In one-to-one exchanges, lipids with higher shape coefficients can be inserted by enzymatic facilitation. To exclude parallel spontaneous incorporation the shape coefficients of the lipids inserted in one-to-one exchanges should be above the threshold for spontaneous incorporation. Accordingly the abscissa in Fig. 3.5 starts at a value of 0.7. If at the same time the headgroups of the inserted lipids are large enough to prevent or at least to retard their redistribution to the inner monolayer one is able to observe the effect of a one-to-one exchange in isolation. Indeed, results have been published (Kuypers et al. 1984) located in both shaded regions of Fig. 3.5.

Conclusions

The results of the above model calculations show the following:

(i) For incorporation of molecules into the outer monolayer without subsequent transbilayer redistribution an almost quantitative interpretation is possible based on the bilayer couple mechanism alone.
(ii) It is not surprising that a red cell shape change due to a change in single-layer based spontaneous curvature has been demonstrated only once (Kuypers et al. 1984). The reason is that only in one-to-one exchanges not accompanied by spontaneous incorporation or secondary transbilayer redistribution such a demonstration appears to be possible.
(iii) The influence of the single-layer spontaneous curvature in shape changes involving a redistribution of inverted truncated conical molecules is strong and must not be neglected in quantitative considerations.

Acknowledgement: I thank Prof. B. Deuticke and Dr. C.W.M. Haest, RWTH Aachen, Germany, for critically reading the manuscript.

References

Beck JS (1978) Relations between membrane monolayers in some red cell shape transformations. J Theor Biol 75:487–501

Bessis M (1972) Red cell shapes an illustrated classification and its rationale. Nouv Rev fr Hématol 12:721–746

Boal DH (1994) Computer simulation of a model network for the erythrocyte cytoskeleton. Biophys J 67:521–529

Božič B, Svetina S, Žekš B, Waugh RE (1992) Role of lamellar membrane structure in tether formation from bilayer vesicles. Biophys J 61:963–973
Canham PB (1970) The minimum energy of bending as a possible explanation of the biconcave shape of the human red blood cell. J Theoret Biol 26:61–81
Deuling HJ, Helfrich W (1976) Red blood cell shapes as explained on the basis of curvature elasticity. Biophys J 16:861–868
Deuticke B (1968) Transformation and restoration of biconcave shape of human erythrocytes induced by amphiphilic agents and changes of ionic environment. Biochim Biophys Acta 163:494–500
Discher DE, Mohandas N (1996) Kinematics of red cell aspiration by fluorescence-imaged microdeformation. Biophys J 71:1680–1694
Discher DE, Boal DH, Boey SK (1998) Simulations of the erythrocyte cytoskeleton at large deformation II Micropipette aspiration. Biophys J 75:1584–1597
Evans EA (1974) Bending resistance and chemically induced moments in membrane bilayers. Biophys J 14:923–931
Evans E, Buxbaum K (1981) Affinity of red blood cell membrane for particle surfaces measured by the extent of particle encapsulation. Biophys J 34:1–12
Evans EA, Skalak R (1979) Mechanics and thermodynamics of biomembranes. CRC Crit Rev Bioeng 3:181–418
Farge E, Devaux PF (1992) Shape changes of giant liposomes induced by an asymmetric transmembrane distribution of phospholipids. Biophys J 61:347–357
Ferrell JE Jr, Lee K-J, Huestis WH (1985) Membrane bilayer balance and erythrocyte shape: A quantitative assessment. Biochemistry 24:2849–2857
Fischer TM (1978) A comparison of the flow behaviour of disk shaped versus elliptic red blood cells (RBC). Blood Cells 4:453–461
Fischer TM (1989) Mechanical hemolysis of cross-bonded red cells in the microcirculation. Int J Microcirc 8:159–176
Fischer TM (1992a) Bending stiffness of lipid bilayers. I. Bilayer couple or single-layer bending? Biophys J 63:1328–1335
Fischer TM (1992b) Is the surface area of the red cell membrane skeleton locally conserved? Biophys J 61:298–305
Fischer TM (1993) Bending stiffness of lipid bilayers. IV. Interpretation of red cell shape change. Biophys J 65:687–692
Fischer TM, Stöhr-Liesen M, Schmid-Schönbein H (1978) The red cell as a fluid droplet: Tank tread-like motion of the human erythrocyte membrane in shear flow. Science 202:894–896
Fischer TM, Haest CWM, Stöhr-Liesen M, Schmid-Schönbein H, Skalak R (1981) The stress-free shape of the red blood cell membrane. Biophys J 34:409–422
Gilligan DM, Bennet V (1993) The junctional complex of the membrane skeleton. Semin Hematol 30:74–83
Haest CWM (1982) Interaction between membrane skeleton proteins and the intrinsic domain of the erythrocyte membrane. Biochim Biophys Acta 694:331–352
Haest CWM, Driessen GK, Kamp D, Heidtmann H, Fischer TM, Stöhr-Liesen M (1980a) Is "deformability" a parameter for the rate of elimination of erythrocytes from the circulation? Pflügers Arch 388:69–73
Haest CWM, Fischer TM, Plasa G, Deuticke B (1980b) Stabilization of erythrocyte shape by a chemical increase in membrane shear stiffness. Blood Cells 6:539–553

Israelachvilli JN, Marcelja S, Horn RG (1980) Physical principles of membrane organization. Quart Rev Biophys 13:121–200

Johnson GJ, Allen DW, Flynn TP, Finkel B, White JG (1980) Decreased survival *in vivo* of diamide-incubated dog erythrocytes. J Clin Invest 66:955–961

Käs J, Sackmann E (1991) Shape transitions and shape stability of giant phospholipid vesicles in pure water induced by area-to-volume changes. Biophys J 60:825–844

Kuypers FA, Roelofsen B, Berendsen W, Op den Kamp JAF, Van Deenen LLM (1984) Shape changes in human erythrocytes induced by replacement of the native phosphatidylcholine with species containing various fatty acids. J Cell Biol 99:2260–2267

Lenormand G, Hénon S, Richert A, Siméon J, Gallet F (2001) Direct measurement of the area expansion and shear moduli of the human red blood cell membrane skeleton. Biophys J 81:43–56

Mohandas N, Evans E (1994) Mechanical properties of the red cell membrane in relation to molecular structure and genetic defects. Annu Rev Biophys Biomol Struct 23:787–818

Rand RP, Fuller NL (1994) Structural dimensions and their changes in a reentrant hexagonal-lamellar transition of phospholipids. Biophys J 66:2127–2138

Raphael RM, Waugh RE (1996) Accelerated interleaflet transport of phosphatidylcholine molecules in membranes under deformation. Biophys J 71:1374–1388

Schmid-Schönbein H, Wells R (1969) Fluid drop-like tranisition of erythrocytes under shear. Science 165:288–291

Schwarz S, Haest CWM, Deuticke B (1999) Extensive electroporation abolishes experimentally induced shape transformations of erythrocytes: a consequence of phospholipid symmetrization? Biochim Biophys Acta 1421:361–379

Sheetz MP, Singer SJ (1974) Biological membranes as bilayer couples. A molecular mechanism of drug-erythrocyte interactions. Proc Natl Acad Sci USA 71:4457–4461

Song J, Waugh RE (1993) Bending rigidity of SOPC membranes containing cholesterol. Biophys J 64:1967–1970

Waugh RE, Agre P (1988) Reductions of erythrocyte membrane viscoelastic coefficients reflect spectrin deficiencies in hereditary spherocytosis. J Clin Invest 81:133–141

Waugh RE, Hochmuth RM (1995) Mechanics and deformability of hematocytes. In: Bronzino JD (ed) The biomedical engineering handbook. CRC Press, Boca Raton, pp 474–486

White JG (1976) Scanning electron microscopy of erythrocyte deformation: The influence of a calcium ionophore A23187. Semin Hematol 13:121–132

Wilbrandt W (1953) Hämolysemechanismen. Schweizerische Medizinische Wochenschrift 43:1023–1027

4 Passive Membrane Permeability for Ions and the Membrane Potential

Ingolf Bernhardt and Erwin Weiss

Arbeitsgruppe Biophysik, Naturwissenschaftlich-Technische Fakultät III, Universität des Saarlandes, Gebäude 6, Postfach 151150, 66041 Saarbrücken, Germany

4.1 Introduction

One characteristic feature of all mammalian cells is the asymmetric and non-equilibrium distribution of inorganic ions across the cell membrane. Such a distribution, established during evolution, is of fundamental importance for the regulation of cellular metabolism as well as for processes of signal transduction and excitation. The maintenance of the intracellular cation concentration as well as the cell volume is based on the balance between active and passive transport processes.

Initially it was postulated that a non-equilibrium distribution of ions in the red blood cell (RBC) is only possible if the cell membrane is impermeable to ions (e.g. as discussed in Gürber 1895). However, van Slyke et al. (1923) found a high permeability of RBC membranes to Cl^-. Later on it was shown directly, using isotopes, that the RBC membrane of different species is also permeable to Na^+ and K^+ (Cohn and Cohn 1939; Dean et al. 1940; Eisenman et al. 1940). The important finding by Harris (1941) showing that the addition of glucose to the cell suspension is important for the maintenance of the Na^+ and K^+ gradients led to the first proposal of the existence of a Na^+/K^+ pump in the cell membrane (Krogh 1946). Systematic investigations of ion transport across the RBC membrane in the 1950s (see e.g., Solomon 1952; Glynn 1956) led to the "pump-leak" concept formulated by Tosteson and Hoffman (1960).

4.2 Specific Transport Systems for Monovalent Cations in the Mammalian Red Cell Membrane

Originally, passive transport was envisaged as a simple electrodiffusive leak that could be described by the classical Goldman flux equation (Goldman 1943). Although the "pump-leak" concept was developed, it already became evident on the basis of the results of ion flux measurements across the RBC membrane carried out in the 1950s that the real situation is much more complicated. For instance, it was shown in RBCs from horse, human, and rat that the K^+ uptake consists of a linear and a saturated component (Shaw 1955; Glynn 1956; Beaugé and Ortiz 1970). Interestingly, a saturable component of the K^+ uptake could be seen also in the presence of ouabain (Glynn 1957).

Beginning with the work of Wiley and Cooper (1974) who first discussed the presence of a cotransport system for Na^+ and K^+ that is sensitive to diuretic drugs (e.g. furosemide) but insensitive to ouabain, a large variety of specific passive, mediated ion transport pathways has been discovered in the RBC membrane. The presently known transport pathways for monovalent cations in the human RBC membrane are presented in Fig. 4.1.

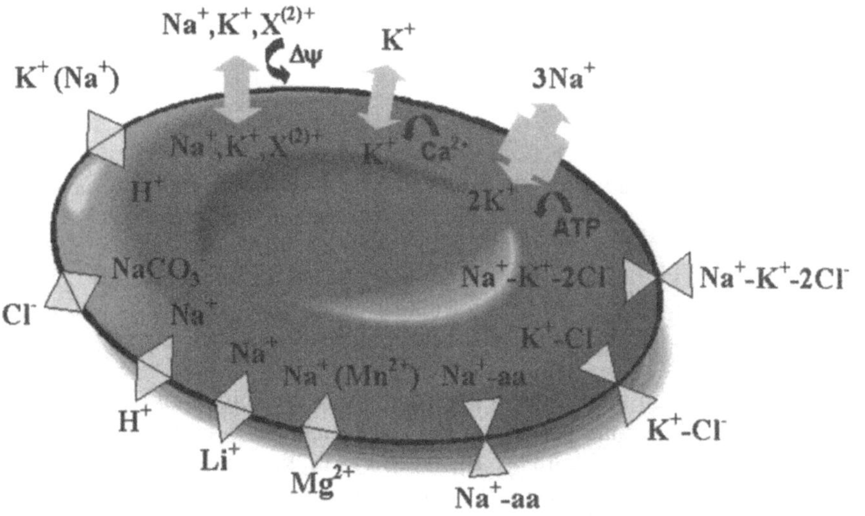

Fig. 4.1. Summary of the principal transport pathways for Na^+ and K^+ in the human RBC membrane: Na^+/K^+ pump, Na^+-K^+-$2Cl^-$ symporter, K^+-Cl^- symporter, Na^+-dependent amino acid (aa) transport (several discrete transporters), $Na^+(Mn^+)/Mg^{2+}$ antiporter, Na^+/Li^+ antiporter, Na^+/H^+ antiporter, $NaCO_3^-/Cl^-$ exchange (via band 3), $K^+(Na^+)/H^+$ antiporter, Ca^{2+}-activated K^+ channel, non-selective voltage-activated cation channel

In addition to the Na^+/K^+ pump (see also Chap. 5), the RBC membrane contains carriers and channels. Among the carriers, there are symporters and antiporters (or exchangers). We will use the terminology "cotransport" for both symport and antiport transport mechanisms, however, one should realise that it is used sometimes in the literature only for symporters. Some of the cotransporters are described and discussed in separate chapters of this book. For Na^+-K^+-$2Cl^-$ symporter, K^+-Cl^- symporter, Na^+-dependent amino acid transport, and Na^+/Mg^{2+} antiporter see Chaps. 8, 9, 12, and 16, respectively. The anion exchanger capnophorin (band 3) that under physiological conditions mediates HCO_3^-/Cl^- exchange can also transport Na^+ or Li^+ (but not K^+, Rb^+, Cs^+) as $NaCO_3^-$ or $LiCO_3^-$ (Funder 1980; Funder and Wieth 1980) and is reviewed in Chap. 11. The Na^+/H^+ and the Na^+/Li^+ antiporter are not considered in a separate chapter of this book but both are mentioned in the context of hypertonicity (see Chap. 25). The $K^+(Na^+)/H^+$ exchanger is discussed in detail in the present chapter since its discovery in the human RBC membrane is directly related to the explanation of the membrane leak over the last century (see below). The presently known results on cation channels in the human RBC membrane are discussed in Chap. 6.

In addition to the transport pathways for monovalent cations shown in Fig. 4.1, present in the RBC membrane of various species, there is a Na^+/Ca^{2+} exchanger in the RBC membrane from species of carnivora (dog, cat, seal, bear, ferret). RBC membranes from these species do not have a Na^+/K^+ pump (e.g. Bernhardt et al. 1988).

It is necessary to consider and eliminate transport via the large variety of specific pathways in the RBC membrane given in Fig. 4.1 before a formal description of the electrodiffusive leak can be attempted. One should realise that in many publications calculations of the cation permeability coefficients have been made without taking into account the existence of these specific transport pathways. In practice, the use of inhibitors for certain pathways as well as the choice of conditions at which the transport via one or more pathways (for which no inhibitors are presently known) is suppressed or not activated represents the most direct approach for eliminating specific mediated ion transporters. In general, such an approach of the elimination of all known specific transport pathways (pumps, carriers, channels) for a certain ion results in the possibility of the measurement of the remaining ion flux that is called "residual" transport of this ion across the membrane. This residual transport in fact can include further, but so far unknown, specific transport pathways, and therefore, should be regarded as different from the true leak ion flux across the membrane. The problem of identifying the membrane permeability coefficients for ions is related directly to the estimation of the electrical conductance of the membrane. A detailed overview in this respect is given by Hoffman (1992).

In general, the problem of how a monovalent cation passes (diffuses) through a biological membrane in the absence of a specific transport system is still unsolved. The obvious model system for ion transport studies, including studies on the electrodiffusive mechanism, appeared to be the RBC, which offered many experimental advantages for ion flux measurements. As one can see from the known transport pathways for monovalent cations presented in Fig. 4.1, it is much easier to measure the residual K^+ transport in these cells in comparison to the residual Na^+ transport because of the greater variety of specific transport systems for Na^+ compared to K^+. Until quite recently, the best methodological approach to measure residual K^+ fluxes involved unidirectional K^+ influx or efflux in media of physiological ionic strength in the presence of ouabain, bumetanide (or furosemide) and EGTA, with chloride replaced by nitrate (or methylsulphate), to suppress the Na^+/K^+ pump, Na^+-K^+-$2Cl^-$ cotransport, Ca^{2+}-activated K^+ channel, and any K^+ flux mediated by the Cl^--dependent K^+-Cl^- cotransport system, respectively. However, one has to take into consideration that under these conditions there are still two transport pathways present in the human RBC membrane, which are not affected. This are the voltage-dependent, non-specific cation channel and the $K^+(Na^+)/H^+$ exchanger. The first is activated at positive transmembrane potentials only (see

Chap. 6) and, therefore, should not play a substantial role at physiological conditions (i.e. negative transmembrane potentials). The contribution of the second pathway, however, is more complicated to analyse (see below). The $K^+(Na^+)/H^+$ exchanger was only recently identified in the human RBC membrane (Richter et al. 1997; Kummerow et al. 2000), and cloned in humans (Numata and Orlowski 2001).

In this context it becomes clear that conditions reported in the literature over the last 50 years assumed to influence the residual fluxes are not necessarily affecting the true leak since the presence of one or more specific transport pathways in these investigations cannot be ruled out. At least five different conditions have been described under which the assumed residual K^+ and Na^+ fluxes (not taking into consideration all existing specific transport pathways) can be altered:

(i) The replacement of extracellular NaCl by sucrose, i.e. at reduced ionic strength of the extracellular solution but at constant isotonic conditions, leads to an enhancement of the (ouabain + bumetanide + EGTA)-insensitive Na^+ and K^+ influx as well as efflux (Denner et al. 1993). This effect is explained in more detail in Sect. 4.3.

(ii) High hydrostatic pressure increases the (ouabain + bumetanide + EGTA)-insensitive Na^+ and K^+ transport (Hall and Ellory 1986). It is interesting to note that the cation fluxes stimulated by low ionic strength (LIS) and by high hydrostatic pressure show similar characteristics, e.g. linear concentration dependence, independence of the anion present, not specific for one monovalent cation, and reduction by cell swelling. In addition, the effect of LIS and high hydrostatic pressure are synergistic (Bernhardt et al. 1987a, 1991).

There are also many reports in the literature discussing a significant effect of membrane stress (caused by cell deformation) and of a defective membrane skeleton on the RBC residual cation transport (e.g. Johnson 1994; Joiner et al. 1995). In addition, there is an interesting report by Passow (1969) who found that human RBCs suspended in solutions of very high NaCl concentrations (up to 2.656 M), i.e. under conditions where the cells are dramatically shrunken, lose about 90% of their intracellular K^+ in less than one minute. At the same time, this K^+ loss is compensated by an enormous Na^+ uptake.

(iii) It was shown that in the presence of the anions salicylate or thiocyanate, the ouabain-insensitive fluxes of Na^+ and K^+ show a paradoxical temperature dependence with a flux minimum at about 20 °C (Wieth 1970). Later on, Blackstock and Stewart (1986) demonstrated that such a minimum also occurred in the presence of ouabain and bumetanide and in chloride media; but in this case it is shifted to 8 °C. If Na^+ is replaced by an organic cation (choline, *N*-methyl-D-glucamine, arginine, L-lysine, trimethylphenylammonium) the paradoxical temperature dependence is more pronounced and the minimum can be seen again at 20 °C. Interestingly, LIS-stimulated (ouabain + bumetanide + EGTA)-insensitive K^+ influx also shows such a paradoxical temperature dependence with a minimum at 8 °C (Bernhardt et al. 1991).

(iv) In various publications it is reported that the formation of disulfide bonds by different oxidative mechanisms results in enhanced leak fluxes for ions and hydrophilic substances (see e.g., Haest et al. 1981; Deuticke et al. 1983). Thus, e.g., the disturbance of the membrane barrier function after 5 mM diamide treatment has been interpreted as the formation of aqueous leaks (Deuticke et al. 1984). On the other hand, Lauf (1988) explained enhanced cation fluxes in sheep RBCs caused by diamide treatment (0–2 mM) as a stimulation of the K^+-Cl^- cotransport. Investigating the diamide-induced increase of the K^+ as well as Na^+ influxes in human RBCs under physiological as well as LIS conditions, Ihrig et al. (1991) were able to show that this effect is neither a result of a stimulation of the K^+-Cl^- cotransport system nor due to electrodiffusion through aqueous pores.

It is also known that organic mercurials (e.g. *p*-chloromercuribenzene sulphonate (PCMBS)), which bind to sulfhydryl groups increase Na^+ and K^+ fluxes across the human RBC membrane (Sutherland et al. 1967; Knauf and Rothstein 1971). However, also in this case it was shown that the PCMBS-induced increase of the K^+ and Na^+ transport is partly chloride-dependent and also involves a K^+/K^+ exchange (Haas and Schmidt 1985).

(v) A number of studies have indicated that membrane expansion with a variety of amphiphilic substances, including charged, zwitterionic, and uncharged molecules, can also modify the residual ion transport (Fortes and Ellory 1975; Isomaa et al. 1986). For drug effects on ion permeability of the RBC membrane see also Deuticke et al. (1990) and Bernhardt et al. (1999).

4.3 The Low Ionic Strength (LIS) Effect

Although it was assumed that under physiological conditions the RBC membrane is impermeable to salts (see above), already Gürber (1904) noticed that RBCs from ox lose nearly all Cl^- and Na^+ after repeatedly washing of the cells in cane sugar solutions. Since he centrifuged the cells up to 28 times, membrane disturbances due to mechanical stress could not be ruled out. Some years later, Bang (1909) demonstrated with intelligent and carefully done haemolysis experiments diluting both NaCl and cane sugar solutions with water that indeed the bovine RBCs lose salt in cane sugar solutions in a relatively short time (0.5 h). These observations were confirmed by Mond (1927) on the basis of direct chemical analysis of Cl^- and K^+ in the extracellular solutions of bovine and pig RBCs suspended in cane sugar solutions with $pH > 8$ (in this paper it was suggested that the RBC membrane is permeable to anion and impermeable to cations at $pH < 8$ but at $pH > 8$ the situation is *visa versa*).

In 1939, Davson showed that there is an increase in net K^+ efflux of a variety of mammalian RBCs including human RBCs in isotonic LIS solutions (NaCl replaced for sucrose). This effect was confirmed by a number of other authors (Wilbrandt 1940; Wilbrandt and Schatzmann 1960; Carolin and Maizels 1965; LaCelle and Rothstein 1966). However, the replacement of NaCl by sucrose results not only in a change of the ionic strength but also in a change of the transmembrane potential (the transmembrane potential of erythrocytes in physiological solutions is the Nernst potential for chloride, see Sect. 4.6). Therefore, attempts have been made to explain the increased K^+ efflux of human erythrocytes solely on the basis of electrodiffusion (Donlon and Rothstein 1969). These authors had to assume a change in the permeability coefficient with respect to the transmembrane potential because the measured increase in K^+ efflux in a solution of low NaCl concentration was much higher than expected from the Goldman flux equation (for equation see Goldman 1943). It is important to emphasize that the Goldman flux equation is derived from the linear relationship:

$$\text{Flux } (J) = \text{Linear coefficient } (L) \times \text{Driving force } (X) \tag{4.1}$$

Thus, L is a constant independent of X. This means that if the permeability coefficient (part of L) depends on the transmembrane potential (part of X), as proposed by Donlon and Rothstein (1969), the Goldman flux equation does not hold under these conditions. To overcome this problem another attempt was made by Bernhardt et al. (1984). They extended the Goldman flux equation by taking into account the inner and outer surface potentials of the RBC. Using this extended Goldman flux equation, it was possible to explain the increase in K^+ efflux observed in human RBCs suspended in LIS media without considering a change in the permeability coefficient. The idea was to calculate the electrodiffusive ion flux taking into account the electrical potential difference existing at the inner and outer membrane surface (this is different from the transmembrane potential that is the difference of the potential from the middle of the cell to the outside infinity), as well as the ion concentration near the membrane surface (that can be calculated using the Boltzmann equation). For an illustration, see Sect. 4.6. However, later it became evident that such an explanation is also not likely to be the case. One reason for this statement is the observation that no enhanced K^+ efflux was seen in bovine RBCs under the same experimental conditions as for human RBCs (Bernhardt 1986; Bernhardt et al. 1987b). The transmembrane potential of bovine RBCs, however, changes by the same amount as that for human RBCs after reduction of the extracellular chloride concentration. In addition, cell electrophoretic measurements made at varying extracellular NaCl concentrations have shown that there is a similar change in the external surface potential of cow RBCs compared with human RBCs (Bernhardt 1986). Furthermore, it can be assumed that the internal surface potentials of cow and human RBCs do not differ significantly, since all of the negatively charged membrane phospholipid, phosphatidylserine, is located in the inner leaflet of the cell membrane (van Dijck et al. 1976).

Comparison of the rate constants of the K^+ efflux from RBCs of different mammalian species in physiological ionic strength solution as well as those of LIS showed that there was a significant increase in the K^+ efflux in LIS solution in human, cat, rat, horse, and rabbit RBCs, whereas no significant change was observed in pig and cow RBCs (Bernhardt et al. 1986; Erdmann et al. 1990).

Although cow RBCs do not show the LIS effect, RBCs from newborn calves show the same increase in the rate constant of K^+ efflux in a LIS solution as human RBCs. This increase depends on the age of the calves; it has a maximal effect one day after birth (earliest time after birth measured) and is diminished 6 weeks after birth of the animals (Bernhardt et al. 1992).

Based on these findings it was necessary to take into account a possible participation of one of the above mentioned specific cation transport pathways (at that time the $K^+(Na^+)/H^+$ exchanger was not known). One should realise that no specific transport inhibitors were used in the experiments until the 1980s. Thus, it has been shown that there is no involvement of a partial flux mediated by the Na^+/K^+

pump or the Ca^{2+}-activated K^+ channel in the increase of the K^+ efflux in LIS media (Erdmann et al. 1990). Using furosemide to block the Na^+-K^+-$2Cl^-$ cotransporter it could be shown that both the K^+ efflux mediated by this cotransporter as well as the remaining residual K^+ efflux (in the presence of furosemide) show a marked enhancement in LIS solutions (Bernhardt et al. 1987b).

With the beginning of the 1990s, the investigations of the LIS effect were extended to measurements of the residual K^+ influx (Bernhardt et al. 1991). Using the influx measuring technology it was possible to extend the characterisation of the LIS effect. An important finding was the demonstration of a marked enhancement of the (ouabain + bumetanide + EGTA)-insensitive K^+ influx (the so called residual K^+ influx, see above) when human RBCs were suspended in LIS solution. This result serves as an additional indication that the transmembrane potential is not responsible for the LIS effect. It is not possible to explain a dramatic increase of the leak K^+ influx of human RBCs by reducing the extracellular NaCl concentration on the basis of the Goldman flux equation (Bernhardt et al. 1991). Under these conditions one would predict a threefold decrease in the unidirectional leak K^+ influx. This is because of a depolarisation of the membrane from the physiological value of about −10 mV to about +50 mV in solution of low NaCl concentration (but see Sect. 4.6). This point was studied in more detail by altering the anion concentration of the extracellular solution and adding an impermeant anion, gluconate or glucuronate. Under these conditions there was only a small increase in the K^+ influx, in marked contrast to sucrose solution, although the transmembrane potentials were equal in both cases (identical low extracellular chloride concentration) (Bernhardt et al. 1991). However, there were first indications that the residual K^+ influx is sensitive to the internal pH (pH_i). Chipperfield and Shennan (1986) and Zade-Oppen et al. (1988) also demonstrated a pH dependence of the residual K^+ transport in human RBCs (see also Sect. 4.4).

Another important point was the demonstration that the K^+-Cl^- cotransport system is not involved in the LIS effect. As the K^+-Cl^- cotransport system in human RBCs is dependent on Cl^- (or Br^-), methylsulphate ($CH_3SO_4^-$) replacement suppresses the K^+ flux via this transport system. It could be shown, however, that the LIS-induced flux is independent of Cl^- (Bernhardt et al. 1991). Volume sensitivity (i.e. activation by cell swelling) is another important characteristic of K^+-Cl^- cotransport (Ellory and Hall 1988). In contrast, the LIS effect decreases slightly on cell swelling. Furthermore, the K^+-Cl^- cotransport system excludes Na^+, whereas the LIS effect is shown equally for Na^+ fluxes (efflux as well as influx) (see Bernhardt et al. 1991).

It can also be concluded that the voltage-activated cation channel (cf. Chap. 6) does not participate in the LIS effect. This channel occurs in human RBCs and in high-potassium-type (HK) but not low-potassium-type (LK) sheep RBCs (Halperin et al. 1989, 1990). In contrast, the effect of an increase of residual K^+ fluxes in a LIS medium was observed in LK but not in HK sheep RBCs (Erdmann et al. 1991). Furthermore, in human RBCs the channel does not discriminate between Na^+, K^+, and Ca^{2+} (Halperin et al. 1989; Kaestner et al. 2000), whereas Ca^{2+} transport is not significantly changed in LIS medium in comparison to physiological solution (Kucherenko and Bernhardt, unpublished results).

In addition to the reported results, the following characteristic features of the residual K^+ (and Na^+) transport could be demonstrated (Bernhardt et al. 1991):

(i) It is known that human RBCs suspended in sucrose solution with the same osmolarity as the physiological NaCl solution (290–300 mosmol/l) have a smaller cell volume. This shrinkage, however, is not responsible for the elevated K^+ influx observed in LIS solutions, since a sucrose medium with an osmolarity of 250 mosmol/l (which leads to about the same cell volume of human RBCs as in physiological ionic strength solution) only slightly influences the residual K^+ transport.

(ii) The enhanced residual K^+ flux observed in LIS solution was reversible when the RBCs were re-suspended in a solution of physiological ionic strength.

(iii) Both Na^+ and K^+ influxes were elevated in the LIS medium by about the same amount. In addition, both fluxes were linear with the concentration showing no signs of saturation over the range studied (0.25–10 mM). In agreement with these findings, Garay et al. (1981) showed that the rate constant of the residual Na^+ and K^+ influxes into media with high $MgCl_2$ concentrations (75 mM) were a linear function of the internal Na^+ or K^+ concentrations. On the other side, Canessa et al. (1986) have shown that the residual K^+ efflux into K^+- and Na^+-free solutions saturates when the intracellular K^+ concentration is increased to 40 mM. However, if high K_m values are a property of the flux, saturation will appear only at higher ion concentrations.

Based on the findings described above, there are several possible explanations for the LIS effect on residual transport. A general explanation might be that lowering ionic strength of the solution might result in altered protein-protein and/or lipid-protein interaction, leading to regions able to allow transmembrane ion movements at instabilities between proteins and/or lipid molecules.

Another possibility is that a specific membrane protein is involved in the LIS-stimulated residual K^+ and Na^+ fluxes. Solomon et al. (1983) have already speculated that the anion transport protein (band 3) could be involved in cation transport. The idea of an involvement of band 3 in the LIS-stimulated residual K^+ and Na^+ fluxes was further developed by Jones and Knauf (1985) finding that the addition of 4,4'-diisothiocyanatostilbene-2,2'-disulphonic acid (DIDS; an inhibitor of anion transport via band 3) to the flux solutions results in a significant decrease of the K^+ efflux in a LIS medium as compared to the same solution without DIDS. In contrast, there is no significant influence of DIDS on K^+ efflux in a physiological ionic strength solution (for discussion of a possible involvement of band 3 see also Sect. 4.4).

An alternative strategy was to explain the LIS-stimulated residual K^+ and Na^+ fluxes on the basis of an electroneutral cation carrier mechanism. In this respect the first attempt was made by Denner et al. (1993). These authors tested various carrier models to fit the experimental data of the K^+ and Na^+ effluxes as well as influxes in solutions of different ionic strength simultaneously. It was possible to describe the residual K^+ and Na^+ fluxes on the basis of a carrier mechanism of competing substrates, however, with modifier sites.

4.4 The $K^+(Na^+)/H^+$ Exchanger

Further detailed theoretical model calculations of possible carrier mechanisms and tracer-kinetic K^+ flux measurements at different pH's of the extracellular solutions lead to the idea that a $K^+(Na^+)/H^+$ exchanger is involved in the LIS-stimulated residual K^+ and Na^+ fluxes (Richter et al. 1997). A fundamental idea in their work was to assume that the local K^+ and Na^+ concentration near the binding site of the hypothetical carrier, i.e. near the membrane surface, are of importance for the carrier-mediated ion fluxes. It is evident that a reduction of the ionic strength of the extracellular solution at a constant negative surface charge density will lead to an enhancement of the absolute value of the negative outer membrane surface potential. This in turn will result in a higher cation concentration near the surface in comparison to the free solution. Combining the corresponding equations describing these effects, i.e. the linearized Gouy-Chapman equation and the Boltzmann equation, and assuming that the ion flux is proportional to the ion concentration (which was demonstrated in separate experiments), a final equation showing that the logarithm of the apparent rate constant of the ion flux is proportional to 1 divided by the square root of the ionic strength is obtained (for more details see Richter et al. (1997)). As a matter of fact, the experimental results of the increased residual K^+ and Na^+ influxes as well as effluxes in dependence on the reduction of the extracellular ionic strength plotted as logarithm of the flux rate constant *vs.* 1/square root of ionic strength showed linear lines for all 4 fluxes (see Fig. 4.2).

In addition to the theoretical investigations based on Na^+ and K^+ flux measurements, it was necessary to demonstrate experimentally a 1:1 relation of the residual, i.e. (ouabain + bumetanide + EGTA)-insensitive, K^+ efflux and the H^+ influx. Therefore, Kummerow et al. (2000) investigated the change of intracellular pH of RBCs under different experimental conditions using the pH-sensitive fluorescent dye 2',7'-bis-(2-carboxylethyl)-5(6)-carboxyfluorescein (BCECF). The net H^+ influx was calculated from intracellular pH (pH_i) measurements taking into account the buffer capacity of haemoglobin (10 mmol/(mmol haemoglobin × pH unit); Dalmark 1975). To convert the fluorescence ratio of BCECF-loaded RBCs into pH_i values, a calibration was carried out by equalizing pH_i and pH_o using the K^+/H^+ ionophore nigericin. In addition to the pH_i and K^+ efflux (Rb-86 as tracer) measurements, the Cl^- efflux (Cl-36 as tracer) from RBCs in LIS media in the absence of extracellular Cl^- was determined.

When human RBCs were suspended in a physiological NaCl solution ($pH_o = 7.4$), the measured pH_i was 7.19 ± 0.04 which remained constant for 30 min. When RBCs were transferred into a LIS solution (NaCl replaced by sucrose) an immediate alkalinization increased the pH_i to 7.70 ± 0.15, which was followed by a slower cell acidification (see Fig. 4.3).

To decide whether the H^+ flux across the RBC membrane depends on transmembrane potential, experiments were carried out in a solution in which Cl^- was replaced by the impermeable anion tartrate, shifting the transmembrane potential to substantial positive values (cf. Lew and Bookchin 1986). The replacement of Cl^- by tartrate led to nearly the same alkaline shift of pH_i as under conditions where the cells were transferred into LIS medium. However, the subsequent acidification was much slower in tartrate compared to LIS medium, since the tartrate

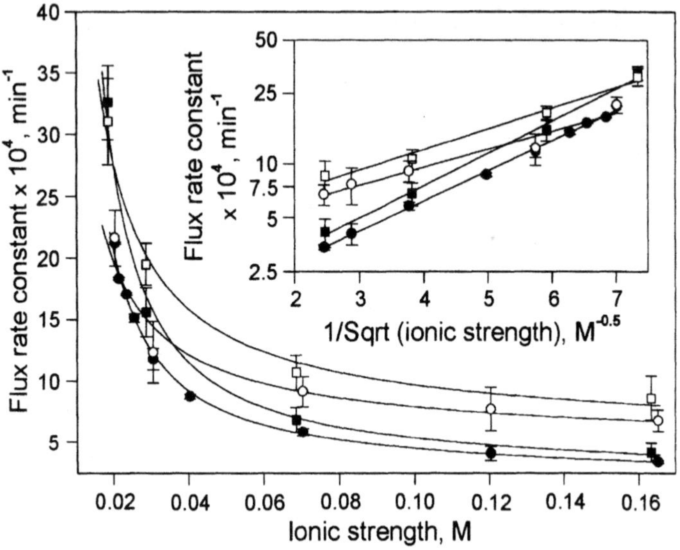

Fig. 4.2. Effect of gradual NaCl or KCl (for Na^+ influx) replacement (0–145 mM) by sucrose (decreased ionic strength, constant osmolarity) on the rate constant of Na^+ (○, □) and K^+ (●, ■) efflux (□, ■) and influx (○, ●) of human RBCs. Na^+ and K^+ fluxes were measured in the presence of 0.1 mM ouabain, bumetanide, EGTA. The insert shows the same experimental results plotted as logarithm *vs.* reciprocal of the square root of ionic strength of the flux solutions. Taken from Richter et al. (1997), for more details see therein

solution was 212 mM, i.e. high ionic strength. Since the transmembrane potential after the rapid alkalinization of the cells is similar in both LIS and tartrate solutions (see also Sect. 4.6), it is evident that the observed H^+ influx was not due to the altered transmembrane potential.

That the effect of immediate alkalinization of cells in LIS media occurs via the anion transport system (band 3) is supported by the following findings (Kummerow et al. 2000): (i) an efflux of about 60 mM Cl^- occurred immediately, accompanied by an increase of pH_i of 0.51 units, an observation consistent with 1:1 Cl^-/OH^- exchange when the buffer capacity of haemoglobin is taken into account; (ii) the Cl^- loss as well as the pH_i increase was strongly inhibited by DIDS and showed the same delayed time course in the presence of the anion transport inhibitors 4,4'-dinitrostilbene-2,2'-disulphonic acid (DNDS) or niflumic acid. In addition, in the seconds immediately following suspension in LIS solution the rapid loss of Cl^- is not accompanied by a significant loss of K^+ from the cells.

The main focus in the investigations of Kummerow et al. (2000), however, was the process of cell acidification after the initial pH_i increase. The movement of H^+ (or H^+ equivalents) was found to increase with decreasing ionic strength of the solution. The comparison of the calculated H^+ influx with the measured unidirectional K^+ efflux at different extracellular ionic strengths showed a correlation with a stoichiometry nearly 1:1. In LIS solution the H^+ influx and the K^+ efflux was

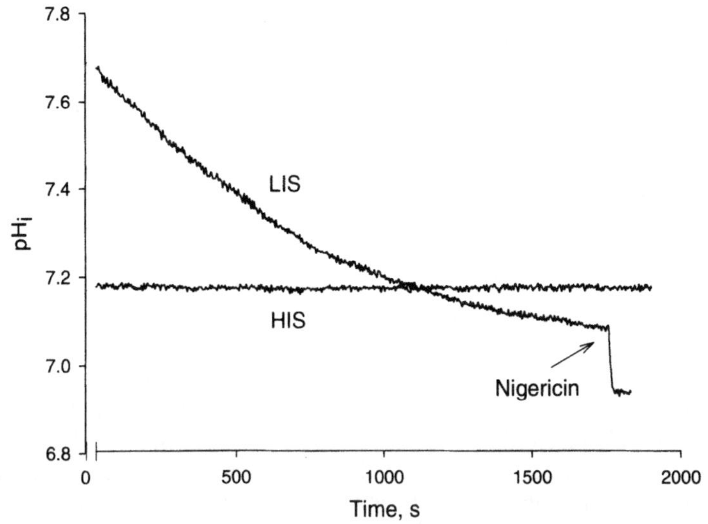

Fig. 4.3. Online records of intracellular pH (pH_i) changes in human RBCs in physiological ionic strength (HIS) and low ionic strength (LIS) solution using the pH-sensitive fluorescent dye BCECF-AM. Nigericin (10 µM) was added after 30 min inducing a K^+ and H^+ equlibrium across the cell membrane. Taken from Kummerow et al. (2000), with permission of the publisher, for more details see therein

108.2 ± 20.4 mmol/(l_{cells}·h) and 98.7 ± 19.3 mmol/(l_{cells}·h), respectively. For bovine and porcine RBCs, in LIS media, H^+ influx and K^+ efflux were of comparable magnitude, but only about 10% of the fluxes observed in human RBCs under LIS conditions. Quinacrine, a known inhibitor of the mitochondrial $K^+(Na^+)/H^+$ exchanger (Garlid et al. 1986), inhibited the K^+ efflux in LIS solution by about 80% and inhibited the H^+ influx nearly completely (demonstrated by pH measurements of cell lysates).

Further evidence for the existence of the $K^+(Na^+)/H^+$ exchanger in the human RBC membranes is provided by the effect of three classical anion transport inhibitors (DIDS, DNDS, niflumic acid) on the K^+ efflux and H^+ influx in LIS solution (Kummerow et al. 2000). These investigations were carried out to see whether the anion exchanger (band 3) is involved in the process of acidification. DIDS and DNDS reduced the K^+ efflux to about 8.1 ± 1.2 mmol/(l_{cells}·h) and 7.7 ± 1.5 mmol/(l_{cells}·h), respectively. Niflumic acid did not block the K^+ efflux significantly. However, niflumic acid induced a delay of the K^+ efflux of the RBCs, comparable with the deceleration induced for the H^+ influx (see above). There might also be a delay time for the K^+ efflux in the presence of DNDS, however, the flux is too small to see such an effect.

When the human RBCs were transferred from the physiological NaCl solution into LIS solution containing the inhibitors, inhibitory effects on the time course of pH_i were found. In the presence of DIDS the pH_i was 6.84 ± 0.05 and constant over 30 min. If, on the other hand, DIDS was added to the cell suspension after the erythrocytes were transferred to the LIS media, resulting in an initial increase of pH_i, a significant inhibition of the acidification process was observed. The acidification of the cells in the solution containing DNDS was also significantly smaller than in LIS solution without the inhibitor. In contrast, the decrease of pH_i in the presence of niflumic acid is much faster than in the presence of DNDS or DIDS and comparable to the acidification in the absence of any anion transport inhibitor. For DNDS and niflumic acid no significant difference in the pH_i decrease, i.e. the H^+ flux, was observed if the inhibitors were present either before or after transferring the cells to the LIS solution.

Alternatives to a $K^+(Na^+)/H^+$ exchanger for explaining the described effects include: (i) a separate channel for K^+ and net H^+ (OH^-) transport via band 3; (ii) K^+/OH^- symport. The first possibility is supported by the model of Lew and Bookchin (1986) assuming that the membrane permeability for K^+ is increased under LIS conditions. Such a raised permeability will inevitably be inferred from flux measurements if only electrodiffusive processes are considered. Should electrodiffusion underlie the LIS response for K^+, efflux of this ion would be predicted to be dependent on transmembrane potential. However, data from previous reports (e.g. Bernhardt et al. (1991)) suggest that LIS-induced fluxes do not show such sensitivity, and support the proposal that an electroneutral $K^+(Na^+)/H^+$ exchanger is more likely. For such a mechanism, it would be unnecessary to invoke a change in membrane permeability to explain the LIS effect. Second, although a K^+/OH^- symport cannot be ruled out, a K^+/H^+ exchange is again more likely. This is based on the apparent dependence of the K^+, Na^+, and H^+ (or OH^-) fluxes on surface potential. The surface potential of the RBC membrane is more negative under LIS conditions than under normal physiological conditions (e.g. Bernhardt 1994). Therefore, in LIS media an increase of the cation concentration and a decrease of the anion concentration near the cell surface will arise. This would probably further lower the transport rate of a cation-anion symport under LIS conditions.

As already mentioned above, Jones and Knauf (1985) have suggested that the LIS-induced cation transport is mediated by band 3, acting as a low-conductivity channel for cations under these conditions. Although the possibility of K^+ transport via band 3 cannot be entirely excluded, the above results, in our opinion, favour a $K^+(Na^+)/H^+$ exchanger separate from the band 3 protein. One reason to exclude a band 3-mediated process has been already indicated, namely the influence of surface potential. Second, the anion transport inhibitor niflumic acid, although able to retard LIS-induced alkalinization, has no effect upon the K^+ efflux and H^+ influx. However, comparison of the effects of the various anion transport inhibitors suggests that DIDS and DNDS directly inhibit the $K^+(Na^+)/H^+$ exchanger. Such an effect is not surprising since it is known that DIDS affects the K^+/H^+ exchanger proposed to operate in *Amphiuma* and trout RBCs (Adorante and Cala 1987; Fievet et al. 1993). Third, quinacrine, a known inhibitor of the $K^+(Na^+)/H^+$ exchanger in the mitochondrial membrane (Garlid et al. 1986), inhibits the K^+ efflux and the H^+ influx in LIS solution but has no effect on anion transport. Fourth, bovine and porcine RBCs have normal band 3 activity, but no measurable fluxes equivalent to the proposed $K^+(Na^+)/H^+$ exchange in human RBCs.

A more complex mechanism involving band 3 would be to assume that Cl^-, leaving the cells immediately after the cells have been transferred into the LIS medium (via Cl^-/OH^- exchange), is re-entering the cells slowly over the next 30 min with hydroxyl ions coming out. In this case, Cl^- could leave the cells slowly together with K^+ (and Na^+). If there are two separate pathways for Cl^- and K^+, both should be dependent on the transmembrane potential, which is not the case (see above). On the other hand, one could think about a role for K^+-Cl^- co-transport (Hall and Ellory 1986). However, as was shown by Richter et al. (1997), the K^+-Cl^- cotransport is not involved in this process. In addition, the same arguments against an involvement of band 3 in the process given above can be used to conclude that such a complex mechanism is less likely than a $K^+(Na^+)/H^+$ exchanger.

4.5 General Consideration of the Residual and Leak Cation Fluxes

Whatever the mechanisms of the residual fluxes in human RBCs, it is evident that there are one or more specific monovalente cation transport pathways involved. One approach is to find inhibitors for these mechanisms before solving the general question of how an ion can move across a biological membrane without using a specific transport pathway, i.e. pump, channel or carrier. Discovering in specific ion transport pathways this way, it will finally lead to the possibility of the measurement of the true leak flux of an ion across a biological membrane. Probably the RBC membrane will serve again as the classical example to solve this problem. Whether or not the Goldman flux equation will then be able to describe the true leak is still a matter of debate. However, in our opinion the Goldman flux equation does not fulfill the requirements to explain the leak ion fluxes. Simply, in an adequate time of ion diffusion there is no stable water-filled pore in the biological membrane that allows free ion diffusion, i.e. without an interaction of the moving ion with the structure of the pore. Therefore, the Goldman flux equation can principally be a helpful tool for ion channel investigations but never for explaining the mechanism of a leak ion flux! Since the leak ion transport on the basis of our present understanding of a biological membrane can be assumed to occur due to fluctuations of membrane constituents at instabilities between proteins and/or lipid molecules, it needs a new mathematical description of this process based on equations describing these biological fluctuations.

The permeability coefficient of an artificial lipid bilayer membrane e.g. for Na^+ and K^+ is assumed to be about 2 orders of magnitude lower than for biological membranes. Such a statement is given in many publications despite the fact that the true leak permeability of biological membranes is not known (see above). Explaining the mechanism of a leak transport of a small cation across a lipid bilayer membrane, one should also realise that there are two different possibilities discussed in the literature. The fundamental approach is the "solubility-diffusion mechanism" based on the classical concept first described by Overton (1899). This mechanism postulates that the permeating ion must first partition into the membrane, diffuse across it, and finally leave the membrane at the other side. It is assumed that the diffusion through the hydrophobic region of the bilayer membrane is the rate-limiting step of the transport process. Although on the basis of the "solubility-diffusion mechanism" it is possible to explain many experimental results of permeability investigations, it does not explain all results. In particular studies with thin lipid bilayer membranes show deviations from the predicted theoretical results. Therefore, another mechanism, the permeation through "transient hydrated pores or water-filled channels" formed by fluctuations in the bilayer membrane has been postulated (e.g. Solomon 1968). For detailed mathematical description as well as for comparison of the two mechanisms the reader is referred to Weiss (1996) and Paula and Deamer (1999). In addition to both mechanisms, it needs a different explanation for the permeability of protons across an artificial or biological membrane. The reason is that the H^+ permeability coefficient is several orders of magnitude higher than expected from the permeability coefficients of other monovalent cations. Therefore, it has been suggested that the complex ion $H_9O_4^+$ (4 water molecules plus a proton) is permeating the lipid bilayer membrane corresponding to the "solubility-diffusion mechanism" (Paula et al. 1996). Alternatively, short-lived transient pores or defects could be present in the lipid bilayer membrane allowing protons but not other cations to be transported through the membrane (Nichols and Deamer 1980; Marrink et al. 1996).

The incorporation of integral membrane proteins, including proteins that have no specific transport function, into artificial phospholipid membranes, in general results in a significant increase of the bilayer permeability for non-electrolytes as well as ions (e.g. van der Steen et al. 1982; van Hoogevest et al. 1983, 1984). If, however, not a single phospholipid but a lipid extract from erythrocyte is used to make protein-containing lipid bilayer vesicles, these vesicles are much less permeable to these solutes (van der Steen et al. 1982). From such findings it has been suggested that packing defects at the lipid-protein interface and/or pores in protein aggregates are of importance for the membrane permeability to non-electrolytes and ions (van der Steen et al. 1982; van Hoogevest et al. 1983, 1984).

Also for RBCs there is some evidence that a modification of the membrane lipids affecting the lipid-protein interaction is of importance for the leak ion flux across the membrane. Using a phospholipid exchange protein, Kuypers et al. (1984) showed that human RBCs were leaky for K^+ if the native phosphatidylcholine was partly replaced by phosphatidylcholine containing arachidonic acid. However, it has to be clarified whether the arachidonic acid (or highly unsaturated fatty acid) content of the membrane phospholipids is really affecting the true leak or whether it is modifying a specific transport pathway. Dwight and Hendry (1995) described a significant increase in the residual, i.e. (ouabain + bumetanide)-insensitive $K^+(Rb^+)$ influx after insertion of non-esterified arachidonic acid into the erythrocytes. To explain the differences of the LIS-stimulated K^+ flux across the RBCs of various mammalian species, an attempt was made to correlate the measured rate constants in solutions of physiological and low ionic strength with data for the membrane phospholipids of the RBCs of these species. Bernhardt et al. (1992) found some evidence that the ionic strength-induced alteration of the residual cation transport (at that time, but see for $K^+(Na^+)/H^+$ exchanger) correlates with the arachidonic acid content of the membrane phospholipids. Another important finding in this respect was that the increase in the K^+ efflux in LIS solution of RBCs from newborn calves depending on the age of the calves (see Sect. 4.3) also correlates with the arachidonic acid content of the membrane phospholipids. It should be mentioned that also for the anion transport protein in the RBC membrane (band 3) of different species, correlations were found between the phosphate influx as well as the bicarbonate-chloride exchange and the arachidonic acid content of the phospholipids of the RBC membrane (Gruber and Deuticke 1973; Lu and Chow 1982).

In addition, there are many reports in the literature demonstrating an effect of a changed cholesterol/phospholipid (C/P) ratio on various membrane transport systems as well as membrane permeability for non-electrolytes and ions of RBCs. However, it remains to be clarified whether the C/P ratio is indeed affecting the leak ion transport. Alternatively, it could be possible that a changed C/P ratio is influencing specific transport pathways only, since many of them have not been identified at the time the investigations were carried out.

4.6 The Transmembrane Potential, Surface Potential, and the Electric Field in the Membrane

Under physiological condition, the net ion movement of K^+ and Na^+ in comparison to Cl^- across the RBC membrane is very small (about 2 orders of magnitude lower). Therefore, the electric transmembrane potential (*Δψ*; cf. Fig. 4.4) can be described as a diffusion potential of K^+ and Na^+ only:

$$\Delta\psi = \frac{RT}{F}\ln\frac{P_K c_K^o + P_{Na} c_{Na}^o}{P_K c_K^i + P_{Na} c_{Na}^i} \tag{4.2}$$

It is identical to the Nernst potential for chloride and protons:

$$\Delta\psi = \frac{RT}{F}\ln\frac{P_{Cl} c_{Cl}^i}{P_{Cl} c_{Cl}^o} = \frac{RT}{F}\ln\frac{P_H c_H^o}{P_H c_H^i} \tag{4.3}$$

In Eqs. (4.2) and (4.3), P_K, c_K, P_{Na}, c_{Na}, P_{Cl}, c_{Cl}, P_H, c_H are the permeability coefficients and concentrations for K^+, Na^+, Cl^-, and H^+, respectively. Symbols *o* and *i* denote outside and inside the membrane. *R* is the gas constant, *F* the Faraday constant, and *T* the absolute temperature. Both equations are based on the following consideration (for detailed mathematical analysis the reader is referred to Kotyk and Janacek 1977):

The calculation of the transmembrane potential is possible only for conditions where the sum of all partial ion fluxes is zero, i.e. for steady-state conditions only! Thus, $\Sigma\, z_i F J_i = 0$, where J_i are the electrodiffusive fluxes of all ion species. Taking into account only K^+, Na^+, and Cl^-, the transmembrane potential can be calculated using the Goldman-Hodgkin-Katz equation (Goldman 1943; Hodgkin and Katz 1949):

$$\Delta\psi = \frac{RT}{F} \ln \frac{P_K c_K^o + P_{Na} c_{Na}^o + P_{Cl} c_{Cl}^i}{P_K c_K^i + P_{Na} c_{Na}^i + P_{Cl} c_{Cl}^o} \tag{4.4}$$

If, however, the Cl^- distribution is in equilibrium, and therefore the sum of the Cl^- influx and efflux is zero, the Cl^- ions can be neglected in the derivation of the Eq. (4.4) of the transmembrane potential. This Cl^- equilibrium exists due to the fact that for RBCs the net Cl^- permeability is about 2 orders of magnitude larger than the K^+ and Na^+ permeabilities.

The value of the transmembrane potential of human RBCs in physiological solution of about –5 to –10 mV was estimated with different techniques such as microelectrodes (Lassen and Sten-Knudsen 1968; Jay and Burton 1969), fluorescent dyes (Hoffman and Larris 1974; Freedman and Hoffman 1979), and measurements of distribution of chloride ions and radioactive lipophilic ions (Funder and Wieth 1966; Donlon and Rothstein 1969; Beaugé 1975; Deutsch et al. 1979). Taking into account the transmembrane potential as well as standard values for intracellular and extracellular Na^+ and K^+ concentrations of human red blood cells under physiological conditions, from Eq. (4.2) it follows that P_K/P_{Na} is about 1.5.

If the situation of the analysis of the transmembrane potential of RBCs in physiological solutions is relatively simple, it changes under non-physiological conditions (except in a situation where the permeability of one ion species is much higher than the permeability of the others, e.g. in the presence of the potassium carrier valinomycin leading to the Nernst potential for K^+). Under non steady-state-conditions as in LIS media, the transmembrane potential changes with time until a new steady-state situation is reached. During this process, the transmembrane potential can be estimated only by taking into account the permeability coefficients as well as the internal and external concentration of K^+, Na^+, and Cl^-. In addition, it should become clear that under such conditions the transmembrane potential cannot be estimated even roughly as the Nernst potential for Cl^- (as assumed in some publications). This is due to the fact that the Cl^- permeability is necessarily no longer 2 orders of magnitude larger than the K^+ and Na^+ permeabilities (this statement holds both under conditions where a relative decrease of the Cl^- permeability or/and a relative increase of the K^+ and/or Na^+ permeability occurs). An illustration of the situation is given below.

Let us assume that in LIS solution we have only 7.5 mM KCl and 0 mM NaCl (NaCl is replaced by sucrose to maintain osmolarity). The internal concentrations for ($K^+ + Na^+$) and for Cl^- are 152 mM and 77 mM, respectively. In addition, for simplification let us assume that $P_K = P_{Na} = P_{Kat}$. With these assumptions it is easy to calculate the transmembrane potential ($\Delta\psi$) assuming quasi-steady-state-conditions in dependence on the relation of P_{Cl} to P_{Kat} (see Table 4.1). For comparison, the Nernst potential for Cl^- (at 37 °C) would be +62.2 mV.

Thus, the transmembrane potential of human RBCs after transferring the cells into a LIS solution clearly depends on the P_{Cl}/P_{Kat} ratio. Assuming a large and immediate increase in cation permeability following transfer to a LIS solution, as many people do on the basis of the classical investigations (see Sect. 4.3) it will be impossible to estimate the transmembrane potential correctly. If, in contrast, one assumes that the increased K^+ and Na^+ fluxes are to a great extent due to the stimu-

Table 4.1. Transmembrane potential ($\Delta\psi$) of human RBCs in LIS solution in dependence on the ratio of the membrane permeability of Cl^- (P_{Cl}) to the membrane permeability of cations (P_{Kat})

P_{Cl}/P_{Kat} ratio		$\Delta\psi$ [mV]
$P_{Cl} >> P_{Kat}$	$(P_{Cl} = 100\ P_{Kat})$	+57.3
$P_{Cl} > P_{Kat}$	$(P_{Cl} = 10\ P_{Kat})$	+32.9
$P_{Cl} = P_{Kat}$		−17.0
$P_{Cl} < P_{Kat}$	$(10\ P_{Cl} = P_{Kat})$	−61.0

lation of the electroneutral $K^+(Na^+)/H^+$ exchanger (see Sect. 4.4) it is possible that in LIS solution the P_{Cl}/P_{Kat} ratio is not changed significantly. However, a more realistic explanation has to be considered. After transferring the cells into a LIS solution, not only the $K^+(Na^+)/H^+$ exchanger but also the non-specific voltage-activated cation channel (see Chap. 6) is activated. Depending on how far the ionic strength of the extracellular solution is decreased, i.e. on how positive the transmembrane potential becomes immediately after transferring the cells into a LIS solution, the non-specific voltage-activated cation channel is activated or not. This view is in agreement with the findings reported by Donlon and Rothstein (1969). These authors described a triphasic increase of K^+ efflux when the extracellular NaCl concentration was reduced (osmolarity compensated by adding sucrose). When the transmembrane potential of human RBCs, calculated as the Nernst potential for Cl^-, was changed from 0 mV to about +40 mV, an increase of K^+ efflux was observed. Changing the transmembrane potential to values higher than +40 mV produced a more pronounced effect, and at transmembrane potentials higher than +170 mV a third phase with a dramatic increase of the K^+ efflux, probably due to the beginning of the electrical breakdown of the membrane, was seen. If a significant opening of the non-specific voltage-activated cation channel is occurring there would be an increased permeability of the RBC membrane to K^+ and Na^+. As a general conclusion, there should be no doubt that the transmembrane potential immediately after the cells are surrounded by the LIS solution is positive. In a very short time interval (some seconds), the protons are redistributed by the Cl^-/OH^- exchanger resulting in an internal pH of about 7.7 (Kummerow et al. 2000). Even calculating the transmembrane potential on the basis of the Nernst distribution for H^+ (assuming a quasi-steady-state for protons), and taking into account an internal and external pH of 7.7 and 7.4, respectively, it results in +18.5 mV. Of course it is evident that this calculation is an underestimation due to the fact that on both sides of the RBC membrane there is a significant buffer capacity (the extracellular solution was buffered in the reported investigations).

The situation is even more complicated if one takes into consideration that the human RBCs lose 70% of their internal Cl^- in the first two minutes in LIS solution, and nearly 90% in 30 min. In addition, the RBCs lose K^+ and take up protons. After 30 min in LIS medium, the internal pH is 7.1, i.e. 0.1 units lower than at the beginning of the experiment (7.2 in physiological solution). Because of these dramatic changes in ion distribution, it is evident that the transmembrane potential cannot be calculated correctly during this process. In addition, during this 30 min process in LIS solution there will be a change of the P_{Cl}/P_{Kat} ratio. This is due to the fact that the open probability of the non-specific voltage-activated cation channel is very low if the transmembrane potential decreases below +30 mV (Kaestner et al. 1999, 2000).

That indeed the transmembrane potential will decrease during the 30 min period, the cells spent in LIS medium, can be concluded from Cl^- efflux measurements. For a 90% Cl^- loss in a 5% haematocrit erythrocyte suspension (cf. Kummerow et al. 2000, the investigations were carried out in a solution with an osmolarity of 200 mosmol/l instead of 250 mosmol/l, i.e. under conditions where the RBC volume in physiological solution is the same as in LIS solution after 15 min), the Nernst potential assuming a new steady-state-equilibrium for Cl^- would be –9.5 mV. For comparison, the Nernst potential for H^+ after 30 min in LIS solution can be calculated as –18.5 mV. In any case, it seems realistic to imagine that the transmembrane potential of the human RBCs after being 30 min in LIS solution is again very close to the original transmembrane potential value in the physiological ionic strength solution. A significant negative transmembrane potential of human RBCs in LIS solution containing DIDS was also reported by Halperin et al. (1989).

It is known that fixed charges are distributed at the inner and outer surface of biological membranes. They are generated by the adsorption and dissociation of ionogenic groups of the surface coat glycoproteins and glycolipids (particularly at the outer membrane surface), as well as membrane phospholipids. The classical description of the electric potential profile near the cell surface is based on the assumption that the fixed charges are homogeneously distributed over the surface (Gouy-Chapman theory). The calculations of the electric potential profile perpendicular to the membrane surface ($\psi(x)$) are based on the Poisson-Boltzmann equation (e.g. Bernhardt 1994). However, at least at the outer membrane surface, the fixed electric charges are spatially distributed in a layer of some nanometers thickness around the cell surface in the glycocalyx and not only at the membrane surface. Data exist for the human RBC showing that about $1.5 \cdot 10^7$ elementary electric charges are distributed in a surface layer of approx. 6 nm thickness and a surface area of about 140 μm^2 (Donath and Pastushenko 1979). Therefore, for the extracellular surface, a fixed space charge density has to be taken into account instead of the surface charge density.

When varying the ionic strength of the solution surrounding the RBCs one has to consider diverse effects caused by changes in the electric potential profile near the cell surface. For example, in a LIS solution, the increase of the absolute value of the (negative) electric potential results in an enhancement of the intra- and intermolecular electrostatic repulsion or attraction of the charged sites. This in turn leads to structural changes of the glycocalyx molecules (glycoproteins, glycolipids) as well as changes in the physical characteristics of the whole glycocalyx (density profile, thickness). For human RBCs it has been calculated from electrophoretic measurements that the glycocalyx of the cells suspended in a LIS solution of 10 mM NaCl increased in thickness from 5.5 (in physiological ionic strength solution) to 12 nm. The distribution of the fixed charges in space and not only at the surface, as well as the fact that there is an increase in thickness of the glycocalyx results in a significant lower absolute value of the (negative) surface potential as compared with the surface potential calculated assuming a charge distribution at the surface only (for details see Bernhardt (1994)).

Depending on the electric potential of a microenvironment of a given site on the membrane surface or of the glycocalyx, local changes in the ion concentration as well as in the pH value have to be considered in accordance with the Boltzmann equation. It should be realised that such changes may occur not only in the perpendicular direction of the membrane but also in the direction of the plane of the membrane. These variations of the local electric potential, i.e. local ion concentrations, are of importance for changes in a large variety of cellular and membrane parameters including membrane transport processes. In general, the ion concentration near the binding site (or entrance site) of the transport molecule (pathway) and not the bulk concentration of the corresponding ion is of importance for the transport itself.

Another important parameter, which can influence membrane constituents and, therefore, modify membrane processes including membrane transport, is the membrane electric field. The electric field inside a biological cell membrane is determined by the gradient of the electric potential in all three dimensions. Focusing on the direction perpendicular to the membrane surface only and assuming zero charge density inside the membrane, the electric field is given by the difference between the actual electric potentials on both sides of the cell membrane divided by the membrane thickness. Therefore, it is evident that the transmembrane potential as well as the inner and outer surface potential is of importance for the electric field inside a cell membrane. The electric field strength in a RBC membrane under physiological conditions can be assumed to be in the order of 10^6 V/m. It is easy to understand that a reduction of the ionic strength of the extracellular solution results in a change of the electric field inside the membrane due to alterations of the transmembrane potential as well as the outer surface potential. The situation is illustrated in Fig. 4.4. In addition, in Fig. 4.4 a situation is given where only the transmembrane potential of the RBC is changed. This can be realised e.g. by a replacement of the extracellular Cl^- by an impermeant anion (cf. Sect. 4.3 and Fig. 4.4).

It has to be mentioned that the constant field approximation is useful only for general discussion of electric field effects in cell membranes. In fact, the electric potential profile across the cell membrane is not linear since electrically charged and polarizable groups are present and, besides, are located in different positions relative to both sides of the membrane surface. This concerns not only the head groups of the membrane phospholipids contributing with a dipole potential (not illustrated in Fig. 4.4) that can be much larger than 100 mV to the overall potential profile in the membrane but also regions of membrane proteins. The electric potential profiles proposed for ion channel proteins may serve as a classical example (e.g. Hille and Schwarz 1978; Jordan 1984).

In addition, a change of the electric potential gradient leads to considerable alterations of the mechanical tension inside the membrane, given by the Maxwell stress, P_e:

$$P_e = \frac{\varepsilon\varepsilon_o (d\psi / dx)^2}{2} \tag{4.5}$$

Here ε is the permittivity, ε_o the dielectric constant, and $d\psi/dx$ the electric potential gradient inside the membrane.

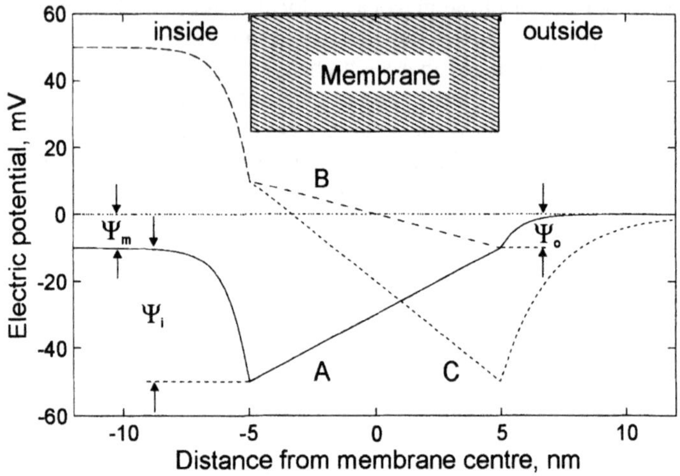

Fig. 4.4. Electric membrane potential profile across the RBC membrane: Ψ_m transmembrane (diffusion) potential, Ψ_o outer surface potential, Ψ_i inner surface potential, *A* physiological ionic strength solution, e.g. sodium chloride containing solution, *B* solution of reduced Cl^- concentration but constant ionic strength (compared to *A*), e.g. sodium tartrate containing solution, *C* solution of low ionic strength (LIS), e.g. sucrose solution. Ψ_m calculated according to Bernhardt 1994, Ψ_o and Ψ_i calculated according to Gouy-Chapman theory, dipole potentials are not considered

The electric field in the cell membrane in the direction of the plane of the membrane depends on the present localization of the membrane proteins and lipids. Of special importance is the distribution of charged and polarized membrane constituents as well as their lateral movement (McLaughlin and Poo 1981). Furthermore, electrogenic pumps and electrodiffusive pathways can contribute to the actual electric field in the membrane (Fromherz 1988).

Considering molecular mechanisms like ion transport, one has to take into account electric field effects on membrane phospholipids and membrane proteins. It is known that the electric field inside a biological membrane influences the mobility and the position of the hydrocarbon chains of the phospholipids as well as the phase transition temperature (Träuble and Eibl 1974; Jähnig 1976; Forsyth et al. 1977). The electric field strength is also affecting the head groups of the phospholipids. An increase in the electrostatic repulsion (possibly due to a reduced ionic strength of the extracellular solution) contributes to the interfacial tension, which in turn can have mechanical consequences (e.g. morphological changes for RBCs) (Coakley and Deeley 1980). The influence of the membrane electric field on membrane proteins can be either direct or indirect via changes in the phospholipid environment of the proteins leading to changes in the lipid-protein interaction. The direct effect of the membrane electric field on membrane proteins is based on mechanisms like charge displacement, dipole reorientation, and dipole induction (e.g. Glaser 1990).

Thus, the transmembrane potential, the outer and inner surface potential, and the electric field strength inside the biological membrane are of fundamental importance influencing and regulating specific ion transport pathways (pumps, channels, carriers). In addition, there should be no doubt that changes of the electrical potential near the membrane surface and inside the membrane have at least two consequences for the leak ion transport. First, these changes result in modifications of the ion concentration at the membrane-solution interface and, therefore, at the local membrane site where the entrance of the ion into the membrane phase occurs. Second, the membrane electric field strength will affect fluctuations of membrane constituents at instabilities between proteins and/or lipid molecules important for the leak ion transport.

4.7 Conclusion

Over the last three decades several discrete carriers and channels transporting monovalent cations through the RBC membrane have been discovered. The mechanism of how an ion can cross a biological membrane when all specific transport pathways (pumps, carriers, channels) for this ion are inhibited (defined as leak transport) is, however, still unclear. Mediated transport, whether via carriers or channels dominates passive membrane permeability. To investigate the true leak fluxes in a biological membrane for a certain ion, all specific transport pathways for this ion have to be identified. Furthermore, for these investigations inhibitors for the specific pathways or conditions at which the transporters are not activated have to be find.

Although there is no doubt that the electric potential profile in the membrane as well as in the space close to the extracellular and intracellular membrane surface is of fundamental importance for regulating membrane-bound processes including membrane transport, it seems unlikely that a leak ion flux can be described by a simple electrodiffusive mechanism. It is more realistic to assume that the leak ion transport mechanism in biological membranes is based on fluctuations of the membrane constituents, and therefore, an adequate mathematical description is needed.

References

Adorante JS, Cala PM (1987) Activation of electroneutral K^+ flux in *Amphiuma* red blood cells by N-ethylmaleimide. Distinction between K/H exchange and KCl cotransport. J Gen Physiol 90:209-227

Bang I (1909) Physiko-chemische Verhältnisse der Blutkörperchen. Biochemische Zeitschrift 16:255-276

Beaugé LA (1975) Non-pumped sodium fluxes in human red blood cells. Evidence for facilitated diffusion. Biochim Biophys Acta 401:95-108

Beaugé LA, Ortiz O (1970) Rubidium, sodium and ouabain interactions on the influx of rubidium in red blood cells. J Physiol 210:519-533
Bernhardt I (1986) Untersuchungen zur Regulation des Ouabain-insensitiven Membrantransports monovalenter Kationen an Erythrozyten. D.Sc. thesis, Humboldt University
Bernhardt I (1994) Alteration of cellular features after exposure to low ionic strength medium. In: Bauer J (ed) Cell electrophoresis. CRC Press, Boca Raton, pp 163-179
Bernhardt I, Donath E, Glaser R (1984) Influence of surface charge and transmembrane potential on rubidium-86 efflux of human red blood cells. J Membrane Biol 78:249-255
Bernhardt I, Erdmann A, Glaser R, Reichmann G, Bleiber R (1986) Influence of lipid composition on passive ion transport of erythrocytes. In: Klein R, Schmitz B (eds) Topics in lipid research. Royal Society of Chemistry, London, pp 243-248
Bernhardt I, Ellory JC, Hall AC (1987a) Alteration to human erythrocyte passive cation permeability induced by decreasing ionic strength. J Physiol 396:47P
Bernhardt I, Erdmann A, Vogel R, Glaser R (1987b) Factors involved in the increase of K^+ efflux of erythrocytes in low chloride media. Biomed Biochim Acta 46:36-40
Bernhardt I, Hall AC, Ellory JC (1988) Transport pathways for monovalent cations through erythrocyte membranes. Studia Biophys 126:5-21
Bernhardt I, Hall AC, Ellory JC (1991) Effects of low ionic strength media on passive human red cell monovalent cation transport. J Physiol 434:489-506
Bernhardt I, Seidler G, Ihrig I, Erdmann A (1992) Species-dependent differences in the influence of ionic strength on potassium transport of erythrocytes: The role of lipid composition. Gen Physiol Biophys 11:287-299
Bernhardt I, Bogdanova AY, Kummerow D, Kiessling K, Hamann J, Ellory JC (1999) Characterisation of the $K^+(Na^+)/H^+$ monovalent cation exchanger in the human red blood cell membrane: Effects of transport inhibitors. Gen Physiol Biophys 18:1-19
Blackstock EJ, Stewart GW (1986) The dependence on external cation of sodium and potassium fluxes across the human red cell membrane at low temperatures. J Physiol 375:403-420
Canessa M, Brugnara C, Cusi D, Tosteson DC (1986) Modes of operation and variable stoichiometry of the furosemide-sensitive Na and K fluxes in human red cells. J Gen Physiol 87:13-142
Carolin DA, Maizels M (1965) Effect of the duration of loading lactose-treated red cells with cations on the rate of subsequent cation flux. J Physiol 179:54-94
Chipperfield AR, Shennan DB (1986) The influence of pH and membrane potential on passive Na^+ and K^+ fluxes in human red blood cells. Biochim Biophys Acta 886:373-382
Coakley WT, Deeley JO (1980) Effects of ionic strength, serum protein and surface charge on membrane movements and vesicle production in heated erythrocytes. Biochim Biophys Acta 602:355-375
Cohn WE, Cohn ET (1939) Permeability of red corpuscles of the dog to sodium ion. Prog Soc Exp Biol Med 41:445-449
Dalmark M (1975) Chloride and water distribution in human red cells. J Physiol 250:65-84
Davson H (1939) Studies on the permeability of erythrocytes. VI. The effect of reducing salt content of the medium surrounding the cell. Biochem J 33:389-401
Dean RB, Noonan TR, Haege L, Fenn WO (1940) see Harris (1941) and reference therein
Denner K, Heinrich R, Bernhardt I (1993) Carrier-mediated residual K^+ and Na^+ transport of human red blood cells. J Membrane Biol 132:137-145

Deuticke B, Poser B, Lütkemeier P, Haest CWM (1983) Formation of aqueous pores in the human erythrocyte membrane after oxidative cross-linking of spectrin by diamide. Biochim Biophys Acta 731:196-210

Deuticke B, Lütkemeier P, Sistemich M (1984) Ion selectivity of aqueous leaks induced in the erythrocyte membrane by crosslinking of membrane proteins. Biochim Biophys Acta 775:150-160

Deuticke B, Grebe R, Haest CWM (1990) Action of drugs on the erythrocyte membrane. In: Harris JR (ed) Blood cell biochemistry, vol 1. Plenum Press, New York, pp 475-529

Deutsch CJ, Holian A, Holian SK, Daniele RP, Wilson DF (1979) Transmembrane electrical and pH gradients across human erythrocytes and human peripheral lymphocytes. J Cell Physiol 99:79-93

Donath E, Pastushenko V (1979) Electrophoretical study of cell surface properties. The influence of the surface coat on the electric potential distribution and on general electrokinetic properties of animal cells. Bioelectrochem Bioenergetics 6:543-554

Donlon JA, Rothstein A (1969) The cation permeability of erythrocytes in low ionic strength media of various tonicities. J Membrane Biol 1:37-52

Dwight JFSJ, Hendry BM (1995) Actions of arachidonic acid on erythrocyte membrane Rb permeability. Clin Chim Acta 238:187-197

Eisenman AJ, Ott L, Smith PK, Winkler AW (1940) A study of the permeability of human erythrocytes to potassium, sodium and inorganic phosphate by the use of radioactive isotopes. J Biol Chem 135:165-173

Ellory JC, Hall AC (1988) Human red cell volume regulation in hypotonic media. Comp Biochem Physiol 90A:533-537

Erdmann A, Bernhardt I, Hermann A, Glaser R (1990) Species-dependent differences in the influence of ionic strength on potassium transport of erythrocytes: The role of membrane fluidity and Ca^{2+}. Gen Physiol Biophys 9:577-588

Erdmann A, Bernhardt I, Pittman SJ, Ellory JC (1991) Low potassium-type but not high potassium-type sheep red blood cells show passive K^+ transport induced by low ionic strength. Biochim Biophys Acta 1061:85-88

Fievet B, Guizouarn H, Pellissier B, Garcia-Romeu F, Motais R (1993) Evidence for a K^+-H^+ exchange in trout red blood cells. J Physiol 462:597-607

Forsyth PA, Marcelja S, Mitchell DJ, Ninham BW (1977) Phase transition in charged lipid membranes. Biochim Biophys Acta 469:335-344

Fortes PAG, Ellory JC (1975) Asymmetric membrane expasion and modification of active and passive cation permeability of human red cells by the fluorescent probe 1-anilino-8-naphtalene sulfonate. Biochim Biophys Acta 413:65-78

Freedman JC, Hoffman JF (1979) The relation between dicarbocyanine dye fluorescence and the membrane potential of human red blood cells set at varying Donnan equilibria. J Gen Physiol 74:187-212

Fromherz P (1988) Spatio-temporal patterns in the fluid-mosaic model of membranes. Biochim Biophys Acta 944:108-111

Funder J (1980) Alkali metal cation transport through the human erythrocyte membrane by the anion exchange mechanism. Acta Physiol Scand 108:31-37

Funder J, Wieth JO (1966) Chloride and hydrogen distribution between human red cells and plasma. Acta Physiol Scand 68:234-245

Funder J, Wieth JO (1980) Passive cation transport by ion pair formation. In: Lassen UV, Ussing HH, Wieth JO (eds) Membrane transport in erythrocytes (Alfred Benzon Symposium). Munksgaard, Copenhagen, pp 520-527

Garlid KD, DiResta DJ, Beavis AD, Martin WH (1986) On the mechanism by which dicyclohexylcarbodiimide and quinine inhibit K^+ transport in rat liver mitochondria. J Biol Chem 261:1529-1535

Garay R, Adragna N, Canessa M, Tosteson DC (1981) Outward sodium and potassium cotransport in human red cells. J Membrane Biol 62:169-174

Glaser R (1990) The influence of membrane electric field on cellular functions. In: Glaser R, Gingell D (eds) Biophysics of the cell surface. Springer, Berlin, pp 173-192

Glynn IM (1956) Sodium and potassium movement in human red cells. J Physiol 134:278-316

Glynn IM (1957) The action of cardiac glycoside on sodium and potassium movements in human red cells. J Physiol 136:148-173

Goldman DE (1943) Potential, impedance and rectification in membranes. J Gen Physiol 27:37-60

Gruber W, Deuticke B (1973) Comparative aspects of phosphate transfer across mammalian erythrocyte membranes. J Membrane Biol 13:19-36

Gürber A (1895) Die Salze des Blutes. Jahresbericht über die Fortschritte der Tierchemie oder der physiologischen und pathologischen Chemie 24:172-175

Gürber A (1904) Salze des Blutes. II. Teil. Salze der Blutkörper. Habilitationsschrift, Kgl. Bayer. Julius-Maximilians-Universität Würzburg

Haas M, Schmidt WF (1985) p-Chloromercuribenzensulfonic acid stimulation of chloride-dependent sodium and potassium transport in human red blood cells. Biochim Biophys Acta 814:43-49

Haest CWM, Kamp D, Deuticke B (1981) Topology of membrane sulfhydryl groups in the human erythrocyte. Demonstration of a non-reactive population in intrinsic proteins. Biochim Biophys Acta 643:319-326

Hall AC, Ellory JC (1986) Evidence for the presence of volume-sensitive KCl transport in 'young' human red cells. Biochim Biophys Acta 85:317-320

Halperin JA, Brugnara C, Tosteson MT, Van Ha T, Tosteson DC (1989) Voltage-activated cation transport in human erythrocytes. Am J Physiol 257:C986-C996

Halperin JA, Brugnara C, Van Ha T, Tosteson DC (1990) Voltage-activated cation permeability in high potassium but not in low potassium red blood cells. Am J Physiol 258:C1169-C1172

Harris JE (1941) The influence of the metabolism of human erythrocytes on their potassium content. J Biol Chem 141:579-595

Hille B, Schwarz W (1978) Potassium channels as multi-ion single-file pores. J Gen Physiol 72:409-442

Hodgkin AL, Katz B (1949) The effect of sodium ions on the electrical activity of the giant axon of the squid. J Physiol 108:37-77

Hoffman JF (1992) Estimates of the electrical conductance of the red cell membrane. In: Bamberg E, Passow H (eds) Progress in cell research, vol. 2. Elsevier, Amsterdam, pp 173-178

Hoffman JF, Laris PC (1974) Determinations of membrane potentials in human and *Amphiuma* red blood cells by means of a fluorescent probe. J Physiol 239:519-552

Ihrig I, Hessel E, Seidler G, Erdmann A, Bernhardt I (1991) Investigation of monovalent cation influxes of diamide-treated human erythrocytes in solutions of different ionic strength. Biochim Biophys Acta 1069:171-174

Isomaa B, Hägerstrand H, Paatero G, Engblom AC (1986) Permeability alterations and antihaemolysis induced by amphiphiles in human erythrocytes. Biochim Biophys Acta 860:510-524

Jähnig F (1976) Electrostatic free energy and shift of the phase transition for charged lipid membranes. Biophys Chem 4:309-318

Jay AWL, Burton AC (1969) Direct measurements of potential difference across the human red blood cell membrane. Biophys J 9:115-121

Johnson RM (1994) Membrane stress increases cation permeability in red cells. Biophys J 67:1876-1881

Joiner CH, Franco RS, Jiang M, Franco MS, Barker JE, Lux SE (1995) Increased cation permeability in mutant mouse red blood cells with defective membrane skeletons. Blood 86:4307-4314

Jones GS, Knauf PA (1985) Mechanism of the increase of the cation permeability of human erythrocytes in low-chloride media. Involvement of the anion transport protein capnophorin. J Gen Physiol 86:721-238

Jordan PC (1984) Effect of pore structure on energy barriers and applied voltage profiles. I. Symmetric channels. Biophys J 45:1091-1100

Kaestner L, Bollensdorff C, Bernhardt I (1999) Non-selective voltage-activated cation channel in the human red blood cell membrane. Biochim Biophys Acta 1417:9-15

Kaestner L, Christophersen P, Bernhardt I, Bennekou P (2000) The non-selective voltage-activated cation channel in the human red blood cell membrane: Reconciliation between two conflicting reports and further characterisation. J Bioelectrochem 52:117-125

Knauf PA, Rothstein A (1971) Chemical modification of membranes. II. Permeation paths for sulfhydryl agents. J Gen Physiol 58:211-223

Kotyk A, Janacek K (1977) Membrane transport. An interdisciplinary approach. Plenum Press, New York

Krogh A (1946) The active and passive exchange of inorganic ions through the surface of living cells and through living membranes generally. Proc R Soc (London) Biol Sci 133:140-200

Kuypers FA, Roelofsen B, Op den Kamp JAF, Van Deenen LLM (1984) The membrane of intact human erythrocytes tolerates only limited changes in the fatty acid composition of its phosphatidylcholine. Biochim Biophys Acta 769:337-347

Kummerow D, Hamann J, Browning JA, Wilkins R, Ellory JC, Bernhardt I (2000) Variations of intracellular pH in human erythrocytes via $K^+(Na^+)/H^+$ exchange under low ionic strength conditions. J Membrane Biol 176:207-216

LaCelle PL, Rothstein A (1966) The passive permeability of red blood cells to cations. J Gen Physiol 50:171-188

Lassen UV, Sten-Knudsen O (1968) Direct measurements of membrane potential and membrane resistance of human red cells. J Physiol 195:681-696

Lauf PK (1988) Thiol-dependent K:Cl transport in sheep red cells. VIII. Activation through metabolically and chemically reversible oxidations by diamide. J Membrane Biol 101:179-188

Lew VL, Bookchin RM (1986) Volume, pH, and ion-content regulation in human red cells: Analysis of transient behavior with an integrated model. J Membrane Biol 92:57-74

Lu Y, Chow EI (1982) Bicarbonate/chloride transport kinetics at 37 °C and its relationship to membrane lipid in mammalian erythrocytes. Biochim Biophys Acta 689:485-489

Marrink SJ, Jähnig F, Berendsen HJC (1996) Proton transport across transient single-file water pores in a lipid membrane studied by molecular dynamic simulations. Biophys J 71:632-647

McLaughlin S, Poo MM (1981) The role of electro-osmosis in the electric-field-induced movement of charged macromolecules on the surface of cells. Biophys J 34:85-93

Mond R (1927) Umkehr der Anionenpermeabilität der roten Blutkörperchen in eine elective Durchlässigkeit für Kationen. Ein Beitrag zur Analyse der Zellmembranen. Pflügers Arch 217:618-630

Nichols JW, Deamer DW (1980) Net proton-hydroxyl permeability of large unilamellar liposomes measured by an acid-base titration technique. Proc Natl Acad Sci USA 77:2038-2042

Numata M, Orlowski J (2001) Molecular cloning and characterization of a novel (Na^+, K^+)/H^+ exchanger localized to the trans-Golgi network. J Biol Chem 276:17387-17394

Overton E (1899) Über die allgemeinen osmotischen Eigenschaften der Zelle, ihre vermutlichen Ursachen und ihre Bedeutung für die Physiologie. Vierteljahreschrift Naturforsch Ges (Zürich) 44:88-114

Paula S, Deamer DW (1999) Membrane permeability barriers to ionic and polar solutes. In: Deamer DW, Kleinzeller A, Fambrough DM (eds) Membrane permeability (Current Topics in Membranes, vol 48). Academic Press, San Diego, pp 77-95.

Paula S, Volkov AG, Van Hoek AN, Haines TH, Deamer DW (1996) Permeation of protons, potassium ions, and small polar molecules through phospholipid bilayers as a function of membrane thickness. Biophys J 70:339-348

Passow H (1969) The molecular basis of ion discrimination in the erythrocyte membrane. In: Tosteson DC (ed) The molecular basis of membrane function. Prentice-Hall, Englewood Cliffs, pp 319-351

Richter S, Hamann J, Kummerow D, Bernhardt I (1997) The monovalent cation "leak" transport in human erythrocytes: An electroneutral exchange process. Biophys J 73:733-745

Shaw TI (1955) K movements in washed erythrocytes. J Physiol 129:464-475

Solomon AK (1952) The permeability of the human erythrocyte to sodium and potassium. J Gen Physiol 36:57-110

Solomon AK (1968) Characterization of biological membranes by equivalent pores. J Gen Physiol 51:335-364

Solomon AK, Chasan B, Dix JA, Lukacovic MF, Toon MR, Verkman AS (1983) The aqueous pore in the red cell membrane: band 3 as a channel for anions, cations, nonelectrolytes, and water. Ann New York Acad Sci 414:97-124

Sutherland RM, Stannard JN, Weed RJ (1967) Involvement of sulfhydryl groups in radiation damage to the erythrocyte membrane. Int J Radiat Biol 12:551-564

Tosteson DC, Hoffman JF (1960) Regulation of cell volume by active cation transport in high and low potassium sheep red cells. J Gen Physiol 44:169-194

Träuble H, Eibl H (1974) Electrostatic effects on lipid phase transitions: Membrane structure and ionic environment. Proc Natl Acad Sci USA 71:214-219

Van der Steen ATM, De Kruijff B, De Gier J (1982) Glycophorin incorporation increases the bilayer permeability of large unilamellar vesicles in a lipid-dependent manner. Biochim Biophys Acta 691:13-23

Van Dijck PWM, Van Zoelen EJJ, Seldenrijk R, Van Deenen LLM, De Gier J (1976) Calorimetric behaviour of individual phospholipid classes from human and bovine erythrocyte membranes. Chem Phys Lipids 17:336-343

Van Hoogevest P, Du Maine APM, De Kruijff B (1983) Characterization of the permeability increase induced by the incorporation of glycophorin in phosphatidylcholine vesicles. FEBS Letters 157:41-45

Van Hoogevest P, Du Maine APM, De Kruijff B, De Gier J (1984) The influence of lipid composition on the barrier properties of band 3-containing lipid vesicles. Biochim Biophys Acta 777:241-252

Van Slyke DD, Wu H, McLean FC (1923) Studies of gas and electrolyte equilibria in the blood. V. Factors controlling the electrolyte and water distribution in the blood. J Biol Chem 56:765-849

Weiss TF (1996) Cellular biophysics, vol 1. The MIT Press, Cambridge, Massachusetts

Wieth JO (1970) Effects of monovalente cations on sodium permeability of human red cells. Acta Physiol Scand 79:76-87

Wilbrandt W (1940) Die Ionenenpermeabilität der Erythrozyten in Nichtleiterlösungen. Pflügers Arch 242:537-556

Wilbrandt W, Schatzmann HJ (1960) Changes in the passive cation permeability of erythrocytes in low electrolyte media. Ciba Found. Study Group 5:34-52

Wiley JS, Cooper RA (1975) Inhibition of cation cotransport by cholesterol enrichment of human red cell membranes. Biochim Biophys Acta 413:425-431

Zade-Oppen AMM, Tosteson DC, Adragna NC (1988) Effects of pH, potential, chloride and furosemide on Na^+ and K^+ effluxes from human red blood cells. J Membrane Biol 103:17-225

5 Na^+/K^+ Pump

John R. Sachs

Department of Medicine, State University of New York at Stony Brook, Stony Brook, NY, 11794-8151, USA

5.1 Introduction

Red cells, especially human red cells, have played a central role in the identification and characterization of the Na^+/K^+ pump. The red cell has been especially useful in defining the cation transport characteristics of the pump. Red cell membranes have played a lesser part in characterizing the $(Na^+ + K^+)$-dependent ATPase, the enzymatic manifestation of the pump. In recent years the availability of partially purified enzyme preparations and enzyme reconstituted into liposomes, and the use of cells in which membrane potentials can be precisely measured, have diminished the importance of the red cell as an experimental model, but there remain questions which might best be answered with its help. The findings obtained using red cells have been confirmed in studies using other experimental models, and, in fact, a finding in another preparation that differs from information obtained with red cells is often treated with suspicion.

5.2 Development of the Concept of Active Transport

In this section, the early history of the contribution of the red cell to the study of the Na^+/K^+ pump will be recounted, and the significance of some of the early findings discussed. Animal cells are characterized by high concentrations of K^+ and low concentrations of Na^+ although they are surrounded by solutions high in Na^+ and low in K^+. Human red cells and the red cells of most other species are also high in K^+ and low in Na^+. In principle, three explanations for the marked difference in the concentrations of the two ions in the intracellular and extracellular solutions can be advanced: 1. the physicochemical properties of Na^+ and K^+ differ on the two sides of the cell membrane so that ions inside the cell are actually at thermodynamic equilibrium with ions in the extracellular solution (e.g., intracellular ions are bound); 2. the cell membrane is impermeable to Na^+ and K^+; 3. there is an energy driven process which maintains the concentration of Na^+ and K^+ inside the cell at a steady state away from equilibrium with the ions outside the cell. By 1941, the first possibility seemed unlikely because of the osmotic properties of cells, especially mammalian red cells; because of the electric conductivity properties of cytoplasm; and because biological macromolecules with the required physicochemical properties were not found.

The second possibility also seemed unlikely by 1941. Muscle was shown to lose K^+ and gain Na^+ during stimulation (Fenn 1936); liver and muscle of rats lost K^+ and gained Na^+ when the animals were fed a K^+-deficient diet (Heppel 1939); and muscles bathed in K^+ free solution lost K^+ and lost Na^+, but regained K^+ and lost Na^+ when reimmersed in solution containing K^+ (Steinbach 1940). Human red cells stored at low temperature lost K^+ and gained Na^+ (Flynn and Maizels 1949), and the cells regained K^+ and lost Na^+ when reinfused (Maizels and Paterson 1940), or when incubated at 37 °C in the presence of glucose (Harris 1941; Danowski 1941). When radioactive isotopes of Na^+ and K^+ became available, both *in vivo* and *in vitro* experiments clearly showed that extracellular cations readily exchange with intracellular cations in red cells and other cells (Cohn and Cohn 1939; Mullins et al. 1941; Hahn and Hevesy 1940). The findings rendered untenable the concept that an impermeable membrane is responsible for the asymmetric distribution of Na^+ and K^+. In 1941 Dean pointed out that the findings require the existence of active transport of Na^+, a Na^+ pump, in the cell membrane (Dean 1941, 1987).

5.3 Early Characterization of the Na⁺/K⁺ Pump in Red Cells

By 1946, the existence of a cation pump was widely accepted, but it was believed that either Na^+ or K^+ is the actively transported species, and the movement of the other ion is coupled indirectly to the movement of the actively transported species, perhaps by means of the membrane potential (Krogh 1946). However, Flynn and Maizels (1949) found that Na^+ efflux from human red cells can occur without K^+ uptake, but K^+ uptake is always associated with an equal or greater loss of Na^+, and Harris and Maizels (1952) showed that external K^+ (K^+_o) increases Na^+ efflux. They then demonstrated that changes in membrane potential did not account for the coupling of the fluxes. Pump rate is stimulated by Na^+ only from the inside (Na^+_i), and by K^+ only from the outside (K^+_o) (Whittam and Ager 1964). The relationship between the efflux of Na^+ through the red cell Na^+/K^+ pump and the internal Na^+ concentration $[Na^+_i]$ is exactly the same as the relation between the K^+ uptake and $[Na^+_i]$, and activation of K^+ influx or Na^+ efflux shows an identical relation to the external K^+ concentration $[K^+_o]$. The evidence that the active transport of Na^+ through the pump is directly coupled to the active transport of K^+ became overwhelming as a result of these findings with red cells.

Harris proposed that coupling is stoichiometric, that coupling is achieved by a carrier which alternately transports Na^+ and K^+, that K^+ carriers are converted to Na^+ carriers by an energy dependent process, and that the coupling ratio is 2 Na^+:1 K^+ (Harris 1954). A more accurate determination of the coupling ratio in red cells led to the conclusion that the ratio of Na^+ efflux to K^+ influx through the pump is 3:2 (Post and Jolly 1957). This is a characteristic of Na^+/K^+ pumps from all sources. The 3:2 ratio means that the pump should be electrogenic, and this too has been shown to be the case for the pump from all sources, including the human red cell (Hoffman et al. 1979).

Active transport requires a source of energy, and early experiments with human red cells demonstrated that the energy for Na^+ transport is derived from glucose (Harris 1941; Harris 1954). Gardos (1954) prepared resealed ghosts of human red cells. When ATP was included in the resealing solution, the ghosts were able to accumulate K^+ through the Na^+/K^+ pump, demonstrating that the proximate source of energy for the pump is ATP. Measurement of the ratio of Na^+ ions lost through the pump to the amount of ATP hydrolyzed showed that each ATP supported the efflux of 3 Na^+ ions (Sen and Post 1964). A pump cycle, then, consists of the coupled efflux of three Na^+ ions, the influx of two K^+ ions, and the hydrolysis of one ATP. It is now known that partially purified enzyme preparations bind 3 Na^+:2 K^+:1 ATP per enzyme protomer.

An important discovery which has been of great importance in the study of the Na^+/K^+ pump, particularly in intact cells such as red cells, was made by Schatzman (1953) who found that cardioactive glycosides and aglycones (such as ouabain) inhibit Na^+ and K^+ movements through the Na^+/K^+ pump. The inhibition is specific for the pump; cation movements through other pathways are not affected (Post et al. 1967). Use of these inhibitors has permitted the separation of pump mediated movements of Na^+ and K^+, defined as that portion of the total movement which is inhibited by ouabain, from the movement of the cations through other transport pathways and from the non-mediated leak.

Shaw (1955) and Glynn (1956), using radioactive K^+, measured K^+ uptake in red cells as a function of $[K^+_o]$, and found that the results were adequately described by:

$$v = V_m/(1 + K_{1/2}/[K^+_o]) + {}^{i}k[K^+_o] \qquad (5.1)$$

which is an explicit expression of the pump-leak hypothesis. v is the total K^+ influx, V_m is the maximal K^+ influx through the pump, $K_{1/2}$ is the concentration of K^+_o at which the pump rate is half maximal, and ${}^{i}k$ is a constant. ${}^{i}k[K^+_o]$ is the K^+ influx through the leak. The relation between the pump rate and external K^+ concentration was adequately described by a rectangular hyperbola similar to the Michaelis-Menten relation. The results showed that the pump is saturable and present in limiting amounts.

Shaw (1955) proposed a model for the pump that was adapted by Glynn (1956). It accounted for these findings and many of the findings described above. The model is shown in Fig. 5.1. The model proposes that Na^+ and K^+ cross the membrane bound to the carriers X or Y, Y carrying Na^+ and X carrying K^+. After release of Na^+ to the outside, Y spontaneously changes to X, which has a high affinity for K^+. K^+ is then released from X at the inside surface, and X is converted to Y as a result of the input of energy from ATP. Form Y has a high affinity for Na^+. The model is a truncated version of the most popular current model of the reaction mechanism of the Na^+/K^+ pump and illustrates three of its most important characteristics: 1) the transport of Na^+ and K^+ occurs consecutively rather than simultaneously (the reaction is ping-pong with respect to Na^+ and K^+ (Cleland 1963)); 2) the change in the affinity of the "carrier" (or transport protein) from high affinity for Na^+ to high affinity for K^+ results from a change in its conformation; and 3) the energy for transport of the ions and change in conformation comes from ATP.

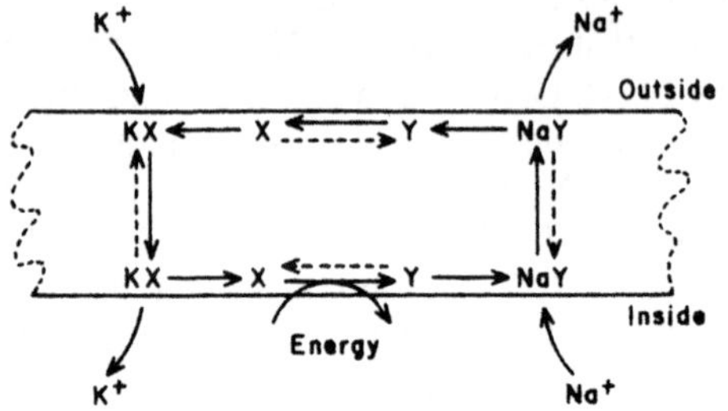

Fig. 5.1. Sequential carrier model for operation of the Na^+/K^+ pump. For description, see text. From Glynn (1956)

5.4 The Enzymatic Basis of the Na^+/K^+ Pump

Skou (1957) reported that membranes of nerves from the walking legs of crabs demonstrated ATPase activity that required the presence of Mg^{2+}, Na^+, and K^+. The ATPase activity required the simultaneous presence of Na^+ and K^+, and the relation between activity and the concentration of the two ions was similar to the relation of the pump rate of red cells and the cytoplasmic concentration of Na^+ and external concentration of K^+. Soon Na^+- and K^+-dependent ATPase activity was demonstrated with red cell membranes (Post et al. 1960). Na^+ stimulated ATPase activity from inside the cells, and K^+ stimulated from outside (Whittam and Ager 1964). The relation between ATPase activity and the internal Na^+ and external K^+ was the same as the relation between the pump rate and the concentrations of the two ions, indicating that ions that activate ATPase are the ions that are transported.

Using partially purified enzyme from other sources, Albers et al. (1963) and Post et al. (1965) showed that the γ phosphate of ATP phosphorylates the ATPase in the presence of Na^+. The phosphate group is covalently bound to the enzyme by an acyl bond as demonstrated by its pH sensitivity and its hydrolysis by hydroxylamine. The phosphorylated intermediate actually consists of two species. ADP, in the presence of Na^+, dephosphorylates one, E_1P, with the formation of ATP. The other, E_2P, is dephosphorylated by water in the presence of K^+, with the formation of phosphate, but not of ADP. E_2P can also be formed from inorganic phosphate in the presence of K^+ and Mg^{2+}, and this phosphoenzyme can also be dephosphorylated by water in the presence of K^+. The acyl bond in E_1P is therefore high energy, and the bond in E_2P low energy. Since the same aspartyl group is phosphorylated in the two species, the difference in the energy of the phosphate bond in E_1P and E_2P must result from some structural difference in the protein. This is the basis for the Albers-Post model of the Na^+/K^+ pump reaction mechanism that, in turn, is the basis for much of the research on the enzyme. A scheme of the Albers-Post mechanism is shown in Fig. 5.2.

Partially purified ATPase preparations also hydrolyze *p*-nitrophenylphosphate (pNPP) in the presence of K^+. This pNPPase activity may represent the step in which E_2P is dephosphorylated in the reaction cycle.

Red cell membranes are not a favorable preparation for study of the Na^+, K^+-ATPase since the specific activity is very low. Ouabain binding studies have shown that there are 500 or fewer copies of the enzyme per red cell (Joiner and Lauf 1978). Nevertheless, it has been shown that phosporylated intermediate can be prepared in the presence of Na^+, Mg^{2+}, and ATP with properties similar to those of phosphorylated intermediate in other preparations (Blostein 1968). As in other preparations, the phosphate is covalently bound by an acyl bond, and is partially dephosphorylated by ADP, but not by water, in the presence of Na^+, and by water, but not by ADP, in the presence of K^+. As in partially purified preparations, the phosphorylated intermediate in red cell membranes is a mixture of E_1P and E_2P. Red cell membranes also demonstrate K^+-dependent pNPPase activity with properties similar to those of purified preparations (Garrahan et al. 1969). The red cell (Na^+ + K^+)-dependent ATPase is not different from the enzyme from other sources.

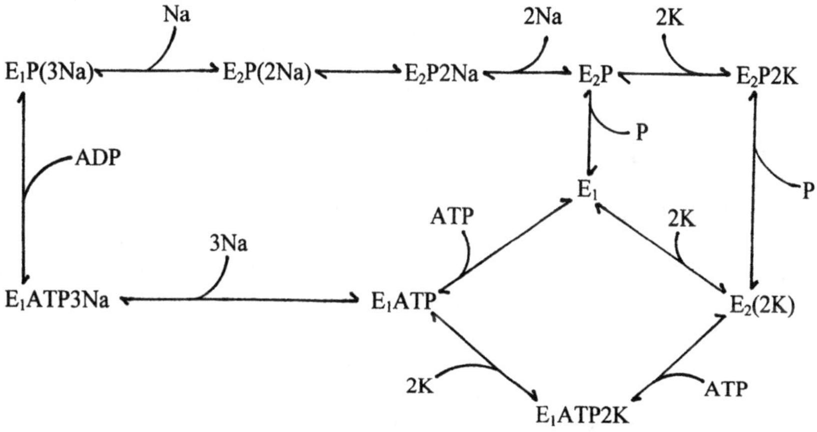

Fig. 5.2. Albers-Post model of Na^+/K^+ reaction mechanism. For description, see text

5.5 Red Cell Characteristics Favorable for the Study of the Na^+/K^+ Pump

Red cells are singularly suited for the study of Na^+ and K^+ transport by the Na^+/K^+ pump. Red cells are readily available. They are robust so that release of intracellular contents due to cell lysis is minimal. There is also little lysis when the cells are washed by repeated sequential suspension in isotonic salt solution and separation of the cells from the solution by centrifugation and aspiration of the supernatant. Since the cells are nearly impermeable to cations at 0 °C, intracellular cations are not lost during washing at that temperature and contamination of the cell fraction by extracellular solutes can be reduced to trivial levels. The cells are large enough to permit the measurement of cation transport with the use of isotopes without altering intracellular concentration. On the other hand, it is also possible to follow cation movements by measuring changes in intracellular and extracellular concentrations.

Intracellular cation concentrations can be readily altered in intact cells by several methods. The simplest, but least effective, is incubation for prolonged periods in a solution approximating the required intracellular cation concentration. The approach to equilibrium is more rapid at higher temperatures, but the cells begin to lose their integrity after prolonged incubation at elevated temperature.

Incubation of cells with *p*-chloromercuribenzene sulphonate (PCMBS) is an effective means for changing intracellular Na^+ and K^+ concentration and for replacing some or all of the inorganic cations with inert organic cations such as choline (Garrahan and Rega 1967; Sachs 1977a). PCMBS combines with sulphhydryl groups at the external surface and increases the permeability of the cells to cations, but does not penetrate the membrane. Reduction of Na^+ or K^+ or both to low levels can be achieved by incubation of the cells at low hematocrit at 4 °C in a solution of appropriate cation concentration containing PCMBS for prolonged periods, e.g., 36 h. Best results are obtained if the extracellular solution is exchanged for fresh solution every 12 h, and if the extracellular solution contains the appropriate concentration of an impermeant solute, such as sucrose, to maintain cell volume constant. At the end of the incubation period, cells are removed from the PCMBS solution, resuspended in a solution containing a sulphhydryl reducing agent such as dithiothreitol, and incubated at 37 °C for about an hour. Impermeability is restored, and the characteristics of the Na^+/K^+ pump are not altered.

An alternative way to achieve the same result is incubation of the cells at low hematocrit and low temperature with the antibiotic nystatin, an ionophore for Na^+ and K^+ (and H_3O^+). Incubation of the cells with solutions containing the antibiotic and the appropriate salt solution alters the internal cation content (Cass and Dalmark 1973). The antibiotic can then be washed away and the impermeability of the altered cells restored. With this method, too, use of an impermeant solute at appropriate concentration maintains cell volume.

Although these techniques permit easy manipulation of the intracellular concentration of small ions, they do not permit manipulation of the content of intracellular metabolites such as ATP and ADP. However, the availability of resealed ghosts allows extensive manipulation of intracellular metabolite concentration. By varying the concentration of metabolites in the resealing solution, and incorporating appropriate enzymes in the ghosts, precise control of the simultaneous concentration of ATP and ADP can be achieved (Glynn and Hoffman 1971). Early experiments were done with ghosts formed from cells hemolyzed at 0 °C in a large volume of hypotonic solution and then resealed at 37 °C in a small volume of isotonic solution containing the appropriate cations and metabolites. Such ghosts retain some of the components of the original cells, although increasing the volume in which the cells are hemolyzed can reduce the amount of such retention.

A more effective method for preparing ghosts utilizes gel filtration (Kaplan 1982). Red cells are run over a sepharose column with a molecular exclusion of about 1 million daltons equilibrated with hypotonic solution at pH 6.0 and maintained at 0 °C. The ghosts that emerge from the column are usually white and contain only the ions present in the hypotonic solution; they are devoid of other inorganic cations, metabolites, hemoglobin and enzymes. Reconstitution at 37 °C in an appropriate solution permits precise definition of the intracellular content of the ghosts.

Finally, when red cells are hemolyzed at high pH and the ghosts then subjected to shear stress, they fragment, and some of the fragments reseal as relatively impermeable vesicles, inside-out with reference to the original red cell membrane, which can be collected by gradient centrifugation (Steck et al. 1970). Such vesicles allow easy access to the intracellular surface of the pump, and fluorescent reporter molecules can be incorporated within the vesicles at the physiological exttracellular surface (Polvani and Blostein 1989).

The utilization of red cells and their derivatives for characterizing the transport properties of the Na^+/K^+ pump depends in large part on the impermeability of the cells to cations and the availability of specific inhibitors such as ouabain to inhibit cation movements through the pump. Transport through the pump is then taken as the difference in the total transport rate in the presence and absence of the inhibitor. Although the permeability of red cell ghosts is significantly higher than that of intact cells, it is usually sufficiently low for significant measurements to be made. Since a large part of ouabain insensitive movement in red cells and ghosts is attributable to the Na^+-K^+-$2Cl^-$ cotransporter and the K^+-Cl^- cotransporter, ouabain insensitive movements can be significantly reduced if measurements are made in nitrate solutions rather than chloride solutions. There is no change in the Na^+/K^+ pump activated movements.

5.6 Transport of Na^+ and K^+ and the Reaction Mechanism of the Na^+/K^+ Pump

Figure 5.2 is a diagram of the currently accepted reaction mechanism of the Na^+/K^+ pump (Sachs 1991). The forward direction is clockwise. Events that occur at the inner surface of the pump are at the bottom of the diagram, and events at the outside at the top. A cycle begins with the addition of one ATP and three Na^+ ions at the inner surface with formation of an E_1 conformation. A large conformational change brings the γ phosphate of ATP in close contact with Asp 369, which is phosphorylated. At the same time, the bound Na^+ ions are occluded (i.e., they are not readily exchanged at either membrane surface). ADP is released and the species $E_1P(3Na)$ is formed where the brackets indicate that the Na^+ ions are occluded. E_1P undergoes a conformational change to E_2P, and one of the occluded Na^+ ions is released to the outside (Na^+_o). The two remaining Na^+ ions in E_2P are deoccluded and become accessible to the outside solution. If K^+ is present at the outside two K^+ ions may replace the two Na^+ ions in E_2P. E_2P is hydrolyzed, and the two K^+ ions are occluded in the dephospho form $E_2(2K)$.

At this point, the reaction mechanism is branched (Moczydlowski and Fortes 1981). In the upper branch, $E_2(2K)$ spontaneously converts to the form E_1 with the release of the two occluded K^+ ions to the inside (K^+_i), and ATP binds to E_1, with high affinity, to form E_1ATP. In the lower branch, ATP binds to $E_2(2K)$ with low affinity, and promotes a conformational change to E_1ATP2K. The K^+ ions are deoccluded and released to the inside. The final result is, again, E_1ATP, ready for the addition of Na^+_i, and another cycle.

5.6.1 Na⁺/K⁺ Exchange

If multiple Na^+ and K^+ ions are transported in each cycle, pump rate *vs.* $[K^+_o]$ or $[Na^+_i]$ should be a sigmoid rather than a rectangular hyperbolic curve. The expectation is borne out if Na^+ efflux or K^+ influx through the pump is measured as a function of Na^+_i (McConaghey and Maizels 1962; Sachs 1970; Garay and Garrahan 1973) or K^+_o (Garrahan and Glynn 1967b; Sachs and Welt 1967a).

The sigmoid relation, though consistent with the belief that multiple ions must be bound during a transport cycle, does not provide independent evidence. Sigmoid curves arise in steady-state kinetic schemes whenever substrate binds at more than one point in a cycle. Although sigmoid curves appear in a mechanism in which more than one substrate molecule (Na^+ or K^+) must be bound before the reaction proceeds, they also arise when a substrate binds to a single site in different branches of a branched chain mechanism.

In the case of K^+ influx, evidence that more than two K^+ ions are transported in each cycle is obtained from studies on the effect of K^+ congeners (Rb^+, Cs^+, Li^+) on the K^+ influx rate measured in Na^+ solutions at very low $[K^+_o]$ (well below the $K_{1/2}$ for activation of K^+ influx) and at increasing concentration of one of the congeners. K^+ influx at first increases, reaches a peak, and then decreases. For a mechanism in which K^+ adds only once, or twice in a branched chain mechanism, once in each branch, monotonic inhibition is expected. However, if there are two or more transport sites that must be filled simultaneously before transport occurs, at very low $[K^+_o]$ there will be many pumps with a single bound K^+ ion: such pumps will not transport K^+. At very low $[K^+_o]$, congeners at low concentration combine with the second site and partner the influx of the associated K^+, so that K^+ influx increases. At higher concentrations, the congener displaces K^+ from both sites and K^+ influx is inhibited. Such competitive activation has been seen in red cells with Rb^+, Cs^+, and Li^+ (Sachs and Welt 1967b).

There is good evidence that Li^+ (Dunham and Senyk 1977) and protons (Polvani and Blostein 1988) can substitute for Na^+ at the inside surface and activate K^+ or Rb^+ influx. Competitive activation of Na^+ efflux by H_3O^+ may occur, and competitive activation by Li^+ has clearly been shown.

The kinetic studies provide convincing independent evidence that more than one K^+ ion and more than one Na^+ ion are transported in each transport cycle.

The effect of Na^+_o on the pump rate is complex. As $[Na^+_o]$ is lowered, the $K_{1/2}$ for K^+_o decreases, and the curve relating pump rate to $[K^+_o]$ becomes less sigmoid. In Na^+-free solution, the curve is either a rectangular hyperbola or is antisigmoid, i.e., at low $[K^+_o]$ it rises more rapidly than a rectangular hyperbola. Dixon plots of the reciprocal of the K^+ influx *vs.* the $[Na^+_o]$ are linear up to about 150 mM Na^+ at intermediate concentrations of K^+. These findings suggest that Na^+ and K^+ bind to the same enzyme form, (i.e., Na^+ and K^+ bind competitively), and, to account for the linearity of the Dixon plot, binding of a single Na^+ ion is sufficient to prevent K^+ transport (Sachs 1977a).

However, the situation is more complicated. At very low [K^+_o], Dixon plots are curved with the concavity directed downward (Cavieres and Ellory 1975) and at very high [K^+_o], Dixon plots are also nonlinear (Hobbs and Dunham 1976) so that the simple competitive model does not fully explain the effect of Na^+_o on the pump rate. Nonlinearity of the Dixon plots led Cavieres and Ellory (1975) to suggest that Na^+ inhibits from a site other than the K^+ transport sites. A model that combines competitive inhibition with allosteric inhibition by Na^+_o at least qualitatively describes the complex interaction of external Na^+ and external K^+ with the pump at all concentrations of K^+ (Sachs 1977a).

Both at the allosteric site, and at the transport sites at which Na^+_o presumably competes with K^+_o to inhibit the pump rate, the concentration at which Na^+_o exerts its effect is much lower than the concentration at which it replaces K^+_o and activates transport (see below). It is not possible to directly compare the affinities since, in the case of the inhibition studies, the affinity determined from kinetic studies is likely to be about the same as the affinity determined in binding measurements, but estimates from transport measurements are kinetic constants that cannot be considered comparable to affinities derived from binding studies. Nevertheless, the possibility that the effect of Na^+_o on the curve relating pump rate to K^+_o concentration is, indeed, an allosteric effect (as described by Cavieres and Ellory (1975)), of Na^+_o at its third binding site, and that competition with K^+_o occurs at that site rather than, or in addition to, competition at transport sites is attractive.

The observation that, in Na^+-free solutions, a plot of pump rate against [K^+_o] is antisigmoid cannot be accommodated by a scheme that requires that both K^+ sites must be filled before transport occurs. Such a model always predicts a sigmoid curve that, however, may be difficult to distinguish from a rectangular hyperbola; it does not predict an antisigmoid curve. Na^+ reduces the affinity for K^+ of the first binding site to be filled, and converts the activation curve to a sigmoid curve.

The antisigmoid curve in the absence of Na^+_o indicates that, at low [K^+_o], binding of a single K^+ ion is sufficient to activate a transport cycle. There is other evidence for this view. If two K^+ ions are always required to initiate a transport cycle, it should be possible to demonstrate competitive activation by K^+ ion congeners at low [K^+_o] even in Na^+-free solutions. In fact, Li^+ does not competitively activate K^+ influx or Cs^+ influx in such circumstances (Sachs 1977a). Support also comes from the complex effect of [K^+_o] on pump inhibition by strophanthidin and tetrapropylanmonium in solutions with and without Na^+ (Kropp and Sachs 1977).

Interaction of Na^+ and K^+ with the ion binding sites at the inner surface is more straightforward. The curve relating pump rate in red cells to [Na^+_i] is sigmoid at all K^+_i concentrations [K^+_i] (Sachs 1970). Garay and Garrahan (1973) proposed that Na^+_i binds randomly and independently to three equivalent cation binding sites each with dissociation constant K_N for Na^+_i and K_K for K^+_i, and derived a rate equation which describes the relation between pump rate and [Na^+_i] and [K^+_i]:

$$v = V_m/(1+(K_N/[Na^+_i])(1 + [K^+_i]/K_K))^3 \tag{5.2}$$

where V_m is the maximum pump rate at saturating Na^+_i and zero K^+_i. As predicted by Eq. (5.2), 1) the pump rate is competitively (with Na^+) inhibited by K^+, and K_N

is a linear function of K^+ over a wide range of [K^+], and 2) at fixed [Na^+], the reciprocal of pump rate is a parabolic function of [K^+]. Rate equations based on Eq. (5.2), therefore, describe the relation between pump rate and [Na^+] and [K^+] remarkably well despite the somewhat unlikely assumptions involved in its derivation. In purified membrane preparations, control of the conformational change between E_1 and E_2 by Na^+ and K^+, measured with fluorescein isothiocyanate as a reporter, can be described by an equivalent binding site mechanism similar to the one described above (Smirnova et al. 1995). In this case, a two equivalent binding site mechanism fit the data equally well. The findings with red cells described above can also be accounted for by a model with three Na^+ binding sites, two of which are binding sites for Na^+ and K^+.

5.6.2 Na^+/Na^+ Exchange

In solutions free of K^+_o, the Na^+/K^+ pump carries out two types of Na^+/Na^+ exchange, i.e., exchange of Na^+_o ions for Na^+_i ions. One type of Na^+/Na^+ exchange is dependent on the presence of intracellular ADP and, with reference to Fig. 5.2, uses the pathway $E_1ATP \longleftrightarrow E_2P$ and its reverse. The second type of Na^+/Na^+ exchange takes place in the absence of ADP, is associated with the hydrolysis of ATP, and takes place by the normal clockwise operation of the transport cycle shown in Fig. 5.2 with Na^+ replacing K^+ at the outside. In K^+-free solutions, the red cell pumps carry out an efflux of Na^+ that takes place without exchange with any extracellular cation by means of a cycle in which E_2P dephosphorylates to E_1 (Fig. 5.2).

When, in red cells containing both ATP and ADP, Na^+ efflux is measured into K^+-free solutions as a function of [Na^+_o], a complex relation is found. Na^+ efflux at first decreases until the [Na^+_o] reaches about 5 mM (Garrahan and Glynn 1976a). As [Na^+_o] continues to rise to about 300 mM, Na^+ efflux increases along a sigmoid curve with a $K_{1/2}$ for Na^+ of about 100 mM (Garay and Garrahan 1973). The Na^+ efflux into solutions free of both Na^+ and K^+ and the Na^+ efflux into high Na^+, K^+-free solutions have the same relation to the [Na^+_i] as the Na^+ efflux into solutions containing K^+_o (Sachs 1970). At high Na^+_o, the Na^+ efflux is accompanied by an equal Na^+ influx (Garrahan and Glynn 1967a). K^+_o inhibits the Na^+ influx in Na^+ solution, and the $K_{1/2}$ at which K^+ inhibits Na^+ influx is the same as the $K_{1/2}$ at which it activates Na^+/K^+ exchange (Sachs 1970). All the Na^+ movements, which occur in K^+ free solutions are inhibited by cardiotonic steroids. There is no doubt that efflux of Na^+ into K^+-free solutions at all concentrations of Na^+_o are carried out by the Na^+/K^+ pump.

Although ATP is required for Na^+/Na^+ exchange, hydrolysis of ATP cannot be detected (Garrahan and Glynn 1967d). ADP is also required, and the Na^+ efflux into high Na^+, K^+-free solution is proportional to the ADP concentration (Glynn and Hoffman 1971). From Fig. 5.2 it can be seen that Na^+/Na^+ exchange that takes place by the bidirectional operation of the sequence E_1ATP to E_2P, should be associated with an exchange of the γ phosphate between ATP and ADP. In red cell ghosts, ATP/ADP exchange has been reported (Kaplan 1982). Oligomycin, which inhibits Na^+/K^+ exchange by blocking the conversion of E_1P to E_2P, also inhibits

Na^+/Na^+ exchange (Garrahan and Glynn 1967d), but enhances ATP/ADP exchange (Blostein 1970), a result which can be predicted from consideration of Fig. 5.2.

Using inside-out vesicles derived from red cell ghosts, Blostein (1983) found that, in solutions free of ATP and of K^+, the Na^+/K^+ pump carries out an approx. 1:1 exchange of internal for external Na^+. The exchange is associated with hydrolysis of ATP, and approx. three Na^+ ions are transported outward for every ATP hydrolyzed. The relation between the ATP dependent Na^+/Na^+ exchange and $[Na^+_i]$ is about the same as the relation between the Na^+/K^+ exchange and $[Na^+_i]$, but, at the outside, high concentrations of Na^+ (50 mM) are necessary to demonstrate the exchange. The $K_{1/2}$ for Na^+ at the outside is not known. It is likely that this exchange results from normal operation of the pump with Na^+ replacing K^+ at the outside surface, but it is not known why the ratio of efflux to influx is 1:1 rather than 3:2, as it is for Na^+/K^+ exchange.

The Na^+ efflux into solutions free of both Na^+ and K^+ is also associated with the hydrolysis of ATP (Glynn and Karlish 1976). Inhibition of the efflux by low $[Na^+_o]$ is associated with inhibition of ATP hydrolysis and with inhibition of the hydrolysis of the phosphorylated intermediate (Lee and Blostein 1980). There are, then, three effects of Na^+_o on Na^+ efflux into K^+-free solutions: 1) activation of ADP dependent Na^+/Na^+ exchange with a high $K_{1/2}$ for Na^+; 2) activation of ATP dependent Na^+/Na^+ exchange with, presumably, a similar high $K_{1/2}$ for Na^+; and 3) inhibition of the Na^+ efflux into Na^+ and K^+ free solution with low $K_{1/2}$ for Na^+. It is tempting to consider that the first two effects take place at the two transport sites and the third at the third Na^+ binding site. It is also tempting to identify the site at which Na^+ inhibits Na^+ efflux with the site at which it allosterically modifies the curve relating K^+ influx to external K^+ concentration.

5.6.3 K^+/K^+ Exchange

The red cell Na^+/K^+ pump also carries out an exchange of K^+_i for K^+_o through clockwise operation of the pathway $E_2P \longleftrightarrow E_1PATP$ and its reverse.

The ratio of K^+ efflux to K^+ influx for the K^+/K^+ exchange is nearly 1:1 (Simons 1974). The $K_{1/2}$ for K^+_i is about 10 mM (Simons 1974), and the $K_{1/2}$ for K^+_o in Na^+ free solutions is 0.17 mM, higher than the $K_{1/2}$ for K^+ for the Na^+/K^+ exchange in the same circumstances (Sachs 1986a).

For a true exchange, an increase in the concentration of substrate at one side of the membrane should increase both V_m and the $K_{1/2}$ for the substrate at the other side by the same proportion. In fact, when K^+_i was increased (Sachs 1981), V_m and $K_{1/2}$ for K^+_o both increased as expected, and the ratio $V_m/K_{1/2}$ remained constant. The K^+/K^+ exchange therefore satisfies this criterion for a true exchange.

Support for the statement that the exchange takes place by the pathway E_1ATP to E_2P comes from an experiment with oligomycin (Sachs 1980). Oligomycin inhibits Na^+/K^+ exchange by stabilizing intermediates to which Na^+ is bound, especially $E_1P(3Na)$. In cells free of Na^+, oligomycin does not inhibit K^+/K^+ exchange. In cells that contain Na^+ at a concentration too low to inhibit K^+/K^+ exchange by itself, addition of oligomycin then results in inhibition. Na^+_i generates intermediates such as $E_1P(3Na)$ which are stabilized by oligomycin, and therefore accumulate, resulting in depletion of the E_2 forms which carry out the K^+/K^+ exchange.

Figure 5.2 predicts that K^+/K^+ exchange requires phosphate, Mg^{2+}, and ATP, and it is known that it does (Glynn et al. 1970). ATP can be replaced by a non-hydrolyzable ATP analog such as AMP-PNP or AMP-PCP (Simons 1975), TNP-ADP or TNP-ATP (Sachs 1988), or ADP (Kaplan 1982).

Inspection of Fig. 5.2 suggests that stimulation by ATP should be biphasic. Since the exchange results in no net concentration changes, addition of ATP at constant $MgPO_4$ at first increases the velocity of the exchange. But, as the concentration of ATP increases further, the concentration of E_1ATP increases with depletion of the other intermediates in the exchange pathway. The velocity of the exchange should peak and then decline. A similar sequence should occur with $MgPO_4$. As $MgPO_4$ increases at constant ATP, the velocity of the exchange increases, but eventually E_2PK accumulates with depletion of the other intermediates, and the exchange velocity peaks and then decreases. Experimentally, however, no decrease in the exchange rate is seen when ATP or $MgPO_4$ are increased to high concentrations in the presence of high fixed concentrations of the other (Simons 1974; Sachs 1981). This observation requires that there is an intermediate at some point in the exchange pathway which simultaneously binds ATP and $MgPO_4$, and it has been advanced as evidence for the belief that the functional unit of the Na^+/K^+ pump is an $(\alpha\beta)_2$ dimer, one protomer of which binds ATP at the same time that the other binds $MgPO_4$. However, it is now clear that the functional unit of the red cell pump is an $\alpha\beta$ protomer (Martin and Sachs 1992; Sachs 1994), The intermediate which simultaneously binds ATP and $MgPO_4$ is an $\alpha\beta$ protomer. Adenosine tetraphosphate, and TNP-adenosine tetraphosphate, in the presence of $MgPO_4$, support K^+/K^+ exchange as well as ATP or TNP-ATP so there must be enough room in the ATP binding site for two extra phosphate groups (Sachs 1988). The 2.6 Å structure of the closely related Ca^{2+}-ATPase, presumably in the E_1 conformation, shows that the nucleotide binding site is separated from the phosphorylation site by about 25 Å (Toyoshima et al. 2000) so that the ability of the Na^+/K^+ pump to bind adenosine tetraphosphate and $MgPO_4$ simultaneously to the same α chain is not surprising.

Na^+_i inhibits the K^+/K^+ exchange (Simons 1974). If the model proposed by Garay and Garrahan (1973), and discussed above, is expanded by the assumption that K^+/K^+ exchange occurs when K^+ is bound to all three sites, then, assuming random binding of Na^+ and K^+, the velocity of the K^+/K^+ exchange (v) is related to $[Na^+_i]$ and $[K^+_i]$ by the relation:

$$v = V_m/(1 + (K_K/[K^+_i])(1 + [Na^+_i]/K_N))^3 \tag{5.3}$$

where V_m is the maximal velocity of the K^+/K^+ exchange at saturating K^+. With the same value of K_N used in evaluating the effect of $[Na^+_i]$ on the Na^+/K^+ exchange, Eq. (5.3) describes several of the characteristics of the K^+/K^+ exchange. At fixed concentrations of K^+_i the reciprocal of K^+/K^+ exchange is a parabolic function of $[Na^+_i]$ (Sachs 1986a), as predicted by Eq. (5.3). Inhibition of the K^+/K^+ exchange by Na^+_i is competitive, and, as required of a competitive inhibitor, sufficiently high concentrations of Na^+_i drive the K^+/K^+ exchange to zero (Sachs 1986a).

The model from which Eq. (5.3) is derived assumes that K^+ must be bound at three binding sites before transport occurs. Since two K^+ ions are transported in each cycle during Na^+/K^+ exchange, it might be expected that K^+/K^+ exchange occurs when K^+ is bound only to the two transport sites. In that case, the third Na^+ binding site would differ from the other two in that it either does not bind K^+, or that K^+ bound to it does not cross the membrane during exchange. Although it is known that K^+ influx equals K^+ efflux during K^+/K^+ exchange, it is not known how many K^+ ions are transported per exchange cycle. Since the influx component of ATP dependent Na^+/Na^+ exchange, which utilizes the same mechanism as the K^+/K^+ exchange, transports three Na^+ ions per cycle, it is possible that three K^+ ions are transported inward during K^+/K^+ exchange. However, support for the belief that the K^+/K^+ exchange involves only two of the three cation binding sites comes from experiments with purified membrane preparations using the electrochromic fluorescent dye RH421 as a reporter. The results showed that other monovalent cations compete with Na^+ at two of the binding sites, but not at the third (Schneeberger and Apell 2001).

Examination of Fig. 5.2 shows that the existence of an Na^+ efflux into Na^+- and K^+-free solution which does not exchange with an external cation through the pathway E_2P to E_1 necessarily implies the presence of an uncoupled K^+ efflux through the same pathway under the same conditions. Such an efflux can be demonstrated (Sachs 1986a). It is inhibited by low concentrations of Na^+_o, as is the comparable Na^+ efflux, and, as expected, by the inhibitors ouabain and vanadate, but not by oligomycin.

5.6.4 The Reaction Mechanism of the Na^+/K^+ Pump

If inward transport of K^+ follows outward transport of Na^+, then it is possible that the same set of cation binding sites carry out both movements by changing relative affinities for Na^+ and K^+ during a transport cycle. If Na^+ and K^+ sites do not coexist, the reaction mechanism of the pump must be ping-pong with respect to Na^+ and K^+; i.e., Na^+ must be released to the outside before outside K^+ adds, and K^+ must be released to the inside before inside Na^+ adds. The reverse does not hold; the reaction mechanism may be ping-pong, and nevertheless sites for Na^+ and K^+ coexist. On the other hand, if the pump mechanism is sequential with respect to Na^+ and K^+; i.e., if both Na^+_i and K^+_o must be simultaneously bound to the pump at some point in the transport cycle, then high affinity sites for Na^+ and K^+ must coexist.

Indications that the reaction mechanism must be ping-pong are many. The enzyme binds or occludes either Na^+ or K^+, but not both, suggesting that binding sites do not coexist. If the enzyme cannot bind both ions simultaneously during a transport cycle, the mechanism must be ping-pong. On the other hand, the design of the binding and occlusion experiments is such that it would be difficult to detect simultaneous binding of both Na^+ and K^+. Results from molecular biology indicate that the same amino acids in the transmembrane helices are involved in the binding of Na^+ and K^+. When high resolution structures of the enzyme become available it might be possible to resolve the question definitively. Meanwhile, suggestions that the mechanism is sequential continue to appear. Such suggestions are often put forward along with the proposal that the functional unit of the enzyme is an $(\alpha\beta)_2$ dimer, and that each monomer acts cooperatively during a transport cycle.

Cleland (1989) has written "because it looks at the reaction while it is taking place, a kinetic study is the ultimate criteria of mechanism". There is a great deal of evidence from experiments with the red cell pump that the mechanism is, indeed, ping-pong.

A ping-pong mechanism of Na^+/K^+ pump can be represented as:

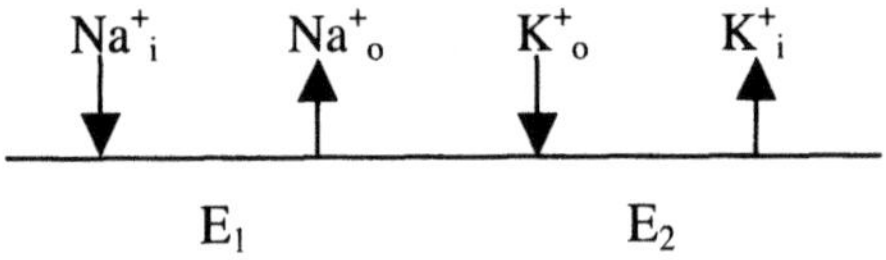

where E_1 and E_2 is enzyme in its two major conformations. For a ping-pong mechanism, measurement of V_m and the $K_{1/2}$ for one of the substrates (Na^+_i or K^+_o) at fixed concentrations of the second and in the absence of products (Na^+_o and K^+_i) shows increases in both parameters as the fixed concentration of the second substrate increases. The ratio $V_m/K_{1/2}$ remains constant. As the fixed concentration of Na^+_i increases in the absence of Na^+_o, the amount of E_2, and therefore V_m, increases, and, because the concentration of E_2 increases, the $[K^+_o]$ necessary to half saturate also increases to the same extent.

When K^+ influx in red cells was measured at a number of fixed $[Na^+_i]$ and varying $[K^+_o]$ in the absence of the products Na^+_o and K^+_i, V_m and the $K_{1/2}$ for K^+_o both increased, but the ratio $V_m/K_{1/2}$ was a sigmoid function of the $[Na^+_i]$ with a $K_{1/2}$ for Na^+_i about the same as that found when pump rate is plotted against the $[Na^+_i]$ (Sachs 1977b). The finding is not consistent with the predictions of a ping-pong mechanism.

However, the situation is complicated by the Na^+ efflux in solutions free of both Na^+ and K^+. Because of the efflux, the Albers-Post mechanism is not strictly ping-pong, but is better represented as:

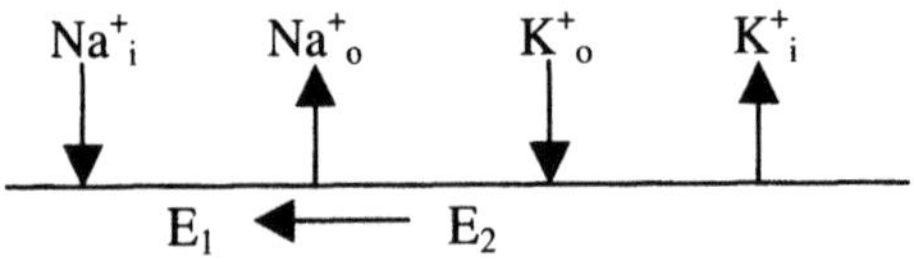

where the arrow between E_2 and E_1 represents Na^+ efflux into Na^+- and K^+-free solution through the pathway E_2P to E_1 (Fig. 5.2). Although increasing the concentration of Na^+_i increases V_m, the increase in the concentration of E_2 is less than the increase in a pure ping-pong mechanism and therefore the increase in the $K_{1/2}$ for K^+_o is less than the increase in V_m. When we derived a rate equation from the mechanism which included the Na^+ efflux into Na^+- and K^+-free solutions, we found that it described the data from the experiments summarized above remarkably well (Sachs 1979). The model exactly described the sigmoid relation between the ratio $V_m/K_{1/2}$ for K^+_o or Cs^+_o and $[Na^+_i]$. Furthermore, when we estimated the magnitude of the Na^+ efflux into Na^+- and K^+-free solutions from the parameters obtained from the measurement of Na^+/K^+ exchange or Na^+/Cs^+ exchange as K^+ or Cs^+ influx, the predicted value was about the same as the value found when Na^+ efflux was directly measured in a Na^+- and K^+-free solution. The sequential model without a Na^+ efflux in the absence of Na^+_o and K^+_o was also able to describe the results, although not as well as the modified ping-pong mechanism. Steady-state kinetic analyses of the overall Na^+/K^+ exchange cannot unambiguously distinguish between a sequential mechanism and a modified ping-pong mechanism, which includes the Na^+ efflux into Na^+- and K^+-free solutions.

Strong support for the ping-pong mechanism comes from the Na^+/Na^+ and K^+/K^+ exchanges provided they are true exchanges, and part of the overall Na^+/K^+ exchange pathway, rather than alternative pathways. If the pump mechanism were sequential rather than ping-pong, so that both internal Na^+ and external K^+ must bind before a transport cycle is completed, and Na^+ must be released to the outside and K^+ to the inside before a new cycle begins:

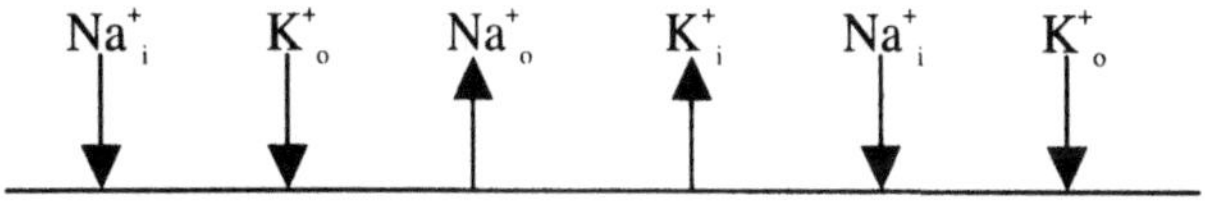

Na^+_i will be necessary for K^+/K^+ exchange to occur, and K^+_o must be present for Na^+/Na^+ exchange to occur. Na^+_i is not necessary for K^+/K^+ exchange, and, in fact, inhibits it (Simons 1974) and K^+_i is not required for Na^+/Na^+ exchange, but inhibits it (Garrahan and Glynn 1967c).

Evidence that the K^+/K^+ exchange is a true exchange was discussed above. There are many reasons for believing the K^+/K^+ exchange shares the pathway of the overall Na^+/K^+ exchange derived from a comparison of the biochemistry of purified Na^+, K^+-ATPase preparations with the properties of the K^+/K^+ exchange (Glynn and Karlish 1982). There are several reasons for accepting the shared pathway of the two exchanges that do not rely on biochemical data.

From Fig. 5.2, one would expect the $K_{1/2}$ for K^+_o of the Na^+/K^+ exchange and the $K_{1/2}$ for K^+_o of the K^+/K^+ exchange would be the same when measured under identical conditions. In fact, the two values differ significantly (Sachs 1980). Evaluation of the scheme in Fig. 5.2 showed that the value of the $K_{1/2}$ for K^+_o of the two exchanges differ because of the pathway E_2P to E_1 that accounts for the efflux of Na^+ and the efflux of K^+ into solutions free of both Na^+ and K^+. These effluxes are inhibited at low concentrations of Na^+_o (Garrahan and Glynn 1967a; Sachs 1986), and when the $K_{1/2}$ for K^+_o of the Na^+/K^+ exchange and K^+/K^+ exchange are measured simultaneously at low $[Na^+_o]$, the two values approach each other. The reaction mechanism of Fig. 5.2 satisfactorily explains a puzzling discrepancy between the Na^+/K^+ exchange and the K^+/K^+ exchange provided the two exchanges share the same pathway.

In a preceding section, it was pointed out that competition between Na^+_i and K^+_i for the same set of binding sites satisfactorily describes the dependence of the Na^+/K^+ exchange and the K^+/K^+ exchange on the concentration of the two ions.

Since K^+/K^+ exchange shares the pathway of the overall Na^+/K^+ exchange, and since it does not require internal Na^+, Na^+ must add to the pump at the inside after K^+ is released.

The situation with the ADP dependent Na^+/Na^+ exchange is less clear. Comparison of the characteristics of the exchange with the biochemistry of the Na^+, K^+-ATPase strongly suggests that the two processes must utilize the same pathway. However, the $K_{1/2}$ for Na^+_o of the exchange does not vary with $[Na^+_i]$, nor does the $K_{1/2}$ for Na^+_i vary with the $[Na^+_o]$ (Garay and Garrahan 1973). The efflux of Na^+ into Na^+- and K^+-free solutions complicates interpretation. Although K^+_o is not required for the Na^+/Na^+ exchange, and, in fact, suppresses it, even at saturating K^+_o the exchange is not completely suppressed (Garrahan and Glynn 1967a). One cannot, therefore, conclude that the two ions compete at the same binding sites.

The pump reaction mechanism can be independently evaluated by determining the characteristics of pump inhibition by reversible inhibitors which combine with specific enzyme forms. Two inhibitors, oligomycin and vanadate, have proven useful. The rationale can be understood intuitively.

Oligomycin reversibly inhibits Na^+/K^+ exchange and Na^+/Na^+ exchange, but not K^+/K^+ exchange unless the cells contain Na^+. Inhibition of Na^+/Na^+ exchange increases with increasing $[Na^+_i]$ and with increasing $[Na^+_o]$. Oligomycin inhibits by combining with an enzyme form that occurs between the addition of Na^+ at the inside and its release at the outside:

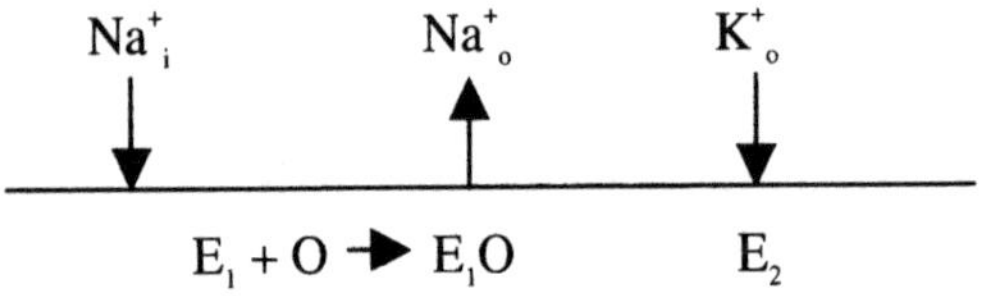

where O is oligomycin. Fractional inhibition by oligomycin is proportional to the steady-state concentration of E_1 as a fraction of all enzyme forms. Since E_1 can be formed by forward or reverse operation of the pump when Na^+_o is present, the fraction of all enzyme forms represented by E_1 is independent of $[K^+_o]$. Therefore at constant $[K^+_o]$, fractional inhibition will also be independent of $[K^+_o]$. At all $[K^+_o]$, the pump rate will be reduced by a constant factor. As a result, V_m decreases, but $K_{1/2}$ for K^+_o remains constant (Sachs 1991). Inhibition in which V_m decreases, but $K_{1/2}$ remains unchanged is noncompetitive.

If Na^+_o is absent, E_1 can be formed only by forward operation of the pump, and the fraction of all enzyme forms represented by E_1 increases with $[K^+_o]$. Fractional inhibition at constant oligomycin concentration increases with increasing $[K^+_o]$. As a result, V_m decreases, but since inhibition is least at the lowest $[K^+_o]$, $K_{1/2}$ for K^+_o also decreases. Inhibition in which both V_m and $K_{1/2}$ decrease is uncompetitive.

Inhibition of pump rate by oligomycin is noncompetitive with K^+_o if Na^+_o is present, but uncompetitive if it is absent. Na^+ is released to the outside before external K^+ adds (Sachs 1981).

Vanadate reversibly inhibits K^+/K^+ exchange, but does not inhibit Na^+/Na^+ exchange. Inhibition of the K^+/K^+ exchange increases with increasing $[K^+_i]$ and increasing $[K^+_o]$ so that it must add between the addition of K^+ at the outside and its release at the inside:

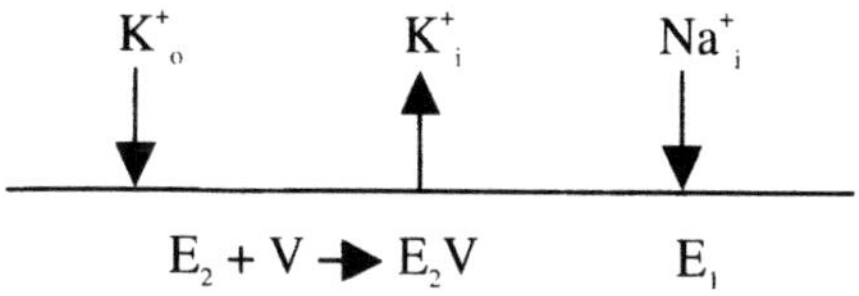

where V is vanadate. Inhibition by vanadate of the Na^+/K^+ exchange is noncompetitive with Na^+_i if K^+_i is present, but uncompetitive if it is absent (Sachs 1986b). From considerations similar to those described above in the case of oligomycin, it can be concluded that K^+ is released to the inside before internal Na^+ adds.

The results from the inhibition studies indicate that the pump mechanism is ping-pong with respect to Na^+ and K^+.

5.7 The Quaternary Structure of the Red Cell Na^+/K^+ Pump

The curve relating Na^+, K^+-ATPase activity to ATP concentration is biphasic both in purified membrane preparations (Neufeld and Levy 1969) and in red blood cells (Glynn and Karlish 1976). As discussed above, biphasic curves arise when the substrate (ATP) interacts with the enzyme at more than one point in the reaction cycle. For some time it was believed that the existence of two points at which ATP adds means that the functional enzyme unit is a dimer of αβ protomers, each with an ATP site, which interact during an enzyme cycle (Glynn and Karlish 1976). This interpretation continues to attract strenuous support (Askari 2000).

Moczydlowski and Fortes (1981) pointed out that, as shown in Fig. 5.2, the reaction mechanism is branched at the joint at which ATP adds. Each branch starts with $E_2(2K)$ in which K^+ is occluded. In the upper branch, E_2 spontaneously charges conformation to E_1, K^+ is deoccluded and released to the inside, and ATP adds to E_1 with high affinity. In the lower pathway, which accounts for the greater part of the flow, ATP binds to $E_2(2K)$ with low affinity, and induces a change in conformation from E_2 to E_1 with deocclusion of K^+ and its release to the inside. This mechanism produces a biphasic ATPase curve using measured values of the rate constants of intermediate steps between E_2K and E_1ATP in the reaction pathway.

When Li^+ replaces K^+, V_m is about half that seen with K^+, and the $K_{1/2}$ for Li^+ is low. Li^+ is not detectably occluded (Post et al. 1972). All of the flow during a transport cycle in which Li^+ replaces K^+ must take place through the upper branch (Fig. 5.2) in which Li^+ is spontaneously released to the inside before ATP adds with high affinity. The curve of ATPase activity as a function of ATP concentration should not be biphasic, but should be represented by a rectangular hyperbola with a very low $K_{1/2}$ for ATP. Figure 5.3 shows the reciprocal of Na^+, K^+-ATPase activity as a function of the reciprocal of the ATP concentration. The downward curvature of the Rb^+ curve results from the biphasic direct plot, and the $K_{1/2}$ for ATP is relatively high. Rb^+ is occluded more strongly than K^+, and most of the flow during a transport cycle takes place through the lower branch (Fig. 5.2). On the other hand, the linear reciprocal plot in the presence of Li^+ results from a direct plot of a rectangular hyperbola and the $K_{1/2}$ for ATP is very low (Sachs 1994). The

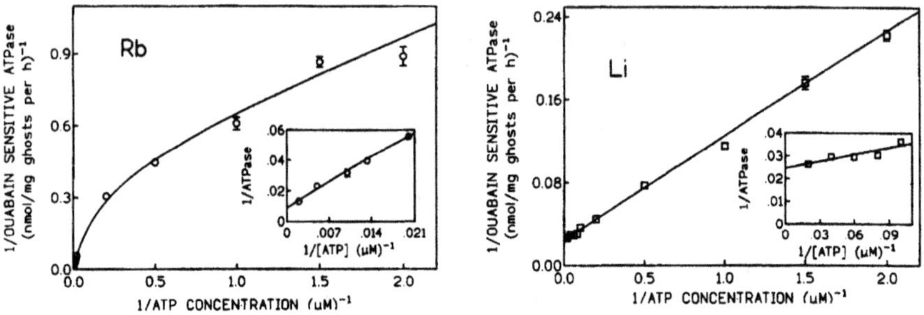

Fig. 5.3. Reciprocal of ouabain-sensitive ATPase activity *vs*. reciprocal of ATP concentration. Measured in solutions in which K^+ was replaced with 24 mM Rb^+ or 80 mM Li^+. The insert is an expanded plot of the results at the highest ATP concentrations. From Sachs (1994)

findings are readily explained by the interpretation of Moczydlowski and Fortes (1981). Na^+ replacing K^+ at the outside is occluded little, if at all, and a plot of ATPase activity as a function of ATP concentration with Na^+ replacing K^+ at the outside is also a rectangular hyperbola with very low $K_{1/2}$ for ATP, but V_m is a very small fraction of V_m with K^+.

There is evidence that interaction between coexisting ATP sites in a functional enzyme unit does not account for the biphasic curve. If coexisting binding sites exist, inhibitors, such as FITC, which react randomly and irreversibly with ATP binding sites, would produce a mixed population of functional units in some of which both ATP sites are blocked, in some neither is blocked, and in others one site is blocked and the other is not. At inhibitor concentrations at which functional units in which one ATP site is blocked and the other not predominate, interaction between ATP sites will be much less than in uninhibited enzyme, and the biphasic curve should be converted toward a rectangular hyperbola. When the measurements were made, the shape of the biphasic curve was not detectably altered, although V_m was reduced to 30% for less of the V_m of uninhibited enzyme. The biphasic curve does not provide evidence for multiple interacting ATP binding sites in each functional unit (Sachs 1994).

In purified membrane preparations, crosslinkers such as Cu and *o*-phenanthroline from αα dimers in a conformationally dependent manner (Periyasamy et al. 1983). αβ protomers are closely packed in purified preparations (12,000–18,000/μ^2) so that it is possible that crosslinking results from casual association of protomers in a densely packed environment rather than from permanent structural dimers. In red cell membranes, αβ protomers are much less dense, 3/μ^2. Cu and *o*-phenanthroline does not crosslink α chains in red cell membranes to each other, but does crosslink them to band 3, the dominant membrane protein, in a conformationally dependent manner (Martin and Sachs 1992). In red cell membranes ankyrin has a high affinity binding site for α chains and attaches them to the cytoskeleton. This may aid in preventing the protomers from forming dimers. In purified membrane preparations, from which the cytoskeleton has been removed, measurement of fluorescence resonance energy transfer between fluorescent ouabain analogs, show that αβ protomers are all close enough to be dimers. However, the microsomes from which the purified membranes are prepared retain significant amounts of cytoskeleton, and in microsomes more than 50% of the protomers are monomers. The molecular turnover number is the same for both preparations so that dimereization does not significantly alter the reaction mechanism (Martin and Sachs 2000). The bulk of the evidence does not support multiple ATP sites as the explanation for the biphasic activation curve, but is consistent with the interpretation of Moczydlowski and Fortes (1982).

5.8 Unresolved Issues

There are several issues about the reaction mechanism of the Na^+/K^+ pump, which arise, at least in part, from experiments with red cells.

5.8.1 Properties of the Third Na^+ Site

Studies with purified membranes indicate that one of three Na^+ ions is different from the other two. Addition of the third Na^+ ion at the inner surface, after the other two Na^+ ions have bound, is electrogenic and causes a conformational change (Schneeberger and Apell 1999). Three Na^+ ions are occluded in oligomycin inhibited enzyme (EP(3Na)), but only two in ouabain inhibited enzyme (E_2P(2Na)) (Jørgensen 1991). Electrophysiological studies indicate that release of one Na^+ ion from E_1P(3Na)) is highly electrogenic, but release of the other two is not (Rakowski et al. 1997). The binding site for one of the Na^+ ions may differ qualitatively from the binding sites for the other two, which alternate in their affinity for Na^+ and K^+ during a transport cycle.

In preceding sections, it was pointed out that external Na^+ at low concentration inhibits Na^+ efflux and K^+ efflux in Na^+ and K^+ free solution, and at higher concentrations activates both ADP-dependent and ATP-dependent Na^+/Na^+ exchange. This suggests Na^+ binding sites with different properties.

There is another observation that suggests qualitatively heterogeneous Na^+ binding sites. Vanadate inhibits (Na^+ + K^+)-dependent ATPase activity by binding to E_2 forming a transition state analog of E_2P. It has long been known that in solutions containing both Na^+ and vanadate the relation between pump rate and [K^+_o] is biphasic (Cantley et al. 1978; Beauge 2000). The effect is attributable to the presence of Na^+_o. Figure 5.4 shows that, when vanadate is present, a plot of transport rate as a function of [K^+_o] is biphasic if the solution contains Na^+, but in Na^+-free solutions, inhibition is constant at all [K^+_o] (Sachs 1987). Na^+_o protects against vanadate inhibition, and protection is abolished by K^+ (and the K^+ congeners, except Li^+). The K^+ transport sites and the site at which Na^+ protects against inhibition cannot be the same since Fig. 5.5 shows that Na^+ protects against inhibition even at [K^+_o] at which K^+_o maximally stimulates Na^+/K^+ exchange, and therefore the K^+ transport sites are saturated. The point is reinforced by Fig. 5.6, which shows the uninhibited pump rate (normalized to the V_m) and fractional inhibition of transport

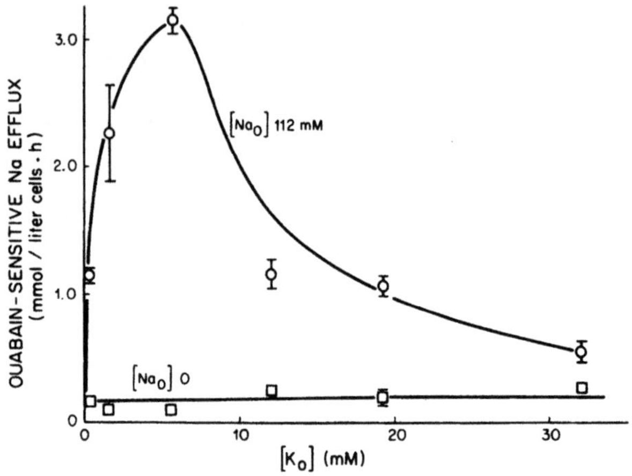

Fig. 5.4. Ouabain-sensitive Na^+ efflux *vs.* $[K^+_o]$ into solutions with and without Na^+. 2.2 μM vanadate was present in all solutions. From Sachs (1987)

rate at two vanadate concentrations as a function of $[K^+_o]$. The $K_{1/2}$ at sites at which $[K^+_o]$ activates K^+ transport is lower than the $K_{1/2}$ at sites at which it protects against vanadate inhibition, and highest at the lowest vanadate concentration. The site at which K^+_o abolishes Na^+ protection is different from the K^+ transport sites. It may be another manifestation of the third Na^+ site.

The estimated affinity of the protective site for both Na^+ and K^+ is high: the estimated $K_{1/2}$ for both Na^+ and K^+ is less than 1 mM. The affinity for Na^+ is comparable to the affinity of the site at which Na^+ inhibits Na^+ and K^+ efflux into Na^+- and K^+-free solution, and the site at which it inhibits dephosphorylation.

The nature of the Na^+ site at which it exerts these effects remains to be elucidated.

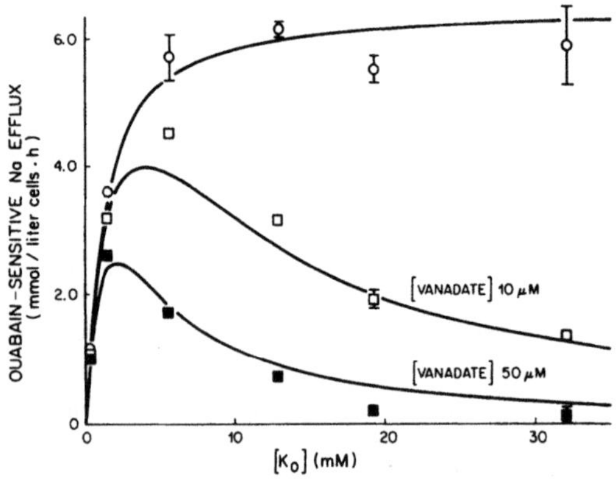

Fig. 5.5. Ouabain-sensitive Na^+ efflux *vs.* $[K^+_o]$. The extracellular solution contained 112 mM Na^+ and the indicated concentrations of vanadate. From Sachs (1987)

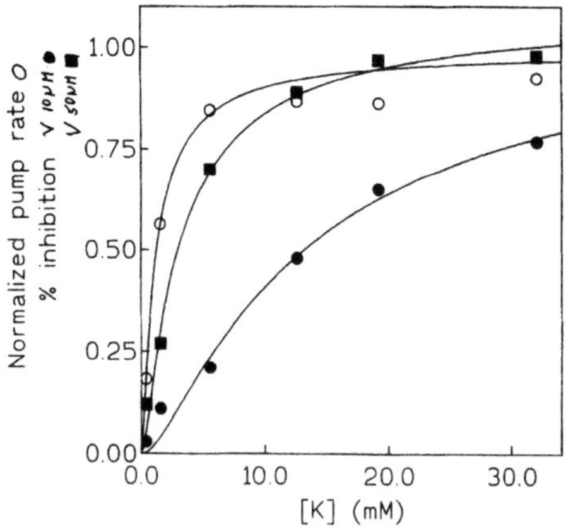

Fig. 5.6. Pump rate normalized to V_m *vs.* $[K^+_o]$ (○), and % inhibition at 50 μM (■) and 10 μM (●) vanadate from the data of Fig. 5.5. From Sachs (1987)

5.8.2 Selectivity of the Cation Binding Sites

Ion binding sites in proteins are made up of carboxyl, carbonyl and hydroxyl oxygens of amino acid side chains, and sometimes peptide backbone carbonyl oxygens, arranged so that they can replace the oxygens of water of hydration of the ions in solution. The distance from the ions to the binding oxygens is 1.4 Å (the oxygen radius) plus the radius of the bound ion, and the net change of the binding site equals that of the bound ion.

At the inside, the binding sites have a high affinity for Na^+, a lower affinity for Li^+ and low affinity for K^+ (measured as an inhibitor of Na^+ binding). The ionic radius of Na^+ is 0.95 Å, of Li^+ 0.53 Å, and of K^+ 1.32 Å. The difference in the sizes of the three ions accounts in a general way for the difference in the affinity of the binding sites, but does not account for the observation that Li^+ can replace Na^+ at the three binding sites and initiate a transport cycle, but K^+ cannot.

At the outside, the sites have high affinities for K^+, Rb^+, and Tl^+, and somewhat lower affinities for NH_4^+ and Cs^+. The radii of K^+, Rb^+, and Tl^+ range from 1.32 Å–1.46 Å, and Cs^+ is somewhat larger, 1.63 Å. Na^+ can substitute for K^+ at the outside sites even though it is much smaller than K^+. The affinity for Na^+ is low and V_m with Na^+ replacing K^+ is 3% or less of the V_m with K^+. Li^+, which is much smaller than Na^+, can also substitute for K^+ with low affinity, but in this case V_m is about 50% of the V_m with K^+. Once again, Na^+ at the outside is qualitatively different from the other cations. Li^+ is not detectably occluded when it activates transport from the outside so that occlusion cannot be an essential part of the ion transport mechanism.

Interest in Li^+ as a substrate for the Na^+/K^+ pump ended a decade ago. In that time, much has been learned about the interaction of ions with the pump. No understanding of the way in which ions activate transport cycles will be complete without an understanding of the way in which Li^+ is recognized and transported.

5.8.3 Na^+_i/H^+_o Exchange and K^+_o/H^+_i Exchange

At pH 6.2, the Na^+/K^+ pump carries out an exchange of K^+_o for H^+_i when Na_i^+ is zero, and an exchange of Na^+_i for H^+_o when K^+_o is zero (Polvani and Blostein 1988). The affinity of the inside sites for H^+ is quite high since H^+ transport can be detected at $[H^+]$ as low as 10^{-8} M. Since the highly homologous H^+/K^+ pump exchanges H^+_i for K^+_o, and since other transporters are known which transport both Na^+ or H^+, it is not surprising that substitution can occur in the Na^+/K^+ pump. Polvani and Blostein (1988) point out that crown ethers, which form crystals with Na^+ and K^+, also form crystals with H_3O^+ (Boyer 1988). In the crown ether crystals, H_3O^+ forms hydrogen bands 2.7 Å long with three oxygen atoms. The distance between the oxygen of the hydronium ion and the coordinating oxygens in an H_3O^+ crystal is about the same as the distance between K^+ and its coordinating oxygens in crystals formed with crown ethers and K^+ salts. The distance is, however, greater than the distance between Na^+ and coordinating oxygens in Na^+ salts. Li^+ also binds to inside sites with a high affinity for Na^+ and outside sites with a high affinity for K^+. It is widely accepted that the binding sites change their affinity for ions as a result of changes in the relative positions of the transmembrane helices that alter the configuration of the coordinating oxygens in the binding site. In other proteins, H^+ transport occurs by the sequential formation and breaking of hydrogen bonds in a defined pathway. It is conceivable that such a mechanism might contribute to the change in affinity of the cation binding sites.

5.8.4 Na^+ Transport Coupled with Anion Transport

The unresolved issues discussed above are related to the characteristics of the ion binding sites. There is a perplexing set of findings that is not related to the binding sites, but has relevance for the reaction mechanism. The efflux of Na^+ into solutions free of both Na^+ and K^+ is usually referred to as the uncoupled efflux since it is not coupled to the influx of another cation. However, this flow in red cells is coupled to the efflux of an anion (Dessing and Hoffman 1990). If the anion is sulfate, the coupling ratio is two Na^+ ions per sulfate so that the efflux is electroneutral. Although Na^+ efflux into Na^+- and K^+-free solutions in red cells of many species displays the anion coupling seen in human red cells, purified enzyme from the kidneys of the same species incorporated into liposomes has markedly different properties (Martin et al. 1994). The uncoupled Na^+ efflux is much less (2–4% of the Na^+/K^+ exchange rate) than the uncoupled efflux in the red cells (15–30% of the Na^+/K^+ exchange rate), and the efflux is not accompanied by an anion. The difference in the properties of purified enzyme incorporated in liposomes and the native enzyme in red cells is unexplained.

5.9 Conclusion

Until recently, study of the Na^+/K^+ pump of red cells was a major source of information about the Na^+, K^+-ATPase, and provided the basis for many of the concepts about its properties. In recent years, however, emphasis on other preparations has eclipsed the red cell as a source of new information. However, as the previous section attempted to demonstrate, there are a number of fundamental questions about the pump that might be answered with the help of red cells. Moreover, it may be possible to use red cells to answer questions about other preparations that are difficult to approach. For instance, to decide whether the kinetic differences between the various α isomers are properties of the proteins themselves, or related to the fact that they predominate in different tissues, Munzer et al. (1992) devised a method for fusing purified enzyme preparations with red cells. Using this system, the kinetic differences between the isoforms were found to persist even when they functioned in the same membrane environment (Munzer et al. 1994). Kidney enzyme incorporated into dog red cells differ from the native red cell pump in that there is very little uncoupled Na^+ efflux.

The chapter has endeavored to provide convincing evidence that the red cell has proven indispensable in the study of Na^+/K^+ pump. Perhaps it will also encourage those interested in this enzyme to consider whether it can continue to resolve outstanding questions.

Acknowledgement: Work from the author's laboratory described in this chapter was supported in part by a grant DK19185 from the USPHS.

References

Albers RW, Kovall GJ, Fahn S (1963) Role of sodium ions in activation of electrophorous electric organ adenosine triphosphatase. Proc Natl Acad Sci USA 50:474-481

Askari A (2000) Significance of potein-protein interactions to Na^+/K^+ ATPase function. In: Taniguchi K, Kaya S (eds) Na^+/K^+ ATPase and related ATPases. Proceedings of the IXth international conference on the Na^+/K^+ ATPase and related ATPases. Elsevier, Amsterdam, pp 17-26

Beaugé L (2000) Reactivity of the Na^+, K^+-ATPase while cycling as Na^+-ATPase. In: Taniguchi K, Kaya S (eds) Na^+/K^+ ATPase and related ATPases. Proceedings of the IXth international conference on the Na^+/K^+ ATPase and related ATPases. Elsevier, Amsterdam, pp 325-332

Blostein R (1968) Relationships between erythrocyte membrane phosphorylation and adenosine triphosphate hydrolysis. J Biol Chem 243:1957-1965

Blostein R (1970) Sodium-activated adenosine triphosphatase activity of the erythrocyte membrane. J Biol Chem 245:270-275

Blostein R (1983) Sodium pump-catalyzed sodium-sodium exchange associated with ATP hydrolysis. J Biol Chem 258:7948-7953

Boyer PD (1988) Bioenergetic coupling to protonmotive force: should we be considering hydronium ion coordination and not group protonation? Trends Biochem Sci 13:5-7

Cantley LC Jr, Cantley LG, Josephson L (1978) A characterization of vanadate interactions with the (Na,K)-ATPase. Mechanistic and regulatory implications. J Biol Chem 253:7361-7368

Cass A, Dalmark M (1973) Equilibrium dialysis of ions in nystatin-treated red cell. Nature New Biol 244:47-49

Cavieres JD, Ellory JC (1975) Allosteric inhibition of the sodium pump by external sodium. Nature 255:338-340

Cleland WW (1963) The kinetics of enzyme-catalyzed reactions with two or more substrates or products. I. Nomenclature and rate equations. Biochim Biophys Acta 67:104-137

Cleland WW (1989) Enzyme kinetics revisited: a commentary on 'The kinetics of enzyme-catalyzed reactions with two or more substrates or products'. Biochim Biophys Acta 1000:209-212

Cohn WE, Cohn ET (1939) Permeability of red corpuscles of the dog to sodium ion. Proc Soc Exp Biol Med 41:445-449

Danowski TS (1941) The transfer of potassium across the human red blood cell membrane. J Biol Chem 139:693-705

Dean RB (1941) Theories of electrolyte equilibrium in muscle. Biol Symp 3:331-348

Dean RB (1987) Reminiscences on the sodium pump. TINS 10:451-454

Cavieres JD, Ellory JC (1975) Allosteric inhibition of the sodium pump by external sodium. Nature 255: 338-340

Dissing S, Hoffman JF (1990) Anion-coupled Na efflux mediated by the human red blood cell Na/K pump. J Gen Physiol 96:167-193

Dunham PB, Senyk O (1977) Lithium efflux through the Na/K pump in human erythrocytes. Proc Natl Acad Sci USA 74:3099-3103

Fenn WO, Cobb DM (1936) Electrolyte changes in muscle during activity. Am J Physiol 115:345-356

Flynn F, Maizels M (1949) Cation control in human erythroytes. J Physiol 110:301-318

Garay RP, Garrahan PJ (1973) The interaction of sodium and potassium with the sodium pump in red cells. J Physiol 231:297-325

Gardos G (1954) Akkumulation der Kaliumionen durch menschliche Blutkörperchen. Acta Physiol Hung 6:191-199

Garrahan PJ, Glynn IM (1967a) The behaviour of the sodium pump in red cells in the absence of external potassium. J Physiol 192:159-174

Garrahan PJ, Glynn IM (1967b) The sensitivity of the sodium pump to external sodium. J Physiol 192:175-188

Garrahan PJ, Glynn IM (1967c) Factors affecting the relative magnitudes of the sodium:potassium and sodium:sodium exchanges catalysed by the sodium pump. J Physiol 192:189-216

Garrahan PJ, Glynn IM (1967d) The stoicheiometry of the sodium pump. J Physiol 192:217-235

Garrahan PJ, Rega AF (1967) Cation loading of red blood cells. J Physiol 193:459-466

Garrahan PJ, Pouchan MI, Rega AF (1969) Potassium activated phosphatase from human red blood cells. The mechanism of potassium activation. J Physiol 202:305-327

Glynn IM (1956) Sodium and potassium movements in human red cells. J Physiol 134:278-310

Glynn IM, Hoffman JF (1971) Nucleotide requirements for sodium-sodium exchange catalysed by the sodium pump in human red cells. J Physiol 218:239-256

Glynn IM, Karlish SJ (1976) ATP hydrolysis associated with an uncoupled efflux of Na through the sodium pump: evidence for allosteric effects of intracellular ATP and extracellular sodium. J Physiol 256:465-496

Glynn IM, Karlish SJD (1982) Conformational changes associated with K^+ transport by the Na^+-K^+-ATPase. In: Martinosi AN (ed) Membranes and transport. Plenum Publishing Corp, New York, pp 529-536

Glynn IM, Lew VL, Lüthi U (1970) Reversal of the potassium entry mechanism in red cells, with and without reversal of the entire pump cycle. J Physiol 207:371-391

Hahn L, Hevesy G (1941) Rate of penetration of ions into erythrocytes. Acta Physiol Scand 3:193-223

Harris JE (1941) The influence of the metabolism of human erythrocytes on their potassium content. J Biol Chem 141:579-595

Harris EJ (1954) Linkage of sodium- and potassium-active transport in human erythrocytes. Symp Soc Exp Biol 8:228-241

Harris EJ, Maizels M (1952) Distribution of ions in suspensions of human erythrocytes. J Physiol 118:40-53

Heppel LA (1939) The electrolytes of muscle and liver in potassium-depleted rats. Am J Physiol 127:385-392

Hobbs AS, Dunham PB (1976) Letter: Evidence for two sodium sites on the external aspect of Na-K pump in human erythrocytes. Nature 260:651-652

Hoffman JF, Kaplan JH, Callahan TJ (1979) The Na:K pump in red cells is electrogenic. Fed Proc 38:2440-2441

Joiner CH, Lauf PK (1978) The correlation between ouabain binding and potassium pump inhibition in human and sheep erythrocytes. J Physiol 283:177-196

Jørgensen PL (1991) Conformational transitions in the alpha-subunit and ion occlusion. Soc Gen Physiol Ser 46:189-200

Kaplan JH (1982) Sodium pump mediated ATP:ADP exchange: the sided effect of sodium and potassium ions. J Gen Physiol 88:915-932

Krogh A (1946) The active and passive exchanges of inorganic ions through the surfaces of living cells and through living membranes generally. Proc Roy Soc (London) Biol Sci 133:140-200

Kropp DL, Sachs JR (1977) Kinetics of the inhibition of the Na-K pump by tetrapropylammonium chloride. J Physiol 264:471-487

Lee KH, Blostein R (1980) Red cell sodium fluxes catalysed by the sodium pump in the absence of K^+ and ADP. Nature 285:338-339

Maizels M, Paterson JH (1940) Survival of stored blood after transfusion. Lancet 2:417-420

Martin DW, Sachs JR (1992) Cross-linking of the erythrocyte Na^+,K^+-ATPase. Chemical crosslinkers induce α subunit-band 3 heterodimers and do not induce α-subunit homodimers. J Biol Chem 267:23922-23929

Martin WH, Richards DE, Marin R, Jack-Hays M, Hoffman JF (1994) Comparative aspects of Na^+/K^+ pump-mediated uncoupled Na^+ efflux in red blood cells and kidney proteoliposomes. Proc Natl Acad Sci USA 91:9881-9885

Martin DW, Marecek J, Scarlata S, Sachs JR (2000) $\alpha\beta$ protomers of Na^+,K^+-ATPase from microsomes of duck salt gland are mostly monomeric: formation of higher oligomers does not modify molecular activity. Proc Natl Acad Sci USA 97:3195-3200

McConaghey PD, Maizels M (1962) Cation exchanges of lactose-treated human red cells. J Physiol 162:485-493

Moczydlowski EG, Fortes PA (1981) Inhibition of sodium and potassium adenosine triphosphatase by 2',3'-O-(2,4,6-trinitrocyclohexadienylidene) adenine nucleotides. Implications for the structure and mechanism of the Na:K pump. J Biol Chem 256:2357-2366

Mullins LJ, Fenn WO, Noonan TR, Haege L (1941) Permeability of erythrocytes to radioactive potassium. Am J Physiol 135:93-101

Munzer JS, Silvius JR, Blostein R (1992) Delivery of ion pumps from exogenous membrane-rich sources into mammalian red blood cells. J Biol Chem 267:5202-5210

Munzer JS, Daly SE, Jewell-Motz EA, Lingrel JB, Blostein R (1994) Tissue- and isoform-specific kinetic behavior of the Na^+,K^+-ATPase. J Biol Chem 269:16668-16676

Neufeld AH, Levy HM (1969) A second ouabain-sensitive sodium-dependent adenosine triphosphate in brain microsomes. J Biol Chem 244:6493-6497

Periyasamy SM, Huang WH, Askari A (1983) Subunit associations of ($Na^+ + K^+$)-dependent adenosine triphosphatase. Chemical cross-linking studies. J Biol Chem 258:9878-9885

Polvani C, Blostein R (1989) Effects of cytoplasmic sodium concentration on the electrogenicity of the sodium pump. J Biol Chem 264:15182-15185

Post RL, Jolly PC (1957) The linkage of sodium, potassium, and ammonium active transport across the human erythrocyte membrane. Biochim Biophys Acta 25:118-128

Post RL, Merritt CR, Kinsolving CR, Allbright CD (1960) Membrane adeosine triphosphatase as a participant in the active transport of sodium and potassium in the human erythrocyte. J Biol Chem 235:1796-1802

Post RL, Sen AK, Rosenthal AS (1965) A phoshorylated intermediate in adenosine triphosphate-depedent sodium + potassium transport across kidney membranes. J Biol Chem 240:1437-1445

Post RL, Albright CD, Dayani K (1967) Resolution of pump and leak components of sodium and potassium ion transport in human erythrocyte. J Gen Physiol 50:1201-1220

Post RL, Hegyvary C, Kume S (1972) Activation by adenosine triphosphate in the phosphorylation kinetics of sodium and potassium ion transport adenosine triphosphatase. J Biol Chem 247:6530-6540

Rakowski RF, Bazanilla P, DeWeer P, Gadsby DC, Holmgren M, Wagg J (1992) Charge translocation by the Na/K pump. Ann New York Acad Sci 834:231-243

Sachs JR (1970) Sodium movements in the human red blood cell. J Gen Physiol 56:322-341

Sachs JR (1977a) Kinetics of the inhibition of the Na-K pump by external sodium. J Physiol 264:449-470

Sachs JR (1977b) Kinetic evaluation of the Na-K pump reaction mechanism. J Physiol 273:489-514

Sachs JR (1979) A modified consecutive model for the Na,K-pump. In: Skou JC, Nøby JG (eds) Na,K ATPase: structure and kinetics. Academic Press, London, pp 463-473

Sachs JR (1980) The order of release of sodium and addition of potassium in the sodium-potassium pump reaction mechanism. J Physiol 302:219-240

Sachs JR (1986a) Potassium-potassium exchange as part of the over-all reaction mechanism of the sodium pump of the human red blood cell. J Physiol 374:221-244

Sachs JR (1986b) The order of addition of sodium and release of potassium at the inside of the sodium pump of the human red cell. J Physiol 381:149-168

Sachs JR (1987) Inhibition of the Na,K pump by vanadate in high-Na solutions. Modification of the reaction mechanism by external Na acting at a high-affinity site. J Gen Physiol 90:291-320
Sachs JR (1988) Phosphate inhibition of the human red cell sodium pump: simultaneous binding of adenosine triphosphate and phosphate. J Physiol 400:545-574
Sachs JR (1991) Successes and failures of the Albers-Post model in predicting ion flux kinetics. Soc Gen Physiol Ser 46:249-266
Sachs JR (1994) The role of (alpha beta) protomer interaction in determining functional characteristics of red cell Na,K-ATPase. Biochim Biophys Acta 1193:199-211
Sachs JR, Welt LG (1967) The concentration dependence of active potassium transport in the human red blood cell. J Clin Invest 46:65-76
Sachs JR, Welt LG (1968) Concentration dependence of active potassium transport in the human red blood cell in the presence of inhibitors. J Clin Invest 47:949-959
Schatzman HJ (1953) Herzglykoside als Hemmstoffe für den aktiven Kalium- und Natrium-Transport durch die Erythrocytenmembran. Helv Physiol Acta 11:346-354
Schneeberger A, Apell HJ (1999) Ion selectivity of the cytoplasmic binding sites of the Na,K-ATPase: I. Sodium binding is associated with a conformational rearrangement. J Membrane Biol 168:221-228
Schneeberger A, Apell HJ (2001) Ion selectivity of the cytoplasmic binding sites of the Na,K-ATPase: II. Competition of various cations. J Membrane Biol 179:263-273
Sen AK, Post RL (1964) Stoichiometry and localization of adenosine triphosphate-dependent sodium and potassium transport in erythrocyte. J Biol Chem 239:345-352
Shaw TI (1955) Potassium transport in washed erythrocytes. J Physiol 129:464-475
Simons TJ (1974) Potassium:potassium exchange catalysed by the sodium pump in human red cells. J Physiol 237:123-155
Simons TJ (1975) The interaction of ATP-analogues possessing a blocked gamma-phosphate group with the sodium pump in human red cells. J Physiol 244:731-739
Skou JC (1957) The influence of cell cations on adenosine triphosphatase from peripheral nerves. Biochim Biophys Acta 23:394-401
Smirnova IN, Lin SH, Faller LD (1995) An equivalent site mechanism for Na^+ and K^+ binding to sodium pump and control of the conformational change reported by fluorescein-5'-isothiocyanate modification. Biochemistry 34:8657-8667
Steck TL, Weinstein RS, Straus JH, Wallach DF (1970) Inside-out red cell membrane vesicles: preparation and purification. Science 168:255-257
Steinbach HB (1940) Sodium and potassium in frog muscle. J Biol Chem 133:695-701
Toyoshima C, Nakasako M, Nomura H, Ogawa H (2000) Crystal structure of the calcium pump of sarcoplasmic reticulum at 2.6 Å resolution. Nature 405:647-655
Whittam R, Ager ME (1964) Vectorial aspects of adenosine-triphosphatase activity in erythrocyte membranes. Biochem J 93:337-348

6 Ion Channels

Poul Bennekou[1] and Palle Christophersen[2]

[1]LCMF, The August Krogh Institute, University of Copenhagen, Universitetsparken 13, 2100 Copenhagen Ø, [2]NeuroSearch A/S, Ballerup, Pederstrupvej 93, 2750 Ballerup, Denmark

6.1 Introduction

Erythrocyte ion channels have been described from several animal species and have been more or less extensively characterized. However, the most comprehensive picture has been established for mammalian erythrocytes, especially human, which will be the focus of this chapter. Among the pathways for passive, conductive transport of inorganic ions, two different cation channels have been extensively characterized at present, an intermediate conductance Ca^{2+}-activated K^+ channel, also known as the Gárdos channel, and a voltage-dependent non-selective cation channel. It should be remembered, however, that transport mediated by these channels was observed decades before ion channels were established facts of life.

Furthermore, a small conductance K^+ channel and an anion-selective channel have been anecdotally observed, but neither are well characterized. In this context, the word channel is used only for mechanisms, which unequivocally have been shown to cause discrete openings and closings in patch-clamp experiments. It should be kept in mind, that channels with very low conductance and/or short open times cannot be observed as such, since the limit for single channel observations – *single channel current* × *open time* – is about 10^{-16} As.

6.2 The Ca^{2+}-Activated K^+ Channel

The existence of a Ca^{2+}-activated K^+ pathway was originally demonstrated by G. Gárdos (Gárdos 1958) who showed that the K^+ efflux from human red cells in suspension, which could be induced by inhibition of glycolysis (Wilbrandt 1937), could be prevented by the presence of calcium chelating agents. In the following years, the original observation was reproduced and extended to ghosts and membrane vesicles (e.g. Knauf et al. 1975; Lew et al. 1982; Yingst and Hoffman 1984), demonstrating that this pathway was a genuine property of the membrane. Pb^{2+}, which had been shown to cause loss of K^+ (Ørskov 1935) was also shown to act as an activator (Riordan and Passow 1971) with the Ca^{2+}- (or Pb^{2+}-) binding site located on the inside of the membrane (Blum and Hoffman 1972).

The question of the electrogenicity of the Ca^{2+}-activated K^+ transport was answered by Lassen and co-workers using intracellular microelectrodes to measure the membrane potential of giant red cells from *Amphiuma means*. During an initial puncture followed by fast withdrawal of the electrode, Ca^{2+} could enter the cell along the leak introduced by the electrode tip before the cell resealed. On repuncture a transient hyperpolarisation, the size of which depended on the potassium equilibrium potential, was observed (Lassen et al. 1974; Lassen et al. 1976). Apart from establishing the conductive nature of the pathway, these experiments demonstrated the reversibility of the activation and the interplay between the channel and the calcium pump in cells with normal metabolism. Use of potential-sensitive fluorescent probes confirmed these results for human red cells (Hoffman and Laris 1974), which are too small for micropuncture (Lassen and Rasmussen 1978).

Although it was generally accepted that the Gárdos pathway was electrogenic, the question with regard to a carrier mechanism *vs.* a channel was still open, due to the complicated kinetics of the transport. An increase of the extracellular K^+ concentration up to 4 mM increased the Ca^{2+}-activated net efflux, whereafter a further increase lead to the expected decrease (Knauf et al. 1975). When no extracellular K^+ was present, the pathway was irreversibly inactivated (Heinz and Passow 1980). However, this irreversible inactivation of the Gárdos channel occurred only in the presence of Ca^{2+}, and extracellular K^+ was thus suggested to protect the channel against Ca^{2+} inactivation (Lepke et al. 1987). The presence of a distinct extracellular K^+ site for activation of the Gárdos permeability was also supported by the blocker quinine, which acted competitively with extracellular K^+ (Reichstein and Rothstein 1981). This type of *trans* effects was taken as evidence for carrier-mediated transport and believed to be unreconcilable with the behaviour of a simple ion channel.

Vestergaard-Bogind and co-workers used protonophore-mediated estimates of the membrane potential, and traditional flux determinations in a Huxley-Hodgkin based approach, to give a phenomenological description of the Ca^{2+}-activated K^+ transport. The transient nature of the Ca^{2+}-activated K^+ conductance in metabolically-fed cells was confirmed, and it was shown that the K^+ conductance oscillates in response to oscillations of intracellular Ca^{2+} (Vestergaard-Bogind and Bennekou 1982). With concomitant determination of the driving force ($V_m - E_K$), the net flux and one of the unidirectional fluxes, it was possible to apply a flux-ratio analysis to the Gárdos-channel. The analysis demonstrated that the unidirectional fluxes could be described by the Ussing flux-ratio equation, when raised to the power of 2.7 (Vestergaard-Bogind et al. 1985). This type of kinetic behaviour, known as 'single file' diffusion, which was first described for the squid axon delayed rectifier (Hodgkin and Keynes 1955), is consistent with a channel pathway containing several ions restricted to concerted jumps. The exponent indicates the (integer) number of ions in the pore, in this case 3 or higher.

The question with regard to the carrier *vs.* channel nature of the Gárdos pathway was finally settled with the introduction of the 'giga-seal' (Hamill et al. 1981) patch-clamp technique. With this technique, genuine single-channel events with a conductance of about 18 pS were observed on ATP-depleted human red cells

(Hamill 1981). The proof for the identity of the Gárdos-pathway and the observed single channels came from studies on excised inside-out patches, which demonstrated that the channel was obligatorily Ca^{2+}-dependent with a half-activation concentration in the range of 1–3 μM (Grygorczyk and Schwarz 1983), in accordance with most estimates of EC_{50} values based on flux experiments. Furthermore, it was shown that Pb^{2+} could also activate the isolated channel. In symmetric, high K^+ solutions the channel was shown to exhibit weak, inward rectification with a single channel conductance (zero-current) of approx. 20 pS. From flux experiments, using tracers, it had been inferred that the Gárdos pathway showed a high selectivity for K^+ over Na^+, and that $^{86}Rb^+$ was usable as tracer for K^+. In single channel studies, only K^+, NH_4^+, Rb^+, and Tl^+ were found to permeate, with a selectivity sequence K^+:Rb^+:NH_4^+ = 1.0:1.4:8.5 determined from bi-ionic reversal potentials and K^+:Tl^+:NH_4^+:Rb^+ = 1.0: ~1:1.7:2.1 from the conductance sequence determined in symmetric solutions (Christophersen 1991). The selectivity sequences conformed to Eisenman's series IV or V for binding to a site of intermediate field strength, as is generally seen for K^+-selective channels (Latorre and Miller 1983). An unexpected observation, however, was the dependence of the gating on the nature of the permeating ion. At a constant concentration of intracellular Ca^{2+}, Rb^+ stabilized the channel in the open conformation (relative to K^+), whereas NH_4^+ had the reverse effect. In bi-ionic experiments the stabilization/destabilization followed the direction of the current, indicating that the gating is influenced by the channel occupancy by specific ions (Christophersen and Bennekou 1990). Similar gating effects of the permeating ions have been described for squid axon delayed rectifiers (Swenson and Armstrong 1981) and BK-channels (Demo and Yellen 1992). As is the case for other K^+ channels, impermeant cations act as voltage-dependent pore blockers of the Gárdos channel. It was found (Stampe and Vestergaard-Bogind 1989) that intracellular Na^+ blocks the channel with an effective valence of up to 1.5, depending on the Na^+ concentration, indicating that several ions participate in the blocking reaction in accordance with the notion of a multi-ion pore. The block was sufficiently potent to be relevant at physiological Na^+ concentrations. In a study of the action of Cs^+ on inside-out patches, Grygorczyk and Schwarz (1985) also found a block with a valence above 1 when this ion was added to the outside. Of the impermeant divalent cations, Ba^{2+} has been found to be an open channel blocker with slow kinetics, while Mg^{2+} seems to act both as a competitive antagonist to Ca^{2+} (Grygorczyk and Schwarz 1985) and possibly also as a fast open channel blocker, since 1 mM Mg^{2+} reduced the single channel conductance about 30% at saturating Ca^{2+}-concentrations (Christophersen and Bennekou 1988). The pronounced anomalous rectification observed in ATP-depleted cells (Vestergaard-Bogind et al. 1987) may be due to a voltage-dependent block by an unidentified intracellular cation.

With regard to the temperature dependence (Grygorczyk 1987), a moderate increase of the single channel conductance with temperature was found (Q_{10} = 1.3), in accordance with a diffusion-like transport through the open channel. The open state probability at a saturating Ca^{2+} concentration, however, showed a pronounced inverse temperature dependence, with an open state probability at 0 °C of 0.8, decreasing to about 0.1 at 35 °C.

A number of chemically diverse organic modulators of the Gárdos channel has been found. Early reports established quinine (Armando-Hardy et al. 1975; Reichstein and Rothstein 1981), oligomycin (Wood 1984), and cetiedil (Berkowitz and Orringer 1982) as inhibitors although with rather low affinities, in the 10–100 μM range. More recently members of the antihypertensive drug-class dihydropyridines have been shown to be potent blockers of the channel (Kaji 1990; Ellory et al. 1992), with nitrendipine having the highest affinity. Another high-affinity blocker is the imidazole antimycotic clotrimazole, with an IC_{50} of 50 nM (Alvarez et al. 1992). Single channel studies have shown that clotrimazole induces long silent periods in channel activity and thus lowers the open state probability, without affecting the single channel conductance (Dunn 1998). An even higher affinity, IC_{50} = 5.4 nM, was found for the peptide charybdotoxin from the scorpion (*Leirus quinquestritus*) venom, which previously had been shown to block the Ca^{2+}-activated maxi K^+ channels from muscle cell T-tubules (Smith et al. 1986), but was shown to block the Gárdos channel too (Castle and Strong 1986; Wolff et al. 1988). From K^+ flux studies the potency of the charybdotoxin-type peptides was found to be charybdotoxin > stichodactylatoxin > margatoxin (Brugnara et al. 1995).

Among the Gárdos channel activators the classical observation is the effect of the antihypertensive drug propranolol. In the presence of calcium, propranolol was found to increase the Ca^{2+}-activated K^+ fluxes (Ekman et al. 1969; Porzig 1975; Gárdos et al. 1975). Propranolol has also been shown to inhibit the calmodulin-dependent Ca^{2+}-ATPase, which could contribute to the response in the intact red cell (Meltzer and Kassir 1983). However, single channel studies on isolated patches revealed that the drug increased the sensitivity of the channels towards Ca^{2+} (Schwarz et al. 1989b) in inside-out patches by an unknown mechanism. The activator of the IK-channel from colonic T84 cells, 1-EBIO (Devor et al. 1996), has been shown to activate the Gárdos-channel too by increasing the Ca^{2+}-sensitivity (Bennekou et al. 1999). The mechanism is unknown in this case as well.

Numerous experiments indicate that redox state and phosphorylation could play a role in the modulation of the Gárdos channel. Binding of calpromotin, a major erythrocyte cytosolic component, which has been identified as a "Thiol Specific Antioxidant" protein, TSA (Kristensen et al. 1999), has been shown to participate in the activation of the channel (Moore et al. 1990, 1991). The activity of the Gárdos channel may also be regulated by protein phosphorylation. Increase of channel activity by application of a phosphorylation promoting mix including theophylline has been demonstrated (Pellegrino and Pellegrini 1998) and it was concluded that the channel was regulated by tightly associated phosphokinase A (PKA) and phosphodiesterase enzymes. The stimulating activity of theophylline has been revealed on the cloned IK-channel too, but the effect could not be ascribed to phosphorylation of the α-subunit, since mutation of the only consensus site for PKA action was without effect (Schrøder et al. 2000).

The number of Gárdos channels per red cell was originally found to be in the same range as the number of Na^+/K^+ pumps, 100–200 (Lew et al. 1982) estimated from the correspondence of the $^{86}Rb^+$ loss from $^{86}Rb^+$-loaded inside-out vesicles responding to Ca^{2+} activation of the channels or to ATP activation of the pumps, under the assumption that the number of vesicles originating from a red cell was greater than the number of transporters present in the same cell. Based on a comparison between macroscopic Ca^{2+}-activated equilibrium effluxes, traced by ^{86}Rb, single channel conductance and open state probability from inside-out patches, assuming constant field behaviour, evidence was found for two cell populations, with 25% of the cells having 11–55 and 75% 1–5 channels (Grygorczyk et al. 1984). It was later shown, however, that the mode of transport was single-file behaviour (Vestergaard-Bogind et al. 1985), where the constant field requirement of independence is not fulfilled. A heterogeneous distribution of K^+ fluxes under net efflux conditions has also been reported (Yingst and Hoffman 1984) at submaximal Ca^{2+} activation, as judged from a resulting non-uniform density distribution. Brugnara et al. (1993), using charybdotoxin-binding found 120 ± 36 extracellular binding sites. Using a number of 150 channels/cell, Bennekou et al. (2000) calculated the expected macroscopic conductance from independent estimates of the single channel conductance in near-physiological gradients of 10 pS (Christophersen 1991) and from the temperature-corrected open state probability (Grygorczyk 1987) and arrived at 85 $\mu S/cm^2$, which is in good agreement with the macroscopic K^+ conductance of 61 $\mu S/cm^2$ measured at saturating Ca^{2+} concentrations (Stampe and Vestergaard-Bogind 1989).

The molecular identity of the Gárdos channel has not yet been established unequivocally. However, the recent cloning, expression and subsequent characterization of the human intermediate conductance Ca^{2+}-activated K^+ channel (Joiner et al. 1997; Ishii et al. 1997; Logsdon et al. 1997; Jensen et al. 1998) from pancreas, placenta and lymphocytes, strongly suggests that the Gárdos channel is identical to the channel encoded by the KCNN4 gene, located at chromosome 19. This channel is about 40–42% homologous at the amino acid level to the small conductance Ca^{2+}-activated K^+ channels (SK1, SK2, SK3) and was therefore originally included in this family and named SK4 by Joiner et al. (1997). However, Ishii et al. (1997) and Jensen et al. (1998) identified this channel as being identical to the Gárdos channels and suggested it as the first member of the hIK channel family. Hydropathy plots of the amino acid sequence conform to the well known 6 TM and P structure of voltage-dependent K^+ channels. In analogy to these, the functional hIK channel is supposed to be formed by 4 identical subunits organized around a central pore. However, in contrast to the K_v channels, hIK does not display any significant voltage dependence of gating. The macroscopic IV curve is weakly inward-rectifying in symmetrical K^+ solutions a feature, which is completely accounted for by the single channel rectification. Under the same circumstances a zero current single channel conductance of about 20 pS is observed, in good agreement with single channel recordings from erythrocytes. Pharmacological evidence also points to the identity of hIK and the Gárdos channel: the classical Gárdos channel blocker and anti-sickling compound cetiedil is a relatively weak blocker of the cloned channel as well. More importantly, the high affinity blockers clotrimazole and nitrendipine also block hIK potently, whereas the selective SK-channel blocker apamin has no effect. Finally, the compound 1-EBIO modulates hIK (Jensen et al. 1998) as well as the Gárdos channel (Bennekou et al. 1999).

The involvement of calmodulin in the activation of the Gárdos channel by Ca^{2+} has been much debated over the years: Pape and Kristensen (1984), found evidence for calmodulin dependence of Ca^{2+}-activated Rb^+ transport in inside-out vesicles, a conclusion that was supported by the inhibitory action of calmodulin antagonists, also found by Hoffman and Yingst (1984). However, Alvarez et al.

(1986) were not able to confirm this finding. Patch-clamp studies using the inside-out configuration demonstrated Ca^{2+}-dependent channel activity without added calmodulin. The general opinion was that the Gárdos channel was itself capable of Ca^{2+} binding! With the cloning of the SK and IK channels a search for the Ca^{2+}-binding site was initiated by systematic mutations of negatively charged amino acid residues exposed to the cytosol. Surprisingly, these mutation programs failed to identify a channel Ca^{2+} binding site. Using the yeast two hybrid system, however, a close interaction between SK channels (Xia et al. 1998), IK channels (Fanger et al. 1999) and calmodulin was revealed. Complementary site-directed mutations in both proteins localized the interaction domain to the proximal part of the C-terminus. Calmodulin was found to be constitutively bound to the channel and to confer allosteric Ca^{2+} sensitivity to the channels. The tight interaction between calmodulin and the channel was proposed to explain the lack of effect of calmodulin antagonists in these studies.

6.3 The Voltage-Dependent Non-Selective Cation Channel

Activation of the voltage-dependent cation channel occurs at positive membrane potentials. Due to the dominating erythrocyte anion conductance relative to cations, the membrane potential is determined by the distribution of anions (in particular Cl^-) and is therefore normally clamped at –12 mV. Strongly positive membrane potentials have classically been achieved by a reduction of the extracellular Cl^- concentration, typically by suspension in isotonic, low salt containing media, such as sucrose, which, however, in buffer-free media also causes an extracellular acidification. Therefore, much debate has centred around the relative effects of the changed membrane potential *vs.* the changed pH gradients and the effect of low ionic strength (see Chap. 4 for a review of the low ion strength experiments).

Following the idea of generation of large positive membrane potentials by a changed anion distribution, Wilbrandt and Schatzmann (1960) pointed out that the electrical potential *per se* might influence the cation permeability. Halperin et al. (1990), using the CCCP-method to estimate the membrane potential, proposed that the salt leak in solutions of low ionic strength was caused by a separate membrane protein, a voltage-dependent cation channel activated at potentials more positive than 40 mV with the conductance increase reaching 1.5 $\mu S/cm^2$ at +50 mV. In contrast to earlier findings (Dawson 1939) they found conductances for Na^+ and K^+ continuing if the cells were repolarized by transfer to a normal Ringer's solution. An important conclusion from these studies was that the voltage-dependent cation permeability increase involved Ca^{2+}. Addition of 1 mM Ca^{2+} to cells depolarized in a Ca^{2+}-free sucrose medium caused a strong hyperpolarization, which could be inhibited by addition of carbocyanine, a Gárdos channel inhibitor. This result was interpreted as a secondary activation of the Gárdos channel following influx of Ca^{2+} through the voltage-dependent pathway. Furthermore, it was shown that the fluxes were blocked by ruthenium red with an IC_{50} of 0.7–1.0 μM. In a subsequent paper (Halperin et al. 1990), the voltage-activated pathway was shown to be present in high K^+ sheep red cells, but not in low K^+ sheep and dog cells, thus extending the observations of Dawson 1939, on the animal species variation. Since this distribution coincides with the distribution of Na^+/K^+ pumps, it was suggested that the voltage-dependent cation pathway was related to the pump protein.

Using the patch-clamp technique on inside-out patches from human red cells, the first direct evidence for a non-selective, voltage-activated cation channel in the human red cell membrane was found (Christophersen and Bennekou 1991). Since the frequency of channel observations increased at high salt concentrations, which is still unexplained, it was characterized using 500 mM salt solutions. The channel did not discriminate between Na^+, K^+, Rb^+ and NH_4^+ and the single channel IV-curve was found to be symmetric with respect to 0 mV driving force, with a main-state, zero-current conductance of 35 pS. The open state probability however, was strongly voltage-dependent, achieving a value of 0.85 at positive potentials of 100 mV and becoming zero at membrane potentials more negative than –20 to –30 mV. The observed gating characteristics were complicated with multiple conductance states and 'gear-shifts'. Based on the selectivity and voltage-dependent open state probability it was proposed that this channel represented the pathway, which was activated in low salt media.

The channel was further described by Bennekou (1993), who found that the presence in the pipette solution of up to 2.5 µM of the non-selective acetylcholine agonist carbachol activated the channel further; with increasing concentrations the activation declined again, possibly due to desensitisation. Furthermore, since nicotine was effective too, the channel was concluded to be of nicotinic subtype. However, the behaviour of the non-selective channel is very different from the cloned nicotinic acetylcholine receptors, especially with regard to the voltage-dependent gating and the complete lack of selectivity towards monovalent cations, indicating that the channel does not belong to the known family of nicotinic acetylcholine receptors. It should also be noted, that carbachol has no effect on the permeability changes for red cells suspended in low chloride media (Kaestner et al. 2000). This could be due to unmasking of the binding site in the inside-out patch configuration.

Based on an estimate of the patch area and the number of channel observations the number of channels in the red cell membrane was assumed to be 300/red cell. Kaestner et al. (1999), working with physiological salt concentrations reported a single channel conductance of 21 pS, but found that the open state probability as function of the membrane potential was right-shifted 30–40 mV relative to earlier observations (Bennekou and Christophersen 1991), with activity observed only at membrane potentials more positive than +20 mV. Based on a cation flux of 25 mmoles/(l_{cells}·h) at +45 mV and a corresponding open state probability of 0.35, the number of channels was estimated to be 0.25/red cell.

The discrepancy with regard to voltage activation was resolved by the finding that the channel activity was strongly dependent on the experimental history. If the channel was activated by an initial holding potential of +100 mV and the open state probability was determined in negative-going steps the activity was found to be increased compared to measurements initiated from negative holding potentials (Bennekou and Christophersen 1992; Kaestner et al. 2000). This shift was shown to be an inherent property of the gating mechanism, rather than the result of irreversible changes, in that the experiment could be repeated on the same patch (channel) after a period at a deactivating (negative) potential. The reason for this hysteresis-like characteristic, which is rare, is not known. It was found that the current-voltage relation for monovalent cations changed from sublinear at concentrations below 150 mM salt to superlinear above, thus being linear at the physiological concentration range. The single channel conductances for Ca^{2+}, Mg^{2+} and Ba^{2+} were found to be 12–18 pS at 75 mM, about the same size as for monovalent cations, the current voltage characteristic however, retained the superlinearity at this concentration. The voltage-dependent gating for divalent cations is reported to show anomalies compared to monovalent ions, in that the open state probability at high positive voltages does not tend to 1. Furthermore, for monovalent cations inward currents are only observable as tail-currents in response to deactivating voltage steps (Christophersen and Bennekou 1991), whereas in experiments with divalent cations, rare gating events can be observed at large negative voltages (Kaestner et al. 2000).

The gene encoding the voltage-dependent cation channel has not been identified! Furthermore, the pharmacology of the channel is essentially unknown and therefore of little use in the establishment of a final correspondence between the channel and the low chloride induced pathway in intact cells. As the preceding discussion has shown, transport systems other than a voltage-dependent cation channel have been proposed as contributing to the cation loss in low Cl^- media. Recently, Bernhardt et al. (1999) have also found evidence for the participation of a $K^+(Na^+)/H^+$ monovalent cation exchanger in the process (see Chap. 4). Despite the complexity, it seems probable, that the voltage-dependent ion channel, which is now routinely observed in patch-clamp studies, is a major contributor to the depolarisation-induced salt loss from intact cells.

6.4 The Low Conductance K^+ Channel

In patch-clamp experiments, a small K^+ channel with a conductance of about 10 pS is infrequently observed compared to the Gárdos channel (Grygorczyk and Schwarz 1983; Bennekou and Christophersen 1988). Whether this channel is a separate pathway or a functional state of the Gárdos channel is not known.

6.5 The Anion-Selective Channel

The conductive pathway for anions, which at least for the major part seems to be mediated by the anion exchange protein, AE1, has been proposed to be channel-like, but has never physically been identified as such. However, an anion-selective channel, with a single channel conductance of 6 pS in a Cl^- medium, and about 50% higher in NO_3^- media has been reported (Passow et al. 1988; Schwarz et al. 1989a). The conductance sequence $Cl^- < NO_3^-$ and the reported sensitivity to anion exchange and conductance blockers corresponds to the data from experiments on intact red cell. However it remains to be established whether this pathway is a separate pathway or a functional mode of the anion exchange protein.

6.6 Whole Cell Patch-Clamp Recordings

Whole cell recordings from human red cells are notoriously difficult to achieve due to their small size and the special shape, which is determined by the cytoskeleton. Furthermore, unequivocal identification of the individual channel(s) contributing to the whole cell current is difficult, due to the lack of gating patterns. However, a few studies have been published recently. Using classical whole cell recording Desai et al. (2000) in a study of malaria-infected human red cells determined the normal human whole cell conductance in K^+-free solutions, excluding the contribution from the Gárdos channel. The ohmic conductance was found be less than 100 pS/cell. In contrast to this observation, Huber et al. (2001) in a study on volume-dependent currents in human red cell ghosts (due to the establishment of the whole-cell configuration, which causes the cytoplasm to diffuse into the pipette), found large non-selective cation currents as well as anion currents not previously described, corresponding to a whole cell conductance of 7.4 nS/cell. Working with malaria-infested cells, Huber et al. (2002) found two anion conductances, differing in voltage dependence and pharmacological profile with regard to inhibitors. However, both types of conductances were redox-dependent, being inhibited by reducing agents. In non-infected cells oxidation induced identical conductances, indicating that the malaria infection activated endogenous pathways. Formenti and Rodighiero (2001) using whole cell permeabilized patches were able to demonstrate the presence of a voltage-activated whole-cell current with properties resembling the voltage-activated cation channel discussed earlier in this chapter. Furthermore, Duranton et al. (2002) have shown that oxidation of human red cells induces a chloride dependent cation conductance for monovalent cations and Ca^{2+}.

Although the correspondence between the single channel work on excised patches, and the pathways found in whole cell/perforated patches and intact cells in suspension, at present is unclear, the importance of this type of experiments is immense, building as they do a bridge between the molecular properties of the isolated channel and the physiology of the intact cell.

6.7 Conclusion

Despite the extensive work done on both the Gárdos channel and the voltage regulated non-specific cation channel the physiological relevance of both remains obscure. Due to the low Ca^{2+} permeability and the large Ca^{2+} pumping capacity of the erythrocyte membrane, combined with the lack of membrane delimited intracellular stores which could be released, the Gárdos channel is thought to remain effectively closed in healthy erythrocytes (for a discussion of sickle cell anaemia, see Chap. 23). The voltage-dependent cation channel is, at the normal erythrocyte membrane potential of −12 mV, believed to be effectively closed too. However, it should be kept in mind that the two channels and the Ca^{2+} pump, in principle, represents a coupled, feed-back regulated system of potential importance for the Ca^{2+} homeostasis and maintenance of the human red cell membrane potential. If the non-specific channel opens (even rarely), the conductances for mono- and divalent ions increase, Ca^{2+} influx may activate nearby Gárdos channels leading to hyperpolarization and closing of the non-specific channels and subsequently the system is restored to the resting state by extrusion of Ca^{2+}. It remains to be established whether this mechanism may operate physiologically.

References

Alvarez J, Garcia-Sancho J, Herreros B (1986) The role of calmodulin on Ca^{2+}-dependent K^+ transport regulation in the human red cell. Biochim Biophys Acta 860:25-34

Alvarez J, Montero M, Garcia-Sancho J (1992) High affinity inhibition of Ca^{2+}-dependent K^+ channels by cytochrome P-450 inhibitors. J Biol Chem 267:11789-11793

Armando-Hardy M, Ellory JC, Ferreira HG, Fleminger S, Lew VL (1975) Inhibition of the calcium-induced increase in the potassium permeability of human red blood cells by quinine. J Physiol 250:32P

Bennekou P (1993) The voltage-gated non-selective cation channel from human red cells is sensitive to acetylcholine. Biochim Biophys Acta 1147:165-167

Bennekou P, Christophersen P (1988) The occurrence of a low conductance channel in the human red cell. Acta Physiol Scand 13:S575

Bennekou P, Christophersen P (1992) A human red cell cation channel showing hysteresis like voltage activation/inactivation. Acta Physiol Scand 146:608

Bennekou P, Rasmussen DE, Kristensen BI (1999) The effect of a channel opener, 1-EBIO, on the human red cell Ca^{2+}-activated K^+-channel. Physiol Research 48, suppl 1:57

Bennekou P, Pedersen O, Møller A, Christophersen P (2000) Volume control in sickle cells is facilitated by the novel anion conductance inhibitor NS1652. Blood 95:1842-1848

Berkowitz LR, Orringer EP (1982) Effects of cetiedil on monovalent cation permeability in the erythrocyte: An explanation for the efficacy of cetiedil in the treatment of sickle cell anemia. Blood Cells 8:283-288

Bernhardt I, Bogdanova AY, Kummerow D, Kiessling K, Hamann J, Ellory JC (1999) Characterization of the $K^+(Na^+)/H^+$ monovalent cation exchanger in the human red blood cell membrane: Effects of transport inhibitors. J Gen Physiol 18:119-137

Blum RM, Hoffman JF (1972) Ca-induced K transport in human red cells: Localization of the Ca-sensitive site to the inside of the membrane. Biochem Biophys Res Commun 46:1146-1153

Brugnara C, De Franceschi L, Alper SL (1993) Ca^{2+}-activated K^+ transport in erythrocytes. J Biol Chem 268:8760-8768

Brugnara C, Armsby CC, De Franceschi L, Crest M, Euclaire MF, Alper SL (1995) Ca^{2+}-activated K^+ channels of human and rabbit erythrocytes display distinctive patterns of inhibition by venom peptide toxins. J Membrane Biol 147:71-82

Castle NA, Strong PN (1986) Identification of two toxins from scorpion (*Leiurus quinquestriatus*) venom which block distinct classes of calcium-activated potassium channel. FEBS Letters 209:117-121

Christophersen P (1991) Ca^{2+}-activated K^+ channel from human erythrocyte membranes: Single channel rectification and selectivity. J Membrane Biol 119:75-83

Christophersen P, Bennekou P (1988) Mg^{2+} affects the single channel conductance of the human red cell K^+-channel, but not rectification. Acta Physiol Scand 132:41A

Christophersen P, Bennekou P (1990) The gating of human red cell Ca^{2+}-activated K^+-channels is strongly affected by the permeant cation species. Biochim Biophys Acta 1030:183-187

Christophersen P, Bennekou P (1991) Evidence for a voltage-gated, non-selective cation channel in the human red cell membrane. Biochim Biophys Acta 1065:103-106

Dawson H (1939) Studies on the permeability of erythrocytes. VI. The effect of reducing the salt concentration in the medium surrounding the cell. Biochem J 33:389-401

Demo SD, Yellen G (1992) Ion effects on gating of the Ca^{2+}-activated K^+ channel correlate with occupancy of the pore. Biophys J 61:639-648

Desai SA, Bezrukov SM, Zimmerberg J (2000) A voltage-dependent channel involved in nutrient uptake by the red blood cells infected with the malaria parasite. Nature 406:1001-1004

Devor DC, Singh AK, Frizzell RA, Bridges RJ (1996) Modulation of Cl^- secretion by benzimidazolones. I. Direct activation of a Ca^{2+}-dependent K^+ channel. Am J Physiol 271:L775-L784

Dunn PM (1998) The action of blocking agents applied to the inner face of Ca^{2+}-activated K^+ channels from human erythrocytes. J Membrane Biol 165:133-143

Duranton SM, Huber SM, Lang F (2002) Oxidation induces a Cl^--dependent cation conductance in human red blood cells. J Physiol 539:847-855

Ekman A, Manninen V, Salminen S (1969) Ion movements in red cells treated with propranolol. Acta Physiol Scand 75:333-344

Ellory JC, Kirk K, Culliford SJ, Nash GB, Stuart J (1992) Nitrendipine is a potent inhibitor of the Ca^{2+}-activated K^+ channel of human erythrocytes. FEBS Letters 296:219-221

Fanger CM, Ghanshani S, Logsdon NJ, Rauer H, Kalman K, Zhou J, Beckingham K, Chandy KE, Cahalan MD, Aiyar J (1999) Calmodulin mediates calcium-dependent activation of the intermediate conductance KCa channel, IKCa1. J Biol Chem 274:5746-5754

Formenti A, Rodighiero S (2001) Whole-cell and perforated-patch recordings of voltage-dependent outward and inward cation currents in human red blood cells. XIII Meeting of the European Association for Red Cell Research. Abstract 72, p 59

Gárdos G (1958) The function of calcium in the potassium permeability of human erythrocytes. Biochim Biophys Acta 30:653-654

Gárdos G, Szász I, Sarkadi B (1975) Mechanism of Ca-dependent K-transport in human red cells. In: Gárdos G, Szász I (eds) Biomembranes: structure and function. North-Holland, Amsterdam, pp 167-180

Grygorczyk R (1987) Temperature dependence of Ca^{2+} activated K^+ currents in the membrane of human erythrocytes. Biochim Biophys Acta 902:159-168

Grygorczyk R, Schwarz W (1983) Properties of the Ca^{2+} activated K^+ conductance of human red cells as revealed by the patch-clamp technique. Cell Calcium 4:499-510

Grygorczyk R, Schwarz W (1985) Ca^{2+}-activated K^+ permeability in human erythrocytes: Modulation of single-channel events. Eur Biophys J 12:57-65

Grygorczyk R, Schwarz W, Passow H (1984) Ca^{2+}-activated K^+ channels in human red cells. Comparison of single-channel currents with ion fluxes. Biophys J 45:693-698

Halperin JA, Brugnara C, Van Ha T, Tosteson DC (1990) Voltage-activated cation permeability in high-potassium but not low-potassium red blood cells. Am J Physiol 258:C1169-C1172

Hamill OP (1981) Potassium channel currents in human red blood cells. J Physiol 319:97-98

Hamill OP, Marty A, Neher E, Sakmann B, Sigworth FJ (1981) Improved patch-clamp techniques for high-resolution current recording from cells and cell-free membrane patches. Pflügers Arch 391:85-100

Heinz A, Passow H (1980) Role of external potassium in the calcium-induced potassium efflux from human red blood cell ghosts. J Membrane Biol 57:119-131

Hodgkin AL, Keynes RD (1955) The potassium permeability of a giant nerve fibre. J Physiol 128:61-88

Hoffman JF, Laris PC (1974) Determinations of membrane potentials in human and *Amphiuma* red blood cells by means of a fluorescent probe. J Physiol 239:519-552

Huber SM, Gamper N, Lang F (2001) Chloride conductance and volume-regulatory nonselective cation conductance in human red blood cell ghosts. Pflügers Arch 441:551-558

Huber SM, Uhlemann AC, Gamper NL, Duranton C, Kremsner PG, Lang F (2002) *Plasmodium falciparum* activates endogenous Cl^- channels of human erythrocytes by membrane oxidation. EMBO J 21:22-30

Ishii TM, Silvia C, Hirsberg B, Bond CT, Adelman JP, Maylie J (1997) A human intermediate conductance calcium-activated potassium channel. Proc Natl Acad Sci USA 94:11651-11656

Jensen BS, Strøbæk D, Christophersen P, Jørgensen TD, Hansen C, Silahtaroglu A, Olesen SP, Ahring PK (1998) Characterization of the cloned human intermediate-conductance Ca^{2+}-activated K^+-channel. Am J Physiol 275:C848-C856

Joiner WJ, Wang LY, Tang MD, Kaczmarek LK (1997) hSK4, a member of a novel subfamily of calcium-activated potassium channels. Proc Natl Acad Sci USA 94:11013-11018

Kaestner L, Bollensdorff C, Bernhardt I (1999) Non-selective voltage-activated cation channel in human red blood cell membrane. Biochim Biophys Acta 1417:9-15

Kaestner L, Christophersen P, Bernhardt I, Bennekou P (2000) The non-selective voltage-activated cation channel in the human red blood cell membrane: Reconciliation between two conflicting reports and further characterization. Biolectrochem 52:117-125

Kaji DM (1990) Nifedipine inhibits calcium-activated K transport in human erythrocytes. Am J Physiol 259:C332-C339

Knauf PA, Riordan JR, Schuhman B, Wood-Guth I, Passow H (1975) Calcium-potassium-stimulated net potassium efflux from human erythrocyte ghosts. J Membrane Biol 25:1-22

Kristensen P, Rasmussen DE, Kristensen BI (1999) Properties of thiol-specific anti-oxidant protein or calpromotin in solution. Biochem Biophys Res Commun 262:127-131

Lassen UV, Rasmussen BE (1978) Use of microelectrodes for measurement of membrane potentials. In: Giebisch G, Tosteson DC, Ussing HH (eds) Membrane transport in biology. Springer-Verlag, Berlin, pp 169-203

Lassen UV, Vestergaard-Bogind B, Bengtson O (1974) Calcium-related hyperpolarization of the *Amphiuma* red cell membrane following micropuncture. J Membrane Biol 18:125-144

Lassen UV, Pape L, Vestergaard-Bogind B (1976) Effect of calcium on the membrane potential of *Amphiuma* red cells. J Membrane Biol 26:51-70

Latorre R, Miller C (1983) Conduction and selectivity in potassium channels. J Membrane Biol 71:11-30

Lepke S, Shields M, Passow H (1987) Irreversible inhibition of Ca^{2+} activated K^{+} channels in the red blood cell membrane by Ca^{2+}. Deutsches Membranforum 368:1262

Lew VL, Muallem S, Seymour CA (1982) Properties of the Ca^{2+}-activated K^{+} channel in one-step inside-out vesicles from human red cell membranes. Nature 296:742-744

Logsdon NJ, Kang J, Togo JA, Christian EP, Ayiar J (1997) A novel gene, hKCa4, encodes the calcium-activated potassium channel in human T lymphocytes. J Biol Chem 272:32723-32726

Meltzer HL, Kassir S (1983) Inhibition of calmodulin-activated Ca^{2+}-ATPase by propranolol and nadol. Biophys Biochem Acta 755:452-456

Moore RG, Plishker GA, Shriver SK (1990) Purification and measurement of calpromotin, the cytoplasmic protein which activates calcium-dependent potassium transport. Biochem Biophys Res Commun 166:146-153

Moore RB, Mankad MV, Shriver SK, Mankad VN, Plishker GA (1991) Reconstitution of Ca^{2+}-dependent K^{+} transport in erythrocyte membrane vesicles requires a cytoplasmic protein. J Biol Chem 266:18964-18968

Pape L, Kristensen BI (1984) A calmodulin activated Ca^{2+}-independent K^{+} channel in human erythrocyte membrane inside-out vesicles. Biochem Biophys Acta 770:1-6

Passow H, Schwarz W, Glibowicka M, Aranibar N, Raida M (1988) Studies of erythroid band 3 protein-mediated anion transport in red blood cells and in *Xenopus* oocytes. In: Palmieri F, Quagliariello E (eds) Molecular basis of biomembrane transport. Elsevier, Amsterdam, pp 121-140

Pellegrino M, Pellegrini M (1998) Modulation of Ca^{2+}-activated K^{+} channels of human erythrocytes endogeneous cAMP-dependent protein kinase. Pflügers Arch 436:749-756

Porzig H (1975) Comparative study of the effects of propranolol and tetracaine on cation movements in resealed human red cell ghosts. J Physiol 249:27-49

Reichstein E, Rothstein A (1981) Effects of Quinine on Ca^{++}-induced K^{+} efflux from human red blood cells. J Membrane Biol 59:57-63

Riordan JR, Passow H (1971) Effects of calcium and lead on potassium permeability of human erythrocyte ghosts. Biochim Biophys Acta 249:601-605

Schrøder RL, Jensen BS, Strøbæk D, Olesen S-P, Christophersen P (2000) Activation of the human intermediate-conductance Ca^{2+}-activated K^{+} channel by methylxanthines. Pflügers Arch 40:809-818

Schwarz W, Grygorczyk R, Hof D (1989a) Recording single-channel currents from human red cells. Methods Enzymol 173:112-122

Schwarz W, Keim H, Fehlau R, Fuhrmann GF (1989b) Modulation of the Ca^{2+}- or Pb^{2+}-activated K^{+}-selective channels in human red cells. I. Effects of propranolol. Biochim Biophys Acta 978:32-36

Smith C, Phillips M, Miller C (1986) Purification of charybdotoxin, a specific inhibitor of the high-conductance Ca^{2+}-activated K^{+}-channel. J Biol Chem 261:14607-14613

Stampe P, Vestergaard-Bogind B (1989) Ca^{2+}-activated K^{+} conductance of the human red cell membrane: Voltage-dependent Na^{+} block of outward-going currents. J Membrane Biol 112:9-14

Swenson RP Jr, Armstrong CM (1981) K^+ channels close more slowly in the presence of external K^+ and Rb^+. Nature 292:427-429

Vestergaard-Bogind B, Bennekou P (1982) Induced oscillations in K^+-conductance and membrane potential of human erythrocytes mediated by ionophore A23187. Biochem Biophys Acta 688:37-44

Vestergaard-Bogind B, Stampe P, Christophersen P (1985) Single-file diffusion through the Ca^{2+}-activated K^+ channel of human red cells. J Membrane Biol 88:67-75

Vestergaard-Bogind B, Stampe P, Christophersen P (1987) Voltage dependence of the Ca^{2+}-activated K^+ conductance of the human red cell membranes is strongly dependent on the extracellular K^+ concentration. J Membrane Biol 95:121-130

Wilbrandt W (1937) A relation between the permeability of the red cell membrane and its metabolism. Trans Faraday Soc 33:956-959

Wilbrandt W, Schatzmann HJ (1960) Changes in the passive cation permeability of erythrocytes in low elelectrolyte media. In: Wolstenholme GEW, O'Connor CM (eds) Ciba Foundation Study Group No. 5. J & A Churchill, London, pp 34-53

Wolff D, Cecchi X, Spalvins A, Canessa M (1988) Charybdotoxin blocks with high affinity the Ca^{2+}-activated K^+ channel of Hb A and Hb S red cells: individual differences in the number of channels. J Membrane Biol 106:243-252

Wood PG (1984) The spontaneous activation of a potassium channel during the preparation of resealed human erythrocyte ghosts. Biochim Biophys Acta 774:103-109

Xia X-M, Fakler B, Rivard A, Wayman G, Johnson-Pais T, Keen JE, Ishii T, Hirschberg B, Bond CT, Lutsenko S, Maylie J, Adelman JP (1998) Mechanism of calcium gating in small conductance calcium-activated potassium channels. Nature 395:503-507

Yingst DR, Hoffman JF (1984) Ca-induced K transport in human red blood cell ghosts containing arsenazo III. Transmembrane interactions of Na, K, and Ca and the relationship to the functioning Na-K pump. J Gen Physiol 83:19-45

Ørskov SL (1935) Untersuchungen über den Einfluss von Kohlensäure und Blei auf die Permeabilität der Blutkörperchen für Kalium und Rubidium. Biochem Z 279:250-261

7 The Swelling-Sensitive Osmolyte Channel

René Motais, Hélène Guizouarn and Franck Borgese

Laboratoire de physiologie des membranes cellulaires, Université de Nice - Sophia Antipolis, UMR 6078 CNRS, Bâtiment Jean Maetz, 284 chemin du Lazaret, 06230 Villefranche sur mer, France

7.1 Introduction

Changes in intracellular or extracellular solute concentration generate a transmembrane osmotic gradient. Animal cell membranes being non-rigid and generally freely permeable to water, any such gradient results in flow of water into or out of the cell inducing cell volume perturbation. Regulation of cellular volume, which is important for cells to function properly, is achieved by activation of membrane transport processes. In response to swelling, cells that undergo a regulatory volume decrease (RVD) adjust their volume through the release of intracellular osmotically active solutes (termed "osmolytes") accompanied by an obligate water loss. "Inorganic osmolytes" (mainly K^+ and Cl^-) play a central role in the RVD response, leaving cells via a K^+-Cl^- cotransport system (see Chap. 9) and/or volume-activated channels (Sarkadi and Parker 1991). However, most cells also contain at high concentrations (tens to hundreds of millimolar) small organic, osmotically active solutes which exert a stabilizing influence on intracellular proteins and are synthesized from metabolic precursors or taken up from the extracellular medium via energy-dependent transport mechanisms. These "organic osmolytes" fall into 3 different classes: amino acids (e.g. taurine), polyols (e.g. sorbitol and myo-inositol) and methylamines (e.g. betaine). For a wide range of cells, organic osmolytes are also released in response to cell swelling (Strange and Jackson 1995; Kirk 1997) and they may account for as much as half of the total RVD (Garcia-Romeu et al. 1991). There is increasing evidence that in a variety of mammalian cell types this release occurs via channels, yet unidentified, having a significant permeability to a wide variety of both charged and uncharged organic solutes (Kirk 1997).

After maturation, mammalian anucleated red blood cells have lost the capacity to regulate their volume although some quiescent regulatory transport systems (e.g. K^+-Cl^- cotransporter) are still present in the membrane. Conversely, avian, amphibian and fish nucleated red blood cells undergo a RVD in response to swelling. The role amino acids or other organic osmolytes play in the avian erythrocyte response has not been studied. However, the total quantity of amino acids in these cells is not large, 9–10 mmoles/l_{cells} (Wheeler and Christensen 1967), so one would not expect them to have a major osmotic role. RVD appears to be achieved essentially via the K^+-Cl^- cotransport (Kregenow 1971; McManus et al. 1985).

By contrast, erythrocytes from fish (elasmobranchs and teleosts) release both amino acids and K^+-Cl^- for volume control. Thus fish red blood cells have proven to be extremely useful as a model system to study the different transport pathways involved in RVD and their coordination in this regulatory process. The relative contribution that organic and inorganic osmolyte pathways make to the process of RVD is different in elasmobranchs and teleosts. Because the cell osmolality in elasmobranchs (e.g. skate) is about 950 mosmol/kg H_2O *vs.* 320 in teleosts and other vertebrates, elasmobranch erythrocytes contain very high levels of amino acids in their cytosol. The major components are taurine and β alanine (Boyd et al. 1977), which are both released from the cell in response to osmotic swelling (Dickman and Goldstein 1990). K^+ is also released but the rate of K^+ loss is tenfold less than the rate of taurine loss. Thus in skate, RVD is due predominantly to the release of amino acids (Dickman and Goldstein 1990). Erythrocytes from teleosts (e.g. trout, eel, flounder) have high levels of amino acids (25–55 mM), mainly taurine (Fugelli and Zachariassen 1976; Fincham et al. 1987; Garcia-Romeu et al. 1991). The loss of taurine in these cells can account for half of total RVD (Garcia-Romeu et al. 1991), the remainder being due to the loss of K^+ and Cl^-.

To achieve RVD, fish erythrocytes activate multiple swelling-sensitive transport systems. One of these is the well-defined K^+-Cl^- cotransporter. But the others, which are only defined as mediating the net loss of a specific solute, share a number of functional properties. The purpose of this chapter is to summarize what is known about these so-called multiple transport pathways. Firstly, we discuss why and how fish red blood cells, depending on swelling conditions, make use or not of organic osmolytes to readjust their volume. Secondly, by reviewing volume-sensitivity, functional properties and regulation of transport systems, then by analysing the molecular approach used to identify the pathway, we present evidence that a single transport pathway having the characteristics of an anion channel mediates the movements of all the structurally unrelated, charged and uncharged intracellular osmolytes. Evidence is given which unambiguously involves the anion exchanger protein (AE1 called band 3 in erythrocytes) in volume regulation.

7.2 Swollen to the Same Extent, Red Cells can Adopt Different Regulatory Patterns

Na^+, K^+, Cl^- and taurine account for up to 90% of trout erythrocyte osmolality. The remaining osmolality is due to the presence of nearly or totally impermeant solutes such as ATP, ADP, PO_4^{2-} and proteins, mainly haemoglobin (Guizouarn and Motais 1999). Because cell volume is regulated by the net loss of osmotically active solutes, RVD of trout erythrocytes can only be achieved by the loss of K^+, Cl^- and taurine down their electrochemical gradients. However, depending on how its volume has been altered, the erythrocyte will or will not involve taurine in its volume correction and will choose between different transport systems to jettison K^+ and Cl^- (Motais et al. 1991; Guizouarn and Motais 1999). As illustrated in Fig. 7.1, when the volume increase is due to a net uptake of salts and osmotically obliged water resulting in an increase in cell electrolyte concentration (e.g. after hormonal stimulation of a Na^+/H^+ exchange or by exposure of the erythrocytes to a solution containing NH_4Cl), the regulatory response involves only a KCl loss mediated by a K^+-Cl^- cotransport (Borgese et al. 1987; Motais et al. 1991).

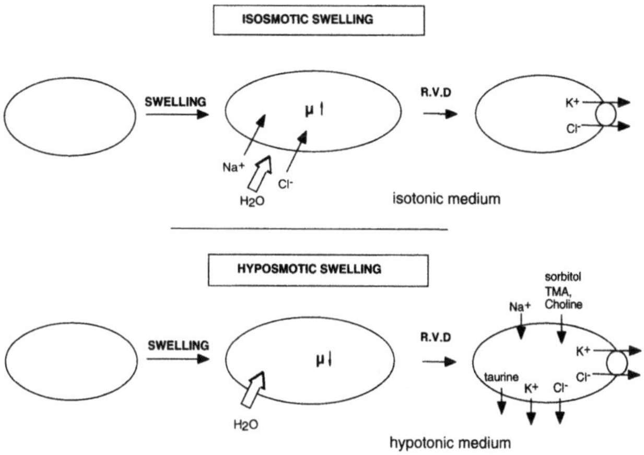

Fig. 7.1. Isosmotic swelling is induced by a net uptake of electrolytes and osmotically obliged water. Hormonal stimulation of Na^+/H^+ exchange or suspension of erythrocytes in an isotonic solution containing NH_4Cl promotes an isosmotic swelling, characterized by a slight increase of the intracellular electrolyte concentration, i.e. of the ionic strength, μ (calculated from the molar concentrations of Na^+, K^+ and Cl^- in cell water as indicated in Motais et al. (1991)). Cells respond by releasing K^+ via a Cl^--dependent pathway (K^+-Cl^- cotransport). **Hyposmotic swelling** is induced by a decrease in extracellular osmolality (or by diffusion of an uncharged solute such as urea), generating an osmotic inflow of water which decreases intracellular ionic strength. The volume regulatory response is then much more complex, involving taurine loss and K^+ loss via two distinct (Cl^--dependent and Cl^--independent) pathways. Other pathways are activated mediating downhill uptake of Na^+ (which counteracts *RVD*) and movements of diverse structurally unrelated compounds (e.g. choline, tetramethylammonium (*TMA*), sorbitol etc.). From Guizouarn and Motais (1999)

Conversely, when a volume increase of the same magnitude occurs as a result of an entry of water into the cell along its chemical gradient ("hyposmotic" swelling) or dragged by an uncharged solute (e.g. urea), leading to dilution of cell electrolytes, RVD occurs by a large loss of taurine (accounting for 50% of the total RVD) and by a loss of KCl. The loss of KCl is then mediated by two different systems: the K^+-Cl^- cotransport (the Cl^--dependent component) and a "Cl^--independent" component. Furthermore such a hyposmotic swelling also activates pathways permeable to diverse cations such as Na^+, choline, or tetramethylammonium (TMA) (Garcia-Romeu et al. 1991). However, although activation of taurine, K^+ and Cl^- transport systems will tend to correct volume changes, the simultaneous activation of other pathways may either partly counteract RVD (net entry of Na^+

down its electrochemical gradient (Garcia-Romeu et al. 1991)) or appear physiologically irrelevant (transport of choline, tetramethylammonium). In several other fish species, a similar hyposmotic regulatory pattern has been described involving both K^+ (Lauf 1983; Cossins et al. 1994; Jensen 1994; Bursell and Kirk 1996) and diverse organic solutes (Fugelli and Thoroed 1986; Fincham et al. 1987; Goldstein and Brill 1991; Kirk et al. 1992; Haynes and Goldstein 1993; Joyner and Kirk 1994; Thoroed and Fugelli 1994).

Thus trout erythrocytes possess multiple swelling-sensitive transport pathways. The fact that they adopt different regulatory patterns after isosmotic and hyposmotic swellings of similar magnitude shows that the cells are responding to more than just simple swelling. Clearly changes in cellular electrolyte concentration are involved in regulation of different pathways and impose the observed regulatory patterns. However, before analysing mechanisms underlying volume-sensitivity and regulation of these pathways, it is relevant to consider the significance of such a complex regulatory process.

7.3 Red Cells will Adopt the Regulatory Pattern that Maintains Cell Homeostasis

The ability of a cell to adopt different regulatory patterns depending on how its volume has been altered becomes physiologically relevant if we consider that great changes in cellular solute and water contents, occurring during RVD, are not expected to drastically alter intracellular pH, cell membrane potential, enzyme activities or other cell processes.

When the volume increase is due to a net uptake of salts resulting in an increase in cell electrolyte concentration, it makes sense that, to recover volume and a normal electrolyte content, the cell specifically extrudes salts in excess and avoids using the organic compound taurine as an osmoregulatory solute. Indeed under these conditions, as experimentally observed, RVD occurs exclusively by a KCl loss mediated by the K^+-Cl^- cotransport (Borgese et al. 1987; Motais et al. 1991).

When swelling is due to an entry of water not accompanied with salts, resulting in a dilution of cell electrolytes, it could be considered that the best way to recover cell volume and electrolyte content is to extrude water in excess by the loss of the organic compound taurine and to avoid the use of electrolytes, already diluted, as osmoregulatory solutes. However, that strategy would lead to a large alteration of the Cl^-_{int}/Cl^-_{ext} ratio: after removal of water in excess, the cell will recover its volume and the concentration of electrolytes in the cell, Cl^-_{int} will be the concentration of cell electrolytes before swelling. But at the same time the external chloride in the hypotonic saline, in which cells are suspended, remains diluted. Thus the ratio Cl^-_{int}/Cl^-_{ext} would be greater after RVD than before and, for a dilution of one third of the external medium, the membrane potential would be shifted from about –22 mV to –11 mV and the intracellular pH from 7.40 to 7.60. Volume recovery also cannot be obtained by a loss of KCl exclusively: this would induce a large decrease in the intracellular chloride pool and would greatly alter the Cl^-_{int}/Cl^-_{ext} ratio, then promoting hyperpolarisation and acidification of the cell. Thus, in response to hypotonic swelling, it makes sense that the cell loses both taurine and electrolytes, as experimentally observed (Garcia-Romeu et al. 1991; Motais et al. 1991; Guizouarn and Motais 1999). However, it means that release of Cl^- and taurine have to be coordinated if variations in membrane potential and cell pH are to be limited. It has been experimentally shown (Fig. 7.2) that over the full time course of RVD the chloride concentration in the cell water remains remarkably constant (Guizouarn et al. 2000).

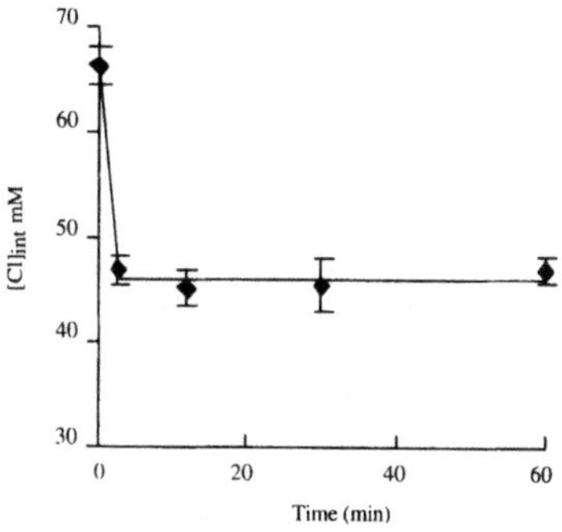

Fig. 7.2. Changes in chloride concentration of cell water induced by suspending erythrocytes in hypotonic medium. Red blood cells were suspended at time 0 in hypotonic saline (215 mosmol/kg H_2O), promoting a rapid dilution of internal chloride from 66.33 ± 1.94 mM to 46.89 ± 1.42 mM. Subsequently, over the full time course of RVD, the intracellular chloride concentration remains constant. n = 21. Mean values ± SEM. From Guizouarn et al. (2000)

In other words, despite variable loss of cell electrolytes and changes in cell volume, the Cl^-_{int}/Cl^-_{ext} ratio remains fixed during RVD and then both membrane potential and cell pH, which depend on this ratio, remain unchanged. When cell volume decreases, it is only possible to keep the chloride concentration in the cell water constant if an equal percentage of the cell chloride pool and the cell water pool are lost simultaneously. However, the amount of water leaving from the cell depends on the release of different osmolytes. Thus, to keep constant chloride concentration in cell water, a strict linkage must exist between the net loss of Cl^- and the net loss of other osmolytes. Indeed by measuring continuously the changes in electrolytes, taurine and water contents we showed that, over the full time course of RVD, cells lost osmotically active solutes with a constant stoichiometry of *1 Cl^-/1 positive charge/2.35 taurine*. This stoichiometric relationship provokes two comments. Firstly, the large amount of taurine released during RVD (50% of all solutes lost) is transported as an electroneutral zwitterion and not as an anion. It is the first time that taurine is demonstrated to move as a zwitterion during volume regulatory decrease. Secondly, it has been shown experimentally that the constancy of cell chloride concentration is not finely tuned by some mechanism able to modulate and adjust the relative ionic and taurine permeabilities, but results essentially from the fact that the swelling-sensitive osmolyte channel constitutively possesses an adequately fixed relative permeability for cations, chloride and taurine. However, as a small but significant fraction of KCl loss occurs via the K^+-Cl^- cotransporter, the parallel contribution of the cotransport to the stoichiometric relationship remains to be defined.

7.4 Regulation of Swelling-Sensitive, Osmolyte Pathways

As discussed above (Fig. 7.1), when cell electrolyte concentration has been increased during swelling (isosmotic swelling) only one regulatory pathway, a K^+-Cl^- cotransport, is activated. Conversely, when cell electrolyte concentration has been decreased (by exposure to hyposmotic medium or when water is dragged by urea) not only the K^+-Cl^- cotransport but also several osmolyte pathways are activated: thus, opening of these osmolyte pathways is triggered by a decrease in cell electrolyte concentration. It is noteworthy, as illustrated in Fig. 7.3, that the K^+-Cl^- cotransport activity is also controlled, but negatively, by a decrease in cell electrolyte concentration (Guizouarn and Motais 1999): during isosmotic swelling, as soon as a slight increase in volume occurs, a strong KCl activation is observed and the KCl activity continues to increase linearly as a function of cell volume. Hyposmotic swelling also induces KCl activation, but under these conditions the efficiency of the K^+-Cl^- cotransport is clearly reduced: the set point is shifted to higher cell volumes and the volume dependence is damped and displays a sigmoidal relationship. This is in agreement with the results obtained for the control of K^+-Cl^- cotransport in dog red blood cells (Parker et al. 1995).

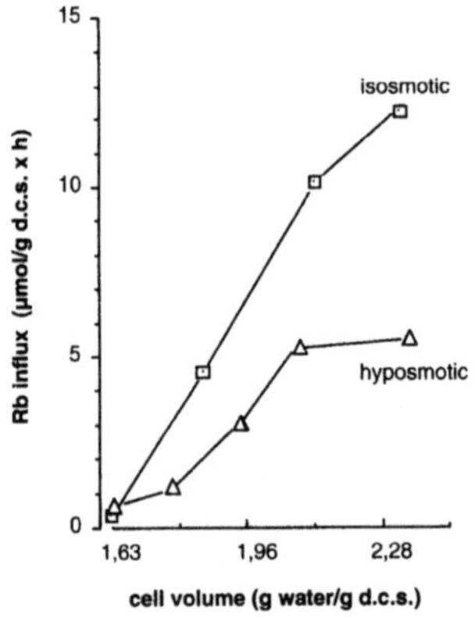

Fig. 7.3. The K^+-Cl^- cotransport activity is reduced by the decrease in cell electrolyte concentration. Cl^--dependent K^+ fluxes as a function of cell volume in erythrocytes swollen isosmotically (□) and hyposmotically (Δ). Data were obtained by subtracting fluxes measured in NO_3^--containing media from those measured in Cl^--containing media. *d.c.s.* = dry cell solids. From Guizouarn and Motais (1999)

7.4.1 The Opening of Osmolyte Pathways is Controlled by Ionic Strength and is Insensitive to Cell Volume Expansion

Clearly the decrease in cell electrolyte concentration is involved in the opening of diverse osmolyte pathways in trout erythrocytes. However, exposure of erythrocytes to a hypotonic medium results in an entry of water which simultaneously induces cell volume expansion and dilution of electrolyte concentration. Thus the relative contribution of volume expansion and electrolyte concentration in activation of pathways must be clarified. Moreover, in skate hepatocytes and C6 glioma cells, it has been shown that activation of the taurine pathway is dependent on intracellular Cl^- (Jackson et al. 1996). Therefore Cl^-, rather than total electrolytes, can be involved in the regulation of these pathways.

To test whether the decrease in intracellular chloride alone is able to open the taurine channel, swellings with ammonium chloride *vs.* ammonium nitrate have been compared (Guizouarn and Motais 1999). It has been observed that the taurine pathway remains inactive after total replacement of Cl^- by NO_3^- in swollen cells. Moreover, the nature of cations contained in cells also does not control pathway activation (Guizouarn and Motais 1999). Thus the concentration of electrolytes (i.e. ionic strength), and not their nature, is the factor controlling activation. Similarly, in skate erythrocytes, the taurine pathway has been recently shown to be insensitive to intracellular chloride concentration and controlled by the decrease in cell ionic strength (Wittels et al. 2000).

Figure 7.4A shows that exposure of trout erythrocytes to hypotonic media of different tonicities activates, as expected, the transport of a series of structurally unrelated compounds such as taurine, sorbitol, Na^+ and K^+ (independently of K^+ mediated by the cotransporter). Note that, first, the magnitude of fluxes increases as a function of electrolyte dilution; second, fluxes of all solutes are very similarly controlled by electrolyte concentration. In this type of experiment, however, dilution of electrolytes and volume expansion are strictly linked (dashed line). The relative contribution of volume expansion and electrolyte dilution in the control of these pathways can be deduced from data illustrated in Fig. 7.4B. In this experiment trout erythrocytes are swollen to the same extent (40% increase in cell water; dashed line) by different treatments (see legend to figure), which induces large variations in cell electrolyte concentration. The figure shows how flux of different solutes varies as a function of electrolyte dilution when cell volume is maintained constant. Despite the fact that the cell volume is greatly increased, no discernible activation is measured when electrolyte concentrations remain equivalent to (dilution factor = 1) or is greater (dilution factor = 0.9) than that of control unswollen erythrocytes. Activation of flux starts when electrolytes are significantly diluted: the set point for activation corresponds to about a 10% dilution (dilution factor = 1.1). Then it increases progressively as a function of electrolyte dilution. A comparison of Fig. 7.4A and B shows that the two curves are very similar: at the same degree of electrolyte dilution, activation is equivalent when the volume increase is still only small (for example 2% Fig. 7.4A) or large (40% Fig. 7.4B). These data clearly show that fluxes are controlled by the changes in cell electrolyte concentration (i.e. ionic strength) but are independent of the degree of cell swelling. Moreover, recent experiments show that activation of taurine fluxes can be induced just

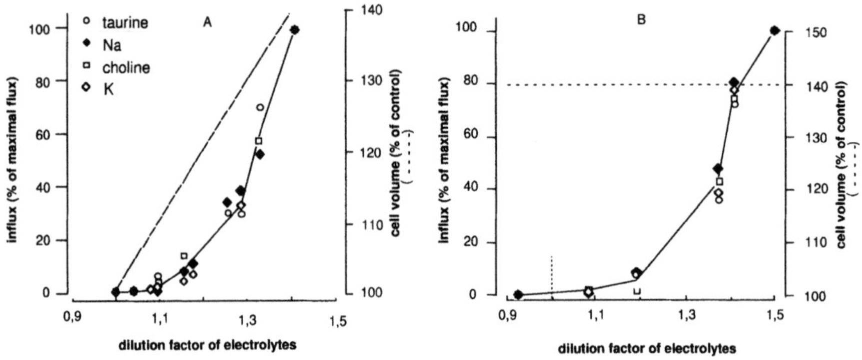

Fig. 7.4. Comparison of taurine, Na^+, choline and Cl^--independent K^+ fluxes as a function of electrolyte dilution. (**A**) After hyposmotic swelling (electrolyte dilution varying as cell volume as indicated by the dashed line). To measure Cl^--independent fluxes, NO_3^- was used as a Cl^- substitute. Cells were suspended in media of different tonicities varying from 320 (isotonic) to 220 mosmol/kg H_2O. In these conditions, as long as external osmolality decreased, cell volume increased and simultaneously electrolytes were progressively diluted. Cell volume, measured as cell water content per gram of dry cell solids, is expressed as percentage of control, unswollen cells. Fluxes are expressed as percentage of fluxes measured when cells are exposed to maximal hypotonicity. Intracellular amounts of electrolytes were measured, their concentrations in cell water (mM) were then calculated. The changes in cell electrolyte concentration are expressed as the dilution factor in comparison with the electrolyte concentration before cells were swollen, i.e. 1. (**B**) When volume expansion is maintained constant. All cells were 40% swollen (dashed line) by suspension in isosmotic, Cl^--free NO_3^- -containing media (320 mosmol/kg H_2O) in which 90 mM $NaNO_3$ was replaced by defined amounts of urea, NH_4NO_3 or urea - NH_4NO_3 mixtures (see Guizouarn and Motais (1999)): penetration of these highly permeable electrolyte and nonelectrolyte osmolytes drags the same amount of water (thus inducing same swelling) but alters intracellular electrolyte concentration differently. Fluxes are expressed as percentage of flux measured at maximal urea concentration, i.e. 180 mM. From Guizouarn and Motais (1999)

by dilution of intracellular electrolytes without any volume expansion (Fig. 7.5).

In summary, in trout erythrocyte changes in intracellular ionic strength alone can open osmolyte pathways and these pathways are clearly insensitive to the degree of volume expansion. In other words activation is characterized by an "ionic strength set point" and not by a "volume set point" which could be modulated by changes in electrolyte concentration as suggested by Emma et al. (1997) for the swelling-activated taurine channel of C6 glioma cells (these data, however, could be interpreted differently: see Guizouarn and Motais (1999)). Recently Wittels et al. (2000) confirms the role of intracellular ionic strength in the activation of the taurine pathway in skate erythrocytes. In these cells, however, a similar volume expansion achieved via hyposmolarity or via diffusion of a nonelectrolyte (urea derivative, ethylene glycol), resulting in the same decrease in ionic strength, induces a rise in taurine efflux greater in hyposmotic than in nonelectrolyte media. Thus in skate, there is not a direct correlation between a decrease in ionic strength

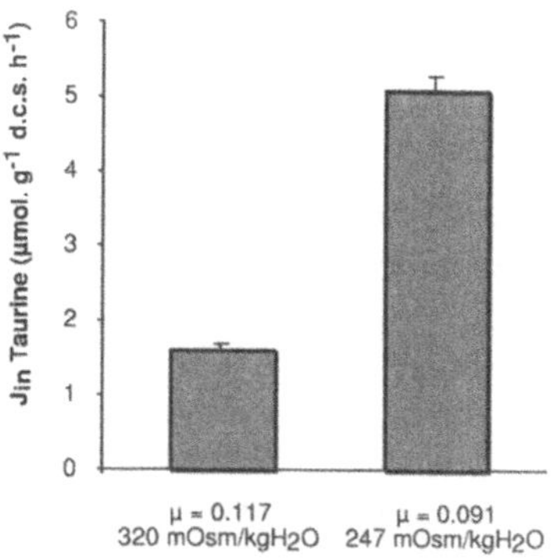

Fig. 7.5. The effect of ionic strength on activation of taurine fluxes in unswollen cells. Valinomycin ($5 \cdot 10^{-7}$ M) was added to trout red blood cells suspended in an isosmotic saline solution (320 mosmol/kg H_2O). Addition of valinomycin for 60 minutes induced a cell shrinkage by the loss of KCl and obliged water. Valinomycin was then washed out (3 rinses with 25 vol. of saline + bovine serum albumin (BSA) 1%, followed by 2 rinses without BSA) and the initial volume was restored by suspending red cells in a hypotonic saline (247 mosmol/kg H_2O) leading to a dilution of intracellular electrolytes and a decrease of ionic strength: μ. Ion and water contents and unidirectional influxes were measured just before and after the hypotonic shock. *d.s.c.* = dry cell solids

and an increase in taurine loss. As suggested by the authors, it may be that in elasmobranch erythrocytes, which have a much higher concentration of intracellular organic osmolytes than trout erythrocytes (osmolality = 940 *vs.* 320 mosmol/kg H_2O), the pathway is under dual control, by both ionic strength and organic osmolyte concentration.

7.4.2 A Single or Several Osmolyte Transport Pathways?

The decrease in intracellular ionic strength induces fluxes of structurally unrelated osmolytes: taurine, sorbitol, Na^+, choline and K^+ (independently of K^+ mediated by the cotransporter). Figure 7.4A and B shows that the dependence of fluxes on ionic strength is strictly identical for all solutes. These data suggest either that the different pathways are regulated in a coordinated fashion or that all solutes move via a common transport pathway (this point will be discussed later).

It is noteworthy that a massive increase in cation fluxes, sensitive to 4,4'-diisothiocyanatostilbene-2,2'-disulphonic acid (DIDS) and other anion exchange inhibitors, occurs when human red cells are suspended in solutions of very low ionic strength. Jones and Knauf (1985) proposed that these cation fluxes could be mediated by AE1. Moreover, Culliford et al. (1995) showed that a low ionic strengh medium induced the activation of organic osmolyte solute transport sharing strong functional similarities with the cation route. It is attractive to think that when red cells are suspended in media of low ionic strength, some changes of protein structure could be responsible for a perturbation altering AE1 selectivity leading to this peculiar solute permeation.

7.4.3 Opening of the Osmolyte Channel in Response to a Gradual and Slow Decrease in Medium Osmolality

In most studies on volume regulation, as in those described above, cells are exposed to large and sudden hypotonic shocks (e.g. 320 to 220 mosmol/kg H_2O). However, it has been shown that renal cells (Lohr and Grantham 1986; Van Driesche et al. 1997) maintain a stable volume over a range of osmolality extending from isotonicity to 167 mosmol/kg H_2O, provided that the rate of osmolality change is very slow. In other words, in these cells, experimentally undetectable volume changes can activate regulatory pathways, illustrating their exquisite sensitivity. By contrast trout erythrocytes do not have the capacity to maintain cell volume constant when exposed to a gradual decrease in osmolality, even at a rate (0.66 mosmol/kg H_2O per min) much slower than those used in studies with renal cells, i.e. 1.5 to 27 mosmol/kg H_2O per min (Godart et al. 1999). No significant activation of the taurine channel occurs before a volume increase of at least 15% is attained, indicating that the ionic strength set point for activation of the channel is similar to that observed with a rapid shock or even is shifted to a lower intracellular ionic strength.

7.5 Swelling-Sensitive Osmolyte Pathways: Association with Band 3 Protein

7.5.1 The Organic Osmolyte Pathway has the Functional Characteristics of an Anion Channel

The swelling-sensitive transport of organic solutes has been characterised in detail by several laboratories in a variety of fish erythrocytes (skate and teleosts). Amino acid, polyol and methylamine efflux pathways share a number of characteristics: the loss of solutes is mediated by passive, Na^+-independent transport mechanisms (Fugelli and Thoroed 1986; Fincham et al. 1987). From kinetics and functional analysis of unidirectional fluxes it has been shown that these pathways (e.g. for taurine) mediate the movement of solutes in both directions across the cell membrane (Haynes and Goldstein 1993; Thoroed and Fugelli 1994). They fail to saturate with increasing substrate concentrations (Kirk et al. 1992; Goldstein and Davis 1994), and to show either *trans*-stimulation (Haynes and Goldstein 1993) or substrate competition (Goldstein and Davis 1994). They cannot discriminate between stereoisomers, e.g. D- and L-alanine (Fincham et al. 1987), D- and L-glucose (Wolowyk et al. 1989). Taken together, these functional characteristics are unlike those of conventional transporters but are those expected of pores or channels. Moreover, in a number of studies the inhibitor-sensitivity of the swelling activated transport of a range of different organic solutes has been compared and shown to be similar. Such data have been interpreted as indicating that in erythrocytes the volume regulatory efflux of a range of structurally unrelated organic solutes is via a common channel of broad specificity (Kirk et al. 1992; Goldstein and Davis 1994). Similarly, in many mammalian cell types, there is increasing evidence that structurally dissimilar organic osmolytes are lost from cells via a common swelling-activated channel. This channel is thought to be anion selective (see Strange et al. 1996; Kirk 1997). Moreover, electrophysiological studies performed in different mammalian cell types have demonstrated a significant permeability of swelling-activated anion channels to organic osmolytes bearing a net negative charge, e.g. the anionic form of taurine (Banderali and Roy 1992; Jackson and Strange 1993; Basavappa et al. 1996; Boese et al. 1996). These results demonstrate, at least, that mammalian swelling-activated anion channels are of sufficient dimensions to accommodate these compounds.

As in mammalian cell types, the swelling-induced transport of all organic osmolytes observed in fish erythrocytes is inhibited by a range of anion channel blockers, e.g. DIDS, niflumic acid, 5-nitro-2-(3-phenylpropylamino)-benzoic acid (NPPB), pyridoxal-5'-phosphate, furosemide. Although such reagents are notoriously non-specific, the data suggest that, in fish erythrocytes, the common osmolyte channel may also be an anion channel (Kirk et al. 1992; Lewis et al. 1996). Two independent arguments reinforce this hypothesis: 1) There is a quantitative agreement between the effect of inhibitors on the transport of organic solutes and their effect on the swelling-activated transport of Cl^- (Lewis et al. 1996). 2) Using the patch-clamp technique, it has been observed that trout erythrocytes, when exposed to a hypotonic medium, activate a reversible Cl^- conductance which is blocked by DIDS (Egee et al. 1997).

7.5.2 The Organic Osmolyte Channel: a Pathway for Cations

In erythrocytes from various fish species, as previously discussed, hyposmotic swelling activates pathways permeable to diverse cations such as Na^+, K^+ (not mediated by the K^+-Cl^- cotransport), choline or TMA (Garcia-Romeu et al. 1991; Joyner and Kirk 1994; Bursell and Kirk 1996). These pathways share a number of functional characteristics with the swelling-sensitive organic osmolyte channel: the failure to saturate with increasing substrate concentrations (Joyner and Kirk 1994), the failure to show *trans*-stimulation and competition (Garcia-Romeu and Motais, unpublished data). Furthermore cation pathways and osmolyte channel are modulated in a similar way: same time course of activation (Joyner and Kirk 1994); strictly similar dependence on the intracellular ionic composition (Fig. 7.4). Lastly they are also pharmacologically indistinguishable (Garcia-Romeu et al. 1991; Joyner and Kirk 1994; Bursell and Kirk 1996; Garcia-Romeu et al. 1996). Note that if the drugs used in these pharmacological studies are anion transport blockers, it has been shown that these inhibitors interfere with the cation transport itself and not through a decrease in Cl^- conductance (Garcia-Romeu et al. 1991).

In summary, the existence of a swelling-activated cation channel coexisting in the membrane with the swelling-sensitive osmolyte channel and subject to the same control mechanisms is a possibility, which cannot be ruled out. However, the hypothesis that the swelling-sensitive osmolyte channel, despite being anion selective, has a significant cation permeability is an alternative and the simplest explanation (Joyner and Kirk 1994; Bursell and Kirk 1996; Guizouarn and Motais 1999; Kiessling et al. 2000).

7.5.3 The Swelling-Sensitive Osmolyte Channel is Associated with Band 3 Protein (AE1)

The anion exchanger band 3 protein (also termed AE1) is a major constituent of the plasma membrane of erythrocytes. The particular sensitivity of the erythrocyte swelling-sensitive osmolyte channel to a wide range of drugs known as potent inhibitors of band 3 protein has prompted the suggestion that band 3 might be involved in volume regulation, either as the transporter or as a regulator of transport (Garcia-Romeu et al. 1991; Goldstein and Brill 1991; Motais et al. 1991, 1992). These pharmacological data are consistent with such an implication of band 3 but they are neither unambiguous nor conclusive. However, a very convincing argument in support of this hypothesis is given by experiments using covalently bound DIDS. The stilbenedisulphonate derivatives DIDS and H_2DIDS are known as potent reversible inhibitors of red blood cell anion exchangers. But when they are in contact with red blood cells for one hour or less, these drugs progressively interact with band 3 protein and then behave as *specific, irreversible inhibitors* of the anion exchanger. Irreversible inhibition is due to the covalent binding of these compounds with a lysine residue of the protein. This residue (human Lys-539) is located close to the extracellular medium in the 5th putative transmembrane helix (Cabantchick and Rothstein 1974; Passow 1986). It has been shown that the covalent binding of DIDS on trout red blood cells blocks both anion exchange (Romano and Passow 1984) and the swelling-induced transport of taurine, cations (Motais et al. 1992; Garcia-Romeu et al. 1996) and polyol (Guizouarn and Motais 1999), strongly indicating that volume-induced movements of solutes are dependent on band 3. In red blood cells, however, the possibility can never be completely ruled out that DIDS binds covalently to a minor protein in addition to band 3, even if on SDS polyacrylamide gel electrophoresis no other protein than band 3 is significantly labelled (Motais et al. 1992).

Other experimental data suggest, much more indirectly, that band 3 could play a role in the volume-stimulated solute efflux during volume expansion. In skate erythrocytes, Goldstein and co-workers showed that a number of biochemical events involving band 3 protein occur rapidly and transiently upon hyposmotically-induced volume expansion. These events, however, could be coincidental and not directly related to RVD. (i) volume expansion leads to the formation of oligomeric forms of the band 3 protein (Musch et al. 1994a); (ii) hypotonic swelling induces a rapid increase in band 3 phosphorylation (Musch et al. 1994b); (iii) an increase in the activities of the band 3 phosphorylating tyrosine kinases p72syk and p56lyn is observed when erythrocytes are volume expanded in hypotonic media (Musch et al. 1999). Conversely, no significant increase occurs when erythrocytes are swollen in NH_4Cl-containing isosmotic medium (Wittels et al. 2000); (iv) hypotonic swelling increases the affinity of ankyrin binding to band 3 (Musch and Goldstein 1996) and decreases the amount of band 4.1 bound to band 3 (Musch and Goldstein 2001).

Taken together, these data are consistent with the view that in fish erythrocytes band 3 protein is either the osmolyte channel or a crucial component of the transport pathway (Motais et al. 1997). Clearly, however, only cloning and functional expression of trout band 3 can produce the proof of its involvement in volume regulation.

7.6 Molecular Identification of the Swelling-Sensitive Osmolyte Channel

When expressed in *Xenopus* oocytes, the trout red blood cell anion exchanger (tAE1) elicits, as expected, an anion exchange activity. But simultaneously tAE1 expression induces several other changes in oocyte membrane permeability, which are expected if tAE1 serves, in erythrocyte, as the route for the volume-regulatory efflux of osmolytes.

(i) tAE1 expression results in the appearance of both an anion conductance, a transport of taurine (Fiévet et al. 1995) and a transport of polyol (Fiévet et al. 1998). Anion conductance and transport of organic osmolytes are strictly related. Moreover, they are similarly affected by AE1 inhibitors. Thus tAE1 is involved in the appearance of an anion channel and the transport of neutral and zwitterion organic solutes.

(ii) tAE1 expression also elicits a large increase in Na^+ and K^+ permeability (Fig. 7.6; Guizouarn et al. 2001). Diffusive in nature, the cation permeability is ouabain-insensitive and unaffected when Cl^- is replaced by NO_3^-. Anion conductance and cation permeability are strictly related and similarly affected by AE1 inhibitors. That could be explained by assuming that Na^+ and K^+ movements are mediated by a cation channel pre-existing in the oocyte membrane (endogenous chanel) and are electrically linked to Cl^- moving via the tAE1-induced channel. However, several arguments refute this simple interpretation:

(a) after treatment with tAE1 inhibitors which for example reduce anion conductance by half, the membrane potential remains practically unchanged (i.e. Cl^- conductive permeability is still much greater than cation permeability). Nevertheless, cation fluxes are reduced by half. Thus the inhibitory effect of drugs like NPPB, niflumic acid, *N*-(4-azido-2-nitrophenyl)-2-amino-ethylsulfonate (NAP-taurine) is not via a decrease in the driving force for diffusional fluxes but is due to a direct action on the cation permeability pathway.

(b) generation in control oocytes of a large anion conductance (e.g. by expression and stimulation of the “Cystic Fibrosis Transmembrane Conductance Regulator” protein (CFTR)) does not induce cation permeability. Thus the observed cation permeability is strictly dependent on expression of tAE1, probably via the anion conductance due to tAE1 expression.

(c) expression of tAE1 induces a large depolarisation of oocytes (from –57 mV to –20 mV). If cation transport occurs by electrodiffusion via an endogenous channel and if the membrane conductance for cation is constant, a change in membrane potential would have an opposite effect on influx and efflux. However, both are greatly increased. Obviously the increase in cation permeability reflects an increase of membrane conductance for cations. Only two possibilities can account for such an increase: either expression of tAE1 results in the appearance of a new cation channel (i.e. the tAE1-induced osmolyte anion channel also functions as a cation channel) or expression of tAE1 activates an endogenous cation channel which is normally quiescent in control oocytes (i.e. tAE1 acts as a regulator for cation transport). In other words, tAE1 is clearly involved in the appearance of cation transport.

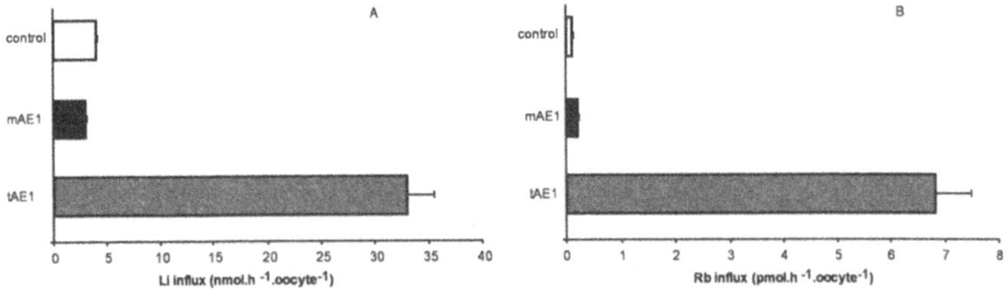

Fig. 7.6. Chloride-independent, ouabain-insensitive Li^+ and Rb^+ influxes in oocytes expressing mouse (*mAE1*) and trout (*tAE1*) anion exchanger. Li^+ influxes (**A**) and Rb^+ influxes (**B**) were measured at day 5 post-cRNA injection, in a Cl^--free, NO_3^--containing saline in presence of ouabain (10^{-4} M). Mean ± SEM; n = 8

(iii) The irreversible covalent binding of H_2DIDS on the oocyte membrane expressing the trout band 3 induces a large decrease of anion conductance but has no significant effect on the membrane potential (Guizouarn et al. 2001). Nevertheless, cation fluxes are strongly inhibited. Thus, as discussed for reversible inhibitors like NPPB and niflumic acid, the inhibitory effect of covalent bound H_2DIDS is not via a decrease in the driving force for diffusional fluxes but is due to a direct action of the drug on the cation pathway. H_2DIDS binds covalently on tAE1 by interacting with the trout lysine 522. Conversion of this Lys to Arg by site-directed mutagenesis prevents the specific binding of the drug. After such a point mutation, Cl^- exchange, anion conductance and cation permeability remain fully unchanged after a treatment normally producing an irreversible binding of H_2DIDS. Nevertheless, all these transports remain sensitive to reversible inhibitors (reversible H_2DIDS, niflumic acid, NPPB) (Guizouarn et al. 2001). These results clearly demonstrate that the protein is directly involved in the channel-forming pathway and is a crucial component of the cation transport system. It is likely that the trout anion exchanger can form an anion, organic osmolyte channel of broad specificity, having a significant cation permeability. The data cannot completely rule out the possibility that H_2DIDS, by interacting with tAE1, inhibits both the channel function of tAE1 and a putative capacity of tAE1 to activate an endogenous cation channel. But then H_2DIDS would have to inhibit *to the same extent* the tAE1-channel function and the tAE1-regulatory function. The simplest hypothesis compatible with expression data seems to be that tAE1 acts as a channel permitting net flow of K^+ and Na^+. It also accounts for the fact that tAE1 expressed in oocytes reproduces all permeability changes observed in hypotonically swollen erythrocytes without assuming that oocyte and erythrocyte possess the same tAE1-regulated, endogenous cation channel.

(iv) Mammalian red blood cells, in contrast to fish red cells, do not regulate their volume and lack a swelling-sensitive osmolyte channel. Band 3 from mouse erythrocyte (mAE1) and tAE1 are highly homologous proteins. However, expression of mAE1 does not result in increased anion conductance, taurine, polyol or cation transport (Fig. 7.6) (Fiévet et al. 1995, 1998; Guizouarn et al. 2001). These data are also consistent with a role for the fish band 3 protein in the volume-regulatory response of erythrocytes.

In summary, when expressed in oocytes, tAE1 generates a permeability change, which is constitutively active and has all the characteristics observed in erythrocytes after hypotonic swelling. Moreover, the specific covalent binding of H_2DIDS to the protein affects transport properties similarly in oocyte and in erythrocyte. Thus it is clear that band 3 protein is unambiguously involved in fish red blood cell volume regulation and probably functions as a swelling-sensitive osmolyte channel of broad specificity. By contrast, the homologous AE1 isofom from mammalian erythrocytes is devoid of such a volume-regulatory function.

The mechanism by which cell ionic strength can activate the band 3-dependent osmolyte channel remains to be defined. It can be envisioned that a change in intracellular ionic strength, by altering cytoskeletal architecture and/or interaction between band 3 and some elements of the cytoskeleton, allows rearrangement or aggregation of band 3 in the membrane (Motais et al. 1991; Guizouarn and Motais 1999). The signalling cascade activated by the reduction in intracellular ionic strength is likely to involve protein phosphorylation (Musch et al. 1994b; Kiessling et al. 2000; Wittels et al. 2000).

7.7 Conclusion

The available data are consistent with the view that the trout red blood cell anion exchanger tAE1 can serve a dual transport role, the well-established Cl^-/HCO_3^- exchange function and a channel function. We know several examples of a specific transport system that similarly may display channel properties, which are revealed under unusual experimental circumstances. Thus a triose phosphate/phosphate exchanger from chloroplasts, when reconstituted into artificial membranes, generates a voltage-dependent anion channel (Schwartz et al. 1994). Some cloned mammalian Na^+ cotransporters, when expressed in oocytes, reveal both cotransport and channel activities (Wright et al. 1996). Palytoxin, by binding to the Na^+, K^+-ATPase, reveals a channel thought to be a fundamental feature of the pump (Gadsby et al. 1993). It remains to be shown whether these unforeseen properties are normally expressed in wild type cells and may serve a regulatory function. It also remains to be shown whether some mammalian homologues of the fish anion exchanger, which are found in cell types regulating their volume, e.g. AE2 and AE3, also possess a dual transport function and may be involved in volume regulation. It is noteworthy that suspension of human red blood cells in solutions of very low ionic strength results in appearance of DIDS-sensitive, diffusional Na^+ and K^+ fluxes, thought to be mediated by AE1 (Jones and Knauf 1985, see also Chap. 4). In other words, even mammalian AE1, when forced into an altered conformation, could display channel-like properties.

Acknowledgement: The authors would like to thank their co-workers Nicole Gabillat and Bernard Pellissier for their excellent technical assistance.

References

Banderali U, Roy G (1992) Anion channels for amino acids in MDCK cells. Am J Physiol 263:C1200-C1207

Basavappa S, Huang CC, Mangel AW, Lebedev DV, Knauf PA, Ellory JC (1996) Swelling-activated amino acids efflux in the human neuroblastoma cell line CHP-100. J Neurophys 76:764-769

Boese SH, Wehner F, Kinne RKH (1996) Taurine permeation through swelling-activated anion conductance in rat IMCD cells in primary culture. Am J Physiol 40:F498-F507

Borgese F, Garcia-Romeu F, Motais R (1987) Control of cell volume and ion transport by β adrenergic catecholamines in erythrocytes of rainbow trout *Salmo gairdneri*. J Physiol 382:123-144

Boyd TA, Cha CJ, Forster RP, Goldstein L (1977) Free amino acids in tissues of the little skate *Raja erinacea* and the stingray *Dasyatis sabina*; effects of environmental dilution. J Exp Zool 199:435-442

Bursell JDH, Kirk K (1996) Swelling-activated K^+ transport via two functionally distinct pathways in eel erythrocytes. Am J Physiol 270:R61-R70

Cabantchick ZI, Rothstein A (1974) Membrane proteins related to anion permeability of human red blood cells. Localisation of disulfonic stilbene binding sites in proteins involved in permeation. J Membrane Biol 15:207-226

Cossins AR, Weaver YR, Lykkeboe G, Nielsen OB (1994) Role of protein phosphorylation in control of K flux pathways of trout red blood cells. Am J Physiol 267:C1641-C1650

Culliford SJ, Bernhardt I, Ellory JC (1995) Activation of a novel organic solute transporter in mammalian red blood cells. J Physiol 489:755-765

Dickman KG, Goldstein L (1990) Cell volume regulation by skate erythrocytes: role of potassium. Am J Physiol 258:R1217-R1223

Egée S, Harvey BJ, Thomas S (1997) Volume-activated DIDS-sensitive whole-cell chloride currents in trout red blood cells. J Physiol 504:57-63

Emma F, McManus M, Strange K (1997) Intracellular electrolytes regulate the volume set point of the organic osmolyte/anion channel VSOAC. Am J Physiol 272:C1766-C1775

Fiévet B, Gabillat N, Borgese F, Motais R (1995) Expression of band 3 anion exchanger induces chloride current and taurine transport: structure-function analysis. EMBO J 14:5158-5169

Fiévet B, Perset F, Gabillat N, Guizouarn H, Borgese F, Ripoche P, Motais R (1998) Transport of uncharged organic solutes in *Xenopus* oocytes expressing red cell anion exchangers (AE1s). Proc Natl Acad Sci USA 95:10996-11001

Fincham DA, Wolowyk MW, Young JD (1987) Volume-sensitive taurine transport in fish erythrocytes. J Membrane Biol 96:45-56

Fugelli K, Zachariassen KE (1976) The distribution of taurine, gamma-aminobutyric acid and inorganic ions between plasma and erythrocytes in flounder (*Platichtys flesus*) at different plasma osmolalities. Comp Biochem Physiol 55A:173-177

Fugelli K, Thoroed SM (1986) Taurine transport associated with cell volume regulation in flounder erythrocytes under anisosmotic conditions. J Physiol 37:245-261

Gadsby DC, Rakowski RF, De Weer P (1993) Extracellular access to the Na,K pump: pathway similar to ion channel. Science 260:100-103

Garcia-Romeu F, Cossins AR, Motais R (1991) Cell volume regulation by trout erythrocytes: characteristics of the transport systems activated by hypotonic swelling. J Physiol 440:547-567

Garcia-Romeu F, Borgese F, Guizouarn H, Fievet B, Motais R (1996) A role for the anion exchanger AE1 (band 3 protein) in cell volume regulation. Cell and Mol Biol 42:985-994

Godart H, Ellory JC, Motais R (1999) Regulatory volume response of erythrocytes exposed to a gradual and slow decrease in medium osmolality. Eur J Physiol 437:776-779

Goldstein L, Brill SR (1991) Volume-activated taurine efflux from skate erythrocytes: possible band 3 involvement. Am J Physiol 260:R1014-R1020

Goldstein L, Davis EM (1994) Taurine, betaine and inositol share a volume-sensitive transporter in skate erythrocyte cell membrane. Am J Physiol 267:R426-R431

Guizouarn H, Motais R (1999) Swelling activation of transport pathways in erythrocytes: effects of Cl^-, ionic strength and volume changes. Am J Physiol 276:C210-C220

Guizouarn H, Motais R, Garcia-Romeu F, Borgese F (2000) Cell volume regulation: the role of taurine loss in maintaining membrane potential and cell pH. J Physiol 523:147-154

Guizouarn H, Gabillat N, Motais R, Borgese F (2001) Multiple transport functions of a red blood cell anion exchanger, tAE1. Its role in cell volume regulation. J Physiol 535:497-506

Haynes JK, Goldstein L (1993) Volume-sensitive amino acid transport in erythrocytes of the little skate, *Raja erinacea*. Am J Physiol 265:R173-R179

Jackson PS, Strange K (1993) Volume-sensitive anion channels mediate swelling-activated inositol and taurine efflux. Am J Physiol 265:C1489-C1500

Jackson PS, Churchwell K, Ballatori N, Boyer JL, Strange K (1996) Swelling-activated anion conductance in skate hepatocytes: regulation by cell Cl and ATP. Am J Physiol 270:C57-C66

Jensen FB (1994) Regulatory volume decrease in carp red blood cells: mechanisms and oxygenation-dependency of volume-activated potassium and amino acid transport. J Exp Biol 198:155-165

Jones JS, Knauf PA (1985) Mechanism of the increase in cation permeability of human erythrocytes in low-chloride media. Involvement of the anion transport protein capnophorin. J Gen Physiol 86:721-738

Joyner SE, Kirk K (1994) Two pathways for choline transport in eel erythrocytes: a saturable carrier and a volume-activated channel. Am J Physiol 267:R773-R779

Kiessling K, Ellory JC, Cossins AR (2000) The relationship between hypotonically-induced taurine and K fluxes in trout red blood cells. Eur J Physiol 440:467-475

Kirk K (1997) Swelling-activated organic osmolyte channels. J Membrane Biol 158:1-16

Kirk K, Ellory JC, Young JD (1992) Transport of organic substrates via a volume-activated channel. J Biol Chem 267:23475-23478

Kregenow FM (1971) The response of duck erythrocytes to non hemolytic hypotonic media. Evidence for a volume controlling mechanism. J Gen Physiol 58:372-395

Lauf PK (1983) Evidence for a chloride-dependent potassium and water transport induced by hyposmotic stress in erythrocytes of the marine teleost, *Opsanus tao*. J Comp Physiol 146:9-16

Lewis RA, Bursell JDH, Kirk K (1996) Anion-selectivity of the swelling-activated osmolyte channel in eel erythrocytes. J Membrane Biol 149:103-111

Lohr JW, Grantham JJ (1986) Isovolumetric regulation of isolated S2 proximal tubules in anisotonic media. J Clin Invest 78:1165-1172

McManus TJ, Haas M, Starke LC, Lytle C (1985) The duck red cell model of volume-sensitive chloride dependent cation transport. Proc New York Acad Sci 456:183-186

Motais R, Guizouarn H, Garcia-Romeu F (1991) Red cell volume regulation: the pivotal role of ionic strength in controlling swelling-dependent transport systems. Biochim Biophys Acta 1075:169-180

Motais R, Fievet B, Borgese F, Garcia-Romeu F (1992) Some functional properties of band 3 protein in nucleated red cells. In: Bamberg E, Passow H (eds) Progress in cell research, vol 2. Elsevier, Amsterdam, pp 253-262

Motais R, Fievet B, Borgese F, Garcia-Romeu F (1997) Association of the band 3 protein with a volume-activated anion and amino acid channel: a molecular approach. J Exp Biol 200:361-367

Musch MW, Goldstein L (1996) High affinity binding of ankyrin induced by volume expansion in skate erythrocytes. J Biol Chem 271:21221-21225

Musch MW, Goldstein L (2001) Modulation of erythrocyte band 4.1 binding by volume expansion. J Exp Zool 289:177-183

Musch MW, Davis EM, Goldstein L (1994a) Oligomeric forms of skate erythrocyte band 3. Effect of volume expansion. J Biol Chem 269:19683-19686

Musch MW, Leffingwell TR, Goldstein L (1994b) Band 3 modulation and hypotonic stimulated taurine efflux in skate erythrocytes. Am J Physiol 255:R65-R74

Musch MW, Hubert EM, Goldstein L (1999) Volume expansion stimulates p52syk and p56lyn in skates erythrocytes. J Biol Chem 274:7923-7928

Parker JC, Dunham PB, Minton AP (1995) Effect of ionic strength on the regulation of Na/H exchange and K-Cl cotransport in dog red blood cells. J Gen Physiol 105:677-699

Passow H (1986) Molecular aspects of band 3 protein-mediated anion transport across the red blood cell membrane. Rev Physiol Biochem Pharmacol 103:62-223

Romano L, Passow H (1984) Characterization of anion transport system in trout red blood cell. Am J Physiol 246:C330-C338

Sarkadi B, Parker JC (1991) Activation of ion transport pathways by changes in cell volume. Biochim Biophys Acta 1071:407-427

Schwarz M, Gross A, Steinkamp T, Flugge UI, Wagner R (1994) Ion channel properties of the reconstituted chloroplast triose phosphate/phosphate translocator. J Biol Chem 269:29481-29489

Strange K, Jackson PS (1995) Swelling-activated organic osmolyte efflux: a new role for anion channels. Kidney Int 48:994-1003

Strange K, Emma F, Jackson PS (1996) Cellular and molecular physiology of volume sensitive anion channels. Am J Physiol 270:C711-C730

Thoroed SM, Fugelli K (1994) The Na-independent taurine influx in flounder erythrocytes and its association with the volume regulatory taurine efflux. J Exp Biol 186:245-268

Van Driessche P, De Smet JL, Allen S, Zizi M, Mountian I (1997) Isovolumetric regulation in a distal nephron cell line, A6. Am J Physiol 272:C1890-C1898

Wheeler KP, Christensen HN (1967) Interdependent fluxes of amino acids and sodium ion in the pigeon red cell. J Biol Chem 242:3782-3788

Wittels KA, Hubert EM, Musch MK, Goldstein L (2000) Osmolyte channel regulation by ionic strength in skate RBC. Am J Physiol 279:R69-R76

Wolowyk MW, Fincham DA, Young JD (1989) The effects of furosemide, piretanide and MK-196 on volume sensitive solute transport in fish erythrocytes. Proc Western Pharmacol Soc 32:309-311

Wright EM, Loo DF, Turk E, Hirayama BA (1996) Sodium cotransporters. Curr Opin Cell Biol 8:468-473

8 Na^+-K^+-$2Cl^-$ Cotransport

Christian Lytle

Division of Biomedical Sciences, University of California, Riverside, Riverside, California, 92521, USA

8.1 Introduction

Since Na^+-K^+-$2Cl^-$ cotransport (NKCC) was first described almost three decades ago (Wiley and Cooper 1974; Geck et al. 1980), numerous studies have contributed to an impressive understanding of this complex transport protein. In recent years, significant advances have been made in discerning the structure, regulation, and physiological functions of electroneutral salt transporters. Here we review salient features of NKCC and then consider its possible roles in red cells. True to their rich heritage as simpler model systems, red cells have proven ideal for exploring the kinetic characteristics and signaling pathways that control this complex transporter. And yet the precise function of NKCC in red cells remains far from certain.

Na^+-K^+-$2Cl^-$ cotransporters are members of an evolutionarily ancient family of electroneutral ion transport proteins that also includes four K^+-Cl^- cotransporter (KCC) isoforms and one Na^+-Cl^- cotransporter (NCC). In mammals, there are two functionally distinct NKCC proteins encoded by separate genes (Payne and Forbush 1995; Mount et al. 1998). The NKCC1 (*Slc12a2*) isoform is expressed in most cells where it participates in the homeostatic control of cell volume and $[Cl^-]$, commonly in coordination with its kinsman KCC. This primal function is enormously amplified in many Cl^--secreting epithelial cells where basolateral NKCC1 units maintain cytosolic $[Cl^-]$ above electrochemical equilibrium in opposition to regulated apical Cl^- efflux (Epstein and Silva 1985; O'Grady et al. 1987; Lytle et al. 1995). Given its ubiquitous distribution, it is probable, though not yet proven, that NKCC1 is the form found in erythroid cells. The second isoform, NKCC2 (*Slc12a1),* is expressed only in the kidney where it mediates apical salt uptake in the cells lining the thick ascending limb and macula densa (Gamba et al. 1994; Payne and Forbush 1994). In the mammalian nephron, there are three distinct variants of NKCC2 produced by alternative splicing of mutually exclusive cassette exons (Payne and Forbush 1994; Igarashi et al. 1995). The spatially segregated expression of these variants in the thick ascending limb has been proposed to account for axial heterogeneity in electrolyte reabsorption along the loop of Henle.

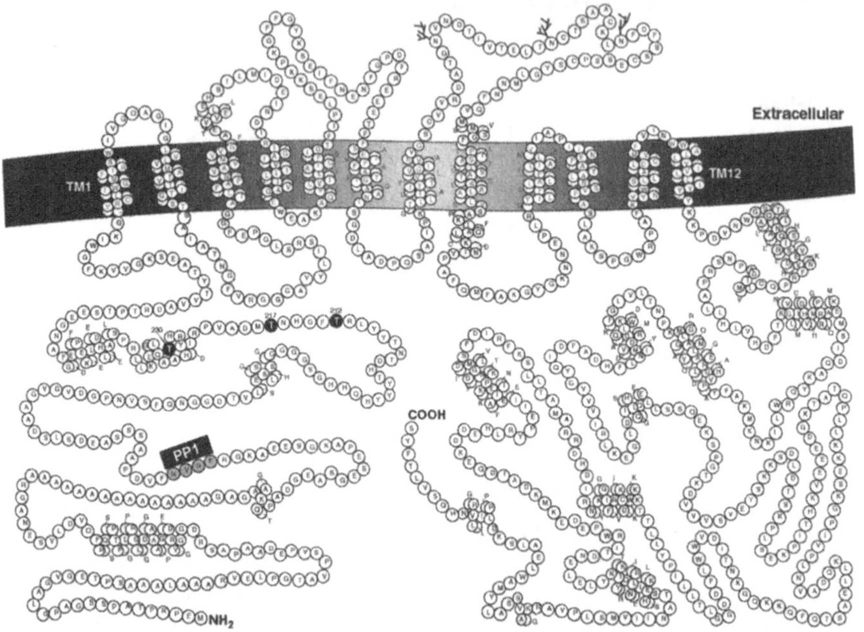

Fig. 8.1. Proposed model of NKCC1 based on hydropathy, membrane insertion, and secondary structure (Payne and Forbush 1995). Human NKCC1 is comprised of 1212 amino acids forming 12 transmembrane domains (*TM1–TM12*), a cytoplasmic amino domain harboring regulatory phosphothreonines (T^{212}, T^{217}, T^{230}) and a docking site for deactivating phosphatase (*PP1*), and a cytoplasmic carboxy domain. Probable extracellular sites of heavy glycosylation are indicated by short branched lines

8.2 The NKCC1 Protein

A general model of mammalian NKCC1 predicated on hydropathy (Payne et al. 1995), membrane insertion (Gerelsaikhan and Turner 2000), and secondary structure is illustrated in Fig. 8.1. The core protein is comprised of ~1200 amino acid residues that form 12 membrane-spanning α-helix units flanking large cytoplasmic amino and carboxy termini. The protein is glycosylated at one or more asparagine residues situated on the extracellular loop between transmembrane segments 7 and 8. The structure and/or amount of attached oligosaccharide varies greatly between different cells, and this heterogeneity is presumed to account for large differences in the mobility and breadth of NKCC homologs on SDS gels before, but not after enzymatic *N*-deglycosylation with peptide-*N*-glycosidase F (Lytle et al. 1995). The importance of glycosylation status on the function and life cycle of NKCC remains uncertain. The membrane-spanning and carboxy domains are well conserved across the phylogenetic spectrum. In contrast, much of the amino domain varies considerably between species, although it contains certain highly conserved sectors that appear to constitute part of the molecular switch that regulates ion translocation (as discussed below). Studies using chimeras and point mutations (Isenring and Forbush 1997; Isenring et al. 1998a,b) have established that the structural determinants of substrate ion binding are contained entirely within the membrane spanning segments. Residues within 3 of the 12 membrane segments exert a particularly strong influence on the transporter's macroscopic affinity for ion cargo (K_m). Thus, the K_m values observed for Na^+ and K^+ depend on structural features within TM2, those for K^+ and Cl^- on TM4, and those for Na^+, K^+ and Cl^- on TM7. Whether these features influence the intrinsic affinity of the ion binding sites or the relative abundance of transporter intermediates available to load a particular ion is uncertain.

NKCC appears to exist in the plasma membrane as a mixed complex of itself and regulatory elements. Evidence that NKCC assembles into stable homodimers has come from chemical crosslinking studies (Moore-Hoon and Turner 2000). Because mutants missing the amino domain likewise dimerize, the determinants for self-assembly must reside within the membrane domain, the carboxy domain, or both. The curious propensity for the carboxy domain to precipitate out of solution in pure form (Lytle, unpublished observations) suggests that it contains the dimerization interface. The existence of docking sites for another NKCC unit or accessory proteins (involved in transport, targeting, or regulation) might explain why numerous regions of the carboxy domain are highly conserved between species (Delpire et al. 1996). The extent to which the individual units of the dimer function independently or cooperatively remains uncertain. Insight should come, once the oligomerization interface is identified, from analyses of expressed NKCC mutants.

There is compelling evidence that a type-1 protein phosphatase (PP1) physically associates with the amino domain of NKCC and modulates cotransport activity (Darman et al. 2001). The association of PP1 with a conserved domain of NKCC (human R^{140}VNFVD) brings the phosphatase into proximity with at least one threonine residue whose phosphorylation is required for ion translocation (human T^{217}). Mutations in this domain that strengthen or weaken PP1 binding dramatically accelerate or retard cotransporter dephosphorylation and shift the relationship between NKCC activity and cytosolic $[Cl]_i$, as discussed below. An intriguing possibility is that the amino domain of NKCC functions as a scaffold for other regulatory elements. Indeed, a protein kinase that phosphorylates the regulatory domain of NKCC *in vitro* can be detected in cytosol-free membranes (Lytle, unpublished observations).

In nucleated red cells, NKCC might also associate with elements of the cortical cytoskeleton. Whereas NKCC is readily solubilized from most mammalian cells by nondenaturing detergents like Triton X-100 at normal ionic strength (Moore-Hoon and Turner 2000), it remains tightly associated with the detergent-stripped duck red cell unless the concentration of salt (KCl) is raised above ~700 mM (Lytle, unpublished observations). Whether there are similar differences in the physical organization of NKCC in the membrane of mammalian red cells remains unexplored.

8.3 NKCC1 Antibodies

The set of tools aiding research on NKCC continues to expand and improve. Among the first selective probes were monoclonal antibodies (J3, J4, J7, J25) raised against protein isolated from the shark rectal gland (Lytle et al. 1992). The epitopes of J3 and J4 are situated within the amino domain near the cluster of regulatory threonines (T^{184}, T^{189}, T^{202} in shark NKCC) and the phosphatase binding site (Darman et al. 2001), whereas J7 recognizes the carboxy domain between residues 1000–1100 (Lytle et al. 1992). Since these J-series antibodies do not recognize mammalian or avian NKCC, there use is limited to studies on shark NKCC. Another monoclonal antibody with a restrictive species preference is Q3, which recognizes only avian NKCC (Torchia et al. 1992). Monoclonal antibodies generated against the carboxy domain of human NKCC have proven useful for detecting and isolating the mammalian cotransporter (Lytle et al. 1995). This panel includes T4, a widely available[1] and highly versatile antibody that recognizes NKCC in animals species from insects to man. In addition, several excellent polyclonal antibodies have been raised against the C-terminus (designated T84 and TEF-S) and the N-terminus (dubbed NT and anti-P-NKCC). Interestingly, all but a few (e.g., J4) of the NKCC antibodies developed thus far greatly favor the chemically- or thermally-denatured form of the protein. Presumably their epitopes are concealed within the native NKCC dimer or blocked either by adjoining regulatory subunits or attached oligosaccharide. Thus, with the notable exception of the J series for shark NKCC, antibodies have been of limited use for studies of the native NKCC complex. A promising alternative approach is to capture native NKCC through its association with type-1 phosphatase, using antibodies against the PP1 catalytic unit or the tightly-binding PP1 inhibitor microcystin (Darman et al. 2001). The advent of new antibodies should simplify biochemical investigations on the cotransport protein, its regulation, and its network of protein interactions. One remarkable new probe is anti-P-NKCC, a polyclonal antibody that selectively recognizes the functional (phosphorylated) form of NKCC (Flemmer and Forbush 1999; Darman et al. 2001). The antibody was raised against a portion of the N-terminal regulatory domain containing phosphorylated T^{212} and T^{217}. In immunoblot assays, antibody binding exhibits a 25-fold preference for the phosphorylated form of NKCC and is linearly correlated with ^{32}P-NKCC over a broad scale.

8.4 The NKCC1 Reaction Cycle

Each complete cycle of cotransport requires all three ion species (Na^+, K^+, and Cl^-) on at least one side of the membrane (Schmidt and McManus 1977a; Geck et al. 1980; Haas et al. 1982). The transporter accepts Li^+ in place of Na^+ but with lower affinity and slower translocation (Lytle et al. 1998). In place of K^+ it accepts Tl^+,

[1] Monoclonal antibody T4 is available to the research community through the Developmental Studies Hybridoma Bank (dshb@uiowa.edu).

Rb^+, NH_4^+, and Cs^+ to decreasing degrees (Schmidt and McManus 1977a; Kinne et al. 1985; Geck and Heinz 1986; Hegde and Palfrey 1992). The only transportable surrogate for Cl^- is Br^-, at least in red cells, although there is evidence that NO_3^- can be secreted via NKCC by the shark rectal gland (Silva and Epstein 1993).

NKCC activity is most simply measured as the decrement of cation influx on adding 10 μM bumetanide or on replacing Cl^-. Typically, the inward transport of K^+ is monitored as the initial linear flux of its congener Rb^+, in either the chemical ($^{85}Rb^+$) or the radioactive ($^{86}Rb^+$) form, in the presence of ouabain to eliminate parallel uptake through the Na^+/K^+ pump. In those cells or experimental conditions in which NKCC dominates passive cation movement, it is convenient to monitor cotransport in both directions by measuring the bumetanide-sensitive loss of K^+ and Na^+ and gain of Rb^+ and Li^+ (Lytle et al. 1998). Transported cations may be varied or removed from the extracellular medium by substitution with choline$^+$, *N*-methyl-D-glucamine$^+$, or tetramethylammonium$^+$ with similar results. Anions such as NO_3^-, I^-, SCN^-, methanesulfonate (MSA^-), and sulfamate (SFA^-) are commonly used as replacements for Cl^-. While not carried by the cotransporter, each of these anions rapidly equilibrates across the red cell membrane via the anion exchanger AE1. However, anions of the lyotropic series (NO_3^-, I^-, and SCN^-) should be avoided in studies of volume-responsive or pH-sensitive transporters, such as NKCC, due to their perturbing effects on protein charge, conformation, and aggregation. In red cells, for example, substitution of Cl^- with NO_3^-, I^-, or SCN^- markedly increases the net negative charge on intracellular protein, causing a loss of permeant anions, a decrease in cell volume, and an acidic shift in cytosolic pH (Payne et al. 1989). Besides hemoglobin, these anions appear to interact with the cotransport protein itself (Turner and George 1988; Hegde and Palfrey 1992) and with intracellular elements that regulate its activity (Lytle, unpublished observations). Use of either MSA^- or SFA^- to replace Cl^- circumvents many if not most of these confounding effects.

Several indirect measures NKCC activity have proven useful in experimental situations where measurements of ion transport are not feasible. One approach is based on evidence that 5-sulfamoylbenzoate loop diuretics bind avidly only to NKCC units that are active and laden with Na^+, K^+, and a single Cl^- ion (Palfrey et al. 1980; Forbush and Palfrey 1983; Haas and Forbush 1986). As such, both the rate and the extent of ion-dependent [^{3}H]bumetanide binding to isolated membranes (Pewitt et al. 1990b) or to living cells (Forbush et al. 1992; Lytle and Forbush 1992, 1996) can serve as sensitive indicator of *transporting* NKCC units. A second approach has been to isolate the core (denatured) cotransport protein and measure its relative degree of phosphorylation. This is most simply accomplished by immunoprecipitating P-NKCC from detergent extracts of ^{32}P-labeled cells. Monoclonal antibody T4 (Lytle et al. 1995) has been commonly used for this purpose, and under denaturing conditions it is capable of capturing P-NKCC with quantitative efficiency (Lytle 1997). When the cotransport protein is isolated from ^{32}P-labeled duck red cells (Lytle 1997) or ferret red cells (Mairbäurl and Lytle 1999), its ^{32}P content closely parallels cotransport activity whatever the nature of the stimulus. The new anti-P-NKCC antibody, which recognizes only the activated form of the cotransporter (Darman et al. 2001), should facilitate concomitant determinations of NKCC abundance and regulatory status (by immunoblot assay or

ELISA), and possibly even spatial distribution (by immunocytochemistry). It must be noted that the validity of all three indirect assays hinges on the premise that NKCC units acquire the ability to transport ions and to bind bumetanide only after they are phosphorylated. This presumption seems correct, at least in simpler cells such as the duck red cell where a strict parallelism between cotransport activity, [^{3}H]bumetanide binding, and fractional NKCC phosphorylation can be demonstrated. In more complex cells, however, an increasing body of evidence suggests that the functional state or cell surface density of NKCC units depends on factors other than phosphorylation (see review by Russell (2000)).

Much of the passive cation traffic across the red cell membrane represents Na^+-K^+-$2Cl^-$ cotransport. Notable exceptions are found in the mature red cells of rabbits, various ungulates (sheep, horse, pig), salamanders (*Amphiuma*), fish (teleosts, elasmobranchs), and certain members of order carnivora (dogs, cats) in which NKCC activity is either negligible or absent. Among other animal red cells, NKCC activity, like overall cation transport capacity, differs widely. The low end of the activity spectrum includes the red cells of many mammals (human, rat, guinea pig), where bumetanide-sensitive K^+ (or $^{86}Rb^+$) influx is slower than ~2 mmol/(l_{cells}·h). Within the human population is found a pronounced variability in NKCC activity, typically ~3-fold (Duhm 1987a), that probably reflects both genetic and environmental factors; this heterogeneity has prompting many investigators to seek correlations between NKCC activity and familial disorders such as essential hypertension (Parker and Berkowitz 1986), though no singular association has emerged (Tepper et al. 1998). Thirty-fold higher rates of NKCC are found in the red cells of the ferret (Flatman 1983, 1989; Hall and Ellory 1985; Mairbäurl and Herth 1996; Flatman and Creanor 1999a). Even higher rates of NKCC are observed in the red cells of certain birds. For example, when duck red cells are stimulated by β-adrenergic catecholamines, cell shrinkage, or hypoxia, the movement of K^+ through the cotransport system can approach 140 mmol/(l_{cells}·h) (Kregenow 1973; Schmidt and McManus 1977b; Haas et al. 1982; Lytle 1998).

The inventory of ion cargo has been established in only a few model cells. The most direct method for defining the stoichiometry of net cotransport is to measure the effluxes of co-ions into media lacking them, or the loss or gain of cell ions in circumstances where the combined ion gradients are unbalanced. In cases in which net ion movements can be accurately determined, such as the Ehrlich ascites tumor cell (Geck et al. 1980), the duck red cell (Haas et al. 1982; Haas and McManus 1985; Lytle et al. 1998), and the ferret red cell (Lytle 1988; Flatman 1989) the stoichiometric ratio of net ion flows is uniformly $1Na^+{:}1K^+{:}2Cl^-$. The only certain exception is the squid axon where the stoichiometry is $2Na^+{:}1K^+{:}3Cl^-$ (Russell 2000). Careful determinations of cell ion and water contents have indicated that the transported fluid is isosmotic with the cell and its environment (Geck et al. 1980; Lytle 1988). Hence, each cycle of quaternary cotransport drives the movement of ~700 molecules of water by osmosis. While this osmotic flow is presumed to occur through aquaporin channels, some water may instead be cotransported along with the ions, as recently proposed for other cotransporters (Zeuthen 1994; Meinild et al. 1998). In human red cells, a ratio of $1Na^+{:}1K^+$ has likewise been reported for net mutually-interdependent cation uptakes (Kracke et al. 1988), as well as for net furosemide-sensitive cation losses (Garay et al. 1981).

A coupling stoichiometry of $1Na^+$:$1K^+$:$2Cl^-$ is consistent with the observed hyperbolic dependence of inward cotransport on external Na^+ and K^+ concentrations and sigmoidal dependence on Cl^-, and with evidence that net cotransport neither affects (Haas et al. 1982) nor is affected by (McManus et al. 1985) the membrane potential. The invariant stoichiometry and electrically neutral character of net cotransport make it evident that translocation of partially loaded or electrically unbalanced intermediate forms is negligible. Minimal "slippage" allows for both efficient ion gradient coupling and electroneutral salt transport.

Although the stoichiometry of net transport is fixed, the ratio of unidirectional fluxes in red cells varies from $2K^+$:$1Na^+$ to $1K^+$:$5Na^+$ depending on whether K^+ or Na^+, respectively, is the predominant cytosolic cation (Duhm 1987a,b; McManus 1987; Lytle 1988; Flatman 1989; Lytle et al. 1998). The excess unidirectional flux of one cotransported cation over the other reveals the presence of a 1:1 K^+/K^+ exchange mode prominent in normal high-K^+ cells and Na^+/Na^+ exchange mode prominent in high-Na^+ cells (Duhm 1987b; McManus 1987; Lytle et al. 1998). As both exchange modes are obligate (unproductive), they contribute to unidirectional, but not net, ion flux. In duck, human, and ferret red cells, the K^+/K^+ and Na^+/Na^+ exchanges are clearly responsible for the observed variation in unidirectional flux stoichiometries.

Several lines of evidence indicate that each cation self-exchange represents a different segment (partial reaction) of the cotransport cycle that occurs when a fully loaded carrier releases part of its cargo, rebinds fresh ions, and returns back across the membrane fully loaded. Thus a full turn of the cycle does not occur, and all that is observed is an unproductive exchange of one internal cation for one external cation. Net cotransport, by definition, occurs when the carrier unloads completely and returns empty. The two exchange modes require different sets of co-ions: K^+/K^+ exchange requires only K^+ and Cl^- outside but all three co-ions inside, whereas Na^+/Na^+ exchange requires all three co-ions outside but only Na^+ inside.

Study of these partial reactions suggested a conceptually simple model of the cotransport reaction (McManus 1987; Lytle et al. 1998). As illustrated in Fig. 8.2, the model advocates translocation of only fully-loaded and empty carriers to account for the invariant stoichiometry of net cotransport. For this scheme to work, the transition rate of the empty carrier (steps 10 and 1 in Fig. 8.2) must be slower than fully loaded one (steps 5 and 6). Thus, the return of the empty form must be a kinetic barrier or rate-limiting step in the cycle. Otherwise, the cation exchange partial reactions would not be apparent and the mechanism would have only one mode of operation, net salt transport. Another aspect of the cotransport reaction that allows for a more simple model is that the ions load and unload in rigid succession. This is made evident by the absolute dependence of each exchange mode on co-ions that are not themselves being exchanged. It is possible to deduce the entire sequence of binding and release from the asymmetrical ion requirements of the exchange partial reactions. For the inward journey, the ions bind on the outside in the sequence Na^+, Cl^-, K^+, Cl^- (steps 2–5). On the inside, the ions are released in the same order (steps 6–9). This first-on, first-off behavior has been called glide symmetry (Stein 1986). In this scheme, K^+/K^+ exchange is represented as step 4 through step 8 and back. The natural predominance of this mode in cells can be explained by the tendency of high intracellular [K^+] to hinder, by the principle of mass action, release of the second Cl$^-$. Na$^+$/Na$^+$ exchange, depicted in Fig. 8.2 as step 2 to step 6 and back, would become prominent in cells rendered high in Na$^+$ since in this circumstance mass action would discourage release of this ion. This explains why Na$^+$/Na$^+$ exchange is the predominant mode in the ferret red cell, which is naturally rich in Na$^+$ and poor in K$^+$.

Although K$^+$/K$^+$ exchange is quite conspicuous in red cells, this mode of NKCC has been difficult to demonstrate in other types of cells. The simplest explanation is that red cells are unusually rich in Cl$^-$. This would promote the outward return of carriers in the exchange mode and make fewer carriers available for net cotransport. In other types of cells in which Cl$^-$ levels are much lower, the tendency would be for the carriers to complete the internal unloading process rather than participate in the exchange. Another possibility is that the transition rate of the empty carrier is faster in epithelial systems, owing to differences in membrane

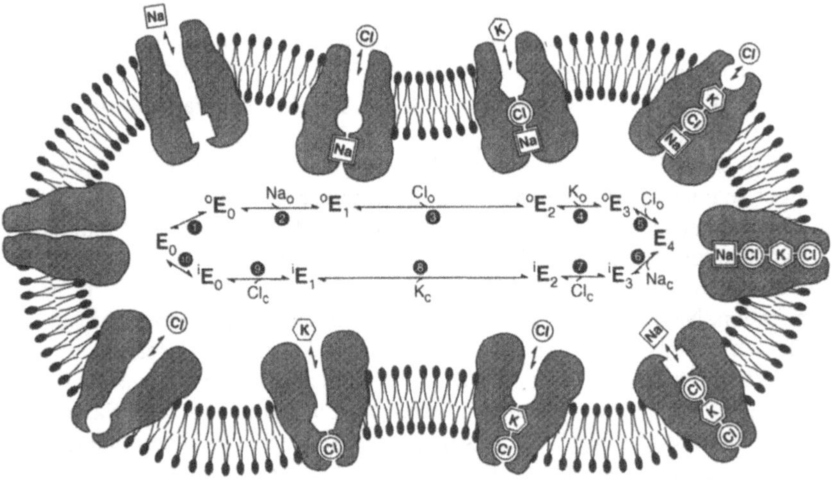

Fig. 8.2. A model of Na$^+$-K$^+$-2Cl$^-$ cotransport. Ten intermediate states of a membrane-spanning transporter are depicted. A specialized ion binding pocket is alternately exposed by a hinge mechanism to external (^oE) and internal (^iE) medium. Association of an ion with its site triggers a minor conformational change that occludes the bound ion and induces a new site to form with a selective affinity for the next ion in the sequence. All binding is assumed to be at equilibrium. When all sites are occupied (E_4), a major conformational change occurs, allowing bound ions to unload on the opposite side of membrane in the order they were bound (first-on, first-off). When carrier is completely unloaded (E_o), the empty form can also undergo a major conformational change, allowing it to begin reloading ions at the opposite side of the membrane, thus facilitating net cotransport. This latter change, however, occurs at a much slower rate than when the carrier is loaded. Partial unloading and reloading with return of the fully loaded carrier, although unproductive, can be detected by use of tracers or surrogate ions. Exchanges are a consequence of a kinetic barrier presented by rate-limiting movement of the empty carrier. The model explains the experimental findings for the net and unidirectional movement of ions. From Lytle et al. (1998), with permission of publisher

composition, NKCC structure, or accessory proteins. Whatever the reason, it is perhaps not surprising that the cotransport reaction is more fully optimized for net movement in salt-transporting epithelial cells (where its primary function is to take up Cl^- ions destined for secretion) than in red cells (where NKCC normally operates at or near thermodynamic equilibrium).

The ordered binding model has been corroborated by kinetic modeling (Miyamoto et al. 1986; Benjamin and Johnson 1997). In fact, a series of differential equations explicitly defining the reaction cycle depicted in Fig. 8.2 could accurately simulate the behavior of NKCC in various cells under different experimental conditions (Benjamin and Johnson 1997). While the details of the cotransport reaction remain far from certain, this simple model continues to offer a conceptual framework for further study.

8.5 Breadth of Physiological Functions

Members of the cation-chloride cotransporter family appear to have evolved from a single ancestral gene already present in primitive eukaryotes (Kaplan et al. 1996). Nature presumably conceived these transporters as mechanisms for moving salt into or out of the cell without perturbing the transmembrane electrical potential. Evidence that NKCC and KCC are reciprocally regulated by a negative feedback mechanism that is modulated by cytosolic volume and Cl^- concentration (as discussed below) suggests that they evolved as part of a primeval homeostatic mechanism for regulating intracellular volume, macromolecular crowding, and ionic strength.

This basic mechanism is amplified in various cells to perform a broad repertoire of specialized functions. Ion uptake via NKCC maintains an outwardly-directed electrochemical gradient for Cl^- that is utilized for various purposes, i.e., in salt-transporting epithelial cells to drive Cl^- efflux through apical membrane channels to accomplish Cl^- secretion (Silva et al. 1977; Haas and Forbush 2000), or in sensory neurons to provide a depolarizing electrical battery for modulating synaptic potentials (Alvarez-Leefmans 1997).

The processes by which animal cells preserve their volume are perhaps the most fundamental of life. Water occupies ~3/4 of the cellular interior and therefore determines cell volume. Cells are encapsulated by a membrane that is, with few exceptions, freely permeable to water and too fragile to resist hydrostatic pressure gradients. Hence, alterations in external or internal solute create osmotic gradients that draw water to or from the cytosol. Although the composition and osmolarity of the extracellular fluid is carefully regulated in mammals, numerous pathological conditions are associated with substantial alterations in plasma osmolarity (McManus et al. 1995; O'Neill 1999). A more common threat to cell volume is from changes in the cellular content of osmolytes as a result of imbalances in uptake and release of solutes, or of normal fluctuations in intracellular metabolism (Sachs 1996; Lang et al. 1998). Increases and decreases in cell hydration compromise all cellular functions because the attendant dilution and concentration of cytosolic solutes affect the interdependent activities of macromolecules, small solutes, and solvent water. Indeed, alterations in the degree to which enzymes and substrates are crowded and confined can have profound effects on the rate and equilibria of their reactions (Minton 1983; Burg 2000).

Over the course of evolution, cells have developed a variety of homeostatic mechanisms for regulating cell volume and the concentrations of intracellular protein and salt. Most cells respond within seconds to abrupt osmotic swelling or shrinkage by activating membrane transporters and metabolic enzymes that remove or add cytoplasmic solutes (Cossins 1991; Parker 1993; McManus et al. 1995; Sachs 1996; Okada 1997). Osmotic shrinkage typically evokes salt uptake via NKCC, Na^+/H^+ exchange (NHE), or Na^+ channels, whereas swelling promotes salt loss via KCC or separate K^+ and Cl^- channels (Sachs 1996; Lang et al. 1998).

8.6 Coordinate Regulation of Volume-Regulatory Transporters

Evidently there are mechanisms that coordinately regulate volume-responsive ion transporters so that they work together effectively. Various transporters are switched on or off at a characteristic cell volume termed the "set point" (Parker 1993; Parker et al. 1995). In many cells, the set points for shrinkage-activated and swelling-activated transporters are the same. For example, when a duck red cell is equilibrated in an isosmotic artificial medium, it adopts a relatively stable fluid volume at which both swelling-activated KCC and shrinkage-activated NKCC are nearly inactive (Haas and McManus 1985; Starke 1989). Osmotic shrinkage or swelling stimulates one cotransport process and further suppress the other, which brings about a flow of ions and water until the volume set point is reestablished. An important feature of this set point is that it is subject to modulation by a complex interplay of signals. In duck red cells, for example, β-adrenergic stimulation via intracellular cAMP (Starke 1989; Palfrey 1994), Mg^{2+} loading (Starke and McManus 1990), deoxygenation (Starke 1989), calyculin A and fluoride (Palfrey 1994) all reprogram the set point to higher volumes and thereby stimulate NKCC in normal cells and suppress KCC in swollen cells. By contrast, increases in cytosolic $[Cl^-]$, Mg^{2+} depletion (Starke and McManus 1990), *N*-ethylmaleimide and staurosporine lower the set point and thereby stimulate KCC in normal cells and inhibit NKCC in shrunken cells (Fig. 8.3).

The observation that various manipulations exert tandem effects on a common set point for swelling- and shrinkage-activated transporters suggests that the two transporters are regulated by the same coordinating mechanism. A similar mechanism appears to coordinate swelling-activated K^+-Cl^- cotransport and shrinkage-activated Na^+/H^+ exchange in dog red cells (Parker 1993) (Fig. 8.3).

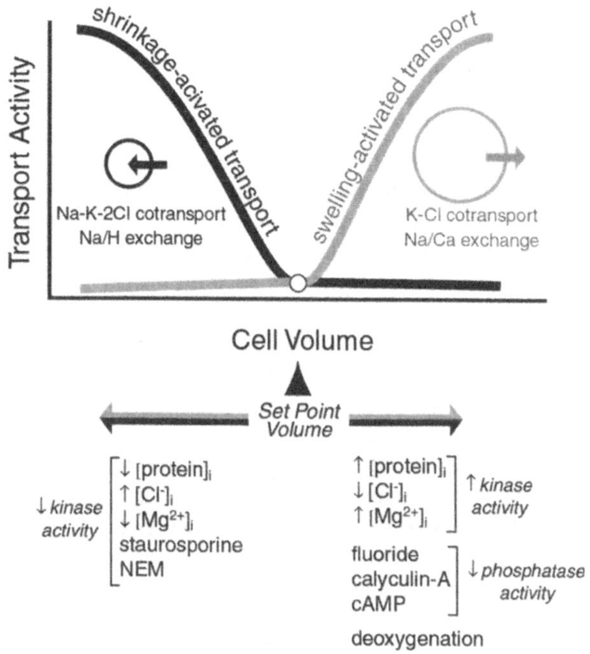

Fig. 8.3. Coordinate regulation of volume-responsive transporters in red cells. Signals of volume change, presumably generated by alterations in $[\text{protein}]_i$, are transmitted to shrinkage-activated NKCC or NHE and swelling-activated KCC or Na^+/Ca^{2+} exchanger by a common mechanism that involves protein phosphorylation. The set point volume at which these transporters are quiescent is modulated by numerous factors. In duck red cells, the protein kinase that directly phosphorylates NKCC is strongly stimulated by cell shrinkage, reduced $[Cl^-]_i$, and elevated $[Mg^{2+}]_i$, and it is inhibited by staurosporine and *N*-ethylmaleimide (*NEM*). The type-1 phosphatase that returns NKCC to its quiescent state is inhibited by fluoride, calyculin A, and elevated $[cAMP]_i$. Deoxygenation upshifts the set point by a mechanism unexplored

8.7 Cell Volume Detection

The mechanisms by which cells detect changes in their fluid volume are poorly understood. There are two schools of thought. One postulates that alterations in membrane or cytoskeletal configuration or tension give rise to regulatory signals (Sachs 1991; Watson 1991; Mills and Mandel 1994). Mechanical forces are known to affect the activity or assembly of several membrane systems, including ion channels (Sachs 1991) and signaling enzymes (Mills et al. 1990; Yoo et al. 1994), although their role in volume control remains unsettled. The other theory contends that the primary signal of volume change is generated by the concentration and dilution of cytosolic protein (Zimmerman and Harrison 1987; Parker 1993). The first evidence favoring this possibility came from ingenious experiments using dog red cells (Parker 1993). When red cell ghosts were resealed with elevated concentrations of protein, be it hemoglobin or albumin, the common set point for swelling-activated KCC and shrinkage-activated NHE was shifted to larger volumes, making the cell behave as if shrunken at normal cell volume (Colclasure and Parker 1991). Likewise, when barnacle muscle cells are perfused with an artificial cytosol containing subphysiological concentrations of protein, they behave as if swollen and mount a volume regulatory decrease, even if factors such as cell volume, ionic strength, membrane stretch, and osmolality are held constant (Summers et al. 1997). These experiments demonstrate that volume-regulatory transporters react to alterations in intracellular protein concentration rather than cell volume *per se*.

The basis for the protein effect appears to be macromolecular crowding (Parker and Colclasure 1992; Parker 1993; Summers et al. 1997; Burg 2000). Within cells, chemical reactions do not occur in dilute solution but in a cytoplasm that contains proteins and other macromolecules which collectively occupy a considerable fraction of the solvent volume (Minton 1983). Indeed, the modern view of the cytoplasm is one of a "densely entangled filament network interpenetrated by a fluid phase crowded with globular macromolecules" (Burg 2000). Because large enzymes and their protein substrates and products are effectively excluded from a sizable fraction of the cytoplasmic solvent volume by other ambient macromolecules, their thermodynamic activity can greatly exceed their simple molar concentration. As the effect depends exponentially on the concentration and size of both the crowded and crowding macromolecules (Minton 1983), subtle changes in cell water content could have powerful effects on protein-protein interactions and on enzyme kinetics and equilibria (Zimmerman and Minton 1993; Garner and Burg 1994). Evidence that cell volume perception depends on the concentration and size of cytosolic macromolecules, but not their chemical nature (Summers et al. 1997), provides strong support for the macromolecular crowding theory of cell volume detection. While not yet proven, the theory remains attractive and plausible. As Parker (1993) has argued, cells may be striving to regulate not volume but cytosolic nonideality, i.e., the physical-chemical conditions influencing macromolecular associations. The controlled formation of protein complexes is a central theme in cellular signal transduction.

Confirmation of the macromolecular crowding theory of cell volume detection will undoubtedly present many technical challenges. The first challenge will be to identify the critical components on the transduction pathway. This might necessitate an intact cell model in which the physical-chemical properties of cytosol are preserved. All previous attempts to delineate this pathway have employed conventional biochemical approaches without much success. Genetic approaches (e.g., yeast two-hybrid screening, mutational analysis) with intact cells could offer obvious practical advantages. Since the cell volume set point remains responsive to macromolecular crowding after intracellular perfusion of barnacle muscle cells (Summers et al. 1997) or after lysis of dog red cells (Colclasure and Parker 1992), all of the necessary signaling components would seem to be tethered to the cell membrane, perhaps in association with the transporter itself.

8.8 Regulation of NKCC

NKCC is dynamically regulated by a complex interplay of signals elicited by cell shrinkage, reduced $[Cl^-]_i$, hypoxia, and hormones (Russell 2000). These signals may blend together in combinatorial or synergistic ways, and the relative importance of one signal over another may vary among different cells (Haas 1994). Most of these signals seem to converge on elements that convey information on cell volume to shrinkage-activated and swelling-activated ion transporters via protein phosphorylation (Fig. 8.3). In avian red cells, four circumstances are known to raise the set point volume below which NKCC activates and above which KCC activates: reduced $[Cl^-]_i$, deoxygenation, elevated $[cAMP]_i$, and elevated $[Mg^{2+}]_i$ (Starke 1989; Starke and McManus 1990; Muzyamba et al. 1999; Lytle and McManus 2002); as a result, each circumstance evokes NKCC activity in normally hydrated cells. As in duck red cells, cell shrinkage and deoxygenation stimulate NKCC in ferret red cells (Mairbäurl and Lytle 1999). The arsenite anion is another powerful stimulus of NKCC in the red cells of ferrets (Flatman and Creanor 1999b), yet it has no effect on those of ducks. One unique feature of NKCC in the human red cell is that it is largely insensitive to osmotic shrinkage (Duhm 1987b).

NKCC itself is a phosphoprotein whose phosphorylation state parallels its activation state (reviewed by Russell 2000). Our studies on the duck red cell have indicated that NKCC becomes active when phosphorylated by a ser/thr protein kinase whose activity is dually responsive to cell shrinkage and reduced $[Cl^-]_i$ (Lytle 1996, 1998). This kinase can be inhibited by staurosporine and *N*-ethylmaleimide. The kinase is functionally opposed by a tonically active type-1 protein phosphatase (Pewitt et al. 1990a; Palfrey 1994) that is anchored directly to the transporter's amino domain (Darman et al. 2001). Our results suggest that this phosphatase is inhibited by cAMP-dependent protein kinase (PKA) via inhibitor-1. Thus, NKCC appears to be regulated *in vivo* through modulation of both the activating kinase (by macromolecular crowding, $[Cl^-]$, and possibly oxygen tension) and the deactivating phosphatase (by PKA via inhibitor-1).

An important part of the molecular switch that controls NKCC activity lies within a short sector of the amino domain. Studies on NKCC in the shark rectal gland, which responds to the same stimuli as its counterpart in the duck red cell, have shown that cotransport activity correlates with phosphorylation at T^{184}, T^{189}, and T^{202} (Darman and Forbush 2002). MALDI-TOF mass analysis has confirmed that these residues are phosphorylated when NKCC is stimulated with calyculin A (Behnke and Forbush 1998). Genetic mutation of T^{189} to either A or D renders NKCC refractory to all stimuli when expressed in HEK-293 cells, even though the mutant transporters are correctly processed to the cell surface and phosphorylated at all remaining sites. While phosphorylation of T^{189} is strictly required and possibly sufficient for ion translocation, additional phosphorylation of T^{184} and T^{202} is needed for optimal sensitivity and full activation. Mutagenesis of T^{184} and/or T^{202} resulted in transporters that were less active and less responsive to activation by intracellular Cl^- depletion (Darman and Forbush 2002). This cluster of threonine residues, which is situated within a region of the N-terminus that is highly conserved in all known NKCC forms, is proposed to form an important regulatory locus. The transport protein is also phosphorylated at serine residues when activated, yet none have been identified. Although NKCC harbors consensus targets for three known kinases (1 PKA site, 8 PKC sites, and 10 casein kinase II sites), there is no evidence that these sites are phosphorylated *in vivo* or that these kinases are required for transduction of the cell volume signal (Lytle 1997).

8.9 Thermodynamic Considerations

The direction and magnitude of net salt cotransport is determined by the vectorial sum of the chemical gradients of the three ions (Haas et al. 1982) as described by the thermodynamic expression:

$$\mu_{net} = \mu_{Na} + \mu_K + 2\mu_{Cl} = RT\ln\frac{[Na]_o \cdot [K]_o \cdot [Cl]_o^2}{[Na]_i \cdot [K]_i \cdot [Cl]_i^2} \qquad (8.1)$$

where the exponents of the concentration terms reflect the coupling ratio and R and T have their usual meanings. Because the cation product ($[Na^+]\cdot[K^+]$) on the outside and the inside are approx. equal in most cells, and because the $[Cl^-]$ term is squared, the main source of free energy available to drive net cotransport (μ_{net}) is the inwardly-directed Cl^- chemical gradient (Cl^- is the majority anion outside all cells whereas it is the minority anion inside most cells). Hence, in most cells, NKCC drives salt and water uptake.

The rare exception may be red cells, which are unusually rich in Cl^- (75–100 mM). This concentration reflects the passive distribution of small anions across the membrane at Donnan electrochemical equilibrium. In human (Duhm 1987a), duck (Haas and McManus 1985), and ferret (Lytle 1988) red cells under physiological conditions, the inward gradients of Na^+ and Cl^- nearly balance the outward gradient of K^+, leaving little if any net driving force to move salt through the system (Haas and McManus 1985).

8.10 Function of NKCC in Red Cells

The roles played by NKCC in red cell function remain uncertain. A fundamental uncertainty is whether salt cotransport is limited by thermodynamic constraints or by kinetic ones. Does the cotransporter operate more or less continuously to maintain an equilibrium at which the ion products ($[Na^+]\cdot[K^+]\cdot[Cl^-]^2$) on the outside and inside are balanced? Or does it operate slightly away from equilibrium and take up salt only when stimulated by cell shrinkage or Cl^- depletion?

Let us consider the first possibility. If we assume that plasma pH and hence the Cl^- distribution ratio remain reasonably constant, the direction of net salt cotransport should depend critically on just two main variables: internal $[Na^+]$ and external $[K^+]$ in human and duck red cells, or internal $[K^+]$ and external $[K^+]$ in ferret red cells. These thermodynamic relationships are graphically illustrated in Fig. 8.4. Over the physiological range of these two variables, shown as a shaded trape-

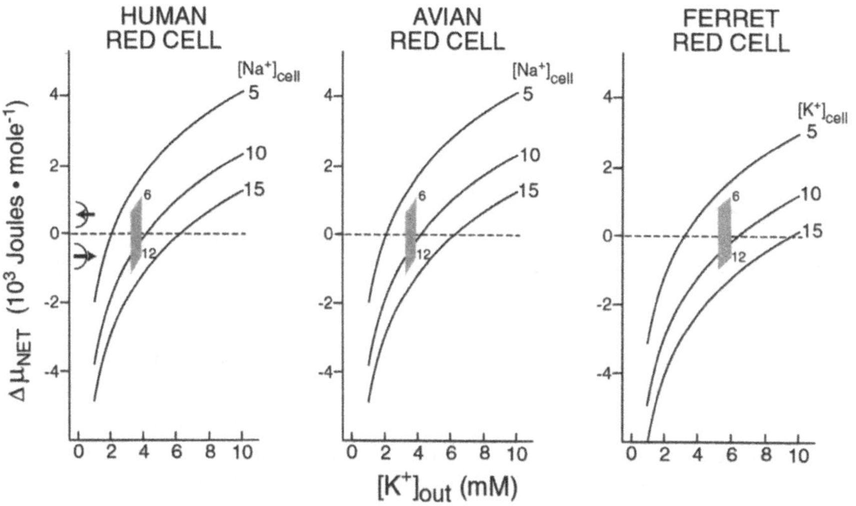

Fig. 8.4. Net driving force ($\Delta\mu_{net}$) on Na^+-K^+-$2Cl^-$ cotransport as a function of physiologically critical variables ($[K^+]_o$ and $[Na^+]_i$ in human and duck red cells or $[K^+]_o$ and $[K^+]_i$ in ferret red cells). $\Delta\mu_{net}$ was computed from ion concentrations measured in freshly drawn red cells and plasma from each species. Arrows denote that the theoretical driving force is inwardly directed when $\Delta\mu_{net}$ is positive and outwardly-directed when $\Delta\mu_{net}$ is negative. The calculations assume that the activity coefficients in the cytosol and plasma do not differ. The shaded trapezoid indicates the estimated working range of normal red cells. Note that in each red cell NKCC operates on the cusp of thermodynamic equilibrium. This delicate balance could be tipped by rather small changes in either $[K^+]_o$ or $[Na^+]_i$, resulting in net salt cotransport in a corrective direction

zoid, the predicted direction of the net driving force changes from inward to outward. The operation of NKCC on the cusp of thermodynamic equilibrium in red cells could enable it to buffer deviations in plasma $[K^+]$. Hyperkalemia should make inward cotransport energetically favorable and allow the circulating red cells to absorb and sequester any excess K^+. Conversely, red cells might donate K^+ to the plasma via outward cotransport during episodes of hypokalemia. Although not as large as skeletal muscle, the circulating red cell compartment could serve as a substantial store of K^+. For example, mobilization of just 25% of the red cell K^+ store could replace all K^+ ions in the extracellular water (~15% body weight). One exception would be the ferret red cell which contains only ~11 mM K^+; although unable to donate much K^+, the ferret red cell could conceivably store it during periods of hyperkalemia. The possibility that red cells help scavenge K^+ released from working muscle has been evaluated in humans (Maassen et al. 1998). During exhaustive handgrip exercise, antecubital venous plasma $[K^+]$ rises to ~8 mM, but almost none of this excess K^+ is taken up by red cells passing through working muscle, despite a favorable gradient for inward ion cotransport. Cation transport in the human red cell is far too slow. The role of NKCC in plasma K^+ buffering could be more important in birds, especially during the extremes of physical exertion and dehydration imposed by high altitude migratory flight (McManus and Schmidt

1978). This might explain why the capacity of the cotransport system is far greater in red cells of birds adept at long-distance migratory flight (e.g., duck) than in red cells of passerine birds (e.g., raven), and why it is almost undetectable in red cells of flightless birds (chicken, emu, rhea). The stresses of migratory flight would presumably trigger the release of catecholamines, which would stimulate NKCC activity and allow the circulating red cells to more rapidly buffer K^+ leaked from working skeletal muscles. Further stimulation of NKCC would occur as the red cells encounter low oxygen tension on traveling through active muscle beds (Muzyamba et al. 1999). Whether red cells actually help buffer plasma K^+, and whether this involves NKCC, remains to be determined.

One variable that affects $[Cl^-]$, and hence the driving force on NKCC, is plasma pH. The equilibrium distribution ratio of Cl^- in red cells is determined by the concentration of impermeant solutes (mainly protein and organic phosphates) and their net negative (Donnan) charge (Hladky and Rink 1977; Payne et al. 1989). Moderate acidemia would neutralize Donnan charge and elevate cell $[Cl^-]$ to levels that might energetically favor outward ion cotransport under normokalemic conditions, perhaps acting to offset the cell swelling caused by acidemia.

An alternative possibility is that the red cell cotransport system operates sufficiently away from thermodynamic equilibrium to move salt when biochemically activated. Work on mammalian red cells has suggested that the net driving force on NKCC switches from inward to outward as reticulocytes mature (Mairbäurl et al. 2000) and that continuous outward cotransport exerts a dehydrating influence on the mature circulating red cell (Duhm and Göbel 1984; Mairbäurl and Hoffman 1992). Thus, NKCC might contribute to the decreases in cell water and volume associated with reticulocyte maturation.

In the duck red cell, by contrast, a small driving force appears to be available *in vivo* to support inward cotransport (Fig. 8.4). This would make it possible for NKCC, acting in concert with KCC, to precisely regulate the salt content of the red cell, and therefore its fluid volume. Two observations support this possibility. First, when duck red cells are incubated in their own plasma and an oxygen atmosphere, they slowly gain KCl and water when exposed to norepinephrine (which activates NKCC) but gradually lose KCl and water if exposed to 10 μM bumetanide (which inhibits NKCC). Second, when duck red cells are removed from the influence of catecholamines in their native plasma and equilibrated in an isosmotic artificial medium, they engage KCC and jettison salt until they adopt and maintain the set point volume, i.e., a water content at which both NKCC and KCC are essentially inactive (Haas and McManus 1985; Starke 1989).

An important feature of the set point volume, i.e., the water content at which NKCC activates and KCC inactivates, is its modulation by $[Cl^-]_i$. This effect is especially powerful in the range of $[Cl^-]$ ordinarily present in most excitable cells (8–20 mM), smooth muscle cells (34–51 mM), and epithelial cells (20–40 mM) (Chipperfield and Harper 2000). Hence, physiological fluctuations of $[Cl^-]_i$ could modulate the volume set point in these cells. The cell would perceive its volume as lower in circumstances where $[Cl^-]_i$ is reduced. What purpose might this serve? This modulation could enable the cell to gauge not only the severity of the volume perturbation but also its underlying cause, so as to implement a corrective strategy that restores not only water content but ionic balance (Strange 1994; O'Neill

1999). Thus, if the cell should shrink isosmotically as a result of KCl loss, the coincident decrease in $[Cl^-]_i$ would potentiate activation of NKCC for a response compensatory for both cell volume and $[Cl^-]_i$. If, on the other hand, the cell should shrink hypertonically as a result of water loss, the increase in $[Cl^-]_i$ would decrease the volume set point and thereby desensitize NKCC to the shrunken state, avoiding additional enrichment of $[Cl^-]_i$. As noted by O'Neill (1999), the tendency for increases in $[Cl^-]_i$ to make the cell perceive itself swollen could explain why many animal cells recover from shrinkage if its cause is Cl^- loss (isosmotic) but not water loss (hypertonic). By influencing cell volume perception, $[Cl^-]$ would also affect the response to swelling; if the mechanism of swelling is isosmotic (due to Cl^- uptake), the increase in $[Cl^-]_i$ would potentiate activation of KCC and compensatory salt loss. If, on the other hand, the mechanism of swelling is hyposmotic swelling (due to water uptake), the decrease in $[Cl^-]_i$ would desensitize KCC to the swollen state and avoid a further fall in $[Cl^-]_i$. It is now evident that this regulatory system amplifies the corrective NKCC response in circumstances where cell shrinkage is caused by Cl^- loss, and the corrective KCC response when cell swelling is caused by Cl^- gain. This would enable the cell to gauge the severity of the offending stress as well as to identify its nature. Thus, volume regulatory transporters appear to be designed with the dual purpose of stabilizing $[Cl^-]_i$ while defending against potentially deleterious increases or decreases in cytosolic ionic strength.

Chloride could play a special role in red cells, which are uniquely rich in Cl^- (~74 mM) and in protein (~8 mM hemoglobin). At this high concentration, the effect of Cl^- on the set point volume is near-maximal and saturating, at least in duck red cells (Lytle, unpublished results). Although its effect is clearly not modulatory in the red cell, high $[Cl^-]_i$ could act in a continuous manner to maintain the set point at a much lower volume. Perhaps this helps counteract the effect of highly concentrated cytosolic protein (hemoglobin) on the set point. If unopposed, inordinate macromolecular crowding might make the red cell seek a swollen state with adverse rheological consequences. Thus, high $[Cl^-]_i$ might create the permissive environment necessary for red cells to package exceptionally high concentrations of soluble protein (hemoglobin). This teleological rationalization assumes that volume-sensing mechanisms based on macromolecular crowding are shared by red cells and other cells alike, which may not be so.

Is the steady-state physiological volume of a cell *in vivo* actually determined by the set point volume at which shrinkage-activated transporters like NKCC and swelling-activated transporters like KCC are quiescent? Our studies of duck red cells support this possibility. When the set point volume is experimentally shifted to higher volumes (by reducing $[Cl^-]_i$) under conditions where the driving force on NKCC is slightly inward (as *in vivo*), duck red cells engage NKCC and gain salt until they adopt and maintain the artificially elevated set point volume (Lytle and McManus 2002). We conclude that duck red cells maintain their internal volume and $[Cl^-]$ at interdependent set points by coordinately regulating separate KCl-loading (NKCC) and KCl-extruding (KCC) transporters.

Acknowledgement: I am grateful to Thomas McManus, Mark Haas, John Payne, and Biff Forbush for insightful discussions and to the National Science Foundation (MCB-9904605) for supporting our red cell studies.

References

Alvarez-Leefmans FJ (1997) Chloride transport: osmotic balance and presynaptic inhibition. In: Rudomin P, Romo R, Mendell L (eds) Presynaptic inhibition and neural control. Oxford Univ Press, New York, pp 50-79

Behnke R, Forbush III B (1998) Stimulatory phosphorylation sites on the shark rectal gland Na-K-Cl cotransporter. FASEB J 12:A1013

Benjamin BA, Johnson EA (1997) A quantitative description of the Na-K-2Cl cotransporter and its conformity to experimental data. Am J Physiol 273:F473-F482

Burg MB (2000) Macromolecular crowding as a cell volume sensor. Cell Physiol Biochem 10:251-256

Chipperfield AR, Harper AA (2000) Chloride in smooth muscle. Prog Biophys Mol Biol 74:175-221

Colclasure GC, Parker JC (1991) Cytosolic protein concentration is the primary volume signal in dog red cells. J Gen Physiol 98:881-892

Colclasure GC, Parker JC (1992) Cytosolic protein concentration is the primary volume signal for swelling-induced [K-Cl] cotransport in dog red cells. J Gen Physiol 100:1-10

Cossins AR (1991) A sense of cell size. Nature 352:667-668

Darman RB, Forbush III B (2002) A regulatory locus of phosphorylation in the N terminus of the Na-K-Cl cotransporter, NKCC1. J Biol Chem 277:37542-37550

Darman RB, Flemmer A, Forbush III B (2001) Modulation of ion transport by direct targeting of protein phosphatase type 1 to the Na-K-Cl cotransporter. J Biol Chem 276:34359-34362

Delpire E, Kaplan MR, Plotkin MD, Hebert SC (1996) The Na-(K)-Cl cotransport family in the mammalian kidney: molecular identification and function(s). Nephrol Dial Transplant 11:1967-1973

Duhm J (1987a) Furosemide-sensitive K (Rb) transport in human erythrocytes: modes of operation, dependence on extracellular and intracellular Na, kinetics, pH dependency, and the effect of cell volume and N-ethylmaleimide. J Membrane Biol 98:15-32

Duhm J (1987b) Modes of furosemide-sensitive K (Rb) transport in human (and rat) erythrocytes: Effects of $[Na]_o$, $[Na]_c$, $[K]_c$, pH, cell volume, and NEM. Fed Proc 46:2383-2385

Duhm J, Göbel BO (1984) Role of the furosemide-sensitive Na^+/K^+ transport system in determining the steady-state Na^+ and K^+ content and volume of human erythrocytes *in vitro* and *in vivo*. J Membrane Biol 77:243-254

Epstein FH, Silva P (1985) Na-K-Cl cotransport in chloride-transporting epithelia. Ann New York Acad Sci 456:187-197

Flatman PW (1983) Sodium and potassium transport in ferret red cells. J Physiol 341:545-557

Flatman PW (1989) Stoichiometry of net sodium and potassium fluxes mediated by the Na-K-Cl cotransport system in ferret red cells. Quart J Exp Physiol 74:939-941

Flatman PW, Creanor J (1999a) Na^+ arsenite: A potent stimulant of Na^+-K^+-$2Cl^-$ cotransport in ferret red cells. J Physiol 518P:118P

Flatman PW, Creanor J (1999b) Stimulation of Na^+-K^+-$2Cl^-$ cotransport by arsenite in ferret erythrocytes. J Physiol 519:143-152

Flemmer AW, Forbush III B (1999) Changes in amount and phosphorylation state of the Na-K-Cl cotransporter (NKCC) in teleost gill during salt adaptation. FASEB J 13:A399

Forbush III B, Palfrey HC (1983) [^{3}H]Bumetanide binding to membranes isolated from dog kidney outer medulla. Relationship to the Na,K,Cl co-transport system. J Biol Chem 258:11787-11792

Forbush III B, Haas M, Lytle C (1992) Na-K-Cl cotransport in the shark rectal gland. I. Regulation in the intact perfused gland. Am J Physiol 262:C1000-C1008

Gamba G, Miyanoshita A, Lombardi M, Lytton J, Lee WS, Hediger MA, Hebert SC (1994) Molecular cloning, primary structure, and characterization of two members of the mammalian electroneutral sodium-(potassium)-chloride cotransporter family expressed in kidney. J Biol Chem 269:17713-17722

Garay RP, Adragna N, Canessa M, Tosteson D (1981) Outward sodium and potassium co-transport in human red cells. J Membrane Biol 62:169-174

Garner MM, Burg MB (1994) Macromolecular crowding and confinement in cells exposed to hypertonicity. Am J Physiol 266:C877-C892

Geck P, Heinz E (1986) The Na-K-2Cl cotransport system. J Membrane Biol 91:97-105

Geck P, Pietrzyk C, Burckhardt BC, Pfeiffer B, Heinz E (1980) Electrically silent co-transport of Na, K, and Cl in Ehrlich cells. Biochim Biophys Acta 600:432-447

Gerelsaikhan T, Turner RJ (2000) Transmembrane topology of the secretory Na^+-K^+-$2Cl^-$ cotransporter NKCC1 studied by *in vitro* translation. J Biol Chem 275:40471-40477

Haas M (1994) The Na-K-Cl cotransporters. Am J Physiol 267:C869-C885

Haas M, Forbush III B (1986) [^{3}H]bumetanide binding to duck red cells. Correlation with inhibition of (Na+K+2Cl) cotransport. J Biol Chem 261:8434-8441

Haas M, Forbush III B (1998) The Na-K-Cl cotransporters. J Bioenergetics Biomembranes 30:161-172

Haas M, Forbush III B (2000) The Na-K-Cl cotransporter of secretory epithelia. Annu Rev Physiol 62:515-534

Haas M, McManus TJ (1985) Effect of norepinephrine on swelling-induced potassium transport in duck red cells: evidence against a volume-regulatory decrease under physiological conditions. J Gen Physiol 85:649-667

Haas M, Schmidt III WF, McManus TJ (1982) Catecholamine-stimulated ion transport in duck red cells. Gradient effects in electrically neutral [Na-K-2Cl] cotransport. J Gen Physiol 80:125-147

Hall AC, Ellory JC (1985) Measurement and stoichiometry of bumetanide-sensitive (2Na:1K:3Cl) cotransport in ferret red cells. J Membrane Biol 85:205-213

Hegde RS, Palfrey HC (1992) Ionic effects on bumetanide binding to the activated Na/K/2Cl cotransporter: Selectivity and kinetic properties of ion binding sites. J Membrane Biol 126:27-37

Hladky SB, Rink TJ (1977) pH equilibrium across the red cell membrane. In: Ellory JC, Lew VL (eds) Membrane transport in red cells. Academic Press, London, pp 115-135

Igarashi P, Van den Heuvel GB, Payne JA, Forbush III B (1995) Cloning, embryonic expression, and alternative splicing of a murine kidney-specific Na-K-Cl cotransporter. Am J Physiol 269:F405-F418

Isenring P, Forbush III B (1997) Ion and bumetanide binding by the Na-K-Cl cotransporter: Importance of transmembrane domains. J Biol Chem 272:24556-24562

Isenring P, Jacoby SC, Chang J, Forbush III B (1998a) Mutagenic mapping of the Na-K-Cl cotransporter for domains involved in ion transport and bumetanide binding. J Gen Physiol 112:549-558

Isenring P, Jacoby SC, Forbush III B (1998b) The role of transmembrane domain 2 in cation transport by the Na-K-Cl cotransporter. Proc Natl Acad Sci USA 95:7179-7184

Kaplan MR, Mount DB, Delpire E, Gamba G, Hebert SC (1996) Molecular mechanisms of NaCl cotransport. Annu Rev Physiol 58:649-668

Kinne R, Koenig B, Hannafin J, Kinne-Saffran E, Scott DM, Zierold K (1985) The use of membrane vesicles to study the NaCl/KCl cotransporter involved in active transepithelial chloride transport. Pflügers Arch 405, suppl 1:S101-S105

Kracke GR, Anatra MA, Dunham PB (1988) Asymmetry of Na-K-Cl cotransport in human erythrocytes. Am J Physiol 254:C243-C250

Kregenow FM (1973) The response of duck erythrocytes to norepinephrine and an elevated extracellular potassium. J Gen Physiol 61:509-527

Lang F, Busch GL, Ritter M, Völkl H, Waldegger S, Gulbins E, Häussinger D (1998) Functional significance of cell volume regulatory mechanisms. Physiol Rev 78:247-306

Lytle C (1988) Na-K-2Cl and K-Cl cotransport in duck red cells. Ph.D. thesis, Duke University, Durham, NC

Lytle C (1997) Activation of the avian erythrocyte Na-K-Cl cotransport protein by cell shrinkage, cAMP, fluoride, and calyculin-A involves phosphorylation at common sites. J Biol Chem 272:15069-15077

Lytle C (1998) A volume-sensitive protein kinase regulates the Na-K-2Cl cotransporter in duck red blood cells. Am J Physiol 274:C1002-C1010

Lytle C, Forbush III B (1992) Na-K-Cl cotransport in the shark rectal gland. II. Regulation in isolated tubules. Am J Physiol 262:C1009-C1017

Lytle C, Forbush III B (1996) Regulatory phosphorylation of the secretory Na-K-Cl cotransporter: modulation by cytoplasmic Cl. Am J Physiol 270:C437-C448

Lytle C, McManus TJ (2002) Coordinate modulation of Na-K-2Cl cotransport and K-Cl cotransport by cell volume and chloride. Am J Physiol 283:C1422-C1431

Lytle C, Xu J-C, Biemesderfer D, Haas M, Forbush III B (1992) The Na-K-Cl cotransport protein of shark rectal gland. I. Development of monoclonal antibodies, immunoaffinity purification, and partial biochemical characterization. J Biol Chem 267:25428-25437

Lytle C, Xu J-C, Biemesderfer D, Forbush III B (1995) Distribution and diversity of Na-K-Cl cotransport proteins: A study with monoclonal antibodies. Am J Physiol 269:C1496-C1505

Lytle C, McManus TJ, Haas M (1998) A model of Na-K-2Cl cotransport based on ordered ion binding and glide symmetry. Am J Physiol 274:C299-C309

Maassen N, Foerster M, Mairbäurl H (1998) Red blood cells do not contribute to removal of K^+ released from exhaustively working forearm muscle. J Appl Physiol 85:326-332

Mairbäurl H, Herth C (1996) Na^+-K^+-$2Cl^-$ cotransport, Na^+/H^+ exchange, and cell volume in ferret erythrocytes. Am J Physiol 271:C1603-C1611

Mairbäurl H, Hoffman JF (1992) Internal magnesium, 2,3-diphosphoglycerate, and the regulation of the steady-state volume of human red blood cells by the Na/K/2Cl cotransport system. J Gen Physiol 99:721-746

Mairbäurl H, Lytle C (1999) Shrinkage and deoxygenation stimulate NKCC independent of Mgi in ferret red blood cells. FASEB J 13:A716

McManus TJ (1987) Na,K,2Cl cotransport: kinetics and mechanism. Fed Proc 46:2378-2381

McManus TJ, Schmidt III WF (1978) Ion and co-ion transport in avian red cells. In: Hoffman JF (ed) Membrane transport processes, vol 1. Raven Press, New York, pp 79-106

McManus TJ, Haas M, Starke LC, Lytle C (1985) The duck red cell model of volume-sensitive chloride-dependent cation transport. Ann New York Acad Sci 456:183-186

McManus ML, Churchwell KB, Strange K (1995) Regulation of cell volume in health and disease. N Engl J Med 333:1260-1266

Meinild AK, Klaerke DA, Loo DDF, Wright EM, Zeuthen T (1998) The human Na^+-glucose cotransporter is a molecular water pump. J Physiol 50:15-21

Mills JW, Mandel LJ (1994) Cytoskeletal regulation of membrane transport events. FASEB J 8:1161-1165

Mills A, Leitsou G, Rabban J, Sumpio B, Ginretz H (1990) Mechanosensitive adenylate cyclase activity in coronary vascular smooth muscle cells. Biochem Biophys Res Commun 171:143-147

Minton AP (1983) The effect of volume occupancy upon the thermodynamic activity of proteins: some biochemical consequences. Mol Cell Biochem 55:119-140

Miyamoto H, Ikehara T, Yamaguchi H, Hosokawa K, Yonezu T, Masuya T (1986) Kinetic mechanism of Na^+, K^+, Cl^- -cotransport as studied by Rb^+ influx into HeLa cells: Effect of extracellular monovalent ions. J Membrane Biol 92:135-150

Moore-Hoon ML, Turner RJ (2000) The structural unit of the secretory Na^+-K^+-$2Cl^-$ co-transporter (NKCCl) is a homodimer. Biochemistry 39:3718-3724

Mount DB, Delpire E, Gamba G, Hall AE, Poch E, Hoover RS, Heber SC (1998) The electroneutral cation-chloride cotransporters. J Exp Biol 201:2091-2102

Muzyamba MC, Cossins AR, Gibson JS (1999) Regulation of Na^+-K^+-$2Cl^-$ cotransport in turkey red cells: The role of oxygen tension and protein phosphorylation. J Physiol 517:421-429

O'Grady SM, Palfrey HC, Field M (1987) Characteristics and functions of Na-K-Cl co-transport in epithelial tissues. Am J Physiol 253:C177-C192

O'Neill WC (1999) Physiological significance of volume-regulatory transporters. Am J Physiol 276:C995-C1011

Okada Y (1997) Volume expansion-sensing outward-rectifier Cl^- channel: fresh start to the molecular identity and volume sensor. Am J Physiol 273:C755-C789

Palfrey HC (1994) Protein phosphorylation control in the activity of volume-sensitive transport systems. In: Strange K (ed) Cellular and molecular physiology of cell volume regulation. CRC Press, Boca Raton, pp 201-214

Palfrey HC, Feit PW, Greengard P (1980) cAMP-stimulated cation cotransport in avian erythrocytes: inhibition by "loop" diuretics. Am J Physiol 238:C139-C148

Parker JC (1993) In defense of cell volume? Am J Physiol 265:C1191-C1200

Parker JC, Berkowitz LR (1986) Genetic variants affecting the structure and function of the human red cell membrane. In: Andreoli TE, Hoffman JF, Fanestil DD (eds) Physiology of membrane disorders. Plenum Press, New York, pp 785-813

Parker JC, Colclasure GC (1992) Macromolecular crowding and volume perception in dog red cells. Mol Cell Biochem 114:9-11

Parker JC, Dunham PB, Minton AP (1995) Effects of ionic strength on the regulation of Na/H exchange and K-Cl cotransport in dog red blood cells. J Gen Physiol 105:677-699

Payne JA, Forbush III B (1994) Alternatively spliced isoforms of the putative renal Na-K-Cl cotransporter are differentially distributed within the rabbit kidney. Proc Natl Acad Sci USA 91:4544-4548

Payne JA, Forbush III B (1995) Molecular characterization of the epithelial Na-K-Cl cotransporter isoforms. Curr Opin Cell Biol 7:493-503

Payne JA, Lytle C, McManus TJ (1989) Osmotic and ionic equilibria in human red cells: effects of foreign anions. Am J Physiol 259:C819-C827

Payne JA, Xu J-C, Haas M, Lytle C, Ward D, Forbush III B (1995) Primary structure, functional expression, and chromosomal localization of the bumetanide-sensitive Na-K-Cl cotransporter in human colon. J Biol Chem 270:17977-17985

Pewitt EB, Hegde RS, Haas M, Palfrey HC (1990a) The regulation of Na/K/2Cl cotransport and bumetanide binding in avian erythrocytes by protein phosphorylation and dephosphorylation. Effect of kinase inhibitors and okadaic acid. J Biol Chem 265:20747-20756

Pewitt EB, Hegde RS, Palfrey HC (1990b) [^{3}H]Bumetanide binding to avian erythrocyte membranes. Correlation with activation and deactivation of Na/K/2Cl cotransport. J Biol Chem 265:14364-14370

Russell JM (2000) Sodium-potassium-chloride cotransport. Physiol Rev 80:211-276

Sachs F (1991) Mechanical transduction by membrane ion channels: a mini review. Mol Cell Biochem 104:57-60

Sachs JR (1996) Cell volume regulation. In: Schultz SG, Andreoli TE, Brown A, Fambrough DM, Hoffman JF, Welsh MJ (eds) Molecular biology of membrane transport disorders. Plenum Press, New York, pp 379-406

Schmidt III WF, McManus TJ (1977a) Ouabain-insensitive salt and water movement in duck red cells. I. Kinetics of cation transport under hypertonic conditions. J Gen Physiol 70:59-79

Schmidt III WF, McManus TJ (1977b) Ouabain-insensitive salt and water movements in duck red cells. II. Norepinephrine stimulation of sodium plus potassium cotransport. J Gen Physiol 70:81-97

Silva P, Epstein FH (1993) Secretion of nitrate by rectal gland of *Squalus acanthias*. Comp Biochem Physiol A 104:255-259

Silva P, Stoff J, Field M, Fine L, Forrest JN Jr, Epstein FH (1977) Mechanism of active chloride secretion by shark rectal gland: role of Na-K-ATPase in chloride transport. Am J Physiol 233:F298-F306

Starke LC (1989) Regulation of [Na-K-2Cl] cotransport and [K-Cl] cotransport in duck red cells. Ph.D. thesis, Duke University, Durham, NC

Starke LC, McManus TJ (1990) Intracellular free magnesium determines the volume regulatory set point in duck red cells. FASEB J 4:A818

Stein WD (1986) Transport and Diffusion Across Cell Membranes. Academic Press, Orlando

Strange K (1994) Are all cell volume changes the same? News Physiol Sci 9:223-228

Summers JC, Trais L, Lajvardi R, Hergan D, Buechler R, Chang H, Peña-Rasgado C, Rasgado-Flores H (1997) Role of concentration and size of intracellular macromolecules in cell volume regulation. Am J Physiol 273:C360-C370

Tepper T, Sluiter WJ, Huisman RM, De Zeeuw D (1998) Erythrocyte Na^+/Li^+ countertransport and Na^+-K^+-$2Cl^-$ co-transport measurement in essential hypertension: Useful diagnostic tools or failure? A meta-analysis of 17 years of literature. Clin Sci 95:649-657

Torchia J, Lytle C, Pon DJ, Forbush III B, Sen AK (1992) The Na-K-Cl cotransporter of duck salt gland: Phosphorylation in response to cAMP-dependent and calcium-dependent secretagoues. J Biol Chem 267:25444-25450

Turner RJ, George JN (1988) Ionic dependence of bumetanide binding to the rabbit parotid Na/K/Cl cotransporter. J Membrane Biol 102:71-77

Watson PA (1991) Function follows form: generation of intracellular signals by cell deformation. FASEB J 5:2013-2019

Wiley JS, Cooper RA (1974) A furosemide-sensitive cotransport of sodium plus potassium in the human red cell. J Clin Invest 53:745-755

Yoo J, Ellis R, Morgan KG, Hai C-M (1994) Mechanosensitive modulation of myosin phosphorylation and phosphatidylinositol turnover in smooth muscle. Am J Physiol 267:C1657-C1665

Zeuthen T (1994) Cotransport of K^+, Cl^- and H_2O by membrane proteins from choroid plexus epithelium of *Necturus maculosus*. J Physiol 478:203-219

Zimmerman SB, Harrison B (1987) Macromolecular crowding increased binding of DNA polymerase to DNA: an adaptive effect. Proc Natl Acad Sci USA 84:1871-1875

Zimmerman SB, Minton AP (1993) Macromolecular crowding: Biochemical, biophysical, and physiological consequences. Annu Rev Biophys Biomol Struct 22:27-65

9 K^+-Cl^- Cotransport in Vertebrate Red Cells

John S. Gibson[1] and J. Clive Ellory[2]

[1]Department of Clinical Veterinary Medicine, University of Cambridge, Madingley Road, Cambridge CB3 0ES, and [2]Laboratory of Physiology, Oxford University, Parks Road, Oxford OX1 3PT, UK

9.1 Introduction

9.1.1 Historical Perspective

The "pump-leak" model of Tosteson and Hoffman (1960) presented an elegant and simple paradigm for regulation of red blood cell (RBC) cation content and volume. It emphasised the importance of a specific, ouabain-sensitive, Na^+/K^+ pump, driven by ATP, as the primary active transport process for maintaining high intracellular K^+, balancing K^+ loss through passive processes. The passive pathways were not defined. For many years, they were thought of as non-specific "leaks" by which ions could travel down their electrochemical gradients across the cell membrane. A second route for mediated Na^+ efflux persisted, however, in cells treated with ouabain, and thereby lacking functional Na^+/K^+ pump activity (Hoffman and Kregenow 1966). In these ouabain-treated cells, movement of K^+ ions was not proportional to the electrochemical gradient for K^+ unless Cl^- was removed, implying coupled ion movements (Funder and Wieth 1967). The second Na^+ "pump" was found to require an outward K^+ gradient (Sachs 1971). It became apparent that Na^+ and K^+ fluxes were dependent on each other (Wiley and Cooper 1974), hence the notion of Na^+-K^+ cotransport. Subsequently, their Cl^--dependence raised the possibility of Na^+-K^+-Cl^- cotransport (Schmidt and McManus 1977; Dunham and Ellory 1980). Later a second passive pathway for K^+ transport, requiring Cl^- but not Na^+, was observed (Dunham and Ellory 1981). These passive pathways were saturable, ion-specific, exhibited *trans*-stimulation, and could be inhibited by furosemide and other compounds – all characteristics of mediated transport. Clearly, the "leaks" were also specific pathways. For some time, it was questioned whether Cl^--dependent K^+ cotransport via Na^+-dependent and Na^+-independent pathways represented alternative transport modes of the same carrier. It was even suggested that both might be functions of the Na^+/K^+ pump, notwithstanding their persistence in the presence of ouabain. Such alterations in the transport mode of the pump are not without precedent, for example palytoxin converts the Na^+/K^+ pump into an ouabain-sensitive ion channel (Tosteson et al. 1991). It is now known, however, that K^+-Cl^- cotransport (here ascribed KCC) is a discrete transporter, and one of several passive or secondary active pathways in the RBC. In this chapter, we discuss further its identity and characteristics in normal vertebrate RBCs, see Chap. 23 for its behaviour in sickle cells.

9.1.2 The LK Sheep Red Cell

Although Kregenow had shown that K^+ and Cl^- efflux occurred in swollen duck RBCs (Kregenow 1971, and later), RBC KCC was first characterised in detail in sheep (Ellory and Dunham 1980; Lauf and Theg 1980; Dunham and Ellory 1981). The RBC K^+ content of sheep is under genetic control (Kerr 1937; Evans 1954), being either high K^+-containing (HK; like RBCs from most other vertebrate species) or low K^+-containing (LK). HK RBCs have a high capacity for Na^+/K^+ pumping and low passive K^+ leak; LK cells have opposite "pump-leak" characteristics with a low pump capacity and high K^+ leak (Tosteson and Hoffman 1960). This dimorphism is inherited in a Mendelian fashion, with a single gene of two alleles, *l* and *m*, encoding for the proteins (Evans and King 1955; Rasmusen and Hall 1966). Mature RBCs from homozygote *ll* and heterozgyote *lm* sheep are LK, those from homozygote *mm* are HK. Interestingly, all immature and precursor RBCs, regardless of the genotype, are HK, with those destined to become LK changing their transport phenotype as they mature. Immunological studies have identified two antigens, L and M, associated with LK and HK cell types, and the anti-L antibody(ies) (raised against LK RBCs inoculated into HK sheep) both stimulated the variant Na^+/K^+ pump in LK cells and inhibited KCC fluxes (Ellory and Tucker 1969; Dunham 1976; Lauf et al. 1977; Smalley et al. 1982). The passive K^+ pathway has since been shown to be the specific Cl^--dependent K^+ pathway, ascribed to the K^+-Cl^- cotransporter. Cl^--dependent K^+ fluxes, consistent with K^+-Cl^- cotransport, have been described subsequently in RBCs from many vertebrate species. The influence of the LK sheep model remains large and much of our understanding of KCC in RBCs comes from work on it. It is frustrating that, notwithstanding the pre-eminence of sheep RBCs in the history of K^+-Cl^- cotransport, the identification of the genes that control the LK/HK dimorphism, the nature of L and M antigens and the mechanism of action of the anti-L antibodies, remain elusive. In addition, species differences in the behaviour of RBC KCC do occur, and the reader should beware of extrapolating from one species to another.

9.1.3 Ionic Requirements and Transporter Kinetics

Like most cation transporters, KCC displays considerable ion selectivity (see Lauf et al. (1992) for references). Rb^+ is able to substitute for K^+, but Na^+ and choline are poorly transported. Br^- is preferred over Cl^-, but many other univalent anions (e.g. I^-, NO_3^-, SCN^-) do not support K^+ transport. Some replacement anions can cause reduction in the isoelectric point of intracellular charges (which reside mainly on Hb), hence an increase in nondiffusible intracellular negative charges, cell shrinkage and intracellular acidification (Payne et al. 1990). These actions often follow the Hofmeister (or lyotropic) series, possibly via protein destabilisation (or the chaotropic effect). On the other hand, impermeable replacements for Cl^- (e.g. glucuronate, sucrose) cause depolarisation and alkalinisation of the cell interior, as well as shrinkage (e.g. Bernhardt et al. 1991; Ellory et al. 1991). Some anions appear to alter the volume set point (Lauf 1991) and there are also reports that replacement of Cl^- with NO_3^- causes ATP reduction, at least in sheep RBCs (Flatman et al. 1996). The best Cl^- substitutes are probably the organic anions, methylsulphate and methylsulphonate (Payne et al. 1990). From the above, however, it will be appreciated that ion replacements are not always neutral, and care must be taken to ensure that the secondary effects of Cl^- substitution *per se* do not alter KCC function indirectly. Unfortunately, in the absence of high affinity, specific inhibitors for KCC – the best being furosemide and bumetanide analogues like H25 (Ellory et al. 1990), or Cl^- channel blockers like 5-nitro-2-(3-phenylpropylamino)-benzoic acid (NPPB) (Cabantchik and Greger 1992) and [(dihydroindenyl)oxy]alkanoic acid (DIOA) (Garay et al. 1989; Vitoux et al. 1989) – anion-substitution remains the method of choice for assessing transporter activity.

Whilst Cl^--dependent K^+ transport is often taken to indicate obligatory coupling of K^+ and Cl^- movements, this is not necessarily the case. Unequivocal linkage of movement is difficult to prove experimentally because of the high Cl^- conductance of the RBC membrane. Uphill K^+ movement, driven by a Cl^- gradient has been demonstrated in human and sheep RBCs under particular experimental conditions (Brugnara et al. 1989b). Consistent with cotransport is a report that K^+ transport via KCC is insensitive to changes in membrane potential, implying an electroneutral process (Kaji 1993). Recently, a 1:1 stoichiometry of K^+ and Cl^- fluxes has been observed in rabbit RBCs whose KCC was stimulated using *N*-ethylmaleimide (NEM) (Jennings and Adame 2001).

Kinetic analysis of the transporter in RBCs from different vertebrates has shown considerable variability in the apparent K_m's for K^+ and Cl^- at the external site (10–100 mM for K^+, typically around 30 mM; 15–84 mM for Cl^-: Lauf et al. 1992, Table 3 therein). It is clear, however, that KCC has a much lower affinity for K^+ than the Na^+/K^+ pump. Stoichiometry is probably 1:1 in RBCs (Kaji 1989; but see Dunham and Ellory 1981). Cotransport can occur in either direction across the membrane, with net movement depending on the chemical gradients for the two transported ions. There are reports that the kinetics of the transporter are asymmetric (which is not unusual for carrier-mediated transporters), with different orders of ion binding at each face (Delpire and Lauf 1991b) and possibly different K_m's and V_{max}'s for influx and efflux (Kaji 1989). Differences in ion selectivity or affinities between the various KCC isoforms (see next section) have not yet been addressed but may also be of considerable physiological importance.

9.1.4 Molecular Identity of the K^+-Cl^- Cotransporter

KCC represents a third subclass of the chloride-cation cotransporter (CCC) superfamily, after the Na^+-Cl^- cotransporter and the Na^+-K^+-$2Cl^-$ cotransporter (Mount et al. 1998). The first Cl^--dependent cotransporter identified molecularly was the thiazide-sensitive Na^+-Cl^- cotransporter (NCC; Gamba et al. 1993), followed soon after by the bumetanide-sensitive Na^+-K^+-$2Cl^-$ cotransporter, NKCC1 (Xu et al.

1994). Cloning was facilitated by the existence of specific or high affinity inhibitors (such as thiazide and bumetanide) and tissues particularly rich in the cotransporter under study (such as urinary bladder of the winter flounder for NCC and dogfish shark rectal gland for NKCC1). No similar advantages exist for KCC, which hindered progress. Finally, a protein mediating K^+-Cl^- cotransport was identified molecularly in the mid-1990s, cloned by homology screening of human ESTs (expressed sequence tags) using expected similarity to NKCC1 as a starting point (Gillen et al. 1996). The first isoform to be cloned, KCC1, has a gene encoding for a protein of 1,085 amino acids, with 12 transmembrane domains, cytoplasmic N- and C-termini, a large extracellular loop between the fifth and sixth transmembrane domains with potential glycosylation sites, and also consensus sites for phosphorylation. The resulting protein when expressed in HEK293 cells has a mass of about 120 kDa, increasing to 150 kDa when glycosylated. As expected from its successful cloning strategy, KCC1 has substantial sequence homology (about 25%) with NKCC1 and with NCC. At least four isoforms exist, KCC1–4. (Note: there is some confusion over the nomenclature: KCC4 was termed KCC3, and *vice versa*, by Mount et al. (1999); here we follow the terminology of Race et al. (1999)). KCC1, KCC3 and KCC4 are all widely expressed in mammalian tissues, though KCC4 is most abundant in kidney and heart (Mount et al. 1999). By contrast, KCC2 is confined to neuronal tissue (Payne et al. 1996; Payne 1997). The RBC isoform is probably KCC1 (Pellegrino et al. 1998), although KCC3 has also been reported in sheep RBCs (Lauf et al. 2001). Limited work has been carried out on the characteristics of these proteins using expression systems such as *Xenopus* oocytes and HEK293 cells. Thus it has been shown that KCC1 and KCC4 are highly volume-sensitive (KCC4 apparently more so than KCC1), and inhibited by furosemide (Gillen et al. 1996; Hiki et al. 1999; Mount et al. 1999), whereas KCC3 is only weakly stimulated by swelling (Race et al. 1999). KCC1 and KCC3 are stimulated by NEM (Gillen et al. 1996; Race et al. 1999), and KCC3 by staurosporine (Race et al. 1999). These isoforms therefore share some of the features of K^+ and Cl^- cotransport in RBCs. The neuronal isoform, KCC2, is volume-insensitive and expression in some neurons changes during development (Clayton et al. 1998; Miles 1999; Delpire 2000). Finally, in endothelial cells, mRNA levels for KCC3 are affected by cytokines, being up-regulated by vascular endothelial growth factor (VEGF) and down-regulated by tumour necrosis factor (TNF) (Hiki et al. 1999). Homologues to KCC have been found in *C. elegans* (Holtzman et al. 1998).

9.2 Physiological Characteristics

9.2.1 Volume Dependence

The RBC in general, and KCC in particular, has been much studied in the context of cellular volume regulation (e.g. Hoffmann and Dunham 1995). In many species, the RBC K^+-Cl^- cotransporter is highly volume-sensitive: thus a few percent increase in cell volume stimulates activity by several fold (Dunham and Ellory

1981). In RBCs from various species (rabbit, fish, horse, human), activation shows a delay before peak activity is achieved. This delay increases with swelling or protein phosphatase inhibition (Jennings and Al-Rohil 1990; Jennings and Schulz 1991), but is decreased or abolished by Mg^{2+} depletion or staurosporine (Dunham 1990 – reported in Lauf et al. 1992; Dunham et al. 1993; Bize and Dunham 1994). By contrast, decrease in KCC activity on shrinkage occurs with no lag phase. These initial observations have led to development of the concept of regulation of KCC activity by protein kinases and phosphatases, as discussed in detail later. Rate of swelling is also important, thus slow anisotonic swelling of trout RBCs increases the set point for activation of KCC and substantially reduces the final activity reached (Godart et al. 1999).

In human, horse, rabbit and other high K^+-containing RBCs, volume sensitivity has been associated with regulatory volume decrease (RVD): when activated, the transporter will mediate K^+ and Cl^- efflux, water will follow osmotically, thus causing shrinkage towards normal cell volume. In a similar way, KCC may be involved in the volume reduction, which occurs as precursor and young RBCs mature. Certainly, activity of the transporter is higher in young cells compared to mature cells (Hall and Ellory 1986b; Brugnara and Tosteson 1987a; Canessa et al. 1987a; Lauf and Bauer 1987). In fact, in the case of human RBCs, on maturation, the K^+-Cl^- cotransporter becomes largely quiescent and refractory to stimuli such as cell swelling. The transporter is still present, however, and can be activated pharmacologically (e.g. by NEM) or by high hydrostatic pressure, implying modification of some of its regulatory apparatus (see later) with cell age. In certain species, the transporter is active at normal cell volume in mature cells, being reduced by shrinkage and further activated on swelling (e.g. horse RBCs – Gibson et al. 1993); in others, it is quiescent without swelling or some other stimulus (e.g. human sickle cells). Clearly, where KCC represents the major passive transporter, maintenance of RBC cation content requires balanced ion movement via Na^+/K^+ pump and KCC. Abnormal or excessive activity of KCC is associated with loss of intracellular solute and reduction in cell volume below normal – which may contribute to the pathology of sickle cell disease (Stuart and Ellory 1988; Ellory et al. 1989b; and see Chap. 23).

Volume change can occur anisotonically, in response to alteration in medium tonicity, or isosmotically, with constant osmotic pressure. Anisotonicity is encountered physiologically, for example in the vasa recta of the renal medulla, and in intestinal and gastric capillaries during water or solute absorption, and also pathologically in a number of conditions. In some species, isosmotic swelling can also occur *in vivo*, e.g. in trout RBCs exposed to adrenaline, via NaCl uptake by beta-NHE and band 3 (Bourne and Cossins 1982; Borgese et al. 1987). *In vitro*, isosmotic volume change can also be induced by treating cells with compounds such as NH_4Cl (NH_3 enters the cell and is converted to NH_4Cl thereby increasing the intracellular solute content and causing swelling; on removal of extracellular NH_4Cl, the reverse occurs and cells shrink). There is evidence that in trout RBCs, isosmotic swelling preferentially activates KCC, while anisotonic swelling by reduction in ionic strength activates a Cl^--independent K^+ pathway (Motais et al. 1991; Guizouarn and Motais 1999). The latter pathway may also be permeable to certain organic osmolytes such as taurine. Whether or not these two pathways represent different transport modes of the same transporter has been much debated (Guizouarn et al. 1993; Berenbrink et al. 1997). Latterly, the role of organic osmolyte pathways in volume regulation of RBCs and other cell types has received considerable attention (Kirk et al. 1992; Kirk and Strange 1998). A teleological argument has been advanced that, when inorganic ions are relatively scarce (i.e. under hypotonic conditions), volume regulation is achieved through preferential loss of organic molecules like taurine (leaving via the Cl^--independent pathway). By contrast, when ionic strength remains high, organic molecules are conserved and regulation is achieved with inorganic ions leaving via KCC.

9.2.2 H^+ Dependence

A second important physiological variable to which KCC responds is $[H^+]$. It has been suggested that changes in pH represent a more important physiological regulator of KCC than volume (Ellory et al. 1989a). Thus *in vivo* the RBC is more likely to encounter pH fluctuations than anisotonicity. Typically, KCC activity displays a bell-shaped response to changes in $[H^+]$ over the physiological range, with peak activity at about pH 7 in many cases, and inhibition at higher or lower values (Brugnara et al. 1986; Zade-Oppen and Lauf 1990; Speake et al. 1997). Recently, however, we have found that the response to $[H^+]$ is more complicated, with considerable stimulation at very alkaline pH's (Drew et al. 2002). The phenomenon is probably due to changes in intracellular $[H^+]$, rather than extracellular (Ellory et al. 1991). These two are normally coupled via the Jacob-Stewart cycle (e.g. Hladky and Rink 1977), but can be dissociated using impermeant anions such as glucuronate which elevate intracellular pH (and depolarise and shrink the cell) at constant extracellular pH (Ellory et al. 1991). Part of the stimulation of KCC as pH is reduced could be due to the concomitant cell swelling. From pH 7.4 to 7.0, however, this is a relatively minor effect (~3% swelling), and further acid-induced swelling would not obviously produce a biphasic response, with inhibition at lower pH values. There is synergism between swelling and H^+ concentration, with reduction in pH sensitising the cell to the effect of swelling (Ellory et al. 1991; Speake et al. 1997). Mg^{2+}-depleted cells or cells treated with inhibitors of protein phosphorylation have an altered response to $[H^+]$, so that activity increases monophasically with pH (Brugnara and Tosteson 1987b; Lauf et al. 1994; Godart and Ellory 1996; Gibson, unpublished observations on horse RBCs). Interestingly, sickle cells, following reduction in cell [Hb] (using nystatin-induced swelling to dilute intracellular contents), and sickle or normal RBC ghosts (which have a very much reduced [Hb] cf. intact cells) also show this monophasic stimulation of KCC with increase in pH (Brugnara et al. 1989a; Vitoux et al. 1999). These observations may be pertinent to a possible role for macromolecular crowding in regulation of KCC (see Sect. 9.3.2). Finally, H^+ can interact with modulators of KCC in another way. Thus isomolar loss of KCl via KCC (or, for that matter, via K^+ and Cl^- channels) may lower intracellular pH. This is not intuitive, and was first proposed following analysis of RBC behaviour using the model of Lew et al. (1991), illustrating the value of this approach. Thus as Cl^- is only one of the complement of intracellular anions, isomolar loss of KCl will reduce $[Cl]_i$, Cl^- will then re-enter the cell via band 3 in exchange for HCO_3^-, hence lowering pH. There is the potential here for positive feedback and volume instability under certain conditions, as discussed by Lew et al. (1991) in relation to "fast-track" dehydration of reticulocytes in sickle cell patients.

9.2.3 Urea

Urea represents a third potential physiological stimulus for K^+-Cl^- cotransport (dog: Parker 1993; sheep: Dunham 1995; human: Kaji and Gasson 1995; horse: Speake and Gibson 1997). Concentrations of urea above about 100 mM stimulate KCC, with the peak response occurring at about 750–1000 mM. Around 500 mM urea is found in renal medulla of humans during antidiuresis (and higher concentrations in certain other mammals), and it can also accumulate in the circulation during renal failure; so this effect may be relevant *in vivo*. The mechanism of the urea effect has received some attention. Thus, urea stimulates following a delay whose magnitude increases with urea concentration (Dunham 1995), reminiscent of activation by swelling. At low concentrations, stimulation by urea is reversed by shrinkage. Urea may increase the stability of the water shell around proteins, and thereby mimic swelling (Dunham 1995). On the other hand, it is a chaotropic agent, considered to destabilise proteins. When used for osmoregulation in lower vertebrates, urea is usually found in combination with other, stabilising or counteracting osmolytes (mainly methylamines: Yancey 1994). At higher concentrations, urea produces a range of additional effects, which are not seen with simple swelling (Speake and Gibson 1997).

9.2.4 Oxygen

KCC in RBCs from many vertebrate species has been shown to respond to oxygen tension (Po_2) – reviewed recently in Gibson et al. (2000). In normal RBCs from many vertebrates, KCC activity is increased as Po_2 is raised. In several mammalian species, the transporter activity responds to physiological Po_2's with a sigmoidal relationship like that of O_2 saturation of Hb (horse: Speake et al. 1997; sheep: Campbell and Gibson 1998; human: Gibson et al. 1998). Thus KCC activity is maximal at arterial Po_2's (> 80 mmHg), half-maximal at the Po_2 of mixed venous blood (around 30–40 mmHg), and refractory at lower Po_2's (< 20 mmHg) found in the capillary beds of highly metabolically active tissues. In trout, however, it has recently been reported that KCC activity responds to changes in Po_2's up to very high values (> 700 mmHg), far in excess of those required to saturate Hb maximally (Berenbrink et al. 2000). These observations are relevant to the mechanism by which O_2 regulates KCC activity. O_2 also modifies the response to other stimuli. Thus when the transporter is inactivated at low Po_2's, it is unable to respond to cell volume or moderate changes in pH (Nielsen et al. 1992; Gibson et al. 1995; Cossins and Gibson 1997). The case of urea is slightly different. At low or moderate concentrations, the transporter is highly O_2-dependent; at high concentrations, however, the requirement for O_2 is progressively lost so that at 1 M urea, the transporter is O_2 insensitive (Speake and Gibson 1997). The history of exposure to O_2 is also important, with the transporter becoming inactivated by prolonged exposure to high Po_2's and requiring a spell at low Po_2's before it can become fully active again (Nielsen et al. 1992; Gibson, unpublished results). Clearly, O_2 can affect the response of KCC to other stimuli, but the rate of the response to O_2, and the effect of cyclical fluctuations in Po_2, such as encountered by the RBC as it circulates *in vivo*, is not known. Interestingly, KCC in sickle cells (containing Hb S) has an abnormal O_2 response – it is no longer inactivated at low Po_2's, rather activity in the complete absence of O_2 is not dissimilar (in fact, consistently about 10% greater) than that at high, arterial Po_2's (Gibson et al. 1998). This behaviour means that KCC efflux may contribute more to sickle cell dehydration, e.g. following stimulation by low pH in hypoxic regions of the circulation, i.e. active muscle beds, or by urea in the relatively hypoxic renal medulla (as discussed in Ellory et al. (1998)). Finally, it should be noted that other membrane transporters (e.g. amino acid transporters: Kiessling et al. 2000; anion transporter AE1: Teti et al. 2002) are also O_2 sensitive, providing evidence for a generalised role for O_2 tension in control of RBC membrane transport, whose importance is only beginning to emerge (e.g. Gibson et al. 2000).

9.2.5 Hydrostatic Pressure

Elevation of hydrostatic pressure also stimulates RBC KCC. The effect has been observed in human, sheep and horse RBCs (human: Hall et al. 1982; sheep: Ellory et al. 1985; horse: Gibson and Hall 1995). It is debatable whether this phenomenon is physiological, since certainly very high pressures (> 10 MPa) are required. High pressures may be found in certain parts of the body, however, during load bearing (i.e. 15 MPa in the vertebral column during postural changes), and will be experienced by certain aquatic vertebrates during deep dives. In addition, the effect has been used as a tool with which to study the mechanism of KCC activation. Stimulation by pressure is additive with that by Mg^{2+} depletion, and inhibited by serine-threonine phosphatase inhibitors, making it likely that again the effect is mediated via the regulatory phosphorylation cascade (Gibson and Hall 1995; Godart and Ellory 1996).

9.2.6 Temperature

Temperature has a marked stimulatory effect on KCC in RBCs from a number of species including human sickle cells (Ellory et al. 1985; Jennings and Al-Rohil 1990; Willis and Anderson 1998; Jennings 1999; Khan and Ellory 2000). It is important here to distinguish between effects of temperature on activation as opposed to the final steady state transport rates. Thus temperature has a much greater effect on swelling- or urea-activated KCC than on that stimulated by NEM (with estimates for activation energy of several fold higher: Lauf 1983; Jennings and Al-Rohil 1990; Khan and Ellory 2000). Its effect therefore appears to be upon the activation step, rather than on the transport process *per se* (Willis and Anderson

1998; Jennings 1999; Khan and Ellory 2000). KCC regulation occurs via a cascade of protein kinases and phosphatases (see Sect. 9.3.1) and it therefore appears that high temperature mediates its effect upon this signalling pathway, e.g. the activating protein phosphatase (Jennings 1999; Khan and Ellory 2000). It is interesting that highly temperature sensitive phosphatases have been reported in other tissues (Mitsui et al. 1994). Even in homeothermic animals, where core temperature is well regulated, small changes, such as 2–3 °C rises during heavy exercise or pyrexia, do occur; the peripheral circulation is also liable to larger temperature fluctuations. The high temperature-sensitivity of the transporter may be of particular significance in sickle cells, where small rises in temperature may elicit cell dehydration, potentially inciting a sickle cell crisis (Ellory et al. 1998).

9.2.7 Bicarbonate

Red blood cell KCC activity is lower when measured in autologous plasma compared to saline (Gibson et al. 1995; Godart et al. 1997). In a series of experiments designed to investigate this phenomenon, it was found that phosphate had a small inhibitory effect (at 5.6 mM, but not 1.1 mM). However, in sickle cells, the greatest inhibition was due to the presence of physiological levels of HCO_3^-/CO_2 buffer. This may be another difference between sickle cells and other RBCs because such inhibition was not present in KCC of horse RBCs incubated with similar levels of bicarbonate (Speake et al. 1997) and there are no other reports of bicarbonate-induced inhibition of KCC. On the other hand, the effect may be relevant to the anion-dependence observed in the response of KCC to volume and other stimuli. The effect may well reflect an anion requirement in the signalling pathway, e.g. bicarbonate may alter the activity of volume sensitive kinases/phosphatases.

9.3 Mechanism of Regulation

9.3.1 Protein Phosphorylation

A major advance in understanding regulation of the RBC KCC came in the early 1990s, when it was found that protein phosphatase inhibitors including orthovanadate, inorganic phosphate, fluoride, okadaic acid and calyculin A inhibit KCC (Jennings and Al-Rohil 1990; Jennings and Schulz 1991; Kaji and Tsukitani 1991; O'Neill 1991). Conversely, similar agents stimulate transporters mediating regulatory volume increase (RVI) like NKCC (see Chap. 8), another example of the reciprocal response of RVD and RVI transporters (see Parker 1994). These observations led to the suggestion that the transporters were modulated by protein phosphorylation, with KCC activated by dephosphorylation (Jennings and Al-Rohil 1990). Stimulation of KCC by swelling could therefore result from an increase in activity of the relevant protein phosphatase (PP), or decreased activity of a protein kinase (PK), or both. As the lag time to full activity increases with swelling, a kinetic argument was used to suggest that stimulation by swelling probably results from inhibition of a kinase, termed the "V"-kinase (Jennings and Al-Rohil 1990). The last decade has seen considerable development from these initial observations: on the identification of the enzymes and their targets; in the elaboration of the phosphorylation transduction pathway to include several pairs of kinase/phosphatase enzymes; and in the mechanism, which couples enzyme activity and the perturbing stimulus (volume, H^+, O_2, etc.).

The main evidence for the involvement of protein phosphorylation is pharmacological (at least for KCC – NKCC is certainly phosphorylated). Inhibitors of serine-threonine phosphorylation (e.g. NEM, staurosporine) stimulate KCC (Lauf 1983; Bize and Dunham 1994), whilst phosphatase inhibitors (like calyculin A and okadaic acid) inhibit (Jennings and Schulz 1991; Kaji and Tsukitani 1991; Starke and Jennings 1993). By contrast, tyrosine kinase inhibitors (like tyrphostin and genistein) decrease KCC activity (Cossins et al. 1994; Flatman et al. 1996); tyrosine phosphatase inhibitors (like vanadate) stimulate. Thus more than one pair of PP/PK enzymes appear to be involved, and of both the serine-threonine and tyrosine types. Combinations of inhibitors have been used to produce pharmacological maps, which are best developed in trout (Cossins et al. 1994) and LK sheep (Dunham et al. 1993; Lauf et al. 1994, 1995; Flatman et al. 1996). Although species differences may exist, the final step in activation of KCC is generally held to be dephosphorylation of serine-threonine residues.

Notwithstanding a role for phosphorylation, the identity of the enzymes involved is unknown, nor is it yet clear whether the transporter *per se* or some regulatory protein is the target for phosphorylation. A major experimental problem is the probable low copy number of the relevant proteins (both target and enzyme) in RBCs. The only molecular information available to date comes from experiments on RBCs of knockout mice lacking the tyrosine kinases *Fgr* and *Hck*. These RBCs show elevated activity of KCC, which remains sensitive to inhibition by okadaic acid but is no longer stimulated by staurosporine or NEM (De Franceschi et al. 1997). This would imply inhibitory tyrosine kinase activity, perhaps regulating the stimulatory protein phosphatase (De Franceschi et al. 1997). There is also some evidence concerning the intracellular location of the regulatory enzymes. Human RBC ghosts become insensitive to PP inhibitors like okadaic acid unless added at the beginning of resealing, suggesting the loss of the conjugate PK during the ghosting procedure (Sachs and Martin 1993, 1999). Stimulation of KCC activity in human ghosts by Mg^{2+}-ATP is abolished by genistein, and, in the absence of ATP, KCC is stimulated by vanadate (Sachs and Martin 1993) – cf. the results from knockout mice (De Franceschi et al. 1997). PP activity, particularly of type 1 but also type 2A, is associated with RBC membranes and has been correlated with KCC activity (Bize et al. 1999, 2000). In summary, the "V"-kinase is probably cytoplasmic, with the conjugate phosphatase probably membrane-bound – these target serine-threonine residues; in addition, there are also membrane-bound tyrosine kinases which stimulate KCC and inhibitory tyrosine phosphatases.

"Phosphorylation clamp" experiments, in which combinations of PK inhibitors (like NEM) followed by PP inhibitors (like calyculin A) are added sequentially to lock the phosphorylation state and transporter activity, are relevant here. Cells treated in this way have KCC activity intermediate between maximal and zero, but are unable to respond to such stimuli as changes in volume, O_2, urea or pressure

(Cossins et al. 1994; Gibson and Hall 1995; Honess et al. 1996; Cossins and Gibson 1997) and have an altered response to H^+. The experiments have been interpreted to indicate that such stimuli act via the phosphorylation cascade, rather than directly on the transporter.

Despite the remaining uncertainties, a major role for protein phosphorylation is envisaged for control of cation-chloride cotransporters. Nevertheless, it is unlikely that changes in protein phosphorylation represent the sole modulators of KCC activity.

9.3.2 Macromolecular Crowding, Ion Concentration or Mechanoreception?

For a transporter responding to any stimulus, the system requires a sensor and a transduction apparatus, to mediate appropriate changes in transport activity. There is no unifying hypothesis to explain the mechanism of volume-sensitive KCC, nevertheless, some important advances have been made. In the case of cell volume regulation, in the majority of situations (an exception being stretch-activated channels), it is not certain to what the cell is responding. It is unlikely to be volume *per se*. Rather, the cell may respond to change in concentration of an intracellular mediator – but obviously amplification must be very high and very sensitive (to a change in volume of only a few percent); to mechanical stimuli, such as membrane stretch or cytoskeletal distortion; or to a combination of these.

Modulation of the common intracellular signals, such as $[Ca^{2+}]_i$, does not seem to be involved directly. On the other hand, seminal experiments with dog RBC ghosts suggest that protein concentration is critical (Colclasure and Parker 1991, 1992). Red blood cells, of course, contain high concentrations of haemoglobin (~5 mM in human). It has been suggested that these affect the regulatory apparatus, possibly affecting the thermodynamic activity of relevant protein kinases and phosphatases – a process termed macromolecular crowding (Minton 1983, 1994; Minton et al. 1992). The identity of the crowding protein did not seem to matter, since albumin was able to replace haemoglobin in these experiments (Colclasure and Parker 1992). Later, Vitoux et al. (1999) using human RBC ghosts found that KCC in white ghosts lacking Hb was unable to respond to volume or pH, whilst re-addition of Hb restored volume dependency, but not the pH response. The Hb isoform was important, thus re-addition of Hb S always produced higher fluxes than Hb A, regardless of the original ghost genotype. It has been suggested that urea mimics the effect of swelling on macromolecular crowding (perhaps by stabilising the surrounding water shell?) and thereby inhibiting the "V"-kinase (Dunham 1995). If urea did stimulate the same pathway as swelling, it would explain why the delay required for full KCC activity increased with urea concentration, as observed for volume, and for its reversal by cell shrinkage. On the other hand, stimulation by high urea concentrations is not reversed by shrinkage, nor calyculin A, and provokes a different Po_2 response, suggesting that other effects also occur (Speake and Gibson 1997). It is possible that higher concentrations inhibit regulatory enzymes directly.

Ion concentrations, as well as that of macromolecules, may also be important. The different responses of trout RBCs to isotonic and anisotonic swelling has already been discussed. In addition, increased ionic strength reduces the volume set point for KCC activation in dog RBCs (Parker et al. 1995). Pre-incubation of cells with lyotropic anions (SCN^-, I^-, NO_3^-) at 37 °C increased KCC in sheep RBCs (Lauf 1991). In both cases, KCC was activated at lower cell volumes. In sickle cells, the ability to respond to swelling, but not to NEM, required permissive concentrations of Cl^-, complete substitution with NO_3^- did not activate KCC (Wilkins et al. 1999). As discussed earlier, bicarbonate appears to inhibit sickle cell KCC (Godart et al. 1997). These various observations suggest an important role for the concentration of inorganic anions in regulation of KCC. In other cell types, a key role for intracellular ionic strength, and especially $[Cl^-]_i$, has also been suggested for volume responses (Cannon et al. 1998; Lang et al. 1998; Voets et al. 1999; Wilkins et al. 1999). Protein kinases, sensitive to either cations (Na^+ and K^+) or anions (e.g. Cl^-) have been documented (e.g. in ovine tracheal epithelium: Muimo et al. 2000), and would provide a link between ion concentration and regulatory protein phosphorylation. An alternative explanation may involve changes in membrane negative surface charges.

Others have argued for a mechanical signal in both ghosts and intact RBCs. Thus the experiments of Sachs and others (Dunham et al. 1993; Sachs and Martin 1993; Ortiz-Carranza et al. 1997) show that KCC remains volume-sensitive in human RBC ghosts or intact sheep RBCs in the absence of changes in macromolecular crowding or in protein phosphorylation (since they lack cytoplasmic macromolecules, Mg^{2+} or ATP). Heating intact rat RBCs to 49 °C, which disrupts the cytoskeleton, increased KCC activity but abolished further stimulation by swelling (Orlov et al. 1989). Polyvalent cations, either inorganic or organic, inhibited volume-activated KCC (Sachs and Martin 1999). Heat-treatment of human ghost preparations (again to 49 °C) also stimulated KCC but did not prevent inhibition by polyamines. These observations have led Sachs and Martin (1999) to suggest that KCC is inhibited by positive charges on spectrin, which is removed on cell swelling, but can be mimicked by polyvalent cations. In this model, KCC is activated if the negative surface charge on the interior face of the membrane increases.

Volume responsiveness involving changes in concentration and/or membrane stretch can be resolved if the transduction pathway involves both modalities of stimulus, perhaps acting in series, as argued by Dunham et al. (1993). Notwithstanding, there is still much that remains obscure concerning regulation of KCC.

9.3.3 Magnesium and Organic Phosphates

$[Mg^{2+}]_i$ can be altered experimentally by using the divalent cation ionophore A23187 and different $[Mg^{2+}]_o$'s. Lowering $[Mg^{2+}]_i$ stimulates KCC in RBCs from a number of different vertebrates including sheep (Lauf and Mangor-Jensen 1984; Lauf 1985), human (Brugnara and Tosteson 1987b), dog (Parker et al. 1990) and rabbit (Jennings 1999). Reintroducing divalent cations inhibited KCC, with the sequence: $Mn^{2+} > Mg^{2+} > Ca^{2+} = Cd^{2+} = Zn^{2+} > Sr^{2+}$ (Lauf and Mangor-Jensen 1984;

Lauf 1985). Most of the inhibitory effect of Mg^{2+} requires the presence of intracellular ATP (Delpire and Lauf 1991a), leading to the hypothesis that Mg^{2+} loading and depletion act via the regulatory phosphorylation cascade. This explanation could account for the loss of the delay for swelling-activated KCC in Mg^{2+}-depleted cells (see Dunham et al. 1993) and the inability of Mg^{2+} to inhibit KCC in NEM-treated LK sheep RBCs (Lauf 1985).

Several lines of evidence suggest that this cannot be the only mechanism by which divalent cations including Mg^{2+} affect KCC. First, other divalent cations like Ca^{2+} and Sr^{2+} inhibit KCC following NEM activation (Lauf 1985). Second, Mg^{2+} is still able to reduce activity even in ghosts lacking cytoplasmic ATP (although a membrane-bound pool may still be present – e.g. Hoffman 1997) and therefore apparently unable to carry out protein phosphorylation. Possibilities include a direct effect on the transporter or via effects on the negative surface charge of the membrane. In addition, Mg^{2+}-ATP is also able to *increase* KCC activity. Thus ATP stimulates KCC both in ghosts and intact RBCs (Delpire and Lauf 1991a; Sachs and Martin 1993). Tyrosine kinase inhibitors prevent this effect, consistent with the presence of a stimulatory tyrosine kinase (Sachs and Martin 1993; Cossins et al. 1994; and Sect. 9.3.1).

Fluctuations in free $[Mg^{2+}]_i$ have also been hypothesised as responsible for the O_2 effect (Canessa et al. 1987b). DeoxyHb binds the principal intracellular Mg^{2+} buffers (mainly 2,3-DPG and ATP in mammals) with greater affinity than oxyHb, thus elevating free $[Mg^{2+}]_i$ from about 0.4 to 0.6 mM in human RBCs (Flatman and Lew 1980). High $[Mg^{2+}]/[Mg^{2+}\text{-ATP}]$ could inhibit KCC via increased activity of inhibitory PK's. There is evidence that this mechanism cannot fully account for O_2-sensitivity of KCC activity, however. Thus, RBCs whose free intracellular $[Mg^{2+}]$ had been clamped using A23187 still show KCC activated by increases in O_2 (Campbell and Gibson 1998; Campbell et al. 1999; Muzyamba et al. 2001). In addition, the magnitude of changes in Mg^{2+} required to elicit changes in KCC activity are far outside the physiological range (Delpire and Lauf 1991a; Campbell and Gibson 1998; Campbell et al. 1999). A role for the cytoskeleton has been suggested (see Sect. 9.4.1).

9.3.4 KCC and Redox Potential

The role of redox reactions in regulation of KCC activity also remains unclear. A number of oxidants stimulate KCC activity whilst reciprocally lowering reduced glutathione ([GSH]) (Olivieri et al. 1993; Lauf et al. 1995). Under these conditions, the transporter is still susceptible to inhibition by the phosphatase inhibitor calyculin A. A fall in [GSH] may therefore inhibit protein phosphorylation but without completely abolishing activity of the relevant protein kinase(s) (Lauf et al. 1995). In dog RBCs, low [GSH] correlate with elevated KCC activity (Fujise et al. 1997). Sickle cells also tend to have lower [GSH], cf. normal RBC, which may contribute to their high KCC activity (Olivieri et al. 1993; Adragna and Lauf 1998). GSH depletion will cause oxidation of protein thiol groups, which may alter KCC activity in a similar manner to that seen with NEM. The identity of any relevant thiols, however, is not known. In addition, regulation may involve other components of the anti-oxidant system of the RBC, such as NADH or NADPH, rather than GSH *per se*. Thus in horse RBCs, stimulation of KCC by oxidants does not correlate simply with depletion of GSH (Muzyamba et al. 2000). Other evidence shows that H_2O_2 stimulates KCC, possibly via stimulation of the regulatory PP (Bize et al. 1998). We speculate that these observations may have particular relevance to control of KCC activity by Po_2, which could act via RBC redox systems to generate a chemical signal (e.g. rise in H_2O_2, fall in NADPH) acting on PP/PK's involved in transporter regulation. Clearly further work is required in this area.

9.4 Future Perspectives

The preceding account documents a number of areas of controversy or uncertainty concerning RBC KCC. In this last section, we highlight several areas whose resolution should yield a greater understanding of its function and regulation.

9.4.1 The Cytoskeleton and Regulation of KCC

The concept that the cytoskeleton plays a role in regulation of KCC came originally from observations that high hydrostatic pressure, which activates KCC (see Sect. 9.2.5), also induces shape changes (Hall and Ellory 1986a). Hereditary defects such as the stomatocytoses are also associated with increased K^+ fluxes (e.g. Tse and Lux 1999; Stewart and Turner 1999). Jennings and Schulz (1990) found no simple association between KCC activity and shape, however, and drugs acting on the cytoskeleton, such as cytochalasin, did not inhibit activation of KCC. Notwithstanding, there is a considerable body of evidence for cytoarchitecture affecting membrane transport. These effects may be highly localised. For example, the work of Hoffman (1997) on ATP pools and Na^+/K^+ pump activity, and the results of Mehta with NPDK (e.g. Muimo et al. 2000), indicate considerable nonhomogeneity of the RBC cytoplasm and membrane.

The RBC cytoskeleton comprises a network of spectrin and actin, anchored to the plasma membrane via band 3 and glycophorins (e.g. Takakuwa 2001). The intracellular C-terminus of band 3 (sometimes termed cdb3) contains binding sites for several important molecules, including glycolytic enzymes, Hb and hemichromes, carbonic anhydrase, and certain organic phosphates (reviewed by Low 1986). It also contains multiple phosphorylation sites. If selected kinases and phosphatases are associated with cytoskeletal elements, anchored via linker proteins to cdb3, reversible binding of Hb to this site could alter phosphorylation of key proteins concerned with transporter regulation. Hb affinity for cdb3 is altered by O_2 tension, with deoxyHb probably having a higher affinity than the oxy form. Such a scheme could account for the O_2 sensitivity of KCC activity described above. In addition, the differential O_2 sensitivity of pink (O_2-sensitive KCC) and white (O_2-insensitive KCC) ghosts (Khan et al. 2000) implies a role of membrane-bound Hb, rather than bulk Hb. Spectroscopic evidence shows that Hb interacts with band 3, and with a different affinity depending on Hb isoform (A, S, C, etc.), but often these experiments have not been carried out under conditions approximating to physiological (i.e. normal ionic concentration, pH, organic phosphate, etc.).

There are a number of possible strategies with which to investigate the role of band 3 and cdb3 in transporter regulation. Thus band 3 deficient RBCs do occur, either through natural mutation (Japanese Black wagyu cattle: Inaba et al. 1996; certain human hereditary spherocytoses: Bruce and Tanner 1999), or experimental genetic manipulation (knockout mice: Southgate et al. 1996). These models provide the opportunity to investigate the correlation between band 3, and/or key cytoskeletal proteins, and KCC activity. The use of antibodies directed to cdb3, or competing peptides, incorporated into RBC ghosts may also help in the elucidation of this effect. Binding studies with physical (NMR, ESR) measurements on band 3-Hb interactions should resolve the issue of membrane-bound Hb. In addition, the ability to culture progenitor erythroid cells *in vitro* and induce differentiation, with sequential expression of components of the cytoskeleton and other proteins, also allows the correlation of these various elements with function and regulation of KCC.

Finally, whilst protein phosphorylation appears to have a major role in control of KCC, its interaction with the cytoskeleton remains unknown. To our knowledge, no-one has shown direct phosphorylation of KCC in response to regulation. NKCC can be directly phosphorylated, but it remains unclear whether this represents the sole or main regulatory event (Lytle 1997; Matthews et al. 1998). The possibility of a regulatory associated protein(s), cf. NHE and ezrin (Shenolikar and Weinman 2001), in modulating members of the CCC super family has not been resolved. A major limitation to the resolution of this problem remains the lack of availability of anti-KCC antibodies for immunoprecipitation studies. Such reagents will give valuable insights into regulation of transporter function.

9.4.2 Diverse Functions of KCC in Red Cells and Other Tissues

The physiological characteristics indicate a number of possible roles for KCC. Red blood cell KCC is notably stimulated by cell swelling, and could contribute to regulatory volume decrease (RVD), protecting cells from osmotic lysis (e.g. Hoffmann and Dunham 1995). It may also be participate in the decrease in cell volume, which occurs during RBC maturation (Lauf and Bauer 1987). These roles would be expected to involve the volume-sensitive isoforms, KCC1 and KCC4. The neuronal isoform, KCC2, by contrast, is volume-insensitive. KCC2, together with another member of the CCC (cation-coupled chloride cotransporter) superfamily, NKCC1, may function to regulate intracellular [Cl^-] ([Cl^-]$_i$) of certain neurons (Miles 1999; Delpire 2000; Vardi et al. 2000). Thus early expression of NKCC1 could elevate [Cl^-]$_i$, later expression of KCC1 acting to reduce it. These events have been correlated with depolarisation (with high [Cl^-]$_i$) or hyperpolarisation (low [Cl^-]$_i$) by neurotransmitters such as GABA, and may be important in neuronal development, plasticity or possibly apoptosis (Miles 1999). KCC3 is also found in endothelial cells, where its expression is up-regulated by vascular endothelial growth factor (VEGF) and down-regulated by tumour necrosis factor (TNF) (Hiki et al. 1999). In addition, in cervical cancer cells (SiHa cells), KCC3 expression correlates with cell cycle and plays a role in cell growth regulation (Shen et al. 2001). We speculate that these characteristics would allow KCC to modulate normal or abnormal growth and division of cells, again possibly acting via changes in $[Cl^-]_i$. Finally, of course, KCC transports K^+ as well as Cl^-. It may therefore function as an extrarenal regulator of plasma $[K^+]$. Direction of net K^+ movement across the RBC membrane may be variable. Thus O_2-sensitivity of KCC would support K^+ sequestration (via the Na^+/K^+ pump) in hypoxic regions of the circulation, K^+ efflux (via KCC) at high Po_2's. Red blood cell KCC may thus play a role in modulating blood vessel tone. In addition, in low K^+-containing RBCs of carnivores (and certain other vertebrates), KCC may have an inwardly directed electrochemical gradient enabling it to buffer plasma K^+ following ingestion of K^+-rich meals. Elucidation of the various roles of the various KCC isoforms, and other similar transporters, represents an exciting area for future research.

Acknowledgement: We thank the Wellcome Trust and Action Research for financial support.

References

Adragna NC, Lauf PK (1998) Role of nitric oxide derivative, in K-Cl activation of low-potassium sheep red blood cells. J Membrane Biol 166:157-167

Berenbrink M, Weaver YR, Cossins AR (1997) Defining the volume dependence of multiple K flux pathways in trout red blood cells. Am J Physiol 272:C1099-C1111

Berenbrink M, Volkel S, Heisler N, Nikinmaa M (2000) O_2-dependent K^+ fluxes in trout red blood cells: the nature of O_2 sensing revealed by the O_2 affinity, cooperativity and pH dependence of transport. J Physiol 526:69-80

Bernhardt I, Hall AC, Ellory JC (1991) Effects of low ionic strength media on passive human red cell monovalent cation transport. J Physiol 434:489-506

Bize I, Dunham PB (1994) Staurosporine, a protein kinase inhibitor, activates K-Cl cotransport in LK sheep erythrocytes. Am J Physiol 266:C759-C770

Bize I, Munoz P, Canessa M, Dunham PB (1998) Stimulation of Membrane serine-threonine phosphatase in erythrocytes by hydrogen peroxide and staurosporine. Am J Physiol 274:C440-C446

Bize I, Guvenc B, Robb A, Buchbinder G, Brugnara C (1999) Serine/threonine protein phosphatases and regulation of K-Cl cotransport in human erythrocytes. Am J Physiol 277:C926-C936

Bize I, Guvenc B, Buchbinder G, Brugnara C (2000) Stimulation of human erythrocyte K-Cl cotransport and protein phosphatase type 2A by N-ethylmaleimide: role of intracellular Mg^{2+}. J Membrane Biol 177:159-168

Borgese F, Garcia-Romeu F, Motais R (1987) Control of cell volume and ion transport by beta-adrenergic catecholamines in erythrocytes of rainbow trout, *Salmo gairdneri*. J Physiol 382:123-144

Bourne PK, Cossins AR (1982) On the instability of K^+ influx in erythrocytes of the rainbow trout *Salmo gairdneri*, and the role of catecholamine hormones in mainitaining *in vivo* influx activity. J Exp Biol 101:93-104

Bruce LJ, Tanner MJ (1999) Erythroid band 3 variants and disease. Ballieres Best Practical Research and Clinical Haematology 12:637-654

Brugnara C, Tosteson DC (1987a) Cell volume, K transport, and cell density in human erythrocytes. Am J Physiol 252:C269-C276

Brugnara C, Tosteson DC (1987b) Inhibition of K transport by divalent cations in sickle erythrocytes. Blood 70:1810-1815

Brugnara C, Bunn HF, Tosteson DC (1986) Regulation of erythrocyte cation and water content in sickle cell anemia. Science 232:388-390

Brugnara C, Van Ha T, Tosteson DC (1989a) Acid pH induces formation of dense cells in sickle erythrocytes. Blood 74:487-495

Brugnara C, Van Ha T, Tosteson DC (1989b) Role of chloride in potassium transport through the K-Cl cotransport system in human red blood cells. Am J Physiol 256:C944-C1003

Cabantchik ZI, Greger R (1992) Chemical probes for anion transporters of mammalian cell Membranes. Am J Physiol 262:C803-C827

Campbell EH, Gibson JS (1998) Oxygen-dependent K^+ fluxes in sheep red cells. J Physiol 506:679-688

Campbell EH, Cossins AR, Gibson JS (1999) Oxygen dependent K^+ influxes in Mg^{2+}-clamped equine red cells. J Physiol 515:431-437

Canessa M, Fabry ME, Blumenfeld N, Nagel RL (1987a) Volume-stimulated, Cl-dependent K efflux is highly expressed in young human red cells containing normal hemoglobin or HbS. J Membrane Biol 97:97-105

Canessa M, Fabry ME, Nagel RL (1987b) Deoxygenation inhibits the volume-stimulated Cl^--dependent K efflux in SS and young AA cells: a cytoplasmic Mg^{2+} modulation. Blood 70:1861-1866

Cannon CL, Basavappa S, Strange K (1998) Intracellular ionic strength regulates the volume sensitivity of a swelling-activated anion channel. Am J Physiol 275:C416-C422

Clayton GH, Owens GC, Wolff JS, Smith RL (1998) Ontogeny of cation-Cl^- cotransporter expression in rat neocortex. Brain Res and Developm Brain Res 109:281-292

Colclasure GC, Parker JC (1991) Cytosolic protein concentration is the primary volume signal in dog red cells. J Gen Physiol 98:881-892

Colclasure GC, Parker JC (1992) Cytosolic protein concentration is the primary volume signal for swelling-induced [K-Cl] cotransport in dog red cells. J Gen Physiol 100:1-10

Cossins AR, Gibson JS (1997) Volume-sensitive transport systems and volume homeostasis in vertebrate red blood cells. J Exp Biol 200:343-352

Cossins AR, Weaver YR, Lykkeboe G, Nielsen OB (1994) Role of protein phosphorylation in control of K flux pathways of trout red blood cells. Am J Physiol 267:C1641-C1650

De Franceschi L, Fumagalli L, Olivieri O, Corrocher R, Lowell CA, Berton G (1997) Deficiency of Src family kinases Fgr and Hck results in activation of erythrocyte K/Cl cotransport. J Clin Invest 99:220-227

Delpire E (2000) Cation-chloride cotransporters in neuronal communication. News Physiol Sci 15:309-312

Delpire E, Lauf PK (1991a) Magnesium and ATP dependence of K-Cl co-transport in low K^+-sheep red blood cells. J Physiol 441:219-231

Delpire E, Lauf PK (1991b) Trans effects of cellular K and Cl on ouabain-resistant Rb(K) influx in low K sheep red blood cells: Further evidence for asymmetry of K-Cl co-transport. Pflügers Arch 419:540-542

Drew C, Ball V, Gibson JS, Ellory JC (2002) H^+ ion dependence of K^+-Cl^- cotransport in normal human red blood cells. J Physiol 539P:6P

Dunham PB (1976) Anti-L serum. Two populations of antibodies affecting cation transport in LK erythrocytes of sheep and goats. Biochim Biophys Acta 443:219-226

Dunham PB (1995) Effects of urea on K-Cl cotransport in sheep red blood cells: evidence for two signals of swelling. Am J Physiol 268:C1206-C1032

Dunham PB, Ellory JC (1980) Chloride activated potassium transport in human erythrocytes. Proc Natl Acad Sci USA 77:1711-1715

Dunham PB, Ellory JC (1981) Passive potassium transport in low potassium sheep red cells: dependence upon cell volume and chloride. J Physiol 318:511-530

Dunham PB, Klimczak J, Logue PJ (1993) Swelling activation of K-Cl cotransport in LK sheep erythrocytes: a three-state process. J Gen Physiol 101:733-766

Ellory JC, Dunham PB (1980) Volume-dependent passive potassium transport in LK sheep red cells. In: Lassen UV, Ussing HH, Wieth JO (eds) Membrane transport in erythrocytes. Munksgaard, Copenhagen, pp 409-427

Ellory JC, Tucker EM (1969) Stimulation of the potassium transport system in low potassium type sheep red cells by a specific antibody reaction. Nature 222:477-478

Ellory JC, Hall AC, Stewart GW (1985) Volume-sensitive cation fluxes in mammalian red cells. Mol Physiol 8:235-246

Ellory JC, Hall AC, Ody SA (1989a) Is acid a more potent activator of KCl co-transport than hypotonicity in human red cells? J Physiol 420:149P

Ellory JC, Player M, Chalder SM, Stuart J (1989b) Rheological effect of activation of the KCl-cotransport pathway in normal and sickle erythrocytes. Clin Hemorheol 9:1009-1016

Ellory JC, Hall AC, Ody SO, Englert HC, Mania D, Lang HJ (1990) Selective inhibitors of KCl cotransport in human red cells. FEBS Letters 262:215-218

Ellory JC, Hall AC, Ody SA, DeFigueiredos CE, Chalder S, Stuart J (1991) KCl cotransport in HbAA and HbSS red cells: activation by intracellular acidity and disappearance during maturation. In: Mangani M, DeFlora A (eds) Red blood cell ageing. Plenum Press, New York, pp 47-57

Ellory JC, Gibson JS, Stewart GW (1998) Pathophysiology of abnormal cell volume in human red cells. Contrib Nephrol 123:220-239

Evans JV (1954) Electrolyte concentrations in red blood cells of British breeds of sheep. Nature 174:931

Evans JV, King JWB (1955) Genetic control of sodium and potassium concentrations in the red blood cells of sheep. Nature 176:171

Flatman PW, Lew VL (1980) Magnesium buffering in intact human red blood cells measured using the ionophore A23187. J Physiol 305:13-30

Flatman PW, Adragna NC, Lauf PK (1996) Role of protein kinases in regulating sheep erythrocyte K-Cl cotransport. Am J Physiol 271:C255-C263

Fujise H, Abe K, Kamimura M, Ochiai H (1997) K^+-Cl^- cotransport and volume regulation in the light and the dense fraction of high K^+ dog red blood cells. Am J Physiol 273:R2991-R2998

Funder J, Wieth JO (1967) Effect of ouabain on glucose metabolism and on fluxes of sodium and potassium of human blood cells. Acta Physiol Scand 71:113-124

Gamba G, Saltzberg SN, Lombardi M, Miyanoshita A, Lytton J, Hediger MA, Brenner BM, Hebert SC (1993) Primary structure and functional expression of a cDNA encoding the thiazide-sensitive, electroneutral sodium-chloride cotransporter. Proc Natl Acad Sci USA 90:2749-2753

Garay RP, Nazaret C, Hannaert PA, Cragoe EJ (1989) Demonstration of a [K^+,Cl^-]-cotransport system in human red cells by its sensitivity to [(dihydroindenyl)oxy]alkanoic acids: regulation of cell swelling and distinction from the bumetanide-sensitive [Na^+,K^+,Cl^-]-cotransport system. Mol Pharmacol 33:696-701

Gibson JS, Hall AC (1995) Stimulation of KCl co-transport in equine erythrocytes by hydrostatic pressure: effect of kinase/phosphatase inhibition. Pflügers Arch 429:446-448

Gibson JS, Ellory JC, Culliford SJ, Fincham DA (1993) Volume-sensitive KCl co-transport and taurine fluxes in horse red blood cells. Exp Physiol 78:685-695

Gibson JS, Godart H, Ellory JC, Staines H, Honess NA, Cossins AR (1995) Modulation of K^+-Cl^- cotransport in equine red blood cells. Exp Physiol 79:997-1009

Gibson JS, Speake PF, Ellory JC (1998) Differential oxygen sensitivity of the K^+-Cl^- cotransporter in normal and sickle human red blood cells. J Physiol 511:225-234

Gibson JS, Cossins AR, Ellory JC (2000) Oxygen-sensitive membrane transporters in vertebrate red cells. J Exp Biol 203:1395-1407

Gillen CM, Brill S, Payne JA, Forbusch IB (1996) Molecular cloning and functional expression of the KCl cotransporter from rabbit, rat and human. J Biol Chem 217:16237-16244

Godart H, Ellory JC (1996) KCl cotransport activation in human erythrocytes by high hydrostatic pressure. J Physiol 491:423-434

Godart H, Dormandy A, Ellory JC (1997) Do HbSS erythrocytes lose KCl in physiological conditions? Brit J Haematol 98:25-31

Godart H, Ellory JC, Motais R (1999) Regulatory volume response of erythrocytes exposed to a gradual and slow decrease in osmolality. Pflügers Arch 437:776-779

Guizouarn H, Motais R (1999) Swelling activation of transport pathways in erythrocytes: effects of Cl^-, ionic strength and volume changes. Am J Physiol 276:C210-C220

Guizouarn H, Harvey BJ, Borgese F, Gabillat N, Garcia-Romeu F, Motais R (1993) Volume-activated Cl^--independent and Cl^--dependent K^+ pathways in trout red blood cells. J Physiol 462:609-626

Hall AC, Ellory JC (1986a) Effect of high hydrostatic pressure on 'passive' monovalent cation transport in human red cells. J Membrane Biol 94:1-17

Hall AC, Ellory JC (1986b) Evidence for the presence of volume-sensitive KCl transport in 'young' human red cells. Biochim Biophys Acta 858:317-320

Hall AC, Ellory JC, Klein RA (1982) Pressure and temperature effects on human red cell cation transport. J Membrane Biol 68:47-56

Hiki K, D'Andrea RJ, Furez J, Crawford J, Woollatt E, Sutherland GR, Vadas MA, Gamble JR (1999) Cloning, characterization and chromosomal location of a novel human K^+-Cl^- cotransporter. J Biol Chem 274:10661-10667

Hladky SB, Rink TJ (1977) pH equilibrium across the red cell membrane. In: Ellory JC, Lew VL (eds) Membrane transport in red cells. Academic Press, London, pp 115-135

Hoffman JF (1997) ATP compartmentalization in human erythrocytes. Curr Opin Hematol 4:112-115

Hoffmann EK, Dunham PB (1995) Membrane mechanisms and intracellular signalling in cell volume regulation. Int Rev Cytol 161:173-262

Hoffman JF, Kregenow FM (1966) The characterization of new energy dependent cation transport processes in red blood cells. Ann New York Acad Sci 137:566-576

Holtzman EJ, Kumar S, Faaland CA, Warner F, Logue PJ, Erickson SJ, Ricken G, Waldman J, Kumar S, Dunham PB (1998) Cloning, characterization, and gene organization of K-Cl cotransporter from pig and human kidney and *C. elegans*. Am J Physiol 275:F550-F564

Honess NA, Gibson JS, Cossins AR (1996) The effects of oxygenation upon the Cl-dependent K flux pathway in equine red cells. Pflügers Arch 432:270-277

Inaba M, Yawata A, Koshino I, Sato K, Takeuchi M, Takakuwa Y, Yawata Y, Kanzaki A, Sakai J, Ban A, Ono K, Maede Y (1996) Defective anion transport and marked spherocytosis with membrane instability caused by hereditary total deficiency of red cell band 3 in cattle due to a nonsense mutation. J Clin Invest 97:1804-1817

Jennings ML (1999) Volume-sensitive K^+/Cl^- cotransport in rabbit erythrocytes. Analysis of the rate-limiting activation and inactivation events. J Gen Physiol 114:743-757

Jennings ML, Adame MF (2001) Direct estimate of 1:1 stoichiometry of K(+)-Cl(-) cotransport in rabbit erythrocytes. Am J Physiol 281:C825-C832

Jennings ML, Al-Rohil N (1990) Kinetics of activation and inactivation of swelling-stimulated K/Cl transport: The volume-sensitive parameter is the rate constant for inactivation. J Gen Physiol 95:1021-1040

Jennings ML, Schulz RK (1991) Okadaic acid inhibition of KCL cotransport: evidence that protein dephosphorylation is necessary for activation of transport by either swelling or N-ethylmaleimide. J Gen Physiol 97:799-817

Kaji DM (1989) Kinetics of volume-sensitive K transport in human erythrocytes: evidence for asymmetry. Am J Physiol 256:C1214-C1223

Kaji DM (1993) Effect of membrane potential on K-Cl transport in human erythrocytes. Am J Physiol 264:C376-C382

Kaji DM, Gasson C (1995) Urea activation of K-Cl cotransport in human erythrocytes. Am J Physiol 268:C1018-C1025

Kaji DM, Tsukitani Y (1991) Role of protein phosphorylation in activation of KCl cotransport in human erythrocytes. Am J Physiol 260:C178-C182

Kerr SE (1937) Studies on the inorganic composition of blood. J Biol Chem 117:227-235

Khan A, Ellory JC (2000) Elevated temperatures enhance KCC1 activity in sickle cells. Bioelectrochemistry 52:127-131

Khan A, Gibson JS, Ellory JC (2000) Oxygen-dependent KCl cotransport in ghosts from normal human red blood cells. J Physiol 527P:38P

Kiessling K, Roberts N, Gibson JS, Ellory JC (2000) A comparison in normal individuals and sickle cell patients of reduced glutathione precursors and their transport between plasma and red cells. Hematol J 1:243-249

Kirk K, Strange K (1998) Functional properties and physiological roles of organic solute channels. Annu Rev Physiol 60:719-739

Kirk K, Ellory JC, Young JD (1992) Transport of organic substrates via a volume-activated channel. J Biol Chem 267:23475-23478

Kregenow FM (1971) The reponse of duck erythrocytes to nonhemolytic hypotonic media. Evidence for a volume-controlling mechanism. J Gen Physiol 58:372-395

Lang F, Busch GL, Ritter M, Volkl H, Waldegger S, Gulbins E, Haussinger D (1998) Functional significance of cell volume regulatory mechanisms. Physiol Rev 78:247-306

Lauf PK (1983) Thiol-dependent passive K/Cl transport in sheep red cells. I. Dependence on chloride and external K^+ (Rb^+) ions. J Membrane Biol 73:237-246

Lauf PK (1985) KCl cotransport: sulfhydryls, divalent cations and the mechanism of volume activation in a red cell. J Membrane Biol 88:1-13
Lauf PK (1991) Foreign anions modulate volume set point of sheep erythrocyte K-Cl cotransport. Am J Physiol 260:C502-C512
Lauf PK, Bauer J (1987) Direct evidence for chloride-dependent volume reduction in macrocytic sheep reticulocyte. Biochem Biophys Res Commun 144:849-855
Lauf PK, Theg BE (1980) A chloride dependent K^+ flux induced by N-ethylmaleimide in genetically low K^+ sheep and goat erythrocytes. Biochem Biophys Res Commun 92:1422-1428
Lauf PK, Mangor-Jensen A (1984) Effects of A23187 and Ca^{2+} on volume- and thiol-stimulated ouabain-resistant K^+Cl^- fluxes in low K^+ sheep erythrocytes. Biochem Biophys Res Commun 125:790-796
Lauf PK, Stiehl BJ, Joiner CH (1977) Active and passive cation transport and L-antigen heterogeneity in low potassium sheep red cells. J Gen Physiol 70:221-242
Lauf PK, Bauer J, Adragna NC, Fujise H, Martin A, Zade-Oppen M, Ryu KH, Delpire E (1992) Erythrocyte K-Cl cotransport: properties and regulation. Am J Physiol 263:C917-C932
Lauf PK, Erdmann A, Adragna NC (1994) K-Cl cotransport, pH and role of Mg in volume-clamped low-K sheep erythrocytes: three equilibrium states. Am J Physiol 266:C95-C103
Lauf PK, Adragna NC, Agar NS (1995) Glutathione removal reveals kinases as common targets for K-Cl cotransport stimulation in sheep erythrocytes. Am J Physiol 269:C234-C241
Lauf PK, Zhang J, Delpire E, Fyffe RE, Mount DB, Adragna NC (2001) K-Cl co-transport: immunocytochemical and functional evidence for more than one KCC isoform in high K and low K sheep erythrocytes. Comp Biochem Physiol A 130:499-509
Lew VL, Freeman CJ, Ortiz OE, Bookchin RM (1991) A mathematical model of the volume, pH and ion content regulation in reticulocytes. J Clin Invest 87:100-112
Low PS (1986) Structure and function of the cytoplasmic domain of band 3: center of erythrocyte membrane-peripheral protein interactions. Biochim Biophys Acta 864:145-167
Lytle C (1997) Activation of the avian erythrocyte Na-K-Cl cotransport protein by cell shrinkage, cAMP, fluoride, and calyculin-A involves phosphorylation at common sites. J Biol Chem 272:15069-15077
Matthews JB, Smith JA, Mun EC, Sicklick JK (1998) Osmotic regulation of intestinal epithelial Na^+-K^+-Cl^- cotransport: role of Cl^- and F-actin. Am J Physiol 274:C697-C706
Miles R (1999) A homeostatic switch. Nature 397:215-216
Minton AP (1983) The effect of volume occupancy upon the thermodynamic activity of proteins: some biochemical consequences. Mol Cell Biochem 55:119-140
Minton AP (1994) Influence of macromolecular crowding on intracellular association reactions: possible role in volume regulation. In: Strange K (ed) Cellular and molecular physiology of cell volume regulation. CRC Press, Boca Raton, pp 181-190
Minton AP, Colclasure GC, Parker JC (1992) Model for the role of macromolecular crowding in regulation of cellular volume. Proc Natl Acad Sci USA 89:10504-10506
Mitsui T, Kitazawa T, Ikebe M (1994) Correlation between high temperature dependence of smooth muscle myosin light chain phosphatase activity and muscle relaxation rate. J Biol Chem 269:5842-5848

Motais R, Guizouarn H, Garcia-Romeu F (1991) Red cell volume regulation: the pivotal role of ionic strength in controlling swelling-dependent transport systems. Biochim Biophys Acta 1075:169-180

Mount DB, Delpire E, Gamba G, Hall AE, Poch E, Hoover RSJ, Hebert SC (1998) The electroneutral cation-chloride cotransporters. J Exp Biol 201:2091-2102

Mount DB, Mercado A, Song L, Xu J, George AL, Delpire E, Gamba G (1999) Cloning and characterization of KCC3 and KCC4, new members of the cation-chloride cotransporter gene family. J Biol Chem 274:16355-16362

Muimo R, Hornickova Z, Riemen CE, Gerke V, Matthews H, Mehta A (2000) Histidine phosphorylation of annexin I in airway epithelia. J Biol Chem 275:36632-36636

Muzyamba MC, Speake PF, Gibson JS (2000) Oxidants and regulation of KCl cotransport in equine red blood cells. Am J Physiol 279:C981-C989

Muzyamba MC, Campbell EH, Rahman T, Gibson JS (2001) Intracellular Mg^{2+} and K^+-Cl^- cotransport in human red blood cells. J Physiol 535P:20P

Nielsen OB, Lykkeboe G, Cossins AR (1992) Oxygenation-activated K^+ fluxes in trout red blood cells. Am J Physiol 263:C1057-C1064

Olivieri O, Bonollo M, Friso S, Girelli D, Corrocher R, Vettore L (1993) Activation of K^+/Cl^- cotransport in human erythrocytes exposed to oxidative agents. Biochim Biophys Acta 1176:37-42

O'Neill WC (1991) Swelling-activated K-Cl cotransport: metabolic dependence and inhibition by vanadate and fluoride. Am J Physiol 260:C308-C315

Orlov SN, Pokudin NI, Kotelevtsev YV, Gulak PV (1989) Volume-dependent regulation of ion transport and membrane phosphorylation in human and rat erythrocytes. J Membrane Biol 107:105-117

Ortiz-Carranza O, Adragna NC, Carnes L, Lauf PK (1997) Two operational models of K-Cl cotransport in low K^+ sheep red blood cells. Cell Physiol Biochem 7:251-263

Parker JC (1993) Urea alters set point volume for K-Cl cotransport, Na-H exchange and Ca-Na exchange in dog red blood cells. Am J Physiol 265:C447-C452

Parker JC (1994) Coordinated regulation of volume-activated transport pathways. In: Strange K (ed) Cellular and molecular physiology of volume regulation. CRC Press, Boca Raton, pp 311-321

Parker JC, McManus TJ, Starke LC, Gitelman HJ (1990) Coordinated regulation of Na/H exchange and K-Cl cotransport in dog red cells. J Gen Physiol 96:1141-1152

Parker JC, Dunham PB, Minton AP (1995) Effects of ionic strength on the regulation of Na/H exchange and K-Cl cotransport in dog red blood cells. J Gen Physiol 105:677-699

Payne JA (1997) Functional characterization of the neuronal-specific K-Cl cotransporter: implications for $[K^+]_o$ regulation. Am J Physiol 273:C1516-C1525

Payne JA, Lytle C, McManus TJ (1990) Foreign anion substitution for chloride in human red blood cells: effect on ionic and osmotic equilibria. Am J Physiol 259:C819-C827

Payne JA, Stevenson TJ, Donaldson LF (1996) Molecular characterization of a putative K-Cl cotransporter in rat brain. J Biol Chem 271:16245-16252

Pellegrino CM, Rybicki AC, Musto S, Nagel RL, Schwartz RS (1998) Molecular identification of erythroid K:Cl cotransporter in human and mouse erythroleukemic cells. Blood Cells, Molecules, and Diseases 24:31-40

Race JE, Makhlouf FN, Logue PJ, Wilson FH, Dunham PB, Holtzman EJ (1999) Molecular cloning and functional characterization of KCC3, a new K-Cl cotransporter. Am J Physiol 277:C1210-C1219

Rasmusen BA, Hall JG (1966) Association between potassium concentration and serological type of sheep red blood cells. Science 151:1551-1552

Sachs JR (1971) Ouabain-insensitive sodium movements in the human red blood cell. J Gen Physiol 57:259-282

Sachs JR, Martin DW (1993) The role of ATP in swelling-stimulated K-Cl cotransport in human red cell ghosts. J Gen Physiol 102:551-573

Sachs JR, Martin DW (1999) Role of polyamine structure in inhibition of K^+-Cl^- cotransport in human red cell ghosts. J Physiol 520:723-735

Schmidt WFR, McManus TJ (1977) Ouabain-insensitive salt and water movements in duck red cells. III. The role of chloride in the volume response. J Gen Physiol 70:99-121

Shen MR, Chou CY, Hsu KF, Liu HS, Dunham PB, Holtzman EJ, Ellory JC (2001) The KCl cotransporter isoform KCC3 can play an important role in cell growth regulation. Proc Natl Acad Sci USA 98:14714-14719

Shenolikar S, Weinman EJ (2001) NHERF: targeting and trafficking membrane proteins. Am J Physiol 280:F389-F395

Smalley CE, Tucker EM, Dunham PB, Ellory JC (1982) Interaction of L antibody with low potassium-type sheep red cells: resolution of two separate functional antibodies. J Membrane Biol 64:167-174

Southgate CD, Chishti AH, Mitchell B, Yi SJ, Palek J (1996) Targeted disruption of the murine erythroid band 3 results in spherocytosis and severe haemolytic anaemia depsite a normal membrane skeleton. Nature Genetics 14:227-230

Speake PF, Gibson JS (1997) Urea-stimulated K-Cl cotransport in equine red blood cells. Pflügers Arch 434:104-112

Speake PF, Roberts CA, Gibson JS (1997) Effect of changes in respiratory blood parameters on equine red blood cell K-Cl cotransporter. Am J Physiol 273:C1811-C1818

Starke LC, Jennings ML (1993) K-Cl cotransport in rabbit red cells: further evidence for regulation by protein phosphatase type 1. Am J Physiol 264:C118-C124

Stewart GW, Turner EJ (1999) The hereditary stomatocytoses and allied disorders: congenital disorders of erythrocyte membrane permeability to Na and K. Ballieres Best Practical Res Clin Haematol 12:707-727

Stuart J, Ellory JC (1988) Rheological consequences of erythrocyte dehydration. Br J Haematol 69:1-4

Takakuwa Y (2001) Regulation of red cell membrane protein interactions: implications for red cell function. Curr Opin Hematol 8:80-84

Teti K, Venza I, Crupi M, Busa M, Loddo S, Romano L (2002) Anion transport in normal erythrocytes, sickle red cells, and ghosts in relation to haemoglobins and magnesium. Arch Biochem Biophys 403:149-154

Tosteson DC, Hoffman JF (1960) Regulation of cell volume by active cation transport in high and low potassium sheep red cells. J Gen Physiol 44:169-194

Tosteson MT, Halperin JA, Kishi Y, Tosteson DC (1991) Palytoxin induces an increase in the cation conductance of red cells. J Gen Physiol 98:969-985

Tse WT, Lux SE (1999) Red blood cell membrane disorders. Blood 104:2-13

Vardi N, Zhang LL, Payne JA, Sterling P (2000) Evidence that different cation chloride cotransporters in retinal neurons allow opposite responses to GABA. J Neurosci 20:7657-7663

Vitoux D, Olivieri O, Garay RP, Cragoe EJ, Galacteros F, Beuzard Y (1989) Inhibition of K^+ efflux and dehydration of sickle cells by [(dihydroindenyl)oxy]alkanoic acid: an inhibitor of the K^+Cl^- cotransport system. Proc Natl Acad Sci USA 86:4273-4276

Vitoux D, Beuzard Y, Brugnara C (1999) The effect of hemoglobin A and S on the volume- and pH-dependence of K-Cl cotransport in human erythrocyte ghosts. J Membrane Biol 167:233-240

Voets T, Droogmans G, Raskin G, Eggermont J, Nilius B (1999) Reduced intracellular ionic strength as the initial trigger for activation of endothelial volume-regulated anion channels. Proc Natl Acad Sci USA 96:5298-5303

Wiley JS, Cooper RA (1974) A furosemide-sensitive cotransport of sodium plus potassium in the human red cell. J Clin Invest 53:745-75

Wilkins RJ, Browning JA, Ellory JC (1999) Cell volume regulation reduced to size. In: Kozlowski R (ed) Chloride channels. Isis Medical Publishing, Oxford, pp 19-23

Willis JS, Anderson GL (1998) Activation of K-Cl cotransport by mild warming in guinea pig red cells. J Membrane Biol 163:193-203

Xu JC, Lytle C, Zhu TT, Payne JA, Benz EJ, Forbush BI (1994) Molecular cloning and functional expression of the bumetanide-sensitive Na-K-Cl cotransporter. Proc Natl Acad Sci USA 91:2201-2205

Yancey PH (1994) Compatible and counteracting solutes. In: Strange K (ed) Cellular and molecular physiology of cell volume regulation. CRC Press, Boca Raton, pp 81-109

Zade-Oppen AM, Lauf PK (1990) Thiol-dependent passive K:Cl cotransport in sheep red blood cells: IX. Modulation by pH in the presence and absence of DIDS and the effect of NEM. J Membrane Biol 118:143-151

10 The Band 3 Protein: Anion Exchanger and Anion-Proton Cotransporter

Sigrid Lepke[1,2], Joachim Heberle[3] and Hermann Passow[1,2]

[1]Max Planck Institut für Biophysik, Heinrich-Hoffmann-Str. 7, 60528 Frankfurt am Main, [2]Institut für Biochemie und Biophysik, Friedrich Schiller Universität Jena, Philosophenweg 12, 07743 Jena, [3]Forschungszentrum Jülich, IBI-2: Strukturbiologie, 52425 Jülich, Germany

10.1 Introduction

The anion transport protein of the red blood cell membrane, the so-called band 3 protein, mediates two distinct processes: anion exchange and anion-proton cotransport. The present chapter deals with the relationship between the two processes.

When the transporter operates in the exchange mode, the transfer of an anion across the membrane is accompanied by the transfer of another anion in the opposite direction such that the electrical neutrality in the two adjacent media is preserved. The process is, however, not only electrically neutral but also "electrically silent". The latter term refers to the fact that the anion exchange takes place without contributing to the electrical conductance of the red cell membrane. This indicates that the exchange involves the binding of the transported anion to the "transfer site" of the protein and a translocation of the bound anion across a rate-determining barrier by one or a series of conformational changes of the non-diffusible transporter. It also shows that the conformational change cannot take place when no anion is bound. Hence the unloaded exchanger is unable to return to that membrane surface, from which it had left together with the bound anion. It can only do so after picking up another anion for the return journey. The process described shows a fairly complex dependence on ion composition and pH of the internal and external media. For our present purpose we confine ourselves to the study of chloride and sulfate transport. We need to remember that for these anions, the relationship between transmembrane flux and concentration shows a superimposition of saturation kinetics and self inhibition. With respect to the pH dependence, we need to consider its acidic branch below a plateau that extends from, approximately, pH 9.5 to 7.1. Below pH 7.1, the rate of chloride equilibrium exchange (i.e. the flux measured with a radioisotope under conditions of no net flow) shows a continuous decrease with an apparent pK value of about 6.0 at 0 °C (Gunn et al. 1975; Funder and Wieth 1976; Wieth and Bjerrum 1982). The rate of sulfate equilibrium exchange passes through a maximum near pH 6.3 (23 °C) before decreasing to negligibly small values. The maximum resides close to the point of intersection of two superimposed dissociation curves with slightly different pK values. This point shifts to a lower pH value if the measurements of sulfate flux are performed in the presence of a large excess of Cl^- over SO_4^{2-} (Schnell et al. 1977).

The anion exchange across the red cell membrane is known for more than a century. The anion-proton cotransport represents a more recent discovery. It was observed by M.L. Jennings that, after elimination of the Jacobs-Parpart cycle (which involves a chloride/bicarbonate exchange) by careful removal of CO_2, a joint net movement of H^+ and SO_4^{2-} (Jennings 1976) and of H^+ and Cl^- (Jennings 1978) across the red cell membrane could be observed. Using typical inhibitors of anion exchange, including stilbene disulphonate derivatives, he was able to show that this cotransport was mediated by the same protein that was responsible for anion exchange. He added an array of further observations (temperature dependence, comparison of the relative rates of transmembrane translocation of various anion species by anion exchange and cotransport, and others, including electrical silence), which strongly supported his conclusion. A striking difference between anion exchange and anion-proton cotransport reported by Jennings consisted in the observation that the rate of H^+-Cl^- cotransport was about one thousand times lower than the rate of Cl^-/Cl^- exchange. In contrast, the H^+-SO_4^{2-} cotransport was found to take place about two orders of magnitude faster than SO_4^{2-}/SO_4^{2-} exchange (Milanick and Gunn 1984).

In the years after the discovery of the anion-proton cotransport, many aspects of H^+-SO_4^{2-} cotransport and of its relationship to the pH dependence of sulfate equilibrium exchange were thoroughly examined, notably by Milanick and Gunn (1982, 1984). In contrast, no effort was made to explore in more detail the original contributions of Jennings on H^+-Cl^- cotransport. In the present chapter the focus will be on investigations of the relationship between anion exchange and anion-proton cotransport with specific emphasis on chloride transport: An attempt will be made to show *that* and *how* the decrease of the rapid Cl^-/Cl^- exchange with decreasing pH is causally related to the emergence of the very much slower H^+-Cl^- cotransport. It will be supplemented by some comments on the relationship to H^+-SO_4^{2-} cotransport. Work in progress on H^+-Cl^- cotransport was reviewed by Passow et al. (1997) and presented at various international meetings, most recently in Boston, 2001 (Passow et al. 2001).

10.2 Measuring Anion-Proton Cotransport

The experimental investigation of anion-proton cotransport is based on the measurement of net proton flows associated with band 3-mediated net anion transfer. This requires the rigorous exclusion of proton flows through other pathways. Most important is the complete absence of carbon dioxide which, catalyzed by intracellular carbonic anhydrase, could lead to the formation of the proton carriers HCO_3^- and H_2CO_3. Jennings (1976, 1978) met this condition by working in the complete absence of carbon dioxide by degassing with N_2. Alternatively, the experiments

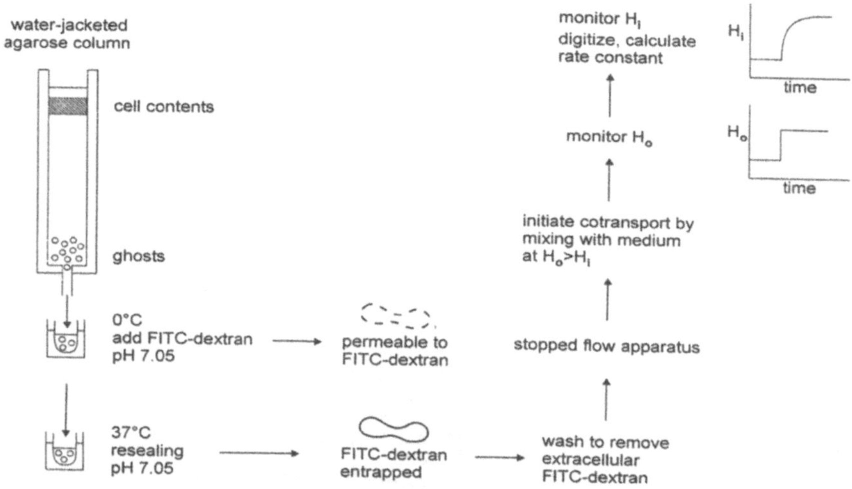

Fig. 10.1. Protocol for preparation of carbonic anhydrase-free red cell ghosts (by an adaptation of the method of Wood (1989)) and for the execution of the transport measurements

can be executed even in the presence of CO_2 in carbonic anhydrase-free human red cell ghosts (Fig. 10.1).

It is possible to prepare such ghosts by hemolyzing red cells on an agarose column equilibrated with hypotonic NaCl solutions. The cell contents including the carbonic anhydrase are released and retarded, while the ghosts move rapidly in the void volume and can be collected separately (Wood 1989). They are collected in media containing NaCl at desired concentrations, an impermeant glycyl-glycine-EDTA buffer (1.25 mM EDTA, 1.875 mM glycylglycine, pH ~ 7), and FITC dextran as an impermeant, fluorescent pH indicator. The absence of carbonic anhydrase is verified by conventional tests, by the absence of inhibition of net anion transport by acetazolamide, and – most sensitively –, by demonstrating the occurrence of the extra- or intracellular acidification by, respectively, the net efflux or influx of H^+-SO_4^{2-}, which is reversed to an alkalinization in the presence of proton carriers other than those involved in anion-proton cotransport (Jennings 1976). The buffer was specifically designed to maintain its buffer capacity constant over the pH range covered in the experiments (Fig. 10.2, lower left). Proton uptake was initiated by mixing in a stopped flow apparatus a small volume of the resealed ghosts containing glycyl-glycine-EDTA-buffered NaCl solution of pH_i ~ 7.05 with a large volume of medium with the same composition as inside the ghosts (160 mM) and varying extracellular proton concentration H_o, or with constant H_o in the external medium (pH 5.6) but varying the Cl^- concentration at $Cl^-_i = Cl^-_o$ (the symbols indicate ion concentrations). Since the buffer capacity inside the ghosts is known and constant throughout the whole time course of each experiment, the changes of proton concentration calculated from the fluorescence readings are a direct measure of the amount of protons taken up at any given time at the external pH established by the initial pH jump in the stopped flow apparatus. The time

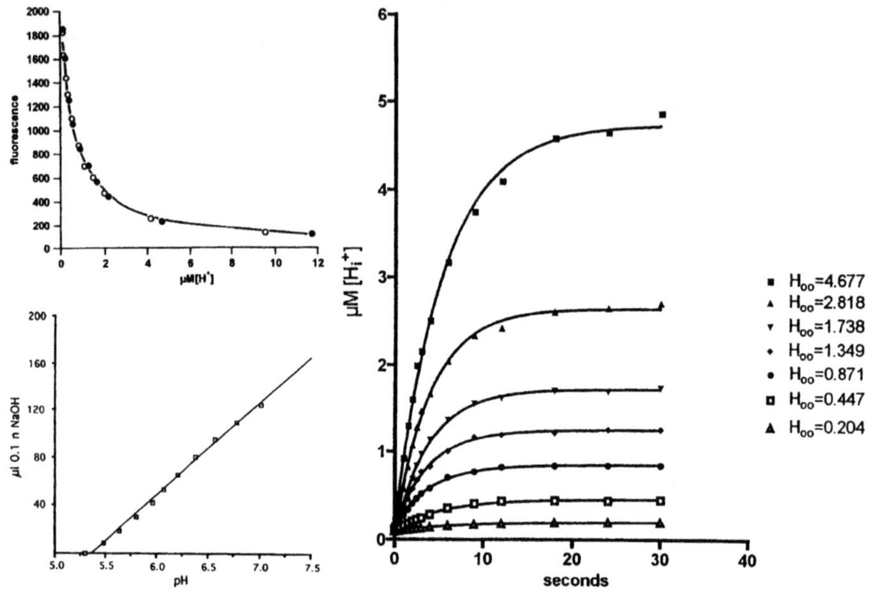

Fig. 10.2. Measuring H^+-Cl^- cotransport in carbonic anhydrase-free, resealed red cell ghosts containing the fluorescent pH indicator FITC-dextran: basic observations and evaluation of H^+-Cl^- cotransport data. Right panel: Intracellular proton concentration (H^+_i) as a function of time (s), calculated from the recorded intracellular FITC-dextran fluorescence changes induced by sudden jumps of the extracellular proton concentration to the values indicated in the figure. The proton concentrations were calculated from the fluorescence readings using a dissociation constant for FITC-dextran of 0.505 mM (pK 6.3). The curves represent fits to the data of the empirical equation: $H^+_i = H^+_{ioo} + (H^+_{io} - H^+_{ioo}) \times \exp(-kt)$. H^+_i, H^+_{io}, H^+_{ioo} indicate, respectively, intracellular proton concentrations at times t, zero, and infinity, in µM. The curves start at the same initial pH of 7.05, i.e. at the pH at which the ghosts had been resealed prior to subjecting them to the pH jump. The rate constants k or the initial slopes of a semilog representation of the data were used as a measure of the initial rate of proton uptake $\{(dn_H/C\ dt)_{(t=0)} = m_{(H)}\}$ which pertains to the initial intracellular proton concentration of 0.1 µM. Since the buffer capacity inside the ghosts (C) is constant, the changes of the *concentration* H^+_i are a direct measure of the *amount* of protons (n_H) taken up at time t: $n_H = C \cdot H^+_i(t)$. – The constancy of the buffer capacity dn_H/dpH over the pH range covered in the experiments is shown at the lower left. The abscissa indicates pH, the ordinate µl of 0.1 N NaOH added to 10 ml buffer (1.25 mM EDTA, 1.875 mM glycylglycine, 160 mM NaCl plus white ghosts, 0.0175 OD at 280 nm, 20 °C). – The upper left shows the applicability of the mass law to the FITC fluorescence entrapped inside the ghosts. Upon excitation at 495 nm, emission was measured at 518 nm (a) in buffered, ghost-free media (closed circles) and (b) in ghosts with entrapped FITC-dextran (open circles). The ghosts were suspended in FITC-dextran-free media buffered to the same pH values as in (a) by the addition of a trace of the proton carrier HN_3, which mediates rapid pH equilibration between external medium and FITC-containing interior of the ghosts. The drawn curve represents a fit to the data with pK = 6.3. The ordinate is normalized with respect to the maximal fluorescence at complete dissociation of FITC-dextran at high pH. Abscissa: proton concentration

course of fluorescence change and hence of proton uptake can be described by a single exponential (Fig. 10.2, right panel), which extrapolates back to the intracellular pH originally established during resealing of the ghosts before initiating the proton uptake in the stopped flow apparatus by the sudden pH jump. This permits the calculation of the initial rate of proton uptake pertaining to the initially existing intracellular pH and the extracellular pH established at the instant of mixing ghosts and extracellular medium (Fig. 10.2).

10.3 H^+-Cl^- Cotransport and its Relationship to Cl^-/Cl^- Exchange

In any cotransport process, the rate of transport should depend on the concentrations of all cotransported substrate species. So far, the dependence of H^+-Cl^- cotransport on chloride concentration had never been studied. Therefore, to confirm the occurrence of H^+-Cl^- cotransport, in the first set of experiments described below, we investigated the dependence of cotransport on chloride concentration at $Cl^-_o = Cl^-_i$ after a pH jump in the external medium from always the same initial pH_o of 7.05 (which is identical to the pH inside the ghosts before initiation of cotransport) to $pH_o = 5.6$. As described above (see Figs. 10.1 and 10.2), after the jump, the latter pH remains constant throughout the whole time course of the experiment and is equal to the intracellular pH, which the equilibration process tends to approach. The rates of net proton uptake increase with increasing Cl^- concentration and tend to approach a saturation value (Fig. 10.3). The data can be fitted to the Hill equation with n = 2 and a half saturation constant of about 200 mM. These findings are compatible with a cotransport process in which Cl^- is translocated jointly with a proton.

The results presented above only obtain over the Cl^- concentration range between 80 and 320 mM. Below and above this range, for unexplored reasons, the time course of cotransport can no longer be described by a single exponential (Fig. 10.4). Nevertheless, even under these experimental conditions, cotransport could always be inhibited with H_2DIDS by more than 95 percent, indicating that the effects were essentially related to band 3.

In a second set of experiments, we studied the pH dependence of H^+-Cl^- cotransport again at equal Cl^- concentrations inside and outside the ghosts, but the chloride concentration was fixed at 0.160 M. The cotransport was initiated by pH jumps in the external medium from pH 7.05 to various levels ranging from pH 5.3 to 6.7. As shown in Figs. 10.1 and 10.2 and described in the preceding section, the pH newly established in the medium after the jump remains constant while the pH inside the ghosts changes until it becomes equal to the pH outside. From the rate constants of the single exponentials fitted to the data points, the initial rates of proton uptake ($m_{(H)}$) pertaining to the internal pH at time t = 0 were calculated as described in the legend to Fig. 10.2. When $m_{(H)}$ is plotted against the external proton concentration, the data can be fitted to saturation curves with Hill coefficients of n = 1 or n = 2. The quality of the fit with n = 2 is slightly better than with n = 1 but the difference is not significant. In either case, the half saturation concentrations

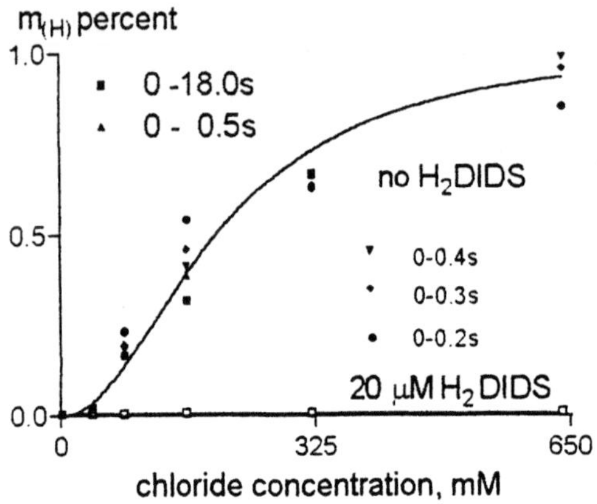

Fig. 10.3. Rate of net proton influx $m_{(H)}$ in the absence and presence of H_2DIDS as a function of Cl^- concentration (mM) at equal Cl^- concentrations inside and outside the ghosts. At all Cl^- concentrations, the proton influx was initiated by the same jump of the external proton concentration from 0.09 to 2.51 μM. The external proton concentration of 0.09 μM is equal to the internal proton concentration at t = 0. The various symbols in the figure indicate $m_{(H)}$ values obtained when the respective uptake curves are truncated after the time intervals indicated in the figure. Although the scatter increases with decreasing number of data points used for the calculation of the $m_{(H)}$ values, no systematic deviation of the calculated initial uptake rates becomes apparent. The data obtained in the absence of H_2DIDS represent a Hill plot fitted with n = 2.0: $m_{(H)} = F\,[Cl^-]^2/(K+[Cl^-]^2)$, where F = 1.03 ± 0.10 was fixed. The half saturation concentration (= sqrt K) amounts to 203 ± 64 mM. In spite of the scatter of the rate constants for each one of the time segments measured at the various chloride concentrations, the Hill plots for the rate constants pertaining to these time segments nearly coincide. This indicates that, at Cl^- concentrations of 80 mM or more, the intracellular pH changes due to cotransport exert no detectable effect on transport rate

are rather similar. Over the whole range of external proton concentrations, proton influx is inhibited by the anion transport inhibitor H_2DIDS to more than 95% (Fig. 10.5).

The results described above confirm and extend, with a different technique, the original observation of Jennings (1978) that after inhibition of the Jacobs-Parpart cycle, pH equilibration between cells and medium is accomplished by chloride-proton cotransport. They shed, in addition, some light on the relationship between H^+-Cl^- cotransport and Cl^-/Cl^- exchange (Fig. 10.6). A comparison of the rate of net proton uptake at constant and equal Cl^- concentrations in- and outside the ghosts (160 mM) as a function of extracellular pH with data on pH dependence of Cl^-/Cl^- exchange taken from the literature (Funder and Wieth 1976) shows that normalized plots of cotransport and equilibrium exchange against pH are nearly mirror images of one another. The curves intersect at pH 5.7–5.9 where both activation of cotransport and inhibition of exchange become approx. half maximal. Since the maximal rates of cotransport and exchange differ by a factor of about 1000 we conclude that lowering the pH transforms band 3 from the rapidly operating anion exchanger to the very much more slowly operating H^+-Cl^- cotransporter. To a first approximation, we stipulate that the sum of exchange and cotransport mediating forms of band 3 is constant and the binding of a proton converts one form into the other. On the basis of our data we are unable to decide, whether or not the proton

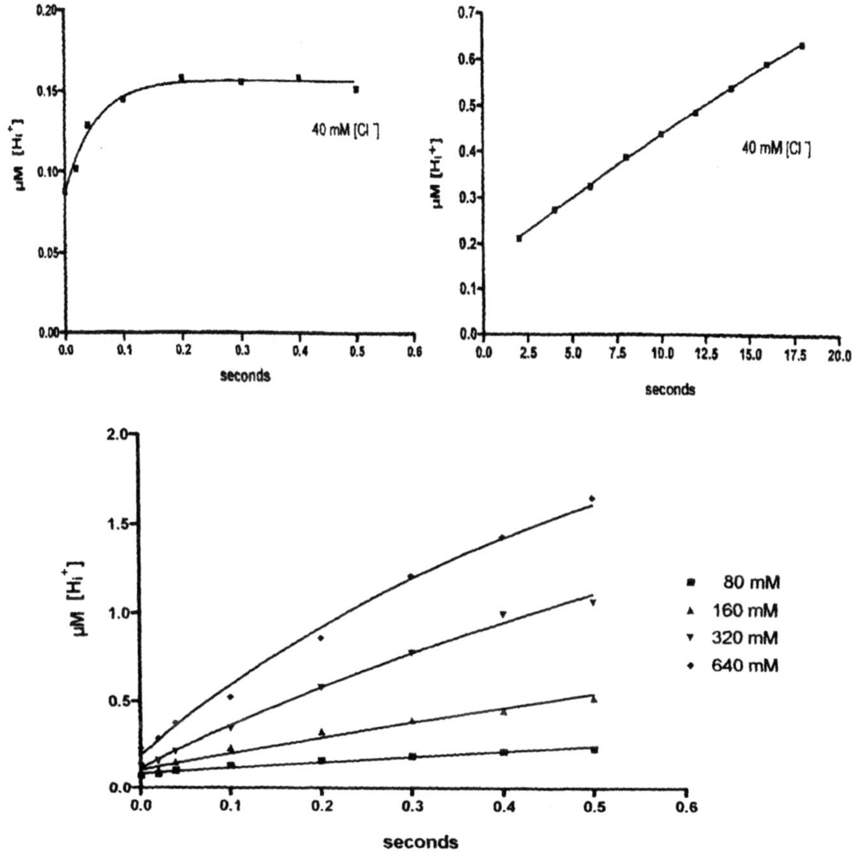

Fig. 10.4. Time course of proton equilibration after jumps of the external proton concentration from 0.09 to 2.51 μM at Cl^- concentrations ranging from 40–640 mM. Ordinate: intracellular proton concentration in μM. Abscissa: time. Lower panel: Fits of single exponentials (with $[H^+_{i\infty}]$ = 2.54 μM) to the data pertaining to chloride concentrations ranging from 80–640 mM. Upper panels: Time course of proton equilibration at 40 mM Cl^-. Left: The rapid initial phase of equilibration. The curve represents a single exponential. The data point at t = 0 was obtained by extrapolation of the fitted curve. Right panel: Continuation of the equilibration process depicted in the middle, covering the time interval from 2–18 s. The curve is not a strictly straight line but represents a fit to the data of the exponential represented in the legend of Fig. 10.2. $[H^+_{i\infty}]$ = 2.54 μM. The initial rate of proton equilibration during the rapid phase depicted in the middle panel can only be guessed. Based on such guess, it appears that at low Cl^- concentration, during the transition from rapid to slow phase, the rate constant of proton uptake decreases about 30-fold

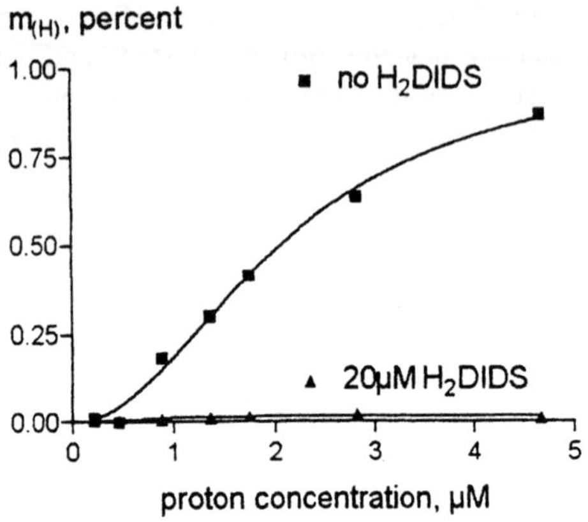

Fig. 10.5. Rate of net proton uptake, $m_{(H)}$, as a function of external proton concentration. The influx was measured in the absence and presence of H_2DIDS at constant Cl^- concentration (160 mM) inside and outside the ghosts after sudden jumps of external proton concentration from 0.1 µM (identical to the internal proton concentration at t = 0) to the external proton concentrations indicated on the abscissa (µM). In the course of the equilibration process, the intracellular proton concentration tends to approach these values. The drawn curves represent Hill plots with n = 2: $m_{(H)} = F \cdot [H^+_{oo}]^2/(K+[H^+_{oo}]^2)$. In the absence of H_2DIDS: F = 7.69 ± 0.41, sqrt(K) = 2.1 ± 0.64 µM, corresponding to pH 5.68. It is also possible to obtain converging fits for n = 1. They are slightly less satisfactory, but the K values obtained are close to that derived from the quadratic expression (K = 1.32 µM, corresponding to pH 5.87, see Fig. 10.7B, lowest panel to the right)

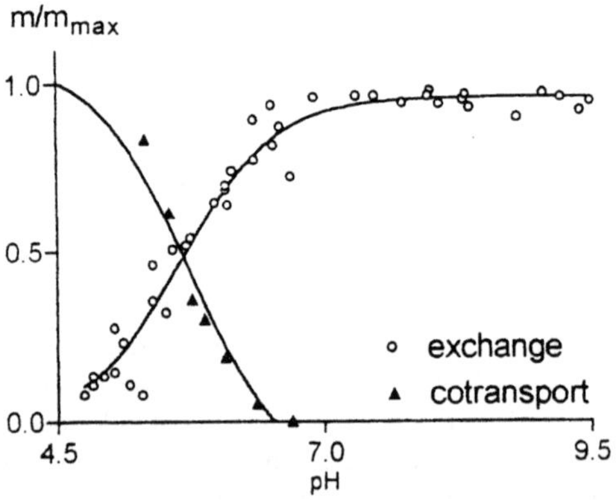

Fig. 10.6. The inverse relationship of the pH dependences of chloride/chloride equilibrium exchange and H^+-Cl^- cotransport as measured at $Cl_i^- = Cl_o^-$ = 160 mM. Data for Cl^-/Cl^- exchange from several sources (notably from Funder and Wieth 1976), slightly modified for adaptation to 20 °C (Müller-Berger et al. 1995). Data for cotransport from the present contribution. Ordinate: rates of exchange or cotransport as fractions of their respective plateau values

that accomplishes this conversion is itself transported or if an additional proton is required to render possible the net Cl^- transfer. Our interconversion hypothesis is supported by our finding that at a given proton gradient across the membrane, increasing the chloride concentration at $[Cl^-_i] = [Cl^-_o]$ enhances the rate of Cl^- uptake, as one would expect for a saturable cotransport process which involves the joint translocation of H^+ and Cl^-.

10.4 Molecular Basis of the Relationship between Anion Exchange and Anion-Proton Cotransport

In a comment on our results, M.L. Jennings suggested to us a hypothesis about the molecular mechanism of the interconversion between exchange and cotransport:

At sufficiently low pH, the binary complex of the deprotonated anion exchanger, loaded with one single Cl ion, combines with one proton and one additional Cl ion. The resulting quaternary complex, consisting of band 3, two Cl ions, and one proton (designated HClTCl) accomplishes the slow cotransport of the proton and the additional Cl^-. The quaternary complex is formed at the expense of the binary complex with the single chloride responsible for the rapid Cl^-/Cl^- exchange (TCl). Thus, the appearance of cotransport at low pH would necessarily lead to the observed inhibition of rapid anion exchange:

HClTCl	← slow →	HClTCl	slow H^+-Cl^- cotransport
|		|	
H + Cl + TCl	← very fast →	TCl + Cl + H	fast Cl^-/Cl^- exchange

We observed that chloride activation of cotransport is saturable and seems to occur with a Hill coefficient of two. In terms of the Jennings hypothesis, one of the Cl ions involved could be identical to the Cl ion that combines with the "transfer site" for anion exchange. The other could possibly combine with the so-called "modifier site" (an idea originally contributed by J. Salhany, but later abandoned by him), the occupancy of which is known to inhibit anion exchange (Dalmark 1975). When the external medium is suddenly acidified in the stopped flow apparatus, an inwardly-directed proton gradient is established. Under its influence, the two chloride ions bound to the transfer and the modifier site, and the proton are jointly transferred across the membrane. At the inner membrane surface, the proton and the two Cl ions dissociate off. The transfer site of the unloaded band 3 picks up one single Cl ion, which rapidly returns via the anion exchange mechanism to the outer surface from where the cotransport had originated. The Cl ion is discharged and replaced by another Cl ion. The exchanger is now ready to pick up another pair of H^+ and Cl^- for another round of slow inwardly-directed cotransport.

To avoid confusion in the further discussion of this model, the transfer site involved in Cl^-/Cl^- exchange will be called "exchange site" and designated by the letter *a*, the modifier site involved in the inhibition of Cl^-/Cl^- exchange mediated by *a*, will be called "cotransport site" and designated by the letter *b*. Finally, the letter *p* designates the proton binding site.

Although the model is basically rather simple, closer inspection reveals a considerable complexity. This is due to the fact that the occupancies of the chloride binding sites *a* and *b* and of the proton binding site *p* may vary with the composition of the internal and external media. The possible permutations are indicated in the diagram (D1):

$$\begin{array}{ccccccccc} \mathrm{TCl} & - & \mathrm{T} & - & \mathrm{ClT} & - & \mathrm{ClTCl} & - & \mathit{ClTClH} \\ | & & | & & | & & | & & | \\ \mathrm{HTCl} & - & \mathrm{HT} & - & \mathrm{HClT} & - & \mathrm{HClTCl} & - & \mathit{HClTClH} \end{array} \qquad \text{(D1)}$$

In this diagram we indicate chloride binding to *a* by placing the symbol Cl to the right of the symbol for the transporter, T, and chloride binding to *b* by placing the symbol Cl to the left of T. Proton binding to site *p* is indicated by placing the symbol H to the left of the chloride bound to the cotransport site (*b*). Using these conventions, the above diagram represents the various states of occupancy of the transporter. It shows that, at each surface, there exist eight different states of occupancy interconnected by mass law equations (indicated by the dashes between the symbols) with eight independent mass law constants. Cotransport is accomplished by the form HClTCl. After its formation in the more acidic medium at the outer membrane surface, it is slowly converted to its inwardly-facing conformer, which discharges the bound proton and the bound chloride to the more alkaline intracellular medium. This transforms HClTCl to TCl, the form that mediates the rapid Cl^-/Cl^- exchange. Thus, the inward-facing TCl is rapidly reoriented and faces again the external medium. There it may pick up another proton and chloride ion for another round of cotransport. At the elevated internal pH, the return of the transporter from inward to outward orientation does never become rate limiting and, for the present purpose, does not require further attention.

The diagram includes, in *italics*, the forms ClTClH and HClTClH. They serve to remind the reader that we are unable to exclude the existence of these forms (see legend to Fig. 10.5 and the accompanying discussion). In what follows, the possible influence of these forms will be neglected. Within the framework of the present hypothesis, for stoichiometrical reasons, a second proton could not be cotransported. It could only exert a modulating effect, which is unlikely to obscure the essential features of the model.

Previous work has demonstrated that, apart from the forms, which we stipulate to be predominantly responsible for the cotransport cycle (TCl and HClTCl) the intermediate forms ClTCl (Dalmark 1975), HTCl, HT, and T (Milanick and Gunn 1986) do in fact occur. Thus, our findings and speculations seem to represent a rather natural extension of the comprehensive previous work of others, which did not include the possible occurrence of H^+-Cl^- cotransport, but was confined to the pH dependence of the intermediates responsible for Cl^-/Cl^- equilibrium exchange.

Due to the various interconnected mass law relationships, the sum of TCl and HClTCl is not necessarily independent of pH; hence the stoichiometry of the relationship between inhibition of anion exchange and emergence of proton-chloride cotransport does not need to be strictly 1:1 over any possible pH and chloride concentration range. As shown in Fig. 10.6, under our experimental conditions the inhibition of Cl^-/Cl^- exchange and activation of H^+-Cl^- cotransport form nearly mirror images of one another. This implies that under these conditions the interconversion of band 3 from anion exchanger (TCl) to anion-proton cotransporter (HClTCl) is the predominant factor. Apparently, the specific numerical values of the various mass law constants are responsible for this fact.

Measuring the dependence of chloride equilibrium exchange on external and internal Cl^- and proton concentration, Milanick and Gunn were not only able to provide convincing evidence for the actual occurrence of the proton-binding forms indicated above. In addition they managed to derive numerical values for a number of mass law constants governing some of the equilibria among the various forms represented in the diagram (D1). They also considered an involvement of the modifier site (= cotransport site in our terminology, see above), but stated that their experimental conditions were limited to a range where their methods were not accurate enough to discover its possible participation in the various equilibria. On the basis of their findings they suggested that the inhibition of Cl^-/Cl^- equilibrium exchange is entirely due to the conversion of the deprotonated, rapidly moving exchanger (TCl) into its immobile protonated forms (HT and HTCl).

The cotransport model in the version presented in the diagram (D1) encompasses, in addition to the Cl^- loaded exchanger TCl and the forms T, HT, and HTCl, further forms, notably HClTCl, which we suppose to be needed for the cotransport of H^+ and Cl^- to take place. Thus, the elucidation of the relationship between exchange and cotransport requires reconciliation, within a single theory, of our cotransport measurements with the measurements of Cl^-/Cl^- exchange by Milanick and Gunn and other authors.

10.5 Chloride Equilibrium Exchange and Chloride-Proton Cotransport: Comparison of Experimental Evidence with Theoretical Predictions

In order to extend the work of Milanick and Gunn (1986) and to take the influence of the formation of the cotransporting band 3 species into account, we derived the equations for the pH dependence of equilibrium exchange and cotransport as outlined in Table 10.1. The derivations are based on the assumption that the dependence on chloride and proton concentration of both processes is determined by the same reaction diagram (D1) with identical numerical values of the various mass law constants (assumption 3 in Table 10.1). The derivations follow conventional patterns. We assume that in anion exchange without net transport, chloride influx and efflux are equal (assumption 1) while for cotransport the difference between the two fluxes has a finite value (assumption 2). We stipulate that the exchange flux is mediated by TCl while the net cotransport is accomplished by HClTCl. We further assume that the sum of all forms in (D1) is constant and independent of pH and chloride concentration (assumption 4, mass balance). By algebraic transformations we obtain equations for equilibrium exchange fluxes (m_{Cl}) and net H^+-Cl^- cotransport (m_H).

We first used the abundant experimental evidence on anion equilibrium exchange to determine the 11 parameter values, which are required to describe the

Table 10.1. Chloride equilibrium exchange and chloride-proton cotransport: theoretical predictions

(1) anion exchange flux without net transport:

$$j_{ex} = \kappa_1 \cdot T_oCl - \kappa_{11} \cdot T_iCl = 0, \text{ hence } \kappa_1 \cdot T_oCl = \kappa_{11} \cdot T_iCl = m_{Cl}$$

(2) anion-proton cotransport: $j_{co} = k_1 \cdot HClT_oCl - k_{11} \cdot HClT_iCl = m_H$

(3) mass law-governed occupancy of two Cl binding sites and one proton binding site at the band 3 protein yields eight different band 3 species. We assume that chloride binding is a cooperative process, where the binding of chloride to the exchange site facilitates the binding of another chloride to the cotransport site. We postulate, therefore, that the occupancy of the cotransport site in the absence of chloride binding to the exchange site is negligible. Thus, the system can be described by 10 mass law constants (K) and one distribution coefficient, q:

$$\frac{T_o \cdot H_o}{HT_o} = K_1 \quad \frac{T_oCl \cdot H_o}{HT_oCl} = K_2 \quad \frac{ClT_oCl \cdot H_o}{HClT_oCl} = K_3 \quad \frac{T_o \cdot Cl_o}{T_oCl} = K_a \quad \frac{T_oCl \cdot Cl_o}{ClT_oCl} = K_b$$

A corresponding set of equations pertains to the inner membrane surface, with the corresponding mass law constants K_{11}, K_{22}, K_{33}, K_{aa}, K_{bb}.

(4) mass balance:

$$T_{tot} = \left\{1 + \frac{H_o \cdot K_a}{K_1 \cdot Cl_o} + \frac{H_o}{K_2} + \frac{H_o \cdot Cl_o}{K_3 \cdot K_b} + \frac{K_a}{Cl_o} + \frac{Cl_o}{K_b}\right\} \cdot T_oCl$$

$$+ \left\{1 + \frac{H_i \cdot K_{aa}}{K_{11} \cdot Cl_i} + \frac{H_i}{K_{22}} + \frac{H_i \cdot Cl_i}{K_{33} \cdot K_{bb}} + \frac{K_{aa}}{Cl_i} + \frac{Cl_i}{K_{bb}}\right\} \cdot T_iCl$$

(5) using (1), $T_iCl = q \cdot T_oCl$ (where $q = \kappa_1/\kappa_{11}$), and solving for T_oCl yields for equilibrium exchange (unidirectional flux):

$$m_{Cl} = \frac{F \cdot Cl_i \cdot Cl_o}{Cl_i \cdot \left\{K_a\left(1+\frac{H_o}{K_1}\right) + Cl_o\left(1+\frac{H_o}{K_2}\right) + \frac{Cl_o^2}{K_b}\left(1+\frac{H_o}{K_3}\right)\right\} + Cl_o \cdot q\left\{K_{aa}\left(1+\frac{H_i}{K_{11}}\right) + Cl_i\left(1+\frac{H_i}{K_{22}}\right) + \frac{Cl_i^2}{K_{bb}}\left(1+\frac{H_i}{K_{33}}\right)\right\}}$$

where $F = \kappa_1 \cdot T_{tot}$

(6) calculating $HClT_oCl$ and $HClT_iCl$ from the mass law relationships (3), equating $Cl_i = Cl_o = Cl$ (corresponding to the conditions under which all cotransport experiments had been conducted), and inserting into (2) yields for cotransport (net flux):

$$m_H = \frac{k_1 \cdot T_{tot} \cdot \frac{Cl^2}{K_b \cdot K_3} \cdot (H_o - H_i)}{K_a\left(1+\frac{H_o}{K_1}\right) + Cl\left(1+\frac{H_o}{K_2}\right) + \frac{Cl^2}{K_b}\left(1+\frac{H_o}{K_3}\right) + q \cdot \left\{K_{aa}\left(1+\frac{H_i}{K_{11}}\right) + Cl\left(1+\frac{H_i}{K_{22}}\right) + \frac{Cl^2}{K_{bb}}\left(1+\frac{H_i}{K_{33}}\right)\right\}}$$

The equations for m_{Cl} and m_H can be simplified on the assumption that H_o/K_1, H_o/K_2, $H_i/K_{33} << 1.0$ and $K_{bb} >> Cl_i^2$. Hence these terms can be neglected (see legend to Table 10.2 for justification).

(7) The derivations can be extended to obtain equations for exchange and cotransport in a system capable of forming band 3 species with two chloride and two proton binding sites, by adding mass law equations that account for the existence of the species T_oH, ClT_oH, ClT_oClH, and $HClT_oClH$ and the corresponding species facing the opposite membrane surface i. Expressions for equilibrium exchange and cotransport are obtained if suitable assumptions are made about which ones of these protonated forms are able to undergo the transport-accomplishing conformational change.

available data in terms of the pertinent equations represented in Table 10.1. We then attempted to apply the numerical values so obtained for the various parameters to fit our data on cotransport: According to the theory, except for a single normalizing (constant) factor, the same set of parameter values should suffice to provide a satisfactory fit to all of them, regardless of the experimental conditions with respect to the concentrations of internal and external protons and chloride ions.

The data on equilibrium exchange were derived from the published work of Funder and Wieth (1978), Knauf and Mann (1986), and Milanick and Gunn (1986). The pertinent figures in the papers of these authors were magnified, the coordinates of the data points were read off, multiple determinations were averaged, and the averages were used for non-linear curve fitting with a suitable statistics program (PRISM). The curves presented in Figs. 10.7A and B, including those on cotransport, were obtained with the same set of fixed parameter values, except for the respective individual normalizing factor. They indicate that the data on anion exchange from several different laboratories can be used to predict our observations on the dependence of cotransport on proton and chloride concentration. Although this does not prove that our theoretical treatment represents a strictly unique result, it shows that it is *compatible* with the best experimental results currently available. It supports our intuition that the inhibition of rapid anion exchange with decreasing pH is *not only* associated with a decrease of TCl at the expense of the formation of HT and HTCl (as postulated by Milanick and Gunn 1986) *but also, and predominantly*, with the emergence of very slow chloride-proton cotransport; it also provides information on a number of details of the kinetics of the cotransport cycle.

As suggested by Jennings, according to the model, the inwardly-directed cotransport is accomplished by binding to T_o of a single proton and two chloride ions. One of the Cl ions tends to combine loosely with the externally-oriented transfer site *a* to form HT_oCl (with mass law constant K_{22} of about 10 μM, as estimated by Milanick and Gunn 1986, Table 4 therein) and the other one (as originally suggested by Salhany) to the externally oriented, Cl^- loaded modifier site *b* to form $HClT_oCl$. The binding of the second chloride ion increases the apparent affinity for proton binding to *p* by almost eight-fold as indicated by the decrease of the mass law constant for proton binding from about 10 μM to 1.32 μM. It also increases cooperatively the strength of Cl^- binding to the cotransport site (= externally-oriented Dalmark's (1975) modifier site) by about the same factor. After the conformational change from $HClT_oCl$ to $HClT_iCl$, H^+-Cl^- is released into the cell interior. This release is facilitated by a decrease of the apparent affinity for proton binding to T_iCl as compared to T_oCl (K_{33} is large, see Table 10.2). In accord with Jennings' suggestion, the ensuing form T_iCl rapidly returns to the conformation T_oCl. The transport cycle is completed. T_oCl is now able to pick up another pair of H^+-Cl^- for another round of cotransport. This process accomplishes a sustained net H^+-Cl^- influx until Donnan equilibrium is attained. It could be slightly modulated by the formation of small amounts of HT_i and HT_iCl, which does not only affect the number of TCl species available for exchange but also of HClTCl species involved in cotransport.

The observations just described emphasize that allosteric relationships between the binding sites of the proton, *p*, and of the second chloride, *b*, exert a considerable influence on the inhibition of the rapid Cl^-/Cl^- exchange mediated by *a*. In other words, the model does not imply that band 3 represents a double barrelled transporter which is capable of independently mediating exchange and cotransport. Upon binding of the proton to *p* and of the second chloride to *b*, the Cl ion bound to the exchange site *a* becomes entrapped and can no longer mediate the rapid transmembrane Cl^-/Cl^- exchange. The entrapped Cl^- is also no longer able to exchange with the chloride in the adjacent media. This exchange is a prerequisite for the line broadening seen in 35-Cl NMR spectroscopy. As was shown by Falke and Chan (1985) chloride binding to the so-called modifier site (site *b* in the present terminology) cannot be observed by NMR measurements. Thus, our observation that lowering the pH immobilizes the Cl ion at the exchange site, *a*, could explain why, with decreasing pH, the 35-Cl NMR signal decreases strictly parallel to the inhibition of Cl^-/Cl^- exchange. It thus supports an earlier hypothesis based on our previous observations on the relationship between the effects of pH on Cl^-/Cl^- exchange and 35-Cl NMR line broadening (see Müller-Berger et al. 1995, Fig. 6 therein).

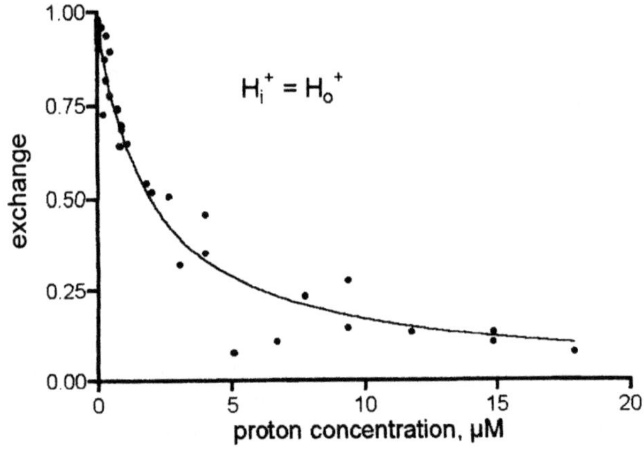

Fig. 10.7. A, B. Fit of the equations in Table 10.1 for Cl^-/Cl^- exchange (m_{Cl}) and H^+-Cl^- cotransport (m_H) to experimental data by a nonlinear curve fitting procedure (PRISM). All fits were performed with the parameters fixed to the values listed in Table 10.2 except for the scaling factors $F = \kappa_i \cdot T_{tot}$ (for equilibrium exchange) or $k_i \cdot T_{tot}$ (for cotransport).

Fig. 10.7A. Dependence of Cl^-/Cl^- exchange on proton concentration at constant and equal chloride concentrations inside and outside the ghosts ($Cl_i = Cl_o = 0.166$ M). The proton concentrations inside and outside the ghosts are varied together such that intracellular and extracellular concentrations are always equal. Data from Funder and Wieth (1976)

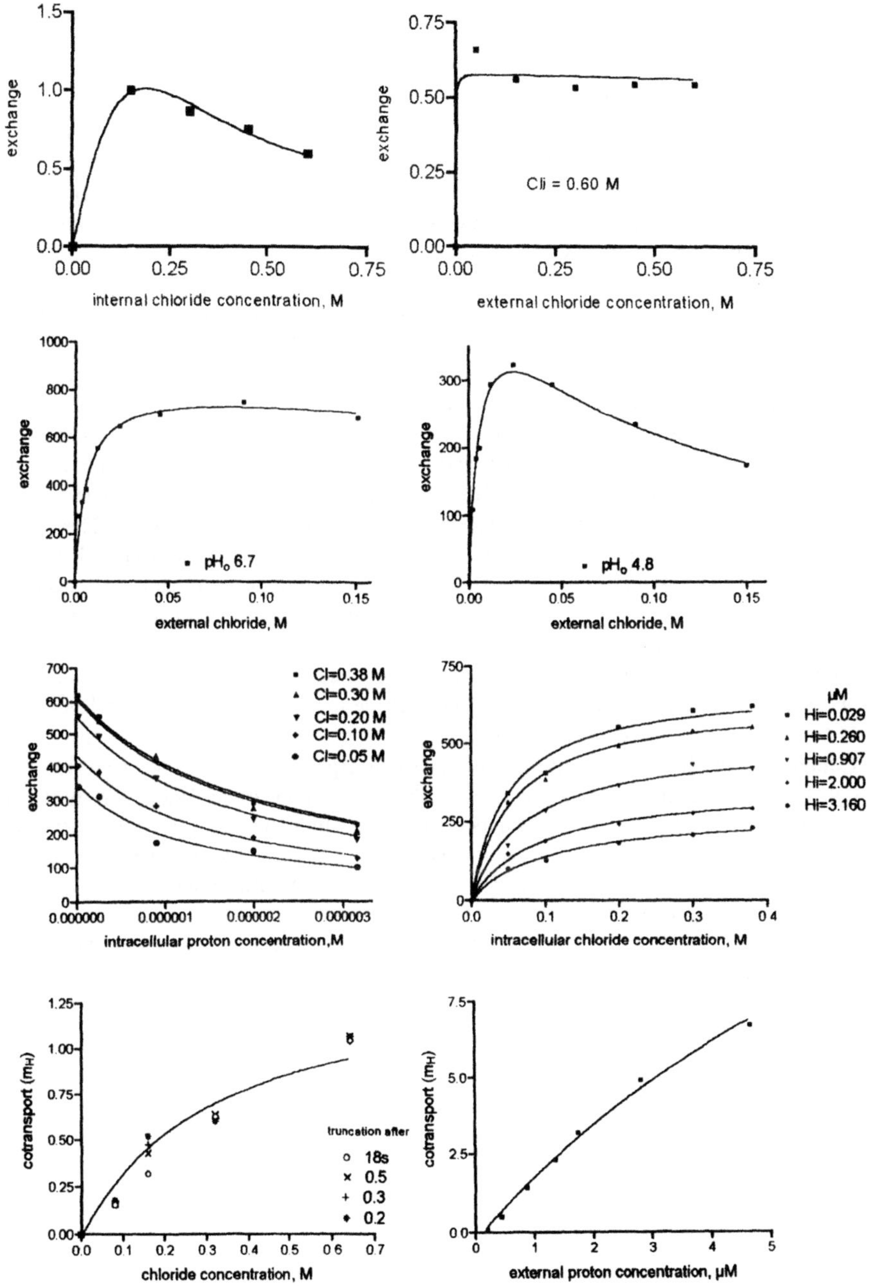
exchange
internal chloride concentration, M
exchange
Cli = 0.60 M
external chloride concentration, M
exchange
pH$_o$ 6.7
external chloride, M
exchange
pH$_o$ 4.8
external chloride, M
Cl=0.38 M
Cl=0.30 M
Cl=0.20 M
Cl=0.10 M
Cl=0.05 M
exchange
intracellular proton concentration,M
μM
Hi=0.029
Hi=0.260
Hi=0.907
Hi=2.000
Hi=3.160
exchange
intracellular chloride concentration, M
cotransport (m$_H$)
truncation after
18s
0.5
0.3
0.2
chloride concentration, M
cotransport (m$_H$)
external proton concentration, μM

Fig. 10.7 A, B. Fit of the equations in Table 10.1 for Cl^-/Cl^- exchange (m_{Cl}) and H^+-Cl^- co-transport (m_H) to experimental data by a nonlinear curve fitting procedure (PRISM). All fits were performed with the parameters fixed to the values listed in Table 10.2 except for the scaling factors $F = \kappa_1 \cdot T_{tot}$ (for equilibrium exchange) or $k_1 \cdot T_{tot}$ (for cotransport).
Fig. 10.7B. First row: Cl^-/Cl^- equilibrium exchange as a function of external chloride concentration at fixed internal (0.6 M) and varied external chloride concentration (right panel), or at fixed external (0.15 M) and varied internal chloride concentration (left panel). In both cases the internal and external pH values are constant and equal to 7.2. The points at the origins of both curves were not measured but indicate that in the absence of Cl^- on either the inner or the outer membrane surface exchange must approach zero. (Data from Knauf and Mann (1986), Figs. 2 and 1, respectively). – **Second row**: Dependence of Cl^-/Cl^- exchange on extracellular chloride concentration at a fixed intracellular chloride concentration of 0.11 M and fixed internal pH of 7.6. Left: external pH 6.7. Right: external pH 4.8. (Data from Milanik and Gunn (1986), Fig. 7). – **Third row**, left: Dependence of Cl^-/Cl^- exchange on intracellular proton concentration at a fixed extracellular chloride concentration of 0.05 M, a fixed extracellular pH of 7.1, and a range of fixed intracellular chloride concentrations indicated in the figure. Right: The same data replotted as a function of intracellular chloride concentration for a range of fixed intracellular proton concentrations. The conditions in the extracellular medium are the same as in the left panel. (Data from Milanik and Gunn (1986), Fig. 2). – **Fourth row,** left: Proton influx as a function of chloride concentration at $Cl_i = Cl_o$, and a fixed proton concentration difference across the membrane; initial intracellular proton concentration 0.089 µM. Right: Proton influx as a function of external proton concentration at a fixed initial internal proton concentration of 0.089 µM and a fixed chloride concentration of 0.16 M inside and outside the cells. For details, see legends to Figs. 10.3 and 10.5, respectively.
Comments: In all cases, the fitting procedure converged with r^2 equal to or greater than 0.98 (r = correlation coefficient), except for the data of Funder and Wieth (1976) and of Knauf and Mann (1986), Fig. 10.7B upper panel, right. In both cases, $r^2 = 0.96$. Note: *None of the individual fits suffices to prove its uniqueness.* Thus, the reader has to relay on his own judgment on the qualities of the fits and the range of experimental conditions covered. It seems obvious, however, that the data from the various sources and obtained under a variety of experimental conditions are at least qualitatively compatible with the predictions of the model

Table 10.2. Numerical values of the parameters used for the evaluation of the equations about H^+-Cl^- cotransport and Cl^-/Cl^- exchange as defined by the model underlying the equations for m_H and m_{Cl} in Table 10.1

Orientation		Orientation	
Outside	Reaction	Inside	Reaction
K_a 0.00765 [1])	$K_a = T_o \cdot Cl_o / T_oCl$	K_{aa} 0.058 [3])	$K_{aa} = T_i \cdot Cl_i / T_iCl_i$
K_b 0.971 [1])	$K_b = T_oCl \cdot Cl_o / ClT_oCl$	K_{bb} large [4])	$K_{bb} = T_iCl_i \cdot Cl_i / ClT_iCl$
K_1 large [2])	$K_1 = T_o \cdot H_o / HT_oCl$	K_{11} 0.84 μM [3])	$K_{11} = T_i \cdot H_i / HT_iCl$
K_2 large [2])	$K_2 = T_oCl \cdot H_o / HT_oCl$	K_{22} 2.00 μM [3])	$K_{22} = T_iCl \cdot H_i / HT_iCl$
K_3 1.32 μM	$K_3 = ClT_oCl \cdot H_o / HClT_oCl$	K_{33} large [1])	$K_{33} = ClT_iCl \cdot H_i / HClT_iCl$
q 0.132	$q = \kappa_1/\kappa_{11}$		

All values taken from Milanick and Gunn (1986), except for q, K_3, and K_{33}. [1]) Unnumbered equation on p. C961, K_a is average of the values for pH 4.8 and 6.7. If not indicated otherwise, all data in M = moles/l. [2]) Large as compared to the respective proton concentrations; an approximation compatible with our results is $K_1 \cong 10$ μM (in Milanick and Gunn (1986), Table 4, p. C961). [3]) From unnumbered equation on p. C960. K_b is slightly higher than given by Milanick and Gunn (0.910 ± 0.390 M) but within their rather wide margin of error. [4]) From the data of Knauf and Mann (1986), $K_{bb} = 0.61$, which is large enough to be compatible with our analysis. This value is also not significantly different from that for K_b = 0.971.

10.6 H^+-SO_4^{2-} Cotransport and its Relationship to H^+-Cl^- Cotransport: General Features

In the experiments described in the preceding pages, cotransport was only measured under conditions where the difference of the proton concentrations inside and outside the red cell ghosts acted as a driving force to induce net movements of H^+-Cl^- across the membrane. However, even in the absence of concentration differences, under conditions of no net flow, and hence under the sole influence of heat movements, an equilibrium exchange of H^+-Cl^- takes place. Under these conditions, each pair of H^+ and Cl^- entering the ghosts is replaced by another one leaving the ghosts. This equilibrium exchange can only be followed by means of radioisotopes. The contribution of H^+-Cl^- cotransport to the isotopically measured exchange of labelled chloride across the red cell membrane is negligibly small and would amount to no more than about 0.1% of the total Cl^-/Cl^- exchange flux (Jennings 1978). This behavior differs from that of the isotopically measured sulfate equilibrium exchange. Most of it is mediated by H^+-SO_4^{2-}/H^+-SO_4^{2-} exchange while the contribution of SO_4^{2-}/SO_4^{2-} exchange can be neglected, except at very high pH values where the supply of protons becomes rate-limiting. Estimates of Milanick and Gunn suggest that under conditions of equilibrium exchange, sulfate cotransport occurs at a rate, which exceeds that of sulfate/sulfate exchange about 200 times. Thus, measuring equilibrium exchange by means of radiosulfate essentially yields information on H^+-SO_4^{2-} equilibrium exchange.

In Fig. 10.8, the curve designated "0.11 M sulfate" depicts the relationship between isotopically measured sulfate equilibrium exchange and pH in the absence of chloride. The nearly bell-shaped curve passes through a maximum. When the curve is decomposed into two superimposed dissociation curves, one related to activation, the other to inhibition of H^+-SO_4^{2-} equilibrium exchange, one obtains two nearly identical apparent pK values of about 6.2. This points to the involvement of two proton binding sites with rather similar apparent pK values.

The picture is dramatically changed (Fig. 10.9) if one measures the pH dependence of H^+-SO_4^{2-} cotransport from an all sulfate medium into cells containing Cl^- in place of SO_4^{2-} (Milanick and Gunn 1984). Since the rapidly penetrating chloride substitutes for the slowly penetrating sulfate, the vast majority of the transfer sites of band 3 are "recruited" to the outer membrane surface. Thus, every band 3 molecule, which delivers a pair of H^+-SO_4^{2-} to the cell interior, returns almost instantaneously to the outer membrane surface to pick up another pair of H^+ and SO_4^{2-} such that the net influx of H^+-SO_4^{2-} becomes the rate-limiting step. If under these conditions, the extracellular pH is varied while the intracellular pH is maintained at an elevated level (e.g. pH 7.2), the maximum disappears. One observes a nearly perfect, monotonically increasing dissociation curve with a single apparent pK of 5.8.

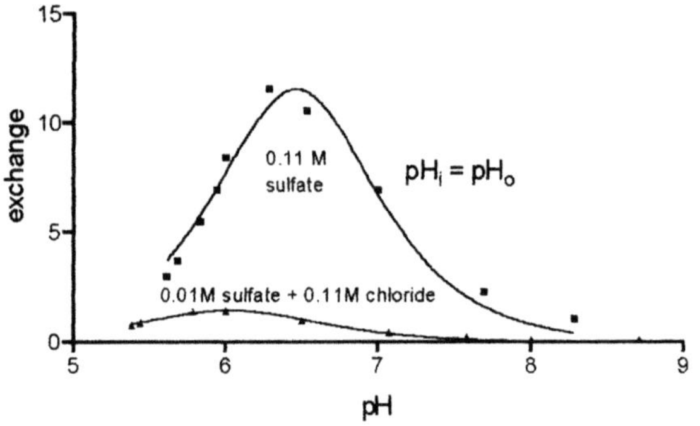

Fig. 10.8. pH dependence of sulfate equilibrium exchange as measured in media of equal composition at both membrane surfaces. 25 °C. Squares: 0.11 M, sulfate, no chloride; triangles: 0.01 M sulfate, 0.11 M Cl^-. Media buffered with 0.023 M phosphate to the pH values indicated on the abscissa. The curves represent fits to the equation presented in the text (data from Schnell et al. 1977). Ordinate: sulfate exchange. Abscissa: pH

Thus, the recruitment of the transfer sites to the outward orientation reveals the uniformity of the population of H^+-SO_4^{2-} cotransporters. It strongly suggests that the maximum seen under conditions where both external and internal pH are varied equally, is due to the existence of at least one inwardly-oriented inhibitory proton binding site. The pK of this site lies below 7.2 and seems to be similar to that of the outwardly-oriented transfer site. This suggestion is further supported by measuring the dependence of sulfate equilibrium exchange as a function of intracellular pH (Jennings 1992). If measured at constant extracellular pH of 7.0, at a saturating sulfate concentration, and in the absence of chloride, the curve relating efflux to pH_i is qualitatively similar to the curve obtained under similar conditions where pH is varied at $pH_i = pH_o$ with a maximum around pH 6.5. It is most likely but not yet proven definitively that the inhibitory site is simply the inwardly-oriented transfer site.

The experimental conditions under which the data in Fig. 10.9 had been obtained are essentially similar to those under which our measurements of the pH dependence of H^+-Cl^- cotransport had been executed. The initial pH inside the red cell ghosts was always about 7 while the extracellular pH was varied at constant chloride concentrations at the inner and outer membrane surface. Thus the rate determining step was always the net H^+-Cl^- entry, just as the H^+-SO_4^{2-} uptake in the experiment of Milanick and Gunn. If, after appropriate normalization, our data are incorporated into their figure (see Fig. 10.9), we observe that the titration curves for H^+-Cl^- and H^+-SO_4^{2-} cotransport nearly coincide. Thus, conditions can be created under which the two processes appear to be subject to identical rate-controlling steps with an apparent pK value of 5.7–5.9. This strongly suggests that the proton cotransported with a sulfate ion is identical to the proton cotransported with the chloride ion.

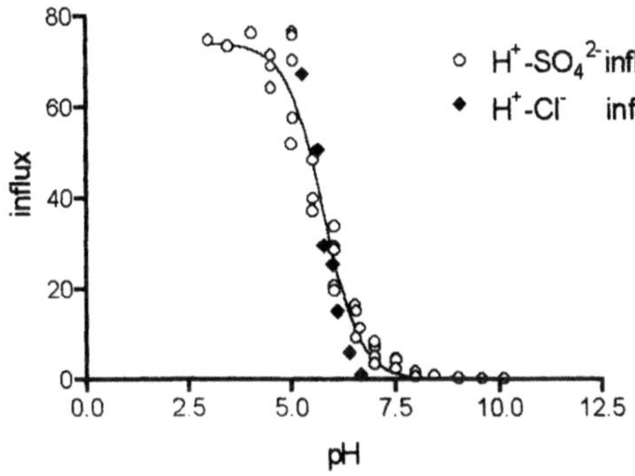

Fig. 10.9. pH dependence of H^+-SO_4^{2-} influx into red cells (open symbols). The cells contained 110 mM Cl, pH 7.2; the external medium consisted of solutions with nearly saturating Na_2SO_4 concentrations (62.5 mM or 125 mM) buffered to the pH values indicated on the abscissa (taken from Milanick and Gunn (1984), Fig. 2). Ordinate: rate of cotransport as a fraction of the rate at the plateau level (= 75%). Filled triangles: H^+-Cl^- influx into the ghosts, data from Fig. 10.5. The experimental conditions are not strictly identical (see legend to Fig. 10.5), but similar enough to demonstrate an approximately equal increase of both processes with decreasing pH and a tendency to approach saturation. The deviations at the foot of the curve may be related in part to the sulfate/sulfate exchange occurring at elevated pH. The drawn curve corresponds to a pK of 5.8

10.7 The Relationship between H^+-Cl^- and H^+-SO_4^{2-} Cotransport: Specific Features

Although the similarities between H^+-Cl^- and H-SO_4^{2-} cotransport are most impressive, a closer examination of the available information suggests specific differences, at least in detail, of the mechanism of cotransport of the two anion species.

Originally it was thought that proton binding to the transfer site of band 3 transformed this site from a chloride binding site into a binding site for the divalent sulfate. This would imply that without an additional proton present, no sulfate binding should be feasible and that any increase of proton concentration would solely affect the K_m value in a Michaelis-Menten representation of the cotransport process. Studies of Schnell et al. (1977) had shown, however, that changes of pH affected both K_m *and* V_{max} of the cotransport process. In their work on the chemical modification of H^+-SO_4^{2-} cotransport (see below), Legrum et al. (1980) designed, therefore, a mathematical model of the cotransport process, which involved the assumption that SO_4^{2-} and H^+ are able to combine independently of one another with specific binding sites and that cotransport would occur when both sites are occupied. At least qualitatively, this model explained the behavior observed by Schnell and coworkers.

Independently, Milanick and Gunn (1984) performed a much more extensive analysis of the molecular mechanism of H^+-SO_4^{2-} cotransport. Their results led them to postulate the operation of the same random-ordered mechanism considered by Legrum et al. (1980). They demonstrated, step by step, that cotransport involves the binding of either one sulfate, or one proton, or both without regard of the sequence in which the binding of the two ions occurs. Within this frame of reference, sulfate binds about nine times less firmly to the deprotonated transporter, but if bound, it increases about nine-fold the apparent affinity for proton binding. Although most of sulfate transfer as measured by radiosulfate takes place by cotransport, Milanick and Gunn could show that, at sufficiently high pH, sulfate as such is also translocated by band 3, although – as mentioned already above – about 200 times more slowly than by H^+-SO_4^{2-} cotransport. This may contribute to the deviation of the curves for the pH dependence of H^+-Cl^- and H^+-SO_4^{2-} cotransport at high pH seen in Fig. 10.9.

In order to appraise the relationship between chloride and sulfate cotransport in more detail, it is useful to recapitulate briefly certain aspects of the well-known competition between the two ion species for common binding sites. We shall first consider measurements under conditions of equilibrium exchange with (i) identical solutions at both membrane surfaces and (ii) under asymmetric conditions, which make it possible to attribute the observed effects to binding sites in one or the other membrane surface.

(i) The data obtained with identical solutions at both surfaces show saturation kinetics with superimposed self-inhibition. The apparent binding constants for sulfate and chloride to the transfer site are definitely lower than the estimates for binding of chloride or sulfate to the inhibitory modifier site (Schnell et al. 1977).

Competition for the common transfer site is most convincingly demonstrated by 35-Cl NMR, which does not recognize chloride binding to the modifier site (Falke and Chan 1985): it shows clearly the displacement of bound chloride by sulfate (Araníbar 1991). Extensive studies of Schnell et al. (1977) on inhibition kinetics yielded K_i values for inhibition of sulfate equilibrium exchange by chloride of about 14 mM, and for chloride equilibrium exchange by sulfate of about 19 mM, both at pH 7.2–7.3. The competition between Cl^- and SO_4^{2-} varies with pH. Lowering the pH favors the inhibition of chloride equilibrium exchange by sulfate and reduces the inhibition of sulfate equilibrium exchange by chloride.

The effects of variations of sulfate and chloride concentrations are, however, not confined to competition for the common transfer site. When measured under equilibrium exchange conditions, the maximum of H^+-SO_4^{2-} cotransport resides near pH 6.25 with approx. equal pK's for the ascending and descending flanks of the approx. bell-shaped curve. When the same measurements are performed in a medium containing a large excess of chloride over sulfate, the curve tends to become asymmetric. The maximum is moved to pH 5.9–6.0 and the pK's for the flanks are shifted to lower values, closer to an average pK of 5.9 (Fig. 10.8). This shows that at least in the absence of ion gradients across the membrane (i.e. under equilibrium exchange conditions) chloride modifies the pK of H^+-SO_4^{2-} equilibrium exchange. The effect cannot simply be related to sulfate displacement by chloride at the common transfer site and suggests the involvement of an additional site, possibly the site responsible for self-inhibition.

Information on self-inhibition of isotopically measured chloride or sulfate movements at equilibrium is less abundant than information on transport activation. Self-inhibition of both chloride and sulfate exchange is facilitated by lowering the pH. However, there seems to exist one characteristic difference: self-inhibition of chloride equilibrium exchange also takes place when the proton concentration is negligible (Dalmark 1975), while self-inhibition of sulfate exchange disappears completely. For sulfate self-inhibition to occur, there seems to exist an absolute requirement for protons. The self-inhibition constant increases from 0.22 M at pH 6.5, to 0.5 at pH 7.3, and becomes immeasurably large at pH 8.5 (Schnell et al. 1977, Fig. 3 and Table 2 therein).

(ii) We turn now to the more specific investigations of the respective mechanisms of chloride and sulfate cotransport. For this purpose we focus on the detailed studies of Milanick and Gunn (1982), who had demonstrated the operation of the random-ordered mechanism of sulfate and proton binding described above (see Sect. 10.6). Their results were essentially based on studies of the interaction between sulfate, chloride, and protons in the presence of an outwardly-directed chloride gradient (0.11 M Cl_i^-, 0.0066 M Cl_o^-), which recruited most of the transfer sites to the outer membrane surface. Two of their results are pertinent for the elucidation of the relationship between anion-proton cotransport and anion/anion exchange.

1. Due to the recruitment, the apparent K_i value for inhibition of chloride flux by sulfate becomes independent of pH at pH > 7.0. Hence in order to inhibit chloride flux, sulfate does not require a proton either bound to itself (to form the ion pair HSO_4^-) nor to the transporter. Thus, in accord with the NMR measurements and the studies of Schnell et al. (1977) on equilibrium exchange described above, chloride and sulfate as such compete for binding to the outwardly-oriented transfer site which is involved in the mediation of both the rapid Cl^-/Cl^- exchange and the very much slower SO_4^{2-}/SO_4^{2-} exchange.

2. External protons inhibit Cl^-/Cl^- exchange at fixed intracellular pH (7.8) and the fixed intra- and extracellular Cl^- concentrations indicated above. This effect is augmented by the addition of sulfate to the external medium. The sulfate-dependent change of the capacity of the protons to inhibit chloride exchange extrapolates to the same value as the value measured at zero sulfate concentration. This confirms that the single proton that inhibits Cl^-/Cl^- exchange in the absence of

sulfate is "the same proton whose inhibition of anion exchange is assisted by SO_4^{2-}" (Milanick and Gunn 1982).

From their observations, Milanick and Gunn draw the conclusion that a single site in the outer membrane surface is sufficient to explain both the chloride exchange inhibition and the activation of sulfate-proton cotransport. Since our results on chloride-proton cotransport strongly suggest that the proton inhibition of Cl^-/Cl^- exchange is due to its conversion to slow H^+-Cl^- cotransport, we are now able to explain the findings of Milanick and Gunn: they are simply due to the fact that one and the same proton is involved in both H^+-Cl^- and H^+-SO_4^{2-} cotransport. Evidently, the proton accompanying the respective anion is not rate-limiting. Hence the nature of the anion seems to determine whether or not exchange is faster or slower than cotransport. Below we shall address this problem making use, among other things, of recent suggestions of P.A. Knauf concerning possible differences of the mode of penetration of halides and oxy-anions (see Chap. 11).

10.8 H^+-SO_4^{2-} Cotransport *vs.* H^+-Cl^- Cotransport: Discussion

H^+-Cl^- cotransport was discussed in terms of a model with one proton binding site and two chloride binding sites, the exchange site *a* and the cotransport site *b*. The question arises whether or not the observations on H^+-SO_4^{2-} cotransport could also be fitted into this model. Some of the more pertinent observations for a discussion of this question are the following:

1. Sulfate displaces chloride from *a*, as suggested by the competition studies and confirmed more directly by 35-Cl NMR measurements.
2. Binding of a proton to *p* leads to an occlusion of chloride bound to *a* and promotes cotransport associated with chloride binding to *b*.
3. Binding of a proton to *p* is facilitated by binding of sulfate to *a* (and *vice versa*).

The findings described would be in accord with the assumption that sulfate bound to *a* is cotransported with the proton bound to *p*. In both cases, *p* could serve as the binding site for the cotransported proton. Since H^+-Cl^- cotransport via *b* and *p* is much slower than Cl^-/Cl^- exchange via *a*, H^+-Cl^- cotransport leads to inhibition of Cl^-/Cl^- exchange. Since H^+-SO_4^{2-} cotransport via *a* and *p* is much faster than SO_4^{2-}/SO_4^{2-} exchange (possibly via *b*, the evidence available does not permit a definitive statement), increasing the proton concentration should lead to a continuous increase of cotransport until a plateau value is reached. This only happens, however, as long as the protonation of a second, normally inwards-facing proton binding site is prevented by maintaining a sufficiently high intracellular pH. However, if internal and external pH is changed equally, the curve passes through a maximum. This could, perhaps, indicate that the orientation of the sulfate-loaded transporter varies with pH, if no special precautions are taken to prevent a re-equilibration across the membrane of the sites *a* and *p* as in the experiment shown in Fig. 10.9.

In plain language, the findings described would seem to be compatible with the view that the exchange site for chloride is the cotransport site for sulfate and *vice versa.* Although this statement is highly conjectural, it is pronounced explicitly since it could possibly contribute to a further development of the ideas of P.A. Knauf concerning different mechanisms of band 3-mediated chloride and sulfate transport (see Chap. 11).

A major problem of this model is posed by the self-inhibition at elevated sulfate concentrations. To stay within its bounds, it would be necessary to assume that the extremely slow sulfate/sulfate exchange takes place through the cotransport pathway for chloride ions. There seems to exist no evidence whatsoever to this effect. Moreover, at high pH, when *p* is fully deprotonated, in contrast to chloride binding to *b*, high sulfate concentrations produce no inhibition of the isotopically measured sulfate flux (Schnell et al. 1977, Fig. 3 therein). Only when the pH is lowered does inhibition result. Thus a proton not involved in cotransport (i.e. a proton bound to a site other than *p*, which may be called *q*) seems to be required to accomplish sulfate binding to the cotransport site for chloride, and thus to produce self inhibition of the isotopically measured sulfate flux. To account for this feature we would need to postulate a second proton binding site. This requirement reminds one that even in H^+-Cl^- cotransport, in addition to the cotransported proton, the binding of a modulating proton to *q* may possibly facilitate chloride binding to the cotransport site *b*.

The reaction diagram (D2) outlines the hypothesis discussed above. It may serve as an example of a hypothesis about possible relationships between sulfate and chloride cotransport. Assuming that the sum of all forms represented in the diagram is constant (mass conservation), that in each surface, the concentrations of the various forms are interrelated by mass law equations, and that the transitions from inward to outward orientation and *vice versa* are confined to TCl and $HTSO_4$, it is possible to derive an equation for H^+-SO_4^{2-} cotransport as a function of chloride, sulfate and proton concentration. For the pH dependence of equilibrium exchange between two media in which the concentrations of the two anion species are equal and constant, and the pH is varied equally at both membrane surfaces, ($pH_i = pH_o$), one obtains for the isotopically measured sulfate flux an equation of the following structure: Flux = $k_1 \cdot T_oSO_4 = A_o \cdot H^+/(A_1 + H^+ + A_2 \cdot H^{+2})$, where the coefficients A_o, A_1, and A_2 are functions of the concentrations of chloride, sulfate and the various mass law constants. The equation has been fitted to the data of Schnell et al. (1977) about the pH dependence of sulfate equilibrium exchange (Fig. 10.8). The fits are quite reasonable. This was to be expected since the structure of the equation presented above is identical to that used by Schnell et al. (1977). However, in our equation, each parameter value is explained in terms of the reaction diagram (D2), while the equation of Schnell et al. (1977) only summarily considers the operation of one transfer and one modifier site. We abstain from presenting the unwieldy equations for the parameters since we were unable to find in the literature the data necessary for their computation. Thus, in the present context, the diagram represents no more than one possible qualitative interpretation of the fragmentary information at hand.

The reaction diagram (D2) is discussed above. Note that it includes four unexplained forms, written in *italics:* $HClT_oClH$, $HClT_iClH$, $HSO_4T_oSO_4H$, and

$HSO_4T_iSO_4H$. They serve to demonstrate the uncertainties with respect to the possible existence of an additional, immobile proton binding site q in the outer and/or inner membrane surface, which possibly modifies rate constants and the various mass law equilibria determining proton and anion binding. The possible existence of $HClT_oClH$ could give rise to the uncertainties concerning the pH dependence of H^+-Cl^- cotransport with a Hill coefficient n = 1.0 (Fig. 10.7A) or n = 2.0 (Fig. 10.5). If n = 2, under our experimental conditions, the effect of the second proton would be confined to the outer membrane surface, since our data were obtained by varying external pH at a fixed intracellular pH. With respect to sulfate, the effect of variations of intracellular pH on an additional binding site seems to be essential. In this case, a binding site exposed to the internal medium would play a role.

$$\begin{array}{ccc}
HClT_oClH & & HClT_iClH \\
| & & | \\
HClT_oCl & & HClT_iCl \\
| & & | \\
ClT_oCl & & ClT_iCl \\
| & & | \\
T_oCl & \leftarrow \text{exch.} \leftarrow & T_iCl \\
| & & | \\
T_o & & T_i \\
| & & | \\
HT_o & & HT_i \\
| & & | \\
HT_oSO_4 & \rightarrow \text{cotr.} \rightarrow & HT_iSO_4 \\
| & & | \\
HSO_4T_oSO_4H & & HSO_4T_iSO_4H
\end{array} \qquad \text{(D2)}$$

The three sites model discussed above gives rise to 16 different permutations for the occupancies of the binding sites a, b, and p with Cl^-, SO_4^{2-} and the proton, plus two mixed forms, HSO_4TCl and $HClTSO_4$. This number is increased if the various dissociation constants differ for inward- and outward orientation of the transporter and if allosteric effects are also involved. Finally, if an additional intracellular proton binding site q also plays a role, the total number of parameters to be determined experimentally becomes even larger. Thus, at the present stage of our knowledge, the discussion of the mechanistic basis of the relationship between cotransport and exchange cannot be expected to yield a definitive result. One can only hope that it represents an invitation to further research.

10.9 Chemical Modification of the Cotransport Process

The functioning of at least some of the amino acid residues involved in cotransport can be affected by the chemical modification of band 3 with dansyl chloride as solubilized by means of cyclodextrin. When red cell ghosts dansylated at pH

6.6 under standard conditions (130 mM NaCl, 1 mM Na_2SO_4, 20 mM EDTA, 37 °C) are subjected to measurements of sulfate equilibrium exchange over the pH range 5.0–8.4, the following observations were made (Fig. 10.10): The bell shaped pH dependence with the maximum at about pH 6 is replaced by a curve which resembles the pH dependence of Cl^-/Cl^- equilibrium exchange. With increasing pH the maximal rate of sulfate movements increases up to a plateau, which is situated slightly above the maximal rate reached in the control with untreated ghosts and extends at least up to pH 8.4 (Legrum et al. 1980). The enhancement of sulfate movements is accompanied by a proportionate inhibition of Cl^-/Cl^- exchange (not shown). This inhibition remains incomplete even though the measurements were performed after adjusting the pH to 8.0, where the sulfate movements have reached the plateau value (Fig. 10.10, curve designated "Dans"). The position of the plateau depends on the conditions existing at the time of dansylation. When dansylation is performed at pH 6.6 and variable Cl^- concentrations up to 320 mM, the effect measured after re-equilibration of the modified ghosts at fixed chloride concentration (160 mM) and pH (8.0) is close to maximal. The changes of sulfate and chloride movements can be augmented even further when during exposure to dansyl chloride a disulphonic acid, APMB (2-(4'-aminophenyl)-6-methylbenzene-thiazol-3',7-disulphonic acid), is present (Fig. 10.10, "Dans+APMB"). The augmented sulfate fluxes seen after removal of APMB and the excess dansyl chloride can still be inhibited by H_2DIDS, some reversibly binding stilbene disulphonates, and APMB, although with somewhat altered K_i values. The findings suggest that the observed effects are due to the modification of at least 3 different sites (Lepke and Passow 1982).

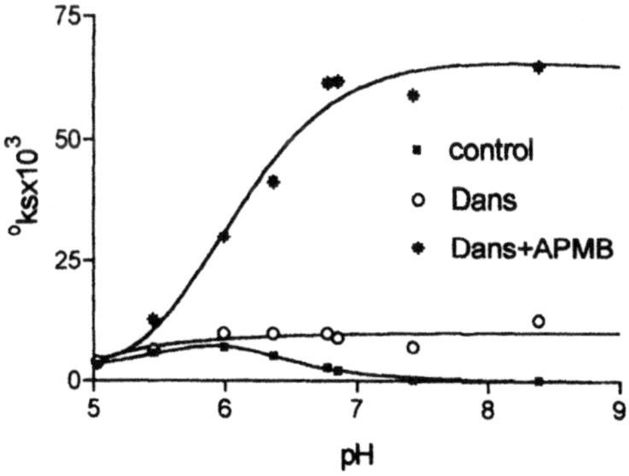

Fig. 10.10. Effect of dansylation in the absence and presence of the disulphonic acid APMB on pH dependence of sulfate equilibrium exchange. Media inside and outside the ghosts: 130 mM NaCl buffered to the pH values indicated on the abscissa. Dansylation at pH 6.6, in the presence of 125 μM dansyl chloride with or without 5.0 mM APMB, at 37 °C, for 30 min. Subsequently, the ghosts are washed to remove APMB and breakdown products of dansyl chloride, and subjected to flux measurements at the pH values indicated on the abscissa, in 130 mM buffered NaCl solution at 30 °C. Ordinate: rate constant for sulfate exchange ^{o}ks, as measured by means of 35-SO_4^{2-}. Abscissa: pH, identical values inside and outside the cells (according to Lepke and Passow 1982)

The potentiation of the effects of dansylation depends on the nature of the disulphonic acid present. A non-penetrating, covalently binding APMB derivative called PENS-Cl produces a maximal effect intermediate between the maximal effects seen with dansyl chloride alone and with dansyl chloride plus APMB (Raida and Passow 1985). The spectacular changes of the anion transport kinetics in the dansylated ghosts were described in terms of changes of sulfate or chloride "movements". This neutral term was used to indicate that no quantitative investigations have been performed to discriminate between effects on anion exchange unaccompanied by protons and on anion exchange accompanied by a cotransported proton. Nevertheless, Berghout et al. (1988) convinced themselves that regardless of whether dansylation was performed in the presence or absence of APMB or PENS-Cl, H^+-SO_4^{2-} cotransport is still feasible, at least over the the pH range 6.8–7.3. Thus, a quantitative comparison of the exchange and cotransport in modified ghosts and control ghosts is still missing and none of the modified amino acid residues (presumably lysine or histidine residues) involved in the alteration of transport kinetics have been identified.

This situation is different with respect to observations of Jennings (1992, 1995) on the effects of treatment of the red cell membrane with Woodward's reagent K (WRK) followed by exposure to borohydride. This treatment selectively converts the carboxyl group of Glu 681 into an OH group. It inhibits both H^+-SO_4^{2-}(Jennings and Al-Rhayiel 1988) and H^+-Cl^- cotransport (Fig. 10.11). The inhibition of cotransport is accompanied by the rise of a rapid electrogenic sulfate-chloride exchange with a 1:1 stoichiometry (Jennings 1992). If measured under conditions where no net transport of protons or anions takes place (equilibrium exchange), the electrogenic process no longer shows the maximum, which is characteristic for the pH dependence of the non-electrogenic H^+-SO_4^{2-} cotransport (Jennings 1995). Although the nature of the modification by Woodwards Reagent K and of the ensuing effects are known in much more detail than those of dansylation, a definitive interpretation has not yet been achieved.

In summary, it is possible to produce two clearly distinct effects on the isotopically measured, band 3-mediated sulfate transport: (1) an inversion of its pH dependence by dansylation without production of electrogenicity and without destroying the capacity to perform cotransport and (2) a conversion by treatment with Woodwards Reagent K plus BH_3 of the isotopically measured fluxes of chloride and sulfate into electrogenic processes that are essentially independent of pH. Effect (1) could possibly constitute an uncoupling of translocation of sulfate bound to site *a*, from proton translocation involving proton binding to *p*. This would enable the sulfate bound to site *a* to exchange across the membrane like chloride. Effect (2) possibly induces an entirely new mode of translocation, which involves a more complete disruption of the physiological transport process than dansylation.

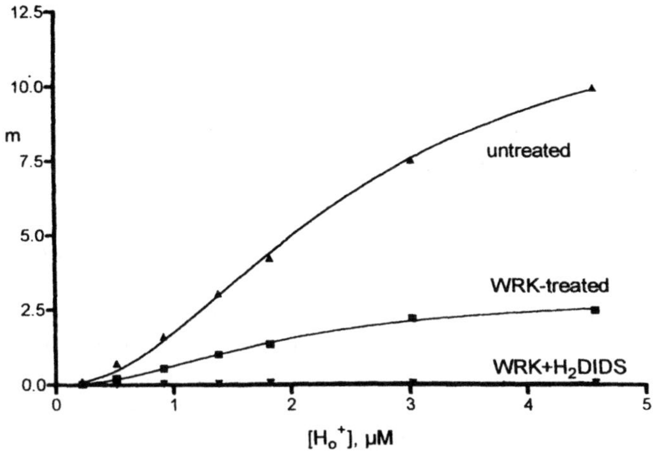

Fig. 10.11. Effect of pretreatment of ghosts with WRK plus $NaBH_4$ on net proton uptake ($m_{(H)}$) due to H^+-Cl^- cotransport at 160 mM NaCl inside and outside the ghosts, initiated by pH jumps from pH 7.0 to 5.6. Hill plots: $m_{(H)} = F \cdot H^{+2}/(K+H^{+2})$.
No WRK treatment: F = 13.0 ± 0.34, sqrt(K) = 2.55 ± 0.65; WRK-treated: F = 2.96 ± 0.08, sqrt(K) = 2.55 ± 0.92; WRK-treated, H_2DIDS present during the influx measurements: F = 0.084 ± 0.006, sqrt(K) = 0.92 ± 0.51. The success of the WRK-$NaBH_4$ treatment was verified by observing a concomitant 3.8-fold increase of rate of sulfate equilibrium exchange if measured at pH 6.37 and a 11.4-fold increase if measured at pH 7.11. Ordinate: cotransport ($m_{(H)}$). Abscissa: proton concentration

10.10 Summary

Band 3-mediated anion transport may take place either by anion/anion exchange or by anion-proton cotransport. Cl^--H^+ cotransport proceeds about 1000 times more slowly than Cl^-/Cl^- exchange, while sulfate-proton cotransport occurs about 200 times faster than SO_4^{2-}/SO_4^{2-} exchange. Hence when measured with radiochloride, almost all of the radioactivity crosses the membrane by anion exchange. When measured with radiosulfate, most of the radioactivity penetrates by H^+-SO_4^{2-} cotransport.

The present chapter is primarily concerned with the relationship between the pH dependence of Cl^-/Cl^- exchange and H^+-Cl^- cotransport. It describes a new method for the measurement of H^+-Cl^- cotransport, using carbonic anhydrase-free red cell ghosts in which the Jacobs-Parpart cycle is unable to operate. We show that the decrease of rapid Cl^-/Cl^- exchange with decreasing pH is associated with an increase of the very much slower H^+-Cl^- cotransport, and that the two processes exhibit the same pK value of 5.7–5.9 (Fig. 10.6). This suggests that the decrease of anion exchange is due to the conversion of band 3 from a rapidly operating Cl^-/Cl^- exchanger to a slowly operating chloride-proton cotransporter.

Based on this assumption, equations for the pH dependence of exchange and cotransport were derived (Table 10.1). Data on the variation of Cl^-/Cl^- exchange with pH were obtained by reevaluation of measurements from several different laboratories. Inserting these data into our equations, we determined the numerical values of the parameters pertinent to exchange (Table 10.2). We then used the values so obtained to predict the rise of H^+-Cl^- cotransport with decreasing external pH as measured at a fixed internal pH. (Fig.10.7 A, B). Our cotransport data could be represented quite well with the apparent pK value of 5.8 derived from the measurements of exchange. This suggests that, in contrast to previous interpretations of the mechanism of the pH dependence of Cl^-/Cl^- exchange, proton binding to band 3 does not simply inactivate the exchange process by reducing the number of deprotonated band 3 molecules, but converts the exchange into the slowly operating H^+-Cl^- cotransport. These findings provide a new aspect for the interpretation of the relationship between the pH dependence of the well known Cl^-/Cl^- exchange and the H^+-Cl^- cotransport first demonstrated to exist by Jennings in 1978.

A comparison of our data on H^+-Cl^- cotransport with those on H^+-SO_4^{2-} cotransport obtained by Milanick and Gunn (1984) yields superimposable dissociation curves with essentially similar pK values in the range 5.7–5.9. This confirms the previously advanced suggestion by these authors that the proton that inhibits Cl^-/Cl^- exchange is the same proton that activates H^+-SO_4^{2-} cotransport. It shows in addition that the binding of the proton does not simply immobilize the chloride binding site as originally assumed, but combines with a site that confers to band 3 the ability to accomplish H^+-Cl^- cotransport. Finally, we demonstrate that the activation of an electrogenic Cl^-/SO_4^{2-} exchange by treatment of band 3 with WRK + borohydride (Jennings and Al-Rhaiyel 1988) leads to an inhibition of H^+-Cl^- cotransport similar to the inhibition of H^+-SO_4^{2-} cotransport described by Jennings and Al-Rhaiyel.

The theoretical treatment of the relationship between anion exchange and anion-proton cotransport is compatible with the assumption that the slow H^+-Cl^- cotransport is mediated by the inhibitory modifier site of band 3 first described by Dalmark (1975). We discuss the relationship between H^+-SO_4^{2-} and H^+-Cl^- cotransport in terms of a tentative model. It stipulates that the proton cotransported with SO_4^{2-} or Cl^- combines with the same mobile binding site, called *p*. The cotransported Cl^- seems to be bound to a site *b*, which, presumably, is identical to Dalmark's modifier site and possibly different from the site *a* involved in cotransport of SO_4^{2-}. This latter site seems to be the same site *a*, which is involved in Cl^-/Cl^- exchange. These suggestions may show some bearing to the different pathways for Cl^- and oxy-anions discussed by Knauf in Chap. 11.

Appendix

To explain the cotransport, we had previously proposed that inside the band 3 protein, a transmembrane array of side chain hydrogen bonds may be involved in the proton translocation. To test this possibility, we used IR spectroscopy to look for the existence of a continuum band at wavenumbers above 1700 cm^{-1} that is typical for arrays of mobile protons in membrane proteins (Zundel and Merz 1984). In search for such band, we studied the pH dependence of infra red spectra of the hydrophobic domain of band 3. We attached a film of trypsinized red cell ghosts, which contain the hydrophobic domain of band 3 as the principal protein constituent. Such ghosts are still able to perform anion exchange like unmodified ghosts (Lepke and Passow 1976; Lepke et al. 1992). These ghosts were attached to the surface of a 45° ZnS trapezoidal internal reflection element (IRE) and difference spectra were obtained after changes of the pH in the supernatant. These measurements were preceded by determinations of linear dichroism, which showed that the attachment by partial drying of the ghosts to the surface of the crystal did not destroy the native structure of the transmembrane helices.

Figure 10.12 depicts the linear dichroism of the infrared spectrum of the trypsinized ghosts. Strongest linear dichroism is exerted by the C=O stretching vibrations of the ester lipids of the lipid bilayer and the amide vibrations of the hydrophobic domain of band 3. The negative dichroism of the ester C=O stretch demonstrates that the lipid molecules are oriented predominantly perpendicular to the surface of the IRE (Ludlam et al. 1996). Thus, the erythrocytes rest, with a smooth surface, flatly on the crystal. The dichroism pattern of the amide bands is typical for a transmembrane alpha-helical protein: positive amide I' and negative amide II dichroism. Accordingly, the amide II' dichroism is also negative. The predominantly alpha-helical structure of the band 3 protein is confirmed by comparison with bacteriorhodopsin (bR, dashed line in Fig. 10.12). The bR resides in large patches of purple membranes oriented parallel to the surface of the IRE. The linear dichroism band pattern is virtually identical to that of the band 3 protein, except for the C=O vibration of the lipid esters, which is missing due to the negligibly small concentration of such lipids in the purple membrane.

The linear dichroism of the trypsinized ghosts clearly demonstrates that the secondary structure of the hydrophobic domain of band 3 remains intact when these ghosts are attached to the surface of the IRE. Figure 10.12 shows difference spectra measured at a range of decreasing pH values against a reference spectrum at pH 7.83. Although the pH values covered the whole range from far above to far below the pK value of the proton flux, no continuum band could be discovered.

In summary, our attempts to obtain structural information by means of FT-IR measurements yielded no unambiguous results. Although in our membrane preparations the helicality of the transmembrane regions was essentially preserved, the continuum band could not be detected. In particular, a large continuum band such as that observed in the F_0F_1 ATPase (Bartl et al. 1995) can definitely be ruled out for the band 3 protein. In contrast, pH titration over the range pH 7–3 in untreated membranes (and after treatment with Woodward's reagent K followed by reduction with BH_4) revealed no significant changes of carboxyl group dissociation in the critical region between pH 5 and 7. These latter spectroscopic results are, however, still rather preliminary since we are not sure that the sensitivity of our technique permits us to detect the pH dependent dissociation or the disappearance by chemical modification of one single carboxyl group out of 20 that reside in the anion transport domain of band 3.

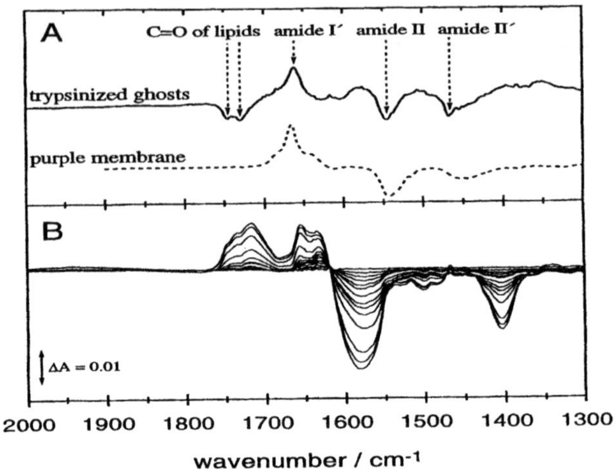

Fig. 10.12. ATR/FT-IR spectra of permeabilized red cell ghosts treated with trypsin to remove cytoskeletal proteins and the hydrophilic domain of band 3. The ghosts were deposited as a film on the surface of the crystal of the photometer and kept under PBS buffer. (**A**) Buffer made up in D_2O. Irradiation with vertical and parallel polarized IR radiation. Ordinate: linear dichroism calculated from the data plotted as a function of wavenumber (abscissa) and compared with linear dichroism of bacteriorhodopsin residing in the purple membrane (dashed line). Amide I' corresponds mainly to the C=O stretch of the protein backbone, amide II is a coupled C=N-H vibration. Amide II without a prime indicates that the replacement of H_2O by D_2O remains incomplete. Amide II' shows the downshift of amide II associated with complete replacement by D_2O. Comparison of the spectra demonstrates that the secondary structure of the hydrophobic domain of band 3 is not influenced by attaching the ghosts to the surface of the internal reflection element. (**B**) pH dependence of IR-absorbance difference spectra of trypsinized ghosts in aqueous buffer. Spectra were recorded successively at pH 7.83, 7.28, 6.84, 6.45, 6.11, 5.88, 5.53, 5.04, 4.77, 4.53, 4.05, 3.43, 3.08, and 2.86. The spectrum at pH 7.83 had been taken as a reference

Acknowledgement: Supported by the Deutsche Forschungsgemeinschaft and the Verband der chemischen Industrie. We are grateful to Professors E. Bamberg (Frankfurt) and G. Büldt (Jülich) for their kind support. We thank Drs. M.L. Jennings and J. Salhany for inspiration and helpful discussions, and Dr. Ph. Knauf for sending us a draft of his stimulating contribution to the present book prior to publication.

References

Araníbar N (1991) The anion exchanger of the human red blood cell membrane. A chloride binding and transport study. Ph.D. thesis, J.W. Goethe Universität, Frankfurt

Bartl F, Decker-Hebestreit G, Altendorf K, Zundel G (1995) The F_o complex of the ATP synthetase of E. coli contains a proton pathway with large proton polarizability caused by collective proton fluctuation. Biophys J 68:104-110

Berghout A, Raida M, Legrum B, Passow H (1988) The effects of dansylation on pH dependence of SO_4 and Cl equilibrium exchange and on H/SO_4 cotransport across the red cell membrane. Biochim Biophys Acta 986:75-84

Dalmark M (1975) Chloride transport in human red cells. J Physiol 230:39-64

Falke JJ, Chan SI (1985) Evidence that anion transport by band 3 proceeds via a ping-pong mechanism involving a single transport site. J Biol Chem 260:9537-9544

Funder J, Wieth JO (1976) Chloride transport in human erythrocytes and ghosts. A quantitative comparison. J Physiol 262:679-698

Gunn RB, Wieth JO, Tosteson DC (1975) Some effects of low pH on chloride exchange in human red blood cells. J Gen Physiol 65:731-749

Jennings ML (1976) Proton fluxes associated with erythrocyte membrane anion exchange. J Membrane Biol 28:187-205

Jennings ML (1978) Characteristics of CO_2-independent pH equilibration in human red blood cells. J Membrane Biol 40:365-391

Jennings ML (1992) Chemical labelling of a glutamate residue involved in proton-sulfate cotransport catalysed by band 3. In: Progress in cell research, vol 2. Elsevier, Amsterdam, pp 73-84

Jennings ML (1995) Rapid electrogenic sulfate-chloride exchange mediated by the chemically modified band 3 in human erythrocytes. J Gen Physiol 105:21-47

Jennings ML, Al-Rhaiyel S (1988) Modification of a carboxyl group that appears to cross the permeability barrier in the red blood cell anion transporter. J Gen Physiol 92:161-178

Legrum B, Fasold H, Passow H (1980) Enhancement of anion equilibrium exchange by dansylation of the red cell membrane. Hoppe-Seyler's Z Physiol Chem 361:1573-1590

Lepke S, Passow H (1976) Effect of incorporated trypsin on anion exchange and membrane proteins in human red blood cell ghosts. Biochim Biophys Acta 455:353-370

Lepke S, Passow H (1982) Inverse effects of dansylation of red cell membrane on band 3 protein-mediated transport of sulphate and chloride. J Physiol 328:27-48

Lepke S, Becker A, Passow H (1992) Mediation of inorganic anion transport by the hydrophobic domain of mouse erythroid band 3 expressed in oocytes of *Xenopus laevis*. Biochim Biophys Acta 1106:13-16

Ludlam CFC, Atkin LT, Liu X-M, Rothman MS, Rath P, Aimoto S, Smith SO, Engelman DM, Rothschild KJ (1996) Fourier transform infrared spectroscopy and site-directed isotope labeling as a probe of local secondary structure in the transmembrane domain of phospholamban. Biophys J 70:1728-1736

Milanick MA, Gunn RB (1982) Proton-sulfate cotransport: Mechanism of H and sulfate addition to the chloride transporter of human red cells. J Gen Physiol 79:87-113

Milanick MA, Gunn RB (1984) Proton-sulfate cotransport: External proton activation of sulfate influx into human red cells. Am J Physiol 247:C247-C259

Milanick MA, Gunn RB (1986) Proton inhibition of chloride exchange: Asynchrony of band 3 proton and anion transport sites. Am J Physiol 250:C955-C969

Müller-Berger S, Karbach D, König J, Araníbar N, Lepke S, Wood PG, Appelhans H, Passow H (1995) Inhibition of mouse erythroid band 3-mediated chloride transport by site-directed mutagenesis of histidine residues and its reversal by second site mutation of Lys 558, the locus of H_2DIDS binding. Biochemistry 34:9325-9332

Passow H (1986) Molecular aspects of band 3 protein-mediated transport across the red blood cell membrane. Rev Physiol Biochem Pharmacol 103:61-203

Passow H, Karbach D, Araníbar N, Liebold K, Wood PG, Lepke S (1997) Studies on the pH dependence of band 3-mediated anion transport. Facts and speculations. In: Mihara K, Hamasaki N (eds) Membrane proteins: Structure, function, and expression control. Fukuoka University Press, Karger, Basel, pp 373-404

Passow H, Lepke S, Heberle J (2001) Inhibition of band 3-mediated Cl^-/Cl^- exchange by pH dependent transition to proton-anion cotransport. Biophys J 80, part 2:46a

Raida M, Passow H (1985) Enhancement of divalent anion transport across the human red cell membrane by the water soluble dansyl chloride derivative 2-(N-piperidine)-ethylamine-naphthyl-5-sulfonylchloride (PENS-Cl). Biochim Biophys Acta 812:624-632

Schnell KF, Gerhardt S, Schöppe-Fredenburg A (1977) Kinetik characteristics of the sulfate self-exchange in human red cells and red cell ghosts. J Membrane Biol 30:319-350

Wieth JO, Bjerrum PJ (1982) Titration of transport and modifier sites in the red cell anion transport system. J Gen Physiol 79:253-282

Wood PG (1989) Preparation of resealable membranes maximally depleted of cytosolic components by gel filtration and tryptic digestion: Resealable white ghosts. Methods Enzymol 173:346-355

Zundel G, Merz H (1984) On the role of hydrogen bonds and hydrogen bonded systems with large proton polarizability for mechanisms of proton activation and conduction in bacteriorhodopsin. In: Bolis CL, Helmreich EJM, Passow H (eds) Information and energy transduction in biological membranes. Alan Liss, New York, pp 153-164

11 Band 3 Mediated Transport

Philip A. Knauf and Prithwish Pal

Department of Biochemistry and Biophysics, University of Rochester Medical Center, Rochester, NY 14642, USA

11.1 Functions of Band 3 and Related Proteins

The band 3 protein is the first member, AE1, of the AE family of anion exchange proteins. The full-length 911 amino acid (Tanner et al. 1988; Lux et al. 1989) version of human AE1 (hAE1) is found primarily, though not exclusively (Papageorgiou et al. 2001), in red blood cells, where it mediates the very tightly coupled exchange of Cl^- for HCO_3^-. This permits HCO_3^- formed from CO_2 in the systemic capillaries by the intracellular enzyme carbonic anhydrase to leave the cell in exchange for Cl^-, thus effectively storing CO_2 in the blood in the form of plasma HCO_3^-, greatly increasing the CO_2 carrying capacity of the blood (Gunn et al. 1989). In the lungs, the process is reversed to regenerate CO_2 that is then exhaled. Although this function is important in facilitating the removal from the body of CO_2 produced by tissue metabolism, it is not essential for life, as animals can survive despite nearly complete inhibition of this activity by disulphonic stilbenes (DS) (Swenson et al. 1993) or knockout of the functional gene for this protein (Peters et al. 1996).

AE1, through its cytoplasmic N-terminal portion, also plays an important structural role in providing a site of membrane attachment for red cell cytoskeletal elements (Low 1986; Low et al. 1991). Mutations that result in reduced expression of functional AE1 in the red cell membrane lead to alterations of red cell shape and mechanical stability, as seen in AE1-knockout mice (Peters et al. 1996) and in haematological abnormalities (Bruce et al. 2000) such as hereditary spherocytosis (Jarolim et al. 1995; Miraglia del Guidice et al. 1997) and Southeast Asian ovalocytosis (SAO) (Schofield et al. 1992b). The N-terminal cytoplasmic portion of AE1 also binds haemoglobin and glycolytic enzymes, while the C-terminal cytoplasmic tail binds carbonic anhydrase II (Vince and Reithmeier 1998, 2000; Vince et al. 2000). Recently it has been suggested that AE1 may also play a role in the transport of NO out of the red cell to cause vascular smooth muscle relaxation under conditions of low tissue oxygen (Pawloski et al. 2001).

A truncated form of AE1, known as kAE1, produced by alternate splicing of the same gene, SLC4A1, is expressed in the kidney collecting ducts at the basolateral side of type A intercalated cells (Alper et al. 1988; Huber et al. 1999). Here it functions to remove HCO_3^- from the intercalated cells in exchange for peritubular Cl^-, thereby contributing to the process of urinary acidification by removing

HCO_3^- that accumulates in intercalated cells as H^+ is transported into the urine (Weinstein 2000). Mutations of AE1 that result in poor expression of functional kAE1 at the cell membrane cause distal renal tubular acidosis (Tanphaichitr et al. 1998; Toye et al. 2002); in some cases red cell anion exchange is unaffected, indicating that the cellular machinery for surface membrane expression of AE1 is different in kidney cells from that in red cell precursors (see Chap. 26). In rats, changes in acid-base status were found to alter the amount of kAE1 mRNA expressed, indicating some capacity for regulation (Huber et al. 1999).

The AE1 family also includes AE2 and AE3, both of which are also expressed as variants of different lengths (Alper 1991; Tanner 1993). AE3 is primarily found in brain neurons and in cardiac cells, where it is probably involved in regulation of intracellular Cl^- as well as cell pH (Sterling and Casey 1999). Cardiac cells also express a form of AE1 that accounts for at least some of the Cl^-/HCO_3^- exchange activity in these cells (Puceat et al. 1995; Richards et al. 1999; Papageorgiou et al. 2001). AE2 is widely distributed in various cells and tissues, where it serves a housekeeping function in cell pH and volume regulation. It is also widely expressed in epithelia, where it functions in transepithelial anion transport.

For the structure of the AE genes and the function of members of the AE family other than red blood cell AE1, band 3, the reader is referred to other chapters in this book (Chap. 18, 26) and other sources (Alper 1991; Tanner 1993). This chapter will focus primarily on a single function of AE1, the one-for-one exchange of halides (such as Cl^-) and oxyanions (such as HCO_3^-), and those parts of the AE1 structure that are involved in this function. Other functions of AE1 are dealt with in Chaps. 3, 4, 10 and 18. Because this topic has been extensively reviewed in the past (Passow 1986; Gunn et al. 1989; Knauf 1989; Knauf and Brahm 1989; Salhany 1990; Cabantchik and Greger 1992; Jennings 1992a,b; Reithmeier 1993), we will concentrate on recent developments, particularly those which give clues about the relation of anion exchange to AE1 structure.

11.2 Basic Characteristics

11.2.1 One-for-One Exchange

AE1 serves as a prime example of a tightly-coupled ion exchanger, or antiport. Evidence for the tight coupling includes the fact that net Cl^- flux, driven by a membrane potential (V_m) that is different from the Cl^- equilibrium potential, V_{Cl}, takes place at a much slower rate than homoexchange of radioactive ^{36}Cl for another chloride isotope or heteroexchange of Cl^- for HCO_3^- (Hunter 1977; Knauf et al. 1977, 1983; Fröhlich 1988).

Further evidence for one-for-one coupling is the very strong dependence of unidirectional Cl^- or HCO_3^- efflux from the red cell on the presence of a suitable exchange partner in the external medium. Unidirectional anion outflow is rapid with Cl^- or HCO_3^- outside, slower with NO_3^-, much slower with F^- or Br^-, two orders of magnitude slower with I^-, and nearly completely inhibited with citrate (Wieth 1979; Knauf 1989). A minimum estimate of the "tightness" of the coupling can be gained by comparing the efflux of radioactive chloride either with chloride outside or with no exchangeable anion present. The flux in the latter case is only 0.03% of that in the former (Knauf 1989), and even this small residual efflux is probably due at least in part to the small amount of bicarbonate present in the medium because of the CO_2 in air. Comparison of V_m-driven Cl^- flux (Knauf et al. 1977) *vs.* Cl^- exchange (Brahm 1977) indicates that the system "slips" and allows flow of Cl^- out without another anion going in less than one time in every 10,000!

Because of this tight coupling, the transport activity of AE1 and other AE transporters can be assayed by the ability of a Cl^- gradient (in > out) to drive HCO_3^- into the cell against an electrochemical gradient (Wieth 1979), thereby raising the internal pH (Kopito et al. 1989; Lindsey et al. 1990; Lee et al. 1991; Tang et al. 1998). Although the tight coupling is very well-established, the reason for this characteristic is not so clear in the case of the red cell, where the primary role of the system is not pH regulation but rather HCO_3^- transport. In the first paper describing radioactive Cl^- exchange, Tosteson (1959) noted that diffusion of Cl^- through pores occupying only 0.1% of the total membrane area could account for the entire flux, so why does the cell make about a million copies of AE1, when fewer channel proteins could accomplish the same rate of anion transport? Two reasons may be suggested: (i) AE1 may have evolved from other AE proteins that served the purpose of pH regulation in other cells, where tight coupling would be essential for function; and (ii) the tight coupling gives the system a much higher selectivity for anions *vs.* cations (over a million-fold) than could be accomplished by a channel, unless that channel had an extraordinarily large positive charge density. Low cation "leak" permeability gives the red cell the ability to regulate its volume with very low Na^+/K^+ pump activity, and thus at a very low metabolic cost that can be supplied by the cell's limited ATP production through anaerobic glycolysis.

11.2.2 Divalent *vs.* Monovalent Anion Transport

As described in Chap. 10, AE1 can be converted by protonation (Gunn 1972) to a carrier for divalent anions (with one more + charge), such as sulphate. The net result is that the monovalent form of the carrier binds a proton and a divalent anion such as sulphate, so the sulphate is cotransported with a proton (Jennings 1976). Thus, AE1 also serves as a model symport system. The binding of sulphate to the external-facing transport site facilitates the binding of protons, and *vice versa* (Milanick and Gunn 1982), and the form with both sulphate and a proton bound is much more readily converted to the inward-facing form than are the partly loaded forms. Both of these features promote the cotransport of H^+ with SO_4^{2-}.

11.2.3 Basic Transport Mechanism – Ping-Pong Model

Most of the available evidence supports the concept that AE1 functions by means of a ping-pong lock-carrier mechanism, in which the protein can be in either of two conformations, Ei, in which the anion binding (transport) site faces the cytoplasm, and Eo, in which the transport site faces the external medium (Figs. 11.1 and 11.2). The change in transport site orientation probably occurs by coordinated movement of gating regions of AE1 (Fig. 11.2), which control the access of the site to the cytoplasm or external medium, just as the gates in a canal lock control access of the boat in the lock to the portions of the canal on either side of the lock.

The idea that the transition state must involve an occluded form (Fig. 11.2 B), in which the transported anion is inaccessible from either side of the membrane, has not been experimentally demonstrated as in the case of the Na^+/K^+ pump (Chap. 5), but it follows from the fact that there is little net diffusion of anions through this system. If one gate did not close before the other opened, then there would be an open channel for anion diffusion, contrary to the evidence mentioned in Sect. 11.2.1 above (see also Chap. 4).

In terms of this model (Fig. 11.1), the physiological exchange of internal Cl^- for external HCO_3^-, for example, which occurs in the lungs as internal HCO_3^- (designated Bi in Fig. 11.1) decreases due to release of CO_2 from the cell into the alveoli, can be described as follows:

(1) with AE1 in the Ei form, a Cl^- ion binds from the inside to form ECli (panel A in Fig. 11.2, where the "boat" represents Cl^-);
(2) while in this form, by chance the gate between the binding site and the inside closes to form an occluded transition state, in which the Cl^- is trapped (panel B in Fig. 11.2);
(3) the gate between the binding site and the outside opens, forming EClo (panel C in Fig. 11.2);
(4) the Cl^- ion is released to the external medium to generate Eo;
(5) an external HCO_3^- ion (Bo) binds to Eo to form EBo;
(6) by another random event, the gate between the binding site and the external medium closes, forming an occluded transition state with HCO_3^- bound;
(7) the gate toward the inside of the binding site opens, forming EBi;
(8) HCO_3^- is released to the cytoplasm to regenerate Ei, thus completing the cycle.

Note that the Cl^- in step 1 may bind and debind tens or hundreds of times (Falke et al. 1985) before on one "lucky" occasion, the gate closes in step 2. Also, with the small gradients of Cl^- and HCO_3^- that are usually present physiologically (about 2 mM), much of the time the system will exchange Cl^- for Cl^- or HCO_3^- for HCO_3^- and even HCO_3i for Clo (Gunn et al. 1989), the opposite of the thermodynamically favoured process! The laws of chance, however, dictate that if the game is played enough times (and each AE1 molecule transports 50,000 ions per second at body temperature (Brahm 1977)), the result will be net exchange of Cli for HCO_3o under the normal conditions in the lung.

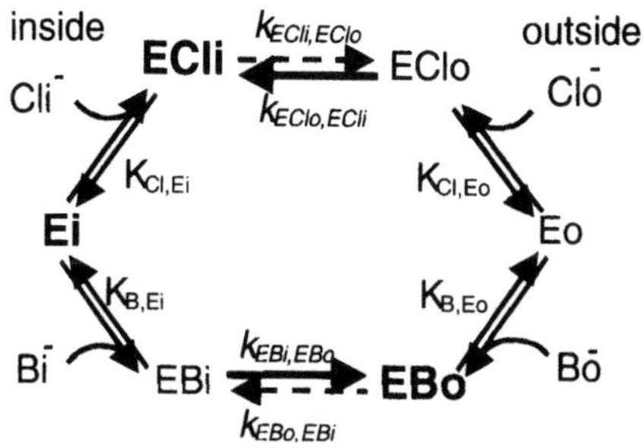

Fig. 11.1. Ping-pong model for Cl^-/HCO_3^- exchange by AE1. B^- represents HCO_3^-. Dissociation constants for binding of anions are designated as, e.g. $K_{Cl,Ei}$, where Cl^- is the anion ligand and *Ei* is the form of AE1 to which it binds. Rate constants are designated as, e.g. $k_{ECli,EClo}$ for the transition from *ECli* to *EClo*. Note that the system is constrained to one-for-one exchange of anions because reorientation of the "empty" forms of AE1, Ei and Eo, is strongly disfavoured. Larger bold type is used to represent the predominant forms of AE1: *Ei* >> *Eo*, *ECli* >> *EClo*, but *EBi* << *EBo*. For the transport steps (*EXi* ↔ *EXo*, where X^- is any anion substrate), the darker solid arrows represent the faster rate constants and the lighter dashed arrows represent the slower rate constants

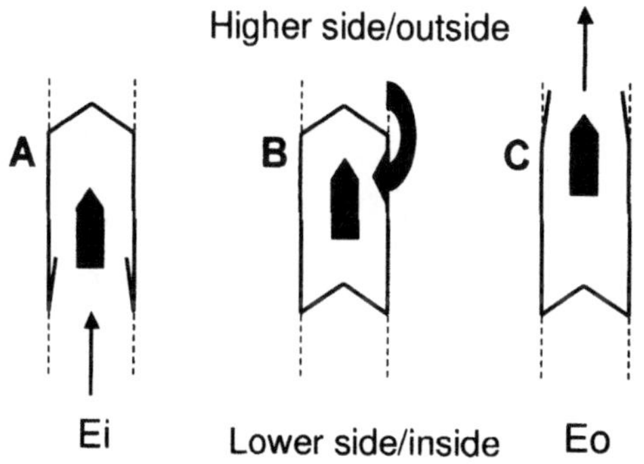

Fig. 11.2. Lock-carrier analogy for AE1. Like a lock in a canal, AE1 can exist in one form (*Ei*, panel **A**) in which the substrate anion (boat) has access to the cytoplasmic side of the membrane (inside, analogous to e.g. the lower water side of the lock). Closing of the gate toward the inside creates an occluded form (panel **B**) in which the anion (boat) is trapped and cannot move to either side. In the case of the canal, this permits adjustment of the water level in the lock - no such "active" process is required for the AE1 anion exchanger. Finally, opening of the gate toward the outside gives rise to a form (*Eo*, panel **C**) in which the anion can diffuse to the outside. As in the canal lock, both gates are never open at the same time, so the movement of the gates must be co-ordinated so that transitions from EXi to EXo always take place via an occluded form. Unlike the canal lock, the AE1 transition from *Ei* to *Eo* or *vice versa* **only** occurs at an appreciable rate when there is an anion in the binding site, and the rate is highly dependent on the kind of anion substrate present

To explain the one-for-one transport characteristic, it is necessary to assume that the transporting conformational change (from Ei to Eo or *vice versa*), designated TCC, can only occur at an appreciable rate if a suitable anion substrate (such as Cl^- or HCO_3^-) is bound at the transport site. The TCC itself is a random event, driven by thermal motion, whose rate is determined by the amount of the free energy increase needed to form the transition state, defined as the highest free energy state on the pathway between the Ei and Eo forms (Fig. 11.3). Binding of certain anions, such as Cl^- or HCO_3^-, must cause changes in AE1 conformation (binding conformational change, BCC) that result in a lowering of this free energy barrier, so that the TCC is facilitated. It is important to emphasize the random nature of this process. One is easily tempted to assume that there is something about the binding of the Cl^- or HCO_3^- ion that "drives" the conformational change, but in fact the anion binding only "permits" a thermally-driven random process to occur at a much higher rate than in its absence.

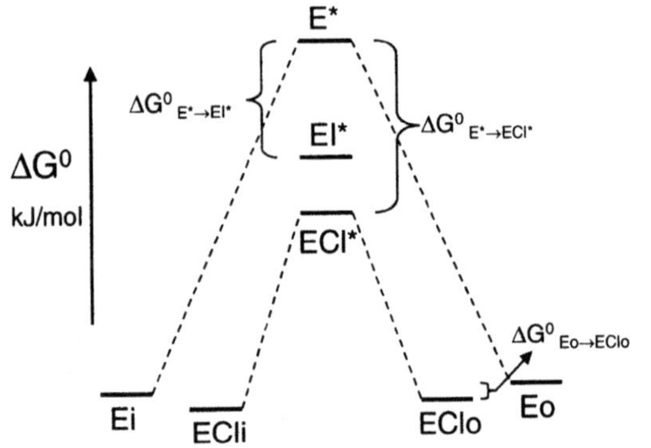

Fig. 11.3. Free-energy diagram for AE1 conformations at 0 °C. Vertal scale indicates free energy in joules/mole - the scale is arbitrary and is only meant to show general aspects of the system. The free energies of *Ei* and *ECli* are slightly lower than those of *Eo* and *EClo*, respectively, so more AE1 molecules will be in the inward-facing (*Ei*) states. The transition state for the *Ei* ↔ *Eo* conversion, *E**, is very high, so this process is highly disfavoured. Binding of a "good" substrate such as Cl^- greatly reduces the free energy, *ECl**, for the *ECli* ↔ *EClo* conversion, thereby increasing the transport rate. This decrease in the free energy of *ECl** compared to *E** may be thought of (Krupka 1989a) as corresponding to a very low dissociation constant, K_{Cl,E^*}, for Cl^- binding to the transition state, because $K_{Cl,E^*} = \exp(\Delta G^0_{E^*\rightarrow ECl^*}/RT)$. This apparent dissociation constant is much lower than the dissociation constants for Cl^- binding to *Ei* or *Eo*, for which the free energy difference, e.g. $\Delta G^0_{Eo\rightarrow EClo}$, is much smaller. Substrate anions that are transported more slowly, such as I^-, do not lower the transition state free energy, *EI**, to such a large extent, and so they have a higher apparent dissociation constant (lower affinity) for the *E** transition state

The binding conformational change (BCC) can, of course, be different for each different anion substrate, and so each substrate can lower the transition state free energy for the TCC to a different extent. "Good" substrates like Cl^- and HCO_3^- lower the free energy to the greatest extent, while "poor" substrates like iodide lower it to a lesser extent (Fig. 11.3). This mechanism can explain why a substrate such as iodide, that has a *higher* affinity for AE1 than Cl^- (half saturation concentrations of about 10 (Dalmark 1976) and 50 mM (Gasbjerg and Brahm 1991), respectively), is transported at a much lower rate. In concept, such a mechanism provides a marvelously clever way of achieving very high substrate specificity without requiring high binding affinity, an important consideration when physiological substrate concentrations are 20–150 mM and when the driving force for the exchange is a concentration difference of only ~2 mM (Gunn et al. 1989). It is likely that this mechanism represents an evolutionary "discovery" that probably underlies the operation of various coupled transport systems, including antiports like AE1 and the Na^+/H^+ exchanger as well as symports like the K^+-Cl^- cotransporter.

Krupka (1989a,b) pointed out that the lower free energy of, e.g., the ECl* transition state as compared to the E* transition state (Fig. 11.3) implies that the transition state has a very low apparent dissociation constant for Cl^-, that is, Cl^- binds with very high affinity. This "dissociation constant" is, of course, only a virtual one, because Cl^- cannot directly dissociate from AE1 in the occluded transition state form (Fig. 11.2), so this process can only be accomplished by a series of steps. For both physiological substrates, these considerations give dissociation constants for the transition states of 1–2 μM! Because Cl^- and HCO_3^- are so different structurally, Krupka argued that it is very unlikely that a single site could bind both of them tightly, and thus that Cl^- and HCO_3^- must bind to separate sites. This argument is not completely convincing because the factors that cause apparently tight binding within an occluded complex (e.g. Fig. 11.2 B) may be quite different from those involved in a direct association-dissociation process, but other evidence also favours this concept (see Sect. 11.6.3).

11.2.4 Evidence for the Ping-Pong Model

Probably the strongest evidence for the fundamental notion of the ping-pong model, that anion exit from the cell and anion entry occur in two separate steps, comes from the half-turnover experiments of Jennings (1982). Red cell ghosts were loaded with very low concentrations of ^{36}Cl, so there were only a few-fold more Cl^- ions inside the ghosts than the number of AE1 molecules in the membrane. Suspension of these ghosts at 0 °C in a medium with sulphate as the only anion resulted in a brief release of a small amount of ^{36}Cl from the ghosts, which then stopped. The total ^{36}Cl efflux was nearly equivalent to the number of AE1 molecules present, as if each AE1 molecule had been able to transport one Cl^- ion outward, after which it became "stuck" in the Eo or ESO_4o form. (Because sulphate is very slowly transported at 0 °C (Ku et al. 1979), the conversion from ESO_4o to ESO_4i is negligibly slow).

Many other types of experiments favour this model, particularly the observation that Cl^- gradients across the membrane can cause a change in the number of AE1 in the Eo form, as detected by the apparent affinity of inhibitors that prefer that form (Furuya et al. 1984; Knauf and Mann 1984; Knauf et al. 1984, 1987b, 2000; Raha et al. 1993). Also, the concept of a single site with alternating access received strong support from Grinstein et al.'s demonstration that reaction of one inhibitor, DIDS, at the external side could prevent binding of another inhibitor, NAP-taurine, at the transport site on the cytoplasmic side (Grinstein et al. 1979). Falke et al. (1984) found that DNDS, that only inhibits anion exchange from the outside (Barzilay and Cabantchik 1979) and that binds preferentially to the Eo form (Fröhlich 1982; Liu et al. 1996c), inhibits Cl^- binding (measured by ^{35}Cl-NMR) at both sides of the membrane, consistent with an alternating site ping-pong model. Jennings (1998) has recently shown that when red cells are preequilibrated in phosphate media with very low (80–250 μM) Cl^-, addition of H_2DIDS to the external medium causes an uphill movement of Cl^- toward the outside, consistent with a ping-pong model in which H_2DIDS recruits AE1 from the ECli form to the Eo form, resulting in each AE1 molecule transporting one Cl^- outward. Moreover, as discussed in Sect. 11.5.5, a cytoplasmically-accessible histidine also exhibits properties indicating that AE1 can be recruited to two different conformations (Izuhara et al. 1989). These are only a few of the examples of "recruiting" phenomena, consistent with the ping-pong alternating site model.

Although the ping-pong model is successful in explaining many phenomena, even the very first study of the dependence of flux on Cl^- concentration showed that the system did not exhibit the hyperbolic saturation behaviour expected for the simplest form of this model (Cass and Dalmark 1973). The decrease in flux at very high $[Cl^-]$ requires at least one additional Cl^- binding site besides the transport site, and was explained by postulating an inhibitory "modifier" site at which Cl^- can bind (with low affinity) and inhibit its own transport (Cass and Dalmark 1973). Other anion substrates exhibit similar self-inhibitory behaviour (Dalmark 1976), and Salhany and co-workers (1983) and others (Hautmann and Schnell 1985) have provided kinetic evidence inconsistent with the simple ping-pong model and demonstrating binding of more than one anion at a time to AE1. Integration of the evidence for additional sites into a structural model for anion transport still remains an area of speculation (see Sect. 11.6.3). Strictly speaking, the recruiting and half-turnover experiments mentioned above only provide evidence for an alternating site model, of which the ping-pong model is the simplest example. They do not rule out other alternating site models, such as the ordered sequential model which involves one state with two anions bound (Salhany 1996).

11.3 Asymmetry of the System

11.3.1 Sidedness of Unloaded Transport Sites

Because Ei and Eo represent two different conformations of a protein, AE1, their free energies can be different (Fig. 11.3). Even a small free energy difference will cause one form to be favoured over the other, so that more of the unloaded transport sites will face either the inside or outside. Unloaded site asymmetry is expressed by a parameter A, which is equal to Eo/Ei with equal concentrations of substrate at the two sides of the membrane. Asymmetry ($A \neq 1$) reveals itself in the different apparent affinities for substrate at each side of the membrane or in the effects of external or internal substrates on the apparent affinities for inhibitors that bind preferentially to either Ei or Eo (Knauf 1979; Fröhlich and Gunn 1986; Knauf and Brahm 1989). Both kinds of evidence agree that the unloaded sites are highly asymmetric, both at 0 °C (Knauf and Brahm 1989; Gasbjerg and Brahm 1991) and at 38 °C (Gasbjerg et al. 1996; Knauf et al. 1996), with an A value of < 0.1, indicating that there are more than 10 times as many AE1 molecules in the Ei form (shown in bold in Fig. 11.1) as in the Eo form.

As first pointed out by Fröhlich and Gunn (1986), the unloaded site asymmetry should not be affected by the kind of substrate anion present, because it represents the effect of a difference in free energy of the *unloaded* forms of AE1. Indeed, measurements have shown that the A value measured with Cl^- or HCO_3^- as substrate are nearly the same, both at 0 °C (Gasbjerg and Brahm 1991) and at 38 °C (Gasbjerg et al. 1996; Knauf et al. 1996).

This asymmetry (Ei >> Eo) leads to an asymmetry in *apparent* affinities for substrates, with the external affinity being higher than the internal (Gunn and Fröhlich 1979). This seems paradoxical if one thinks in terms of mass action, because there is less of AE1 in the Eo form, so one might suppose that it would be more difficult to form complexes with external anions, e.g. EClo. The paradox is resolved, however, when one realizes that the apparent external dissociation constant for Cl^-, $K_{1/2o}$, is the concentration of external Cl^- at which half of the AE1 molecules are in the Eo form (Fröhlich and Gunn 1986). If this form is disfavoured by having a higher free energy, then [Clo] must be reduced to very low values to recruit half of the AE1 molecules to Eo. Thus, $K_{1/2o}$ is very low and the corresponding apparent affinity is very high.

11.3.2 Effects of Chloride on Sidedness of Transport Sites

The asymmetry ratio for Cl^- loaded sites, A_{Cl}, is related to the free energy difference between ECli and EClo. A_{Cl} is equal to EClo/ECli, and thus $A_{Cl} = k_{Cli,EClo}/k_{EClo,ECli}$ (Fig. 11.1). In principle, this ratio can be quite different from A for the unloaded forms. At present, there is no way of determining A_{Cl} precisely. Estimates can be made based on ^{35}Cl-NMR measurements of the dissociation constant for Clo binding to Eo, under conditions where the ECli ↔ EClo transition is prevented by a noncompetitive inhibitor. This assumes that the inhibitor does not affect the affinity, an assumption that cannot be adequately tested. Data with eosin-5-maleimide (EM) as inhibitor indicate that $A_{Cl} \sim A$ (Liu et al. 1996a). Recent refinements give A_{Cl} of about 0.14 (Kennedy, Wu, and Knauf, unpublished data), suggesting that the Cl^- loaded sites are also largely in the inward-facing form (Fig. 11.1), but that their asymmetry may be somewhat smaller than that of the unloaded sites.

11.3.3 Effects of Bicarbonate on Sidedness of Transport Sites

The asymmetry factor for bicarbonate (B)-loaded sites, A_B = EBo/EBi, is determined by the free energy difference between EBi and EBo, and thus may be different from A and A_{Cl}. The only method so far for estimating this asymmetry involves comparisons of the rates of heteroexchange of Cli for Bo and Bi for Clo. If A_B is similar to A_{Cl}, the rates will be similar, but if $A_B >> A_{Cl}$, Bi/Clo exchange will be faster than Cli/Bo exchange. Experiments at both 38 °C (Wieth and Brahm 1980) and 0 °C (Knauf et al. 2002b) show pronounced asymmetry in this direction; quantitative analysis using a model (Weinstein 2000) based on Cli/Clo and Bi/Bo exchange data (Gasbjerg and Brahm 1991; Gasbjerg et al. 1996; Knauf et al. 1996) indicates that $A_B \geq 10$ (Knauf 2002; Knauf et al. 2002b), so far more bicarbonate-loaded sites face the outside than the inside, the opposite of the asymmetry seen for unloaded or Cl^- loaded sites (Fig. 11.1). As shown in Fig. 11.1, Bi/Clo exchange (anticlockwise in Fig. 11.1) involves the larger anion translocation rate constants for both Cl^- and B^- (solid lines), while Cli/Bo exchange (clockwise) involves the smaller rate constants (dashed lines), giving rise to asymmetry in the heteroexchange flux rates.

11.3.4 Consequences of Asymmetry for Fluxes under Physiological Conditions

The surprising observation of different asymmetry with Cl^- or B^- as substrate raises the question of the possible physiological significance of this characteristic. Although the heteroexchange flux asymmetry is as much as 5-fold at 0 °C, this is seen under extremely asymmetric conditions, with all Cl^- at one side and all B^- at the other, or *vice versa*. In systemic and lung capillaries, where physiological Cl^-/B^- exchange takes place, the gradients of B^- that drive this process are very small, usually < 2 mM (Gunn et al. 1989). Calculations for these conditions indicate that the bicarbonate net fluxes in the tissues and lungs take place at nearly the same rate but in opposite directions (Knauf 2002). Thus, the strongly asymmetric system performs physiological Cl^-/B^- exchange no differently than would a completely symmetric system.

The other major function of AE1 is transport of HCO_3^- (formed from CO_2) out through the basolateral membranes of intercalated cells in the kidney outer medullary collecting duct. Here the asymmetric system shows a very slight advantage over a completely symmetric one (Knauf 2002), but the advantage is so slight (about 5%) that it seems unlikely to have exerted significant selection pressure favouring asymmetry. Thus, the asymmetry is probably an accidental consequence arising because two different protein conformations are unlikely to have precisely the same free energy, and because such differences do not significantly affect physiological function.

An important corollary of the opposite asymmetries with Cl^- and B^- bound is that the Ei and Eo forms have opposite *selectivities* (relative affinities) for Cl^- *vs.* B^- (Knauf et al. 2002b). The Ei form prefers Cl^- by ≥ 5-fold and the Eo form prefers B^- by about 10-fold. Thus the characteristic of side-dependent changes in substrate selectivity, critically important for ATP-driven pumps like the Na^+, K^+-ATPase and often thought to be a special characteristic of such active transporters, is also seen in this non-energy-requiring passive ion exchange system.

Another implication of the asymmetry is that AE1 can be recruited preferentially toward either the EBo or ECli forms in B^- or Cl^- media respectively, thus providing an experimental means for altering the sidedness of transport sites for studies of the structural effects of such changes. Other evidence indicates that with iodide (Knauf et al. 1986) or sulphate (Jennings 1995) bound, there are approx. equal numbers of inward- and outward-facing forms of AE1. This provides further opportunities to alter AE1 conformation by substrate replacement.

11.4 Structure of AE1

Proteolysis and gene sequencing studies have revealed that AE1 consists of two distinct structural and functional domains divided by sites for cytoplasmic trypsin (Lys360) and chymotrypsin (Tyr359) cleavage. The N-terminal 40 kDa domain (1–360) is located within the cytoplasm (cdAE1) and contains the sites of cytoskeleton, haemoglobin and glycolytic enzyme binding (Low 1986). The membrane-integrated 55 kDa C-terminal domain (361–911) (mdAE1) is necessary and sufficient for the obligatory exchange of anions (Kopito and Lodish 1985).

11.4.1 The Cytoplasmic Domain (cdAE1)

The cytoplasmic domain provides an anchor through which the cytoskeleton is linked to the membrane, thus playing an important role in stabilising the membrane in the face of mechanical stress. Mutations that affect this function cause abnormalities of red cell shape and/or premature lysis (Schofield et al. 1992b; Bruce et al. 1993; Jarolim et al. 1994; Miraglia del Guidice et al. 1997). Recently, a 2.8 Å resolution crystal structure (cdAE1x) has been published (Zhang et al. 2000) that defines most of the cdAE1 (residues 1–356) which, like AE1 as a whole (Casey and Reithmeier 1991), forms a dimer (Appell and Low 1981; Colfen et al. 1996). Unfortunately, the cdAE1 was crystallised at an unphysiologically low pH (4.8), and there is evidence that this structure differs in important ways from that at normal pH (Low et al. 1984; Doyle et al. 1998). Nevertheless, the crystal structure is useful for revealing the nature of the monomer-monomer contacts in the AE1 dimer. In particular, the crystal structure shows near 2-fold symmetry around an axis at the dimer interface.

The exact position of the cdAE1 relative to the membrane is also not clear. The part of cdAE1 that was crystallised contains residues 1–379, but residues 357–379 were not resolved, possibly because this region is flexible. Hydropathy analysis suggests that AE1 does not enter the membrane until about residue Val405 (Fig. 11.4). The region around Val405 seems to be important, because a common mutation with a deletion of residues 400 to 408 (Schofield et al. 1992b), Southeast Asian ovalocytosis (SAO), inhibits anion exchange and has effects on cytoskeletal contacts and cell rigidity (Moriyama et al. 1992; Schofield et al. 1992a). Because of the 48 amino acid gap between the cdAE1x crystal structure and the probable membrane entry site, the relationship of cdAE1x to the membrane surface is unknown. Since it is highly likely that the 2-fold axis of symmetry (z-axis) is perpendicular to the membrane surface, it can be estimated that the cross-sectional thickness of cdAE1x is about 50 Å along the z-axis (Zhang et al. 2000).

11.4.2 Structure of the Membrane Domain Revealed by Electron Microscopy

A high-resolution structure of the membrane-integrated portion of AE1 (mdAE1) has proved to be elusive due to the difficulties in procuding well ordered crystals of AE1 from detergent micelles. However, Wang et al. (1993, 1994) and Wang (1994) have succeeded in reconstituting the protein into lipid vesicles to obtain two-dimensional crystals that were studied by electron microscopy and image reconstitution. Deglycosylated mdAE1 crystallised as dimers in two different crystal forms - sheets and tubes - and the negatively stained crystals diffracted to give an image at a very low (20 Å) resolution. A 3D map was reconstituted from a tilt series of electron micrographs of the crystalline sheets. At this resolution, details like individual α-helices are not observed, but the overall shape of the protein dimer and a few interesting structural features are seen. The dimer is 100 Å long, 60 Å wide and 80 Å thick. It consists of a basal domain, which is membrane embedded, and two upper protrusions, probably on the cytosolic side of the membrane, that form a stain-filled 'canyon' at the dimer interface. The basal domain is roughly 40 Å thick and its cross sectional area is consistent with a 12- or 14-helix model of the protein (see Sect. 11.4.3). The 'canyon' facing the cytosol was proposed to represent the anion access channels (Wang 1994). However, in order to accommodate the known anion substrates, the access channel need only be approx. 9 Å in diameter (Giebel and Passow 1960), and so might not be visible in a 20 Å-resolution structure.

11.4.3 Topology of the AE1 Membrane Domain

In absence of a high resoltuion structure of AE1, effort has been made to understand the topology of the protein within the membrane. Hydropathy plots (Wood 1992; Reithmeier et al. 1996), vectorial chemical labeling, and proteolysis (Jennings 1992a) were used to determine the topology model of the protein. Recently, molecular biology techniques have provided further insights into the topology issue. Casey and co-workers (Tang et al. 1998; Fujinaga et al. 1999; Tang et al. 1999) have generated a series of unique single-cysteine mutants that can be expressed in HEK 293 cells and have investigated the accessibility of each engineered cysteine to biotin maleimide (BMA). Only cysteines outside the apolar part of the membrane bilayer are able to react with BMA (Tang et al. 1998). External membrane impermeant reagents, e.g. Lucifer Yellow Iodoacetamide (LYIA) are used to block BMA reaction, thereby distinguishing extracellular and cytosolic residues. Reithmeier and co-workers (Popov et al. 1997, 1999) have utilised a scanning *N*-glycosylation mutagenesis method where a consensus sequence for *N*-glycosylation (Asn-Xaa-Ser/Thr) or a part of the loop containing the endogenous Asn site (the extracellular (EC) 4 loop; see below and Fig. 11.4) is inserted into various putative loop regions of AE1. The degree of *N*-glycosylation at the mutated sites can be used to fix the boundaries of the extracellular loop regions and hence the topology of the protein (Popov et al. 1997).

These studies indicate that mdAE1 traverses the bilayer about 14 times, forming transmembrane (TM) segments (Fig. 11.4) that are probably α-helical in nature, as indicated by the high helical content of AE1 from circular dichroism studies.

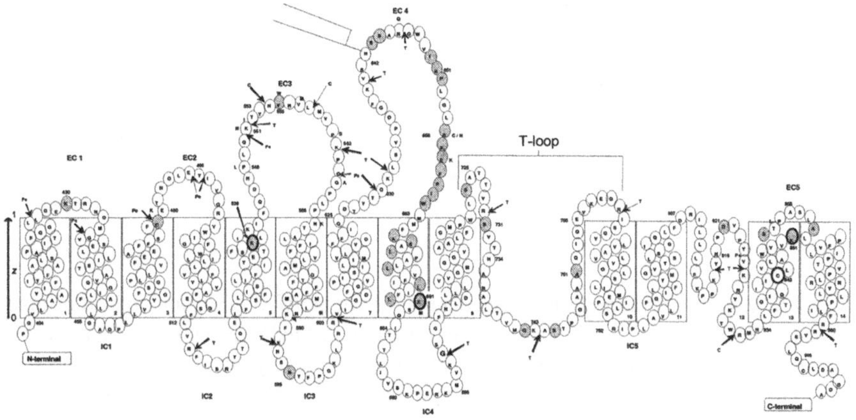

Fig. 11.4. Topology model of mdAE1. N-terminal cytoplasmic domain residues 1–402 and residues 892–911 (C-terminus) are not shown. The transmembrane (TM) segments are enclosed in rectangular boxes (except for TM 12) and identified by the TM number at the bottom right hand corner. The branched lines at Asn642 represent the N-linked oligosaccharide. Proteolytic cleavage sites are indicated by arrows and the enzymes are abbreviated as follows: *Pa*, papain; *Pe*, pepsin; *T*, trypsin; *C*, chymotrypsin. Shaded residues indicate those that are accessible to chemical probes from the outside (see Table 11.1)

Description of the AE1 topology is facilitated by defining a term, the 'z-coordinate' or 'z', which specifies the location of various sites with respect to their distance from the cytoplasmic side of the lipid bilayer (left edge of Figs. 11.4 and 11.5). For a site at the lipid head-group region on the cytoplasmic side, z is defined as zero, while $z = 1$ at the external head-group region; so $z = 0.5$ indicates a position halfway through the bilayer.

11.4.4 The First Eight TM Segments

The topologies of the first eight transmembrane segments are clearly defined and the different methods yield mainly consistent results. The AE1 protein enters the membrane from the cytoplasmic side at Val405. The boundaries of EC loops 1–4 have been defined by a variety of methods such as *N*-glycosylation insertional mutagenesis (Popov et al. 1997, 1999), chemical accessibility (at Lys430 – see Table 11.1 and Fig. 11.4) (Cobb and Beth 1990; Cobb et al. 1990), proteolysis (Okubo et al. 1994), and through presence of natural mutations (e.g. R432W,

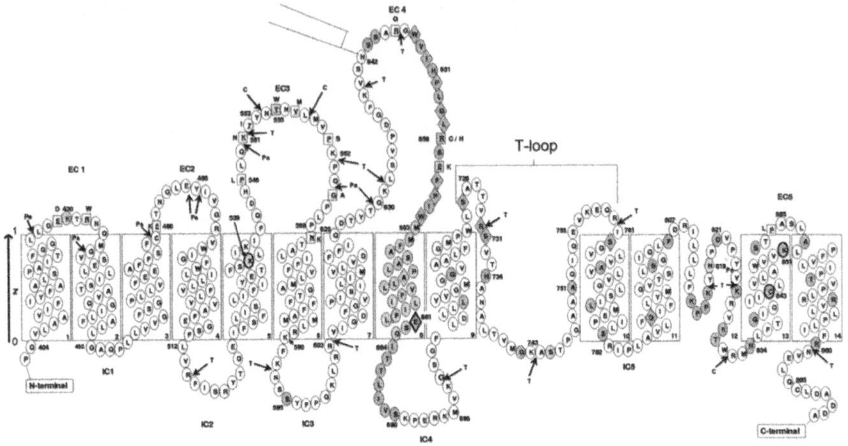

Fig. 11.5. Map of natural and site directed mutations in mdAE1. *Gray*: amino acid residues that have been studied by site directed mutagenesis methods. *Diamond*: residues where function of AE1 is abolished (less than 15% of wild type) by site directed mutagenesis. *Hexagon*: loss of function at these sites due to mutation to Gln is recovered by a 'second site' mutation K539N. *Square and underlined*: Sites for natural AE1 mutations that produce antibodies against red blood cells but do not affect the function. Single letter amino acid codes near these residues indicate the nature of the mutations (also see Table 11.2)

E429D, E480K) that cause antibody production against intact red cells in humans (see Chap. 26) and which must therefore be extracellular (see Fig. 11.5).

The *N*-glycosylation site at Asn642 serves as an endogenous marker placing this region in the extracellular domain (EC 4) (Tanner et al. 1988). Scanning cysteine mutagenesis studies (Tang et al. 1998) have indicated an interesting feature of this loop proposed to extend from Gln625 to Met663. Residues Ser643–Met663 are consistent with an extracellular location. However, the extent of LYIA labelling gradually decreased from Arg656 to Met663. Also the segment closest to the glycosylation site, Ser643–Leu655 (excluding His651) showed less labelling by LYIA than expected (Fujinaga et al. 1999). This was explained by proposing that this region has a folded structure whereby Arg656–Met663 forms a funnel that acts as an extended anion access channel, with the Ser643–Leu655 sequence supporting the funnel from underneath.

TM 8, containing the important residue Glu681, which appears to cross the permeability barrier (see Sect. 11.5.2), is defined by the sequence Met664–Gln683, in which all substituted cysteines are poorly labeled by BMA. Cysteines substituted from residues Ile684 to Ser690 are again labeled by BMA, but not by LYIA, showing that they are located on the cytoplasmic side (Fujinaga et al. 1999). Thus, Glu681 is located three residues or about 5 Å from the internal side of the lipid bilayer (assuming a helical pitch of 1.5 Å per residue), giving it a z value of ~0.15.

11.4.5 Remaining C-Terminal Segments; Controversial Topologies

Topological assignment of the remaining C-terminal portion of the protein has not been straightforward. One of the major controversies has been over the location of the residues from Ala726–Ile761, the so-called 'T-loop' (trypsin-sensitive loop) region. The centre of this region is proposed to be located on the cytoplasmic side due to cleavage by cytoplasmic trypsin at Lys743 (Jennings et al. 1986; Kuma et al. 2002). Although trypsin sealed inside resealed ghosts and trypsin treatment of inside-out vesicles produced the cleavage fragment, no cleavage could be achieved even after extensive external trypsin treatment of intact red cells. However, Fujinaga et al. (1999) found that cysteines substituted at Ser731, Gly742, Ser745 and Ala751 were accessible to both BMA and LYIA, indicating that these residues are accessible from the outside. Popov et al. (1997) created a novel glycosylation site with the mutant K743N and inserted glycosylation signals into various points in the T-loop region. *N*-glycosylation was observed in these mutants when expressed by cell-free translation in a rabbit reticulocyte lysate in the presence of microsomal membranes, but the levels of glycosylation were lower than that of the endogenous site at Asn642. In addition, glycosylation was abolished when similar mutants were expressed in HEK 293 cells (Popov et al. 1999), except for the mutant with an insertion after Gln754.

To reconcile these conflicting pieces of evidence, Fig. 11.4 shows the T-loop folding back into the membrane with both the N- and C-terminal ends at the extracellular side. Such a topology would agree with evidence suggesting that His734 is accessible to protons from the cytoplasmic side and that it interacts with Glu681 (see Sect. 11.5.5). A purely extracellular location of the T-loop ($z > 1$) would put the His734 residue too far away from Glu681 (at $z \sim 0.15$) to allow any meaningful interaction. In addition, as can be seen from Fig. 11.5, the natural antibody producing AE1 mutations (squares) can be found on all the large putative extracellular loops like EC 3 and EC 4 and even on some short loops like EC 1 and EC 5. While this is not conclusive evidence, it would be strange if an extracellularly exposed loop with 36 residues did not have a single antigenic site.

This model could still accommodate the observed external accessibility of the substituted cysteines, since the folded loop could form a part of an aqueous channel that allows entry of bulky hydrophilic molecules like BMA and LYIA. Part of this loop could also be located at the permeability barrier and thereby exposed to either the cytoplasm or the extracellular medium depending on the conformational changes associated with the transport cycle (Sect. 11.6.2). Cytoplasmic trypsin cleavage could result so long as Lys743 is accessible to the cytoplasmic side of the membrane in at least some AE1 conformations. This model would not explain the ability of the topologically extracellular oligosaccharyl transferase enzymes to reach the glycosylation site in the K743N mutants. However, since *N*-glycosylation is observed mainly in the cell-free translation systems, it could be the result of a transient exposure of the region to the lumen of the endoplasmic reticulum during protein biosynthesis, giving rise to an aberrant form of the protein. Such aberrant forms would be degraded by cellular proteases when expressed in the HEK 293 cells, explaining the absence of glycosylation in this system.

Beyond the T-loop, in the direction of the C-terminus, it is likely that there are two TM helices (10 and 11) with a very short intracellular loop (IC 5) before we come to another peptide segment, Asp807–His834 (TM 12 in Fig. 11.4), that shows a peculiar topology. In this region, cysteines substituted at Lys814 and Thr830 are unreactive with BMA while that at Asp821 is accessible to BMA as well as extracellular LYIA (Fujinaga et al. 1999). A labelling study with an impermeant, positively charged carbodiimide, on the other hand, indicates that Asp821 is accessible from the cytoplasm (Erickson 1997). The Bric132 monoclonal antibody recognises its epitope, Phe813–Tyr824 only in leaky erythrocytes (Wainwright et al. 1990), suggesting a cytoplasmic location, but the antibody labelling is weak. A glycosylation signal inserted at Asp821 is not *N*-glycosylated, which also argues against an extracellular location (Popov et al. 1999). It is possible that this region spans the membrane with Asp821 adjacent to an aqueous channel. Interestingly, Asp821 is accessible to the negatively charged LYIA from the outside and a positively charged carbodiimide from the inside. This parallels studies by Pan and Cherry (1998), who showed that positively charged quenchers are able to reach a fluorescent label on Lys430 only from the cytoplasmic side, while negatively charged ones do so only from the extracellular side. The concentration of positive charges in this sequence – as many as six positive charges (seven if we consider His) as compared to only two negative charges – might indicate that it forms part of a channel through which external anionic substrates and probes might diffuse to sites deep within AE1. This region has been represented as an extended helix in Fig. 11.4.

Cysteine accessibility scanning of the protein sequence from Trp831–Gln884 indicates that there are two TM helices joined by a very short extracellular loop (Fujinaga et al. 1999). The accessibility of A855C to BMA (Fig. 11.4) and the presence of the blood group antigen P854L (Fig. 11.5) confirm the extracellular location of this short loop.

The residues from Arg880 onward form the C-terminal tail often referred to as the third domain of AE1. It is located in the cytoplasm and recent evidence indicates that it binds to carbonic anhydrase II (CAII) under physiological conditions. This interaction has been proposed to create a 'metabolon', in which the physical proximity of CAII, which hydrates CO_2 to form HCO_3^-, to AE1, which transports HCO_3^-, may enable the red blood cell to store CO_2 as plasma bicarbonate more efficiently (Vince and Reithmeier 1998, 2000).

11.5 From Structure to Function: Information from Mutations and Chemical Labelling

Various lysine, arginine, histidine, glutamate, and other residues have been implicated as being important for transport function and/or as sites for reaction with anion transport inhibitors. A list of such residues along with their chemical modifiers and proposed functions is given in Table 11.1. The roles of some of these residues were further investigated using site directed mutagenesis (Table 11.2). To avoid confusion, in the ensuing discussion we will refer to mutations of amino acids in both mouse and human AE1 in terms of the number of the corresponding amino acid in *human* AE1.

Table 11.1. Some important chemical modification and proteolysis sites

Residue[#]	Location	Importance (reference)	Effect* [section]	Cons[^]
Tyr359/ Lys 360	Cytoplasm	Internal cleavage site for chymotrypsin and trypsin respectively; boundary between cytoplasmic and membrane domains	(+) [11.4.1]	+
Lys430	EC 1	Chemical labeling by EM and reductive methylation from the external side	(–) [11.5.4]	+
Cys479	TM 3 (near EC 2)	Unreactive to maleimides (EM) but reacts with PCMBS in intact cells; called 'cryptic cysteine' (Ojcius and Solomon 1988)	(+)	+
Lys539	TM 5	H_2DIDS rapid covalent reaction site	(–) [11.5.3]	+
Lys542	TM 5	Possible alternate site for DIDS (not H_2DIDS) reaction at neutral pH when Lys539 is unavailable (Bartel et al. 1989)	(–)	+
Lys551/ Lys562	EC 3	Sites for intermonomer crosslinking with BSSS in AE1 dimer	(+) [11.4.6]	AE1s
Lys590	TM 6 (near IC 3)	Reaction site for phenylisothiocyanate; Possible site for photoaffinity labeling by FAzNPPB (Branchini et al. 1995)	(–)	hAEs
Gln630	EC 4	Extracellular cleavage by papain (Jennings et al. 1984)	(–)	+
Asn642	EC 4	Endogenous *N*-glycosylation site	(+) [11.4.4]	AE1s
Glu681	TM 8	Can be modified by WRK/BH_4^- from extracellular side	Proton binding site [11.5.2]	+
His734	T-loop	Principal site for labeling by DEPC, accessible to protons from cytoplasm	(–) [11.5.5]	+
Lys743	T-loop	Internal trypsin cleavage site (low yield)	(+) [11.4.6]	hAEs
Cys843	TM 13	Acylation (principally palmitoylation) site	(+) [11.5.7]	hAEs
Lys851	TM 13	PLP binding and H_2DIDS crosslinking in intact cells	(–) [11.5.3]	Met in AE2
Cys885	C-terminus	Internal NEM reactive site (Rao and Reithmeier 1979)	(+) [11.5.7]	hAEs

* (–) Indicates that labeling or proteolysis at that site causes inhibition of anion exchange. (+) No known effect on function; [Section in which details and further references will be found]; ^ '+' Indicates conserved in all known AE1, AE2 and AE3; # Residues in bold have been studied using site-directed mutagenesis.

Table 11.2. Effects of site directed mutation of certain residues in human AE1

Location	Mutation*	Effects^ (reference) and/or [section]
EC 1	K430S/G	(+) Only reversible EM binding (1) [11.5.4]
EC 2	R490K/C/T	No surface expression detected (2) [11.5.6]
TM 5	K539N	(+) Irreversible PLP inhibition not affected, DNDS affinity decreased ~4× (3) [11.5.4]
TM 8	E681Q/T/R/K /G	Abolishes Cl^-/Cl^- or Cl^-/SO_4^{2-} or SO_4^{2-}/SO_4^{2-} transport except E699Q enhances SO_4^{2-}/SO_4^{2-} transport [11.5.3]
T-loop	H734S	Some transport observed and significant reduction of inactivation rate by DEPC (4) [11.5.5]
TM 13	K851M	(+) Irreversible PLP inhibition not affected, DNDS affinity decreased ~12× (3) [11.5.4]
Double Mutants:		
TM 5/TM 13	K539M-K851M	(+) Only reversible inhibition by PLP/DNDS affinity decreased ~80×; K_m for Cl^- binding decreased ~4×; No H_2DIDS irreversible inhibition (4) [11.6.1]
TM 5/T-loop	K539N-R730K	(–) (3)
EC 2/T-loop	R490K-R730K	No surface expression (3)
TM 5/IC 4	K539N-H703Q	AE activity > H703Q (less than wt) (5)
TM 5/TM 12	K539N-H819Q/R	(K539N/H819Q) AE > H819Q (less than wt) (K539N/H819R) AE ~ H819R (5)
TM 5/TM 12	K539N-H834Q	AE comparable to wt levels (5) [11.6.1]

* Mutations are all represented by hAE1 number; (for the mouse amino acid numbers see Wood (1992)); ^ (+) indicates normal anion exchange (AE) function, (–) indicates AE function lost; References: (1) Passow et al. 1992, (2) Karbach et al. 1998, (3) Wood et al. 1992, (4) Müller-Berger et al. 1995a, (5) Müller-Berger et al. 1995b

11.5.1 Interpretation of Mutation and Chemical Modification Data

Before discussing the data, it is important to emphasize some problems in interpreting chemical modification and mutation data in AE1. Considerable evidence, primarily gathered by Salhany and co-workers (Salhany et al. 1987, 1990, 1993, 1994; Batenjany et al. 1993), demonstrates that there are strong allosteric interactions not only within the membrane domain of AE1 but also between membrane and cytosolic domains and between AE1 monomers, so alteration of one part of AE1 can have effects on distant parts of the protein. Because of this, the fact that mutation of an amino acid affects function does not permit one to conclude that this amino acid normally plays a role in the function. Many mutations alter the degree of expression of AE1 in the membrane. Even if such mutations are excluded, more subtle structural effects are always possible. Chemical probes have the additional problem that they can inhibit by affecting parts of AE1 different from the site of chemical reaction. In contrast, the observation that mutation or chemical modification of an amino acid has *no* effect on a function still provides strong evidence that that residue is not essential.

In the following discussion, we will try to interpret the experimental data in terms of the simplest models, which usually involve proximity or steric interactions. The reader should keep in mind, however, that such conclusions are very tentative in the absence of more precise information about such a highly cooperative molecule as AE1.

11.5.2 Carboxyl Group Modification: Putative Identification of Proton Binding Sites

As mentioned above (Sect. 11.2.2), AE1 can also serve as a cotransporter of H^+ and SO_4^{2-} ions. Strangely, there is more information about the binding site for the only known positively charged substrate, H^+, than for any anionic substrate. Jennings and Al-Rhaiyel (1988) discovered that the anionic carboxyl-reactive reagent, Woodward's reagent K (WRK), inhibits anion transport in intact red cells when added to the extracellular medium. If the reaction product is then cleaved with borohydride (hereafter designated W-B treatment) to convert the WRK-reactive R-COO^- to the corresponding alcohol, R-CH_2OH, Cl^- transport is still inhibited, but sulphate transport increases dramatically at neutral pH, where sulphate exchange normally is slow compared to more acidic conditions. This shift in pH dependence is seen for both internal and external pH, suggesting that the WRK-reactive COO^- group is accessible from either side of the membrane and thus that it might be the site that binds protons and translocates them across the permeability barrier.

These effects occur under conditions where the W-B modification is highly specific for one carboxyl group of AE1, Glu681 (Jennings and Anderson 1987; Jennings and Smith 1992). Because there is topological evidence that Glu681 is located near the cytoplasmic side of the membrane (Fig. 11.4) near $z = 0.15$ (Tang et al. 1998), this implies that the impermeant polar reagent, WRK, can diffuse to a site deep within the membrane, so there must be an aqueous channel within AE1 that extends from the outside nearly to the cytoplasmic surface of the bilayer.

Normally, exchange of a proton (H^+) together with a SO_4^{2-} anion for one Cl^- results in zero net charge transfer. After W-B treatment, however, Cl^-/SO_4^{2-} exchange becomes electrogenic (Jennings 1995), resulting in flow of one negative charge across the membrane with the SO_4^{2-}. This is exactly what would be expected if the SO_4^{2-} normally crosses the membrane together with a proton bound to Glu681. Release of the proton together with the SO_4^{2-} generates the form of AE1 with one less positive charge appropriate for transport of the monovalent anion Cl^- in the opposite direction. W-B converts the R-COO^- proton binding site to the alcohol, with the H atom much more tightly bound, so H^+ cannot be released into the medium with the SO_4^{2-}. The net result is transport of 2 negative charges in exchange for the single negative charge of the Cl^- flowing in the opposite direction. The dependence of SO_4^{2-}/Cl^- exchange on V_m fits best with a model in which most of the charge transfer occurs during the Cl^- transport step, with flow of Cl^- being accompanied by 2 positive charges.

Experiments with site-directed mutations of Glu681 (Glu699 in mAE1) expressed in *Xenopus* oocytes (Chernova et al. 1997) support the concept that it is the proton transport site. Substitution of Q, T, R, K, or G for Glu681 abolishes Cl^- efflux in exchange for either Cl^- or SO_4^{2-}. All with the exception of Q (Gln) also abolish SO_4^{2-}/SO_4^{2-} exchange. Compared to wild type AE1, the Gln mutant shows increased SO_4^{2-}/SO_4^{2-} exchange, and even greater disulphonic stilbene (DS)-sensitive $SO_4^{2-}{}_i/Cl^-_o$ exchange. The latter exhibits a 1:1 $Cl^-:SO_4^{2-}$ stoichiometry and is electrogenic. In a separate study (Sekler et al. 1995), replacement of Glu681 with Gln or Ala greatly diminished the pH dependence of sulphate efflux over the pH range from 5.5 to 11.5. These results are remarkably parallel to the effects of W-B treatment in intact human red cells, as expected since substitution of glutamine for glutamate removes the negative charge of the $R–COO^-$ group and prevents it from translocating a proton, as does W-B treatment.

The seemingly conservative substitution of Asp for Glu681 inhibits Cl^- exchange (Müller-Berger et al. 1995a) and alters the pK for its inhibition by protons (see Chap. 10). Because Glu and Asp differ by only a single $–CH_2–$ group in the side chain, this suggests that the AE1 structure near Glu681 is sufficiently rigid that a small change in position of a negative charge greatly alters its function.

11.5.3 Amino Group Modification by Disulphonic Stilbenes

DS isothiocyanates, such as DIDS, SITS, and their derivatives, covalently react with amino groups of lysine. The correlation between their selective reaction with band 3 and inhibition of anion exchange provided the first evidence that band 3 was the anion exchange protein (Cabantchik and Rothstein 1974; Lepke et al. 1976; Ship et al. 1977). Measurements of reversible inhibition of anion transport with more slowly-reactive (H_2DIDS) or non-covalently-reactive (DNDS) derivatives revealed that Cl^- reduces the potency of the stilbenes as reversible inhibitors of anion exchange in the manner predicted if Cl^- and DS compete for the external-facing transport site (Shami et al. 1978; Fröhlich 1982). Because the amino-reactive DS bind reversibly prior to covalent reaction (affinity labelling) (Rakitzis et al. 1978), it was thought that the DS could be used to identify the transport site in AE1 (Shami et al. 1978).

Relation of DS-Labelled Sites to the Transport Site

Although Passow et al. (1980b) pointed out very early that allosteric interactions between Cl^- and an inhibitor could masquerade as direct competition, the first experimental indication that the DS site might not be the transport site came from the observation that mutation of Lys539 and Lys851 (Table 11.2), the sites with which H_2DIDS reacts in AE1 (Jennings and Passow 1979; Bartel et al. 1989; Okubo et al. 1994), does not inhibit Cl^- exchange (Wood et al. 1992). Dix et al. (1986) showed that the reversibly-binding fluorescent DS, DBDS, binds to AE1 in unsealed ghost membranes by a 2-step mechanism, in which there is an initial binding of the DS to AE1 to form a first complex (1), AE1•DS, followed by a conformational change to form a second complex (2), AE1–DS*:

$$AE1 + DS \underset{k_{-1}}{\overset{k_1}{\rightleftarrows}} AE1 \bullet DS \underset{k_{-2}}{\overset{k_2}{\rightleftarrows}} AE1 - DS^*$$

Rapid kinetic measurements of the biexponential time course of DS binding enabled Salhany and co-workers (Salhany et al. 1995; Salhany 1996, 1999, 2001b; Salhany and Schopfer 2001) to determine the rate constants for each step of the 2-step mechanism more precisely. The only significant effect of Cl^- is to increase k_{-2} by over 15-fold. An increase in k_{-2} is also seen when resealed ghosts are exposed to increased extracellular Cl^- (Salhany et al. 1994). Because there is no effect on k_1, the second-order rate constant for initial binding of DBDS, it is clear that Cl^- does not compete directly with DBDS for binding, but rather exerts an allosteric effect on DBDS binding kinetics (Salhany et al. 1995).

The effect of Cl^- on the DBDS k_{-2} is saturable (Salhany et al. 1994), but with a much higher half-saturation constant (266 mM) than that for Cl^- transport (~50 mM, (Gasbjerg and Brahm 1991)), perhaps because of the effect of DBDS on the Cl^- affinity. For the monovalent anions, except iodide, the relative affinities as measured by effects on DBDS parallel their apparent affinities for transport (Salhany 1999). The data clearly show that DBDS and substrate anions form a ternary complex.

H_2DIDS and DIDS exhibit similar 2-step reversible binding prior to covalent reaction (Salhany et al. 1995; Salhany 1996; Salhany and Schopfer 2001). Although DIDS shows single-exponential kinetics under conditions of low temperature and low concentration (Janas and Janas 2000), this is consistent with the 2-step model, under conditions where the first step of the reaction is rate-limiting (Salhany and Schopfer 2001). The constants obtained by these experiments agree well with the measured effects of Cl^- on DBDS binding in intact cells (Salhany 1996), but predict the opposite effects of Cl^- on equilibrium H_2DIDS binding from those observed in intact cells (Shami et al. 1978). Despite these discrepancies, the evidence strongly suggests that the DS and Cl^- transport sites are different.

This conclusion is supported by evidence that addition of sulphate in the presence of Cl^- increases DBDS binding, contrary to what would be expected if both sulphate and Cl^- compete with DBDS (Aranibar et al. 1994). Furthermore, although DNDS binds with highest affinity to the Eo (outward-facing, empty) form, it can also bind with lower affinity to the Ei form, contrary to the prediction if it competed with Clo for binding to the transport site (Knauf et al. 1993a).

Covalent Reaction of DS

The H_2DIDS reaction sites provide an excellent example of the use of mutations to test hypotheses about the action of chemical modifiers. From a careful analysis of the pH dependence of the reaction rates, Kampmann et al. (1982) proposed that H_2DIDS reacts with two sites: one with an unusually low pK, ~11.3 (Lys539), and one with a much higher pK, ~9.9 (Lys851). At neutral pH, the reaction takes place preferentially at the low pK site, because isothiocyanates such as H_2DIDS react with the uncharged form of the amino group. Mutation of Lys539 caused a dramatic increase in the pK of the H_2DIDS covalent reaction, exactly as expected if the low-pK amino group is ablated and reaction only takes place with the high-pK group (Kietz et al. 1991). The pK of Lys539 measured by H_2DIDS reaction is considerably lower than that measured for the same Lys by reaction with FDNB (≥ 8.5) (Ruffing et al. 1996) or DIDS (Schopfer and Salhany 1995), possibly because of changes in AE1 structure induced by H_2DIDS. The FDNB value is most likely to represent the unperturbed pK, because FDNB does not bind reversibly to AE1 before covalent reaction (Ruffing et al. 1996).

The rate of covalent reaction of DIDS with AE1, which probably takes place after formation of reversible complex 2, is greatly increased by various anions (Schopfer and Salhany 1995), including Cl^- and nitrate. Surprisingly, citrate has an effect greater than Cl^- and the dissociation constant for citrate stimulation of the reaction is similar to that for Cl^-. This contrasts with the inability of citrate to compete effectively for the transport site (Liu et al. 1996b), and the lack of a citrate effect on k_{-2} for the reversible binding of DBDS or H_2DIDS (Salhany et al. 1994, 1995), so the binding site responsible for the covalent reaction effect is probably not the transport site.

Geographic Location

The geographic location of the DS site has been studied by measuring fluorescence resonance energy transfer (FRET) between BIDS, a covalently reactive fluorescent stilbene, and fluorescent labels attached to the cysteines in the cytoplasmic domain of AE1, Cys317 and Cys201 (Rao et al. 1979). These measurements indicated a distance of 34–42 Å, which, when combined with a measurement of 40 Å as the distance between the cytoplasmic domain cysteines and the inner face of the lipid bilayer (Thevenin et al. 1996), would put the DS sites very deep within AE1, at a z-coordinate position ≤ 0.1, near Glu681. More recent FRET measurements (Knauf et al. 2002a), however, together with ^{35}Cl-NMR evidence that reaction of Glu681 with WRK does not prevent DIDS binding (Bahar et al. 1999), relocates the DS sites to a $z \geq 0.5$. This location would also fit better with the proposed location of the H_2DIDS-reactive sites, Lys539 and Lys851, near the external surface at $z \sim 1$ (Fig. 11.4). Also, DS attached to macromolecules with relatively short spacer arms (4–8 atoms) can bind AE1 (Pimplikar and Reithmeier 1986) and can inhibit anion exchange from the external side (Eidelman et al. 1991).

Further insight into the parts of AE1 that are affected by DS binding comes from the effects of the reversible DS, DNDS, on the ability of biotin maleimide to label cysteines introduced into EC 3, EC 4, and the T-loop of AE1 (Tang and Casey 1999). Significant inhibition of biotinylation was seen for cysteines substituted for Gly565 (in EC 3, Fig. 11.4), Ser643, Pro652, Leu655, Arg656, Ser657, Ile661, Try662, Met663 (all in EC 4), and Gly742 (in the T-loop), indicating that these parts of AE1 are affected, possibly allosterically, by DS binding. The effect on Gly742 is consistent with the fact that H_2DIDS binding protects (Jennings et al. 1986) internal trypsin cleavage of Lys743 (Sect. 11.4.5). It is somewhat surprising that residues in EC loops are affected by DS binding, as well as Gly742 and Lys743 that are located cytoplasmically, but this is perhaps to be expected in light of the cooperative interactions between the various parts of the membrane domain of AE1.

11.5.4 Other Amino-Reactive Probes

In contrast to the DS, the covalent modification of Lys430 (Cobb and Beth 1990) by EM (which is mutually exclusive with disulphonic stilbene binding (Macara and Cantley 1981a)) does allow Cl^- binding, as observed by ^{35}Cl line broadening in NMR studies (Liu et al. 1995). Mutation of this residue did not inhibit anion transport (Passow et al. 1992), even though reaction with EM causes nearly complete inhibition of Cl^- exchange (Liu and Knauf 1993). This, together with the fact that EM and analogous compounds that are unreactive can inhibit anion exchange reversibly (Knauf et al. 1993b), indicates that the inhibition is due to interaction with parts of AE1 other than Lys430.

Pan and Cherry (1995) found that external iodide quenches the eosin triplet fluorescence in a manner suggesting that iodide binds to a site with characteristics somewhat similar to, but not identical with, the transport site. Further experiments showed that negatively-charged quenchers were effective from the external side, but that positively-charged quenchers acted from the cytoplasmic side (Pan and Cherry 1998), supporting an earlier observation that Cs^+ quenches from the cytoplasmic side (Macara et al. 1983). These data suggest that the relatively hydrophobic ring moiety of eosin may insert itself between parts of AE1 that form the permeability barrier, in a manner analogous to that proposed to explain transport by AE1 of anionic molecules with hydrophobic moieties (Knauf 1986), so that it is accessible to quenchers from either side of the barrier. Because Lys430 is almost certainly physically close to the outside ($z \sim 1$) (Fig. 11.4), the quenching by polar reagents added to the cytoplasmic side shows that there is a deep aqueous channel that extends from the cytoplasmic side of AE1 almost to $z = 1$.

Although another amino-reactive reagent, pyridoxal-5-phosphate (PLP), is a substrate for AE1, the effects on Cl^- binding as measured by NMR (Salhany 2001a) are insufficient to explain the degree of inhibition on the basis of blockage of the transport site by PLP. This probe provides further evidence that reaction with Lys539 or Lys851 (Kawano et al. 1988; Wood et al. 1992) can block anion exchange, but does not help in identifying the transport site.

11.5.5 Histidine Modification

Inhibition of phosphate transport by the histidine-selective reagent DEPC (Izuhara et al. 1989) was found to increase with increasing pH. Most of this effect was due to changes in internal pH, indicating that the DEPC-reactive His is accessible to H^+ from the cytoplasmic side. Binding of DNDS from the outside greatly reduced the ability of DEPC to inhibit anion exchange by reacting at this inward-facing site. Conversely, reaction with DEPC greatly reduced labelling by extracellular H_2DIDS. Binding of Cl^- to DS-sensitive sites, as measured by ^{35}Cl-NMR, was also inhibited by DEPC (Hamasaki et al. 1990). Since the DS bind preferentially to the Eo conformation of AE1 (Fröhlich 1982; Furuya et al. 1984), the His seems to react preferentially with cytoplasmic DEPC when AE1 is in the Ei form, trapping AE1 in a form that binds neither external Cl^- nor DS. These data, however, do not prove that the DEPC-reactive His crosses the permeability barrier, and other evidence argues against this possibility (Sect. 11.6.1).

The possible role of particular His residues in anion exchange and DEPC inhibition was further explored by site-directed mutagenesis (Table 11.2). Mutation of His703, His819 and His834 to Gln (hexagons in Fig. 11.5) greatly reduced Cl^- transport (Müller-Berger et al. 1995b). Surprisingly, mutation of Lys539 to Asn caused partial or complete restoration of function in these mutants, indicating (probably allosteric) interactions between Lys539 and these various histidines, despite the fact that they are located far apart in the primary sequence (Fig. 11.5).

None of the His mutants above caused a decrease in DEPC reaction rate, but mutation of His734 to Ser, which showed less transport inhibition, caused the DEPC reaction rate to decrease to about 1/3 of its value in wild type AE1 (Müller-Berger et al. 1995b). This indicates that His734 is the principal site for DEPC inhibition of anion exchange.

The H734S mutation also shifted the external pH dependence of Cl^- transport from a pK of 5.8 to 6.9 (Müller-Berger et al. 1995a), very similar to the shift (5.8 to 6.7) observed when Glu681 is mutated to Asp. The evidence that His734 is accessible to H^+ only from the *cytoplasmic* side, but that mutations at Glu681 and His734 have similar effects on the pK for *external* H^+ titration of Cl^- transport (Müller-Berger et al. 1995a), can perhaps best be explained by postulating that in the wild type and both mutants the external proton titration is always occurring at a carboxyl group at position 681 (Glu or Asp). Because of the shorter side chain, substitution of Asp for Glu681 could move the negative charge farther from the positive charge of His734, which would tend to raise the pK. Replacement of His734 with Ser would remove this interaction (Müller-Berger et al. 1995a), thereby causing the Glu681 pK to increase just as when Asp replaces Glu681.

11.5.6 Arginine Modification

The functional importance of arginine residues was first indicated by Wieth and Bjerrum's (1982) observation that Cl^- exchange is reversibly inactivated at very high external pH with a pK ~ 12. In agreement with this hypothesis, transport was irreversibly inhibited by covalent reaction with the arginine-selective reagent, phenylglyoxal (PG). Although PG binds reversibly to AE1 to inhibit Cl^- exchange even at 0 °C, the irreversible inhibition takes place by a second-order reaction and seems to be unrelated to the reversible binding (Wieth et al. 1982). The reaction rate increases greatly with increasing external pH (pHo), indicating that the PG-reactive site is accessible to H^+ from the outside. At high pHo the major labelling site is in the region of AE1 from Asn554 to the C-terminus (Bjerrum et al. 1983). Although there are several Arg in this region, the fact that site-directed mutation of Arg730 to Lys or Glu greatly inhibits Cl^- exchange (Karbach et al. 1998) suggests that this is the functionally essential arginine. Many other Arg are also possible, but some (e.g. Arg871, Arg879) can be eliminated because site-directed mutation data indicate that they are not essential (Fig. 11.4 and Table 11.2).

When PG modification is done at lower pHo, the effects are even more complicated, including under some conditions acceleration of anion transport (Gartner et al. 1997), and involving at least one additional reactive site.

11.5.7 Cysteine Modification

AE1 has five endogenous cysteines, three of them in the membrane domain. None these can be labelled from the external side by maleimide reagents. Although some of the cysteines are quite well-conserved in the AE protein family, anion exchange remains intact after removal of all of the cysteines by site-directed mutation (Casey et al. 1995), so none is essential. Cys843 is fatty acid-acylated, but this is not required for anion exchange (Kang et al. 1994). The absence of essential cysteines is useful in permitting specific site-directed labelling in mutants with Cys substituted at various positions in the cysteineless mutant of AE1.

11.5.8 Functional Importance of Serine and Threonine

Neither of these amino acids has been chemically modified, but the presence of many Ser and Thr in the TM segments of AE1 (Fig. 11.4), as well as the hydrophilic character of their side chain OH, has led to the suggestion that they might line the channels through which anions gain access to the transport site from either membrane surface (Reithmeier et al. 1989). Some support for the functional importance of Ser and Thr comes from their proximity to the bound Cl^- in the crystal structures of both halorhodopsin (Kolbe et al. 2000) and bacterial Cl^- channels (Dutzler et al. 2002). In AE1, the relatively conservative mutation of certain serines, such as Ser725 and Ser731 in the T-loop and Ser762 in TM 10, to Cys greatly decreases anion exchange activity, indicating their probable importance.

11.5.9 Protein Flexibility and Cross-Linking

AE1 undergoes conformational changes during its transport cycle from an inward facing to an outward facing form and also between the anion loaded and unloaded form of the protein. Information on the flexibility of different parts of the protein comes from cross-linking studies with single cysteine mutants, in which the cross-linking reagent can join two AE1 monomers by linking them at one specific position. The mutants T431C (EC 1), Y486C (EC 2), Y555C (EC 3), G565C (EC 3) and A751C (T-loop) were all cross-linked efficiently both by direct disulphide bonding (zero-distance cross-link) as well as by maleimide cross linkers with spacer arms. This indicates these regions are flexible enough to allow the residues in adjacent monomers to come as close as 7.5 Å (disulphide-bond distance) or as far apart as 16 Å (length of the spacer arm). On the other hand R656C can be zero-distance cross-linked but none of the maleimide linkers work, indicating that this region is relatively inflexible, as might be expected for its proposed function as part of an external anion access funnel (Tang et al. 1998).

It is somewhat surprising that –S–S– cross-linking, that should restrict molecular motion, does not inhibit anion exchange for the cysteine-substituted mutants at positions Gly565 and Arg656, positions where DNDS binding alters BMA labelling. This suggests that these parts of AE1 do not undergo much movement during the transport process, and that the effect of DNDS on labelling could be due either to steric hindrance by DNDS or to alterations in the position of other parts of AE1 nearby. Similarly, the lack of effect of crosslinking at position Thr431 on anion exchange suggests that the part of AE1 near the EM binding site at Lys430 does not undergo major positional changes during transport, and in particular that significant rotational movements of TM helices 1 and 2 probably do not occur. The lack of effects of crosslinking at Gly565 on transport argue somewhat against movements of TM 6, but similar evidence for Arg656 does not preclude movement of TM 8, because Arg656 is located quite far from the likely beginning of TM 8 (Fig. 11.4).

11.5.10 Sensing of Changes in AE1 Conformation by Chemical Probes

Understanding the mechanism of AE1 at the molecular level requires that we know what conformational changes occur in the protein when various substrate anions are bound (binding conformational change, BCC), and what happens when the system changes from the inward to the outward-facing state (transporting conformational change, TCC). Even in the absence of a high-resolution crystal structure, inhibitory probes can provide some information about these changes, if their binding and/or reaction is affected by substrate binding and/or the change from Ei to Eo or *vice versa*. Indeed, many probes have been found to exhibit this property (Knauf 1979, 1989; Passow et al. 1980a; Jennings 1992b; Knauf et al. 1993b, 2000), even probes whose ability to inhibit (defined by the concentration that gives half-inhibition, the IC_{50}) is not strongly affected by substrate concentration (that is, noncompetitive inhibitors in the most general sense).

Inhibitors that prefer one conformation, say the outward-facing form, can recruit AE1 molecules toward that form. For example an inhibitor that binds solely to outward-facing forms would recruit all of AE1 to the outward-facing form, but the distribution (in Cl^- media) between Eo and EClo will be affected by the degree to which the inhibitor prefers Eo over EClo or *vice versa* (Knauf et al. 2000). Thus, unless the dissociation constants for all forms of AE1 are known, even an ideal "recruiting" inhibitor will cause changes in the fraction of empty *vs.* substrate-bound sites, so the TCC caused by the inhibitor is accompanied by a BCC.

Reversible Inhibitors

Two kinds of inhibitors bind preferentially to the outward-facing forms of AE1. The first kind is typified by the disulphonic stilbenes, discussed in Sect. 11.5.3, whose affinity is very strongly reduced by binding of substrates. The second class of conformation-sensitive reversible inhibitors, which includes EM (Knauf et al. 1993b), niflumic (Cousin and Motais 1979; Knauf and Mann 1984) and flufenamic (Knauf et al. 1989) acids, NAP- and NIP-taurine (Knauf et al. 1984, 1987a; Knauf and Spinelli 1995), and the oxonol dye WW781 (Knauf et al. 2000), is not strongly affected by binding of substrates. The effects of [Cl^-] on the IC_{50} are much smaller than would be predicted for competitive inhibition, but the binding affinity for Eo is greater than that for Ei by factors of 3.5 to > 12 (Knauf et al. 1993b, 2000). Substrate binding does have some effect – the slowly-transported iodide reduces binding affinity for Eo by factors of 1.4 to 6.5 for various inhibitors – but this is far less than expected if the inhibitors were binding to the transport site. Thus, these inhibitors can help to identify parts of AE1 other than the anion binding site that are affected by the TCC and can also be useful for recruiting AE1 to outward-facing forms.

Once the dissociation constants for binding to Eo and Ei have been determined, these inhibitors can also be used to test the value of unloaded site asymmetry, A (Sect. 11.3.1), against the predictions of the ping-pong model (Fröhlich and Gunn 1986). Data with flufenamic acid at 38 °C (Knauf et al. 1996) and with WW781 at 0 °C (Knauf et al. 2000) agree well with the hypothesis that A << 1. For WW781, the data also agree with the indications from ^{35}Cl-NMR that A_{Cl} << 1, providing further support for the ping-pong model with these asymmetry values.

The oxonols such as WW781, like the DS (Sect. 11.5.3), bind to AE1 reversibly by a 2-step mechanism, but they do not react covalently. Both binding steps are strongly affected by the TCC and, to a lesser extent, by substrate binding (Knauf et al. 2000), providing evidence that WW781 binds to a part of AE1 that is involved in the transport-related conformational changes. Another oxonol, diBA(5)C4, has an extremely high inhibitory potency, with an IC_{50} < 1 nM (Knauf et al. 1995). Recent data show that the fluorescence increase upon binding of a related oxonol, diTBA, to AE1 is strongly affected by the kind of anion present (Knauf et al. 2002c), with bicarbonate giving enhanced fluorescence and iodide giving lower fluorescence relative to Cl^-. Bicarbonate greatly reduces k_{-2} for diTBA, as compared to its value in Cl^- or citrate media, providing direct evidence for substrate-induced changes in AE1 conformation (Sect. 11.3.3).

Covalent Inhibitors

An advantage of covalently-reactive inhibitors is that their site of reaction with AE1 can be identified, so effects of the TCC or BCC can be related to specific sites in the protein. For example, Passow and co-workers (Ruffing et al. 1996) have shown that 1-fluoro-2,4-dinitrobenzene (FDNB) reacts with Lys539 in resealed red cell ghosts by a second-order reaction, so FDNB should sense the conformation of AE1 unperturbed by prior reversible binding. Unfortunately, FDNB also reacts at another site, where it alters the reactivity of Lys539 in Cl^- media. Kinetic analysis to account for the latter effect showed that in Cl^- media, the reaction rate of Lys539 was about 15 times higher than in sulphate media (Ruffing et al. 1996). Although Lys539 is not essential for transport, these data provide evidence that substrate binding alters the reactivity of Lys539, possibly because of movements of Lys539 itself or of adjacent parts of AE1 that affect the interaction of Lys539 with FDNB. Other probes that affect anion exchange also modify the

FDNB reactivity of Lys539 (Passow et al. 1980a), possibly because of effects on AE1 structure similar to those caused by Cl^- and sulphate.

11.6 Model for Anion Exchange

At our present stage of knowledge, any proposed molecular mechanism for anion exchange by AE1 must be highly speculative. Nevertheless, we will attempt to provide an outline of some of the general features of such a model.

11.6.1 Residues that Cross the Permeability Barrier (PB) and/or Bind Substrates

We begin with the idea that Glu681 is likely to change positions (sidedness) relative to the permeability barrier (PB) as a result of the Ei ↔ Eo change that occurs during anion transport (transporting conformational change, TCC). The most straightforward interpretation of the W-B chemical modification and site-directed mutation data (Sect. 11.5.2) is that in normal AE1 the negative charge of the Glu681 $-COO^-$ crosses the permeability barrier together with a Cl^- ion and two positive charges, for net neutral transport, whereas when the negative charge of Glu681 is removed, Cl^- transport with 2 positive charges leads to net transport of one positive charge (Fig. 11.6).

What are likely candidates for the two positive charges? Data on the effects of alkaline extracellular pH on Cl^- exchange indicate that there are two titratable sites (Bjerrum 1992), one with a very high pK (> 11) (Bjerrum 1992; Liu et al. 1996b) that is probably an arginine guanidino group, and another with a lower pK (9.3–9.4) that has been proposed to be an amino (Liu et al. 1996b) or guanidino (Bjerrum 1992) group. Titration of the lower pK group converts the system to the "alkaline state" (Bjerrum 1992), characterised by a much lower external affinity for anions, somewhat lower rate constant for outward translocation of ECli (Liu et al. 1996b), and an unloaded site asymmetry value, A > 1, the opposite of the situation at lower pH (Sect. 11.3.1). Titration of the higher pK group (Wieth and Bjerrum 1982) or reaction with the arginine-selective reagent phenylglyoxal (Wieth et al. 1982) abolishes transport, as well as Cl^- binding measured by ^{35}Cl-NMR (Falke and Chan 1986a; Passow et al. 1997). The fact that replacement of Arg730 with either Gln or Lys profoundly inhibits transport (Karbach et al. 1998) suggests that Arg730 may be the high pK positive charge that is critical for anion binding to the transport site (Sect. 11.5.6).

Some insights into the nature of anion binding at the transport site may be gained from the high resolution crystal structures of the archaebacterial light-driven Cl^- pump, halorhodopsin (Kolbe et al. 2000) (Protein Data Bank (PDB) 1E12), and of bacterial ClC chloride channels (Dutzler et al. 2002). Halorhodopsin may be a particularly good analogy, because its dissociation constant for Cl^- binding is about 100 mM, not far from the value for Cl^- binding to AE1. In Halorhodopsin, the Cl^- ion is stabilised by interaction with the protonated Schiff base,

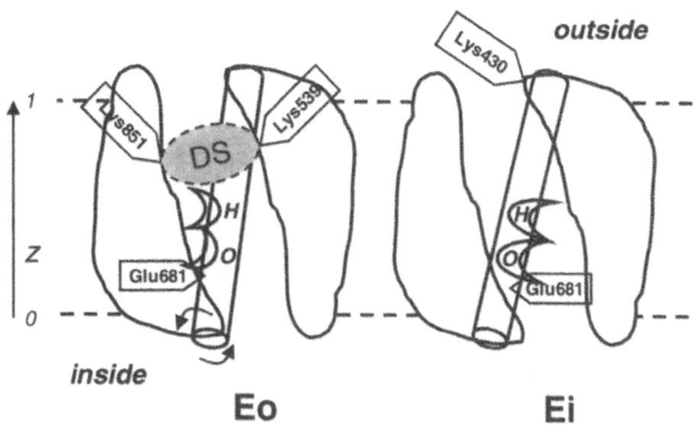

Fig. 11.6. Schematic model for the transporting conformational change in AE1. The model is shown with two distinct binding sites for anions, one (*H*) primarily for halide anions such as Cl^-, and one (*O*) primarily for oxyanions such as bicarbonate, sulphate, and nitrate. These sites display negative cooperativity so that binding to one decreases the ability of anions to bind to the other. When such binding occurs, e.g. of halides to the *O* (wrong) site, the conformational change from EXi to EXo or *vice versa* is inhibited. The *O* site is shown as being toward the inside (cytoplasmic side) of the bilayer, near the *Glu681* residue that is the proton binding site for H^+-SO_4^{2-} cotransport. The *H* site is shown farther toward the middle of the membrane, on the basis that it seems to be located far from *Glu681*. When AE1 is in the Eo conformation, external anion substrates gain access to the transport sites by diffusing through an aqueous channel that extends deep into the membrane, formed by parts of TM 5, 8, 13, 14, and probably TM 12 and the T-loop as well. Disulphonic stilbenes (*DS*) bind to a site located nearer to the extracellular side of the membrane than the transport sites or *Glu681* and, when bound, can form covalent links with *Lys851* and *Lys539*. This blocks transport, either by restricting access to the transport sites or, more likely, by preventing the transporting conformational change (TCC).
For simplicity, the TCC is shown schematically as taking place by means of rotation of a single cylindrical portion of AE1, but the TCC more likely involves a co-ordinated movement of several parts of AE1. This TCC results in simultaneous change in position of both anion binding sites, together with *Glu681* and probably two positively charged sites, that may be guanidino groups of Arg (not shown), one of which is probably Arg730. The TCC from ***Eo*** to ***Ei*** also dramatically decreases the affinity for *DS* and oxonols (not shown). In the resulting ***Ei*** form, both transport sites are accessible to cytoplasmic anions via an aqueous channel extending from the cytoplasmic side of AE1. Positively-charged quenchers can reach part of eosin maleimide (EM) attached to *Lys430* near the outside, so this channel must extend deep within AE1. His734 (not shown) is accessible from the inside and probably interacts with *Glu681*; reaction of His734 with DEPC locks the system in the ***Ei*** form and inhibits *DS* binding, as well as Cl^- binding and transport

as well as with an Arg, which is ion paired with a Glu. However, the guanidino group of the Arg is located quite far (7.4 Å) from the bound Cl^-, which accepts H-bonds from two tightly bound water molecules (one of which lies between Cl^- and the Arg) and from a Ser. Three other Ser, a Trp, a Val, and a Thr also help to stabilise the bound Cl^-. Calculations indicate that all of these interactions are important, not just the charge interaction with the guanidino group, despite the fact that the Arg was earlier thought to be the Cl^- binding site on the basis of site-directed mutations that abolish Cl^- transport. The Cl^- binding site in halorhodopsin is thus far more complex and less symmetrical than, for example, the K (Rb) ion binding sites in the crystal structure of the potassium channel from *Streptomyces lividans* (PDB 1BL8), where backbone *O* atoms replace the ion's water of hydration (Doyle et al. 1998).

The crystal structures (PDB 1KPL and 1KPK) of two bacterial ClC Cl^- channels (Dutzler et al. 2002) also show complex Cl^- binding sites, with binding stabilised by partial charges rather than direct interaction with Arg or Lys side chains. The binding is favoured by interactions with backbone amide nitrogens and side-chains of Tyr and Ser, as well as many hydrophobic residues. Cl^- binding is also enhanced by the orientation of the positive ends of three α-helix dipoles toward the binding site, a possibility originally suggested by Reithmeier et al. (1989) on the basis of similar architecture in the *Salmonella typhimurium* sulphate-binding protein (Pflugrath and Quiocho 1988). The Cl^- binding site in the bacterial ClC is formed by the juxtaposition of two regions of the polypeptide that, like the two halves of the water channel protein, aquaporin (Preston and Agre 1991), are structural repeats oriented in an antiparallel fashion so as to bring α-helices from each half together with the positive ends of their dipoles facing each other near the Cl^- binding site. Aqueous vestibule regions extend inward toward the Cl^- ion from either side of the membrane, but a Glu blocks the access of Cl^- to the external aqueous region. Dutzler et al. (2002) suggest that this residue may act as a gate and that Cl^- ions, that stabilise the open conformation of the channel (Pusch et al. 1995), may do so by competing with and displacing the Glu from its position in the crystal structure. It is tempting to speculate that Glu681 in AE1 might play a similar role, whereby its displacement by binding of Cl^- or HCO_3^- would alter AE1 conformation so as to reduce the transition state free energy for the TCC.

The probable structural complexity of the Cl^- binding site makes it even more difficult to identify the lower pK alkaline-titratable group of AE1. Could it be His734, in view of the evidence that His734 interacts with Glu681 (Sect. 11.5.5)? This seems unlikely because His734, that is probably the primary site of DEPC reaction (Müller-Berger et al. 1995b), is accessible to H^+ from the inside but not the outside, and the pK for the DEPC-reactive His is only about 6.6 (Izuhara et al. 1989), far lower than the pK (> 9) for the alkaline titratable site.

Since His734 is unlikely, the lower pK alkaline-titratable group is probably either an amino group (Liu et al. 1996b) or else a guanidino group in a very hydrophobic environment (Bjerrum 1992). In either case, because of the absence of rapid H^+ efflux at alkaline external pH values between the lower and higher pK values (Wieth and Bjerrum 1982) (where the low pK site would be protonated on the cytoplasmic side, but deprotonated on the outside, so each transport cycle would cause net efflux of a proton), it is very unlikely that the low pK group crosses the permeability barrier (Bjerrum 1992) during the TCC. How, then, does the extra positive charge get across? The positive charge could be a guanidino group of Arg with a high pK or, alternately, the sum of the effects of reorientation of various partial charges or dipoles during the TCC. The latter model could also explain why slightly less than two unit positive charges are transferred together with Cl^- and Glu681 (Jennings 1995).

11.6.2 Evidence for Anion Access Channels in AE1

External Access Channel

The fact that Glu681 is located near the cytoplasmic side of the lipid bilayer (Sect. 11.4.4), together with the observation that the bulky hydrophilic reagent, WRK, can reach Glu681 from the outside to react with it (Sect. 11.5.2), demonstrates that there must be an aqueous channel within AE1 leading from the outside nearly to the cytoplasmic side of the lipid bilayer (Fig. 11.6). Consistent with this idea, some inhibitors of anion exchange act as if they restrict access of Cl^- to the transport site, but do not compete directly with Cl^- (Falke and Chan 1986b). The observation that reaction of WRK with Glu681 does not prevent external Cl^- or DIDS binding to Eo (Bahar et al. 1999) indicates that, if Cl^- and DIDS enter AE1 through the same access channel as WRK, the Cl^- and DIDS binding sites must be located toward the extracellular side of Glu681, i.e. at $z > 0.15$. This is reinforced by data indicating that reversible DNDS binding or prior H_2DIDS reaction does prevent WRK from reacting with Glu681 (Jennings and Anderson 1987), as expected if the DS site is located toward the outside ($z > 0.15$) of Glu681 (Fig. 11.6). From the effects of V_m on the ability of sulphate to act as a competitive inhibitor, Jennings et al. (1990) concluded that external sulphate moves through about 10–15% of the transmembrane V_m on the way to its binding site. This would fit with the idea that the sulphate binding site (*O* in Fig. 11.6) lies deep within the membrane (at $z < 0.5$), but that it can be reached from the outside through a relatively low-resistance aqueous channel. Further evidence for the concept that there is an access channel from the outside for negatively-charged ions and probes comes from the fact that external DS inhibitors attached through a spacer arm to bulky hydrophilic molecules are only effective if the spacer is > 3 atoms long, but anthranilic acid derivatives are effective even with 2-atom spacers, indicating that the sites of action for these two classes of inhibitors lie at different depths within an access channel for anions (Eidelman et al. 1991).

More specific information suggesting that part of TM 8 forms the anion access channel comes from the effects of PCMBS (*p*-chloromercuribenzene sulphonate), a hydrophilic SH reagent that crosses the membrane through the anion exchange system (Knauf and Rothstein 1971), on Cl^-/HCO_3^- exchange in HEK 293 cells expressing various single-cysteine mutants of AE1 (Tang et al. 1999). Anion exchange was inhibited by PCMBS in mutants with introduced cysteines at Ala666, Ser667, Leu669, Leu673, Leu677, and Leu680 (Fig. 11.4). If TM 8 is modelled as an α-helix, these residues would lie on the same side of the helix as Glu681, so they may line the external anion access channel. The presence of some hydrophobic residues (Leu, Ala) may favour fast movement of anions through the channel – a similar feature in halorhodopsin (Kolbe et al. 2000) is known as the "greasy edge". Hydrophobicity of this surface of the access channel may explain why biotin maleimide fails to react with Cys substituted at these sites (Tang et al. 1998).

PCMBS had no effect on cysteines at positions beyond Leu680, as expected if Glu681 lies near the innermost part of the external access channel.

A different negatively-charged SH reagent, MTSES ((2-sulphonatoethyl)-methanethiosulphonate), inhibited anion exchange in mutants with cysteine at Ser852 and Ala858 (Fujinaga et al. 1999), suggesting that these residues toward the external side of TM 13 and TM 14 (Fig. 11.4) face an aqueous region near or at a functionally important site. They lie near Lys851, the site of H_2DIDS cross-linking to Lys539, and thus must be within about 15 Å of TM 5, indicating that TM helices 5, 8, and 13/14 may form parts of an aqueous channel extending from the external side into AE1. The T-loop and TM region 12 may also be involved, on the basis that they can be labelled by LYIA (Fujinaga et al. 1999) (Fig. 11.4).

Cytoplasmic Access Channel

A membrane-permeant positively-charged reagent, MTSEA (2-(aminoethyl)-methanethiosulphonate), inhibited anion exchange by reacting with cysteines substituted for Ile684 and Ile688 (Tang et al. 1999), suggesting that these residues might line an aqueous access channel extending into AE1 from the cytoplasm. The channel appears to extend quite far into the membrane because benzene sulphonate inhibitors attached to bulky macromolecules require an 11-atom spacer to be fully effective (Eidelman et al. 1991). The depth of the channel is most clearly indicated by the observation that positively-charged quenchers, such as Cs^+ (Macara et al. 1983) and positively-charged nitroxide compounds (Pan and Cherry 1998), can quench fluorescence of externally-reacted EM (Sect. 11.5.4), which labels Lys430 (Cobb and Beth 1990) that is almost certainly located near the external side of the lipid bilayer ($z \sim 1$, Sect. 11.4.5, Fig. 11.6).

The fact that AE1 has an anion-permeable channel extending from the outside to Glu681 (at $z \sim 0.15$), as well as an aqueous channel from the cytoplasmic side that extends all the way to Lys430 (at $z \sim 1$), has important structural and mechanistic implications (Fig. 11.6). In particular, it means that the permeability barrier cannot be thought of as residing at a particular depth (z-position) within the membrane, but rather that a permeability barrier between two aqueous compartments within AE1 extends at least from $z = 0.15$ to $z = 1$. As in the Shaker K^+ channel S4 segment, which is postulated to form a barrier between two aqueous regions that extend into the protein from opposite sides of the membrane (Cha et al. 1999; Glauner et al. 1999; Bezanilla 2000), such a structure implies that sites at different z positions can simultaneously cross the permeability barrier during the TCC. Aqueous regions in the halorhodopsin trimer adjacent to the anion binding regions have been suggested to raise the dielectric constant near the binding sites, thus facilitating anion binding (Kolbe et al. 2000). The aqueous channels in AE1 might play a similar role.

11.6.3 Separate Binding Sites for Halides and Oxyanions

Abundant data show that a ping-pong model with only one binding site cannot explain many features of the system, such as "self-inhibition" by substrates at high concentrations, which requires a second, "modifier", site at which anion binding is inhibitory (Sect. 11.2.3). Recently, considerable evidence has accumulated to support Krupka's (Krupka 1989a) suggestion, based on theoretical arguments (Sect. 11.2.3), that the halide (e.g. Cl^-) and oxyanion (e.g. HCO_3^-) substrates bind to different sites (designated *H* and *O* respectively in Fig. 11.6), at least if sulphate may be taken as an example of an oxyanion. External sulphate binding to Eo raises the pK of the proton binding site (probably Glu681) from 5 to 6 (Milanick and Gunn 1982). This apparent increase in proton binding affinity could be explained by electrostatic attraction between the negatively-charged sulphate and the proton, but the fact that iodide binding has the *opposite* effect on proton affinity cannot be explained electrostatically if halides and oxyanions bind to the same site (Milanick and Gunn 1986). Also, the effects of raising pH on external sulphate and iodide affinities do not fit quantitatively with an electrostatic model in which both anions bind at the same position in AE1 (Liu et al. 1996c). The dissociation constant for Cl^- binding measured by competition with nitrate is far lower than that measured in the absence of nitrate (Galanter and Labotka 1991). Since nitrate is a close analogue of bicarbonate, it could be that nitrate recruits AE1 toward more outward-facing states, as does bicarbonate (Sect. 11.3.3), thus increasing Cl^- affinity by binding to a site (*O*) separate from the primary Cl^- binding site (*H*).

Interactions with inhibitors provide further evidence for different sites. Chloride but not sulphate decreases the rate of irreversible inactivation of anion exchange by phenylglyoxal at alkaline pH (Wieth et al. 1982). Chloride but not sulphate greatly enhances the inhibitory potency of the reversible inhibitor, dipyridamole (Renner et al. 1988; Legrum and Passow 1989). FDNB reaction with Lys539 is about 15 times faster in Cl^- than in sulphate media, indicating that Lys539 conformation is altered by binding of these different anion substrates (Ruffing et al. 1996). DBDS binding affinity is increased by addition of chloride to sulphate media (Aranibar et al. 1994), a paradoxical result that cannot be explained by binding to a single site. Low pH titration (probably of Glu681) interferes with Cl^- binding as measured by ^{35}Cl-NMR (Müller-Berger et al. 1995a), but must still permit sulphate binding, because sulphate transport increases at low pH. ^{35}Cl-NMR data indicate that WRK reaction with Glu681 prevents binding of external sulphate, but not of iodide, as measured by their ability to compete with Cl^- for binding (Kennedy, Wu, Law, Schultz and Knauf, unpublished data). Recent work by Passow and Lepke (Passow et al. 2001) strongly suggests that at low pH proton transport is accompanied by simultaneous transport of *two* Cl^- ions (see Chap. 10), which would require two distinct binding sites. The different effects of papain treatment on transport of various organic and/or amphipathic anions has been explained in terms of two sites (Voswinkel et al. 2001), although it is not clear that these are the *H* and *O* sites.

All of these observations fit the hypothesis that oxyanion substrates, such as sulphate, bicarbonate, and nitrate, bind to a different place in AE1 than do halides such as Cl^-, Br^-, I^- and F^- (Fig. 11.6). In this model, there must be strong negative cooperative interactions between the two sites to explain the apparent competition between oxyanion and halide substrates. Self-inhibition is easily explained by proposing that binding of a second anion at high concentrations to the "wrong" site, so that both sites are occupied, greatly inhibits transport (Macara and Cantley

1981b; Krupka 1989a). This model is supported by the fact that the taurine derivatives, NAP- and NIP-taurine, that compete with sulphate (Fröhlich and Gunn 1987) but not Cl^- (Knauf et al. 1978), also compete with iodide for binding to the "*modifier*" site (*not* the substrate site), where iodide inhibits Cl^- (and I^-) transport (Knauf and Spinelli 1995). These data could indicate that the taurine derivatives and iodide inhibit halide transport by binding to the oxyanion binding site (the "wrong" site for halides), where sulphate (or bicarbonate) normally binds prior to transport (Krupka 1989a).

If, for some substrate anions, binding of one anion to its transport site can occur even when the other binding site is occupied, this model could help to reconcile the evidence for formation of a ternary complex of two substrates with AE1 under some conditions (Salhany and Rauenbuehler 1983) with the idea that the mechanism of AE1 is basically ping-pong (Jennings 1982). The two-site model proposed here can be viewed as a form of the ordered sequential model, which has two simultaneously-translocated anion binding sites, only one of which is usually "loaded" during the translocating conformational change, that Salhany and coworkers (Salhany and Rauenbuehler 1983) proposed to explain these data. The two-site model would also help to explain how the change from Ei to Eo alters the selectivity for Cl^- *vs.* HCO_3^- (Knauf et al. 2002b) – with two different sites for oxyanions (*O*) and halides (*H*), the conformational change might be expected to have different effects on their affinities. Moreover, two anion binding sites in AE1 would help to rationalise data indicating that AE2, whose membrane portion is highly homologous to AE1 (Demuth et al. 1986; Wood 1992), operates by a simultaneous, rather than a ping-pong mechanism (Restrepo et al. 1989, 1991, 1992), in which two anions bind from one side to form a ternary complex before the translocation event occurs.

A significant advantage of this model is that the self-inhibition phenomenon, rather than being an unexplained anomaly, is seen as a natural consequence of having one binding site for halides and one for oxyanions, neither of which is perfectly selective for its preferred class of substrates. The evidence for this model, however, is far from conclusive – in particular far more data are needed with bicarbonate and other oxyanions to see if their binding site is really different from that of the halide anions.

11.6.4 Structural Nature of the Transporting Conformational Change (TCC)

Because regions of AE1 that are at least tens of Angstroms apart, such as Glu681 and the Cl^-, DIDS, EM, and oxonol binding sites, are *all* affected by the TCC, it seems highly unlikely that the TCC involves only small-scale movements of a few side chains on the TM helices of AE1. Calculations based on such a "thin-barrier basket inversion" model, involving simultaneous introversion of residues Glu681, Lys826, and Arg730, together with a substrate anion, show some behaviour consistent with observation, but fail to explain why the TCC is so strongly disfavoured without an anion substrate bound (Ramakrishnan and Busath 2002).

In view of the evidence for large-scale "breathing" movements in proteins, together with data indicating that AE1 is a highly cooperative structure, it seems more likely that the TCC may involve helix rotations or equivalent motions that simultaneously alter the positions of various parts of AE1 located at different levels (z-values) within the membrane thickness (Fig. 11.6). Such a model harks back to a helix rotation model proposed many years ago by Tanford (1982) for the active transport Ca^{2+}-ATPase. Also, the only structure of an ion exchange protein at sufficiently high resolution to see individual TM helices, that of a bacterial Na^{+}/H^{+} exchanger (Williams 2000), shows several helices that are tipped or bent. Such bends could act as gates to regulate access to intramembrane regions of the protein, depending on the angular orientation of particular TM helices. A helix rotation model would explain why alterations of parts of AE1 located at various z-coordinate levels, such as mutation of Glu681 (at $z \sim 0.15$) to Gln (Chernova et al. 1997) or mutation of Trp648, Ile650, Pro652, Gly654, Leu655, Phe659, Pro660, or Trp662 (all at $z > 1$) to Cys (Tang et al. 1998), all inhibit Cl^- transport. Similarly, binding of DIDS, EM, or oxonols (all to sites with $z > 0.5$) are able to prevent the TCC, as does reaction with WRK (with Glu681 at $z \sim 0.15$).

A helix rotation model, similar to the kinds of models that have been proposed to explain changes in sidedness of positive charges located at different z-coordinate positions along the voltage-sensing S4 segment of voltage-activated K^+ channels (Cha et al. 1999; Glauner et al. 1999), would also help in understanding how the relatively low-energy binding of certain anions, such as Cl^- and HCO_3^-, can lower the transition state free energy for the TCC. Binding of these ions, even binding to separate but interacting sites, could change the position of side chains or even backbone atoms to remove steric or electrostatic restrictions to the helix rotations involved in the TCC. Furthermore, as in the original Tanford (1982) model, rotational movements of one or more helices could result in closing of the gate toward one side of the membrane before the gate toward the other side opens (Fig. 11.2), an essential feature that prevents an ion exchanger from acting like an ion channel (Sect. 11.2.3).

11.6.5 Relation to AE1 Structure

Although the known mutations in AE1 are by no means evenly distributed throughout the membrane domain (Table 11.2), it is probably worth noting that, with the exception of the SAO deletion of Ala400 to Ala408 at the beginning of TM 1, all of the mutations that abolish anion exchange are located between the end of TM 7 and the C-terminus. Fragments of AE1 lacking TM 6 and TM 7 expressed in *Xenopus* oocytes still exhibit Cl^- transport, albeit at a lower level than when TM 6 and 7 are included, indicating that these TM segments are not essential (Groves et al. 1998). Although chemical reaction with parts of the membrane domain N-terminal to TM 7 can inhibit transport, indicating that these parts of AE1 are near to or interact with functionally-important regions, it seems that the most important parts of AE1 for transport function lie in the region after TM 7 which, paradoxically, is the least well-understood part in terms of structure (Sect. 11.4.5). This may not just be a coincidence – the parts of AE1 that move during the TCC must be structurally labile, that is, able to exist in different conformations that do not differ much in free energy. Even minor structural alterations, such as introduction of single cysteines or of glycosylation sites, may alter this labile structure and thus give confusing topological information. If these parts of AE1 lie next to the aqueous access channels, the problem of structural determination is exacerbated, because vectorial labelling reagents may penetrate deep into AE1 via these channels, so sites geographically near one side of the membrane can react with labels added from the opposite side. Thus, the very part of AE1 whose structure may be most important is the part that is most refractory to usual methods for assessing membrane protein topology. It is perhaps worth noting that the bacterial ClC Cl^- channel crystal structures (Dutzler et al. 2002) also reveal a very complex folding pattern that defied attempts at topological analysis by the usual techniques.

11.6.6 Relation to Structure and Mechanism of Other Membrane Transport Proteins

The lock-carrier model (Fig. 11.2) bears an obvious relation to the structure of gated ion channels, the difference being that there are *two* gates that operate synchronously so that ions can never diffuse freely through the entire length of the channel. The model we propose is also similar to that proposed for voltage-gated K^+ channels in having internal aqueous regions that are separated by a barrier that extends through the membrane (at various z-values).

Interestingly, the crystal structures of two bacterial anion channels (Dutzler et al. 2002) show that they are homodimers, with a 2-fold axis of symmetry perpendicular to the membrane, as proposed for AE1. Furthermore, each monomer has an independent Cl^- transport pathway. The gating and conductance properties of Cl^- channels are taken as evidence for two anion binding sites per channel monomer, each with a different anion selectivity, that interact in the processes of gating and transport of Cl^- (Pusch et al. 1995). Finally, it is worth noting that for two of the non-bacterial Cl^- channels, that probably have similar structural elements to the crystallised channels, the two channels in a dimer can be simultaneously gated (Miller and White 1984; Saviane et al. 1999). This implies strong interactions between the monomers in a dimer. It is not too hard to imagine that evolutionary modifications of such a coupled gating mechanism could transform a Cl^- channel into an anion exchanger in which coordinated movement of the gates creates a "channel" that is never fully open for diffusion, but that has the occluded transition state properties needed for a coupled ion exchanger (Sect. 11.2.3). Although it does not seem likely that the monomers in the AE1 dimer operate in a coupled fashion to exchange anions, because one monomer in a dimer can be inactive without preventing transport by the other monomer (Jennings and Gosselink 1995), coupling of the movements of two gates at either side of the anion binding sites could just as easily occur within the structure of a single monomer.

The idea that conformational changes in AE1 can result in occlusion of ion binding sites (preventing diffusion to the site from either side of the membrane) receives support from the crystallographic data. In the crystal structure of halorhodopsin, Cl^- is in an occluded state, with access to one side blocked by an arginine and access to the other requiring opening of a channel. In the bacterial Cl^- channel, a glutamate blocks access toward the extracellular side – clearly major conformational changes must occur during the halorhodopsin pumping cycle and the Cl^- channel gating process.

The AE1 system shows surprising similarity to active ion pumps, like the Na^+, K^+- or Ca^{2+}-ATPases, in that the change in sidedness of the ion binding sites is accompanied by a change in anion selectivity for Cl^- *vs.* HCO_3^- (Sect. 11.3.3). This is an essential property of ion pumps, but is also compatible with the function of this non-energy-requiring ion exchange system (Sect. 11.3.4).

Finally, the concept that AE1 ion selectivity is related to the ability of particular substrates to decrease the transition state free energy for the TCC (Fig. 11.3) is common to all biological catalysts, which act by lowering the transition state free energy of specific reactions, thereby increasing the rate of these reactions. The only difference in the case of AE1 and other transport proteins is that they catalyse a change in sidedness, rather than a chemical change in the substrate.

These similarities are to be expected in light of the evolutionary development of biological catalysts. Structural elements that prove "useful" (in the sense of conferring a selective advantage) for various functions are combined and then selected to serve new purposes. This has benefits not only for the organism, but also for the researcher. It means that information and insights about structure-function relationships from one system, including relatively simple ones like AE1, halorhodopsin, and the bacterial Cl^- channels, may be useful in generating hypotheses about the molecular mechanism of more complex systems of physiological importance.

References

Alper SL (1991) The band 3-related anion exchanger (AE) gene family. Annu Rev Physiol 53:549-564

Alper SL, Kopito RR, Libresco SM, Lodish HF (1988) Cloning and characterization of a murine band 3-related cDNA from kidney and from a lymphoid cell line. J Biol Chem 263:17092-17099

Appell KC, Low PS (1981) Partial structural characterization of the cytoplasmic domain of the erythrocyte membrane protein, band 3. J Biol Chem 256:11104-11111

Aranibar N, Ostermeier C, Legrum B, Ruterjans H, Passow H (1994) Influence of stilbene disulfonates on the accessibility of the substrate-binding site for anions in the band 3 protein (AEB1). Renal Physiol Biochem 17:187-189

Bahar S, Gunter CT, Wu C, Kennedy SD, Knauf PA (1999) Persistence of external chloride and DIDS binding after chemical modification of Glu-681 in human band 3. Am J Physiol 277:C791-C799

Bartel D, Hans H, Passow H (1989) Identification by site-directed mutagenesis of Lys-558 as the covalent attachment site of H_2DIDS in the mouse erythroid band 3 protein. Biochim Biophys Acta 985:355-358

Barzilay M, Cabantchik ZI (1979) Anion transport in red blood cells. III. Sites and sidedness if inhibition by high-affinity reversible binding probes. Membrane Biochem 2:297-322

Batenjany MM, Mizukami H, Salhany JM (1993) Near-UV circular dichroism of band 3. Evidence for intradomain conformational changes and interdomain interactions. Biochemistry 32:663-668

Bezanilla F (2000) The voltage sensor in voltage-dependent ion channels. Physiol Rev 80:555-592

Bjerrum PJ (1992) The human erythrocyte anion transport protein, band 3. Characterization of exofacial alkaline titratable groups involved in anion binding/translocation. J Gen Physiol 100:301-339

Bjerrum PJ, Wieth JO, Borders CL Jr (1983) Selective phenylglyoxalation of functionally essential arginyl residues in erythrocyte anion transport protein. J Gen Physiol 81:453-484

Brahm J (1977) Temperature-dependent changes of chloride transport kinetics in human red cells. J Gen Physiol 70:283-306

Branchini BR, Murtiashaw MH, Eckman EA, Egan LA, Alfano CV, Stroh JG (1995) Modification of chloride ion transport in human erythrocyte ghost membranes by photoaffinity labeling reagents based on the structure of 5-nitro-2-(3-phenylpropylamino)-benzoic acid (NPPB). Arch Biochem Biophys 318:221-230

Bruce LJ, Kay MMB, Lawrence C, Tanner MJA (1993) Band 3 HT, a human red-cell variant associated with acanthocytosis and increased anion transport, carries the mutation Pro-868-Leu in the membrane domain of band 3. Biochem J 293:317-320

Bruce LJ, Wrong O, Toye AM, Young MT, Ogle G, Ismail Z, Sinha AK, McMaster P, Hwaihwanje I, Nash GB, Hart S, Lavu E, Palmer R, Othman A, Unwin RJ, Tanner MJA (2000) Band 3 mutations, renal tubular acidosis and south-east asian ovalocytosis in Malaysia and Papua New Guinea: Loss of up to 95% band 3 transport in red cells. Biochem J 350:41-51

Cabantchik ZI, Greger R (1992) Chemical probes for anion transporters of mammalian cell membranes. Am J Physiol 262:C803-C827

Cabantchik ZI, Rothstein A (1974) Membrane proteins related to anion permeability of human red blood cells. I. Localization of disulfonic stilbene binding sites in proteins involved in permeation. J Membrane Biol 15:207-226

Casey JR, Reithmeier RAF (1991) Analysis of the oligomeric state of band 3, the anion transport protein of the human erythrocyte membrane, by size exclusion high performance liquid chromatography. Oligomeric stability and origin of hetrogeneity. J Biol Chem 266:15726-15737

Casey JR, Ding Y, Kopito RR (1995) The role of cysteine residues in the erythrocyte plasma membrane anion exchange protein, AE1. J Biol Chem 270:8521-8527

Cass A, Dalmark M (1973) Equilibrium dialysis of ions in nystatin-treated red cells. Nature 244:47-49

Cha A, Snyder GE, Selvin PR, Bezanilla F (1999) Atomic scale movement of the voltage-sensing region in a potassium channel measured via spectroscopy. Nature 402:809-813

Chernova MN, Jiang L, Crest M, Hand M, Vandorpe DH, Strange K, Alper SL (1997) Electrogenic sulfate/chloride exchange in *Xenopus* oocytes mediated by murine AE1 E699Q. J Gen Physiol 109:345-360

Cobb CE, Beth AH (1990) Identification of the eosinyl-5-maleimide reaction site on the human erythrocyte anion-exchange protein: overlap with the reaction sites of other chemical probes. Biochemistry 29:8283-8290

Cobb CE, Lin H, Beth AH (1990) Identification of the eosin-5-maleimide and eosin-5-iodoacetamide reaction sites on human erythrocyte band 3. Biophys J 57:96a

Colfen H, Harding SE, Boulter JM, Watts A (1996) Hydrodynamic examination of the dimeric cytoplasmic domain of the human erythrocyte anion transporter, band 3. Biophys J 71:1611-1615

Cousin JL, Motais R (1979) Inhibition of anion permeability by amphiphilic compounds in human red cell: evidence for an interaction of niflumic acid with the band 3 protein. J Membrane Biol 46:125-153

Dalmark M (1976) Effects of halides and bicarbonate on chloride transport in human red blood cells. J Gen Physiol 67:223-234

Demuth DR, Showe LC, Ballantine M, Palumbo A, Fraser PJ, Cioe L, Rovera G, Curtis PJ (1986) Cloning and structural characterization of a human non-erythroid band 3-like protein. EMBO J 5:1205-1214

Dix JA, Verkman AS, Solomon AK (1986) Binding of chloride and a disulfonic stilbene transport inhibitor to red cell band 3. J Membrane Biol 89:211-223

Doyle DA, Cabral JM, Pfuetzner RA, Kuo A, Gulbis JM, Cohen SL, Chait BT, MacKinnon R (1998) The structure of the potassium channel: molecular basis of K^+ conduction and selectivity. Science 280:69-77

Dutzler R, Campbell EB, Cadene M, Chait BT, MacKinnon R (2002) X-ray structure of a ClC chloride channel at 3.0 Å reveals the molecular basis of anion selectivity. Nature 415:287-294

Eidelman O, Yanai P, Englert HC, Lang HG, Greger R, Cabantchik ZI (1991) Macromolecular conjugates of transport inhibitors: new tools for probing topography of anion transport proteins. Am J Physiol 29:C1094-C1103

Erickson HK (1997) Cytoplasmic disposition of aspartate 821 in anion exchanger from human erythrocytes. Biochemistry 36:9958-9967

Falke JJ, Chan SI (1986a) Molecular mechanisms of band 3 inhibitors. 1. Transport site inhibitors. Biochemistry 25:7888-7894

Falke JJ, Chan SI (1986b) Molecular mechanisms of band 3 inhibitors. 2. Channel blockers. Biochemistry 25:7895-7898

Falke JJ, Pace RJ, Chan SI (1984) Direct observation of the transmembrane recruitment of band 3 transport sites by competitive inhibitors. A ^{35}Cl NMR study. J Biol Chem 259:6481-6491

Falke JJ, Kanes KJ, Chan SI (1985) The kinetic equation for the chloride transport cycle of band 3. A ^{35}Cl and ^{37}Cl NMR study. J Biol Chem 260:9545-9551

Fröhlich O (1982) The external anion binding site of the human erythrocyte anion transporter: DNDS binding and competition with chloride. J Membrane Biol 65:111-123

Fröhlich O (1988) The "tunneling" mode of biological carrier-mediated transport. J Membrane Biol 101:189-198

Fröhlich O, Gunn RB (1986) Erythrocyte anion transport: the kinetics of a single-site obligatory exchange system. Biochim Biophys Acta 864:169-194

Fröhlich O, Gunn RB (1987) Interactions of inhibitors on anion transporter of human erythrocyte. Am J Physiol 252:C153-C162

Fujinaga J, Tang X-B, Casey JR (1999) Topology of the membrane domain of human erythrocyte anion exchange protein, AE1. J Biol Chem 274:6626-6633

Furuya W, Tarshis T, Law F-Y, Knauf PA (1984) Transmembrane effects of intracellular chloride on the inhibitory potency of extracellular H_2DIDS. Evidence for two conformations of the transport site of the human erythrocyte anion exchange protein. J Gen Physiol 83:657-681

Galanter WL, Labotka RJ (1991) The binding of nitrate to the human anion exchange protein (AE1) studied with ^{14}N nuclear magnetic resonance. Biochim Biophys Acta 1079:146-151

Gartner EM, Liebold K, Legrum B, Fasold H, Passow H (1997) Three different actions of phenylglyoxal on band 3 protein-mediated anion transport across the red blood cell membrane. Biochim Biophys Acta 1323:208-222

Gasbjerg PK, Brahm J (1991) Kinetics of bicarbonate and chloride transport in human red cell membranes. J Gen Physiol 97:321-350

Gasbjerg PK, Knauf PA, Brahm J (1996) Kinetics of bicarbonate transport in human red blood cell membranes at body temperature. J Gen Physiol 108:565-575

Giebel O, Passow H (1960) Die Permeabilität der Erythrocytenmembran für organische Anionen. Zur Frage der Diffusion durch Poren. Pflügers Arch 271:378-388

Glauner KS, Mannuzzu LM, Gandhi CS, Isacoff EY (1999) Spectroscopic mapping of voltage sensor movement in the *Shaker* potassium channel. Nature 402:813-817

Grinstein S, McCulloch L, Rothstein A (1979) Transmembrane effects of irreversible inhibitors of anion transport in red blood cells. J Gen Physiol 73:493-514

Groves JD, Wang L, Tanner MJA (1998) Complementation studies with co-expressed fragments of human red cell band 3 (AE1): the assembly of the anion transport domain in *Xenopus* oocytes and a cell-free translation system. Biochem J 332:161-171

Gunn RB (1972) A titratable carrier model for both mono- and di-valent anion transport in human red blood cells. In: Rorth M, Astrup P (eds) Oxygen affinity of hemoglobin and red cell acid-base status. Munksgaard, Copenhagen, pp 823-827

Gunn RB, Fröhlich O (1979) Asymmetry in the mechanism for anion exchange in human red blood cell membranes. Evidence for reciprocating sites that react with one transported anion at a time. J Gen Physiol 74:351-374

Gunn RB, Fröhlich O, King PA, Shoemaker DG (1989) Anion transport. In: Agre P, Parker JC (eds) Red blood cell membranes: structure, function, clinical implications. Marcel Dekker, New York, pp 563-596

Hamasaki N, Izuhara K, Okubo K, Omachi A, Kelps RA (1990) Inhibition of chloride binding to the anion transport site by diethylpyrocarbonate modification of band 3. J Membrane Biol 116:87-91

Hautmann M, Schnell KF (1985) Concentration dependence of the chloride selfexchange and homoexchange fluxes in human red cell ghosts. Pflügers Arch 405:193-201

Huber S, Asan E, Jons T, Kerscher C, Puschel B, Drenckhahn D (1999) Expression of rat kidney anion exchanger 1 in type A intercalated cells in metabolic acidosis and alkalosis. Am J Physiol 277:F841-F849

Hunter MJ (1977) Human erythrocyte anion permeabilities measured under conditions of net charge transfer. J Physiol 268:35-49

Izuhara K, Okubo K, Hamasaki N (1989) Conformational change of band 3 protein induced by diethyl pyrocarbonate modification in human erythrocyte ghosts. Biochemistry 28:4725-4728

Janas T, Janas U (2000) Reversible DIDS binding to band 3 protein in human erythrocyte membranes. Mol Membr Biol 17:109-115

Jarolim P, Rubin HL, Liu SC, Cho MR, Brabec V, Derick LH, Yi SJ, Saad STO, Alper S, Brugnara C, Golan DE, Palek J (1994) Duplication of 10 nucleotides in the erythroid band 3 (AE1) gene in a kindred with hereditary spherocytosis and band 3 protein deficiency (band 3 ^{PRAGUE}). J Clin Invest 93:121-130

Jarolim P, Rubin HL, Brabec V, Chrobak L, Zolotarev AS, Alper SL, Brugnara C, Wichterle H, Palek J (1995) Mutations of conserved arginines in the membrane domain of erythroid band 3 lead to a decrease in membrane-associated band 3 and to the phenotype of hereditary spherocytosis. Blood 85:634-640

Jennings ML (1976) Proton fluxes associated with erythrocyte membrane anion exchange. J Membrane Biol 28:187-205

Jennings ML (1982) Stoichiometry of a half-turnover of band 3, the chloride transport protein of human erythrocytes. J Gen Physiol 79:169-185

Jennings ML (1992a) Anion transport proteins. In: Seldin DW, Giebisch G (eds) The kidney: physiology and pathophysiology, 2nd edn. Raven Press, New York, pp 503-535

Jennings ML (1992b) Cellular anion transport. In: Seldin DW, Giebisch G (eds) The kidney: physiology and pathophysiology, 2nd edn. Raven Press, New York, pp 113-145

Jennings ML (1995) Rapid electrogenic sulfate-chloride exchange mediated by chemically modified band 3 in human erythrocytes. J Gen Physiol 105:21-47

Jennings ML, Al-Rhaiyel S (1988) Modification of a carboxyl group that appears to cross the permeability barrier in the red blood cell anion transporter. J Gen Physiol 92:161-178

Jennings ML, Anderson MP (1987) Chemical modification and labeling of glutamate residues at the stilbenedisulfonate site of human red blood cell band 3 protein. J Biol Chem 262:1691-1697

Jennings ML, Gosselink PG (1995) Anion exchange protein in southeast asian ovalocytes: heterodimer formation between normal and variant subunits. Biochemistry 34:3588-3595

Jennings ML, Passow H (1979) Anion transport across the erythrocyte membrane, *in situ* proteolysis of band 3 protein, and cross-linking of proteolytic fragments by 4,4'-diisothiocyano dihydrostilbene-2,2'-disulfonate. Biochim Biophys Acta 554:498-519

Jennings ML, Smith JS (1992) Anion-proton cotransport through the human red blood cell band 3 protein: role of glutamate 681. J Biol Chem 267:13964-13971

Jennings ML, Adams-Lackey M, Denney GH (1984) Peptides of human erythrocyte band 3 protein produced by extracellular papain cleavage. J Biol Chem 259:4652-4660

Jennings ML, Anderson MP, Monaghan R (1986) Monoclonal antibodies against human erythrocyte band 3 protein, localization of proteolytic cleavage sites and stilbenedisulfonate-binding lysine residues. J Biol Chem 261:9002-9010

Jennings ML, Allen M, Schulz RK (1990) Effect of membrane potential on electrically silent transport: potential-independent translocation and asymmetric potential-dependent substrate binding to the red blood cell anion exchange protein. J Gen Physiol 96:991-1012

Jennings ML, Whitlock J, Shinde A (1998) Pre-steady state transport by erythrocyte band 3 protein: uphill countertransport induced by the impermeant inhibitor H_2DIDS. Biochem Cell Biol 76:807-813

Kampmann L, Lepke S, Fasold H, Fritzsch G, Passow H (1982) The kinetics of intramolecular cross-linking of the band 3 protein in the red blood cell membrane by 4,4'-diisothiocyano dihydrostilbene-2,2'-disulfonic acid (H_2DIDS). J Membrane Biol 70:199-216

Kang D, Karbach D, Passow H (1994) Anion transport function of mouse erythroid band 3 protein (AE1) does not require acylation of cysteine residue 861. Biochim Biophys Acta 1194:341-344

Karbach D, Staub M, Wood PG, Passow H (1998) Effect of site-directed mutagenesis of the arginine residues 509 and 748 on mouse band 3 protein-mediated anion transport. Biochim Biophys Acta 1371:114-122

Kawano Y, Okubo K, Tokunaga F, Miyata T, Iwanaga S, Hamasaki N (1988) Localization of the pyridoxal phosphate binding site at the COOH-terminal region of erythrocyte band 3 protein. J Biol Chem 263:8232-8238

Kietz D, Bartel D, Lepke S, Passow H (1991) pH-dependence of inhibition by H_2DIDS of mouse erythroid band 3-mediated Cl^- transport in *Xenopus* oocytes. The effect of oligonucleotide-directed replacement of Lys-558 by an Asn residue. Biochim Biophys Acta 1064:81-88

Knauf PA (1979) Erythrocyte anion exchange and the band 3 protein: transport kinetics and molecular structure. Curr Topics Membrane Trans 12:249-363

Knauf PA (1986) Anion transport in erythrocytes. In: Andreoli TE, Hoffman JF, Fanestil DD, Schultz SG (eds) Physiology of membrane disorders, 2nd edn. Plenum Press, New York, pp 191-220

Knauf PA (1989) Kinetics of anion transport. In: Raess BU, Tunnicliff G (eds) The red cell membrane. Humana Press, Clifton, NJ, pp 171-200

Knauf PA (2002) Asymmetry of the AE1 anion exchange system: implications for modeling the physiological rates of chloride-bicarbonate exchange. In: Layton HE, Weinstein AM (eds) Membrane transport and renal physiology, vol. 129 of IMA volumes in mathematics and its applications. Springer-Verlag, New York, pp 85-100 (in press)

Knauf PA, Brahm J (1989) Functional asymmetry of the anion exchange protein, capnophorin: effects on substrate and inhibitor binding. In: Fleischer S, Fleischer B (eds) Methods in enzymology, vol. 173. Academic Press, New York, pp 432-453

Knauf PA, Mann N (1984) Use of niflumic acid to determine the nature of the asymmetry of the human erythrocyte anion exchange system. J Gen Physiol 83:703-725

Knauf PA, Rothstein A (1971) Chemical modification of membranes. II. Permeation paths for sulfhydryl agents. J Gen Physiol 58:211-223

Knauf PA, Spinelli LJ (1995) NIP- and NAP-taurine bind to external modifier site of AE1 (band 3), at which iodide inhibits anion exchange. Am J Physiol 269:C410-C416

Knauf PA, Fuhrmann GF, Rothstein S, Rothstein A (1977) The relationship between anion exchange and net anion flow across the human red blood cell membrane. J Gen Physiol 60:363-386

Knauf PA, Ship S, Breuer W, McCulloch L, Rothstein A (1978) Asymmetry of the red cell anion exchange system. Different mechanisms of reversible inhibition by N-(4-azido-2-nitrophenyl)-2-aminoethylsulfonate (NAP-taurine) at the inside and outside of the membrane. J Gen Physiol 72:607-630

Knauf PA, Law F-Y, Marchant PJ (1983) Relationship of net chloride flow across the human erythrocyte membrane to the anion exchange mechanism. J Gen Physiol 81:95-125

Knauf PA, Law F-Y, Tarshis T, Furuya W (1984) Effects of the transport site conformation on the binding of external NAP-taurine to the human erythrocyte anion exchange system: evidence for intrinsic asymmetry. J Gen Physiol 83:683-701

Knauf PA, Mann N, Brahm J, Bjerrum PJ (1986) Asymmetry in iodide affinities of external- and internal-facing red cell anion transport sites. Fed Proc 45:1005-1005

Knauf PA, Mann NA, Kalwas JE, Spinelli LJ, Ramjeesingh M (1987a) Interactions of NIP-taurine, NAP-taurine, and Cl^- with the human erythrocyte anion exchange system. Am J Physiol 253:C652-C661

Knauf PA, Mann NA, Spinelli LJ (1987b) Effects of transport site conformation and anion binding on the affinity of the human red blood cell anion transport protein for niflumic acid (NA). Biophys J 51:566a

Knauf PA, Spinelli LJ, Mann NA (1989) Flufenamic acid senses conformation and asymmetry of human erythrocyte band 3 anion transport protein. Am J Physiol 257:C277-C289

Knauf PA, Ries EA, Romanow LA, Bahar S, Szekeres ES (1993a) DNDS (4,4'-dinitrostilbene-2,2'-disulfonate) does not act as a purely competitive inhibitor of red blood cell band 3-mediated anion exchange. Biophys J 64:307a

Knauf PA, Strong NM, Penikas J, Wheeler RB Jr, Liu SJ (1993b) Eosin-5-maleimide inhibits red cell Cl^- exchange at a noncompetitive site that senses band 3 conformation. Am J Physiol 264:C1144-C1154

Knauf PA, Law F-Y, Hahn K (1995) An oxonol dye is the most potent known inhibitor of band 3-mediated anion exchange. Am J Physiol 269:C1073-C1077

Knauf PA, Gasbjerg PK, Brahm J (1996) The asymmetry of chloride transport at 38°C in human red blood cell membranes. J Gen Physiol 108:577-589

Knauf PA, Raha NM, Spinelli LJ (2000) The noncompetitive inhibitor WW781 senses changes in erythrocyte anion exchanger (AE1) transport site conformation and substrate binding. J Gen Physiol 115:159-173

Knauf PA, Law F-Y, Leung T, Atherton SJ (2002a) Relocation of the disulfonic stilbene (DS) sites of band 3 (AE1) on the basis of fluorescence lifetime measurements. Biochemistry, submitted:

Knauf PA, Law F-Y, Leung T, Gehret AU, Perez ML (2002b) Substrate-dependent reversal of AE1 anion transport site orientation. Proc Natl Acad Sci 99:10861-10864

Knauf PA, Leung T, Law F-Y, Li X (2002c) Oxonol dye binding reports substrate-induced changes in human red blood cell anion exchanger AE1. Biophys J 82:570a

Kolbe M, Besir H, Essen L-O, Oesterhelt D (2000) Structure of the light-driven chloride pump halorhodopsin at 1.8 Å resolution. Science 288:1390-1396

Kopito RR, Lodish HF (1985) Primary structure and transmembrane orientation of the murine anion exchange protein. Nature 316:234-238

Kopito RR, Lee BS, Simmons DM, Lindsey AE, Morgans CW, Schneider K (1989) Regulation of intracellular pH by a neuronal homolog of the erythrocyte anion exchanger. Cell 59:927-937

Krupka RM (1989a) Role of substrate binding forces in exchange-only transport systems: II. Implications for the mechanism of the anion exchanger of red cells. J Membrane Biol 109:159-171

Krupka RM (1989b) Role of substrate binding forces in exchange-only transport systems: I. Transition-state theory. J Membrane Biol 109:151-158

Ku CP, Jennings ML, Passow H (1979) A comparison of the inhibitory potency of reversibly acting inhibitors of anion transport on chloride and sulfate movements across the human red blood cell membrane. Biochim Biophys Acta 553:132-141

Kuma H, Shinde AA, Howren TR, Jennings ML (2002) Topology of the anion exchange protein AE1: the controversial sidedness of lysine 743. Biochemistry 41:3380-3388

Lee BS, Gunn RB, Kopito RR (1991) Functional differences among nonerythroid anion exchangers expressed in a transfected human cell line. J Biol Chem 266:11448-11454

Legrum B, Passow H (1989) Inhibition of inorganic anion transport across the human red blood cell membrane by chloride-dependent association of dipyridamole with a stilbene disulfonate binding site on the band 3 protein. Biochim Biophys Acta 979:193-207

Lepke S, Fasold H, Pring M, Passow H (1976) A study of the relationship between inhibition of anion exchange and binding to the red blood cell membrane of 4,4'-diisothiocyano stilbene-2,2'-disulfonic acid (DIDS) and its dihydro derivative (H_2DIDS). J Membrane Biol 29:147-177

Lindsey AE, Schneider K, Simmons DM, Baron R, Lee BS, Kopito RR (1990) Functional expression and subcellular localization of an anion exchanger cloned from choroid plexus. Proc Natl Acad Sci USA 87:5278-5282

Liu SJ, Knauf PA (1993) Lys-430, site of irreversible inhibition of band 3 Cl^- flux by eosin-5-maleimide, is not at the transport site. Am J Physiol 264:C1155-C1164

Liu D, Kennedy SD, Knauf PA (1995) ^{35}Cl nuclear magnetic resonance line broadening shows that eosin-5-maleimide does not block the external anion access channel of band 3. Biophys J 69:399-408

Liu D, Kennedy SD, Knauf PA (1996a) Source of transport site asymmetry in the band 3 anion exchange protein determined by NMR measurements of external Cl^- affinity. Biochemistry 35:15228-15235

Liu SJ, Law F-Y, Knauf PA (1996b) Effects of external pH on substrate binding and on the inward chloride translocation rate constant of band 3. J Gen Physiol 107:271-291

Liu SJ, Ries EA, Knauf PA (1996c) Effects of external pH on binding of external sulfate, 4,4'-dinitro-stilbene-2,2'-disulfonate (DNDS), and chloride to the band 3 anion exchange protein. J Gen Physiol 107:293-306

Low PS (1986) Structure and function of the cytoplasmic domain of band 3: center of erythrocyte membrane-peripheral protein interactions. Biochim Biophys Acta 864:145-167

Low PS, Westfall MA, Allen DP, Appell KC (1984) Characterization of the reversible conformational equilibrium of the cytoplasmic domain of erythrocyte membrane band 3. J Biol Chem 259:13070-13076

Low PS, Willardson BM, Mohandas N, Rossi M, Shohet SB (1991) Contribution of the band 3-ankyrin interaction to erythrocyte membrane mechanical stability. Blood 77:1581-1586

Lux SE, John KM, Kopito RR, Lodish HF (1989) Cloning and characterization of band 3, the human erythrocyte anion-exchange protein (AE1). Proc Natl Acad Sci USA 86:9089-9093

Macara IG, Cantley LC (1981a) Interactions between transport inhibitors at the anion binding sites of the band 3 dimer. Biochemistry 20:5095-5105

Macara IG, Cantley LC (1981b) Mechanism of anion exchange across the red cell membrane by band 3: interactions between stilbenedisulfonate and NAP-taurine binding sites. Biochemistry 20:5695-5701

Macara IG, Kuo S, Cantley LC (1983) Evidence that inhibitors of anion exchange induce a transmembrane conformational change in band 3. J Biol Chem 258:1785-1792

Milanick MA, Gunn RB (1982) Proton-sulfate co-transport. Mechanism of H^+ and sulfate addition to the chloride transporter of human red blood cells. J Gen Physiol 79:87-113

Milanick MA, Gunn RB (1986) Proton inhibition of chloride exchange: asynchrony of band 3 proton and anion transport sites? Am J Physiol 250:C955-C969

Miller C, White MM (1984) Dimeric structure of single chloride channels from *Torpedo* electroplax. Proc Natl Acad Sci USA 81:2772-2775

Miraglia del Guidice E, Vallier A, Maillet P, Perrotta S, Cutillo S, Iolascon A, Tanner MJA, Delaunay J, Alloisio N (1997) Novel band 3 variants (Bands 3 Foggia, Napoli I and Napoli II) associated with hereditary spherocytosis and band 3 deficiency: status of the D38A polymorphism within the EPB3 locus. Br J Haematol 96:70-76

Moriyama R, Hiroshi I, Lombardo CR, Van Dort H, Low PS (1992) Structural and functional characterization of band 3 from southeast asian ovalocytes. J Biol Chem 267:25792-25797

Müller-Berger S, Karbach D, Kang D, Aranibar N, Wood PG, Ruterjans H, Passow H (1995a) Roles of histidine 752 and glutamate 669 in the pH dependence of mouse band 3 protein-mediated anion transport. Biochemistry 34:9325-9332

Müller-Berger S, Karbach D, König J, Lepke S, Wood PG, Appelhans H, Passow H (1995b) Inhibition of mouse erythroid band 3-mediated chloride transport by site-directed mutagenesis of histidine residues and its reversal by second site mutation of Lys 558, the locus of covalent H_2DIDS binding. Biochemistry 34:9315-9324

Ojcius DM, Solomon AK (1988) Sites of p-chloromercuribenzenesulfonate inhibition of red cell urea and water transport. Biochim Biophys Acta 942:73-82

Okubo K, Kang D, Hamasaki N, Jennings ML (1994) Red blood cell band 3: lysine 539 and lysine 851 react with the same H_2DIDS (4,4'-diisothiocyanodihydrostilbene-2,2'-disulfonic acid) molecule. J Biol Chem 269:1918-1926

Pan R, Cherry RJ (1995) Evidence that eosin-5-maleimide binds close to the anion transport site of human erythrocyte band 3: a fluorescence quenching study. Biochemistry 34:4880-4888

Pan R, Cherry RJ (1998) The eosin-5-maleimide binding site on human erythrocyte band 3: investigation of membrane sidedness and location of charged residues by triplet state quenching. Biochemistry 37:10238-10245

Papageorgiou P, Shmukler BE, Stuart-Tilley AK, Jiang L, Alper SL (2001) AE anion exchangers in atrial tumor cells. Am J Physiol 280:H937-H945

Passow H (1986) Molecular aspects of band 3 protein-mediated anion transport across the red blood cell membrane. Rev Physiol Biochem Pharmacol 103:61-223

Passow H, Fasold H, Gartner EM, Legrum B, Ruffing W, Zaki L (1980a) Anion transport across the red cell membrane and the conformation of the protein in band 3. Ann New York Acad Sci 341:361-383

Passow H, Kampmann L, Fasold H, Jennings M, Lepke S (1980b) Mediation of anion transport across the red blood cell membrane by means of conformational changes of the band 3 protein. In: Lassen UV, Ussing HH, Wieth JO (eds) Membrane transport in erythrocytes. Munksgaard, Copenhagen, pp 345-367

Passow H, Lepke S, Wood PG (1992) Exploration of the mechanism of mouse erythroid band 3-mediated anion exchange by site-directed mutagenesis. Prog Cell Res 2:85-98

Passow H, Karbach D, Aranibar N, Liebold K, Wood PG, Lepke S (1997) Studies on the pH dependence of band 3-mediated anion transport. Facts and speculations. In: Hamasaki N, Mihara K (eds) Membrane proteins: structure, function and expression control. Kyushu University Press, Fukuoka, pp 373-404

Passow H, Lepke S, Heberle J (2001) Inhibition of band 3-mediated Cl/Cl exchange by pH-dependent transition to proton-anion cotransport. Biophys J 80:46a

Pawloski JR, Hess DT, Stamler JS (2001) Export by red blood cells of nitric oxide bioactivity. Nature 409:622-626

Peters LL, Shivdasani RA, Liu SC, Hanspal M, John KM, Gonzalez JM, Brugnara C, Gwynn B, Mohandas N, Alper SL, Orkin SH, Lux SE (1996) Anion exchanger 1 (band 3) is required to prevent erythrocyte membrane surface loss but not to form the membrane skeleton. Cell 86:917-927

Pflugrath JW, Quiocho FA (1988) The 2Å resolution structure of the sulfate-binding protein involved in active transport in *Salmonella Typhimurium.* J Mol Biol 200:163-180

Pimplikar SW, Reithmeier RAF (1986) Affinity chromatography of band 3, the anion transport protein of erythrocyte membranes. J Biol Chem 261:9770-9778

Popov M, Li J, Reithmeier RAF (1999) Transmembrane folding of the human erythrocyte anion exchanger (AE1, band 3) determined by scanning and insertional N-glycosylation mutagenesis. Biochem J 339:269-279

Popov M, Tam LY, Li J, Reithmeier RAF (1997) Mapping the ends of transmembrane segments in a polytopic membrane protein. Scanning N-glycosylation mutagenesis of extracytosolic loops in the anion exchanger, band 3. J Biol Chem 272:18325-18332

Preston GM, Agre P (1991) Isolation of the cDNA for erythrocyte integral membrane protein of 28 kilodaltons: member of an ancient channel family. Proc Natl Acad Sci USA 88:11110-11114

Puceat M, Korichneva I, Cassoly R, Vassort G (1995) Identification of band 3-like proteins and Cl^-/HCO_3^- exchange in isolated cardiomyocytes. J Biol Chem 270:1315-1322

Pusch M, Ludewig U, Rehfeldt A, Jentsch T (1995) Gating of the voltage-dependent chloride channel ClC-0 by the permeant anion. Nature 373:527-531

Raha NM, Spinelli LJ, Knauf PA (1993) WW781, a potent reversible inhibitor of red cell Cl^- flux, binds to band 3 by a two-step mechanism. Am J Physiol 265:C521-C532

Rakitzis ET, Gilligan PJ, Hoffman JF (1978) Kinetic analysis of the inhibtion of sulfate transport in human red blood cells by isothiocyanates. J Membrane Biol 41:101-115

Ramakrishnan V, Busath DD (2002) An inverting basket model for AE1 transport. J Theor Biol 215:215-226

Rao A, Reithmeier RAF (1979) Reactive sulfhydryl groups of the band 3 polypeptide from human erythrocyte membranes. J Biol Chem 254:6144-6150

Rao A, Martin P, Reithmeier RAF, Cantley LC (1979) Location of the stilbene disulfonate binding site of the human erythrocyte anion-exchange system by resonance energy transfer. Biochemistry 18:4505-4516

Reithmeier RAF (1993) The erythrocyte anion transporter (band 3). Curr Opin Struct Biol 3:515-523

Reithmeier RAF, Pirraglia CA, Lieberman DM, Casey JR, Anderson WF (1989) Towards crystallization of band 3. In: Hamasaki N, Jennings ML (eds) Anion transport protein of the red blood cell membrane. Elsevier, Amsterdam, pp 213-223

Reithmeier RAF, Chan SL, Popov M (1996) Structure of the erythrocyte band 3 anion exchanger. In: Konings WN, Kaback HR, Lolkema JS (eds) Handbook of biological physics. Elsevier, Amsterdam, pp 281-309

Renner M, Dietl M, Schnell KF (1988) Chloride mediated inhibition of the phosphate and the sulfate transport by dipyridamole in human erythrocyte ghosts. FEBS Letters 238:77-81

Restrepo D, Kozody DJ, Spinelli LJ, Knauf PA (1989) Cl-Cl exchange in promyelocytic HL-60 cells follows simultaneous rather than ping-pong kinetics. Am J Physiol 257:C520-C527

Restrepo D, Cronise BL, Snyder RB, Spinelli LJ, Knauf PA (1991) Kinetics of DIDS inhibition of HL-60 cell anion exchange rules out ping-pong model with slippage. Am J Physiol 260:C535-C544

Restrepo D, Cronise BL, Snyder RB, Knauf PA (1992) A novel method to differentiate between ping-pong and simultaneous exchange kinetics and its application to the anion exchanger of the HL60 cell. J Gen Physiol 100:825-846

Richards SM, Jaconi ME, Vassort G, Puceat M (1999) A spliced variant of AE1 gene encodes a truncated form of band 3 in heart: the predominant anion exchanger in ventricular myocytes. J Cell Sci 112:1519-1528

Ruffing W, Gärtner E-M, Lepke S, Legrum B, Passow H (1996) Transport-related conformational states of the band 3 protein: probing with 1-fluoro-2,4-dinitrobenzene. Cell Mol Biol 42:1097-1118

Salhany JM (1990) Erythrocyte band 3 protein. CRC Press, Boca Raton

Salhany JM (1996) Allosteric effects in stilbenedisulfonate binding to band 3 protein (AE1). Cell Mol Biol 42:1065-1096

Salhany JM (1999) Anion binding characteristics of the band 3 / 4,4-dibenzamidostilbene-2,2-disulfonate binary complex: evidence for both steric and allosteric interactions. Biochem Cell Biol 77:543-549

Salhany JM (2001a) Mechanistic basis for site-site interactions in inhibitor and substrate binding to band 3 (AE1): evidence distinguishing allosteric from electrostatic effects. Blood Cells, Molecules, and Diseases 27:901-912

Salhany JM (2001b) Stilbenedisulfonate binding kinetics to band 3 (AE1): relationship between transport and stilbenedisulfonate binding sites and role of subunit interactions in transport. Blood Cells, Molecules, and Diseases 27:127-134

Salhany JM, Rauenbuehler PB (1983) Kinetics and mechanism of erythrocyte anion exchange. J Biol Chem 258:245-249

Salhany JM, Schopfer LM (2001) Kinetic mechanism of DIDS binding to band 3 (AE1) in human erythrocyte membranes. Blood Cells, Molecules, and Diseases 27:844-849

Salhany JM, Rauenbuehler PB, Sloan RL (1987) Evidence for multisite allosteric interactions on the band 3 monomer. Biochem Biophys Res Commun 143:959-964

Salhany JM, Sloan RL, Cordes KA (1990) *In situ* cross-linking of human erythrocyte band 3 by bis(sulfosuccinimidyl) suberate. Evidence for ligand modulation of two alternate quaternary forms: covalent band 3 dimers and noncovalent tetramers formed by the association of two covalent dimers. J Biol Chem 265:17688-17693

Salhany JM, Cordes KA, Schopfer LM (1993) Kinetics of conformational changes associated with inhibitor binding to the purified band 3 transporter. Direct observation of allosteric subunit interactions. Biochemistry 32:7413-7420

Salhany JM, Sloan RL, Cordes KA, Schopfer LM (1994) Kinetic evidence for ternary complex formation and allosteric interactions in chloride and stilbenedisulfonate binding to band 3. Biochemistry 33:11909-11916

Salhany JM, Sloan RL, Cordes KA, Schopfer LM (1995) Quantitative analysis of the kinetics of stilbenedisulfonate binding to band 3. Int J Biochem 27:953-964

Saviane C, Conti F, Pusch M (1999) The muscle chloride channel ClC-1 has a double-barreled appearance that is differentially affected in dominant and recessive myotonia. J Gen Physiol 113:457-468

Schofield AE, Reardon DM, Tanner MJ (1992a) Defective anion transport activity of the abnormal band 3 in hereditary ovalocytic red blood cells. Nature 355:836-838

Schofield AE, Tanner MJA, Pinder JC, Clough B, Bayley PM, Nash GB, Dluzewski AR, Reardon DM, Cox TM, Wilson RJM, Gratzer WB (1992b) Basis of unique red cell membrane properties in hereditary ovalocytosis. J Mol Biol 223:949-958

Schopfer LM, Salhany JM (1995) Characterization of the stilbenedisulfonate binding site on band 3. Biochemistry 34:8320-8329

Sekler I, Lo RS, Kopito RR (1995) A conserved glutamate is responsible for ion selectivity and pH dependence of the mammalian anion exchangers AE1 and AE2. J Biol Chem 270:28751-28758

Shami Y, Rothstein A, Knauf PA (1978) Identification of the Cl^- transport site of human red blood cells by a kinetic analysis of the inhibitory effects of a chemical probe. Biochim Biophys Acta 508:357-363

Ship S, Shami Y, Breuer W, Rothstein A (1977) Synthesis of tritiated 4,4'-diisothiocyano-2,2'-stilbene disulfonic acid ($[^3H]_2$DIDS) and its covalent reaction with sites related to anion transport in human red blood cells. J Membrane Biol 33:311-323

Sterling D, Casey JR (1999) Transport activity of AE3 chloride/bicarbonate anion-exchange proteins and their regulation by intracellular pH. Biochem J 344:221-229

Swenson ER, Grønlund J, Ohlsson J, Hlastala MP (1993) *In vivo* quantitation of carbonic anhydrase and band 3 protein contributions to pulmonary gas exchange. J Appl Physiol 74:838-848

Tanford C (1982) Simple model for the chemical potential change of a transported ion in active transport. Proc Natl Acad Sci USA 79:2882-2884

Tang X-B, Casey JR (1999) Trapping of inhibitor-induced conformational changes in the erythrocyte membrane anion exchanger AE1. Biochemistry 38:14565-14572

Tang X-B, Fujinaga J, Kopito R, Casey JR (1998) Topology of the region surrounding Glu^{681} of human AE1 protein, the erythrocyte anion exchanger. J Biol Chem 273:22545-22553

Tang X-B, Kovacs M, Sterling D, Casey JR (1999) Identification of residues lining the translocation pore of human AE1, plasma membrane anion exchange protein. J Biol Chem 274:3557-3564

Tanner MJA (1993) Molecular and cellular biology of the erythrocyte anion exchanger (AE1). Semin Hematol 30:34-57

Tanner MJA, Martin P, High S (1988) The complete amino acid sequence of the human erythrocyte membrane anion-transport protein deduced from the cDNA sequence. Biochem J 256:703-712

Tanphaichitr VS, Sumboonnanonda A, Ideguchi H, Shayakul C, Brugnara C, Takao M, Veerakul G, Alper SL (1998) Novel AE1 mutations in recessive distal renal tubual acidosis. Loss-of-function is rescued by glycophorin A. J Clin Invest 102:2173-2179

Thevenin BJ, Bicknese SE, Verkman AS, Shohet SB (1996) Distance between CYS-201 in erythrocyte band 3 and the bilayer measured by single-photon radioluminescence. Biophys J 71:2645-2655

Tosteson D (1959) Halide transport in red blood cells. Acta Physiol Scand 46:19-41

Toye AM, Bruce LJ, Unwin RJ, Wrong O, Tanner MJA (2002) Band 3 Walton, a C-terminal deletion associated with distal renal tubular acidosis, is expressed in the red cell membrane but retained internally in kidney cells. Blood 99:342-347

Vince JW, Reithmeier RAF (1998) Carbonic anhydrase II binds to the carboxyl terminus of human band 3, the erythrocyte Cl^-/HCO^{3-} exchanger. J Biol Chem 273:28430-28437

Vince JW, Reithmeier RAF (2000) Identification of the carbonic anhydrase II binding site in the Cl^-/HCO_3^- anion exchanger AE1. Biochemistry 39:5527-5533

Vince JW, Carlsson U, Reithmeier RAF (2000) Localization of the Cl^-/HCO_3^- anion exchanger binding site to the amino-terminal region of carbonic anhydrase II. Biochemistry 39:13344-13349

Voswinkel S, Haest CWM, Deuticke B (2001) Complex effects of papain on function and inhibitor sensitivity of the red cell anion exchanger AE1 suggest the presence of different transport subsites. J Membrane Biol 179:205-221

Wainwright D, Mawby WJ, Tanner MJA (1990) The membrane domain of the human erythrocyte anion transport protein. Epitope mapping of a monoclonal antibody defines the location of a cytoplasmic loop near the C-terminus of the protein. Biochem J 272:265-268

Wang DN (1994) Band-3 protein - structure, flexibility and function. FEBS Letters 346:26-31

Wang DN, Kuhlbrandt W, Sarabia VE, Reithmeier RAF (1993) Two-dimensional structure of the membrane domain of human band 3, the anion transport protein the erythrocyte membrane. EMBO J 12:2233-2239

Wang DN, Sarabia VE, Reithmeier RAF, Kuhlbrandt W (1994) Three-dimensional map of the dimeric membrane domain of the human erythrocyte anion exchanger, band 3. EMBO J 13:3230-3235

Weinstein AM (2000) A mathematical model of the outer medullary collecting duct of the rat. Am J Physiol 279:F24-F45

Wieth JO (1979) Bicarbonate exchange through the human red cell membrane determined with [^{14}C] bicarbonate. J Physiol 294:521-539

Wieth JO, Bjerrum PJ (1982) Titration of transport and modifier sites in the red cell anion transport system. J Gen Physiol 79:253-282

Wieth JO, Brahm J (1980) Kinetics of bicarbonate exchange in human red cells - physiological implications. In: Lassen UV, Ussing HH, Wieth JO (eds) Membrane transport in erythrocytes. Munksgaard, Copenhagen, pp 467-482

Wieth JO, Bjerrum PJ, Borders CL Jr (1982) Irreversible inactivation of red cell chloride exchange with phenylglyoxal, an arginine-specific reagent. J Gen Physiol 79:283-312

Williams KA (2000) Three-dimensional structure of the ion-coupled transport protein NhaA. Nature 403:112-115

Wood PG (1992) The anion exchange proteins: homology and secondary structure. Prog Cell Res 2:325-338

Wood PG, Muller H, Sovak M, Passow H (1992) Role of Lys 558 and Lys 869 in substrate and inhibitor binding to the murine band 3 protein: a study of the effects of site-directed mutagenesis of the band 3 protein expressed in the oocytes of *Xenopus laevis*. J Membrane Biol 127:139-148

Zhang D, Kiyatkin A, Bolin JT, Low PS (2000) Crystallographic structure and functional interpretation of the cytoplasmic domain of erythrocyte membrane band 3. Blood 96:2925-2933

12 Amino Acid Transport

J. Clive Ellory[1], Pawel Swietach[1] and John S. Gibson[2]

[1]Laboratory of Physiology, Oxford University, Parks Road, Oxford OX1 3PT, and [2]Department of Clinical Veterinary Medicine, University of Cambridge, Madingley Road, Cambridge CB3 0ES, UK

12.1 Introduction

All cells need a supply of amino acids, and membrane transporters for their uptake are fundamental membrane transport proteins. Unlike the sugars, nucleotides or other metabolically important nutrients, the twenty-three or so amino acids are more diverse in size, charge and structure, making it necessary to have several transporters with different substrate specificities. The pioneering work of Christensen (e.g. Christensen 1985; Christensen and Kilberg 1987) used tracer flux studies, with paradigm substrates and inhibitors to define the number and specificity of amino acid transporters in mammalian cells. Although very successful, this approach was limited by the low affinities and overlapping specificities of the systems it defined, and in particular the paucity of high-affinity specific inhibitors prevented successful biochemical characterisation by classical membrane protein isolation procedures. A limited number of paradigm substrates (MeAIB (*N*-methylisobutyric acid); BCH (2-aminonorbornane 2-carboxylic acid)) proved useful in identifying specific systems. Molecular biological techniques, and in particular expression cloning, have revolutionised the identification of membrane proteins, and now the identity and properties of several large families of amino acid transporters are established (Palacin et al. 1998; Bode 2001; Chillaron et al. 2001). Particular attention has been paid to neuronal transporters from the point of view of neurotransmitter uptake and synthesis, and epithelia in the context of renal and intestinal absorption, but the main systems for amino acid uptake into most types of mammalian cells are now well-characterised.

Mammalian red blood cells (RBCs) represent a special problem for identifying transporters, since they lack a nucleus and protein synthetic machinery, and are therefore not amenable to RT-PCR and other molecular biological techniques for identifying amino acid transporter expression. Some studies (Lannigan et al. 1986; Vadgama et al. 1991; Bading et al. 1996; Boyd et al. 2000) have used erythrogenic cell lines (K562 cells, Friend mouse erythroleukaemic cells), and it is possible that plasma-derived colony forming units (CFUs) (Suwanmanee et al. 2002) could play a role in the future. In addition avian and lower vertebrate RBCs are nucleated, and therefore can provide mRNA and DNA preparations. Antibodies raised against specific epitopes on identified amino acid transporters are starting to become available, and provide a secure means of identifying specific transport systems (Ohkame et al. 2001; Chen et al. 2002; Marie et al. 2002; Pow et al. 2002). However, kinetic and inhibitor information derived from the properties of identified transporters expressed in *Xenopus* oocytes, or cell lines, remains the principal route for defining the number, identity and properties of the RBC amino acid transporters (Palacin et al. 1998).

12.2 Why do Red Cells Need Amino Acid Transporters?

As cells lacking the machinery for protein synthesis, it may not be obvious why mature RBCs need to transport amino acids. One reason stems from the early synthetic requirements of RBC precursors. The transporters represent a relic from pre-reticulocyte stages of the RBC, when functioning in the bone marrow these cells had an absolute requirement for amino acid uptake for protein synthesis, leading to high activity of a variety of specific transporters (Christensen and Kilberg 1987). There has been some work on the elimination of membrane transporters on reticulocyte maturation, which indicates that particular amino acid transporters are selectively retained, whilst others are eliminated (Benderoff et al. 1978; Blostein and Grafova 1987; Johnstone et al. 1987). Another important role for amino acid transporters in RBC precursor cells is to export basic amino acids derived from histone metabolism during reticulocyte maturation. Failure to remove these amino acids leads to the accumulation of 20–30 mM of arginine, ornithine (derived from arginine by arginase) and lysine in sheep RBCs with a transporter defect, altering the alkali metal cation content significantly (Young et al. 1975a).

An essential biochemical requirement for amino acid transporters is to provide the precursor amino acids for reduced glutathione (GSH) biosynthesis. As a tripeptide, GSH is synthesised from glycine, cysteine and glutamate in two steps (Fig. 12.1). With an intracellular GSH level of 3 mM, and a turnover of 1–3 days, there is a requirement for uptake of the three relevant amino acids. Transport of cysteine and glycine is straightforward via defined transporters (ASC, L for cysteine; gly, L, and band 3 for glycine) but mediated transport of glutamate is not found in human RBCs. Rather, glutamine uptake via system N, followed by hydrolysis to glutamate intracellularly, has been suggested to provide the glutamate for GSH biosynthesis (Ellory et al. 1983) although this has been questioned since no glutaminase activity was detected by NMR (King and Kuchel 1985). In fact, the very low level of glutamate uptake may be sufficient for GSH synthesis and may still be mediated, via band 3 or another transport system. In certain cells and tissues, the gamma-glutamyl transferase system has been shown to play a role in amino acid uptake for GSH synthesis (Orlowski and Meister 1973). However, this does not occur in human RBCs (Young et al. 1975b).

As well as a likely source for GSH synthesis, the uptake of glutamine is necessary in the synthesis of NAD from nicotinic acid, another essential part of anti-oxidant defence in RBCs. Recently this has been highlighted in work on homozygous sickle (Hb SS) RBCs, where striking increases in amino acid transport have been observed (Niihara et al. 1997, 1998). There are also a number of animal species with transport modifications, including dog RBCs with very high glutamate transport rates leading to raised GSH levels (Maede et al. 1982), and sheep RBCs lacking transporters which lowers intracellular GSH levels (Young et al. 1975a).

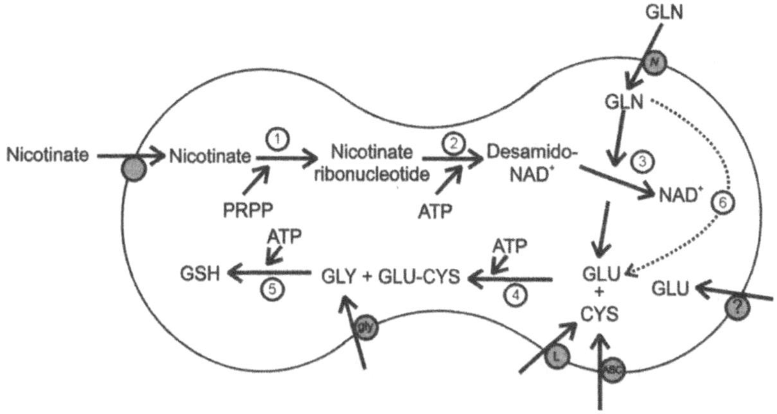

Fig. 12.1. Two important biochemical pathways requiring uptake of amino acids (glutamine or glutamate, glycine, cysteine) into human RBCs. *1.* Nicotinate phosphoribosyl transferase; *2.* NAD$^+$ pyrophosphorylase; *3.* NAD$^+$ synthetase; *4.* γ-glutamylcysteine synthetase; *5.* GSH synthetase; *6.* Glutaminase (?)

An old idea (Elwyn et al. 1972), that has recently been receiving increasing support is that RBCs participate in whole body metabolism of amino acids, either promoting interorgan transport, or acting as a pool or reserve of particular amino acids (Suliman et al. 1999). Measurements of plasma and RBC levels of amino acid concentrations have provided variable data, but for certain amino acids, particularly lysine, arginine and cysteine, there is good evidence for intracellular pools being significant. The important role of arginine as a nitric oxide (NO) precursor, has promoted considerable interest in its transport via system y^+ and y^+L (Deves and Boyd 1998). Red cells have made an important contribution to research on arginine transport. They have been used as a model system to characterise the transporters, but have also been considered as playing a physiological role in buffering plasma availability of arginine (Mendes Ribeiro et al. 2001).

Chapter 17 identifies a further role for amino acid transport with the uptake of divalent cation-histidine complexes, particularly driving zinc and copper transport.

In the RBCs of non-mammalian species, and specifically in the nucleated RBCs of birds and fishes, intracellular amino acids are accumulated for defending cell volume, participating in regulatory volume decrease (RVD) as osmolytes able to be eliminated after cell swelling (Garcia-Romeu et al. 1991). Principal amino acids include glycine, aspartate, and the sulphonic amino acid analogue taurine. This system is probably a relic of adaptation to life in seawater, at a higher osmolality than normal vertebrate plasma. The pathway involved in amino acid loss has analogues in many mammalian cells, and is known variously as VRAC, VSOAC, etc.

(Lang et al. 1998). Although it is likely to be present in human RBCs (Culliford et al. 1995), its role in RVD is replaced by KCl transport via KCC1 in this case.

Red cells infected with intraerythrocytic parasites, such as malaria and babesia, have greatly enhanced rates of amino acid transport, necessary both for the uptake of essential substrates, and the export of breakdown products of proteolysis. This topic is addressed in Chap. 24, but the system(s) involved show several similarities to the VRAC, VSOAC pathways mentioned above.

Finally, amino acids may be taken up "accidentally", as minor substrates on other transport pathways. Although showing low affinities compared with the dedicated substrates, rates can nevertheless be significant and may participate in exchange fluxes driving preferred substrates into the cell. Transport via band 3 of glycine, sarcosine, etc. (Young et al. 1981; King and Gunn 1991), and transport of aromatic amino acids via system T are examples of this behaviour (Zhou et al. 1992).

12.3 A Comment on Methodology

Traditionally, RBCs have provided a simple model system for membrane transport studies, involving incubation with radioisotopically labelled substrate, and separation by centrifugation with a washing procedure (Young and Ellory 1982). More rarely amino acid concentrations are estimated chemically, enzymatically or by NMR (King and Kuchel 1985). Aside from the obvious caveats of radioisotope purity and linearity of time courses, there are two important considerations when measuring amino acid transport in RBCs. The first concerns contamination of RBC preparations with platelets and white cells. Unless these other cell types are removed by filtration/absorption there is a high possibility of enzyme activity (glutaminase, proteases) and high transport rates by transporters not found in RBCs (e.g., for acidic amino acids) biasing the results. The second recognises that reticulocytosis, and the age of the RBC population may affect transport rates. For example, there is a five-fold increase in glutamine uptake into RBCs with patients with 9% reticulocytes (Niihara et al. 1997). Other factors such as phosphorylation of transporters, oxygen-sensitivity and effects of cell metabolism should also be considered.

12.4 Transport System Nomenclature

Nomenclature of the amino acid transporters has proved a difficult area. In 1984 (Bannai et al. 1984) the principal workers in the field at that time suggested a system broadly based on common practice. Systems were given acronymic letters, with capital letters implying secondary active transporters i.e., with Na^+-dependence, and lower case being facilitated diffusion systems. Thus ASC was the name for a Na^+-dependent transporter for alanine, serine, cysteine, whilst y^+ was a sodium-independent system for cationic amino acids. There were many exceptions (e.g., L and T being upper case but Na^+-independent) but the principle was established. With the advent of cloning, several systems were found to be proteins characterised previously and named for a different function. Thus y^+, which was named CAT-1 or -2 on cloning, turned out to be identical with ecoR, a previously identified viral receptor protein (Wang et al. 1991). A final complication came later when it was found to be heterodimeric, with a second variable subunit cloned previously as a surface histocompatability marker 4F2hc or CD98 (Palacin et al. 1998). There were also prefixes for the species in which the transporter was expressed i.e. mCAT from mouse (Chillaron et al. 2001). In the present chapter, Table 12.1 presents the simplest name based on the original strategy for identifying discrete transport systems. Other names associated with similar or identical systems cloned in other tissues are also given.

12.5 Identification of Red Cell Amino Acid Transport Systems

Until good specific antibodies become available, and because PCR is not possible with mammalian RBCs, the traditional strategy of tracer flux studies with paradigm substrates and selective inhibitors remains the most effective means of identifying RBC amino acid transporters. Ion substitution studies, with Na^+ or Cl^- replacement by K^+ or NO_3^- can help identify secondary active transporters, (e.g., ASC, N and gly), particularly if there are differences in the effectiveness of the replacement cation, e.g. Li^+ substitution for Na^+ normally results in transport inhibition, but for y^+L, and system A in other tissues. Li^+ still supports transport. There is a particular lack of specific tightly bound inhibitors, and no parallels for e.g. ouabain on the Na^+/K^+ pump, or cytochalasin on the glucose transporter exist. Only the nonspecific SH-group reagents such as *p*-chloromercuribenzene sulphonate (PCMBS) or *N*-ethylmaleimide (NEM) have been used successfully as non-competitive inhibitors (Young 1980; Ellory et al. 1982; Deves et al. 1993) to separate systems (e.g., y^+, y^+L). More frequently, competitive inhibition by other amino acids or amino acid analogues has been used to define specific transport systems, e.g. αAB or serine on cysteine flux via ASC (Young and Ellory 1977). The increasing availability of antibodies to transporters means Western blotting, and even functional studies with antibodies in resealed ghosts (cf. antibodies to the Na^+/K^+ pump (Glynn et al. 1974)) may be useful in the future. Table 12.1 reports the paradigm substrates identifying the transporters found in RBCs.

12.6 Kinetic Studies

Kinetic studies have been carried out with two separate goals. As outlined above, kinetic data, i.e. experimentally determined K_m, K_i and V_{max} values play an important role in the process of identifying the expression and activity of particular transporters. If substrates can be shown to be true competitive inhibitors, with K_i values equal to K_m values under appropriate conditions, this strengthens the identity of particular transporters. This approach has been used in particular by Deves to separate system y^+ and y^+L (Deves and Boyd 1998; Deves et al. 1998). Taken further, kinetic experiments have been used to separate exchange, from net, fluxes in particular using stringent conditions such as zero *trans* or infinite *cis* (Stein 1986). It is apparent that both net fluxes and homo- and heteroexchange occur via both facilitated diffusion (L, T) and Na^+-dependent secondary active transport type carriers (ASC, gly). (In the general case, the best-studied example of a facilitated diffusion carrier is band 3, where the heteroexchange of bicarbonate and chloride has obvious functional importance, and the homoexchange of chloride may be a structural consequence of the transporter function.) For the amino acid transporters, heteroexchange of a rare substrate (e.g., cysteine) for a common substrate (e.g., alanine) again has functional significance. It is homoexchange, which might compete functionally with "useful" net transport that raises questions about the physiological consequences of such a mechanism. This suggests a further use of kinetic studies, in the assessment of transport of a particular amino acid under physiological conditions.

Table 12.1. The principal amino acid (aa) transporters in human red cells

System	Synonyms/ proteins	Amino acid selectivity	Ionic dependency	Comments	Red cell references
ASC	ASCT 1–2	Small neutral (ala, ser, cys)	Na^+	Na^+-independent analogs in sheep (C) and horse (asc) red cells	Al-Saleh and Wheeler 1982 Ellory et al. 1985
gly	GLYT 1–2	Glycine (gly, sarc)	Na^+, Cl^-	Important in avian red cells	Ellory et al. 1981
L	LAT 1–2	Large neutral (leu)	None	High capacity	Rosenberg 1981
T	(?) OATP 2–3	Aromatic (tyr, try, phe)	None	physiological substrate (T_3)	Rosenberg et al. 1980 Zhou et al. 1992
y^+	CAT 1–4	Cationic (lys, arg, orn)	None	Heterodimeric complex	Deves and Boyd 1998
y^+L	y^+LAT	Cationic/neutral (arg, leu)	Neutral aa plus Na^+	High affinity for cationic aa	Deves et al. 1998
N	SN 1–2	Amine/N-ring (gln, asn, his)	Na^+ (H^+)		Niihara et al. 1997
glu	GLAST EAAT	Anionic (glu, asp)	Na^+ (K^+, H^+)	Important in dog, controversial in humans	Sato et al. 2000

12.7 System gly

Vidaver and Shepherd (1968) first characterised RBC transport of glycine via system gly in avian erythrocytes, a cell type where it has a high capacity. More recently there has been much general interest in the transport of glycine and GABA in neuronal cell types from the point of view of their role as neurotransmitters (Herdon et al. 2001). A large family of amino acid transporters has been characterised (GLYT1–2 being the relevant glycine transporters; GAT1–3; BGTI; TAUT representing transporters for other amino acids which act as neurotransmitters or neuromodulators (GABA, betaine, taurine)) (Palacin et al. 1998). The transporter designated gly is present in human RBCs, and represents a significant route for glycine uptake, accounting for about 30% of influx at physiological concentrations. The transporter is unusual in requiring both Na^+ and Cl^- ions for function. It is also very narrow in its substrate requirements, transporting only glycine or sarcosine (Ellory et al. 1981). The need for glycine for GSH biosynthesis makes glycine an important substrate in terms of RBC amino acid transport.

12.8 Band 3

The chloride:bicarbonate transporter band 3 (AE1) has also been shown to transport a variety of organic anions, and some amino acids, principally glycine and serine. This protein is present at a high copy number (10^6) in the RBC membrane, and therefore can mediate significant fluxes even of molecules, which have a low affinity, and are transported at much lower rates than the inorganic anions, which represent the preferred substrates. Band 3 is a member of the AE family of transporters, of which there are at least three principal isoforms (see Chap. 11). Functionally, the stilbene derivatives DIDS and SITS have been used to define fluxes via band 3, the DIDS-sensitive component of glycine transport being taken as mediated by band 3. The sulphur-containing amino acid analogue taurine is important in cell volume regulation in several mammalian tissues (heart, retina) and the RBCs of lower vertebrates, and has been proposed to involve a variant of band 3 in its transport in trout RBCs (Fievet et al. 1995).

12.9 The Heterodimeric Amino Acid Transporters (L, asc, y^+L)

The most striking advance in understanding mammalian amino acid transporters has come from the body of work identifying a family of heteromeric transporters, composed of two distinct polypeptides, heavy chain and light chain, linked by a disulphide bridge (Kanai and Endou 2001; Palacin et al. 2001). The heavy subunit is either (more usually) the cell-surface antigen 4F2hc (also named CD98), or rBAT. These molecules are type II membrane glycoproteins, with a single transmembrane domain, an intracellular N-terminus, and a large extracellular domain. The light subunit is much more variable. Twenty-two different (including species' variants) light subunits have been identified so far, and it is this subunit which conveys the substrate specificity on the membrane transporter. The principal transporters are given in Table 12.2, where systems y^+L, L, asc, X_c^- and the epithelial transporter b_o^+ are listed. The association of subunits to form discrete transporters has been confirmed by oocyte coexpression, followed by transport studies (radiosiotope fluxes or electrophysiology). These methods are quite sensitive, and superior to e.g. Western blotting due to low titre and nonspecific antibodies, and a low abundance of transporters in the membrane. For example, a vesicular distribution study estimated the copy number for the y^+ transporter to be four hundred molecules per RBC (Tse et al. 1991), a level difficult to identify by Western blotting. Alternative strategies are to study reticulocytes or K562 cells, although the latter did not prove useful in characterising y^+L (Boyd et al. 2000). The same paper was also unsuccessful in identifying 4F2hc in RBCs, emphasising the difficulties of using RBCs in molecular biological studies.

The identification of a large number of isoforms and family members points to considerable diversity in this family of transporters. The recent identity of asc-2, with an antibody in mouse erythrocytes (Chairoungdua et al. 2001), which represents another heterodimeric transporter with a different, unidentified heavy chain suggests that further members of this family will be found.

Table 12.2. Heteromeric amino acid transporters

System	Heavy chain	Light chain	Human chromosome
y^+L	4F2hc*	y^+LAT-1	14q11.2
y^+L	4F2hc	y^+LAT-2	16q22.1
L	4F2hc	LAT-1	16q24.3
L	4F2hc	LAT-2	14q11.2
asc	4F2hc	asc-1	19q12-13
X_c^-	4F2hc	xCT	4q28-q32
b_o^+	rBAT	b_o^+AT	19q12-13

*also known as CD98hc

12.9.1 System L

This system is a Na^+-independent transporter with a high transport rate in human RBCs. System L prefers large neutral amino acids i.e. leucine. In other tissues it consists of 4F2hc with LAT-1 or LAT-2 light chain. It is not stereospecific, showing a 3.5-fold lower affinity for D-leucine *vs.* L-leucine, but a twofold greater V_{max}. It is one of the transporters to have been extensively studied kinetically, and performs exchange as well as net fluxes (Rosenberg 1981). As a high capacity, relatively low affinity transporter it is less susceptible to inhibition by competing substrates at plasma levels, and carries a significant fraction of e.g., cysteine (Fig. 12.2), although there is also a more specific, high affinity Na^+-dependent system (ASC).

12.9.2 System y^+L

First identified as carrying a fraction of lysine influx into human RBCs with an unusually high affinity for a facilitated diffusion, rather than secondary active

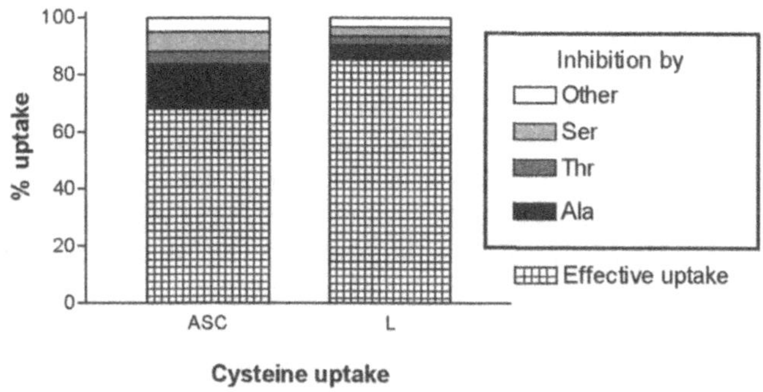

Fig. 12.2. Cysteine uptake into human red blood cells at plasma concentrations of substrate plus competing amino acids

transport system, this transporter is now known to represent 4F2hc linked to a different light chain (y^+LAT-1 or -2) (Kanai and Endou 2001). The transporter can take either cationic amino acids, or substitute Na^+ plus neutral amino acid (e.g., leucine) at the substrate binding site. This separates the system from the other cationic amino acid transporter y^+, which has a lower affinity for cationic amino acids, and will not accept Na^+ plus a neutral amino acid (Deves and Boyd 1998). Additionally y^+L, but not y^+, is not susceptible to inhibition by NEM, allowing the separation of these two transporters in flux studies. As can be seen in Fig. 12.3, y^+L does not carry a significant fraction of arginine into RBCs at physiological amino acid levels. Interest in both y^+ and y^+L comes from the role of arginine as a nitric oxide precursor, although the RBC probably does not contribute to NO production *in vivo*, due to NO binding to haemoglobin.

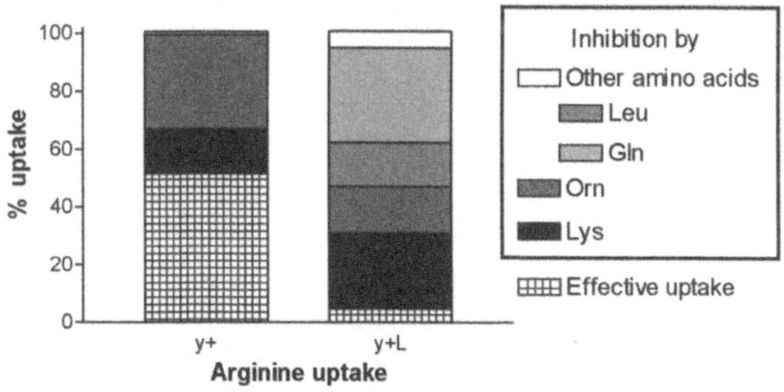

Fig. 12.3. Arginine uptake into human red blood cells at plasma concentrations of substrate plus competing amino acids. The presence of leu and gln (offset in key) is only significant for y^+L, although in both cases all plasma amino acids were tested

12.10 System ASC

In contrast to the possibility of RBC glutamate transport via EAAT mentioned above, ASCT1 and 2, two other members of the superfamily to which the EAAT belong, are directly relevant to human RBCs. The transporter characterized as ASC (for ala, ser, cys; gly is not transported, and large neutral amino acids are poor substrates) was identified as a high affinity Na^+-dependent transporter in RBCs by Al-Saleh and Wheeler (1982). This transporter has a 1:1 stoichiometry for Na^+ and amino acids, but also operates in the exchange mode. It has high affinity for cysteine, (15 μM) and acts as the physiological route for cysteine entry in human RBCs (Young et al. 1980). The concept of heteroexchange, exchanging a rare amino acid (cysteine) for a common one (alanine) even in a Na^+-dependent manner is probably physiologically relevant in this case. ASC is particularly susceptible to inhibition by SH-group inhibitors NEM or PCMBS (Young et al. 1980). Variants of ASC exist in other species' RBCs, in particular the low affinity system (designated C) found in sheep RBCs (Young et al. 1975) and the triallelic variants (asc etc.) found in horse RBCs (Fincham et al. 1987).

12.11 System y^+

This cationic amino acid transporter was identified early on in human and rabbit RBCs (Antonioli and Christensen 1969; Gardner and Levy 1972; Young et al. 1980). It is Na^+-independent, but has a relatively high affinity for cationic amino acids. The system is electrogenic, and the transporter has been studied electrophysiologically following oocyte expression. In fact there are at least four members of a family of transporters designated CAT which expresses y^+ activity. The various isoforms CAT1–4 are distributed in specific tissues with differing K_m values for arginine and lysine which reflect physiological function (Palacin et al. 1998; Hammermann et al. 2001).

12.12 System N

An early study identified a Na^+-dependent uptake of glutamine in human RBCs consistent with properties of system N (Ellory and Osotimehin 1983), a transporter, which has recently received considerable attention, particularly in muscle and liver, where its role in metabolism has been emphasized (Bode 2001). More recently, (Niihara et al. 1997, 1998) the presence of system N type activity in human RBCs with a K_m for glutamine of 0.4 mM, and a V_{max} of 0.36 mmol/(l_{cells}·h) has been demonstrated. The transport rate was increased in reticulocytosis and in Hb SS patients, with additional changes in K_m in the latter group. In other tissues where glutamine transport occurs via N or A system, these two systems have been associated with members of a transporter gene family with three system A clones (SAT1–3) and two system N clones (SN1–2) being identified (Nakanishi et al.

2001). Although the RBC transporter has not yet been cloned, further flux studies should resolve whether it is SN1, since the latter tolerates a Li^+ for Na^+ substitution, and also transports a proton (Bode 2001).

12.13 System T

System T was first studied in human RBCs as the transporter for aromatic amino acids, including the acronymic tryptophan (Rosenberg et al. 1980). Subsequently, the preferred substrates for this transporter were shown to be the thyroid hormones, and in particular T_3 (Zhou et al. 1992). Nevertheless, transport of tryptophan and phenylalanine on this transporter is significant, and has been discussed in the context of introducing such amino acids into Hb SS cells pharmacologically to stabilise Hb S in sickle cell disease. The kinetic evidence for system T has been presented in some detail, and it shows a moderate affinity for its narrow substrate range of amino acids (tyr, trp, phe).

12.14 System glu

Glutamate transport is particularly important in the nervous system, and Na^+- (and K^+, H^+ and anion) dependent transporters have been studied in detail (Levy et al. 1998). The potency of glutamate as an excitatory neurotransmitter means that it must be maintained extracellularly at very low levels. Neuronal tissue therefore uses an inward Na^+ gradient, and outward K^+ and H^+ gradient via the secondary active transporter to rapidly take up glutamate. The nomenclature of glutamate transporters is confusing but there are five members of the family primarily associated with glutamate transport (EAAT1–5, which are also synonymous with GLAST and EAAC). All EAAT isoforms are found in nervous tissue, and several of these isoforms are neuronal or even retina specific (Chillaron et al. 2001). A high affinity Na^+-dependent transporter for glutamate and aspartate is found in carnivore RBCs, but not in human cells (Sato et al. 2000). Interestingly, a genetic trait in certain dogs leads to greatly enhanced transport of glutamate, with increased intracellular GSH production and alterations in cell Na^+ levels (Maede et al. 1982). Thus although this superfamily of amino acid transporters is important overall, the EAAT transporters are not relevant to human RBCs.

12.15 Physiological Transport Rates for Amino Acids

Human RBCs possess at least nine discrete amino acid transporters, with widely different affinities for various amino acids. Also, there are at least twenty three relevant substrate amino acids in plasma, at varying concentrations, making it a difficult task to assess substrate interactions and identify the magnitude of competitive inhibition by one substrate on another. It is likely that a high-affinity inhibitor present at low concentrations in plasma can have a more important effect than a higher concentration of another amino acid, which has a high K_i value for the transport system. Our recent approach to this has been to use published values for K_i and K_m on well-characterised human RBC amino acid transporters to develop a matrix to calculate interactions between substrates. We have chosen two examples, cysteine and arginine to present in this calculation (Figs. 12.2 and 12.3). The effects of amino acids at their concentrations in normal plasma on cysteine uptake via systems ASC or L, and arginine uptake via y^+ or y^+L, are given in Figs. 12.2 and 12.3.

For cysteine uptake via ASC, the principal competing amino acids are alanine, threonine and serine, but at their plasma concentrations these only inhibit by about one third, with cysteine transport remaining at 68% of the level in the absence of other amino acids, i.e. plasma-free conditions. On system L, a low affinity, high capacity transport system, the presence of other amino acids also makes surprisingly little difference to cysteine transport. In this case, at plasma levels of amino acids, the transporter is operating far below the K_m values for relevant amino acids, making it kinetically insensitive. Fluxes are nearly linear with cysteine concentration, and represent a very low (2%) fraction of V_{max}. This will mean that, although system L has a high transport capacity (V_{max}), the actual transport of cysteine as a substrate at its plasma concentration via system L will be small.

The other two examples (illustrated in Fig. 12.3) concern arginine transport via systems y^+ and y^+L. Here the narrower specificity of system y^+ for cationic amino acids means that significant competition from lysine, ornithine and cystine for arginine transport is expected. In fact modelling shows that arginine transport into RBCs in plasma via system y^+ occurs at about half the rate it would be in the total absence of other (competing) amino acids, confirming y^+ as an effective transporter for arginine from plasma. More surprising is the result for transport via the high affinity system y^+L. This transporter will carry either cationic amino acids, or a Na^+ ion and neutral amino acid (Deves and Boyd 1998). As seen in Fig. 12.3, there is very significant competition for uptake via y^+L from glutamine, as well as lysine and ornithine. Coupled to the effects of (neutral amino acids + Na^+), e.g., the 20% inhibition by leucine in this case, the system is not functioning for arginine transport under physiological conditions, operating at 6% of the value under amino acid-free conditions.

This first approach to trying to understand amino acid uptake from plasma is not completely rigorous, since it only considers inhibition of zero *trans* influx, and does not investigate heteroexchange, and *trans* acceleration. Data are available for intracellular amino acid concentrations, and for some systems (L, ASC) it should be possible in the future to model exchange as well as net fluxes. The idea that, for example exchange of intracellular alanine for extracellular cysteine is more important that net (zero *trans*) cysteine uptake has always been an argument for the existence of the exchange mode of operation of carriers. A further conclusion from the present modelling, is that transport with a low affinity, on a very high capacity transporter (cf., band 3 "accidental" substrates) may be physiologically robust i.e., insensitive to competition from other amino acids, in providing an uptake route for key amino acids. It is a paradox that such accidental or nonspecific transport may be more important physiologically than high affinity low capacity transporters designed to transport particular substrates selectively with high affinities.

Finally, the dependence of amino acid transporters on oxygen tension may also be relevant physiologically. Limited work has been carried out in this area. It is known, however, that many RBC membrane transporters are affected by oxygen tension (Gibson et al. 2000). System ASC and gly are stimulated by oxygenation in normal and Hb SS RBCs (Kiessling et al. 2000). Kinetic effects differ: system ASC has a lower K_m on oxygenation with no change in V_{max}, whilst for system gly K_m is unaltered but capacity is elevated. We speculate that such changes may be relevant in interorgan transport of amino acids, or in the provision of precursors for GSH and NADH synthesis, given the role for GSH and NADH in protection against peroxidation. The effect of oxygen on other amino acid transporters awaits investigation.

12.16 Alterations of Amino Acid Transport in Disease States

Red blood cells represent the easiest cell type to sample and analyse in various disease states (although platelets and leucocytes are also used, with more difficulty). There have therefore been a number of studies of amino acid transport in RBCs from patients with a variety of clinical conditions including renal and heart failure (Mendes Ribeiro et al. 2001), septic shock (Reade et al. 2002), starvation (Pico et al. 1992) and sickle cell disease (Kiessling et al. 2000; see also Chap. 27). A particular problem lies in distinguishing disease-related changes, from alterations due to reticulocytosis or changes in RBC ageing. There is considerable work identifying enhanced amino transport activity in reticulocytes, and selective shedding of amino acid transporters during reticulocyte maturation (Blostein and Grafova 1987; Johnstone et al. 1987). Nevertheless, there is also good evidence for important changes in amino acid transport in mature Hb SS RBCs, after correcting for effects of young cell populations.

Another justification for studying amino acid transport in RBCs during disease states is to reflect general transport changes which will also occur in other tissues. For example, the systemic effects of uraemic toxins could be partially mediated by membrane transport inhibition, and RBCs may represent a convenient paradigm to show this. In particular, transport of important amino acids such as arginine as an NO synthesis precursor, and cysteine as a limiting substrate for GSH synthesis may be relevant.

Malaria was omitted from the list above, since the role of the parasite in modifying membrane transport, including transport of amino acids is considered in detail in another chapter (Chap. 24). Nevertheless, the fact that digestion of haemoglobin is not an adequate source of all amino acids for parasite nutrition, emphasizes a need for enhanced amino acid transport in parasitized cells. Additionally, there is a potential role for amino acid transporters in volume regulation in infected RBCs, where export of breakdown products (i.e., amino acids) of haemoglobin is necessary to counteract cell swelling.

12.17 Conclusion

In conclusion, amino acids should be recognized as important substrates to be transported across the cell membrane for the overall functioning of the red blood cell. It is ironic that molecular biology, which has made great contributions to our understanding of the different classes and isoforms of transporters present in many cells is necessarily of limited value in the anuclear RBC. However, the study of RBC amino acid transport will ultimately derive much benefit from such advances when a library of antibodies is available to allow characterisation of individual transporters in the human RBC membrane. Specific amino acid transporters play a role in RBC metabolism and redox state and are also altered in disease states.

References

Al-Saleh EAS, Wheeler KP (1982) Transport of neutral amino acids in human erythrocytes. Biochim Biophys Acta 772:337-346

Antonioli JA, Christensen HN (1969) Cationic amino acid transport in the rabbit reticulocyte Na^+-dependent inhibition of Na^+-independent transport. J Biol Chem 244:1497-1504

Bading JR, Kan-Mitchell J, Conti PS (1996) System A amino acid transport in cultured human tumor cells: implications for tumor imaging with PET. Nucl Med Biol 23:779-786

Bannai SO, Christensen HN, Vadgama JV, Ellory JC, Englesberg E, Guidotti GG, Gazzola GC, Kilberg MS, Lajtha A, Sacktor B, Sepulveda FV, Young JD, Yudilevich D, Mann G (1984) Amino acid transport systems. Nature 311:308

Benderoff S, Blostein R, Johnstone RM (1978) Changes in amino acid transport during red cell maturation. Membr Biochem 1:89-106

Blostein R, Grafova E (1987) Characteristics of membrane transport losses during reticulocyte maturation. Biochem Cell Biol 65:869-875

Bode BP (2001) Recent molecular advances in mammalian glutamine transport. J Nutr 131:S2475-S2485

Boyd CA, Deves R, Laynes R, Kudo Y, Sebastio G (2000) Cationic amino acid transport through system y+L in erythrocytes of patients with lysinuric protein intolerance. Pflügers Arch 439:513-516

Chairoungdua A, Kanai Y, Matsuo H, Inatomi J, Kim DK, Endou H (2001) Identification and characterization of a novel member of the heterodimeric amino acid transporter family presumed to be associated with an unknown heavy chain. J Biol Chem 276:49390-49399

Changani KK, Fincham DA, Kilgour L, Tucker EM, Young JD (1990) Amino acid transport systems in sheep reticulocytes. Biochem Soc Trans 18:646-647

Chen W, Aoki C, Mahadomrongkul V, Gruber CE, Wang GJ, Blitzblau R, Irwin N, Rosenberg PA (2002) Expression of a variant form of the glutamate transporter GLT1 in neuronal cultures and in neurons and astrocytes in the rat brain. J Neurosci 22:2142-2152

Chillaron J, Roca R, Valencia A, Zorzano A, Palacin M (2001) Heteromeric amino acid

transporters: biochemistry, genetics, and physiology. Am J Physiol 281:F995-F1018
Christensen HN (1985) On the strategy of kinetic discrimination of amino acid transport systems. J Membrane Biol 84:97-103
Christensen HN, Kilberg MS (1987) Amino acid transport across the plasma membrane: role of regulation in interorgan flows. In: Yudilevich DL, Boyd CAR (eds) Amino acid transport in animal cells. Manchester University Press, Manchester, pp 10-46
Culliford SJ, Bernhardt I, Ellory JC (1995) Activiation of a novel organic solute transporter in mammalian red blood cells. J Physiol 489:755-765
Deves R, Boyd CA (1998) Transporters for cationic amino acids in animal cells: discovery, structure, and function. Physiol Rev 78:487-545
Deves R, Angelo S, Chavez P (1993) N-ethylmaleimide discriminates between two lysine transport systems in human erythrocytes. J Physiol 468:753-766
Deves R, Angelo S, Rojas AM (1998) System y+L: the broad scope and cation modulated amino acid transporter. Exp Physiol 83:211-220
Ellory JC, Osotimehin B (1983) Glutamine uptake in human erythrocytes. J Physiol 348:48P
Ellory JC, Jones SEM, Young JD (1981) Glycine transport in human erythrocytes. J Physiol 320:403-422
Ellory JC, Preston RL, Osotimehin B, Young JD (1983) Transport of amino acids for glutathione biosynthesis in human and dog red cells. Biomed Biochim Acta 42:S48-S52
Elwyn DH, Launder WJ, Parikh HC, Wise EM (1972) Roles of plasma and erythrocytes in interorgan transport of amino acids in dogs. Am J Physiol 222:1333-1342
Fievet B, Gabillat N, Borgese F, Motais R (1995) Expression of band 3 anion exchanger induces chloride current and taurine transport: structure-function analysis. EMBO J 14:5158-5169
Fincham DA, Mason DK, Paterson JY, Young JD (1987) Heterogeneity of amino acid transport in horse erythrocytes: a detailed kinetic analysis of inherited transport variation. J Physiol 389:385-409
Garcia-Romeu F, Cossins AR, Motais R (1991) Cell volume regulation by trout erythrocytes: characteristics of the transport systems activated by hypotonic swelling. J Physiol 440:547-567
Gardner JD, Levy AG (1972) Transport of dibasic amino acids by human erythrocytes. Metabolism 21:413-431
Gibson JS, Cossins AR, Ellory JC (2000) Oxygen-sensitive membrane transporters in vertebrate red cells. J Exp Biol 203:1395-1407
Glynn IM, Karlish SJ, Cavieres JD, Ellory JC, Lew VL, Jorgensen PL (1974) The effects of an antiserum to Na^+, K^+-ATPase on the ion-transporting and hydrolytic activities of the enzyme. Ann New York Acad Sci 242:357-371
Hammermann R, Brunn G, Racke K (2001) Analysis of the genomic organization of the human cationic amino acid transporters CAT-1, CAT-2 and CAT-4. Amino Acids 21:211-219
Herdon HJ, Godfrey F, Brown AM, Coulton S, Evans JR, Cairns WJ (2001) Pharmacological assessment of the role of the glycine transporter GlyT-1 in mediating high-affinity glycine uptake by rat cerebral cortex and cerebellum synaptosomes. Neuropharmacology 41:88-96
Johnstone RM, Adam M, Hammond JR, Orr L, Turbide C (1987) Vesicle formation during reticulocute maturation. Association of plasma membrane activities with released vesicles (exosomes). J Biol Chem 262:9412-9420

Kanai Y, Endou H (2001) Heterodimeric amino acid transporters: molecular biology and pathological and pharmacological relevance. Curr Drug Metab 2:339-354

Kiessling K, Roberts N, Gibson JS, Ellory JC (2000) A comparison in normal individuals and sickle cell patients of reduced glutathione precursors and their transport between plasma and red cells. Hematol J 1:243-249

King GF, Kuchel PW (1985) Assimilation of alpha-glutamyl-peptides by human erythrocytes. A possible means of glutamate supply for glutathione synthesis. Biochem J 227:833-842

King PA, Gunn RB (1991) Glycine transport by human red blood cells and ghosts: evidence for glycine anion and proton cotransport by band 3. Am J Physiol 261:C814-C821

Lang F, Busch GL, Ritter M, Volkl H, Waldegger S, Gulbins E, Haussinger D (1998) Functional significance of cell volume regulatory mechanisms. Physiol Rev 78:247-306

Lannigan DA, Knauf PA, Macara IG (1986) Relationship of the decreases in protein synthesis and intracellular Na^+ during friend murine erythroleukemic cell differentiation. J Biochem 261:14430-14436

Levy LM, Warr O, Attwell D (1998) Stoichiometry of the glial glutamate transporter GLT-1 expressed inducibly in a Chinese hamster ovary cell line selected for low endogenous Na^+-dependent glutamate uptake. J Neurosci 18:9620-9628

Maede V, Inaba M, Taniguchi N (1982) Increase of Na-K ATPase activity, glutamate and aspartate uptake in dog erythrocytes associated with hereditary high accumulation of GSH, glutamate, glutamine, and aspartate. Blood 61:493-499

Marie H, Billups D, Bedford FK, Dumoulin A, Goyal RK, Longmore GD, Moss SJ, Attwell D (2002) The amino terminus of the glial glutamate transporter GLT-1 interacts with the LIM protein Ajuba. Mol Cell Neurosci 19:152-164

Mendes Ribeiro AC, Brunini TM, Ellory JC, Mann GE (2001) Abnormalities in L-arginine transport and nitric oxide biosynthesis in chronic renal and heart failure. Cardiovasc Res 49:697-712

Nakanishi T, Kekuda R, Fei YJ, Hatanaka T, Sugawara M, Martindale RG, Leibach FH, Prasad PD, Ganapathy V (2001) Cloning and functional characterization of a new subtype of the amino acid transport system N. Am J Physiol 281:C1757-C1768

Niihara Y, Zerez CR, Akiyama DS, Tanaka KR (1997) Increased red cell glutamine availability in sickle cell anemia: demonstration of increased active transport, affinity, and increased glutamate level in intact red cells. J Lab Clin Med 130:83-90

Niihara Y, Zerez CR, Akiyama DS, Tanaka KR (1998) Oral L-glutamine therapy for sickle cell anemia: I. Subjective clinical improvement and favorable change in red cell NAD redox potential. Am J Hematol 58:117-121

Ohkame H, Masuda H, Ishii Y, Kanai Y (2001) Expression of L-type amino acid transporter (4F2hc) in liver tumor lesions of rat models. J Surg Oncol 78:265-271

Orlowski M, Meister A (1970) The gamma-glutamyl cycle: a possible transport system for amino acids. Proc Natl Acad Sci USA 67:1248-1255

Palacin M, Estevez R, Bertran J, Zorzano A (1998) Molecular biology of mammalian plasma membrane amino acid transporters. Physiol Rev 78:969-1054

Pico C, Pons A, Palou A (1992) Regulation of rat erythrocyte L-glutamine, L-glutamate and L-lysine uptake by short-term starvation. Int J Biochem 24:1731-1735

Pow DV, Sullivan R, Reye P, Hermanussen S (2002) Localization of taurine transporters, taurine, and (3)H taurine accumulation in the rat retina, pituitary, and brain. Glia 37:153-168

Reade MC, Clark MF, Young JD, Boyd CA (2002) Increased cationic amino acid flux through a newly expressed transporter in cells over producing nitric oxide from patients with septic shock. Clin Sci 102:645-650

Rosenberg R, Young JD, Ellory JC (1980) L-tryptophan transport in human red blood cells. Biochim Biophys Act 598:375-384

Rosenberg R (1981) L-leucine transport in human red blood cells. A detailed kinetic analysis. J Membrane Biol 62:79-93

Sato K, Inaba M, Suwa Y, Matsuu, A, Hikasa Y, Ono K, Kagota K (2000) Inherited defects of sodium-dependent glutamate transport mediated by glutamate/aspartate transporter in canine red cells due to a decreased level of transporter protein expression. J Biol Chem 275:6620-6627

Stein WD (1986) Transport and diffusion across cell membranes. Academic Press, London

Suliman ME, Divino Filho JC, Barany P, Anderstam B, Lindholm B, Bergstrom J (1999) Effects of high-dose folic acid and pyridoxine on plasma and erythrocyte sulfur amino acids in hemodialysis patients. J Am Soc Nephrol 10:1287-1296

Suwanmanee T, Sierakowska H, Lacerra G, Svasti S, Kirby S, Walsh CE, Fucheroen S, Kole R (2002) Restoration of human beta-globin gene expression in murine and human IVS2-654 thalassemic erythroid cells by free uptake of antisense oligonucleotides. Mol Pharmacol 62:545-553

Tse CM, Fincham DA, Ellory JC, Young JD (1991) Use of membrane vesicles to estimate the numbers of system y+ and system L amino acid transporters in human erythrocytes. Biochem J 277:565-568

Vadagama JV, Chan MN, Wu JM (1991) Differential expression of amino acid transport systems A and ASC during erythroleukemia cell differentiation. Am J Physiol 260:C392-C399

Vidaver GA, Shepherd SL (1968) Transport of glycine by haemolyzed and restored pigeon red blood cells. J Biol Chem 243:6140-6150

Wang H, Kavanaugh MP, North RA, Kabat D (1991) Cell-surface receptor for ecotropic murine retroviruses is a basic amino-acid transporter. Nature 352:729-731

Young JD (1980) Effects of thiol-reactive agents on amino acid transport by sheep erythrocytes. Biochim Biophys Acta 602:661-672

Young JD, Ellory JC (1977) Red cell amino acid transport. In: Ellory JC, Lew VL (eds) Membrane transport in red cells. Academic Press, London, pp 301-325

Young JD, Ellory JC (1982) Flux measurements. In: Ellory JC, Young JD (eds) Red cell membranes – a methodological approach. Academic Press, London, pp 119-134

Young JD, Ellory JC, Tucker EM (1975a) Amino acid transport defect in glutathione-deficient sheep erythrocytes. Nature 254:156-157

Young JD, Ellory JC, Wright PC (1975b) Evidence against the participation of the γ-Glutamyltransferase-γ-Glutamylcyclotransferase pathway in amino acid transport by rabbit erythrocytes. Biochem J 152:713-715

Young JD, Jones SEM, Ellory JC (1980) Amino acid transport in human and sheep erythrocytes. Proc Roy Soc London 209:355-375

Young JD, Jones SE, Ellory JC (1981) Amino acid transport via the red cell anion transport system. Biochim Biophys Acta 645:157-160

Zhou Y, Samson M, Francon J, Blondeau JP (1992) Thyroid hormone concentrative uptake in rat erythrocytes. Involvement of the tryptophan transport system T in countertransport of tri-iodothyronine and aromatic amino acids. Biochem J 281:81-86

13 Equilibrative Nucleoside Transport Proteins

James D. Young[1], Sylvia Y. M. Yao[2], Carol E. Cass[3] and Stephen A. Baldwin[4]

Membrane Protein Research Group, Departments of [1,2]Physiology and [3]Oncology, University of Alberta, and [3]Cross Cancer Institute, Edmonton, Alberta T6G 2H7, Canada, and the [4]School of Biochemistry and Molecular Biology, University of Leeds, Leeds, LS2 9JT, UK

13.1 Introduction

In human and other mammalian cells, plasma membrane transport of nucleosides is brought about by members of the concentrative (Na^+-dependent) nucleoside transporter (CNT) and equilibrative (Na^+-independent) nucleoside transporter (ENT) families (Hyde et al. 2001; Ritzel et al. 2001). CNTs have been described primarily in specialized epithelia, whereas ENTs occur in most, possibly all, cell types and tissues, including red blood cells (RBCs).

Three mammalian ENT isoforms (ENT1, ENT2 and ENT3) have been identified (Hyde et al. 2001). Human (h) and rat (r) ENT1 (formerly the *es* nucleoside transport process) and ENT2 (formerly the *ei* nucleoside transport process) are broadly selective for purine and pyrimidine nucleosides and are distinguished by a difference in sensitivity to inhibition by nitrobenzylthioinosine (NBMPR), h/rENT2 being NBMPR-insensitive. They also differ in sensitivity to inhibition by the coronary vasodilator drugs dipyridamole, dilazep and draflazine (hENT1 > hENT2 > rENT1 = rENT2) and by the ability of the ENT2 isoform to transport nucleobases in addition to nucleosides. Mouse orthologs of human and rat ENT1 and ENT2 have also been identified (mENT1, mENT2), as well as human and mouse versions of a third mammalian ENT family member, ENT3. hENT3 and mENT3 may have a role in nucleoside transport across intracellular membranes (Hyde et al. 2001).

Nucleoside uptake by RBCs is an important component of adenosine homeostasis in blood plasma (Beutler 2001) and is the obligatory first step of RBC intracellular nucleoside metabolism (Murray 1971). Roles of nucleoside transport and metabolism in RBCs revealed by studies of genetic variation in human and sheep RBCs include maintenance of adenine nucleotide pools (via adenosine salvage) and utilization of plasma inosine (and adenosine) in addition to glucose as a source of metabolic energy (Murray 1971; Young 1978; Tucker and Young 1988; Beutler 2001). Pig RBCs, which are impermeable to glucose, rely exclusively on inosine as their energy substrate (Young et al. 1985). *In vivo* pharmacological blockade of nucleoside transport activity in these cells using a soluble pro-drug form of NBMPR leads to a rapid decline in cellular ATP (t½ < 1 h) and corresponding elevation of intracellular ADP and AMP (Young et al. 1986).

Historically, RBCs have played a central role in functional studies of equilibrative nucleoside transport mechanisms, beginning in 1971 with the original demonstration in human RBCs of carrier-mediated nucleoside transport activity and the parallel discovery of the first high-affinity nucleoside transport inhibitor, nitrobenzylthioguanosine (NBTGR) (Oliver and Paterson 1971; Paterson and Oliver 1971). More recently, RBCs have also played a central role in molecular studies of nucleoside transporter proteins, culminating in 1997 with the cDNA cloning of human and other mammalian ENT family members. This chapter describes the research that led to these and subsequent molecular advances in nucleoside transport biology, focussing on the involvement of RBCs and the contributions of our laboratories to the discovery process. Earlier reviews provide good in depth coverage of functional studies of RBC and other nucleoside transport processes (Jarvis and Young 1983; Gati and Paterson 1989; Cass 1995; Griffith and Jarvis 1996).

13.2 Biochemical Studies of Red Cell Nucleoside Transport Proteins

Red blood cells from four species (human, pig, rat and sheep) have played a prominent role in biochemical studies of nucleoside transport proteins. While human and pig RBCs transport nucleosides exclusively by *es* (ENT1), rat RBCs exhibit both *es* (ENT1) and *ei* (ENT2) functional activity (Jarvis and Young 1986). Sheep are polymorphic with respect to RBC nucleoside transport activity, some animals having nucleoside transport-positive RBCs, others having RBCs deficient in nucleoside transport activity (Young 1978; Tucker and Young 1988). Approximate numbers of *es* (ENT1) nucleoside transporters per RBC in the four species, determined by measurements of high-affinity NBMPR binding, are 10^4 (human), $5{\cdot}10^3$ (pig), $4{\cdot}10^2$ (rat) and 20 (nucleoside transport-positive sheep).

Evidence that high-affinity NBMPR binding to RBC membranes represents a specific interaction with *es* nucleoside transporters includes proportionality between NBMPR binding and transport inhibition, competition by nucleoside permeants, and absence of sites from the membranes of nucleoside transport-deficient sheep RBCs (Cass et al. 1974; Cass and Paterson 1976; Jarvis and Young 1980a).

13.2.1 Identification of the Red Cell Nucleoside Transporter as a Band 4.5 Protein

As described elsewhere in this volume, human RBC membranes possess large numbers of AE1 Cl^-/HCO_3^- anion-exchange and GLUT1 glucose transport proteins ($\sim 10^6$ and $2{\cdot}10^5$ copies per cell, respectively). AE1, representing the major integral membrane protein of RBC membranes, corresponds to band 3 on SDS/polyacrylamide gels (average apparent M_r 95,000), while GLUT1 is a component of band 4.5 (average apparent M_r 55,000). Detergent-solubilized preparations of human RBC membranes depleted of extrinsic membrane proteins consist largely of band 3 with smaller amounts of band 4.5 and some other proteins. In the first molecular cloning of a mammalian transport protein, Kopito and Lodish (1985) used a band 3-specific polyclonal antibody to screen a reticulocyte cDNA library and isolate cDNA containing the complete coding sequence of mouse AE1. Correspondence between the biochemical properties of band 3 and the predicted amino acid sequence of the encoded protein was used to confirm the identity of AE1 as the RBC Cl^-/HCO_3^- anion-exchange transporter. Band 4.5-specific polyclonal antibodies, prepared from band 4.5 material separated from band 3 by ion-exchange chromatography on DEAE-cellulose, were subsequently used to clone the first mammalian glucose transport protein (human GLUT1) from a HepG2 cDNA library (Mueckler et al. 1985). The identity of hGLUT1 as the RBC glucose transporter was confirmed by fast atom bombardment mapping and gas-phase sequencing.

Efforts to provide biochemical information on the *es* nucleoside transport process began in human and nucleoside transport-positive sheep RBCs with solubilization studies of high-affinity NBMPR-binding sites (Jarvis and Young 1980b). This was followed by the demonstration that NBMPR-binding activity was enriched in preparations of human band 4.5 proteins (Jarvis and Young 1981). These studies implicating band 4.5 proteins in nucleoside transport were given impetus by the demonstration that [^{3}H]NBMPR also functioned as a photoaffinity probe, and by the finding that the resulting radiolabelled *es* nucleoside transporter co-migrated with [^{3}H]cytochalasin B photoaffinity-labelled glucose transporter in the band 4.5 region of SDS/polyacrylamide gels (Wu et al. 1983a,b; Young et al. 1983). The specificity of photo-incorporation was established through competition studies with NBTGR and nucleoside permeants, and by the absence of labelling in membranes prepared from nucleoside transport-deficient sheep RBCs. The ability of band 4.5 proteins to mediate NBMPR-sensitive nucleoside transport when reconstituted into proteoliposomes provided direct evidence for their role in nucleoside permeation (Tse et al. 1985b).

NBMPR photoaffinity-labelling studies in other mammalian cell types and tissues demonstrated that putative *es* nucleoside transport protein(s) with similar characteristics were widely distributed (e.g., Almedia et al. 1984; Shi et al. 1984; Wu and Young 1984). Photoaffinity-labelling studies also established that the *es* transporters of human and pig RBCs differed substantially in size (Wu et al. 1983b). In human RBC membranes, NBMPR-photolabelled polypeptides migrated on SDS/polyacrylamide gels as a broad band of average apparent M_r 55,000, which could be converted by deglycosylation with endoglycosidase-F to a species that migrated as a sharp band of apparent M_r 45,000 (Kwong et al. 1986). By comparison, the photolabelled pig RBC *es* transporter migrated with apparent M_r values of 64,000 and 57,000 before and after deglycosylation, respectively (Kwong et al. 1986). The photolabelled transporter from rat RBCs and other tissues also migrated as a larger protein (apparent M_r 62,000) (Shi et al. 1984; Wu and Young 1984; Jarvis and Young 1986). In this case, however, deglycosylation produced a protein that co-migrated with the deglycosylated human *es* transporter, suggesting that the initial difference in apparent M_r was a consequence of more extensive glycosylation of the rat protein.

13.2.2 Production of Nucleoside Transporter Antibodies

To generate antibody probes, it is first necessary to obtain a suitable immunogen. Pig RBCs were chosen as the source because they are relatively rich in *es* nucleoside transporters and because, unlike human RBCs, they lack a glucose transporter (the major constituent of human band 4.5 preparations). After removal of extrinsic membrane proteins, integral membrane proteins solubilized in the nonionic detergent octyl glucoside were fractionated by gradient-elution ion-exchange chromatography on DEAE-cellulose to yield a preparation of band 4.5 membrane proteins (Kwong et al. 1987). Although by no means pure, this material proved to be sufficiently enriched in nucleoside transporters to enable the successful preparation of transporter-specific monoclonal antibodies (Good et al. 1987). The resulting antibodies (11C4) recognized polypeptides of a single species of apparent M_r 64,000 on Western blots of both adult and neonatal pig RBC membranes (Good et al. 1987; Craik et al. 1988). Antibody specificity was confirmed by the ability to immunoprecipitate [^{3}H]NBMPR-photolabelled protein (Good et al. 1987). As described in the next section, these antibodies provided a means to further purify the pig RBC *es* transporter. It was not possible to raise *es* transporter-specific antibodies using human RBC band 4.5 membrane proteins as immunogen, because the transporter represented only a very minor component of this fraction (see below).

13.2.3 Purification of the Human and Pig Red Cell Nucleoside Transporters

In human RBCs, efforts to isolate the *es* nucleoside transporter were hampered by its low membrane abundance relative to the glucose transporter (10^4 and $2 \cdot 10^5$ copies per cell, respectively), and by the complication that glucose and nucleoside transporters co-purified during ion-exchange chromatography (Jarvis and Young 1981). Because of this, nucleoside transporter represented ~5% of purified human band 4.5 preparations. This technical difficulty was overcome by separating the two band 4.5 proteins on a column of immobilized glucose transporter antibodies (Kwong et al. 1988). The resultant preparation was demonstrated by its NBMPR binding content to be at least 60% pure, and the absence of glucose transporter was confirmed by cytochalasin B binding assays and reconstitution of transport activity into proteoliposomes. Polyclonal antibodies raised in rabbits against this material proved to be *es* transporter-specific: on Western blots of human RBC membranes they labelled a single band of apparent M_r 55,000 that was identical in mobility to material photolabelled by NBMPR (Kwong et al. 1992). In immunocytochemical investigations, these antibodies recognized immunologically crossreactive *es* nucleoside transporters at the syncytiotrophoblast brush border membrane surface in sections of human placenta (Barros et al. 1995). As described below, placenta was subsequently used as the tissue from which to clone the human *es* transporter.

In parallel with these studies, further purification of the *es* transporter from pig RBC band 4.5 proteins was achieved by adsorption of the nucleoside transporter onto a column of immobilized 11C4 monoclonal antibodies (Beaumont et al.

1995). This resulted in the removal of several unbound band 4.5 proteins, yielding a purified nucleoside transport protein that migrated as a broad band of apparent M_r 64,000 on SDS/polyacrylamide gels with a pattern of silver staining exactly matching that given by 11C4 monoclonal antibodies on Western blots (Beaumont et al. 1995).

When subjected to gas-phase sequence analysis, the purified human protein generated 21 residues of amino-terminal amino acid sequence (Beaumont et al. 1995). The first 11 residues were hydrophilic in character and were followed by a segment of hydrophobic residues typical of a membrane-spanning segment in an integral membrane protein. Sequence analysis of the purified pig protein also generated 21 residues of amino-terminal sequence. The human and pig sequences were identical, apart from a single conservative substitution in the hydrophilic region (Beaumont et al. 1995). We therefore concluded that the two sequences obtained by alternative positive and negative purification procedures in two different species belonged to orthologs of the same protein. As described in a following section, this sequence information was used to isolate a human placental cDNA encoding the full length *es* nucleoside transport protein (hENT1).

13.2.4 Tryptic Cleavage Studies

The human nucleoside transporter antibodies were also used in combination with NBMPR-photolabelling, trypsin cleavage and endoglycosidase-F digestion to provide molecular information on the sites of NBMPR and carbohydrate attachment to the purified human RBC *es* nucleoside transporter (Kwong et al. 1993). It was established that the site(s) of *N*-glycosylation were very close to one end of the protein, and that the site of NBMPR-photolabelling was within 16 kDa of that site.

13.3 cDNA Cloning of ENT Nucleoside Transport Proteins

13.3.1 cDNA Cloning and Heterologous Expression of Recombinant Human and Rat ENT1

The sequence of the amino-terminal 21 residues of the human RBC *es* nucleoside transporter, determined by gas-phase sequencing of the purified protein, was TTSHQPQDRYKAV?LIFFMLG, where "?" indicates an unidentified residue (Beaumont et al. 1995; Griffiths et al. 1997a). A specific DNA encoding residues 8 to 15 of the sequence was amplified from a human placental cDNA library using a 5' degenerate primer corresponding to residues 1–7 (plus an additional amino-terminal methionine) and a 3' degenerate primer corresponding to residues 16–21. Sequencing of this PCR product showed that it contained the amino acids expected for positions 8 to 15 in the amino-terminal sequence of the transporter, and identified position 14 as tryptophan. Database searching showed the presence of an identical nucleotide sequence in a ~250 bp EST from human Burkitt's lymphoma. With this information, it was possible to amplify a human placental cDNA product containing the complete coding sequence of the transporter by nested PCR. Assignment of the translation initiation site of the open reading frame was based on its resemblance to the consensus sequence described by Kozak (1989), and by exact correspondence between the deduced amino-terminal sequence and that determined for the purified RBC protein, except that the latter lacked the initial methionine residue.

The encoded 456-residue protein (M_r 50,259), designated hENT1, was predicted to contain 11 transmembrane (TM) segments connected by short (≤ 16 residue) hydrophilic regions, with the exception of the loops linking TMs 1 and 2, and TMs 6 and 7, which were predicted to contain 41 and 66 residues, respectively. These large loops contained the three potential *N*-glycosylation sites in the protein (one in the first loop, and two in the second). The previous observation by Kwong et al. (1993) that the human RBC nucleoside transporter was glycosylated very close to one end of the protein suggested that only the first of the three sites was glycosylated and led to the putative topography illustrated in Fig. 13.1. When produced in *Xenopus* oocytes, recombinant hENT1 exhibited uridine transport activity with all of the characteristics expected of an *es*-type transporter, including (i) cross-inhibition by both pyrimidine and purine nucleosides and anticancer nucleoside drugs, and (ii) inhibition by NBMPR, dipyridamole and dilazep. The apparent

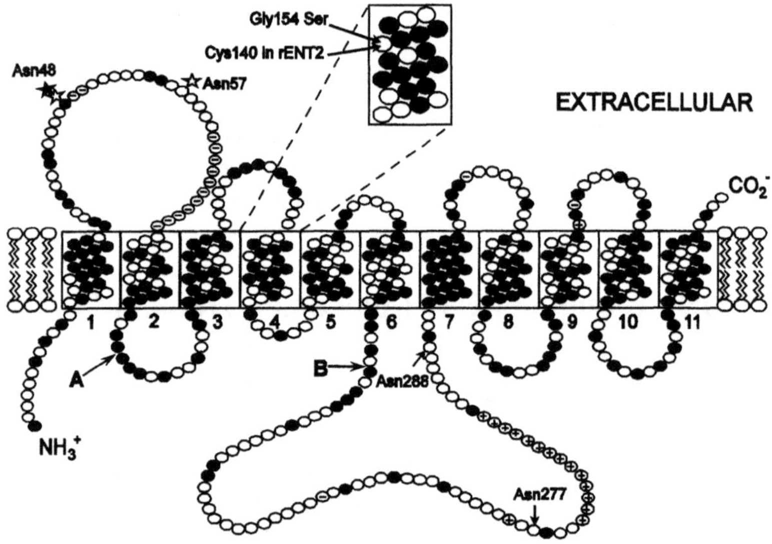

Fig. 13.1. Membrane topology model of hENT1 and hENT2. Potential membrane-spanning α-helices are *numbered,* and glycosylation sites in hENT1 and hENT2 are indicated by *solid* and *open stars*, respectively. Residues identical in the two proteins are shown in *solid circles*. Residues corresponding to insertions in the sequence of hENT1 or hENT2 are indicated by *circles* containing "–" and "+" signs, respectively. Splice sites used for construction of hENT1/hENT2 chimeras are represented by *arrows A* and *B*. The same splice sites were utilized in construction of rENT1/rENT2 chimeras. As well, an additional rENT1/rENT2 spice site was introduced in the middle of the loop linking TMs 4 and 5

K_m value for uridine influx was 0.2 mM and the apparent K_i for NBMPR inhibition was 2 nM (Griffiths et al. 1997a). Consistent with the apparently ubiquitous cell and tissue distribution of *es* transport activity, hENT1 mRNA was broadly distributed, and cDNA encoding a protein identical to hENT1 was subsequently isolated from human erythroleukemia (K562) cells (Boleti et al. 1997). The hENT1 gene was localized to chromosome 6p21.1-p21.2 (Coe et al. 1997).

The rat homolog of hENT1, isolated from a rat jejunal cDNA library (Yao et al. 1997), was a 457-residue protein that was 78% identical in sequence to hENT1. Consistent with earlier photoaffinity labelling studies suggesting that rENT1 was more heavily glycosylated than hENT1, the predicted extracellular loop between TMs 1 and 2 of rENT1 contained three potential sites of *N*-glycosylation (rather than the single site of hENT1). Consistent also with previous functional studies of native rat *es*, recombinant rENT1 exhibited NBMPR-sensitive uridine transport activity when produced in *Xenopus* oocytes (uridine apparent K_m ~ 0.2 mM), but was unaffected by high concentrations of dipyridamole and dilazep (Yao et al. 1997).

13.3.2 cDNA Cloning and Heterologous Expression of Recombinant Human and Rat ENT2

In database searches, it was found that the carboxyl-terminal two thirds of the hENT1 sequence showed strong similarity (~44% identity) to 36 kDa human and mouse HNP36 proteins, which are delayed-early proliferative response gene products (Williams and Lanahan 1995). These HNP36 proteins were reported to have a nucleolar location and were predicted to have only 8 TMs. However, the nucleotide sequence upstream of the assigned start codon in human HNP36 cDNA included an open reading frame that was 49% identical in predicted amino acid sequence to the amino-terminal region (TMs 1 and 2) of hENT1. This suggested that a continuous full-length open reading frame might encode a second mammalian equilibrative nucleoside transporter isoform. The hypothesis was tested by PCR amplification, resulting in the isolation of rENT2 cDNA from rat jejunum (Yao et al. 1997) and, shortly thereafter, hENT2 cDNA from human placenta (Griffiths et al. 1997b). Production of the encoded 456 residue proteins (designated rENT2 and hENT2, respectively) in *Xenopus* oocytes revealed uridine transport activity with characteristics expected of rat and human *ei* (Griffiths et al. 1997b). Thus, both were NBMPR-insensitive. The human protein exhibited intermediate sensitivity to inhibition by dipyridamole, dilazep and draflazine, while the rat protein, like rENT1 was unaffected by vasodilator drugs. Apparent K_m values for uridine influx (0.2–0.3 mM) were similar to those for hENT1 and rENT1 (Griffiths et al. 1997b), although K_m values for guanosine, cytidine and the deoxycytidine analog gemcitabine were substantially higher (Mackey et al. 1999; Ward et al. 2000). Human ENT2 was also independently cloned from cultured HeLa cells by expression selection in a nucleoside transport-deficient human leukaemia cell line (Crawford et al. 1998). Human ENT2 and rENT2 share the predicted 11 TM topology shown for hENT1 in Fig. 13.1 and have three and two potential *N*-glycosylation sites, respectively, in the loop linking TMs 1 and 2. Consistent with the broad distribution of *ei* transport activity, hENT2 message was found in many different cells and tissues and in variable levels, with the greatest quantities in skeletal muscle. The gene for hENT2 has been localized to chromosome 11q13 (Williams et al. 1997).

While hENT1 and rENT1 are nucleoside-specific, we have recently established through expression studies in *Xenopus* oocytes and the yeast *Saccharomyces cerevisiae* that recombinant hENT2 and rENT2 also transport nucleobases (Yao et al. 2002). Therefore, while there is clear functional evidence that many cells and tissues (including RBCs) also possess nucleobase-specific uptake mechanisms (Griffith and Jarvis 1996; De Koning and Diallinas 2000), hENT2 and rENT2 represent the first identified mammalian nucleobase transport proteins. This dual capability of the ENT2 isoform to transport both nucleosides and nucleobases provides a possible rationale why many cell types co-express both *es* and *ei* functional activities.

Like the human and rat ENT1 isoforms and members of the CNT family of concentrative nucleoside transporters, hENT2 and rENT2 transport a broad range of pyrimidine and purine nucleoside analogs used in cancer chemotherapy (e.g., Hyde et al. 2001; Ritzel et al. 2001). In addition, they exhibit a greater ability than ENT1 to transport antiviral 3'-deoxy-nucleoside drugs such as AZT, ddC and ddI (Yao et al. 2001a).

13.3.3 Other ENT Family Members

Mammals

Other mammalian ENTs that have been cloned include mouse ENT1 and mouse ENT2 (Kiss et al. 2000). In addition, database searches have recently identified human and mouse versions of a third mammalian ENT isoform (Hyde et al. 2001). hENT3 and mENT3 cDNAs were isolated from placenta and kidney, respectively, and encode 475 and 474 residue proteins, respectively, with 30–33% identity to human, rat and mouse ENT1 and ENT2. The transport characteristics of hENT3 and mENT3 have not yet been established. Both, however, appear to have a wide tissue distribution and have the characteristic ENT membrane topology of 11 predicted TMs. One striking difference is that the ENT3 hydrophilic amino-terminal region proceeding TM1 is much longer than in ENT1/2. This region contains dileucine motifs consistent with targeting to intracellular membranes and suggests that ENT3 might function as an intracellular transporter. One ENT that has not yet been cloned is that corresponding to the pig RBC transporter.

Hagfish (an Ancient Marine Pre-Vertebrate)

Hagfish are jawless eel-shaped animals that diverged from the main line of vertebrate evolution ~550 million years ago. They are the most ancient extant craniate (the subphylum that also includes humans) and represent the transition between early chordates and the first true vertebrates. Hagfish are highly conserved and represent a unique research resource in molecular studies of vertebrate evolution. Functional studies of nucleoside transport in hagfish RBCs have shown that they possess an *ei*-type mechanism (Fincham et al. 1991). We have recently cloned a novel hagfish CNT family member (Loewen et al. 1999; Ritzel et al. 2001) and are currently also searching for hagfish ENTs.

Other Eukaryotes

ENTs represent a relatively recent protein family in evolutionary terms. They are widely distributed in lower eukaryotes but, unlike CNTs, have not yet been found in prokaryotes (Hyde et al. 2001; Ritzel et al. 2001). To date, ENTs have been identified in insects, nematodes, apicomplexan protozoans (including *Plasmodium*), kinetoplastid protozoans, yeast and plants. A number of the protozoan ENTs are known or predicted to be proton symporters (Hyde et al. 2001).

13.4 Molecular Properties of Recombinant Mammalian ENT Proteins

13.4.1 ENT Membrane Topology

The topology of hENT1, rENT1, hENT2 and rENT2, predicted by hydropathy analysis of their amino acid sequences using the algorithm of Eisenberg et al. (1984) has, as shown in Fig. 13.1, 11 putative TMs with an amino-terminus that is intracellular and a carboxyl-terminus that is extracellular. Other features of the topology are the large loops linking TMs 1 and 2 and TMs 6 and 7. The former is predicted to be extracellular and contains one of three hENT1 *N*-glycosylation acceptor sites. The latter is proposed to be intracellular and contains the remaining two glycosylation acceptor sites. We have recently combined various computer predictions of membrane topology with glycosylation scanning mutagenesis to provide an in depth analysis of the membrane topology of hENT1 and other ENT family members (Sundaram et al. 2001a).

Application of the TMHMM (Sonnhammer et al. 1998), TMAP (Persson and Argos 1994) and PHDhtm algorithms (Rost et al. 1996) to the sequences of hENT1 and 33 other mammalian, insect, nematode, protozoan, yeast and plant ENT family members led to the prediction of an 11 TM topology common to all of the proteins. For hENT1, the locations of the 11 TMs were almost identical to those shown in Fig. 13.1. Additional support for the locations of the proposed TMs was provided by the observation that both insertions and deletions were present in the aligned sequences of each of the proposed loops linking the TMs, except for that between TMs 4 and 5. The predicted shortness of that loop, the fact that it never naturally has an insertion or deletion, and the observation that it contains a highly conserved proline residue, suggests that it plays an important structural role in the ENT family.

Glycosylation scanning mutagenesis in *Xenopus* oocytes was used to determine experimentally the orientation of seven key topographical landmarks of recombinant hENT1 membrane architecture (the amino-terminus, two exofacial loops in the amino-terminal half of the protein (including that between TMs 1 and 2), the central cytoplasmic loop, two exofacial loops in the carboxyl-terminal half of the protein, the carboxyl terminus) and, in so doing, define the overall topology of the protein as a whole. The results validated the topology model of hENT1 shown in Fig. 13.1. They were also consistent with fusion-protein and synthetic anti-peptide antibody studies in RBCs and isolated cardiac myocytes that independently established the intracellular location of the central cytoplasmic loop between TMs 6 and 7. Previous peptide mapping experiments with the native *es* transporter in human RBCs and rat liver membranes found a common trypsin cleavage site situated in an intracellular domain at the approximate centre of both proteins (Kwong et al. 1993). It is likely that the site is located in this loop.

The glycosylation scanning mutagenesis studies also confirmed Asn^{48} in the loop between TMs 1 and 2 as the only endogenous site of *N*-linked glycosylation in wild-type hENT1. The location of this glycosylation site near the amino-terminus of the protein is, as discussed earlier, consistent with enzymic cleavage studies in RBCs (Kwong et al. 1993). The aglyco form of hENT1 (hENT1/N48Q) was functional in *Xenopus* oocytes (Sundaram et al. 2001a) but when expressed in yeast, exhibited a reduced binding affinity for NBMPR (Vickers et al. 1999).

13.4.2 Chimeric Studies

Differences in transport inhibition between vasoactive drug-sensitive hENT1 and vasoactive drug-insensitive rENT1 and between NBMPR-sensitive rENT1 and NBMPR-insensitive rENT2 have been exploited in chimeric studies of recombinant transporters produced in *Xenopus* oocytes to reveal structural domains involved in inhibitor binding (Sundaram et al. 1998, 2001b). These studies of chimeras produced with various hENT1/rENT1 and rENT1/rENT2 combinations, identified the TM 3–6 region of the transporter as the major site of both dipyridamole and NBMPR binding (splice site locations are shown in Fig. 13.1). Similarly, insertion of TMs 3–6 from rENT2 (which transports nucleobases as well as nucleosides) into rENT1 (which does not transport nucleobases) resulted in acquisition of nucleobase transport activity (Yao et al. 2002). Within the TM 3–6 region, TMs 3–4 and TMs 5–6 both contributed to NBMPR binding, while TMs 5–6 were sufficient to confer nucleobase transport capability (Yao et al. 2002). The significance of these observations in relation to other recent studies of the ENT translocation pore is discussed below.

13.4.3 Identification of an Exofacial Cysteine Residue within the Rat ENT2 Translocation Pore

PCMBS (*p*-chloromercuriphenyl sulphonate) is a membrane-impermeant reagent that reacts selectively and reversibly with protein cysteine residues. It acts as an inhibitor of *es* nucleoside transport activity in RBCs from various species (human, rat, sheep), but only in unsealed membranes or inside-out vesicles where PCMBS has access to the interior membrane surface (Jarvis and Young 1982, 1986; Tse et al. 1985a). In intact cells and right-side-out membrane vesicles, the organomercurial has no effect. In contrast, *ei* nucleoside transport activity in rat RBCs shows sensitivity to PCMBS inhibition in intact cells (Jarvis and Young 1986). Loss of transport function was prevented by extracellular uridine, suggesting that PCMBS caused transport inhibition by reacting with exofacial cysteine residue(s) within or closely adjacent to the substrate-translocation channel of the transporter. In experiments comparing the effects of PCMBS on recombinant hENT1, rENT1, hENT2 and rENT2 produced individually in *Xenopus* oocytes, it was found that rENT2 was uniquely sensitive to inhibition by PCMBS added to the extracellular medium (Yao et al. 2001b). In agreement with the earlier rat RBC experiments, extracellular uridine was able to protect recombinant rENT2 against PCMBS inhibition.

Human ENT1 and hENT2 each contain 10 cysteine residues, whereas rENT1 and rENT2 contain 15 and 11, respectively. Of the 11 cysteine residues in rENT2, only one (Cys^{140}) is not conserved in one or more of the other ENTs. This cysteine residue, which is located in the outer half of TM 4, is therefore the only candidate residue that could be responsible for the unique exofacial sensitivity of recombinant rENT2 (and rat RBC *ei*) to inhibition by PCMBS. Involvement of Cys^{140} in PCMBS inhibition of rENT2 was tested experimentally by site-directed mutagenesis. Conversion of rENT2 Cys^{140} into serine (the residue at this position in rENT1) gave a functional mutant protein (rENT2/C140S) with uridine transport activity similar to wild-type recombinant rENT1 and rENT2. Unlike wild-type rENT2, however, the mutant protein was no longer PCMBS-sensitive (Yao et al. 2001b).

Therefore, these experiments identify Cys^{140} as the cysteine residue responsible for the unique sensitivity of rENT2 to inhibition by extracellular PCMBS. This residue is located in the outer half of a transmembrane helix (TM4) that is within a region of the transporter (TMs 3–6) implicated independently by chimeric studies to be involved in transport inhibition of ENT proteins by NBMPR and dipyridamole (Sundaram et al. 1998, 2001b). It has also been established recently that mutation of Gly^{154} in hENT1 to the corresponding residue in hENT2 converts the transporter from NBMPR-sensitive to NBMPR-insensitive (Hyde et al. 2000). The latter residue (Gly^{154}) occupies the same position in TM 4 of hENT1 as Cys^{140} in rENT2 (Fig. 13.1). The finding that binding of PCMBS to rENT2 Cys^{140} is prevented by uridine suggests that this residue and the helix to which it belongs (TM 4) form part of the nucleoside-translocation channel of the transporter. The observation from chimeric studies that TMs 5 and 6 are involved in nucleobase transport (Yao et al. 2002) provides additional evidence that parts of the TM 3–6 structural domain contribute to the translocation channel of the transporter, and argues against the possibility that uridine protects Cys^{140} against PCMBS inhibition through long-range, substrate-induced conformational changes.

The finding that parts of the nucleoside/nucleobase transport machinery colocalize with the sites of NBMPR and dipyridamole binding in TMs 3–6 provides a structural basis to interpret previous functional and molecular studies of native RBC *es* nucleoside transporters. Kinetic and ligand binding experiments, for example, have shown that NBMPR (a substrate analog) and vasoactive drugs function as competitive inhibitors of zero-*trans* nucleoside influx and interact with the transporter at common or overlapping exofacial sites (Jarvis et al. 1982; Jarvis 1986; Agbanyo et al. 1988). Also, as discussed previously, studies of the human

RBC transporter have established that the site of covalent attachment of NBMPR is within 16 kDa of the site of *N*-glycosylation (now identified as Asn^{48} in the loop between TMs 1 and 2).

Since PCMBS has access only to the external membrane surface, and since uridine protection against PCMBS inhibition occurs under conditions where there is minimal intracellular accumulation of substrate, the identification of Cys^{140} as the residue in rENT2 responsible for PCMBS inhibition provides validation for the predicted membrane architecture of TM 4 and adjacent helices.

13.5 Conclusions

Just as RBCs played a central role in the cDNA cloning and characterization of the first members of the AE and GLUT families of membrane transport proteins, the studies reviewed in this chapter reveal an equally important role of RBCs in the molecular identification, cDNA cloning and characterization of members of the ENT protein family. In addition to mammals, ENTs are present in a diverse range of eukaryotes including insects, nematodes, protozoa, yeast and plants. The finding that all these proteins have a common membrane architecture sets the stage for detailed studies of ENT structure and function.

In their topology, ENTs are strikingly similar to members of the GLUT family of sugar transporters, although the latter are predicted to have an additional TM at the carboxyl-terminus. For example, the human glucose transporter GLUT1 has a cytoplasmic amino-terminus, a large glycosylated loop connecting TMs 1 and 2, and a large cytoplasmic loop between TMs 6 and 7. The finding from biochemicall studies that the human RBC glucose and nucleoside transporters were both band 4.5 polypeptides and the fact that the two transporters co-migrated during ion-exchange chromatography on DEAE cellulose fuelled speculation that the two proteins might be related. Although ENTs and GLUTs, which belong to the major facilitator superfamily of proteins (Pau et al. 1998), show no obvious sequence similarities, they do share some functional parallels. For example, in human RBCs, which possess both ENT1 and GLUT1, the potent sugar transport inhibitors cytochalasin B and phloretin weakly inhibit uridine fluxes, while conversely the nucleoside transport inhibitor dipyridamole weakly inhibits sugar transport (Plagemann and Woffendin 1987). Moreover, adenosine inhibits hexose transport in these cells in a concentration-dependent manner (Plagemann and Woffendin 1987). It has also been shown that 8-azidoadenosine photoaffinity labels the glucose transporter (Jarvis et al. 1986). These observations suggest similarities between the substrate/inhibitor binding sites of the two transporters. It remains to be seen whether these functional and topological similarities reflect similar helix-packing arrangements and molecular mechanisms in the two groups of transporters, and a possible common evolutionary origin.

Acknowledgement: This work was supported by the Canadian Institutes of Health Research, the Alberta Cancer Board, the Natural Sciences and Engineering Research Council of Canada, the Wellcome Trust and the Medical Research Council of the United Kingdom.

CEC holds the Canada Research Chair in Oncology at the University of Alberta. JDY is a Heritage Scientist of the Alberta Heritage Foundation for Medical Research.

References

Agbanyo FR, Cass CE, Paterson ARP (1988) External location of sites on pig erythrocyte membranes that bind nitrobenzylthioinosine. Mol Pharmacol 33:332-337

Almedia AR, Jarvis SM, Young JD, Paterson ARP (1984) Photoaffinity labelling of the nucleoside transporter of cultured mouse lymphoma cells. FEBS Letters 176:444-448

Barros LF, Yudilevich DL, Jarvis SM, Beaumont N, Young JD, Baldwin SA (1995) Immunolocalisation of nucleoside transporters in human placental trophoblast and endothelial cells: Evidence for multiple transporter isoforms. Eur J Physiol 429:394-399

Beaumont N, Baldwin SA, Cass CE, Young JD (1995) Antibodies as probes of nitrobenzylthioinosine-sensitive nucleoside transporters. In: Belardinelli L, Pelleg A (eds) Adenosine and adenine nucleotides: from molecular biology to integrative physiology. Kluwer Academic Publishers, Norwell, MA, pp 55-60

Beutler E (2001) Energy metabolism and maintenance of erythrocytes. In: Beutler E, Coller BS, Lichtman MA, Kipps TJ, Seligsohn U (eds) Williams hematology, 6th edn. McGraw-Hill, New York, pp 319-332

Boleti H, Coe I, Baldwin SA, Young JD, Cass CE (1997) Molecular identification of the equilibrative NBMPR-sensitive (es) nucleoside transporter and demonstration of an equilibrative NBMPR-insensitive (ei) transport activity in human erythroleukemia (K - 562) cells. Neuropharmacology 36:1167-1179

Cass CE (1995) Nucleoside transport. In: Georgopadakou NH (ed) Drug transport in antimicrobial and anticancer chemotherapy. Marcel Dekker, New York, pp 404-451

Cass CE, Paterson ARP (1976) Nitrobenzylthioinosine binding sites in the erythrocyte membrane. Biochim Biophys Acta 419:285-294

Cass CE, Gaudette LA, Paterson ARP (1974) Mediated transport of nucleosides by human erythrocytes. Specific binding of the inhibitor nitrobenzylthioinosine to nucleoside transport sites in the erythrocyte membrane. Biochim Biophys Acta 345:1-10

Coe IR, Griffiths M, Young JD, Baldwin SA, Cass CE (1997) Assignment of the human equilibrative nucleoside transporter (hENT1) to 6p21.1-p21.1. Genomics 45:459-460

Craik JD, Good AH, Gottschalk R, Jarvis SM, Paterson ARP, Cass CE (1988) Identification of glucose and nucleoside transport proteins in neonatal pig erythrocytes using monoclonal antibodies against band 4.5 polypeptides of adult human and pig erythrocytes. Biochem Cell Biol 66:839-852

Crawford CR, Patel DH, Naeve C, Belt JA (1998) Cloning of the human equilibrative, nitrobenzylmercaptopurine riboside (NBMPR)-insensitive nucleoside transporter *ei* by functional expression in a transport-deficient cell line. J Biol Chem 273:5288-5293

De Koning H, Diallinas G (2000) Nucleobase transporters. Mol Membrane Biol 75:75-94

Eisenberg D, Schwarz E, Komaromy M, Wall R (1984) Analysis of membrane and surface protein sequences with the hydrophobic moment plot. J Mol Biol 179:125-142

Fincham DA, Wolowyk MW, Young JD (1991) Nucleoside uptake by red blood cells from a primitive vertebrate, the Pacific hagfish (*Eptatretus stouti*), is mediated by a nitrobenzylthioinosine-insensitive transport system. Biochim Biophys Acta 1069:123-126

Gati WP, Paterson ARP (1989) Nucleoside transport. In: Agre P, Parker JC (eds) Red blood cell membranes. Marcel Dekker, New York, pp 635-661

Good AH, Craik JD, Jarvis SM, Kwong FYP, Young JD, Paterson ARP, Cass CE (1987) Characterisation of monoclonal antibodies that recognise band 4.5 polypeptides associated with nucleoside transport in pig erythrocytes. Biochem J 244:749-755

Griffith DA, Jarvis SM (1996) Nucleoside and nucleobase transport systems of mammalian cells. Biochim Biophys Acta 1286:153-181

Griffiths M, Beaumont N, Yao SYM, Sundaram M, Boumah CE, Davies A, Kwong FYP, Coe I, Cass CE, Young JD, Baldwin SA (1997a) Cloning of a human nucleoside transporter implicated in the cellular uptake of adenosine and chemotherapeutic drugs. Nature Medicine 3:89-93

Griffiths M, Yao SYM, Abidi F, Phillips SEV, Cass CE, Young JD, Baldwin SA (1997b) Molecular cloning of a nitrobenzylthioinosine (NBMPR)-insensitive (ei) equilibrative nucleoside transporter from human placenta. Biochem J 328:739-743

Hyde RJ, Abidi F, Griffiths M, Yao SYM, Sundaram M, Phillips SEV, Cass CE, Young JD, Baldwin SA (2000) Probing the structure/function relationships of human equilibrative nucleoside transporters using site-directed mutagenesis. Drug Dev Res 50:38

Hyde RJ, Cass CE, Young JD, Baldwin SA (2001) The ENT family of eukaryote nucleoside and nucleobase transporters: recent advances in the investigation of structure/function relationships and the identification of novel isoforms. Mol Membrane Biol 18:53-63

Jarvis SM (1986) Nitrobenzylthioinosine-sensitive nucleoside transport system: mechanism of inhibition by dipyridamole. Mol Pharmacol 30:659-665

Jarvis SM, Young JD (1980a) Nucleoside transport in human and sheep erythrocytes: evidence that nitrobenzylthioinosine binds specifically to functional nucleoside transport sites. Biochem J 190:377-383

Jarvis SM, Young JD (1980b) Solubilisation of the nucleoside translocation system from human and nucleoside-permeable sheep erythrocytes. FEBS Letters 117:33-36

Jarvis SM, Young JD (1981) Extraction and partial purification of the nucleoside transport system from human erythrocytes based on the assay of nitrobenzylthioinosine binding activity. Biochem J 194:331-339

Jarvis SM, Young JD (1982) Nucleoside translocation in sheep reticulocytes and in erythrocytes from newborn lambs. A proposed molecular model for the nucleoside transporter. J Physiol 324:47-66

Jarvis SM, Young JD (1983) Nucleoside transport in animal cells. Bioscience Reports 3:309-322

Jarvis SM, Young JD (1986) Nucleoside transport in rat erythrocytes. Two components with differences in sensitivity to inhibition by nitrobenzylthioinosine and *p*-chloromercuriphenyl sulphonate. J Membrane Biol 93:1-10

Jarvis SM, McBride D, Young JD (1982) Erythrocyte nucleoside transport: asymmetrical binding of nitrobenzylthioinosine to nucleoside permeation sites. J Physiol 324:31-46

Jarvis SM, Young JD, Wu JSR, Belt JA, Paterson ARP (1986) Photoaffinity labelling of the human erythrocyte glucose transporter with 8-azidoadenosine. J Biol Chem 261:11077-11085

Kiss A, Farah K, Kim J, Garriock RJ, Drysdale TA, Hammond JR (2000) Molecular cloning and functional characterization of inhibitor-sensitive (mENT1) and inhibitor-resistant (mENT2) equilibrative nucleoside transporters from mouse brain. Biochem J 352:363-372

Kopito RR, Lodish HF (1985) Primary structure and transmembrane orientation of the murine anion exchange protein. Nature 316:234-238

Kozak M (1989) The scanning model for translation: an update. J Cell Biol 108:229-241

Kwong FYP, Baldwin SA, Scudder PR, Jarvis SM, Choy MYM, Young JD (1986) Erythrocyte nucleoside and sugar transport, endo-beta-galactosidase and endoglycosidase-F digestion of partially-purified human and pig transporter proteins. Biochem J 240:349-356

Kwong FYP, Tse CM, Jarvis SM, Choy YM, Young JD (1987) Purification and reconstitution studies of the nucleoside transporter from pig erythroctyes. Biochim Biophys Acta 904:105-116

Kwong FYP, Davies A, Tse CM, Young JD, Henderson PJF, Baldwin SA (1988) Purification of the human erythrocyte nucleoside transporter by immunoaffinity chromatography. Biochem J 255:243-249

Kwong FYP, Fincham HE, Davies A, Beaumont N, Henderson PJF, Young JD, Baldwin SA (1992) Mammalian nitrobenzylthioinosine-sensitive nucleoside transport proteins: immunological evidence that transporters differing in size and inhibitor-specificity share sequence homology. J Biol Chem 267:21954-21960

Kwong FYP, Wu JSR, Fincham HE, Davies A, Henderson PJF, Baldwin SA, Young JD (1993) Enzymic cleavage as a probe of the molecular structures of mammalian equilibrative nucleoside transporters. J Biol Chem 268:22127-22134

Loewen SK, Ng AML, Yao SYM, Cass CE, Baldwin SA, Young JD (1999) Identification of amino acid residues responsible for the pyrimidine and purine nucleoside specificities of human concentrative Na^+-nucleoside cotransporters hCNT1 and hCNT2. J Biol Chem 274:24475-24484

Mackey JR, Yao SYM, Smith KM, Karpinski E, Baldwin SA, Cass CE, Young JD (1999) Gemcitabine transport mediated by recombinant plasma membrane mammalian nucleoside transporters expressed in *Xenopus* oocytes. J Natl Cancer Inst 91:1876-1881

Mueckler M, Caruso C, Baldwin SA, Panico M, Blench I, Morris HR, Allard WJ, Lienhard GE, Lodish HF (1985) Sequence and structure of a human glucose transporter. Science 229:941-945

Murray AW (1971) The biological significance of purine salvage. Annu Rev Biochem 40:881-826

Oliver JM, Paterson ARP (1971) Nucleoside transport. I. A mediated process in human erythrocytes. Can J Biochem 49:262-270

Paterson ARP, Oliver JM (1971) Nucleoside transport. II. Inhibition by *p*-nitrobenzylthioinosine and related compounds. Can J Biochem 49:271-274

Pau SS, Paulsen IT, Saier MH (1998) Major facilitator superfamily. Microbiol Mol Biol 62:1-34

Persson B, Argos P (1994) Prediction of transmembrane segments in proteins utilising multiple sequence alignments. J Mol Biol 237:182-192

Plagemann PG, Woffendin C (1987) Purine and pyrimidine transport and permeation in human erythrocytes. Biochim Biophys Acta 899:295-301

Ritzel MWL, Ng AML, Yao SYM, Graham K, Loewen SK, Smith KM, Hyde RJ, Karpinski E, Cass CE, Baldwin SA, Young JD (2001) Recent molecular advances in studies of the concentrative Na^+-dependent nucleoside transporter (CNT) family: identification and characterization of novel human and mouse proteins (hCNT3 and mCNT3) broadly selective for purine and pyrimidine nucleosides (system *cib*). Mol Membrane Biol 18:65-72

Rost B, Fariselli P, Casadio R (1996) Topology prediction for helical transmembrane proteins at 86% accuracy. Prot Sci 5:1704-1718

Shi MM, Wu JSR, Lee CM, Young JD (1984) Nucleoside transport. Photoaffinity labelling of high affinity nitrobenzylthioinosine binding sites in rat and guinea pig lung. Biochem Biophys Res Commun 118:594-600

Sonnhammer ELL, Von Heijne G, Krogh A (1998) A hidden Markov model for predicting transmembrane helices in protein sequences. In: Glasgow J, Littlejohn T, Major F, Lathrop R, Sankoff D, Sensen C (eds) Proceedings of the sixth international conference on intelligent systems for molecular biology. AAAI Press, Menlo Park, CA, pp 175-182

Sundaram M, Yao SYM, Ng AML, Griffiths M, Cass CE, Baldwin SA, Young JD (1998) Chimeric constructs between human and rat equilibrative nucleoside transporters (hENT1 and rENT1) reveal hENT1 structural domains interacting with coronary vasoactive drugs. J Biol Chem 273:21519-21525

Sundaram M, Yao SYM, Ingram JC, Berry ZA, Abidi F, Cass CE, Baldwin SA, Young JD (2001a) Topology of a human equilibrative, nitrobenzylthioinosine (NBMPR)-sensitive nucleoside transporter (hENT1) implicated in the cellular uptake of adenosine and anti-cancer drugs. J Biol Chem 276:45270-45275

Sundaram M, Yao SYM, Ng AML, Cass CE, Baldwin SA, Young JD (2001b) Equilibrative nucleoside transporters: mapping regions of interaction for the substrate analogue nitrobenzylthioinosine (NBMPR) using rat chimeric proteins. Biochemistry 40:8146-8151

Tse CM, Wu JSR, Young JD (1985a) Evidence for the asymmetrical binding of p-chloromercuriphenyl sulphonate to the human erythrocyte nucleoside transporter. Biochim Biophys Acta 88:316-324

Tse CM, Belt JA, Jarvis SM, Paterson ARP, Wu JS, Young JD (1985b) Reconstitution studies of the human erythrocyte nucleoside transporter. J Biol Chem 260:3506-3511

Tucker EM, Young JD (1988) Genetic control of red cell nucleoside transport and its association with the B blood group locus and nucleoside phosphorylase activity in sheep. Biochem Genet 26:489-501

Vickers MF, Mani R, Sundaram M, Hogue DL, Young JD, Baldwin SA, Cass CE (1999) Functional production and reconstitution of the human equilibrative nucleoside transporter (hENT1) in *Saccharomyces cerevisiae*: Interaction of inhibitors with recombinant hENT1 and a glycosylation-defective derivative (hENT1/N48Q). Biochem J 339:21-32

Ward JL, Sherali A, Mo Z-P, Tse CM (2000) Kinetic and pharmacological properties of cloned human equilibrative nucleoside transporters, ENT1 and ENT2, stably expressed in nucleoside transporter-deficient PK15 cells. ENT2 exhibits a low affinity for guanosine and cytidine but a high affinity for inosine. J Biol Chem 275:8375-8381

Williams JB, Lanahan AA (1995) A mammalian delayed-early response gene encodes HNP36, a novel, conserved nucleolar protein. Biochem Biophys Res Commun 213:325-333

Williams JB, Rexer B, Sirripurapu S, Suthra J, Goldstein R, Phillips JA, Haley LL, Sait SN, Shows TB, Smith CM, Gerhard DS (1997) The human HNP36 gene is localized to chromosome 11q13 and produces alternate transcripts that are not mutated in multiple endocrine neoplasia, type 1 (MEN I) syndrome. Genomics 42:325-330

Wu JS, Young JD (1984) Photoaffinity labelling of nucleoside transport proteins in plasma membranes isolated from rat and guinea pig liver. Biochem J 220:499-506

Wu JS, Jarvis SM, Young JD (1983a) The human erythrocyte nucleoside and glucose transporters are both band 4.5 membrane polypeptides. Biochem J 214:995-997

Wu JS, Kwong FYP, Jarvis SM, Young JD (1983b) Identification of the erythrocyte nucleoside transporter as a band 4.5 polypeptide. Photoaffinity labelling studies using nitrobenzylthioinosine. J Biol Chem 258:13745-13751

Yao SYM, Ng AML, Muzyka WR, Griffiths M, Cass CE, Baldwin SA, Young JD (1997) Molecular cloning and functional characterisation of NBMPR-sensitive (*es*) and NBMPR-insensitive (*ei*) equilibrative nucleoside transporter proteins (rENT1 and rENT2) from rat tissues. J Biol Chem 272:28423-28430

Yao SYM, Ng AML, Sundaram M, Cass CE, Baldwin SA, Young JD (2001a) Transport of antiviral 3'-deoxy-nucleoside drugs by recombinant human and rat equilibrative, NBMPR-insensitive (ENT2) nucleoside transporter proteins produced in *Xenopus* oocytes. Mol Membrane Biol 18:161-167

Yao SYM, Sundaram M, Chomey EG, Cass CE, Baldwin SA, Young JD (2001b) Identification of Cys140 in helix 4 as an exofacial cysteine residue within the substrate-translocation channel of rat equilibrative nitrobenzylthioinosine (NBMPR)-insensitive nucleoside transporter rENT2. Biochem J 353:387-393

Yao SYM, Sundaram M, Vickers MF, Chomey EG, Cass CE, Baldwin SA, Young JD (2002) Functional and molecular characterization of nucleobase transport by recombinant human and rat ENT1 and ENT2 equilibrative nucleoside transport proteins produced in *Xenopus* oocytes. Chimeric constructs reveal a role for the ENT2 helix 5-6 region in nucleobase translocation. J Biol Chem 277:24938-24948

Young JD (1978) Nucleoside transport in sheep erythrocytes: genetically controlled transport variation and its influence on erythrocyte ATP concentrations. J Physiol 277:325-339

Young JD, Jarvis SM, Robins MJ, Paterson ARP (1983) Photoaffinity labelling of the human erythrocyte nucleoside transporter by N^6-(*p*-azidobenzyl) adenosine and nitrobenzylthioinosine. Evidence that the transporter is a band 4.5 polypeptide. J Biol Chem 258:2202-2208

Young JD, Paterson ARP, Henderson PJF (1985) Nucleoside transport and metabolism in erythrocytes from the Yucatan miniature pig. Evidence that inosine functions as an *in vivo* energy substrate. Biochim Biophys Acta 842:214-224

Young JD, Jarvis SM, Clanachan AS, Henderson JF, Paterson ARP (1986) Nitrobenzylthioinosine - an *in vivo* inhibitor of pig erythrocyte energy metabolism. Am J Physiol 251:C90-C94

14 Glucose Transport

Richard J. Naftalin

Physiology Division, King's College London, Guy's Campus, London SE1 1UL, UK

14.1 Introduction

Since I last wrote a review on this topic 25 years ago (Naftalin and Holman 1977) the dominant advance in our understanding of sugar transport has come from the publication of the amino acid sequence of the glucose transporters (GLUTs) with the 2D topology of the transporter in the membrane (Mueckler et al. 1985). Corroboration of the details of this 2D structure has been a major and continuing effort. Unfortunately, there has not been a commensurate increase in understanding of the mechanism of sugar transport. Although some advance in our perceptions of the mechanism of sugar transport has occurred, until recently, this has only been tenuously related to the transporter structure. Real advances can only come from knowledge of the 3D structural changes in GLUT, which accompany transport.

14.2 Specificity of Sugar Transporters

Unravelling the GLUT1 sequence led quickly to discovery of the main isoforms of GLUTs. The primary sequences of the five most common isoforms have 50–75% homology (Baldwin 1993; Bell et al. 1993; Gould and Holman 1993). The hydropathy plots of all these isoforms can almost be superimposed. GLUTs 1–5 are all considered to contain 12 transmembrane amphilic helical regions. The glucose transporter has been shown to be part of a major facilitator superfamily (MFS), which is one of the two largest families of membrane transporters. All permeases of the MFS possess either 12 or 14 putative transmembrane alpha-helical spanning domains and all appear to have undergone gene duplication so that the N-terminal half of the molecule is similar to the C-terminal half (Pao et al. 1998; Saier et al. 2000).

The main differences in sugar specificity of GLUTs lie between the relative affinities for glucose and fructose. GLUT1, 3 and 4 have a higher affinity for D-glucose than D-fructose, whereas GLUT2 and 5 have relatively high affinities for fructose. GLUT5 has a higher absolute affinity for fructose than for glucose. GLUT1 is the most widely distributed of all the GLUTs, and is particularly prevalent in erythrocytes and brain endothelium. GLUT3 has a higher affinity (K_m ~ 1–

2 mM at 20 °C) for glucose than GLUT1 (K_m ~ 2–3 mM at 20 °C) and is found in brain astrocytes and neuronal cells. GLUT2 has a lower affinity for glucose (K_m ~ 10–15 mM) but a wider range of affinities, in that it can also transport D-fructose (K_m ~ 15–30 mM). Cytochalasin B (K_i ~ 100 nM) inhibits GLUTs 1, 2, 3 and 4, forskolin (K_i ~ 100 nM) inhibits GLUTs 1, 3 and 4, but not GLUT2 and neither cytochalasin B, nor forskolin inhibit GLUT5. GLUT2 is found mainly in liver, baso-lateral membranes of intestine and in pancreatic islet cells of some species (but not human) (Bell et al. 1993; Gould and Holman 1993; Baldwin 1993).

The low affinity of GLUT2 is thought to be advantageous where cells are exposed to high glucose concentration, e.g. in cells exposed to the portal venous drainage. The higher specificity for fructose of GLUT5 is utilized in tissues exposed to high concentrations of this sugar, mainly in the apical membranes of intestinal enterocytes and in the male reproductive tract. GLUT4 has a high affinity for glucose and is responsive to insulin in adipocytes and muscle tissues. On exposure to low insulin concentrations, insulin-sensitive adipocytes and muscle cells respond by insertion of GLUT4 into the plasma membranes. This increased rate of GLUT insertion is not confined to GLUT4 alone, as increased amounts of GLUT1 are inserted albeit to a lesser extent and rate than GLUT4 (~4:1) in adipocytes (Calderhead et al. 1990) and 3T3-L1 cells (Kozka et al. 1991; Yang et al. 1993). However, as will be shown below, insertion of GLUT4 confers an advantage over GLUT1 in terms of enhanced sugar transport.

14.3 Net *vs.* Exchange Transport of Glucose

Glucose transport in erythrocytes via GLUT1 is characterised as an equilibrative process. At equilibrium, the concentrations of transported sugars in the intracellular and extracellular solutions are the same. There is a faster maximal rate of exchange transport than of net transport for the well-transported sugars, e.g. D-glucose, 2-deoxy D-glucose (2-dGlc), 3-*O*-methyl glucoside (3-OMG), galactose and mannose in GLUT1. The maximal rate, V_m of equilibrium exchange transport of glucose (measured with equal concentrations of sugar on both sides, using the rate of equilibration of labelled sugar present initially only in one compartment) is faster than the maximal rate of net exit flux by a factor of 2 and entry flux by a factor of 5–10 (Levine et al. 1965; Lacko et al. 1973; Lieb and Stein 1974; Wheeler 1986; Lowe and Walmsley 1986; Stein 1986; Cloherty et al. 1996). The K_m for equilibrium exchange flux is 3–10 fold higher for net uptake; i.e. for net uptake of glucose at 25 °C, the K_m is typically 2–3 mM; whereas for equilibrium exchange flux, it is ~10–12 mM (Fig. 14.1, Table 14.1).

Two other processes relate to accelerated exchange transport: infinite *trans* uptake of labelled hexose into a cytosolic compartment containing a fixed high concentration of transported hexose. The K_m of this process is similar to that for net glucose (2–3 mM), but the maximal rate of infinite *trans* uptake is similar to that of equilibrium exchange (Lacko et al. 1973).

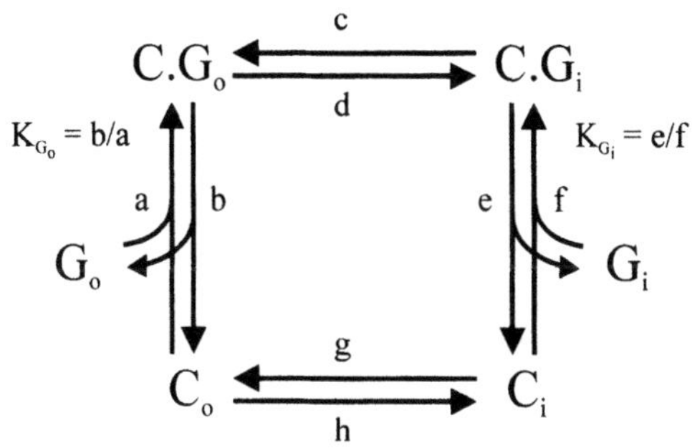

Fig. 14.1. Diagram of single cycle symmetrical or asymmetrical alternating carrier transporter. If glucose binding to the exposed sites of the transporter is much faster than the rate of transport then the dissociation constants: 1) $K_{Go} = b/a$ and $K_{Gi} = e/f$; 2) The asymmetric transporter requires $K_{Gi} > K_{Go}$; 3) The law of microscopic reversibility requires that rate products *adeg* = *bhfc*. Assuming that exchange rates of sugar movement via processes *c* and *d* are symmetrical, then $K_i / K_o = h/g$

Table 14.1. The variation of V_m and K_m during the different transport modes at 0 °C of 3-OMG, D-glucose and 2-dGlc

Experiment	Parameter	3-OMG	D-glucose	2-dGlc
Zero *trans* entry	K_m	0.38 ± 0.13	0.46 ± 0.09	0.59 ± 0.27
	V_m	0.18 ± 0.02	0.15 ± 0.01	0.29 ± 0.04
Zero *trans* exit	K_m	4.35 ± 0.62		
	V_m	1.62 ± 0.10		
Equilibrium exchange	K_m	22.62 ± 6.17		
	V_m	9.17 ± 3.44		
Infinite *trans* entry	K_m	1.57 ± 0.10	2.19 ± 0.36	0.76 ± 0.14
	V_m	5.62 ± 0.11	12.19 ± 0.88	8.55 ± 0.79
Infinite *cis* entry	K_m^a	0.66 ± 0.15		
	K_m^b	0.47 ± 0.11		
	K_m^c	3.35 ± 0.62		

K_m in mM, V_m in mmol/(l_{cells}·min). *a* = intracellular 3-OMG which reduced influx by a half; *b* = obtained from integrated rate equation; *c* = the intracellular 3-OMG concentration which stimulated 3-OMG entry half maximally. From Cloherty et al. (1996). Note the apparent asymmetry of net transport exit K_m exceeds entry K_m of 3-OMG by 11.4-fold whereas V_m exit exceeds V_m net entry by 9-fold. Note the much higher V_m (accelerated exchange) and K_m (accelerated exchange) than of net flux. Also, that V_m of net exit rate of 2-dGlc is ~2 times that of D-glucose, but anomalously, V_m (accelerated exchange) of D-glucose is 1.5 times higher than V_m (accelerated exchange) of 3-OMG (see text). Using the nomenclature in Fig. 14.1. V_m (equilibrium exchange) $\alpha(1/c + 1/d)$, whereas V_m (zero *trans* net efflux) $\alpha(1/c + 1/h)$. Where α is a proportionality coefficient.

The other phenomenon related to accelerated exchange flux is *counterflow* (Wilbrandt 1972). *Counterflow* occurs when a high concentration (50–100 mM) of unlabelled transported hexose is loaded into the cytosolic compartment and a low (< 4 mM) concentration of labelled hexose is placed in the external solution. The labelled sugar flows into the cell and accumulates to a higher "concentration" inside the cell than outside. The high concentration of unlabelled sugar in the cell compartment reduces the specific activity of labelled intracellular and so labelled sugar exit is retarded. The overshoot of accumulated radioactivity within the cell results from accelerated inflow of labelled sugar in exchange for sugar of much lower specific activity. The specific activity of the sugar (dpm/mole glucose) in the internal compartment rises continuously, without overshooting, until it matches that in the external compartment. Full equilibration of specific activity may be delayed because of compartmentalization of sugars within the cytosolic compartment. This results in a lower overshoot than would be predicted. (Naftalin et al. 1985; Cloherty et al. 1995; Zottola et al. 1995).

Counterflow illustrates that exit flux (µmoles/($ml_{cell\ water}$·s)) of unlabelled sugar, which can exceed entry flux of labelled sugar by more than an order of magnitude, does not retard sugar inflow. Non-interference between entry and exit streams implies either that sugar molecules flow through the same channel sequentially, or through separate antiparallel channels simultaneously or sequentially. Acceleration of exchange implies that the presence of *transported ligand* on the *trans* side increases the mobility of the *cis-trans* flow of the alternate transported ligand.

Since the rate of return of the empty carrier *h* (Fig. 14.1) is the same for all sugars and for equilibrium exchange $c = d$ and these are similar for 3-OMG and mannose at 25 °C in rat (Naftalin and Rist 1994) it follows that V_m net flux should be similar for both mannose and 3-OMG and yet in fact V_m 3-OMG is 4 times faster than for mannose. Several hypotheses, account for accelerated exchange. The alternating mobile carrier model explains some, but not all of the phenomena out above.

14.4 Membrane Topology of GLUTs

The original hypothesis of a 12 transmembrane (12 TM) helical conformation of all GLUTs (Mueckler et al. 1985) based on hydropathy plots of the primary structure of GLUTs has been largely corroborated (Fig. 14.2).

14.5 Evidence for the 12 TM Helix Model for GLUTs

14.5.1 Trypsinolysis

The basis of this corroboration is as follows: trypsin, when added to the outside of cells, has no effect on glucose transport. It results in a single break, which is without effect on the transporter function (Masiak and LeFevre 1977). On the other hand, introduction of trypsin into the cytosolic compartment of erythrocyte ghosts, results in a rapid loss of glucose transport activity in the reconstituted ghosts and to loss of cytochalasin B binding capacity (Cairns et al. 1984). Analysis of the peptides released by trypsin digestion studies show that the linker segment 207–

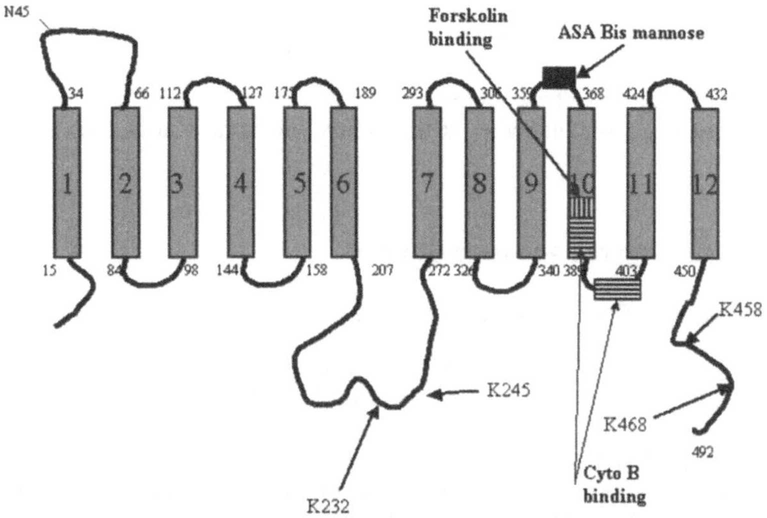

Fig. 14.2. Diagram showing the 2D 12 transmembrane helices of GLUTs (Muecker et al. 1985). The positions of extracellular and cytoplasmic inter-TM helical linkers, sites of trypsinolysis and binding sites of covalent inhibitors are shown

271 between TM 6 and TM 7 is cytoplasmic. This linker segment only becomes accessible to trypsin when the cell membranes are broken (Gibbs et al. 1988; Haspel et al. 1988). Two sites within segment 207–271 are susceptible to trypsinolysis, Arg 232 and Lys 245. The other cytoplasmic segment of the transporter attacked by trypsin is the C-terminal fragment from 451–492. This contains a further two sites at Arg 458 and Arg 468 (Baldwin 1993).

These studies confirm that two hypothetical intracellular segments of the 12 TM model of GLUT1 are intracellular.

14.5.2 Antibody Studies

Antibodies have been developed against synthesised peptides corresponding to specific regions along the entire length of GLUTs. Antibodies, which bind to regions predicted by the 12 TM model to face toward the cytoplasm segments between TM 6 and 7 amino acids 231–246 and 240–255 and the C-terminal peptide between 450–467 and 460–477 also bind to the native protein in resealed ghosts. The antibodies binding to the cytoplasmic segment between TM 6 and 7 also inhibit cytochalasin B binding to depleted erythrocyte membranes in a dose-dependent fashion. Another study using antibodies to identify structural features of GLUT1 showed that antibody binding to the C-terminal peptide residues 480–492 bound to the inside of permeabilized red cells but not to impermeant cells. Antibodies raised against the N-terminal peptide 1–12 did not bind to the native protein (Haspel et al. 1988; Davies et al. 1990).

However, the antibodies to the C-terminal segment had no effect on cytochalasin B binding. None of the antibodies, predicted to bind to the linker regions joining the TM helices at their external faces bind to the native GLUT1 protein. This may be because these epitopes are protected by phospholipid interactions from interacting with antibodies at the external membrane surface (Davies et al. 1990).

14.5.3 Scanning Mutagenesis Studies

Glycosylation Scanning Mutagenesis

Natural glycosylation occurs at aspargine (Asn) residues. Only one naturally occurring site of *N*-glycosylation occurs in GLUT1 at Asn^{46}. The 12 TM model predicts that Asn^{46} is in the extracellular linker region between TM 1 and TM 2. Introduction of Asn's by mutations at all the putative extramembranous linker regions between the transmembrane helices of GLUT1 permit new glycosylation sites to be introduced. Only those sites facing the extramembranous compartment are glycosylated as this side is exposed to glycosylating enzymes in the endoplasmic reticulum. Expression of the mutant proteins in *Xenopus* oocytes demonstrated that when Asn mutants were inserted in the linker segments between TM 1–2, 3–4, 5–6, 7–8, 9–10 and 11–12 glycosylation occurred. This was confirmed by subjecting the extracted protein from immunoprecipitated intracellular oocyte membrane to endoglycosidase H digestion on the SDS-PAGE gels. The enzyme induced deglycosylations caused a shift in the position of the protein on the gel whereas no change was observed with unglycosylated protein. All the hypothetical exofacial mutants were glycosylated; whereas, none of the mutants facing towards the cytoplasm were glycosylated.

It could be argued that insertion of glycosylation sites could distort the GLUT topology in the oocyte expression system, i.e. by flipping the resultant mutated protein segment to an abnormal exposed extracellular loop. However, expression of the glycosylation mutants in the oocytes showed that at least some of these constructs were biologically active, thus "proving" that their topology must be correct. Since none of the "endofacial" mutants were glycosylated, this is strong evidence supporting of the 2D topology of the 12 TM model (Hresko et al. 1994).

Cysteine Scanning Mutagenesis Studies

Substitution of all the cysteine residues in GLUT1 with serine gives a functional cysteineless mutant transporter protein when expressed in *Xenopus* oocytes. The only difference is that the alkylating sulphydryl reagents inhibit glucose transport across GLUT1, whereas the cysteineless mutant is not inhibited (Due et al. 1995). This result has some important implications for studies on GLUT dimerization (see below).

Cysteine residues can readily be substituted for any residue in the GLUT mRNA and the mutants expressed in oocytes. Systematic expression of cysteine substitutions within the cysteineless mutant GLUT1 transporter in a single or multiple TM regions in parallel with site-directed chemical modification of the mutated cysteine residues, using impermeant sulphydryl modifiers, e.g. *p*-chloromercuribenzene sulphonate (PCMBS) or *N*-ethylmaleimide (NEM) gives information regarding the accessibility of any particular residue to the extracellular or intracellular solution.

These labour intensive methods have yielded useful information about the accessibility of helices to alkylating reagent present in the extracellular solution (Mueckler and Makepeace 1997, 1999). They indicate that a hydrated pore, or cleft exists within the transporter. The residues on the helical facets lining the pore are readily alkylated by PCMBS, whereas antipodal residues facing the hydrophobic exterior of the transporter are not readily alkylated. The 12 TM model predicts that TMs 3, 5, 7, 8, and 11 helices are amphipathic, i.e. the hydrophilic part faces towards a pore and the hydrophobic part faces outwards towards the membrane.

14.5.4 Cysteine Scanning Mutagenesis of TM's 7, 10 and 11 Evidence for Two Hydrophilic Pores Traversing GLUT1

It has recently been suggested that there are two hydrophilic pores traversing GLUT1. The evidence for this was obtained by a 3D reconstruction of the GLUT1 using best fit conjectures and homologies with similar proteins (Zuniga et al. 2001) (Fig. 14.3). There is good corroborative experimental evidence supporting a two-pore structure. Cysteine scanning mutagenesis shows that entire exofacial surface of helix TM 7 is accessible to the external solvent as judged by inhibition studies of glucose transport with the sulphydryl-directed reagent PCMBS (Hruz and Mueckler 1999). Whereas, analysis of TM segments 10 and 11 show that in the exofacial region of TM's 10 and 11, the solvent is only accessible to a single helix face. (Hruz and Mueckler 2000; Mueckler and Makepeace 2002) (Fig.14.3).

These studies support the notion that GLUT1 has a hydrophilic core surrounded by a ring of TM segments which have mainly hydrophilic residues facing towards the pore and which are accessible to alkylating reagents from the extracellular side. The residues near to the cytoplasmic surface are inaccessible to the extracellular inhibitors. This finding supports the view that a conformational change has to occur to effect net transport across the membrane (Mueckler and Makepeace 1999).

Large Cytoplasmic Loop

Alanine scanning mutagenesis and the introduction of insertions and deletions into the large 65 residue long cytoplasmic loop between TMs 6 and 7 showed that progressive truncation of the loop caused a progressive decrease in the rate of 2-dGlc uptake from 33% (delta 236–241) to 4% (delta 216–262). Insertion of hexaalanine into the deletion (236–241) restored the rate of 2-dGlc uptake to 73% of wild type GLUT1. These findings demonstrate that the cytoplasmic loop has a major role to play in controlling the rate of glucose transport possibly in controlling the distance separating TM's 6 and 7. Unfortunately there are no data on exchange *vs.* net flow rates (Monden et al. 2001).

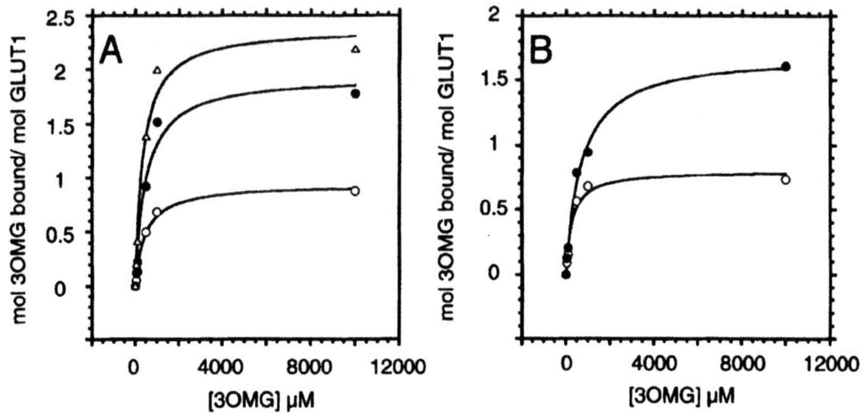

Fig. 14.3. Model of the effects of ATP on 3-OMG association with GLUT1 in red cell membranes (**A**) and purified GLUT1 (**B**). From Heard et al. (2000). (Thanks to Prof A Carruthers, Univ. Worcester, Mass., USA for permission to reproduce this figure.) The figure illustrates how tetramerization can lead to formation of an occlusion of a compartment which can accommodate 2–3 moles 3-OMG/mole GLUT1. 3-OMG binding activity of erythrocytes, erythrocyte ghosts, and purified GLUT1. (**A**) Red blood cells (•) and red blood cell ghosts resealed in the presence (Δ) and absence (○) of 4 mM ATP were assayed for 3-OMG binding by equilibrium 3-OMG space analysis. Ordinate: moles of 3-OMG bound per mole of GLUT1. Abscissa: micromolar concentration of 3-OMG. This plot summarizes at least four separate experiments made in triplicate. The curves drawn through the points were computed by nonlinear regression assuming that equilibrium 3-OMG binding is described by the expression [3-OMG bound] = B_{max}[3-OMG]/{$K_{d(app)}$ + [3-OMG]}, where B_{max} is 3-OMG binding capacity and $K_{d(app)}$ is that [3-OMG] where [3-OMG bound] = 0.5 B_{max}. The curves have the following constants: RBC, B_{max} = 1.93 ± 0.15 mol of 3-OMG bound/mol of GLUT1 and $K_{d(app)}$ = 475 ± 126 μM 3-OMG; ghosts + ATP, B_{max} = 2.38 ± 0.16 mol of 3-OMG bound/mol of GLUT1 and $K_{d(app)}$ = 344 ± 83 μM 3-OMG; ghosts - ATP, B_{max} = 0.94 ± 0.4 mol of 3-OMG bound/mol of GLUT1 and $K_{d(app)}$ = 458 ± 67 μM 3-OMG. (**B**) Purified GLUT1 was assayed for 3-OMG binding in the presence (•) and absence (○) of 4 mM ATP. Ordinate: moles of 3-OMG bound per mole of GLUT1. Abscissa: micromolar concentration of 3-OMG. This plot summarizes at least four separate experiments made in triplicate. The curves drawn through the points were computed by nonlinear regression assuming that equilibrium binding is described by the expression [3-OMG bound] = B_{max}[3-OMG]/{$K_{d(app)}$ + [3-OMG]} and have the following constants: GLUT1 + ATP, B_{max} = 1.71 ± 0.06 mol of 3-OMG bound/mol of GLUT1 and $K_{d(app)}$ = 689 ± 76 μM 3-OMG; GLUT1 - ATP, B_{max} = 0.80 ± 0.05 mol of 3-OMG bound/mol of GLUT1 and $K_{d(app)}$ = 269 ± 69 μM 3-OMG

14.5.5 Covalent Linkage of Inhibitors

Bis-mannose inhibitors

The impermeant selective photolabel 2-N^4-(1-azi-2,2,2-trifluoroethyl)benzolyl-1,3-bis-(D-mannosyloxy)-2-propylamine (ATB-BMPA) binds to externally facing epitopes mainly to the external face of TM helix 7–9, although secondary effects resulting from conformational changes can affect ATB-BMPA binding, e.g. mutagenesis on Pro 385 in TM 10 reduces ATB-BMPA binding (Tamori et al. 1994). Bis-mannose covalent inhibitors are used to monitor the number of transporters present in the membrane domain (Palfreyman et al. 1992; Yang et al. 1992; Nishimura et al. 1993; Doege et al. 1998; Koumanov et al. 1998).

Cytochalasin B and Forskolin

Cytochalasin B binds to the inside of GLUT1 (Fig. 14.2). The evidence for this is based on kinetics, as it acts as a competitive inhibitor of glucose exit (Basketter and Widdas 1978; Krupka and Devés 1978) and on photolabelling studies. Trp 388 and Trp 412 in TM 10 and TM 11 are preferentially labelled by exposing the purified protein to labelled cytochalasin B (Baldwin 1993). Mutation of both Trp 388 and 412 to Leu abolished photolabelling with cytochalasin B and reduced cytochalasin B-dependent inhibition of glucose transport by 30% (Inukai et al. 1994). Mutation of Trp 388 alone in TM 10 caused 30% decease in transport activity. Apart from demonstrating that cytochalasin B binds to Trp 388 and 412, these findings show that other regions, unsusceptible to photolabelling, also must be involved in cytochalasin B interactions. Forskolin also inhibits glucose transport in red cells (Sergeant and Kim 1985). Photolabelling studies show that forskolin binds to an adjacent site to cytochalasin B (Wadzinski et al. 1990).

14.5.6 Conformational Change Resulting from Ligand Interactions

Study of the localization of labelling of GLUT with covalent ligands is closely linked with conformational changes induced by ligand interactions with the transporter. Both transported and non-transported ligand interactions alter the accessibility of thiol groups and other reactive species within GLUT1 to alkylating reagents, such as FDNB or NEM. Binding of labelled ATB-BMPA to the external surface of GLUT, or maltose protects against proteolytic cleavage of GLUT1, either by thermolysin (Holman and Rees 1987), or trypsin (King et al. 1991). Transportable ligands, e.g. glucose, or methylglucoside that bind to the cytoplasmic surface accelerate cleavage, whereas, 4,6-*O*-ethylidene glucose (EG), which binds preferentially to the outside site (Barnett et al. 1973; Basketter and Widdas 1978), retards the rate of thermolysin proteolysis. The non-transportable inhibitor, cytochalasin B, which binds to the cytoplasmic side, does not protect GLUT1 against thermolysin-induced proteolysis (Holman and Rees 1983), however, it abolishes both the methylglucoside activation of proteolysis and the EG-dependent protection against proteolysis, i.e. inside ligands increase proteolysis whereas outside ligands reduce it.

Similar results have been obtained with GLUT4 (Yano and May 1993). EG protects against trypsin and thermolysin digestion of GLUT4 and cytochalasin B prevents EG protection (King et al. 1991). However, at low trypsin concentrations cytochalasin B, in contrast to its effects on GLUT1, was found to accelerate proteolysis. Glucose and phenyl β-D-glucopyranoside, which bind preferentially to the inside face of the transporter (Barnett et al. 1973) also accelerate trypsinolysis of GLUT4. These studies indicate that ligand binding alters the conformation of GLUTs. The conformational changes vary; with GLUT1, ligand binding to the outside induces conformational changes that prevent access of proteolytic agents to the inside sites. In contrast, ligand interactions with GLUT4 increase accessibility of the cytoplasmic side of the transporter to proteolytic agents. These differences in GLUT response to ligand induced changes in accessibility to proteolytic agents may correlate with the retardive effects on glucose exit of the C-terminal chain interactions with the N-terminal chain in GLUT1 and the absence of retardive effects of C-terminal chain interactions in GLUT4 (Oka et al. 1990; Muraoka et al. 1995; Dauterive et al. 1996).

14.5.7 Biophysical Studies Showing α Helicity of the Transmembrane Strands

The 12 TM model (Mueckler et al. 1985) requires that the majority of the TM strands are α helical. Supportive evidence for this is based both on spectroscopic studies, circular dichroism (CD) and Fourier Transform Infrared (FTIR) (Alvarez at al. 1987; Chin et al. 1987) indicating that 70% of the native protein is α helical. Contrary reports (Fischbarg et al. 1993) suggest that a substantial fraction of the protein exists in the form of β-sheets. However, most of this β-sheet conformation must exist in the extramembranous domains, since studies on the protein situated within the membrane show that the majority of this is α helical.

Fischbarg et al. (1993) have suggested that GLUT1 consists not of the accepted 12α helical TM domains, but of 16 TM β-barrel porin structure, surrounding an aqueous channel through which glucose can be transported. The only piece of experimental evidence supporting this view is a specific antibody to the peptide Ile 386–Ala 405 increases uptake of 2-dGlc into GLUT1 or GLUT4 expressed in *Xenopus laevis* oocytes. The increased sugar uptake arises from a decrease in the K_m for uptake from 8.6 to 2.8 mM in the case of GLUT1 and similarly decreases with GLUT4. V_m is unaffected by the antibody. According to the 12 TM structure, the strand 386–405 is expected to bind only to cytoplasmic facing strand joining TM 10 and 11. Hence, it is anomalous that the antibody should affect sugar transport when added to the external solution. No effect of the antibody was observed when the antibody was injected into the oocytes. The 16 TM model, predicts that the sequence 386–405 is present in a strand facing the external solution.

The Fischbarg model has some obvious demerits, the most important being that the data from glycosylation scanning mutagenesis of Hresko et al. (1994) plainly contradicts the prediction that the 386–405 face the extracellular solution. Also, the bulk of the β-sheet formation is in the extramembranous domains of GLUT1 so the membranous domains must mainly consist of α helix (Chin et al. 1987; Zeng et al. 1996).

However, the β-barrel model provoked necessary discussion with regard to the 12 TM structure in relation to function.

There is a need for a synthesis of the wealth of detailed information relating 2D structure of GLUTs to a 3D model. Although no crystallographic structure of the glucose transporter, or any of the super-family of 12 TM transporters currently exists, structures of homologous protein sequences can be used as templates to predict the 3D structures. Templates have been obtained from a mechanosensitive channel derived from mycobacteria with 10 TM segments and from 6 TM segments of aquaporin, which is distantly related to GLUT (Dywer 2001).

Another 3D model of GLUT, not based on any template structures has been derived from fundamental protein structural principles (Zeng et al. 1996). Both these 3D model of GLUT share the important property that transporter is barrel-like structure surrounding a central hydrophilic pore, through which glucose permeates. These simulated 3D protein structures retain the 12 TM α helical structure predicted by Mueckler et al. (1985) and demonstrate that the adoption of a β-sheet structure is not a necessity for pore formation.

14.6 Structural-Functional Studies on GLUTs

Molecular biological techniques have been exploited to study the structural basis of accelerated exchange.

14.6.1 Functional Differences between the Isoforms GLUT1 and GLUT4

The response to insulin in adipocytes and 3T3 cells is to increase the amount of GLUT4 and GLUT1 in the cell membrane. This increase in GLUTs at the membrane surface is thought to account for the insulin-dependent augmentation of glucose transport. Differences in glucose transport between the two GLUT isoforms could account for the marked increase in turnover rate of glucose transport following insulin stimulation.

Glucose transport in adipocytes, unlike in erythrocytes, does not exhibit accelerated exchange transport (Taylor and Holman 1981; Wheeler 1984). The absence of accelerated exchange in GLUT4 has been confirmed several times, e.g. when expressed in liposomes containing extracted GLUT4 (Wheeler et al. 1998), or in eukaryote cells in culture (Palfreyman et al. 1992), or *Xenopus* oocytes (Nishimura et al. 1993; Dauterive et al. 1996), or COS-7 cells (Wandel et al. 1996) and yeast (Kasahara and Kasahara 1997), whereas accelerated exchange is observed when GLUT1 is expressed in all these systems.

Comparison of the transport parameters of GLUT1 and 4 in oocytes, or cultured cells is not a simple matter. A major problem is determining how many transporter molecules are present in the cell membrane. Immunoelectrophoresis of whole cell extracts, or of isolated membranes with anti-GLUT immunoglobulins, does not discriminate between non-functional transporters, present in the cytosol or close to membrane surface and functional transporters inserted in the membrane. The transport turnover, i.e. the moles of sugar transported per min per transporter, may be underestimated, as the reference number of functional transporters may be exaggerated. An improvement on this methodology is to use a selective photoactivatable ligand, e.g. ATB-BMPA, which binds specifically and covalently to the external surface of the transporter. Bis-mannose derivatives have been used successfully to label GLUTs from the external surface using UV light to activate the linkage (Baldwin 1993; Gould and Holman 1993; Koumanov et al. 1998). This method circumvents the problem of errors generated by non-functioning transporters attached to plasma membrane in the estimates of the density of functioning transporters in the cell membrane. As long as each GLUT isoform binds ATB-BMPA with equal affinity and there is only a small amount of non-specific binding of label, the method will be accurate.

Estimating the turnover rate of transporters inserted in liposomes, in theory should avoid the problems of non-functioning transporters in expression systems. However, transporters may be sensitive to the kind of lipid in which they are reconstituted. The GLUTs may have mixed orientation or be occluded inside the liposomal lumen. So, in practice this methodology is no more reliable than the cell expression systems at determining the activity of GLUTs (Wheeler et al. 1998).

Although comparison between different GLUT isoforms using the available expression systems is inaccurate, it is still possible to observe differences between the transport parameters. The K_m of 3-*O*-methyl-D-glucoside (3-OMG) transport, across GLUT1 for accelerated exchange is ~10 mM (Table 14.1); whereas, the K_m for exchange of 3-OMG transport across GLUT4 is around 2–3 mM.

The consensus view is that the maximal turnover numbers of GLUT1 and GLUT4 isoforms in equilibrium exchange mode are similar $\sim 1.7 \cdot 10^4$ molecules/min (Palfreyman et al. 1992; Nishimura et al. 1993; Dauterive et al. 1996). In the physiological sugar concentration range 3–5 mM, net uptake, via GLUT1 is 5–10-fold slower than exchange uptake. Since there is no difference between the rate of exchange and net flux in GLUT4, it follows that the unidirectional uptake of labelled sugar into muscle, or adipose tissue, is ~10 times faster via GLUT4 than GLUT1. Unidirectional flux across GLUT4 does not have to be activated by high concentrations of cytosolic sugar to be transported at the highest rate, whereas the accelerated rate of unidirectional sugar entry across GLUT1 only obtains when the transporter is activated by high concentrations of cytosolic sugar. This property of GLUT4 seems to have been unregarded in consideration of the higher activity of this isoform in insulin-stimulated tissue.

14.6.2 Structure-Functional Comparisons of GLUTs 1 and 4

The amino acid sequences of GLUT1 and GLUT4 differ mainly in the C-terminal cytoplasmic domain. Chimeras of GLUT1 and GLUT4, in which different lengths of the primary sequence of one of the isoforms are substituted at a similar position in the other, have been used to investigate the region in GLUTs that affects accelerated exchange in *Xenopus* oocytes. When the C-chain of GLUT4 replaced the C-terminal chain in GLUT1, GLUT1 lost its capacity to perform accelerated exchange. Substitution of the C-terminal sequence of GLUT1 into GLUT4 resulted in accelerated exchange in GLUT4 (Dauterive et al. 1996). These findings indicate that the C-terminal interaction of GLUT1 with some other sequences on the inner face of the transporter slows the turnover rate of net transport.

Other studies have suggested that the C-terminal sequence of GLUT1 interacts with the cytoplasmic N-terminal sequence (Kozka et al. 1991). Net glucose entry into GLUT4 may occur without any conformational change involving retardation of glucose entry. This view is supported by the similarities in K_m's of net flux and exchange flux in GLUT4, compared to with the wide disparity in the values 2–3 mM (zero *trans* net) and 20 mM (equilibrium exchange) for 3-OMG glucose transport in GLUT1 (Table 14.1).

Inter-chain reactions on the cytoplasmic side of GLUT1 leading to retardation of net flux can only be a partial explanation for accelerated exchange. Interactions occurring in the outward facing regions of GLUT1 also must play a role in acceleration of exchange. Glucose exit is accelerated both by transported sugars in the external solutions (Miller 1968a,b; Lacko et al. 1973; Eilam and Stein 1974) and by low concentrations of maltose (< 250 μM) binding to high affinity external sites. As maltose is not transported across the membrane, it cannot act directly with the inward facing sites (Hamill et al. 1999; Sultzman and Carruthers 1999; and see below).

14.6.3 Structural Basis of Specificity Differences between GLUTs

GLUTs 1, 3 and 4, have a high affinity for glucose, 2-dGlc, 3-OMG and maltose and almost zero affinity for fructose, whereas GLUTs 2 and 5, have low affinities for 2-deoxyglucose, maltose, glucose, in comparison with GLUTs 1, 3 and a relatively high affinity for fructose.

The differences in affinity between these two types of transporter have been investigated by two methodologies; chimeras of GLUT2/GLUT3 (Arbuckle et al. 1996; Wu et al. 1998), or GLUT1/GLUT2 (Noel and Newgard 1997), or GLUT5/GLUT3 (Buchs et al. 1998), have been constructed containing different segment lengths of the alternate isoform substituted at appropriate positions in the other isoform. GLUT selectivity for 3-OMG, 2-dGlc, or maltose has been compared with that of fructose. The changes in K_m for transport, of high affinity sugars, which typically vary from 1–3 mM for GLUTs 1 and 3 to > 10 mM for GLUTs 2 and 5, were monitored. The K_m for fructose is too high to be monitored accurately so the change in rates of fructose transport at a fixed fructose concentration are usually observed. Arbuckle et al. (1996) measured changes in fructose K_m. These studies revealed that the key differences for fructose selectivity lie within TM helices 7–8 of GLUTs 2 and 5. Substitution of these TM helices for the equivalents in GLUT1 or 3 reduced selectivity for fructose. Sequences in TM segments 9–12 are responsible for the glucose affinity within GLUT5. Almost no selectivity for the external site of the transporter exists within TM helices 1–6 and substitution of TM 12 does not affect selectivity.

A more precise methodology for investigating selectivity differences between GLUTs 2 and 3 has been adopted recently (Seatter et al. 1998). This involves study of TM helix 7. GLUTs 1, 3, 4 with high affinity for deoxyglucose, contain the sequence QLS (Gln-Leu-Ser) (residues 276–279 in GLUT1). GLUTs 2 and 5, with low affinity for deoxyglucose, do not contain this motif. Substitution of the alternate HVA (His-Val-Ala) motif in GLUT3 increased fructose transport in the mutant GLUT3 to a higher rate than in the native GLUT2 and increased sensitivity to fructose inhibition of 2-dGlc transport. Substitution of QLS for HLV in GLUT2 reduced the K_m for deoxyglucose transport from 11 mM to 3 mM.

These findings suggest that the HLV motif acts as one of the discriminants selecting for fructose or glucose. However, this motif in TM 7 cannot be the only site for fructose binding, since fructose affinity of the substituted HVA on GLUT3 (173 ± 51 mM) is much less than it is in GLUT2 (73 ± 6 mM). Glucosamine is transported 3–4-fold slower in GLUTs 1–4 than D-glucose as has been shown in *Xenopus* oocytes.

Glucosamine has a higher affinity for GLUT2 (K_m = 0.8 ± 0.1 mM) than glucose 17–20 mM. Glucosamine uptake by liver cells has been shown to be exclusively via GLUT2 (Uldry et al. 2002).

14.7 Problems Relating to Glucose Transport

14.7.1 Problem 1: Is the Carrier a One Mobile or Two Fixed Site Transporter?

The alternating carrier model makes several unfulfilled kinetic predictions (Miller 1968a,b; Naftalin 1971; Naftalin and Rist 1994; Cloherty et al. 1996a) (see also Table 14.1). The conventional explanation for accelerated exchange being faster than net flux is that the loaded carrier form moves faster than the unloaded carrier. Hence, zero *trans* net exit results in depletion of carrier sugar complex from the *cis* side with carrier accumulation at the *trans* side. The slow rate of return of the empty carrier is the rate-controlling step in the net exit cycle. Consequently, addition of sugar to the *trans* (external solution) leads both to return of ligand to the inside on the fast tract available to the ligand carrier complex and augmentation of the proportion of inward facing carriers. Miller (1968a,b) was the first to explore the implications of this analysis quantitatively. A clear prediction of the alternating carrier theory is that sugars that move faster across the carrier during net flux should also have the fastest rates of exchange and show the greatest differences between net and exchange flux. Miller found, contrary to theory, that slower moving sugars were accelerated more in the homo-, or hetero-exchange mode than the faster homo-exchanges of glucose. This was confirmed (Naftalin 1971).

These anomalous results were considered unsound because they did not account for the retardation of exchange due to competition from unlabelled sugar entering the cytosol (Eilam and Stein 1974). Additionally, K_m's of exchange flux are relatively high (30–50 mM). The concentrations used to determine V_m (100 mM) were considered to be too low to give an accurate estimate of V_m (accelerated exchange). Because it is technically difficult to measure exchange fluxes in human erythrocytes due to the very high rates of this process, the question as to whether there is a real discrepancy remained unresolved for twenty years. Rat erythrocytes contain only about 0.1% of the number (~500/cell) of GLUT1 that human erythrocytes have (~500000/cell). Glucose flux occurring in 1 second in human erythrocytes requires ~10 min in rat erythrocytes. Monitoring exchange fluxes in rat erythrocytes permits a straightforward means of testing whether there is any anomaly in the exchange predictions of the single site carrier.

The maximal rate of mannose influx at 24 °C in rat erythrocytes is only about 20% of that of 3-OMG, or 2-dGlc. However homo and hetero-exchange rates of all three sugars are similar. The alternating carrier model predicts that the rates of homo-exchange of mannose or hetero-exchange with mannose should be slower than the rates of homo exchange with 3-OMG (Naftalin and Rist 1994). Cloherty et al. (1996) undertook a similar series of exchange experiments in human erythrocytes at 0 °C (Table 14.1). They compared the net entry of 3-OMG, D-glucose and 2-dGlc. The V_m (exchange) of D-glucose is 1.5 times faster than that of 2-dGlc. The V_m (exchange) of 3-OMG is only half that of D-glucose, although the V_m (zero *trans* uptake) of 3-OMG slightly exceeds that of D-glucose. These findings confirm that both qualitatively and qualitatively exchange fluxes of sugars are discordant with the alternating carrier predictions, as Miller (1968a,b) first showed.

Another divergence of alternating carrier theory from experimental results is the finding that the V_m (zero *trans*) of various sugars differ widely. If sugar transport is rate limited by the rate of return of the empty carrier, the V_m's of net transport and their temperature coefficients (activation energies) should all be similar (LeFevre and Marshall 1958). This is evidently not the case from the examples already quoted. At temperatures in the range 25–30 °C, the V_m's of glucose and mannose are similar; below 25 °C, the maximal rate of glucose zero *trans* net exit exceeds that of mannose, whereas above 35 °C the V_m (zero *trans*) net exit of mannose exceeds that of glucose. The activation energy of mannose is twice than of D-glucose (Naftalin and Rist 1994). Since net exit fluxes of mannose and glucose have widely differing activation energies, there must be differences in the process(es), which rate-limit their passage across the transporter.

The two-site model (Naftalin and Rist 1994; Naftalin 1997; Naftalin et al. 2001) does not require that sugar transport is rate limited by protein movement, instead ligand movements across the transporter are determined by the extent of conformational changes in the transporter that they induce (Fig. 14.4). This version of the two-site model postulates that net transport occurs by ligands jumping to a vacant site more readily through a more open transporter than a closed one. Glucose, which induces a larger conformational change than mannose, or sorbose, has a lower activation energy for transport. The more recent version of Carruthers' dimer 2-site model (Hamill et al. 2000) is another variant of the alternating carrier, which like the simpler model (Fig. 14.1) requires synchronous movements of ligands and transporter protein rather than relative sugar mobility within the transporter (Figs. 14.1 and 14.4).

Studies of high pressure on the kinetics of glucose transport in human erythrocytes indicate that glucose-specific conformational changes induce hydration changes in GLUT, which increase the activation volume of the transporter (Naftalin et al. 2001). These findings are more consistent with the current view that glucose flows through a central hydrated pore within GLUT (Mueckler et al. 1985; Gould and Holman 1993; Zang et al. 1996; Muecker and Makepeace 1999; Dwyer 2001) (Figs. 14.5 and 14.6).

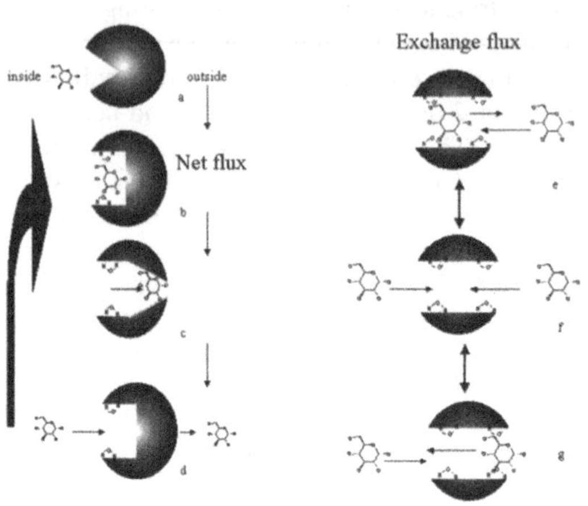

Fig. 14.4. Diagram showing the substates of a two-site glucose transporter. Specific D-glucose binding causes a conformational change, (*a, b*) which admits water to the site. The water competes with ligand and both lowers the affinity, increases the mobility and reduces the activation energy of net glucose movement across the transporter (*c, d, b*). The presence of glucose in both inside and outside compartments induces bilateral hydration of the transporters, (*e*), this raises the mobility of glucose further (*f, h*) thus permitting exchange transport to be faster than net flux

Once hydration has occurred, the ensuing sugar exchanges occur via metastable transporter sub-states. No further conformation change is required as the ligands move within the pore relative to the protein and not with the protein as required by the alternating carrier theory (Fig. 14.4). The alternating carrier requires intramolecular unfolding and refolding movements in the sub-millisecond time domain synchronous with each transport cycle. As net flux in GLUT1 is slowed by interaction of the C-terminal peptide with other cytoplasmic strands, probably in the N-terminal half of the transporter (Cope et al. 1994; Dauterive et al. 1996), it is likely that the net flux transport process involves large scale conformational changes in the cytoplasmic domains of GLUT1.

Fluorescence relaxation studies indicate that much of the conformational change in GLUT occurs more slowly ($\tau \sim$ 1–5 ms) (Gorga and Lienhardt 1981; Lowe and Walmsley 1987; Appleman and Lienhardt 1989; Janoshazi and Solomon 1993; May and Beechem 1993; Sultzman and Carruthers 1999). However, large scale protein folding has a time constant, $\tau \sim$ 1–5 ms (Talaga et al. 2000), this is too slow to accommodate the turnover rates ~1000 s^{-1}. Thus it would seem more probable that conformational change in GLUT is a recognition response (Janoshazi and Solomon 1993) to initial contact with ligand and also a gating process. However, maintenance of the net flux process requires no further large-scale conformational changes. The two site transport process, in which ligands move through a pore gated by a hydration step would better fit with these slow activation steps than the alternating carrier which requires small scale oscillations in the 1–10 μs time domain.

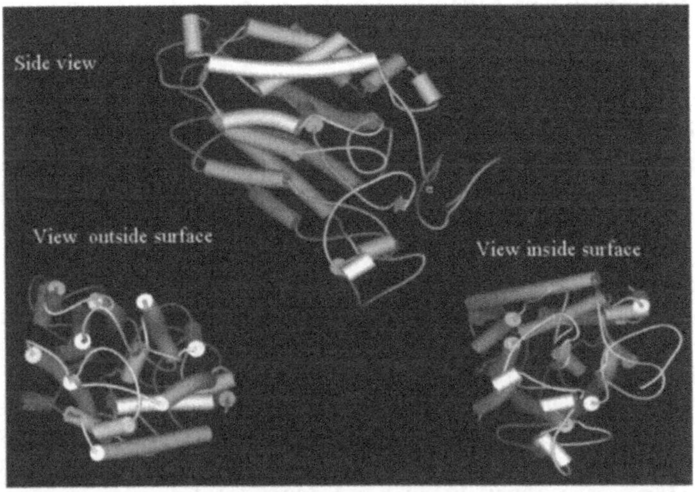

Fig. 14.5. Schematic views of 3D model of GLUT1 (Zuniga et al. 2001). These shows three views from the outside, side and inside. The schematic view clearly demonstrates that there are two possible hydrophilic porous regions in the protein monomer with TM 7 lying diagonally between the two regions. This explains the high accessibility of all facets of TM 7 in comparison with neighbour TM's 10 and 11, which only have one facet accessible to soluble alkylating reagents like PCMBS. These new views were obtained using Weblab Viewer Lite software obtainable as a freeware software from Accelrys Inc

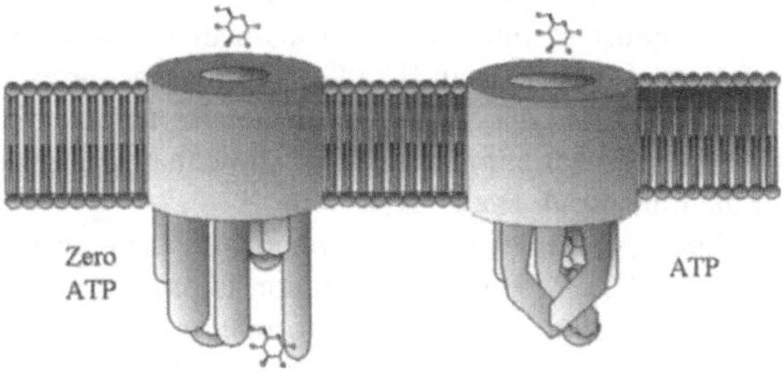

Fig. 14.6. The GLUT transporter with cytoplasmic tentacles. ATP causes partial occlusion of glucose within the cytoplasmatic linker regions. A synthesis of the models of Cloherty et al. (1995), Levine et al. (1998), Heard et al. (2000) (Fig. 14.4) and Dwyer (2001) and Zuniga et al. (2001) (Fig. 14.5) with current understanding of the role of C-terminal peptide and the Walker domains in retardation of net flux. The cytoplasmic processes bend following interaction with ATP. This creates an enclosure, which increases resistance to glucose dissociation from the cytoplasmic surface of the transmembrane pore

14.7.2 Problem 2: Can Ligands Bind to Inside and Outside Sites Simultaneously?

The alternating carrier hypothesis predicts that saturation of the transporter with high affinity ligand in the outside surface of the transporter should cause nearly all the transporter to migrate to the outside and deplete the inside of inward facing transporters, thereby preventing cytochalasin B from binding to the inside. The apparent competition between maltose, or phloretin and cytochalasin B appears to confirm that there is only one site available for ligand binding. Devés and Krupka (1981) and Gorga and Lienhardt (1981) showed that non-transported ligands, which bind exclusively to the external sites on transporter, e.g. maltose and phloretin, or the transported sugar with high affinity for the external site EG prevents cytochalasin B binding to the inner sites on the transporter. Cytochalasin B binds exclusively to the inside of the transporter (Basketter and Widdas 1978; Devés and Krupka 1978; Griffin et al. 1982; Holman and Rees 1987; Chin et al. 1987; King et al. 1991; Carruthers and Helgerson 1989, 1991; May and Beechem 1993; Inukai et al. 1994; Due et al. 1995b; Wandel et al. 1996). It behaves as a non-competitive inhibitor of glucose entry (Basketter and Widdas 1978; Devés and Krupka 1978).

Another view is that it acts as a channel blocker rather than as a competitor to substrate binding (May and Beechem 1993). These results are taken as very strong evidence in support of the alternating carrier.

However, contrary evidence now exists favouring the two fixed-site hypothesis for transport. External ligand binding of maltose results in allosteric interactions across the carrier. These reduce the affinity of the inward facing sites to cytochalasin B (Carruthers and Helgerson 1987). Stopped flow spectrophotometry of tryptophan fluorescence quenching by cytochalasin B in isolated GLUT1, or in depleted erythrocyte ghost membranes has shown that external maltose reduces the rate of association of cytochalasin B to GLUT1 and increases the rate of cytochalasin B dissociation from GLUT1, i.e. competes for net association (Sultzman and Carruthers 1999). Allosteric interactions do show that two ligand-binding sites on either side of the transporter can coexist.

This allosteric interaction between the two sides of the carrier appears to be a property of dimerised GLUT. When dimerization is prevented by treatment with dithiothreitol, which prevents –S–S– bridge formation then allosteric interaction between cytochalasin B and maltose is lost. A question remains as to whether dimerisation can occur in cysteineless mutants of GLUT (Due et al. 1995a; Zottola et al. 1995; Coderre et al. 1995; Hamill et al. 1999).

14.7.3 Problem 3: Is GLUT1 an Asymmetric Transporter?

Glucose transport across the erythrocyte transporter is operationally asymmetrical, in that the initial rate of net entry into cells that are glucose-free (zero *trans* influx) is slower by a factor of 2–5-fold than zero *trans* net exit (see below). The K_m for zero *trans* net uptake of glucose at 25 °C is lower than for exit by a similar factor, i.e. 2–3 mM *vs.* 8–12 mM, although with integrated rate equation methods of estimating net flux parameters the K_m for net exit are higher (15–25 mM) (Hankin et al. 1972; Naftalin et al. 1985).

Glucose transport across GLUT1 is passive; at steady state, glucose concentrations in the bathing solutions on either side of the membrane are equal. Comparison of net exit with entry kinetics shows that the affinity at the export site is lower than at the import site and the maximal exit rate is higher than the rate of asymmetric net glucose entry. The alternating carrier proposes that a closed single cyclical process can explain net glucose transport. The single cycle model requires that the ratio of the Haldane's relationship $V_m(\text{exit})/K_m(\text{exit}) = V_m(\text{entry})/K_m(\text{entry})$ (Hankin et al. 1972; Naftalin and Holman 1977; Baker and Naftalin 1979; Naftalin et al. 1985; Lieb and Stein 1986; Carruthers 1990, 1991; Cloherty et al. 1996). For a single cyclical process, this is a restatement of the *law of microscopic reversibility* (Fig. 14.1). However, *microscopic reversibility* applies only to mechanisms involving single cycles. If the transport process incorporating more than one transport cycle with parallel branches, as in the two-site transport model, then the Haldane relationships do not apply to the ratios V_m/K_m for inward and outward sugar flows (Fig. 14.4).

There are two major reasons for rejecting the simple asymmetric carrier model. The model requires a single low affinity site at the export surface of the transporter. The affinity of the internal (export site) of a simple asymmetric transporter should be independent of the method of monitoring glucose exit. Slow association or dissociation of glucose with, or from low affinity glucose binding sites within the cytoplasmic compartment could alter the kinetics of both entry and exit across the transporter.

If bound and free glucose coexist in the cytoplasmic compartment, then monitoring the initial few seconds of exit, (initial rates approach) would give a more accurate estimate of the membrane flux parameters than one involving measurements over a prolonged period, i.e. the integrated rate approach. If glucose is not compartmentalized within the cytosol, then an integrated rate equation taking proper account of the cell volume changes, will give similar estimates of K_m and V_m at the export site as the initial rate method (Hankin et al. 1972; Lieb and Stein 1974; Ginsburg and Ram 1975; Naftalin et al. 1985; Carruthers 1990; Cloherty et al. 1995, 1996a).

If glucose is compartmentalized within the cytosol, then the estimates of intracellular glucose obtained on the basis of a uniform distribution within the cytosol will be in error because the free glucose will leave the cell before the bound glucose. Once the free sugar has drained from the cell, estimates of free intracellular glucose concentration (expressed per ml of total cell water) and the K_m for exit will be exaggerated.

During net uptake, entry of free sugar into the membrane-associated compartment precedes binding. The K_m measured at the inside site obtained by infinite *cis* entry concentrations of intracellular sugar may be lower than for net exit (Baker and Naftalin 1979; Naftalin et al. 1985; Cloherty et al. 1995, 1996a) (Table 14.1).

The high affinity export site observed during infinite *cis* entry (Hankin et al. 1972; Baker and Naftalin 1979; Ginsburg and Ram 1975; Cloherty et al. 1995) and low affinity export site, observed during zero *trans* net exit, support the view that asymmetry of the glucose transport process in human erythrocytes is due mainly to a slow binding process, independent of the transport process, rather than to an intrinsic asymmetry.

A key question remains: what is the extrinsic cause of the asymmetry? The first candidate was haemoglobin. This provides a viscous microenvironment at the cytoplasmic surface of the cell membrane. Cloherty et al. (1995, 1996a) observed that the time course of net uptake and net exit of 3-OMG, or glucose is bi-exponential. This bi-exponential process time course is also found in erythrocyte ghosts denuded of haemoglobin and in K562 cells, which do not express haemoglobin and in K562, which have been induced to synthesise haemoglobin. Thus, glucose binding to haemoglobin cannot be considered as a likely cause of asymmetric glucose transport.

Bi-exponential uptake is consistent with slow filling and exchange between serial cytoplasmic compartments. The membrane compartment is estimated to be ~30% of the cytoplasmic volume of ultracentrifuge-compacted human red cell membranes. More recent studies have identified this cytoplasmic sugar-binding site as being integral with GLUT1. Studies of 3-OMG binding to erythrocyte ghosts or GLUT1 incorporated into proteoliposomes show that the amount of sugar bound is directly related to the amount of GLUT1 present. Binding of 3-OMG to GLUT1 is monitored using proteoliposomes pelleted with a high-speed centrifuge (1,222,000 × g). 3-OMG also binds as a function of ATP concentration. ATP increases the amount of bound 3-OMG from 0.8 ± 0.05 mol of 3-OMG bound/mole GLUT1 in its absence to 1.71 ± 0.06 mol of 3-OMG bound/mole GLUT1 at 4 mM. It is thought that ATP increases 3-OMG binding only to tetrameric forms of GLUT1. The evidence for this is based mainly on the effects of dithiothreitol, which causes breakdown of the tetramers and prevents ATP dependent increases in 3-OMG binding (Heard et al. 2000) (Fig. 14.4). These latest results from Carruthers' laboratory appear to solve the problem of asymmetry and maybe others too (see below).

Asymmetry appears to be the result of extrinsic binding of glucose to protein. The amount and affinity of this extrinsic glucose binding is dependent on ATP interactions with GLUT, possibly as a result of aggregation of the protein (Fig. 14.7). The extrinsic protein may coincide with the cytoplasmic domains of GLUT, i.e. the linkers between TM 6 and 7, and the N- and C-terminal peptides, which may occlude the export site of the hydrated pore in the presence of ATP (Figs. 14.4–14.7).

14.8 ATP Interactions with Sugar Transport

14.8.1 ATP Effects on Human GLUT1

The effects of ATP on glucose transport are currently unresolved and contentious. Early studies in which cell ATP concentration was reduced using raised Ca^{2+} and A23187 (Hebert and Carruthers 1986) may not be reliable as high Ca^{2+} concentration can trigger proteolysis. Manipulation of erythrocytes to produce ghosting to alter ATP may damage the cytoskeleton and dislodge ATP binding proteins,

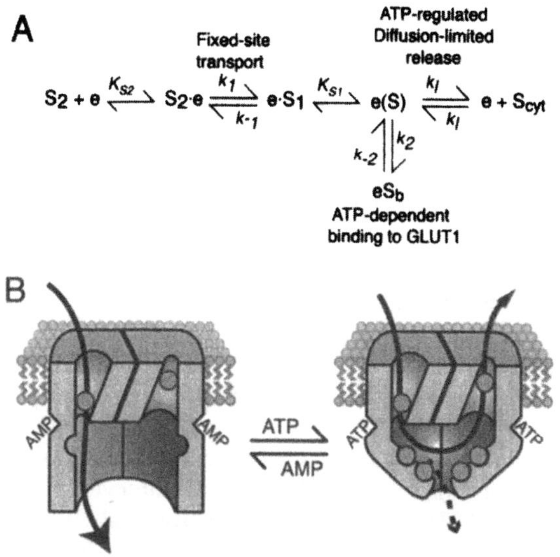

Fig. 14.7. Carruthers' model of GLUT tetramer interaction with ATP and glucose at the cytosolic surface. (**A**) is a schematic model of the asymmetric glucose fluxes across GLUT1 and (**B**) is a diagram showing the same. From Heard et al. (2000)

thereby exposing sites to ATP, which normally may be occluded (Liu et al. 1995; Bunn et al. 1999). Thus the data from these experiments should be viewed with some caution.

The experiments of Helgerson et al. (1989) (Table 14.2) show that radioactively labelled 3-OMG uptakes into resealed ghosts at 0 °C are affected by ATP. Similar results were reported with the effects of non-hydrolysable ATP analogues on 3-OMG uptake into ghosts at 0 °C, namely AMP-PCP, AMP-PNP and ATPγS all stimulated uptake compared with controls (Heard et al. 2000).

Table 14.2. Effects of ATP on kinetics of 3-OMG entry into human erythrocytes

Condition ATP [mM]	Zero *trans* uptake 3-OMG K_m [mM]	Zero *trans* uptake 3-OMG V_m [μmole/(l·min)]
0	7.10 ± 1.80	841 ± 191
4	0.42 ± 0.11	110 ± 15

from Helgerson et al. (1989)

On the other hand Kaloyianni and Baker (1998) using metabolically depleted cells to reduce ATP observed that the rate of glucose exit at 30 °C was slower in ATP depleted cells than fresh cells and the K_m at the external surface was raised slightly from 2.5 mM in normal cells to 3.0 mM in depleted cells. Additionally, they showed that the K_i of cytochalasin B inhibition of glucose exit from human erythrocytes decreased from 0.85 ± 0.09 μM in normal cells to 0.51 ± 0.06 μM in ATP depleted cells. In contrast, Cloherty et al. (1996) reported no observable effect of ATP on the K_i of cytochalasin B inhibition on 3-OMG uptakes into human erythrocyte ghosts. Our own results (Afzal et al. 2001, 2002) are similar to those of Kaloyianni and Baker (1998) in that raising ATP from 0–2 mM increases the K_i of cytochalasin B on D-glucose exit from erythrocyte ghosts at 20 °C.

According to Zottola et al. (1995) and Hamill et al. (1999) tetramerization of GLUT1 leads to loss of half the cytochalasin B binding and to doubling of 3-OMG binding to GLUT1 (Heard et al. 2000) (Fig. 14.4).

14.8.2 ATP on Glucose Transport in Avian Red Cells

The glucose transporter in avian erythrocytes is closely homologous to human GLUT1; nevertheless avian glucose transport differs markedly from in human erythrocytes. Net transport of 3-OMG is virtually absent from fresh avian erythrocytes. Following cell depletion of ATP with metabolic inhibitors, the V_m of net uptake increased by more than 40-fold. In avian erythrocyte ghosts, uptake is reduced when cytosolic ATP is replenished (Cloherty et al. 1996b). In contrast to net uptake, exchange uptake of 3-OMG is present at a relatively high level in both ATP replete and depleted avian erythrocytes. Depletion of ATP has no significant effect on this high rate of exchange.

There are a number of explanations for this suppression of net transport by ATP, amongst which are sequestration of the transporters within the cytosol, or suppression of the transport activity at the cell surface. Since the exchange transport is unaffected by ATP and net transport activity of 3-OMG in ghosts can be suppressed by ATP, it is considered that transport at the cell surface is suppressed by action of ATP on the transporter. It seems likely that ATP regulates net glucose transport in avian erythrocytes in a similar, but more exaggerated way to its effect on human erythrocytes (Hamill et al. 2000). ATP binds to the inside facing sites and results in occlusion of the transporter, perhaps by near apposition of two or more sites. Occlusion of the transporter would reduce net transport but leave exchange transport unaffected, as the occluding site continues to exchange with the cytoplasmic sugar (Figs. 14.4, 14.5 and 14.7 and Table 14.2).

14.8.3 Structural Basis for ATP Interaction with GLUT1

ATP reduces the V_m and K_m of 3-OMG transport at 0 °C by an allosteric mechanism. Only AMP antagonises ATP, most other metabolic intermediates were inactive, although non-metabolizable ATP analogues, e.g. AMP-PNP could substitute for ATP (Hebert and Carruthers 1986; Carruthers and Helgerson 1989; Levine et al. 1998). Several studies have shown that ATP binding to GLUT1 is localized to 2 main sites. Photolabelling studies have shown that 8 azidoadenosine 5'-[γ-^{32}P] trisphosphate is immunoprecipitated with an antiserum raised against the C-terminal peptide of GLUT1. After trypsinolysis of GLUT, the peptide products show that the major sites of label incorporation are residues 213–269, which lie within the cytoplasmic segment between TM 6 and TM 7 and 457–492 which is within the C-terminal peptide also facing towards the cytosol (Fig. 14.2). These two binding sites are close to two of the three Walker ATP sequences. Walker I, GXXXXGK is present in several ATPases and GTPases (Fry et al. 1986). It is found in position 111–118, which in the 12 TM model lies at the outside of TM 3, and consequently is unlikely to binding cytosolic ATP. Walker II, KXVXK has homology with adenylate kinase and protein kinases. Walker III is present between 332 and 338, which lies in the cytoplasmic segment joining TM's 8 and 9 (Carruthers and Helgerson 1989). Levine et al. (1998) identified GLUT1 residues 301–364 as the ATP binding portion. This includes the cytoplasmic domain between TM 8 and 9 and a large portion of TM 9.

Levine et al. (1998) claim that ATP binding to GLUT1 is mainly to the tetrameric form of the transporter and only ATP modulates this form of the transporter. The evidence for this is that the reducing agent dithiothreitol DTT, prevents labelling of GLUT1 by 8-azidoadenosine 5'-[γ-^{32}P] trisphosphate and also prevents the ATP dependent modulation of 3-OMG uptake into erythrocyte ghosts (Levine et al. 1998) (Fig. 14.1).

Mutational analysis by substituting Gly111→Ala in Walker ATP binding domain I or Gly336→Ala Walker domain III GLUT shows that neither of expressed mutants GLUT1 has marked differences in net transport parameters from controls. There is an absence in mutants of any differences between equilibrium exchange and net flux K_m and V_m (Liu et al. 2001), which indicates that the mutants have lost their capacity for accelerated exchange.

14.9 Drug Interactions with Glucose Transport

This area of research is rapidly developing and may ultimately lead to new insights about the mechanism of glucose transport.

14.9.1 Barbiturates

Barbiturates are known to inhibit glucose transport in red cells (El Barbary et al. 1995; Honkanen et al. 1995; Haspel et al. 1999; Naftalin and Arain 1999) and across the blood brain barrier (Gjedde and Rasmussen 1980; Lowry and Fillenz 2001). Barbiturates are amphipathic drugs, which have a higher efficacy in hydrophobic environments. It has been suggested that the drug binds to hydrophobic regions in the interstices of the protein folds and thereby reduces, or prevents conformational changes induced by substrate binding (Dickenson et al. 1995). The affinity of barbiturates for GLUT inhibition varies widely, thiol substitution raises the affinity for the glucose transporter, e.g. thiamylal has a 10 times higher affinity than sodium pentobarbital. However, most of the K_i's of barbiturates action on GLUT exceed by a wide margin ~10 times their IC_{50} for induction of anaesthesia. It is considered that the main site of action of barbiturates in induction of anaesthesia is on $GABA_A$ receptors (Study and Barker 1981).

Recent work has suggested that pentobarbital acts as a mixed inhibitor of net glucose exit (Naftalin and Arain 1999) possibly by preventing the glucose-specific conformational change required to facilitate its passage via the transporter (Naftalin et al. 2002).

14.9.2 Glucose Transporter Deficiency Syndrome (GTDS) in Relation to Drug Action

More than 70 patients have been identified with *glucose transporter deficiency syndrome* since 1991. The diagnostic features of the disease are a low glucose concentration in the CSF (hypoglycorrhachia) in the presence of normoglycaemia (CSF/blood glucose ratio < 0.4). Clinical features include seizures, developmental delay, acquired microcephaly, hypotonia, and a complex motor disorder with elements of ataxia, dystonia, and spasticity. The GLUT1 defect is also present in erythrocytes. Glucose uptake studies in red cells of GTDS patients show a slowed glucose uptake usually without any significant alteration in glucose affinity for the transporter. Other more exact ways of identifying the disease are by molecular analysis of the GLUT1 gene. Several heterozygous mutations result in GLUT1 haploinsufficiency. As ketone bodies serve as an alternative fuel for the developing brain, treatment with a ketogenic diet of the affected children reduces the clinical symptoms (Klepper and Voit 2002). Children with *glucose transporter protein syndrome* (GTPS; Klepper et al. 1999b), may particularly at hazard from barbiturates and xanthine drugs, like caffeine, which further diminish the low rate of glucose and dehydroascorbate transport in these patients (Klepper et al. 1998, 1999a).

14.9.3 Steroids, Flavones and Isoflavones

Oestrogens and some phytoestrogens, like genistein exert strong direct effects on GLUT1. Genistein and oestradiol inhibit glucose, 2-dGlc and dehydroascorbate (DHA), the oxidised form of ascorbate uptake into an a number of cultured cells, in particular Chinese Hamster Ovary (CHO) cells over expressing GLUT1 and human erythrocytes (Vera et al. 2001). Since, genistein competitively displaces cytochalasin B binding from GLUT it is evident that it binds to the inside site. Interestingly, only isoflavones that inhibit tyrosine kinases at their ATP binding site inhibit glucose transport. Thus, whilst genistein inhibits both tyrosine kinase and glucose transport daidzein does neither (Vera et al. 2001).

ATP antagonises oestradiol inhibition of glucose exit from human red cell ghosts. The anti-oestrogens, tamoxifen and faslodex competitively reverse these effects of oestradiol on glucose transport. However these effects of anti-oestrogens are only evident in the presence of ATP (Afzal et al. 2001, 2002). These results indicate that there is a pseudo-oestrogen binding site at the inner surface of GLUT1, which interacts both with ATP binding and glucose transport. Four sequences in the cytoplasmic surface of GLUT1 have close homologies with the ligand binding domain in the human oestrogen receptor β (ESR-2). One sequence is adjacent to the Walker II ATP-binding motif between transmembrane spans 6 and 7 (GLUT 225–229); another contains the Walker ATP-binding motif III,

GLUT (421–423). Mapping of these regions onto a 3D template of GLUT indicates that an oestrogen binding site lies between His 336, Arg 326 and Glu 327. These groups have similar topology to those anchoring the head and tail OH groups of the oestradiol in ESR-2 and are suitably sited at the cytoplasmic entrance to a hydrophilic pore traversing GLUT to act as a competitive inhibitor to glucose export (Afzal et al. 2002). Whether this is of physiological significance in relation to the non-genomic effects of oestrogens and phytoestrogens remains to be discovered.

14.9.4 The L-Type Channel Antagonists

Nimodipine, nifedipine, verapamil, and diltiazem all inhibit glucose transport in a dose-dependent manner. Nimodipine is the most potent L-type Ca^{2+} channel blocker of glucose transport. T- and N-type channel antagonists do not affect glucose transport. Glucose appears to act competitively with the drug interactions at the external surface as raising the concentration of glucose in the medium reduced the inhibition by the drugs. The inhibitor effects of the drugs were unaffected by the presence or absence of Ca^{2+} in the medium (Ardizzone et al. 2002).

14.9.5 Antipsychotic drugs

Treatment of schizophrenics with some antipsychotic drugs, e.g. risperidone and clozapine, and structural analogs of clozapine has been associated with an increased incidence of hyperglycemia and new-onset type 2 diabetes. Some of these drugs also inhibit glucose transport in rat pheochromocytoma (PC12) cells. Risperidone and clozapine inhibit glucose transport in a dose-dependent fashion with IC_{50}'s estimated to be 35 and 20 μM, respectively. The clozapine metabolite, desmethylclozapine, is more potent than clozapine. They inhibition of glucose transport is non-competitively antagonized by including 2-deoxyglucose (5 mM) in the uptake medium. These studies suggest that the drugs may block glucose accumulation directly at the level of the glucose transporter (GLUT) protein in cells derived from both peripheral and brain tissue (Ardizonne et al. 2001).

14.9.6 Actions of HIV Protease Inhibitors on GLUTs

The HIV protease inhibitors (Pi) inhibit glucose uptake via GLUT4. The sensitivity to inhibition by indinavir among the Glut isoforms assayed in the *X. laevis* oocyte system was Glut4 >> Glut2 > Glut3 > Glut1 approximate to Glut8. Indinavir non-competitively inhibits zero *trans* uptake of 2-dGlc in insulin-stimulated primary rat adipocytes K_i = 15 μM. Indinavar raised glucose and insulin levels in rats. Use of Pis has been associated with lipodystrophy, hyperlipidemia, and insulin resistance. These data suggest that these complications may be directly related to the drug interactions with GLUT4 on muscle and adipose tissue. To some extent there appears to be some similarity between the diabetogenic effects of indinivir and those antipsychotic drugs.

14.10 Conclusions

Much of the function of GLUT1 is determined by interactions within cytoplasmic segments, particularly the N- and C-terminal chains (Cope et al. 1994), and the linkers between TM 6 and 7 and between TM 8 and 9 (Nishimura et al. 1993; Dauterive et al. 1996).

These effects can be summarised as follows:

- The interactions between GLUT1 C-terminal and other cytoplasmic structures retard net glucose entry (Oka et al. 1990; Nishimura et al. 1993; Dauterive et al. 1996).
- Cytoplasmic structures form ATP-dependent cytoplasmic glucose binding sites (Hebert and Carruthers 1986; Carruthers and Helgerson 1989; Helgerson et al. 1989; Cloherty et al. 1996b; Levine et al. 1998; Heard et al. 2000; Liu et al. 2001; Vera et al. 2001). These structures prevent net flux in avian erythrocytes and retard both entry and exit in human erythrocytes causing asymmetric glucose transport (Cloherty et al. 1996b; Hamill et al. 1999; Sultzman et al. 1999; Heard et al. 2000).
- These cytoplasmic structures may be sites of drug and inhibitor binding, e.g. genistein and cytochalasin B (Afzal et al. 2001, 2002; Liu et al 2001; Vera et al. 2001).
- GLUT1 has long been considered to have a hydrophilic pore. The 3D simulation of GLUT3 (Dwyer 2001) suggest that the cytoplasmic segments of GLUT cover the export site of hydrated pore through which glucose traverses the transporter and at least partially occlude the exit in the presence of ATP (Figs. 14.3, 14.5 and 14.6).
- Recent support for the two channel hypothesis has been found. Glucose transport across a mutant GLUT1 expressed in *Xenopus* oocytes with cRNA derived from a patient with glucose transporter deficiency syndrome in which threonine 310 in TM 8 is replaced with the more bulky and hydrophobic isoleucine (mutant T310I) is reduced by 87%, whereas osmotic water permeability associated with the mutant GLUT is unaffected. 10 mM glucose and L-glucose inhibit the osmotic permeability of wild type GLUT1, but leave the water transport across the mutant unaffected. Interestingly 5 mM L-glucose increases the osmotic permeability of the mutant GLUT (Iserovich et al. 2002). These experimental observations may suggest that there are separate channels for glucose and water permeation. However, the possibility that water and glucose share the same pathway cannot be excluded by this data. The kinetic implications of a parallel pathway for glucose transport have not been considered yet and in any case, it is unclear whether the alternate pathway is a transport route for glucose permeation.

Acknowledgement: My thanks to Prof. Anthony Carruthers, Prof. Goeffrey Holman and Dr. Timothy Simons for their critical reading of this manuscript and helpful suggestions.

References

Afzal I, Cunningham P, Naftalin RJ (2001) Interactions of ATP, 17β-estradiol, genistein and the anti-estrogens, ICI 182780 (Faslodex™) and 4-OH tamoxifen with human erythrocyte glucose transporter, GLUT1. Proceedings, Pharmaconference 2001, Interlaken, organizer: MA Hediger

Afzal I, Cunningham P, Naftalin RJ (2002) Interactions of ATP, oestradiol, genistein and the anti-oestrogens, faslodex (ICI 182780) and tamoxifen, with the human erythrocyte glucose transporter, GLUT1. Biochem J 365:707-719

Alvarez J, Lee DC, Baldwin SA, Chapman D (1987) Fourier transform infrared spectroscopic study of the structure and conformational changes of the human erythrocyte glucose transporter. J Biol Chem 262:3502-3509

Appleman JR, Lienhard GE (1989) Kinetics of the purified glucose transporter - direct measurement of the rates of interconversion of transporter conformers. Biochemistry 28:8221-8227

Arbuckle MI, Kane S, Porter LM, Seatter MJ, Gould GW (1996) Structure-function analysis of liver type (GLUT2) and brain type (GLUT3) glucose transporters: Expression of chimeric transporters in *Xenopus* oocytes suggests an important role for putative transmembrane helix 7 in determining substrate selectivity. Biochemistry 35:16519-16527

Ardizzone TD, Bradley RJ, Freeman AM, Dwyer DS (2001) Inhibition of glucose transport in PC12 cells by the atypical antipsychotic drugs risperidone and clozapine, and structural analogs of clozapine. Brain Res 923:82-90

Ardizzone TD, Lu XH, Dwyer DS (2002) Calcium-independent inhibition of glucose transport in PC-12 and L6 cells by calcium channel antagonists. Am J Physiol 283:C579-C586

Baker GF, Naftalin RJ (1979) Evidence for multiple operational affinities for D-glucose inside the human erythrocyte membrane. Biochim Biophys Acta 550:474-484

Baker GF, Rogers HJ (1972) Effects of psychotropic drugs on the erythrocyte permeability to glucose and ethylidene glucose. Biochem Pharmacol 21:1871-1878

Baker GF, Widdas WF (1973) Asymmetry of the facilitated transfer system for hexoses in human red cells and the simple kinetics of two component model. J Physiol 231:143-165

Basketter DA, Widdas WF (1978) Asymmetry of the hexose transfer system in human erythrocytes: Comparison of the effects of cytochalasin B, phloretin and maltose as competitive inhibitors. J Physiol 278:389-401

Baldwin SA (1993) Mammalian passive glucose transporters - members of a ubiquitous family of active and passive transport proteins. Biochim Biophys Acta 1154:17-49

Barnett JE, Holman GD, Munday KA (1973) Structural requirements for binding to the sugar transport system of the human erythrocyte. Biochem J 131:211-221

Bell GI, Burant CF, Takeda T, Gould GW (1993) Structure and function of mammalian facilitative sugar transporters. J Biol Chem 268:19161-19164

Buchs AE, Sasson S, Joost HG, Cerasi E (1998) Characterisation of GLUT5 domains responsible for fructose transport. Endocrinology 139:827-831

Bunn RC, Jensen MA, Reed BC (1999) Protein interactions with the glucose transporter binding protein GLUT1CBP that provide a link between GLUT1 and the cytoskeleton. Mol Biol Cell 10:819-832

Cairns MT, Elliot DA, Scudder PR, Baldwin SA (1984) Proteolytic and chemical dissection of the human erythrocyte glucose transporter. Biochem J 221:179-188

Calderhead DM, Kitagawa K, Tanner LI, Holman GD, Lienhard GE (1990) Insulin regulation of the 2 glucose transporters in 3T3-L1 adipocytes. J Biol Chem 265:13800-13808

Carruthers A (1990) Facilitated diffusion of glucose. Physiol Rev 70:1135-1176

Carruthers A (1991) Mechanisms for the facilitated diffusion of substrates across cell membranes. Biochemistry 30:3898-3907

Carruthers A, Helgerson AL (1989) The human erythrocyte sugar transporter is also a nucleotide binding protein. Biochemistry 28:8337-8346

Carruthers A, Helgerson AL (1991) Inhibition of sugar transport produced by ligand binding at opposite sides of the membrane. Evidence for simultaneous occupancy of the carrier by maltose and cytochalasin B. Biochemistry 30:3907-3915

Chin JJ, Jung EK, Chen V, Jung CY (1987) Structural basis of erythrocyte glucose transporter function in proteoliposome vesicles: Circular dichroism measurements. Proc Natl Acad Sci USA 84:4113-4116

Cloherty EK, Sultzman LA, Zottola RJ, Carruthers A (1995) Net sugar transport is a multistep process - evidence for cytosolic sugar binding-sites in erythrocytes. Biochemistry 34:15395-15406

Cloherty EK, Heard KS, Carruthers A (1996a) Human erythrocyte sugar transport is incompatible with available carrier models. Biochemistry 35:10411-10421

Cloherty EK, Diamond DL, Heard KS, Carruthers A (1996b) Regulation of GLUT1-mediated sugar transport by an antiport/uniport switch mechanism. Biochemistry 35:13231-13239

Coderre PE, Cloherty EK, Zottola RJ, Carruthers A (1995) Rapid substrate translocation by the multi-subunit, erythroid glucose-transporter requires subunit associations but not cooperative ligand-binding. Biochemistry 34:9762-9773

Cohen RM, Nordahl TE, Semple WE, Andreason P, Litman RE, Pickar D (1997) The brain metabolic patterns of clozapine and fluphenazine treated patients with schizophrenia during a continuous performance task. Arch Gen Psychiatry 54:481-486

Cope DL, Holman GD, Baldwin SA (1994) Domain assembly of the Glut1 glucose-transporter. Biochem J 300:291-294

Dauterive R, Laroux S, Bunn RC, Chaisson A, Sanson T, Reed BC (1996) C-terminal mutations that alter the turnover number for 3-*O*-Methylglucose transport by GLUT1 and GLUT4. J Biol Chem 271:11414-11421

Davies A, Ciardeli TL, Leinhard GE, Boyle JM, Whetton AD, Baldwin SA (1990) Site-specific antibodies as probes of the topology and function of the human erythrocyte glucose transporter. Biochem J 266:799-808

Devés R, Krupka RM (1978) Cytochalasin B and the kinetics of inhibition of biological transport. A case of asymmetric binding to the glucose carrier. Biochim Biophys Acta 510:339-348

Dickinson R, Lieb WR, Franks NP (1995) The effects of temperature on the interactions between volatile general-anesthetics and a neuronal nicotinic acetylcholine-receptor. Br J Pharmacol 116:2949-2956

Doege H, Schürmann A, Ohnimus H, Monser V, Holman GD, Joost H (1998) Serine-294 and threonine-295 in the exofacial loop domain between helices 7 and 8 of glucose transporters (GLUT) are often involved in the conformational alterations during the transport process. Biochem J 329:289-293

Due DA, Cook JA, Fletcher SJ, Zhi-Chao Q (1995a) A "cysteineless" GLUT1 glucose transporter has normal function when expressed in *Xenopus* oocytes. Biochem Biophys Res Commun 208:590-596

Due DA, Zhi-Chao Q, Thomas JM, Buchs A, Powers A, May JM (1995b) Role of the C-terminal tail of the GLUT1 glucose transporter in its expression and function in *Xenopus laevis* oocytes. Biochemistry 34:5462-5471

Dwyer DS, Liu Y, Bradley RJ (1999) Dopamine receptor antagonists modulate glucose uptake in rat pheochromocytoma (PC12) cells. Neurosci Letters 274:151-154

Dwyer DS (2001) Model of the 3-D structure of the GLUT3 glucose transporter and molecular dynamics simulation of glucose transport. Proteins: Structure, Function and Genetics 42:531-541

Eilam Y, Stein WD (1974) A simple resolution of the kinetic anomaly in the exchange of different sugars across the membrane of the human red cell. Biochim Biophys Acta 266:161-173

El-Barbary A, Fenstermacher JD, Haspel HC (1996) Barbiturate inhibition of GLUT-1 mediated hexose transport in human erythrocytes exhibits substrate dependence for equilibrium exchange but not unidirectional sugar flux. Biochemistry 35:15222-15227

Fischbarg J, Min C, Czegledy F, Li J, Iserovich P, Kuang KY, Hubbard J, Garner M, Rosen OR, Golde DW, Vera JC (1993) Evidence that facilitative glucose transporters may fold as β-barrels. Proc Natl Acad Sci USA 90:11658-11662

Fry DC, Kuby SA, Mildvan AS (1986) ATP-binding site of adenylate kinase: Mechanistic implications of homology with ras-encoded p21, F1-ATPase, and other nucleotide binding proteins. Proc Natl Acad Sci USA 83:907-911

Gaut ZN (1974) Influence of various substance which induce and inhibit aggregation on the uptake of deoxyglucose by human blood platelets. J Pharmacol Exp Therap 190:180-186

Gibbs AF, Chapman D, Baldwin SA (1988) Proteolytic dissection as a probe of conformational changes in the human erythrocyte glucose transport protein. Biochem J 256:421-427

Ginsburg H, Ram D (1975) Zero trans and equilibrium exchange efflux and infinite trans uptake of galactose by human erythrocytes. Biochim Biophys Acta 382:396-376

Gjedde A, Rasmussen M (1980) Pentobarbital anesthesia reduces blood-brain glucose transfer in the rat. J Neurochem 35:1382-1387

Gorga F, Lienhard GE (1981) Equilibria and kinetics of ligand binding to the human erythrocyte glucose transporter: Evidence for an alternating conformation model for transport. Biochemistry 20:5108-5113

Gould GW, Holman GD (1993) The glucose-transporter family - structure, function and tissue-specific expression. Biochem J 295:329-341

Griffin JF, Rampal AL, Jung CY (1982) Inhibition of glucose transport in human erythrocytes by cytochalasins: A model based on diffraction studies. Proc Natl Acad Sci USA 79:3759-3763

Hamill S, Cloherty EK, Carruthers A (1999) The human erythrocyte sugar transporter presents two sugar import sites. Biochemistry 38:16974-16983

Hankin BL, Lieb WR, Stein WD (1972) Rejection criteria for the asymmetric carrier and their application to glucose transport in the human red blood cell. Biochim Biophys Acta 288:114-126

Haspel HC, Rosenfeld MG, Rosen MO (1988) Characterisation of antisera to a synthetic carboxyl-terminal peptide of the glucose transporter protein. J Biol Chem 263:398-403

Haspel HC, Stephenson KN, Davies-Hill T, El-Barbary A, Lobo JF, Croxen RL, Mougrabi W, Koehler-Stec EM, Fenstermacher JD, Simpson IA (1999) Effects of barbiturates on facilitative glucose transporters are pharmacologically specific and isoform selective. J Membrane Biol 169:45-53

Heard KS, Fidyk N, Carruthers A (2000) ATP-dependent substrate occlusion by the human erythrocyte sugar transporter. Biochemistry 39:3005-3014

Hebert DN, Carruthers A (1986) Direct evidence for ATP modulation of sugar-transport in human erythrocyte ghosts. J Biol Chem 261:93-99

Helgerson AL, Carruthers A (1987) Equilibrium ligand binding to the human erythrocyte sugar transporter. J Biol Chem 262:5464-5475

Helgerson AL, Hebert DN, Naderi S, Carruthers A (1989) Characterisation of two independent modes of action of ATP on human erythrocyte sugar transport. Biochemistry 28:6410-6417

Holman GD, Rees WD (1987) Photolabelling of the hexose transporter at external and internal sites: fragmentation patterns and evidence of conformational change. Biochim Biophys Acta 897:395-405

Honkanen RA, McBath H, Kushmerick C, Callender GE, Scarlata SF, Fenstermacher JD, Haspel HC (1995) Barbiturates inhibit hexose transport in cultured mammalian cells and human erythrocytes and interact directly with purified GLUT-1. Biochemistry 34:535-544

Hresko RC, Kruse M, Strube M, Mueckler M (1994) Topology of the Glut1 glucose transporter deduced from glycosylation scanning mutagenesis. J Biol Chem 269:20482-20488

Hruz PW, Mueckler MM (1999) Cysteine-scanning mutagenesis of transmembrane segment 7 of the GLUT1 glucose transporter. J Biol Chem 274:36176-36180

Hruz PW, Mueckler MM (2000) Cysteine-scanning mutagenesis of transmembrane segment 11 of the GLUT1 facilitative glucose transporter. Biochemistry 39:9376-9372

Hruz PW, Murata H, Qiu HJ, Mueckler MM (2002) Indinavir induces acute and reversible peripheral insulin resistance in rats. Diabetes 51:937-942

Inukai K, Asano T, Katagiri H, Ana IM, Funaki M, Ishihara H, Tsukuda K, Kikuchi M, Yazaki Y, Oka Y (1994) Replacement of both tryptophan residues at 388 and 412 completely abolished cytochalasin-B photolabeling of the GLUT1 glucose-transporter. Biochem J 302:355-361

Iserovich P, Wang D, Ma L, Yang H, Zuniga FA, Pascual JM, Kuang KY, De Vivo DC, Fischbarg J (2002) Changes in glucose transport and water permeability resulting from the T310I pathogenic mutation in Glut1 are consistent with two transport channels per monomer. J Biol Chem 277:30991-30997

Janoshazi A, Solomon AK (1993) Initial steps of alpha-D-glucose and beta-D-glucose binding to intact red cell membranes. J Membrane Biol 132:167-178

Joost HG, Thorens B (2001) The extended GLUT-family of sugar/polyol transport facilitators: nomenclature, sequence characteristics, and potential function of its novel members. Mol Membrane Biol 18:247-256

Kaloyianni M, Baker GF (1998) The effect of ATP-depletion on the inhibition of glucose exits from human red cells. Biochim Biophys Acta 1369:295-303

Kasahara T, Kasahara M (1997) Characterisation of rat GLUT4 glucose transporter expressed in the yeast *Saccharomyces cerevisiae*: comparison with GLUT1 glucose transporter. Biochim Biophys Acta 1324:111-119

King APJ, Tai P-KK, Carter-Su C (1991) Cytochalasin B interferes with conformational changes of the human erythrocyte glucose transporter induced by internal and external sugar binding. Biochemistry 30:11546-11553

Klepper J, Voit T (2002) Facilitated glucose transporter protein type 1 (GLUT1) deficiency syndrome: impaired glucose transport into brain - a review. Eur J Pediatrics 161:295-304

Klepper J, Vera JC, De Vivo DC (1998) Deficient transport of dehydroascorbic acid in the glucose transporter protein syndrome. Ann Neurol 44:286-287

Klepper J, Fischbarg J, Vera JC, Wang D, De Vivo DC (1999a) GLUT-1 deficiency: barbiturates potentiate haploin sufficiency *in vitro*. Pediatr Res 46:677-683

Klepper J, Wand D, Fischbarg J, Vera JC, Jarjour IT, O'Driscoll KR, De Vivo DC (1999b) Defective glucose transport across brain tissue barriers: a newly recognized neurological syndrome. Neurochem Res 24:587-594

Kozka IJ, Clark AE, Holman GD (1991) Chronic treatment with insulin selectively down-regulates cell-surface Glut4 glucose transporters in 3T3-L1 adipocytes. J Biol Chem 266:11726-11731

Koumanov F, Yang J, Jones AE, Hatanaka Y, Holman GD (1998) Cell-surface biotinylation of GLUT4 using bis-mannose photolabels. Biochem J 330:1209-1215

Lachaal M, Rampal AL, Lee W, Shi Y-W, Jung CY (1996) GLUT1 transmembrane glucose pathway affinity labelling with a transportable D-glucose diazirine. J Biol Chem 271:5225-5230

Lacko L, Wittke B, Geck P (1973) The temperature dependence of exchange transport of glucose in human erythrocytes. J Cell Physiol 82:213-318

LeFevre PG, Marshall JK (1958) Conformational specificity in a biological sugar transport system. Am J Physiol 194:333-337

Levine M, Oxender DL, Stein WD (1965) The substrate-facilitated transport of the glucose carrier across the human erythrocyte membrane. Biochim Biophys Acta 109:151-163

Levine KB, Cloherty EK, Fidyk NJ, Carruthers A (1998) Structural and physiological determinants of human erythrocyte sugar transport regulation by adenosine triphosphate. Biochemistry 37:12221-12232

Lieb WR, Stein WD (1974) Testing and characterising the simple carrier. Biochim Biophys Acta 373:178-196

Liu H, Xiong S, Shi Y, Samuel SJ, Lachaal M, Jung CY (1995) ATP-sensitive binding of a 70 kDa cytosolic protein to the glucose transporter in rat adipocytes. J Biol Chem 270:7869-7875

Liu Q, Vera JC, Peng H, Golde DW (2001) The predicted ATP-binding domains in the hexose transporter GLUT1 critically affect transporter activity. Biochemistry 40:7874-7881

Lowe AG, Walmsley AR (1986) The kinetics of glucose transport in human red cells. Biochim Biophys Acta 857:146-154

Lowe AG, Walmsley AR (1987) A single half-turnover of the glucose carrier of the human erythrocyte. Biochim Biophys Acta 903:547-550

Lowry JP, Fillenz M (2001) Real time monitoring of brain energy metabolism *in vivo* using microelectrochemical sensors: the effects of anesthesia. Bioelectrochem 54:39-47

Masiak SJ, LeFevre PG (1977) Glucose transport inhibition by proteolytic degradation of the human erythrocyte membrane. Biochim Biophys Acta 465:371-377

May JM, Beechem JM (1993) Monitoring conformational change in the human erythrocyte glucose carrier - use of a fluorescent probe attached to an exofacial carrier sulfhydryl. Biochemistry 32:2907-2915

May JM, Qui ZC, Beechem JM (1993) Tryptic digestion of the human erythrocyte glucose transporter - effects on ligand-binding and tryptophan fluorescence. Biochemistry 32:9524-9531

Miller DM (1968a) The kinetic of selective biological transport. III erythrocyte monosaccharide transport data. Biophys J 8:1329-1338

Miller DM (1968b) The kinetics of selective biological transport. IV Assessment of three carrier systems using the erythrocyte-monosaccharide transport data. Biophys J 8:1339-1352

Monden I, Olsowski A, Krause G, Keller K (2001) The large cytoplasmic loop of the glucose transporter GLUT1 is an essential structural element for function. Biol Chem 382:1551-1558

Mueckler M, Makepeace C (1997) Identification of an amino acid residue that lies between the exofacial vestibule and exofacial substrate-binding site of the Glut1 sugar permeation pathway. J Biol Chem 272:30141-30146

Mueckler M, Makepeace C (1999) Transmembrane segment 5 of the Glut1 glucose transporter is an amphipathic helix that forms part of the sugar permeation pathway. J Biol Chem 274:10923-10926

Mueckler M, Makepeace C (2002) Analysis of transmembrane segment 10 of the Glut1 glucose transporter by cysteine-scanning mutagenesis and substituted cysteine accessibility. J Biol Chem 277:3498-3503

Mueckler M, Caruso C, Baldwin SA, Panico M, Blench I, Morris HR, Allard WJ, Leinhard GE, Lodish HF (1985) Sequence and structure of a human glucose transporter. Science 229:941-945

Muraoka A, Hashiramoto M, Clark AE, Edwards LC, Sakura H, Kadowaki T, Holman GD, Kasuga M (1995) Analysis of the structural features of the C-terminus of Glut1 that are required for transport catalytic activity. Biochem J 311:699-704

Murata H, Hruz PW, Mueckler M (2002) Indinavir inhibits the glucose transporter isoform Glut4 at physiologic concentrations. AIDS 16:859-863

Naftalin RJ (1971) The role of unstirred layers in control of sugar movements across red cell membranes. Biochim Biophys Acta 211:65-78

Naftalin RJ (1997) Evidence from studies of temperature-dependent changes of D-glucose, D-mannose and L-sorbose permeability that different states of activation of the human erythrocyte hexose transporter exist for good and bad substrates. Biochim Biophys Acta 1328:13-29

Naftalin RJ, Arain M (1999) Interactions of sodium pentobarbital with D-glucose and L-sorbose transport in human red cells. Biochim Biophys Acta 1419:78-88

Naftalin RJ, Holman GD (1977) Transport of sugars in human red cells. In: Ellory JC, Lew VL (eds) Membrane transport in red cells. Academic Press, New York, pp 257-300

Naftalin RJ, Rist RJ (1994) Re-examination of hexose exchanges using rat erythrocytes - evidence inconsistent with a one-site sequential exchange model, but consistent with a 2-site simultaneous exchange model. Biochim Biophys Acta 1191:65-78

Naftalin RJ, Smith PM, Roselaar SE (1985) Evidence for non-uniform distribution of D-glucose within human red cells during net exit and counterflow. Biochim Biophys Acta 820:235-249

Naftalin RJ, Afzal I, Browning J, Wilkins RJ, Ellory JC (2002) Effects of high-pressure on glucose transport and hydration interactions with the human erythrocyte glucose transporter. J Membrane Biol 186:113-129

Nishimura H, Pallardo FW, Seidner GA, Vannucci S, Simpson IA, Birnbaum MJ (1993) Kinetics of GLUT1 and GLUT4 glucose transporters expressed in *Xenopus* oocytes. J Biol Chem 268:8514-8520

Noel LE, Newgard CB (1997) Structural domains that contribute to substrate specificity in facilitated glucose transporters are distinct from those involved in kinetic function: studies with GLUT-1/GLUT-2 chimeras. Biochemistry 36:5465-5475

Oka Y, Asano T, Shibasaki Y, Lin JL, Tsukuda K, Katagiri H, Akanuma Y, Takaku F (1990) C-terminal truncated glucose transporter is locked into an inward facing form without transport activity. Nature 345:550-553

Olsowski A, Monden I, Krause G, Keller K (2000) Cysteine-scanning mutagenesis of helices 2 and 7 in GLUT1 identifies an exofacial cleft in both transmembrane segments. Biochemistry 39:2469-2474

Palfreyman RW, Clark AE, Denton RM, Holman GD, Kozka IJ (1992) Kinetic resolution of the separate GLUT1 and GLUT4 glucose transport activities in 3T3-L1 cells. Biochem J 284:275-281

Pao SS, Paulsen IT, Saier MH (1998) Major facilitator superfamily. Microbiol Mol Biol Rev 62:1-34

Saier MH (2000) Families of transmembrane sugar transport proteins. Mol Microbiol 35:699-710

Seatter MJ, De La Rue SA, Porter LM, Gould GW (1998) QLS motif in transmembrane helix VII of the glucose transporter family interacts with the C-1 position of D-glucose and is involved in substrate selection at the exofacial binding site. Biochemistry 37:1322-1326

Sergeant S, Kim HD (1985) Inhibition of 3-O-methylglucose transport in human erythrocytes by forskolin. J Biol Chem 260:14677-14682

Stein WD (1986) Transport and diffusion across cell membranes. Academic Press, New York

Study RE, Barker JL (1981) Diazepam and (--)-pentobarbital: fluctuation analysis reveals different mechanisms for potentiation of gamma-aminobutyric acid responses in cultured central neurons. Proc Natl Acad Sci USA 78:7180-7184

Sultzman LA, Carruthers A (1999) Stop-flow analysis of cooperative interactions between GLUT1 sugar import and export sites. Biochemistry 38:6640-6650

Talaga DS, Lau WL, Roder H, Tang J, Jia Y, De Grado WF, Hochstrasser RM (2000) Dynamics and folding of single two-stranded coiled-coil peptides studied by fluorescent energy transfer confocal microscopy. Proc Natl Acad Sci USA 9:13021-13026

Tamori Y, Hashiramoto M, Clark AE, Mori H, Muraoka A, Kadowaki T, Holman GD, Kasuga M (1994) Substitution at Pro(385) of Glut1 perturbs the glucose transport function by reducing conformational flexibility. J Biol Chem 269:2982-2986

Taylor LP, Holman GD (1981) Symmetrical kinetic parameters of 3-O-methyl D-glucose transport in adipocytes in the presence and absence of insulin. Biochim Biophys Acta 642:325-335

Uldry M, Ibberson M, Hosokawa M, Thorens B (2002) GLUT2 is a high affinity glucosamine transporter. FEBS Letters 524:199-203

Vera JC, Reyes AM, Velasquez FV, Rivas CI, Zhang RH, Strobel P, Slebe JC, Nunez-Alarcon J, Golde DW (2001) Direct inhibition of the hexose transporter GLUT1 by tyrosine kinase inhibitors. Biochemistry 40:777-790

Wadzinski BE, Shanahan MF, Seamon KB, Ruoho AE (1990) Localization of the forskolin photolabelling site within the monosaccharide transporter of human erythrocytes. Biochem J 272:151-158

Wandel S, Buchs A, Schürmann A, Summers SA, Powers AC, Shanahan MF, Joost H (1996) Glucose transport activity and ligand binding (cytochalasin B, IAPS-forskolin) of chimeric constructs of GLUT2 and GLUT4 expressed in COS-7-cells. Biochim Biophys Acta 1284:56-62

Wheeler TJ (1986) The kinetics of glucose transport in human erythrocytes: zero transport and infinite trans efflux at 0 °C. Biochim Biophys Acta 862:387-398

Wheeler TJ, Cole D, Hauck MA (1998) Characterization of glucose transport activity reconstituted from heart and other tissues. Biochim Biophys Acta 1414:217-230

Wilbrandt W (1972) Coupling between simultaneous movements of carriers. J Membrane Biol 10:357-366

Wu L, Fritz JD, Powers AC (1998) Different functional domains of GLUT2 glucose transporter are required for glucose affinity and substrate specificity. Endocrin 139:4205-4212

Yang J, Clark AE, Kozka IJ, Cushman SW, Holman GD (1992) Development of an intracellular pool of glucose transporters in 3T3-L1 cells. J Biol Chem 267:10393-10399

Yano Y, May JM (1993) Ligand induced conformational changes modify proteolytic cleavage of adipocytes insulin-sensitive glucose transporter. Biochem J 295:183-188

Zeng H, Parthasarathy R, Rampal AL, Jung CY (1996) Proposed structure of putative glucose channel in GLUT 1 facilitative glucose transporter. Biophys J 70:14-21

Zottola RJ, Cloherty EK, Coderre PE, Carruthers A (1995) Glucose transporter function is controlled by transporter oligomeric structure - a single, intramolecular disulfide promotes Glut1 tetramerization. Biochemistry 34:9734-9747

15 Calcium Homeostasis in Normal and Abnormal Human Red Cells

Teresa Tiffert[1], Robert M. Bookchin[2] and Virgilio L. Lew[1]

[1]Physiological Laboratory, University of Cambridge, Cambridge CB2 3EG, UK, and [2]Department of Medicine, Albert Einstein College of Medicine, Bronx, New York 10461, USA

15.1 Introduction

The calcium homeostasis of normal human red cells may appear deceptively simple. Mature red cells lack Ca^{2+}-accumulating organelles and Ca^{2+} signalling functions. Their total calcium content and Ca^{2+} permeability (P_{Ca}) are extremely low, and they have minimal cytoplasmic Ca^{2+} buffering capacity compared to other cell types. One property red cells share with all prokaryots and eukaryots is a low cytoplasmic ionized Ca^{2+} concentration ($[Ca^{2+}]_i$). The low $[Ca^{2+}]_i$ represents the balance between passive Ca^{2+} influx and active Ca^{2+} extrusion. Passive Ca^{2+} influx is through low capacity transport pathways with carrier properties (Ferreira and Lew 1977; Lew and Ferreira 1978; Desai et al. 1991a). Active Ca^{2+} extrusion is mediated by a large-capacity (high-V_{max}), ATP-fuelled P-type pump (PMCA), expressed with genomic variants 1 and 4 (Schatzmann 1966, 1982; Schatzmann and Vincenzi 1969; Rega and Garrahan 1986; Wang et al. 1992; Carafoli 1992, 1997). Thus, as with the resting $[Ca^{2+}]_i$ level in other cell types, the physiological low-$[Ca^{2+}]_i$ condition of red cells is probably maintained within very narrow margins throughout their life-span (~120 days). There have been claims that the Ca^{2+} permeability, and hence $[Ca^{2+}]_i$, may increase in physiological conditions, transiently during shear stress in the microcirculation (Larsen et al. 1981), or terminally, as part of red cell programmed death (Lew and Beaugé 1979; Romero and Romero 1999a), and the evidence for and against will be discussed below. The suggestion that high intracellular Ca^{2+} must necessarily lead to early demise of the red cells, when tested *in vivo*, proved wrong (Bookchin et al. 1985a,b).

Elevations of total red cell calcium *in vivo* were documented only in specific pathological conditions: in *P. falciparum* malaria, sickle cell anaemia and thalassaemia. Thus, the abnormalities of red cell Ca^{2+} homeostasis identified so far involve malaria infection and the genetic haemoglobinopathies with prevalence in malaria endemic regions.

Many red cell abnormalities, reversible and irreversible, have been attributed to elevated $[Ca^{2+}]_i$. The abnormalities with the potential to cause the largest alterations in cell metabolism, volume and ion content are those associated with substantial increases in P_{Ca}, resulting in activation of the PMCA and of the Ca^{2+}-sensitive K^+ channels (IK1, (Gardos 1958; Lew and Ferreira 1978)). Most of the purported ancillary effects, such as high-$[Ca^{2+}]_i$ mediation of cell senescence, activation of Ca^{2+}-sensitive proteases (Wang et al. 1989; Glaser et al. 1994), cytoskeletal cross-linking (Lorand et al. 1976, 1978, 1979), stimulatory or inhibitory effects on PMCA-mediated ATPase activity of glucose (Davis et al. 1985; Gonzalez Flecha et al. 1990, 1999), thyroid hormones (Davis et al. 1989), haem radicals (Leclerc et al. 1988, 1991) and other agents were documented in ghosts, lysates, broken membrane preparations or isolated targets. As discussed below, in the few instances where such effects were tested directly *in vivo* or *in vitro* using intact red cells, the results did not agree with those hypothesised or observed in cell-disrupted preparations. In this chapter, therefore, we emphasize research performed with intact red cells.

Homeostasis is about integrated function. To understand red cell Ca^{2+} homeostasis, we must take into account all of the components regulating $[Ca^{2+}]_i$, and determine how they interact in the red cells. The "red cell organ" of an organism is comprised of its entire red cell mass, so that red cell function in general has to be examined in the context of the behaviour of cell populations, with an eye on variations within populations. As shown below, such an approach is necessary to understand red cell Ca^{2+} homeostasis, Ca^{2+}-induced pathology, and a large body of experimental work in the field.

The scope of the subject may be summed up as follows. Ca^{2+} moves across the red cell plasma membrane by passive influx through low-capacity leak pathways and by active extrusion via the PMCAs. Pump-leak turnover maintains $[Ca^{2+}]_i$ at least three orders of magnitude below the level of extracellular Ca^{2+} ($[Ca^{2+}]_o$). The PMCA is a multi-regulated pump (Carafoli 1992; Wang et al. 1992) whose activity and properties may be modulated by dimerization (Kosk-Kosicka and Bzdega 1988; Kosk-Kosicka et al. 1990) and by intracellular factors such as calmodulin or calpain. There is a dynamic equilibrium between free and bound calcium (Ferreira and Lew 1976; Tiffert and Lew 1997a) and between free cytosolic Ca^{2+} and Ca^{2+} contained within endocytic inside-out vesicles (EIOVs) in equilibrium with calcium-phosphate precipitates (Lew et al. 1985; Bookchin et al. 1985c, 1986). Population variations result from constitutive differences in each of these red cell components and in the enzymes regulating their nucleotide metabolism. Pathological Ca^{2+} homeostasis occurs when the balance between the homeostatic components is disrupted, either uniformly across the entire red cell population, or heterogeneously, involving some cells more than others.

15.2 Methodological Considerations

A variety of methods, some specific to red cells, have been applied in studies of red cell Ca^{2+} homeostasis. The measurements obtained have been variously integrated within specific experimental protocols to deliver information on total cell calcium contents, physiological $[Ca^{2+}]_i$ levels, passive and active Ca^{2+} fluxes and cytoplasmic Ca^{2+} buffering. Specific methodologies were also developed to study the effects of Ca^{2+} loads on nucleotide metabolism, on population variations in Ca^{2+} transport, and on the *in vivo* survival of Ca^{2+}-loaded red cells. Important tools or procedures which have contributed much to our current knowledge include the use of acetoxy-methyl (AM) derivatives of Ca^{2+} chelators which could be incorporated non-disruptively into the cells, the use of Ca^{2+} ionophores, and a variety of manoeuvres used to inhibit the plasma membrane Ca^{2+} pump. In evaluating the information gained with such tools it is important to bear in mind some critical properties and limitations.

15.3 Use of Incorporated Ca^{2+} Chelators

The use of AM compounds to introduce impermeant solutes into cells was pioneered by Tsien (1981) in human red cells with the non-disruptive incorporation of Ca^{2+} chelators. AM compounds are highly hydrophobic, are usually dissolved in dimethyl sulfoxide (DMSO) and added to cell suspensions from concentrated stock solutions. The AM compound immediately partitions into the cell plasma membrane. Esterases hydrolyse the AM ester bonds liberating formaldehyde. Intracellular esterases release the free chelator into the cytoplasm whereas extracellular esterases (PI-anchored actylcholinesterase, e.g.) presumably release free chelator to the external medium, reducing the yield of cell-incorporated chelator. Initial expectations that the formaldehyde released would diffuse away without toxic effects on the cells, and that the dissociation constant of the incorporated compound would remain the same as in free solution, were not generally fulfilled. All the formaldehyde released by AM hydrolysis remains trapped within the cell where it causes a functional blockade of the glyceraldehyde-3-phosphate dehydrogenase by NAD depletion, leading to an irreversible fall in ATP at a rate determined by ongoing ATP-consuming processes (Orringer and Mattern 1976; Tiffert et al. 1984; García-Sancho 1985). The glycolytic block can be prevented by the presence of pyruvate in the incubation media, to maintain physiological NAD/NADH ratios.

The apparent Ca^{2+} dissociation constants (K_D) of two incorporated Ca^{2+} chelators, BAPTA and benz2, were found to be increased six to eight fold over the K_D values in free solution (Tiffert and Lew 1997a), a shift attributed to low affinity binding of the incorporated chelators to haemoglobin (Hb). The distribution of incorporated chelators among and within individual red cells was investigated using fura2 and quantitative fluorescence microscopy (Lew et al. 1993). Hb quenched the fura2 fluorescence yield by a factor of 100. The intracellular distribution of fura2 was uniform, but the population variation in fura2 load was Gaussian, with a coefficient of variation of about 20%. Therefore, besides major quenching, there are relatively large cell to cell variations in fura2 load. Although in need of confirmation, the most serious and invalidating change in fura2 properties, attributed to the high intracellular Hb concentration, was the profound distortion in the spectral properties of fura2, with almost complete loss of Ca^{2+} sensitivity, described by Blackwood et al. (1997). Thus, estimates of red cell $[Ca^{2+}]_i$ from fura2 fluorescence imaging may be seriously flawed and will not be discussed further here.

15.4 Use of Ca^{2+} Ionophores

Ca^{2+} ionophores (A23187, ionomycin), used in combination with $^{45}Ca^{2+}$, are particularly powerful tools in studies of red cell Ca^{2+} homeostasis. Firstly, the ionophore allows precise control of the membrane Ca^{2+} permeability, and the induced permeability is uniform in all the cells of a suspension (Lew and Simonsen 1980; Simonsen and Lew 1980; Simonsen et al. 1982). Secondly, because the endogenous calcium pools of red cells are so minute (see below), the mean total calcium content of the cells ($[Ca_T]_i$) can be accurately estimated from the tracer specific activity together with the measured total cell $^{45}Ca^{2+}$ content. Thirdly, it is extremely easy to maintain cell volume, pH, ion content (other than calcium) and membrane potential within physiological ranges regardless of the size of the Ca^{2+} load induced by ionophores (Ferreira and Lew 1976; Flatman and Lew 1977; Lew and Brown 1979; Lew and García-Sancho 1989). A23187 transports Ca^{2+} as well as other divalent cations (Mg^{2+}, Co^{2+}) in a strict M^{2+}:$2H^+$ exchange mode (Pressman 1976). Intracellular Mg^{2+} ($[Mg^{2+}]_i$) may thus be kept at any desired level simply by setting extracellular Mg^{2+} ($[Mg^{2+}]_o$) so that $[Mg^{2+}]_o = [Mg^{2+}]_i/([H^+]_i/[H^+]_o)^2 = r^2$. Since in red cells the combined operation of the anion exchanger and CO_2 shunt ensures rapid equilibration of Cl^-, HCO_3^- and H^+ across the membrane, the proton ratio will remain unchanged after ionophore addition as long as the anion ratio is clamped. Changes in intracellular Cl^- concentrations in Ca^{2+}-loaded cells may be elicited by IK1 channel activation driving a net loss of KCl and water when the cells are suspended in low-K^+ media. This may be easily prevented by using selective IK1 channel blockers such as charybdotoxin or clotrimazole (Alvarez et al. 1992; Tiffert et al. 2000a), or by setting the external K^+ concentration at electrochemical equilibrium with intracellular K^+, thus removing the driving gradients for cell dehydration by net KCl loss.

15.5 Inhibition of the Ca^{2+} Pump

With fresh, substrate-fed red cells, no significant net $^{45}Ca^{2+}$ gain is observed unless the Ca^{2+} pump is inhibited. Thus, the passive Ca^{2+} transport pathways of the red cell membrane have been investigated by measuring unidirectional Ca^{2+} fluxes with the Ca^{2+} pump inhibited. No specific inhibitors of the PMCA have yet been found (Tiffert et al. 2001). Three strategies have been used to inhibit the pump: depleting the cells of ATP, poisoning the pump with vanadate, or loading the cells with Ca^{2+} chelators to keep $[Ca^{2+}]_i$ below pump-activating levels. Rapid ATP depletion may be induced by incubation with combinations of metabolic inhibitors (iodoacetamide, iodoacetic acid, formaldehyde) and metabolic substrates (inosine, adenosine, glucose) (Lew 1971a); the substrate speeds up depletion by consuming ATP while its synthesis is blocked by the inhibitor. These treatments rapidly reduce the intracellular ATP concentration from mM to µM levels but no further, because of continued generation of ATP from large reserves of 2,3-DPG (Glynn et al. 1971). These low (µM) ATP levels are sufficient to sustain significant residual Ca^{2+} pump fluxes by Ca^{2+}-saturated pumps (Harbak and Simonsen 1987; Alvarez et al. 1988). Thus, the contribution of residual pump fluxes must be considered when interpreting early studies of passive Ca^{2+} permeability in ATP-depleted red cells (Ferreira and Lew 1977; Lew and Ferreira 1978). The observed tendency of the Ca^{2+} influx curves to level off in time may result from true self-limiting effects of internal Ca^{2+} on its own permeability, from limited but progressive pump activation as $[Ca^{2+}]_i$ increases, from an heterogeneous distribution of passive Ca^{2+} permeabilities in the cell population, or from a combination of these factors. Admixture of tetrathionate with ATP-depleting cocktails have been shown to induce more pronounced ATP depletion (Alvarez et al. 1988) but no passive Ca^{2+} permeability studies have yet been reported under these conditions. Orthovanadate, a non-specific P-ATPase inhibitor (Cantley et al. 1977; Bond and Hudgins 1978; Barrabin et al. 1980; Schatzmann et al. 1986), produces irreversible inhibition of the red cell Ca^{2+} pump in less than 15 s, with a $K_{1/2}$ of about 3 μM (Tiffert and Lew 2001). With 1 mM vanadate in a red cell suspension, the pump is inhibited by about 99.7% (Tiffert and Lew 1997a). A $[Ca^{2+}]_i$-saturated pump with a V_{max} of 15 mmol/(l_{cells}·h) may still sustain residual Ca^{2+}-extrusion fluxes of about 50 μmol/(l_{cells}·h), enough to oppose normal passive Ca^{2+} influx of comparable magnitude.

When Ca^{2+} influx was measured in chelator-loaded cells, Ca^{2+} uptake was linear for at least 90 min in substrate-fed cells (Lew et al. 1982), and for at least 5 h in ATP-depleted cells (Tiffert et al. 1993). Thus, under conditions in which $[Ca^{2+}]_i$ elevation is precluded and residual Ca^{2+}-pump mediated fluxes are effectively suppressed, the linearity of passive Ca^{2+} influx is preserved. To resolve the nature of the apparent self-limiting effects of Ca^{2+} on its own permeability it will be necessary to allow the build-up of $[Ca^{2+}]_i$ with the Ca^{2+} pump effectively inhibited.

15.6 Calcium Homeostasis in Physiological Conditions

15.6.1 Total and Total Exchangeable Calcium Content of Red Cells

The total calcium content of red cells, including calcium in any form, was measured by atomic absorption spectroscopy (Harrison and Long 1968; Dunn 1974; Bookchin and Lew 1980). Because of its low value in normal human red cells from freshly drawn blood compared with ambient calcium in laboratory chemicals and solutions, special treatments of all washing and processing solutions were needed to remove contaminant traces of calcium. After washing the cells with chelator-containing solutions to remove mobilizable calcium bound to the external membrane surface, the total calcium content was reduced from about 15–20 μmol/l_{cells} to about 5 μmol/l_{cells}. Of this total, less than 1 μmol/l_{cells} could be extracted by incubating the cells in the presence of A23187 and external Ca^{2+} chelators, suggesting that the rest is non-exchangeable calcium covalently bound to cell components (Bookchin and Lew 1980).

It is important to distinguish between the total calcium content of the cells, as measured by atomic absorption, from the functionally available calcium. Because the non-exchangeable calcium (which cannot be extracted by an external chelator, when the red cells are permeabilized to Ca^{2+} by the ionophore) is apparently functionally unavailable, we will exclude it from further consideration here. Thus, in the following, the expression $[Ca_T]_i$ will refer to the total exchangeable calcium, which is functionally available. Part of this $[Ca_T]_i$ is in the "free, ionized" form, expressed here as "$[Ca^{2+}]_i$", and part is bound to cytoplasmic buffers in the red cells or precipitated as calcium phosphates within cytoplasmic endocytic inside-out vesicles.

Is it possible that the real total Ca^{2+} content of the cells *in vivo* is higher, or much higher, than what the above measurements suggest? Could it be that during the necessary washing steps in Ca^{2+}-free media the powerful Ca^{2+} pump empties an existing cytoplasmic Ca^{2+} pool, and that what is measured is a left-over from higher $[Ca_T]_i$ *in vivo*? Earlier observations argue against such a possibility. The temperature-dependence of the Ca^{2+} pump is very steep below 3 °C ($Q_{10} > 30$ (García-Sancho and Lew 1988a)). Ca^{2+}-loaded cells washed in ice-cold Ca^{2+}-free solutions retained all the loaded Ca^{2+} (García-Sancho and Lew 1988a). Measurements made after washing freshly drawn red cells in ice-cold media showed an exchangeable calcium pool of less than 1 $\mu mol/l_{cells}$ (Bookchin and Lew 1980). Thus, the very low $[Ca_T]_i$ measured in fresh normal red cells probably represents their *in vivo* state in the circulation. It is important to bear in mind that the minute pool of exchangeable Ca^{2+} represents a mean estimate for the cell population. This imposes a very restrictive limit on the existence of cells with elevated $[Ca_T]_i$, on the size of their fraction, and on the magnitude of their $[Ca_T]_i$ increase.

The results of the $[Ca_T]_i$ measurements may offer clues about possible variations in $[Ca_T]_i$ in circulating red cells, such as whether periodic oxygenation-deoxygenation or shearing forces in the microcirculation may generate transient $[Ca_T]_i$ increases. The measurements of red cell $[Ca_T]_i$ were made with venous blood samples, processed as described above to ensure conservation of their original $[Ca^{2+}]_i$. Therefore, the near calcium-free condition represents the true $[Ca_T]_i$ content of red cells in the veins. As described below, deoxygenation causes a mild inhibition of the V_{max} of the Ca^{2+} pump but has no effect on the Ca^{2+} permeability. Thus, oxygenated cells from arterial blood are not likely to have higher $[Ca_T]_i$ than cells from venous blood. The occurrence of transient $[Ca_T]_i$ elevations in the microcirculation cannot be investigated directly.

15.6.2 The Physiological $[Ca^{2+}]_i$ Level

Because of the methodological limitations considered above, the physiological red cell $[Ca^{2+}]_i$ level remains uncertain. The powerful quenching and apparent spectral distortion caused by haemoglobin would seem to preclude the use of fluorescent signalling to obtain accurate estimates of red cell $[Ca^{2+}]_i$. Thus, previous attempts to estimate $[Ca^{2+}]_i$ in intact red cells using fluorescent indicators will not be considered further here.

Red cell $[Ca^{2+}]_i$ was estimated by two methods, each using incorporated, non-fluorescent, Ca^{2+} chelators. In one experimental design, the incorporated chelator (benz2) was used to generate a measurable Ca^{2+} pool within the cells (Lew et al. 1982; Muallem et al. 1985; Etzion et al. 1993; Tiffert et al. 1993). The chelator-loaded cells were suspended in autologous plasma and Ca^{2+} influx was followed by tracer ($^{45}Ca^{2+}$) until a pump-leak steady-state was established. $[Ca^{2+}]_i$ was then estimated from the measured chelator load and from the total cell Ca^{2+} content. In the other experimental design, red cells were loaded with fluorinated (^{19}F) Ca^{2+} chelators (FBAPTA or QuinMF), and $[Ca^{2+}]_i$ estimated by ^{19}F nuclear magnetic resonance spectroscopy (Murphy et al. 1986, 1987). The original values reported by both methods were very similar, between 10 and 30 nM for the benz2 loaded cells and 21 ± 2 nM for the NMR estimates in QuinMF loaded cells. However, the $[Ca^{2+}]_i$ estimates by both methods depend critically on the value assigned to the Ca^{2+} dissociation constants of the incorporated chelators; as noted before, their value *in situ* was recently found to be substantially higher than originally assumed or estimated (Tiffert and Lew 1997b). Thus, the true physiological $[Ca^{2+}]_i$ range is likely to be higher than the original estimates. Until these issues are resolved, it appears reasonable to consider the physiological $[Ca^{2+}]_i$ level in human red cells as below 100 nM, probably within the range 30 to 60 nM.

15.6.3 Cytoplasmic Ca^{2+} Buffering

Cytoplasmic Ca^{2+} buffering was investigated by measuring the equilibrium distributions of $^{45}Ca^{2+}$ between the red cells and medium at different external Ca^{2+} concentrations. Ca^{2+} permeabilization and rapid, full Ca^{2+} equilibration was induced by high concentrations of the ionophore A23187, with the Ca^{2+} pump effectively inhibited either by marked ATP depletion induced with the combined use of inosine, iodoacetamide and tetrathionate, or by maximal inhibitory vanadate concentrations in cells with normal ATP contents (Tiffert and Lew 1997a). Similar results were obtained with both pump-inhibition methods, indicating that cytoplasmic Ca^{2+} buffering was not affected by ATP depletion. The $^{45}Ca^{2+}$ measurements provided values for $[Ca_T]_i$ and $[Ca^{2+}]_o$ at each $[Ca^{2+}]_o$ level.

With the ionophore-induced Ca^{2+}-equilibration method, the following relations apply (Ferreira and Lew 1976; Lew and García-Sancho 1989). At each $[Ca^{2+}]_i$ level, $[Ca_T]_i$ is the sum of $[Ca^{2+}]_i$ and of the intracellular concentration of bound Ca^{2+}, $[CaB]_i$, so that $[Ca_T]_i = f_w[Ca^{2+}]_i + [CaB]_i$, where $[Ca_T]_i$ and $[CaB]_i$ are expressed in units of $\mu mol/l_{cells}$, and f_w represents the fraction of cell water in units of $l_{cell\ water}/l_{cells}$. At ionophore-induced equilibrium, $[Ca^{2+}]_i$ is computed from $[Ca^{2+}]_i = r^2 [Ca^{2+}]_o$, where r represents the proton concentration ratio, so that $[CaB]_i = [Ca_T]_i - f_w r^2[Ca^{2+}]_o$. The concentration of intracellular bound Ca^{2+} may thus be obtained for each $[Ca^{2+}]_i$ level from measured $[Ca_T]_i$ and $[Ca^{2+}]_o$, in experimental conditions which ensure the constancy of f_w and r.

The plots of $[Ca_T]_i$ *vs.* $[Ca^{2+}]_i$ showed that the cytoplasmic Ca^{2+} buffering behaviour of fresh, normal human red cells can be adequately described by a two-buffer equation of the form:

$$[Ca_T]_i = (1/\alpha)\,[Ca^{2+}]_i + B\,([Ca^{2+}]_i/([Ca^{2+}]_i + K_B)) \qquad (15.1)$$

where α represents a large capacity, low affinity Ca^{2+} buffer, mostly haemoglobin, and B and K_B describe a low capacity, medium affinity buffer, probably the combined Ca^{2+} binding properties of calmodulin and calpromotin (Cheung 1982;

Plishker et al. 1986, 1992; Moore et al. 1990; Tiffert and Lew 1997a). α was measured with much higher precision than was possible for B and K_B; their respective ranges were: for α, 0.32 to 0.37; for B, 20 to 160 $\mu mol/l_{cells}$; and for K_B, 4 to 50 μM (Tiffert and Lew 1997a). For most practical purposes, this description may be simplified as follows. For $[Ca_T]_i$ above ~100 $\mu mol/l_{cells}$, $[Ca^{2+}]_i$ may be estimated from $[Ca^{2+}]_i \sim 0.3\ [Ca_T]_i$, i.e., the free Ca^{2+} concentration in the red cell cytoplasm is about 30% of the total calcium content of the cell. For $[Ca_T]_i$ below ~100 $\mu mol/l_{cells}$, $[Ca^{2+}]_i$ may be estimated from $[Ca^{2+}]_i \sim 0.2\ [Ca_T]_i$; nearer physiological $[Ca^{2+}]_i$ levels, below 1 μM, $[Ca^{2+}]_i \sim 0.1\ [Ca_T]_i$. Thus, the measured total exchangeable calcium content of the cells, less than 1 $\mu mol/l_{cells}$, also imposes an upper limit on the physiological $[Ca^{2+}]_i$ level of less than 100 nM (10% of $[Ca_T]_i$).

15.7 Passive Ca^{2+} Transport

The difficulties of measuring passive Ca^{2+} influx in the presence of incomplete Ca^{2+} pump inhibition (see Sect. 15.5) were overcome by the use of chelator-loaded cells suspended in autologous plasma. The results yielded values of about 50 $\mu mol/(l_{cells}\cdot h)$ for the physiological Ca^{2+} influx (Lew et al. 1982; Tiffert et al. 1993). This flux was influenced by a variety of intra- and extracellular factors. ATP depletion reduced Ca^{2+} influx by about 30%. At similar $[Ca^{2+}]_o$ levels, Ca^{2+} influx was about 30% higher in plasma than in plasma-like saline. With ATP-depleted red cells incubated in saline media containing vanadate, to ensure more complete pump inhibition, the reported value for Ca^{2+} influx (Desai et al. 1989) was similar to that found earlier for red cells incubated in plasma (Lew et al. 1982). However, the red cells used for the saline incubations had been stored for weeks in a preservative, and the linear slope of Ca^{2+} influx as a function of $[Ca^{2+}]_o$ was much larger than that observed in fresh cells. The nature of the metabolism-dependent component of the influx, and of the plasma factor(s) responsible for the plasma-saline differences, are presently unknown. Ca^{2+} influx was reduced at high external K^+ concentrations (Lew 1971a; Ferreira and Lew 1977), highly increased in low (Na^+ + K^+), choline or sucrose media (Ferreira and Lew 1977), competitively inhibited by external Sr^{2+} (Desai et al. 1989) and non-competitively blocked by external Co^{2+} (Tiffert et al. 1984). Ca^{2+} influx as a function of $[Ca^{2+}]_o$ increased along a curve best described by the sum of a rectangular hyperbola and a linear component (Ferreira and Lew 1977; Tiffert et al. 1984; McNamara and Wiley 1986; Desai et al. 1991a). The reported $K_{1/2}$ of the saturating component ranged from 0.8 to 1.3 mM. With fresh red cells (Tiffert et al. 1984), or in blood-bank cells up to 6-days old (Ferreira and Lew 1977) the linear component was hardly detectable, but with aged blood-bank cells (Romero and Whittam 1971) or cells kept at 70% haematocrit and 4 °C in a preservative medium (Desai et al. 1991b), it was highly increased. With ATP-depleted red cells, the unidirectional fluxes of $^{45}Ca^{2+}$ were accelerated by the presence of cold Ca^{2+} on the *trans*-membrane side (Ferreira and Lew 1977) and similar *trans*-membrane effects were also reported in ATP-depleted, vanadate-poisoned cells for Ca^{2+} and Sr^{2+} (Desai et al. 1991b).

These results suggest that a substantial fraction of the passive Ca^{2+} influx is carrier-mediated.

15.8 The Ca^{2+} Pump and the Physiological Pump-Leak Turnover of Ca^{2+}

The physiological pump-leak turnover of Ca^{2+} is about 50 μmol/(l_{cells}·h) (Lew et al. 1982; Desai et al. 1991b). The PMCA is a multi-regulated pump (Scharff et al. 1982; Schatzmann 1983; Rega and Garrahan 1986; Kosk-Kosicka and Bzdega 1988, 1990; Wang et al. 1992; Guerini et al. 1998) and it is unclear which of its multiple configurations is predominantly expressed and operational in the intact, fresh red cell. This uncertainty arises because the maximal Ca^{2+} extrusion rate of the Ca^{2+}- and ATP-saturated pump (V_{max}) is so high relative to the pump-leak turnover rate, about 200- to 500-fold, that even its low Ca^{2+} affinity configurations may comfortably accommodate a Ca^{2+} efflux of 50 μmol/(l_{cells}·h) at physiological $[Ca^{2+}]_i$ levels (within the uncertainty range of 30–100 nM). The major alternative pump configurations consist of high Ca^{2+} affinity dimers, and monomers with and without calmodulin bound to its specific C-terminal domain, rendering high and low Ca^{2+}-affinity kinetics, respectively. Additional variation may arise from potentially different kinetic behaviours of the two genomic variants expressed in human red cells, and from the possible coexistence of some or all of these functional configurations, in stable proportions or in dynamic exchange.

In chelator-loaded red cells with sub-physiological $[Ca^{2+}]_i$, the kinetics of pump-mediated Ca^{2+} extrusion clearly follows a sigmoid curve, with the efflux proportional to $([Ca^{2+}]_i)^2$ (Lew et al. 1982). It remains unclear whether these kinetics result from a single dominant pump configuration or from a complex assembly of pump states. Exploration of Ca^{2+} pump kinetics of red cells with $[Ca^{2+}]_i$ levels above the physiological range requires Ca^{2+} loading of the cells, and this is considered next.

15.9 The V_{max} of the Ca^{2+} Pump

The V_{max} of the Ca^{2+} pump in intact red cells is usually measured by the Co^{2+}-exposure method of Dagher and Lew (1988): The cells are exposed to a high concentration of Ca^{2+} ionophore A23187 in the presence of external $^{45}Ca^{2+}$ so that the ionophore-mediated Ca^{2+} influx far exceeds the V_{max} of the pump. In such conditions, $^{45}Ca^{2+}$ rapidly approaches the equilibrium distribution expected with a pure ionophore-membrane system, i.e. $[Ca^{2+}]_i/[Ca^{2+}]_o = ([H^+]_i/[H^+]_o)^2$. One or two minutes after ionophore addition, Co^{2+} is added to the medium in excess of external Ca^{2+}. Co^{2+} instantly blocks all ionophore-mediated Ca^{2+} influx and exposes the uphill extrusion of Ca^{2+} by the pump. As discussed below, speed is necessary to prevent significant ATP depletion, and the whole procedure can be completed in less than five minutes. Ca^{2+} extrusion is uphill because it alters the Ca^{2+} distribution away from an equilibrium condition, and is pump-mediated because it is inhibited by ATP depletion or vanadate.

Co^{2+} is also transported into the cells by the ionophore (Tiffert et al. 1984; Brown and Simonsen 1985). That Co^{2+}, *per se*, does not interfere with pump-mediated Ca^{2+} fluxes was demonstrated by comparing the Ca^{2+} extrusion patterns elicited by Co^{2+} addition or by ionophore removal (repeated washes in ice-cold media containing albumin); the two patterns were essentially identical (Pereira et al. 1993). Within the cells Co^{2+} acts like a Mg^{2+} substitute for the Ca^{2+} pump (Raftos and Lew 1994), and does not compete with Ca^{2+} for binding sites on the α-buffer (Pereira et al. 1993). With the original method, the secondary effects of elevated $[Ca^{2+}]_i$ on cell volume, pH and ion content (derived from IK1 channel activation and ionophore-mediated Mg^{2+} permeabilization) were prevented by setting the external K^+ and Mg^{2+} concentrations at electrochemical equilibrium with the intracellular concentrations of K^+ and Mg^{2+}. Volume and pH changes could also easily be prevented by the use of IK1 channel inhibitors, enabling V_{max} measurements in low-K^+ media (Tiffert et al. 2000a).

Typically, the linear V_{max} Ca^{2+} extrusion rate continued until $[Ca_T]_i$ was reduced to about 50–100 μmol/(l_{cells}·h) (Fig. 15.1). Below these $[Ca_T]_i$ levels the rate of extrusion was progressively reduced, following an apparent pattern of Ca^{2+} desaturation. As shown below, this Ca^{2+} extrusion pattern reflects the distribution of V_{max} values in the cell population, and cannot be used to derive the kinetic parameters of the pump.

15.10 Effects of Deoxygenation and pH_i on the V_{max} of the Ca^{2+} Pump

Full deoxygenation of fresh red cells reduced the V_{max} of the Ca^{2+} pump by 18–32% and increased the fraction of ionized Ca^{2+} in cell water (α) by 34–74% (Tiffert et al. 1993). Deoxygenation generated large alkaline pH_i shifts, due to the increased protonation of deoxygenated haemoglobin. The magnitude of the pH shifts could be explained if deoxygenation increased the isoelectric point of the intracellular haemoglobin by 0.2–0.4 units, consistently higher than the value (0.15) reported for pure haemoglobin solutions (Bunn and McDonough 1974; Bunn and Forget 1986). The increased value of α, reflecting decreased Ca^{2+} buffering capacity on the α-buffer, could be explained if Ca^{2+} and H^+ ions shared common binding sites on haemoglobin, the main component of the α-buffer.

Could pump inhibition result from deoxygenation-induced pH effects? The V_{max} of the Ca^{2+} pump was found to be inhibited by intracellular acidification and by medium alkalinization (Tiffert et al. 1993). Operation of the pump as a Ca^{2+}/H^+ or $Ca^{2+}/2H^+$ exchanger (Gassner et al. 1988) explains the inhibitory effect of medium alkalinization. On the other hand, intracellular alkalinization at constant external pH had no effect on the V_{max} of the pump (Tiffert et al. 1993). Therefore, the mild inhibitory effect of deoxygenation on the pump V_{max} could not be attributed to the cell pH change accompanying deoxygenation. Tiffert et al. (1993) suggested that the inhibition might result from a deoxy-induced increase in the $[Mg^{2+}]_i$/[ATP] ratio within the cells, above the optimal value in the oxygenated state.

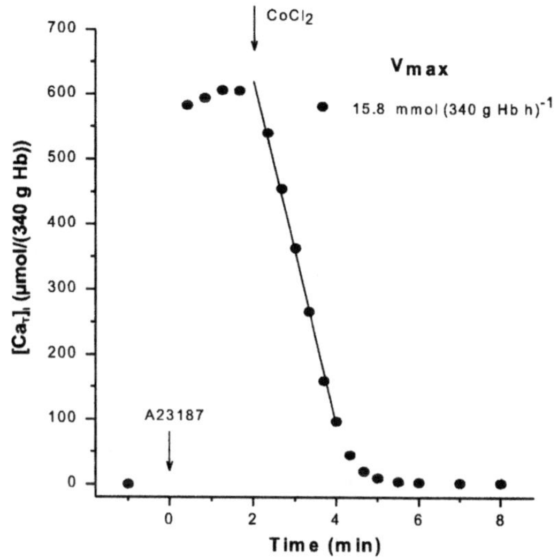

Fig. 15.1. Typical result obtained with the Co^{2+}-exposure method. A brief ionophore-induced total Ca^{2+} load of about 500 $\mu mol/l_{cells}$ is usually extruded by the Ca^{2+} pump in 2–3 min. The curve describing the fall in $[Ca_T]_i$ with time after Co^{2+} addition reports the mode in which the Ca^{2+} pump extrudes the Ca^{2+} load. The initial linear slope after Co^{2+} addition reflects the V_{max} of the Ca^{2+} pump. The V_{max} was found to vary between donors, and to a lesser extent between red cell samples obtained from the same donor on different days, with about 90% of the values falling within a range of 8 to 20 $mmol/(l_{cells} \cdot h)$

15.11 Elevated Cell Ca^{2+} in Experimental Conditions

15.11.1 Secondary Effects of Elevated $[Ca^{2+}]_i$

The most powerful effects of experimentally induced $[Ca^{2+}]_i$ elevations are mediated through activation of IK1 channels and Ca^{2+} pumps. Depending on the experimental conditions, IK1 channel activation may have profound effects on red cell homeostasis, whereas Ca^{2+} pump activation affects primarily the nucleotide metabolism of the cells.

On Red Cell Homeostasis

In normal red cells, the electrodiffusional K^+ permeability is less than two orders of magnitude below that of Cl^- (Hunter 1971, 1977). Full activation of the IK1 channels sets the K^+ permeability at least one order of magnitude above that of Cl^-. With red cells suspended in plasma or low-K^+ plasma-like media, the immediate effect of IK1 channel activation is hyperpolarization of the membrane potential

(Lassen 1977) from its physiological level, about –10 mV, towards the K^+ equilibrium potential, about –90 mV. This drives Cl^- out of the cells resulting in a net loss of K^+ and Cl^- at a rate limited by the Cl^- permeability. The net loss of KCl drives an osmotic loss of cell water, and the cells become progressively dehydrated. Because the original intracellular concentration of Cl^- is lower than that of K^+, the balanced loss of K^+ and Cl^- causes dilution of intracellular Cl^- and an increase in the $[Cl^-]_o/[Cl^-]_i$ concentration ratio. This increase, in turn, drives a net flux of H^+ into the cells via the powerful Jacobs-Stewart mechanism (Jacobs and Stewart 1947), causing cell acidification and medium alkalinization (Lew and Bookchin 1986; Freeman et al. 1987; Lew et al. 1991). As the cells dehydrate, haemoglobin is concentrated, and its osmotic coefficient increases exponentially, forcing osmotic water retention within the cells, thus limiting the extent of cell dehydration. The net result is cell dehydration with the loss of an alkaline, hypertonic effluent containing KCl and KOH. A new equilibrium is reached, with the cells markedly dehydrated, the pH differential highly increased, and the membrane potential set at a hyperpolarized Donan potential.

All these effects can be accelerated by partial or full replacement of Cl^- by more permeant anions such as thiocyanate or nitrate (García-Sancho and Lew 1988a). Alternatively, they can be prevented or blocked by the use of specific IK1 channel inhibitors, or by setting the external K^+ concentration at levels found to prevent net K^+ fluxes.

On the Nucleotide Metabolism of the Cells

The Ca^{2+}:ATP stoichiometry of the PMCA is 1:1 (Schatzmann 1973; Clark and Carafoli 1983; Dagher and Lew 1988), so that red cells with typical V_{max} values of 15 or 20 mmol/(l_{cells}·h) will hydrolyse ATP at these same rates. Since the mean glycolytic capacity of glucose- or inosine-fed red cells is only about 5 mmol/(l_{cells}·h) (Whittam 1964; McManus 1967), full pump activation in Ca^{2+}-loaded cells, will progressively deplete ATP, as was demonstrated (Arese et al. 1981; Till et al. 1981; Almaraz et al. 1988; Dagher and Lew 1988; García-Sancho and Lew 1988b).

Red cells have two extremely powerful enzymes: adenylate kinase, which catalyses the reversible conversion of two ADP molecules to ATP and AMP, and AMP deaminase, which catalyses the irreversible conversion of AMP to IMP. When saturated, both enzymes can operate at rates exceeding 1 mol/(l_{cells}·h), far higher than the range of Ca^{2+}-ATPase activities expected from the Ca^{2+} pump. The combined operation of the Ca^{2+} pump, adenylate kinase and AMP deaminase has two opposing effects – they buffer the net fall in ATP, while irreversibly converting ATP to IMP (Lew 1971a,b; Almaraz et al. 1988), by the following mechanisms: Hydrolysis of ATP by the pump elevates the ADP concentration. The adenylate kinase converts ADP back to ATP and in the process generates AMP. The AMP activation kinetics of the AMP deaminase is sigmoid (Askari and Franklin 1965; Askari 1966; Askari and Rao 1968), and a small increase in AMP concentration is sufficient to initiate the irreversible conversion to IMP. Ca^{2+} has a direct stimulatory effect on the AMP deaminase, so that the rate of conversion to IMP is increased in Ca^{2+}-loaded cells (Almaraz et al. 1988). As ATP falls, the Ca^{2+} pump becomes progressively inhibited and the depletion rate slows. After about 30–60 min of maximal pump activation, a steady state is reached in which glycolytic ATP production, also limited by the reduced ATP level, matches the rate of ATP hydrolysis by the pump; at this time, ATP has been reduced from its normal level of about 1.5 mM to a new stable level of about 200 $\mu mol/l_{cells}$, with the difference appearing as IMP. Irreversible IMP accumulation may also be elicited, independently of Ca^{2+}, by any of the rapid ATP depletion procedures described above (Lew 1971a,b).

Thus Ca^{2+} loads trigger profound, irreversible changes in the nucleotide metabolism of the cells. An interesting potential application of the results discussed above is that IMP may be used to trace events which caused elevated red cell Ca^{2+} *in vivo*. The IMP concentration in normal red cells is undetectable. Thus, any measurable IMP level may indicate that abnormal events of increased Ca^{2+} permeability in the circulation have occurred.

On Red Cell Survival and Ageing

Ca^{2+} was often thought to be implicated in red cell ageing and removal (Lew and Beaugé 1979; Bookchin et al. 1985b; Clark 1988a,b). Romero and Romero (1997, 1999a,b) and Romero et al. (1997) have recently elaborated on this hypothesis suggesting that during red cell ageing there is a progressive increase in Ca^{2+} permeability and in $[Ca^{2+}]_i$, IK1 channels and proteases become activated altering cell homeostasis and causing Ca^{2+} pump proteolysis with further increases in $[Ca^{2+}]_i$, finally leading to the removal of old cells by macrophages in the spleen.

The main problem with the above hypothesis is the absence of any direct evidence. The effect of elevated $[Ca^{2+}]_i$ on red cell survival *in vivo* was tested directly only once, in rabbits (Bookchin et al. 1985a). Like human red cells, rabbit red cells have a powerful Ca^{2+} pump activity and Ca^{2+}-sensitive K^+ channels capable of dehydrating the cells rapidly if fully activated. Contrary to the prevailing notion described above, when rabbit red cells were loaded *in vitro* with high Ca^{2+} levels, and were substantially dehydrated and ATP depleted at the time of re-infusion, they showed normal survival *in vivo*.

15.11.2 Effects of Physiological Agents on the Performance of the Ca^{2+} Pump in Intact Red Cells

Over the last three decades, a variety of molecules normally present in cells and blood were found to have stimulatory or inhibitory effects on the Ca^{2+}-ATPase, eliciting claims on their potential physiological roles in controlling cell Ca^{2+} homeostasis. These include hormones such as T4 (Davis et al. 1989), glucose (Davis et al. 1985; Gonzalez Flecha et al. 1990, 1993, 1997, 1999; Castello et al. 1994) and ferriprotoporphyrin IX (FP) (Leclerc et al. 1988, 1991). Even under the most stringent experimental conditions that had been described for the studies with isolated membranes or purified Ca^{2+}-ATPase preparations, however, in intact red cells the V_{max} of the Ca^{2+} pump was essentially unaffected by glucose or FP (Tiffert et al. 2000b; Raftos et al. 2001). Although these apparent discrepancies remain to be explained, they show that observations on Ca^{2+}-ATPase activity in fragmented cells or extracts cannot be extrapolated to Ca^{2+} pump behaviour in the intact red cells.

15.11.3 Population Response to Increased Ca^{2+} Influx

Fresh red cells exposed to sudden ionophore-mediated Ca^{2+} influx, at rates submaximal for the Ca^{2+} pump, showed complex responses (Ferreira and Lew 1976; Lew and Ferreira 1978). At the onset of influx, whether elicited by Ca^{2+} addition in the presence of ionophore or by ionophore addition in the presence of external Ca^{2+}, there was a rapid but transient surge in $[Ca_T]_i$ (designated the initial "peak") followed by a rapid decline towards a pump-leak steady state, sometimes along a peculiar wavy pattern. With increasing Ca^{2+} influx, the peaks became less apparent and the pump-leak $[Ca_T]_i$ levelled off at progressively higher values. When ^{42}K was added to the cell suspensions during the pump-leak steady states, to provide estimates of the Ca^{2+} sensitivity of the IK1 channels, the change in ^{42}K uptake with increasing steady state $[Ca_T]_i$ was surprising: instead of progressing towards the same ^{42}K equilibrium distribution at faster rates, the tracer uptake curves levelled off at progressively higher levels (Lew and Ferreira 1976). At face value the results suggested that the Ca^{2+} sensitivity varied with the ionophore concentration, but the ^{42}K tracer distribution pattern appeared to indicate that the higher the Ca^{2+} influx, the more cells ^{42}K was entering, as if the mean measured $[Ca_T]_i$ was distributed very heterogeneously among the cells. This possibility did not appear likely at the time particularly since it had been clearly demonstrated that heterogeneity could not have arisen from uneven ionophore distributions (Simonsen et al. 1982).

The question became accessible to direct scrutiny with a method developed *ad hoc* by García Sancho and Lew (1988a), the details of which will not be reviewed here. The picture which finally emerged from the investigation that followed (Almaraz et al. 1988; García-Sancho and Lew 1988b,c) may be summed up as follows. A sudden and uniform increase in Ca^{2+} influx of less than the mean V_{max} of the Ca^{2+} pump causes a rapid initial rise in $[Ca^{2+}]_i$ in all the cells. After a few seconds, a fraction of cells starts pumping out the Ca^{2+} they have gained, down to negligible levels (low Ca^{2+}, "L-cells"), whereas the remaining cells continue to gain Ca^{2+} until, after about 30 min, they approach $[Ca^{2+}]_i$ values near the equilibrium distribution of Ca^{2+} set by the ionophore (high-Ca^{2+}, "H-cells"). The proportion of L- and H-cells is entirely determined by the value of the experimentally set Ca^{2+} influx: the higher the Ca^{2+} influx, the higher the proportion of H-cells. At very high Ca^{2+} influx, all cells become H-cells. L-cells conserve their normal ATP levels whereas H-cells become progressively and irreversibly ATP depleted, with ATP converted to IMP by the mechanism described above. Thus, at all low and intermediate Ca^{2+} influx values the measured $[Ca_T]_i$ reports the mean of two distinct cell subpopulations with extremely heterogeneous Ca^{2+} contents. As a good approximation, it was observed that mean $[Ca_T]_i \sim f_H[Ca_T]_H$, where f_H is the fraction of H-cells and $[Ca_T]_H$ is the measured total Ca^{2+} content of the H-cells, i.e., essentially all the measured $[Ca_T]_i$ is in the H-cells.

The peculiar results obtained with the ^{42}K tracer flux measurements in Ca^{2+}-loaded cells may now easily be explained. The $[Ca^{2+}]_i$ level within L-cells is below the activation threshold of the IK1 channels, whereas that within H-cells activates them to saturation. Thus, at any given Ca^{2+} influx, the ^{42}K tracer entered only H-cells. This explains the different ^{42}K distribution levels observed, and why the results had no bearing on the Ca^{2+} sensitivity of the IK1 channels, as originally interpreted (Lew and Ferreira 1976).

What could be the mechanism of this heterogeneous population response to a uniform increase in Ca^{2+} influx? Let us assume, for the sake of argument, that the measured pump V_{max} represents the mean of a broad normal distribution. For any given experimentally-induced Ca^{2+} influx, there will be cells with pump V_{max} values above and below the set influx. Those cells whose pump V_{max} is below the set influx will gain Ca^{2+} continuously and become H-cells; those with V_{max} above will be able to sustain a pump-leak balance with minimal elevations in $[Ca^{2+}]_i$ if the kinetic modality of the pump is of a high Ca^{2+} affinity (L-cells). The fact that all H-cells become depleted of ATP may be explained by the observed stimulatory effect of high $[Ca^{2+}]_i$ on the AMP deaminase (Almaraz et al. 1988). As the H-cells gain Ca^{2+} the stimulated AMP deaminase activity will irreversibly convert ATP to IMP even if the glycolytic capacity of the H-cell had been able to match the rate of ATP consumption by the pump. How L-cells maintain normal ATP levels is more difficult to explain. For instance, let us assume that we are inducing a Ca^{2+} influx of 10 mmol/(l_{cells}·h) in a cell suspension with a measured mean pump V_{max} of 10 mmol/(l_{cells}·h), and that after 30 min 50% of the cells are H and 50% are L. The 50% L-cells would have to be those whose pump V_{max} is equal or higher than 10 mmol/(l_{cells}·h) since they have to sustain pump-leak turnover fluxes of Ca^{2+} of 10 mmol/(l_{cells}·h). But L-cells will also need a glycolytic capacity equal to or higher than 10 mmol/(l_{cells}·h) to maintain ATP constant when the pump is consuming ATP at that rate. Although the mean glycolytic rate in the cell population is about 5 mmol/(l_{cells}·h), this value may also represent the mean of a broad distribution. Under these circumstances, the H or L fate of Ca^{2+} permeabilized cells could be determined not only by the V_{max} of the pump but also by the glycolytic V_{max} of each cell, whichever is limiting. But the commitment of cells to H or L fates at each Ca^{2+} influx takes place within the first 2–3 min of Ca^{2+} permeabilization (García-Sancho and Lew 1988b). Within this time period, there would appear to be no mechanisms for the Ca^{2+} pumps to deplete the cytoplasmic pool of ATP and become H-cells due to a limiting glycolytic V_{max}. On the other hand, if there were submembrane microdomains at pump-mediated hydrolytic sites, reactions between Ca^{2+}, membrane bound glycolytic enzymes and soluble glycolytic intermediates and substrates in those domains might induce local ATP depletion (Proverbio et al. 1988).

Additional evidence for variation in pump V_{max} values in populations of normal red cells was provided by Tiffert et al. (1988, 2001). In one set of experiments the Ca^{2+} permeability was gradually increased using low ionophore concentrations at low external Ca^{2+} levels. When Ca^{2+} influx exceeded 0.7 mmol/(l_{cells}·h) variable fractions of cells became dehydrated. Progressively higher Ca^{2+} influx generated higher fractions of dehydrated cells, never higher rates of dehydration in all the cells. Since the Ca^{2+} influx induced by the ionophore was uniform in all the cells, the simplest explanation of this all-or-none dehydration pattern is that at each Ca^{2+} influx level, differences in Ca^{2+} pump V_{max} generated cell subpopulations with different $[Ca^{2+}]_i$ levels, below and above the Ca^{2+} activation threshold of the IK1 channels. In the more recent set of experiments (Tiffert et al. 2001) red cells were loaded with Ca^{2+} using ionophore. After Co^{2+} addition, as Ca^{2+} was being extruded, the rate at which the Ca^{2+} loaded cells ceased dehydrating was monitored using the profile migration method developed by Raftos et al. (1996, 1997). The results showed that cells ceased dehydrating over a broad range of times, much wider than observed when Ca^{2+} was rapidly extracted by addition of chelators in excess of Ca^{2+} in the external medium, supporting the view of a broad distribution of pump V_{max} values among red cells.

This variation of pump V_{max} values in the cell population invalidates the use of the Co^{2+} exposure method for estimating the Ca^{2+} activation kinetics of the Ca^{2+} pump. The tail end of the Ca^{2+} extrusion curves after Co^{2+} addition simply reflects the dispersion of times at which different cells empty their Ca^{2+} loads. This explains why the Ca^{2+} extrusion curves begin to "desaturate" at such high $[Ca_T]_i$ levels, usually when $[Ca_T]_i$ is over 50 $\mu mol/l_{cells}$, and $[Ca^{2+}]_i$ would be expected to be above 10 μM.

The distribution of glycolytic capacity among red cells or of possible heterogeneity arising from Ca^{2+} effects on glycolytic rate, remain unexplored fields. Brown and Johnston (1983) reported that elevated $[Ca^{2+}]_i$ stimulates lactic acid production with a biphasic Ca^{2+} concentration dependence, but there has been no follow-up on this pioneering work.

The nature of the $[Ca_T]_i$ peaks described at the beginning of this section (during Ca^{2+} loading at rates below the pump V_{max}) remains controversial. The most elegant explanation, provided by Scharff et al. (1983) and Scharff and Foder (1986), is that the native state of the Ca^{2+} pump in the intact cell is that of low Ca^{2+} affinity, low V_{max} monomers, and that Ca^{2+} entry triggers the $[Ca^{2+}]_i$-dependent association of calmodulin to the monomers, converting the pumps to a high Ca^{2+} affinity, high V_{max} state responsible for the declining phase of the peak. Alternative explanations are that the peaks result from very brief local inhomogeneities of Ca^{2+} or ionophore concentration on addition of these agents to cell suspensions, transiently exposing some cells to higher Ca^{2+} influx (Ferreira and Lew 1976), or that sudden pump activation depletes a local sub-membrane pool of ATP, transiently inhibiting the pumps (Perdomo and Lew, unpublished). Transient inhomogeneities are appealing because the peaks are not consistently observed or easily reproducible in parallel conditions, and vary considerably in different experimental conditions.

15.12 Elevated Cell Ca^{2+} in Pathological Conditions

Since all currently identified pathological conditions with elevated red cell Ca^{2+} are directly or indirectly related to malaria, it seems appropriate to consider first the altered Ca^{2+} homeostasis of malaria-infected red cells. Other abnormalities of malaria-infected red cells are discussed in Chap. 24. Almost all the available information on malaria infection in human red cells derives from experiments with

Plasmodium falciparum. Therefore, the material revised in the next section focuses on *P. falciparum* infected red cells (pRBCs). In the subsequent sections we review our current knowledge about Ca^{2+} dysfunction in sickle cell anaemia and thalassaemia, two hereditary haemoglobinopaties whose persistence in the present human genetic pool results from the selective survival of heterocygote carriers in *P. falciparum* endemic regions.

15.12.1 Malaria

Ca^{2+} is essential for parasite invasion and growth. During the intraerythrocytic cycle of *P. falciparum* parasites there is a dramatic increase in the total Ca^{2+} content of parasitised red cells (pRBCs) (Bookchin et al. 1980; Tanabe et al. 1982; Kramer and Ginsburg 1991; Adovelande et al. 1993) but the time-course of the Ca^{2+} gain is unknown. The elevated Ca^{2+} appears to be fully localised in the parasite (Shalev et al. 1981; Leida et al. 1981; Tanabe et al. 1982; Tanabe 1990; Adovelande et al. 1993), sequestered within two functionally distinct pools. Early observations that one of the pools was discharged by treatment with vanadate led to the suggestion that it represented an endoplasmic reticulum (ER)-like store, a conclusion supported by the presence of a gene encoding a SERCA P-type pump in the parasite genome (Garcia et al. 1996; Passos and Garcia 1997). The other pool was found to be discharged by treatment with protonophores and proton pump inhibitors, suggesting the existence of acidic vacuole-like compartments; this possibility is supported by the presence in the parasite genome of genes encoding a presumptive Ca^{2+}/H^{+} exchanger similar to the yeast *HUM1* gene product (Pozos et al. 1996), and also of subunits of a V-type ATPase (Passos and Garcia 1997; Passos and Garcia 1998). In more recent work, Alleva and Kirk (2001), using fura2-loaded parasites retained within saponin-haemolysed hosts, confirmed the existence of those pools and provided important additional insights on their origin, regulation and interactions. They showed that Ca^{2+} release from the ER-like store could be elicited by cyclopropiazonic acid, an inhibitor of plant and animal ER stores, but not by thapsigargin, a more specific inhibitor of animal ER stores. This suggested that during the intraerythrocytic stage, the ER of *P. falciparum* is like that of plants and other lower eukaryotes. They also showed that Ca^{2+} could be discharged from the acidic store by nigericin and by NH_4^+, and that the acidic store could act as a reservoir for Ca^{2+} released from the ER. The response of the stores to increasing external Ca^{2+} concentrations was also different, in that the ER took up more Ca^{2+} whereas the acidic store did not. The physiological "extracellular" environment of the parasite has the low normal Ca^{2+} concentration of the host red cell cytosol. Alleva and Kirk (2001) showed that in the saponin-lysed pRBC preparation the parasite can sustain cytoplasmic Ca^{2+} concentrations of ~100 nM in nominally Ca^{2+}-free media. When the external Ca^{2+} concentration was 1 mM the cytosolic free Ca^{2+} concentration within the parasite was ~700 nM. It was surprising that the acidic store would take up Ca^{2+} discharged from the ER, but not Ca^{2+} supplied from the extracellular medium. This behaviour is reminiscent of that observed in animal cells between ER and mitochondria (Pinton et al. 1998; Montero et al. 2000); mitochondria take up Ca^{2+} released from ER stores at specific locations, representing microdomains of much higher Ca^{2+} concentrations than in the bulk of the cytosol. Further research will be needed to establish whether such microdomains exist between ER and acidic stores in the parasite.

How does the malaria parasite gain Ca^{2+}: from the external medium directly, or through the host? Throughout the intraerythrocytic phase the malaria parasite is enclosed within the so-called parasitophorous vacuolar membrane (PVM), formed during the process of invasion. Connecting ducts were suggested as a possible direct path for Ca^{2+} between the plasma and the space between the PVM and the parasite plasma membrane (PPM) (Pouvelle et al. 1991; Haldar 1994; Garcia 1999). However, because the PVM is highly permeable to small solutes and ions (Desai et al. 1993; Desai and Rosenberg 1997; Ginsburg and Kirk 1998; Kirk 2001), the Ca^{2+} within the PVM space would also be expected to gain free access to the host cell cytoplasm, with the consequent rise in $[Ca^{2+}]_i$; this rise in $[Ca^{2+}]_i$ has not been observed. Therefore, most parasite Ca^{2+} gain is likely to be through the host. In principle, the parasite might extract its Ca^{2+} from the tiny but continuously replenished normal host Ca^{2+} pool (Staines et al. 1999a; Kirk 2001). Alternatively, the parasite may induce elevated $[Ca^{2+}]_i$ levels in the host, by increasing the host Ca^{2+} permeability and/or partially inhibiting the PMCA. There is one report of partial PMCA inhibition (Tanabe et al. 1983; Tanabe 1990) and an abundant early literature reporting large increases in host cell Ca^{2+} permeability in murine and human malaria infections (Bookchin et al. 1980; Tanabe et al. 1982, 1983; Krungkrai and Yuthavong 1983; Krishna and Ng 1989; Kramer and Ginsburg 1991; Desai et al. 1996). However, recent work showed that PMCA function is normal in intact pRBCs with mature *P. falciparum* parasites (Tiffert et al. 1999, 2000a) and also that the Ca^{2+} permeability of glucose-fed pRBCs remains at its normal low level throughout parasite development (Staines et al. 1999a,b). We found that the previously noted high Ca^{2+} permeability pathway is only expressed late in parasite development (30–35 h post-invasion), well past the critical stage at which external Ca^{2+} is required for growth (see following section on *Ca^{2+} Requirement for Malaria Parasite Invasion and Growth*). Furthermore, this pathway is activated only in conditions of glucose-deprivation (named DICAP, for depletion-induced calcium permeability). Thus, it appears that neither PMCA inhibition nor high Ca^{2+} permeability contribute to parasite Ca^{2+}-store filling during early development. However, the available evidence cannot rule out the occurrence of brief, transient Ca^{2+} permeabilization episodes during these stages.

Ca^{2+} Requirement for Malaria Parasite Invasion and Growth

With *in vitro* cultures, Ca^{2+} is required in the medium for invasion (Johnson et al. 1980; Raventos Suarez et al. 1982; Wasserman et al. 1982), but it remains controversial whether the Ca^{2+} target is external or intraerythrocytic. Wasserman et al. (1990) and Olaya and Wasserman (1991) suggested that red cell Ca^{2+} entry signals a cytoskeletal reorganization presumed necessary for membrane invagination and fusion during invasion. However, assays of invasion efficiency with either Ca^{2+} chelator-filled red cell ghosts (Dluzewski et al. 1983; McCallum-Deighton and Holder 1992) or intact red cells loaded with the Ca^{2+} chelator Indo-1 (McCallum-Deighton and Holder 1992) showed no detectable differences from chelator-free controls. Wasserman and Chaparro (1996) pointed out shortcomings in that evidence, and showed that invasion efficiency may be reduced when red cells loaded with high levels of the Ca^{2+} chelators quin2 or BAPTA, were suspended in low Ca^{2+} media. Furthermore, the inhibition of glycolysis accompanying the intracellular release of 4 moles of formaldehyde per mole of incorporated free chelator (Tiffert et al. 1984) was not considered in the original experiments with chelator loaded ghosts and cells. Thus it is still unclear whether the external Ca^{2+} requirement reflects a role for Ca^{2+} influx during invasion. Further complexity emerged from the observation of a small $^{45}Ca^{2+}$ gain by recently invaded RBCs (Wasserman et al. 1990). It is difficult to envisage retained $^{45}Ca^{2+}$ within a host red cell with normal Ca^{2+} pump activity unless such $^{45}Ca^{2+}$ is contained within invading parasites. Merozoites use a myosin (Pfmyo-A) motor, which presumably requires Ca^{2+} for gliding motility during internalization (Pinder et al. 1998) raising the possibility of merozoites carrying the $^{45}Ca^{2+}$ gained by pRBCs. Recently, Rozmajzl et al. (2001) sequenced, and functionally characterized a Ca^{2+}-transport-ATPase (PfATP4) from *P. falciparum*, and Caldas and Wasserman (2001) provided cytochemical evidence for the presence of Ca^{2+}-ATPase activity in the PPM throughout all its developmental stages, supporting the idea that merozoites and early ring parasite stages may actively transport Ca^{2+} inwards from low Ca^{2+} concentrations in the environment.

Normal parasite development following invasion also has an absolute requirement for external Ca^{2+}, from early ring stage until about 26 h post-invasion (Raventos Suarez et al. 1982; Wasserman et al. 1982). In the absence of external Ca^{2+}, morphological abnormalities become apparent after about 16 h post-invasion (Wasserman et al. 1982), and by 40 h development is fully arrested, with abnormal trophozoites and no schizonts. In contrast, chelation of external Ca^{2+} after 26 h has no effect on parasite development, schizogony and merozoite release (Raventos Suarez et al. 1982; Wasserman et al. 1982). Therefore, sufficient Ca^{2+}-loading of parasite stores must occur within 26 h post-invasion to sustain all subsequent parasite functions.

Possible DICAP Functions

What could be the function of DICAP, the depletion-induced calcium permeability pathway? Given that parasites have acquired sufficient Ca^{2+} for subsequent development by 26 h post-invasion, the role of the late DICAP pathway is far from clear. *In vivo*, most pRBCs are sequestered in the microcirculation by the time the DICAP pathway is expressed (Staines et al. 1999a). It was suggested (Staines et al. 1999a) that DICAP might play a role in promoting parasite apoptosis in circumstances where its continued survival would threaten the host. DICAP activation may occur if the density of sequestered pRBCs in some small vessels becomes high enough to reduce local blood flow, allowing the highly active parasite metabolism to deplete the local glucose supply. The increased uptake of Ca^{2+} could then trigger an apoptotic sequence leading to the rapid demise of both parasite and host cell, ultimately limiting damage to the host organism, by helping to unblock critical blood vessels. An alternative suggestion was that DICAP represented a capacitative Ca^{2+} channel, activated when the SERCA-type Ca^{2+} pump becomes inhibited by substrate starvation eliciting ER Ca^{2+} depletion (Staines et al. 1999b).

15.12.2 Sickle Cell Anaemia

The mechanism of sickle cell dehydration is discussed in Chap. 23. Here we focus exclusively on the altered Ca^{2+} homeostasis of sickle cells. The link between deoxy-Hb S polymerization and sickle cell dehydration was a mystery until Eaton et al. (1973) and Palek (1973), independently discovered that the total Ca^{2+} content of sickle cells was highly increased. These findings immediately pointed to the possible involvement of Ca^{2+}-sensitive IK1 channels in sickle cell dehydration. Moreover, the elevated Ca^{2+} content of sickle cells was found to be unevenly distributed, with more total Ca^{2+} contained within progressively denser cells (Eaton et al. 1973; Palek 1973; Bookchin et al. 1981). However, many hurdles and puzzles had to be confronted and resolved before the role of Ca^{2+} was fully clarified.

Sickling-Induced Increase in Ca^{2+} Permeability

The first question concerned the mechanism by which Ca^{2+} entered the red cells. Tosteson's pioneering work had established a link between deoxy-Hb S polymerization and increased membrane permeability, though the increased permeability was then only documented for Na^+ and K^+ (Tosteson et al. 1952, 1955; Tosteson 1955). The question of whether sickling also increases the Ca^{2+} permeability was investigated in red cells from heterozygote (sickle cell trait) and homozygotes sickle cell anaemia subjects (Eaton et al. 1973; Palek 1973, 1977; Bookchin and Lew 1981) and, more recently, in different density subpopulations of sickle cells (Bookchin et al. 1991). The results showed that, as with monovalent ions and with Mg^{2+} (Ortiz et al. 1990), sickling increased the Ca^{2+} permeability of the red cell membrane and, at physiological $[Ca^{2+}]_o$ levels, the Ca^{2+} influx increased from its normal value around 50 $\mu mol/(l_{cells} \cdot h)$ to up to ~300 $\mu mol/(l_{cells} \cdot h)$ (Etzion et al. 1993). The measured increase in each case was a mean value. Because, as shown below, each sickling episode generates a distribution of Ca^{2+} permeabilities among sickle cell populations, in some cells Ca^{2+} influx must be much higher than the measured mean. The sickling-induced increase in Ca^{2+} permeability persisted for the duration of the sickling episode and was fully reversible (Lew et al. 1997). The extent of Ca^{2+} permeabilization on deoxygenation was similar for the different density-separated fractions of sickle cells – the light, reticulocyte-rich; – the normal density, discocyte-rich; and – the dense fraction, rich in irreversible sickle cells (Bookchin et al. 1991). These results showed that the transient permeability pathway generated by sickling (P_{sickle}) could carry Ca^{2+}, but how high levels of $[Ca_T]_i$ were retained within the red cells remained a mystery. The puzzle was confounded by findings that the increased $[Ca_T]_i$ in sickle cells could be fully extracted by exposure to ionophore and external EGTA, and yet failed to activate Ca^{2+} pumps or IK1 channels, leading to mistaken suggestions of sickling-induced dysfunction of these transporters (Bookchin and Lew 1980; Lew and Bookchin 1980).

Endocytic Inside-Out Vesicles (EIOVs)

One critical experiment provided a powerful clue that Ca^{2+} was compartmentalized within the cells (Lew and Bookchin 1980). When sickle cells were suspended in a Ca^{2+}-free medium and the uptake of ^{42}K tracer measured as a function of time, addition of a small concentration of the ionophore A23187 produced a transient increase in ^{42}K influx which could be inhibited by quinine, an inhibitor of IK1 channels (Lew and Ferreira 1978); this suggested that the ionophore could somehow make the intracellular Ca^{2+} accessible to IK1 channels while Ca^{2+} was being slowly released from the cells. The notion than red cells could contain Ca^{2+}-accumulating vesicles was so alien at the time that no mention of such possibility was acceptable in the original paper. The idea gained acceptance only after conclusive evidence was provided, by a variety of methods (Lew et al. 1985; Bookchin et al. 1985c, 1986; Rubin et al. 1986; Williamson et al. 1986). Serial section analysis of sickle and normal red cells revealed fully enclosed vesicular compartments, more abundant in the sickle cells (Lew et al. 1985). X-ray microanalysis of the elemental composition of vesicles and cytosol showed that the elevated Ca^{2+} was fully localized to the vesicles; the Na^+/K^+ ratio in the vesicles was reversed relative to the cytosol suggesting an endocytic origin to the vesicles, with retention of the high Na^+/K^+ ratio prevailing in plasma. The vesicles also contained high concentrations of phosphate, suggesting that its passive entry into the vesicles from the red cell cytoplasm led to formation of hydroxyapatite precipitates. In ghosts generated by disruption of the red cell plasma membrane, while conserving the integrity of at least some of the internal vesicles, it was possible to demonstrate powerful (ATP + Mg^{2+})-dependent Ca^{2+} accumulation. Finally, the use of a specific antibody against the inner aspect of the red cell Ca^{2+} pump in histochemical studies conclusively confirmed the endocytic origin of the vesicular compartment (Williamson et al. 1992). Furthermore, thapsigargin failed to inhibit the vanadate-sensitive, ATP-dependent Ca^{2+} uptake by membrane-disrupted ghosts from sickle cell anaemia red cells thus ruling out contributions from residual endoplasmic reticulum-derived organelles (Muoma and Lew, unpublished). Thus, Ca^{2+} compartmentalization in EIOVs accounts for the failure of the elevated $[Ca_T]_i$ of sickle cells to activate their Ca^{2+} pumps and IK1 channels, with no implications for the functional state of these transporters. Early observations (Holroyde and Gardner 1970) had indicated that in splenectomised subjects endocytic vesicles accumulate within red cells, suggesting that these form spontaneously in the circulation and are normally pitted from the cells during passage through the spleen. The loss of splenic function in sickle cell anaemia patients in early childhood may explain the relative abundance of endocytic inside-out vesicles in their red cells.

Unlike normal red cells, sickle cells exposed to ionophore-induced Ca^{2+} loads do not fully extrude the Ca^{2+} after washing away the ionophore indicating that their vesicles retain a substantial fraction of the load (Bookchin and Lew 1980). The Ca^{2+} trapped within the EIOVs of sickle cells thus represents a cumulative record of increased Ca^{2+} permeability events in the circulation. This fully explains the origin of the high $[Ca_T]_i$ content of sickle cells. Although increased Ca^{2+} permeability through P_{sickle} may be sufficient to explain the high Ca^{2+} content of sickle cells, contributions from other sources, such as shear-induced increases in permeability, cannot be excluded.

The presence of EIOVs within normal red cells with negligible total Ca^{2+} contents suggests either that shear-induced, age-related increases in Ca^{2+} permeability (described above) do not occur in the circulation, or that a high turnover of vesicles in subjects with normal splenic function precludes the persistence of vesicles with increased Ca^{2+} contents. An investigation of the $[Ca_T]_i$ of red cells from splenectomized subjects may help settle this point.

As noted above, the demonstration of high phosphate concentrations within the sickle cell EIOVs together with their high calcium indicates that calcium phosphate precipitates form within the EIOVs. The presence of such precipitates can provide an explanation for their accumulation of Ca^{2+}. For otherwise, simple biophysical considerations suggest that endocytic vesicles could not retain the accumulated Ca^{2+} for extended periods of time. When the Ca^{2+} permeability of the red cell is increased, with the consequent rise in $[Ca^{2+}]_i$, the vesicles will accumulate Ca^{2+}. When the normal permeability is recovered, the Ca^{2+} pumps will rapidly restore the cytosolic Ca^{2+} concentration to its normal level by both extrusion through the PMCA and further intra-vesicular accumulation. Since the vesicles are derived from the plasma membrane by endocytosis, they may be expected to have the same intrinsic membrane permeability of the plasma membrane (*P*). But their surface area to volume ratio is much higher than that of the cell; hence, the rate constant for Ca^{2+} leak from the vesicles, *k*, proportional to *P*/radius, could be extremely fast. Simple calculations indicate that the pump-leak steady state $[Ca^{2+}]$ in the vesicles would have to be much lower than in the extracellular medium, i.e., no Ca^{2+} accumulation. Thus, for Ca^{2+} accumulation to occur in the absence of rapid formation of phosphate precipitates, either the frequency of Ca^{2+} permeabilization episodes would have to overcome the discharge rate set by *k*, or Ca^{2+}-phosphate precipitates can rapidly form to trap Ca^{2+} within the vesicles, or the intrinsic Ca^{2+} permeability of the membrane of the Ca^{2+}-accumulating vesicles is actually lower than that of the plasma membrane. This may occur if the surface density of Ca^{2+} permeability pathways was very low, raising the probability of forming vesicles with fewer permeability pathways per unit membrane area.

Finally, let us consider the significance of the fact that the densest sickle cell fraction, mostly irreversibly sickled cells, also has the highest total Ca^{2+} content. These cells represent a relatively young subpopulation, with most of the cells dehydrated by a fast track path (Bookchin et al. 1991; Lew et al. 1991). This path combines dehydration by pH_i-stimulated K^+-Cl^- cotransporters and Ca^{2+}-sensitive K^+ channels (see Chap. 23). Although K^+-Cl^- cotransport clearly contributes to their rapid dehydration at the reticulocyte stage, thus generating relatively dense reticulocytes, their high Ca^{2+} content, despite their young age, attests to a selective vulnerability to frequent Ca^{2+} permeabilization events *in vivo*. These may derive exclusively from P_{sickle} or also from shear effects, perhaps increased by the marked rigidity of these dense cells.

Population Distribution and the Nature of P_{sickle}

When a sample of sickle cells of similar density is fully deoxygenated, is the Ca^{2+} permeability through P_{sickle} increased uniformly in all the cells? This question was addressed in sickle cell discocytes using a P_{sickle}-dependent dense cell formation method designed *ad hoc* (Lew et al. 1997). The method reported the fraction of cells whose $[Ca^{2+}]_i$ was high enough to activate IK1 channels during each deoxygenation pulse ("activated cells"). The results showed that each deoxygenation pulse generated between 10 and 45% activated cells at physiological $[Ca^{2+}]_o$, reflecting donor variation. After removal of activated cells from a previous pulse, similar fractions of activated cells were generated by subsequent pulses, indicating random determination rather than the response of a specific vulnerable cell subpopulation. Thus, deoxygenation generated a stochastic P_{sickle} distribution in the cell population. The fraction of activated cells increased monotonically with $[Ca^{2+}]_o$, and the pattern observed indicated that in a small fraction of cells Ca^{2+} entry through P_{sickle} was high enough to elicit near-maximal IK1 channel activation. Consistent with the stochastic nature of P_{sickle}, repeated deoxygenation-reoxygenation pulses led to progressive dense cell formation whereas single long pulses caused only an early single density shift to a steady level.

What could be the possible mechanism behind this behaviour of P_{sickle}? P_{sickle} arises from reversible interactions between Hb S and membrane components, which either generate a new transport pathway or modify a pre-existing one. Based on known interactions between the central cavity of the Hb tetramer and band 3 protein (Walder et al. 1984), it was suggested that the nucleation of the deoxy-Hb S polymer might occur at that interaction site. If so, oligomeric aggregates of band 3 dimers might form, which could function as a large ion channel (Lew et al. 1997), a suggestion consistent with the observed inhibitory effect of stilbenes on P_{sickle} (Joiner 1990, 1993). The stochastic behaviour discussed above could be explained if the number of permeabilizing contacts between deoxy-Hb S polymers and the membrane in each deoxygenation pulse were small, random and variable from pulse to pulse in each cell, and if the permeabilizing structure remained stable for the duration of each deoxygenation episode (Lew et al. 1997). At this time, the nature of P_{sickle} remains one of the major open questions concerning the pathophysiology of sickle cell dehydration.

15.12.3 Thalassaemia

Red cells from patients with β-thalassaemia intermedia (β-TI) have increased $[Ca_T]_i$, higher in red cells from splenectomized patients, but their Ca^{2+} permeability and physiological $[Ca^{2+}]_i$ levels were within normal values (Bookchin et al 1988). Unlike findings with sickle cells, most of the elevated cell-associated calcium was found within the lighter density cells (presumed younger), and about half of the total calcium content of the cells was not mobilizable by ionophore treatment in the presence of external chelators, suggesting that it could be firmly bound to remnants of organelles and normoblast nuclei. Serial electronmicroscopical studies on ghosts derived from controls and splenectomized β-TI patients showed increased numbers of fully enclosed vesicles (relative to ghosts from normal red cells), and many large vesicles with structured electron-dense material, particularly in the red cells from splenectomized patients. Ionophore-induced Ca^{2+} loads were extruded poorly from β-TI red cells, with higher $[Ca_T]_i$ retention in red cells from splenectomized patients. The results suggested that the elevated total calcium content of β-TI red cells is contained within residual organelles as well as within endocytic inside-out vesicles, both in covalently bound and extractable forms. Unlike red cells from β-TI, those from β-thalassaemia major (β-TM) seem to have an increased Ca^{2+} permeability, probably related to the greater immaturity of their membrane (Wiley 1981). It is uncertain, however, whether an increased K^+ permeability observed with β-TM red cells (Nathan and Gunn 1966; Nathan et al. 1969) is related to their elevated Ca^{2+} permeability. Except for these uncertainties concerning a possible role of Ca^{2+} in β-TM, the elevated $[Ca_T]_i$ content of red cells from thalassaemia patients appears to be solely an epiphenomenon resulting from the presence of residual organelles and endocytic vesicles, and, unlike sickle cell anaemia, it does not seem to play a relevant role in the pathophysiology of the disease.

Acknowledgement: We wish to thank the Wellcome Trust (UK) for funds.

References

Adovelande J, Bastide B, Délèze J, Schrével J (1993) Cytosolic free calcium in *Plasmodium falciparum*-infected erythrocytes and the effect of verapramil:a cytofluorimetric study. Exp Parasitol 76:247-258

Alleva LM, Kirk K (2001) Calcium regulation in the intraerythrocytic malaria parasite *Plasmodium falciparum*. Mol Biochem Parasitol 117:121-128

Almaraz L, García-Sancho J, Lew VL (1988) Calcium-induced conversion of adenine nucleotides to inosine monophosphate in human red cells. J Physiol 407:557-567

Alvarez J, García-Sancho J, Herreros B (1988) All or one cell responses of Ca^{2+}-dependent K channels elicited by calcium or lead in human red cells can be explained by heterogeneity of agonist distribution. J Membrane Biol 104:129-138

Alvarez J, Montero M, García-Sancho J (1992) High affinity inhibition of Ca^{2+}-dependent K^+ channels by cytochrome P-450 inhibitors. J Biol Chem 267:11789-11793

Arese P, Bosia A, Pescarmona GP, Till U (1981) The connection between ionophore-mediated Ca^{2+}-movements and intermediary metabolism in human red cells. II. Site and mode of glycolytic activation during Ca^{2+}-loading. Cell Calcium 2:509-524

Askari A (1966) Modifying effects of anions on the alkali-cation-activated AMP deaminase of human erythrocyte. Mol Pharmacol 2:518-525

Askari A, Franklin JE Jr (1965) Effects of monovalent cations and ATP on erythrocyte AMP deaminase. Biochim Biophys Acta 110:162-173

Askari A, Rao SN (1968) Regulation of AMP deaminase by 2,3-diphosphoglyceric acid: a possible mechanism for the control of adenine nucleotide metabolism in human erythrocytes. Biochim Biophys Acta 151:198-203

Barrabin H, Garrahan PJ, Rega AF (1980) Vanadate inhibition of the Ca^{2+}-ATPase from human red cell membranes. Biochim Biophys Acta 600:796-804

Blackwood AM, Sagnella GA, Markandu ND, MacGregor GA (1997) Problems associated with using Fura-2 to measure free intracellular calcium concentrations in human red blood cells. J Hum Hypertens 11:601-604

Bond GH, Hudgins PM (1978) Kinetics of inhibition of red cell membrane Ca-ATPase by pentavalent vanadium. Fed Proc 313:abstract

Bookchin RM, Lew VL (1980) Progressive inhibition of the Ca pump and Ca:Ca exchange in sickle red cells. Nature 284:561-563

Bookchin RM, Lew VL (1981) Effect of a 'sickling pulse' on calcium and potassium transport in sickle cell trait red cells. J Physiol 312:265-280

Bookchin RM, Lew VL, Nagel RL, Raventos C (1980) Increase in potassium and calcium transport in human red cells infected with *Plasmodium falciparum in vitro.* J Physiol 312:65P

Bookchin RM, Raventos C, Lew VL (1981) Abnormal vesiculation and calcium transport by "one-step" inside-out vesicles from sickle cell anemia red cells. Comparisons with transport by intact cells. In: Brewer GJ (ed) The red cell: Fifth Ann Arbor conference. Alan R Liss, New York, pp 163-182

Bookchin RM, Roth EF Jr, Lew VL (1985a) Near-normal circulatory survival of rabbit red cells exposed to high levels of Ca and ionophore *in vitro.* Blood 66:220-223

Bookchin RM, Lew VL, Roth EF Jr (1985b) Elevated red cell calcium: Innocent bystander or kiss of death? In: Eaton JW (ed) Cellular and molecular aspects of aging: The red cell as a model. Alan R Liss, New York, pp 369-375

Bookchin RM, Ortiz OE, Somlyo AV, Somlyo AP, Sepulveda MI, Hockaday A, Lew VL (1985c) Calcium-accumulating inside-out vesicles in sickle cell anemia red cells. Trans Assoc Am Phys 98:10-20

Bookchin RM, Ortiz OE, Hockaday A, Sepulveda MI, Lew VL (1986) The state of calcium in sickle cell anemia red cells. In: Bader H, Gietzen K, Rosenthal J, Rudel R, Wolf HU (eds) Intracellular calcium regulation. Manchester University Press, Wolfeboro NH, pp 327-334

Bookchin RM, Ortiz OE, Shalev O, Tsurel S, Rachmilewitz EA, Hockaday A, Lew VL (1988) Calcium transport and ultrastructure of red cells in beta-thalassemia intermedia. Blood 72:1602-1607

Bookchin RM, Ortiz OE, Lew VL (1991) Evidence for a direct reticulocyte origin of dense red cells in sickle cell anemia. J Clin Invest 87:113-124

Brown AM, Johnston MJ (1983) Stimulation of lactate production by Ca^{2+} ionophore A23187 in inosine-fed human red cells. J Physiol 341:63P

Brown AM, Simonsen LO (1985) Intracellular cobalt buffering by intact human red cells - evidence for time-dependent and metabolism-dependent changes. J Physiol 361:26P

Bunn HF, Forget BG (1986) Hemoglobin: molecular, genetic and clinical aspects. WB Saunders Company, Philadelphia

Bunn HF, McDonough M (1974) Asymmetrical hemoglobin hybrids. An approach to the study of subunit interactions. Biochemistry 13:988-993

Caldas ML, Wasserman M (2001) Cytochemical localisation of calcium ATPase activity during the erythrocytic cell cycle of *Plasmodium falciparum.* Int J Parasitol 31:776-782

Cantley LC Jr, Josephson L, Warner R, Yanagisawa M, Lechene C, Guidotti G (1977) Vanadate is a potent (Na, K)-ATPase inhibitor found in ATP derived from muscle. J Biol Chem 252:7421-7423

Carafoli E (1992) The Ca^{2+} pump of the plasma membrane. J Biol Chem 267:2115-2118

Carafoli E (1997) Plasma membrane calcium pump: structure, function and relationships. Basic Res Cardiol 92, suppl 1:59-61

Castello PR, Caride AJ, Gonzalez Flecha FL, Fernandez HN, Rossi JP, Delfino JM (1994) Identification of transmembrane domains of the red cell calcium pump with a new photoactivatable phospholipidic probe. Biochem Biophys Res Commun 201:194-200

Cheung WY (1982) Calmodulin: an overview. Fed Proc 41:2253-2257

Clark MR (1988a) Senescence of red blood cells: progress and problems. Physiol Rev 68:503-554

Clark MR (1988b) Calcium extrusion by high-density human red blood cells. Blood Cells 14:119-137

Clark A, Carafoli E (1983) The stoichiometry of the Ca^{2+}-pumping ATPase of erythrocytes. Cell Calcium 4:83-88

Dagher G, Lew VL (1988) Maximal calcium extrusion capacity and stoichiometry of the human red cell calcium pump. J Physiol 407:569-586

Davis FB, Davis PJ, Nat G, Blas SD, MacGillivray M, Gutman S, Feldman MJ (1985) The effect of *in vivo* glucose administration on human erythrocyte Ca^{2+}-ATPase activity and on enzyme responsiveness *in vitro* to thyroid hormone and calmodulin. Diabetes 34:639-646

Davis PJ, Davis FB, Lawrence WD (1989) Thyroid hormone regulation of membrane Ca2(+)-ATPase activity. Endocr Res 15:651-682

Desai SA, Rosenberg RL (1997) Pore size of the malaria parasite's nutrient channel. Proc Natl Acad Sci USA 94:2045-2049

Desai SA, Schlesinger PH, Krogstad DJ (1989) Passive Ca^{2+} fluxes across the erythrocyte membrane suggest a carrier model. Biophys J 55:143a

Desai SA, Schlesinger PH, Krogstad DJ (1991a) Physiologic rate of carrier-mediated Ca^{2+} entry matches active extrusion in human erythrocytes. J Gen Physiol 98:349-364

Desai SA, Schlesinger PH, Krogstad DJ (1991b) Physiologic rate of carrier-mediated Ca^{2+} entry matches active extrusion in human erythrocytes. J Gen Physiol 98:349-364

Desai SA, Krogstad DJ, McCleskey EW (1993) A nutrient-permeable channel on the intraerythrocytic malaria parasite. Nature 362:643-646

Desai SA, McCleskey EW, Schlesinger PH, Krogstad DJ (1996) A novel pathway for Ca^{2+} entry into *Plasmodium falciparum*-infected blood cells. Am J Trop Med Hyg 54:464-470

Dluzewski AR, Rangachari K, Wilson RJ, Gratzer WB (1983) A cytoplasmic requirement of red cells for invasion by malarial parasites. Mol Biochem Parasitol 9:145-160

Dunn MJ (1974) Red blood cell calcium and magnesium: effects upon sodium and potassium transport and cellular morphology. Biochim Biophys Acta 352:97-116

Eaton JW, Skelton TD, Swofford HS, Koplin CE, Jacob HS (1973) Elevated erythrocyte calcium in sickle cell disease. Nature 246:105-106

Etzion Z, Tiffert T, Bookchin RM, Lew VL (1993) Effects of deoxygenation on active and passive Ca^{2+} transport and on the cytoplasmic Ca^{2+} levels of sickle cell anemia red cells. J Clin Invest 92:2489-2498

Ferreira HG, Lew VL (1976) Use of ionophore A23187 to measure cytoplasmic Ca buffering and activation of the Ca pump by internal Ca. Nature 259:47-49

Ferreira HG, Lew VL (1977) Passive Ca transport and cytoplasmic Ca buffering in intact red cells. In: Ellory JC, Lew VL (eds) Membrane transport in red cells. Academic Press, London, pp 53-91

Flatman PW, Lew VL (1977) Use of ionophore A23187 to measure and to control free and bound cytoplasmic Mg in intact red cells. Nature 267:360-362

Freeman CJ, Bookchin RM, Ortiz OE, Lew VL (1987) K-permeabilized human red cells lose an alkaline, hypertonic fluid containing excess K over diffusible anions. J Membrane Biol 96:235-241

Garcia CR (1999) Calcium homeostasis and signaling in the blood-stage malaria parasite. Parasitol Today 15:488-491

Garcia CRS, Dluzewski AR, Catalani LH, Burting R, Hoyland J, Mason WT (1996) Calcium homeostasis in intraerythrocytic malaria parasites. Eur J Cell Biol 71:409-413

García-Sancho J (1985) Pyruvate prevents the ATP depletion caused by formaldehyde or calcium-chelator esters in the human red cell. Biochim Biophys Acta 813:148-150

García-Sancho J, Lew VL (1988a) Detection and separation of human red cells with different calcium contents following uniform calcium permeabilization. J Physiol 407:505-522

García-Sancho J, Lew VL (1988b) Heterogeneous calcium and adenosine triphosphate distribution in calcium-permeabilized human red cells. J Physiol 407:523-539

García-Sancho J, Lew VL (1988c) Properties of the residual calcium pools in human red cells exposed to transient calcium loads. J Physiol 407:541-556

Gardos G (1958) The function of calcium in the potassium permeability of human erythrocytes. Biochim Biophys Acta 30:653-654

Gassner B, Luterbacher S, Schatzmann HJ, Wüthrich A (1988) Dependence of the red blood cell calcium pump on the membrane potential. Cell Calcium 9:95-103

Ginsburg H, Kirk K (1998) Membrane transport in the malaria-infected erythrocyte. In: Sherman IW (ed) Malaria: Parasite biology, pathogenesis, and protection. ASM Press, Washington DC, pp 219-232

Glaser T, Schwarz-Benmeir N, Barnoy S, Barak S, Eshhar Z, Kosower NS (1994) Calpain (Ca^{2+}-dependent thiol protease) in erythrocytes of young and old individuals. Proc Natl Acad Sci USA 91:7879-7883

Gonzalez Flecha FL, Bermudez MC, Cedola NV, Gagliardino JJ, Rossi JP (1990) Decreased Ca2(+)-ATPase activity after glycosylation of erythrocyte membranes *in vivo* and *in vitro*. Diabetes 39:707-711

Gonzalez Flecha FL, Castello PR, Caride AJ, Gagliardino JJ, Rossi JP (1993) The erythrocyte calcium pump is inhibited by non-enzymic glycation: studies *in situ* and with the purified enzyme. Biochem J 293:369-375

Gonzalez Flecha FL, Castello PR, Gagliardino JJ, Rossi JP (1997) Structural characterization of the glycation process of the plasma membrane calcium pump. Ann New York Acad Sci 834:126-128

Gonzalez Flecha FL, Castello PR, Gagliardino JJ, Rossi JP (1999) Molecular characterization of the glycated plasma membrane calcium pump. J Membrane Biol 171:25-34

Guerini D, Garcia-Martin E, Zecca A, Guidi F, Carafoli E (1998) The calcium pump of the plasma membrane: membrane targeting, calcium binding sites, tissue-specific isoform expression. Acta Physiol Scand, suppl 643:265-273

Haldar K (1994) Ducts, channels and transporters in *Plasmodium*-infected eryhtrocytes. Parasitology Today 10:393-395

Harbak H, Simonsen LO (1987) Residual Ca pump activity in vanadate-inhibited and in ATP-depleted human red cells. J Physiol 390:95P

Harrison DG, Long C (1968) The calcium content of human erythrocytes. J Physiol 199:367-381

Holroyde CP, Gardner FH (1970) Acquisition of autophagic vacuoles by human erythrocytes. Physiological role of the spleen. Blood 36:566-575

Hunter MJ (1971) A quantitative estimate of the non-exchange-restricted chloride permeability of the human red cell. J Physiol 218:49P-50P

Hunter MJ (1977) Human erythrocyte anion permeabilities measured under conditions of net charge transfer. J Physiol 268:35-49

Jacobs MH, Stewart DR (1947) Osmotic properties of the erythrocyte. XII. Ionic and osmotic equilibria with a complex external solution. J Cell Comp Physiol 30:79-103

Johnson JG, Epstein N, Shiroishi T, Miller LH (1980) Factors affecting the ability of isolated *Plasmodium knowlesi* merozoites to attach to and invade erythrocytes. Parasitol 80:539-550

Joiner CH (1990) Deoxygenation-induced cation fluxes in sickle cells: II. Inhibition by stilbene disulfonates. Blood 76:212-220

Joiner CH (1993) Cation transport and volume regulation in sickle red blood cells. Am J Physiol 264:C251-C270

Kirk K (2001) Membrane transport in the malaria-infected erythrocyte. Physiol Rev 81:495-537

Kosk-Kosicka D, Bzdega T (1988) Activation of the erythrocyte Ca^{2+}-ATPase by either self-association or interaction with calmodulin. J Biol Chem 263:18184-18189

Kosk-Kosicka D, Bzdega T (1990) Effects of calmodulin on erythrocyte Ca^{2+}-ATPase activation and oligomerization. Biochemistry 29:3772-3777

Kosk-Kosicka D, Bzdega T, Johnson JD (1990) Fluorescence studies on calmodulin binding to erythrocyte Ca^{2+}-ATPase in different oligomerization states. Biochemistry 29:1875-1879

Kramer R, Ginsburg H (1991) Calcium transport and compartment analysis of free and exchangeable calcium in *Plasmodium falciparum*-infected red blood cells. J Protozool 38:594-601

Krishna S, Ng LL (1989) Cation metabolism in malaria-infected red cells. Exp Parasitol 69:402-406

Krungkrai J, Yuthavong Y (1983) Enhanced Ca^{2+} uptake by mouse erythrocytes in malarial (*Plasmodium berghei*) infection. Mol Biochem Parasitol 7:227-235

Larsen FL, Katz S, Roufogalis BD, Brooks DE (1981) Physiological shear stresses enhance the Ca^{2+} permeability of human erythrocytes. Nature 294:667-668

Lassen UV (1977) Electrical potential and conductance of the red cell membrane. In: Ellory JC, Lew VL (eds) Membrane transport in red cells. Academic Press, London, pp 137-172

Leclerc L, Vasseur C, Bursaux E, Marden M, Poyart C (1988) Inhibition of membrane erythrocyte (Ca^{2+} + Mg^{2+})-ATPase by hemin. Biochim Biophys Acta 946:49-56

Leclerc L, Marden M, Poyart C (1991) Inhibition of the erythrocyte (Ca^{2+} + Mg^{2+})-ATPase by nonheme iron. Biochim Biophys Acta 1062:35-38

Leida MN, Mahoney JR, Eaton JW (1981) Intraerythrocytic plasmodial calcium metabolism. Biochem Biophys Res Commun 103:402-406

Lew VL (1971a) On the ATP dependence of the Ca^{2+}-induced increase in K^+ permeability observed in human red cells. Biochim Biophys Acta 233:827-830

Lew VL (1971b) Effect of ouabain on the Ca^{2+}-dependent increase in K^+ permeability in depleted guinea-pig red cells. Biochim Biophys Acta 249:236-239

Lew VL, Beaugé LA (1979) Passive cation fluxes in red cell membranes. In: Giebisch G, Tosteson DC, Ussing HH (eds) Transport across biological membranes, vol II. Springer-Verlag, Berlin, pp 85-115

Lew VL, Bookchin RM (1980) A Ca^{2+}-refractory state of the Ca^{2+}-sensitive K^{+} permeability mechanism in sickle cell anemia red cells. Biochim Biophys Acta 602:196-200

Lew VL, Bookchin RM (1986) Volume, pH and ion content regulation in human red cells: analysis of transient behavior with an integrated model. J Membrane Biol 92:57-74

Lew VL, Brown AM (1979) Experimental control and assessment of free and bound calcium in the cytoplasm of intact mammalian red cells. In: Ashley CC, Campbell AK (eds) Detection and measurement of free Ca in cells, Elsevier/North-Holland, Amsterdam, pp 423-432

Lew VL, Ferreira HG (1976) Variable Ca sensitivity of a K-selective channel in intact red-cell membranes. Nature 263:336-338

Lew VL, Ferreira HG (1978) Calcium transport and the properties of a calcium-activated potassium channel in red cell membranes. In: Kleinzeller A, Bronner F (eds) Current Topics in Membranes and Transport, vol 10. Academic Press, New York, pp 217-277

Lew VL, García-Sancho J (1989) Measurement and control of intracellular calcium in intact red cells. In: Fleischer S, Fleischer B (eds) Methods in enzymology, vol 173, Biomembranes, part T: Cellular and subcellular transport: Eukaryotic (nonepithelial) cells. Academic Press, San Diego, pp 100-112

Lew VL, Simonsen LO (1980) Ionophore A23187-induced calcium permeability of intact human red blood cells. J Physiol 308:60P

Lew VL, Tsien RY, Miner C, Bookchin RM (1982) Physiological $[Ca^{2+}]_i$ level and pump-leak turnover in intact red cells measured using an incorporated Ca chelator. Nature 298:478-481

Lew VL, Hockaday A, Sepulveda MI, Somlyo AP, Somlyo AV, Ortiz OE, Bookchin RM (1985) Compartmentalization of sickle cell calcium in endocytic inside-out vesicles. Nature 315:586-589

Lew VL, Freeman CJ, Ortiz OE, Bookchin RM (1991) A mathematical model on the volume, pH and ion content regulation of reticulocytes. Application to the pathophysiology of sickle cell dehydration. J Clin Invest 87:100-112

Lew VL, Etzion Z, Bookchin RM, DaCosta R, Väänänen H, Sassaroli M, Eisinger J (1993) The distribution of intracellular calcium chelator (Fura-2) in a population of intact human red cells. Biochim Biophys Acta 1148:152-156

Lew VL, Ortiz-Carranza OE, Bookchin RM (1997) Stochastic nature and red cell population distribution of the sickling-induced Ca^{2+} permeability. J Clin Invest 99:2727-2735

Lorand L, Weissmann LB, Epel DL, Bruner-Lorand J (1976) Role of intrinsic transglutaminase in the Ca^{2+}-mediated crosslinking of erythrocyte proteins. Biochemistry 75:4479-4481

Lorand L, Siefring GE Jr, Lowe-Krentz L (1978) Formation of gamma-glutamyl-epsilon-lysine bridges between membrane protein by a Ca^{2+}-regulated enzyme in intact erythrocytes. J Supramol Struct 9:427-440

Lorand L, Siefring GE Jr, Lowe-Krentz L (1979) Enzymatic basis of membrane stiffening in human erythrocytes. Semin Hematol 16:65-74

McCallum-Deighton N, Holder AA (1992) The role of calcium in the invasion of human erythrocytes by *Plasmodium falciparum*. Mol Biochem Parasitol 50:317-324

McManus TJ (1967) Comparative biology of red cells. Fed Proc 26:1821-1826

McNamara MK, Wiley JS (1986) Passive permeability of human red blood cells to calcium. Am J Physiol 250:C26-C31

Montero M, Alonso MT, Carnicero E, Cuchillo-Ibanez I, Albillos A, Garcia AG, Garcia-Sancho J, Alvarez J (2000) Chromaffin-cell stimulation triggers fast millimolar mitochondrial Ca^{2+} transients that modulate secretion. Nature Cell Biol 2:57-61

Moore RB, Plishker GA, Shriver SK (1990) Purification and measurement of calpromotin, the cytoplasmic protein which activates calcium-dependent potassium transport. Biochem Biophys Res Commun 166:146-153

Muallem S, Miner C, Seymour CA (1985) The nature of the Ca^{2+}-pump defect in the red blood cells of patients with cystic fibrosis. Biochim Biophys Acta 819:143-147

Murphy E, Levy L, Berkowitz L, Orringer E, Gabel S, London R (1986) Nuclear magnetic resonance measurement of cytosolic free calcium levels in human red blood cells. Am J Physiol 251:C496-C504

Murphy E, Berkowitz LR, Orringer E, Levy L, Gabel SA, London RE (1987) Cytosolic free calcium level in sickle red blood cells. Blood 69:1469-1474

Nathan DG Gunn RB (1966) Thalassemia: the consequences of unbalanced hemoglobin synthesis. Am J Med 41:815-830

Nathan DG, Stossel TB, Gunn RB, Zarkowsky HS, Laforet MT (1969) Influence of hemoglobin precipitation on erythrocyte metabolism in alpha and beta thalassemia. J Clin Invest 48:33-41

Olaya P, Wasserman M (1991) Effect of calpain inhibitors on the invasion of human erythrocytes by the parasite *Plasmodium falciparum*. Biochim Biophys Acta 1096:217-221

Orringer EP, Mattern WD (1976) Formaldehyde-induced hemolysis during chronic hemodialysis. New Eng J Med 294:1416-1420

Ortiz OE, Lew VL, Bookchin RM (1990) Deoxygenation permeabilizes sickle cell anaemia red cells to magnesium and reverses its gradient in the dense cells. J Physiol 427:211-226

Palek J (1973) Calcium accumulation during sickling of haemoglobin S red cells. Blood 42:988-1000

Palek J (1977) Red cell calcium content and transmembrane calcium movements in sickle cell anemia. J Lab Clin Med 89:1365-1374

Passos APD, Garcia CRS (1997) Characterization of Ca^{2+} transport activity associated with a non-mitochondrial calcium pool in the rodent malaria parasite *P. chabaudi*. Biochem Mol Biol Int 42:919-925

Passos APD, Garcia CRS (1998) Inositol 1,4,5-trisphosphate induced Ca^{2+} release from chloroquine-sensitive and -insensitive intracellular stores in the intraerythrocytic stage of the malaria parasite *P. chabaudi*. Biochem Biophys Res Commun 245:155-160

Pereira AC, Samellas D, Tiffert T, Lew VL (1993) Inhibition of the calcium pump by high cytosolic Ca^{2+} in intact human red cells. J Physiol 461:63-73

Pinder JC, Fowler RE, Dluzewski AR, Bannister LH, Lavin FM, Mitchell GH, Wilson RJ, Gratzer WB (1998) Actomyosin motor in the merozoite of the malaria parasite, *Plasmodium falciparum*: implications for red cell invasion. J Cell Sci 111:1831-1839

Pinton P, Brini M, Bastianutto C, Tuft RA, Pozzan T, Rizzuto R (1998) New light on mitochondrial calcium. Biofactors 8:243-253

Plishker GA, White PH, Cadman ED (1986) Involvement of a cytoplasmic protein in calcium-dependent potassium efflux in red blood cells. Am J Physiol 251:C535-C540

Plishker GA, Chevalier D, Seinsoth L, Moore RB (1992) Calcium-activated potassium transport and high molecular weight forms of calpromotin. J Biol Chem 267:21839-21843

Pouvelle B, Spiegel R, Hsiao L, Howard RJ, Morris RL, Thomas AP, Taraschi TF (1991) Direct access to serum macromolecules by intraerythrocytic malaria parasites. Nature 353:73-75

Pozos TC, Sekler I, Cyert MS (1996) The product of HUM1, a novel yeast gene, is required for vacuolar Ca^{2+}/H^{+} exchange and is related to mammalian Na^{+}/Ca^{2+} exchangers. Mol Cell Biol 16:3730-3741

Pressman BC (1976) Biological applications of ionophores. Annu Rev Biochem 45:501-530

Proverbio F, Shoemaker DG, Hoffman JF (1988) Functional consequences of the membrane pool of ATP associated with the human red blood cell Na/K pump. Prog Clin Biol Res 268A:561-567

Raftos JE, Lew VL (1994) Dependence of the plasma membrane Ca^{2+} pump on intracellular Mg^{2+}, investigated in intact human red cells. J Physiol 475:97P

Raftos JE, Bookchin RM, Lew VL (1996) Distribution of chloride permeabilities in normal human red cells. J Physiol 491:773-777

Raftos JE, Bookchin RM, Lew VL (1997) Measurement of the distribution of anion-exchange function in normal human red cells. J Physiol 499:17-25

Raftos JE, Edgley A, Bookchin RM, Etzion Z, Lew VL, Tiffert T (2001) Normal Ca^{2+} extrusion by the Ca^{2+} pump of intact red blood cells exposed to high glucose concentrations. Am J Physiol 280:C1449-C1454

Raventos Suarez C, Bookchin RM, Lew DJ, Nagel RL, Lew VL (1982) Ca^{2+} is required for reinfection of red cells by *Plasmodium falciparum*. Fed Proc 41:1121

Rega AF, Garrahan PJ (1986) The Ca^{2+} pump of plasma membranes. CRC Press, Boca Raton

Romero PJ, Romero EA (1997) Differences in Ca^{2+} pumping activity between subpopulations of human red cells. Cell Calcium 21:353-358

Romero PJ, Romero EA (1999a) The role of calcium metabolism in human red blood cell ageing: a proposal. Blood Cells, Molecules, and Diseases 25:9-19

Romero PJ, Romero EA (1999b) Effect of cell ageing on Ca^{2+} influx into human red cells. Cell Calcium 26:131-137

Romero PJ, Whittam R (1971) The control by internal calcium of membrane permeability to sodium and potassium. J Physiol 214:481-507

Romero PJ, Romero EA, Winkler MD (1997) Ionic calcium content of light dense human red cells separated by Percoll density gradients. Biochim Biophys Acta 1323:23-28

Rozmajzl PJ, Kimura M, Woodrow CJ, Krishna S, Meade JC (2001) Characterization of P-type ATPase 3 in *Plasmodium falciparum*. Mol Biochem Parasitol 116:117-126

Rubin E, Schlegel RA, Williamson P (1986) Endocytosis in sickle erythrocytes: A mechanism for elevated intracellular Ca^{2+} levels. J Cell Physiol 126:53-59

Scharff O, Foder B (1986) Delayed activation of calcium pump during transient increases in cellular Ca^{2+} concentration and K^{+} conductance in hyperpolarizing human red cells. Biochim Biophys Acta 861:471-479

Scharff O, Foder B, Skibsted U (1982) Two states of Ca^{2+}-ATPase and activation of Ca^{2+}-pump of human erythrocytes. Acta Physiol Lat Amer 32:93-95

Scharff O, Foder B, Skibsted U (1983) Hysteretic activation of the Ca^{2+} pump revealed by calcium transients in human red cells. Biochim Biophys Acta 730:295-305

Schatzmann HJ (1966) ATP-dependent Ca^{++}-extrusion from human red cells. Experientia 22:364-365

Schatzmann HJ (1973) Dependence on calcium concentration and stoichiometry of the calcium pump in human red cells. J Physiol 235:551-569

Schatzmann HJ (1982) The plasma membrane calcium pump of erythrocytes and other animal cells. In: Carafoli E (ed) Membrane transport of calcium. Academic Press, London, pp 41-108

Schatzmann HJ (1983) The red cell calcium pump. Annu Rev Physiol 45:303-312

Schatzmann HJ, Vincenzi FF (1969) Calcium movements across the membrane of human red cells. J Physiol 201:369-395

Schatzmann HJ, Luterbacher S, Stieger J, Wuthrich A (1986) Red blood cell calcium pump and its inhibition by vanadate and lanthanum. J Cardiovasc Pharmacol, suppl 8:S33-S37

Shalev O, Leida MN, Hebbel RP, Jacob HS, Eaton JW (1981) Abnormal erythrocyte calcium homeostasis in oxidant-induced hemolytic disease. Blood 58:1232-1235

Simonsen LO, Gomme J, Lew VL (1982) Uniform ionophore A23187 distribution and cytoplasmic calcium buffering in intact human red cells. Biochim Biophys Acta 692:431-440

Simonsen LO, Lew VL (1980) The correlation between ionophore A23187 content and calcium permeability of ATP-depleted human red blood cells. In: Lassen UV, Ussing HH, Wieth JO (eds) Membrane transport in erythrocytes. Munksgaard, Copenhagen, pp 208-212

Staines HM, Chang W, Ellory JC, Tiffert T, Kirk K, Lew VL (1999a) Passive Ca^{2+} transport and Ca^{2+}-dependent K^{+} transport in *Plasmodium falciparum*-infected red cells. J Membrane Biol 172:13-24

Staines HM, Chang W, Ellory JC, Tiffert T, Kirk K, Lew VL (1999b) A novel Ca^{2+} transport pathway in *Plasmodium falciparum* infected red cells. J Physiol 520P:19P

Tanabe K (1990) Ion metabolism in malaria-infected erythrocytes. Blood Cells 16:437-449

Tanabe K, Mikkelsen RB, Wallach DFH (1982) Calcium transport of *Plasmodium chabaudi*-infected erythrocytes. J Cell Biol 93:680-689

Tanabe K, Mikkelsen RB, Wallach DF (1983) Transport of ions in erythrocytes infected by plasmodia. Ciba Found Symp 94:64-73

Tiffert T, Lew VL (1997a) Cytoplasmic Ca^{2+} buffers in intact human red cells. J Physiol 500:139-154

Tiffert T, Lew VL (1997b) Apparent Ca^{2+} dissociation constant of Ca^{2+} chelators incorporated non-disruptively into intact human red cells. J Physiol 505:403-410

Tiffert T, Lew VL (2001) Kinetics of inhibition of the plasma membrane calcium pump by vanadate in intact human red cells. Cell Calcium 30:337-342

Tiffert T, García-Sancho J, Lew VL (1984) Irreversible ATP depletion caused by low concentrations of formaldehyde and of calcium-chelator esters in intact human red cells. Biochim Biophys Acta 773:143-156

Tiffert T, Spivak JL, Lew VL (1988) Magnitude of calcium influx required to induce dehydration of normal human red cells. Biochim Biophys Acta 943:157-165

Tiffert T, Etzion Z, Bookchin RM, Lew VL (1993) Effects of deoxygenation on active and passive Ca^{2+} transport and cytoplasmic Ca^{2+} buffering in normal human red cells. J Physiol 464:529-544

Tiffert T, Staines HM, Ellory JC, Lew VL (1999) Conserved Ca^{2+} function in *Plasmodium falciparum*-infected red cells. J Physiol 521P:72P

Tiffert T, Staines HM, Ellory JC, Lew VL (2000a) Functional state of the plasma membrane Ca^{2+} pump in *Plasmodium falciparum*-infected human red blood cells. J Physiol 525:125-134
Tiffert T, Lew VL, Perdomo D, Ginsburg H (2000b) Ferriprotoporphyrin IX does not affect the Ca^{2+} pump in intact erythrocytes. J Membrane Biol 175:107-113
Tiffert T, Daw N, Perdomo D, Lew VL (2001) A fast and simple screening test to search for specific inhibitors of the plasma membrane calcium pump. J Lab Clin Med 137:199-207
Till U, Petermann H, Wenz I, Arese P (1981) The connection between ionophore-mediated Ca^{2+}-movements and intermediary metabolism in human red cells. I. Relationships between Ca^{2+}-loading, ATP-consumption and glycolytic flux. Cell Calcium 2:495-507
Tosteson DC (1955) The effects of sickling on ion transport II. The effect of sickling on sodium and cesium transport. J Gen Physiol 39:55-67
Tosteson DC, Shea E, Darling RC (1952) Potassium and sodium of red blood cells in sickle cell anemia. J Clin Invest 31:406-411
Tosteson DC, Carlsen E, Dunham ET (1955) The effects of sickling on ion transport I. Effect of sickling on potassium transport. J Gen Physiol 39:31-53
Tsien RY (1981) A non-disruptive technique for loading calcium buffers and indicators into cells. Nature 290:527-528
Walder JA, Chatterjee R, Steck TL, Low PS, Musso GF, Kaiser ET, Rogers PH, Arnone A (1984) The interaction of hemoglobin with the cytoplasmic domain of band 3 of the human erythrocyte membrane. J Biol Chem 259:10238-10246
Wang KKW, Roufogalis BD, Villalobo A (1989) Calpain I activates Ca^{2+} transport by the reconstituted erythrocyte Ca^{2+} pump. J Membrane Biol 112:233-245
Wang KKW, Villalobo A, Roufogalis BD (1992) The plasma membrane calcium pump: a multiregulated transporter. Trends in Cell Biol 2:46-52
Wasserman M, Alarcón C, Mendoza PM (1982) Effects of Ca^{++} depletion on the asexual cell cycle of *Plasmodium falciparum*. Am J Trop Med Hyg 31:711-717
Wasserman M, Vernot JP, Mendoza PM (1990) Role of calcium and erythrocyte cytoskeleton phosphorylation in the invasion of *Plasmodium falciparum*. Parasitol Res 76:681-688
Wasserman M, Chaparro J (1996) Intraerythrocytic calcium chelators inhibit the invasion of *Plasmodium falciparum*. Parasitol Res 82:102-107
Whittam R (1964) Transport and diffusion in red blood cells. Edward Arnold, London
Wiley JS (1981) Increased erythrocyte cation permeability in thalassemia and conditions of marrow stress. J Clin Invest 67:917-922
Williamson P, Rubin L, Schlegel R (1986) Internal Ca^{2+} containing vesicles in sickle erythrocytes. J Cell Biol 99:430a
Williamson P, Puchulu E, Penniston JT, Westerman MP, Schlegel RA (1992) Ca^{2+} accumulation and loss by aberrant endocytic vesicles in sickle erythrocytes. J Cell Physiol 152:1-9

16 Magnesium Transport

Peter W. Flatman

Membrane Biology Group, Division of Biomedical and Clinical Laboratory Sciences, The University of Edinburgh, Hugh Robson Building, George Square, Edinburgh EH8 9XD, Scotland, UK

16.1 Introduction

Mg^{2+} is a vital constituent of living things. Once considered a trace mineral it is now recognised as a major cellular component being the second most abundant intracellular cation in mammalian tissue (Walser 1967; Wacker 1980). Beneficial and medicinal uses of Mg^{2+} salts have been described since the sixteenth century (Aikawa 1963; Wacker 1980) and there have been many studies revealing the importance of Mg^{2+} in animal and human health (Wacker 1980; Smetana 1997).

Mg^{2+} plays a special role in red cell physiology as its intracellular free ionized concentration ($[fMg^{2+}]_i$) fluctuates when haemoglobin is oxygenated or deoxygenated. $[fMg^{2+}]_i$ is lowest in oxygenated cells and highest in deoxygenated cells (Gupta et al. 1978; Flatman 1980; Campbell and Gibson 1998; Campbell et al. 1999). Thus $[fMg^{2+}]_i$ is a messenger that relays information about the oxygenation state of haemoglobin and may regulate metabolism and transport in the cell. $[fMg^{2+}]_i$ is thought to play a particularly important role in cell volume regulation through its effects on Na^+-K^+-$2Cl^-$ cotransport, K^+-Cl^- cotransport and Na^+/H^+ antiport (Jennings and Schulz 1990; Mairbäurl and Hoffman 1992; Parker 1993; Flatman 1993). Long-term regulation of $[fMg^{2+}]_i$ by membrane transport may thus be of particular interest in red cells and recent evidence suggests that changes in Mg^{2+} transport may play a part in several pathophysiological conditions.

16.2 Magnesium Homeostasis in Red Cells

Red cell $[fMg^{2+}]_i$ is controlled by two main processes, transport across plasma membrane (and intracellular membranes, if present) and binding (buffering) by intracellular ligands, most of which are metabolic intermediates. In principal, $[fMg^{2+}]_i$ could be controlled entirely by regulating buffering, but for effective long-term homeostasis, any Mg^{2+} influx across the plasma membrane must be balanced by an equal efflux, indicating the need for Mg^{2+} transport systems. These may also be necessary to compensate for dramatic changes in Mg^{2+} buffering capacity that may occur, for instance following metabolic depletion. The eventual recovery of buffering capacity must then also be accompanied by an appropriate change in Mg^{2+} transport and content if $[fMg^{2+}]_i$ is to be maintained within physiological limits.

There is evidence in humans and mice that $[fMg^{2+}]_i$ is controlled by several genes (Henrotte 1993). These determine not only the type and membrane surface density of Mg^{2+} transporter, but also cytoplasmic Mg^{2+}-buffering, and the levels of circulating transport-regulating factors, and the ability to respond to these factors.

Regulated Mg^{2+} uptake and efflux is clearly necessary during growth and division of stem cells and during red cell maturation so a variety of Mg^{2+} transporters may be present in their membranes. It is unlikely that all these systems are necessary in mature cells, some may be retained for specific purposes, and some may persist because of insufficient pressure for their removal. Thus red cells from different species may have different ways of transporting Mg^{2+}.

16.3 Red Cell Magnesium Content

Red cell Mg^{2+} contents for a variety of species are given in Table 16.1. In mammals the values range from less than 1 $mmol/l_{cells}$ in cattle to 6 $mmol/l_{cells}$ in pigs, with the majority lying between 2 and 3 $mmol/l_{cells}$. Thus the Mg^{2+} content of red cells is lower than in most other tissues, probably because they do not have nuclei and ribosomes that bind substantial Mg^{2+}. Red cell Mg^{2+} content is genetically determined. In human males, both total and ionized $[Mg^{2+}]$ are lower in HLA-B35$^+$ red cells than in HLA-B35$^-$ cells, whereas the Mg^{2+} content of red cells in HLA-B38$^+$ males is higher than that in HLA-B38$^-$ males. Interestingly, no such correlations could be found in red cells from women (Santarromana et al. 1989; Henrotte et al. 1990; Henrotte 1993). In mice it has been possible to breed strains with low or high red cell Mg^{2+} contents (Henrotte 1993).

16.3.1 Distribution of Magnesium between Bound and Free Forms

About 90% of Mg^{2+} inside red cells is not free and ionized but is bound to a variety of ligands (mainly polyanions, for instance organic phosphates) (Flatman and Lew 1980). These ligands depend on the particular species, but in most mammalian red cells include ATP and 2,3-bisphosphoglycerate (BPG). There may also be some weak binding to haemoglobin (Raftos et al. 1999). The ability of these ligands to bind Mg^{2+} depends on the oxygenation state of haemoglobin. Deoxygenation causes a change in the arrangement of the polypeptide chains in haemoglobin resulting in increased binding of ATP and BPG. As there is now less of these ligands available to bind Mg^{2+} in the cytoplasm, $[fMg^{2+}]_i$ increases. The reverse occurs during oxygenation and $[fMg^{2+}]_i$ falls. Early models of this process (Bunn et al. 1971; Gerber et al. 1973) predicted very large fluctuations in $[fMg^{2+}]_i$ but this prediction was not borne out by measurements of $[fMg^{2+}]_i$ using ^{31}P-NMR or null point titration in the presence of A23187 (Gupta et al. 1978; Flatman 1980). More recent models that take into account the many interactions between ligands, and the changes in pH that accompany changes in oxygenation, fit the data better (Mulquiney and Kuchel 1997). If it is also assumed that Mg^{2+} binds to haemoglobin and to additional low affinity sites on ATP and BPG, then the fit is excellent, not only for red cells under normal metabolic conditions, but also under a variety of other conditions (Raftos et al. 1999).

Table 16.1. Mg^{2+} content of red cells

Species	Mg^{2+} content [$mmol/l_{cells}$]	Reference
Man	2.64 (range 1.84–3.12, n = 16)	Valberg et al. 1965
	2.55 (range 1.75–3.15, n = 22)	Charbon and Hoekstra 1962
	2.10 (range 1.55–3.25, n = 349 males), varies with HLA phenotype	Henrotte 1993
Monkey	2.08 (range 1.57–2.27, n = 4)	Valberg et al. 1965
Rat	2.31 (range 2.05–2.57, n = 16)	Valberg et al. 1965
	2.95 ± 0.22 (S.E.M., n = 12)	Günther et al. 1990
Mouse	2.5, strains of mice have been bred with high and low Mg^{2+} contents	Henrotte 1993
	2.20 ± 0.06 (S.E.M., n = 6)	Büttner et al. 1998
Guinea pig	3.20 (range 2.6–4.0, n = 4)	Charbon and Hoekstra 1962
	2.36 ± 0.06 (S.E.M., n = 6)	Büttner et al. 1998
Ground squirrel	2.23 ± 0.21 (S.E.M., n = 4)	Marjanovic et al. 1993
Hamster	2.80 (mean, n = 5)	Miseta et al. 1993
Cat	2.80 (range 2.5–3.1, n = 3)	Charbon and Hoekstra 1962
	1.80 ± 0.05 (S.E.M., n = 6)	Büttner et al. 1998
Dog	3.20 (range 2.26–4.11, n = 4)	Valberg et al. 1965
	2.91 (range 2.3–3.2, n = 6)	Charbon and Hoekstra 1962
	1.82 ± 0.09 (S.E.M., n = 6)	Büttner et al. 1998
Ferret	3.01 ± 0.12 (S.E.M., n = 10)	Flatman and Andrews 1983
Rabbit	3.90 ± 0.32 (S.E.M., n = 6)	Ginsburg et al. 1962
	3.72 (range 3.35–4.40, n = 4)	Valberg et al. 1965
	3.15 ± 0.14 (S.E.M., n = 6)	Büttner et al. 1998
Horse	3.19 (range 3.0–3.6, n = 4)	Charbon and Hoekstra 1962
	2.26 ± 0.10 (S.E.M., n = 6)	Büttner et al. 1998
	2.55 ± 0.07 (S.E.M., n = 6)	Campbell et al. 1999
Pig	5.79 (range 5.3–6.7, n = 3)	Charbon and Hoekstra 1962
	4.00 ± 0.18 (S.E.M., n = 6)	Büttner et al. 1998
Deer	5.33 (Mean, n = 5)	Miseta et al. 1993
Cattle	1.18 (0.7–1.35, n =3)	Charbon and Hoekstra 1962
	0.50 ± 0.04 (S.E.M., n = 6)	Büttner et al. 1998
Sheep	0.99 ± 0.05 (S.E.M., n = 6)	Büttner et al. 1998
	1.29 ± 0.08 (S.E.M., n = 5)	Campbell and Gibson 1998
Chicken	3.26 ± 0.11 (S.E.M., n = 6)	Büttner et al. 1998
Duck	4.41 (range 3.28–6.19, n = 3)	Valberg et al. 1965
	3.85 ± 0.13 (S.E.M., n = 6)	Büttner et al. 1998
Turkey	4.92 (mean, n = 5)	Miseta et al. 1993
Pigeon	8.44 (mean, n = 3)	Miseta et al. 1993
Trout	4.70 ± 0.39 (S.E.M., n = 6)	Büttner et al. 1998

16.3.2 Levels of Free Ionized Magnesium

$[fMg^{2+}]_i$ is about 0.4 mM in oxygenated human red cells rising to 0.65 mM in fully deoxygenated cells (Gupta et al. 1978; Flatman 1980). Values for other species are given in Table 16.2. Those obtained by null point titration are higher than values obtained by NMR, even when the measurements are made in the same laboratory. $[fMg^{2+}]_i$ obtained by NMR is critically dependent on the value chosen for the apparent MgATP dissociation constant (K_{MgATP}). Values used in the past (about 50 μM) may have been too low, as recent measurements suggest values close to 80 μM (Zhang et al. 1997, Malon et al. 2002). When this K_{MgATP} is used, $[fMg^{2+}]_i$ determined by NMR is very similar to values obtained by null point titration. Similar values have recently been obtained using ion selective electrodes on concentrated red cell lysates (Malon et al. 2002). Interestingly, significant changes in $[fMg^{2+}]_i$ have been reported in sheep red cells despite these cells containing virtually no BPG, and having a haemoglobin with a low affinity for polyanions (Campbell and Gibson 1998). Thus, $[fMg^{2+}]_i$ in oxygenated red cells lies between 0.3 and 0.7 mM.

Surprisingly few measurements have been made of $[fMg^{2+}]$ in extracellular fluids ($[fMg^{2+}]_o$), and most of these have been on human plasma. It is difficult to measure because of competition between Mg^{2+} and Ca^{2+} for binding to plasma proteins and interference from Ca^{2+} and other materials, including silicone oil (syringe and needle lubricant) and heparin, with the detection systems. Initially $[fMg^{2+}]_o$ was determined by chemical methods on ultrafiltrates of human plasma, and values of about 0.5 mM were obtained (Nordbö 1939; Walser 1961; Heaton 1967). More recently, Mg^{2+}-selective electrodes have been used (Altura and Altura 1997; Elin 1997). Current electrodes show less interference from Ca^{2+} and H^+ than previous ones but this interference is still significant and must be compensated for by simultaneous measurements of ionized $[Ca^{2+}]$ and pH. Values of 0.62 ± 0.01 mM (mean ± S.E.M., $n = 61$) were obtained for $[fMg^{2+}]_o$ in a recent study of healthy human adults (Altura and Altura 1997).

For human red cells $[fMg^{2+}]_i$ is below electrochemical equilibrium both in oxygenated and deoxygenated cells. With $[fMg^{2+}]_o$ at approx. 0.6 mM and a membrane potential (E) of −10 mV, $[fMg^{2+}]_i$ would be about 1.3 mM at equilibrium. Thus passive entry of Mg^{2+} into human red cells is favoured throughout their life span. Data for red cells from other species is incomplete. However, unless a far smaller proportion of plasma Mg^{2+} is free in these animals, $[fMg^{2+}]_i$ is probably below electrochemical equilibrium in many mammalian red cells.

16.3.3 Changes in Magnesium Content with Cell Age

After density-fractionation of human red cells by prolonged centrifugation, cells in the top 10% fraction had the highest Mg^{2+} content whereas those in the bottom fraction had the lowest (Bernstein 1959; see also Chap. 29). As the top fraction contained reticulocytes and young cells whereas the densest fraction was thought to contain the oldest cells (but see Clark 1988), it was suggested that cell Mg^{2+} content falls as cells get older. This idea is supported by the observation that the

Table 16.2. Free ionized [Mg^{2+}] in red cells ([fMg^{2+}]$_i$)

Species	[fMg^{2+}]$_i$ oxygenated cells [mM], mean ± S.E.M., (n)	[fMg^{2+}]$_i$ deoxygenated cells [mM], mean ± S.E.M., (n)	Method (Reference)
Human	0.25 ± 0.07	0.67 ± 0.15	NMR (Gupta et al. 1978)
	0.39 ± 0.01 (15)	0.59 ± 0.03 (7)	Null point (Flatman 1980)
	0.24 ± 0.01 (7)		NMR (Bock et al. 1985)
	0.31 ± 0.01 (12)		Null point (Bock and Yusuf 1988)
	0.20 ± 0.01 (14)		NMR (Geven et al. 1991)
	0.55 ± 0.03 (14)		Null point (Geven et al. 1991)
	0.38 ± 0.04	0.69 ± 0.06	Fluorescent dye (Barbul et al. 1999)
Rat	0.22 ± 0.01 (6)		NMR (Büttner et al. 1998)
	0.32 – 0.38		Null point (Günther et al. 1990)
Guinea-pig	0.25		NMR (Jelicks et al. 1989)
Ferret	0.65 ± 0.03 (4)		Null point (Flatman 1987)
Horse	0.45 ± 0.04 (4)	0.68 ± 0.03 (4)	Null point (Campbell et al. 1999)
Pig	0.22 ± 0.01 (6)		NMR (Büttner et al. 1998)
Cattle	0.20 ± 0.01 (6)		NMR (Büttner et al. 1998)
Sheep	0.39 ± 0.05 (5)	0.52 ± 0.04 (5)	Null point (Campbell and Gibson 1998)

Mg^{2+} content of human and rabbit red cells taken from donors with high reticulocyte counts have the highest Mg^{2+} contents (Ginsburg et al. 1962). Data from patients with reticulocytosis suggest that human reticulocytes contain two or three times the Mg^{2+} content of the average mature cell. Similarly, the Mg^{2+} content of rabbit red cells rises from a normal value of 3.9 mmol/l_{cells} to 14.2 mmol/l_{cells} in samples containing 85% reticulocytes following treatment of the rabbits with phenylhydrazine. Other experimental approaches also suggest that reticulocytes contain more (total) Mg^{2+} than mature cells. In order to produce a model that explains the time course of ^{28}Mg equilibration between plasma and human red cells it was necessary to assume that reticulocytes contain about 50% more Mg^{2+} than mature cells, and that cells lose Mg^{2+} progressively with a half life of 100 days (Watson et al. 1979, 1980). These data suggest that reticulocytes have a high Mg^{2+} content and that Mg^{2+} is lost from cells as they age in the circulation. Interestingly, measurements of $[fMg^{2+}]_i$ in guinea-pig reticulocytes and mature red cells by NMR show that $[fMg^{2+}]_i$ rises on maturation from 0.15 to 0.25 mM (Jelicks et al. 1989), perhaps indicating a change in Mg^{2+} buffering as well as transport during maturation.

However, there is an observation that does not fit with the idea that Mg^{2+} content falls with cell age. When human red cell Mg^{2+} content was measured in cells separated on a Percoll density gradient (Mairbäurl and Hoffman 1992), the Mg^{2+} content of the top, light fraction (12% of total cells) was 2.31 mmol/l_{cells} and this rose to 2.73 mmol/l_{cells} in the densest (15%) cells. The reason for the discrepancy between this and earlier work has still to be resolved.

16.4 Red Cell Membrane Magnesium Permeability

^{28}Mg uptake by rat, cat and dog red cells reaches about 60% equilibration in 9 hours at 38 °C (Rogers 1961). Similar rates of equilibration were observed *in vitro*, or *in vivo* following intravenous injection of isotope. On the other hand, uptake by human red cells only reaches about 5% equilibration and uptake into bovine red cells was barely detectable after 9 hours incubation (Rogers 1961). Further *in vitro* studies failed to detect any significant uptake of ^{28}Mg by human red cells incubated for 24 hours at 37 °C and also failed to detect uptake by rabbit reticulocytes or mature cells incubated for 18 hours at 40 °C (Ginsburg et al. 1962). *In vivo* ^{28}Mg uptake by human red cells has been observed over a 5 day period following injection of the isotope into the circulation of healthy volunteers (Watson et al. 1979). In view of the *in vitro* observations it was suggested that most ^{28}Mg uptake into human red cells occurs during the early stages of erythropoiesis (Watson et al. 1979). ^{28}Mg then appears in the circulation in reticulocytes which have little capacity to exchange ^{28}Mg further. Thus red cells from rats, cats and dogs are at least ten times more permeable to Mg^{2+} than those from humans, rabbits and cattle.

16.5 Is Active Magnesium Transport Needed?

Red cells should gradually accumulate Mg^{2+} as they age (as long as their buffering capacity does not fall), because there is a prevailing inward electrochemical gradient for Mg^{2+} and because the cells have a finite permeability to Mg^{2+}. The fact that the opposite is observed, and red cell Mg^{2+} content falls with age (but see Mairbäurl and Hoffman 1992) suggests that cells possess an active Mg^{2+} transporter that can move Mg^{2+} out of the cell against its electrochemical gradient. Much effort has been put into trying to identify this transporter. Although this is the most likely explanation for how $[fMg^{2+}]_i$ is regulated in a flexible and responsive way, there is an alternative way of keeping $[fMg^{2+}]_i$ low.

The thermodynamic analysis above applies to cells with a normal complement of metabolic intermediates, and hence Mg^{2+} buffering capacity, surrounded by plasma at pH 7.4. However, under some conditions, for instance when deoxygenated cells are exposed to acidic plasma (causes cytoplasmic Mg^{2+} buffering capacity to fall), or when the membrane depolarises, $[fMg^{2+}]_i$ may rise above electrochemical equilibrium ($[fMg^{2+}]_i$ must rise above 1.3 mM or E must rise above +5 mV). If these changes coincide with a rise in membrane Mg^{2+} permeability, then Mg^{2+} could leave cells passively. Thus cell Mg^{2+} content may be kept low by minimizing Mg^{2+} permeability when the electrochemical gradient is inward (minimizes Mg^{2+} gain), and by increasing permeability during conditions, which favour loss. Cells are repeatedly exposed to conditions that favour loss throughout their life span. If such a mechanism does exist, Mg^{2+} transporters in mature red cells may not need the capacity for active transport. In a sense, the energy to keep $[fMg^{2+}]_i$ low would come from the heart which pushes the cell from place to place.

16.6 Magnesium Transport Mechanisms

In the last few years significant progress has been made in understanding Mg^{2+} transport in a variety of tissues, and particularly in prokaryotic and yeast cells using molecular biological techniques (Moncrief and Maguire 1999; Graschopf et al. 2001). These studies now provide a framework to help us understand what may be happening in red cells. Mg^{2+} can move through channels, on Na^+/Mg^{2+} antiporters, on a Mg^{2+}/H^+ antiporter, on Mg^{2+}-anion (Cl^- or bicarbonate) symports, perhaps in exchange for Ca^{2+}, or on a Mg^{2+}-pump (Flatman 1991a; Beyenbach 2000; Quamme and de Rouffignac 2000; Rasgado-Flores and Gonzalez-Serratos 2000; Romani and Scarpa 2000; Schweigel and Martens 2000). In many tissues, Na^+/Mg^{2+} antiport, plays an important role in magnesium homeostasis and may be responsible for keeping $[fMg^{2+}]_i$ below electrochemical equilibrium. There may be a family of these antiporters with properties to suit particular physiological and developmental needs. In squid axon this transporter may also carry Cl^- and K^+ (Rasgado-Flores et al. 1994).

16.7 Methods for Investigating Magnesium Transport in Red Cells

Under normal conditions, Mg^{2+} fluxes in red cells are small, so they are difficult to detect as changes in cell Mg^{2+} content even when cells are suspended in media containing high $[Mg^{2+}]$. Fluxes are therefore usually measured as an efflux into an initially Mg^{2+}-free medium. This is a condition where there is little back flux to complicate analysis, and the small increase in $[Mg^{2+}]_o$ can be resolved from background noise. It is vital to correct accurately for cell lysis in these experiments, as Mg^{2+} can accumulate in the medium more rapidly by this route than by transport. However, most experiments on Mg^{2+} transport in red cells involve the use of cells heavily loaded with Mg^{2+}. Under these conditions fluxes are large and can be easily detected as changes in cell or medium Mg^{2+} content.

The first experiments on Mg^{2+} transport in red cells involved use of red cells with high Mg^{2+} contents (Dunn 1974). These cells were obtained either from uraemic patients or by loading normal cells. Loading was achieved by incubating cells in media containing *p*-chloromercuribenzene sulphonate (PCMBS) and a high $[Mg^{2+}]$, followed by incubation in a medium containing dithiothreitol (DTT), and substrates to restore energy and reducing power, to terminate the effects of PCMBS. This method has been widely used by others with minor modification (cysteine instead of DTT). It has now been largely replaced by the use of the ionophore A23187 to increase Mg^{2+} permeability. At a concentration of between 2 and 10 μM A23187 can be used to control red cell Mg^{2+} content precisely and selectively (Flatman and Lew 1977, 1980). The ionophore is then removed by incubating and washing cells in large volumes of media containing about 1 mg/ml bovine serum albumin (BSA) at 37 °C. Although this technique removes all the ionophore from Mg^{2+}-depleted cells it does not always remove all the ionophore from cells loaded with high Mg^{2+} levels. The ionophore may form intracellular precipitates with Mg^{2+} that leach ionophore into the membrane (Flatman and Simonsen, unpublished results). Resealed red cell ghosts have also been used when it was necessary to load cells with Mg^{2+} and simultaneously change the levels of other cell constituents.

There are a number of problems interpreting data from Mg^{2+} loaded cells. Residual effects of the loading method may present problems, for instance, significant fluxes may occur through residual ionophore, or because the membrane remains perturbed after PCMBS or hypotonic lysis. Heavy Mg^{2+} loading may activate transport systems that are quiescent under more physiological conditions so these studies may focus on transporters that normally contribute little to normal Mg^{2+} homeostasis. Finally, the loading method may alter the preferred ion that accompanies Mg^{2+} to balance charge across the membrane. However, when used carefully the method provides valuable information on Mg^{2+} transport processes.

16.8 Magnesium Transport in Human Red Cells

Mg^{2+} efflux from human red cells into a Mg^{2+}-free medium containing physiological levels of Na^+ and K^+ is barely detectable. Total Mg^{2+} efflux is 7.3 ± 2.8 μmol/(l_{cells}·h) (n = 14 subjects) (Féray and Garay 1986), with a Na^+-dependent component of 4.6 ± 1.2 μmol/(l_{cells}·h) (n = 4) (Lüdi and Schatzmann 1987). From these studies the Na^+-independent flux can be calculated as about 3 μmol/(l_{cells}·h), yielding a Mg^{2+} permeability (P_{Mg}) of $2 \cdot 10^{-10}$ cm/s ($[fMg^{2+}]_i$ = 0.4 mM, E = −10 mV). A similar value can be calculated for the Na^+-independent flux measured in A23187-treated cells (Lüdi and Schatzmann 1987), and a value of $4 \cdot 10^{-10}$ cm/s is obtained for the quinidine-resistant flux in PCMBS-treated cells (Féray and Garay 1986). This indicates that A23187 can be effectively removed, but that treatment with PCMBS may leave membrane slightly more permeable to Mg^{2+}. P_{Mg} is similar to red cell Na^+ and K^+ permeabilities and is larger than the ground permeability of membrane lipids to small cations (Lew and Beaugé 1979) and thus probably represents Mg^{2+} transport pathways in addition to the Na^+-dependent route. Much of this Mg^{2+} may move through a non-selective cation channel that is also thought to account for a significant portion of the leak fluxes of Na^+, K^+ and Ca^{2+} (Kaestner et al. 2000). Using $P_{Mg} = 2 \cdot 10^{-10}$ cm/s it can be calculated that human red cells ($[fMg^{2+}]_o$ = 0.6 mM, $[fMg^{2+}]_i$ = 0.4 mM, E = −10 mV) should take up Mg^{2+} from plasma at a rate of about 7 μmol/(l_{cells}·h), and this must be balanced by an equivalent efflux to prevent accumulation of Mg^{2+}.

Using this permeability it is also possible to calculate that Mg^{2+} uptake should be easily detected in human red cells incubated in media containing more than 10 mM Mg^{2+}. However, Mg^{2+} uptake has not been detected in fresh human red cells even when cells are incubated in media containing 108 mM Mg^{2+}, indicating either that the efflux mechanism responds rapidly to increased influx, or that P_{Mg} falls as $[Mg^{2+}]_o$ rises. Although it is difficult to detect any uptake of Mg^{2+} by untreated human red cells, significant uptake can be seen in cells loaded with Mg^{2+} by the PCMBS method (Lüdi and Schatzmann 1987).

Total efflux from untreated cells has an activation energy of 53 kJ/mol measured over the temperature range of 5 to 40 °C (Féray and Garay 1986), is 20–30% stimulated by increasing or decreasing the pH of the medium by one unit (in the presence of 4,4'-diisothiocyanatostilbene-2,2'-disulphonic acid (DIDS) to minimise pH induced changes to cell volume and Cl^- distribution), is mildly stimulated by making the medium hypertonic and markedly stimulated by hypotonicity (Féray and Garay 1986). The data suggest that cell swelling activates a Mg^{2+} transporter.

16.8.1 Sodium-Dependent Magnesium Transport

Mg^{2+} efflux into a Mg^{2+}-free medium is markedly stimulated by loading cells with Mg^{2+} (Dunn 1974; Féray and Garay 1986; Lüdi and Schatzmann 1987), and is inhibited by raising $[Mg^{2+}]_o$ (Dunn 1974). Efflux can be divided into two components, one requiring external Na^+, and one that is independent of external Na^+ and is probably stimulated by ATP depletion (Frenkel et al. 1989). Most of the increased efflux seen following Mg^{2+} loading is due to stimulation of the Na^+-dependent mechanism. Kinetic analysis of these data show that the maximal rate of Na^+-dependent efflux is very variable amongst donors (40–500 µmol/(l_{cells}·h), (Féray and Garay 1986)). It is higher in cells from HLA-B35⁺ males perhaps partially explaining the low Mg^{2+} content and $[fMg^{2+}]_i$ in these cells (Féray et al. 1989).

Efflux is half maximally stimulated when total $[Mg^{2+}]_i$ is about 4–5 mmol/l_{cells} (Féray and Garay 1986). When expressed as a function of $[fMg^{2+}]_i$, calculated from published Mg^{2+} buffer curves (Flatman and Lew 1980), the data are well described by the Hill equation, yielding an apparent affinity of about 0.9–1.3 mM, and Hill coefficient, $h = 2.4$ (Lüdi and Schatzmann 1987). Similar values for affinity and h were obtained following recalculation of earlier data (Flatman 1991a; Féray and Garay 1986). These data suggest that more than one Mg^{2+} ion binds at the inner site to promote transport. External Na^+ activates efflux with a $K_{0.5}$ of 16–21 mM (Féray and Garay 1986; Lüdi and Schatzmann 1987). Na^+-dependent Mg^{2+} efflux is unaffected by replacing Cl^- in the cell and medium by nitrate (Féray and Garay 1986).

16.8.2 Inhibitors of Sodium-Dependent Magnesium Transport

The transporter is highly selective for Na^+ and is not activated by K^+, Li^+, Rb^+, Cs^+ and choline (Féray and Garay 1986). Na^+-dependent efflux is inhibited by quinidine (IC_{50} = 50 µM), Mn^{2+} (IC_{50} = 0.5–1.0 mM), amiloride (IC_{50} = 0.4 mM) imipramine (IC_{50} = 25 µM) and other tricyclic antidepressant drugs, but is not affected by ouabain (0.1 mM), vanadate (0.1 mM), furosemide (1 mM), bumetanide (20 µM), DIDS (10 µM) or 4-acetamido-4′-isothiocyanatostilbene-2,2′-disulphonic acid (SITS, 30 µM) (Féray and Garay 1986, 1988; Lüdi and Schatzmann 1987; Frenkel et al. 1989; Günther and Vormann 1989a). Non-tricyclic antidepressants (Trazodone, Mianserin) and phenothiazines are weak inhibitors (Féray and Garay 1988). External Ca^{2+} (10 mM), Sr^{2+} (1 mM), Ba^{2+} (1 mM), and phloretin (0.5 mM) cause only minor inhibition (Féray and Garay 1986).

16.8.3 Interactions between Sodium and Magnesium across the Membrane

Initial studies showed that Mg^{2+} efflux is not affected by changing internal $[Na^+]$ ($[Na^+]_i$, or $[K^+]_i$, which was used as replacement) over a wide range (Féray and Garay 1986). This holds when $[Na^+]_o$ is saturating, and indicates that internal Na^+ and Mg^{2+} do not compete with each other for transport. However, further analysis (Lüdi and Schatzmann 1987) showed that as $[Na^+]_i$ was raised, it appeared to compete with external Na^+, shifting the activation curve to the right (higher $[Na^+]_o$), but did not affect the maximal rate of transport. In addition, there is an interaction between the effects of internal Mg^{2+} and external Na^+ (Frenkel et al. 1989). As $[Mg^{2+}]_i$ rises, not only does the rate of Mg^{2+} efflux increase but higher $[Na^+]_o$ is needed to activate the flux. Analysis of the kinetics suggests that the true affinity of the external Na^+ site (no internal Na^+ or Mg^{2+}) is about 8 mM, and the affinity of the internal Mg^{2+} site is about 1 mM (Frenkel et al. 1989). The interactions between Na^+ and Mg^{2+} across the membrane strongly suggest that Na^+ and Mg^{2+} are exchanged by the transporter.

16.8.4 How Many Sodium Ions are Exchanged with Each Magnesium Ion?

Thermodynamic analysis of the Na^+ and Mg^{2+} gradients across the human red cell membrane show that movement of a single Na^+ ion into the cell should provide sufficient energy to drive the efflux of a Mg^{2+} ion under normal circumstances. It requires about 3 kJ to move 1 mol Mg^{2+} ions out of the cell under physiological conditions and about 8 kJ can be obtained by allowing 1 mol Na^+ ions into the cell. Active Mg^{2+} efflux has been observed in cells under conditions where 9 kJ were required to drive efflux of a mol of Mg^{2+} (Lüdi and Schatzmann 1987) and over 9 kJ of energy were available from a single mol of Na^+ entering the cell. Similarly, analysis of curves relating Na^+-dependent Mg^{2+} efflux to $[Na^+]_o$ suggest that a single Na^+ ion is exchanged for each Mg^{2+} ion ($h = 1$; Frenkel et al. 1989). However, it has been very difficult to measure the stoichiometry of exchange directly because inhibitors of Na^+/Mg^{2+} antiport are non-specific and inhibit a wide range of other Na^+ transporters possibly resulting in an overestimate of the number of Na^+ ions exchanged. A stoichiometry of $3Na^+:1Mg^{2+}$ was obtained using imipramine or quinidine to inhibit transport in cells that had also been treated with ouabain and bumetanide to reduce Na^+ movement through other possible imipramine- or quinidine-sensitive routes (Féray and Garay 1988). However, as Mg^{2+} fluxes are small, the presence of other minor Na^+ transporters could have distorted the analysis.

16.8.5 Can Magnesium Efflux Occur against an Electrochemical Gradient?

Na^+-dependent Mg^{2+} efflux can occur against an electrochemical gradient for Mg^{2+}. Cells loaded to a $[fMg^{2+}]_i$ of 1.25 mM (PCMBS method) lose Mg^{2+} when incubated in media containing 5 mM Mg^{2+}, but gain Mg^{2+} when 2 mM amiloride is added to the solution (Lüdi and Schatzmann 1987). Efflux seen in the absence of amiloride is clearly against an electrochemical gradient. The amiloride-sensitive efflux (about 40 $\mu mol/(l_{cells} \cdot h)$) was found to be constant in media containing 1–20 mM Mg^{2+}. In these experiments there was a large inward Na^+ gradient that may have provided energy for Mg^{2+} efflux. However, there is evidence that the Na^+ gradient may not provide energy for Mg^{2+} transport in human red cells. If Mg^{2+} were transported on a simple Na^+/Mg^{2+} antiport reversing the Na^+ gradient should reverse the direction of Mg^{2+} transport and result in Mg^{2+} uptake on the system. Despite many careful experiments, it has not been possible to reverse the direction of the Na^+-dependent Mg^{2+} transporter in human cells (Lüdi and Schatzmann 1987; Schatzmann 1993). The earlier experiments (Lüdi and Schatzmann 1987) were carried out with cells that had a low $[Mg^{2+}]_i$ and this may have reduced the efficiency of the transporter (Mg^{2+} may be required for reactions that phosphorylate the transporter). Later experiments used cells loaded with $[Mg^{2+}]_i$ above normal, but the same result was obtained (Schatzmann 1993). In addition, abolition of the Na^+ gradient (cells loaded with 110 mM Na^+), only resulted in a halving of the Na^+-dependent Mg^{2+} efflux, rather than complete abolition of transport (Frenkel et al. 1989). It seems that the transporter can move large amounts of Mg^{2+} out of the cell using relatively small outward gradients but moves little, if any Mg^{2+} into the cell, despite large inward gradients (this is true whether 1, 2 or 3 Na^+ ions are exchanged for Mg^{2+} on the transporter) (Schatzmann 1993).

16.8.6 Role of ATP

Human red cell membranes have high levels of Mg^{2+}-ATPase activity that cannot be ascribed to the Na^+ or Ca^{2+} pumps, or other known functions. The possibility that part of this activity is associated with a Mg^{2+} pump has been explored in purified membranes (Lüdi and Schatzmann 1987). No activity fulfilling the requirements of a Mg^{2+} pump could be found in membranes unless cell lysate was added. The level of this activity could not be quantified, but may have been sufficient to drive Mg^{2+} transport at the observed rates (this would have needed less than 2% of the total Mg^{2+}-ATPase activity). Thus it has not yet been possible to demonstrate the stoichiometric relationship between Mg^{2+} transport and ATP hydrolysis that would be indicative of an ATP-fuelled Mg^{2+} pump. In addition, the finding that Mg^{2+} efflux is not affected by vanadate indicates that the Mg^{2+} transporter is not a P-class ATPase like the Na^+, Ca^{2+} or H^+ pumps (Frenkel et al. 1989). However, ATP clearly does play a role in Mg^{2+} transport. ATP depletion by treating cells with iodoacetamide or prolonged starvation inhibits Na^+-dependent Mg^{2+} efflux (Féray and Garay 1986; Lüdi and Schatzmann 1987). By adding ATP to red cell ghosts it was possible to show that 0.15 mM ATP half maximally activates the transporter. Non-hydrolysable ATP analogues like AMP-PCP are ineffective (Frenkel et al. 1989). These findings are consistent with Mg^{2+} transport being regulated by phosphorylation, as seen with Na^+/Mg^{2+} antiport in squid axon (DiPolo and Beaugé 1988), though the possibility that ATP provides energy for transport still exists. Here, ATP hydrolysis would be used to reduce the inner Na^+ affinity to such a low level that the transporter is essentially irreversible and unresponsive to the Na^+ gradient (Lüdi and Schatzmann 1987; Frenkel et al. 1989; Schatzmann 1993).

16.8.7 Sodium-Independent Magnesium Efflux

Heavily Mg^{2+} loaded (using A23187) human red cells lose significant amounts of Mg^{2+} when incubated in a sucrose medium (Günther and Vormann 1989a,b; 1990). The rate of Mg^{2+} efflux is about twice that seen in Na^+ containing media (Günther and Vormann 1989a). Efflux is inhibited by adding Cl^- to the medium. K^+, Li^+ and choline salts are equally effective, whereas NaCl causes less inhibition (Günther and Vormann 1989b). 30 μM SITS reduces the flux by about 50%, whereas furosemide (1 mM) and bumetanide (0.1 mM) are without effect (Günther and Vormann 1989b). Amiloride (1 mM) inhibits about 70% of efflux (Günther and Vormann 1989b), and a variety of amiloride derivatives are even more potent (Günther and Vormann 1990).

These data have been interpreted as showing that Mg^{2+} can leave human red cells by a channel, and that the charge moved across the membrane is balanced by efflux of Cl^-. SITS and external Cl^- inhibit movement through the Cl^- pathway and thus inhibit Mg^{2+} efflux, whereas amiloride and its derivatives probably inhibit the Mg^{2+} pathway itself. Additional support for this model is the finding that the inhibition of Mg^{2+} efflux into sucrose by 40 μM DIDS is overcome by adding the membrane permeant cation, tetraphenylphosphonium (TPP) to the medium (Günther and Vormann 1990). This ion is thought to take over the role of charge compensation from Cl^- ions.

Red cells also lose large amounts of K^+ when incubated in sucrose (see Chap. 4). It was suggested that K^+ and Mg^{2+} leave the cells via different routes as efflux of these ions show different sensitivities to inhibition by amiloride analogues (Günther et al. 1989). However, as K^+ fluxes are 30–40 times larger than Mg^{2+} fluxes, it seems possible that some K^+ flows through the same path as Mg^{2+} and that there are additional routes for K^+ efflux. In these circumstances an alternative explanation is possible. Incubation of human red cells in sucrose causes the membrane potential to become large and positive (Macey et al. 1978) and this activates a non-selective cation channel that has been shown to permit the movement of Na^+, K^+ and Ca^{2+} ions across the membrane (Halperin et al. 1989). Mg^{2+} probably flows through this route as well (Kaestner et al. 2000). The potent inhibitory effect of SITS and DIDS on Mg^{2+} efflux may not only be due to the inhibition of charge compensation by Cl^-, but also be due to a reduction in sucrose-induced depolarization. This would in turn reduce channel activation and the driving force on Mg^{2+} exit. The role this transporter plays in normal red cell Mg^{2+} homeostasis has still to be established.

16.8.8 Changes in Magnesium Transport in Disease

Changes in $[fMg^{2+}]_i$ may play a role in altering cation transport in red cells from patients with sickle cell disease (SS) (Joiner et al. 1998; see also Chap. 23). As discussed earlier, deoxygenation causes a rise in $[fMg^{2+}]_i$ due to increased binding of ATP and BPG to deoxyhaemoglobin, but $[fMg^{2+}]_i$ remains below electrochemical equilibrium with $[fMg^{2+}]_o$ in deoxygenated cells. In SS red cells $[fMg^{2+}]_i$ may rise to such an extent on deoxygenation that the electrochemical gradient for Mg^{2+} reverses (becomes outward), and Mg^{2+} is lost from these cells because deoxygenation also increases the membrane's permeability to Mg^{2+} (Ortiz et al. 1990b). Density-fractionation of cells showed that SS cells with normal density and morphology (discocytes) have a normal Mg^{2+} content whereas those with high density (mainly irreversibly sickled cells) have a low Mg^{2+} content (Bookchin et al. 1987). Deoxygenation was found to increase the membrane permeability to Mg^{2+} in SS cells (but not in normal, AA cells), no matter what their density. However, the densest cells were also found to have a low BPG content. As expected $[fMg^{2+}]_i$ in these dense cells was higher than in AA red cells, and this level increased markedly on deoxygenation to levels where the Mg^{2+} gradient was outward (Ortiz et al. 1990b). It has also been found that there is a large loss of Mg^{2+} from reticulocyte-rich SS cells incubated at pH 7 for 30–60 minutes. This loss is reduced by amiloride, quinine and by replacing Cl^- by nitrate (Ortiz et al. 1990a). Thus acidosis can substantially stimulate Mg^{2+} loss from SS reticulocytes facilitating the generation of irreversibly sickled cells.

Increased Na^+-dependent Mg^{2+} transport has been seen in red cells from patients with hypertension (Picado et al. 1994; see also Chap. 25) and renal failure (Vormann et al. 1996). In the latter case this may be due to an increased level of a circulating factor in patients with renal failure as transport is reduced by haemodialysis, and as heat inactivated plasma from these patients stimulates Mg^{2+} transport in red cells from normal controls. Na^+-dependent Mg^{2+} transport is also significantly increased in red cells from patients with cystic fibrosis showing severe symptoms (Vormann et al. 1996). Again it has been suggested that transport stimulation may be due to the presence of a factor in the plasma of these patients. Red cells from severely depressed patients, particularly those with psychomotor retardation, have high Mg^{2+} contents (Widmer et al. 1995b, 1998). The maximal rate of Na^+-dependent Mg^{2+} efflux from these red cells was higher than that seen in controls. Mood improvement was correlated with a further steady increase in Mg^{2+} transport rate and normalization of red cell magnesium content (Widmer et al. 1995a).

16.9 Magnesium Transport in Ferret Red Cells

Net Mg^{2+} fluxes can be measured in ferret red cells without the need for loading, enabling the study of Mg^{2+} transport under more physiological conditions, and at the same time providing an opportunity to study the consequences of loading (and in particular the method of loading) on the properties of the transporters examined (Flatman and Smith 1990, 1991, 1996).

16.9.1 Magnesium Efflux from Ferret Red Cells

Ferrets are members of the order carnivora, and their red cells have a high Na^+ (96 $mmol/l_{cells}$) and low K^+ (3.9 $mmol/l_{cells}$) content. There is little, if any Na^+ pump activity in mature red cells, and yet these cells volume regulate and have similar survival times to other mammalian red cells (Flatman and Andrews 1983). These cells contain about 3 $mmol/l_{cells}$ Mg^{2+} of which about 0.65–0.73 mM is free and ionized (Flatman 1987). $[fMg^{2+}]_o$ in plasma has not been determined, but the total Mg^{2+} concentration (0.97 mM, (Flatman and Andrews 1983)) is similar to that of other mammalian plasmas. With E = –10 mV (Flatman and Smith 1991), $[fMg^{2+}]_o$ would have to be 0.3 mM for Mg^{2+} to be at electrochemical equilibrium. If a similar proportion of Mg^{2+} is free in ferret, as in human plasma, $[fMg^{2+}]_o$ is likely to be

0.5–0.6 mM. Thus $[fMg^{2+}]_i$ is likely to be below electrochemical equilibrium in oxygenated cells.

Mg^{2+} efflux can be accurately determined in ferret red cells incubated in a Mg^{2+}-free medium (Flatman and Smith 1990). The flux, 41 µmol/(l_{cells}·h), is about 10 times larger than efflux in human red cells and has an activation energy of about 25 kJ/mol at temperatures between 25 and 40 °C. Efflux appears to have at least two components. About 70% is inhibited by a variety of external ions and inhibitors whereas the remaining 30% is unaffected. External divalent cations (1 mM), Ca^{2+}, Mn^{2+}, Ni^{2+} and Co^{2+}, partially inhibit efflux with Co^{2+} being the most potent (IC_{50} = 0.3 mM). Similarly, quinidine (IC_{50} = 0.2 mM), amiloride (IC_{50} = 0.5 mM), and imipramine (0.5 mM had maximal effect) partially inhibit efflux. Efflux is not affected by ouabain (0.5 mM), lanthanum (0.05 mM), bumetanide (0.5 mM), α-cyano-4-hydroxycinnamate (an inhibitor of lactate and pyruvate transport, 0.3 mM), *S*-(*p*-nitrobenzyl)-6-thioinosine (an inhibitor of nucleotide transport, 10 nM). Interestingly, 0.1 mM SITS stimulates efflux by about 23%, possibly because it depolarizes the membrane potential to about 0 mV. This is consistent with Mg^{2+} efflux through a channel or $1Na^+/1Mg^{2+}$ antiport (Flatman and Smith 1990, 1991). The residual Mg^{2+} efflux seen in the presence of a variety of inhibitors (about 10–20 µmol/(l_{cells}·h) probably represents movement on a transporter as P_{Mg} (4–8·10^{-10} cm/s) is far higher than the permeability of membrane lipids to small cations.

Mg^{2+} efflux from ferret red cells is not via a P-class ATPase. Depletion of cell ATP (from 0.64 to 0.02 mmol/l_{cells}) by incubating cells in media containing 2-deoxyglucose stimulates Mg^{2+} efflux by 50–60%, most likely due to an increase in $[fMg^{2+}]_i$ when the cytoplasmic Mg^{2+} buffer capacity fell (Flatman and Smith 1990). Vanadate (up to 0.2 mM) has no effect on efflux, though it causes cell ATP content to increase (Flatman and Smith 1991). These experiments suggest that ATP does not provide energy for Mg^{2+} transport, though they do not rule out the possibility that phosphorylation regulates transport as proteins remain phosphorylated for many hours after ATP depletion in ferret red cells (Flatman 1991b).

Incubation of ferret red cells in media containing choline or *N*-methyl-D-glucamine in place of Na^+ results in an initial fall in Mg^{2+} efflux to about 5 µmol/(l_{cells}·h) which is maintained for about 30 min, followed by a large rise in Mg^{2+} efflux. This rise is associated with significant loss of Na^+ and cell shrinkage, both of which can be reduced by adding 10 µM bumetanide to the medium. Bumetanide inhibits Na^+-K^+-$2Cl^-$ cotransport, a major route for Na^+ loss, without directly affecting Mg^{2+} transport (Flatman and Smith 1990). In the presence of bumetanide cell Na^+ content and volume remain stable for at least an hour, during which time it is possible to measure Mg^{2+} fluxes reproducibly and with precision. Reducing $[Na^+]_o$ from 145 to 10 mM stimulates Mg^{2+} efflux by about 80% but further reduction causes inhibition. At very low $[Na^+]_o$ (average < 1 mM over flux period), Mg^{2+} uptake is seen. Uptake is usually observed in the presence of bumetanide as this prevents accumulation of Na^+ in the medium. However, it has also been seen in the absence of bumetanide with cells that lost Na^+ unusually slowly, so that $[Na^+]_o$ remained low (Flatman and Smith 1990). This Mg^{2+} uptake is against a large electrochemical gradient ($[fMg^{2+}]_o$ = 2 µM, $[fMg^{2+}]_i$ = 0.65 mM, E = –10 mV), indicating that the transporter is capable of active Mg^{2+} transport. Mg^{2+}

uptake at low $[Na^+]_o$ can be prevented by adding EDTA to the medium, which keeps $[fMg^{2+}]_o$ at sub-micromolar levels, or by adding 1 mM Co^{2+} to the medium.

Thus there is strong evidence for a reversible Na^+/Mg^{2+} antiport in ferret red cells that is capable of uphill Mg^{2+} transport. This later point was more rigorously tested in experiments where $[fMg^{2+}]_o$ was set at levels (0.1–0.2 mM), that were well below electrochemical equilibrium with $[fMg^{2+}]_i$, and yet high enough to minimize changes in $[fMg^{2+}]_o$ while Mg^{2+} was being taken up by the cells. Membrane potential was carefully monitored and found to remain at –10 mV throughout. Mg^{2+} uptake was 1.1 mmol/(l_{cells}·h) when $[Na^+]_o$ averaged about 0.4 mM, and was 0.12 mmol/(l_{cells}·h) when $[Na^+]_o$ averaged 10 mM (to minimize changes in internal and external $[Na^+]$) (Flatman and Smith 1991). Thermodynamic analysis of active Mg^{2+} transport under a number of different conditions suggests that a single Na^+ ion is exchanged for each Mg^{2+} ion, and that the Na^+ gradient can provide all the energy required for Mg^{2+} transport. Determination of the $[Na^+]_o$ at which Mg^{2+} efflux becomes Mg^{2+} influx with only micromolar concentrations of $[fMg^{2+}]_o$, is strongly indicative of stoichiometry. Using the equation:

$$\frac{[fMg^{2+}]_o}{[fMg^{2+}]_i} = \left(\frac{[Na^+]_o}{[Na^+]_i}\right)^q \cdot \exp\left(\frac{E \cdot F \cdot (2-q)}{R \cdot T}\right) \tag{16.1}$$

where: q is the number of Na^+ ions exchanged with each Mg^{2+} ion, $[fMg^{2+}]_i$ = 0.65 mM, $[fMg^{2+}]_o$ = 2 μM, $[Na^+]_i$ = 135 mM, and E = –10 mV, it can be shown that reversal of Mg^{2+} transport with stoichiometries of 1, 2 or 3 Na^+ per Mg^{2+}, is predicted to occur at $[Na^+]_o$ of 0.6, 7 and 17 mM, respectively. The finding that average $[Na^+]_o$ must be reduced below 1 mM during the flux period to see influx is consistent with $1Na^+:1Mg^{2+}$. Solution of Eq. (16.1) using values found under physiological conditions ($[fMg^{2+}]_i$ = 0.65 mM, $[fMg^{2+}]_o$ = 0.5–0.6 mM, $[Na^+]_i$ = 135 mM, $[Na^+]_o$ =150 mM, E = –10 mV) yields q = 1.0–1.4 for oxygenated ferret red cells (preliminary experiments indicate q < 1 in deoxygenated cells, Flatman, unpublished results). In addition, $[fMg^{2+}]_i$ calculated assuming that 3 or 2 Na^+ ions are exchanged for each Mg^{2+} ion (0.25–0.48 mM) is lower than the value measured in oxygenated cells. Taken together these data suggest that $1Na^+/1Mg^{2+}$ antiport can maintain the physiological level of $[fMg^{2+}]_i$ in ferret red cells.

16.9.2 Magnesium Uptake

Mg^{2+} uptake is observed when cells are incubated in media with normal $[Na^+]_o$ if $[Mg^{2+}]_o$ is increased above 1 mM. Some of this flux may be through reversed Na^+/Mg^{2+} antiport and some through another pathway. However, uptake is much greater when $[Na^+]_o$ is reduced (in the presence of bumetanide) (Flatman and Smith 1991). At 5 mM $[Na^+]_o$ Mg^{2+} influx rises to a maximum rate of 0.9 mmol/(l_{cells}·h) as $[Mg^{2+}]_o$ is increased (apparent $K_{0.5}$ = 0.37 mM). The Na^+-dependence of Mg^{2+} uptake was measured as a function of $[Mg^{2+}]_o$. Mg^{2+} uptake was half-maximally active at $[Na^+]_o$ of 17, 22 and 62 mM when $[Mg^{2+}]_o$ was 2, 5, and 10 mM, respectively. These interactions between Na^+ and Mg^{2+} are consistent with uptake by reversed Na^+/Mg^{2+} antiport, a view supported by the finding that uptake is inhibited by the same agents that inhibit Mg^{2+} efflux, including amiloride, quinidine and imipramine. SITS has no effect on Mg^{2+} uptake in 5 mM $[Na^+]_o$ whereas vanadate stimulates uptake (50% at 0.05 mM and 200% at 0.2 mM). Incubation of cells with 2-deoxyglucose to reduce cell ATP level to about 0.02 $mmol/l_{cells}$ reduced Mg^{2+} uptake, possibly due to the rise in $[fMg^{2+}]_i$ and the effects of ATP on membrane Mg^{2+} permeability.

16.9.3 Transport in Magnesium-Loaded Cells

Mg^{2+} efflux has been measured in ferret red cells loaded with Mg^{2+} either by the A23187 method or by incubating cells in a buffered isotonic medium containing 5 mM Na^+, 5 mM Mg^{2+} and bumetanide (Flatman and Smith 1996). These methods load cells with similar amounts of Mg^{2+}, however, efflux from these cells have different properties.

The Mg^{2+} content of cells loaded with Mg^{2+} by the A23187 method falls back to the initial pre-load level when the cells are incubated in a Mg^{2+}-free medium, indicating that cells regulate their $[Mg^{2+}]_i$, and that the membranes have not simply become leaky to Mg^{2+}. Efflux rate increases steeply when $[fMg^{2+}]_i$ increases above 0.9 mM and is dependent on the presence of external Na^+. Reduction of $[Na^+]_o$ inhibits efflux, and if reduced enough causes the transporter to reverse leading to additional Mg^{2+} loading (if the medium also contains Mg^{2+}). This uptake of Mg^{2+} can occur against an electrochemical gradient. Both efflux and influx of Mg^{2+} in Mg^{2+}-loaded cells is inhibited by amiloride. Thus loading ferret red cells with Mg^{2+} markedly stimulates Na^+/Mg^{2+} antiport. A23187 can also be used to deplete cells of Mg^{2+}. These cells take up Mg^{2+} at rates comparable to those seen in cells with normal Mg^{2+} contents. Thus physiological $[Mg^{2+}]_i$ in ferret red cells is not simply maintained by a transporter that switches off when this level is achieved.

The Mg^{2+} content of ferret red cells loaded with Mg^{2+} by exposure to low $[Na^+]_o$ also returns to pre-load levels when cells are incubated in Mg^{2+}-free media. However, at each $[fMg^{2+}]_i$ fluxes are about 4 times larger than those seen in cells loaded using A23187. Amiloride has no detectable effect on efflux from these cells. It seems that a new transporter is revealed in these cells that obscures the antiporter. This transporter is activated by something that occurs during the loading procedure and which requires a high cell Mg^{2+} content. Simply exposing ferret red cells to low $[Na^+]_o$ under conditions where $[Mg^{2+}]_i$ does not change, has no effect on Mg^{2+} fluxes compared to controls (Flatman and Smith 1996).

16.10 Magnesium Transport in Rodent Red Cells

Rodent red cells provide an attractive model in which to investigate Mg^{2+} transport. These cells have some of the highest rates of Mg^{2+} transport seen in red cells. In addition, selective breeding and genetic manipulation of mice (and perhaps rats) could provide powerful techniques to investigate transport properties and regulation. Two strains of mice, one with high Mg^{2+} levels (MGH mice), and one with low Mg^{2+} levels (MGL mice) have been bred (Henrotte and Franck 1990; Henrotte

1993). Work on these strains suggest that the Mg^{2+} levels in red cells, kidneys, bone and plasma may be controlled by the same set of genes. Rodent red cells may also provide a robust and reproducible assay for factors that affect Mg^{2+} transport. For instance, it was shown that rat and human plasma contain factors that potently inhibit Mg^{2+} loss from rat or mice red cells (Henrotte and Franck 1990).

16.10.1 Rat Red Cells

Rat red cells take up ^{28}Mg at least ten times more quickly than human red cells. About 20–30% of Mg^{2+} in rat red cells exchanges rapidly with ^{28}Mg (within two hours), the rest exchanging much more slowly (Rogers and Mahan 1959; Rogers 1961). Mg^{2+} efflux from non-loaded rat red cells incubated in a Mg^{2+}-free medium is about 0.1–0.2 mmol/(l_{cells}·h) (Féray and Garay 1987; Franck et al. 1999; Ebel and Günther 1999). This efflux is partially inhibited by amiloride, quinidine and replacement of external Na^+ by choline, but is not affected by SITS (Ebel and Günther 1999). The Na^+-dependent component of efflux is reversibly inhibited by ATP depletion (prolonged starvation) (Büttner et al. 1998b). However, most work on rat red cells has involved use of cells that had been either lightly Mg^{2+}-loaded (10–50% increase in Mg^{2+} content) by overnight incubation in high $[Mg^{2+}]$ media (Féray and Garay 1987), or heavily Mg^{2+}-loaded using A23187 (Günther and Vormann 1989a; Günther et al. 1990b). Efflux from these cells has similar properties to those described above suggesting that rat cells tolerate well the loading procedure (Féray and Garay 1987; Günther and Vormann 1989a, 1990; Günther et al. 1990b). The Na^+-dependent component of efflux is half maximally activated non-co-operatively (h = 1) by 11–60 mM $[Na^+]_o$ (increases with $[Mg^{2+}]_i$) but is activated by $[Mg^{2+}]_i$ in a co-operative manner (h = 2.4). It is inhibited by quinidine (IC_{50} = 0.4 mM), amiloride and ATP depletion (Féray and Garay 1987; Günther et al. 1990b). Despite Hill plots suggesting that a single Na^+ ion is exchanged for each Mg^{2+} ion, direct measurements of changes in cell Na^+ content suggest 2 Na^+ ions are exchanged (Günther et al. 1990b). The Mg^{2+}content of Mg^{2+}-loaded cells returns to the initial pre-load level suggesting that activity of the Mg^{2+} transporter is gated by $[Mg^{2+}]_i$, the transporter inactivating when the physiological $[Mg^{2+}]_i$ is achieved (Günther et al. 1990b). In Na^+-free media, Mn^{2+} ($K_{0.5}$ = 35 μM) and Co^{2+} activate Mg^{2+} efflux. The Mn^{2+}-dependent efflux is also inhibited by quinidine and ATP depletion. Mn^{2+} uptake can be shown to accompany Mg^{2+} efflux. It appears that the Na^+/Mg^{2+} antiport in rat red cells can also transport Mn^{2+} giving rise to a $1Mg^{2+}/1Mn^{2+}$ exchange in the absence of external Na^+ (Féray and Garay 1987). Similar results were reported for heavily Mg^{2+}-loaded cells (Günther et al. 1990a), in this case the sensitivities of Na^+-dependent Mg^{2+} efflux and Mn^{2+}-dependent Mg^{2+} efflux to inhibition by a range of amiloride derivatives were shown to be identical.

Although quinidine and imipramine inhibit a component of Mg^{2+} efflux from rat red cells they do not affect the strophanthidin-resistant ATPase activity measured in their membranes. Amiloride and its derivatives inhibit some of this ATPase (Günther et al. 1990b). Thus, not even in rat red cells with their high Mg^{2+} transport rates, has it been possible to find a stoichiometric relationship between ATP hydrolysis and Mg^{2+} transport rate.

Rat red cells, whether Mg^{2+} loaded or not, lose Mg^{2+} into a sucrose medium (Günther and Vormann 1989a, 1990; Ebel and Günther 1999). As in other red cells this loss is partially inhibited by SITS and amiloride, but unusually it is also inhibited by quinidine, which was also found to inhibit Na^+-independent Mg^{2+} efflux from lightly loaded cells (Féray and Garay 1987). Studies using inhibitors of choline transport suggest that in the absence of external Na^+ some Mg^{2+} efflux may occur on a choline transporter (Ebel et al. 2002).

By loading rat red cells with Mg^{2+} as well as with Na^+ it is possible to show that Na^+/Mg^{2+} antiport is reversible (Günther and Vormann 1995). Reversal of the Na^+ gradient in these cells, reverses the direction of Mg^{2+} movement to produce Mg^{2+} influx. This influx is inhibited by quinidine (IC_{50} = 0.1 mM), amiloride (IC_{50} = 0.3 mM) and imipramine (IC_{50} = 0.15 mM), and is accompanied by the loss of twice as much Na^+. It is reduced by ATP depletion and by the presence of Ca^{2+}, Ni^{2+}, Co^{2+} and lanthanum in the medium.

Mg^{2+} transport has also been examined in red cells from rats made Mg^{2+}-deficient by feeding them a low Mg^{2+} diet (Günther et al. 1990c). Red cell $[fMg^{2+}]_i$ in these animals was 0.22 mM compared to 0.38 mM in controls, whereas Mg^{2+} efflux into Na^+, choline or sucrose media was about 30% higher than controls. It was suggested that Mg^{2+} deficiency causes a rise in the internal Mg^{2+} affinity of the transport systems, explaining both the low $[fMg^{2+}]_i$ and high flux in loaded cells.

Plasma factors affect Mg^{2+} transport in rat red cells. By comparing the effects of plasma on the efflux of Mg^{2+} and Mn^{2+}, it has been suggested that plasma factors cause the Na^+/Mg^{2+} antiporter to adopt a conformation that favours Mg^{2+}/Mg^{2+} exchange over Na^+/Mg^{2+} exchange (Henrotte et al. 1997). Mg^{2+} efflux is inhibited by glucose (Franck et al. 1999). Efflux into a medium containing Na^+ but no glucose is about 30% higher than in the presence of 5 mM D-glucose, D-mannose, D-fructose or 2-deoxyglucose. Inhibition of Mg^{2+} efflux is due to inhibition of the Na^+-independent component, and is partially offset by stimulation of the Na^+-dependent component. In heavily Mg^{2+}-loaded cells, glucose stimulates Mg^{2+} efflux as the majority of this flux is Na^+-dependent. Blocking red cell glucose transport with cytochalasin B or phloretin prevents the glucose-mediated inhibition suggesting that glucose must enter the cell to have its effects.

16.10.2 Hamster and Guinea-Pig Red Cells

Amiloride-sensitive Na^+ (^{22}Na) influx into hamster red cells is stimulated 40-fold by loading cells with Mg^{2+} (A23187 method) (Willis et al. 1992; Xu and Willis 1994). This is accompanied by efflux of Mg^{2+} that depends on $[Na^+]_o$ and is inhibited by amiloride (IC_{50} = 0.3 mM) suggesting that hamster red cells have a powerful Na^+/Mg^{2+} antiport. Maximal fluxes in the range of 40 mmol/(l_{cells}·h) for Na^+ and 20 mmol/(l_{cells}·h) for Mg^{2+} were found, suggesting a stoichiometry of $2Na^+$:$1Mg^{2+}$. Hill plots of amiloride-sensitive Na^+ uptake as a function of $[Mg^{2+}]_i$ also suggest a 2:1 stoichiometry (h = 1.8), and give an apparent internal Mg^{2+} affinity of 2.1 mM. Plots of amiloride-sensitive Mg^{2+} efflux as a function of $[Mg^{2+}]_i$ also gave h = 1.8

and an apparent Mg^{2+} affinity of 2.3 mM. The data suggest that more than one Mg^{2+} ion is involved in regulating transport inside the cell. The apparent external Na^+ affinity for these processes is about 105–107 mM.

Amiloride-sensitive Na^+ (^{22}Na) uptake appears to be exclusively by Na^+/Mg^{2+} antiport in hamster red cells, thus providing a simple robust assay of transporter activity (Xu and Willis 1994). $^{22}Na^+$ uptake is inhibited in ATP-depleted cells, suggesting the system is ATP-dependent (IC_{50} = 0.09 mM), but it is not affected by phorbol esters, vanadate, okadaic acid or cAMP. It is stimulated by hypotonicity, inhibited by external Mg^{2+} and Mn^{2+} but unaffected by Ca^{2+}.

Attempts to reverse the transporter failed. Amiloride-sensitive ^{22}Na efflux is not activated by high $[Mg^{2+}]_o$. Measurements of ^{22}Na efflux from Na^+-loaded cells, incubated in the presence or absence of external Na^+ and Mg^{2+}, indicate that the transporter is incapable of Na^+/Na^+ exchange (Xu and Willis 1994).

Guinea-pig red cells may also have a powerful Na^+/Mg^{2+} antiport as loading these cells with Mg^{2+} (A23187 method) strongly stimulates (up to about 40 mmol/(l_{cells}·h)) amiloride-sensitive Na^+ (^{22}Na) influx (Zhao and Willis 1993). Mg^{2+} efflux from these Mg^{2+}-loaded cells requires external Na^+ and may be weakly inhibited by amiloride.

16.11 Magnesium Transport in Avian Red Cells

Mg^{2+} is rapidly lost from chicken red cells that have been loaded with Mg^{2+} using A23187 (Günther et al. 1984). The Mg^{2+} content of these cells falls back to the initial, pre-load level, but does not fall below this level, even if the cells are incubated for long periods in a medium that is Mg^{2+}-free or contains the Mg^{2+} chelator EDTA (Günther et al. 1984). It was suggested that the transporter is gated by $[Mg^{2+}]_i$, becoming inactive when physiological $[Mg^{2+}]_i$ is achieved. Replacement of external Na^+ by K^+, Li^+ or choline inhibits efflux, suggesting involvement of a Na^+-dependent transporter. The Na^+-dependent component is half-maximally activated by 25 mM Na^+ (Günther and Vormann 1985a). The apparent internal Mg^{2+} affinity of the transporter is about 3.5 mM (Günther et al. 1984). Transport (in Na^+ media) is inhibited by amiloride (IC_{50} = 0.6 mM), quinidine (0.5 mM), Mn^{2+} and metabolic inhibitors (of glycolysis or oxidative phosphorylation). It is not affected by ouabain (0.1 mM), PCMBS (0.1 mM), *N*-ethylmaleimide (NEM, 1 mM), furosemide (1 mM), acetazolamide (1 mM), DIDS (0.1 mM), SITS (0.03 mM), trifluoperazine (0.2 mM), phloretin (0.1 mM), phloridzin (1 mM), external Ca^{2+} and catecholamines (Günther et al. 1984; Günther and Vormann 1985a, 1986, 1989a).

Approximately 2 mol of Na^+ were taken up for each mol of Mg^{2+} lost from loaded cells, suggesting transport on a $2Na^+/1Mg^{2+}$ antiport (Günther et al. 1984). However, curves relating Mg^{2+} loss to $[Na^+]_o$ are more consistent with a single Na^+ ion being exchanged for each Mg^{2+} ion (Günther and Vormann 1985a). In addition, it was shown that reducing the transmembrane Na^+ gradient by treating cells with PCMBS ($[Na^+]_i$ rose 5-fold) did not affect the rate of Mg^{2+} loss. This was taken to indicate that although the transporter is Na^+-dependent, it does not obtain energy from the Na^+ gradient.

Studies with ^{28}Mg confirm these findings. They reveal a Na^+-dependent Mg^{2+} transporter that is inhibited by amiloride and stimulated by internal Mg^{2+} (Günther and Vormann 1987). In the absence of external Na^+, amiloride-sensitive Mg^{2+}/Mg^{2+} exchange becomes apparent, and this reaction is also found in cells that have a normal Mg^{2+} content. These data suggest that the transporter in chicken red cells is similar to that in rat, with the exception that it cannot transport Mn^{2+} (Günther et al. 1990a; Büttner et al. 1998a).

It has proved difficult to reverse the Mg^{2+} transporter in chicken red cells (Günther and Vormann 1985b). Mg^{2+} uptake could not be detected when Mg^{2+}-depleted cells (previously incubated with A23187 and EDTA) were incubated for two hours in physiological saline, cell culture media or homologous plasma ($[fMg^{2+}]_o$ about 0.5 mM), but this may have been due to the low medium $[Mg^{2+}]$.

Loading chicken red cells with Mg^{2+} not only increases Mg^{2+} efflux but also phosphorylation of a 230 kDa membrane protein and a 14 kDa cytosolic protein (Günther and Vormann 1986), both of which are not significantly phosphorylated in control cells. Thus amiloride may inhibit transport by preventing phosphorylation of proteins that regulate Mg^{2+} transport, or the transporter itself, perhaps by inhibiting a protein kinase. This is probably not a protein kinase C (phorbol esters do not affect Mg^{2+} transport), a Ca^{2+}-calmodulin dependent kinase (trifluoperazine is without effect), nor a protein kinase A (catecholamines are without effect), but may be a casein kinase (Günther and Vormann 1986).

Mg^{2+} is also lost from Mg^{2+}-loaded chicken red cells incubated in a sucrose medium (Günther and Vormann 1989a). This loss is slower than that seen with human or rat cells under similar conditions, and slower than the flux into Na^+-containing media. It is not inhibited by quinidine, nor by amiloride and its derivatives, but is slightly reduced by SITS and by isosmotically substituting sucrose by choline Cl^- (Günther and Vormann 1990; Günther et al. 1989). K^+ efflux, which also occurs into sucrose, is strongly inhibited by amiloride derivatives suggesting that Mg^{2+} and K^+ leave the cell through different routes (Günther et al. 1989) (but see above).

Mg^{2+} efflux from Mg^{2+}-loaded chicken red cells incubated in media comprising mainly of either NaCl, sucrose or choline Cl^- is partially inhibited by homologous plasma and by BSA (IC_{50} = 0.8 g/l) (Günther and Vormann 1991).

16.12 Conclusions

Red cells from a wide variety of species display at least two mechanisms of Mg^{2+} transport when they are incubated in media of normal ionic strength. In all cells examined, there is a component of Mg^{2+} efflux that requires external Na^+. This component is inhibited by quinidine, amiloride and imipramine, but not by vanadate. The transporter is activated by $[Mg^{2+}]_i$, not only as substrate, but also at regulatory sites, so activity is low or perhaps absent at physiological $[Mg^{2+}]_i$. This has led to the concept that Mg^{2+} transport is gated by $[Mg^{2+}]_i$ (Vormann and Günther 1993). The transporter is probably regulated by phosphorylation involving ATP, but it is not clear whether it can obtain energy for transport from ATP.

Amiloride may inhibit transport by interfering with transporter phosphorylation. Transporter activity is affected by factors circulating in the plasma. An additional Na^+-independent route also exists. This is rather poorly defined, but is necessary to explain rates of Mg^{2+} transport far in excess of those expected for simple diffusion through the membrane lipids. It may be a channel or uniport, and it too, may be gated by $[Mg^{2+}]_i$. High rates of Mg^{2+}efflux are seen when red cells are incubated in sucrose. Properties of this efflux suggest that it does not involve the Na^+-dependent route, but may represent either activation of a latent transporter capable of carrying Mg^{2+}, or stimulation of the Mg^{2+} channel or uniport.

In red cells from several species, the Na^+-dependent Mg^{2+} transporter has many of the characteristics required of a Na^+/Mg^{2+} antiporter. However, Na^+ transport on the putative antiport has only been clearly demonstrated in rodent red cells. In ferret, the system is reversible, can move Mg^{2+} against an electrochemical gradient using energy from the Na^+ gradient, and probably exchanges a single Na^+ ion for each Mg^{2+} ion. In human, the Na^+-dependent Mg^{2+} transporter can move Mg^{2+} against an electrochemical gradient, and in Mg^{2+}-loaded rat cells the system can reverse to give Mg^{2+} influx. The stoichiometry of the putative antiporter is not clear. Hill plots suggest $1Na^+:1Mg^{2+}$ in human, ferret, rat and chicken red cells, and this idea is supported by the finding that transport is sensitive to membrane potential in ferret cells. However, direct measurements of changes in cell Na^+ content suggest 2 or even 3 $Na^+:1Mg^{2+}$ in rat, chicken and human cells. These estimates are very tentative until more selective inhibitors of the transporter are available. As yet, it has not been possible to show that the Na^+-dependent Mg^{2+} transporters in various red cells are the same. The fact that the transporter cannot be reversed in human cells may indicate that a different transporter is present or may simply result from difficulties working with a system only capable of mediating very small fluxes.

What would be the consequences of Na^+/Mg^{2+} antiport working at equilibrium in red cells? An antiport operating with a 1:1 stoichiometry would explain the maintenance of physiological $[Mg^{2+}]_i$ in red cells from ferrets and perhaps other carnivores. These cells have a high $[Na^+]_i$ and therefore a small Na^+ gradient across their membranes. However, in red cells with a more usual low Na^+ content, the antiport would reduce $[Mg^{2+}]_i$ to less than 0.1 mM which would compromise the operation of other transport systems and metabolism. If the stoichiometry were 2 or 3 to 1, then $[Mg^{2+}]_i$ would fall to levels (< 1 μM) where the cells could not survive. Clearly, if the antiporter is operating in these cells, it must be in conjunction with a channel or uniport, so that $[Mg^{2+}]_i$ is regulated by a pump-leak system. Given the low rates of ^{28}Mg equilibration in red cells, the fluxes through these systems are very small under normal circumstances. A possible advantage of the system is that $[Mg^{2+}]_i$ does not slavishly follow $[Na^+]_i$, and, within limits, can be regulated independently of $[Na]_i$ in cells where this concentration is low, and consequently subject to (significant fractional) variation.

The possibility that Mg^{2+} is transported by a Ca^{2+} transporter, particularly the Na^+/Ca^{2+} antiporter has been frequently examined (Flatman and Lew 1980; Flatman 1991a), and the idea has gained ground following the recent demonstration that expression of the Na^+/Ca^{2+} antiporter (from canine heart, NCX1, or rat brain, NCX3) in fibroblasts not only increases Ca^{2+} transport but Mg^{2+} transport as well

(Tashiro et al. 2000). It is clear that Mg^{2+} transport on a Na^+/Ca^{2+} antiporter cannot explain all Mg^{2+} transport, however, it may explain some. For instance Na^+/Ca^{2+} antiport has been demonstrated in ferret red cells (Milanick 1989) with their high Mg^{2+} transport rates, but is not present in human red cells (Ferreira and Lew 1977; Desai et al. 1991) with their low Mg^{2+} transport rates. An alternative, attractive possibility, is that some transporters, previously described as Na^+/Ca^{2+} antiporters (often on the basis of experiments with very high internal Ca^{2+} levels), are in fact Mg^{2+} transporters. Thus some Na^+-dependent Mg^{2+} transporters may be part of what is currently called the Na^+/Ca^{2+} antiporter family. Alteration of relatively few amino acids may change the divalent cation specificity of these proteins. Recently, it has been suggested that some Na^+-dependent Mg^{2+} transporters may be part of the Cl^--dependent cation cotransport super-family (CCC) (Rasgado-Flores and Gonzalez-Serratos 2000). Perhaps the elusive Mg^{2+} transporter is already known, but under a different name. Experiments are currently underway to investigate this hypothesis.

References

Aikawa JK (1963) The role of magnesium in biologic processes. Charles C Thomas, Springfield

Altura BT, Altura BM (1997) Ionized magnesium measurements in serum, plasma and whole blood in health and disease. In: Smetana R (ed) Advances in magnesium research: 1. Magnesium in cardiology. John Libbey, London, pp 538-546

Barbul A, Zipser Y, Nachles A, Korenstein R (1999) Deoxygenation and elevation of intracellular magnesium induce tyrosine phosphorylation of band 3 in human erythrocytes. FEBS Letters 455:87-91

Bernstein RE (1959) Alterations in metabolic energetics and cation transport during aging of red cells. J Clin Invest 38:1572-1586

Beyenbach KW (2000) Renal handling of magnesium in fish: from whole animal to brush border membrane vesicles. Front Biosci 5:d712-d719

Bock JL, Yusuf Y (1988) Further studies on alterations in magnesium binding during cold storage of erythrocytes. Biochim Biophys Acta 941:225-231

Bock JL, Wenz B, Gupta RK (1985) Changes in intracellular Mg adenosine triphosphate and ionized Mg^{2+} during blood storage: detection by ^{31}P nuclear magnetic resonance spectroscopy. Blood 65:1526-1530

Bookchin RM, Ortiz OE, Lew VL (1987) Red cell magnesium content and permeability in sickle cell anemia. Clin Res 35:651A

Bunn HF, Ransil BJ, Chao A (1971) The interaction between erythrocyte organic phosphates, magnesium ion, and hemoglobin. J Biol Chem 246:5273-5279

Büttner S, Günther T, Schäfer A, Vormann J (1998a) Magnesium metabolism in erythrocytes of various species. Magnes - Bull 20:101-109

Büttner S, Günther T, Vormann J (1998b) Reversible inactivation of Na^+/Mg^{2+} antiport in rat erythrocytes by glucose starvation. Magnes - Bull 20:11-14

Campbell EH, Gibson JS (1998) Oxygen-dependent K^+ fluxes in sheep red cells. J Physiol 506:679-688

Campbell EH, Cossins AR, Gibson JS (1999) Oxygen-dependent K^+ influxes in Mg^{2+}-clamped equine red blood cells. J Physiol 515:431-437

Charbon GA, Hoekstra MH (1962) Mineral content in plasma and blood cells of various species. Acta Physiol Pharmacol Neerlandica 10:209-214

Clark MR (1988) Senescence of red cells: progress and problems. Physiol Rev 68:503-554

Desai SA, Schlesinger PH, Krogstad DJ (1991) Physiologic rate of carrier-mediated Ca^{2+} entry matches active extrusion in human erythrocytes. J Gen Physiol 98:349-364

DiPolo R, Beaugé L (1988) An ATP-dependent Na^+/Mg^{2+} countertransport is the only mechanism for Mg extrusion in squid axons. Biochim Biophys Acta 946:424-428

Dunn MJ (1974) Red blood cell calcium and magnesium: effects upon sodium and potassium transport and cellular morphology. Biochim Biophys Acta 352:97-116

Ebel H, Günther T (1999) Characterization of Mg^{2+} efflux from rat erythrocytes non-loaded with Mg^{2+}. Biochim Biophys Acta 1421:353-360

Ebel H, Hollstein M, Günther T (2002) Role of the choline exchanger in Na^+-independent Mg^{2+} efflux from rat erythrocytes. Biochim Biophys Acta 1559:135-144

Elin RJ (1997) Evaluating the role of ionized magnesium in laboratory and clinical practice. In: Smetana R (ed) Advances in magnesium research: 1. Magnesium in cardiology. John Libbey, London, pp 525-531

Ferreira HG, Lew VL (1977) Passive Ca transport and cytoplasmic Ca buffering in intact red cells. In: Ellory JC, Lew VL (eds) Membrane transport in red cells. Academic, London, pp 53-91

Féray JC, Franck G, Garay R, Henrotte JG (1989) Inter-individual differences in red cell Mg^{2+} contents are related to the activity of a Na^+:Mg^{2+} exchanger. Possible relationship with HLA-associated genetic factors. Magnes Res 2:124

Féray JC, Garay R (1986) An Na^+-stimulated Mg^{2+}-transport system in human red blood cells. Biochim Biophys Acta 856:76-84

Féray JC, Garay R (1987) A one-to-one Mg^{2+}:Mn^{2+} exchange in rat erythrocytes. J Biol Chem 262:5763-5768

Féray JC, Garay R (1988) Demonstration of a Na^+ : Mg^{2+} exchange in human red cells by its sensitivity to tricyclic antidepressant drugs. Naunyn-Schmiedebergs Arch Pharmacol 338:332-337

Flatman PW (1980) The effect of buffer composition and deoxygenation on the concentration of ionized magnesium inside human red blood cells. J Physiol 300:19-30

Flatman PW (1987) The effects of calcium on potassium transport in ferret red cells. J Physiol 386:407-423

Flatman PW (1991a) Mechanisms of magnesium transport. Annu Rev Physiol 53:259-271

Flatman PW (1991b) The effects of metabolism on Na^+-K^+-Cl^- co-transport in ferret red cells. J Physiol 437:495-510

Flatman PW (1993) The role of magnesium in regulating ion transport. In: Birch NJ (ed) Magnesium and the cell. Academic Press, London, pp 197-216

Flatman PW, Andrews PLR (1983) Cation and ATP content of ferret red cells. Comp Biochem Physiol 74A:939-943

Flatman P, Lew VL (1977) Use of ionophore A23187 to measure and to control free and bound cytoplasmic Mg in intact red cells. Nature 267:360-362

Flatman PW, Lew VL (1980) Magnesium buffering in intact human red blood cells measured using the ionophore A23177. J Physiol 305:13-30

Flatman PW, Smith LM (1990) Magnesium transport in ferret red cells. J Physiol 431:11-25

Flatman PW, Smith LM (1991) Sodium-dependent magnesium uptake by ferret red cells. J Physiol 443:217-230

Flatman PW, Smith LM (1996) Magnesium transport in magnesium-loaded ferret red blood cells. Pflügers Arch 432:995-1002

Franck G, Wdzieczak-Bakala J, Henrotte JG (1999) Modulating role of glucose on magnesium transport in rat erythrocytes. J Nutr Biochem 10:433-437

Frenkel EJ, Graziani M, Schatzmann HJ (1989) ATP requirement of the sodium-dependent magnesium extrusion from human red blood cells. J Physiol 414:385-397

Gerber G, Berger H, Jänig GR, Rapoport SM (1973) Interaction of haemoglobin with ions. Quantitative description of the state of magnesium, adenosine 5'-triphosphate, 2,3-bisphosphoglycerate, and human haemoglobin under simulated intracellular conditions. Eur J Biochem 38:563-571

Geven WB, Vogels-Mentink GM, Willems JL, Van Os CH, Hilbers CW, Joordens JJM, Rijksen G, Monnens LAH (1991) ^{31}P Nuclear magnetic resonance and zero-point titration compared for measurement of free magnesium concentration in erythrocytes. Clin Chem 37:2076-2080

Ginsburg S, Smith JG, Ginsburg FM, Reardon JZ, Aikawa JK (1962) Magnesium metabolism of human and rabbit erythrocytes. Blood 20:722-729

Graschopf A, Stadler JA, Hoellerer MK, Eder S, Sieghardt M, Kohlwein SD, Schweyen RJ (2001) The yeast plasma membrane protein Alr1 controls Mg^{2+} homeostasis and is subject to Mg^{2+}-dependent control of its synthesis and degradation. J Biol Chem 276:16216-16222

Gupta RK, Benovic JL, Rose ZB (1978) The determination of the free magnesium level in the human red blood cell by ^{31}P NMR. J Biol Chem 253:6172-6176

Günther T, Vormann J (1985a) Mg^{2+} efflux is accomplished by an amiloride-sensitive Na^+/Mg^{2+} antiport. Biochem Biophys Res Commun 130:540-545

Günther T, Vormann J (1985b) Removal and reuptake of intracellular magnesium. Magnes - Bull 7:66-69

Günther T, Vormann J (1986) Probable role of protein phosphorylation in the regulation of Mg^{2+} efflux via Na^+/Mg^{2+} antiport. Magnes - Bull 8:307-309

Günther T, Vormann J (1987) Characterization of Na^+/Mg^{2+} antiport by simultaneous $^{28}Mg^{2+}$ influx. Biochem Biophys Res Commun 148:1069-1074

Günther T, Vormann J (1989a) Characterization of Mg^{2+} efflux from human, rat and chicken erythrocytes. FEBS Letters 250:633-637

Günther T, Vormann J (1989b) Na^+-independent Mg^{2+} efflux from Mg^{2+}-loaded human erythrocytes. FEBS Letters 247:181-184

Günther T, Vormann J (1990) Characterization of Na^+-independent Mg^{2+} efflux from erythrocytes. FEBS Letters 271:149-151

Günther T, Vormann J (1991) Inhibition of Mg^{2+} efflux from erythrocytes by serum albumin. Magnes - Bull 13:82-84

Günther T, Vormann J (1995) Reversibility of Na^+/Mg^{2+} antiport in rat erythrocytes. Biochim Biophys Acta 1234:105-110

Günther T, Vormann J, Förster R (1984) Regulation of intracellular magnesium by Mg^{2+} efflux. Biochem Biophys Res Commun 119:124-131

Günther T, Vormann J, Cragoe EJ, Höllriegl V (1989) Characterization of Na^+-dependent and Na^+-independent Mg^{2+} efflux from erythrocytes by amiloride derivatives. Magnes - Bull 11:103-107

Günther T, Vormann J, Cragoe EJ (1990a) Species-specific Mn^{2+}/Mg^{2+} antiport from Mg^{2+}-loaded erythrocytes. FEBS Letters 261:47-51

Günther T, Vormann J, Höllriegl V (1990b) Characterization of Na^{+}-dependent Mg^{2+} efflux from Mg^{2+}-loaded rat erythrocytes. Biochim Biophys Acta 1023:455-461

Günther T, Vormann J, Höllriegl V (1990c) Concentration of intracellular free Mg^{2+} and Mg^{2+} efflux from magnesium-deficient erythrocytes. Magnes - Bull 12:43-46

Halperin JA, Brugnara C, Tosteson MT, Ha TV, Tosteson DC (1989) Voltage-activated cation transport in human erythrocytes. Am J Physiol 257:C986-C996

Heaton FW (1967) The determination of ionized magnesium in serum and urine. Clin Chim Acta 15:139-144

Henrotte JG (1993) Genetic regulation of cellular magnesium content. In: Birch NJ (ed) Magnesium and the cell. Academic Press, London, pp 177-195

Henrotte JG, Franck G (1990) Plasma factor(s) inhibiting cell magnesium efflux rapidly determined with a biological assay. Clin Chem 36:1859-1860

Henrotte JG, Pla M, Dausset J (1990) HLA- and H-2-associated variations of intra- and extracellular magnesium content. Proc Natl Acad Sci USA 87:1894-1898

Henrotte JG, Santarromana M, Féray JC, Franck G, Garay R (1997) Increased Mn^{2+} and decreased Mg^{2+} efflux from rat erythrocytes by plasma. In: Smetana R (ed) Advances in magnesium research: 1. Magnesium in cardiology. John Libbey, London, pp 514-517

Jelicks LA, Weaver J, Pollack S, Gupta RK (1989) NMR studies of intracellular free calcium, free magnesium and sodium in the guinea pig reticulocyte and mature red cell. Biochim Biophys Acta 1012:261-266

Jennings ML, Schulz RK (1990) Swelling-activated KCl cotransport in rabbit red cells: flux is determined mainly by cell volume rather than shape. Am J Physiol 259:C960-C967

Joiner CH, Jiang M, Fathallah H, Giraud F, Franco RS (1998) Deoxygenation of sickle red blood cells stimulates KCl cotransport without affecting Na^{+}/H^{+} exchange. Am J Physiol 274:C1466-C1475

Kaestner L, Christophersen P, Bernhardt I, Bennekou P (2000) The non-selective voltage-activated cation channel in the human red blood cell membrane: reconciliation between two conflicting reports and further characterisation. Bioelectrochem 52:117-125

Lew VL, Beaugé L (1979) Passive cation fluxes in red cell membranes. In: Giebisch G, Tosteson DC, Ussing HH (eds) Membrane transport in biology, vol. 2. Transport across single biological membranes. Springer, Berlin, pp 81-115

Lüdi H, Schatzmann HJ (1987) Some properties of a system for sodium-dependent outward movement of magnesium from metabolizing human red blood cells. J Physiol 390:367-382

Macey RI, Adorante JS, Orme FW (1978) Erythrocyte membrane potentials determined by hydrogen ion distribution. Biochim Biophys Acta 512:284-295

Mairbäurl H, Hoffman JF (1992) Internal magnesium, 2,3-diphosphoglycerate, and the regulation of the steady-state volume of human red blood cells by the Na/K/2Cl cotransport system. J Gen Physiol 99:721-746

Malon A, Wagner B, Bulska E, Maj-Zurawska M (2002) Comparison of the potentiometric, ^{31}P NMR, and zero-point titration methods of determining ionized magnesium in erythrocytes. Anal Biochem 302:220-223

Marjanovic M, Gregory C, Ghosh P, Willis JS, Dawson MJ (1993) A comparison of the effects of temperature on phosphorous metabolites, pH and Mg^{2+} in human and ground squirrel red cells. J Physiol 470:559-574

Milanick MA (1989) Na-Ca exchange in ferret red blood cells. Am J Physiol 256:C390-C398

Miseta A, Bogner P, Berényi E, Kellermayer M, Galambos C, Wheatley DN, Cameron IL (1993) Relationship between cellular ATP, potassium, sodium and magnesium concentrations in mammalian and avian erythrocytes. Biochim Biophys Acta 1175:133-139

Moncrief MBC, Maguire ME (1999) Magnesium transport in prokaryotes. J Biol Inorg Chem 4:523-527

Mulquiney PJ, Kuchel PW (1997) Model of the pH-dependence of the concentrations of complexes involving metabolites, haemoglobin and magnesium ions in the human erythrocyte. Eur J Biochem 245:71-83

Nordbö R (1939) Bestimmung der Magnesiumionenkonzentration im Ultrafiltrat von Blutserum. Skand Archiv Physiol 81:265-268

Ortiz OE, Lew VL, Bookchin RM (1990a) A sodium- and chloride-dependent magnesium efflux in sickle (SS) reticulocytes is highly stimulated by acid pH. Blood 76:14A

Ortiz OE, Lew VL, Bookchin RM (1990b) Deoxygenation permeabilizes sickle cell anaemia red cells to magnesium and reverses its gradient in the dense cells. J Physiol 427:211-226

Parker JC (1993) In defense of cell volume? Am J Physiol 265:C1191-C1200

Picado MJ, De la Sierra A, Aguilera MT, Coca A, Urbano-Márquez A (1994) Increased activity of the Mg^{2+}/Na^{+} exchanger in red blood cells from essential hypertensive patients. Hypertension 23:987-991

Quamme GA, De Rouffignac C (2000) Epithelial magnesium transport and regulation by the kidney. Front Biosci 5:d694-d711

Raftos JE, Lew VL, Flatman PW (1999) Refinement and evaluation of a model of Mg^{2+} buffering in human red cells. Eur J Biochem 263:635-645

Rasgado-Flores H, Gonzalez-Serratos H (2000) Plasmalemmal transport of magnesium in excitable cells. Front Biosci 5:d866-d879

Rasgado-Flores H, Gonzalez-Serratos H, DeSantiago J (1994) Extracellular Mg^{2+}-dependent Na^{+}, K^{+}, and Cl^{-} efflux in squid giant axons. Am J Physiol 266:C1112-C1117

Rogers TA (1961) The exchange of radioactive magnesium in erythrocytes of several species. J Cell Comp Physiol 57:119-121

Rogers TA, Mahan PE (1959) Exchange of radioactive magnesium in the rat. Proc Soc Exp Biol Med 100:235-239

Romani A, Scarpa A (2000) Regulation of cellular magnesium. Front Biosci 5:d720-d734

Santarromana M, Delepierre M, Féray JC, Franck G, Garay R, Henrotte JG (1989) Correlation between total and free magnesium levels in human red blood cells. Influence of HLA antigens. Magnes Res 2:281-283

Schatzmann HJ (1993) Asymmetry of the magnesium sodium exchange across the human red cell membrane. Biochim Biophys Acta 1148:15-18

Schweigel M, Martens H (2000) Magnesium transport in the gastrointestinal tract. Front Biosci 5:d666-d677

Smetana R (1997) Advances in magnesium research: 1. Magnesium in cardiology. John Libbey, London

Tashiro M, Konishi M, Iwamoto T, Shigekawa M, Kurihara S (2000) Transport of magnesium by two isoforms of the Na^{+}-Ca^{2+} exchanger expressed in CCL39 fibroblasts. Pflügers Arch 440:819-827

Valberg LS, Card RT, Paulson EJ, Szivek J (1965) The metal composition of erythrocytes in different species and its relationship to the lifespan on the cells in the circulation. Comp Biochem Physiol 15:347-359

Vormann J, Günther T (1993) Magnesium transport mechanisms. In: Birch NJ (ed) Magnesium and the cell. Academic Press, London, pp 137-155

Vormann J, Günther T, Magdorf K, Rob P (1996) Changed erythrocyte magnesium transport in patients with cystic fibrosis and renal failure. J Physiol 493:66P-67P

Wacker WEC (1980) Magnesium and man. Harvard University Press, Cambridge

Walser M (1961) Ion association. VI. Interactions between calcium, magnesium, inorganic phosphate, citrate and protein in normal human plasma. J Clin Invest 40:723-730

Walser M (1967) Magnesium metabolism. Ergebn Physiol 59:185-296

Watson WS, Hilditch TE, Horton PW, Davies DL, Lindsay R (1979) Magnesium metabolism in blood and the whole body in man using ^{28}Magnesium. Metabolism 28:90-95

Watson WS, Lyon TDB, Hilditch TE (1980) Red cell magnesium as a function of cell age. Metabolism 29:397-399

Widmer J, Féray JC, Bovier P, Hilleret H, Raffin Y, Chollet D, Gaillard JM, Garay R (1995a) Sodium-magnesium exchange in erythrocyte membranes from patients with affective disorders. Neuropsychobiol 32:13-18

Widmer J, Henrotte JG, Raffin Y, Bovier P, Hilleret H, Gaillard JM (1995b) Relationship between erythrocyte magnesium, plasma electrolytes and cortisol, and the intensity of symptoms in major depressed-patients. J Affect Disord 34:201-209

Widmer J, Henrotte JG, Raffin Y, Mouthon D, Chollet D, Stepanian R, Bovier P (1998) Relationship between blood magnesium and psychomotor retardation in drug-free patients with major depression. Eur Psychiatry 13:90-97

Willis JS, Xu W, Zhao Z (1992) Diversities of transport of sodium in rodent red cells. Comp Biochem Physiol 102A:609-614

Xu W, Willis JS (1994) Sodium transport through the amiloride-sensitive Na-Mg pathway of hamster red cells. J Membrane Biol 141:277-287

Zhang W, Truttmann AC, Lüthi D, McGuigan JAS (1997) Apparent Mg^{2+}-adenosine 5-triphosphate dissociation constant measured with Mg^{2+} macroelectrodes under conditions pertinent to ^{31}P NMR ionized magnesium determinations. Anal Biochem 251:246-250

Zhao Z, Willis JS (1993) Cold activation of Na influx through the Na-H exchange pathway in guinea pig red cells. J Membrane Biol 131:43-53

17 Trace Metal Transport

Niall M. Horn, Alan L. Thomas and Fiona Oakley

School of Biological Sciences, University of Southampton, Bassett Crescent East, Southampton SO16 7PX, UK

17.1 Introduction

The term "trace metals" is used here to include both those transition metals, which are essential for cell function and those, which do not have a known biological role but which produce toxic effects by interacting with biological systems. The metals discussed are zinc, copper, cobalt, nickel, chromium, manganese, iron and cadmium. A common property of these metals is the ability to form coordination compounds with amino acids with very high affinity. Conventionally, the affinity of the metal for the amino acid ligand is expressed as a log stability constant (the association constant for the forward reaction). In many cases the metal can form stable complexes with multiple ligands; for example zinc can associate with one histidine to form a *mono* complex which can then associate with a second histidine to form a *bis* complex. Mixed complexes such as Zn(cys-his) can also exist in biological systems. The distribution of the metal between the various forms is an equilibrium and will vary with the relative concentrations of metal and ligand. The stability constants for these interactions vary over a wide range. Values for the log stability constant for the formation of $Ni(cys)_2$ *in vitro* are approx. 20, compared with that for $Mn(his)_2$ of 5.8. This behaviour affects the availability of trace metals for uptake into cells and tissues from plasma. For example, approx. 98% of plasma zinc is incorporated in α_2-macroglobulin (non-labile) or complexed to serum albumin (labile). Virtually all the remaining zinc is complexed to amino acids, mainly histidine and cysteine (Giroux and Henkin 1972). The free ionic metal concentration is therefore extremely low (the only direct measurements published are those of Magneson et al. (1987) in horse plasma, where free zinc activity was found to be approx. $2 \cdot 10^{-10}$ M, which is compatible with the value predicted from computer simulations). This distribution is shown diagrammatically (Fig. 17.1). For all other metals discussed here, except manganese, the free ionic activities in plasma are predicted to be even lower than for zinc. Even in the case of manganese the activity will be much lower than for divalent cations such as calcium and magnesium. The extremely low ionic metal activities of trace metals in extracellular fluids should be kept in mind when evaluating the relevance of experiments on carriers conducted *in vitro* at millimolar concentrations. The fact that a significant proportion of the trace metals will be complexed to amino acids such as histidine

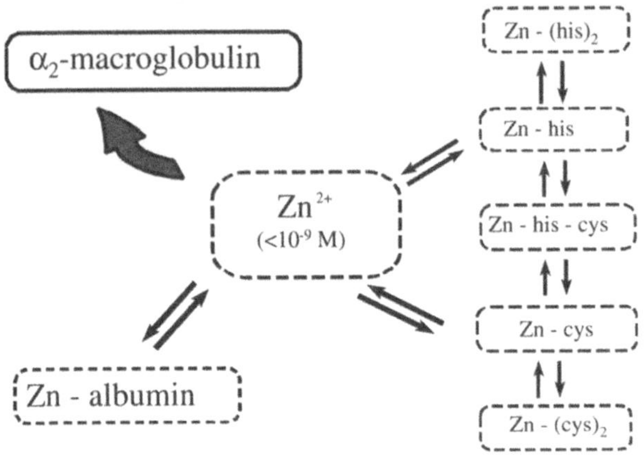

Fig. 17.1. The distribution of zinc in plasma. Bi-directional arrows indicate an equilibrium distribution. Free ionic zinc is in equilbrium with that bound to serum albumin and to ligands such as amino acids, with histidine and cysteine being the most important. The solid arrow to α_2-macroglobulin indicates that zinc incorporated in this (and similar) protein(s) is not labile

and cysteine suggests that these complexes might provide an important substrate for metal uptake into cells.

The high affinity for amino acid ligands will also ensure that the intracellular free ionic activities of trace metals within erythrocytes are very low. Direct measurement is not practical, but the erythrocytic free intracellular zinc concentration has been measured by a zero-flux method (Simons 1990). It was found to be $2.4 \cdot 10^{-11}$ M. It would be expected that the values for metals such as copper and nickel will be even lower owing to their greater affinities for amino acids. Many trace metals have redox behaviour and can promote free radical formation thus there is an obvious biological advantage in maintaining low ionic activities both inside and outside cells. The high affinities of trace metals for amino acids also determine their biological roles. There is little evidence at present for an extracellular role for the free hydrated metal ions, with the exception of synaptically-released zinc as a modulator of transmission in specific brain regions (for review see Frederickson and Moncrieff (1994)). Similarly, a role for intracellular free ions appears problematical except in those cells that can synthesise metallothionein, where a wide range of ions is capable of acting on metal-responsive elements and initiating metallothionein gene transcription. In general, the trace elements have biological roles at structural or catalytic regions in enzymes or other proteins.

In view of the complex speciation of trace metals in plasma, and the fact that many of the erythrocyte uptake systems studied appear to have broad substrate specificity, it is easier to review metal uptake mechanisms by substrate type rather than by individual metal. In many cases, the full range of possible substrates has not been investigated, so that absence of data for uptake of a particular metal species by an identified system should not be taken to mean that it is not a substrate for that system.

17.2 Transport Systems

17.2.1 As Free Metal (Hydrated Ions)

Membrane transport systems for transition metal ions have been described in a range of prokaryotic and eukaryotic cells. The P-type ATPase family of transporters are of great interest because of their role in Wilson's and Menke's diseases (for review see Harris (2000)). The DMT/DCT/Nramp (Divalent Metal Transporter/Divalent Cation Transporter/Natural resistance associated macrophage protein) family have been reviewed by Rolfs and Hediger (1999). The cation diffuser family (for review see Paulsen and Saier (1997)), and the ZIP family (for review see Gaither and Eide (2000)) are also widely-distributed. So far, evidence for the presence of any of these systems in the mature erythrocyte is lacking, but some or all of them may play an important role during the development of erythroid cells and their precursors.

Iron

The uptake of free (non-transferrin bound) iron into erythroid cells has been investigated by Morgan and co-workers (Morgan 2001). It was found that Fe(II) could be transported into reticulocytes by a saturable high-affinity system ($K_m \sim 0.2\ \mu M$) which disappeared on maturation into erythrocytes. This appeared to be mediated by a carrier corresponding to DMT1. This conclusion was supported by the finding that iron uptake was impaired in reticulocytes from rats and mice having mutations affecting DMT1. In addition, a low-affinity transporter with a K_m of approx. 15 μM is present in both reticulocytes and erythrocytes. This uptake was believed to involve the Na^+/Mg^{2+} antiport.

An earlier study using rabbit and rat reticulocytes and rat erythrocytes (Savigni and Morgan 1998) reported that both the high-affinity and low-affinity iron transport systems could also transport manganese, zinc, cobalt and nickel with similar affinity to that for iron and that all transported metals appeared to compete with each other. In addition, the uptake characteristics varied according to whether the medium had low ionic strength (which favoured the high-affinity system) or had low sodium concentration (which favoured the low-affinity system).

Zinc

Zinc transport in mammalian cells has been comprehensively reviewed by Reyes (1996).

Van Wouwe et al. (1990) reported that the average adult contains 28 mmoles of zinc, of which the majority is stored in the tissues with only 0.5 mmoles of zinc remaining in the circulation. They calculated that of the 0.5 mmoles of circulatory zinc, 0.015 mmoles are bound to α_2-macroglobulin, 0.03 mmoles are bound to the plasma protein albumin and the remaining 0.45 mmoles are carried in the erythrocytes. As erythrocytic zinc binds to the plasma membrane as well as to haemoglobin, other proteins and small molecules, Van Wouwe et al. (1990) proposed that the less stable erythrocytic zinc complexes might exchange with the plasma zinc fraction. They calculated the intracellular exchangeable erythrocytic zinc pool to be 5 μmoles of zinc per litre of packed cells, with a half-life for exchange of 7 hours. Naber et al. (1994) suggested that 10% of total erythrocytic zinc in rats, and 3.5% in humans, could act as an intracellular pool of zinc that is rapidly exchangeable with plasma. Ohno et al. (1985) used atomic absorption spectrometry to measure the total zinc content of human erythrocyte lysate and found that 92.4% of erythrocytic zinc is contained in carbonic anhydrase and superoxide dismutase (SOD), and is therefore not freely exchangeable.

Rifkind (1983) reviewed earlier studies on erythrocyte zinc exchange. Subsequently, a number of studies on zinc exchange across the erythrocyte membrane have been published, although the contribution made by transport of free ionic (non-complexed) zinc under physiological conditions is still not clear (uptake mechanisms depending on the formation of extracellular zinc complexes are discussed later). Van Wouwe et al. (1990) investigated the steady state kinetics of zinc exchange between human erythrocytes and a medium containing 7.6 μM zinc and 40 g/l human serum albumin (HSA) to chelate the zinc. The results of the experiments showed a rapid "uptake" component, which was thought to be extracellular binding, followed by a steady state zinc exchange. They also showed that 100 μM histidine promoted zinc uptake into erythrocytes, and that the addition of citric acid or prednisolone (25 mg/l) decreased zinc uptake. Ouabain, iodoacetate, adrenaline, dexamethasone and pH variation did not produce any significant effects on zinc uptake.

Naber et al. (1994) studied the effects of temperature on ^{65}Zn uptake in human erythrocytes, in the presence of 40 g/l HSA. Zinc uptake decreased with decreasing temperature; at 4 °C the uptake was reduced to 12.5% of that observed at 37 °C. The uptake could be divided into 3 phases: an initial rapid uptake phase (0–30 min), a second rapid uptake phase (30 min–5.5 h) and a slow quasi-linear phase (> 5.5 h). The initial rapid uptake phase was attributed to membrane binding and was shown to reach equilibrium within the first 30 minutes (however, as the erythrocytes were washed in ice-cold saline the extent of the surface binding may have been underestimated). The second rapid phase was thought to represent an exchange of labile intracellular zinc with the ^{65}Zn in the incubation medium. The third phase was attributed to a slower exchange between extracellular ^{65}Zn and compartmentalised zinc. Naber et al. (1994) proposed that the uptake occurred via an active process, although this conclusion appeared to be based solely on the temperature dependence rather than demonstration of a specific requirement for ATP or the presence of a sodium gradient. Dekok et al. (1993) studied the uptake of ^{65}Zn into rat erythrocytes *in vitro* under near-physiological conditions (i.e. in the presence of albumin to maintain a low free ionic zinc concentration). They reported a high affinity carrier (K_m of 0.2 nM), which transported zinc across the erythrocyte membrane and concluded that the *in vitro* zinc uptake rate was dependent on the zinc status of the donor rat. The actual mechanism was not discussed.

Mandolfino et al. (1994) compared zinc uptake in human and eel erythrocytes from a medium without metal buffering ligands using atomic absorption spectrophotometry. In both cases uptake was directly related to external zinc concentration up to 500 μM, with no evidence of saturation. Uptake rate in eel cells was approximately half that measured in human cells. More interestingly, the zinc uptake in eel cells was reduced by approx. 40% following pretreatment with trypsin, whereas uptake in human cells was not affected. Eel cells also showed a component accounting for about 36% of total uptake which was resistant to the red cell anion exchange inhibitor DIDS (4,4'-diisothiocyanostilbene-2,2'-disulphonic acid) whereas, in human cells, DIDS blocked almost all uptake. The pH dependence of eel cell zinc uptake was different from that in human cells. Uptake in eel cells increased progressively towards pH 8.1, whereas that in human cells showed a bell-shaped relationship, which was maximal at pH 7.4. The nature of the DIDS-resistant pathway in eel cells deserves further investigation. It should be remembered that fish erythrocytes are nucleated and that their membrane transporters may not have degenerated to the same extent as in mature human erythrocytes. These findings suggest that comparative studies on trace metal transport may be fruitful.

Agte et al. (1997) studied the effects of nicotinic acid supplementation on the absorption and utilisation of zinc and iron and the *in vitro* zinc uptake by human erythrocytes obtained from either fasting or non-fasting humans. They showed that the addition of 8 μM nicotinic acid or NADP to the medium, increased ^{65}Zn uptake into human erythrocytes. The mechanism was not investigated further.

Simons (1991) reported the presence of a saturable zinc efflux from human erythrocytes, which was dependent on the extracellular calcium ion concentration. Zinc could be transported outwards against its electrochemical gradient provided calcium was present in the external medium. It was proposed that a zinc/calcium exchanger could mediate the efflux of zinc at the expense of an inward calcium gradient maintained by the calcium pump.

Manganese

Manganese is an essential component of a number of enzymes important at the nucleated stages of erythropoiesis, including pyruvate carboxylase and oxaloacetate decarboxylase. Erythrocytes also contain manganese-SOD (superoxide dismutase). In spite of the clear importance of manganese for cellular functions, there has been a tendency to use manganese flux measurements as a convenient model for the investigation of magnesium or iron transport mechanisms rather than considering manganese transport as important in its own right.

Weed and Rothstein (1960) studied the uptake of ^{54}Mn into mature human erythrocytes, using a saline-Tris incubation medium. As in many similar studies, the cells were washed with cold saline before counting which may not have removed all of the surface-bound metal, so that apparent uptake does not necessarily represent internalisation. At Mn^{2+} concentrations below $5{\cdot}10^{-4}$ M, uptake was directly related to extracellular concentration with no sign of saturation. Uptake was not energy-dependent, and was only reduced by cobalt, calcium and magnesium when present at very high concentrations. They concluded that influx and efflux of manganese was by passive diffusion. Their findings are difficult to relate to subsequent studies. Chau et al. (1996) described three saturable mechanisms for manganese uptake in erythroid cells. Reticulocytes showed receptor-mediated Mn-transferrin complex uptake. In addition, reticulocytes expressed a high-affinity ionic manganese uptake system with a K_m of 0.4 μM, which was inhibited by metabolic inhibitors, and a low-affinity ionic uptake mechanism with a K_m of 20 μM. The high-affinity system was suggested to be cell surface expression of a transporter responsible for manganese export from the endosome after release from transferrin. Erythrocytes showed only a low-affinity ionic uptake with a K_m of 50 μM. This system was inhibited by valinomycin and amiloride, and was more prominent in media having a low sodium concentration. It was suggested that the low-affinity system involved the Na^+/Mg^{2+} antiporter. As discussed above, the mechanisms for ionic manganese uptake appear to correspond to those described for iron and other metals. Feray and Garay (1997) also described a Mg^{2+}/Mn^{2+} exchange in rat erythrocytes which had similar properties to the previously-described Na^+/Mg^{2+} antiporter. Manganese efflux was stimulated by external sodium, and was ATP-dependent. In the absence of external sodium, manganese could stimulate magnesium efflux.

Lucacui et al. (1997) used an electron paramagnetic resonance technique to study manganese transport across human erythrocyte membranes. A sigmoid relationship between concentration and rate was obtained over the range 0.5–20 mM, with apparent cooperativity at low extracellular manganese concentrations (up to 5 mM) and competition for the transport site at higher concentrations. Vanadate-treated cells, and cells stored overnight in plasma, showed a large increase in maximum influx rate. This was compatible with the presence of an ATP-dependent manganese export mechanism. In vanadate-treated cells, influx was reduced by nifedipine, suggesting that uptake might also occur through calcium channels.

Cadmium

Although cadmium has no known biological role in humans, it is a ubiquitous pollutant, and the average daily intake by Americans was reported to be as high as 200–500 μg (Schroeder et al. 1967). It is both toxic and carcinogenic.

Limited evidence for transport of ionic (non-complexed) cadmium into erythrocytes is available. Garty et al. (1986) described cadmium uptake into rat erythrocytes, which was passive and not competitively inhibited by equimolar zinc. NEM (*N*-ethylmaleimide) treatment decreased uptake, suggesting involvement of sulfhydryl groups. Nguyen and Chien (1988) discussed a range of possible mechanisms for cadmium uptake into human erythrocytes. They found evidence for both saturable and non-saturable components. The saturable component had a K_m of 60 μM, and was inhibited by zinc and apparently stimulated by copper. The non-saturable component was observed at high cadmium concentrations. The actual mechanisms involved were not clear.

Mukhopadhyay et al. (1991) described cadmium uptake into human erythrocytes, which was not active and not inhibited by extracellular zinc. Lou et al. (1991) reported cadmium uptake in the absence of added bicarbonate that was not blocked by DIDS, implying that it did not involve the anion exchanger. Similarly, Horn and Thomas (1996) reported that cadmium uptake into human erythrocytes in the absence of added bicarbonate and histidine was not blocked by DIDS, suggesting the presence of a mechanism for the uptake of ionic cadmium. These observations invite further investigation.

Chromium

^{51}Chromium has been extensively used to label red cells in clinical studies involving measurement of whole body red cell volume, but little attention has been directed towards the mechanisms of cellular uptake. Although chromium has limited biological functions its metabolism is of considerable interest because of the wide range of toxic effects it can produce.

Coogan et al. (1990) studied uptake of hexavalent chromium into human and rat erythrocytes using ^{51}Cr labelled K_2CrO_4 (final concentration 50 μM) in a medium without metal buffering ligands, giving a very high free ionic activity. Accumulation was linear over 120 minutes. However, since the extracellular metal concentration in these experiments was not varied, it is not possible to confirm whether the uptake mechanism was saturable. Uptake was greater in human than rat erythrocytes. Uptake into rat erythrocytes *in vivo* was also studied over longer periods (up to 7 days) following oral and intravenous exposure. Erythrocyte chromium content remained relatively stable over 7 days following intravenous injection, whereas there was a significant decline following oral administration. Erythrocyte uptake of trivalent chromium (from $CrCl_3$) 60 min after an intravenous injection was not significantly different from that following K_2CrO_4 injection. This suggests that either the uptake mechanism does not differentiate between valency states, or that the metal reaches the cell in the same valency state irrespective of the form in which it is injected. Overall, no conclusions can be drawn regarding the actual cellular mechanisms involved.

Cobalt and Copper

Although a considerable amount of data exists on the transport of ionic cobalt and copper in nucleated cells (for reviews, see Paulsen and Saier (1997), Nelson (1999), Peña et al. (1999), Harris (2000)), there is at present no information on possible uptake mechanisms in erythrocytes. The intracellular buffering of cobalt in human erythrocytes was found to be time and metabolism dependent (Brown and Simonsen 1985).

17.2.2 Transport by Band 3 as an Anionic Complex

The erythrocyte anion exchanger (band 3) can bind and translocate a wide range of substrates including amino acids (Young et al. 1981) and anionic complexes of transition metals as well as mediating HCO_3^- and Cl^- exchange. So far, this mechanism has been most clearly established for zinc and cadmium, but its relevance to other metals would repay further investigation. For example, it has also been reported that ABH blood group antigens can act as recognition sites for stimulation of a DIDS-sensitive Ca^{2+} influx in human erythrocytes (Engelmann et al. 1991).

Zinc

Kruckeberg and Brewer (1978) first suggested a link between bicarbonate ions and zinc transport. Alda Torrubia and Garay (1989) measured the initial zinc uptake rates in human erythrocytes using atomic absorption spectrometry. The zinc uptake was inhibited by DIDS and was dependent on the presence of both external HCO_3^- and Cl^- ions. They suggested that transport occurred in the form of a monovalent $[Zn(HCO_3)_2Cl]^-$ complex. Kalfakakou and Simons (1990) reported that ionic zinc could be transported across the erythrocyte membrane by the anion exchanger as part of an anionic complex with bicarbonate ions. The uptake was dependent on bicarbonate ions in the external medium, and was DIDS-sensitive. As DIDS irreversibly blocks all of the erythrocyte cation-coupled HCO_3^- transporters with the exception of the K^+-HCO_3^- cotransporter, they concluded that ionic zinc uptake into human erythrocytes could be mediated by the anion exchanger. They also reported that zinc uptake could be promoted by the formation of a neutral complex of zinc with salicylate or thiocyanate ions. The influx of both zinc and lead could be changed by altering the concentration of zinc, lead or bicarbonate or by altering the pH, supporting the conclusion that uptake occurs via a transporter. Efflux was not observed by any of these mechanisms.

Cadmium

Lou et al. (1991) measured cadmium uptake in human red cells using atomic absorption spectrometry. Almost all (96%) of the observed cadmium uptake was shown to be inhibited by DIDS and was resistant to ouabain and amiloride. The DIDS-sensitive cadmium uptake required the presence of bicarbonate and chloride ions in the external medium suggesting that the anion exchanger can mediate cadmium uptake into human red cells. Bicarbonate ions produced a biphasic effect as low concentrations of bicarbonate stimulated uptake while high concentrations of bicarbonate ions caused inhibition. It was concluded that at low concentrations of bicarbonate ions a monovalent cadmium bicarbonate complex is formed which is capable of being translocated across the red cell membrane, but increasing the bicarbonate concentration leads to the production of a non-translocatable complex. The observed cadmium uptake was not affected by the presence of potassium, magnesium or calcium in the external medium. The presence of physiological concentrations of phosphate in the external media produced a small inhibitory effect, which may have been due to chelation of the cadmium by the phosphate.

They concluded that the anion exchanger provided a major transport mechanism for cadmium uptake into the human red cell, and suggested that the transported species were monovalent cadmium bicarbonate complexes, possibly $[Cd(OH)(HCO_3)_2]^-$ and $[Cd(OH)(HCO_3)Cl]^-$.

17.2.3 Uptake Dependent on Formation of Metal-Amino Acid Complexes

To date, most work in this area has focussed on the amino acid stimulation of zinc uptake into erythrocytes, but it is becoming apparent that a wide range of metals can also act as substrates, and that such mechanisms may also be present in other erythroid cells.

The first report suggesting a link between amino acid complexes and erythrocyte zinc uptake was by Sivarama Sastry et al. (1960). They found that 1 mM histidine inhibited the uptake of ^{65}Zn by 90%, and explained this result in terms of saturation of a carrier, which binds zinc-histidine complexes. It is much more likely that the reported result was caused by chelation of free zinc by histidine and hence inhibition of other ionic zinc-dependent pathways. Subsequently, Agarwal et al. (1981) investigated amino acid stimulated ^{65}Zn uptake into human erythrocytes. They reported that glycine did not affect ^{65}Zn uptake into human erythrocytes, in contrast to L-histidine (0–200 μM), which increased zinc uptake in the presence of albumin. They proposed that zinc enters erythrocytes as part of an amino acid complex. Van Wouwe et al. (1990) also reported that 100 μM histidine promoted zinc uptake into human erythrocytes.

In 1992, Aiken et al. reported that stimulation of zinc uptake into rat erythrocytes by millimolar concentrations of histidine was stereospecific (L-his >> D-his), dose-dependent, DIDS-resistant and temperature-dependent. Horn and Thomas (1993) reported similar stimulation of ^{60}Co uptake into rat erythrocytes by histidine. In this, and all subsequent studies by this group, extracellular free ionic zinc was kept low by buffering ligands such as BSA (bovine serum albumin) or ADA (*N*-[2-acetamido]-2-iminodiacetic acid) as appropriate. Cell pellets were spun down through EDTA-saline to minimise surface binding before counting. Horn et al. (1995) showed that histidine and cysteine both gave a stereospecific stimulation of zinc uptake in rat erythrocytes, but that stimulation only occurred in human erythrocytes with histidine and not with cysteine. Kinetic studies showed better correlation between uptake rate and calculated $Zn(his)_2$ (*bis*) complex concentration than Zn(his) or free ionic Zn.

Horn and Thomas (1996) reported that the uptake of cadmium into human erythrocytes was stimulated by histidine in a similar fashion to that observed with zinc, and competition occurred between zinc and cadmium histidine complexes. Again the concentration of the $metal(his)_2$ complex appeared to be rate-limiting. Zinc and cadmium *bis*-histidine complexes were found to have very similar affinities for metal uptake, suggesting that the uptake mechanism discriminated poorly between the metal within the complex.

Using longer incubation times Horn and Thomas (1997) reported that zinc uptake into human erythrocytes was stimulated by lower, more physiological (50 μM

~ 1 mM), concentrations of histidine. At these low histidine concentrations relatively less $Zn(his)_2$ complex is formed compared to Zn(his) complex. These data do not preclude the possibility that the *mono*-complex may also be a substrate for the transporter. Horn and Thomas (1997) also reported that copper could competitively inhibit zinc uptake at high histidine concentrations where formation of the *bis*-complexes is forced; as with cadmium the calculated inhibition constants for zinc and copper *bis*-histidine complexes were similar suggesting poor discrimination between metals. Horn et al. (1998) presented the results of a series of human erythrocyte zinc uptake experiments where the relative proportions of ionic zinc, *mono* and *bis* histidine complexes were manipulated by independently varying the concentrations of histidine and the buffering ligands (BSA or ADA). Again the zinc uptake correlated best with the calculated *bis*-complex concentration irrespective of the buffering ligands used. Oakley et al. (1998) reported competition for histidine-stimulated ^{65}zinc uptake by zinc, nickel, cobalt and manganese. The calculated inhibition constants (expressed as *bis*-complex concentrations) were similar for zinc, nickel and cobalt. However, the manganese *bis*-histidine complex inhibited zinc uptake at much lower concentrations, suggesting that the affinity of the transporter for manganese *bis*-histidine complexes was considerably higher. It was also found that histidine would promote ^{54}Mn uptake in a similar manner to that found for zinc, cadmium and cobalt, and that the $Mn(his)_2$ concentration giving half-maximal stimulation of uptake was much lower than for the other metals, confirming that the uptake site has a higher affinity for manganese-histidine complexes.

In more recent experiments Oakley, Horn and Thomas (data to be published) have investigated zinc uptake in a human erythroleukaemic cell line (HEL 92.1.7). A metal/histidine complex dependent uptake system was found which had similar properties to that described for the erythrocyte; it was stereospecific, saturable, appeared to prefer the *bis* complex and had similar affinities for zinc, nickel, and cobalt with a higher affinity for manganese. The rate of zinc uptake per cell was greater in the HEL cells compared to the human erythrocyte and the affinity for all metal *bis*-histidine complexes was higher in the HEL cells than in the erythrocyte. The HEL cells differed from the human erythrocytes in that cysteine was able to stimulate zinc uptake. These differences in uptake rate, affinity and response to cysteine may simply reflect the differing age of the transporter in the two cell systems. Oakley and co-workers have found that the injection of total HEL mRNA into *Xenopus* oocytes causes the expression of a histidine dependent metal uptake system with very similar properties to those seen in the HEL cells and human erythrocytes. The affinity of the transporter for all of the metal complexes tested more closely resembles that seen in the erythrocyte than in the HEL cell. The *Xenopus* expression system has the advantage that the only native zinc uptake system present in the oocyte requires the metal to be present as part of a complex with the protein vitellogenin (Montozori et al. 1995). Experiments are in progress to enable the binding site for the metal histidine complex to be isolated and sequenced.

There are a number of published studies that describe the effects of histidine on copper uptake in nucleated cells, and it is important to compare these with the erythrocyte histidine-stimulated metal uptake described above. Schmitt et al.

(1983) reported that isolated rat hepatocytes took up copper from copper histidine solutions having a 1:1 molar ratio, and that the uptake was saturable and not active. They concluded that ionic copper was probably the transported species, since the K_m for copper uptake was similar with or without histidine. Competition by cadmium, zinc, cobalt and manganese was demonstrated (the histidine concentration being kept equimolar with the total metal). However, the effects of differences in the relative stability constants for histidine complex formation were not considered. A subsequent study using copper *bis*-histidine containing ^{64}Cu and ^{3}H histidine (Darwish et al. 1984) appeared to confirm the suggestion that the copper was internalised without concomitant uptake of histidine. The overall conclusion was that histidine acted by mobilising copper from transport proteins such as albumin, and made it available for uptake as free ionic copper.

Other studies have emphasised the interaction between the amino acids complexed to the copper and the actual metal uptake site. Hartter and Barnea (1988) studied copper uptake into rat hypothalamic tissue slices, and described two uptake processes with different characteristics. A "low-affinity" site had an apparent K_m of 40 µM, and a large excess of histidine over copper inhibited uptake. A "high-affinity" site had a K_m of 6 µM; uptake was stimulated by excess histidine. However, in both cases the uptake in the presence of histidine was greater than that without, indicating that it was actually facilitating the uptake mechanism. However, a subsequent paper (Katz and Barnea 1990) reported that the earlier suggestion of two uptake systems had to be reconsidered. No significant difference in ability to stimulate uptake between L- and D-histidine was found in this preparation. In addition, they found that histidine in excess of the stoichiometric concentration for copper complexation inhibited uptake, suggesting competition for a histidine-binding site on the uptake mechanism. Barnea and Katz (1990) used dual labelling of histidine-copper complexes to study the relationship between histidine and copper uptake kinetics, and concluded that the uptake processes were separate, implying dissociation of the complex at the level of the membrane.

McArdle et al. (1988) reported that histidine stimulated copper uptake in mouse hepatocytes. The effect was dose-dependent, with a maximal effect around 400 µM histidine (i.e. with a large excess of histidine over copper which would favour the formation of *bis*-complexes). A range of amino acids appeared capable of stimulating copper uptake. Mas and Sarkar (1992) concluded that histidine was capable of facilitating copper uptake into human trophoblast cells by mobilising albumin-bound copper. However, the difference in uptake compared to that from free ionic copper or ceruloplasmin-bound copper was not significant. Bingham and McArdle (1994) reported concentration-dependent copper uptake from $Cu(his)_2$ complexes in rat liver plasma membrane vesicles and cultured rat hepatocytes. In both preparations, NEM inhibited uptake, suggesting that transport involved interaction with sulfhydryl groups. Hilton et al. (1995) reported concentration-dependent copper uptake into human placental microvillar vesicles from a medium containing $Cu(his)_2$. Evidence was provided for the involvement of a ceruloplasmin receptor. It was considered unlikely that this receptor corresponded to the various "ceruloplasmin receptors" reported from erythrocytes, but it appeared to have more similarity with that from liver endothelium.

Overall, it is not clear how closely the erythrocytic Zn(his)$_2$ uptake mechanism can be identified with the other Cu(his)$_2$ uptake mechanisms so far described. Essential distinguishing features of the erythrocyte system appear to be the extreme specificity for L-histidine and the lack of any inhibition of metal uptake in the presence of a very large excess of histidine. The effect is not inhibited by NEM. It is also clear that the presence of metal-histidine complexes promotes uptake independently of the free metal concentration. There are some similarities with the copper-histidine uptake so far described; for example with the exception of manganese, the affinities for the metal(his)$_2$ complexes in erythrocytes, HEL cells and HEL RNA-injected oocytes are broadly similar to the copper *bis*-histidine values reported from nucleated cells.

17.3 Conclusions

When considering the metal transporter systems present in the mature erythrocyte it must be remembered that the mature cell's requirements for metals are very limited and that these transporters were inserted into the membrane when the cell was nucleated and actively synthesizing new protein. The red cell's requirements for divalent metals will of course include iron for the synthesis of haemoglobin and zinc for the synthesis of carbonic anhydrase but there are many other zinc containing enzymes involved in normal cellular function (e.g. alcohol dehydrogenase, DNA polymerase and protein kinase C). In the early stages of erythropoiesis the erythrocyte shares progenitor cells with a number of other blood cells and these will require divalent metals for normal cell function especially, in the case of leukocytes, manganese and copper for the formation of superoxide dismutase (SOD). The transporters that remain in the mature erythrocyte are remnants of those present in erythroid precursor cells and there will therefore be fewer of them and their properties may be changed by the loss of sub-units or factors, which modulate their activity. This may explain the observed differences in the rate of transport and the affinity for metal-amino acid complexes between HEL cells and mature erythrocytes.

Given the number of different forms in which trace metals can exist *in vivo* it is not surprising that cells have developed a number of transport systems. In the first section we described mechanisms, which have been proposed for the transport of free ionic trace metals but given the very low concentrations of free ionic metal, which exist *in vivo*, the physiological significance of these systems in erythrocytes is doubtful. The mechanisms involving the transport of metal complexes via the anion exchanger (Kalfakakou and Simons 1990) are likely to play a significant role in the mature erythrocyte and may well be the main route by which trace metals are taken up from plasma *in vivo*. We have described an amino acid facilitated transport system, which depends on the formation of *bis*-histidine or, in some cases, *bis*-cysteine metal complexes. In the mature erythrocyte this system can function at near physiological amino acid concentrations (normal arterial histidine concentration is about 100 μM although concentrations may well be higher in portal blood) but its affinity is more suited to much higher amino acid concentrations.

However, in the HEL cells we have found that the K_m for metal *bis*-amino acid complexes is such that it would be quite feasible for these to play a role in metal uptake in the developing cell. The uptake rate per cell is also greater in the HEL cell than in the erythrocyte presumably because there are a greater number of transporters present in the membrane. In the HEL cell, unlike the mature human erythrocyte, we also found that cysteine stimulated metal uptake. This was also true of the rat erythrocyte. These differences may simply reflect the different ages of the various cells but are important because it has been suggested that at plasma concentrations a large proportion of amino acid-complexed metal is complexed with cysteine rather than histidine or is in the form of a metal(cys-his) complex (Giroux and Henkin 1972).

In both the HEL cell and the erythrocyte we have found much higher affinities for Mn-*bis*-histidine complexes than for those with other metals. Manganese has a lower affinity for histidine than the other trace metals and thus the plasma concentrations of *bis*-histidine complexes will be very low. It may be that the higher manganese affinity of the transporter has evolved to offset this or that the transporter is essentially a manganese transporter that can be persuaded to transport other ions.

The mature erythrocyte may have little requirement for trace metals for metabolic purposes but it does have the ability to take up and store significant quantities of metal. Much of this is bound to haemoglobin so that the intracellular metal concentrations remain low. This property may be important in buffering plasma metal concentrations over the relatively short term providing a reservoir to replenish body stores when they are depleted or, more importantly, as an acute system to "mop up" excessive plasma concentrations during periods of acutely raised concentrations. In this regard the red cell may be an important defence against acute toxicity.

Acknowledgement: Our own work described above was supported by the Pauline Thomas Memorial Trust and The Leverhulme Trust.

References

Agarwal RP, Kosman H, Henkin RI (1981) Control of zinc uptake in human erythrocytes. Fed Proc 40:856, #3555

Agte VV, Paknikar KM, Chiplonkar SA (1997) Effect of nicotinic acid on zinc and iron metabolism. Biometals 10:271-276

Aiken SP, Horn NM, Saunders NR (1992) L-histidine stimulates uptake of zinc by rat erythrocytes *in vitro*. J Physiol 445:69-80

Alda Torrubia JO, Garay R (1989) Evidence for a major route for zinc uptake in human red blood cells: $[Zn(HCO_3)_2Cl]^-$ influx through the $[Cl^-/HCO_3^-]$ anion exchanger. J Cell Physiol 138:316-322

Barnea A, Katz BM (1990) Uptake of ^{67}copper complexed to ^{3}H-histidine by brain hypothalamic slices: evidence that dissociation of the complex is not the only factor determining ^{67}copper uptake. J Inorg Biochem 40: 81-93

Bingham MJ, McArdle HJ (1994) A comparison of copper uptake by liver plasma membrane vesicles and uptake by isolated cultured rat hepatocytes. Hepatology 20:1024-1031

Brown AM, Simonsen LO (1985) Intracellular cobalt buffering by intact human red cells - evidence for time-dependent and metabolism-dependent changes. J Physiol 361:P26

Chau ACG, Stonnell LM, Savingi DL, Morgan EH (1996) Mechanisms of manganese transport in rabbit erythroid cells. J Physiol 493:99-112

Coogan TP, Squibb KS, Motz J, Kinney PL, Costa M (1990) Distribution of chromium within cells of the blood. Toxicol Appl Pharmacol 108:157-166

Darwish HM, Cheney JC, Schmitt RC, Ettinger MJ (1984) Mobilisation of copper(II) from plasma components and mechanism of hepatic copper transport. Am J Physiol 246:G72-G79

Engelmann B, Schumacher U, Duhm J (1991) Role of ABH blood group antigens in the stimulation of a DIDS-sensitive Ca^{2+} influx pathway in human erythrocytes by *Ulex europaeus* agglutinin I and a monoclonal anti-A_1 antibody. Biochim Biophys Acta 1091:261-269

Feray JC, Garay R (1987) A 1:1 Mg^{++}-Mn^{++} exchange in rat erythrocytes. J Biol Chem 262:5763-5768

Frederickson CJ, Moncrieff DW (1994) Zinc-Containing Neurones. Brain Signals 3:127-139

Garty M, Bracken WM, Klaassen CD (1986) Cadmium uptake by rat red-blood-cells. Toxicology 42:111-119

Giroux EL, Henkin RI (1972) Competition for zinc among serum albumin and amino acids. Biochim Biophys Acta 273:64-72

Gunshin H, Mackenzie B, Berger UV, Gunshin Y, Romero MF, Boron WF, Nussberger S, Gollan JH, Hediger MA (1997) Cloning and characterisation of a mammalian proton-coupled metal-ion transporter. Nature 388:482-488

Harris ED (2000) Cellular copper transport and metabolism. Annu Rev Nutr 20:291-310

Hartter DE, Barnea A (1988) Brain tissue accumulates ^{67}copper by two ligand-dependent saturable processes. J Biol Chem 263:799-805

Hilton M, Spenser DC, Ross P, Ramsey A, McArdle HJ (1995) Characterisation of the copper uptake mechanism and isolation of the ceruloplasmin receptor/copper transporter in human placental vesicles. Biochim Biophys Acta 1245:153-160

Horn NM, Thomas AL (1996) Interactions between the histidine stimulation of cadmium and zinc influx into human erythrocytes. J Physiol 496:711-718

Horn NM, Thomas AL (1997) The effect of copper on zinc uptake into human erythrocytes. J Physiol 499:26P

Horn NM, Thomas AL, Thompkins JD (1995) The effect of histidine and cysteine on zinc influx into rat and human erythrocytes. J Physiol 489:73-80

Horn NM, Oakley F, Thomas AL (1998) Histidine stimulated metal uptake into human erythrocytes. J Physiol 513:50P

Kalfalfakou V, Simons TJB (1990) Anionic mechanisms of zinc uptake across the human red cell membrane. J Physiol 421:485-497

Katz BM, Barnea A (1990) The ligand specificity for uptake of complexed copper-67 by brain hypothalamic tissue is a function of copper concentration and copper:ligand molar ratio. J Biol Chem 265:2017-2021

Kruckeberg WC, Brewer GJ (1978) The mechanism and control of human erythrocyte zinc uptake. Med Biol 56:5-10

Lou M, Garay R, Alda JD (1991) Cadmium uptake through the anion exchanger in human red blood cells. J Physiol 443:123-136

Lucacui CM, Dragu C, Copaescu L, Morariu VV (1997) Manganese transport through human erythrocyte membrane. An EPR study. Biochim Biophys Acta 1328:90-98

Magneson GR, Pulveringthal JM, Ray WJ (1987) The concentrations of free Mg^{++} and free Zn^{++} in equine blood plasma. J Biol Chem 262:11140-11148

Mandolfino M, Cimino G, Scuteri A, Cordi R, Romano L (1994) Some aspects of zinc transport across eel (*Anguilla anguilla*) red blood cell membranes. Cell Biol Int 18:279-288

Mas A, Sarkar B (1992) Uptake of ^{67}Cu by isolated human trophoblast cells. Biochim Biophys Acta 1135:123-128

McArdle HJ, Gross SM, Danks DM (1988) Uptake of copper by mouse hepatocytes. J Cell Physiol 136:373-378

Montozori M, Falchuk KH, Vallee BL (1995) Vitellogenin and lipovitellin: zinc proteins of *Xenopus laevis* oocytes. Biochemistry 34:10851-10858

Morgan EH (2001) Mechanism of iron transport into rat erythroid cells. J Cell Physiol 186:193-200

Mukhopadhyay S, Ghosh N, Chattopadhyay D, Mukhopadhyay S, Bhattacharya DK, Chatterjee GC (1989) Uptake of cadmium by isolated human red blood cells. Indian J Exp Biol 27:69-71

Naber THJ, Van der Hamer CJA, Van der Broek WJM, Roelofs H (1994) Zinc exchange by blood-cells in nearly physiological standard conditions. Biol Trace Elem Res 46:29-50

Nelson N (1999) Metal ion transporters and homeostasis. EMBO J 18:4361-4371

Nguyen QH, Chien PK (1989) Cadmium uptake kinetics in human erythrocytes. Biol Trace Elem Res 22:119-129

Oakley F, Horn NM, Hubbard A, Thomas AL (1998) The effects of heavy metals on histidine-stimulated ^{65}Zn uptake in human erythrocytes. J Physiol 511:38P

Ohno H, Doi K, Yamamura K, Yamashita S, Iuzuka S, Taniguchi N (1985) A study of zinc distribution in erythrocytes of normal humans. Blut 50:113-116

Paulsen IT, Saier MH (1997) A novel family of ubiquitous heavy metal ion transporters. J Membrane Biol 156:99-103

Peña MO, Lee J, Thiele DJ (1999) A delicate balance: Homeostatic control of copper uptake and distribution. J Nutr 129:1251-1260

Reyes JD (1996) Zinc transport in mammalian cells. Am J Physiol 270:C407-C410

Rifkind JM (1983) Interaction of zinc with erythrocytes. In: Sigel H, Sigel A (eds) Metal ions in biological systems. Marcel Dekker, New York, pp 275-315

Savigni DL, Morgan EH (1998) Transport mechanisms for iron and other transition metals in rat and rabbit erythroid cells. J Physiol 508:837-850

Schmitt RC, Darwish HM, Cheney JC, Ettinger MJ (1983) Copper transport by isolated rat hepatocytes. Am J Physiol 244:G183-G191

Simons TJB (1990) Intracellular free zinc and zinc buffering in human red blood cells. J Membrane Biol 123:63-71

Simons TJB (1991) Calcium dependant zinc efflux in human red blood cells. J Membrane Biol 123:73-82

Sivarama Sastry K, Viswanathan L, Ramaiah A, Sarma PS (1960) Studies on the binding of ^{65}Zn by equine erythrocytes *in vitro*. Biochem J 74:561-567

Van Wouwe JP, Veldhuizen M, Jeroen JM, Van Den Hamer CJA (1990) *In vitro* exchangeable erythrocytic zinc. Biol Trace Elem Res 25:57-69

Weed RI, Rothstein J (1960) The uptake of divalent manganese ion by mature normal human red blood cells. J Gen Physiol 44:301-313

Young JD, Jones SEM, Ellory JC (1981) Amino acid transport via the red cell anion transport system. Biochim Biophys Acta 645:157-160

18 Monocarboxylate and other Organic Anion Transport

Andrew P. Halestrap

Department of Biochemistry, School of Medical Sciences, University of Bristol, Bristol BS8 1TD, UK

18.1 Introduction

Organic anions comprise a wide range of compounds including naturally occurring metabolites such as taurine, mono-, di- and tricarboxylates, bile acids and oxidised glutathione, as well as numerous clinically important drugs. However, in purely quantitative terms, by far the most important organic anion is lactate, since this metabolite, or more accurately lactic acid, is the end product of glycolysis which produces two lactic acids for every glucose metabolised. It is clear that some mechanism must enable the efflux of lactic acid from the cell if it is not to accumulate and inhibit glycolytic ATP production, and it is the transporter for lactate and related monocarboxlates that will be the focus of the present chapter. However, red blood cells (RBCs) do possess at least two members of the ATP-dependent multispecific organic-anion transporter family that are capable of transporting glucuronides, mercapturates and oxidised glutathione (Heijn et al. 1992; Saxena and Henderson 1996). The reader is referred elsewhere for a review of this transporter family (Sekine et al. 2000). Organic anions can also be transported by swelling-activated osmolyte channels (see Chap. 7) and anion-like channels activated under conditions of low ionic strength (Culliford et al. 1995).

Unlike the situation in most cells, the absence of mitochondria in mammalian RBCs makes them totally dependent on glycolysis to supply the ATP required for the maintenance of ionic homeostasis and structural integrity. The availability of RBCs in large quantities, coupled with their stability, well characterised plasma membrane and limited metabolism, have made them the experimental system of choice for characterising the plasma membrane transport of lactate and other monocarboxylates. Initially it was thought that lactate might cross the plasma membrane purely by non-ionic diffusion of the free acid since it was known that monocarboxylates can penetrate biological membranes by this mechanism (Bakker and Van Dam 1974; Aubert and Motais 1975; Walter and Gutknecht 1984). Conclusive proof for the presence of a specific transport mechanism only came with the discovery in this laboratory that lactate and pyruvate transport into human RBCs could be strongly inhibited by aromatic C2 substituted monocarboxylates such as α-cyano-4-hydroxycinnamate (CHC) (Halestrap and Denton

1974). Subsequent work from several laboratories, but especially our own and that of Deuticke (see Deuticke 1982, 1989), have characterised this transport process in some detail and demonstrated that three pathways are involved.

18.2 Three Pathways for Monocarboxylate Transport across the Plasma Membrane of Red Cells

The presence of parallel, multiple pathways for the transport of L-lactate and pyruvate across the plasma membrane of human RBCs was first proposed as a result of kinetic and inhibitor studies in this laboratory (Halestrap 1976) and subsequently more detailed analysis using a range of inhibitors was performed in Deuticke's laboratory (see Halestrap 1976; Deuticke 1982, 1989). At "physiological" concentrations, about 90% of monocarboxylate transport occurs via a specific CHC-sensitive proton-linked monocarboxylate transporter (MCT) that is also powerfully inhibited by organomercurial thiol reagents such as *p*-chloromercuribenzene sulphonate (PCMBS) and mersalyl (Deuticke et al. 1978; Leeks and Halestrap 1979; Deuticke 1982). Monocarboxylate anion exchange is also mediated by the 'classical' inorganic anion exchanger (AE1 or band 3) that can be totally inhibited by the stilbene disulphonates such as 4,4'-diisothiocyanatostilbene-2,2'-disulphonic acid (DIDS) and 4,4'-dinitrostilbene-2,2'-disulphonic acid (DNDS). These reagents totally inhibit monocarboxylate/chloride exchange across the RBC membrane with the same affinity ($K_i \ll 1$ μM) as chloride/bicarbonate exchange (Halestrap 1976) whereas organomercurial reagents such as PCMBS, are without effect. Thus, band 3 provides a mercurial resistant DIDS-sensitive pathway for monocarboxylate transport, which in the human RBC represents about 5% of transport under physiological conditions. It should be noted that CHC inhibits both band 3 and MCT-mediated monocarboxylate transport. The remaining component of transport that is insensitive to CHC, DIDS and PCMBS, represents diffusion of the non-dissociated acid.

The characteristics of each of the three mechanisms of monocarboxylate transport into the RBC will be considered in turn. The greatest emphasis will be placed on the proton-linked monocarboxylate transporter since this is quantitatively the most important mechanism for the transport of physiologically important monocarboxylates such as lactate and pyruvate. However, it is important to realise that the relative contributions of the three transport mechanisms depends markedly on the monocarboxylate under investigation, its concentration, the pH (which affects the concentration of undissociated acid), temperature and the species from which the RBC was obtained (Deuticke 1982).

18.3 Non-Ionic Diffusion of Monocarboxylates

Many monocarboxylates can penetrate biological membranes by non-ionic diffusion of the undissociated acid (Bakker and Van Dam 1974; Aubert and Motais

1975; Deuticke et al. 1978; Walter and Gutknecht 1984). Whether this system contributes to the transfer of a given anion in a particular membrane could not be predicted from its chemical structure and had to be tested experimentally. Qualitative investigations identified acetate, propionate and substituted monocarboxylates of high lipid solubility, as being capable of permeating through a lipid phase, but lower pK values will reduce their apparent permeability at a given pH. Thus non-ionic diffusion of chloro- and fluoro-substituted acetates (pK_a < 3.0) and 2-chloropropionate (pK_a = 2.88) could not be detected, whilst 3-chloropropionate (pK_a = 4.0) can permeate in this manner. Non-ionic diffusion of physiologically relevant hydroxyl- and oxo-substituted fatty acids (e.g. lactate, pyruvate, 2-hydroxybutyrate, acetoacetate, glycerate, and glyoxylate) were difficult to investigate quantitatively. However it was clear that they could all penetrate the RBC membrane to some extent by ionic diffusion, although for pyruvate (pK_a = 2.5), lactate (pK_a = 3.86) and its C_2-homologue, glycolate, this process is very slow. Free diffusion was found to be non-saturable and to have the pH dependency predicted for the diffusion of the undissociated acid (whose concentration depends on both its pK_a and pH).

18.4 Band 3-Mediated Transport of Monocarboxylates

A wide range of mono- and divalent inorganic anions as well as dicarboxylates and certain monovalent organic anions including monocarboxylates are substrates for the anion exchanger, band 3 (see Chap. 11 and Deuticke (1980)). Although this transporter is only capable of mediating the exchange of one anion for another, net uptake of monocarboxylic acid is possible in the presence of bicarbonate. The monocarboxylate anion enters by exchange with an intracellular bicarbonate anion, which associates with a proton outside the cell to form carbonic acid. This is dehydrated to yield CO_2 which then diffuses back into the RBC where, catalysed by carbonic anhydrase, it is rehydrated to H_2CO_3 that dissociates to form HCO_3^- and a proton (Aubert and Motais 1975; Halestrap 1976). Band 3 was shown to mediate the transport of lactate, pyruvate, and halogenated acids in this manner, only when their concentrations were substantial (> 20 mM), since K_m values for such substrates are very high K_m (> 250 mM). However, the large amount of band 3 in the RBC membrane means that at elevated (non-physiological) concentrations of these monocarboxylates, transport rates by this pathway can be appreciable (Halestrap 1976). The structural requirements of monocarboxylates for acceptance by band 3 have not been well characterised, but there appear to be relatively few constraints (Halestrap 1976; Deuticke 1982). Certainly, the inorganic anion exchanger cannot distinguish between the D- and L-enantiomers of lactate (Deuticke 1977). Thus bovine and sheep RBCs that lack a PCMBS sensitive component to lactate exchange also lack stereo-selectivity for lactate transport (Deuticke 1982). Band 3-mediated monocarboxylate transport is inhibited by the wide range of established inhibitors of this anion exchanger many of which, such as phloretin, CHC and niflumic acid also inhibit the specific monocarboxylate transporter (Halestrap 1976; Deuticke 1989). Non-specific amino-reactive reagents (e.g. pyridoxal phosphate and fluorodinitrobenzene) and thiol reagents (e.g. maleimides) can also inhibit both transport pathways (Sachs et al. 1975; Deuticke 1977, 1982).

18.5 The Specific Monocarboxylate Transporter (MCT1)

18.5.1 Kinetics Properties

The specific monocarboxylate transporter of RBCs (now known as MCT1) was originally thought to catalyse only exchange of one carboxylate anion for another (Halestrap 1976). However, this erroneous conclusion was a consequence of the presence of high concentrations of intracellular L-lactate, which exchanged with the added lactate or pyruvate in preference to net uptake of the monocarboxylic acid (Leeks and Halestrap 1978). When intracellular lactate is removed, extracellular alkalinisation accompanies monocarboxylate uptake and there is a one to one stoichiometry between monocarboxylate and proton transport. This is reflected in the ability of a pH gradient to drive transport, with equilibrium being reached when $[\text{Lactate}]_{in}/[\text{Lactate}]_{out} = [H^+]_{out}/[H^+]_{in}$ (Dubinsky and Racker 1978; Leeks and Halestrap 1978). Detailed kinetic analyses of the carrier of human RBCs have been performed in this laboratory (Halestrap 1976; Leeks and Halestrap 1978) and those of other workers (Rice and Steck 1976, 1977; Anderson et al. 1978; Dubinsky and Racker 1978; Regen and Tarpley 1978; Deuticke 1982; De Bruijne et al. 1983, 1985). Four different experimental protocols have been used to obtain these kinetic data, and there is generally very good agreement in the results obtained. Zero *trans* influx data were obtained using the initial rate of transport of the substrate into a cell whose internal substrate concentration is zero. Zero *trans* efflux kinetics were determined using cells preloaded with substrate incubated with a medium without added substrate. Equilibrium exchange was measured by allowing cells to equilibrate with a known concentration of substrate before addition of radiolabelled substrate and following the uptake of tracer into the cell. Finally *trans*-stimulation kinetics were determined with cells preloaded with radiolabelled substrate and measuring the stimulation of substrate efflux following addition of an external substrate.

These studies have shown that in the human RBC, the K_m for L-lactate is between 4 (pH 6.5) and 13 mM (pH 7.5) for net influx or efflux, the value increasing as the pH is raised on the side the substrate is added. Increasing the pH on the *trans* side increases the V_{max} for transport, whilst having little effect on the K_m values (Dubinsky and Racker 1978; Regen and Tarpley 1978; Leeks and Halestrap 1978; Deuticke et al. 1982; De Bruijne et al. 1983). The K_m values appear to reflect the K_s (dissociation constant of the carrier-substrate complex) since the K_i values (dissociation constant of the carrier-inhibitor complex) for pyruvate and L-lactate as inhibitors of the transport of other monocarboxylates are the same as their K_m values as substrates (Halestrap 1976). Exchange K_m values appear to be somewhat higher than those for net flux, but this is compensated for by a substantial increase in V_{max} (Deuticke 1982). Such behaviour is a common feature of carrier kinetics and is usually taken to indicate that the conformational change involved in moving the substrate binding site from one side of the membrane to the other is slow in the absence of substrate (Deves and Krupka 1979; Deuticke 1989; Stein 1989).

Several schemes have been proposed to account for the observed kinetics of substrate transport (see Deuticke 1982, 1989). However, the most detailed kinetic studies involved the effects of pH and the competitive inhibitor glycolate on L-lactate transport (De Bruijne et al. 1983, 1985) and suggest the mechanism shown in Fig. 18.1. An ordered binding to the carrier of H^+ (K_m = 0.2 μM, equivalent to a pK of 6.7 for the accepting group) followed by lactate (K_m = 6.7 mM) precedes a conformational change, which translocates the substrate across the membrane. The lactate and then the proton are released and the carrier undergoes a conformational change to return the substrate binding site to the external surface. This latter step is rate limiting for substrate transport.

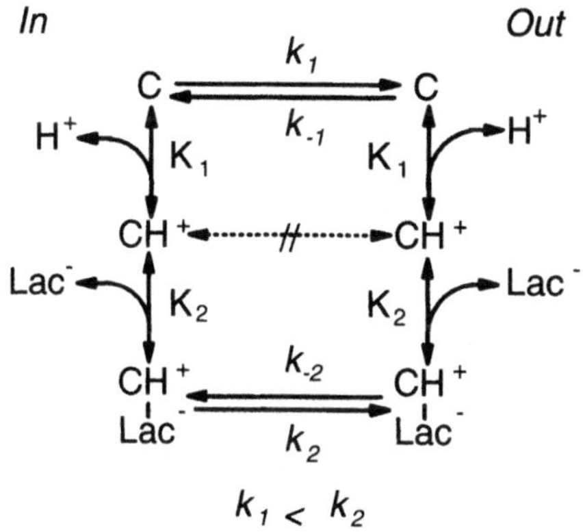

Fig. 18.1. Proposed kinetic scheme for the monocarboxylate carrier of the RBC derived from De Bruijne et al. (1983, 1985) as described in the text. *Lac*⁻ represents the lactate anion, whilst *k* and *K* represent rate and equilibrium constants, respectively

Rates of monocarboxylate transport into human RBCs are too fast to allow kinetics to be determined accurately at physiological temperatures using radiotracer techniques. However, activation energies of about 140 kJ/mol (< 15 °C) and 80 kJ/mol (> 15 °C) have been determined for the human RBC (Halestrap 1976; Dubinsky and Racker 1978; Leeks and Halestrap 1978; Deuticke 1982). This predicts a value for the maximal rate of lactate transport under physiological conditions of about 120 μmol/(ml_{cells}·h), which is considerably greater than the rate of lactate output by a RBC under similar conditions (1.5–2 μmol/(ml_{cells}·h)) (Deuticke 1982). Thus it seems unlikely that lactate efflux from the normal human RBC ever limits glycolysis, and in smaller animals such as rats, rabbits and guinea-pigs, the much higher activities of the monocarboxylate transporter make rate limitation even less likely (Deuticke et al. 1978; Poole and Halestrap 1988, 1992, 1993). It is now known that these differences in MCT activity reflect differences in the amount of transporter protein (MCT1) expressed (Poole and Halestrap 1992). Interestingly, cow and sheep RBC can survive without any specific monocarboxylate transporter and rely solely on diffusion and band 3 mediated lactic acid efflux. It is thus of interest to ask why some RBCs have such high transporter activity. Is the carrier merely an evolutionary remnant, or does it have some function? One possibility is that the carrier allows rapid equilibration of tissue lactic acid into the RBCs thus giving a significant increase in the effective blood volume for lactic acid removal from producing to utilising tissues (Deuticke 1982; Roth 1991). In this context it has been reported that an individual who appeared to have an impaired RBC lactate transporter experienced severe muscle fatigue after only slight exersion (Fishbein 1986), although this might also reflect a decrease in the expression of MCT1 in the muscle plasma membrane. Of course species differences may have no functional significance at all. Thus large species differences are also seen in the amount of glucose transporter but in this case the transporter is present in much higher concentrations in human RBCs than rat RBCs (Widdas 1988).

18.5.2 Substrate Specificity

Direct measurements of the kinetics monocarboxylate transport into RBCs have been made only with those substrates that are available in radiolabelled form, such as L- and D-lactate, pyruvate, β-hydroxybutyrate and glycolate (Halestrap 1976; Anderson et al. 1978; Deuticke et al. 1978). For a wider range of monocarboxylates less direct approaches, such as competitive *cis*-inhibition, or *trans*-stimulation of lactate transport, have been used to assess the specificity of the transporter (Halestrap 1976; Deuticke 1980; Poole and Halestrap 1990). *Trans*-stimulation of lactate transport by other substrates occurs because exchange is faster than net transport on the carrier (see above) and provides evidence that the substrate is actually transported, rather than merely being bound but not transported (inhibitory). In contrast, *cis*-inhibition demonstrates that the monocarboxylate binds at the same site as lactate and pyruvate, but gives no information on whether or not the substrate is transported. There is good agreement between different laboratories on the substrate specificity which is similar in human and rat RBCs, although the carrier in rat RBCs has a higher affinity for most substrates (Halestrap 1976; Deuticke 1980, 1989; Edlund and Halestrap 1986; Poole et al. 1990). Following confirmation that the protein responsible for monocarboxylate transport in RBCs is MCT1 (see below), we were able to make a more detailed characterisation of its substrate specificity in Ehrlich-Lettré tumour cells (a mouse breast tumour cell line) using the fluorescent intracellular pH-sensitive dye 2',7'-bis-(2-carboxyethyl)-5(6)-carboxyfluorescein (BCECF) (Carpenter and Halestrap 1994). This dye allowed the pH of the cytosol to be continuously monitored and thus the rate of acidification accompanying proton-linked monocarboxylate uptake could be determined. With appropriate calibration, initial rates of transport of any monocarboxylate could be determined, even at physiological temperatures. Data obtained with this technique are in good agreement with those obtained for rat RBCs using the radiotracer technique. Yet further confirmation of these data have been provided by expression of MCT1 in *Xenopus* oocytes and measurement of transport kinetics by radiotracer, pH electrode and fluorescence techniques (Bröer et al. 1998; Manning Fox et al. 2000).

Table 18.1 presents a summary of the K_m values of MCT1 for a wide range of monocarboxylates. These data and additional data from Deuticke's laboratory (Deuticke 1980, 1982, 1989) enable the following conclusions to be drawn about the substrate specificity of the monocarboxylate transporter present in RBCs:

- Unbranched aliphatic monocarboxylates from C-2 to C-5 are readily transported whilst formate is a very poor substrate and bicarbonate is not transported.
- Monocarboxylates substituted at C-2 with oxo, chloro, bromo, cyano, hydroxy, azido or thiol groups are transported as are those substituted at C-3, although with lower affinity. Oxo, halo, and cyano substitution usually enhances the affinity of the carrier for the substrate relative to the corresponding fatty acid, but amino- and amido-substitution at C-2 is not tolerated. Monocarboxylates dihalogenated at C-2 are good substrates, but the transporter accepts trihalogenation much less readily. For example, the transporter can transport trifluoroacetate (Xu and Kuchel 1993), but trichloroacetate and tribromoacetate are inhibitory (Deuticke 1982).
- The carrier is stereoselective for L- over D-lactate (> 10-fold) and D- over L-3-hydroxybutyrate (approx. 2-fold). This steroselectivity may apply only to hydroxy-substituted monocarboxylates since L- and D-2-chloropropionate are equally good substrates.
- Glyoxylate, unlike pyruvate and 2-oxobutyrate, is a very poor substrate for the transporter. One explanation for this would be if the transporter binds oxo-acids only when they are in their enol forms since glyoxylate is incapable of keto-enol tautomerisation, whereas for pyruvate about 1% is in the enol form (see Denton and Halestrap 1979; Halestrap et al. 1980). This explanation receives further support from the ability of α-cyanocinnamate derivatives, which are enol-pyruvate analogues, to inhibit transport (see below).
- Branching of the hydrocarbon chain at C-2, as in 2-hydroxy-2-methyl-propionate and 2-oxo-3-methyl-butyrate lowers or abolishes affinity for the transporter, although substrate size is not critical within certain limits, since acids bearing bulky substitutes (e.g. iodo- or dichloro-groups) have no lesser affinity than their unsubstituted congeners.
- Dicarboxylates, tricarboxylates and sulphonates are not transported.

These observations suggest that the substrate binding site of the transporter has the following features (Deuticke 1980, 1982; Poole and Halestrap 1993):

- A single positive charge responsible for binding carboxyl groups (dicarboxylates not accepted).
- A single hydrogen acceptor group, which allows binding of substrates that carry a single H-donor group such as –OH (dihydroxy- and amino-substitutions are not compatible).
- One or more hydrogen donor groups interacting with electronegative substituents of the substrate (carbonyl-, azido-, and cyano-substitution are compatible). Donor and acceptor groups are probably very close to the positive charge inter-

acting with the carboxylate group, since substituents in position 3 are much less readily accepted than at position 2.

- A hydrophobic region, able to bind apolar substrate domains that may be of crucial importance to conformational change and translocation.
- No adequate negatively charged binding site for a positively charged substituent (amino and *N*-substituted amino acids are not accepted).
- Steric constraints that limit binding and transport of branched chain acids and those > 5 carbons.

Table 18.1. K_m values of MCT1 for a range of monocarboxylates

	Substrate	Red cells K_i [μM]	Tumour cells K_m [μM]	*Xenopus* oocytes K_m [μM]
C1	Formate	>100	>100	
	Bicarbonate	>100		
C2	Acetate	3.5	3.7	
	Glyoxylate	70	63.3	
	Oxamate	17	49.2	
	Glycolate	6.6	14.6	
	Chloroacetate		0.7	
	Dichloroacetate		0.6	
C3	Propionate	1.5		
	Pyruvate	0.6	0.7	1.0
	L-Lactate	2.2	4.5	3.5
	D-Lactate	51	27.5	>60
	D-2-Cl-Propionate	0.9	0.7	
	L-2-Cl-Propionate	1.0	0.8	
	Fluoropyruvate		9.0	
C4	2-Oxobutyrate	0.2	0.2	
	3-Oxobutyrate	3.5	5.5	
	D/L-2-Hydroxybutyrate	1.3	2.6	
	D-3-Hydroxybutyrate		10.1	
	L-3-Hydroxybutyrate		11.4	
	4-Hydroxybutyrate	7.2	7.7	
	2-Hydroxy-2-methyl-propionate	>100	>100	
C5	2-Oxo-3-methyl-butyrate			1.25
C6	2-Oxo-4-methyl-pentanoate	1.3		0.7

For rat RBCs K_i values are taken from Poole and Halestrap (1993) and are used in place of K_m values as explained in the text. Additional data for the $K_{0.5}$ of a range of monocarboxylates acting as *trans*-stimulators of lactate tranpsort may be found in Deuticke (1989). K_m values for MCT1 in tumour cells are taken from Carpenter and Halestrap (1994) and for MCT1 expressed in *Xenopus* oocytes from Bröer et al. (1998).

18.5.3 Inhibitors

Inhibitors have played an indispensable part in studies of monocarboxylate transport, from the discovery of carrier-mediated transport (Halestrap and Denton 1974), to dissection of multiple transport pathways (Deuticke et al. 1982) and finally, identification of the carrier protein in RBCs (see below). As with investigations on substrate specificity, it has been possible to confirm and extend data obtained with RBCs by using Ehrlich-Lettré tumour cells (Carpenter and Halestrap 1994) and *Xenopus* oocytes expressing MCT1 (Bröer et al. 1998). Inhibitors of MCT1 can be grouped into two general classes: those, which bind reversibly and those which result in covalent modification of the transporter and hence irreversible inhibition.

Reversible Inhibitors

A summary of reversible inhibitors of the transporter and their efficacies is given in Table 18.2. They can be divided into three categories:

- Bulky or aromatic monocarboxylates, which as substrate analogues are competitive inhbitors of the carrier. These include 2-oxo-4-methylpentanoate and phenylpyruvate, which accumulate in maple syrup urine disease and phenylketonuria, respectively (Halestrap et al. 1974; Halestrap 1976). Derivatives of α-cyanocinnamate are the most potent aromatic monocarboxylate inhibitors (Halestrap and Denton 1974; Halestrap 1976; Poole and Halestrap 1990). The K_i value for inhibition of the transporter by α-cyano-4-hydroxycinnamate (CHC) is dependent upon the pH gradient (Leeks and Halestrap 1978). For RBCs incubated at pH 7.4 in citrate buffer (pH_i ~ 8.0), the K_i is 60–90 μM, whereas in saline medium (pH_i ~ 7.2), the K_i is 400–500 μM. The lower K_i caused by the pH gradient across the plasma membrane (alkaline-inside) does not reflect intracellular accumulation of CHC, and appears to be due to "recruitment" of the unloaded carrier to an exofacial conformation.
- A range of amphiphilic compounds of widely divergent structure act as potent inhibitors ($K_{0.5}$ ~ 1–10 μM). These include bioflavanoids such as quercetin and phloretin and anion transport inhibitors such as 5-nitro-2-(3-phenylpropylamino)-benzoate (NPPB) and niflumic acid; a more complete list is given by Deuticke (1989). However, these compounds are not specific for MCT1 and also inhibit band 3 and other membrane transport processes (Soulinna et al. 1975; Wheeler and Hinkle 1981; Deuticke 1982). Isobutylmethyl xanthine (IBMX: $K_{0.5}$ = 0.3 mM) is perhaps the most specific of this class of inhibitor, since it inhibits the monocarboxylate transporter under conditions where band 3 is unaffected (Anderson et al. 1978; Deuticke 1982).
- Some 4,4'-substituted stilbene-2,2'-disulphonates act as reversible inhibitors of MCT1 in RBCs, although they do so with much lower affinity than they inhibit band 3 mediated transport (Poole and Halestrap 1991; Poole et al. 1991). Indeed 4,4'-diisothiocyanatostilbene-2,2'-disulphonate (DIDS), 4,4'-dibenzamidostilbene-2,2'-disulphonate (DBDS) and *N,N,N',N'*-tetrabenzyl-4,4'-diaminostilbene-2,2'-disulphonate (TbenzDS) are some of the most effective inhibitors

of the monocarboxylate transporter with K_i values of 40, 22 and 2 μM, respectively (Poole et al. 1991). They cause instantaneous inhibition by binding reversibly to an exofacial site on the transporter, in competition with monocarboxylate substrates. In the case of TbenzDS, it seems probable that the high affinity of the inhibitor reflects its extreme hydrophobicity. This causes it to partition into the membrane, producing a high local concentration in the vicinity of the substrate binding site (Poole et al. 1991). Other stilbene disulphonates, such as those substituted with nitro, amino or azido groups, are much less effective (Poole et al. 1991). Inhibition by the isothiocyanate derivatives DIDS and its dihydo-derivative (H_2DIDS) eventually becomes irreversible on prolonged incubation (30 min at 37 °C), reflecting covalent modification of the transporter (Poole and Halestrap 1991).

Irreversible Inhibitors

MCT1 can be inhibited by a range of thiol and amino reagents. Transport is particularly sensitive to inhibition by organomercurial thiol reagents such as PCMBS and mersalyl (Deuticke et al. 1978; Leeks and Halestrap 1978; Carpenter and Halestrap 1994; Bröer et al. 1998). Inhibition occurs rapidly and at micromolar concentrations, although the exact concentration dependence varies with the experimental system (species of RBC, concentration of cells etc). This may reflect the binding of the reagent to a number of extracellular thiols of which those on MCT1 are only a fraction (Deuticke et al. 1978). Lactate transport in human RBCs is inhibited by other thiol reagents such as dithiols including 5,5'-dithiobis-(2-nitrobenzoic acid) (DTNB), but only poorly by maleimides (Halestrap 1976; Deuticke et al. 1978; Deuticke 1986). Of the maleimides, the hydophobic *N*-phenylmaleimide is more effective than *N*-ethylmaleimide (Halestrap and Poole 1989).

A wide variety of treatments that are known to modify amino-groups cause irreversible inhibition of the monocarboxylate transporter. These include treatment of RBCs with fluorodinitrobenzene (FDNB), esters of *N*-hydroxysulphosuccinimide, pyridoxal phosphate/borohydride, fluorescein isothiocyanate, isobutyllactyl anhydride, and reductive methylation with formaldehyde/borohydride (Halestrap 1976; Deuticke et al. 1978; Donovan and Jennings 1986). As noted above, the stilbene disulphonates DIDS and H_2DIDS, which possess amino-reactive isothiocyanate groups, are also capable of causing irreversible inhibition of the lactate transporter (Jennings and Adams-Lackey 1982; Poole and Halestrap 1991). These inhibitors first bind reversibly to the carrier (see above) before a group accessible from within this site (presumably a lysine residue) is modified upon prolonged incubation (Poole and Halestrap 1991). Since some of the amino-reactive compounds mentioned above are essentially membrane impermeant (e.g. pyridoxal phosphate, DIDS, H_2DIDS), an exofacial lysine is implicated as the target for these treatments. This residue, when protonated, may be involved in the binding of the monocarboxylate anion. There is little information on the effects of modification of other types of amino acid side chain on monocarboxylate transport. However, the arginine-specific reagent phenylglyoxal causes partial inhibition of the transporter (Donovan and Jennings 1986; Poole and Halestrap 1993).

Table 18.2. Reversible inhibitors of MCT1

Inhibitor	Red cells $K_{0.5}$ [μM]	Tumour cells $K_{0.5}$ [μM]
α-Cyanocinnamate analogues		
α-Cyanocinnamate	400	1700
α-Cyano-4-hydroxycinnamate	27	166
α-Fluorocinnamate	660	
α-Cyano-β-(1-phenylindol-3-yl)-acrylate (UK5099)	20	8.1
Stilbene disulphonates*		
DIDS	53	434
H_2-DIDS	400	
SITS	132	1180
DBDS	36	215
DNDS	1120	>3000
TBenzDS	1.5	
pNO_2DBDS	9.8	
Monocarboxylates		
Phenylpyruvate	0.6	
2-Oxo-4-methyl-pentanoate (α-ketoisocaproate)	1.3	
Other Inhibitors		
Phloretin	1.4	5.1
Quercetin	2.0	
NPPB		9.0
Niflumic acid	45	6.1
IBMX	200	288

For rat RBCs K_i values are taken from Poole and Halestrap (1993) and those for tumour cells from Carpenter and Halestrap (1994). *For stilbene disulphonates, abbreviations are given in the text.

18.6 Identification, Cloning and Sequencing of the Red Cell Monocarboxylate Carrier

Identification of the protein responsible for monocarboxylate transport into RBCs was a difficult task, largely because of the absence of specific high affinity inhibitors of the process. Rat, rabbit and guinea-pig RBCs were used for most of this work, since they have a far higher capacity for monocarboxylate transport than human RBCs (Deuticke et al. 1978; Poole and Halestrap 1993). Although several groups used a variety of covalent labeling techniques in their attempts to identify the transporter (Jennings and Adams-Lackey 1982; Donovan and Jennings 1985, 1986; Deuticke 1989), unequivocal identification was ultimately achieved in this laboratory using protection from covalent labeling of the protein with DIDS by specific inhibitors of the transporter such as CHC. Separation of the labeled protein by SDS-PAGE showed it to be about 45 kDa (Poole and Halestrap 1992). In other experiments, the unlabeled protein was partially purified following detergent solubilisation and transport activity measured after reconstitution into proteoliposomes (Poole and Halestrap 1988, 1992). Active fractions were enriched in a 45 kDa protein, and when membrane proteins from DIDS-treated RBCs were taken through the same purification procedure, this protein was found to be DIDS-labeled (Poole and Halestrap 1994). N-terminal sequencing showed it to be identical to a putative transporter of unknown function that had been previously cloned by Goldstein and Brown (Kim et al. 1992). These workers had identified a mutation of the wild-type protein that enhanced mevalonate uptake into Chinese Hamster Ovary (CHO) cells. In subsequent expression studies, they demonstrated that this protein catalysed inhibitor-sensitive monocarboxylate transport and named it MCT1 (Kim-Garcia et al. 1994a). Expression of MCT1 in *Xenopus laevis* oocytes has subsequently confirmed that the properties of MCT1 are the same as the monocarboxylate transporter of RBCs (Bröer et al. 1997, 1998). MCT1 from human, rat and mouse and have now been cloned and share about 95% sequence identity with the CHO MCT1 (Kim-Garcia et al. 1994b; Jackson et al. 1995; Takanaga et al. 1995; Carpenter et al. 1996; Koehler Stec et al. 1998). The protein is expressed in most tissues as might be expected for a transporter that is required for glycolytic function (Kim-Garcia et al. 1994a; Price et al. 1998; Halestrap and Price 1999).

Human MCT1 has been mapped to chromosome band 1p13.2-p12 (Kim-Garcia et al. 1994b). Interrogation of the human genome database has enabled the genomic structure of MCT1 to be predicted and this is shown in Fig. 18.2. Inspection of the EST database shows that two mRNA transcripts are produced, one lacking the upstream exon in the 5'UTR, implying the presence of two alternative promoters. A similar situation occurs with MCT3, another member of the MCT family (see below) found exclusively in the retinal pigment epithelium, where promoter use is developmentally regulated (Yoon and Philp 1998). There is no evidence for alternative splicing in the 3'UTR region although it has been proposed that its large size (1.6 kb) may play a role in post-transcriptional regulation of expression in some tissues (Jackson et al. 1997; Halestrap and Price 1999).

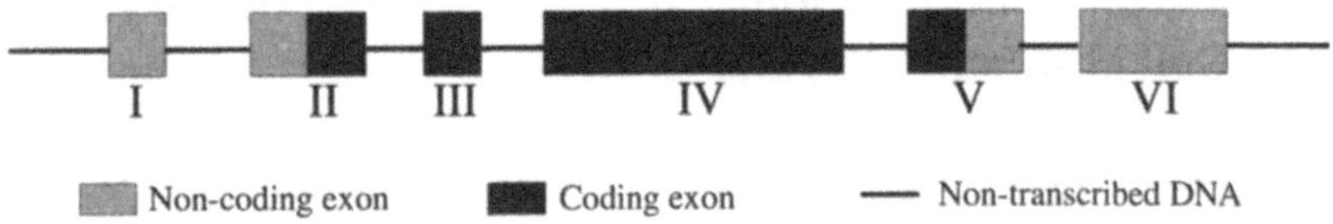

Fig. 18.2. The proposed gene structure of MCT1 derived from the MCT1 genomic sequence

18.7 MCT1 is a Member of a Family of Proton-Linked Monocarboxylate Transporters

Characterisation of the substrate and inhibitor specificity of monocarboxylate transport into a range of mammalian cells, including liver, heart and skeletal muscle, suggested that a family of MCT isoforms with distinct properties must exist. This has now been confirmed in this and other laboratories by the cloning and sequencing of nine mammalian members of the MCT family (see Price et al. 1998; Halestrap and Price 1999). Each isoform has a characteristic tissue distribution at both the mRNA and protein level suggesting distinctive roles, but only MCT1 appears to be expressed in RBCs. Of particular interest, MCT4 represents the major isoform of white muscle fibres (glycolytic), whereas MCT1 predominates in heart and red muscle fibres (oxidative) and is the major isoform in most tissues (Wilson et al. 1998; Halestrap and Price 1999). In contrast MCT2 is expressed in fewer tissues, but is found in those requiring high affinity uptake of lactate or pyruvate such as liver and kidney (both gluconeogenic), or sperm tails and neuronal synapses (lactate and pyruvate oxidation) (Halestrap and Price 1999). Recent studies both here and in Bröer's laboratory have enabled detailed characterisation of MCT1, MCT2 and MCT4 expressed in *Xenopus* oocytes (Bröer et al. 1998, 1999; Dimmer et al. 2000; Manning Fox et al. 2000). K_m values of these isoforms for L-lactate are 5 mM, 0.7 mM and 28 mM, respectively, and for pyruvate the values are 1 mM, 0.1 mM and 150 mM. These values are consistent with the proposed roles of the different isoforms suggested by their distinctive patterns of expression (Bröer et al. 1998, 1999; Price et al. 1998; Dimmer et al. 2000; Halestrap and Price 1999; Manning Fox et al. 2000). The high K_m of MCT4 for L-lactate and pyruvate is especially intriguing and provides an explanation for why lactic acid accumulates in muscle during exercise and leads to fatigue as the intracellular pH drops. This probably represents an important physiological constraint whereby the fatigue prevents exercise continuing to the point at which lactic acid would overload the buffering capacity of the blood and cause damaging systemic acidosis (Juel and Halestrap 1999; Manning Fox et al. 2000). MCT3 has been expressed in a thyroid epithelial cell line and yeast and shown to transport lactate and pyruvate, but detailed characterisation has not been reported (Yoon et al. 1997; Grollman et al. 2000).

To date, we have not been successful in measuring increased transport of any monocarboxylate into *Xenopus* oocytes transfected with MCT5, MCT6 or MCT8. It remains to be established whether this reflects a substrate specificity that excludes the wide range of monocarboxylates we have tested or the lack of an appropriate ancillary protein to chaperone the MCT to the plasma membrane (see below). However, a recent report has shown that a newly identified member of the MCT family (TAT1) catalyses the Na^+-independent transport of aromatic amino acids such as tryptophan, tyrosine and phenylalanine with low-affinity (K_m values: ~5 mM) (Kim et al. 2001). Genomic sequence analysis has also allowed the identification of MCT family members in *Drosophila melanogaster* (6 members), *C. elegans* (6 members) and *S. cerevisiae* (4 members) although none of these has been functionally characterised (Halestrap and Price 1999). In fact in yeast it has been established that none of the MCT family members are required for lactate uptake which is mediated by another transporter, JEN1 (Casal et al. 1999; Halestrap and Price 1999).

18.8 Topology of MCT1 and Other Members of the MCT Family

Human MCT1 is a 499-residue protein which, like many other transporters, is predicted to have 12 transmembrane domains (TMDs), with cytosolic C- and N-termini of 55 and 20 residues, respectively, and a large cytosolic loop (about 60 residues) between TMDs 6 and 7. This predicted topology, which is illustrated in Fig. 18.3, is shared by all members of the MCT family and is characteristic of many membrane transporters including the RBC glucose transporter GLUT1 (Gould and Holman 1993; Henderson 1993). However, unlike GLUT1, there are no potential sites for *N*-linked glycosylation in human MCT1. Even in Chinese hamster MCT1, where a glycosylation site is present, it is not used, probably because the loop within which it is contained (between TMD 1 and TMD 2) is too short (Carpenter et al. 1996). We have used specific proteolytic cleavage of leaky ghosts and intact RBCs in conjunction with antibodies raised to peptides located in different regions of the protein to confirm that the actual structure of MCT1 in the RBC membrane matches the predicted topology (Poole et al. 1996). Thus both the N- and C-terminal sequence and the loop between TMD 6 and TMD 7 could not be cleaved by proteases in intact RBCs, but could be cleaved in leaky ghosts. The only external loop that could be cleaved by proteases added to intact RBCs was that between TMD 11 and TMD 12. The greatest sequence conservation between members of the MCT family is found in the putative transmembrane regions, and the shorter loop regions between them. In contrast the hydrophilic regions of the sequences (the N-terminal residues preceding TMD 1, the loop region between

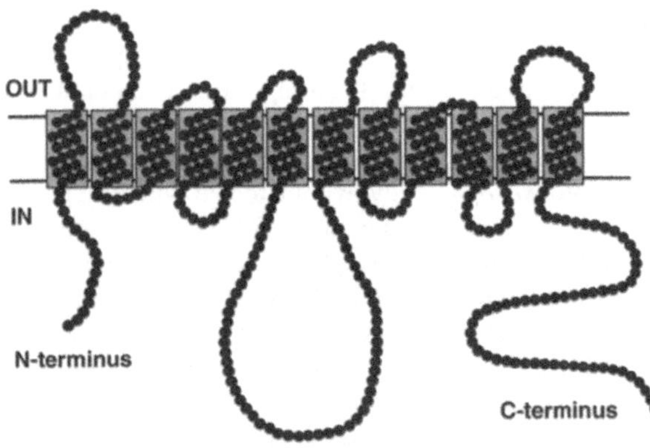

Fig. 18.3. Proposed structure of MCT1 in the RBC membrane derived from topology predictions and proteolytic cleavage studies (Poole et al. 1996)

TMDs 6 and 7, and the C-terminal residues succeeding TMD 12) show little conservation (Halestrap and Price 1999). Such divergent hydrophilic regions are a common feature of 12 TMD transporter family sequences (Saier 1994), and it is unlikely that these regions are directly involved in transport. Rather they may be critical for other aspects of function such as substrate specificity or regulation of transport activity.

18.9 Structure Function Relationships in MCT1

Three approaches have been used to investigate what residues are critical for MCT1 function: chemical modification of specific amino acid side chains, sequence comparison between members of the MCT family and site-directed mutagenesis. As already noted above, reagents that attack cysteine residues (e.g. PCMBS), lysine residues (e.g. DIDS) and arginine residues (phenylglyoxal) can all inhibit transport by MCT1. Since MCT4 is also inhibited by this reagent (Dimmer et al. 2000; Manning Fox et al. 2000) but MCT2 is not (Garcia et al. 1995; Bröer et al. 1999), comparison of the sequences of these isoforms might reveal the cysteine residue that PCMBS attacks. Furthermore, both DIDS and PCMBS are impermeant and thus must react with extrafacial cysteine and lysine groups of MCT1. There is a cysteine present in the extracellular loop between TMDs 3 and 4 of both MCT4 (Cys_{108}) and MCT1 (Cys_{106}) that is lacking in MCT2 (Ser_{106}). However, when this cysteine was mutated to alanine in either MCT1 or MCT4 sensitivity to PCMBS was maintained (Meredith, Manning Fox and Halestrap, unpublished data). Similarly, inhibition of MCT1 by stilbene disulphonates such as DIDS is believed to involve binding to extracellular lysine groups close to the substrate binding site (Halestrap and Price 1999). Since stilbene disulphonates also inhibit MCT2 but not MCT4, it seems probable that DIDS may lack an extracellular lysine residue present in MCT1 and MCT2 but not MCT4. The obvious candidate is Lys_{413} in MCT1 and Lys_{404} in MCT2, which is replaced by a methionine (Met_{385}) in MCT4. This conclusion is further strengthened by our demonstration that changing Lys_{413} to a glutamine in MCT1 prevents covalent modification and irreversible inhibition of MCT1 by DIDS (Meredith et al. 1999).

Sequence comparisons between the members of the MCT family highlight some common features that may be important in relating structure to function (see Halestrap and Price 1999). In TMDs 1–6 there are fifteen residues that are absolutely conserved across all 9 mammalian sequence, and many more positions where only conservative substitutions occur. Two highly conserved regions particularly stand out: the sequence with consensus **[D/E]G[G/S][W/F][G/A]W** which traverses the lead into TMD 1, and the consensus **YF**x**K[R/K][R/L]**x**L**A**x[G/A]**x**A**x**AG** which leads into TMD 5 (residues shown in bold are conserved in all of the sequences, residues in square parentheses indicate alternative amino acids, residues that are in normal type are the consensus amino acid at that position and x represents any amino acid). Even when the non-mammalian MCT family members are included there is still conservation of these sequences motifs. It has been proposed that in twelve TMD transporters, the N-terminal domains are more important for energy coupling (via H^+ or Na^+ translocation), membrane insertion, and/or correct structure maintenance, whereas the C-terminal domains may be more important for the determination of substrate specificity (Saier 1994). Inhibitor and photo-affinity labelling studies on sugar transporters (Gould and Holman 1993) and the characteristics of chimeras constructed with regions from different glucose transporter isoforms (Arbuckle et al. 1996) are consistent with this hypothesis. Since MCTs are monocarboxylate proton co-transporters, it might be predicted that the amino acid residues responsible for proton binding and translocation are in TMDs 1–6. A possible candidate residue for this role would be the only conserved aspartate/glutamate in the N-terminal half of the protein, which is at the start of TMD 1 (Asp_{15} of human MCT1). However, site-directed mutagenesis to change this residue to an asparagine was without effect on transport activity. In contrast even the conservative change of Asp_{302} to Glu in TMD 8 of rat MCT1 led to total loss of lactate transport activity (Rahman et al. 1999).

There is also evidence for the C-terminal half of MCT1 being involved in substrate specificity. Thus conversion of Phe_{360} in TMD 10 of Chinese Hamster MCT1 to Cys enables MCT1 to transport mevalonate whilst reducing its ability to transport a lactate and pyruvate (Kim-Garcia et al. 1994a; Rahman et al. 1999). Furthermore, we have shown that the binding site of MCT1 for DIDS, which acts at or near the substrate binding site (Poole and Halestrap 1992), is in the C-terminal half of the transporter (Poole et al. 1996; Wilson et al. 1999). A positively charged group that binds the COO^- anion is a feature that is likely to be present in all MCTs. In MCT1 arginine probably fulfils this role as it does in lactate dehydrogenase (Hart et al. 1987), since the arginine-specific reagent phenyglyoxal inhibits MCT1-mediated lactate transport (Donovan and Jennings 1986). An arginine in TMD 8 (Arg_{313} of human MCT1) is conserved in all the putative MCTs from higher eukaryotes, except MCT5 and also in all but one of the *S. cerevisiae* and *C. elegans* sequences (Halestrap and Price 1999). Site-directed mutagenesis of this residue to a threoneine greatly reduces the affinity of MCT1 for lactate with transport of 0.1 mM L-lactate being < 3% of control values and not saturating even at 50 mM (Rahman et al. 1999). Another R/K residue that is totally conserved in all sequences is located within the relatively highly conserved region **YF**x**K**[**R/K**][**R/L**]x**L**A**x**[**G/A**] x**A**x**AG** between TMDs 4 and 5 mentioned above. However, since the N-terminal half of the molecule is not thought to be involved in substrate binding, this residue is unlikely to be involved in binding the carboxylate group.

18.10 MCT1 is Tightly Associated with the Cell Surface Glycoproteins GP70 (Embigin) and CD147 (Basigin)

Our endeavours to purify MCT from RBCs were never totally successful because two proteins of 43 and 70 kDa co-purified. When attempts were made to separate these two proteins, transport activity was lost (Poole and Halestrap 1992). Labeling of the transporter with DIDS revealed a cross-linked product that ran at about

130 kDa on SDS-PAGE (Poole and Halestrap 1997). This cross-linked product was purified and antibodies raised against it which, when used to develop Western blots of untreated rat RBC membranes revealed two bands of 43 kDa and 70 kDa. An additional band of 130 kDa appeared in membranes from RBCs pretreated with DIDS. These bands corresponded to MCT1, a glycoprotein of 70 kDa and the cross-linked product. The 70 kDa protein was purified and shown to contain *N*-linked carbohydrate, the apparent core molecular mass being about 40 kDa. Amino acid sequencing showed the protein to be the rat equivalent of the membrane spanning mouse teratocarcinoma glycoprotein GP-70, also known as Embigin. This protein is a member of a small family of closely related proteins of the immunoglobulin superfamily that includes CD147 (also known as OX-47, EMMPRIN, HT7 and basigin) and SDR1 (Ozawa et al. 1988; Miyauchi et al. 1995). These are all predicted to have common structural features including immunoglobulin-like domains in the extracellular region, a short C-terminal cytoplasmic tail, and a single transmembrane domain (Fossum et al. 1991). CD147 is found in most cells (Paterson et al. 1987; Kasinrerk et al. 1992; Seulberger et al. 1992; Biswas et al. 1995), but, as shown in Fig. 18.4, is almost totally absent in blood cells from most species, except rabbits, where it was first identified by Jennings and colleagues as a protein reactive against the "4D4" monoclonal antibody that was raised during studies on the monocarboxylate transporter (Donovan and Jennings 1985). Indeed it was originally thought that this protein might be the transporter, but later studies showed that carrier activity could be separated from the 4D4 antigen (Poole and Halestrap 1993). Thus it is possible that in most cells it is CD147 that fulfils the same function as GP-70 does in rat RBCs. Indeed, we have found that CD147 is always expressed in cells expressing MCT1 and in the same location (Kirk et al. 2000; Zhao et al. 2001).

This led us to consider whether GP70 and CD147 might be required for the proper expression of MCTs, since several other transporters, channels and recep-

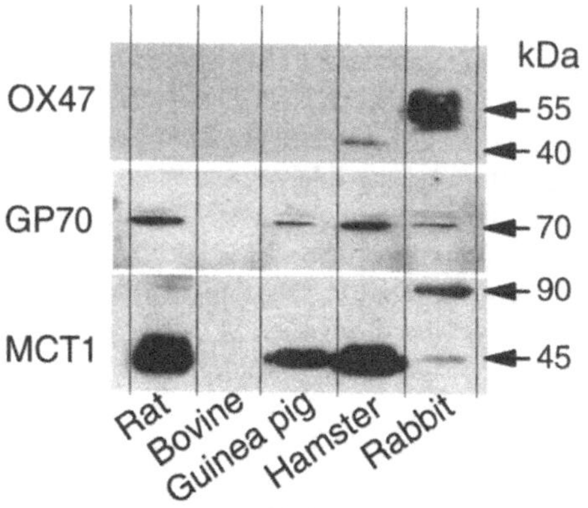

Fig. 18.4. The presence of *MCT1*, CD147 (*OX47*) and *GP70* in the plasma membrane of RBCs from various species as revealed by Western blotting with specific antibodies. Note that for rabbit a significant proportion of MCT1 runs as a 90 kDa dimer. (Data are previously unpublished results of Wilson and Halestrap)

tors require both monotopic and polytopic components for expression and function, including amino acid transporters (Kanai et al. 1998; Mastro Berardino et al. 1998; Pfeiffer et al. 1999; Torrents et al. 1999), band 3 (Bruce et al. 1994), fatty acid transporters (Abumrad et al. 1998), the TrkB receptor (Kafitz et al. 1999) and the calcitonin-like receptor (McLatchie et al. 1998). In collaboration with Neil Barclay's laboratory (Kirk et al. 2000) we demonstrated that MCT1 and MCT4 co-immunoprecipitated with CD147 in a variety of cultured cells lines, whilst MCT2 did not. Furthermore, only when CD147 was co-expressed with either MCT1 or MCT4 did both proteins reach the plasma membrane and lead to an increase in lactate transport activity. Without co-expression of CD147, MCT1 and MCT4 accumulated in the perinuclear region, probably in the Golgi apparatus and endoplasmic reticulum. MCT1 can be expressed in *Xenopus* oocytes without co-transfection with CD147, because these cells synthesise an endogenous CD147 homologue (Kirk et al. 2000). We have confirmed this by co-injection of *Xenopus* oocytes with MCT1 cRNA and an antisense cRNA to a *Xenopus* homologue of CD147. This reduced expression of MCT1 activity, but expression could be restored by co-injection of GP70 cRNA (Meredith and Halestrap 2000).

These data imply that CD147 or GP70 act as a chaperone for MCT1 and MCT4 translocation to the plasma membrane. Once at the plasma membrane, both proteins appear to remain closely associated as shown by the DIDS cross-linking experiments. We have confirmed this in cultured cell lines by using cross-linking antibodies against CD147 to redistribute the CD147 to one end of the cell as a cap. MCT1 and MCT4 were found to redistribute into the caps with CD147, whilst GLUT1 and MCT2 did not (Kirk et al. 2000). These data, like the immunoprecipitation data, imply that MCT2 does not associate with CD147, and this was confirmed by the inability to target MCT2 to the plasma membrane when co-expressed with CD147 (Kirk et al. 2000). The relatively poor expression of MCT2 in *Xenopus* oocytes when compared to MCT1 and MCT4 is also consistent with this observation. Thus MCT2 (and indeed the other MCT isoforms) probably require alternative ancillary proteins.

It appears to be the transmembrane (TM) domain of CD147 that is important for its interaction with MCT1 since CD2/CD147 chimeras only facilitate MCT1 expression if they contain the TM domain of CD147 rather than that of CD2 (Kirk et al. 2000). A shared feature of the TM region of members of the CD147 family is a centrally-positioned Glu (Fossum et al. 1991) which is an unusual feature for a protein with a single TM domain (Green 1991). Other single-pass proteins with a TM charge include FcεRIα, FcγRIII (CD16) and the natural killer cell 'Killer Activatory Receptors', each of which engages in a functionally important lateral association with another membrane protein (Lanier et al. 1989, 1998).

18.11 Monocarboxylate Transport into Malaria Infected Red Cells

When a RBC becomes infected with a malarial parasite, the metabolic turnover of the cell increases dramatically as the parasite grows and reproduces (Sherman

1979; Van der Jagt et al. 1990). This requires a large increase in the passage of nutrients and waste products across the RBC plasma membrane, and there is considerable evidence that the permeability properties of the membrane are substantially altered (see Chap. 24). Since the parasite is almost entirely glycolytic, the requirement for glucose and the production of L-lactate is particularly increased and may be some 40–100 times greater than in a non-infected cell (Sherman 1979; Pfaller et al. 1982; Van der Jagt et al. 1990). Thus rates of lactic acid efflux can reach values as high as 200 μmol/(ml_{cells}·h). However the maximal activity of the lactate carrier in control human RBCs is only 120 μmol/(ml_{cells}·h) at 37 °C, which is inadequate to cope with the increased rates of lactate transport required. Thus it would seem probable that a new transport pathway must be induced, and work in this laboratory (Cranmer et al. 1995) and that of Ginsburg (Kanaani and Ginsburg 1991) have confirmed this. The rate of L-lactate, D-lactate and pyruvate transport are all greatly increased, the effect being most pronounced for D-lactate which is a very poor substrate for the monocarboxylate carrier of the non-infected cells. The stimulated uptake of lactate into the infected RBC is inhibited by α-cyano-4-hydroxycinnamate and DBDS and represents both enhanced transport across the plasma membrane of the RBC and the parasite (Cranmer et al. 1995). Recent work from Kiaran Kirk's laboratory using saponin-derived parasites (Elliott et al. 2001) has confirmed that the latter transport mechanism is proton-linked transport as was implied by our earlier data (Cranmer et al. 1995).

18.12 Conclusions

Although there are three mechanisms by which monocarboxylates such as lactate may enter the RBC, it is the specific proton-linked monocarboxylate transporter MCT1 that predominates. The especially large quantities of MCT1 found in the plasma membrane of rodent and rabbit RBCs provided an ideal starting material for the identification and purification of MCT1 and ultimately led on to the identification of a new family of monocarboxylate transporters with distinct properties, tissue distribution and regulation. Thus, as has been the case with other transporters, studies on the RBC have provided the key to unlock a whole new area of transporter research.

Acknowledgement: Work in the Author's laboratory has been generously funded by the Medical Research Council, The Wellcome Trust and The British Heart Foundation. The enormous contribution of past and present members of this laboratory are gratefully acknowledged. Particular thanks go to Dr. Robert Poole whose exceptional skill and perseverance led, amongst many other achievements, to the identification and purification of MCT1 and its ancillary protein, Embigin, from the plasma membrane of RBCs.

References

Abumrad N, Harmon C, Ibrahimi A (1998) Membrane transport of long-chain fatty acids: evidence for a facilitated process. J Lipid Res 39:2309-2318

Anderson BL, Tarpley HL, Regen DM (1978) Characterisation of β-hydroxybutyrate transport in rat erythrocytes and thymocytes. Biochim Biophys Acta 508:525-538

Arbuckle MI, Kane S, Porter LM, Seatter MJ, Gould GW (1996) Structure-function analysis of liver-type (GLUT2) and brain-type (GLUT3) glucose transporters: Expression of chimeric transporters in *Xenopus* oocytes suggests an important role for putative transmembrane helix 7 in determining substrate selectivity. Biochemistry 35:16519-16527

Aubert L, Motais R (1975) Molecular features of organic anion permeabiliity in ox red blood cells. J Physiol 246:159-179

Bakker EP, Van Dam K (1974) The movement of monocarboxylic acids across phospholipid membranes: evidence for an exchange diffusion between pyruvate and other monocarboxylate ions. Biochim Biophys Acta 339:285-289

Biswas C, Zhang Y, DeCastro R, Guo H, Nakamura T, Kataoka H, Nabeshima K (1995) The human tumor cell-derived collagenase stimulatory factor (renamed EMMPRIN) is a member of the immunoglobulin superfamily. Cancer Res 55:434-439

Bröer S, Rahman B, Pellegri G, Pellerin L, Martin JL, Verleysdonk S, Hamprecht B, Magistretti PJ (1997) Comparison of lactate transport in astroglial cells and monocarboxylate transporter 1 (MCT 1) expressing *Xenopus laevis* oocytes - Expression of two different monocarboxylate transporters in astroglial cells and neurons. J Biol Chem 272:30096-30102

Bröer S, Schneider HP, Bröer A, Rahman B, Hamprecht B, Deitmer JW (1998) Characterization of the monocarboxylate transporter 1 expressed in *Xenopus laevis* oocytes by changes in cytosolic pH. Biochem J 333:167-174

Bröer S, Bröer A, Schneider H-P, Stegen C, Halestrap AP, Deitmer JW (1999) Characterisation of the high-affinity monocarboxylate transporter MCT2 in *Xenopus laevis* oocytes. Biochem J 341:529-535

Bruce LJ, Groves JD, Okubo Y, Thilaganathan B, Tanner MJ (1994) Altered band 3 structure and function in glycophorin A- and B-deficient (MkMk) red blood cells. Blood 84:916-22

Carpenter L, Halestrap AP (1994) The kinetics, substrate and inhibitor specificity of the lactate transporter of Ehrlich-Lettre tumour cells studied with the intracellular pH indicator BCECF. Biochem J 304:751-760

Carpenter L, Poole RC, Halestrap AP (1996) Cloning and sequencing of the monocarboxylate transporter from mouse Ehrlich Lettre tumour cell confirms its identity as MCT1 and demonstrates that glycosylation is not required for MCT1 function. Biochim Biophys Acta 1279:157-163

Casal M, Paiva S, Andrade RP, Gancedo C, Leao C (1999) The lactate-proton symport of *Saccharomyces cerevisiae* is encoded by JEN1. J Bacteriol 181:2620-2623

Cranmer SL, Conant AR, Gutteridge WE, Halestrap AP (1995) Characterization of the enhanced transport of L- and D-lactate into human red blood cells infected with *Plasmodium falciparum* suggests the presence of a novel saturable lactate proton cotransporter. J Biol Chem 270:15045-15052

Culliford SJ, Bernhardt I, Ellory JC (1995) Activation of a novel organic solute transporter in mammalian red blood cells. J Physiol 489:755-765

De Bruijne AW, Vreeberg H, Van Steveninck J (1983) Kinetic analysis of L-lactate transport in human erythrocytes via the monocarboxylate-specific carrier system. Biochim Biophys Acta 732:562-568

De Bruijne AW, Vreeberg H, Van Steveninck J (1985) Alternative substrate inhibition of L-lactate transport via the monocarboxylate-specific carrier system in human erythrocytes. Biochim Biophys Acta 812:841-844

Denton RM, Halestrap AP (1979) Regulation of pyruvate metabolism in mammalian tissues. Essays Biochem 15:37-47

Deuticke B (1977) Properties and structural basis of simple diffusion pathways in the erythrocyte membrane. Rev Physiol Biochem Pharmacol 78:1-97

Deuticke B (1980) Kinetic properties and substrate specificity of the monocarboxylate carrier in the human erythrocyte membrane. In: Lassen UV, Ussing HH, Weith JO (eds) Membrane transport in erythrocytes (Alfred Benzon Symposium 14). Munksgaard, Copenhagen, pp 539-551

Deuticke B (1982) Monocarboxylate transport in erythrocytes. J Membrane Biol 70:89-103

Deuticke B (1986) The role of membrane sulphydryls in passive, mediated transport processes and for the barrier function of the erythrocyte membrane. Membrane Biochem 6:309-326

Deuticke B (1989) Monocarboxylate transport in red blood cells - kinetics and chemical modification. Methods Enzymol 173:300-329

Deuticke B, Beyer E, Forst B (1982) Discrimination of three parallel pathways of L-lactate transport in the human erythrocyte membrane by inhibitors and kinetic properties. Biochim Biophys Acta 684:96-110

Deuticke B, Rickert I, Beyer E (1978) Stereoselective, SH-dependent transfer of lactate in human erythrocytes. Biochim Biophys Acta 507:137-155

Deves R, Krupka RM (1979) A simple experimental approach to the determination of carrier transport parameters for unlabelled substrate analogues. Biochim Biophys Acta 556:524-532

Dimmer KS, Friedrich B, Lang F, Deitmer JW, Bröer S (2000) The low-affinity monocarboxylate transporter MCT4 is adapted to the export of lactate in highly glycolytic cells. Biochem J 350:219-227

Donovan JA, Jennings ML (1985) Membrane polypeptide in rabbit erythrocytes associated with the inhibition of L-lactate transport by a synthetic anhydride of lactic acid. Biochemistry 24:561-564

Donovan JA, Jennings ML (1986) N-Hydroxysulphosuccinimido active esters and the L-(+)-lactate transport protein in rabbit erythrocytes. Biochemistry 25:1538-1545

Dubinsky WP, Racker E (1978) The mechanism of lactate transport in human erythrocytes. J Membrane Biol 44:25-36

Edlund GL, Halestrap AP (1986) Transport of L-lactate into isolated rat hepatocytes and erythrocytes at low temperatures. Biochem Soc Trans 14:1190-1191

Elliott JL, Saliba KJ, Kirk K (2001) Transport of lactate and pyruvate in the intraerythrocytic malaria parasite, *Plasmodium falciparum*. Biochem J 355:733-739

Fishbein WN (1986) Lactate transporter defect: a new disease of muscle. Science 234:1254-1256

Fossum S, Mallett S, Barclay AN (1991) The MRC OX-47 antigen is a member of the immunoglobulin superfamily with an unusual transmembrane sequence. Eur J Immunol 21:671-679

Garcia CK, Brown MS, Pathak RK, Goldstein JL (1995) cDNA cloning of MCT2, a second monocarboxylate transporter expressed in different cells than MCT1. J Biol Chem 270:1843-1849

Gould GW, Holman GD (1993) The glucose transporter family - structure, function and tissue-specific expression. Biochem J 295:329-341

Green NM (1991) Biological membranes. The semiotics of charge. Nature 351:349-50

Grollman EF, Philp NJ, McPhie P, Ward RD, Sauer B (2000) Determination of transport kinetics of chick MCT3 monocarboxylate transporter from retinal pigment epithelium by expression in genetically modified yeast. Biochemistry 39:9351-9357

Halestrap AP (1976) Pyruvate and lactate transport into human erythrocytes. Evidence for the involvement of the chloride carrier and a chloride independent carrier. Biochem J 156:193-207

Halestrap AP, Denton RM (1974) Specific inhibition of pyruvate transport in rat liver mitochondria and human erythrocytes by α-cyano-4-hydroxycinnamate. Biochem J 138:313-316

Halestrap AP, Poole RC (1989) The transport of pyruvate and lactate across mitochondrial and plasma membranes. In: Hamasaki N, Jennings ML (eds) Anion transport protein of the red blood cell membrane. Elsevier, Amsterdam, pp 73-86

Halestrap AP, Price NT (1999) The proton-linked monocarboxylate transporter (MCT) family: structure, function and regulation. Biochem J 343:281-299

Halestrap AP, Brand MD, Denton RM (1974) Inhibition of mitochondrial pyruvate transport by phenylpyruvate and α-keto-isocaproate. Biochim Biophys Acta 367:102-108

Halestrap AP, Scott RD, Thomas AP (1980) Mitochondrial pyruvate transport and its hormonal regulation. Int J Biochem 11:97-105

Hart KW, Clarke AR, Wigley DB, Waldman AD, Chia WN, Barstow DA, Atkinson T, Jones JB, Holbrook JJ (1987) A strong carboxylate-arginine interaction is important in substrate orientation and recognition in lactate dehydrogenase. Biochim Biophys Acta 914:294-298

Heijn M, Oude Elferink RP, Jansen PL (1992) ATP-dependent multispecific organic anion transport system in rat erythrocyte membrane vesicles. Am J Physiol 262:C104-C110

Henderson PJF (1993) The 12-transmembrane helix transporters. Curr Opin Cell Biol 5:708-721

Jackson VN, Price NT, Halestrap AP (1995) cDNA cloning of MCT1, a monocarboxylate transporter from rat skeletal muscle. Biochim Biophys Acta 1238:193-196

Jackson VN, Price NT, Carpenter L, Halestrap AP (1997) Cloning of the monocarboxylate transporter isoform MCT2 from rat testis provides evidence that expression in tissues is species-specific and may involve post-transcriptional regulation. Biochem J 324:447-453

Jennings ML, Adams-Lackey M (1982) A rabbit erythrocyte membrane protein associated with L-lactate transport. J Biol Chem 257:12866-12871

Juel C, Halestrap AP (1999) Lactate transport in skeletal muscle - role and regulation of the monocarboxylate transporter. J Physiol 517:633-642

Kafitz KW, Rose CR, Thoenen H, Konnerth A (1999) Neurotrophin-evoked rapid excitation through TrkB receptors. Nature 401:918-921

Kanaani J, Ginsburg H (1991) Transport of lactate in *Plasmodium-falciparum*-infected human erythrocytes. J Cell Physiol 149:469-476

Kanai Y, Segawa H, Miyamoto K, Uchino H, Takeda E, Endou H (1998) Expression cloning and characterization of a transporter for large neutral amino acids activated by the heavy chain of 4F2 antigen (CD98). J Biol Chem 273:23629-23632

Kasinrerk W, Fiebiger E, Stefanova I, Baumruker T, Knapp W, Stockinger H (1992) Human leukocyte activation antigen M6, a member of the Ig superfamily, is the species homologue of rat OX-47, mouse basigin, and chicken HT7 molecule. J Immunol 149:847-854

Kim CM, Goldstein JL, Brown MS (1992) cDNA cloning of MEV, a mutant protein that facilitates cellular uptake of mevalonate, and identification of a point mutation responsible for its gain in function. J Biol Chem 267:23113-23121

Kim DK, Kanai Y, Chairoungdua A, Matsuo H, Cha SH, Endou H (2001) Expression cloning of a Na^+-independent aromatic amino acid transporter with structural similarity to H^+/Monocarboxylate Transporters. J Biol Chem 276:17221-17228

Kim-Garcia C, Goldstein JL, Pathak RK, Anderson RGW, Brown MS (1994a) Molecular characterization of a membrane transporter for lactate, pyruvate, and other monocarboxylates - Implications for the Cori cycle. Cell 76:865-873

Kim-Garcia C, Li X, Luna J, Francke U (1994b) cDNA cloning of the human monocarboxylate transporter 1 and chromosomal location of the SLC16A1 locus to 1p13.2-p12. Genomics 23:500-503

Kirk P, Wilson MC, Heddle C, Brown MH, Barclay AN, Halestrap AP (2000) CD147 is tightly associated with lactate transporters MCT1 and MCT4 and facilitates their cell surface expression. EMBO J 19:3896-3904

Koehler Stec EM, Simpson IA, Vannucci SJ, Landschulz KT, Landschulz WH (1998) Monocarboxylate transporter expression in mouse brain. Am J Physiol 38:E516-E524

Lanier LL, Yu G, Phillips JH (1989) Co-association of CD3 zeta with a receptor (CD16) for IgG Fc on human natural killer cells. Nature 342:803-805

Lanier LL, Corliss B, Wu J, Phillips JH (1998) Association of DAP12 with activating CD94/NKG2C NK cell receptors. Immunity 8:693-701

Leeks DR, Halestrap AP (1978) Chloride-independent transport of pyruvate and lactate across the erythrocyte membrane. Biochem Soc Trans 6:1363-1366

Leeks DR, Halestrap AP (1979) Pyruvate and lactate transport across the cell membrane of the erythrocyte and other cells. In: Quagliariello E, Papa S, Palmieri F, Slater EC, Siliprandi N (eds) Function and molecular aspects of biomembrane transport. Elsevier-North Holland, Amsterdam, pp 427-430

Manning Fox JE, Meredith D, Halestrap AP (2000) Characterisation of human monocarboxylate transporter 4 substantiates its role in lactic acid efflux from skeletal muscle. J Physiol 529:285-293

Mastroberardino L, Spindler B, Pfeiffer R, Skelly PJ, Loffing J, Shoemaker CB, Verrey F (1998a) Amino-acid transport by heterodimers of 4F2hc/CD98 and members of a permease family. Nature 395:288-291

McLatchie LM, Fraser NJ, Main MJ, Wise A, Brown J, Thompson N, Solari R, Lee MG, Foord SM (1998) RAMPs regulate the transport and ligand specificity of the calcitonin-receptor-like receptor. Nature 393:333-339

Meredith D, Halestrap AP (2000) OX47 (basigin) may act as a chaperone for expression of the monocarboxylate transporter MCT1 at the plasma membrane of *Xenopus laevis* oocytes. J Physiol 526P:23P

Meredith D, Roberts M, Halestrap AP (1999) Both K290 and K413 are essential for DIDS covalent modification of the rat proton-linked monocarboxylate (lactate) transporter MCT1 expressed in *Xenopus laevis* oocytes. J Physiol 517P:25P

Miyauchi T, Jima F, Igakura T, Yu S, Ozawa M, Muramatsu T (1995) Struture of the mouse basigin gene, a unique member of the immunoglobulin superfamily. J Biochem (Tokyo) 118:717-724

Ozawa M, Huang RP, Furukawa T, Muramatsu T (1988) A teratocarcinoma glycoprotein carrying a developmentally regulated carbohydrate marker is a member of the immunoglobulin gene superfamily. J Biol Chem 263:3059-3062

Paterson DJ, Jefferies WA, Green JR, Brandon MR, Corthesy P, Puklavec M, Williams AF (1987) Antigens of activated rat T lymphocytes including a molecule of 50,000 Mr detected only on CD4 positive T blasts. Mol Immunol 24:1281-1290

Pfaller MA, Krogstad DJ, Parquette AR, Nguyen-Dinh P (1982) *Plasmodium falciparum*: Stage -specific lactate production in synchronized cultures. Exp Parasitol 54:391-396

Pfeiffer R, Rossier G, Spindler B, Meier C, Kuhn L, Verrey F (1999a) Amino acid transport of y^{+}L-type by heterodimers of 4F2hc/CD98 and members of the glycoprotein-associated amino acid transporter family. EMBO J 18:49-57

Poole RC, Halestrap AP (1988) Reconstitution of the L-lactate carrier from rat and rabbit erythrocyte plasma membranes. Biochem J 254:385-390

Poole RC, Halestrap AP (1990) Inhibition and labelling of the erythrocyte lactate transporter by stilbene disulphonates. Biochem Soc Trans 18:1245-1246

Poole RC, Halestrap AP (1991) Reversible and irreversible inhibition, by stilbenedisulphonates, of lactate transport into rat erythrocytes - identification of some new high-affinity inhibitors. Biochem J 275:307-312

Poole RC, Halestrap AP (1992) Identification and partial purification of the erythrocyte lactate transporter. Biochem J 283:855-862

Poole RC, Halestrap AP (1993) Transport of lactate and other monocarboxylates across mammalian plasma membranes. Am J Physiol 264:C761-C782

Poole RC, Halestrap AP (1994) N-Terminal protein sequence analysis of the rabbit erythrocyte lactate transporter suggests identity with the cloned monocarboxylate transport protein MCT1. Biochem J 303:755-759

Poole RC, Halestrap AP (1997) Interaction of the erythrocyte lactate transporter (monocarboxylate transporter 1) with an integral 70-kDa membrane glycoprotein of the immunoglobulin superfamily. J Biol Chem 272:14624-14628

Poole RC, Cranmer SL, Halestrap AP, Levi AJ (1990) Substrate and inhibitor specificity of monocarboxylate transport into heart cells and erythrocytes - further evidence for the existence of 2 distinct carriers. Biochem J 269:827-829

Poole RC, Cranmer SL, Holdup DW, Halestrap AP (1991) Inhibition of L-lactate transport and Band-3-mediated anion transport in erythrocytes by the novel stilbenedisulphonate N,N,N',N'-tetrabenzyl-4,4'-diaminostilbene-2,2'-disulphonate (TBenzDS). Biochim Biophys Acta 1070:69-76

Poole RC, Sansom CE, Halestrap AP (1996) Studies of the membrane topology of the rat erythrocyte H^{+}/lactate cotransporter (MCT1). Biochem J 320:817-824

Price NT, Jackson VN, Halestrap AP (1998) Cloning and sequencing of four new mammalian monocarboxylate transporter (MCT) homologues confirms the existence of a transporter family with an ancient past. Biochem J 329:321-328

Rahman B, Schneider HP, Bröer A, Deitmer JW, Bröer S (1999) Helix 8 and helix 10 are involved in substrate recognition in the rat monocarboxylate transporter MCT1. Biochemistry 38:11577-11584

Regen DM, Tarpley HL (1978) Effects of pH on β-hydroxybutyrate transport in rat erythrocytes and thymocytes. Biochim Biophys Acta 508:539-550

Rice WR, Steck TL (1976) Pyruvate flux into resealed ghosts from human erythrocytes. Biochim Biophys Acta 433:39-53

Rice WR, Steck TL (1977) Pyruvate transport into inside-out vesicles icolated from human erythrocyte membranes. Biochim Biophys Acta 468:305-317

Roth DA (1991) The sarcolemmal lactate transporter - transmembrane determinants of lactate flux. Med Sci Sports Exerc 23:925-934

Sachs JR, Knauf PA, Dunham PB (1975) Transport through red cell membranes. In: Surgenor, DMcN (ed) The red blood cell, vol 2. Academic Press, New York, pp 613-703

Saier MH (1994) Computer-aided analyses of transport protein sequences - gleaning evidence concerning function, structure, biogenesis, and evolution. Microbiol Rev 58:71-93

Saxena M, Henderson GB (1996) MOAT4, a novel multispecific organic-anion transporter for glucuronides and mercapturates in mouse L1210 cells and human erythrocytes. Biochem J 320:273-281

Sekine T, Cha SH, Endou H (2000) The multispecific organic anion transporter (OAT) family. Pflügers Arch 440:337-350

Seulberger H, Unger CM, Risau W (1992) HT7, Basigin, gp42 and OX-47 - many names for one developmentally regulated immunoglobulin-like surface glycoprotein on blood-brain barrier endothelium, epthelial tissue barriers and neurons. Neurosci Letters 140:93-97

Sherman IW (1979) Biochemistry of *Plasmodium* (malarial parasites). Microbiol Rev 43:453-495

Soulinna EM, Buchsbaum RN, Racker E (1975) The effect of flavonoids on aerobic glycolysis and growth of tumor cells. Cancer Res 35:1865-1872

Stein WD (1989) Kinetics of transport - analyzing, testing, and characterizing models using kinetic approaches. Methods Enzymol 171:23-62

Takanaga H, Tamai I, Inaba S, Sai Y, Higashida H, Yamamoto H, Tsuji A (1995) cDNA cloning and functional characterization of rat intestinal monocarboxylate transporter. Biochem Biophys Res Commun 217:370-377

Torrents D, Mykkanen J, Pineda M, Feliubadalo L, Estevez R, De Cid R, Sanjurjo P, Zorzano A, Nunes V, Huoponen K, Reinikainen A, Simell O, Savontans ML, Aula P, Palacin M (1999) Identification of SLC7A7, encoding y^{+}LAT-1, as the lysinuric protein intolerance gene. Nat Genet 21:293-296

Van der Jagt DL, Hunsaker LA, Campos NM, Baack BR (1990) D-Lactate production in erythrocytes infected with *Plasmodium falciparum*. Mol Biochem Parasitol 42:277-284

Walter A, Gutknecht J (1984) Monocarboxylic acid permeation through lipid bilayer membranes. J Membrane Biol 77:255-264

Wheeler TJ, Hinkle PC (1981) Kinetic properties of the reconstituted glucose transporter from human erythrocytes. J Biol Chem 256:8907-8914

Widdas WF (1988) Old and new concepts of the membrane transport of glucose in cells. Biochim Biophys Acta 947:385-404

Wilson MC, Heddle C, Kirk P, Roberts M, Meredith D, Barclay AN, Brown MH, Halestrap AP (1999) Co-localization of immunoglobulin superfamily member CD147 with a monocarboxylate transporter (MCT1). J Physiol 517P:58P

Wilson MC, Jackson VN, Heddle C, Price NT, Pilegaard H, Juel C, Bonen A, Montgomery I, Hutter OF, Halestrap AP (1998) Lactic acid efflux from white skeletal muscle is catalyzed by the monocarboxylate transporter isoform MCT3. J Biol Chem 273:15920-15926

Xu ASL, Kuchel PW (1993) Characterisation of erythrocyte transmembrane exchange of trifluoroacetate using F-19-NMR - evidence for transport via the monocarboxylate transporter. Biochim Biophys Acta 1150:35-44

Yoon H, Philp NJ (1998) Genonmic structure and developmental expression of the chicken moncarboxylate transporter MCT3 Gene. Exp Eye Res 67:417-424

Yoon HY, Fanelli A, Grollman EF, Philp NJ (1997) Identification of a unique monocarboxylate transporter (MCT3) in retinal pigment epithelium. Biochem Biophys Res Commun 234:90-94

Zhao C, Wilson MC, Schuit F, Halestrap AP, Rutter GA (2001) Expression and distribution of lactate/monocarboxylate transporter isoforms in pancreatic islets and the exocrine pancreas. Diabetes 50:361-366

19 Water Permeability

Joseph Browning and Robert Wilkins

Laboratory of Physiology, Oxford University, Parks Road, Oxford, OX1 3PT, UK

19.1 Introduction

The plasma membrane of animal cells constitutes a selectively permeable barrier, the properties of which determine intracellular solute composition (e.g., Stein 1990). Characterising these properties has been a longstanding objective of biologists, and it is now recognised that a large array of membrane proteins regulates the exchange of ions and solutes between the cell and its extracellular environment. The passage of water across the lipid bilayer can also occur and, as the chief solvent for body fluids, this movement is axiomatic to life. The characterisation of water movement across cell membranes has relied heavily on the use of the erythrocyte as a paradigm (e.g., Rich et al. 1967). We now know that the swelling and subsequent haemolysis observed when erythrocytes are suspended in hypotonic media predominantly reflects diffusion of water through protein channels, termed aquaporins. In this chapter, we will consider the evidence that supports this assertion, and review the structure and function of the aquaporin family. The importance of aquaporin expression for erythrocyte function will also be discussed, both in terms of its conventional role, and, in the light of recent evidence, as a channel which can facilitate gas carriage.

19.2 Water Diffusion through Lipids

Even simple planar bilayers are permeable to water, so it is clear that diffusion through lipid can occur (Cass and Finkelstein 1967). In this situation, water molecules rapidly dissolve in the lipid phase and then pass slowly across the barrier by simple diffusion. This diffusional water permeability (P_d) is temperature-sensitive, dependent on membrane lipid composition (reduced by increased cholesterol content, for example) and has a high activation energy (E_a), typically 10–20 kcal/mol (Finkelstein and Cass 1967; Petersen 1980). When an osmotic gradient is applied across the bilayer, the water permeability recorded under these conditions (P_f, the osmotic water permeability) is not significantly different from P_d (Träuble 1971).

The fact that water can diffuse across bilayers was, for some time, sufficient reason to discount the existence of specific protein pathways. Nevertheless, a mounting body of evidence dating back to the 1950s indicated that simple diffusion through lipid cannot always account for water permeation, especially in cells where water fluxes are particularly rapid, such as the erythrocyte and certain epithelial cells (Dempster et al. 1992).

19.3 Evidence for Discrete Permeability Pathways

Initial support for the existence of water pathways came from the consideration of the ratio of P_f to P_d. It became recognised that the P_f:P_d ratio offers an indication of the properties of a given membrane (e.g., Parisi and Bourguet 1983). We can summarise this work with three scenarios. For highly porous membranes with large pores, which permit laminar flow, P_f is considerably larger than P_d. For non-porous membranes, P_f:P_d approximates to 1, indicating that water movement occurs solely by solubility diffusion. Where the ratio is only to a small extent greater than 1, the existence of narrow pores is indicated (Dempster et al. 1992). Hence, addition of pore-forming, water-conducting antibiotics to planar bilayers such as nystatin results in an increased P_f:P_d ratio of between 3 and 5 (Holz and Finkelstein 1970).

For biological membranes, P_f:P_d is often greater than 1, and the E_a for water movement is reduced to below 5 kcal/mol, consistent with diffusion of water in solution, at a rate some 60,000 times faster than through lipid (e.g., Berry 1985). For erythrocytes, experiments using tracer ^3H_2O revealed that the half-time for diffusional exchange of water is extremely rapid (of the order of 4 ms), and giving P_d values of approx. $4{\cdot}10^{-3}$ cm/s (Conlon and Outhred 1972). However, measurements of water movement across the erythrocyte membrane in response to an osmotic perturbation have also been made, using light scattering as an index of erythrocyte volume, and provided values of P_f which are considerably higher, at around $2.5{\cdot}10^{-2}$ cm/s (Macey and Farmer 1970; Terwilliger and Solomon 1981). The resultant P_f:P_d ratio of approx. 6 is therefore incompatible with solely diffusive water movement, and is indicative of an aqueous permeation pathway across the erythrocyte membrane.

The observation that, by various chemical and physical means, it is possible to inhibit water movement across the membrane provides further support for the existence of water channel proteins in erythrocytes. Mercurial compounds such as *p*-chloromercuribenzene sulphonate (PCMBS), which reversibly bind protein sulphydryl groups were shown in the early 1970s to reversibly reduce P_f (such that P_f:P_d became 1) while increasing E_a to levels found for artificial bilayers (Naccache and Sha'afi 1974). Twenty years later, radiation inactivation techniques were employed to inhibit proximal tubule osmotic water permeability and to identify a membrane protein of approx. 30 kDa molecular weight responsible for water transport (Van Hoek et al. 1991). These two approaches to inhibiting water permeability have allowed the relative contributions of P_f and P_d to water movement across the erythrocyte membrane to be defined, and indicate that approx. 90% of the water flux passes through water channels (Brahm 1982). In contrast more recent measurements by Lahajnar and co-workers using NMR techniques has indicated a channel-to-lipid diffusional water permeability ratio of about 1:1 (Lahajnar et al. 2000).

A variety of proteins have, at different times, been proposed as candidate water transport pathways. The CFTR Cl^- channel and the glucose carrier GLUT have been reported to confer a degree of water permeability in *Xenopus* oocytes and mouse macrophages (Loike et al. 1993; Schreiber et al. 1997) while in erythrocytes there is some evidence that abundant band 3 anion exchanger (AE1) present in the membrane can mediate water fluxes (Solomon et al. 1983). None of these pathways, however, has properties which are entirely consistent with those outlined in the foregoing discussion, and the high permeability of the erythrocyte membrane to water instead reflects expression of another distinct protein. The discovery of aquaporin water channels in the early 1990s has, to a great extent, solved the paradoxes surrounding water permeability in a variety of eukaryotic cell types. Their initial isolation, properties and patterns of expression, particularly with regard to erythrocytes, are considered in the remainder of this review.

19.4 Molecular Identification of Water Channels

Aquaporin-1, the first water channel to be identified at the molecular level, was isolated serendipitously from human erythrocyte membranes as a 28 kDa protein by investigators attempting to purify the 32 kDa subunit of the human Rhesus blood group polypeptides. Biochemical similarities between the Rhesus protein and aquaporin-1 led to their co-purification in spite of the fact that there is no structural or sequence homology between them. Initially, the function of the 28 kDa protein was unknown, although it was shown to be an integral membrane protein and present in erythrocytes at a high copy number (typically 120,000–160,000 copies per cell) as well as in rat kidney, and it was referred to simply as 28 kDa (Denker et al. 1988). Furthermore, two forms of the protein, of differing molecular weights, could be detected – antibodies raised to 28 kDa reacted on immunoblots of erythrocyte membranes with 28 kDa as well as with a region of higher, albeit diffuse, molecular weight 35–60 kDa – this second, heavier protein or proteins was named high molecular weight 28 kDa (HMW-28 kDa) and was shown to be an *N*-glycosylated form of 28 kDa. Immunocytochemistry of human kidney with anti-28 kDa demonstrated that the protein was widely distributed in the apical brush border membranes of proximal tubule cells, and it was initially thought to play a structural role in linking the membrane skeleton to the lipid bilayer (Denker et al. 1988).

However, sequencing of the first 35 N-terminal amino acid residues of 28 kDa demonstrated that it shares significant sequence homology with major intrinsic protein of lens (MIP26), a protein previously known to form a channel through which lens fibre cells reabsorb interstitial fluid (Pao et al. 1991). Furthermore, 28 kDa was shown to be a non-covalent assembly of four subunits (Smith and Agre 1991); these and other physical similarities between it and other membrane channels led to it being designated CHIP28 (for **ch**annel forming **i**ntegral **p**rotein of 28 kDa) by Agre and co-workers. The subsequent cloning and sequencing of

CHIP28 from a human bone marrow cDNA library was performed using oligonucleotide primers designed from the known N-terminus primary sequence of the protein (Preston and Agre 1991). The deduced complete primary amino acid sequence was used to determine several key structural features of the protein (explained in detail in later discussion).

The cloning of CHIP28 was crucial to its identification as a water channel. Expression experiments, in which *Xenopus* oocytes were injected with CHIP28 mRNA, allowed definitive functional characterisation of the protein product. Oocytes injected with CHIP28 mRNA exhibited significantly increased water permeability, assessed by measuring the relative time taken for lysis to occur following transfer of transfected and non-transfected oocytes from an isotonic into a hypotonic solution (Preston et al. 1992); non-transfected (control) oocytes demonstrate a low permeability to water. This increased water permeability could be inhibited by mercurial sulphydryl reagents, previously identified as inhibitors of water channels.

Permeability to other small ions and solutes was not increased by transfection with CHIP28. Based on these data, CHIP28 was proposed to be a functional unit of membrane water channels (Preston et al. 1992), or, less likely, given its multiple transmembrane domains, an endogenous regulator of water channels. In the light of these findings, CHIP28 was given the name aquaporin-1 (AQP1), which reflects its properties more accurately, and the family of proteins of which it is a member was called aquaporins (AQPs) by the members of the laboratory who first identified it as a water channel (Agre et al. 1993).

19.5 The Aquaporin Family

Although CHIP28 (AQP1) was the first member of the aquaporin family to be identified as such, the existence of MIP26 (major intrinsic protein) from mammalian eye lens had been known for some time prior to the purification of AQP1, although at that time it was not recognised as a member of the AQP family (Gorin et al. 1984). The identification and cloning of AQP1, and the realisation that it shares significant sequence homology with MIP, led to the designation of this new family of related proteins as AQP and MIP was accordingly renamed AQP0 to reflect its status as the first identified AQP water channel. The aquaporin family of proteins is widely distributed across different species and homologues have been identified in most major taxonomic groupings. Presently, approx. 160 members have been described, many of which demonstrate a very high specificity for water. Other members of the AQP family also conduct other ions and molecules, such as glycerol and urea, and have been termed 'aquaglyceroporins' (Fig. 19.1). Here discussion is limited to those mammalian proteins known to exhibit a particularly high permeability to water.

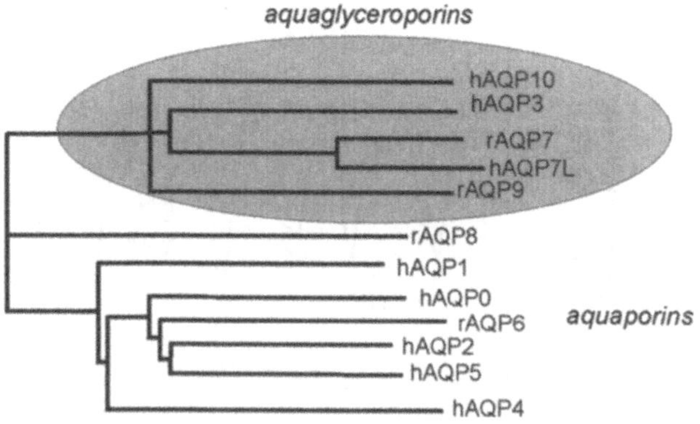

Fig. 19.1. Schematic dendrogram for phylogentic comparison of mammalian aquaporins. Two clusters are identified: channels exclusive for water permeation – aquaporins – and those also permitting the movement of solutes such as glycerol – the aquaglyceroporins. After Borgnia et al. (1999)

19.6 Structure-Function Relationships

Insights into the structure and function of AQP proteins have been made possible by the purification of protein in large quantities. The availability of AQP1, in particular, has triggered rapid advances in our understanding of the relation between structure and function of AQP proteins (e.g., see Walz et al. 1994; Cheng et al. 1997). Primary sequence analysis of AQP1 reveals 269 amino acid residues with a total molecular weight of 28 kDa. Hydropathy analysis of the primary sequence indicates that its constituent residues are predominantly hydrophobic suggesting that they lie largely within the plane of the lipid bilayer itself (Fig.19.2). Six membrane-spanning domains are normally evident, and this has emerged as the consensus view (Fig. 19.3). Antibody labelling of the amino and carboxy terminii confirms that both are located intracellularly and the orientation of the protein has been further confirmed by vectorial proteolysis studies.

The primary sequence of AQP1, like its homologues, contains an internal repeat with the N- and C-terminal halves of the protein sharing significant sequence homology. This is presumed to arise from the duplication of an ancient parent gene, a theory, which is supported by phylogenetic analysis of the aquaporin family (Park and Saier 1996; Zardoya and Villalba 2001). The two halves of the protein are orientated at 180° to each other. The regions of closest homology are in loops B and E, which contain residues that are predicted to line the water pore itself, and each contain an asparagine-proline-alanine (NPA) motif (Fig. 19.4).

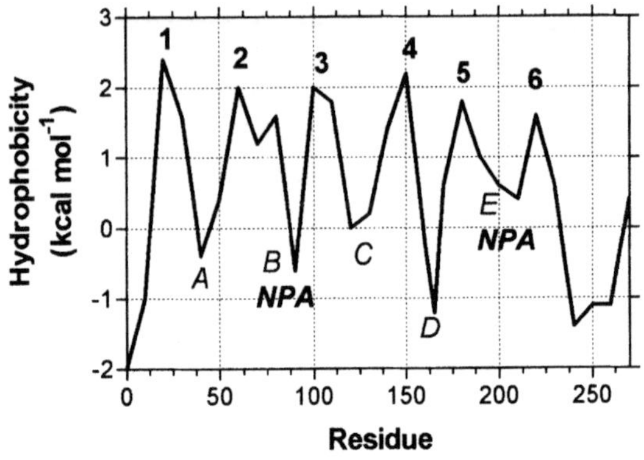

Fig. 19.2. Simplified hydrophobicity plot of AQP1. Membrane spanning regions 1–6 and loops *A–E* are indicated, together with the two *NPA* motifs which define the pore. After Heymann and Engel (1999)

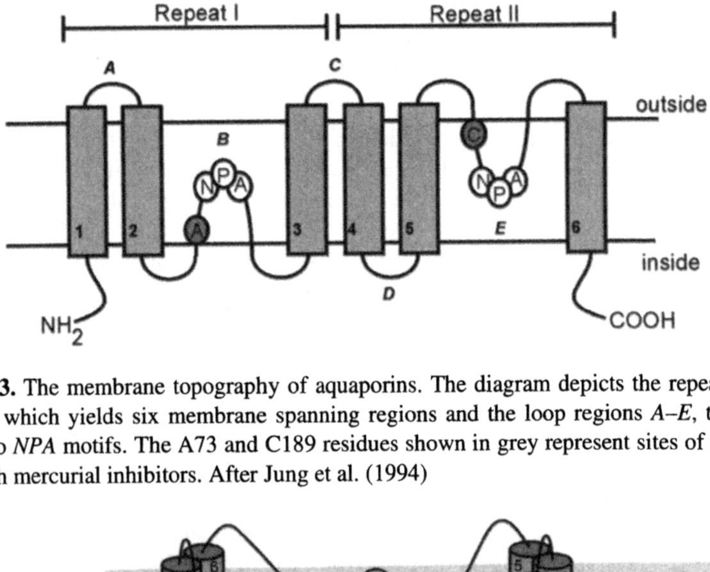

Fig. 19.3. The membrane topography of aquaporins. The diagram depicts the repeated sequence, which yields six membrane spanning regions and the loop regions *A–E*, together with two *NPA* motifs. The A73 and C189 residues shown in grey represent sites of interaction with mercurial inhibitors. After Jung et al. (1994)

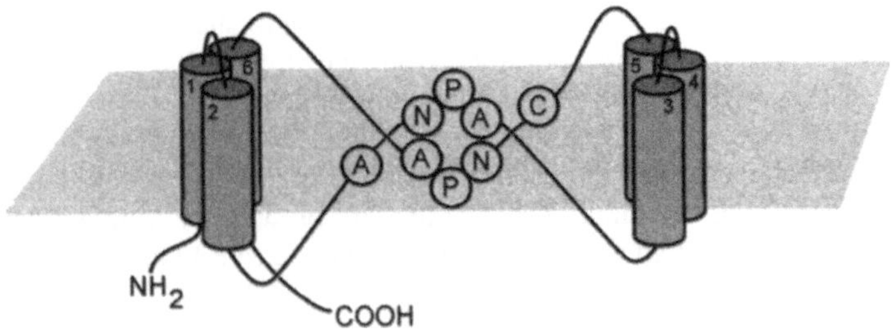

Fig. 19.4. The 'hourglass' model of AQP1. The diagram shows the way in which the repeated sequence orientates within the membrane, with the two *NPA* motifs united to define the pore. After Jung et al. (1994)

A number of residues within the primary sequence of AQP1 can be associated with specific functions of the protein. Most water channels studied to date are inhibitable by $HgCl_2$, an agent used previously as a diuretic but now withdrawn because of its nephrotoxic effects (Ventura et al. 2001). Mercurials are known to modify cysteine residues and mutagenesis studies in which each of the native cysteine residues in AQP1 were systematically replaced by serine revealed Cys 189 to be the mercury-sensitive residue (Zhang et al. 1993). This finding strongly suggests that this amino acid, which is located in loop E, forms part of the aqueous channel. The significant hydrophobicity of loop E, and its homologue in the N-terminal half of the protein, loop B, has led to the development of the 'hourglass' model for the structure of the AQP1 protein in which loops B and E each penetrate partway into the lipid bilayer and overlap with each other forming a continuous aqueous pore (Verkman and Mitra 2000). In this arrangement the Asn-Pro-Ala motifs of loops B and E are juxtaposed. Ala 73, located within loop B, is also required for mercury sensitivity, further strengthening this hypothesis (Fig. 19.2). A number of studies employing electron cryocrystallography of AQP1, as well as radiation inactivation and site-directed mutagenesis have indicated that AQP1 protein oligomerizes within the lipid bilayer to form tetramers. Each subunit, however, is believed to contain a functional pore (reviewed by Verkman and Mitra (2000)).

More recent studies, in which crystals of AQP1 have been studied by X-ray crystallography at resolutions up to 2.2 Å, have shed further light on the structure-function relationships of this protein (reviewed by Sui et al. (2001)). The pore of the channel is indeed hourglass-like in shape, confirming conclusions drawn from earlier primary sequence studies, and consists of a flared extracellular region, a narrow selectivity filter and a second, flared, intracellular region. The pore itself is amphipathic, with about half of the residues lining it being hydrophobic and the other half hydrophilic. Regions binding water molecules within the selectivity filter enable water to pass through the channel, although these regions are few in number (four in total). This low number prevents ions from traversing AQP1, which require a greater number of coordinating sites to render their passage energetically favourable. Furthermore, certain of the aquaporins are exclusively permeable to water (AQP 0, 1, 2, 4, 5, 6 and 8) while others exhibit greater permeability to glycerol and other small, uncharged solutes (AQP 3, 7, 9, 10; Yang et al. 1997). X-ray crystallography studies reveal that the key structural difference between these two classes of protein lies in the size of the pore, rather than the nature of the residues lining it.

A complex glycosylated form of the protein with a molecular weight of between 40 and 60 kDa can also be detected in red cell membranes by protein immunoblots. The physiological function of this modification is unclear, although it is most likely related to membrane trafficking of the protein. The extent and level of glycosylation varies between different tissues and it appears that only one of the four subunits forming a functional tetramer is glycosylated.

19.7 Aquaporins and Red Cells

As already noted, it has long been recognised that erythrocytes demonstrate a constitutively high permeability to water and that this permeability is sensitive to the application of sulphydryl-modifying mercurial agents. As a direct consequence of this, erythrocytes were a paradigm system for many early investigations of channel mediated water movement across biological membranes. The initial identification of CHIP28 from erythrocytes, and the subsequent detailed molecular and functional characterisation of the aquaporin family has led to the realisation that erythrocytes express at least two isoforms of AQP, namely 1 and 3 (Roudier et al. 1998; Yang et al. 2001). Of these, AQP1 is by far the most abundant and there are approx. 120,000 to 160,000 copies of this protein per cell. AQP3 is resistant to the actions of mercurial agents and this, therefore, explains the small mercury-insensitive component of erythrocyte water permeability.

AQP1 is sensitive to mercurial agents and accounts for most of the water permeability (Yang et al. 2001). In addition to being permeable to water, AQP3 demonstrates a high permeability to glycerol and urea and explains the high facilitative permeability of erythrocytes to these small, uncharged solutes. It must, however, be remembered that specific, facilitative urea transporters exist in erythrocyte membranes which also contribute to the high urea permeability of this cell type (Smith and Rousselet 2001). It is thought that the high concentrations of urea present within the renal medulla mean that red cells require a high permeability to this compound during their passage through the renal medulla in order to negate the osmotic effects which arise from its high concentration.

Opinions vary as to why red blood cells require such a high permeability to water. One possibility is that, during their passage through the renal medulla, erythrocytes are routinely exposed to osmolarities greater than 1 M and are required to unload and load water very rapidly. However, one significant line of evidence contradicts this view. Chromosomal location of AQP1 to 7p14 reveals that it is coincident with the well recognised Colton blood group antigens Co^a and Co^b at 7p (Daniels 1999). Immunoprecipitation with anti-Colton antisera reveals that both of these antigens reside on AQP1 and that they arise from a polymorphism of residue 45 (alanine or valine; Daniels 1999). Certain individuals completely lack both Colton antigens arising from deletions, frameshift mutations and missense mutations in the AQP1 gene. Such individuals, although rare, demonstrate no clinical abnormalities and no significant erythrocyte abnormality, apart from a significantly reduced osmotic water permeability. Presumably, therefore, other AQP isoforms are able to compensate for lost function in the absence of AQP1. Interestingly, in contrast to humans, mice lacking AQP1 demonstrate a significantly impaired ability to produce a concentrated urine when dehydrated (Yang et al. 2001). The residual osmotic permeability to water of erythrocytes from Colton null individuals presumably arises from the compensating presence of AQP3, which is moderately permeable to water. Until recently, therefore, it has been difficult to arrive at a convincing explanation as to why normal erythrocytes must possess such significant AQP1 mediated permeability to water. The answer may lie in results from recent studies, which demonstrate that AQP1, in common with certain other water channels, is additionally permeable to CO_2 (Cooper and Boron

1998). Oocytes, injected with mRNA coding for AQP1, exhibited a significantly increased rate of intracellular acidification following exposure to extracellular CO_2, compared to non-transfected controls (Cooper and Boron 1998; Nakhoul et al. 1998). This can be attributed to AQP1 mediating the inward flux of CO_2, which is then hydrated in the cytosol to form H^+ and HCO_3^-. Similar results have been obtained using AQP1 reconstituted into phospholipid vesicles (Ramesh Prasad et al. 1998; Yang et al. 2000). A significant membrane permeability to CO_2 is likely to be of importance given the crucial role of erythrocytes in facilitating the carriage of CO_2 in the blood from peripheral tissues to the lungs. CO_2, produced by metabolically active tissues, diffuses into the blood and from there into red blood cells where it is hydrolysed to form H^+ and HCO_3^- in a reaction catalysed by carbonic anhydrase. HCO_3^- formed in this way accounts for 60% of total CO_2 transported through the venous system from peripheral tissues to the lungs. Evidently, the membrane of erythrocytes must be permeable to CO_2 in order for this to occur; it is assumed that the permeability of AQP1 to CO_2 increases the efficiency of this process. Indeed, Reithmeier and co-workers (Sterling et al. 2001) have described the structural and functional interaction of carbonic anhydrase II and AE1 in erythrocytes, which they have termed a 'transport metabolon'. It may be the case that AQP1 forms the third component of such a metabolon. Nevertheless, the hypothesis that AQP1 has a role in CO_2 permeation is weakened by the apparent lack of effect of the absence of AQP1 in Colton-null individuals. Either erythrocyte permeability to CO_2 through aquaporin proteins is not essential for the efficient carriage and excretion of CO_2, or other AQP isoforms are able to maintain a significant permeability to CO_2 in the absence of AQP1, thereby maintaining the efficiency of erythrocyte mediated CO_2 gas carriage.

19.8 Summary

In summary, in this chapter we have considered the recent work, which has elucidated the molecular basis of the high water permeability, which has been recognised in red blood cells for over fifty years. While the relation between the structure and function of the AQP proteins is now well understood, the precise roles played by the channel in red blood cell physiology await further clarification.

Acknowledgement: Work in the authors' laboratory is supported by the Arthritis Research Campaign, UK.

References

Agre P, Sasaki S, Chrispeels MJ (1993) Aquaporins: a family of water channels. Am J Physiol 265:F461

Berry CA (1985) Characteristics of water diffusion in the rabbit proximal convoluted tubule. Am J Physiol 249:F729-F738

Borgnia M, Nielsen S, Engel A, Agre P (1999) Cellular and molecular biology of the aquaporin water channels. Annu Rev Biochem 68:425-458

Brahm J (1982) Diffusional water permeability of human erythrocytes and their ghosts. J Gen Physiol 79:791-819

Cass A, Finkelstein A (1967) Water permeability of thin lipid membranes. J Gen Physiol 50:1765-1784

Cheng A, Van Hoek AN, Yeager M, Verkman AS, Mitra AK (1997) Three-dimensional organization of a human water channel. Nature 387:627-630

Conlon T, Outhred R (1972) Water diffusion permeability of erythrocytes using an NMR technique. Biochem Biophys Acta 288:354-361

Cooper GJ, Boron WF (1998) Effect of PCMBS on CO_2 permeability of *Xenopus* oocytes expressing aquaporin 1 or its C189S mutant. Am J Physiol 275:C1481-C1486

Daniels G (1999) Functional aspects of red cell antigens. Blood Rev 13:14-35

Dempster JA, Van Hoek AN, Van Os VH (1992) The quest for water channels. News Physiol Sci 7:175-176

Denker BM, Smith BL, Kuhajda FP, Agre P (1988) Identification, purification, and partial characterisation of a novel Mr 28,000 integral membrane protein from erythrocytes and renal tubules. J Biol Chem 263:15634-15642

Finkelstein A, Cass A (1967) Effect of cholesterol on the water permeability of thin lipid membranes. Nature 216:717-718

Gorin MB, Yancey SB, Cline J, Revel JP, Horwitz J (1984) The major intrinsic protein (MIP) of the bovine lens fiber membrane: characterization and structure based on cDNA cloning. Cell 39:49-59

Heymann JB, Engel A (1999) Aquaporins: Phylogeny, structure and physiology of water channels. News Physiol Sci 14:187-194

Holz R, Finkelstein A (1970) The water and nonelectrolyte permeability induced in thin lipid membranes by the polyene antibiotics nystatin and amphotericin B. J Gen Physiol 56:125-145

Jung JS, Preston GM, Smith BL, Guggino WB, Agre P (1994) Molecular structure of the water channel through aquaporin CHIP. The hourglass model. J Biol Chem 269:14684-14654

Lahajnar H, Macek P, Zupancic I (2000) Supression of red cell diffusional water permeability by lipophilic solutes. Bioelectrochem 52:179-185

Loike J, Cao L, Kuang K, Vera JC, Silverstein SC, Fischbarg J (1993) Role of facilitative glucose transporters in diffusional water permeability through J774 cells. J Gen Physiol 102:897-906

Macey RI, Farmer RE (1970) Inhibition of water and solute permeability in human red cells. Biochim Biophys Acta 211:104-106

Naccache P, Sha'afi RI (1974) Effect of PCMBS on water transfer across biological membranes. J Cell Physiol 83:449-456

Nakhoul NL, Davis BA, Romero MF, Boron WF (1998) Effect of expressing the water channel aquaporin-1 on the CO_2 permeability of *Xenopus* oocytes. Am J Physiol 274:C543-C548

Pao GM, Wu LF, Johnson KD, Hofte H, Chrisrppel MJ, Sweet G, Sandal NN, Saier MH (1991) Evolution of the MIP family of integral membrane transport proteins. Mol Microbiol 5:33-37

Parisi M, Bourguet J (1983) The single file hypothesis and the water channels induced by antidiuretic hormone. J Membrane Biol 71:189-193

Park JH, Saier MH (1996) Phylogenetic characterization of the MIP family of transmembrane channel proteins. J Membrane Biol 153:171-180

Petersen DC (1980) Water permeation through the lipid bilayer membrane. Test of the liquid hydrocarbon model. Biochem Biophys Acta 600:666-677

Prasad GV, Coury LA, Finn F, Zeidel ML (1998) Reconstituted aquaporin 1 water channels transport CO_2 across membranes. J Biol Chem 273:33123-33126

Preston GM, Agre P (1991) Isolation of the cDNA for erythrocyte integral membrane protein of 28 kilodaltons: member of an ancient channel family. Proc Natl Acad Sci USA 88:11110-11114

Preston GM, Carroll TP, Guggino WB, Agre P (1992) Appearance of water channels in *Xenopus* oocytes expressing red cell CHIP28 protein. Science 256:385-387

Ramesh Prasad GV, Coury LA, Finn F, Zeidel ML (1998) Reconstituted aquaporin-1 water channels transport CO_2 across membranes. J Biol Chem 273:33123-33126

Rich GT, Sha'afi RI, Barton TC, Solomon AK (1967) Permeability studies on red cell membranes of dog, cat, and beef. J Gen Physiol 50:2391-2405

Roudier N, Verbavatz JM, Maurel C, Ripoche P, Tacnet F (1998) Evidence for the presence of aquaporin-3 in human red blood cells. J Biol Chem 273:8407-8412

Schreiber R, Greger R, Nitschke R, Kunzelmann K (1997) Cystic fibrosis transmembrane conductance regulator activates water conductance in *Xenopus* oocytes. Pflügers Arch 434:841-847

Smith BL, Agre P (1991) Erythrocyte Mr 28,000 transmembrane protein exists as a multisubunit oligomer similar to channel proteins. J Biol Chem 266:6407-6415

Smith CP, Rousselet G (2001) Facilitative urea transporters. J Membrane Biol 183:1-14

Solomon AK, Chasan B, Dix JA, Lukacovic MF, Toon MR, Verkman AS (1983) The aqueous pore in the red cell membrane: band 3 as a channel for anions, cations, nonelectrolytes and water. Ann New York Acad Sci 414:97-124

Stein WD (1990) Channels, carriers and pumps. Academic Press, San Diego

Sterling D, Reithmeier RA, Casey JR (2001) A transport metabolon. Functional interaction of carbonic anhydrase II and chloride/bicarbonate exchangers. J Biol Chem 276: 47886-47894

Sui H, Han B-G, Lee JK, Walian P, Jap BK (2001) Structural basis of water-specific transport through the AQP1 water channel. Nature 414:872-878

Terwilliger TC, Solomon AK (1981) Osmotic water permeability of human red cells. J Gen Physiol 77:549-570

Träuble H (1971) The movement of molecules across lipid membranes: A molecular theory. J Membrane Biol 4:193-208

Van Hoek AN, Hom ML, Luthjens LH, De Jong MD, Dempster JA, Van Os CH (1991) Functional unit of 30 kDa for proximal tubule water channels as revealed by radiation inactivation. J Biol Chem 266:16633-16635

Ventura HO, Mehra MR, Young JB (2001) Treatment of heart failure according to William Stokes: the enchanted mercury. J Card Fail 7:277-282

Verkman AS, Mitra AK (2000) Structure and function of aquaporin water channels. Am J Physiol 278:F13-F28

Walz T, Smith BL, Zeidel ML, Engel A, Agre P (1994) Biologically active two-dimensional crystals of aquaporin CHIP. J Biol Chem 269:1583-1586

Yang B,Verkman AS (1997) Water and glycerol permeabilities of aquaporins 1-5 and MIP determined quantitatively by expression of epitope-tagged constructs in *Xenopus* oocytes. J Biol Chem 272:16140-16146

Yang B, Fukuda N, Van Hoek A, Matthay MA, Ma T, Verkman AS (2000) Carbon dioxide permeability of aquaporin-1 measured in erythrocytes and lung of aquaporin-1 null mice and in reconstituted proteoliposomes. J Biol Chem 275:2686-2692

Yang B, Ma T, Verkman AS (2001) Erthythrocyte water permeability and renal function in double knockout mice lacking aquaporin-1 and aquaporin-3. J Biol Chem 276:624-628

Zardoya R, Villalba S (2001) A phylogenetic framework for the aquaporin family in eukaryotes. J Mol Evol 52:391-404

Zhang R, Van Hoek AN, Biwersi J, Verkman AS (1993) A point mutation at cysteine 189 blocks the water permeability of rat kidney water channel CHIP28k. Biochemistry 32:2938-2941

20 Gas Transport

Mikko Nikinmaa

Department of Biology, University of Turku, FIN-20014 Turku, Finland

20.1 Introduction

The erythrocyte plays a decisive role in the gas transport of vertebrates. First, more than 90% of oxygen transported from the respiratory epithelia to the tissue capillaries is transported by haemoglobin, contained within the erythrocyte. Second, erythrocytic carbonic anhydrase is required for effective carbon dioxide transport. One of the most important factors modifying gas transport is the erythrocytic microenvironment. The intraerythrocytic environment can be modulated by membrane transport. The present chapter, therefore, discusses the role of membrane transport in the control of oxygen and carbon dioxide transport in different vertebrate groups.

20.2 Basic Principles of Gas Transport

20.2.1 Oxygen Transport: Factors Affecting Haemoglobin Function

Owing to its intracellular location, the function of vertebrate haemoglobin can be regulated by changing the intracellular environment within the cell. Furthermore, the oxygen capacity of blood can be rapidly changed by liberation and sequestration of erythrocytes in storage organs, and the flow of erythrocytes to different capillary beds can be modulated by selective opening and closing of precapillary sphincters and arterio-venous anastomoses (Nikinmaa and Salama 1998). Although these aspects of erythrocyte function are taken for granted, none of them are possible in many invertebrates with high molecular weight respiratory pigments dissolved in plasma.

The haemoglobin of most vertebrates is a tetrameric globular molecule consisting of two α and two β chains all capable of binding one molecule of oxygen (for recent reviews, see Perutz (1990); Hsia (1998); Jensen et al. (1998a)). Exceptions to this generalization are the haemoglobins of agnathans (hagfish and lampreys) which are in monomer-oligomer equilibrium in physiological conditions (Nikinmaa et al. 1995; Fago and Weber 1998; Qiu et al. 2000), the haemoglobin of a snake, which appears to be in dimer-tetramer equilibrium (Bonafe et al. 1999), and some haemoglobins of amphibians, reptiles, birds and embryonic mammals, the haemoglobin tetramers of which can aggregate to higher polymers (Riggs 1998).

The number of amino acids in the α chain is slightly smaller than in the β chain – in man 141 and 146, respectively. The polypeptide chains consist of α helical segments (labelled from A to H) and shorter non-helical segments. The oxygen-binding haem group is buried in a hydrophobic pocket between E and F helices of each polypeptide chain and is covalently bound to a histidine residue (His F8), which is present in all haemoglobins studied (Perutz 1990). The conformation and microenvironment of the haem pocket forms the basis for the reversible oxygen binding by haemoglobin.

The oxygen binding properties of haemoglobin molecule can be described using the oxygen equilibrium curve, which relates the oxygen saturation of haemoglobin to the oxygen tension (Fig. 20.1). The oxygen affinity of haemoglobin is given by the position of the equilibrium curve, and in physiological conditions is influenced by the intrinsic oxygen affinity of haemoglobin, its sensitivity to heterotrophic ligands – both of which are determined by the globin chain structure, and the concentration of haemoglobin and heterotrophic ligands within the cell.

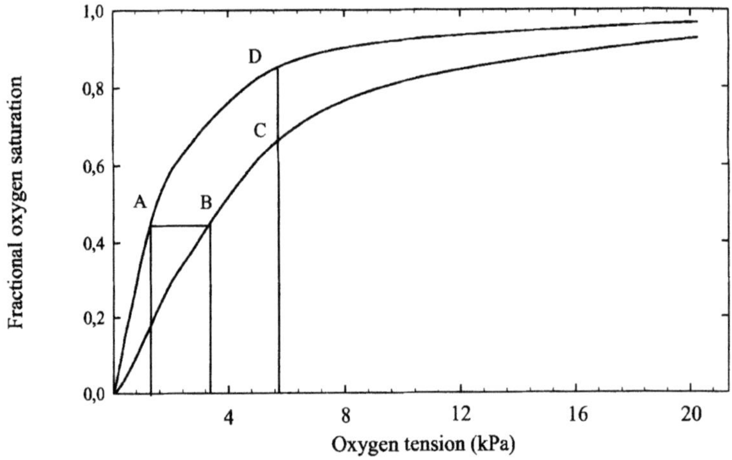

Fig. 20.1. Oxygen equilibrium curves of vertebrate haemoglobins with fractional oxygen saturation on the ordinate and blood oxygen tension on the abscissa. An increase in cellular organic phosphate concentration, a decrease in cell volume or a decrease in intraerythrocytic pH shift the oxygen equilibrium curve to the right. As long as the arterial oxygen saturation can be maintained, a rightward shift is beneficial, because the same amount of oxygen is given up in tissue capillaries at a high oxygen tension, speeding up diffusion (*A–B*). However, in deep hypoxia, a leftward shift of the oxygen equilibrium curve will significantly increase the arterial oxygen saturation (*C–D*)

Physiologically the most important factors affecting the haemoglobin oxygen affinity are cellular haemoglobin concentration (red cell volume), intraerythrocytic pH, and concentrations of organic phosphates, mainly 2,3-diphosphoglycerate (2,3-DPG) in mammals, inositol pentaphosphate (IPP) in birds, ATP in reptiles and amphibians, and ATP and GTP in fish (Nikinmaa 1990). In addition, several other anions may be important in modifying the haemoglobin oxygen affinity such as HCO_3^- ions in the case of crocodilians (Bauer et al. 1981) and chloride ions (Fronticelli et al. 1984). An increase in erythrocyte organic phosphate concentration (and concentration of other anions) decreases the haemoglobin oxygen affinity both by direct binding to haemoglobin and by influencing the erythrocyte pH (Wood and Johansen 1973). A decrease in erythrocyte pH decreases the haemoglobin oxygen affinity via the Bohr effect, and a decrease in erythrocyte volume decreases the haemoglobin oxygen affinity by increasing the likelihood of interactions between haemoglobin and organic phosphates and by favouring the aggregation of haemoglobin molecules.

20.2.2 Factors Affecting Carbon Dioxide Transport and Excretion in Blood

Metabolically produced carbon dioxide acts as a weak acid in aqueous solutions and must thus be excreted at the same rate as it is produced to avoid acid-base imbalances. Carbon dioxide is carried in blood from the site of production to the excretory sites, mainly the respiratory epithelium. Only a small fraction (< 5%) of carbon dioxide is carried as molecular CO_2, owing to the relatively high pH of plasma, and the low solubility of CO_2 in plasma (Boutilier et al. 1984). The proportion of carbon dioxide carried as carbamino compounds varies between vertebrate groups, being high – 10–15% of total – in animals in which the terminal amino groups of haemoglobin are available for carbamino formation, and reduced in species in which the N-terminal amino acids are acetylated (Klocke 1987; Tufts and Perry 1998). Most of the excreted carbon dioxide is, however, carried as HCO_3^-, distributed in the plasma and erythrocytes. Since the proportion of total carbon dioxide as HCO_3^- increases with increasing pH, the total carbon dioxide carrying capacity also increases with increasing pH.

The uncatalysed hydration/dehydration reactions between carbon dioxide and HCO_3^- are too slow to allow significant excretion of carbon dioxide during the residence time of blood in the respiratory epithelia (Edsall 1969; Geers and Gros 2000). Thus, catalysis by the enzyme carbonic anhydrase is required for effective carbon dioxide excretion. Modelling of carbon dioxide transport has indicated that the cytoplasmic carbonic anhydrase of erythrocytes is the most important component of effective carbon dioxide excretion, although extra- and intracellular carbonic anhydrase activity is also present in several other tissues (Maren 1967; Geers and Gros 2000; Henry and Swenson 2000). The activity of the enzyme is very high within the erythrocytes – in most cases much higher than that theoretically required for virtually maximal carbon dioxide excretion during capillary transit time at rest (Roughton 1964). The enzyme present in most vertebrate erythrocytes is the high-activity isoform carbonic anhydrase II (Geers and Gros 2000). However, in agnathans and elasmobranchs (Maren et al. 1980) the high affinity isoform is missing, and only carbonic anhydrase isoform I, with a specific activity of approx. 10% of that of isoform II, is present. Interestingly, the activity of carbonic anhydrase II is markedly reduced (or absent) also in hypoxia-tolerant mammals such as the mole rat, beluga whale and llama (Yang et al. 1998, 2000).

To enable the participation of erythrocytic carbonic anhydrase in carbon dioxide excretion, it is required that extracellular carbon dioxide has access to the intracellular compartment. The permeability of erythrocyte membrane to molecular carbon dioxide is very high, approx. 1 cm/s, and does not limit carbon dioxide transport in any situation. Since the permeability for HCO_3^- across artificial phospholipid vesicles is six orders of magnitude lower than that for carbon dioxide, in the absence of facilitation of HCO_3^- transport only the erythrocytic compartment is able to participate in the CO_2/HCO_3^- interconversion and, thus, in carbon dioxide excretion. This is the case for carbon dioxide excretion in agnathans (hagfish and lampreys). The HCO_3^- and Cl^- permeability of the erythrocyte membrane of these animals is very low owing to the lack of functional anion exchange pathway (Ohnishi and Asai 1985; Nikinmaa and Railo 1987; Ellory et al. 1987; Tufts and Boutilier 1989, 1990; Peters et al. 2000) – the half-time for chloride equilibration across the erythrocyte membrane in the lamprey *Lampetra fluviatilis* is over 2 h (Nikinmaa and Railo 1987). In other vertebrates, the anion exchange protein (band 3) is the most common membrane protein – generally with more than 10^6 copies/cell – and decreases the time constant for chloride equilibration across the erythrocyte membrane to values around 0.1–0.5 s in different species from teleosts to man (Jensen and Brahm 1995; Jensen et al. 1998b). Since the residence time of blood in respiratory epithelia ranges from 0.3 to 3 s (Nikinmaa 1990), the presence of anion exchange pathway enables the utilization of plasma HCO_3^- in carbon dioxide excretion.

Since the hydration/dehydration reactions between carbon dioxide and HCO_3^- involve the release/uptake of protons, the rate of both uncatalyzed and catalyzed reactions depends on the availability of buffers for proton uptake and release. The non-bicarbonate buffering capacity of human erythrocytes is approx. 60 mmol/(l·pH unit) (Siggaard-Andersen 1974), that of the teleost, *Tinca tinca*, only slightly above 20 mmol/(l·pH unit) (Jensen 1986), and that of lampreys approx. 40 mmol/(l·pH unit) (Tufts and Boutilier 1989; Nikinmaa and Mattsoff 1992) – all these values are higher than the non-bicarbonate buffering values in plasma (approx. 10 mmol/(l·pH unit)). In terms of carbon dioxide transport, however, the oxylabile hydrogen binding and release – Haldane effect – is more important than the buffering capacity as such (Brauner and Randall 1996). In the tissues, the production of HCO_3^- from carbon dioxide within the erythrocytes is facilitated by a simultaneously occurring proton uptake by haemoglobin as oxygen is released. Conversely, in the respiratory epithelia the conversion of HCO_3^- to carbon dioxide is facilitated by the proton release by haemoglobin upon oxygenation. The largest Haldane effects are observed in lampreys and teleost fish (Jensen 1989; Nikinmaa 1997). On the basis of the present data it appears that among vertebrates the Haldane effect is smallest in dogfish (Jensen 1989).

20.3 Control of Red Cell pH

20.3.1 Control of Red Cell pH in the Absence of Significant Secondarily Active Cation or Proton Transport

The equilibration of Cl^- and HCO_3^- across the erythrocyte membrane is generally rapid in vertebrates with the exception of the agnathans. The Cl^-/HCO_3^- exchange is driven by the chemical potential differences of the ions across the membrane, and, for this reason, the direction of net transport is determined by the ion having the steeper gradient. Net transport ceases when the distribution ratios for the two ions across the membrane are the same. The intra- and extracellular HCO_3^- activities are coupled to the respective H^+ activities via the hydration/dehydration reactions between HCO_3^- and carbon dioxide. These reactions are catalysed by the enzyme carbonic anhydrase in the intracellular compartment (Maren 1967; Geers and Gros 2000), but generally occur at the uncatalysed rate at the extracellular compartment, making the extracellular reactions rate-limiting for acid equilibration across the erythrocyte membrane. The approx. half-times for uncatalyzed HCO_3^- dehydration are 10–15 s at 37 °C, 1 min at 15 °C and 3–4 min at 5 °C (Heming 1984). The extracellular hydration/dehydration reactions are, however, catalysed by carbonic anhydrase in the plasma in elasmobranchs (Wood et al. 1994) and by extracellularly facing carbonic anhydrase in the capillary endothelium in the lungs and muscles of mammals (Crandall and O'Brasky 1978). As long the transport of acid equivalents via other pathways is significantly slower than the extracellular reactions between carbon dioxide and HCO_3^-, acid equilibration occurs mainly via the Cl^-/HCO_3^- exchange (Fig. 20.2).

When acid equivalents are transported mainly by the Cl^-/HCO_3^- exchange, the major factors influencing erythrocyte pH are changes in the charge and concentration of the intracellular polyions, mainly haemoglobin and organic phosphates. In addition to affecting the pH gradient across erythrocyte membrane, the charge and concentration of intracellular polyions affects erythrocyte volume. As examples, the cellular NTP (ATP and GTP) concentrations of fish erythrocytes decrease in hypoxic conditions. NTPs are virtually impermeable polyanions, and a decrease in their concentration results in a net influx of permeable (univalent) anions, Cl^- and HCO_3^-, to maintain electroneutrality. Consequently, the HCO_3^- distribution ratio (inside/outside) increases and the H^+ distribution ratio (inside/outside) – and thus pH gradient – decreases (Hladky and Rink 1977). The increase in intracellular pH causes a leftward shift of the oxygen equilibrium curve. Simultaneously, the increase in the concentration of permeable, osmotically active anions causes a net influx of water, and cells swell. Notably, hydrolysis of ATP to ADP does not decrease the concentration (or charge) of intracellular impermeable polyions, and, therefore, the decrease in ATP (NTP) levels in hypoxic conditions can only contribute to the intracellular pH and cell volume if the total pool of nucleotides within the erythrocytes decreases. Such a decrease in total nucleotide concentration during *in vivo* hypoxia has been documented (Tetens 1987). However, the mechanisms for nucleotide breakdown or extrusion in hypoxic conditions are unclear at present time.

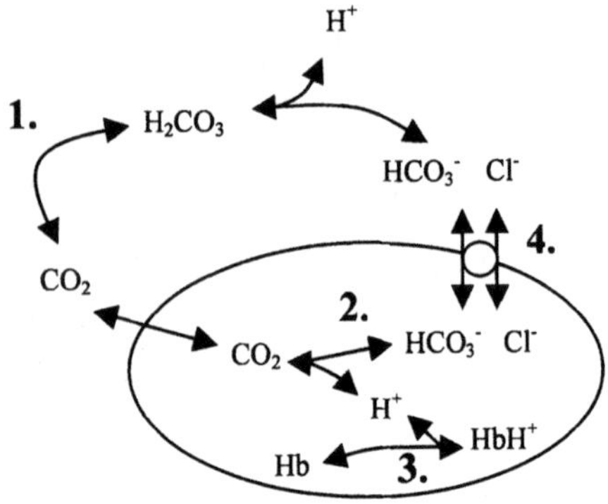

Fig. 20.2. Control of erythrocyte pH in the absence of secondarily active cation or proton transport. Extracellular H^+ loads are transferred to the intracellular compartment via the Jacobs-Stewart cycle (Jacobs and Stewart 1942) in the following fashion. Extracellular protons and HCO_3^- are dehydrated to carbon dioxide and water (*1*). The rate of extracellular dehydration reaction depends on the availability of extracellular carbonic anhydrase. Carbon dioxide diffuses rapidly to the intracellular compartment and is hydrated to HCO_3^- and protons in a reaction catalysed by intracellular carbonic anhydrase, present in all vertebrates (*2*). The protons formed are largely buffered by haemoglobin (*3*), and, thus, intracellular pH changes depend on the buffering capacity of haemoglobin. The HCO_3^- formed can be exchanged for chloride via the anion exchange pathway (*4*), completing the cycle

The charge of intracellular polyions changes as a result of either extracellular pH changes or changes in oxygen partial pressure. A decrease in extracellular pH will be transmitted to the intracellular compartment because of the coupling of HCO_3^- and H^+ activities. A decrease in the intracellular pH will decrease the negative charge of haemoglobin and organic phosphates, Cl^- and HCO_3^- enter the cell to maintain electroneutrality. Consequently, the pH gradient between intracellular and extracellular compartments is reduced and the cells swell (Hladky and Rink 1977; Nikinmaa 1992). Because oxyhaemoglobin is more acidic than deoxyhaemoglobin within the physiological pH range (German and Wyman 1937), deoxygenation of blood will both increase blood pH (because of the proton uptake of haemoglobin) and decrease the negative charge on haemoglobin protein with the net consequence that erythrocyte pH increases.

20.3.2 Effect of Na^+/H^+ Exchange on Red Cell pH

The presence of Na^+/H^+ exchange has been demonstrated in the erythrocytes of most vertebrate groups from lampreys to mammals, including man (Escobales and Canessa 1986). In lampreys, the amiloride-sensitive sodium transport system (Nikinmaa et al. 1986; Tufts 1992) is mainly stimulated by acidification (Virkki and Nikinmaa 1994), and by a decrease in oxygen tension (Virkki et al. 1998). In teleost fish, the major stimuli activating the transport system are osmotic shrinkage (Romero et al. 1996), β-adrenergic stimulation (Nikinmaa and Huestis 1984) and acidification (Weaver et al. 1999). In some cases, as in flounder, the cells respond to all three stimuli (Weaver et al. 1999), in others, as in the eel and rainbow trout, the response is virtually restricted to activation either by osmotic shrinkage (eel) or by catecholamines (trout; Romero et al. 1996). In most cases studied, the activity of Na^+/H^+ exchange is increased with deoxygenation, regardless of the primary stimulus causing increased fluxes (Motais et al. 1987; Weaver et al. 1999), it is even possible that the activation caused by the primary stimulus is only observed at reduced oxygen tension (Salama and Nikinmaa 1988). In the red cells of amphibians, Na^+/H^+ exchange is stimulated by osmotic shrinkage or acidification, as in *Amphiuma* (Cala 1985; Cala et al. 1988), or by β-adrenergic stimulation as in *Rana* species (Rudolph and Greengard 1980; Kaloyianni and Rasidaki 1996). Hypoxia increases the rate of Na^+/H^+ exchange at least in *Rana* (Kaloyianni and Rasidaki 1996). In mammals, osmotic shrinkage activates the Na^+/H^+ exchange at least in rabbit and dog erythrocytes (Parker 1983; Jennings et al. 1986).

Because the sodium gradient across the erythrocyte membrane favours net proton efflux, the Na^+/H^+ exchange tends to increase intracellular pH (Fig. 20.3). In the case of lamprey erythrocytes which lack a functional anion exchange pathway, the amiloride-sensitive H^+ transport pathway, Na^+/H^+ exchange, appears to be the major pathway for transport of acid equivalents across the erythrocyte membrane (Nikinmaa et al. 1986). As a result, the intraerythrocytic pH can be significantly higher than the extracellular one at normal physiological plasma pH values (Nikinmaa 1986).

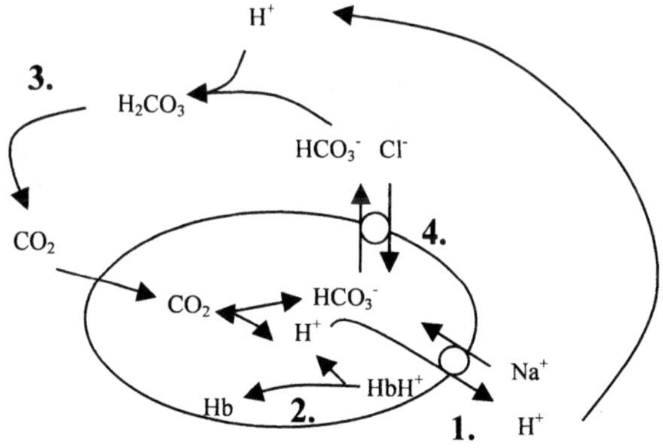

Fig. 20.3. Effect of Na^+/H^+ exchange activation on erythrocyte pH. Increased rate of Na^+/H^+ exchange (*1*) causes net proton extrusion with the consequence that the cellular pH tends to increase. The observed pH increase depends on the buffering capacity of haemoglobin. Haemoglobin also liberates protons to function as substrate for the exchanger (*2*). Net proton excretion occurs as long as the rate of Na^+/H^+ exchange is faster than the rate-limiting step of passive proton equilibration, the extracellular dehydration of HCO_3^- and protons to carbon dioxide (*3*). The rate of extracellular carbonic acid dehydration initially depends mainly on the accumulation of protons in the extracellular compartment and is, therefore, influenced both by the Na^+/H^+ exchange rate and the extracellular buffering capacity. After the initial stages, no further pH changes in the intracellular compartment occur, but net sodium and chloride influx continues via the Na^+/H^+ exchanger and the anion exchanger (*4*)

In teleost fishes, the activation of Na^+/H^+ exchange, especially by catecholamines, can increase erythrocyte pH even in the presence of a functional anion exchange pathway (Nikinmaa 1982a). Any reduction in the oxygen content of arterial blood causes a liberation of catecholamines into the blood stream (Perry et al. 1991). Catecholamines bind to the β-adrenergic receptors on the membrane. It appears that the number of receptors increases in hypoxic conditions (Marttila and Nikinmaa 1988); internalized receptors are probably recycled to the cell membrane (Reid et al. 1991; Reid and Perry 1991). Binding of catecholamines to β-adrenergic receptors activates adenylate cyclase, and cAMP levels increase, causing activation of the Na^+/H^+ exchange (Mahé et al. 1985). Also the accumulation of cAMP depends on the oxygen tension in some species (Salama and Nikinmaa 1990). The Na^+/H^+ exchange causes a net influx of Na^+ and efflux of H^+, displacing H^+ from electrochemical equilibrium during the initial minutes of activation. This is possible even though anion exchange is functional, since the turnover rate of Na^+/H^+ exchange approaches the rate of extracellular uncatalyzed hydration/dehydration reactions between HCO_3^- and carbon dioxide. Thus, a large number of protons can be extruded via the Na^+/H^+ exchanger, and accumulate in the extracellular compartment, before the carbonic acid formed from the accumulated H^+ and HCO_3^- is dehydrated to carbon dioxide to a significant extent. After the initial minutes, the H^+ and HCO_3^- movements are coupled, and no further changes in erythrocyte pH occur, although Na^+ and Cl^- continue to accumulate within the cell, and osmotically obliged water follows causing further erythrocyte swelling (for details see Nikinmaa and Boutilier (1996); Nikinmaa and Salama (1998)).

Several factors influence the extent of intracellular and extracellular pH changes observed after adrenergic stimulation. First, the activity of adrenergically stimulated Na^+/H^+ exchange is influenced both by acidification and oxygen tension (Motais et al. 1987): in rainbow trout maximal activity is observed in hypoxic conditions at around (extracellular) pH 7.3. Second, there are marked differences between species both in the maximal activity of the transporter and its pH- and oxygen sensitivity (Salama and Nikinmaa 1989). Third, for a given activity of the Na^+/H^+ exchanger, the observed intra- and extracellular pH changes depend on the buffering capacity, and buffers, present. The buffering capacity of teleost haemoglobins, the major buffers of the intracellular compartment, is smaller than in most other vertebrates, whereby greater changes in erythrocyte pH occur for a given rate of Na^+/H^+ exchange than in, e.g., mammals (Nikinmaa 1997). With regard to the extracellular compartment, pronounced decreases in pH are observed in carbon dioxide-HCO_3^- buffered medium (Motais et al. 1989), since the proton disequilibrium can fully develop owing to the slow extracellular hydration/dehydration reactions between HCO_3^- and carbon dioxide. Since the extracellular non-bicarbonate buffering capacity of plasma is small, this is also the case *in vivo*, as demonstrated in hypoxic rainbow trout (Thomas and Hughes 1982). With an increase in extracellular non-bicarbonate buffering capacity, the effects of Na^+/H^+ exchanger activation are progressively attenuated (Nikinmaa and Boutilier 1996).

Whereas the role of the Na^+/H^+ exchange in the control of erythrocyte pH has been clearly demonstrated in lampreys and teleost fish, the data on other vertebrate erythrocytes are much more fragmentary. Presently, it is known that Na^+/H^+ exchange makes a contribution to the steady-state pH of the amphibian *Amphiuma*

(Tufts et al. 1987; Cala et al. 1988). However, since the buffering capacity of haemoglobins in elasmobranchs and air-breathing vertebrates is greater than, and the measured activities of the Na^+/H^+ exchanger smaller than in teleost fish, the contribution of Na^+/H^+ exchange to the control of erythrocyte pH is probably smaller than in teleost fish (Nikinmaa 1997).

Since the function of Na^+/H^+ exchange is associated with a net influx of Na^+, its activation leads to cell swelling (e.g. Nikinmaa 1982). The volume changes are maximal when the Na^+/H^+ exchanger is coupled to anion exchange (e.g. Nikinmaa 1989).

20.3.3 Effect of K^+-Cl^- Cotransport on Red Cell pH

At high oxygen tensions, K^+-Cl^- cotransport is activated in erythrocytes of several species (Gibson et al. 2000). The efflux of K^+ and Cl^- causes a decrease in cell volume and is utilized in volume regulation of many erythrocyte types (Cossins and Gibson 1997). Since Cl^- movements are coupled to HCO_3^- movements via the anion exchanger in most species, net loss of Cl^- causes a reduction of cellular HCO_3^- concentration and consecutive increase in the pH gradient across the erythrocyte membrane (i.e. decrease in intracellular pH; Fig. 20.4). To date experimental measurements of changes in erythrocyte pH owing to the increased K^+-Cl^- cotransport activity have not been made. However, marked extracellular alkalinization has been observed in nitrite-treated and hypotonically treated carp erythrocytes, in which K^+-Cl^- cotransport is activated (Jensen 1990, 1995). This suggests that the intracellular compartment is, indeed, acidified. In addition

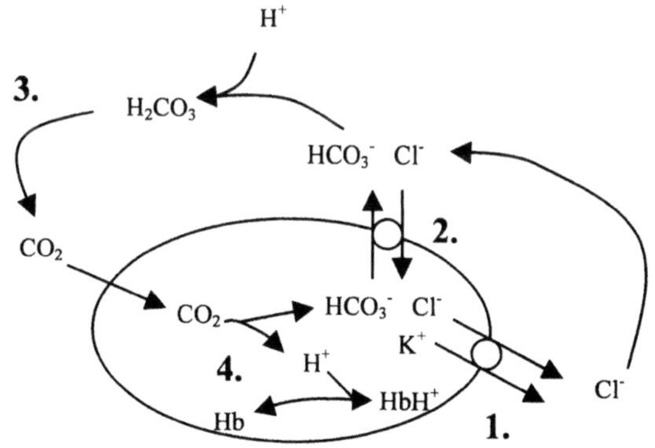

Fig. 20.4. Effect of K^+-Cl^- cotransport activation on the erythrocyte pH. The net efflux of chloride via the K^+-Cl^- cotransport pathway (*1*) generates a disequilibrium for the anion exchanger (2), leading to a net efflux of HCO_3^-. Since basic equivalents are removed from the intracellular compartment, the cell contents are acidified. The degree of acidification depends on the buffering capacity of haemoglobin. Further, the extracellular HCO_3^- will be dehydrated to carbon dioxide and water with the consequent passive acid-base equilibration (*3–4*), whereby the pH changes are limited to the initial stages of K^+-Cl^- cotransport activation

to the effect of pH, K^+-Cl^- cotransport at elevated oxygen tensions causes a reduction of erythrocyte volume to a level far below that expected on the basis of the oxygenation-dependent change in the charge of haemoglobin.

20.4 Roles of pH Changes in Regulation of Oxygen Transport

Although the effect of protons on haemoglobin function varies between species, in virtually all vertebrates a decrease in erythrocyte pH causes a decrease in haemoglobin oxygen affinity. With the help of the Bohr effect, oxygen delivery in tissues can be accomplished while maintaining a large oxygen partial pressure gradient between capillary blood and oxygen-requiring structures in the tissues. The capillary pH of rapidly metabolizing tissues decreases because of the production of metabolic acids and carbon dioxide. Consequently, the oxygen equilibrium curve is shifted to the right, and more oxygen is given up from haemoglobin at any given partial pressure.

Changes in the capillary pH are also caused by oxygenation/deoxygenation reactions of haemoglobin. Thus, during deoxygenation of haemoglobin in the capillaries the capillary pH increases, because deoxyhaemoglobin is a weaker acid than oxyhaemoglobin. It has been suggested that oxygen delivery to the tissues would be compromised by the proton uptake upon deoxygenation, since the net pH changes in the capillary blood are diminished. However, the acidification caused by carbon dioxide and metabolic protons, and the alkalinization caused by deoxygenation differ in one significant respect: whereas carbon dioxide and metabolic protons will acidify blood at constant oxygen saturation, deoxygenation-induced alkalinization is part of the haemoglobin mechanism, and protons will be released as soon as the oxygen saturation of haemoglobin increases to its original level. Thus, pH at constant oxygen saturation is not affected by the oxygenation-dependent proton uptake and release, and, therefore, the haemoglobin oxygen affinity cannot be affected either (Nikinmaa 1997).

Whereas the passive changes in erythrocyte pH in the capillary bed tend to improve oxygen unloading, the function of the Na^+/H^+ exchanger is to improve oxygen loading in the respiratory epithelia. In lamprey erythrocytes, the activation of Na^+/H^+ exchange by acidification and by deoxygenation makes the system well suited to safeguard oxygen transport. Whenever the oxygen affinity of intracellular haemoglobin is decreased by a decrease in intracellular pH or oxygen saturation of haemoglobin decreased because of hypoxia, the Na^+/H^+ exchange will tend to increase erythrocyte pH thus shifting the oxygen equilibrium curve to the left and facilitating oxygen loading (Nikinmaa et al. 1995). In fact, it appears that the high pH of lamprey erythrocytes is required to ensure that the association-dissociation reactions of haemoglobin, required for effective oxygen loading and unloading (Perutz 1990; Nikinmaa et al. 1995), occur in intact erythrocytes. If the erythrocyte pH of *Lampetra fluviatilis* were similar to that of, e.g., rainbow trout or man, the haemoglobin would become only 70% saturated in the respiratory epithelium at 100% air-saturated water (Nikinmaa 1993a).

The adrenergically activated Na^+/H^+ exchanger of teleost fish also functions to facilitate oxygen loading in gills in adverse conditions. *In vitro* oxygen equilibrium curves after adrenergic stimulation indicated that catecholamines increase the haemoglobin oxygen affinity of rainbow trout (Nikinmaa 1983), and *in vivo* injections of adrenaline caused an increase in the arterial oxygen content (Nikinmaa 1982b). The response is present also in physiologically relevant situations. Catecholamine concentrations of fish blood increase during physical disturbance and hypoxia to levels adequate to cause significant activation of the Na^+/H^+ exchange (Nikinmaa 1992). Consequently, the erythrocyte pH is maintained at a constant level despite pronounced plasma acidification in physically disturbed animals (Nikinmaa et al. 1984; Primmett et al. 1986), and may markedly increase in hypoxic animals (Nikinmaa et al. 1987; Tetens and Christensen 1987) – if the animals are treated with the adrenergic antagonist, propranolol, before the disturbance or hypoxic exposure, there is a significant drop in the erythrocyte pH as compared to non-treated animals. Consequently, the arterial oxygen content of hypoxic and physically disturbed control animals is higher than that of animals treated with propranolol.

In contrast, if K^+-Cl^- cotransport contributes to regulation of oxygen transport, its effect is to lower pH and decrease haemoglobin oxygen affinity. The possible physiological significance of this reduction has not been investigated in detail, but could facilitate oxygen unloading at the arterial end of the capillary bed in environmental hyperoxia during which the K^+-Cl^- cotransport is maximally activated (Nikinmaa and Salama 1998).

20.5 Red Cell Volume and Haemoglobin Oxygen Affinity

Since perturbations of the steady-state volume are usually associated with a change in the pH gradient across the erythrocyte membrane (Hladky and Rink 1977; Nikinmaa 1992), direct effects of erythrocyte volume on haemoglobin oxygen affinity are difficult or impossible to separate from the simultaneous effects of pH on haemoglobin function. A separation between the effects of erythrocyte volume and pH can be made for the function of lamprey haemoglobin. A decrease in mean cellular haemoglobin concentration in intact erythrocytes increases the haemoglobin oxygen affinity significantly in lamprey without changes in intraerythrocytic pH (Airaksinen and Nikinmaa 1995). Thus, the decrease in the mean cellular haemoglobin concentration, observed in hypoxic lampreys (Nikinmaa and Weber 1984), will increase the haemoglobin oxygen affinity. The increase in erythrocytic volume of lampreys in hypoxic conditions may be associated with the increased activity of Na^+/H^+ exchange, also observed with decreasing oxygen tension in lampreys (Virkki et al. 1998). It is critical for the influence of erythrocyte volume on haemoglobin oxygen affinity in hypoxic lampreys that the cells remain at the elevated volume throughout the hypoxic period. Notably, the regulatory volume decrease, observed after swelling lamprey erythrocytes (Nikinmaa et al. 1993), is markedly inhibited by hypoxic conditions (Virkki et al. 1998), mainly because of the inhibition of the swelling-activated chloride transport pathway by hypoxic conditions.

20.6 Interactions between Membrane Transport and Regulation of Haemoglobin Oxygen Affinity by Organic Phosphates

The concentrations of the major organic phosphates modifying haemoglobin oxygen affinity, i.e., 2,3-DPG, ATP and GTP, are normally metabolically regulated. The 2,3-DPG concentration of mammalian erythrocytes is influenced by erythrocyte pH. An increase in pH tends to increase the concentration of glycolytic intermediates such as 1,3-diphosphoglycerate with a consequent activation of 2,3-DPG synthase and increase in 2,3-DPG concentration (Duhm and Gerlach 1971; Lenfant et al. 1972). The erythrocyte ATP concentration is influenced by the ratio between energy production and consumption. As an example, when the activity of the sodium pump is markedly increased as a result of increased intracellular sodium concentration in fish erythrocytes, the cellular ATP concentration is reduced (Ferguson et al. 1989). In general, however, the factors adjusting the ATP concentration are poorly known. It appears that adenosine A_2 and β-adrenergic receptors may be involved in the modification of ATP levels in developing bird embryos (Dragon et al. 1999). Adrenergic stimulation also affects erythrocyte ATP concentration in fish (Nikinmaa 1983), but the physiological role of this effect is unclear at present (Nikinmaa and Salama 1998). With regard to the high GTP concentration of fish and IPP concentration of bird erythrocytes, the regulatory mechanisms involved in the modification of the levels remain unknown.

The effects of organic phosphates on haemoglobin oxygen affinity can be influenced by complex formation with other intracellular constituents. For example, both ATP and GTP are complexed with magnesium, whereby their effects on the oxygen equilibrium curves are reduced (Bunn et al. 1971; Weber 1978). Thus, the transport of Mg^{2+} (Günther 1993) across the erythrocyte membrane may indirectly influence haemoglobin function.

20.7 Patterns of Carbon Dioxide Transport and Excretion in Vertebrates

In agnathans (hagfish and lampreys), the major feature influencing carbon dioxide transport and excretion is the lack of a functional anion exchange pathway (Fig. 20.5). For this reason, most of the carbon dioxide excreted in the gills stems from erythrocytic HCO_3^- (Nikinmaa and Mattsoff 1992; Tufts et al. 1992). Carbon dioxide produced in the tissues diffuses down its concentration gradient into the erythrocyte, where carbonic anhydrase facilitates its conversion to HCO_3^-. The protons produced in the reaction are largely taken up by haemoglobin, which is simultaneously deoxygenated. In the gills, the sequence of events is reversed. Carbon dioxide diffuses down its concentration gradient from the erythrocytes into the water shifting the equilibrium of the reaction catalyzed by carbonic anhydrase towards carbon dioxide formation. Simultaneously protons are liberated from haemoglobin upon oxygenation, which also facilitates the reaction catalyzed by carbonic anhydrase. Although this pattern of carbon dioxide excretion is limited to agnathans as a group, it is possible even for mammals to survive without a functional anion exchange pathway, as indicated by the cattle with hereditary total deficiency of red cell band 3 (Inaba et al. 1996).

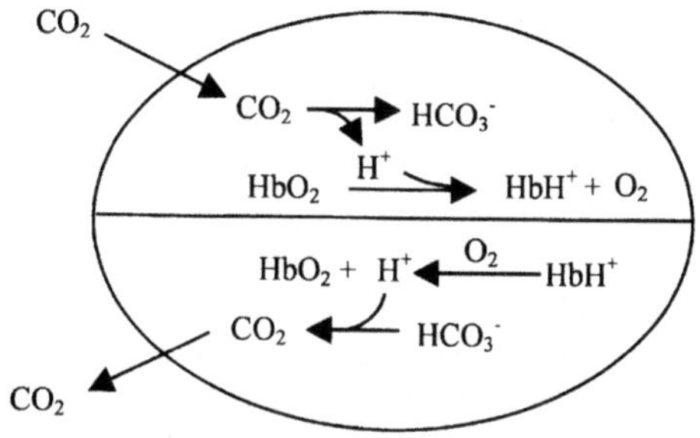

Fig. 20.5. Carbon dioxide excretion in agnathans, which lack a functional anion exchange pathway. In tissues carbon dioxide enters the cell and is converted to HCO_3^- and H^+. Protons are taken up by haemoglobin whereas HCO_3^- remains in the cell. Its accumulation depends on the intracellular pH. In the gills the sequence of events is reversed

In hagfish carbon dioxide excretion is inefficient for several reasons. First, the pH of hagfish erythrocytes is low (at an extracellular pH of 7.6 the erythrocyte pH of *Myxine glutinosa* is approx. 6.9 (Tufts and Boutilier 1990)). As a consequence, the concentration of HCO_3^- at equilibrium at any given carbon dioxide tension is low. Second, the Haldane effect of haemoglobin is small precluding significant uptake and release of protons from the catalyzed reaction in the tissue and gills thus reducing the rates of hydration and dehydration reactions between HCO_3^- and carbon dioxide (and influencing the final equilibrium reached). Third, the haematocrit value of hagfish blood is small – only 8–15% (Nikinmaa 1997).

In lampreys, on the other hand, the effectiveness of carbon dioxide transport appears to be similar to that of other active vertebrates despite the fact that plasma is not available for carbon dioxide transport. The efficiency of carbon dioxide transport is based on two factors. First, the erythrocyte pH is maintained at a high value by Na^+/H^+ exchange (Nikinmaa 1986; Virkki and Nikinmaa 1994), increasing the cellular HCO_3^- stores. Second, the Haldane effect of lamprey erythrocytes is pronounced: the erythrocyte pH increases by 0.3–0.4 units upon deoxygenation (Ferguson et al. 1992; Nikinmaa and Mattsoff 1992). Consequently, the difference in the total carbon dioxide content between arterial and venous erythrocytes is large. However, although carbon dioxide transport by lamprey blood is as effective as that in vertebrates possessing the anion exchanger, there is one significant disadvantage in terms of acid-base regulation. Because the major buffer of blood, haemoglobin, is isolated from buffering non-bicarbonate acid loads in the plasma, marked fluctuations of extracellular pH are observed during acid-base disturbances (Mattsoff and Nikinmaa 1988).

Starting from teleost and elasmobranch fish, vertebrate erythrocytes possess the anion exchange pathway, which enables the transfer of HCO_3^- across the erythrocyte membrane during physiologically relevant time scale, and makes the utilization of plasma HCO_3^- in carbon dioxide excretion possible (Fig. 20.6). Thus, carbon dioxide produced in the tissues diffuses down its concentration gradient into the erythrocyte, where carbonic anhydrase facilitates its conversion to HCO_3^-. The production of HCO_3^- generates a disequilibrium for the anion exchanger, and a net efflux of HCO_3^- and influx of Cl^- occurs. Furthermore, the protons produced in the reaction are largely taken up by haemoglobin, which is simultaneously deoxygenated. Removal of both end products of the catalyzed reaction enables continued hydration of carbon dioxide in the tissues. In the respiratory epithelia, the sequence of events is reversed. Carbon dioxide diffuses down its concentration gradient from the erythrocytes into the water shifting the equilibrium of the reaction catalyzed by carbonic anhydrase towards carbon dioxide formation. Simultaneously protons are liberated from haemoglobin upon oxygenation, and HCO_3^- enters the cell via the anion exchange pathway, further facilitating the reaction catalyzed by carbonic anhydrase.

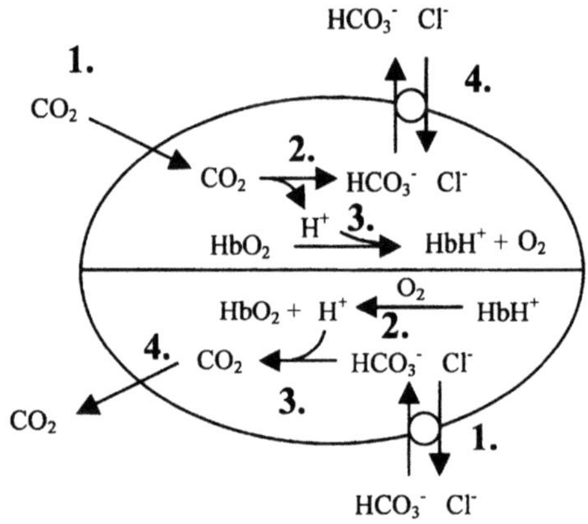

Fig. 20.6. Pattern of carbon dioxide excretion in most vertebrates. Carbon dioxide produced in the tissues diffuses into the erythrocyte (*1*) and is converted to HCO_3^- and protons in the reaction catalysed by carbonic anhydrase (*2*). The protons produced are largely taken up by haemoglobin upon deoxygenation (*3*), and the HCO_3^- ions produced are exchanged for chloride via the anion exchange pathway (*4*). In the respiratory epithelia the sequence of events is reversed. HCO_3^- enters the cell in exchange for chloride (*1*). Protons are released from haemoglobin upon oxygenation (*2*). HCO_3^- and protons form carbon dioxide in the reaction catalysed by carbonic anhydrase (*3*), and carbon dioxide diffuses from the cells to the environment (*4*)

Although the general pattern of carbon dioxide transport is similar in teleost and elasmobranch fish and tetrapods, there are several differences. At the prevailing body temperatures of several teleost fishes, the half-time for Cl^- (and HCO_3^-) equilibration via the anion exchange is similar to the residence time of blood in gills (Jensen and Brahm 1995; Tufts and Perry 1998). In these species the relative role of plasma as compared to erythrocyte HCO_3^- in carbon dioxide excretion is diminished. Interestingly, it appears that the species with relatively slow anion exchange rate have the highest Haldane effects (Jensen 1989; Jensen and Brahm 1995), whereby efficiency of carbon dioxide excretion is increased (Nikinmaa 1993b). In general, the buffering capacity of teleost fish haemoglobins is smaller, but the Haldane effect larger, than that of either elasmobranch fish or tetrapods (Jensen 1989). The lower buffering capacity will tend to reduce the efficiency of hydration/dehydration reactions between carbon dioxide and HCO_3^-, but this reduction is compensated by the large Haldane effect, which maximizes the buffering upon deoxygenation and oxygenation. In fact, it appears that the strategy of low buffering capacity and large Haldane effect (as in teleosts) is more effective in terms of carbon dioxide transport than large buffering capacity and a small Haldane effect (as in elasmobranch fish) – the conversion of radioactive plasma HCO_3^- to carbon dioxide was more effective in the teleost fish studied than in the dogfish *Scyliorhinus canicula* (Perry et al. 1996). Notably, the plasma carbonic anhydrase of elasmobranch fish (Perry et al. 1996) does not appear to play a significant role in the carbon dioxide excretion (Swenson and Maren 1987).

In alligators and crocodiles, the significant oxygen-linked binding of HCO_3^- to haemoglobin (Perutz et al. 1981) is a major factor influencing carbon dioxide transport (Fig. 20.7). The bulk of carbon dioxide taken up in the tissues is carried into the respiratory epithelium as HCO_3^- bound to haemoglobin (Jensen et al. 1998b): in the tissue capillaries HCO_3^- binds to haemoglobin upon deoxygenation and is released in the lung when haemoglobin is oxygenated, and consequently dehydrated to carbon dioxide which diffuses into alveolar space. Despite the fact

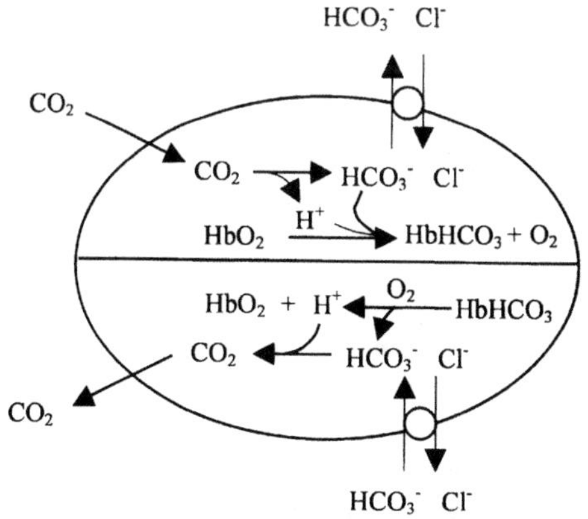

Fig. 20.7. Carbon dioxide excretion in alligators and crocodiles with pronounced bicarbonate binding to haemoglobin. Although the anion exchanger is functional, most of the carbon dioxide excreted stems from oxylabile HCO_3^- binding to haemoglobin. In tissue capillaries HCO_3^- formed binds to haemoglobin upon deoxygenation and is liberated in the lungs upon oxygenation

that alligators do not require an anion exchange pathway for effective carbon dioxide transport, the Cl^-/HCO_3^- exchange is rapid, with a time constant of approx. 200 ms (Jensen et al. 1998b). It is probable that the role of anion exchange in this case is to enable the buffering of extracellular acid (protons produced during burst activity) or base (after feeding during which a marked alkaline tide occurs in crocodilians). However, the lack of extracellular carbonic anhydrase limits the rate of buffering (Jensen et al. 1998b).

In mammals, carbonic anhydrase activity is present in the endothelium of capillaries (facing plasma) both in the lungs and in the muscle (Geers and Gros 2000). Modelling of the role of this extracellular carbonic anhydrase in carbon dioxide transport has indicated that it has only a minor effect. In contrast, acid-base equilibration in the extracellular compartment is speeded up by the presence of extracellular carbonic anhydrase in muscle and lungs (Geers and Gros 2000).

In conclusion, erythrocytic carbonic anhydrase is required for effective carbon dioxide transport in all cases. However, the requirement for anion exchange pathway in carbon dioxide transport can be circumvented by a high erythrocyte pH and large oxylabile proton binding of haemoglobin as in lampreys, or by oxylabile HCO_3^- binding to haemoglobin as in alligators and crocodiles. Effective anion exchange is required for effective buffering of extracellular acid loads by the major intracellular buffer, haemoglobin. In addition, the presence of extracellular carbonic anhydrase either in plasma or in the capillary walls of muscles and lung epithelia speeds up the acid-base equilibria.

References

Airaksinen S, Nikinmaa M (1995) Effect of haemoglobin concentration on the oxygen affinity of intact lamprey erythrocytes. J Exp Biol 198:2393-2396

Bauer C, Forster M, Gros G, Mosca A, Perrella M, Rollema HS, Vogel D (1981) Analysis of bicarbonate binding to crocodilian hemoglobin. J Biol Chem 256:8429-8435

Bonafe CF, Matsukuma AY, Matsuura MS (1999) ATP-induced tetramerization and cooperativity in hemoglobin of lower vertebrates. J Biol Chem 274:1196-1198

Boutilier RG, Heming TA, Iwama GK (1984) Appendix: physicochemical parameters for use in fish respiratory physiology. In: Hoar WS, Randall DJ (eds) Fish physiology, vol XA. Academic Press, New York, pp 403-430

Brauner CJ, Randall DJ (1996) The interaction between oxygen and carbon dioxide movements in fishes. Comp Biochem Physiol A 113:83-90

Bunn HF, Ransil BJ, Chao A (1971) The interaction between erythrocyte organic phosphates, magnesium ion and hemoglobin. J Biol Chem 246:5273-5279

Cala PM (1985) Volume regulation by *Amphiuma* red blood cells: strategies for identifying alkali metal/H^+ transport. Fed Proc 44:2500-2507

Cala PM, Anderson SE, Cragoe EJ Jr (1988) Na-H exchange-dependent cell volume and pH regulation and disturbances. Comp Biochem Physiol 90A:551-555

Cossins AR, Gibson JS (1997) Volume-sensitive transport systems and volume homeostasis in vertebrate red blood cells. J Exp Biol 200:343-352

Crandall ED, O'Brasky JE (1978) Direct evidence for participation of rat lung carbonic anhydrase in CO_2 reactions. J Clin Invest 62:618-622

Dragon S, Carey C, Martin K, Baumann R (1999) Effect of high altitude and *in vivo* adenosine/beta-adrenergic receptor blockade on ATP and 2,3BPG concentrations in red blood cells of avian embryos. J Exp Biol 202:2787-2795

Duhm J, Gerlach E (1971) On the mechanism of the hypoxia induced increase of 2,3-diphosphoglycerate in erythrocytes. Pflügers Arch 326:254-269

Edsall JT (1969) Carbon dioxide, carbonic acid, and bicarbonate ion: physical properties and kinetics of interconversion. In: Forster RE, Edsall JT, Otis AB, Roughton FJW (eds) CO_2: chemical biochemical and physiological aspects. NASA, Washington DC, pp 15-27

Ellory JC, Wolowyk MW, Young JD (1987) Hagfish (*Eptatretus stouti*) erythrocytes show minimal chloride transport activity. J Exp Biol 129:377-383

Escobales N, Canessa M (1986) Amiloride-sensitive Na^+ transport in human red cells: evidence for a Na/H exchange system. J Membrane Biol 90:21-28

Fago A, Weber RE (1998) Hagfish haemoglobins. In: Jorgensen JM, Lomholt JP, Weber RE, Malte H (eds) The biology of hagfishes. Chapman and Hall, London, pp 321-333

Ferguson RA, Tufts BL, Boutilier RG (1989) Energy metabolism in trout red cells: consequences of adrenergic stimulation *in vivo* and *in vitro*. J Exp Biol 143:133-147

Ferguson RA, Sehdev N, Bagatto B, Tufts BL (1992) *In vitro* interactions between oxygen and carbon dioxide transport in the blood of the sea lamprey (*Petromyzon marinus*). J Exp Biol 173:25-41

Fronticelli C, Bucci E, Orth C (1984) Solvent regulation of oxygen affinity in haemoglobin. Sensivity of bovine haemoglobin to chloride ions. J Biol Chem 259:10841-10844

Geers C, Gros G (2000) Carbon dioxide transport and carbonic anhydrase in blood and muscle. Physiol Rev 80:681-715

German B, Wyman J (1937) The titration curves of oxygenated and reduced hemoglobin. J Biol Chem 117:533-550

Gibson JS, Cossins AR, Ellory JC (2000) Oxygen-sensitive membrane transporters in vertebrate red cells. J Exp Biol 203:1395-1407

Günther T (1993) Mechanisms and Regulation of Mg^{2+} Efflux and Mg^{2+} Influx. Miner Electrolyte Metab 19:259-265

Heming TA (1984) The role of fish erythrocytes in transport and excretion of carbon dioxide. Ph.D. thesis, University of British Columbia, Vancouver, Canada

Henry RP, Swenson ER (2000) The distribution and physiological significance of carbonic anhydrase in vertebrate gas exchange organs. Respir Physiol 121:1-12

Hladky SB, Rink TJ (1977) pH equilibrium across the red cell membrane. In: Ellory JC, Lew VL (eds) Membrane transport in red cells. Academic Press, London, pp 115-135

Hsia CC (1998) Respiratory function of hemoglobin. N Engl J Med 338:239-247

Inaba M, Yawata A, Koshino I, Sato K, Takeushi M, Takakuwa Y, Manno S, Yawata Y, Kanzaki A, Sakai J-I, Ban A, Ono K-I, Maede Y (1996) Defective anion transport and marked spherocytosis with membrane instability caused by hereditary total deficiency of red cell band 3 in cattle due to a nonsense mutation. J Clin Invest 97:1804-1817

Jennings ML, Douglas SM, McAndrew PE (1986) Amiloride-sensitive sodium-hydrogen exchange in osmotically shrunken rabbit red blood cells. Am J Physiol 251:C32-C40

Jensen FB (1986) Pronounced influence of Hb-O_2 saturation on red cell pH in tench blood *in vivo* and *in vitro*. J Exp Zool 238:119-124

Jensen FB (1989) Hydrogen ion equilibria in fish haemoglobins. J Exp Biol 143:225-234

Jensen FB (1990) Nitrite and red cell function in carp: control factors for nitrite entry, membrane potassium ion permeation, oxygen affinity and methaemoglobin formation. J Exp Biol 152:149-166

Jensen FB (1995) Regulatory volume decrease in carp red blood cells: Mechanisms and oxygenation-dependency of volume-activated potassium and amino acid transport. J Exp Biol 198:155-165

Jensen FB, Brahm J (1995) Kinetics of chloride transport across fish red blood cell membranes. J Exp Biol 198:2237-2244

Jensen FB, Fago A, Weber RE (1998a) Hemoglobin structure and function. In: Perry SF, Tufts BL (eds) Fish respiration. Academic Press, San Diego, pp 1-40

Jensen FB, Wang T, Jones DR, Brahm J (1998b) Carbon dioxide transport in alligator blood and its erythrocyte permeability to anions and water. Am J Physiol 274:R661-R671

Kaloyianni M, Rasidaki A (1996) Adrenergic responses of *R. ridibunda* red cells. J Exp Zool 276:175-185

Klocke RA (1987) Carbon dioxide transport. In: Farhi LE, Tenney SM (eds) Handbook of physiology, The respiratory system, vol IV: Gas exchange. Am Physiol Soc, Bethesda Maryland, pp 173-197

Lenfant C, Bellingham AJ, Detter JC (1972) Physiological factors influencing the hemoglobin affinity for oxygen. In: Rorth M, Astrup P (eds) Oxygen affinity of hemoglobin and red cell acid base status (Alfred Benzon Symposium IV). Munksgaard, Copenhagen, pp 736-747

Mahé Y, Garcia-Romeu F, Motais R (1985) Inhibition by amiloride of both adenylate cyclase activity and the Na^+/H^+ antiporter in fish erythrocytes. Eur J Pharmacol 116:199-206

Maren TH (1967) Carbonic anhydrase: chemistry, physiology, and inhibition. Physiol Rev 47:595-781

Maren TH, Friedland BR, Rittmaster RS (1980) Kinetic properties of primitive vertebrate carbonic anhydrases. Comp Biochem Physiol 67B:69-74

Marttila ONT, Nikinmaa M (1988) Binding of β-adrenergic antagonists ^{3}H-DHA and ^{3}H-CGP 12177 to intact rainbow trout (*Salmo gairdneri*) and carp (*Cyprinus carpio*) red blood cells. Gen Comp Endocrinol 70:429-435

Mattsoff L, Nikinmaa M (1988) Effects of external acidification on the blood acid-base status and ion concentrations of lamprey. J Exp Biol 136:351-361

Motais R, Garcia-Romeu F, Borgese F (1987) The control of Na^+/H^+ exchange by molecular oxygen in trout erythrocytes. A possible role of hemoglobin as a transducer. J Gen Physiol 90:197-207

Motais R, Fievet B, Garcia-Romeu F, Thomas S (1989) Na^+-H^+ exchange and pH regulation in red blood cells: role of uncatalyzed H_2CO_3 dehydration. Am J Physiol 256:C728-C735

Nikinmaa M (1982a) Effects of adrenaline on red cell volume and concentration gradient of protons across the red cell membrane in the rainbow trout, *Salmo gairdneri*. Mol Physiol 2:287-297

Nikinmaa M (1982b) The effects of adrenaline on the oxygen transport properties of *Salmo gairdneri* blood. Comp Biochem Physiol 71A:353-356

Nikinmaa M (1983) Adrenergic regulation of haemoglobin oxygen affinity in rainbow trout red cells. J Comp Physiol B 152:67-72

Nikinmaa M (1986) Red cell pH of lamprey (*Lampetra fluviatilis*) is actively regulated. J Comp Physiol B 156:747-750

Nikinmaa M (1990) Vertebrate red blood cells. Springer, Berlin

Nikinmaa M (1992) Membrane transport and the control of haemoglobin-oxygen affinity in nucleated erythrocytes. Physiol Rev 72:301-321

Nikinmaa M (1993a) Haemoglobin function in intact *Lampetra fluviatilis* erythrocytes. Respir Physiol 91:283-293

Nikinmaa M (1993b) Blood CO_2 transport in fish with reduced rate of red cell anion exchange. In: Scheid P (ed) Respiration in health and disease. Funktionsanalyse biol Systeme 23:261-266

Nikinmaa M (1997) Oxygen and carbon dioxide transport in vertebrate erythrocytes: An evolutionary change in the role of membrane transport. J Exp Biol 200:369-380

Nikinmaa M, Boutilier RG (1996) Adrenergic control of red cell pH, organic phosphate concentrations and haemoglobin function in teleost fish. In: Heisler N (ed) Mechanisms of systemic regulation: Respiration and circulation. Springer, Berlin, pp 107-133

Nikinmaa M, Huestis WH (1984) Adrenergic swelling in nucleated erythrocytes: cellular mechanisms in a bird, domestic goose, and two teleosts, striped bass and rainbow trout. J Exp Biol 113:215-224

Nikinmaa M, Mattsoff L (1992) Effects of oxygen saturation on the CO_2 transport properties of *Lampetra* red cells. Respir Physiol 87:219-230

Nikinmaa M, Railo E (1987) Anion movements across lamprey (*Lampetra fluviatilis*) red cell membrane. Biochim Biophys Acta 899:134-136

Nikinmaa M, Salama A (1998) Oxygen transport in fish. In: Perry SF, Tufts BL (eds) Fish respiration. Academic Press, New York, pp 141-184

Nikinmaa M, Weber RE (1984) Hypoxic acclimation in the lamprey, *Lampetra fluviatilis*: Organismic and erythrocytic responses. J Exp Biol 109:109-119

Nikinmaa M, Cech JJ Jr, McEnroe M (1984) Blood oxygen transport in stressed striped bass (*Morone saxatilis*): role of beta-adrenergic responses. J Comp Physiol B 154:365-369

Nikinmaa M, Kunnamo-Ojala T, Railo E (1986) Mechanisms of pH regulation in lamprey (*Lampetra fluviatilis*) red blood cells. J Exp Biol 122:355-367

Nikinmaa M, Cech JJ Jr, Ryhänen E-L, Salama A (1987) Red cell function of carp (*Cyprinus carpio*) in acute hypoxia. Exp Biol 47:53-58

Nikinmaa M, Tufts BL, Boutilier RG (1993) Volume and pH regulation in agnathan erythrocytes - comparisons between the hagfish, *Myxine glutinosa*, and the lampreys, *Petromyzon marinus* and *Lampetra fluviatilis*. J Comp Physiol B 163:608-613

Nikinmaa M, Airaksinen S, Virkki LV (1995) Haemoglobin function in intact lamprey erythrocytes: Interactions with membrane function in the regulation of gas transport and acid-base balance. J Exp Biol 198:2423-2430

Ohnishi ST, Asai H (1985) Lamprey erythrocytes lack glycoproteins and anion transport. Comp Biochem Physiol 81B:405-407

Parker JC (1983) Volume-responsive sodium movements in dog red blood cells. Am J Physiol 244:C324-C330

Perry SF, Fritsche R, Kinkead R, Nilsson S (1991) Control of catecholamine release *in vivo* and *in situ* in the Atlantic cod (*Gadus morhua*) during hypoxia. J Exp Biol 155:549-566

Perry SF, Wood CM, Walsh PJ, Thomas S (1996) Fish red blood cell carbon dioxide transport *in vitro*: A comparative study. Comp Biochem Physiol A 113:121-130

Perutz M (1990) Mechanisms of cooperativity and allosteric regulation in proteins. Cambridge University Press, Cambridge, UK
Perutz MF, Bauer C, Gros G, Leclercq F, Van de Casserie C, Schnek AG, Braunitzer G, Friday AE, Joysey KA (1981) Allosteric regulation of crocodilian haemoglobin. Nature 291:682-684
Peters T, Forster RE, Gros G (2000) Hagfish (*Myxine glutinosa*) red cell membrane exhibits no bicarbonate permeability as detected by (18)O exchange. J Exp Biol 203:1551-1560
Primmett DRN, Randall DJ, Mazeaud M, Boutilier RG (1986) The role of catecholamines in erythrocyte pH regulation and oxygen transport in rainbow trout (*Salmo gairdneri*) during exercise. J Exp Biol 122:139-148
Qiu Y, Maillett DH, Knapp J, Olson JS, Riggs AF (2000) Lamprey hemoglobin - Structural basis of the Bohr effect. J Biol Chem 275:13517-13528
Reid SD, Perry SF (1991) The effects and physiological consequences of raised levels of cortisol on rainbow trout (*Oncorhynchus mykiss*) erythrocyte β- adrenoreceptors. J Exp Biol 158:217-240
Reid SD, Moon TW, Perry SF (1991) Characterization of β-adrenoreceptors of rainbow trout (*Oncorhynchus mykiss*) erythrocytes. J Exp Biol 158:199-216
Riggs AF (1998) Self-association, cooperativity and supercooperativity of oxygen binding by hemoglobins. J Exp Biol 201:1073-1084
Romero M, Guizouarn H, Pellissier B, Garcia-Romeu F, Motais R (1996) The erythrocyte Na^+/H^+ exchangers of eel (*Anguilla anguilla*) and rainbow trout (*Oncorhynchus mykiss*): a comparative study. J Exp Biol 199:415-426
Roughton FJW (1964) Transport of oxygen and carbon dioxide. In: Fenn WO, Rahn H (eds) Handbook of physiology, Respiration, vol I. Am Physiol Soc, Washington DC, pp 767-825
Rudolph SA, Greengard P (1980) Effects of catecholamines and prostaglandin E_1 on cyclic AMP, cation fluxes and protein phosphorylation in the frog erythrocyte. J Biol Chem 255:8534-8540
Salama A, Nikinmaa M (1988) The adrenergic responses of carp (*Cyprinus carpio*) red cells: effects of PO_2 and pH. J Exp Biol 136:405-416
Salama A, Nikinmaa M (1989) Species differences in the adrenergic responses of fish red cells: studies on whitefish, pikeperch, trout and carp. Fish Physiol Biochem 6:167-173
Salama A, Nikinmaa M (1990) Effect of oxygen tension on catecholamine-induced formation of cAMP and on swelling of carp red blood cells. Am J Physiol 259:C723-C726
Siggaard-Andersen O (1974) The acid-base status of the blood, 4th edn. Munksgaard, Copenhagen
Swenson ER, Maren TH (1987) Roles of gill and red cell carbonic anhydrase in elasmobranch HCO_3^- and CO_2 excretion. Am J Physiol 253:R450-R458
Tetens V (1987) Regulation of blood O_2 affinity during acute hypoxic exposure of rainbow trout, *Salmo gairdneri*: organismal and cellular processes. Ph.D. thesis, Aarhus University, Denmark
Tetens V, Christensen NJ (1987) Beta-adrenergic control of blood oxygen affinity in acutely hypoxia exposed rainbow trout. J Comp Physiol B 157:667-675
Thomas S, Hughes GM (1982) A study of the effects of hypoxia on acid-base status of rainbow trout blood using an extracorporeal blood circulation. Respir Physiol 49:371-382
Tufts BL (1992) Sodium-dependent pH regulation in sea lamprey (*Petromyzon marinus*) red blood cells. Can J Zool 70:411-416

Tufts BL, Boutilier RG (1989) The absence of rapid chloride/bicarbonate exchange in lamprey erythrocytes: implications for CO_2 transport and ion distributions between plasma and erythrocytes in the blood of *Petromyzon marinus*. J Exp Biol 144:565-576

Tufts BL, Boutilier RG (1990) CO_2 transport properties of the blood of a primitive vertebrate *Myxine glutinosa*. Exp Biol 48:341-347

Tufts BL, Perry SF (1998) Carbon dioxide transport and excretion. In: Perry SF, Tufts BL (eds) Fish respiration. Academic Press, San Diego, pp 229-281

Tufts BL, Nikinmaa M, Steffensen JF, Randall DJ (1987) Ion exchange mechanisms on the erythrocyte membrane of the aquatic salamander, *Amphiuma tridactylum*. J Exp Biol 133:329-338

Tufts BL, Bagatto B, Cameron B (1992) *In vivo* analysis of gas transport in arterial and venous blood of the sea lamprey *Petromyzon marinus*. J Exp Biol 169:105-119

Virkki LV, Nikinmaa M (1994) Activation and physiological role of Na^+/H^+ exchange in lamprey (*Lampetra fluviatilis*) erythrocytes. J Exp Biol 191:89-105

Virkki LV, Salama A, Nikinmaa M (1998) Regulation of ion transport across lamprey (*Lampetra fluviatilis*) erythrocyte membrane by oxygen tension. J Exp Biol 201:1927-1937

Weaver YR, Kiessling K, Cossins AR (1999) Responses of the Na^+/H^+ exchanger of European flounder red blood cells to hypertonic, beta-adrenergic and acidotic stimuli. J Exp Biol 202:21-32

Weber RE (1978) Functional interaction between fish hemoglobin, erythrocyte nucleoside triphosphates and magnesium. Acta Physiol Scand 102:20A-21A

Wood SC, Johansen K (1973) Organic phosphate metabolism in nucleated red cells: influence of hypoxia on eel HbO_2 affinity. Neth J Sea Res 7:328-338

Wood CM, Perry SF, Walsh PJ, Thomas S (1994) HCO_3^- dehydration by the blood of an elasmobranch in the absence of a Haldane effect. Respir Physiol 98:319-337

Yang H, Nevo E, Tashian RE (1998) Unexpected expression of carbonic anhydrase I and selenium-binding protein as the only major non-heme proteins in erythrocytes of the subterranean mole rat (*Spalax ehrenbergi*). FEBS Letters 430:343-347

Yang D, Hewett-Emmett D, Tashian RE (2000) Absence or reduction of carbonic anhydrase II in the red cells of the beluga whale and llama: Implications for adaptation to hypoxia. Biochem Gen 38:241-252

21 'The Hereditary Stomatocytosis and Allied Conditions': Inherited Disorders Na^+ and K^+ Transport

Gordon W. Stewart

Department of Medicine, Rayne Institute, University College London, University Street, London WC1E 6JJ, UK

21.1 Introduction

The diseases under consideration here can be collected under the title, 'The Hereditary Stomatocytoses and Allied Disorders', an expression coined by Dacie (1985). These diseases are defined (in our lab) on the basis of abnormal membrane permeability to Na^+ and K^+. It is characteristic that intracellular $[Na^+]$ is high, $[K^+]$ is low, while isotopic studies show increases in both the passive and active, Na^+/K^+ pump-mediated modalities. The situation is consistent with an increased 'leak', and a part compensation by increased Na^+ and K^+ pumping. Where tested, Li^+ and Rb^+ permeabilities are increased, but not Ca^{2+}. The fundamental functional defect is a bidirectional increase in univalent cation permeability and it is reasonable to suppose that this cation 'leak' compromises the osmotic stability of the cell, shortening its lifespan in the circulation. For reasons that are beginning to be understood, this ionic abnormality is associated with the appearance of a particular cell type on the film known as a stomatocyte, as shown in Fig. 21.1. Several recent reviews are available (Stewart 1993; Lux and Palek 1995; Delaunay et al. 1999; Stewart and Turner 1999).

The name 'hereditary stomatocytosis' is now something of an anachronism. There is no doubt about the 'hereditary' part of the designation, for all these conditions are inherited as autosomal dominants. However, it has become clear that there exist stomatocytic conditions *without* an ion 'leak'; and that there *leaky* conditions, albeit very mild, *without* stomatocytes on the film. But nevertheless, no crisper description exists.

21.2 Normal Red Cell Cation Transport

Normal red cell cation transport is covered elsewhere in this book (see Chaps. 4–9, 15–17). Suffice is to say here that there is a ouabain-inhibited Na^+/K^+ pump and a 'leak' process that behaves as simple diffusion (in NaCl media at least) and a number of other transporters which are either silent or mediate only exchange fluxes in normal red cells. For K^+, these include the Na^+-K^+-$2Cl^-$ cotransporter, which can be inhibited by loop diuretics, while for Na^+, this cotransport system is also important, in addition to the Na^+/Na^+ exchange system, whose activity cannot be eliminated by any high-affinity inhibitor. Thus for the purposes of the work here, the 'leak' is what is left after both the Na^+/K^+ pump and Na^+-K^+-$2Cl^-$ cotransport are inhibited by ouabain and bumetanide, respectively. This residual fraction can include a component of loop-diuretic-resistant K^+-Cl^- cotransport, but this is normally small in human red cells.

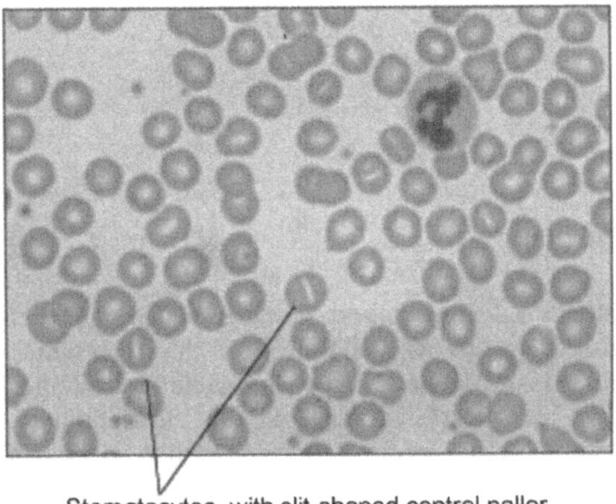

Fig. 21.1. Wright-Giemsa stained blood film from patient with a form of hereditary stomatocytosis showing stomatocytes with slit-shaped central pallor

21.3 History; Nosology

The first description of hereditary stomatocytosis can be attributed to Lock et al. (1961), who described the case of an infant with a haemolytic condition, which resembled hereditary spherocytosis *except* in the morphology, which showed mouth-shaped cells rather than simple spheres (Fig. 21.1). They called these cells 'stomatocytes'. Although these workers recognised the osmotic problems of these cells, it was American workers who first found the gross abnormality in handling of Na^+ and K^+ that these cells show (Zarkowsky et al. 1968): flux rates are about 10–40 times normal. This condition, now known as 'overhydrated HSt' (OHSt), remains the most severe of the collection.

It gradually became apparent that there were variations on this original theme. Miller et al. (1965) found two new forms in Rochester, NY. In one case, which he called 'cryohydrocytosis', the cells showed a striking tendency to swell and lyse if left overnight in the refrigerator. This introduced the idea of temperature as a diagnostic factor and we will return to this later. His second family showed a state of cellular *de*hydration, which contrasted with the overhydration of the original cases (Miller et al. 1971). The cation defect was much less severe but the haemolysis was similar. This 'dehydrated' condition ('dehydrated hereditary stomatocytosis', DHSt), also known as 'hereditary xerocytosis', has emerged as the single commonest form of these conditions. It is identical to a condition originally described as 'hyperphosphatidylcholine haemolytic anaemia' (HPCHA) (Jaffe and Gottfried 1968; Lane et al. 1990).

In the late 1970s, a family was recognised that presented with 'pseudohyperkalaemia', artefactually high plasma [K^+] due to a temperature-dependent net loss of K^+ from the red cells on storage at room temperature (Stewart et al. 1979; Stewart and Ellory 1985). Haematologically trivial, in that there was no haemolysis, and not showing any stomatocytes on the film, this condition presented via the chemical pathology service. It was not immediately obvious that there was any connection between this condition and the haemolytic anaemias described above. The cells in this case were mildly dehydrated. Other families with the same combination of temperature-dependent pseudohyperkalaemia and essentially normal haematology were described (Meenaghan et al. 1985; James and Stansbie 1987; Vantyghen et al. 1991). Like Miller's 'cryohydrocytosis', these cases showed a temperature-dependent mutation.

Other clinical variants have been listed elsewhere (Stewart and Turner 1999). Many of these can be distinguished on the basis of temperature effects (see Sect. 21.4).

In 1980, Lande and others demonstrated that in OHSt and one form of cryohydrocytosis, a membrane protein, later called 'stomatin' after these diseases, was missing from the membrane (Lande et al. 1982). This abnormality has been confirmed by others in other very severely leaky cases (Eber et al. 1989; Stewart et al. 1992). Purification of the protein (Stewart et al. 1992) and sequencing of the gene did not, unfortunately, lead to either a mutation or a molecular explanation: the stomatin gene was normal in the patients and the sequence was quite anonymous (Wang et al. 1992).

Aspects of all these features covered here will now be considered in more detail.

21.4 Temperature Effects

The first temperature-related effect in this group of conditions was Miller's recognition of 'cryohydrocytosis'. These cells show a very marked tendency to lyse under refrigerator conditions. Later, this group identified a 'familial pseudohyperkalaemia' condition in which the red cells leak K^+ into the plasma when stored at room temperature. To investigate this phenomenon we made measurements of K^+ flux as a function of temperature in these red cells. It emerged that inhibition of the Na^+/K^+ pump by ouabain and the Na^+-K^+-$2Cl^-$ cotransport system by a loop diuretic (bumetanide) isolated a residual flux, which is usually a good measure of the 'leak' to K^+. In *normal* cells, this residual 'leak' flux shows a U-shaped temperature dependence with a minimum at 8 °C (Stewart et al. 1980) (Fig. 21.2). Wieth

(1970) had already shown that in SCN^- or salicylate media, the ouabain-insensitive fluxes of Na^+ and K^+ in normal red cells showed a minimum at 23 °C (Fig. 21.2, panel D).

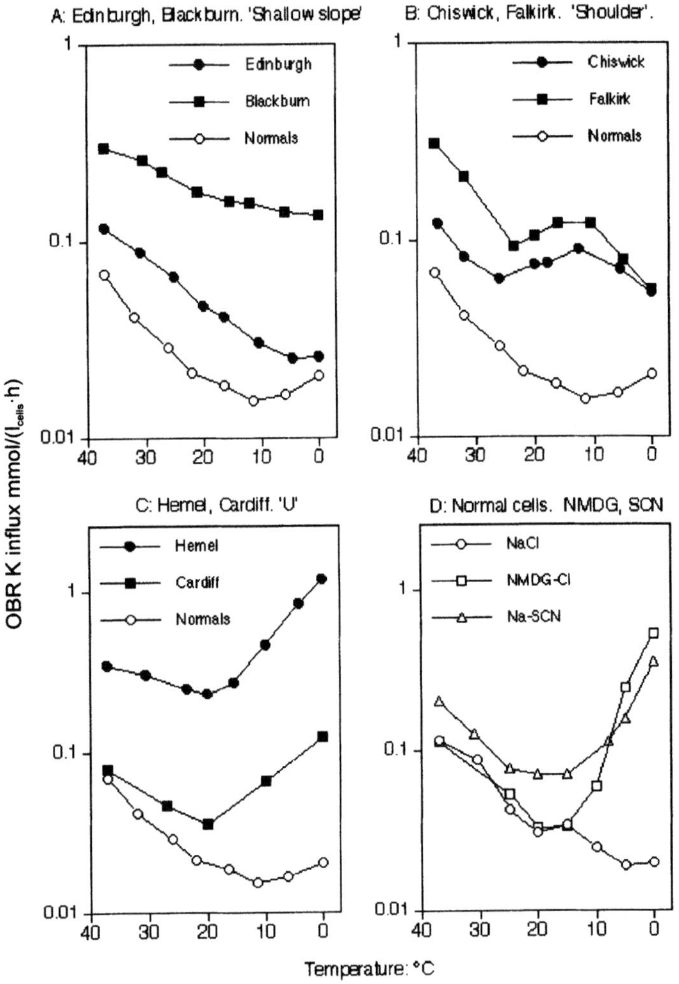

Fig. 21.2. Temperature dependence of passive 'leak' (ouabain + bumetanide)-resistant K^+ influx in different stomatocytosis pedigrees (panels **A–C**) and in normal cells in which Na^+ is substituted by $NMDG^+$ or Cl^- substituted by SCN^-. In all panels, filled symbols denote patients and open symbols, controls. K^+ ($^{86}Rb^+$) influx was measured in the following medium (mM): K^+, 5; Na^+, 145; Cl^-, 150; MOPS, 15 (pH 7.4 at 20 °C); glucose, 5; ouabain, 0.1; bumetanide, 0.1; except in panel **D**, where Na^+ was substituted by *N*-methyl-D-glucamine (NMDG) and Cl^- by SCN^-, where required. Cases are referenced as follows: Edinburgh, (Stewart and Ellory 1985); Blackburn, (Coles et al. 1999b); Chiswick, Falkirk, (Haines et al. 2001a); Cardiff (unpublished); Hemel, (Coles et al. 1999a)

Wieth had not used loop diuretics and the presence of the Na^+-K^+-$2Cl^-$ cotransport had masked the minimum at 8 °C in chloride media (Stewart et al. 1980). We later found that if the Na^+ was replaced by an organic cation such as choline or $NMDG^+$, then a similar effect could be obtained (Blackstock and Stewart 1986) (Fig. 21.2, panel D). These dual effects of anions and cations can be explained on the basis of the lyotropic series: that ranking of ions, related to the ratio of size to charge in different ions, which influences very many biological processes. It was first observed to influence the solubility of starch molecules, but it also influences lipid packing and is used to precipitate proteins with ammonium sulphate and DNA with sodium and magnesium salts (Hunter 1986). But what the actual target of these lyotropic variations in the red cell is, whether it is the packing of lipids or a channel protein, is unknown. It is attractive to assume that it is a lipid effect, given the vast knowledge on the effect of temperature on lipid bilayer organisation (Zimmer and Schirmer 1973; Trueman et al. 2000), but it is not necessarily the mechanism here. An effect on the physical state of a channel protein is equally possible.

On this background of what can happen in normal cells we can look at genetic abnormalities seen in these leaky cell types (Fig. 21.2). It turns out that in Miller's cryohydrocytosis conditions, in which the cells lyse severely at low temperatures, the temperature profile shows a U-shaped pattern (Fig. 21.2, panel C), exactly identical to that seen in normal red cells suspended in media low in the lyotropic series. There is a minimum at 23 °C: at 0 °C, the cells are more leaky than they are at 37 °C, while the Na^+/K^+ pump is effectively silenced. The 'leak' is unopposed and the osmotic consequences of the cation 'leak' become immediately evident: the cells swell and lyse (Coles et al. 1999a) and become spherocytic (Haines et al. 2001b).

Variations in temperature dependence do not stop here. Another pedigree shows a shallow slope variant (Coles et al. 1999b), similar to the original Edinburgh family (Stewart et al. 1979) (Fig. 21.2, panel A), while two more show a shoulder distribution of flux (Haines et al. 2001a) (Fig. 21.2, panel B). Currently, we can distinguish in UK pedigrees 8 different curves: some of these are essentially parallel to each other and vary only in position on the ordinate (roughly reflecting haematological severity), while among many there is a clear qualitative difference. Four general forms emerge: normal (minimum at 8 °C); U-shaped (minimum at 23 °C), shallow slope, shoulder (maximum at 12 °C), monotonic, with variations up and down the ordinate.

21.5 Clinical Aspects

The patients with the most severe variant of these conditions, OHSt, can suffer from a moderately severe, non-transfusion-dependent, chronic haemolytic anaemia, showing pigment gallstones and the occasional crisis precipitated by infection and requiring transfusion. The other conditions are all less severe, and those in which the cation 'leak' is negligible at body temperature are virtually asymptomatic.

The biggest clinical problem occurs in OHSt and DHSt. If splenectomised, these patients exhibit in adult life a marked tendency to venous thrombosis, manifesting as deep or superficial venous thromboses and in some cases portal vein thrombosis (Stewart et al. 1996; Coles et al. 1999b; Perel et al. 1999; Grootenboer et al. 2000). Thrombotic pulmonary hypertension can occur, but can be treated either with anticoagulation or hypertransfusion if necessary (B. Smith, personal communication). This post-splenectomy thrombotic problem does not occur in the British cryohydrocytosis pedigrees (Coles et al. 1999a; Haines et al. 2001b).

Pseudohyperkalaemia, the artefactual finding of a high plasma potassium level due to a temperature dependent 'leak' of K^+ from the red cells on storage, which occurs in some of these pedigrees and can be the only sign of a mild variant, can cause inconvenience for patients and clinicians until the artefactual nature of the hyperkalaemia is recognised. Pseudohyperkalaemia can also occur in the very severe variants in which there is no gross temperature abnormality, simply as a result of high turnover of cations and energy depletion in the test tube (M.C. Chetty, G.W. Stewart, unpublished). Inappropriate treatment for what is taken to be true hyperkalaemia can be instituted (Vantyghen et al. 1991).

In some European families, a state of transient neonatal oedema has been observed in DHSt patients (Entazami et al. 1996; Grootenboer et al. 1998, 2000). This resolves with time. The cause is unknown.

21.6 Genetic Mapping

The heterogeneity in the clinical phenotype of these conditions is reflected in locus heterogeneity on mapping. The dehydrated conditions, including the original familial pseudohyperkalaemia' pedigree, map to a locus on chromosome 16 (Carella et al. 1998; Iolascon et al. 1999) but other conditions do not (Carella et al. 1999). No gene is yet identified. There is no obvious candidate within any of the regions.

21.7 The Stomatocytic Red Cell

The stomatocytic red cell, although associated in the cases here with the presence of a cation 'leak', is not always leaky. Stomatocytes can be found as an acquired feature in liver disease (Lux and Palek 1995). We have investigated one case with stomatocytosis in association with large platelets where there was no cation 'leak' (Stewart et al. 1987). Ducrou and Kimber (1969) were the first to describe this association and we have recently identified a further 4 families (D.C. Rees, A. O'Marcaigh, S.N. Jowitt, J.C. Kendra, G.W. Stewart, unpublished). The stomatocyte can be regarded as one of the relatively few shape changes that the red cell can undergo. There is excellent evidence that it reflects a situation in which the inner leaflet of the lipid bilayer is expanded at the expense of the outer (Sheetz and Singer 1974). If the outer is expanded at the expense of the inner, then an echinocyte is formed. If surface area is lost, then the cell takes on a spherical appearance. The precise reason why cells in these leaky conditions have this stomatocytic shape has never been demonstrated. It is unlikely to result from the cation 'leak' as such, because when control over ion content is lost completely, the cell simply swells and becomes a 'macrospherocyte' ('macro' to distinguish it from the spherocyte seen in hereditary spherocytosis, which results from loss of cell surface area due to nibbling in the spleen (Lux and Palek 1995)).

21.8 The Stomatin Protein

In the absence of a mutant gene from mapping, the best molecular clue to these conditions remains the lack of stomatin in the cells. When first purified and the gene cloned (Stewart et al. 1992), this protein was anonymous, and no mutation has been found in the human gene in these conditions. A murine knock-out is without phenotype (Zhu et al. 1999) and it is now clear that it is exceedingly unlikely that a mutation in this gene causes this condition, although the tiny pedigrees with the stomatin-deficient phenotype do not allow definitive mapping. Nevertheless, a major membrane protein is missing and it must have something to do with the pathophysiology. A number of homologues have been found in other organisms and stomatin can now be regarded as the prototype of a family of proteins found throughout biology (Tavernarakis et al. 1999). Two stomatin homologues in the nematode *C. elegans* are especially informative. Recessive mutations in the first to be cloned, *mec*-2, cause a defect in sensory function (Huang et al. 1995). *Mec*-2 is one of 18 genes that can cause this same 'hypoaesthetic' phenotype. It is of interest here that three of these other genes (*mecs*-4, 6 and 10) code for subunits of an ion channel, itself homologous to the mammalian 'epithelial Na^+ channel' (Canessa et al. 1993), but although there is genetic evidence from *C. elegans* that the protein products of *mec*-2 and these channel genes are functionally related, there is no definite evidence on exactly *how*. A second *C. elegans* stomatin homologue is *unc*-1. *Unc*-1 is closely sequence-homologous to stomatin and mutations in this gene cause an abnormality in sensitivity to volatile anaesthetics (Rajaram et al. 1998, 1999b). *Unc*-1 is connected with a 'degenerin' gene, a further member of the ENaC family (Rajaram et al. 1999a). In the rat, the sensory organ of the whisker is the so-called 'vibrissal' organ and stomatin and ENaC proteins are co-expressed in the transductive 'lanceolate' cells of this organ, confirming an anatomical, if not a functional, relationship between these proteins in a mammal (Fricke et al. 2000).

Although these *C. elegans* examples do not themselves directly display the point, one common theme that emerges from the study of all the stomatin homologues in biology is a role in systems pertaining to the expression of proteins at the external membrane surface. Curiously, the ENaC channel is itself in part controlled by surface exposure: the renal tubular 'leak' to Na^+ which is the flaw in the hypertensive syndrome, Liddle's syndrome, is attributed to a *delay* in the *removal* of the channel protein from the membrane. In some cases of HSt, there is an overt defect in an unphysiological process of ATP-induced endocytic vesiculation

(Mentzer et al. 1978), but exactly how this relates to the pathophysiology of these conditions is currently uncertain and will be discussed below.

The recent demonstration that stomatin is associated with the (sphingomyelin + cholesterol)-rich lipid 'raft' (Snyers et al. 1999) focuses attention on this structure of the cell membrane as a likely focus for these diseases. 'Rafts' exist in red cells but their function is not well understood. In other cell types, 'rafts' act as centres for signalling or endocytosis (Hooper 1999; Kurzchalia and Parton 1999; Cary and Cooper 2000; Ilangumaran et al. 2000; Simons and Toomre 2000). The caveolus, an invaginated dimple in the endothelial cell membrane, is a 'raft'-related structure (Anderson 1998). The drug chlorpromazine, which makes red cells stomatocytic (Schrier et al. 1992), also promotes the formation of caveolae (Hueck et al. 2000).

21.9 Possible Pathogenic Mechanisms for the Stomatocytosis

This finding sets the scene to try to understand the stomatocytoses at the cellular level. We seek to explain why there might exist a cell that leaks Na^+ and K^+, has a stomatocytic shape abnormality, and might (in severe cases) lack the 'raft' protein stomatin. We will accept that the 'stomatocyte' reflects a state of inward membrane bending secondary to some disruption in the balance of lipid distribution between inner and outer leaflets. Somewhere there probably has to be a protein-based ion channel that is functioning abnormally, either because it itself is mutated, or because it is being improperly trafficked, or because it is 'mounted' in an abnormal lipid environment.

One possible explanation is that there is a mutant leaky channel, and that the stomatocytosis (and in severe cases the loss of stomatin) is the *result* of the 'leak'. (It cannot be the currently well-understood ENaC channel genes, for these are expressed elsewhere and their leaky mutants are characterised (Hansson et al. 1995): but this is a large gene family.) If the osmotic regulation mechanism of a red cell are disturbed and intracellular $[Na^+]$ is allowed to increase and $[K^+]$ decrease, then the red cell simply blows up into a sphere and bursts. It does not become a stomatocyte. A more sophisticated mechanism must be invoked to explain why these cells assume this shape. It could be that the cell responds to the 'leak', by trying to get rid of the channel by a process of endocytosis, which would include inward membrane bending. This could be based on 'rafts' (to which channels can be localised (Martens et al. 2001)) and hence the stomatin could be lost as part of a 'sacrifice'. There is an interesting precedent in *C. elegans.* In worms in which an ENaC channel homologue is mutated and leaky, pathological studies reveal whorls of invaginated neuronal plasma membrane within the cell (Hall et al. 1997). Whether or not these whorls contain a stomatin homologue is not known.

Another possibility which could unite these concepts is that it is not the leaky channel which comes first but a defect in its trafficking, which could occur via a 'raft'-based endocytic mechanism. A number of channels are controlled in this way (Hamm-Alvarez and Sheetz 1998). Red cells are not renowned for endocytosis, but it could be taking place on a small scale during erythropoiesis. The stomatocytosis could reflect the inward bending of continued 'attempts' to get rid of the channel from the plasma membrane. This is consistent with the absence of ATP-dependent vesiculation in the most severe forms of the disease, which might reflect an inability to pinch off an inwardly directed 'bud' of membrane from the plasma membrane. Whether or not this decidedly *in vitro* phenomenon has physiological relevance is in doubt, but this mechanism is exactly consistent with the action of mutant dynamin genes on the function of ENaC subunits expressed in *Xenopus* oocytes (Shimkels et al. 1997). These dynamin mutants, which act as dominants, paralyse endocytosis and *prevent* the *removal* of the channel from the membrane, allowing it to function constitutively, or 'leak'. Such a *blockage* in endocytosis offers no immediate explanation for the absence of stomatin: indeed, one might expect to see an excess of proteins to be evident on SDS-PAGE analysis of membrane proteins, but no such excess is evident.

Other possibilities can be suggested. The 'rafts' themselves could be of abnormal lipid composition due to mishandling by as-yet poorly understood shepherding proteins for lipids. A disorder of lipid handling is reported in DHSt (Shohet et al. 1971). This might fit neatly with many of the abnormal temperature effects, which might be explained on the basis of an abnormality in lipid fluidity in a 'raft'. Papahadjopoulos et al. (1973) showed that membrane permeability of vesicles underwent a maximum at a temperature close to the transition temperature of a lipid bilayer. Theoretical calculations have been made which illustrate how a maximum in permeability can occur close to a phase transition as the result of mismatching between the packing geometry of adjacent lipid domains of different physical states (Corvera et al. 1992). The pattern illustrated by Papahadjopoulos is strikingly similar to the maximum in K^+ permeability at about 12 °C observed by ourselves in the Chiswick and Falkirk pedigrees in the UK (Haines et al. 2001a). This kind of physical interpretation is consistent with the observations that the temperature dependence of Na^+, K^+ permeability of the red cell membrane can be modified by variation of the composition of the external medium according to the lyotropic series (Wieth 1970; Blackstock and Stewart 1986). Variations in temperature dependence induced by such lyotropic variations exactly mimic the behaviour of cryohydrocytosis cells (Coles et al. 1999a; Haines et al. 2001b). Nevertheless, lyotropic effects are of promiscuous application in biology and a direct action on the physical structure of an ion channel protein is not impossible: the 'leak' is confined to univalent cations (Li^+, Na^+, K^+, Rb^+) and does not affect, as far as we know, divalent cations or sugars.

In conclusion, the stomatocytoses represent a rare and diverse group of congenital haemolytic anaemias in which the plasma membrane leaks Na^+ and K^+, frequently in a temperature dependent fashion. They represent a series of diseases in which Na^+ and K^+ transport is disturbed, possibly due to an abnormality associated with the lipid 'raft'. They represent a mutant model for salt handling in the human.

Acknowledgement: This work was funded by Action Research and the Sir Jules Thorn Trust.

References

Anderson RG (1998) The caveolae membrane system. Annu Rev Biochem 67:199-225

Blackstock EJ, Stewart GW (1986) The dependence on external cation of sodium and potassium fluxes across the human red cell membrane at low temperatures. J Physiol 375:403-420

Canessa CM, Horisberger JD, Rossier BC (1993) Epithelial sodium channel related to proteins involved in neurodegeneration. Nature 361:467-70

Carella M, Stewart GW, Ajetunmobi JF, Perrotta S, Grootenboer S, Tchernia G, Delaunay J, Totaro A, Zelante L, Gasparini P, Iolascon A (1998) Genomewide search for dehydrated hereditary stomatocytosis (hereditary xerocytosis): Mapping of locus to chromosome 16 (q23-qter). Am J Human Genetics 63:810-816

Carella M, Stewart GW, Ajetunmobi JF, Schettini F, Delaunay J, Iolascon A (1999) Genetic heterogeneity of hereditary stomatocytosis syndromes showing pseudohyperkalaemia. Haematol 84:862-863

Cary L, Cooper J (2000) Molecular switches in lipid rafts. Nature 404:945-946

Coles SE, Chetty MC, Ho MM, Nicolaou A, Kearney JW, Wright SD, Stewart GW (1999a) Two British families with variants on the 'cryohydrocytosis' form of hereditary stomatocytosis. Br J Haematol 105:1055-1065

Coles SE, Ho MM, Chetty MC, Nicolaou A, Stewart GW (1999b) Hereditary stomatocytosis with marked pseudohyperkalaemia. Br J Haematol 104:275-283

Corvera E, Mouritsen OG, Singer MA, Zuckermann MJ (1992) The permeability and the effect of acyl-chain length for phospholipid bilayers containing cholesterol: theory and experiment. Biochim Biophys Acta 1107:261-270

Dacie JV (1985) The haemolytic anaemias, vol. 1. The hereditary haemolytic anaemias, part 1, 2nd edn. Churchill, Edinburgh

Delaunay J, Stewart G, Iolascon A (1999) Hereditary dehydrated and overhydrated stomatocytosis: recent advances. Curr Opin Hematol 6:110-114

Ducrou W, Kimber RJ (1969) Stomatocytes, haemolytic anaemia and abdominal pain in Mediterranean migrants. Some examples of a new syndrome. Med J Australia ii:1087-1091

Eber SW, Lande WM, Iarocci TA, Mentzer WM, Hohn P, Wiley JS, Schroter W (1989) Hereditary stomatocytosis: consistent association with an integral membrane protein deficiency. Br J Haematol 72:452-455

Entazami M, Becker R, Mensen H, Marcinkowki M, Versmold H (1996) Xerocytosis with concomitant intrauterine ascites: first description and therapeutic approach. Blood 90:5392-5393

Fricke B, Lints R, Stewart G, Drummond H, Dodt G, Driscoll M, During MV (2000) Epithelial Na^+ channels and stomatin are expressed in rat trigeminal mechanosensory neurones. Cell Tissue Res 299:327-334

Grootenboer S, Schischmanoff PO, Cynober T, Rodrigue JC, Delaunay J, Tchernia G, Dommergues JP (1998) A genetic syndrome associating dehydrated hereditary stomatocytosis, pseudohyperkalaemia and perinatal oedema. Br J Haematol 103:383-386

Grootenboer S, Schischmanoff PO, Laurendeau I, Cynober T, Tchernia G, Dommergues JP, Dhermy D, Bost M, Varet B, Snyder M, Ballas SK, Ducot B, Babron MC, Stewart GW, Gasparini P, Iolascon A, Delaunay J (2000) Pleiotropic syndrome of dehydrated hereditary stomatocytosis, pseudohyperkalemia, and perinatal edema maps to 16q23-q24. Blood 96:2599-605

Haines P, Crawley C, Chetty M, Jarvis H, Coles S, Fisher J, Nicolaou A, Stewart G (2001a) 'Familial pseudohyperkalaemia Chiswick': a novel congenital thermotropic variant of K and Na transport across the human red cell membrane. Br J Haematol 112:469-474

Haines P, Jarvis H, King S, Noormohamed F, Chetty M, Fisher J, Hill P, Nicolaou A, Stewart G (2001b) Two further British families with the 'cryohydrocytosis' form of hereditary stomatocytosis. Br J Haematol 113:932-937

Hall D, Gu G, Garcia-Anoveros J, Gong L, Chalfie M, Driscoll M (1997) Neuropathology of degenerative cell death in *Caenorhabditis elegans*. J Neurosci 17:1033-1045

Hamm-Alvarez S, Sheetz M (1998) Microtubule-dependent vesicle transport: modulation of channel and transporter activity in liver and kidney. Physiol Rev 78:1109-1129

Hansson JH, Nelson-Williams C, Suzuki H, Schild L, Shimkets R, Lu Y, Canessa C, Iwasaki T, Rossier B, Lifton RP (1995) Hypertension caused by a truncated epithelial sodium channel gamma subunit: genetic heterogeneity of Liddle syndrome. Nat Genet 11:76-82

Hooper NM (1999) Detergent-insoluble glycosphingolipid/cholesterol-rich membrane domains, lipid rafts and caveolae (review). Molec Membrane Biol 16:145-156

Huang M, Gu G, Ferguson E, Chalfie M (1995) A stomatin-like protein necessary for mechanosensation in *C. elegans*. Nature 378:292-295

Hueck IS, Hollweg HG, Schmid-Schönbein GW, Artmann GM (2000) Chlorpromazine modulates the morphological macro- and microstructure of endothelial cells. Am J Physiol 278:C873-C878

Hunter R (1986) Foundations of colloid science, 1. edn. Clarendon Press, Oxford

Ilangumaran S, He HT, Hoessli DC (2000) Microdomains in lymphocyte signalling: beyond GPI-anchored proteins. Immunol Today 21:2-7

Iolascon A, Stewart GW, Ajetunmobi JF, Perrotta S, Delaunay J, Carella M, Zelante L, Gasparini P (1999) Familial pseudohyperkalemia maps to the same locus as dehydrated hereditary stomatocytosis (hereditary xerocytosis). Blood 93:3120-3123

Jaffe ER, Gottfried EL (1968) Hereditary nonspherocytic hemolytic disease associated with an altered phospholipid composition of the erythrocytes. J Clin Invest 47:1375-1388

James DR, Stansbie D (1987). Familial pseudohyperkalaemia: inhibition of erythrocyte K efflux at 4 °C by quinine. Clin Sci 73:557-560

Kurzchalia TV, Parton RG (1999) Membrane microdomains and caveolae. Curr Opin Cell Biol 11:424-431

Lande WM, Thiemann PW, Mentzer WM (1982) Missing band 7 membrane protein in two patients with high Na, low K erythrocytes. J Clin Invest 70:1273-1280

Lane PA, Kuypers FA, Clark MR, Andrews DA, Wagner GM, Butikofer P, Shapiro AD, Chiu DTY, Lubin BH, Mentzer WC (1990) Excess of red cell membrane proteins in hereditary high-phosphatidylcholine hemolytic anemia. Am J Hematol 34:186-192

Lock SP, Sephton-Smith R, Hardisty RM (1961) Stomatocytosis: a hereditary haemolytic anomaly associated with haemolytic anaemia. Br J Haematol 7:303-314

Lux S, Palek J (1995) Disorders of the red cell membrane. In: Handin R, Lux S, Stossel T (eds) Blood: Principles and practice of hematology. JB Lippincott, Philadelphia, pp 1701-1818

Martens JR, Sakamoto N, Sullivan SA, Grobaski TD, Tamkun MM (2001) Isoform-specific localization of voltage-gated K^+ channels to distinct lipid raft populations. Targeting of kv1.5 to caveolae. J Biol Chem 276:8409-8414

Meenaghan M, Follett GF, Brophy PJ (1985) Temperature sensitivity of potassium flux into red blood cells in the familial pseudohyperkalaemia syndrome. Biochim Biophys Acta 821:72-78

Mentzer WC, Lam GK, Lubin BH, Greenquist A, Schrier SL, Lande W (1978) Membrane effects of imidoesters in hereditary stomatocytosis. J Supramol Struc Cell Biochem 9:275-288

Miller G, Townes PL, Macwhinney JB (1965) A new congenital hemolytic anaemia with deformed erythrocytes (?"stomatocytes") and remarkable susceptibility of erythrocytes to cold hemolysis *in vitro*. Pediatrics 35:906-915

Miller DR, Rickles FR, Lichtman MA, La Celle PL, Bates J, Weed RI (1971) A new variant of hereditary hemolytic anemia with stomatocytosis and erythrocyte cation abnormality. Blood 38:184-203

Papahadjopoulos D, Jacobson K, Nir S, Isac T (1973) Phase transitions in phospholipid vesicles. Fluorescence polarization and permeability measurements concerning the effect of temperature and cholesterol. Biochim Biophys Acta 311:330-348

Perel Y, Dhermy D, Carrere A, Chateil JF, Bondonny JM, Micheau M, Barbier R (1999) Portal vein thrombosis after splenectomy for hereditary stomatocytosis in childhood. Eur J Pediatr 158:628-630

Rajaram S, Sedenskey M, Morgan P (1998) *Unc*-1: a stomatin homologue controls sensitivity to volatile anesthetics in *C. elegans*. Proc Natl Acad Sci USA 95:8761-8766

Rajaram S, Spangler TL, Sedensky MM, Morgan PG (1999a) A stomatin and a degenerin interact to control anesthetic sensitivity in *Caenorhabditis elegans*. Genetics 153:1673-1682

Rajaram S, Spangler TL, Sedensky MM, Morgan PG (1999b) UNC-1 and a sodium channel affect sensitivity to volatile anesthetics in *C. elegans*. Genetics 153:1673-1682

Schrier SL, Zachowski A, Devaux PF (1992) Mechanisms of amphipath-induced stomatocytosis in human erythrocytes. Blood 79:782-786

Sheetz M, Singer SJ (1974) Biological membranes as bilayer couples: a molecular mechanism of drug-erythrocyte interactions. Proc Natl Acad Sci USA 71:4457

Shimkels R, Lifton R, Canessa C (1997) The activity of the epithelial sodium channel is regulated by clathrin-mediated endocytosis. J Biol Chem 272:25537-25541

Shohet SB, Livermore BM, Nathan DG, Jaffe ER (1971) Hereditary hemolytic anemia associated with abnormal membrane lipids: mechanism of accumulation of phosphatidylcholine. Blood 38:445-456

Simons K, Toomre D (2000) Lipid rafts and signal transduction. Nature Reviews: Molec Cell Biol 1:31-39

Snyers L, Umlauf E, Prohaska R (1999) Association of stomatin with lipid-protein complexes in the plasma membrane and the endocytic compartment. Eur J Cell Biol 78:802-812

Stewart GW (1993) The membrane defect in hereditary stomatocytosis. In: Tanner MJA, Anstee DJ (eds) Red cell membrane antigens. Bailliere-Tyndall, London, pp 371-400

Stewart GW, Ellory JC (1985) A family with mild xerocytosis showing increased cation permeability at low temperatures. Clin Sci 69:309-319

Stewart GW, Turner EJ (1999) The hereditary stomatocytoses and allied disorders: congenital disorders of erythrocyte membrane permeability to Na and K. Bailliere's Best Practice Res Clin Haematol 12:707-728

Stewart GW, Corrall RJM, Fyffe JA, Stockdill GM, Strong JA (1979) Familial pseudohyperkalaemia. A new syndrome. Lancet ii:175-177

Stewart GW, Ellory JC, Klein RA (1980) Increased human red cell cation passive permeability below 12 °C. Nature 286:403-404

Stewart GW, O'Brien H, Morris SA, Owen JS, Lloyd JK, Amess JAL (1987) Stomatocytosis, abnormal platelets and pseudo-homozygous hypercholesterolaemia. Eur J Haematol 38:376-380

Stewart GW, Hepworth-Jones BE, Keen JN, Dash BCJ, Argent AC, Casimir CM (1992) Isolation of cDNA coding for a ubiquitous membrane protein deficient in high Na, low K stomatocytic erythrocytes. Blood 79:1593-1601

Stewart GW, Amess JAL, Eber S, Kingswood C, Lane P, Smith B, Mentzer WC (1996) Thrombo-embolic disease after splenectomy for hereditary stomatocytosis. Br J Haematol 93:303-310

Tavernarakis N, Driscoll M, Kyrpides NC (1999) The SPFH domain: a universal motif in membrane-associated proteins implicated in regulating targeted protein turnover. Trends Biochem Sci 24:425-427

Trueman RJ, Tiku PE, Caddick MX, Cossins AR (2000) Thermal thresholds of lipid restructuring and delta(9)-desaturase expression in the liver of carp (*Cyprinus carpio L.*). J Exp Biol 203:641-650

Vantyghen MC, Dagher G, Doise B, Lallau G, Vandewoulle B, Racadot A, Lefebre J (1991) Pseudo-hyperkaliemie. A propos d'une observation familiale. Ann Endocrin 52:104-108

Wang D, Turetsky T, Perrine S, Johnson RM, Mentzer WC (1992) Further studies on RBC membrane protein 7.2B deficiency in hereditary stomatocytosis. Blood 80, suppl 1:275a

Wieth JO (1970) Paradoxical temperature dependence of sodium and potassium fluxes in human red cells. J Physiol 207:563-580

Zarkowsky HS, Oski FA, Shaafi R, Shohet SB, Nathan DG (1968) Congenital hemolytic anemia with high sodium, low potassium red cells. Studies of membrane permeability. New Engl J Med 278:573-581

Zhu Y, Paszty C, Turetsky T, Tsai S, Kuypers FA, Lee G, Cooper P, Gallagher PG, Stevens ME, Rubin E, Mohandas N, Mentzer WC (1999) Stomatocytosis is absent in "stomatin"-deficient murine red blood cells. Blood 93:2404-2410

Zimmer G, Schirmer H (1973) Viscosity changes of erythrocyte membrane and membrane lipids at transition temperature. Biochim Biophys Acta 345:314-320

22 Metabolic Disorders

Paolo Arese and Evelin Schwarzer

Department of Genetics, Biology and Biochemistry, University of Torino, Medical School, Via Santena 5 bis, 10126 Torino, Italy

22.1 Introduction

The human red blood cell (RBC) is a simplified cell extremely well-fitted to comply with its functions. During its life span of 120 ± 4 days, 120 miles of travel and $1.7 \cdot 10^5$ circulatory cycles the RBC encounters and successfully copes with a number of perilous situations, of external and internal origin. Externally, shear stress, turbulence, and passage across narrow capillaries and splenic slits are met by a resilient, mechanically rugged yet flexible and highly deformable membrane envelope and underlying cytoskeleton. The external cell surface is non-immunogenic and non-adhesive, to avoid adhesion to endothelia and escape close checks by spleen, liver and bone marrow macrophages, ready to phagocytose any cell showing even subtle membrane alterations. Internally, the potentially dangerous high concentration of oxygen and the continuous formation of oxidant species during the oxygenation-deoxygenation cycles of haemoglobin (Hb) are counteracted by powerful metabolic systems of detoxication and protection.

The RBC is unable to synthesize enzymes or structural protein to substitute for damaged molecules. To counterbalance this inability, the RBC is equipped with a vast excess of enzymes and metabolic potency. Table 22.1 shows a collection of quantitative data on metabolic fluxes, enzyme activities and metabolite levels and briefly summarizes RBC metabolism. It is of note that at the end of RBC life span, its enzyme activities, ATP and other crucial metabolites are still present in sufficient amounts, and rheological and mechanical properties of the cell are apparently intact and do not justify RBC death. What determines the RBC demise is not its fragmentation but its rapid transformation into a non-self cell, recognized as such by the phagocytic system and removed. This transformation is due to rather subtle modifications of the RBC membrane and expression of neo-antigenic sites or molecules that lead to opsonization or direct recognition by phagocytes.

A number of inherited or acquired RBC defects accelerate or magnify the normal process of senescence and induce enhanced and precocious removal of chronologically young, yet biologically and functionally old cells. The term "haemolytic anaemia" usually adopted to indicate enhanced RBC removal and inbalance between formation and loss of RBCs is misleading, because in most cases RBCs do not lyse but are removed by phagocytosis.

Table 22.1. Selected metabolic fluxes, enzyme activities and metabolite levels of glycolysis, hexose monophosphate shunt and anti-oxidant defense system in human erythrocytes

Metabolic fluxes of fresh RBCs in plasma, 46% haematocrit at 37 °C [μmoles/(ml$_{cells}$·h)]	Maximum activity of energy metabolism enzymes in fresh RBC at 37 °C [μmoles/(ml$_{cells}$·h)]	Maximum activity of anti-oxidant defense enzymes in fresh RBCs at 37 °C [μmoles/(ml$_{cells}$·h)]	Steady-state levels of metabolites in fresh RBCs [nmoles/ml$_{cells}$]
Glucose consumption: 1–2	Hexokinase: 37		Glucose-6-phosphate: 28 NADPH: 32 NADP: 1.4
Lactate production: 2–4	G6PD: 173		Lactate: 932 Pyruvate: 53
HMS: 0.08–1.8	Glucosephosphate isomerase: 1265		
DPG shunt: 0.4–1.6	Glyceraldehyde-3-P dehydrogenase: 4700		Glyceraldehyde-3-P: 4
	Phosphoglycerate kinase: 6656		ATP: 1350 ADP: 216 AMP: 21
	Bisphosphoglyceromutase: 99		DPG: 4171
	Pyruvate kinase: 312	Catalase: $3.2 \cdot 10^6$	
	Na^+, K^+-ATPase: 2.5	Glutathione peroxidase: 640	GSH: 2200–2500 GSSG: 3.6
	Mg^{2+}, Ca^{2+}-ATPase: 10–20	Glutathione reductase: 140–216	
		Superoxide dismutase: 46900	
		NADH-met-Hb reductase: 395	

Data recalculated and adapted from Grimes (1980) and Beutler (2001a,b).

Glucose is the normal energy source of the RBC. Glucose can be transformed to 2 molecules lactate via glycolysis (Embden-Meyerhof pathway), or oxidatively via the oxidative section of HMS to ribulose-5-phosphate. Ribulose-5-phosphate can reenter glycolysis as fructose-6-phosphate and glyceraldehydes-3-phosphate via the nonoxidative section of the hexose monophosphate shunt. 2,3-bisphosphoglycerate (DPG) can be generated by diversion of triosephosphate through the Rapoport-Luebering shunt. Glutathione cycling from oxidized to reduced form is dependent upon NADPH generation. NADH is utilized to reduce met-Hb to Hb. Net output of glycolysis is 2 moles of ATP per mole of glucose metabolized. The rate of glucose utilization is limited by the hexokinase and phosphofructokinase reactions. Both enzymes have a relatively high optimum pH. Therefore RBC glycolysis is very pH-sensitive and is stimulated by a rise in pH. The main Embden-Meyerhof pathway has two important branching points: in the first one the product of hexokinase, glucose-6-phosphate can be diverted to the HMS by G6PD. Under normal steady-state conditions, 92% of glucose is metabolized along the Embden-Meyerhof pathway and 8% along the HMS. Under oxidant conditions as much as 90% of glucose can be metabolized along the HMS. In the second branching point, the product of phosphoglycerate kinase, 1,3-diphosphoglycerate, can be diverted to the Rapoport-Luebering shunt by 2,3-diphosphoglycerate mutase to produce DPG. As much as 40% of glucose is diverted through this shunt under normal conditions. Five intermediates are particularly important:*

In the following, we will first briefly summarize the kind of (mostly) oxidative insults that impact on and damage the membrane, physiologically transforming a viable RBC into a senescent cell, and we will describe the protective mechanisms that counteract those insults. We will then describe the membrane modifications that accumulate during the RBC life eventually leading to phagocytic removal, and outline the connections between RBC metabolism and membrane status. Finally, we will show how a congenital error of intermediary metabolism such as deficiency of glucose-6-phosphate dehydrogenase (G6PD) or a generalized non-haematological disease such as diabetes may accelerate the process of senescence causing RBC destruction.

22.2 To Live Perilously: the Life of a Normal Red Cell

Since oxygen can only bind to iron in the ferrous form, the ability of Hb to transport oxygen is dependent on the reduced state of iron in the haem moiety of deoxyhaemoglobin. During the binding of oxygen to form oxyhaemoglobin, one electron is partially transferred from iron to the bound oxygen forming a ferric-superoxide anion complex. During deoxygenation the shared electron is normally returned to the iron when oxygen is released. However, a fraction of the electrons remain and will transform the leaving oxygen into a superoxide anion radical (superoxide or O_2^- for short). In this process, iron is left in the ferric state and Hb is transformed into met-Hb. The autoxidation of Hb occurs spontaneously and transforms 0.5–3% of Hb into met-Hb per day. In addition to this physiological process, met-Hb may be generated by endogenous oxidants, such as hydrogen peroxide (H_2O_2), nitric oxide (NO) and hydroxyl radicals ($OH^•$), formed by granulocytes/monocytes or endothelial cells (for review, see Carrell et al. (1975); Rice-Evans and Baysal (1987)).

Exogenous oxidants, such as drugs or food components may also produce met-Hb directly or via some of their metabolites. Both superoxide and met-Hb are potentially dangerous to the RBC membrane. Met-Hb is unable to bind oxygen and increases the oxygen affinity of the accompanying haems in the Hb tetramer impairing the overall oxygen delivery. Furthermore, met-Hb is the first step in the formation of harmful hemichromes. Superoxide is easily transformed into the potent oxidant H_2O_2 by superoxide dismutase, abundantly present in the RBC. Under normal conditions, the RBC is able to reduce met-Hb back to ferrous-deoxyhaemoglobin by the NADH-dependent met-Hb reductase. NADH is generated during glycolysis and electrons are transferred from NADH to ferrihaem of

* ATP, the primary energy intermediate is essential to ensure the active cation transport across the membrane, to preserve RBC shape and flexibility, to affect the position of the oxygen dissociation curve; DPG, in association with pH and HCO_3^-/CO_2 modulates the position of the oxygen dissociation curve; NADH is the substrate for metHb reductase to keep Hb in a functionally competent state; NADPH is the substrate for glutathione reductase to regenerate GSH after oxidant insults; GSH is the main protector of thiol groups and scavenger of oxides, peroxides and oxidant radicals, and detoxicant of foreign compounds.

met-Hb in a complex reaction involving FAD, cytochrome b5 reductase and cytochrome b5. H_2O_2 is detoxified by catalase and by reduced glutathione (GSH) peroxidase (GSH-POX). Catalase is now believed to be very important in cellular antioxidant defence (Gaetani et al. 1996). The catalase molecule is a tetramer of 60 kDa subunits with four ferrihaem groups per tetramer. H_2O_2 is broken down to water and oxygen with the stoichiometry:

$$2\,H_2O_2 \rightarrow 2\,H_2O + O_2$$

in a rapid two-step reaction:

Reaction 1: Protein-Fe^{3+}-OH + H_2O_2 → Protein-Fe^{3+}-O-OH + H_2O
(complex I)
Reaction 2: Protein-Fe^{3+}-O-OH + H_2O_2 → Protein-Fe^{3+}-OH + H_2O + O_2

In the presence of oxidants complex I is transformed into the inactive complex II (not shown). Interestingly, catalase is effectively protected against inactivation by NADPH, found to be tightly bound to catalase in the ratio of four molecules of NADPH per catalase tetramer. In case of insufficient NADPH generation, catalase is inactivated and thus the anti-oxidant defence of the RBC impaired (Kirkman et al. 1987). The GSH-POX peroxide detoxication route is also dependent of efficient supply of NADPH, the reductant cofactor in the GSH-regenerating reaction catalyzed by glutathione reductase:

$$\text{Oxidized glutathione} + 2\,\text{NADPH} + H^+ \rightarrow 2\,\text{GSH} + 2\,\text{NADP}^+$$

Therefore, metabolically produced reduced coenzymes play essential roles in antioxidant RBC defence: NADH, produced by the glycolytic pathway is the main reductant of met-Hb to Hb, while NADPH, produced by the first two reactions of the hexose monophosphate shunt (HMS), is the co-substrate of GSH-POX and the protector of catalase efficiency (for review, see Carrell et al. (1975); Rice-Evans and Baysal (1987); Shinar and Rachmilewitz (1993); Stern (1989); Winterbourn (1990)).

22.3 The Damaging Role of Hemichromes, Free Haem and Free Iron

In normal, unstressed RBCs spontaneous production of met-Hb on autoxidation of Hb and conversion of met-Hb to its normal functional state are in balance. Any pathological situation, which increases the turnover of this cycle whether increased oxidative stress or impaired antioxidant defences will enhance production of met-Hb and generation of active oxygen species. Enhanced formation of met-Hb leads to formation of hemichromes. Hemichromes are ferric Hb derivatives with characteristic absorption and electron spin resonance spectra in which the sixth haem ligand is either the distal histidine, in which case the process is reversible ("reversible hemichromes") or another aminoacid residue from the globin ("irreversible hemichromes"). Hemichrome formation depends on the amount of met-Hb formed and is accelerated by oxidants such as superoxide or H_2O_2 that enhance the formation of met-Hb. The damaging activity of hemichromes derives from their potential role as generators of powerfully oxidant hydroxyl radical via a so-called iron-catalyzed Haber-Weiss reaction:

Reaction 1: $O_2^- + Fe^{3+} \rightarrow O_2 + Fe^{2+}$
Reduction of Fe^{3+} by O_2^- to form Fe^{2+} and provide the "Fenton reagent" $(H_2O_2 + Fe^{2+})$

Reaction 2: $H_2O_2 + Fe^{2+} \rightarrow OH^- + OH^\bullet + Fe^{3+}$
Iron-catalyzed Haber-Weiss reaction with production of hydroxyl radical $OH^\bullet$

Reaction 3: $O_2^- + H_2O_2 \rightarrow OH^- + OH^\bullet$
Non-catalyzed Haber-Weiss reaction less likely to occur in the cell.

In the Haber-Weiss reaction 1, O_2^- produced continuously by univalent reduction of oxygen would reduce ferri-hemichrome to ferro-hemichrome. In reaction 2, the "Fenton reagent" ferro-hemichrome would catalyse decomposition of H_2O_2 to hydroxyl radical and regenerate ferri-hemichrome. In this sequence, the "Fenton reagent" ferro-hemichrome generates hydroxyl radical upon reaction with H_2O_2. Hydroxyl radical is an extremely reactive oxidizing radical that will react with most biomolecules at diffusion-controlled rates. For example, hydroxyl radical may abstract hydrogen from unsaturated fatty acids and start lipoperoxidation. Fe^{2+} itself is able to cycle between the ferric and ferrous state and given a continuing flux of H_2O_2 and superoxide provides the opportunity for potentially unlimited generation of hydroxyl radical. Evidence that iron of hemichromes is capable of cycling and facilitating superoxide/H_2O_2-driven hydroxyl radical generation was provided by Hebbel et al. (1982).

Hemichromes are not the only hydroxyl radical generators: free haem that may easily detach from met-Hb, or even free iron both facilitate hydroxyl radical generation. Despite the fact that iron is bound tightly to Hb, it may be released from haem under specific circumstances, like oxidant insult, yielding a source of free, reactive iron. A second mechanism by which hemichromes exert a damaging effect is due to their liberation of the free oxidized haem (haemin) moiety. Oxidized Hb or hemichromes easily lose haem that due to its hydrophobic properties readily associates with lipids of the RBC membrane. RBCs of all pathologic conditions characterized by increased susceptibility to oxidation such as sickle cell anaemia, thalassemia and G6PD deficiency contain increased amounts of haemin in their membranes (for review, see Chiu and Lubin (1989); Chiu et al. (1989)). Reports by several authors indicate that at micromolar concentration haemin can induce potassium leak, decrease osmotic fragility, cause RBC swelling, interact with spectrin, actin and protein 4.1, mediate dissociation of membrane skeletal proteins, and destabilize the RBC membrane. Haemin is an extremely potent catalyst of peroxide-induced lipid peroxidation in RBC membrane, and oxidizes thiol groups of membrane proteins dose-dependently (for review, see Chiu and Lubin (1989)).

Peroxidation of membrane lipids, most notably poly-unsaturated fatty acids arachidonic and linoleic acid, generates a wide array of molecules, such as various types of lipid hydroperoxides (e.g., hydroxy-eicosatetraenoic acids, HETEs, and hydroxy-octadecaenoic acids, HODEs); secondary lipid peroxidation products deriving from further decomposition of lipid hydroperoxides (e.g., malondialdehyde and other hydroxyaldehydes such as 4-hydroxynonenal, HNE); volatile alkanes and conjugated diene hydroperoxides. Interestingly, hemichromes and met-Hb were also found to generate prostanoids and hydroxysteroids via a non-enzymatic iron-porphyrin catalysis. Lipid derivatives of oxidant attack, most notably malondialdehyde, exert a number of detrimental effects on RBCs: they can damage membrane structure with formation of membrane pores, increased potassium leak and altered water permeability; polymerize membrane components and decrease cell deformability; cross-link membrane proteins; enhance IgG binding and complement activation; finally, they may enhance exposure of phosphatidylserine (PS) on the outer cell surface (Chiu et al. 1989).

Lastly, oxidant activity against membrane components can be exerted by free non-haem iron. Free non-haem iron was found to accumulate in pathologic conditions such as sickle cell anaemia and thalassemia (Repka et al. 1993). Most likely, free iron enhances the oxidant activity of free haem and hemichromes by enhancing the formation of hydroxyl radicals. In addition to its direct effects on the RBC membrane, haemin was found to inhibit many RBC cytosolic enzymes including aldolase, pyruvate kinase, 6-phosphogluconate dehydrogenase and pyrimidine 5'-nucleotidase.

In conclusion, hemichromes, free haem and non-haem iron all contribute to generate oxygen-derived radicals of different potency that may hit and possibly damage any component of the RBC membrane or cytoplasm and produce a variety of functional impairments. The type of damage will depend on the intensity and specific potency of the oxidant bullet, on the sensitivity of the target, and on the functional role of the target (for review, see Carrell et al. (1975); Hebbel (1985); Rice-Evans and Baysal (1987); Stern (1989); Winterbourn (1990); Shinar and Rachmilewitz (1993)).

A very important feature of hemichrome/haem/iron damage is its non-random occurrence in space. Specific location of attack and propagation are high-affinity binding sites for both haem and hemichromes localized on the long cytoplasmic domain of band 3. This domain is considered a major organizing centre of RBCs and an anchoring site for a number of membrane-associated proteins, such as protein 4.2, 4.1, glyceraldehyde 3-phosphate dehydrogenase, phosphofructokinase, aldolase, Hb, hemichromes and protein tyrosine kinase (p72syk) (Low 1986). Detailed information on organization of the peripheral protein complex on the cytoplasmic domain of band 3 has been provided by the recently determined crystallographic structure (Zhang 2000).

The highly damaging properties of hemichromes result from their tight association with the cytoplasmic domain of band 3. This association propagates beyond the stoichiometric unit complex between the two moieties as it creates a polymer of macromolecular dimensions, promoting clustering of band 3 in the membrane (Waugh et al. 1987). Several authors (for review, see Arese et al. (1992); Kay et al. (1988), Lutz (1990)) have proposed a mechanism for recognition and phagocytic removal of senescent or damaged RBCs based on oxidative clustering of band 3. Oxidative damage to Hb and formation of hemichromes leads to the association between hemichromes and the cytoplasmic domain of band 3 (Low 1986; Zhang 2000). These clusters show enhanced affinity for normally circulating anti-band 3 antibodies and these in turn activate the complement system. It has been shown that less than 1% oligomerized band 3 was sufficient to elicit deposition of autologous anti-band 3 IgG. These few molecules were able to induce generation of large amounts of complement fragments via activation of the alternative pathway. The ultimate signal for phagocytic removal is complement C3 molecules deposited on the RBC surface in their active form C3b (for review, see Lutz (1990)). This model was substantiated by converging evidence showing co-clustering of aggregated band 3 and hemichromes and induction of phagocytosis *in vivo* and *in vitro*, and was verified in different *in vivo* situations (favic crisis, thalassemic patients, canine senescence model) and *in vitro* models. Interestingly, in a canine RBC senescence model closely resembling human RBC senescence (Rettig et al. 1999), senescent RBC were found to contain abruptly elevated membrane-bound denatured globin and seven-fold higher amounts of membrane-associated autologous IgG. In parallel, band 3 was significantly less restrained indicating substantial disruption of the membrane skeleton.

The band 3-clustering model is not necessarily exclusive of other explanations of senescent/damaged cell recognition. Thus opsonization of a cryptic galactosyl residue or recognition of deglycosylated, cleaved or oxidized cell surface protein by autologous IgG may also allow the more avid bivalent IgG attachment. Loss of phospholipid asymmetry and recognition of exposed phospholipids normally hidden such as PS by scavenger receptors might also be associated with band 3 clustering-mediated recognition. Recently, the latter mechanism has been emphasized as a likely mechanism of major importance for removal of senescent RBCs (Kiefer and Snyder 2000). These authors point out that calmodulin, the calcium-responsive protein that activates the plasma membrane Ca^{2+}-ATPase and the flipase, the inward aminophospholipid translocase that governs phospholipid asymmetry in the membrane, are both targets of oxygen free radicals. Kiefer and Snyder also point out that the intracellular calcium level increases with RBC age. Calcium is an important activator of the protease calpain, able to proteolyze the calcium extrusion pump and thus starting a negative cycle of increased influx and decreased efflux of calcium. High calcium levels also activate phospholipid scramblase and promote microvesicle shedding, both inducing loss of phospholipid asymmetry. The PS/scavenger receptor-dependent removal mechanism is certainly attractive and fits with increases in calcium levels observed *in vivo* in favic crises (De Flora et al. 1985; Turrini et al. 1985) and *in vitro* together with enhanced erythrophagocytosis. Support for this view is provided by recent observations in thalassemics (Kuypers et al. 1998). Data indicate that in four splenectomized patients with Hb E/beta-thalassemia and in two splenectomized patients with beta-thalassemia intermedia a varying percentage of RBCs showed the presence of PS on the outside of the cell. Interestingly, areas of PS exposure coincided with aggregation of globin and localization of Heinz bodies. The same coincidence was noted by others in sickle cell anaemia (Liu et al. 1996). The PS exposure hypothesis, however, is open to criticism: first, because treatment of RBCs with oxidants and with phenylhydrazine, a strong oxidant and inducer of Heinz bodies and phagocytosis did not expose PS; second, because patients with Hb H, Hb H/CS and beta-thalassemia intermedia that did not show PS exposure (see above) had aggregated band 3, deposition of complement and IgG and were intensely phagocytosed by human monocytes. In conclusion, the attractive PS exposure hypothesis is presently supported by few and contradictory data and cannot be accepted as the prime mechanism of non-self transformation and phagocytic recognition.

22.4 Pathophysiological Consequences of Oxidative Damage

Oxidative damage outlined above represents normal events during RBC life. In the course of its life in the circulation, damage cannot be completely repaired due to the progressive decrease in activity of defensive enzymes and accumulate until reaching a threshold value. As it has been shown in the dog RBC senescence model, a precipitous increase in protein aggregates occurs at the end of the RBC life-span accompanied by deposition of removal markers. Homeostatic mechanisms that cannot be discussed here keep a precise balance between RBC removal and erythropoiesis, such that a daily demise of 1% old RBCs is exactly matched by entry into circulation of a corresponding number of young RBCs.

A number of genetically-determined or acquired conditions may enhance the formation of oxidant compounds or impair antioxidant defence. In both cases the consequence is accelerated senescence and removal, and increased haemolysis. If the number of removed cells or the kinetics of RBC removal exceeds the erythropoietic capacity, anaemia is the consequence. In quantitative terms, according to the intensity of oxidative stress and the degree of inactivation of the anti-oxidant defences, anaemia can be imperceptible, acute and dramatically severe, or chronic.

Based on well-established evidence from various groups, it is reasonable to assume that the same events that induce removal of normal RBCs are present in a magnified and accelerated form in a series of genetic RBC defects affecting metabolism or haemoglobin structure or synthesis, and in metabolic disorders not primarily involving RBCs but secondarily impacting on RBCs functionality and survival. An incomplete listing of genetic RBC defects affecting intermediary metabolism and anti-oxidant defence would include genetically determined deficiency of glycolytic enzymes (hexokinase, phosphofructokinase, pyruvate kinase, etc.), hexose monophosphate shunt (HMS) enzymes (G6PD, 6-phosphogluconate dehydrogenase), anti-oxidant defence enzymes (superoxide dismutase, glutathione peroxidase, glutathione reductase, met-Hb reductase). Hb synthesis defects may consider the various thalassemias; Hb molecular defects would include sickle cell anaemia, ovalocytosis, unstable haemoglobins, possibly Hb C and Hb E; metabolic disorders or generalized diseases impacting on RBCs include diabetes, alcoholism and cancer (for review, see Jacobasch (2000); Beutler (2001c); Weatherall (2001)).

In the following, we will concentrate on two model examples. We will consider deficiency of G6PD as an example of a genetic defect accompanied by insufficient anti-oxidant defence, and diabetes, as an example of a general disease with distinct RBC damages. We will show that in both cases although through different mechanisms, the final outcome is specific damage to the membrane, leading to accelerated RBC destruction and anaemia.

22.5 G6PD Deficiency

22.5.1 Structure of Normal and Mutant G6PD

G6PD catalyzes the first and pace-making reaction of the HMS by which glucose-6-phosphate (G6P) is oxidized to 6-phosphogluconolactone with production of NADPH. The HMS converts glucose into pentoses necessary for a variety of synthetic reactions, and produces reducing power in the form of NADPH. In RBCs the production of NADPH is essential for the protection of the cell against oxidative stress. A major role of NADPH is to maintain GSH at a ratio grater than 500:1 over the oxidized form GSSG. GSH plays a vital role in antioxidant defence by reacting with H_2O_2 and organic peroxides, as well as maintaining thiol groups of Hb and other proteins and enzymes in the reduced state. H_2O_2 can be also detoxified by catalase which is also stabilized by tightly bound NADPH. The human G6PD monomer is 515 aminoacids long with a molecular weight of about 59 kDa. The enzyme is active in RBCs in a tetrameric or dimeric form. At physiological pH, the dimeric form is predominant. Each monomer has two domains: the N-terminal domain (residues 27–200) with a dinucleotide binding site (residues 38–44) and a second domain with a nine-barrel anti-parallel sheet. The two domains are linked by a α-helix containing the conserved eight-residue substrate binding site (residues 198–206). The stability of these quaternary structures is important for activity.

G6PD deficiency is the most common human enzyme defect, being present in probably more than 300 million people. The gene encoding G6PD is located on the long arm of the X chromosome, consists of 13 exons spanning approx. 18 kb. The coding sequence of the gene is 1545 bp long, coding for the 515 amino acids of the G6PD primary sequence. A male inheriting the defective X chromosome carrying the deficient gene from his mother is a deficient hemizygote. Female homozygotes inherit a defective X chromosomes from each parent. Female heterozygotes inherit only one defective X chromosome. Thus, they have a double population of normal and deficient RBCs ("RBC mosaic"). Due to a varying amount of active X chromosome left, G6PD activity in heterozygotes varies considerably from almost normal to severely deficient. G6PD is one of the most polymorphic proteins known.

Close to 400 variants have been characterized by traditional biochemical methods. The identification of a specific mutation in the gene drastically lowered the number of genetic variants. About 90 different mutations have been detected in all coding regions and in the promoter. Most of these are single amino acid substitutions, and mutations associated with severe chronic anaemia are preferentially localized in the exons 9, 10 and 11. Distinct domains in exon 10 with highly conserved amino acid sequences correspond to the NADP- and G6P-binding sites. In contrast to Hb genes, deletions in the G6PD genes are rare. The most widespread normal variant (B normal variant) is present in Caucasians, Asians and 80% of Africans. The second most widespread normal variant (A variant) is frequent in Africans. G6PD mutants are categorized in four classes: class I, very low residual activity "malign" mutants causing chronic nonspherocytic anaemia; class II, very low (less than 1%) to low (less than 10%) activity mutants with no clinical symptoms unless oxidatively stressed; class III, mutants with slightly decreased activity and no clinical symptoms even under stress; class IV, mutants with increased activity. Widespread low activity mutants are class II mutants: G6PD Mediterranean, frequent in populations facing the Mediterranean and also in Iraq, Iran, Saudi Arabia; G6PD A$^-$, frequent in Africans; G6PD Canton, frequent in Southern China and Taiwan; G6PD Markham, frequent in Papua New Guinea. Essential genetic, molecular and biochemical data on most frequent variants are given in Table 22.2A (for review, see Arese and De Flora (1990); Beutler (1990, 1992, 2001c); Vuillamy et al. (1992); Luzzatto and Mehta (1995); Mason (1996); Jacobasch (2000); Mehta et al. (2000)).

22.5.2 Intracellular Regulation of G6PD in Normal and G6PD-Deficient Red Cells

The intracellular regulation of normal and mutant G6PD has been thoroughly investigated (Yoshida 1973; Gaetani et al. 1974; Kirkman and Gaetani 1986). It is now clear that a large part of control of G6PD activity is mediated by other RBC proteins and is not merely inherent to the intrinsic properties of G6PD itself. Normal G6PD works within the RBC at a very limited fraction of its potential activity (for review, see Arese and De Flora (1990)). Whatever is the cause for increased demand for reducing equivalents, be it oxidative stresses acting on the NADP/NADPH system or on the glutathione system, the result will be the enhanced saturation of G6PD by NADP and the increased intracellular activities of G6PD and HMS. Failure of the G6PD-deficient RBC to cope with acute (class II and III variants) or chronic (class I variants) loss of reducing equivalents, such as NADPH and GSH, is the starting event that eventually results in RBC damage and destruction. Kirkman and Gaetani (1986) have reported inconsistencies between the observed and the kinetically expected activities of intraerythrocytic G6PD. Such inconsistencies were both quantitative (remarkable restraint of intracellular G6PD activity) and qualitative (sigmoid kinetics of intracellular G6PD *vs.* the Michaelian kinetics of the purified enzyme). Further analysis of these inconsistencies revealed that they are due to extensive binding of the erythrocytic NADP to NADPH diaphorase and of NADPH to catalase (Kirkman and Gaetani 1986). The resulting unavailability of NADP and NADPH to G6PD within the RBC was calculated to explain a part of the discrepancy between observed and expected activity of intracellular G6PD. The presence of NADPH tightly bound to RBC catalase and the requirement of NADPH for preventing or reversing inactivation of catalase by H_2O_2 indicated that both catalase and glutathione peroxidase are effective in removing H_2O_2 in human RBCs (Gaetani et al. 1996). Dependence of both H_2O_2-scavenging enzymes on NADPH makes the G6PD-deficient RBC particularly susceptible to oxidative haemolysis.

Studies performed with intact G6PD-deficient RBCs show that these cells utilize glucose through the HMS at a rate similar to that of normal RBCs. On the other hand, in the presence of oxidizing agents, HMS in normal RBCs is stimulated several fold, while in G6PD-deficient RBCs the stimulation was minimal (for review, see Arese and De Flora (1990)) (Table 22.2B). This behaviour can be explained because in the normal cell, almost all NADP is present in the reduced form, so that G6PD is operating at a very low rate, the available NADP being less than 5% of NADP. By contrast, in G6PD-deficient RBCs the NADPH/NADP ratio is reversed, NADPH being approx. 30% of total NADP (Arese and De Flora 1990). Under these conditions HMS is already operating at near maximal rate and is not susceptible to any further stimulation (Table 22.2B).

Table 22.2. Molecular and kinetic data of normal and mutant G6PD

(**A**) Molecular characteristics of normal and most frequent mutant G6PD

Variant	Nucleotide substitution	Amino acid substitution
A	376 A→G	126 Asn→Asp
A^- (African)	202 G→A	68 Val→Met
Mediterranean	563 G→T	188 Ser→Phe

(**B**) Main kinetic parameters of normal and Mediterranean G6PD, and comparison of *in vitro vs. in vivo* behaviour of normal and Mediterranean G6PD

Mutant	Activity at 37 °C [μmoles/(ml_{cells}·h)]	K_m G6P [μM]	K_m NADP [μM]	K_i NADPH [μM]	K_i ATP [μM]
Normal B, *in vivo*	0.15 (100%)	[28]	[0.8–1.4]	[32]	[1350]
Normal B, *in vitro*	170 (100%)	50–70	2.9–4.4	6.7	1030
Mediterranean, *in vivo*	0.12 (80%)	[30–40]	[63]	[20.4]	[1100]
Mediterranean, *in vitro*	1.7–8.5 (1–5%)	11.5	2–2.5	9.5	1000

Activities *in vitro* refer to G6PD activities measured in the haemolysate under optimal conditions. Activities *in vivo* refer to HMS fluxes measured as $^{14}CO_2$ output from 1-^{14}C-glucose in fresh intact RBCs. Numbers in square brackets are steady-state levels of metabolites assayed in fresh RBCs and given as nmoles/ml RBCs. Note that HMS flux *in vivo* in normal and mutant G6PD is approx. 0.8% and approx. 7% of maximal activity of G6PD measured under optimal conditions, respectively. Data recalculated and adapted from Yoshida (1973).

22.5.3 Oxidant-Induced Haemolysis

We will discuss in some detail the acute haemolysis caused by fava bean ingestion in G6PD-deficient individuals, because it is the best-studied and most frequent natural model of oxidant damage, and it bears strong analogies to other haemolytic crises caused by oxidant drugs or a variety of oxidant chemicals. We consider therefore the favic crisis as a paradigm for oxidant damage to the human RBC.

22.5.4 The Favic Crisis

The favic crisis (favism) is an acute haemolytic syndrome occurring in severely

G6PD-deficient individuals after consumption of *Vicia faba major* (broad beans). The association of widespread consumption of fava beans and high incidence of low-activity G6PD variants is typical of the Mediterranean area, the Middle East and Southern China. G6PD deficiency is a sex-linked trait with marked variability of the phenotypic expression in the heterozygous female. Accordingly, the great majority of favism cases occur in males hemizygous for the deficiency. Females are not exempt, depending on the proportion of enzyme-deficient RBCs in relation to the total RBC mass. The first symptoms of oxidant- or fava-bean-induced haemolysis are generalized weakness to severe lethargy, nausea and vomiting, headache, lumbar or abdominal pain and fever. After a delay of variable duration (few hours to 48 hours) jaundice appears and may reach a very intense stage. It is accompanied by enlargement of spleen and liver (Sansone et al. 1958; Luzzatto and Mehta 1995). Haemoglobinuria begins several hours after fava intake and may continue for several days.

Drug-induced, and specifically primaquine-induced, haemolytic crisis in G6PD-deficient subjects is almost identical with favism. The earliest effect of primaquine or fava bean ingestion was the decrease in RBC GSH, which started few hours after intake, remained at the lowest value for 24–36 hours and came slowly back to normal values after about 6 days. Increase in methaemoglobin was delayed by about 20–24 hours and more short-lived with respect to the fall in GSH levels. Heinz bodies appeared at the same time as haemolysis began and were no longer visible as haemolysis stopped. In the densest RBC fraction isolated during early favism, extremely low GSH levels, down to less than 10% of non-haemolytic G6PD-deficient controls have been noted. Anaemia may be extremely severe and life-threatening. During severe attacks, RBC counts as low as $0.5 \cdot 10^6$ RBCs per μl have been measured. Primaquine haemolysis and favism are self-limited and followed by a refractory phase. Self-limitation means that haemolysis stops after a time varying from 6 to 12 days and does not resume even if further oxidant challenge is applied. The current interpretation of this behaviour is that haemolytic susceptibility is related to the age of the RBC. Only old RBCs containing the lowest activity of G6PD are destroyed, while younger RBCs or reticulocytes, which have higher G6PD levels, resist the challenge. A repeated insult would meet only young, more resistant RBCs and this would explain the refractory period.

22.5.5 Toxic Components of Fava Beans and their Haemolytic Activity

Fava beans are rich in two glucosidic compounds, vicine and convicine, which generate the redox aglycones divicine (2,6-diamino-4,5-dihydroxypyrimidine) and isouramil (6-amino-2,4,5-trihydroxypyrimidine) upon hydrolysis of the beta-glucosidic bond (Fig. 22.1). Fava beans also contain high amounts of ascorbate and varying amounts of L-DOPA glucoside (Mager et al. 1980; Arese and De Flora 1990).

Divicine and isouramil share common effects and similar mechanism of action with other oxidant drugs (Mager et al. 1980; Arese et al. 1981; Arese and De Flora

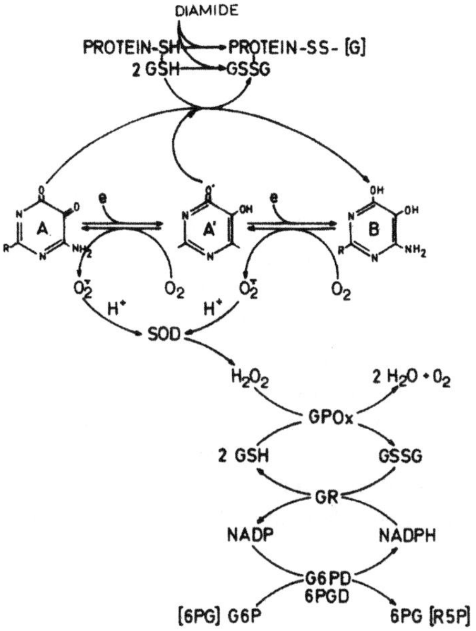

Fig. 22.1. Role of G6PD in antioxidant defence. Oxidant compounds are frequently quinones (*A*) that oxidize GSH or protein –SH groups. Quinones in the figure are fava bean pyrimidine derivatives taken as an example. Quinone-generating drugs such as menadione, primaquine, doxorubicin, alloxan etc. behave similarly. In the process, quinone (*A*) is reduced to the corresponding hydroquinone (*B*) while 2 GSH or protein-SH are oxidized to GSSG or protein mixed disulfide (Protein-SS-[G]). The hydroquinone (*B*) is autooxidized in the presence of molecular oxygen back to the quinone (*A*) via a semiquinone radical intermediate (*A'*). During unielectronic autoxidation superoxide is generated. Superoxide dismutates spontaneously or by superoxide dismutase (SOD) catalysis to H_2O_2. H_2O_2 is reduced to water by GSH-POX or by catalase (not shown), whereby 2 GSH are oxidized to GSSG. GSSG is reduced back to 2 GSH by NADPH-dependent glutathione reductase. NADPH is produced by G6PD and 6-P-gluconate (6PG) dehydrogenase. Product of the latter reaction is ribulose-5-P (R5P). From Arese and De Flora (1990), with permission

1990). In a cell-free system two mechanisms of autoxidation of divicine have been proposed (Fig. 22.1).

First mechanism: Reduced divicine is oxidized to the semiquinoid free-radical form by the one-electron reduction of molecular oxygen to superoxide. The semiquinoid free radical reacts rapidly with GSH or other thiol groups to form GSSG. Superoxide may directly oxidize GSH to GSSG or dismutate to H_2O_2. Redox cycling between oxidized divicine/reduced divicine and GSH/GSSG is expected to occur leading to continuous and suprastoichiometric oxidation of GSH.

Second mechanism: Divicine can produce strongly oxidant hydroxyl radicals in the presence of transition metal ions by a mechanism analogous to the site-specific, metal-mediated Haber-Weiss reaction discussed before.

In summary, the semiquinoid divicine radical, superoxide, H_2O_2 and other aggressive oxygen products oxidize GSH and other thiols, and form ferrylhaemoglobin and methaemoglobin. The effect of oxidants in whole RBCs is different from that described for a cell-free system. Addition of divicine to whole G6PD-deficient RBCs was followed by end-point kinetics of GSH oxidation, with a stoichiometry of approx. one mol oxidized GSH per mol divicine. One reason for the difference with the cell-free system resides in the powerful oxidant scavenging systems of the RBC. In the G6PD-deficient RBCs no regeneration of GSH occurs even after several hours incubation in presence of glucose. In normal RBCs, divicine and isouramil rapidly oxidized GSH to the same extent as in G6PD-deficient RBCs, but initial levels of GSH were rapidly re-established. As a consequence of unbalanced production of oxidants and defence by GSH, incubation of G6PD-deficient RBCs with 1 mM divicine for 12–24 hours produced large polypeptide aggregates unable to migrate into the separating gel (Turrini et al. 1997). Bidimensional electrophoresis after reduction identified those aggregates as consisting largely of band 3 protein and to a minor extent of spectrin. Heinz bodies were formed in a time-dependent and dose-dependent manner (Turrini et al. 1997).

22.5.6 Oxidant Stress and Calcium Homeostasis

In normal RBCs, the inwardly directed gradient of ionized calcium from about 0.1 μmolar (intracellular) to 2500 μmolar (plasma) is maintained by active calcium extrusion by an extremely powerful high affinity Ca^{2+}-ATPase. Basal extruding capacity is 10–20 μmoles $Ca^{2+}/(ml_{cells} \cdot h)$. RBC senescence and conditions with increased oxidant stress and/or reduced anti-oxidant defenses, such as favism, thalassemia, sickle cell disease, Hb C diseases, are accompanied by increased intracellular calcium levels. The mechanism of calcium increase is probably not uniform. It seems to reside in the presence of endocytic, inside-out vesicles containing a high level of calcium in sickle-cells and thalassemic RBCs, while it is due to oxidant-mediated damage of the calcium-extruding pump in favism and RBC senescence, with maximal extrusion capacity reduced by half in aged RBCs.

Keeping ionized calcium low is vital to RBCs. A rise in internal calcium increases cellular rigidity, changes normal morphology into rigid crenated sphere, inhibits the Na^+/K^+ pump and leads to opening of the Gárdos channel, loss of K^+, Cl^- and water ending with profound morphologic alterations (for review, see Romero and Romero (1999)). Raised cell calcium for longer periods of time activates calpain, the main endogenous protease of RBCs, to cleave and degrade the Ca^{2+} pump, the main calpain target. A second calpain target is band 3. Its limited proteolysis might induce band 3 clustering or exposure of antigenic sites and therefore precipitate macrophage recognition and phagocytic removal.

A different way by which increased calcium may influence RBC removal is calcium-elicited loss of asymmetry in phospholipid distribution and exposure of PS to the outer face of the membrane. There appears to be a direct correlation between RBC density and age, intracellular calcium, the amount of PS on the cell surface and the extent of RBC binding and phagocytosis by macrophages. Exposure of PS also mediates increased adhesion of RBCs to endothelial cells, found to be mediated by CD36 scavenger receptor, and to thrombospondin, an adhesive protein able to form intercellular bridges to subendothelial matrix (for review, see Kuypers (1998) and Kiefer and Snyder (2000)).

22.5.7 Predominantly Extravascular Haemolysis in Oxidatively Stressed G6PD-Deficient Red Cells

A crucial question in oxidant-induced haemolysis such as favism concerns the type of haemolysis. Apparently massive haemoglobinuria and increased haemoglobinemia would indicate predominatly intravascular (for example mechanical or osmotic) haemolysis. The share of intravascular haemolysis in the favic crisis has been determined by balancing RBC loss from circulation and appearance of free haem or haemoglobin in urine and blood plasma (Arese and De Flora 1990). The results obtained by follow-up of seven favic crises demonstrate a low share, not exceeding 20% as a mean value, of intravascular haemolysis. It has been estimated that the same share is present in the physiologic removal of senescent RBCs (Garby and Noyes 1959).

In accordance with the model utilized to explain senescent removal, the membrane alterations observed *in vivo* during the favic crisis were superimposable to changes observed in senescent RBCs (Fig. 22.2). In summary, RBC membranes isolated from favic patients contained elevated amounts of complexes between IgG and the complement fragment C3b/C3c; the complex could be cleaved with hydroxylamine, which hydrolyzes primarily ester bonds, and liberated IgG exclusively bound to separated band 3, indicating that autologous anti-band 3 IgG specifically reacted to the aggregated band 3-complement complexes. The relative amount of the complexes correlated with the severity of the crisis and decreased to normal values within 4 to 5 days after fava ingestion. Additional evidence obtained from early favic crises indicated extensive clustering of band 3 (up to 30% of band 3 was aggregated to higher than 4,000 kDa molecular mass clusters) and deposition of hemichromes to the RBC membranes. In parallel, severely damaged RBCs isolated from early crises had extensive cross-bonding (Fischer et al. 1985) and very low reduced glutathione levels (down to 25% of normal levels), and were intensely phagocytosed by adherent human monocytes (up to 9-fold compared to normal RBCs) (Turrini et al. 1997).

In conclusion, the favic crisis and other types of oxidant haemolyses such as those caused by oxidant drugs or chemicals, may be considered as massive and rapid senescence of a large fraction of the RBC population. Those pathologically "senescent" RBCs are removed similarly as in normal RBC senescence, but the rate of the process was magnified up to 15-fold. Indeed, losses of as many as 20% per day of the total RBC mass can frequently be observed in massive favic crises (Arese and De Flora 1990).

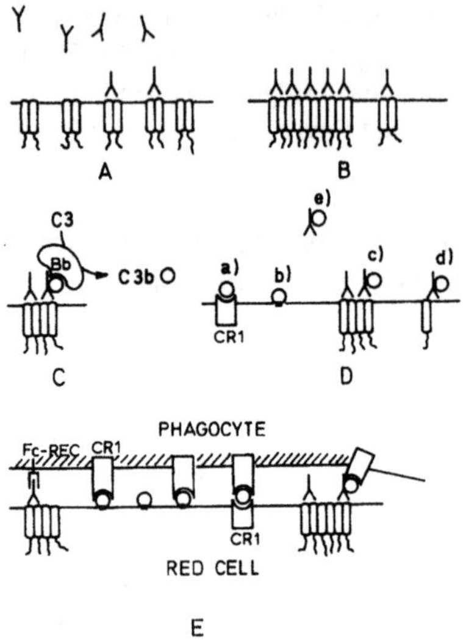

Fig. 22.2. Sequence of events after oxidant damage of the RBC leading to opsonization with anti-band 3 IgG and complement, recognition and removal by the phagocyte. (**A**) Normal or G6PD-deficient RBC before oxidation. Band 3 is not aggregated and only a few molecules of anti-band 3 IgG are bound. (**B, C**) After oxidant insult, aggregation ("clustering") of band 3 enhances binding of anti-band 3 IgG and activates the alternative pathway of complement. Clustered band 3 traps spontaneously generated complement fragments C3b. The binding prevents C3b from inactivation and allows formation of Bb convertase (Bb) that converts more C3 to C3b. (**D**) C3b can bind to complement receptor CR1 (*a*), bind covalently to RBC membrane (*b*), to band 3 clusters (*c*), to monovalently bound anti-band 3 IgG (*d*), or to other IgG (*e*). (**E**) Recognition of opsonized RBC occurs predominantly via CR1 or Fc receptor of the phagocyte. From Arese and De Flora (1990), with permission

22.6 Red Cell Alterations in Diabetes Mellitus

Diabetes mellitus is a metabolic disorder characterized by high blood glucose. Two general types of diabetes mellitus (diabetes for short) are recognized: type I and type II. Type I patients are totally dependent on exogenous insulin, while type II patients can be treated with dietary measures and oral medication. In general, type I diabetics tend to be young and lean ("juvenile onset"), while type II diabetics tend to be older and obese. While insulin and other oral anti-diabetics can control many types of diabetes, numerous complications are common. Peripheral vascular disease primarily originating from endothelial disfunctions is common in diabetes and the likely origin of other severe complications such as blindness, coronary heart disease, renal failure, and infections (King et al. 1994). RBC alterations are frequent in diabetes, and are interesting for several reasons: firstly, the diabetic RBC is a good model to study glycation of proteins and dissect functional impairments in relation to glycation. Second, oxidant damage that accompanies glycation of RBC proteins has similar consequences to those described in oxidatively damaged RBCs, namely accelerated senescence, enhanced phagocytic removal and shortened lifespan; thirdly, metabolic and membrane structure modifications induce increased rigidity, enhanced adhesion, reduced oxygen delivery to tissues and may thus play a role in development of severe diabetic complications such as peripheral vascular disease.

The lifespan of RBC is shortened in diabetes mellitus (Jones and Peterson 1981). The presence in the circulation of a younger RBC population is indicated by several parameters typical of young erythrocytes, such as higher levels of RBC creatine. This situation suggests accelerated removal of altered or precociously senescent RBCs. In the following we will show that enhanced removal of RBCs might be due to a similar sequence of biochemical events that bring about acute anaemia in G6PD deficiency as outlined before.

22.6.1 Glycation of Proteins in Diabetes

Chronically elevated blood levels of glucose is the hallmark of diabetes mellitus. The aldehyde group of the open form of glucose and other monosaccharides or monosaccharide derivatives reacts non-enzymatically with free amino-groups of a wide range of proteins throughout the body, including intracellular proteins such as Hb and lens crystallins, RBC and endothelial membrane proteins, and extracellular proteins such as collagen. Initially a reversible Schiff-base adduct is formed. Successively, following a so-called Amadori-rearrangement the Schiff-base is transformed in a irreversible step into a ketoamine that cannot be split. Over time, these products undergo a series of re-arrangements (similar to nonenzymatic "browning reactions" known as Maillard reactions in the food chemistry) and eventually build up the so-called advanced glycation end products (AGE). AGE are heterocycles with diverse chemical structures that have been partially identified (Njoroge and Monnier 1989). AGE compounds share a yellow-brown color, characteristic fluorescence, the propensity to form crosslinks, generation of oxygen intermediates and interaction with particular receptors on cellular surfaces (Vlassara 1994). One important intermediate, identified as 1-alkyl-2-formyl-3,4-diglycosylpyrrole (AFGP), is considered to play a role in the generation of oxygen radicals.

AGE continue to accumulate intra- and extra-cellularly as a function of time, glucose concentration, number of amino groups, their pK-value, accessibility of amino groups to glucose, nearby presence of protonated amino groups such as those of arginine or histidine. Since the extent of AGE accumulation is determined by the integral of glucose concentration over time, the process occurs over long periods of time *in vivo* where concentrations of glucose rarely exceed 20 mM. Glycation of proteins is a random process, and most proteins isolated from a cellular lysate were glycated to some extent. Inhibitory or damaging effects on e.g. enzymes or structural proteins would depend on the importance of a specific amino residue glycated in the process. In addition to blockage of amino groups, exposure of proteins to glucose leads to protein oxidative damage. This is due to spontaneous autoxidation of glucose and generation of oxygen free radicals in the presence of small amounts of the transition metals iron and copper. Hence one potential consequence of high levels of intracellular glycosylated proteins in diabetic RBCs could be the *de novo* generation of oxygen free radicals leading to membrane oxidative damage (Hunt et al. 1988). A proposed mechanism by which glucose or glycated proteins peroxidize polyunsaturated lipids, thus connecting protein glycation and lipoperoxidation is the following: small amounts of lipoperoxides present in RBCs may react with the enol form of glucose in the open form producing glucose radicals and alkoxy radical. A similar reaction yielding glycoprotein radicals and the alkoxy radical would occur between glycated protein and lipoperoxides. These species can initiate further peroxidative chain reactions, which result in the rapid increase in peroxidation rate, accelerating autocatalysis of lipid peroxidation by activating hydroperoxide decomposition (Hicks et al. 1988). In conclusion, glycated proteins and intermediate compounds on the way to formation of AGE and AGE themselves may produce reactive oxygen species.

22.6.2 Functional Effects of Red Cell Protein Glycation

Due to their long lifetime in circulation and their high permeability to monosaccharides, RBC proteins undergo intense glycation. Even in normal subjects with normal glycemia, glycated Hb may attain up to 4% of total Hb, and much higher levels, up to 12% of total Hb in diabetics. Haemoglobin is glycated predominantly at the valine residue of the N-terminal β-chain and, secondarily, at the epsilon-amino group of lysine on the α- and β-chains and at the N-terminal α-chain, leading to four glycated Hb (named Hb $A1_{a1}$, Hb $A1_{a2}$, Hb $A2_a$, Hb $A3_a$). Glycated Hbs are more negatively charged and have increased affinity to oxygen. Their level is routinely utilized to monitor the average concentration of glucose over a long period of time (6–8 weeks) (Higgins and Bunn 1981).

In view of the prevalence of oxidative damage, the activity of anti-oxidant enzymes, and the levels of anti-oxidant scavengers in RBCs and plasma have received attention. In human diabetes, a drastic decrease in activity of RBC superoxide dismutase (down 97% in diabetes type II), significant decreases in the level of reduced glutathione (down 20% and down 30% in type II and type I diabetes, respectively) and remarkable increases (up 61% to up 91%) in plasma lipoperoxides indicate that glycation processes are accompanied by oxidant damages to RBCs (for review, see Oberley (1988)).

A significant decrease in GSH and decreased GSH/GSSG ratio have been reported in diabetic patients RBCs: both indicate lower antioxidant defence of diabetic RBCs. The impairment of the glutathione level and redox status is the consequence of several factors: firstly, glycation of the glutathione-synthesizing enzyme gamma-glutamylcysteine synthetase and of the thiol transport system decreased GSH synthesis (Yoshida et al. 1995). Second, diabetes activates the polyol pathway and increases the reduction of glucose to sorbitol through the NADPH-dependent aldose reductase. This enhances NADPH consumption and diminishes its availability for glutathione reductase and regeneration of GSH (De Mattia et al. 1994). Other RBC enzymes are also glycated, such as D-aminolevulinate dehydratase, a key enzyme in porphyrin biosynthesis, ($Na^+ + K^+$)-dependent ATPase, Ca^{2+}-

dependent ATPase (Rajeswari et al. 1992) and 2,3-bisphosphoglycerate mutase (Fujita et al. 1998). AGE have been identified on the surface of RBCs. All major RBC proteins (spectrin, band 3, band 4.1) were extensively glycated. The latter was found to be 20–40% glycated (for review, see Chappey et al. (1997)).

The only detailed study on human diabetes causing alterations in RBC membrane proteins has been performed on spectrin (Schwartz et al. 1991). The main results are reported here because they may be extended by analogy, to other exposed RBC proteins. A first important conclusion of the study was the demonstration of spectrin oxidation indicating significant oxidative losses of cysteine, methionine (down 14%) and histidine (down 38%). A second indication was the correlation between glycation of Hb, taken as an indicator of protein glycation, and oxidative damage to spectrin. A third important result was the correlation between the two first modifications and decreased deformability measured with the ektacytometer. Taken together these data indicate that protein glycation is accompanied by oxidative damage and rheological modifications.

Despite structural heterogeneity of AGEs, there appear to exist only a preferential AGE receptor on a variety of cells, named RAGE. RAGE is a newly identified member of the immunoglobulin superfamily of predicted molecular mass of approx. 55 kDa with immunoglobulin-like domains in the extracellular region, a single putative transmembrane domain and a short cytosolic tail. Detailed discussion of RAGE is beyond the scope of this chapter. However, it should be mentioned that it is not a scavenger receptor and that binding with its ligand AGE is followed by generation of oxidant stress, induction of haem oxygenase type I and activation of the transcription factor NF-kB. Thus, adherence of AGE-enriched RBCs to endothelial RAGE is considered to be an important element in vascular damage, and the origin of several pathophysiological complications of diabetes (for review, see Schmidt et al. (2000)).

22.6.3 Increased Phagocytosis of Diabetic Red Cells: Possible Mechanisms

Shortened RBC lifespan and juvenile characteristics of circulating RBCs in diabetics are indicative of accelerated RBC removal. A comprehensive model for RBC removal in diabetes is only tentative and includes AGE as a non-immunologic recognition signal for phagocytosis. A specific component of AGE, [2-furoyl-4-(5)-(2-furanyl-)1-H-imidazole]-hexanoic acid (FFI-HA), was recognized by a receptor present on macrophages, also able to recognize low-density lipoproteins, mannose/fucose and formaldehyde-treated proteins. It has been shown (Vlassara et al. 1987) that both synthetic and naturally formed AGE adducts attached to the surface of intact RBCs induced binding and ingestion of these cells by macrophages. Human monocytes were incubated with FFI-coupled RBCs, IgG-opsonized RBCs or controls. Erythrophagocytosis of FFI-RBCs was significantly higher than that of controls and almost as high as that of IgG-opsonized RBCs. Phagocytosis of FFI-RBCs was competitively inhibited by BSA coupled with FFI. In a murine model of diabetes, RBCs were phagocytosed more intensely than normal RBCs when exposed to murine macrophages. This phagocytosis was competitively inhibited (down 70%) by addition of excess BSA coupled with FFI. The *in vivo* half-life of labelled mouse FFI-coupled RBCs injected into syngeneic mice was reduced to 7 days compared to a half-life of 20 days for the controls. It is conceivable that other removal signals are present on diabetic RBCs, similar to those known to operate in senescent or oxidatively damaged RBCs. This assumption is based on the signs of oxidative stress clearly present in RBCs in human diabetes. Secondly, membrane protein aggregates certainly including band 3 together with spectrin have been shown in RBC membranes isolated from diabetic individuals (Pescarmona et al. 1982). Thirdly, all situations with clear evidence of oxidative damage are associated with band 3 clustering and band-3/complement dependent phagocytosis. To date, no similar studies have been performed in human diabetics. Therefore, the proposed mechanisms to explain increased RBC removal in diabetes are likely but still hypothetical.

22.6.4 Role of Red Cell Alterations in Diabetes Complications

The most common cause of mortality and morbidity in diabetes involves damage to the vasculature. Vascular complications affect micro- and macrovasculature in the retina, renal glomeruli and multiple sites in the microvessels, causing retinopathy, nephropathy and diabetic atherosclerosis. It is beyond the scope of this chapter to discuss some of the theories put forward in recent years to link hyperglycemia with complications, namely the sorbitol hypothesis (based upon increased conversion of glucose to sorbitol in the cells) and the protein kinase C and diacylglycerol theory (based on modifications of protein kinase C in diabetes) (Chappey et al. 1997). By contrast, protein glycation and consequences thereof may play additional roles beyond enhanced phagocytosis and anaemia discussed before.

It has been shown that modified diabetic RBCs have enhanced adhesivity to endothelia and may thus induce endothelial damage and microvascular flow disturbances (McMillan and Utterback 1981). Enhanced adhesivity derives from the presence of AGE on the surface of diabetic RBCs, and increased exposure of PS, promoted by increased intra-erythrocytic calcium in diabetes (Closse et al. 1999). Increased adhesion of RBCs is statistically correlated with extent and severity of the vascular lesions and with the level of glycated Hb. The molecular basis of the interaction has been shown to be mediated by AGE present on the RBC membrane and by the AGE receptor RAGE. Interaction of oxidant-radical producer AGE with RAGE on endothelial cells would induce localized oxidant stress perturbing multiple cellular properties such as induction of cytokines, increased vascular permeability and activation of intracellular pathways not traditionally associated with other scavenger receptors (for review, see Chappey et al. (1997)).

22.7 Concluding Remarks

Based on two distinct examples we have discussed how impaired RBC metabolism may negatively impact on RBC functionality. We have selected a genetic metabolic defect, G6PD deficiency, and a primarily non-erythrocytic generalized metabolic disease with RBC consequences, because of their high prevalence and practical importance in medical practice. Besides their practical importance, both conditions are also classical subjects for genetic studies, protein modification studies, and analysis of cell-to-cell relationships. The parameter of functionality chosen here for discussion is certainly of paramount importance for the RBCs: the duration of its own life. We have shown that a healthy, non-immunogenic, asymmetric, non-glycated, flexible and resilient surface is essential for the RBC well-being and survival, and that, in turn, such an orderly façade is the manifestation of proper metabolism, sufficient anti-oxidant defence, and very low internal calcium. We have tried to show how the internal metabolic machinery and external appearance are connected. The two conditions discussed here indicate that the macrophage, the final executor of RBC demise, is prepared to accept an array of death signals: clustered band 3 and deposited anti-band 3 and complement fragments, predominant in oxidative damage, AGE in diabetes, possibly exposed PS in both. Other signals may also play supplementary roles, such as specific oligosaccharides recognized by macrophage mannose receptors.

Essentially the same mechanisms discussed here for G6PD deficiency may apply to other common RBC mutations of genetic origin affecting several hundred million human beings characterized by disturbancies of Hb synthesis or structure. The thalassemias, sickle-cell anaemia, Hb C, possibly Hb E and a host of less frequent unstable Hbs, are all characterized by increased production of oxidant radicals and unbalanced anti-oxidant mechanisms. All of them are in jeopardy of being phagocytically removed following similar paths as those discussed here for G6PD deficiency. All of them are connected by the red thread of providing efficient protection against *P. falciparum* malaria (Weatherall 1996).

References

Arese P, De Flora A (1990) Pathophysiology of hemolysis in glucose-6-phosphate dehydrogenase deficiency. Semin Hematol 27:1-40

Arese P, Bosia A, Naitana A, Gaetani S, Pescarmona GP (1981) Effect of divicine and isouramil on red cell metabolism in normal and G6PD-deficient (Mediterranean variant) subjects. Possible role in the genesis of favism. In: Brewer G (ed) The red cell, 5th Ann Arbor conference. Alan R Liss, New York, pp 725-744

Arese P, Mannu FR, Megow D, Turrini F (1992) Band 3 and erythrocyte removal. In: Bamberg E, Passow H (eds) Progress in cell research, vol 2. Elsevier, Amsterdam, pp 229-238

Beutler E (1990) The genetics of glucose-6-phosphate dehydrogenase deficiency. Semin Hematol 27:137-164

Beutler E (1992) The molecular biology of G6PD variants and other red cell enzyme defects. Annu Rev Med 43:47-59

Beutler E (2001a) Composition of the erythrocyte. In: Beutler E, Lichtman MA, Coller BS, Kipps TJ, Seligsohn U (eds) Williams hematology, 6th edn. McGraw-Hill, New York, pp 289-293

Beutler E (2001b) Energy metabolism and maintenance of erythrocytes. In: Beutler E, Lichtman MA, Coller BS, Kipps TJ, Seligsohn U (eds) Williams hematology, 6th edn. McGraw-Hill, New York, pp 319-332

Beutler E (2001c) Glucose-6-phosphate dehydrogenase deficiency and other red cell enzyme abnormalities. In: Beutler E, Lichtman MA, Coller BS, Kipps TJ, Seligsohn U (eds) Williams hematology, 6th edn. McGraw-Hill, New York, pp 527-545

Carrell RW, Winterbourn CC, Rachmilewitz EA (1975) Activated oxygen and haemolysis. Br J Haematol 30:259-264

Chappey O, Dosquet C, Wautier M-P, Wautier J-L (1997) Advanced glycation end products, oxidant stress and vascular lesions. Eur J Clin Invest 27:97-108

Chiu D, Lubin B (1989) Oxidative hemoglobin denaturation in RBC destruction: The effect of hemin on red cell membranes. Semin Hematol 26:128-135

Chiu D, Kuypers F, Lubin B (1989) Lipid peroxidation in human red cells. Semin Hematol 26:257-276

Closse C, Dachary-Prigent J, Boisseau MR (1999) Phosphatidylserine-related adhesion of human erythrocytes to vascular endothelium. Br J Haematol 107:300-302

De Flora A, Benatti U, Guida L, Forteleoni G, Meloni T (1985) Favism: Disordered erythrocyte calcium homeostasis. Blood 66:294-297

De Mattia G, Laurenti O, Bravi C, Ghiselli A, Iuliano L, Balsano F (1994) Effect of aldose reductase inhibition on glutathione redox status in erythrocytes of diabetic patients. Metabolism 43:965-968

Fischer TM, Meloni T, Pescarmona GP, Arese P (1985) Membrane cross bonding in red cells in favic crisis. Br J Haematol 59:159-169

Fujita T, Suzuki K Tada T, Yoshihara Y, Hamaoka R, Taniguchi N (1998) Human erythrocyte bisphosphoglycerate mutase: inactivation by glycation *in vivo* and *in vitro*. J Biochem 124:1237-1244

Gaetani GD, Parker JC, Kirkman HN (1974) Intracellular restraint: a new basis for the limitation in response to oxidative stress in human erythrocytes containing low-activity variants of glucose-6-phosphate dehydrogenase. Proc Natl Acad Sci USA 71:3584-3587

Gaetani GF, Ferraris AM, Rolfo M, Mangerini R, Arena S, Kirkman HN (1996) Predominant role of catalase in the disposal of hydrogen peroxide within human erythrocytes. Blood 87:1595-1599

Garby L, Noyes WD (1959) Studies on hemoglobin metabolism. I. The kinetic properties of the plasma hemoglobin pool in normal man. J Clin Invest 38:1479-1483

Grimes AJ (1980) Human red cell metabolism. Blackwell Sci Pub, Oxford

Hebbel RP, Eaton JW, Balasingam A, Steinberg MH (1982) Spontaneous oxygen radical generation by sickle erythrocytes. J Clin Invest 70:1253-1259

Hebbel RP (1985) Auto-oxidation and a membrane-associated "Fenton reagent": A possible explanation for development of membrane lesions in sickle erythrocytes. Clinics in Haematol 14:129-140

Hicks M, Delbridge L, Yue DK, Reeve TS (1988) Catalysis of lipid peroxidation by glucose and glycosylated collagen. Biochem Biophys Res Commun 151:649-655

Higgins PJ, Bunn HF (1981) Kinetic analysis of the nonenzymatic glycosylation of hemoglobin. J Biol Chem 256:5204-5208

Hunt JV, Dean RT, Wolff SP (1988) Hydroxyl radical production and autoxidative glycosylation: glucose autoxidation as the cause of protein damage in the experimental glycation model of diabetes mellitus and ageing. Biochem J 256:205-212

Jacobasch G (2000) Biochemical and genetic basis of red cell enzyme deficiencies. Bailliére's Clin Haematol 13:1-20

Jones RL, Peterson CM (1981) Hematologic alterations in diabetes mellitus. Am J Med 70:339-352

Kay MMB, Bosman GJCG, Johnson GJ, Beth AH (1988) Band-3 polymers and aggregates, and hemoglobin precipitates in red cell aging. Blood Cells 14:275-289

Kiefer CR, Snyder LM (2000) Oxidation and erythrocyte senescence. Curr Opin Hematol 7:113-116

King GL, Shiba T, Oliver J, Inoguchi Y, Bursell SE (1994) Cellular and molecular abnormalities in the vascular endothelium of diabetes mellitus. Annu Rev Med 45:179-188

Kirkman HN, Gaetani GF (1986) Regulation of glucose-6-phosphate dehydrogenase in human erythrocytes. J Biol Chem 261:4033-4038

Kirkman HN, Galiano S, Gaetani GF (1987) The function of catalase-bound NADPH. J Biol Chem 262:660-666

Kuypers FA (1998) Phospholipid asymmetry in health and disease. Curr Opin Hematol 5:122-131

Kuypers FA, Yuan J, Snyder LM, Kiefer CR, Bunyaratvej A, Fucharoen S, Ma L, Styles L, De Jong K, Schrier SL (1998) Membrane phospholipid asymmetry in human thalassemia. Blood 91:3044-3051

Liu SC, Yi SJ, Mehta JR, Nichols PE, Ballas SK, Yacono PW, Golan DE, Palek J (1996) Red cell membrane remodeling in sickle cell anemia: sequestration of membrane lipids and proteins in Heinz bodies. J Clin Invest 97:29-36

Low PS (1986) Structure and function of the cytoplasmic domain of band 3: Center of erythrocyte membrane-peripheral protein interactions. Biochim Biophys Acta 864:145-167

Lutz HU (1990) Erythrocyte clearance. In: Harris JR (ed) Blood cell biochemistry, vol 1. Erythroid cells. Plenum Press, New York, pp 81-120

Luzzatto L, Mehta A (1995) Glucose-6-phosphate dehydrogenase deficiency. In: Scriver RC, Beaudet AL, Sly WS, Valle D (eds) The metabolic and molecular bases of inherited disease, vol 3, 7th edn. McGraw-Hill, New York, pp 3367-3398

Mager J, Chevion M, Glaser G (1980) Favism. In: Liener IE (ed) Toxic constituents of plant foodstuffs, 2nd edn. Academic Press, San Diego, pp 265-294

Mason PJ (1996) New insights into G6PD deficiency. Brit J Haematol 94:585-591

McMillan DE, Utterback NG (1981) Impaired flow properties of diabetic erythrocytes. Clin Hemorheol 1:147-152

Mehta A, Mason PJ, Vulliamy TJ (2000) Glucose-6-phosphate dehydrogenase deficiency. Bailliére's Clin Haematol 13:21-38

Mohandas N, Morrow JS (2000) Plasma membrane dynamics and organization. In: Hoffman R, Benz EJ Jr, Shattil SJ, Furie B, Cohen HJ, Silberstein LE, McGlave P (eds) Hematology. Basic principles and practice. Churchill Livingstone, New York, pp 40-48

Njoroge FG, Monnier VM (1989) The chemistry of the Maillard reaction under physiological conditions: a review. Progr Clin Biol Res 304:85-107

Oberley LW (1988) Free radicals and diabetes. Free Radic Biol Med 5:113-124

Pescarmona GP, Bosia A, Ghigo D (1982) Shortened red cell life span in diabetes: mechanisms of hemolysis. In: Weatherall DJ, Fiorelli G, Gorini G (eds) Advances in red cell biology. Raven Press, New York, pp 391-396

Rajeswari P, Natarajan R, Nadler JL, Kumar D, Kalra VK (1992) Glucose induces lipid peroxidation and inactivation of membrane-associated ion-transport enzymes in human erythrocytes *in vivo* and *in vitro*. J Cell Physiol 149:100-109

Repka T, Shalev O, Reddy R, Yuan J, Abrahamov A, Rachmilewitz EA, Low PS, Hebbel RP (1993) Nonrandom association of free iron with membranes of sickle and beta-thalassemia erythrocytes. Blood 82:3204-3210

Rettig MP, Low PS, Gimm JA, Wang J, Christian JA (1999) Evaluation of biochemical changes during *in vivo* erythrocyte senescence in the dog. Blood 93:376-384

Rice-Evans C, Baysal E (1987) Iron-mediated oxidative stress in erythrocytes. Biochem J 244:191-196

Romero PJ, Romero EA (1999) The role of calcium metabolism in human red blood cell ageing: A proposal. Blood Cells Molec Dis 25:9-19

Sansone G, Piga AM, Segni G (1958) Il Favismo. Minerva Medica Editrice, Torino

Schmidt AM, Yan SD, Yan SF, Stern DM (2000) The biology of the receptor for advanced glycation end products and its ligands. Biochim Biophys Acta 1498:99-111

Schwartz RS, Madsen JW, Rybicki AC, Nagel RL (1991) Oxidation of spectrin and deformability defects in diabetic erythrocytes. Diabetes 40:710-708

Shinar E, Rachmilewitz EA (1993) Haemoglobinopathies and red cell membrane function. Baillière's Clin Haematol 6:357-369

Stern A (1989) Drug-induced oxidative denaturation in red blood cells. Semin Hematol 26:301-306

Turrini F, Naitana A, Mannuzzu L, Pescarmona GP, Arese P (1985) Increased red cell calcium, decreased calcium extrusion and altered membrane proteins during fava bean hemolysis in G6PD-deficient (Mediterranean variant) individuals. Blood 66:302-305

Turrini F, Mannu F, Giribaldi G, Arese P (1997) Biochemistry of favism. In: Greene LS, Danubio ME (eds) Adaptation to malaria. The interaction of biology and culture. Gordon and Breach Publishers, Cambridge, pp 33-56

Vlassara H (1994) Recent progress on the biologic and clinical significance of advanced glycation end products. J Lab Clin Med 124:19-30

Vlassara H, Valinsky J, Brownlee M, Cerami C, Nishimoto S, Cerami A (1987) Advanced glycosylation endoproducts on erythrocyte cell surface induce receptor-mediated phagocytosis by macrophages. J Exp Med 166:539-549

Vulliamy T, Mason P, Luzzatto L (1992) The molecular basis of glucose-6-phosphate dehydrogenase deficiency. Trends Genet 8:138-143

Waugh SM, Walder JA, Low PS (1987) Partial characterization of the copolymerization reaction of erythrocyte membrane band 3 with hemichrome. Biochemistry 26:1777-1783

Weatherall DJ (1996) Host genetics and infectious diseases. Parasitology 112:S23-S29

Weatherall DJ (2001) The thalassemias. In: Beutler E, Lichtman MA, Coller BS, Kipps TJ, Seligsohn U (eds) Williams hematology, 6th edn. McGraw-Hill, New York, pp 547-580

Winterbourn CC (1990) Oxidative denaturation in congenital haemolytic anemias: The unstable anemias. Semin Hematol 27:41-50

Yoshida A (1973) Hemolytic anemia and G6PD deficiency. Science 179:532-537

Yoshida K, Hirokawa J, Tagami S, Kawakami Y, Urata Y, Kondo T (1995) Weakened cellular scavenging activity against oxidative stress in diabetes mellitus: regulation of glutathione synthesis and efflux. Diabetologia 38:201-210

Zhang D, Kiyatkin A, Bolin JT, Low PS (2000) Crystallographic structure and functional interpretation of the cytoplasmic domain of erythrocyte membrane band 3. Blood 96:2925-2933

23 Sickle Cell Disease

Carlo Brugnara

Department of Laboratory Medicine, Children's Hospital Boston, Harvard Medical School, 300 Longwood Avenue, Boston, MA 02115, USA

23.1 Introduction

Sickle cell disease is a classic example of a genetic disorder of hemoglobin structure and function. The mutation from glu to lys at the β 6 position of the hemoglobin molecule (Hb S) results in the formation of Hb S aggregates (polymerization) under conditions of reduced oxygen tension. Polymerization of Hb S leads to the cellular deformation known as erythrocyte sickling. The clinical manifestations of this disease are unique, because they combine those of a chronic hemolytic anemia with those of an acute and chronic vaso-occlusion. The erythrocyte membrane is an important factor in both the premature demise of the circulating sickle erythrocytes and the plugging of capillaries and small blood vessels, resulting from the interaction of sickle erythrocytes with vascular endothelium. One of the distinguishing features of sickle cell disease is the presence of erythrocyte dehydration. Erythrocyte dehydration plays an important role in cell sickling, since the polymerization tendency of Hb S is greatly enhanced when the cellular Hb S concentration is increased.

We will review the evidence in sickle erythrocytes for the possible mechanisms of membrane damage, the abnormalities in membrane function, the abnormalities in membrane transport, and finally, therapeutic approaches based on the inhibition of sickle cell dehydration.

23.2 Mechanisms of Sickle Red Cell Membrane Damage

23.2.1 Oxidation

Hb S is more likely to undergo auto-oxidation and is more unstable than normal Hb A (Asakura et al. 1974; Hebbel et al. 1988; Harrington 1998). An initial electrostatic interaction between Hb S and the inner membrane is followed by formation of metahemoglobin (metHb), Hb denaturation, with formation of hemichromes, and release of free heme. The presence of free heme promotes lipid peroxidation, with formation of lipid hydroperoxide, further promoting release of free iron from heme and additional oxidative damage (Sugihara and Hebbel 1992).

The precipitation of Hb S and derived products on the inner surface of the membrane and the associated oxidative damage lead to the formation of aggregates called Heinz bodies, which are clusters of integral transmembrane proteins, such as band 3 and ankyrin. Autologous immunoglobulins (IgG) bind to clustered band 3, and accelerate the removal of erythrocytes from the circulation (Schlüter and Drenckhahn 1986; Waugh et al. 1986; Liu et al. 1996). SH moieties of membrane protein cysteine residues are another target of free radicals and oxidants. Several membrane proteins of sickle erythrocytes show abnormal oxidation of protein thiols (Rank et al. 1985).

23.2.2 Loss of Membrane

Erythrocyte sickling is associated with release of membrane vesicles and a net loss of membrane surface area (Allan et al. 1981, 1982; Allan and Raval 1983). The vesicles derive from membrane spicules formed during sickling, which uncouple from the underlying cytoskeletal network (Liu et al. 1991, 1996).

23.2.3 Binding of Hb S to the Inner Membrane

Increased deposition of hemoglobin on the inner surface of the membranes is seen in sickle erythrocytes, especially in irreversible sickled cells (ISC) (Fischer et al. 1975; Lessin et al. 1978; Shaklai et al. 1981). Hemoglobins bind to the inner surface of the cell membrane with a selectivity sequence (Hb C > Hb A2 > Hb S > Hb A > Hb F) proportional to their relative positive charge (Fischer et al. 1975; Friedman 1981; Eisinger et al. 1982, 1984; Reiss et al. 1982). An important site of interaction between Hb and the cell membrane is the cytoplasmic domain of band 3: half of the band 3 molecules of each erythrocyte have Hb S bound to their cytoplasmic domains, with a preferential binding of deoxy-hemoglobins (Walder et al. 1984; Liu et al. 1996).

23.3 Red Cell Membrane Structural Abnormalities in Sickle Cell Disease

23.3.1 Altered Cytoskeleton

Several studies have demonstrated functional abnormalities in either the cytoskeleton or the cytoskeleton/integral membrane interaction of sickle erythrocytes (Platt 1994). Sickle erythrocytes maintaining a deformed shape after exposure to oxygen or conditions known to reverse Hb S polymerization are defined as irreversibly sickled cells (ISC) and exhibit an irreversible deformation of the spectrin-actin lattice and altered spectrin dimer-dimer association (Bertles and Döbler 1969; Lux et al. 1976; Liu et al. 1993). Ankyrin function is also abnormal in sickle erythrocytes (Platt et al. 1985). Sickle erythrocytes inside-out vesicles (IOV) show a marked decrease of band 3 binding to ankyrin, with normal binding to protein 4.1 (Platt and Falcone 1995). The mobility of band 3 and glycophorin in the sickle cell membrane is abnormally reduced, indicating irreversible alterations in the interactions with the underlying cytoskeleton (Corbett and Golan 1993). Band 3 can dissociate from the cytoskeleton and become part of the membrane in the spicule of sickle erythrocytes (Liu et al. 1991).

23.3.2 Altered Membrane Lipid Composition

The lipid bilayer of the red cell membrane exhibits a characteristic asymmetry in phospholipid (PL) organization (Zwaal and Schroit 1997). The outer monolayer consists of phosphatidylcholine (PC) and sphingomyelin (SM), while the inner monolayer consists of the aminophospholipids phosphatidylethanolamine (PE) and phosphatidylserine (PS). Although no PS is seen in the outer membrane leaflet, 20% of PE is found on the outside of the membrane. Phosphoinositides are found mostly (~70%) in the inner monolayer (Kuypers et al. 1994). Loss of lipid asymmetry and exposure of PE and PS on the outer layer of the membrane can occur *in vitro* with deoxygenation and sickling (Chiu et al. 1981; Lubin et al. 1981, 1986; Choe et al. 1986; De Jong et al. 2001b). Deoxygenation increases outer leaflet content of PE from 12 to 40% and of PS from 0 to 20% of the total membrane PE and PS, respectively (Blumenfeld et al. 1991; Wang et al. 1993). Cells exposing PS have a decreased survival in the circulation and are removed prematurely (De Jong et al. 2001a).

23.4 Red Cell Membrane Transport Abnormalities in Sickle Cell Disease

Movement of cations (Na^+, K^+, and Ca^{2+}) and/or anions (Cl^-, HCO_3^-) mediated by specific transport systems determines the volume and hydration state of erythrocytes. A distinguishing feature of sickle cell disease is heterogeneity in the distribution of erythrocyte volume and water content (Fabry and Nagel 1982). Based on the level of anemia and associated reticulocytosis, each patient presents a variable fraction of cells with increased volume and normal or low hemoglobin concentration (representing reticulocytes and very young cells). Based on still unknown factors, which however must be related to the capability of sickle cells to lose K^+, Cl^-, and water, the blood of patients with sickle cell disease contains dense, dehydrated erythrocytes and dense reticulocytes (Fabry et al. 1984b; Mohandas et al. 1986a, 1989; Mohandas and Ballas 1994). Erythrocyte dehydration is an important factor in the pathogenesis of sickle cell disease. Polymerization of Hb S is uniquely dependent (to the 20th–30th power) on the cellular concentration of Hb S: the increased cellular Hb concentration of dehydrated sickle erythrocytes markedly enhances their tendency to form polymer and sickle (Eaton and Hofrichter 1987, 1990). This contingent of dense, dehydrated cells varies both from patient to patient (ranging from 0 to 40%) and with time in the same patient. Presence of alpha thalassemia is associated with a reduced number of dense cells (Embury et al. 1984). However, there is no correlation between proportion of dense cells and frequency of crises (Fabry et al. 1984b; Noguchi et al. 1985; Ballas 1991). This is most likely due to the fact that dense cells have been shown to decrease concomitantly with crisis (Fabry et al. 1984a; Lawrence et al. 1985; Lawrence and Fabry 1987), or to increase in the initial phase of a crisis and subsequently decrease (Ballas and Smith 1992). It is generally believed that dense cells play an important role in the pathogenesis of vaso-occlusion in sickle cell disease via at least two possible modalities:

(a) increased polymerization and sickling (Eaton and Hofrichter 1987, 1990; Noguchi et al. 1989),

(b) reduced deformability leading to abnormal rheology and increased adhesion to endothelium (Hebbel et al. 1985; Kaul et al. 1989, 1994; Dong et al. 1992; Morris et al. 1993; Hasegawa et al. 1995; Hiruma et al. 1995a,b).

The seminal work of Tosteson et al. (1952, 1955) and Tosteson (1955) demonstrated alteration of ion content and transport in sickle erythrocytes. Tosteson's work showed that deoxygenation and sickling could induce changes in monovalent cation (Na^+ and K^+) permeability. Identification of the mechanisms responsible for dehydration of sickle erythrocytes and of specific inhibitors is the necessary prerequisite for prevention of dehydration *in vivo* (Joiner 1993). It has been estimated that a 1–4 g/dl decrease in the intracellular hemoglobin S concentration (MCHC) could reduce polymerization in homozygous Hb S disease to the levels seen in the much less clinically severe Hb S/β^+-thalassemia (Sunshine et al. 1978).

Three major pathways have been implicated in the dehydration of sickle erythrocytes (see also Fig. 23.1):

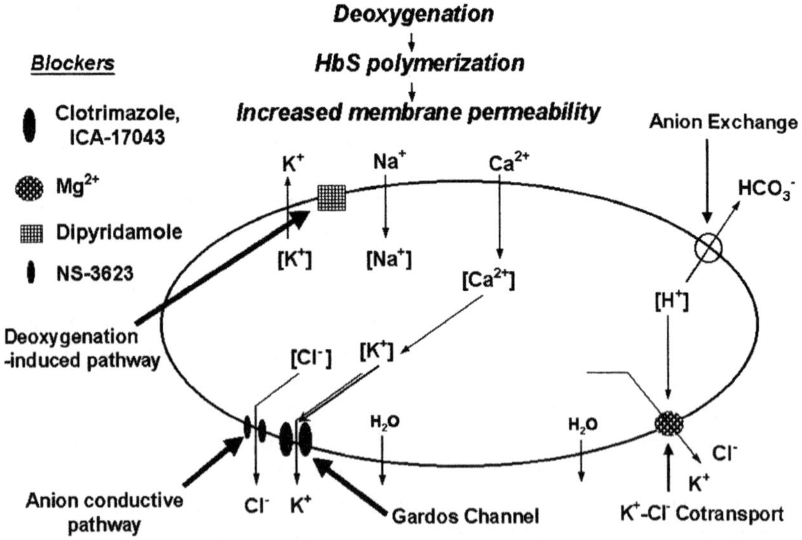

Fig. 23.1. Mechanisms involved in sickle cell dehydration and their inhibitors

23.4.1 K^+-Cl^- Cotransport

The erythrocyte K^+-Cl^- cotransport (KCC) is a member of the chloride-cation cotransporters (CCC) family. Most of the K^+-Cl^- cotransport activity in erythrocytes can be attributed to KCC1 (Gillen et al. 1996) although recent evidence suggests possible contributions of other isoforms, such as KCC3 and KCC4 (Race et al. 1999; Mercado et al. 2000). K^+ transport via erythrocyte KCC is blocked by phosphatase inhibitors (okadaic acid or calyculin A), an indication that the active form of the transporter is dephosphorylated, probably via membrane-associated protein phosphatases (PP1 and PP2A); a still unidentified kinase is responsible for phosphorylation and inactivation (Jennings and Al-Rohil 1990; Jennings and Schulz 1991; Kaji and Tsukitani 1991; Bize et al. 1999, 2000). The functional properties of K^+-Cl^- cotransport are discussed in detail in Chap. 9. There have been multiple reports of activation of K^+-Cl^- cotransport in homozygous sickle (SS) erythrocytes (Brugnara et al. 1986; Canessa et al. 1986; Joiner 1993). When SS cells are isosmotically swollen or are exposed to acid pH, K^+-Cl^- cotransport promotes K^+ and Cl^- loss with cell dehydration (Brugnara et al. 1989). The activity of K^+-Cl^- cotransport is highest in the least-dense, reticulocyte-rich fractions of sickle blood (Brugnara et al. 1986; Canessa et al. 1987). Within sickle reticulocytes, a pool of cells shows an enhanced capability to dehydrate via K^+-Cl^- cotransport (Etzion et al. 1996), and probably accounts for the fast-track dehydration, hypothesized by Lew et al. (1991). Fabry et al. (1991) have identified a similar fraction of reticulocytes/young cells and shown that they also have decreased Hb F cells content. Transferrin receptor (TfR) positivity has been shown to identify the more immature fraction of sickle reticulocytes and characterize their volume regulatory properties (Franco et al. 1994). Dense, TfR^+ reticulocytes have a much greater K^+-Cl^- cotransport activity compared with lighter reticulocytes (Franco et al. 1995). It has been proposed that TfR^+ cells could reach a state of intermediate dehydration via K^+-Cl^- cotransport, and then further dehydrate due to polymerization, sickling and activation of the Gardos channel (Franco et al. 1997).

Activation of K^+-Cl^- cotransport is not an exclusive property of sickle erythrocytes, but has also been observed in other hemoglobinopathies, such as homozygous Hb C (CC) disease (Brugnara et al. 1985; Canessa et al. 1986; Brugnara 1989). A common feature of all Hb variants associated with K^+-Cl^- cotransport activation is their relatively positive charge. Mutations in the β 6 or β 7 position of Hb are strongly associated with increased K^+-Cl^- cotransport (Olivieri et al. 1992). There is evidence for abnormal functional regulation of K^+-Cl^- cotransport in CC erythrocytes (Brugnara 1989; Canessa et al. 1994). A transgenic mouse model expressing human Hb C exhibits K^+-Cl^- cotransport activation and erythrocyte dehydration (Fabry et al. 1998). In Hb SC disease, there is a characteristic erythrocyte dehydration compared with AS or AC erythrocytes (Fabry et al. 1982). K^+-Cl^- cotransport is highly expressed in SC erythrocytes and is responsible for their microcytosis and dehydration (Canessa et al. 1986; Lawrence et al. 1991).

The sensitivity of K^+-Cl^- cotransport to oxygen has been studied in normal and sickle erythrocytes. Deoxygenation inactivated K^+-Cl^- cotransport in normal erythrocytes and rendered it refractory to stimulation by acidification, urea or cell swelling (Gibson et al. 1998). There was no appreciable inhibition of K^+-Cl^- cotransport by deoxygenation in sickle erythrocytes and urea could still stimulate K^+-Cl^- cotransport independently of oxygen tension (Culliford et al. 1998). Studies from another group have also shown stimulation of K^+-Cl^- cotransport during deoxygenation (Joiner et al. 1998). This stimulation has been linked to a generalized dephosphorylation of erythrocyte membrane proteins, which is also observed during deoxygenation (Fathallah et al. 1995).

It is likely that every time sickle erythrocytes are exposed to a relatively hypoxic/acidic environment, such as in the spleen, or to high concentrations of urea, such as in the kidney, or conditions which promote oxidative membrane damage, activation of K^+-Cl^- cotransport ensues with K^+ loss and dehydration.

23.4.2 Ca^{2+}-Activated K^+ Channel (Gardos Pathway)

Properties of the erythrocyte Ca^{2+}-activated K^+ channel (Gardos channel) have been described in Chap. 6. Work by Glader and Nathan (1978) demonstrated that generation of irreversibly sickled cells under conditions of ATP depletion was dependent on the presence of external Ca^{2+} and an outwardly directed K^+ gradient. Several reports have demonstrated that there is a Ca^{2+}-dependent dehydration when sickle erythrocytes are deoxygenated (Horiuchi and Asakura 1987; Horiuchi et al. 1988; Ohnishi et al. 1989; Brugnara et al. 1993; McGoron et al. 2000). The deoxygenation-induced cell dehydration is a probabilistic event: the cells which have the greatest increase in intracellular Ca^{2+} and activation of the Gardos channel are those that will undergo dehydration (Lew et al. 1997). Activation of the Gardos pathway and cell dehydration during deoxygenation is stochastic in nature (Lew et al. 1997), since a relatively constant fraction (10–45%) of cells dehydrates in a Ca^{2+}-dependent fashion during each deoxygenation cycle. Ca^{2+} entry into erythrocytes increases with cell sickling (Rhoda et al. 1990; Bookchin et al. 1991; Etzion et al. 1993). Deoxygenation increases free intracellular Ca^{2+} 2- to 3-fold from the normal range of 9.4–11.4 nM to a range of 21.8–31.7 nM (Etzion et al. 1993). These values are close to the threshold for Gardos channel activation, which is around 40 nM of free Ca^{2+} in normal cells (Tiffert et al. 1988). This Ca^{2+} permeabilization could be localized to discrete areas of the cell membrane that experience the greatest degree of deformation (Mohandas et al. 1986b).

Normal total erythrocyte Ca^{2+} content varies between 0.9 and 2.8 $\mu mol/l_{cells}$ (Engelmann and Duhm 1987). Initial studies on sickle erythrocytes reported increased total cell Ca^{2+}, with values ranging from 110 to 300 $\mu mol/l_{cells}$ (Eaton et al. 1973; Palek 1977). Subsequent studies with the Ca^{2+}-sensitive dye fura2 produced estimates for free cytoplasmic Ca^{2+} levels of 18–30 $nmol/l_{cells}$ for sickle erythrocytes, similar to normal controls (Rhoda et al. 1990). NMR studies of normal and sickle erythrocytes have also yielded similar values for ionized cytoplasmic Ca^{2+} (21 ± 2 *vs.* 18 ± 2 nmol/l) (Murphy et al. 1987). Ca^{2+} pump activity in oxygenated sickle erythrocytes is not significantly different from normal control cells (Etzion et al. 1993). The elevated total Ca^{2+} of sickle erythrocytes is the result of a compartmentalization into cytoplasmic vesicles, as first demonstrated by electron probe X-ray analysis of cryosections (Lew et al. 1985) and subsequently confirmed by the use of chlortetracycline as a Ca^{2+}-sensitive probe (Rubin et al. 1986).

These vesicles, which are derived from the plasma membrane, contain integral membrane proteins, including the Ca^{2+}-ATPase, in an inside-out configuration. The Ca^{2+}-ATPase pumps Ca^{2+} from the cytoplasm into the vesicle, and creates a very high internal concentration of Ca^{2+} in the vesicle with a normal cytoplasmic free Ca^{2+} (Lew et al. 1985; Williamson et al. 1992). Considerable heterogeneity exists in the extent of Gardos channel-mediated dehydration within various subpopulations of sickle erythrocytes, with higher degrees of activation in deoxygenated reticulocytes (Lew et al. 2002).

Deoxygenation of sickle erythrocytes has been shown to lead to a nonspecific dephosphorylation of membrane proteins, which can be prevented by inhibiting protein phosphatases with okadaic acid or stimulating protein kinase C (PKC) with phorbol 12-myristate 13-acetate (PMA) (Fathallah et al. 1995; Merciris et al. 1998). PKC can regulate Ca^{2+} pump and Gardos channel activity in deoxygenated sickle cells (Fathallah et al. 1997). Recent evidence indicates that deoxygenation of sickle erythrocytes specifically stimulates Syk tyrosine protein kinase, and diminishes membrane-associated protein tyrosine phosphatase activity (Merciris et al. 2001). In addition to deoxygenation, the Gardos channel of normal and sickle erythrocytes can be activated by PGE_2 (Li et al. 1996), endothelins (Rivera et al. 1999) and chemokines (Rivera et al. 2002). Gardos channel stimulation by endothelin takes place via a protein kinase C dependent pathway (Rivera et al. 1999). This new potential modality of cell dehydration is particularly relevant since endothelin levels are abnormally elevated at baseline in patients with sickle cell disease (Rybicki and Benjamin 1998; Werdehoff et al. 1998) and increase even further with acute chest syndrome or vaso-occlusive crisis (Hammerman et al. 1997). Sickled cells can induce production of endothelin from endothelial cells, thus leading to a vicious sickle of vaso-occlusion and endothelial damage (Phelan et al. 1995). Arginine supplementation has been shown to reduced dehydration via the Gardos channel in transgenic sickle mice, suggesting a possible role for nitric oxide (NO) in the regulation of this pathway (Romero et al. 2002).

23.4.3 Deoxygenation-Induced Cation Fluxes

In 1952 Tosteson et al. showed that deoxygenation of SS cells was associated with K^+ loss and Na^+ gain (Tosteson et al. 1952). Subsequently, Tosteson et al. (1955) and Tosteson (1955) demonstrated this to be a diffusional pathway. The functional properties of the deoxygenation-induced fluxes have been elucidated mostly by Joiner (1987, 1990) and Joiner et al. (1988, 1993, 1995). The inward Na^+ flux and outward K^+ flux are balanced with no net changes in ion content and cell volume (Berkowitz and Orringer 1985). The fluxes depend on external (and internal) pH, and reach a maximum value at pH 7.4–7.5, with inhibition $<$ pH 7 and $>$ pH 8. The deoxygenation-induced fluxes are not selective for Li^+, Na^+, K^+, Rb^+ or Cs^+, depend linearly on the concentration of Na^+ or K^+; and are sensitive to changes in membrane potential (Joiner et al. 1993). All these properties are suggestive of diffusional ion movement, restricted to monovalent or divalent cations (Clark and Rossi 1990). The deoxy Na^+ and K^+ fluxes are activated when they are chloride independent and O_2 drops below 40–50 mmHg (Joiner et al. 1988). The deoxygenation-induced Na^+ and K^+ fluxes are reduced by the anion exchange inhibitor DIDS, while the morphologic sickling is unaffected (Joiner 1990). If anion transport via the anion exchanger is inhibited by other means, the deoxygenation-induced fluxes are unaffected. The presence of external Ca^{2+} (or other divalent cations) inhibits deoxy Na^+ influx more than K^+ efflux resulting in an unbalance between the two fluxes and net K^+ loss (Bookchin et al. 1991; Joiner et al. 1995). This net K^+ loss is not sensitive to inhibitors of the Gardos channel (Joiner 1993).

The Na^+/K^+ pump may also play a role in cell dehydration. Earlier work in normal human red cells has indicated that activation of the pump in conditions with very high Na^+ content leads to cell dehydration (Clark et al. 1981). *In vitro* evidence for erythrocyte dehydration mediated by the activation of the Na^+/K^+ pump has been provided for sickle cells and hereditary xerocytosis (Joiner et al. 1986). In sickle erythrocytes, the increase cell Na^+ induced by deoxygenation activates the Na^+/K^+ pump, which could lead to cell dehydration due to the 3 Na^+_{out}/2 K^+_{in} stoichiometry of the transporter. However, the reduction in K^+ content far exceeds the observed increase in cell Na^+ and the integrated red cell model developed by Lew and Bookchin (1986) predicts that this mechanism could account for only a small fraction of sickle cell dehydration (Lew et al. 1991).

One important determinant of the deoxygenation induced Na^+ and K^+ movements is the morphologic deformation of the red cell induced by Hb polymerization and sickling. Slow, progressive deoxygenation produces the greatest degree of morphologic sickling (Asakura and Mayberry 1984; Horiuchi et al. 1988). Conditions leading to marked morphology changes are associated with the largest increase in Na^+ and K^+ permeability, suggesting that the deformation induced by polymerization plays a role in increasing Na^+ and K^+ permeability (Mohandas et al. 1986).

The increased cell Na^+ of dense sickle cells (Clark et al. 1978b) is most likely due to the combination of decreased Na^+/K^+ pump activity and increased Na^+ permeability. Potassium transport mediated by the Na^+/K^+ pump is abnormally decreased in ISCs, while ATPase activity is not (Clark et al. 1978a). This has been attributed to an increased intracellular Mg^{2+}/phosphate ratio. When this ratio is decreased to normal values, the transport activity of the Na^+/K^+ pump is normalized (Ortiz et al. 1990). The progressive increase in cell Na^+ observed in the terminal phase of dense cells results in a progressive swelling of these cells, with cells moving into lighter density fractions (Bookchin et al. 2000). This rehydration process of dense cells has been shown to take place via a bumetanide-sensitive pathway which mediates Na^+ entry (Etzion et al. 1996; Holtzclaw et al. 2002).

23.5 Therapeutic Approaches Based on Inhibition of Sickle Cell Dehydration

23.5.1 Inhibition of Gardos Channel

The imidazole antimycotic clotrimazole (CLT) has also been shown to be a specific inhibitor of the human red cell Gardos channel (Alvarez et al. 1992; Brugnara et al. 1993). Studies in normal human volunteers have shown that oral clotrimazole administration induces *ex vivo* inhibition of the erythrocyte Gardos channel (Brugnara et al. 1995; Rifai et al. 1995). Studies in a transgenic sickle mouse model (SAD mouse) have shown that oral CLT administration induces blockade of the Gardos channel and reduces dehydration *in vivo* (De Franceschi et al. 1994). Studies in patients with sickle cell disease have also shown that oral CLT administration results in inhibition of the Gardos channel and reduction in cell dehydration (Brugnara et al. 1996). Adverse side effects include mild/moderate dysuria in all subjects, likely due to urinary excretion of CLT metabolites, and a reversible liver cellular toxicity in two subjects treated with CLT at the dosage of 30 mg/kg body weight/d. Studies based on the chemical structure of CLT metabolites have led to the identification of several new chemical entities, which have similar or greater potency than CLT in inhibiting the Gardos channel, but no effect on cytochrome P450, due to the absence of imidazole residues. One of these compounds, ICA-17043, has been shown to be 10-fold more potent blocker of the Gardos channel than CLT. ICA-17043 prevents dehydration of sickle cells *in vivo* in the transgenic SAD mouse model and is currently undergoing Phase I/II clinical studies (Rigdon et al. 2000; Stocker et al. 2000). Administration of a single dose of ICA-17043 to normal volunteers (dosage range 25–200 mg) resulted in plasma levels, which persisted for up to 10 days and were associated with significant inhibition of the Gardos channel (Rigdon et al. 2000). These results indicate that a high degree of Gardos channel block can be achieved *in vivo* with ICA-17043.

23.5.2 Inhibition of K^+-Cl^- Cotransport

Mg^{2+} is an important negative modulator of K^+-Cl^- cotransport and thus of sickle cell dehydration. Sickle erythrocytes have a reduced Mg^{2+} content (Olukoga et al. 1990; Ortiz et al. 1990; De Franceschi et al. 1997), which is most likely due to net loss of Mg^{2+} during sickling (Ortiz et al. 1990). Sickle erythrocytes exhibit a markedly increased activity of the Na^+/Mg^{2+} exchanger, which could also contribute to their reduced total Mg^{2+} content (Feray et al. 1988; De Franceschi et al. 1997). Oral Mg^{2+} supplementation has been considered as a potential therapy for inhibition of K^+-Cl^- cotransport and dehydration *in vivo*. Studies in diabetic patients have shown that oral Mg-pidolate increases erythrocyte Mg^{2+} content (Borella et al. 1993; Paolisso et al. 1994). In the SAD mouse, oral Mg^{2+} supplementation normalized erythrocyte Mg^{2+} and K^+ contents, and reduced K^+-Cl^- cotransport activity and cell density (De Franceschi et al. 1996). Four weeks of treatment with Mg-pidolate at a dose of 0.6 mEq/kg/d induced an increased erythrocyte Mg^{2+} and K^+ content, and decreased K^+-Cl^- cotransport in 10 patients with sickle cell disease. Mild, transient diarrhea was the only reported side effect (De Franceschi et al. 1997). A subsequent study used oral Mg-pidolate for 6 months in 20 patients with sickle cell disease, confirming the beneficial effects on red cell dehydration and demonstrating a 58% reduction in the number of painful days (De Franceschi et al. 2000). Two randomized, placebo controlled, cross-over studies are currently evaluating the effects of long-term Mg^{2+} supplementation in adult and pediatric patients with sickle cell disease and with SC disease.

23.5.3 Inhibition of Anion Permeability

A new series of high affinity blockers of Cl^- conductance has been produced (Bennekou et al. 2000). The prototype compound, NS1652, reversibly blocked human erythrocyte chloride conductance (IC_{50} = 0.6 μM, 95%) *in vitro* without directly affecting the Gardos channel or the K^+-Cl^- cotransport activity. *In vitro*, deoxygenation induced K^+ loss was also reduced by NS1652. Another Cl^- conductance inhibitor (NS3623) was tested *in vivo* in SAD mice (Bennekou et al. 1999, 2001). Hematocrit increased and MCHC decreased in all dosing groups, with a concomitant increase in cell Na^+ and K^+ contents. A selective loss of the densest red cell population (above 1.1 g/cm) was observed in conjunction with a shift from a high proportion of sickled to well-hydrated discoid erythrocytes. These data indicate that Cl^- conductance blockers might represent a new therapeutic approach for preventing sickle cell dehydration.

23.5.4 Inhibition of Deoxygenation-Induced Fluxes

The deoxygenation-induced fluxes are inhibited by dipyridamole (Persantin®) at concentrations similar to those achievable *in vivo* (Joiner et al. 1997). This drug is widely used and has limited toxicity, thus representing a potential new therapy for sickle cell disease.

In conclusion, studies on ion permeability and volume control of sickle erythrocytes have produced important advances not only in our understanding of membrane physiology but also in the development of new therapies for sickle cell disease. Some of these potential new therapies are now being tested in clinical trials.

References

Allan D, Raval P (1983) Some morphological consequences of uncoupling the lipid bilayer from the plasma membrane skeleton in intact erythrocytes. Biomed Biochim Acta 42:11-12

Allan D, Limbrick AR, Thomas P, Westerman MP (1981) Microvesicles from sickle erythrocytes and their relation to irreversible sickling. Br J Haematol 47:383-390

Allan D, Limbrick AR, Thomas P, Westerman MP (1982) Release of spectrin-free spicules on reoxygenation of sickled erythrocytes. Nature 295:612-613

Alvarez J, Montero M, Garcia-Sancho J (1992) High affinity inhibition of Ca^{2+}-dependent K^+ channels by cytochrome P-450 inhibitors. J Biol Chem 267:11789-11793

Asakura T, Mayberry J (1984) Relationship between morphologic characteristics of sickle cells and method of deoxygenation. J Lab Clin Med 104:987-994

Asakura T, Onishi T, Friedman S, Schwartz E (1974) Abnormal precipitation of oxyhemoglobin S by mechanical shaking. Proc Natl Acad Sci USA 71:1594-1598

Ballas SK (1991) Sickle cell anemia with few painful crises is characterized by decreased red cell deformability and increased number of dense cells. Am J Hematol 36:122-130

Ballas SK, Smith ED (1992) Red blood cell changes during the evolution of the sickle cell painful crisis. Blood 79:2154-2163

Bennekou P, De Franceschi L, Brugnara C, Christophersen P, (1999) Specific blockers of chloride conductance, NS1652 and NS3623, reduce sickle cell dehydration *in vitro* and *in vivo*. Blood 94:677a

Bennekou P, Pedersen O, Moller A, Christophersen P (2000) Volume control in sickle cells is facilitated by the novel anion conductance inhibitor NS1652. Blood 95:1842-1848

Bennekou P, De Franceschi L, Pedersen O, Lian L, Asakura T, Evans G, Brugnara C, Christophersen P (2001) Treatment with NS3623, a novel Cl-conductance blocker, ameliorates erythrocyte dehydration in transgenic SAD mice: A possible new therapeutic approach for sickle cell disease. Blood 97:1451-1457

Berkowitz LR, Orringer EP (1985) Passive sodium and potassium movements in sickle erythrocytes. Am J Physiol 249:C208-C214

Bertles JF, Döbler J (1969) Reversible and irreversible sickling: a distinction by electron microscopy. Blood 33:884-898

Bize I, Güvenç B, Robb A, Buchbinder G, Brugnara C (1999) Serine-threonine protein phosphatases and regulation of K-Cl cotransport in human erythrocytes. Am J Physiol 277:C926-C936

Bize I, Güvenç B, Brugnara C (2000) Stimulation of human erythrocyte K-Cl cotransport and protein phosphatase type 2A by N-ethylmaleimide: role of intracellular Mg^{++}. J Membrane Biol 177:159-168

Blumenfeld N, Zachowski A, Galacteros F, Beuzard Y, Devaux PF (1991) Transmembrane mobility of phospholipids in sickle erythrocytes: effect of deoxygenation on diffusion and asymmetry. Blood 77:849-854

Bookchin RM, Ortiz OE, Lew VL (1991) Evidence for a direct reticulocyte origin of dense red cells in sickle cell anemia. J Clin Invest 87:113-124

Bookchin RM, Etzion Z, Sorette M, Mohandas N, Skepper JN, Lew VL (2000) Identification and characterization of a newly recognized population of high-Na^+, low-K^+, low-density sickle and normal red cells. Proc Natl Acad Sci USA 97:8045-8050

Borella P, Ambrosini G, Concari M, Bargellini A (1993) Is magnesium content in erythrocytes suitable for evaluating cation retention after oral physiological supplementation in marginally magnesium-deficient subjects? Magnesium Res 6:149-153

Brugnara C (1989) Characteristics of the volume- and chloride-dependent K transport in human erythrocytes homozygous for hemoglobin C. J Membrane Biol 111:69-81

Brugnara C, Kopin AS, Bunn HF, Tosteson DC (1985) Regulation of cation content and cell volume in erythrocytes from patients with homozygous hemoglobin C disease. J Clin Invest 75:1608-1617

Brugnara C, Bunn HF, Tosteson DC (1986) Regulation of erythrocyte cation and water content in sickle cell anemia. Science 232:388-390

Brugnara C, Van Ha T, Tosteson DC (1989) Acid pH induces formation of dense cells in sickle erythrocytes. Blood 74:487-495

Brugnara C, De Franceschi L, Alper SL (1993) Inhibition of Ca^{2+}-dependent K^+ transport and cell dehydration in sickle erythrocytes by clotrimazole and other imidazole derivatives. J Clin Invest 92:520-526

Brugnara C, Armsby CC, Sakamoto M, Rifai N, Alper SL, Platt O (1995) Oral administration of clotrimazole and blockade of human erythrocyte Ca^{++}-activated K^+ channel: the imidazole ring is not required for inhibitory activity. J Pharmacol Exper Therap 273:266-272

Brugnara C, Gee B, Armsby CC, Kurth S, Sakamoto M, Rifai N, Alper SL, Platt OS (1996) Therapy with oral clotrimazole induces inhibition of the Gardos channel and reduction of erythrocyte dehydration in patients with sickle cell disease. J Clin Invest 97:1227-1234

Canessa M, Spalvins A, Nagel RL (1986) Volume-dependent and NEM-stimulated K^+,Cl^- transport is elevated in oxygenated SS, SC and CC human red cells. FEBS Letters 200:197-202

Canessa M, Fabry ME, Blumenfeld N, Nagel RL (1987) Volume-stimulated, Cl^--dependent K^+ efflux is highly expressed in young human red cells containing normal hemoglobin or HbS. J Membrane Biol 97:97-105

Canessa M, Romero JR, Lawrence C, Nagel RL, Fabry ME (1994) Rate of activation and deactivation of K:Cl cotransport by changes in cell volume in hemoglobin SS, CC and AA red cells. J Membrane Biol 142:349-362

Chiu D, Lubin B, Roelofsen B, Van Deenen LLM (1981) Sickled erythrocytes accelerate clotting *in vitro*: an effect of abnormal membrane lipid asymmetry. Blood 58:398-401

Choe HR, Schlegel RA, Rubin E, Williamson P, Westerman MP (1986) Alteration of red cell membrane organization in sickle cell anemia. Br J Haematol 63:761-773

Clark MR, Morrison CE, Shohet SB (1978a) Monovalent cation transport in irreversibly sickled cells. J Clin Invest 62:329-337

Clark MR, Unger RC, Shohet SB (1978b) Monovalent cation composition and ATP and lipid content of irreversibly sickled cells. Blood 51:1169-1178

Clark MR, Guatelli JC, White AT, Shohet SB (1981) Studies of the dehydrating effect of the red cell Na^+/K^+ pump in nystain-treated cells with varying Na^+ and water content. Biochim Biophys Acta 646:422-432

Clark MR, Rossi ME (1990) Permeability characteristics of deoxygenated sickle cells. Blood 76:2139-2145

Corbett JD, Golan DE (1993) Band 3 and glycophorin are progressively aggregated in density fractionated sickle and normal red blood cells: evidence from rotational and lateral mobility studies. J Clin Invest 91:208-217

Culliford SJ, Ellory JC, Gibson JS, Speake PF (1998) Effects of urea and oxygen tension on K flux in sickle cells. Pflügers Arch 435:740-742

De Franceschi L, Saadane N, Trudel M, Alper SL, Brugnara C, Beuzard Y (1994) Treatment with oral clotrimazole blocks Ca^{2+}-activated K^+ transport and reverses erythrocyte dehydration in transgenic SAD mice. A model for therapy of sickle cell disease. J Clin Invest 93:1670-1676

De Franceschi L, Beuzard Y, Jouault H, Brugnara C (1996) Modulation of erythrocyte potassium chloride cotransport, potassium content, and density by dietary magnesium intake in transgenic SAD mouse. Blood 88:2738-2744

De Franceschi L, Bachir D, Galacteros F, Tchernia G, Cynober T, Alper S, Platt O, Beuzard Y, Brugnara C (1997) Oral magnesium supplements reduce erythrocyte dehydration in patients with sickle cell disease. J Clin Invest 100:1847-1852

De Franceschi L, Bachir D, Galacteros F, Tchernia G, Cynober T, Neuberg D, Beuzard Y, Brugnara C (2000) Oral magnesium pidolate: effects of long-term administration in patients with sickle cell disease. Br J Haematol 108:248-289

De Jong K, Emerson RK, Butler J, Bastacky J, Mohandas N, Kuypers FA (2001a) Short survival of phosphatidylserine-exposing red blood cells in murine sickle cell anemia. Blood 98:1577-1584

De Jong K, Larkin SK, Styles LA, Bookchin RM, Kuypers FA (2001b) Characterization of the phosphatidylserine-exposing subpopulation of sickle cells. Blood 98:860-867

Dong C, Chadwick RS, Schechter AN (1992) Influence of sickle hemoglobin polymerization and membrane properties on deformability of sickle erythrocytes in the microcirculation. Biophys J 63:774-783

Eaton WA, Hofrichter J (1987) Hemoglobin S gelation and sickle cell disease. Blood 70:1245-1266

Eaton WA, Hofrichter J (1990) Sickle cell hemoglobin polymerization. Adv Prot Chem 40:263-279

Eaton J, Skelton T Swofford H, Kolpin C, Jacob H (1973) Elevated erythrocyte calcium in sickle cell disease. Nature 246:105-106

Eisinger J, Flores J, Salhany JM (1982) Association of cytosol hemoglobin with the membrane in intact erythrocytes. Proc Natl Acad Sci USA 79:408-412

Eisinger J, Flores J, Bookchin RM (1984) The cytosol-membrane interface of normal and sickle erythrocytes. Effect of hemoglobin deoxygenation and sickling. J Biol Chem 259:7169-7177

Embury SH, Clark MR, Monroy G, Mohandas N (1984) Concurrent sickle cell anemia and alpha-thalassemia. Effect on pathological properties of sickle erythrocytes. J Clin Invest 73:116-123

Engelmann B, Duhm J (1987) Intracellular calcium content of human erythrocytes: relation to sodium transport systems. J Membrane Biol 98:79-87

Etzion Z, Tiffert T, Bookchin RM, Lew VL (1993) Effects of deoxygenation on active and passive Ca^{2+} transport and on the cytoplasmic Ca^{2+} levels of sickle cell anemia red cells. J Clin Invest 92:2489-2498

Etzion Z, Lew VL, Bookchin RM (1996) K(86Rb) transport heterogeneity in the low-density fraction of sickle cell anemia red blood cells. Am J Physiol 271:C1111-C1121

Fabry ME, Nagel RL (1982) Heterogeneity of red cells in the sickler: a characteristic with practical clinical and pathophysiological implications. Blood Cells 8:9-15

Fabry ME, Kaul DK, Raventos-Suarez C, Chang H, Nagel RL (1982) SC erythrocytes have an abnormally high intracellular hemoglobin concentration. Pathophysiological consequences. J Clin Invest 70:1315-1319

Fabry ME, Benjamin L, Lawrence C, Nagel RL (1984a) An objective sign in painful crisis in sickle cell anemia: the concomitant reduction of high density red cells. Blood 64:559-563

Fabry ME, Mears JG, Patel P, Schaefer-Rego K, Carmichael LD, Martinez G, Nagel RL (1984b) Dense cells in sickle cell anemia: the effects of gene interaction. Blood 64:1042-1046

Fabry ME, Romero JR, Buchanan ID, Suzuka SM, Stamatoyannopoulos G, Nagel RL, Canessa M (1991) Rapid increase in red blood cell density driven by K:Cl cotransport in a subset of sickle cell anemia reticulocytes and discocytes. Blood 78:217-225

Fabry ME, Romero JR, Suzuka SM, Gilman JG, Feeling-Taylor A, Smithies O, Nagel RL (1998) Hemoglobin C in transgenic mice with full mouse globin knockouts. Blood 92:330a

Fathallah H, Coezy E, De Neef RS, Hardy-Dessources MD, Giraud F (1995) Inhibition of deoxygenation-induced membrane protein dephosphorylation and cell dehdyration by phorbol esters and okadaic acid in sickle cells. Blood 86:1999-2007

Fathallah H, Sauvage M, Romero JR, Canessa M, Giraud F (1997) Effects of PKC alpha activation on Ca^{2+} pump and K(Ca) channel in deoxygenated sickle cells. Am J Physiol 273:C1206-C1214

Feray JC, Beuzard Y, Garay RP (1988) Na/Mg exchange in sickle cells. VI International Magnesium Symposium 22a

Fischer S, Nagel RL, Bookchin RM, Roth EF Jr, Tellez-Nagel I (1975) The binding of hemoglobin to membranes of normal and sickle erythrocytes. Biochim Biophys Acta 375:422-433

Franco RS, Barker-Gear R, Miller MA, Williams SM, Joiner CH, Rucknagel DL (1994) Fetal hemoglobin and potassium in isolated transferrin receptor-positive dense sickle reticulocytes. Blood 84:2013-2020

Franco RS, Palascak M, Thompson H, Joiner CH (1995) KCl cotransport activity in light *versus* dense transferrin receptor-positive sickle reticulocytes. J Clin Invest 95:2573-2580

Franco RS, Thompson H, Palascak M, Joiner CH (1997) The formation of transferrin receptor-positive sickle reticulocytes with intermediate density is not determined by fetal hemoglobin content. Blood 90:3195-3203

Friedman MJ (1981) Hemoglobin and the red cell membrane: increased binding of polymorphic hemoglobins and measurment of free radicals in the membrane. The red cell, 5th Ann Arbor Conference. Alan R Liss, New York, pp 519-531

Gibson JS, Speake PF, Ellory JC (1998) Differential oxygen sensitivity of the K^+-Cl^- cotransporter in normal and sickle human red blood cells. J Physiol 511:225-234

Gillen CM, Brill S, Payne JA, Forbush B III (1996) Molecular cloning and functional expression of the K-Cl cotransporter from rabbit, rat, and human. A new member of the cation-chloride cotransporter family. J Biol Chem 271:16237-16244

Glader BE, Nathan NG (1978) Cation permeability alteration during sickling: relationship to cation composition and cellular dehydration of irreversibly sickled cells. Blood 51:983-989

Hammerman SI, Kourembanas S, Conca TJ, Tucci M, Brauer M, Farber HW (1997) Endothelin-1 production during the acute chest syndrome in sickle cell disease. Am J Resp Crit Care Med 156:280-285

Harrington JP (1998) Alteration of redox stability of hemoglobins A and S by biological buffers. Comp Biochem Physiol 119B:305-309

Hasegawa S, Hiruma H, Uyesaka N, Noguchi CT, Schechter AN, Rodgers GP (1995) Filterability of mixtures of sickle and normal erythrocytes. Am J Hematol 50:91-97

Hebbel RP, Schwartz RS, Mohandas N (1985) The adhesive sickle erythrocyte: cause and consequences of abnormal interactions with endothelium, monocytes/macrophages, and model membranes. Clin Haematol 14:141-161

Hebbel RP, Morgan WT, Eaton JW, Hedlund BE (1988) Accelerated autoxidation and heme loss due to instability of sickle hemoglobin. Proc Natl Acad Sci USA 85:237-241

Hiruma H, Noguchi CT, Uyesaka N, Hasegawa S, Blanchette-Mackie EJ, Schechter AN, Rodgers GP (1995a) Sickle cell rheology is determined by polymer fraction--not cell morphology. Am J Hematol 48:19-28

Hiruma H, Noguchi CT, Uyesaka N, Schechter AN, Rodgers GP (1995b) Contributions of sickle hemoglobin polymer and sickle cell membranes to impaired filterability. Am J Physiol 268:H2003-H2008

Holtzclaw JD, Jiang M, Yasin Z, Joiner CH, Franco RS (2002) Rehydration of high-density sickle erythrocytes *in vitro*. Blood 100:3017-3025

Horiuchi K, Asakura T (1987) Formation of light irreversibly sickled cells during deoxygenation-oxygenation cycles. J Lab Clin Med 110:653-660

Horiuchi K, Ballas SK, Asakura T (1988) The effect of deoxygenation rate on the formation of irreversibly sickled cells. Blood 71:46-51

Jennings ML, Al-Rohil N (1990) Kinetics of activation and inactivation of swelling stimulated K/Cl transport. The volume sensitive parameter is the rate constant for inactivation. J Gen Physiol 95:1021-1040

Jennings ML, Schulz RK (1991) Okadaic acid inhibition of K-Cl cotransport. J Gen Physiol 97:799-818

Joiner CH (1987) Studies on the mechanism of passive cation fluxes activated by deoxygenation of sickle cells. Prog Clin Biol Res 240:229-235

Joiner CH (1990) Deoxygenation-induced cation fluxes in sickle cells: II. Inhibition by stilbene disulfonates. Blood 76:212-220

Joiner CH (1993) Cation transport and volume regulation in sickle red blood cells. Am J Physiol 264:C251-C270

Joiner CH, Platt OS, Lux SE (1986) Cation depletion by the sodium pump in red cells with pathologic cation leaks. Sickle cells and xerocytes. J Clin Invest 78:1487-1496

Joiner CH, Dew A, Ge DL (1988) Deoxygenation-induced cation fluxes in sickle cells: relationship between net potassium efflux and net sodium influx. Blood Cells 13:339-358

Joiner CH, Morris CL, Cooper ES (1993) Deoxygenation-induced cation fluxes in sickle cells. III. Cation selectivity and response to pH and membrane potential. Am J Physiol 264:C734-C744

Joiner CH, Jiang M, Franco RS (1995) Deoxygenation-induced cation fluxes in sickle cells. IV. Modulation by external calcium. Am J Physiol 269:C403-C409

Joiner CH, Claussen W, Yasin Z, Franco RS (1997) Dipyridamole inhibits *in vitro* deoxygenation-induced cation fluxes in sickle red blood cells at the membrane concentration achievable in blood. Blood 90:125a

Joiner CH, Jiang M, Fathallah H, Giraud F, Franco RS (1998) Deoxygenation of sickle red blood cells stimulates KCl cotransport without affecting Na^+/H^+ exchange. Am J Physiol 274:C1466-C1475

Kaji DM, Tsukitani Y (1991) Role of protein phosphatase in activation of KCl cotransport in human erythrocytes. Am J Physiol 260:C176-C180

Kaul DK, Fabry ME, Nagel RL (1989) Microvascular sites and characteristics of sickle cell adhesion to vascular endothelium in shear flow conditions: pathophysiological implications. Proc Natl Acad Sci USA 86:3356-3360

Kaul DK, Chen D, Zhan J (1994) Adhesion of sickle cells to vascular endothelium is critically dependent on changes in density and shape of the cells. Blood 83:3006-3017

Kuypers FA, Van den Berg JJM, Lubin BH (1994) Membrane lipids and vesiculation. In: Embury SH, Hebbel RP, Mohandas N, Steinberg MH (eds) Sickle cell disease: Basic principles and clinical practice. Raven Press, New York, pp 139-152

Lawrence C, Fabry ME (1987) Objective indices of sickle cell painful crisis: decrease in RDW and percent dense cells and increase in ESR and fibrinogen. Prog Clin Biol Res 240:329-336

Lawrence C, Fabry ME, Nagel RL (1985) Red cell distribution width parallels dense red cell disappearance during painful crises in sickle cell anemia. J Lab Clin Med 105:706-710

Lawrence C, Fabry ME, Nagel RL (1991) The unique red cell heterogeneity of SC disease: crystal formation, dense reticulocytes, and unusual morphology. Blood 78:2104-2112

Lessin LS, Kurantsin-Mills J, Wallas C, Weems H (1978) Membrane alterations in irreversibly sickled cells: hemoglobin-membrane interactions. J Supramolec Struct 9:537-554

Lew VL, Bookchin RM (1986) Volume, pH, and ion-content regulation in human red cells: analysis of transient behavior with an integrated model. J Membrane Biol 92:57-74

Lew VL, Hockaday A, Sepulveda MI, Somlyo AP, Somlyo AV, Ortiz OE, Bookchin RM (1985) Compartmentalization of sickle-cell calcium in endocytic inside-out vesicles. Nature 315:586-589

Lew VL, Freeman CJ, Ortiz OE, Bookchin RM (1991) A mathematical model of the volume, pH, and ion content regulation in reticulocytes. Application to the pathophysiology of sickle cell dehydration. J Clin Invest 87:100-112

Lew VL, Ortiz OE, Bookchin RM (1997) Stochastic nature and red cell population distribution of the sickling-induced Ca^{2+} permeability. J Clin Invest 99:2727-2735

Lew VL, Etzion Z, Bookchin RM (2002) Dehydration response of sickle cells to sickling-induced Ca^{++} permeabilization. Blood 99:2578-2585

Li Q, Jungmann V, Kiyatkin A, Low PS (1996) Prostaglandin E_2 stimulates a Ca^{2+}-dependent K^+ channel in human erythrocytes and alters cell volume and filterability. J Biol Chem 271:18651-18656

Liu SC, Derick LH, Zhai S, Palek J (1991) Uncoupling of the spectrin-based skeleton from the lipid bilayer in sickled red cells. Science 252:574-576

Liu SC, Derick LH, Palek J (1993) Dependence of the permanent deformation of red blood cell membranes on spectrin dimer-tetramer equilibrium: implication for permanent membrane deformation of irreversibly sickled cells. Blood 81:522-528

Liu SC, Scott JY, Mehta JR, Nichols PE, Ballas SK, Yacono PW, Golan DE, Palek J (1996) Red cell membrane remodeling in sickle cell anemia: sequestration of membrane lipids and proteins in Heinz bodies. J Clin Invest 97:29-36

Lubin B, Chiu D, Bastacky J, Roelofsen B, Van Deenen LLM (1981) Abnormalities in membrane phospholipid organization in sickled erythrocytes. J Clin Invest 67:1643-1649

Lubin B, Kuypers F, Chiu D (1989) Lipid alterations and cellular properties of sickle red cells. Ann New York Acad Sci 565:86-95

Lux SE, John KM, Karnovsky MJ (1976) Irreversible deformation of the spectrin-actin lattice in irreversibly sickled cells. J Clin Invest 58:955-963

McGoron AJ, Joiner CH, Palascak MB, Claussen WJ, Franco RS (2000) Dehydration of mature and immature sickle red blood cells during fast oxygenation/deoxygenation cycles: role of KCl cotransport and extracellular calcium. Blood 95:2164-2168

Mercado A, Song L, Vazquez N, Mount DB, Gamba G (2000) Functional comparison of the K-Cl cotransporters KCC1 and KCC4. J Biol Chem 275:30326-30334

Merciris P, Hardy-Dessources MD, Sauvage M, Giraud F (1998) Involvement of deoxygenation-induced increase in tyrosine kinase activity in sickle cell dehydration. Pflügers Arch 436:315-322

Merciris P, Hardy-Dessources MD, Giraud F (2001) Deoxygenation of sickle cells stimulates Syk tyrosine kinase and inhibits a membrane tyrosine phosphatase. Blood 98:3121-3127

Mohandas N, Ballas SK (1994) Erythrocyte density and heterogeneity. In: Embury SH, Hebbel RP, Mohandas N, Steinberg MH (eds) Sickle cell disease: Basic principles and clinical practice. Raven Press, New York, pp 195-203

Mohandas N, Kim YR, Tycko DH, Orlik J, Wyatt J, Groner W (1986a) Accurate and independent measurement of volume and hemoglobin concentration of individual red cells by laser light scattering. Blood 68:506-513

Mohandas N, Rossi ME, Clark MR (1986b) Association between morphologic distortion of sickle cells and deoxygenation-induced cation permeability increase. Blood 68:450-454

Mohandas N, Johnson A, Wyatt J, Croisille L, Reeves J, Tycko D, Groner W (1989) Automated quantitation of cell density distribution and hyperdense cell fraction in RBC disorders. Blood 74:442-447

Morris CL, Rucknagel DL, Joiner CH (1993) Deoxygenation-induced changes in sickle cell-sickle cell adhesion. Blood 81:3138-3145

Murphy E, Berkowitz LR, Orringer E, Levy L, Gabel SA, London RE (1987) Cytosolic free calcium in sickle red blood cells. Blood 69:1469-1474

Noguchi CT, Dover GJ, Rodgers GP, Serjeant GR, Antonarakis SE, Anagnou NP, Higgs DR, Wheatherall DJ, Schechter AN (1985) Alpha thalassemia changes erythrocyte heterogeneity in sickle cell disease. J Clin Invest 75:1632-1639

Noguchi CT, Rodgers GP, Schechter AN (1989) Intracellular polymerization. Disease severity and therapeutic predictions. Ann New York Acad Sci 565:75-82

Ohnishi ST, Katagi H, Katagi C (1989) Inhibition of the *in vitro* formation of dense cells and of irreversibly sickled cells by charybdotoxin, a specific inhibitor of calcium-activated potassium efflux. Biochim Biophys Acta 1010:199-203

Olivieri O, Vitoux D, Galacteros F, Bachir D, Blouquit Y, Beuzard Y, Brugnara C (1992) Hemoglobin variants and activity of the (K^+,Cl^-) cotransport system in human erythrocytes. Blood 79:793-797

Olukoga AO, Adewoye HO, Erasmus RT, Adedoyin MA (1990) Erythrocyte and plasma magnesium in sickle-cell anaemia. East Afr Med J 67:348-354

Ortiz OE, Lew VL, Bookchin RM (1990) Deoxygenation permeabilizes sickle cell anaemia red cells to magnesium and reverses its gradient in the dense cells. J Physiol 427:211-226

Palek J (1977) Red cell calcium content and transmembrane calcium movements in sickle cells anemia. J Lab Clin Med 89:1365-1374

Paolisso G, Scheen A, Cazzolino D, Di MG, Varricchio M, D'Onofrio F, Lefebvre PJ (1994) Changes in glucose turnover parameters and improvement of glucose oxidation after 4 week magnesium administration in elderly noninsulin-dependent (type II) diabetic patients. J Clin End Metab 78:1510-1514

Phelan M, Perrine SP, Brauer M, Faller DV (1995) Sickle erythrocytes, after sickling, regulate the expression of the endothelin-1 gene and protein in human endothelial cells in culture. J Clin Invest 96:1145-1151

Platt OS (1994) Membrane proteins. In: Embury SH, Hebbel RP, Mohandas N, Steinberg MH (eds) Sickle cell disease: Basic principles and clinical practice. Raven Press, New York, pp 125-137

Platt OS, Falcone JF (1995) Membrane protein interactions in sickle red blood cells: evidence of abnormal protein 3 function. Blood 86:1992-1998

Platt OS, Falcone JF, Lux SE (1985) Molecular defect in the sickle erythrocyte skeleton. Abnormal spectrin binding to sickle inside-our vesicles. J Clin Invest 75:266-271

Race JE, Makhlouf FN, Logue PJ, Wilson FH, Dunham PB, Holtzman EJ (1999) Molecular cloning and functional characterization of KCC3, a new K-Cl cotransporter. Am J Physiol 277:C1210-C1219

Rank BH, Carlsson J, Hebbel RP (1985) Abnormal redox status of membrane-protein thiols in sickle erythrocytes. J Clin Invest 75:1531-1537

Reiss GH, Ranney HM, Shaklai N (1982) Association of hemoglobin C with erythrocyte ghosts. J Clin Invest 70:946-952

Rhoda MD, Apovo M, Beuzard Y, Giraud F (1990) Ca^{2+} permeability in deoxygenated sickle cells. Blood 75:2453-2458

Rifai N, Sakamoto M, Law T, Platt O, Mikati M, Armsby CC, Brugnara C (1995) HPLC measurement, blood distribution, and pharmacokinetics of oral clotrimazole, potentially useful antisickling agent. Clin Chem 41:387-391

Rigdon G, Stocker JW, McNaughton-Smith G, Hull H, Leese P (2000) A phase 1 safety, pharmacokinetic, and pharmacodynamic study of ascending, single, oral doses of ICA-17043, a novel Gardos channel inhibitor for the treatment of sickle cell disease. Blood 96:21b

Rivera A, Rotter A, Brugnara C (1999) Endothelins activate Ca^{2+}-gated K^{+} channels via endothelin B receptors in CD-1 mouse erythrocytes. Am J Physiol 227:C746-C754

Rivera A, Jarolim P, Brugnara C (2002) Modulation of Gardos channel activity by cytokines in sickle erythrocytes. Blood 99:357-363

Romero JR, Suzuka SM, Nagel RL, Fabry ME (2002) Arginine supplementation of sickle transgenic mice reduces red cell density and Gardos channel activity. Blood 99:1103-1108

Rubin E, Schlegel RA, Williamson P (1986) Endocytosis in sickle erythrocytes: a mechanism for elevated intracellular Ca^{2+} levels. J Cell Physiol 126:53-59

Rybicki AC, Benjamin LJ (1998) Increased levels of endothelin-1 in plasma of sickle cell anemia patients. Blood 92:2594-2596

Schlüter K, Drenckhahn D (1986) Co-clustering of denatured hemoglobin with band 3: Its role in binding of autoantibodies against band 3 to abnormal and aged erythrocytes. Proc Natl Acad Sci USA 83:6137-6141

Shaklai N, Sharma VS, Ranney HM (1981) Interaction of sickle cell hemoglobin with the erythrocyte membrane. Biochemistry 78:65-68

Stocker JW, De Franceschi L, McNaughton-Smith G, Brugnara C (2000) A novel Gardos channel inhibitor, ICA-17043, prevents red blood cell dehdyration *in vitro* and in a mouse model (SAD) of sickle cell disease. Blood 96:486a

Sugihara T, Hebbel RP (1992) Exaggerated cation leak from oxygenated sickle red blood cells during deformation: evidence for a unique leak pathway. Blood 80:2374-2378

Sunshine HR, Hofrichter J, Eaton WA (1978) Requirements for therapeutic inhibition of sickle hemoglobin gelation. Nature 275:238-240

Tiffert T, Spivak JL, Lew VL (1988) Magnitude of calcium influx required to induce dehydration of normal human red cells. Biochim Biophys Acta 943:157-165

Tosteson DC (1955) The effects of sickling on ion transport. II. The effect of sickling on sodium and cesium transport. J Gen Physiol 39:55-67

Tosteson DC, Shea E, Darling RC (1952) Potassium and sodium of red blood cells in sickle cell anemia. J Clin Invest 31:406-411

Tosteson DC, Carlsen E, Dunham ET (1955) The effects of sickling on ion transport. I. Effect of sickling on potassium transport. J Gen Physiol 39:31-53

Walder JA, Chatterjee R, Steck TL, Low PS, Musso GF, Kaiser ET, Rogers PH, Arnone A (1984) The interaction of hemoglobin with the cytoplasmic domain of band 3 of the human erythrocyte membrane. J Biol Chem 259:10238-10246

Wang RH, Philipps G, Medof ME, Mold C (1993) Activation of the alternative complement pathway by exposure of phosphatidylethanolamine and phosphatidylserine on erythrocytes from sickle cell disease patients. J Clin Invest 92:1326-1335
Waugh SM, Willardson BM, Kannan R, Labotka RJ, Low PS (1986) Heinz bodies induce clustering of band 3, glycophorin, and ankyrin in sickle cell erythrocytes. J Clin Invest 78:1155-1160
Werdehoff SG, Moore RB, Hoff CJ, Fillingim E, Hackman AM (1998) Elevated plasma endothelin-1 levels in sickle cell anemia: relationships to oxygen saturation and left ventricular hypertrophy. Am J Hematol 58:195-199
Williamson P, Puchulu E, Penniston JT, Westerman MP, Schlegel RA (1992) Ca^{2+} accumulation and loss by aberrant endocytic vesicles in sickle erythrocytes. J Cell Physiol 152:1-9
Zwaal RFA, Schroit AJ (1997) Pathophysiologic implications of membrane phospholipid asymmetry in blood cells. Blood 89:1121-1132

24 The Membrane Physiology of the 'Malaria-Infected' Red Cell

Kiaran Kirk and Kevin J. Saliba

School of Biochemistry and Molecular Biology, Faculty of Science, Australian National University, Canberra A.C.T. Australia 0200, Australia

24.1 Introduction

In comparison with the metabolic requirements of many cells, those of the human erythrocyte are decidedly modest. The mature red cell lacks intracellular organelles. It does not synthesise DNA, RNA or proteins and its energetic requirements are met by a low level of glycolysis which, under normal conditions, consumes just 1–2 mmol/(l_{cells}·h) of glucose (Murphy 1959; Phillips and Mendershausen 1975).

The situation changes dramatically following invasion of an erythrocyte by a malaria parasite. The malaria parasite is a unicellular eukaryote (genus *Plasmodium*), which infects its vertebrate host via the bite of a female *Anopheles* mosquito. On entering the blood-stream the parasites make their way first to the liver where they multiply before re-entering the circulation and invading the erythrocytes of their host, typically one parasite per erythrocyte. The intraerythrocytic malaria parasite is a voracious creature, with an extensive complement of biosynthetic machinery. Within the 2–3 day period of its occupancy of an erythrocyte the parasite synthesises sufficient DNA, RNA and proteins to produce approx. 20 new parasites, each capable of invading a new erythrocyte.

The high level of metabolic activity required to support the rapid growth and replication of the intraerythrocytic malaria parasite presents the infected cell with a number of physiological challenges relating to the uptake of essential nutrients and ions, and the expulsion of metabolic wastes. The parasite meets these by modifying the membrane transport properties of the erythrocyte membrane, by modulating the permeability of the 'parasitophorous vacuole membrane' (PVM) in which it is enclosed, and by having a range of transporters and channels controlling the flux of solutes across the parasite plasma membrane, as well as a mechanism for the endocytotic ingestion, and subsequent digestion, of the erythrocyte cytosol.

The major focus of this chapter is on the nature, roles and consequences of the parasite-induced modifications to the physiological properties of the infected erythrocyte membrane. The membrane transport properties of the parasite, and those of the PVM are also considered briefly.

24.2 The Intraerythrocytic Phase of the Parasite Life Cycle

Figure 24.1 illustrates the different stages of the asexual intraerythrocytic phase of the life-cycle of *Plasmodium falciparum*, the most virulent of the four species of malaria parasite infectious to humans.

The *P. falciparum* 'merozoite' gains entry into the human erythrocyte by an endocytosis-like process that leaves the intracellular parasite enclosed within the PVM (Fig. 24.1). In the hours immediately following invasion (the so-called 'ring' stage) the intracellular parasite seemingly lies dormant. However, from around 12–16 h post-invasion there is a progressive increase in metabolic and biosynthetic activity within the infected cell as the parasite enters the 'trophozoite' stage. There is extensive synthesis of proteins, RNA and DNA, a situation that contrasts markedly with that in normal erythrocytes. Like its host erythrocyte, the parasite is reliant on glycolysis for the production of ATP; however, its energetic requirements far exceed that of its relatively inert host cell, and as the parasite matures the rate of glucose utilisation and lactic acid production by the parasitised cell increases to up to 100 times the rate in the uninfected erythrocyte (Pfaller et al. 1982; Roth 1990; Van der Jagt et al. 1990).

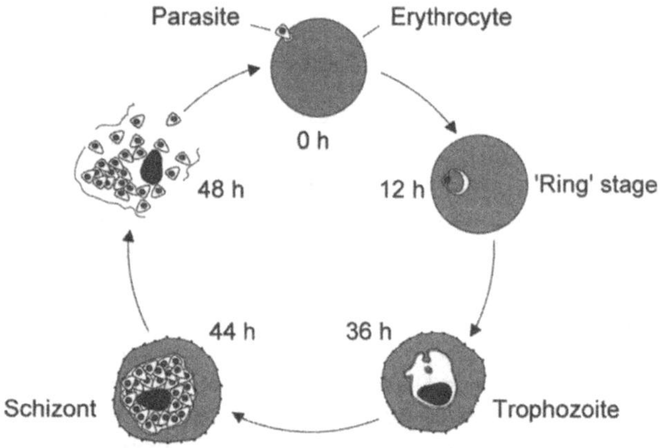

Fig. 24.1. Schematic representation of the intraerythrocytic phase of the malaria-parasite's life-cycle, illustrating the different stages

As it grows, the intracellular parasite endocytoses portions of the erythrocyte cytoplasm. This takes place at a specialised region of the parasite surface, referred to as the 'cytostome'. The resulting haemoglobin-filled vesicles fuse with an internal acidic 'digestive vacuole' in which the proteins of the host cell cytosol are digested to small peptides (Kamchonwongpaisan et al. 1997; Kolakovich et al. 1997; Rosenthal and Meshnick 1998), which serve as a source of amino acids for the parasite. The haem, liberated by the breakdown of haemoglobin, is potentially toxic to the parasite and is disposed of through a process of crystallisation within the digestive vacuole (Pagola et al. 2000). It has also been proposed that haem is broken down within the parasite cytosol (Ginsburg et al. 1998, 1999), although the extent to which this process actually takes place has recently been questioned (Egan et al. 2002).

By 36 h post-invasion, the parasite occupies approx. one third the total volume of the host cell, remaining enclosed within the PVM. There are pronounced changes in the morphology of the infected cell, which is transformed from the smooth biconcave disk of the normal erythrocyte to an irregularly shaped cell, the surface of which becomes covered with numerous small electron-dense protrusions known as 'knobs'. The knobs are the site of localisation of a number of parasite-derived proteins, which play a central role in the dual phenomena of cytoadherence (i.e. binding of infected cells to the endothelial cells lining the capillaries of the brain and other organs, as well as to uninfected erythrocytes) and antigenic variation (Deitsch and Wellems 1996). Figure 24.2 shows a scanning electronmicrograph of a parasitised erythrocyte, alongside several uninfected cells.

Fig. 24.2. Scanning electronmicrograph showing a *P. falciparum*-infected human erythrocyte together with several normal, uninfected human erythrocytes. The infected cell is misshapen and is covered with small protrusions known as 'knobs'. The micrograph was provided by Prof. David Ferguson, University of Oxford

From quite early in the period of occupancy of an erythrocyte by the parasite there appears in the erythrocyte cytosol a variety of tubular and vesicular membrane structures. These are thought to extend out from the PVM and are collectively referred to as the 'tubovesicular membrane' (TVM; Elmendorf and Haldar 1993) or 'tubulovacuolar' (Elford et al. 1995) network. As the parasite develops there are also significant changes in the ionic composition of the infected erythocyte cytosol. In particular, there is a decrease in the normally high concentration of K^+ and an increase in the normally low concentration of Na^+ (Ginsburg et al. 1986; Lee et al. 1988; Staines et al. 2001), with the result that by the time the parasite reaches the mature trophozoite stage the concentrations of Na^+ and K^+ in the erythrocyte cytosol are probably closer to those in the plasma than to those in a normal cell cytosol (Staines et al. 2001). The maturation of the parasite is also accompanied by an increase in the Ca^{2+} content of the infected cell, with most of the Ca^{2+} being restricted to compartments within the intracellular parasite (see Chap. 15).

Some 40–44 h after the initial invasion, the late-stage *P. falciparum* trophozoite enters the 'schizont' stage at which point it subdivides to produce ~20 daughter merozoites. These are released some 48 h after invasion, and each is capable of invading another erythrocyte, thereby continuing the intraerythrocytic cycle. The mechanism of merozoite release has been the subject of a number of recent studies. It had traditionally been thought that the release of merozoites occurs simultaneously with lysis of the host erythrocyte. However, there is evidence that the merozoites escape their host cell via a mechanism that does not necessarily cause haemolysis (Winograd et al. 1999; Salmon et al. 2001). A number of different mechanisms have been proposed that might account for these observations and these have been discussed (Lew 2001) and disputed (Trager 2002).

24.3 Transport Characteristics of the Parasitised Red Cell Membrane

At approx. the same time that the intraerythrocytic parasite begins to increase its metabolic and biosynthetic activity (i.e. 12–16 h after invasion) there begins a gradual increase in the permeability of the host erythrocyte membrane to a range of small molecules and ions. This increase has been demonstrated and characterised in *P. falciparum*-infected human erythrocytes using a range of different techniques, including both the haemolysis of parasitised cells suspended in isoosmotic solutions of solutes to which the erythrocyte membrane normally has a low permeability (and in which uninfected erythrocytes are stable) and radioactive tracers.

Solutes for which an increased rate of transport across the parasitised erythrocyte membrane has been demonstrated include monosaccharide sugars and other small 'polyols' (Ginsburg et al. 1985; Kirk et al. 1994, 1996), amino acids (Elford et al. 1985; Ginsburg et al. 1985; Kirk et al. 1994, 1999), peptides (Atamna and Ginsburg 1997; Saliba and Kirk 1998), nucleosides (Gero and Wood 1991; Kirk et al. 1994; Upston and Gero 1995), various monocarboxylates (Cranmer et al. 1995; Kirk et al. 1994; Saliba et al. 1998), small quaternary ammonium compounds (Staines et al. 2000), and monovalent inorganic anions (Kirk et al. 1993, 1994; Cranmer et al. 1995) and cations (Staines et al. 2001).

A range of pharmacological agents have been shown to block the parasite-induced increase in the erythrocyte membrane permeability. These include compounds best known as inhibitors of anion transport pathways (e.g. furosemide, niflumic acid, 5-nitro-2-(3-phenylpropylamino)-benzoic acid (NPPB) and various derivatives thereof (Kirk et al. 1994; Kirk and Horner 1995a)) as well as a number of other compounds, including glibenclamide (Kirk et al. 1993) and various quaternary ammonium compounds (Staines et al. 2000). Quantitative comparisons of the effect of such compounds on the transport of a range of structurally-unrelated solutes into parasitised erythrocytes has led to the hypothesis that most of the increased flux of low molecular weight solutes across the plasma membrane of infected erythrocytes is via pathways of a single type. These 'New Permeation Pathways' (NPP) show the general characteristics of anion-selective channels. The rate of transport of monovalent anions is orders of magnitude higher than that of similarly sized monovalent cations; nevertheless, the 'trickle' of the inorganic cations Na^+ and K^+ via these pathways is sufficient to constitute a substantial 'leak' of these ions across the erythrocyte plasma membrane and underlies the profound change that occurs in the concentration of Na^+ and K^+ in the erythrocyte cytosol as the intraerythrocytic parasite matures (Staines et al. 2001). Transport via the NPP has been shown (for those solutes for which it has been investigated) to be non-saturable up to millimolar concentrations, and to be non-stereoselective (Kirk 2001).

Although the transport characteristics of the NPP have been described in some detail, their molecular identity remains unknown. It is yet to be clearly established whether the pathways are native to the erythrocyte membrane (activated and/or modified following parasitisation) or synthesised and inserted into the erythrocyte membrane by the intracellular parasite, although recent data may cast some light on this issue (see Sect. 24.4).

In addition to the induction of NPP in the parasitised erythrocyte membrane there is evidence, in some systems, for changes in the rates of transport via various endogenous erythrocyte transporters. For example, in human erythrocytes infected *in vitro* with *P. falciparum* the Na^+/K^+ pump undergoes a progressive increase in activity from 18–30 h post-invasion, followed by a progressive decrease from 36 h (Staines et al. 2001). The initial increase may be attributed to the NPP-induced changes in the Na^+ and K^+ concentrations in the infected erythrocyte cytosol; the cause of the subsequent decrease is less clear. Fig. 24.3 illustrates the progressive increase in the activity of the NPP and modulation of the activity of the Na^+/K^+ pump as the intraerythrocytic parasite matures, as well as showing the predicted consequences for the concentrations of Na^+ and K^+ in the erythrocyte cytosol.

There are a number of reports of parasitised erythrocytes taken from animals infected *in vivo* with different species of malaria showing, in addition to increased permeability via NPP (e.g. Staines and Kirk 1998; Staines et al. 2002), enhanced activity of transporters with the same characteristics as those of normal cells. For example in both *P. knowlesi*-infected monkeys (Ancelin et al. 1991) and *P. vinckei*-infected mice (Staines and Kirk 1998), parasitised erythrocytes show a dramatic (10–20 fold) increase in the V_{max} for the uptake of choline via a pathway with the same K_m (and pharmacological characteristics) as the native erythrocyte choline transporter. The origin of this increase is unclear.

In addition to these reasonably well characterised changes in the transport of solutes into erythrocytes following parasitisation there is evidence for other, less well understood, alterations in the permeability/barrier properties of the infected cell membrane. There is mounting evidence that, unlike the normal erythrocyte, the *Plasmodium*-infected erythrocyte is capable of taking up macromolecules from the extracellular medium (reviewed in Kirk 2001), as well as taking up smaller molecules such as lucifer yellow (M_r = 443; Kirk et al. 1999) and, under conditions of ATP depletion, the divalent cation Ca^{2+} (Staines et al. 1999), via pathways discrete from the NPP. The nature and physiological roles of the pathways involved are not well understood.

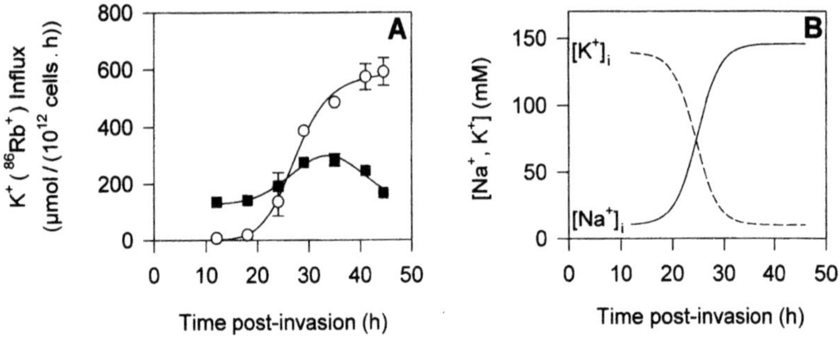

Fig. 24.3. Perturbation of the pump/leak balance for Na^+ and K^+ in the cytosol of the *P. falciparum*-infected human erythrocyte. (**A**) Flux of the K^+ congener $^{86}Rb^+$ via the erythrocyte Na^+/K^+ pump (closed squares) and via the 'New Permeation Pathways' (NPP; open circles) induced by *P. falciparum* in the erythrocyte membrane as the intraerythrocytic parasite matures. (**B**) Variation of $[Na^+]$ (solid line) and $[K^+]$ (broken line) in the cytosol of the parasitised erythrocyte, predicted to occur as a result of the time-dependent transport perturbations illustrated in (**A**). The predictions were made by incorporating the estimated flux of Na^+ and K^+ via the parasite-induced NPP, as well as the altered activity of the Na^+/K^+ pump, into the 'integrated red cell model' of Lew and Bookchin (Lew and Bookchin 1986). Both (**A**) and (**B**) are adapted from (Staines et al. 2001)

24.4 Electrophysiological Characteristics of the Parasitised Red Cell Membrane

The flux of ions via membrane channels may be characterised in detail using variations of the patch-clamp technique. An early review by Cabantchik (1989) refers to unpublished data (from Stutzin and Cabantchik) suggesting the presence of a voltage-dependent, phloridzin-sensitive ion channel in the membrane of human erythrocytes infected with *P. falciparum*. Desai et al. (1996) subsequently presented cell-attached patch-clamp measurements on intact parasitised erythrocytes. In fewer than 10% of the parasitised cells (and in none of the uninfected cells) tested there was a seemingly novel channel activity. However, in each case, the cell lysed before the channel could be characterised in any detail.

More recently there has been a series of more comprehensive studies of the electrophysiological characteristics of erythrocytes infected with the malaria parasite. In the first such study, Desai et al. (2000) reported that in human erythrocytes infected with mature, trophozoite-stage *P. falciparum* parasites the whole-cell current is 150-fold larger than that of uninfected erythrocytes. The increased current was attributed primarily to the presence in the parasitised membrane of a small conductance (< 10 pS), inwardly rectifying anion channel, present at an estimated 1000 copies per cell and showing complex gating behaviour. The selectivity and pharmacological properties of the increased whole-cell currents across the infected erythrocyte membrane showed a remarkable resemblance to those reported previously (on the basis of radiotracer flux and haemolysis experiments) for the NPP induced by the parasite in the infected cell. Like the NPP, the whole cell currents were anion-selective, with a permeability sequence Cl^- > lactate > glutamate >> Na^+, K^+. The order of permeation of monovalent inorganic anions deduced from the whole-cell experiments (SCN^- > I^- > Br^- > Cl^-) is similar to that seen in tracer and haemolysis experiments for the effects of these different anions on the rate of permeation through the NPP of monovalent cations (Kirk and Horner 1995b; Staines et al. 2000). Furthermore the sensitivity of the currents to a range of pharmacological agents matches that of the NPP very closely, with both pathways being blocked by NPPB, furosemide, niflumate, glibenclamide and phloridzin.

Somewhat different results were obtained in a subsequent study by Huber et al. (2002a) who, in whole-cell recordings of *P. falciparum*-infected human erythrocytes, identified *two* discrete anion conductances, differing from one another both in their inhibitor-sensitivity and in their voltage-dependence. One was outwardly rectifying and the other inwardly rectifying. The conductances were diminished on treatment of the parasitised cells with reducing agents, and the same manoeuver was shown to slow the rate of haemolysis of parasitised cells suspended in an isosmotic sorbitol solution. In the same study it was shown that similar anion conductances could be induced in uninfected erythrocytes by exposing them to oxidizing agents, and that oxidative stress also induced haemolysis of uninfected cells suspended in an isosmotic sorbitol solution. On the basis of these observations it was postulated that the NPP induced in infected cells are endogenous erythrocyte channels, activated in response to the oxidative stress to which the host cell is subjected by the intracellular parasite. Of the two anion conductances characterized it was actually the outwardly rectifying conductance which showed the closest similarity to the NPP in terms of its pharmacological profile. A preliminary report of a differential effect of different polyols on the outwardly rectifying conductance in infected cells (Huber et al. 2002b) is also consistent with the hypothesis that the channels underlying this conductance are permeable to small organic solutes of the sort known to permeate the NPP.

The same group has reported the presence in uninfected human erythrocytes of an oxidation-induced cation conductance (Duranton et al. 2002) and have presented preliminary evidence that this conductance is activated in *P. falciparum*-infected cells (Tanneur et al. 2002). The conductance shows the same alkali metal cation selectivity as has been reported for the NPP (i.e. Cs^+ > K^+ > Na^+ > Li^+; Staines et al. 2001), as well as showing an anion-dependence reminiscent of the anion dependence of the permeation of both organic and inorganic cations via the NPP.

The notion of the parasite activating endogenous erythrocyte ion channels is consistent with the results of a study of erythrocytes from *P. gallinaceum*-infected chickens (Thomas et al. 2001) in which the 5–40-fold increase in the whole-cell conductance following parasitisation was attributed to enhanced activity of three endogenous channels that are present, but largely or wholly inactive, in uninfected cells – two non-selective cation (NSC) channels and an anion-selective channel. Nevertheless, as has been discussed elsewhere (Ginsburg 2002; Huber et al.

2002c), there are significant differences between the findings of Huber and co-workers and those reported previously by others, that are yet to be resolved.

In another recent paper Egée et al. (2002) have described the presence in *P. falciparum*-infected human erythrocytes of anion channels showing a low, linear conductance. Channels with the same properties could be activated in uninfected human erythrocytes either by the combination of protein kinase A and ATP, or by membrane stretch, raising the possibility that either one of these mechanisms might be involved in the activation of the channels in infected cells. The properties of the channel were similar, though not identical to, those of the parasite-induced channel described by Desai et al. (2000). In a very small proportion (< 5%) of excised inside-out patch-clamp experiments on uninfected cells a second anion channel, showing outward rectification, was observed. But whereas Huber et al. (2002a) observed an outwardly rectifying current in a majority of *P. falciparum*-infected erythrocytes and have postulated that the enhanced permeability of infected cells to small organic solutes is attributable to the channels underlying this current, Egée et al. (2002) report that the outwardly rectifying channel was "never observed in infected cell patches". They, like Desai et al. (2000), attributed the increased conductance of the parasitised erythrocyte membrane to a single channel type.

In summary, it is clear from the spate of recent electrophysiological studies of erythrocytes infected with the malaria parasite that the membrane of the parasitised erythrocyte has a much higher conductance than that of uninfected erythrocytes. The number and nature of channels underlying this increased conductance is less clear. There is some evidence that the channel activity observed in infected erythrocytes is attributable to the activation of endogenous (normally quiescent) erythrocyte channels and a number of different mechanisms of channel activation have been proposed. The channel originally characterised by Desai et al. (2000), and showing a close (though not exact) resemblance to that described by Egée et al. (2002) does share many characteristics with the NPP responsible for the increased permeability of the parasitised erythrocyte membrane to a wide range of low molecular weight solutes. However, the relationship between this channel and the multiple conductances reported by Huber and colleagues to be active in the membrane of parasitised erythrocytes (Huber et al. 2002a; Tanneur et al. 2002) is yet to be clarified.

24.5 Roles and Consequences of the Altered Permeability of the Infected Red Cell Membrane

The high metabolic and biosynthetic activity of the intraerythrocytic malaria parasite necessitates a substantial traffic of small molecules and ions into and out of the infected cell. For some nutrients the transport capabilities of the host cell membrane are sufficient to meet the parasite's requirements (e.g. the transport of glucose across the human erythrocyte membrane is substantially faster than the utilisation of this substrate by the parasite (see Krishna et al. 2000; Kirk 2001)). For others, transport via the endogenous transport systems of the host erythrocyte is probably too slow to meet the parasite's needs, and in at least a few cases (e.g. the essential amino acid, glutamate (Elford et al. 1985; Lauer et al. 1997; Kirk et al. 1999) and the essential vitamin, pantothenic acid (Saliba et al. 1998)) the erythrocyte simply lacks the appropriate transporters. For such compounds the NPP provide the major (and perhaps sole) route of entry into the parasitised cell (see e.g. Fig. 24.4).

Compounds produced as a by-product of the parasite's high metabolic activity and released from the parasitised cell (e.g. lactic acid, and the amino acids generated as a result of the breakdown of haemoglobin within the parasite) again cross the erythrocyte membrane via a combination of endogenous erythrocyte transporters and the parasite-induced pathways. In the case of lactate it has been argued that its high rate of production by the parasite exceeds the normal lactate transport capability of the erythrocyte membrane (mediated by a combination of a H^+-coupled monocarboxylate transporter, the 'band 3' anion exchanger, and the process of simple diffusion; Poole and Halestrap 1993) and that the efflux of lactate via the NPP (Kanaani and Ginsburg 1991; Kirk et al. 1994; Cranmer et al. 1995) is important in preventing the potentially toxic build-up of lactic acid within the infected cell (Poole and Halestrap 1993).

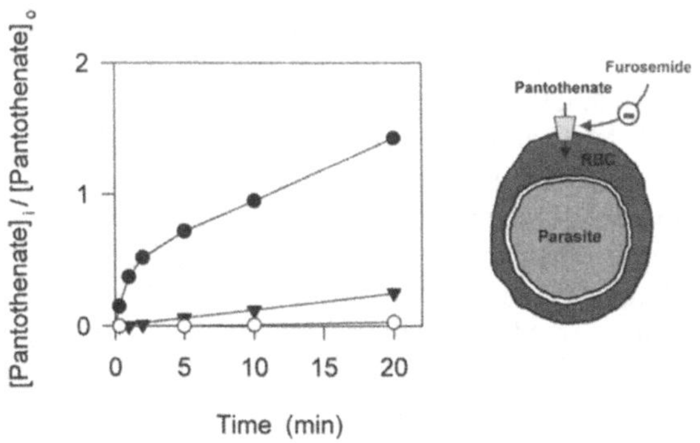

Fig. 24.4. Time-courses for uptake of the essential vitamin pantothenate into uninfected human erythrocytes (open circles) and *P. falciparum*-infected human erythrocytes suspended in the presence (closed triangles) and absence (closed circles) of the NPP inhibitor furosemide (0.1 mM). There is negligible uptake of pantothenate into uninfected cells. By contrast, the parasitised cells take up pantothenate in a biphasic manner. The initial phase corresponds to the entry of the vitamin into the infected cell via the NPP. The second phase represents the accumulation of the phosphorylated derivatives of pantothenate within the parasitised cell. Note that in the presence of furosemide the influx of pantothenate into the parasitised cell becomes rate-limiting for pantothenate metabolism. Adapted from Saliba et al. (1998)

The strong anion-selectivity of the NPP might be expected to ensure that the erythrocyte membrane potential remains at, or close to, the Cl^- potential (i.e. E_{Cl}), as is the case in normal uninfected erythrocytes. It restricts the movement of the major inorganic cations (Na^+ and K^+) sufficiently to prevent the infected erythrocyte undergoing osmotic lysis before the new generation of parasites are ready to emerge. Nevertheless, the small but significant cation permeability does allow sufficient influx of Na^+ into and sufficient efflux of K^+ out of the infected cell to cause a profound change in the Na^+ and K^+ concentrations in the erythrocyte cytosol (Fig. 24.3), and, therefore, in the concentration gradients for these two ions across the parasite plasma membrane. The significance of this and, in particular, the role of these gradients in the transport of solutes across the parasite plasma membrane and in modulating the electrical potential across this membrane is, as yet, unclear.

24.6 The Parasitophorous Vacuole and Parasite Membranes

In comparison with the transport properties of the host erythrocyte membrane, those of the PVM enclosing the intracellular parasite are not well understood. The 'cell-attached' patch clamp technique, applied to parasites freed from their host erythrocyte using either digitonin or a brief electrical pulse administered to the host cell membrane via the patch pipette, has been used to obtain single channel recordings, attributed to a high conductance (140 pS) channel in the PVM (Desai et al. 1993). The channel is permeable to a range of unrelated anions and cations (lysine, Tris, Ca^{2+}, Mg^{2+} and glucuronate), with little ability to discriminate between them. It is present at high density and is open more than 98% of the time, thus rendering this membrane highly permeable to small solutes.

Following on from their initial patch-clamp study Desai et al. (1993) obtained electrical recordings of a channel reconstituted into planar lipid bilayers from *P. falciparum* parasites 'freed' from their host erythrocytes using digitonin (Desai and Rosenberg 1997). The channel showed essentially the same electrophysiological properties as that observed in the patch-clamp study of the PVM and it was proposed, on this basis, to be the same channel. Using this system, evidence was obtained that the channel can pass macromolecules of up to 1400 Da, and has an estimated pore diameter of ~23 Å.

The electrophysiological data have led to the view that the PVM enclosing the intraerythrocytic malaria parasite serves as a 'molecular sieve', allowing the free passage (by diffusion through the PVM channels) of low molecular weight ions, nutrients and metabolic wastes. The recent finding that in parasitised cells in which the erythrocyte membrane (but not the PVM) is permeabilised by the pore-froming protein streptolysin O, a 'membrane impermeant' biotin derivative (~600 Da) gains access to parasitophorous vacuole proteins (Nyalwidhe et al. 2002) is consistent with this view. The exchange of solutes between the erythrocyte cytosol and the parasitophorous vacuole in which the parasite resides may be facilitated by the invagination of regions of the PVM into the erythrocyte cytosol to form the membranous 'tubovesicular' structures observed in fluorescence and electron micrographs of parasitised cells. But whether these structures are always contiguous with, and have the same permeability properties as, the PVM is not entirely clear.

In contrast to the PVM, the parasite plasma membrane is capable of maintaining substantial ion gradients (e.g., Saliba and Kirk 1999; Alleva and Kirk 2001) and the flux of solutes across this membrane is, as in any other cell, via a range of channels and transporters which act in concert to mediate nutrient uptake, ion homeostasis, and the disposal of metabolic wastes. A V-type H^+-ATPase in the parasite plasma membrane serves to extrude H^+, thereby generating a significant inward H^+ electrochemical gradient (Saliba and Kirk 1999; Hayashi et al. 2000). There is also immunological evidence for a H^+-pumping pyrophosphatase in the parasite plasma membrane (Luo et al. 1999; McIntosh et al. 2001), though there is not yet any direct evidence for a role for this protein in transporting H^+ across this membrane. The essential vitamin pantothenate is taken up via a low-affinity H^+-coupled uptake mechanism (Fig. 24.5; Saliba and Kirk 2001). Glucose (Kirk et al. 1996; Woodrow et al. 1999, 2000; Krishna et al. 2000) and nucleosides (Upston and Gero 1995; Carter et al. 2000; Parker et al. 2000; Rager et al. 2001) are taken up via equilibrative transporters. Metabolically-derived lactate escapes from the parasite via a H^+-coupled monocarboxylate transporter (Kanaani and Ginsburg 1991; Cranmer et al. 1995; Elliott et al. 2001), and ATP synthesised by the parasite is provided to the host erythrocyte in exchange for ADP, via an ATP/ADP exchanger (Kanaani and Ginsburg 1989; Choi and Mikkelsen 1990). Fig. 24.6 provides an overview of the various transport pathways present on the host erythrocyte membrane, the PVM, or the parasite plasma membrane, and for which there are at least some functional data available.

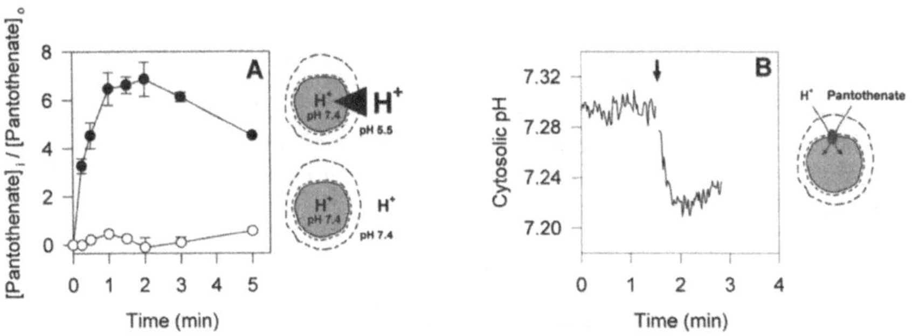

Fig. 24.5. Evidence for the transport of the essential vitamin pantothenate into the intracellular *P. falciparum* trophozoite via a H^+-coupled transporter. (**A**) H^+-driven accumulation of radiolabelled pantothenate in parasites 'isolated' from their host cell using a saponin permeabilisation technique and depleted of ATP to prevent the conversion of pantothenate to phosphopantothenate (the first step in the enzymatic production of coenzyme A). The radiolabelled pantothenate accumulates to high levels within the ATP-depleted parasites in the presence (closed circles), but not in the absence (open circles), of an inward H^+ gradient. (**B**) Pantothenate-induced influx of H^+ into isolated parasites. Pantothenate (20 mM) was added at the point shown by the arrow. The extracellular pH (and the pH of the added pantothenate solution) was 7.1. The pH of the parasite cytosol was measured using the fluorescent pH indicator, 2',7'-bis-(2-carboxyethyl)-5-(and-6)-carboxyfluorescein (BCECF). Both (**A**) and (**B**) are adapted from Saliba and Kirk (2001)

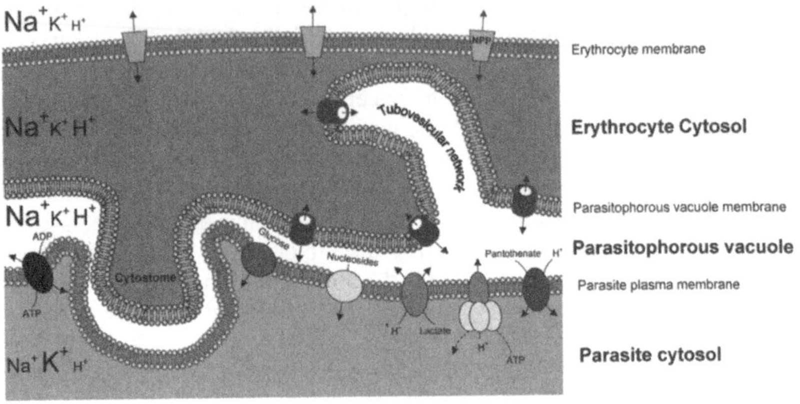

Fig. 24.6. Schematic representation of pathways involved in the transport across the three membranes separating the interior of the intraerythrocytic parasite from the extracellular solution. The figure shows only those pathways for which there are functional data available and which have been demonstrated to operate in the membranes in which they are shown. The relative sizes of the Na^+, K^+, and H^+ are intended to provide a (qualitative) indication of the transmembrane gradient for each ion

Within the parasite there are a number of membrane-bound organelles including a nucleus, a single mitochondrion, a Ca^{2+}-containing endoplasmic reticulum, the 'apicoplast' (a plastid-like organelle of unknown function; McFadden et al. 1996) and an acidic 'digestive vacuole' which is thought to be the site of breakdown of the proteins of the host cell cytosol, ingested by endocytosis. The digestive vacuole may or may not be equivalent to the 'acidocalcisome' an acidic intracellular Ca^{2+} store (Docampo and Moreno 2001), which takes up Ca^{2+} released from the endoplasmic reticulum, thereby preventing its loss from the cell (Alleva and Kirk 2001; see Chap. 15).

24.7 Chemotherapeutic Opportunities

The malaria parasite has demonstrated the ability to become either partially or fully resistant to most of the antimalarial drugs presently available and there is an urgent need to identify new drug targets and to develop new antimalarial drug strategies.

The membrane transport mechanisms discussed in this chapter offer a number of potential chemotherapeutic opportunities. To the extent that they serve as the major routes for the uptake of essential nutrients, the disposal of potentially toxic metabolic wastes and for the flux of inorganic ions, they may be targets in their own right. Although the notion of targeting the parasite's nutrient uptake mechanisms is an attractive one, there have been relatively few studies on the parasite's response to the deprivation of particular nutrients and there is little known about either the metabolic plasticity of the parasite (i.e. the extent to which the requirements for particular compounds or particular pathways can be circumvented) or the nature of the parasite's response (i.e. whether deprivation of particular nutrients is parasitistatic or parasiticidal). In the *in vivo* situation, anything that slows the growth and proliferation of the parasite, without unduly compromising the cells of the host, might be expected to be beneficial. However, it is largely unclear to what extent the transport mechanisms discussed here are 'rate-limiting' for particular metabolic pathways, to what extent these pathways might be rate-limiting for parasite growth, and to what extent the transport mechanisms would need to be inhibited to actually impact on parasite growth.

Irrespective of the feasibility of exploiting transport mechanisms as drug targets, some such mechanisms do offer pathways by which cytotoxic compounds might be targeted into the intracellular parasite. There is evidence that at least some compounds toxic to the intraerythrocytic malaria parasite rely primarily on the parasite-induced NPP in the erythrocyte membrane to gain entry into the parasitised cell. These include the peptidomimetic protease inhibitor, pepstatin A (Saliba and Kirk 1998), pentamidine (Stead et al. 2001) and a range of cytotoxic L-nucleosides (Gero and Hall 1997; Brown et al. 1999). Other antiplasmodial agents permeate the membranes of the parasitised erythrocyte via a process of simple diffusion; however, their uptake into and distribution within the infected cell may be influenced by e.g. the ionic composition (in particular the $[H^+]$) of subcellular compartments. A detailed understanding of the processes considered here therefore offers substantial benefits in terms of understanding and enhancing the antiplasmodial effects of antimalarial drugs.

Acknowledgement: Work in the authors' laboratory was supported by the Australian NH&MRC, the ARC and the Ramaciotti Foundations.

References

Alleva LM, Kirk K (2001) Calcium regulation in the intraerythrocytic malaria parasite *Plasmodium falciparum*. Mol Biochem Parasitol 117:121-128

Ancelin ML, Parant M, Thuet MJ, Philippot JR, Vial H (1991) Increased permeability to choline in simian erythrocytes after *Plasmodium knowlesi* infection. Biochem J 273:701-709

Atamna H, Ginsburg H (1997) The malaria parasite supplies glutathione to its host cell - investigation of glutathione metabolism in human erythrocytes infected with *Plasmodium falciparum*. Eur J Biochem 250:670-679

Brown DM, Netting AG, Chun BK, Choi Y, Chu CK, Gero AM (1999) L-Nucleoside analogues as potential antimalarials that selectively target *Plasmodium falciparum* adenosine deaminase. Nucleosides Nucleotides 18:2521-2532

Cabantchik ZI (1989) Altered membrane transport of malaria-infected erythrocytes: a possible pharmacologic target. Blood 74:1464-1471

Carter NS, Ben Mamoun C, Liu W, Silva EO, Landfear SM, Goldberg DE, Ullman B (2000) Isolation and functional characterization of the PfNT1 nucleoside transporter gene from *Plasmodium falciparum*. J Biol Chem 275:10683-10691

Choi I, Mikkelsen RB (1990) *Plasmodium falciparum*: ATP/ADP transport across the parasitophorous vacuolar and plasma membranes. Exp Parasitol 71:452-462

Cranmer SL, Conant AR, Gutteridge WE, Halestrap AP (1995) Characterization of the enhanced transport of L- and D-lactate into human red blood cells infected with *Plasmodium falciparum* suggests the presence of a novel saturable lactate proton cotransporter. J Biol Chem 270:15045-15052

Deitsch KW, Wellems TE (1996) Membrane modifications in erythrocytes parasitized by *Plasmodium falciparum*. Mol Biochem Parasitol 76:1-10

Desai SA, Rosenberg RL (1997) Pore size of the malaria parasite's nutrient channel. Proc Natl Acad Sci USA 94:2045-2049

Desai SA, Krogstad DJ, McCleskey EW (1993) A nutrient-permeable channel on the intraerythrocytic malaria parasite. Nature 362:643-646

Desai SA, McCleskey EW, Schlesinger PH, Krogstad DJ (1996) A novel pathway for Ca^{2+} entry into *Plasmodium falciparum*-infected blood cells. Am J Trop Med Hyg 54:464-470

Desai SA, Bezrukov SM, Zimmerberg J (2000) A voltage-dependent channel involved in nutrient uptake by red blood cells infected with the malaria parasite. Nature 406:1001-1005

Docampo R, Moreno SN (2001) The acidocalcisome. Mol Biochem Parasitol 114:151-159

Duranton C, Huber SM, Lang F (2002) Oxidation induces a Cl^--dependent cation conductance in human red blood cells. J Physiol 539:847-855

Egan TJ, Combrinck JM, Egan J, Hearne GR, Marques HM, Ntenteni S, Sewell BT, Smith PJ, Taylor D, Van Schalkwyk DA, Walden JC (2002) Fate of haem iron in the malaria parasite *Plasmodium falciparum*. Biochem J 365:343-347

Egée S, Lapaix F, Decherf G, Staines HM, Ellory JC, Doerig C, Thomas SL (2002) A stretch-activated anion channel is up-regulated by the malaria parasite *Plasmodium falciparum*. J Physiol 542:795-801

Elford BC, Haynes JD, Chulay JD, Wilson RJ (1985) Selective stage-specific changes in the permeability to small hydrophilic solutes of human erythrocytes infected with *Plasmodium falciparum*. Mol Biochem Parasitol 16:43-60

Elford BC, Cowan GM, Ferguson DJ (1995) Parasite-regulated membrane transport processes and metabolic control in malaria-infected erythrocytes. Biochem J 308:361-374

Elliott JL, Saliba KJ, Kirk K (2001) Transport of lactate and pyruvate in the intraerythrocytic malaria parasite, *Plasmodium falciparum*. Biochem J 355:733-739

Elmendorf HG, Haldar K (1993) Secretory transport in *Plasmodium*. Parasitol Today 9:98-102

Gero AM, Hall ST (1997) *Plasmodium falciparum*: transport of enantiomers of nucleosides into Sendai-treated trophozoites. Exp Parasitol 86:228-231

Gero AM, Wood AM (1991) New nucleoside transport pathways induced in the host erythrocyte membrane of malaria and *Babesia* infected cells. Adv Exp Med Biol 309A:169-172

Ginsburg H (2002) Oxidative permeabilization? Trends Parasitol 18:346

Ginsburg H, Kutner S, Krugliak M, Cabantchik ZI (1985) Characterization of permeation pathways appearing in the host membrane of *Plasmodium falciparum* infected red blood cells. Mol Biochem Parasitol 14:313-322

Ginsburg H, Handeli S, Friedman S, Gorodetsky R, Krugliak M (1986) Effects of red blood cell potassium and hypertonicity on the growth of *Plasmodium falciparum* in culture. Z Parasitenkd 72:185-199

Ginsburg H, Famin O, Zhang J, Krugliak M (1998) Inhibition of glutathione-dependent degradation of heme by chloroquine and amodiaquine as a possible basis for their antimalarial mode of action. Biochem Pharmacol 56:1305-1313

Ginsburg H, Ward SA, Bray PG (1999) An integrated model of chloroquine action. Parasitol Today 15:357-360

Hayashi M, Yamada H, Mitamura T, Horii T, Yamamoto A, Moriyama Y (2000) Vacuolar H^+-ATPase localized in plasma membranes of malaria parasite cells, *Plasmodium falciparum*, is involved in regional acidification of parasitised erythrocytes. J Biol Chem 275:34353-34358

Huber SM, Uhlemann AC, Gamper NL, Duranton C, Kremsner PG, Lang F (2002a) *Plasmodium falciparum* activates endogenous Cl^- channels of human erythrocytes by membrane oxidation. EMBO J 21:22-30

Huber SM, Duranton C, Uhlemann AC, Kremsner PG, Lang F (2002b) Anion and organic osmolyte channels of human erythrocytes infected with *Plasmodium falciparum*. Pflügers Arch 443, suppl:S164

Huber SM, Uhlemann AC, Gamper NL, Duranton C, Lang F, Kremsner PG (2002c) Oxidative permeabilization? Response from S.M. Huber et al. Trends Parasitol 18:346-347

Kamchonwongpaisan S, Samoff E, Meshnick SR (1997) Identification of hemoglobin degradation products in *Plasmodium falciparum*. Mol Biochem Parasitol 86:179-186

Kanaani J, Ginsburg H (1989) Metabolic interconnection between the human malarial parasite *Plasmodium falciparum* and its host erythrocyte. Regulation of ATP levels by means of an adenylate translocator and adenylate kinase. J Biol Chem 264:3194-3199

Kanaani J, Ginsburg H (1991) Transport of lactate in *Plasmodium falciparum*-infected human erythrocytes. J Cell Physiol 149:469-476

Kirk K (2001) Membrane transport in the malaria-infected erythrocyte. Physiol Rev 81:495-537

Kirk K, Horner HA (1995a) In search of a selective inhibitor of the induced transport of small solutes in *Plasmodium falciparum*-infected erythrocytes: effects of arylaminobenzoates. Biochem J 311:761-768

Kirk K, Horner HA (1995b) Novel anion dependence of induced cation transport in malaria-infected erythrocytes. J Biol Chem 270:24270-24275

Kirk K, Horner HA, Spillett DJ, Elford BC (1993) Glibenclamide and meglitinide block the transport of low molecular weight solutes into malaria-infected erythrocytes. FEBS Letters 323:123-128

Kirk K, Horner HA, Elford BC, Ellory JC, Newbold CI (1994) Transport of diverse substrates into malaria-infected erythrocytes via a pathway showing functional characteristics of a chloride channel. J Biol Chem 269:3339-3347

Kirk K, Horner HA, Kirk J (1996) Glucose uptake in *Plasmodium falciparum*-infected erythrocytes is an equilibrative not an active process. Mol Biochem Parasitol 82:195-205

Kirk K, Staines HM, Martin RE, Saliba KJ (1999) Transport properties of the host cell membrane. In: Transport and trafficking in the malaria-infected erythrocyte (Novartis Foundation Symposium 226). Wiley, Chichester, pp 55-73

Kolakovich KA, Gluzman IY, Duffin KL, Goldberg DE (1997) Generation of hemoglobin peptides in the acidic digestive vacuole of *Plasmodium falciparum* implicates peptide transport in amino acid production. Mol Biochem Parasitol 87:123-135

Krishna S, Woodrow CJ, Burchmore RJ, Saliba KJ, Kirk K (2000) Hexose transport in asexual stages of *Plasmodium falciparum* and kinetoplastidae. Parasitol Today 16:516-521

Lauer SA, Rathod PK, Ghori N, Haldar K (1997) A membrane network for nutrient import in red cells infected with the malaria parasite. Science 276:1122-1125

Lee P, Ye Z, Van Dyke K, Kirk RG (1988) X-ray microanalysis of *Plasmodium falciparum* and infected red blood cells: effects of qinghaosu and chloroquine on potassium, sodium, and phosphorus composition. Am J Trop Med Hyg 39:157-165

Lew VL (2001) Packaged merozoite release without immediate host cell lysis. Trends Parasitol 17:401-403

Lew VL, Bookchin RM (1986) Volume, pH, and ion-content regulation in human red cells: analysis of transient behaviour with an integrated model. J Membrane Biol 92:57-74

Luo S, Marchesini N, Moreno SN, Docampo R (1999) A plant-like vacuolar H^+-pyrophosphatase in *Plasmodium falciparum*. FEBS Letters 460:217-220

McFadden GI, Reith ME, Munholland J, Lang-Unnasch N (1996) Plastid in human parasites. Nature 381:482

McIntosh MT, Drozdowicz YM, Laroiya K, Rea PA, Vaidya AB (2001) Two classes of plant-like vacuolar-type H^+-pyrophosphatases in malaria parasites. Mol Biochem Parasitol 114:183-195

Murphy JR (1959) Erythrocyte metabolism. II. Glucose metabolism and pathways. J Lab Clin Med 55:286-302

Nyalwidhe J, Baumeister S, Hibbs AR, Tawill S, Papakrivos J, Völker U, Lingelbach K (2002) A non-permeant biotin derivative gains access to the parasitophorous vacuole in *Plasmodium falciparum*-infected erythrocytes permeabilized with streptolysin O. J Biol Chem 277:40005-40011

Pagola S, Stephens PW, Bohle DS, Kosar AD, Madsen SK (2000) The structure of malaria pigment beta-haematin. Nature 404:307-310

Parker MD, Hyde RJ, Yao SY, McRobert L, Cass CE, Young JD, McConkey GA, Baldwin SA (2000) Identification of a nucleoside/nucleobase transporter from *Plasmodium falciparum*, a novel target for anti-malarial chemotherapy. Biochem J 349:67-75

Pfaller MA, Krogstad DJ, Parquette AR, Nguyen Dinh P (1982) *Plasmodium falciparum*: stage-specific lactate production in synchronized cultures. Exp Parasitol 54:391-396

Phillips GB, Mendershausen PB (1975) Glucose consumption by red cells of diabetic patients and normal subjects. Effect of ethanol. Clin Chim Acta 61:175-182

Poole RC, Halestrap AP (1993) Transport of lactate and other monocarboxylates across mammalian plasma membranes. Am J Physiol 264:C761-C782

Rager N, Mamoun CB, Carter NS, Goldberg DE, Ullman B (2001) Localization of the *Plasmodium falciparum* PfNT1 nucleoside transporter to the parasite plasma membrane. J Biol Chem 276:41095-41099

Rosenthal PJ, Meshnick SR (1998) Hemoglobin processing and the metabolism of amino acids, heme and iron. In: IW Sherman (ed) Malaria: Parasite biology, pathogenesis, and protection. Am Soc Microbiol, Washington, pp 145-158

Roth E (1990) *Plasmodium falciparum* carbohydrate metabolism: a connection between host cell and parasite. Blood Cells 16:453-460

Saliba KJ, Kirk K (1998) Uptake of an antiplasmodial protease inhibitor into *Plasmodium falciparum*-infected human erythrocytes via a parasite-induced pathway. Molec Biochem Parasitol 94:297-301

Saliba KJ, Kirk K (1999) pH regulation in the intracellular malaria parasite, *Plasmodium falciparum*: H^+ extrusion via a V-type H^+-ATPase. J Biol Chem 47:33213-33219

Saliba KJ, Kirk K (2001) H^+-coupled pantothenate transport in the intracellular malaria parasite. J Biol Chem 276:18115-18121

Saliba KJ, Horner HA, Kirk K (1998) Transport and metabolism of the essential vitamin pantothenic acid in human erythrocytes infected with the malaria parasite *Plasmodium falciparum*. J Biol Chem 273:10190-10195

Salmon BL, Oksman A, Goldberg DE (2001) Malaria parasite exit from the host erythrocyte: a two-step process requiring extraerythrocytic proteolysis. Proc Natl Acad Sci USA 98:271-276

Staines HM, Kirk K (1998) Increased choline transport in erythrocytes from mice infected with the malaria parasite *Plasmodium vinckei vinckei*. Biochem J 334:525-530

Staines HM, Chang W, Ellory JC, Tiffert T, Kirk K, Lew VL (1999) Passive Ca^{2+} transport and Ca^{2+}-dependent K^+ transport in *Plasmodium falciparum*-infected red cells. J Membrane Biol 172:13-24

Staines HM, Rae C, Kirk K (2000) Increased permeability of the malaria-infected erythrocyte to organic cations. Biochim Biophys Acta 1463:88-98

Staines HM, Ellory JC, Kirk K (2001) Perturbation of the pump-leak balance for Na^+ and K^+ in malaria-infected erythrocytes. Am J Physiol 280:C1576-C1587

Staines HM, Godfrey EM, Lapaix F, Egée S, Thomas SL, Ellory JC (2002) Two functionally distinct organic osmolyte pathways in *Plasmodium gallinaceum*-infected chicken red blood cells. Biochim Biophys Acta 1561:98-108

Stead AM, Bray PG, Edwards IG, DeKoning HP, Elford BC, Stocks PA, Ward SA (2001) Diamidine compounds: selective uptake and targeting in *Plasmodium falciparum*. Mol Pharmacol 59:1298-306

Tanneur V, Duranton C, Lang F, Huber SM (2002) Increased cation conductance in *Plasmodium falciparum*-infected red blood cells. Pflügers Arch 443, suppl:S231

Thomas SL, Egée S, Lapaix F, Kaestner L, Staines HM, Ellory JC (2001) Malaria parasite *Plasmodium gallinaceum* up-regulates host red blood cell channels. FEBS Letters 500:45-51

Trager W (2002) On the release of malaria merozoites. Trends Parasitol 18:60-61

Upston JM, Gero AM (1995) Parasite-induced permeation of nucleosides in *Plasmodium falciparum* malaria. Biochim Biophys Acta 1236:249-258

Van der Jagt DL, Hunsaker LA, Campos NM, Baack BR (1990) D-lactate production in erythrocytes infected with *Plasmodium falciparum*. Mol Biochem Parasitol 42:277-284

Winograd E, Clavijo CA, Bustamante LY, Jaramillo M (1999) Release of merozoites from *Plasmodium falciparum*-infected erythrocytes could be mediated by a non-explosive event. Parasitol Res 85:621-624

Woodrow CJ, Penny JI, Krishna S (1999) Intraerythrocytic *Plasmodium falciparum* expresses a high affinity facilitative hexose transporter. J Biol Chem 274:7272-7277

Woodrow CJ, Burchmore RJ, Krishna S (2000) Hexose permeation pathways in *Plasmodium falciparum*-infected erythrocytes. Proc Natl Acad Sci USA 97:9931-9936

25 Hypertension

Sergei N. Orlov

Laboratory of Pathophysiology of Ion Transport Disorders, Research Centre, University of Montreal Hospital, CHUM, 3850 rue St.-Urbain, Montreal, PQ, H2W 1T8, Canada, and Laboratory of Physico-Chemistry of Biomembranes, Faculty of Biology, Lomonosov Moscow State University, Moscow 119899, Russia

25.1 Introduction

In 1973, Jones presented data on increased ^{42}K and ^{36}Cl turnover in aortic strips from spontaneously hypertensive rats (SHR) developed by Okamoto and Aoki as an experimental model of human primary or essential hypertension. These results raised at least 2 questions. First, are these abnormalities caused by altered permeability of plasma membrane for monovalent ions? Second, are these abnormalities linked to the pathogenesis of primary hypertension, or are they a consequence of long-term blood pressure elevation? The complex pattern of monovalent ion compartmentalisation in extra- and intracellular spaces complicated the use of aortic strips for analysis of the first issue. Because cell culture techniques had not been developed at that time, we decided to design an appropriate methodological approach involving erythrocytes. Erythrocytes from SHR, normotensive Wistar-Kyoto (WKY) and Sprague-Dawley rats were randomly investigated in our pilot experiments. To our surprise, we observed that the rate of ^{22}Na efflux was increased in SHR compared to normotensive rats (Postnov et al. 1975). In the next 2 decades, a dozen laboratories were engaged in the further characterisation of ion transport across the plasma membrane in primary hypertension. I will not attempt to cover all these studies but will, instead, offer an overview of key findings on: (i) the identification of ion transporters causing elevated erythrocyte membrane permeability for monovalent cations in primary hypertension; (ii) the relative impact of inheritance and environment on abnormal ion transport; (iii) approaches used to identify the molecular and genetic determinants of these abnormalities.

25.2 Transporters Involved in Abnormal Monovalent Ion Handling in Experimental Models of Primary Hypertension

25.2.1 Na^+/K^+ Pump

The α1-isoform of the Na^+/K^+ pump expressed in rodent erythrocytes possesses an extremely low affinity for ouabain compared to other mammalian species. In initial studies with 50 μM of this compound, we did not observe any differences in the rate of ouabain-sensitive ^{22}Na efflux between erythrocytes from SHR and normotensive rats (Postnov et al. 1976). The rate of ouabain-sensitive ATP hydrolysis was investigated to evaluate maximal activity of the Na^+/K^+ pump. These studies as well as analyses of the number of [^3H]-ouabain-binding sites did not reveal any systematic alterations in SHR (for review, see Postnov and Orlov (1985)).

Normotensive Brown-Norway (BN.lx) rats have served to generate recombinant inbred strains from SHR×BN.lx F_2 hybrids. Using saponin-treated erythrocytes, we observed that the maximal activity of the Na^+, K^+-ATPase is 20–30% lower in BN.lx rats than in SHR (Orlov et al. 1991). Milan hypertensive strains (MHS) developed by Bianchi and co-workers possess pathophysiological markers of abnormal renal function similar to those revealed in essential hypertension. The values of ouabain-sensitive ^{86}Rb uptake were 15–25% lower in MHS compared to normotensive MNS (Bianchi et al. 1985). The same differences between MHS and MNS were observed in the study of ouabain-sensitive ^{22}Na efflux (Orlov et al. 1988).

To study the implication of enhanced salt consumption in the maintenance of elevated blood pressure, inbred strains of Dahl salt-sensitive (SS-JR) and salt-resistant (SR-JR) rats were generated. No differences were observed by Zicha and Duhm (1990) under detailed analysis of ouabain-sensitive Rb^+ uptake and Na^+ extrusion in erythrocytes from young and adult SS-JR and SR-JR. McCormick et al. (1989) also failed to find altered ouabain-sensitive ^{86}Rb uptake despite significant $[Na^+]_i$ elevation in SS-JR erythrocytes. In contrast, Canessa et al. (1993) reported that the rate of ouabain-sensitive ^{22}Na efflux is non-altered in SS-JR rats, whereas ouabain-sensitive ^{86}Rb influx is decreased by ~50%. It was suggested that stoichiometry of the Na^+/K^+ pump is increased in SS-JR rats compared to normal 3:2 value in SR-JR rats. It should be underlined, however, that total content of K^+ and Na^+ rather than the values of their exchangeable fractions were used to calculate the specific radioactivity of intracellular ^{86}Rb and ^{22}Na in this study. Using an other approach, we failed to observe any differences in the stoichiometry of the Na^+/K^+ pump between SS-JR and congenic rats constructed by replacing α1 Na^+/K^+ pump loci with this from MNS (Orlov et al. 2001a).

Several laboratories have demonstrated elevated production of circulating endogenous ouabain-like substances (EOLS) in volume-expanded forms of hypertension. The data on the regulation of the Na^+/K^+ pump by EOLS are outside the scope of this chapter and are considered elsewhere (De Wardener 1996; Hamlyn et al. 1996).

25.2.2 Na^+-K^+-$2Cl^-$ Cotransport

Na^+-K^+-$2Cl^-$ cotransport provides electroneutral symport of Na^+, K^+ and Cl^- (see Chap. 8). In 1982, De Mendonca et al. reported that the activity of the carrier was decreased in Na^+-loaded erythrocytes from SHR. In contrast, our group and other research teams observed that furosemide/bumetanide-sensitive ^{22}Na and ^{86}Rb fluxes were elevated in SHR, in MHS and in a substrain of SHR with severe hypertension and a predisposition to stroke (SHRSP) (for review, see Orlov et al. (1999a)). Several hypotheses have been raised to explain this discrepancy (Postnov and Orlov 1985; Orlov et al. 1999a):

1. Treatment of erythrocytes with SH-group reactive compounds used to elevate intracellular Na^+ content (De Mendonca et al. 1982), can affect Na^+-K^+-$2Cl^-$ cotransporter properties.
2. The rate of Na^+-K^+-$2Cl^-$ cotransport under baseline conditions depends not only on its maximal activity but also on intracellular content of Na^+ and Cl^- which modulate Na^+-K^+-$2Cl^-$ cotransport by a negative feedback mechanism. Several laboratories reported altered Na^+-K^+-$2Cl^-$ cotransport affinity for Na^+_i in MHS and SHR (Rosati et al. 1988; Ferrari et al. 1991). Treatment of erythrocytes in high Na-phosphate medium used for elevation of $[Na^+]_i$ also affects $[Cl^-]_i$.
3. In several studies, the values of specific intracellular ^{22}Na activity were not corrected for differential content of exchangeable sodium in erythrocytes of normotensive and hypertensive rats.
4. Garay and co-workers demonstrated that erythrocyte Na^+-K^+-$2Cl^-$ cotransport is inhibited by endogenous circulating compounds; their content is increased in hypertensive rats and is affected by salt intake (Alvarez-Guerra et al. 1998).

25.2.3 Na^+/H^+ Exchange

Similarly to nucleated cells, cytoplasm acidification as well as isosmotic or hyperosmotic shrinkage of rat erythrocytes transiently activates ^{22}Na influx. Because this increment of ^{22}Na influx was accompanied by equimolar H^+ efflux, we suggested involvement of the Na^+/H^+ exchanger (Orlov et al. 1989a,b). Two methodological approaches have been taken to investigate the activity of this carrier in erythrocytes subjected to intracellular acidification ($\Delta\mu_{H+}$-induced Na^+/H^+ exchange). We utilised amiloride known to be a non-selective inhibitor of all cloned isoforms of Na^+/H^+ exchangers and observed that the amiloride-sensitive component of H^+ efflux is increased by 40–50% in erythrocytes from SHR but is non-altered in MHS (Orlov et al. 1989a). Canessa and co-workers estimated the activity of this carrier as a difference of the initial rate of Na^+ accumulation at $pH_o = 8.0$ and 6.0. They did not observe any differences in SHR and MHS but note that this transporter was increased in SS-JR kept on a high-salt diet (Pontremoli et al. 1992). It should be underlined, however, that both methods have several limitations. First, under analysis of the H^+ efflux rate, precise estimation of the extracellular buffer capacity is complicated due to the much higher buffer capacity of the cytoplasm. This comment is especially important because extracellular buffer capacity is affected by amiloride (Orlov et al. 1998). Second, mammalian erythrocytes are highly abundant with anion exchanger that leads to rapid dissipation of $\Delta\mu_{H+}$, even in the presence of inhibitors of the carrier and carbonic anhydrase. This peculiarity complicates precise measurement of the initial rate of Na^+ and H^+ fluxes mediated by Na^+/H^+ exchange.

To further evaluate Na^+/H^+ exchange, we studied erythrocytes shrunken in hyperosmotic medium. These experiments did not reveal any differences between SHR and WKY (Orlov et al. 1987). In contrast to $\Delta\mu_{H+}$- and hyperosmotic shrinkage-induced Na^+/H^+ exchange, the increment of H^+ efflux in isosmotically-shrunken rat erythrocytes was completely abolished by 10 μM 5-(*N*-ethyl-*N*-isopropyl)-amiloride (EIPA) (unpublished data). Comparative analysis of EIPA-sensitive ion fluxes in isosmotically-shrunken erythrocytes from normotensive and hypertensive rats is currently progressing in our laboratory.

25.2.4 Other Ion Transport Pathways

Quenched activity of Na^+/H^+ exchange under baseline conditions suggests that the enhanced rate of steady-state $^{22}Na/^{23}Na$ exchange in erythrocytes from SHR is mainly caused by Na^+-K^+-$2Cl^-$ cotransport. Side-by-side with abnormalities of Na^+-K^+-$2Cl^-$ cotransport in SHR, several researchers have reported elevated values of Na^+ and K^+ fluxes measured in the presence of inhibitors of this carrier (Zicha 1993). Increased (ouabain + furosemide)-resistant Na^+ and K^+ fluxes were also observed in erythrocytes from SS-JR compared to SR-JR (Zicha and Duhm 1990). These components of ion fluxes, termed "passive permeability", can be mediated by simple diffusion of monovalent cations across the plasma membrane lipid bilayer or/and by non-identified transporters. Bernhardt with co-workers reported that (ouabain + bumetanide)-resistant ion fluxes are elevated in low ionic strength media and are partially mediated by non-selective ion channels (Kaestner et al. 1999) and the $Na^+(K^+)/H^+$ exchanger (Kummerow et al. 2000; see also Chap. 4). We observed that attenuation of ionic strength by equimolar substitution of $NaNO_3$ with sucrose decreases (ouabain + bumetanide)-resistant ^{86}Rb influx in rat erythrocytes, and the $NaNO_3$-dependent component is 3- to 4-fold higher in SHR compared to normotensive rats (Orlov et al. 1991). However, because of the relatively low concentration of ouabain used in this study (0.2 mM), the impact of ionic strength on Na^+/K^+ pump activity cannot be excluded.

25.3 Ion Transporters in Essential Hypertension

Similarly to SHR, the rate of steady-state $^{22}Na/^{23}Na$ exchange is increased in patients with essential hypertension (Postnov et al. 1977). Analysis of Na^+/K^+ pump activity did not reveal any systematic alteration (Orlov et al. 1999a). Considering this, we will focus on the functional properties of Na^+-K^+-$2Cl^-$ cotransport, Na^+/Li^+ countertransport and Na^+/H^+ exchange.

25.3.1 Na^+-K^+-$2Cl^-$ Cotransport

In studies on limited number of patients with essential hypertension, we observed that the rate of bumetanide-sensitive ^{86}Rb influx in erythrocytes was increased by ~40% (Postnov et al. 1988). However, only slight and statistically non-significant elevation of Na^+-K^+-$2Cl^-$ cotransport activity was demonstrated in our subsequent investigation performed on 40 patients with essential hypertension (Orlov et al. 1989a). A modest increment of diuretic-sensitive ion fluxes in erythrocytes from patients with essential hypertension has been reported also by several research teams (Adragna et al. 1982; Tuck et al. 1984, 1987; Imagawa and Arakawa 1986; Cusi et al. 1991; Cacciafesta et al. 1994), whereas other groups have not found any differences (Smith et al. 1984; Delva et al. 1985) or even reported decreased activity of this carrier (Garay et al. 1980; Canessa 1989). Apart from the methodological problems mentioned in Sect. 25.1.2, the features listed below should be taken into account when analysing these data. First, a decreased affinity of Na^+-K^+-$2Cl^-$ cotransport for intracellular Na^+ and the modulation of this parameter by salt intake have been noted in erythrocytes from hypertensive patients (De la Sierra et al. 1989; Canessa et al. 1991). Second, in contrast to inbred rat strains, humans possess wide variability in Na^+-K^+-$2Cl^-$ cotransport activity. Several research groups reported that plasma lipids, salt intake, and plasma renin activity contribute to this variability (for more details, see Zicha et al. (1999)). Third, gender and race affect erythrocyte ion transporters. Thus, in French-Canadians with essential hypertension, the rate of bumetanide-sensitive ^{86}Rb influx is 40% higher in men compared to age-, body mass index- and plasma lipid-matched women (344 ± 21 and 243 ± 20 $\mu mol/(l_{cells} \cdot h)$, respectively, $p < 0.0009$) (Orlov et al. 1999b). In contrast to Caucasians, we did not observe any impact of gender on bumetanide-sensitive ^{86}Rb influx in erythrocytes of African-Americans with essential hypertension while the mean value of Na^+-K^+-$2Cl^-$ cotransport activity was decreased by 40–60% compared to Caucasians (Orlov et al. 2001b). Fourth, essential hypertension is clearly a polygenic disease (Hamet al. 1998). The set of mutated genes can be different in different individuals, which is consistent with the mosaic hypothesis on the pathogenesis of essential hypertension originally proposed by Page (1967). Thus, only some patients with essential hypertension can share abnormalities of ion transporters observed in different experimental models of this disease. Indeed, in contrast to a unimodal distribution of erythrocyte Na^+-K^+-$2Cl^-$ cotransporter among normotensive Caucasians, distribution is bimodal in essential hypertension, and only 20% of patients with low plasma renin activity present an elevated activity of the carrier (Cusi et al. 1991; Cacciafesta et al. 1994).

25.3.2 Na^+/Li^+ Countertransport *vs.* Na^+/H^+ Exchange

Canessa et al. (1980) showed that the activity of Na^+/Li^+ countertransport measured as a rate of Na^+_o-dependent Li^+ efflux, was increased by 30–50% in patients with essential hypertension. With the exception of a few publications, these results were confirmed in numerous laboratories and are reviewed elsewhere (Huot and Aronson 1991; Hardman and Lant 1996). The lack of specific inhibitors of erythrocyte Na^+/Li^+ countertransport complicates the analysis of its physiological significance as well as cloning of cDNA encoding this carrier. It was proposed, however, that Na^+/Li^+ countertransport is a functional mode of operation of the ubiquitous isoform of Na^+/H^+ exchanger, NHE1 (Aronson 1982). Later on, several researchers, including our own team, observed an elevated rate of $\Delta\mu_{H+}$-induced Na^+/H^+ exchange in erythrocytes from patients with essential hypertension measured as indicated in Sect. 25.1.3 (Postnov et al. 1988; Orlov et al. 1989a; Alonso et al. 1993; Delva et al. 1993, 1996). However, detailed analysis contradicted this attractive hypothesis. Indeed, we observed that the ratio of absolute values of Na^+/Li^+ countertransport and $\Delta\mu_{H+}$-induced Na^+/H^+ exchange in human, rat and rabbit erythrocytes are different by 2 orders of magnitude (10:1:170 and 1:5:1, respectively). At 1 mM, amiloride inhibits $\Delta\mu_{H+}$-induced Na^+/H^+ exchanger in human erythrocytes but does not affect Na^+/Li^+ countertransport. And, finally, ATP depletion inhibits $\Delta\mu_{H+}$-induced Na^+/H^+ exchange but does not affect Na^+/Li^+ countertransport (Orlov et al. 1998).

25.4 Are Abnormalities of Red Cell Ion Transporters in Primary Hypertension Genetically Determined?

The positive answer to this question is supported by several observations. First, abnormal monovalent ion transport was demonstrated in young SHR and MHS before the development of chronic hypertension (Postnov et al. 1976; Bianchi et al. 1985) as well as in young offspring of patients with essential hypertension (Adragna et al. 1982; Woods et al. 1982). Second, abnormalities of ion transport observed in SHR erythrocytes are absent in experimental models of secondary hypertension such as Goldblatt and deoxycorticosterone acetate (DOCA)/salt-induced hypertension (Orlov and Postnov 1981) as well as in patients with secondary (symptomatic) forms of hypertension (Liusov et al. 1983; Imagawa and Arakawa 1986). Third, in an elegant study by Bianchi et al. (1985), (MNS×MHS) F_1 hybrids were subjected to X-ray irradiation and bone marrow transplantation from parental strains. Despite of the same extracellular milieu, Na^+-K^+-$2Cl^-$ cotransport in erythrocytes from F_1 hybrids with bone marrow from MHS and MNS was the same as in donor strains. Fourth, in the Caucasian twin study, the heritability estimate (h^2) for the V_{max} of Na^+-K^+-$2Cl^-$ cotransport was 0.84 (Lewiier and Canessa 1985). In 86 Italian families, erythrocyte Na^+-K^+-$2Cl^-$ cotransport was not correlated between spouses sharing the same environment, but there was a correlation between the average value of two parents and their offspring ($h^2 \sim 50\%$) (Cusi et al. 1991). Similar results were obtained in a Caucasian population from Boston (Canessa et al. 1986). Correlation analysis in French-Canadian families showed that sibling resemblance of Na^+/Li^+ countertransport and Na^+-K^+-$2Cl^-$ cotransport was higher in men than in women (Orlov et al. 1999b).

25.5 Role of Ion Transporters Expressed in Red Cells in Blood Pressure Regulation

A priori, erythrocytes could be involved in blood pressure regulation via modulation of their rheological properties. However, there is no evidence that blood microviscosity is affected under modulation of erythrocyte ion transport. Moreover, it has been shown that transplantation of bone marrow from MHS to (MNS×MHS) F_1 hybrids alters the activity of erythrocyte ion transporter but does not affect blood pressure in the recipients (Bianchi et al. 1985).

In the early 1980s, we hypothesised that erythrocytes can be viewed as a marker of widespread ion transport abnormalities in primary hypertension, including organs known to be major determinants of long-term maintenance of elevated blood pressure (Postnov and Orlov 1984, 1985). Considering this, we assumed that ion transport abnormalities are a consequence of mutations within genes whose functional manifestation is independent of a tissue-specific intracellular milieu. During the last 2 decades, this hypothesis has been examined extensively. These studies show that abnormal activity of Na^+-K^+-$2Cl^-$ cotransporter and Na^+/H^+ exchanger can be involved in long-term maintenance of elevated blood pressure via its functional manifestation in renal epithelial cells (Fig. 25.1) and smooth muscle cells (Fig. 25.2) (for more details, see Davis et al. (1993); Golovina and Blaustein (1997); Orlov et al. (1999a)).

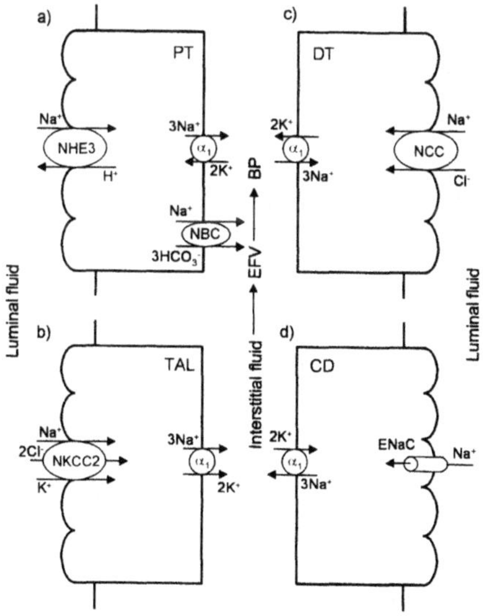

Fig. 25.1. Major ion transporters of the proximal tubule (**a**), thick ascending limb of Henle's loop (**b**), distal tubule (**c**) and collecting duct (**d**) involved in the regulation of extracellular fluid volume (*EFV*) and blood pressure (*BP*). *α1* ubiquitous isoform of the Na^+/K^+ pump; *NBC* Na^+-HCO_3^- cotransporter; *NHE3* renal-specific isoform of the Na^+/H^+ exchanger; *NKCC2* renal-specific isoform of the Na^+-K^+-$2Cl^-$ cotransporter; *NCC* Na^+-Cl^- cotransporter; *ENaC* epithelial Na^+ channels

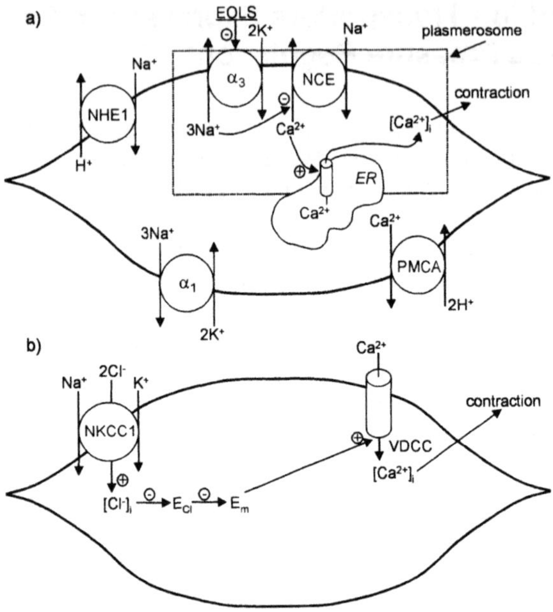

Fig. 25.2. Mechanisms of involvement of the Na^+/Ca^{2+} exchanger (*NCE* - **a**) and Na^+-K^+-$2Cl^-$ cotransporter (*NKCC1* - **b**) in the regulation of vascular smooth muscle cell function. *NHE1* - Na^+/H^+ exchanger; *α1* and *α3* isoforms of the Na^+/K^+ pump; *PMCA* plasma membrane Ca^{2+} pump; *ER* endoplasmic reticulum; E_{Cl} and E_m equilibrium chloride and plasma membrane potential, respectively; *VDCC* voltage-dependent Ca^{2+} channels

25.6 Evidence for the Involvement of Ion Transport Abnormalities in the Pathogenesis of Hypertension

The key role of ion transporters in the long-term maintenance of elevated blood pressure is supported by studies performed in Mendelian (monogenic) forms of hypertension. With an elegant combination of genetic analysis of selected pedigrees, molecular biology and cell physiology approaches, Lifton with co-workers discovered single point mutations in the β- and γ-subunits of ENaC in patients with Liddle's syndrome, leading to sustained activation of this transporter expressed in kidney (Fig. 25.1d) (Shimkets and Lifton 1996). This research team also demonstrated that Mendelian forms of hypotension are caused by mutations leading to partial inactivation of renal the Na^+-K^+-$2Cl^-$ cotransporter NKCC2 (Bartter's syndrome) and Na^+-Cl^- cotransporter (Gitelman's syndrome) (Simon and Lifton 1996) (Fig. 25.1b and c).

The first evidence implicating the ubiquitous isoforms of ion transporters in the pathogenesis of primary hypertension was obtained in F_2 hybrids of hypertensive and normotensive rats. These studies demonstrated that erythrocyte Na^+-K^+-$2Cl^-$ cotransport correlates positively with blood pressure ($r = 0.520$, $p = 0.0009$ and $r = 0.509$, $p = 0.0004$ for MHS×MNS and SHR×WKY F_2 hybrids, respectively) (Bianchi et al. 1985; Kotelevtsev et al. 1987). To further assess the role of the ubiquitous isoform of Na^+-K^+-$2Cl^-$ cotransporter, mutant mice with disrupted NKCC1 gene were produced. As predicted, ion fluxes mediated by Na^+-K^+-$2Cl^-$ cotransport were absent in erythrocyte (Delpire et al. 1999) as well as in basal membrane of gastrointestinal epithelium (Flagella et al. 1999) from homozygous $NKCC1^{-/-}$ mice. Mean blood pressure measured in anaesthetised $NKCC1^{+/+}$, $NKCC1^{+/-}$ and $NKCC1^{-/-}$ mice was 92, 78 and 69 mmHg, respectively (Flagella et al. 1999). Viewing these results collectively, it may be assumed that increased NKCC1 activity in VSMC (Fig. 25.2b) contributes to the maintenance of elevated blood pressure.

In contrast to Na^+-K^+-$2Cl^-$ cotransport, a slight correlation was observed between erythrocyte $\Delta\mu_{H+}$-induced Na^+/H^+ exchange and blood pressure in SHR×WKY F_2 hybrids ($r = 0.196$, $p = 0.06$) (Orlov et al. 1991) that probably reflects methodological difficulties in the quantification of this carrier considered in Sect. 25.1.3. It was reported that NHE1 overexpression in transgenic mice attenuates urinary salt and water excretion and salt-induced increment of blood pressure (Kuro-o et al. 1995). However, in contrast to NHE3 (Fig. 25.1a), mechanisms of the regulation of kidney function by NHE1 remain unknown.

Analysis performed in F_2 hybrids showed that quantitative trait loci (QTL) on chromosome 2 containing the gene encoding α1-Na^+, K^+-ATPase contributes to the 20–30 mmHg increment of blood pressure in SS-JR rats subjected to a high-salt diet (Deng et al. 1994). However, further studies in transgenic rats containing α1-Na^+, K^+-ATPase gene from MNS did not confirm its involvement in salt-induced hypertension (Orlov et al. 2001a).

25.7 Molecular Determinants of Abnormal Ion Transport in Hypertension

Abnormal activity of ion transporters in primary hypertension can be caused by: (i) altered content of their functional units in the plasma membrane, (ii) mutations within genes encoding these transporters, or (iii) mutations within genes encoding their regulators. Both Northern and Western blot analysis in SHR and patients with essential hypertension demonstrated the lack of systematic alteration of the total content of the ubiquitous isoforms of ion transporters studied so far (for review, see Orlov et al. (1999a)). Linkage analysis performed with F_2 hybrids of normotensive and hypertensive rats and their recombinant inbred strains revealed more than 30 QTL for elevated blood pressure and markers for cardiovascular and renal complications of chronic hypertension (Hamet et al. 1998). At least 10 of them contain genes encoding ion transporters, including α1-Na^+, K^+-ATPase, NHE1 and NKCC1 (Orlov et al. 1999a). Herrera and Ruiz-Opazo (1990) reported that adenine-1079 to thymidine transversion in α1-Na^+, K^+-ATPase cDNA derived from SS-JR kidney leads to G276L substitution and attenuation of ouabain-sensitive ^{86}Rb influx in injected *X. laevis* oocytes. However, these results have not been confirmed in other laboratories (Simonet et al. 1991; Harris et al. 2000). To search mutations within NHE1, we isolated mRNA from smooth muscle cells.

Neither single-stranded conformational polymorphism analysis nor sequencing of the 456–2421 fragment revealed any mutations in the coding region of NHE1 cDNA from SHR and SHRSP (Orlov et al. 2000). Thus, mutations within genes encoding regulators rather than ion transporters themselves should be considered as the most feasible source of the abnormal ion transport in primary hypertension. Data on the most potent regulators of NKCC1 and NHE1 are considered below:

Membrane lipids. Abnormal structure of the erythrocyte membrane lipid bilayer in primary hypertension was detected with hydrophobic fluorescent dyes (Montenay-Garestier et al. 1981; Orlov and Postnov 1982; Orlov et al. 1982) as well as by ESR spectroscopy (Gulak et al. 1986). Numerous research teams observed correlation between erythrocyte membrane microviscosity and ion transporter activity (Zicha et al. 1999). However, neither molecular determinants of abnormal membrane structure nor its causal relationship with NKCC1 and NHE1 has been established.

Cytoskeleton. The first evidence of altered erythrocyte cytoskeleton organisation in SHR came from comparison of its annealing by scanning microcalorimetry (Gulak et al. 1984). Modulation of the cytoskeleton network under preparation of erythrocyte inside-out vesicles abolished differences in Na^+-K^+-$2Cl^-$ cotransport revealed in intact cells from MHS and MNS erythrocytes (Ferrari et al. 1991, 1992). In erythrocytes, the cytoskeleton network is formed by heterodimers of α- and β-spectrin and actin bundles that are assembled by means of a set of anchor proteins, including α-, β- and γ-adducin. A few years ago, an immunochemical difference of erythrocyte adducin between MHS and MNS led to subsequent identification of missense mutations in cDNA encoding all 3 isoforms of this protein (Bianchi et al. 1994; Tripodi et al. 1996, 1997). It was shown that transfection of epithelial cells with mutated isoforms of adducin evokes a 40% increase in the V_{max} of the α1-Na^+, K^+-ATPase (Tripodi et al. 1996). However, nothing is known about the role of these mutations in the regulation of NKCC1 and NHE1. Moreover, we did not observe any effect of annealing-induced disruption of spectrin network on the baseline activity of Na^+-K^+-$2Cl^-$ cotransport and the $\Delta\mu_{H+}$-induced Na^+/H^+ exchanger in rat erythrocytes (Orlov et al. 1993).

Cell volume. The hypothesis on the role of cell volume in the altered activity of erythrocyte ion transporters in primary hypertension is based on 3 major findings. First, the shrinkage of rat erythrocytes as well as smooth muscle cells from the rat aorta is accompanied by activation of Na^+-K^+-$2Cl^-$ cotransport and Na^+/H^+ exchange (Orlov et al. 1989b, 1992). Second, erythrocyte volume is slightly decreased in SHR (Sen et al. 1972), MHS (Bianchi et al. 1985) and patients with essential hypertension (Postnov et al. 1988). Third, cell swelling triggered by hyposmotic medium abolished the differences in volume and Na^+-K^+-$2Cl^-$ cotransport between MHS and MNS erythrocytes (Ferrari et al. 1987). It should be underlined, however, that in contrast to rat erythrocytes, neither Na^+-K^+-$2Cl^-$ cotransport nor Na^+/H^+ exchange is affected by hyperosmotic shrinkage in human red blood cells (Orlov et al. 1989b). We failed to observe any effect of cell volume perturbation on Na^+/Li^+ countertransport (Orlov et al. 1998).

Protein kinase C. In human erythrocytes, activation of protein kinase C with 4β-phorbol 12-myristate 13-acetate (PMA) leads to modest cell shrinkage and activation of Na^+-K^+-$2Cl^-$ cotransport and the $\Delta\mu_{H+}$-induced Na^+/H^+ exchanger (Postnov et al. 1988). However, PMA does not affect these parameters in rat erythrocytes (unpublished data) as well as Na^+/Li^+ countertransport in human cells (Postnov et al. 1988).

25.8 Conclusion and Future Directions

Data summarised in this chapter show that altered maximal activity and affinity for Na^+_i of the Na^+-K^+-$2Cl^-$ cotransporter as well as ion fluxes mediated by the lipid bilayer or non-identified transporters contribute to the enhanced permeability of erythrocytes in experimental and clinical forms of primary hypertension. In patients with essential hypertension, enhanced permeability of erythrocytes for Na^+ is also mediated by Na^+/Li^+ countertransport. Both spontaneously hypertensive rats and patients with essential hypertension show augmented activity of the Na^+/H^+ exchanger. However, this carrier is quenched in erythrocytes under baseline conditions and its relationship with Na^+/Li^+ countertransport has not been proven. Abnormal activity of erythrocyte ion transporters in primary hypertension is genetically determined and is not a consequence of chronic elevation of blood pressure. This hypothesis is supported by co-segregation of Na^+-K^+-$2Cl^-$ cotransport with blood pressure in F_2 hybrids of spontaneously hypertensive and normotensive rats. The possible co-segregation of ion transporters with the markers of cardiovascular and renal complications of this disease has not been studied yet.

In contrast to Mendelian forms of secondary hypertension and hypotension caused by single point mutations within genes encoding renal specific isoforms of monovalent ion transporters, the molecular basis of abnormal ion transport in primary hypertension remains unknown. Data obtained during the last few years suggest that abnormal activity of ion carriers in primary hypertension is caused by mutations within genes encoding their intracellular regulators. Altered organisation of the lipid bilayer and cytoskeleton network, decreased cell volume and augmented activity of protein kinase C have been proposed as major determinants of abnormal activity of monovalent ion transporters in primary hypertension. It should be underlined, however, that dozen of genes contribute to the maintenance of membrane/cytoskeleton interaction and cell volume adjustment. As an alternative approach, we proposed positional cloning within QTL for the abnormal activity of ion carriers. Because of completion of the human genome project, this approach seems to be viable for affected sib-pairs from populations of a homogenous ethnic origin and is currently progressing in our laboratory.

Acknowledgement: I would like to acknowledge Prof. Y.V. Postnov for my 15-year training in the field of pathophysiology of hypertension and Prof. P. Hamet who urged me to apply genetic approaches in these studies. The preparation of this manuscript was supported by a grant from the Kidney Foundation of Canada. The technical assistance of Monique Poirier, the editorial help of Ovid Da Silva and the secretarial skills of Ginette Dignard are appreciated.

References

Adragna NC, Canessa ML, Solomon H, Slater E, Tosteson DC (1982) Red cell lithium-sodium countertransport and sodium-potassium cotransport in patients with essential hypertension. Hypertension 4:795-804

Alonso A, Arrazola A, Garciandia A, Esparza N, Gomez-Alamillo C, Diez J (1993) Erythrocyte anion exchanger activity and intracellular pH in essential hypertension. Hypertension 22:348-356

Alvarez-Guerra M, Nazaret C, Garay RP (1998) The erythrocyte Na,K,Cl cotransporter and its circulating inhibitor in Dahl salt-sensitive rats. J Hypertens 16:1499-1504

Aronson PS (1982) Red-cell sodium-lithium countertransport and essential hypertension. N Engl J Med 307:317

Bianchi G, Ferrari P, Trizio D, Ferrandi M, Torielli L, Barber BR, Polli E (1985) Red blood cell abnormalities and spontaneous hypertension in the rat. A genetically determined link. Hypertension 7:319-325

Bianchi G, Tripodi G, Casari G, Salardi S, Barber BR, Garcia R, Leoni P, Torielli L, Cusi D, Ferrandi M (1994) Two point mutations within the adducin genes are involved in blood pressure variation. Proc Natl Acad Sci USA 91:3999-4003

Cacciafesta M, Ferri C, Carlomagno A, De Angelis C, Scuteri A, Guidoni L, Luciani AM, Rosi A, Viti V, Santucci A (1994) Erythrocyte Na-K-Cl cotransport activity in low renin essential hypertensive patients. A ^{23}Na nuclear magnetic resonance study. Am J Hypertens 7:151-158

Canessa ML (1989) The Na-K-Cl cotransport in essential hypertension: cellular functions and genetic environment interactions. Int J Cardiol 25, suppl 1:S37-S45

Canessa M, Adragna N, Solomon HS, Connolly TM, Tosteson DC (1980) Increased sodium-lithium counter-transport in red cells of patients with essential hypertension. N Engl J Med 302:772-776

Canessa M, Bize I, Spalvins A, Falkner B, Katz S (1986) Na-K-C1 cotransport and Na pump in red cells of young blacks and blood pressure response to salt loading. J Clin Hypertens 2:101-108

Canessa M, Morgan K, Goldszer R, Moore TJ, Spalvins A (1991) Kinetics abnormalities of the red blood cell sodium-proton exchange in hypertensive patients. Hypertension 17:340-348

Canessa M, Romero JR, Ruiz-Opazo N, Herrera VL (1993) The alpha 1 Na^+-K^+ pump of the Dahl salt-sensitive rat exhibits altered Na^+ modulation of K^+ transport in red blood cells. J Membrane Biol 134:107-122

Cusi D, Fossali E, Piazza A, Tripodi G, Barlassina C, Dozzoli E, Vezzoli G, Stella P, Soldati L, Bianchi G (1991) Heritability estimate of erythrocyte Na-K-Cl cotransport in normotensive and hypertensive families. Am J Hypertens 4:725-754

Davis JPL, Chipperfield AR, Harper AA (1993) Accumulation of intracellular chloride by (Na-K-Cl) cotransport in rat arterial muscle is enhanced in deoxycorticosterone acetate (DOCA)/salt hypertension. J Mol Cell Cardiol 25:233-237

De la Sierra, Coca A, Aguilera MT, Ingelmo M, Urbano-Marquez A (1989) Clinical profiles and erythrocyte Na^+ transport abnormalities of four major types of primary hypertension in Spain. Kidney Int 36:114-119

De Mendonca M, Knorr A, Grichois ML, Ben Ishay D, Garay RP, Meyer P (1982) Erythrocytic sodium ion transport systems in primary and secondary hypertension of the rat. Kidney Int, suppl 11:S69-S75

De Wardener HE (1996) Sodium transport inhibitors and hypertension. J Hypertens 14:S9-S18

Delpire E, Lu J, England R, Dull C, Thorne T (1999) Deafness and imbalance associated with inactivation of the secretory Na-K-2Cl co-transporter. Nat Genet 22:192-195

Delva P, De Gasperi M, Degan M, Covi G, Lechi A (1985) Kinetics of bumetanide-sensitive Na^+-K^+ co-transport in erythrocytes of essential hypertensive patients. Clin Sci 69:607-611

Delva P, Pastori C, Provoli E, Degan M, Arosio E, Montesi G, Steele A, Lechi A (1993) Erythrocyte Na^+-H^+ exchange activity in essential hypertensive and obese patients: role of excess body weight. J Hypertens 11:823-830

Delva P, Pastori C, Degan M, Montesi G, Lechi C, Steele A, Lechi A (1996) Erythrocyte Na^+-H^+ exchanger kinetics and Na^+-Li^+ countertransport activity in essential hypertensive patients. Eur J Clin Invest 26:64-70

Deng AY, Dene H, Rapp JP (1994) Mapping of a quantitative trait locus for blood pressure on rat chromosome 2. J Clin Invest 94:431-436

Ferrari P, Ferrandi M, Torielli L, Canessa M, Bianchi G (1987) Relationship between erythrocyte volume and sodium transport in the Milan hypertensive rat and age-dependent changes. J Hypertens 5:199-206

Ferrari P, Torielli L, Cirillo M, Salardi S, Bianchi G (1991) Sodium transport kinetics in erythrocytes and inside-out vesicles from Milan rats. J Hypertens 9:703-711

Ferrari P, Torielli L, Salardi S, Rizzo A, Bianchi G (1992) $Na^+/K^+/Cl^-$ cotransport in resealed ghosts from erythrocytes of the Milan hypertensive rats. Biochim Biophys Acta 1111:111-119

Flagella M, Clarke LL, Miller ML, Erway LC, Giannella RA, Andringa A, Gawenis LR, Kramer J, Duffy JJ, Doetschman T, Lorenz JN, Yamoah EN, Cardell EL, Shull GE (1999) Mice lacking the basolateral Na-K-2Cl cotransporter have impaired epithelial chloride secretion and are profoundly deaf. J Biol Chem 274:26946-26955

Garay RP, Dagher G, Pernollet MG, Devynck MA, Meyer P (1980) Inherited defect in a Na^+-K^+ cotransport system in erythrocyte from essential hypertensive patients. Nature 284:281-283

Golovina VA, Blaustein MP (1997) Spatially and functionally distinct Ca^{2+} stores in sarcoplasmic and endoplasmic reticulum. Science 275:1643-1648

Gulak PV, Orlov SN, Pokudin NI, Postnov YV, Litvinov IS, Orlov NJ, Shnyrov VL (1984) Microcalorimetry and electrophoresis of the erythrocyte membrane of spontaneously hypertensive rats. J Hypertens 2:81-84

Gulak PV, Orlov SN, Postnov YV, Kiladze IS, Grigorian GL, Tabagua MI, Kajushin LP (1986) Investigation of membrane proteins in rat erythrocytes in spontaneous hypertension by means of spin-label technique. J Hypertens 4:313-317

Hamet P, Pausova Z, Adarichev S, Adaricheva K, Tremblay J (1998) Hypertension: genes and environment. J Hypertens 16:397-418

Hamlyn JM, Hamilton BP, Manunta P (1996) Endogenous ouabain, sodium balance and blood pressure: a review and a hypothesis. J Hypertens 14:151-167

Hardman TC, Lant AF (1996) Controversies surrounding erythrocyte sodium-lithium countertransport. J Hypertens 14:695-70

Harris EL, Kelly G, Barnard R (2000) Neither GH nor SS/JR rats have a A1079T transition in the $\alpha 1Na^+$, K^+-ATPase gene (abstract). Genetics of Experimental and Human Hypertension – Symposium of the International Society of Hypertension. Toledo, Ohio, USA, pp 19-20

Herrera VLM, Ruiz-Opazo N (1990) Alteration of Na^+, K^+-ATPase ^{86}Rb influx by a single amino acid substitution. Science 249:1023-1026

Huot SJ, Aronson PS (1991) Na^+/H^+ exchanger and its role in essential hypertension and diabetes mellitus. Diabetes Care 14:521-535

Imagawa M, Arakawa K (1986) Sodium-potassium cotransport in hypertension and hypotension. Japan Heart J 27:191-204

Jones AW (1973) Altered ion transport in vascular smooth muscle from spontaneously hypertensive rats. Influence of aldosterone, norepinephrine and angiotensin. Circ Res 33:563-572

Kaestner L, Bollensdorff C, Bernhardt I (1999) Non-selective voltage-activated cation channel in the human red blood cell membrane. Biochim Biophys Acta 1417:9-15

Kotelevtsev YV, Orlov SN, Pokudin NI, Agnaev VM, Postnov YV (1987) Genetic analysis of inheritance of rates of Na,K-cotransport calcium concentration in erythrocytes and blood pressure of F_2 hybrids of spontaneously hypertensive and normotensive rats. Bull Exp Biol Med 103:518-520

Kummerow D, Hamann J, Browning JA, Wilkins R, Ellory JC, Bernhardt I (2000) Variations of intracellular pH in human erythrocytes via $K^+(Na^+)/H^+$ exchange under low ionic strength conditions. J Membrane Biol 176:207-216

Kuro-o M, Hanaoka K, Hiroi Y, Noguchi T, Fujimori Y, Takewaki S, Hayasaka M, Katoh H, Miyagishi A, Nagai R (1995) Salt-sensitive hypertension in transgenic mice overexpressing Na^+- proton exchanger. Circ Res 76:148-153

Lewiier FI, Canessa M (1985) Red cell transport studies in adult twins. Am J Hum Genet 36:172S

Liusov VA, Postnov II, Orlov SN, Riazhskii GG (1983) Differences in the rates of Na/Li-countertransport in the erythrocyte membranes of hypertensive and renal hypertensive patients. Kardiologiia 23:24-26

McCormick CP, Hennessy JF, Rauch AL, Buckalew VM Jr (1989) Erythrocyte sodium concentration and ^{86}Rb uptake in weanling Dahl rats. Am J Hypertens 2:604-609

Montenay-Garestier T, Aragon I, Devynck MA, Meyer P, Helene C (1981) Evidence for structural changes in erythrocyte membranes of spontaneously hypertension rats. A fluorescence polarization study. Biochem Biophys Res Commun 100:660-665

Orlov SN, Postnov II (1981) Erythrocyte membrane permeability of sodium in different types of chronic experimental hypertension. Kardiologiia 21:33-36

Orlov SN, Postnov YV (1982) Calcium binding and membrane fluidity in essential and renal hypertension. Clin Sci 63:281-284

Orlov SN, Gulak PV, Litvinov IS, Postnov Y (1982) Evidence of altered structure of the erythrocyte membrane in spontaneously hypertensive rats. Clin Sci 63:43-45

Orlov SN, Pokudin NI, Postnov Y (1987) Na^+/H^+ exchange in erythrocytes of spontaneously hypertensive rats. J Hypertens 5:S281-S283

Orlov SN, Pokudin NI, Kotelevtsev YV, Postnov YV (1988) Characterization of structural and functional organization of erythrocyte membrane in three strains of spontaneously hypertensive rats. Kardiologiia 28:57-63

Orlov SN, Postnov IY, Pokudin NI, Kukharenko VY, Postnov YV (1989a) Na^+-H^+ exchange and other ion-transport systems in erythrocytes of essential hypertensives and spontaneously hypertensive rats: a comparative analysis. J Hypertens 7:781-788

Orlov SN, Pokudin NI, Kotelevtsev YV, Gulak PV (1989b) Volume-dependent regulation of ion transport and membrane phosphorylation in human and rat erythrocytes. J Membrane Biol 107:105-107

Orlov SN, Petrunyaka VV, Pokudin NI, Kotelevtsev YV, Postnov YV, Kunes J, Zicha J (1991) Cation transport and adenosine triphophatase activity in rat erythrocytes: a comparison of spontaneously hypertensive rats with the normotensive Brown-Norway strain. J Hypertens 9:977-982

Orlov SN, Resink TJ, Bernhardt J, Bühler FR (1992) Volume-dependent regulation of sodium and potassium fluxes in cultured vascular smooth muscle cells: dependence on medium osmolality and regulation by signalling systems. J Membrane Biol 129:199-210

Orlov SN, Kolosova IA, Cragoe EJ, Gurlo TG, Mongin AA, Aksentsev SL, Konev SV (1993) Kinetics and peculiarities of thermal inactivation of volume-induced Na/H exchange, Na,K,2Cl cotransport and K,Cl cotransport in rat erythrocytes. Biochim Biophys Acta 1151:186-192

Orlov SN, Kuznetsov SR, Pokudin NI, Tremblay J, Hamet P (1998) Can we use erythrocytes for the study of the activity of the ubiquitous Na^+/H^+ exchanger (NHE-1) in essential hypertension? Am J Hypertens 11:774-783

Orlov SN, Adragna NC, Adarichev VA, Hamet P (1999a) Genetic and biochemical determinants of abnormal monovalent ion transport in primary hypertension. Am J Physiol 276:C511-C536

Orlov SN, Pausova Z, Gossard F, Gaudet D, Tremblay J, Kotchen T, Larochelle P, Hamet P (1999b) Sibling resemblance of erythrocyte ion transporters in French-Canadian sibpairs affected with essential hypertension. J Hypertens 17:1859-1865

Orlov SN, Adarichev VA, Devlin AM, Maximova NV, Sun YL, Tremblay J, Dominiczak AF, Postnov YV, Hamet P (2000) Increased Na^+/H^+ exchanger isoform 1 activity in spontaneously hypertensive rats: lack of mutations within the coding region of NHE1. Biochem Biophys Acta 1500:169-180

Orlov SN, Dutil J, Hamet P, Deng AY (2001a) Replacement of α1-Na-K-ATPase of Dahl rats by Milan rats lowers blood pressure but does not affect its activity. Physiol Genomics 7:171-177

Orlov SN, Pausova Z, Gossard F, Gaudet D, Tremblay J, Grim CE, Kotchen JM, Kotchen TA, Cowley AW, Larochelle P, Hamet P (2001b) Na^+,K^+,Cl^- cotransport is decreased in African American *vs.* French-Canadian hypertensives: lack of impact of gender and plasma lipids. J Hypertens 19, suppl 2:S46

Page IH (1967) The mosaic theory of arterial hypertension – its interpretation. Perspect Biol Med 10:325-333

Pontremoli R, Spalvins A, Menachery A, Torrelli L, Canessa M (1992) Red cell sodium-proton exchange is increased in Dahl salt-sensitive hypertnsive rats. Kidney Int 42:1355-1362

Postnov YV, Orlov SN (1984) Cell membrane alteration as a source of primary hypertension. J Hypertens 2:1-6

Postnov YV, Orlov SN (1985) Ion transport across plasma membrane in primary hypertension. Physiol Rev 65:904-945

Postnov YV, Orlov SN, Shevchenko AS (1975) Alteration of membrane permeability of erythrocytes in rats with spontaneous hypertension. Kardiologiia 15:88-92

Postnov YV, Orlov SN, Gulak PV, Shevchenko AS (1976) Altered permeability of the erythrocyte membrane for sodium and potassium in spontaneously hypertensive rats. Pflügers Arch 365:257-263

Postnov YV, Orlov SN, Shevchenko AS, Adler AM (1977) Altered sodium permeability, calcium binding and Na-K-ATPase activity in red blood cell membrane in essential hypertension. Pflügers Arch 371:263-270

Postnov YV, Kravtsov GM, Orlov SN, Pokudin NJ, Postnov IY, Kotelevtsev YV (1988) Effect of protein kinase C activation on cytoskeleton and cation transport in human erythrocytes. Reproduction of some membrane abnormalities revealed in essential hypertension. Hypertension 12:267-273

Rosati C, Meyer P, Garay R (1988) Sodium transport kinetics in erythrocytes from spontaneously hypertensive rats. Hypertension 11:41-48

Sen S, Hoffmann GC, Stowe NI, Smeby RR, Bumpus FM (1972) Erythrocytes in spontaneously hypertensive rats. J Clin Invest 51:710-714

Shimkets RA, Lifton RP (1996) Recent advances in the molecular genetics of hypertension. Curr Opin Nephrol Hypertens 5:162-165

Simon DB, Lifton RP (1996) The molecular basis of inherited hypokalemic alkalosis: Bartter's and Gitelman's syndromes. Am J Physiol 271:F961-F966

Simonet L, St Lezin E, Kurtz TW (1991) Sequence analysis of the alpha 1 Na^+,K^+-ATPase gene in the Dahl salt- sensitive rat. Hypertension 18:689-693

Smith JB, Ash KO, Hunt SC, Hentschel WM, Sprowell W, Dadone MM, Williams RR (1984) Three red cell sodium transport systems in hypertensive and normotensive Utah adults. Hypertension 6:159-166

Tripodi G, Valtorta F, Torielli L, Chieregatti E, Salardi S, Trusolino L, Menegon A, Ferrari P, Marchisio PC, Bianchi G (1996) Hypertension-associated point mutations in the adducin α and β subunits affect actin cytoskeleton and ion transport. J Clin Invest 97:2815-2822

Tripodi G, Szpirer C, Reina C, Szpirer J, Bianchi G (1997) Polymorphism of gamma-adducin gene in genetic hypertension and mapping of the gene to rat chromosome 1q55. Biochem Biophys Res Commun 237:685-689

Tuck ML, Gross C, Maxwell MH, Brickman AS, Krasnoshtein G, Mayes D (1984) Erythrocyte Na^+,K^+ cotransport and Na^+,K^+ pump in black and caucasian hypertensive patients. Hypertension 6:536-544

Tuck ML, Corry DB, Maxwell M, Stern N (1987) Kinetic analysis of erythrocyte Na^+-K^+ pump and cotransport in essential hypertension. Hypertension 10:204-211

Woods JW, Falk RJ, Pittman AW, Klemmer PJ, Watson BS, Namboodiri K (1982) Increased red-cell sodium-lithium countertransport in normotensive sons of hypertensive parents. N Engl J Med 306:593-595

Zicha J (1993) Red cell ion transport abnormalities in experimental hypertension. Fundam Clin Pharmacol 7:129-141

Zicha J, Duhm J (1990) Kinetics of Na^+ and K^+ transport in red blood cells of Dahl rats. Effects of age and salt. Hypertension 15:612-627

Zicha J, Kunes J, Devynck MA (1999) Abnormalities of membrane function and lipid metabolism in hypertension: a review. Am J Hypertens 12:315-331

26 Disorders of Band 3

Petr Jarolim

Department of Pathology, Brigham and Women's Hospital, Harvard Medical School, 75 Francis Street, Boston MA 02115, USA

26.1 Introduction

Band 3 is the most abundant protein of the red blood cell (RBC) membrane and plays multiple structural and functional roles. It is therefore no surprise that numerous mutations in the band 3 gene have been described and found to be associated with several erythroid phenotypes. In addition, band 3 is expressed in renal collecting ducts and a distinct group of band 3 mutations have been linked to the phenotype of distal renal tubular acidosis. In the following paragraphs, we will segregate band 3 defects according to the primarily affected tissue (erythroid *vs.* non-erythroid) and to the degree of band 3 deficiency caused by the mutations.

26.2 Erythroid Phenotypes with Partial Deficiency of Band 3

26.2.1 Autosomal Dominant Spherocytosis with Band 3 Deficiency

Most of the band 3 defects have been found in patients with a specific subset of hereditary spherocytosis (HS). In the majority of cases of HS – up to 60% of all kindred studied – the RBC membranes are deficient in α and β spectrins, the main components of the membrane skeleton, or in ankyrin, which links the spectrin–based membrane skeleton to the membrane via interaction with both spectrin and band 3. However, a distinct subset of patients with HS have a deficiency in band 3 of up to 35% as detected by SDS-PAGE (Jarolim et al. 1992a; Savvides et al. 1993). Patients with this subset of HS generally have a mild to moderate clinical course and almost always have characteristic pincered cells in the peripheral blood smear prior to splenectomy (Jarolim et al. 1996). More than one half of these mutations are nonsense and frameshift mutations; the rest are missense mutations that primarily affect highly conserved amino acids (Table 26.1) (Jarolim et al. 1994a, 1995). The mutations are scattered throughout the band 3 molecule.

Table 26.1. Band 3 mutations in autosomal dominant hereditary spherocytosis

Band 3	Mutation			Reference
Foggia	deletion	54-55→frameshift	-C	Delgiudice et al. 1997
Kagoshima	deletion	56→frameshift	-A	Kanzaki et al. 1997
Hodonin	nonsense	81Trp→Stop	TGG→TGA	Jarolim et al. 1996
Cape Town	missense	90Glu→Lys	GAG→AAG	Bracher et al. 2001
Napoli I	insertion	99-100→frameshift	+T	Delgiudice et al. 1997
Fukayama I	deletion	112-113→frameshift	-AC	Kanzaki et al. 1997
Nachod	splicing	117-121GTVLL deleted	-3, C→A (ivs 5)	Jarolim et al. 1996
Fukuoka	missense	130Gly→Arg	GGA→AGA	Kanzaki et al. 1997
Osnabruck I, Lyon	nonsense	150Arg→Stop	CGA→TGA	Eber et al. 1996; Alloisio et al. 1996
Worcester	insertion	170-172→frameshift	+G	Jarolim et al. 1996
Fukuyama II	insertion	183→frameshift	+A	Kanzaki et al. 1997
Campinas	splicing	203→frameshift	+1, G-T (ivs 8)	Lima et al. 1997
Bohain	deletion	241→frameshift	-T	Dhermy et al. 1997
Princeton	insertion	273-275→frameshift	+C	Jarolim et al. 1996
Boston	missense	285Ala→Asp	GCT→GAT	Jarolim et al. 1996
Noirterre	nonsense	330Gln→Stop	CAG→TAG	Jenkins et al. 1996
Bruggen	deletion	419→frameshift	-C	Eber et al. 1996
Benesov	missense	455Gly→Glu	GGG→GAG	Jarolim et al. 1996
Bicêtre II	deletion	456→frameshift	-G	Dhermy et al. 1997
Pribram	splicing	477→frameshift	-1, G→A (ivs 12)	Jarolim et al. 1996
Coimbra	missense	488Val→Met	GTG→ATG	Alloisio et al. 1997
Bicêtre I	missense	490Arg→Cys	CGC→TGC	Dhermy et al. 1997
Evry	deletion	492→frameshift	-T	Dhermy et al. 1997
Milano	duplication	498 23AA insertion	69bp duplication	Bianchi et al. 1997
Dresden	missense	518Arg→Cys	CGC→TGC	Eber et al. 1996
Smichov	deletion	616→frameshift	-C	Jarolim et al. 1996
Trutnov	nonsense	628Tyr→Stop	TAC→TAA	Jarolim et al. 1996
Hobart	deletion	646-647→frameshift	-G	Jarolim et al. 1996
Osnabruck II	deletion	663-664→frameshift	-ATG	Eber et al. 1996
Most	missense	707Leu→Pro	CTG→CCG	Jarolim et al. 1996
Okinawa	missense	714Gly→Arg	GGG→AGG	Kanzaki et al. 1997
Prague II	missense	760Arg→Gln	CGG→CAG	Jarolim et al. 1995
Hradec Kralove	missense	760Arg→Trp	CGG→TGG	Jarolim et al. 1995
Chur	missense	771Gly→Asp	GGC→GAC	Maillet et al. 1995
Napoli II	missense	783Ile→Asn	ATC→AAC	Delgiudice et al. 1997
Jablonec	missense	808Arg→Cys	CGC→TGC	Jarolim et al. 1995
Nara	missense	808Arg→His	CGC→CAC	Kanzaki et al. 1997
Prague I	insertion	822→frameshift	+CACCCAGATC	Jarolim et al. 1994a
Birmingham	missense	834His→Pro	CAC→CCC	Jarolim et al. 1996
Philadelphia	missense	837Thr→Met	ACG→ATG	Jarolim et al. 1996
Tokyo	missense	837→Ala	ACG→GCG	Iwase et al. 1998
Prague III	missense	870Arg→Trp	CGG→TGG	Jarolim et al. 1995
Vesuvio	deletion	984→frameshift	-C	Perrotta et al. 1999

Characterization of multiple mutations leading to autosomal dominant HS with band 3 deficiency clarified the pathogenesis of one subset of HS and pointed to several amino acids critical for co-translational insertion of band 3 into the plasma membrane. We studied the expression of the mutant band 3 allele in spherocytes from two unrelated patients and found that in both cases the mutant allele was absent from the membrane (Jarolim et al. 1995). In a separate study, we found that overexpresssion of band 3 Prague in *Xenopus* oocytes reduced the accumulation of wild type band 3 at the oocytes surface. This inhibition occurred in the absence of detectable heteromer formation between band 3 Prague and wild type band 3. In addition, co-injection of mutant and wild type band 3 cRNAs led to dose-dependent inhibition of Cl^- uptake by the oocytes (Chernova et al. 1995).

Using HEK 293 cells, Quilty and Reithmeier (2000) transiently expressed seven missense HS mutations (L707P, R760Q, R760W, R808C, H834P, T837M, and R870W) constructed in full-length AE1 cDNA. Using immunofluorescence, cell surface biotinylation, and pulse-chase labeling, they demonstrated that all mutants exhibited defective cellular trafficking from the endoplasmic reticulum to the plasma membrane, most likely due to misfolding of the mutant protein.

26.2.2 Autosomal Dominant Spherocytosis with Protein 4.2 Deficiency

Two mutations in one band 3 cDNA allele, K56E and P327R, were detected in a patient with spherocytic hemolytic anemia whose erythrocytes had a partial (29% ± 5%) deficiency of protein 4.2, a major RBC protein that interacts with band 3 and ankyrin. Binding of normal ankyrin to patient's inside-out vesicles was normal, whereas the predicted maximal binding capacity of patient's inside-out vesicles for band 4.2 was decreased. Since the K56E substitution underlies the asymptomatic band 3 Memphis polymorphisms, either the P327R substitution itself or in combination with the band 3 Memphis polymorphism causes the abnormal binding of protein 4.2 to band 3 and results in the spherocytic phenotype (Jarolim et al. 1992b).

26.2.3 Autosomal Recessive Spherocytosis with Protein 4.2 Deficiency

Marked deficiency of protein 4.2 (12% of normal) and spherocytic hemolytic anemia was found to be associated with a homozygous abnormality in the cytoplasmic domain of band 3. Sequencing revealed homozygosity for substitution E40K. The authors concluded that the region of band 3 containing Glu40 is involved in interactions with protein 4.2 (Rybicki et al. 1993).

26.3 Erythroid Phenotypes with Complete Absence of Band 3

26.3.1 Severe Human Autosomal Recessive Hemolytic Anemia

Ribeiro et al. (2000) recently described an absence of band 3 associated with the mutation Coimbra (V488M) in the homozygous state and with severe hereditary spherocytosis in an infant. A hydropic, severely anemic baby was delivered by cesarean section. The baby was resuscitated and initially kept alive with respiratory assistance and hypertransfusion therapy. Cord blood smears revealed erythroblastosis, poikilocytosis, and RBCs with stalk-like elongations. Band 3 and protein 4.2 were absent; spectrin, ankyrin, and glycophorin A were significantly reduced. Renal tubular acidosis was detected in the infant at 3 months of age; nephrocalcinosis appeared shortly thereafter. The child was maintained alive by regular blood transfusions and daily bicarbonate supplements. The mechanism by which the V488M mutation prevents incorporation of band 3 into the RBC membrane remains to be elucidated.

26.3.2 Mouse Band 3 Mutants and Knockouts

The phenotype of the human V488M homozygote is reminiscent of that of mice with a complete knockout of band 3 described earlier by two groups (Peters et al. 1996; Southgate et al. 1996). RBCs lacking band 3 spontaneously shed membrane vesicles and tubules, leading to severe spherocytosis and hemolysis, but the levels of the major components of the red cell membrane skeleton, the synthesis of spectrin in mutant erythroblasts, and the skeletal architecture were normal or nearly normal. The authors concluded that band 3 does not regulate the assembly of the RBC membrane skeleton *in vivo* but is essential for membrane stability. They postulated that stabilization is achieved through band 3-lipid interactions and that loss of these interactions is a key pathogenic event in hereditary spherocytosis. Interestingly, band 3 knockout led to the complete absence of glycophorin A from the erythrocyte membrane (Hassoun et al. 1998a).

Only 5% to 10% of the band 3 null mice survive the neonatal period. Hassoun et al. (1998b) thus studied the cause of death in these animals. Histopathologic examination of the mice revealed evidence of thrombosis, including large thrombotic lesions in the heart, subcapsular necrotic areas in the liver, and large vein thrombi. Increased external exposure of phosphatidylserine was found in the band 3 null erythrocytes. The authors concluded that the high mortality among band 3 null mice may be related to a hypercoagulable state, which appears to originate from changes in the phospholipid composition of the membrane that lead to exposure of phosphatidylserine in the outer leaflet. These findings further support the notion that band 3 stabilizes the RBC membrane by interacting with surrounding lipid molecules.

A spontaneous mutation in inbred mouse, *wan*, in the band 3 gene results in a premature stop codon at amino acid 45. Band 3 is absent in homozygous *wan* erythrocytes, producing a phenotype of severe hereditary spherocytosis. In a striking contrast to the two targeted band 3 gene disruptions, all homozygous *wan* mice die within 72 hours of birth. Peters et al. (2001) described a quantitative trait locus for mean cell volume on mouse chromosome 12 that significantly modifies the band 3 null HS phenotype and identified erythroid β spectrin as a candidate gene.

26.3.3 Severe Spherocytic Hemolytic Anemia in a Steer

Inaba and coworkers (1996) studied a steer whose RBCs completely lacked band 3. The animal was homozygous for a nonsense mutation in both band 3 alleles. RBCs were deficient in spectrin, ankyrin, protein 4.2, and actin, resulting in a distorted and disrupted membrane skeletal network with decreased density. The RBC membranes, like those of the band 3 knockout mice, were extremely unstable and showed the loss of surface area in several distinct ways, such as invagination, vesiculation, and extrusion of microvesicles, leading to the formation of spherocytes. The animal was moderately acidotic. These findings confirmed the essential role of band 3 for membrane stability and acid-base balance. However, the AE1-null steer appeared to be somewhat less severely affected (Inaba et al. 1996) when compared to humans (Ribeiro et al. 2000) and mice (Peters et al. 1996; Southgate et al. 1996) with band 3 null phenotypes.

26.3.4 Phenotype of Congenital Dyserythropoietic Anemia in Zebrafish

Using a chemical mutagenesis screen, Paw et al. (2000) studied mutations associated with abnormal hematopoietic development of zebrafish. Mutant *retsina* fish had normal embryonic erythropoiesis until 72 hours post-fertilization, when they developed a profound anemia with differentiation arrest at the proerythroblast stage with binuclearity. Positional cloning identified association of *retsina* with mutations in the band 3 gene, and sequencing revealed two inactivating band 3 gene mutations in each of the two *retsina* alleles. Immunofluorescent staining of normal zebrafish erythroid cells showed exclusion of band 3 from the cleavage furrow and enrichment around the polar regions, suggesting that band 3 is unlikely to be directly involved in the formation of the contractile ring. In mutant erythroid cells, although the cleavage furrow could initiate properly, further ingression was inhibited by chromosome masses that failed to segregate to the poles. The authors concluded that band 3 might play a critical role in spindle microtubule-cell cortex interaction important in anaphase spindle elongation and cellular division in developing proerythroblasts.

Paw et al. (2000) also analyzed erythroblasts from the murine band 3 null mutants and found similar binuclearity, although at a lower frequency than that observed in zebrafish. Interestingly, although glycosylation of band 3 is markedly decreased in a similar human phenotype, the congenital dyserythropoietic anemia type II (HEMPAS), no mutations of band 3 gene have been described in this disorder and the content of band 3 was normal or only moderately decreased (Jarolim et al. 1994b). Decreased anion transport activity of band 3 and formation of band 3 clusters in HEMPAS were also reported (De Franceschi et al. 1998).

26.4 Other Erythroid Phenotypes Caused by Band 3 Mutations

26.4.1 Southeast Asian Ovalocytosis

Southeast Asian ovalocytosis (SAO) is characterized by oval RBCs often marked by the presence of one or two transverse ridges or a longitudinal slit. The condition is widespread in certain ethnic groups of Malaysia, Papua New Guinea, the Philippines, and Indonesia (Lie-Injo 1965; Amato and Booth 1977; Serjeantson et al. 1977; Holt et al. 1981; Fix et al. 1982). The early studies of ovalocyte properties reported numerous functional abnormalities, such as increased RBC rigidity (Mohandas et al. 1984; Saul et al. 1984), decreased osmotic fragility (Honig et al. 1971), increased thermal stability (Castelino et al. 1981; Kidson et al. 1981), resistance to shape change by echinocytogenic agents (Castelino et al. 1981; Saul et al. 1984; Liu et al. 1990), and a reduced expression of many RBC antigens (Booth et al. 1977). The most remarkable feature of ovalocytes was their resistance to *in vitro* invasion by several strains of malarial parasites, including *Plasmodium falciparum* and *Plasmodium knowlesi* (Castelino et al. 1981; Kidson et al. 1981; Hadley et al. 1983). It also was reported that in areas of endemic malaria the RBCs of subjects with SAO contained reduced numbers of intracellular parasites *in vivo* (Cattani et al. 1987).

The finding of a tight linkage of the abnormal proteolytic digest of erythrocyte band 3 protein to the SAO phenotype (Liu et al. 1990) led to the identification of the underlying mutation. All carriers of the SAO phenotype are heterozygotes with one normal band 3 allele and the other containing two mutations in *cis*: (i) deletion of 27 nucleotides encoding amino acids 400–408 at the boundary of the cytoplasmic and membrane domains of band 3 and (ii) substitution K56E (Jarolim et al. 1991; Mohandas et al. 1992; Schofield et al. 1992a). The K56E substitution corresponds to an asymptomatic band 3 protein polymorphism known as band 3 Memphis (Mueller and Morrison 1977; Yannoukakos et al. 1991; Jarolim et al. 1992c). Subsequent studies associated the SAO phenotype with a tighter binding of band 3 to ankyrin (Liu et al. 1990), increased tyrosine phosphorylation of band 3 (Jones 1984; Jones et al. 1991), inability to transport sulfate anions (Schofield et al. 1992b; Tanner et al. 1992), and restricted lateral and rotational mobility of band 3 in the membrane (Liu et al. 1990; Tilley et al. 1991; Mohandas et al. 1992). Using crosslinking with bis[sulfosuccinimido]suberate, Jennings and Gosselink (1995) demonstrated that normal band 3 and SAO band 3 can form both homo- and heterooligomers in the RBC membrane.

Although initial studies suggested that SAO erythrocytes are refractory to invasion *in vitro* by the malarial parasite *P. falciparum*, a later study showed that fresh SAO cells were invaded by the parasites but became resistant to invasion on storage because intracellular depletion of ATP was more rapid than normal

(Dluzewski et al. 1992). Bruce et al. (1999) demonstrated that, when stored in the cold, SAO RBCs are considerably more leaky to sodium and potassium than are normal RBCs. This leakage was much less marked when the cells were stored at 25 °C or 37 °C. Incubation of cold-stored SAO RBCs at 37 °C did not restore sodium and potassium to normal levels. The increased leakiness of SAO RBCs also was observed for calcium ions, taurine, mannitol and sucrose. The authors suggested that SAO RBCs exhibit a cold-sensitive cation leak and that many of the previously reported abnormalities of SAO may be artifacts caused by the typical storage of RBCs in the cold.

The SAO mutation was the first example of a defect of an integral membrane protein leading to RBC membrane rigidity, which has previously been attributed only to the properties of the membrane skeleton. The mechanism leading to the increased rigidity most likely involves the propensity of mutant band 3 to form higher oligomers. These oligomers were detected by size-exclusion chromatography and visualized in the ovalocyte membrane by freeze-fracture electron microscopy (Liu et al. 1995). Formation of higher oligomers and their tighter binding to the membrane skeleton lead to the marked reduction in lateral mobility of band 3 (Liu et al. 1990; Tilley et al. 1991; Mohandas et al. 1992). However, even these results may be affected by the previous cold storage.

SAO is an ideal RBC phenotype that protects from severe forms of malaria while only minimally affecting its carriers. Heterozygotes for the SAO band 3 allele are hematologically almost normal and do not develop severe, mostly cerebral, forms of malaria. In contrast to sickle cell anemia, in which homozygotes suffer from a debilitating disease, no SAO homozygotes have been described to date, suggesting that homozygosity for the SAO allele is incompatible with life at early stages of embryonic development (Liu et al. 1994).

26.4.2 Hereditary Acanthocytosis

Bruce at al. (1993) reported a substitution P868L in the last putative transmembrane segment of band 3 that was associated with higher apparent molecular weight of band 3. Heterozygous RBCs had a markedly acanthocytic phenotype and an increased anion transport activity. The variant allele was designated band 3^{HT} (High Transport). On the basis of differential sensitivity of three different stilbenedisulfonates in their reaction with band 3^{HT}, Salhany et al. (1995) have suggested that P868L produces a localized conformational change in the C-terminal subdomain of band 3 and that structural change in the vicinity of amino acid 868 influences a turnover step in the transport cycle.

26.4.3 Blood Group Antigens Carried by Band 3

About a decade ago, during the study of the role of band 3 in the pathogenesis of HS and SAO and in chloride-bicarbonate exchange, several polymorphisms were found in the putative external loops of band 3 that were not co-inherited with the abnormal RBC phenotypes. One of them, P854L, eventually was found to underlie the Di^a/Di^b blood group polymorphism (Spring et al. 1992). P854L was found in *cis* with K56E (band 3 Memphis) in the so-called band 3 Memphis variant II allele, which displays a three times higher binding of the anion exchange inhibitor H_2DIDS in addition to decreased electrophoretic mobility (Hsu and Morrison 1985). Blood group antigens Di^a and Di^b thus became the first characterized blood group antigens carried by band 3 and the basis of the newly characterized Diego blood group system (system 010 in the ISBT classification of blood group antigens). Subsequently, through the work of several groups, an additional 18 polymorphisms in the putative external loops of band 3 were found and assigned to antigens of the Diego blood group system. The overview of these antigens and corresponding band 3 polymorphisms is shown in Table 26.2.

Antigens of the Diego blood group system have been found in the putative 1st, 2nd, 3rd, 4th and 7th external loops of band 3, thus confirming some of the theoretical predictions of the 14-transmembrane-segment model. So far, no mutations have been found in the putative 5th and 6th loops. Interestingly, the Wr^b antigen is expressed only when both glycophorin A and band 3 protein are present in the RBC membrane, while the Wr^a blood group is not linked to the glycophorin A locus. This result helped to map a site of band 3-glycophorin A interaction in the RBC membrane.

Amino acid sequences of external loops are considerably less conserved in evolution than are sequences of the transmembrane segments. With some exceptions, such as for the Bp^a antigen, most amino acid mutations in the RBC antigens are not conserved. Not surprisingly, all RBCs carrying these polymorphisms that have been studied so far had normal expression of band 3 and normal influx of radiolabeled sulfate, suggesting that polymorphisms in external loops of band 3 do not affect band 3 expression and its anion-exchange function (Jarolim et al. 1997a,b; 1998a).

26.5 Non-Erythroid Phenotypes Caused by Band 3 Mutations

In addition to being expressed in the erythroid cell, band 3 is expressed in the basolateral membrane of α intercalated cells of renal collecting ducts, where it participates in urine acidification. So far, several mutations in the membrane domain have been described that apparently interfere with this process and lead to either dominant or recessive distal renal tubular acidosis (dRTA) (Shayakul and Alper 2000).

26.5.1 Autosomal Dominant Distal Renal Tubular Acidosis

The first mutation causing autosomal dominant dRTA was described in 1997. Since then, additional mutations have been characterized in various populations (Table 26.3). While they are exceedingly rare in caucasian populations, band 3-associated dRTA is more commonly found in Southeast Asia, frequently in fami-

Table 26.2. Overview of antigens of the Diego blood group system

ISBT number	Blood group	Amino acid	Mutation	Band 3 ecto-loop	Sulfate influx[+]	Reference
010001	Di^a	854	Pro→Leu	7	N	Spring et al. 1992; Bruce et al. 1994; Jarolim et al. 1994c
010002	Di^b	854	Pro[++]	7	N	Spring et al. 1992; Bruce et al. 1994; Jarolim et al. 1994c
010003	Wr^a	658	Glu→Lys	4	N	Bruce et al. 1995
010004	Wr^b	658	Glu[*]	4	N	Bruce et al. 1995
010005	Wd^a	557	Val→Met	3	N	Jarolim et al. 1997a
010006	Rb^a	548	Pro→Leu	3	N	Jarolim et al. 1997a
010007	WARR	552	Thr→Ile	3	N	Jarolim et al. 1997b
010008	ELO	432	Arg→Trp	1	N	Jarolim et al. 1998a
010009	Wu	565	Gly→Ala	3	N	Zelinski et al. 1998; Jarolim et al. 1998a
010010	Bp^a	569	Asn→Lys	3	N	Jarolim et al. 1998a
010011	Mo^a	656	Arg→His	4	N	Jarolim et al. 1998a
010012	Hg^a	656	Arg→Cys	4	N	Jarolim et al. 1998a
010013	Vg^a	555	Tyr→His	3	N	Jarolim et al. 1998a
010014	Sw^a	646	Arg→Gln	4		Zelinski et al. 1997a; Zelinski et al. 2000
010015	BOW	561	Pro→Ser	3	N	Jarolim et al. 1998a
010016	NFLD	429 561	Glu→Asp Pro→Ala	3		Zelinski et al. 1997b
010017	Jn^a	566	Pro→Ser	3		Poole et al. 1997
010018	KREP	566	Pro→Ala	3		Poole et al. 1997
010019[**]	Tr^a	551	Lys→Asn	3	N	Jarolim et al. 1997a
010020	Fr^a	480	Glu→Lys	2	N	McManus et al. 2000
010021	SW1	646	Arg→Trp	4		Zelinski et al. 1996; Zelinski et al. 2000

[+], *N* normal influx of radiolabeled sulfate as measured in the author's laboratory. *Blank fields* flux studies not performed. [++]*Di*[b] corresponds to the wild-type band 3 with proline in position 854; [*]*Wr*[b] epitope is formed by interaction of the segment of wild-type band 3 around glutamic acid 658 and the adjacent molecule of glycophorin A; [**] Only one subject heterozygous for *Tr*[a] has been studied, therefore the ISBT number has been reserved but has not been assigned.

lies with Southeast Asian ovalocytosis. The mutations are associated with abnormal sulfate flux and/or altered band 3 glycosylation. It remains to be elucidated how this particular subset of band 3 mutations leads to renal tubular acidosis. Potential mechanisms include, among others, (i) loss of basolateral localization of band 3 resulting in a shunt in the unidirectional flow of hydrogen ions from plasma into urine via intercalated cells, (ii) changes in the structure of the kidney isoforms of the mutant band 3 or in its interactions with other proteins affecting the band 3-mediated anion exchange, or (iii) complete loss of expression of the kidney isoforms of the mutant protein.

Table 26.3. Band 3 mutations in autosomal dominant distal renal tubular acidosis

Mutation	Reference
589Arg→Cys	Bruce et al. 1997
589Arg→His	Bruce et al. 1997; Jarolim et al. 1998b; Karet et al. 1998
589Arg→Ser	Karet et al. 1998
613Ser→Phe	Bruce et al. 1997
858Ala→Asp	Bruce et al. 2000
Δ901–911*	Karet et al. 1998

* Premature termination of translation due to duplication of 13 nucleotides ranging from second base of codon 896 to second base of codon 900.

Quilty et al. (2002) examined the effect of the R589H mutation on the expression of AE1 and the truncated kidney form (kAE1) in transfected HEK 293 cells. AE1, AE1 R589H and kAE1 were present at the cell surface, while kAE1 R589H was located primarily intracellularly. Co-expression of kAE1 R589H reduced the cell surface expression of kAE1 and AE1 by dominant-negative effect due to heterodimer formation. The mutant AE1 and kAE1 bound to an inhibitor affinity resin suggesting they were not grossly misfolded. Other mutations at R589 also prevented the formation of the cell surface form of kAE1 indicating this conserved arginine residue is important for proper trafficking. The authors concluded that the R589H mutation creates a severe trafficking defect in kidney AE1 but not in erythroid AE1.

Not surprisingly, renal tubular acidosis was found in the child who completely lacked band 3 described earlier in this chapter and in the steer with complete band 3 deficiency. Interestingly, in a study of 10 patients from seven unrelated families with previously characterized band 3 mutations leading to autosomal dominant HS, in both examined family members, mutant band 3 Pribram that encodes a terminally truncated band 3 was associated with an incomplete dRTA (Rysava et al. 1997).

26.5.2 Autosomal Recessive Distal Renal Tubular Acidosis

Both known mutations, G701D and ΔV850 (Tanphaichitr et al. 1998; Vasuvattakul et al. 1999; Bruce et al. 2000), have been described in Southeast Asian population, many in families with SAO. While G701D was found in homozygotes and is clearly autosomal recessive, ΔV850 was described only in *trans* with the SAO allele, which is not functional in anion exchange, and can therefore be characterized as pseudo-dominant. SAO itself did not lead to dRTA. RBC morphology was normal in G701/N, Δ850/N, A858D/N and, Δ850/Δ850 band 3 genotypes. In some individuals, G701D was linked in *cis* to additional mutations M31T and K56E (band 3 Memphis); however, both of these substitutions appeared to be linked polymorphisms without an effect on band 3 abundance and function in erythrocytes (Tanphaichitr et al. 1998).

The pathogenesis of recessive dRTA due to AE1 mutations is better understood than for dominant dRTA associated with AE1 (Tanphaichitr et al. 1998; Bruce et al. 2000). G701D is a complete loss-of-function mutation when expressed in *Xenopus* oocytes. However, consistent with normal AE1 function in patient red cells, co-expression in oocytes of glycophorin A fully rescues G701D anion transport function. Glycophorin A rescues almost as well the loss-of-function phenotype shown by ΔV850. Absence of glycophorin A in kidney explains renal phenotype with normal RBC phenotype. Glycophorin A rescue of G701D is impaired when SAO is coexpressed with G701D (Shayakul et al. 1999).

26.6 Conclusion

This brief summary of band 3 gene defects illustrates the remarkable variability of morphological, biochemical, functional and clinical phenotypes. Abnormal erythrocyte morphology includes spherocytosis, ovalocytosis and acanthocytosis. Nonerythroid phenotypes include primarily distal renal tubular acidosis of dominant or recessive inheritance and of varying degrees of severity. The band 3 content of RBC membranes ranges from normal for antigens of the Diego blood group system and for distal renal tubular acidosis to the complete absence in the targeted gene disruption in mice or in the naturally occurring null phenotypes in humans, steer, and mice. Changes in anion exchange function vary from up-regulation in acanthocytosis and in some cases of distal renal tubular acidosis to a complete loss of function of the SAO allele. During the last decade, numerous band 3 mutations have been characterized and their association with several phenotypes has been established. Nevertheless, much more needs to be learned about phenotypes associated with band 3 mutations, such as the Southeast Asian ovalocytosis and its protective effect against cerebral malaria, congenital dyserythropoietic anemia and the potential role of band 3 in cell division, and distal renal tubular acidosis and the role of band 3 in its pathogenesis.

References

Alloisio N, Maillet P, Carre G, Texier P, Vallier A, Baklouti F, Philippe N, Delaunay J (1996) Hereditary spherocytosis with band 3 deficiency. Association with a nonsense mutation of the band 3 gene (allele Lyon), and aggravation by a low-expression allele occurring in trans (allele Genas). Blood 88:1062-1069

Alloisio N, Texier P, Vallier A, Ribeiro ML, Morle L, Bozon M, Bursaux E, Maillet P, Goncalves P, Tanner MJA, Tamagnini G, Delaunay J (1997) Modulation of clinical expression and band 3 deficiency in hereditary spherocytosis. Blood 90:414-420

Amato D, Booth PB (1977) Hereditary ovalocytosis in Melanesians. Papua New Guinea Medical Journal 20:26-32

Bianchi P, Zanella A, Alloisio N, Barosi G, Bredi E, Pelissero G, Zappa M, Vercellati C, Baronciani L, Delaunay J, Sirchia G (1997) A variant of the EPB3 gene of the anti-Lepore type in hereditary spherocytosis. Br J Haematol 98:283-288

Booth PB, Serjeantson S, Woodfield DG, Amato D (1977) Selective depression of blood group antigens associated with hereditary ovalocytosis among Melanesians. Vox Sanguinis 32:99-110

Bracher NA, Lyons CA, Wessels G, Mansvelt E, Coetzer TL (2001) Band 3 Cape Town (E90K) causes severe hereditary spherocytosis in combination with band 3 Prague III. Br J Haematol 113:689-693

Bruce LJ, Kay MMB, Lawrence C, Tanner MJA (1993) Band 3 HT, a human red-cell variant associated with acanthocytosis and increased anion transport, carries the mutation Pro-868→Leu in the membrane domain of band 3. Biochem J 293:317-320

Bruce LJ, Anstee DJ, Spring FA, Tanner MJA (1994) Band 3 Memphis variant II. Altered stilbene disulfonate binding and the Diego (Di^a) blood group antigen are associated with the human erythrocyte band 3 mutation Pro^{854}→Leu. J Biol Chem 269:16155-16158

Bruce LJ, Ring SM, Anstee DJ, Reid ME, Wilkinson S, Tanner MJA (1995) Changes in the blood group Wright antigens are associated with a mutation at amino acid 658 in human erythrocyte band 3: A site of interaction between band 3 and glycophorin A under certain conditions. Blood 85:541-547

Bruce LJ, Cope DL, Jones GK, Schofield AE, Burley M, Povey S, Unwin RJ, Wrong O, Tanner MJA (1997) Familial distal renal tubular acidosis is associated with mutations in the red cell anion exchanger (band 3, AE1) gene. J Clin Invest 100:1693-1707

Bruce LJ, Ring SM, Ridgwell K, Reardon DM, Seymour CA, Vandort HM, Low PS, Tanner MJA (1999) Southeast Asian ovalocytic (SAO) erythrocytes have a cold sensitive cation leak: implications for *in vitro* studies on stored SAO red cells. Biochim Biophys Acta 1416:258-270

Bruce LJ, Wrong O, Toye AM, Young MT, Ogle G, Ismail Z, Sinha AK, McMaster P, Hwaihwanje I, Nash GB, Hart S, Lavu E, Palmer R, Othman A, Unwin RJ, Tanner MJ (2000) Band 3 mutations, renal tubular acidosis and South-East Asian ovalocytosis in Malaysia and Papua New Guinea: loss of up to 95% band 3 transport in red cells. Biochem J 350:41-51

Castelino D, Saul A, Myler P, Kidson C, Thomas H, Cooke R (1981) Ovalocytosis in Papua New Guinea: dominantly inherited resistance to malaria. Southeast Asian J Trop Med Pub Health 12:549-555

Cattani JA, Gibson FD, Alpers MP, Crane GG (1987) Hereditary ovalocytosis and reduced susceptibility to malaria in Papua New Guinea. Trans Royal Soc Trop Med Hyg 81:705-709

Chernova MN, Jarolim P, Palek J, Alper SL (1995) Overexpression of AE1 Prague, but not of AE1 SAO, inhibits wild-type AE1 trafficking in *Xenopus* oocytes. J Membrane Biol 148:203-210

De Franceschi L, Turrini F, Delgiudice EM, Perrotta S, Olivieri O, Corrocher R, Mannu F, Iolascon A (1998) Decreased band 3 anion transport activity and band 3 clusterization in congenital dyserythropoietic anemia type II. Exp Hematol 26:869-873

Delgiudice EM, Vallier A, Maillet P, Perrotta S, Cutillo S, Iolascon A, Tanner MJA, Delaunay J, Alloisio N (1997) Novel band 3 variants (bands 3 Foggia, Napoli I and Napoli II) associated with hereditary spherocytosis and band 3 deficiency: Status of the D38A polymorphism within the EPB3 locus. Br J Haematol 96:70-76

Dhermy D, Galand C, Bournier O, Boulanger L, Cynober T, Schismanoff PO, Bursaux E, Tchernia G, Boivin P, Garbarz M (1997) Heterogenous band 3 deficiency in hereditary spherocytosis related to different band 3 gene defects. Br J Haematol 98:32-40

Dluzewski AR, Nash GB, Wilson RJM, Reardon DM, Gratzer WB (1992) Invasion of hereditary ovalocytes by *Plasmodium falciparum in vitro* and its relation to intracellular ATP concentration. Mol Biochem Parasitol 55:1-7

Eber SW, Gonzalez JM, Lux ML, Scarpa AL, Tse WT, Dornwell M, Herbers J, Kugler W, Ozcan R, Pekrun A, Gallagher PG, Schroter W, Forget BG, Lux SE (1996) Ankyrin-1 mutations are a major cause of dominant and recessive hereditary spherocytosis. Nat Genet 13:214-218

Fix AG, Baer AS, Lie-Injo LE (1982) The mode of inheritance of ovalocytosis/elliptocytosis in Malaysian Orang Asli families. Hum Genet 61:250-253

Hadley T, Saul A, Lamont G, Hudson DE, Miller LH, Kidson C (1983) Resistance of Melanesian elliptocytes (ovalocytes) to invasion by *Plasmodium knowlesi* and *Plasmodium falciparum* malaria parasites *in vitro*. J Clin Invest 71:780-782

Hassoun H, Hanada T, Lutchman M, Sahr KE, Palek J, Hanspal M, Chishti AH (1998a) Complete deficiency of glycophorin A in red blood cells from mice with targeted inactivation of the band 3 (AE1) gene. Blood 91:2146-2151

Hassoun H, Wang Y, Vassiliadis J, Lutchman M, Palek J, Aish L, Aish IS, Liu SC, Chishti AH (1998b) Targeted inactivation of murine band 3 (AE1) gene produces a hypercoagulable state causing widespread thrombosis *in vivo*. Blood 92:1785-1792

Holt M, Hogan PF, Nurse GT (1981) The ovalocytosis polymorphism on the Western border of Papua New Guinea. Hum Biol 53:23-34

Honig GR, Lacson PS, Maurer HS (1971) A new familial disorder with abnormal erythrocyte morphology and increased permeability of the erythrocytes to sodium and potassium. Pediatric Res 5:159-165

Hsu L, Morrison M (1985) A new variant of the anion transport protein in human erythrocytes. Biochemistry 24:3086-3090

Inaba M, Yawata A, Koshino I, Sato K, Takeuchi M, Takakuwa Y, Manno S, Yawata Y, Kanzaki A, Sakai J, Ban A, Ono K, Maede Y (1996) Defective anion transport and marked spherocytosis with membrane instability caused by hereditary total deficiency of red cell band 3 in cattle due to a nonsense mutation. J Clin Invest 97:1804-1817

Iwase S, Ideguchi H, Takao M, Horiguchiyamada J, Iwasaki M, Takahara S, Sekikawa T, Mochizuki S, Yamada H (1998) Band 3 Tokyo: Thr(837)→Ala(837) substitution in erythrocyte band 3 protein associated with spherocytic hemolysis. Acta Haematol 100:200-203

Jarolim P, Palek J, Amato D, Hassan K, Sapak P, Nurse GT, Rubin HL, Zhai S, Sahr KE, Liu SC (1991) Deletion in erythrocyte band 3 gene in malaria-resistant Southeast Asian ovalocytosis. Proc Natl Acad Sci USA 88:11022-11026

Jarolim P, Brabec V, Ballas SK, Prchal JT, Poon M-C, Castleberry R, Arnold D, Coetzer TL, Liu S-C, Palek J (1992a) Biochemical heterogeneity of the hereditary spherocytosis syndrome. Abstracts of the 24th Congress of the International Society of Haematology, London, p 35

Jarolim P, Palek J, Rubin HL, Prchal JT, Korsgren C, Cohen CM (1992b) Band 3 Tuscaloosa: Pro^{327}→Arg^{327} substitution in the cytoplasmic domain of erythrocyte band 3 protein associated with spherocytic hemolytic anemia and partial deficiency of protein 4.2. Blood 80:523-529

Jarolim P, Rubin HL, Zhai S, Sahr KE, Liu SC, Mueller TJ, Palek J (1992c) Band 3 Memphis: a widespread polymorphism with abnormal electrophoretic mobility of erythrocyte band 3 protein caused by substitution AAG→GAG (Lys→Glu) in codon 56. Blood 80:1592-1598

Jarolim P, Rubin HL, Liu S-C, Cho MR, Brabec V, Derick LH, Yi SJ, Saad STO, Alper S, Brugnara C, Golan DE, Palek J (1994a) Duplication of 10 nucleotides in the erythroid band 3 (AE1) gene in a kindred with hereditary spherocytosis and band 3 protein deficiency (band 3^{PRAGUE}). J Clin Invest 93:121-130

Jarolim P, Brabec V, Chrobak L, Alper SL, Brugnara C, Corbett JD, Cho MR, Golan DE (1994b) Decreased band 3 content, sulfate flux, and band 3 fractional mobility in congenital dyserythropoietic anemia. Blood 84, suppl 1:6a

Jarolim P, Rubin HL, Moulds JM (1994c) Molecular characterization of the Diego blood group antigen. Blood 84, suppl 1:237a

Jarolim P, Rubin HL, Brabec V, Chrobak L, Zolotarev AS, Alper SL, Brugnara C, Wichterle H, Palek J (1995) Mutations of conserved arginines in the membrane domain of erythroid band 3 protein lead to a decrease in membrane-associated band 3 and to the phenotype of hereditary spherocytosis. Blood 85:634-640

Jarolim P, Murray JL, Rubin HL, Taylor WM, Prchal JT, Ballas SK, Snyder LM, Chrobak L, Melrose WD, Brabec V, Palek J (1996) Characterization of 13 novel band 3 gene defects in hereditary spherocytosis with band 3 deficiency. Blood 88:4366-4374

Jarolim P, Murray JL, Rubin HL, Smart E, Moulds JM (1997a) Blood group antigens Rb^a, Tr^a, and Wd^a are located in the third ectoplasmic loop of erythrocyte band 3 protein. Transfusion 37:607-615

Jarolim P, Murray JL, Rubin HL, Coghlan G, Zelinski T (1997b) A Thr552→Ile substitution in erythroid band 3 gives rise to the Warrior blood group antigen. Transfusion 37:398-405

Jarolim P, Rubin HL, Zakova D, Storry J, Reid ME (1998a) Characterization of seven low incidence blood group antigens carried by erythrocyte band 3 protein. Blood 92:4836-4843

Jarolim P, Shayakul C, Prabakaran D, Jiang LW, Stuarttilley A, Rubin HL, Simova S, Zavadil J, Herrin JT, Brouillette J, Somers MG, Seemanova E, Brugnara C, Guaywoodford LM, Alper SL (1998b) Autosomal dominant distal renal tubular acidosis is associated in three families with heterozygosity for the R589H mutation in the AE1 (band 3) Cl^-/HCO_3^- exchanger. J Biol Chem 273:6380-6388

Jennings ML, Gosselink PG (1995) Anion exchange protein in southeast Asian ovalocytes: Heterodimer formation between normal and variant subunits. Biochemistry 34:3588-3595

Jenkins PA, Abou-Alfa GK, Dhermy D, Bursaux E, Feo C, Scarpa AL, Lux SE, Garbarz M, Forget BG, Gallagher PG (1996) A nonsense mutation in the erythrocyte band 3 gene associated with decreased mRNA accumulation in a kindred with dominant hereditary spherocytosis. J Clin Invest 97:373-380

Jones GL (1984) Red cell membrane proteins in Melanesian ovalocytosis - autophosphorylation and proteolysis. Proc Australian Biochem Soc 16:34-34

Jones GL, McLemore-Edmundson H, Wesche D, Saul A (1991) Human erythrocyte band 3 has an altered N-terminus in malaria-resistant Melanesian ovalocytosis. Biochim Biophys Acta 1096:33-40

Kanzaki A, Takezono M, Kaku M, Yawata A, Ozcan R, Kugler W, Eber SW, Yawata Y (1997) Molecular and genetic characteristics in Japanese patients with hereditary spherocytosis: Frequent band 3 mutations and rarer ankyrin mutations. Blood 90, suppl 1:6b

Karet FE, Gainza FJ, Gyory AZ, Unwin RJ, Wrong O, Tanner MJA, Nayir A, Alpay H, Santos F, Hulton SA, Bakkaloglu A, Ozen S, Cunningham MJ, Dipietro A, Walker

WG, Lifton RP (1998) Mutations in the chloride-bicarbonate exchanger gene AE1 cause autosomal dominant but not autosomal recessive distal renal tubular acidosis. Proc Natl Acad Sci USA 95:6337-6342

Kidson C, Lamont G, Saul A, Nurse GT (1981) Ovalocytic erythrocytes from Melanesians are resistant to invasion by malaria parasites in culture. Proc Natl Acad Sci USA 78:5829-5832

Lie-Injo LE (1965) Hereditary ovalocytosis and haemoglobin E-ovalocytosis in Malayan aborigines. Nature 208:1329-1331

Lima PRM, Gontijo JAR, Defaria JBL, Costa FF, Saad STO (1997) Band 3 Campinas: A novel splicing mutation in the band 3 gene (AE1) associated with hereditary spherocytosis, hyperactivity of Na^{+}/Li^{+} countertransport and an abnormal renal bicarbonate handling. Blood 90:2810-2818

Liu SC, Zhai S, Palek J, Golan DE, Amato D, Hassan K, Nurse GT, Babona D, Coetzer T, Jarolim P, Zaik M, Borwein S (1990) Molecular defect of the band 3 protein in Southeast Asian ovalocytosis. N Engl J Med 323:1530-1538

Liu SC, Jarolim P, Rubin HL, Palek J, Amato D, Hassan K, Zaik M, Sapak P (1994) The homozygous state for the band 3 protein mutation in Southeast Asian ovalocytosis may be lethal. Blood 84:3590-3591

Liu SC, Palek J, Yi SJ, Nichols PE, Derick LH, Chiou SS, Amato D, Corbett JD, Cho MR, Golan DE (1995) Molecular basis of altered red blood cell membrane properties in southeast Asian ovalocytosis: Role of the mutant band 3 protein in band 3 oligomerization and retention by the membrane skeleton. Blood 86:349-358

Maillet P, Vallier A, Reinhart WH, Wyss EJ, Ott P, Texier P, Baklouti F, Tanner MJA, Delaunay J, Alloisio N (1995) Band 3 Chur: A variant associated with band 3-deficient hereditary spherocytosis and substitution in a highly conserved position of transmembrane segment 11. Br J Haematol 91:804-810

McManus K, Lupe K, Coghlan G, Zelinski T (2000) An amino acid substitution in the putative second extracellular loop of RBC band 3 accounts for the Froese blood group polymorphism. Transfusion 40:1246-1249

Mohandas N, Lie-Injo LE, Friedman M, Mak JW (1984) Rigid membranes of Malayan ovalocytes: a likely genetic barrier against malaria. Blood 63:1385-1392

Mohandas N, Winardi R, Knowles D, Leung A, Parra M, George E, Conboy J, Chasis J (1992) Molecular basis for membrane rigidity of hereditary ovalocytosis - a novel mechanism involving the cytoplasmic domain of band-3. J Clin Invest 89:686-692

Mueller TJ, Morrison M (1977) Detection of a variant of protein 3, the major transmembrane protein on the human erythrocyte. J Biol Chem 252:6573-6576

Paw BH, Zhou Y, Li R, Pratt SJ, Davidson AJ, Trede NS, Brownlie A, Donovan A, Liao EC, Drejer A, Peters LL, Gwynn B, Lux SE, Zon LI (2000) Cloning of the zebrafish retsina blood mutation: mutations in erythroid band 3 result in dyserythropoiesis and cytokinesis defects. Blood 96, suppl 1:440a

Perrotta S, Polito R, Conte ML, Nobili B, Cutillo S, Delgiudice EM, Nigro V, Iolascon A, Amendola G (1999) Hereditary spherocytosis due to a novel frameshift mutation in AE1 cytoplasmic COOH terminal tail: Band 3 Vesuvio. Blood 93:2131-2132

Peters LL, Shivdasani RA, Liu SC, Hanspal M, John KM, Gonzalez JM, Brugnara C, Gwynn B, Mohandas N, Alper SL, Orkin SH, Lux SE (1996) Anion exchanger 1 (band 3) is required to prevent erythrocyte membrane surface loss but not to form the membrane skeleton. Cell 86:917-927

Peters LL, Andersen SG, Gwynn B, Li R, Lux SE, Churchill GA (2001) A QTL on mouse chromosome 12 modifies the band 3 bull phenotype: β spectrin is a candidate gene. Blood 98, suppl 1:437a

Poole J, Hallewell H, Bruce L, Tanner MJA, Zupanska B, Kusnierz-Alejska G (1997) Identification of two new Jn(a+) individuals and assignment of Jn^a to erythrocyte band 3. Transfusion 37, suppl:90S

Quilty JA, Reithmeier RA (2000) Trafficking and folding defects in hereditary spherocytosis mutants of the human red cell anion exchanger. Traffic 1:987-998

Quilty JA, Li J, Reithmeier AF (2002) Impaired trafficking of distal renal tubular acidosis mutants of the human kidney anion exchanger (kAE1). Am J Physiol 282:F810-F820

Ribeiro ML, Alloisio N, Almeida H, Gomes C, Texier P, Lemos C, Mimoso G, Morle L, Bey-Cabet F, Rudigoz RC, Delaunay J, Tamagnini G (2000) Severe hereditary spherocytosis and distal renal tubular acidosis associated with the total absence of band 3. Blood 96:1602-1604

Rybicki AC, Qiu JJH, Musto S, Rosen NL, Nagel RL, Schwartz RS (1993) Human erythrocyte protein 4.2 deficiency associated with hemolytic anemia and a homozygous ^{40}glutamic acid→lysine substitution in the cytoplasmic domain of band 3 (Band $3^{Montefiore}$). Blood 81:2155-2165

Rysava R, Tesar V, Jirsa M, Brabec V, Jarolim P (1997) Renal tubular acidosis associated a band 3 gene mutation. Nephrol Dial Transplant 12:1869-1873

Salhany JM, Schopper LM, Kay MMB, Gamble DN, Lawrence C (1995) Differential sensitivity of stilbenedisulfonates in their reaction with band 3 HT (Pro-868→Leu). Proc Natl Acad Sci USA 92:11844-11848

Saul A, Lamont G, Sawyer WH, Kidson C (1984) Decreased membrane deformability in Melanesian ovalocytes from Papua New Guinea. J Cell Biol 98:1348-1354

Savvides P, Shalev O, John KM, Lux SE (1993) Combined spectrin and ankyrin deficiency is common in autosomal dominant hereditary spherocytosis. Blood 82:2953-2960

Schofield AE, Tanner MJA, Pinder JC, Clough B, Bayley PM, Nash GB, Dluzewski AR, Reardon DM, Cox TM, Wilson RJM, Gratzer WB (1992a) Basis of unique red cell membrane properties in hereditary ovalocytosis. J Mol Biol 223:949-958

Schofield AE, Reardon DM, Tanner MJA (1992b) Defective anion transport activity of the abnormal band-3 in hereditary ovalocytic red blood cells. Nature 355:836-838

Serjeantson S, Bryson K, Amato D, Babona D (1977) Malaria and hereditary ovalocytosis. Hum Genet 37:161-167

Shayakul C and Alper SL (2000) Inherited renal tubular acidosis. Curr Opin Nephrol Hypertens 9:541-546

Shayakul C, Jarolim P, Alper SL (1999) Autosomal dominant distal renal tubular acidosis and the AE1 gene. Am J Kidney Dis 33:1194-1197

Southgate CD, Chishti AH, Mitchell B, Yi SJ, Palek J (1996) Targeted disruption of the murine erythroid band 3 gene results in spherocytosis and severe haemolytic anaemia despite a normal membrane skeleton. Nat Genet 14:227-230

Spring FA, Bruce LJ, Anstee DJ, Tanner MJA (1992) A red cell band 3 variant with altered stilbene disulphonate binding is associated with the Diego (Di^a) blood group antigen. Biochem J 288:713-716

Tanner MJA, Bruce L, Groves JD, Martin PG, Schofield AE (1992) The defective red cell anion transporter (band 3) in hereditary South East Asian ovalocytosis and the role of glycophorin A in the expression of band 3 anion transport activity in *Xenopus* oocytes. Biochem Soc Trans 20:542-546

Tanphaichitr VS, Sumboonnanonda A, Ideguchi H, Shayakul C, Brugnara C, Takao M, Veerakul G, Alper SL (1998) Novel AE1 mutations in recessive distal renal tubular acidosis - Loss-of-function is rescued by glycophorin A. J Clin Invest 102:2173-2179

Tilley L, Nash GB, Jones GL, Sawyer WH (1991) Decreased rotational diffusion of band 3 in melanesian ovalocytes from Papua New Guinea. J Membrane Biol 121:59-66

Vasuvattakul S, Yenchitsomanus PT, Vachuanichsanong P, Thuwajit P, Kaitwatcharachai C, Laosombat V, Malasit P, Wilairat P, Nimmannit S (1999) Autosomal recessive distal renal tubular acidosis associated with Southeast Asian ovalocytosis. Kidney Int 56:1674-1682

Yannoukakos D, Vasseur C, Driancourt C, Blouquit Y, Delaunay J, Wajcman H, Bursaux E (1991) Human erythrocyte band 3 polymorphism (band 3 Memphis): characterization of the structural modification (Lys56→Glu) by protein chemistry methods. Blood 78:1117-1120

Zelinski T, Mckeown I, McAlpine PJ, Philipps S, Coghlan G (1996) Assignment of the gene(s) governing Froese and Swann blood group polymorphism to chromosome 17q. Transfusion 36:419-420

Zelinski T, Rusnak A, McManus M, Coghlan G (1997a) A G→A mutation in the anion exchanger 1 gene (AE1) accounts for the Swann blood group polymorphism. Transfusion 37, suppl:89S

Zelinski T, Pongoski J, Coghlan G (1997b) The low incidence erythrocyte antigen NFLD is associated with membrane protein band 3. Transfusion 37, suppl:90S

Zelinski T, McManus K, Punter F, Moulds M, Coghlan G (1998) A Gly_{565}→Ala substitution in human erythroid band 3 accounts for the Wu blood group polymorphism. Transfusion 38:745-748

Zelinski T, Rusnak A, McManus K, Coghlan G (2000) Distinctive Swann blood group genotypes: molecular investigations. Vox Sang 79:215-218

27 Amino Acid Transport in Disease

A. Cláudio Mendes Ribeiro and Tatiana M. C. Brunini

Departamento de Farmacologia e Psicobiologia, Instituto de Biologia, Av. 28 de Setembro 87, CEP 20551-030, Rio de Janeiro, Brazil

27.1 Introduction

Amino acid uptake and metabolism by cells plays an important role in regulating and maintaining cellular function. In most mammalian cells, amino acids are a major source of energy and the products of their metabolism influence many physiological and pathological processes. Many pathophysiological alterations observed in clinical syndromes can only be elucidated at the cellular or molecular level. Membrane transport is essential for physiological functioning of cells, and alterations in transport may help explain several pathological changes in disease states. Red blood cells (RBCs) lack a nucleus and mitochondria and have lost the ability to synthesize protein, however, they are resilient and capable of extraordinary changes in shape, and their haemoglobin efficiently transports and delivers oxygen. The adequate functioning and longevity of erythrocytes depend on a satisfactory relationship between the cell membrane and metabolism (Hillman and Finch 1996). In the absence of mitochondria, RBCs have a limited ability to metabolize amino and fatty acids, and their energy supply is derived largely from the breakdown of glucose. The specific metabolic characteristics of RBCs provide a particular need for amino acid transport that is not entirely clear. Red cells synthesize glutathione and therefore require transport systems capable of supplying amino acid precursors of this molecule, namely cysteine, glutamate and glycine. Glutathione protects RBCs from exogenous oxidants and reactive electrophiles (Tunnicliff 1994). Indeed, in Finnish Landrade sheep, a genetically lowered glutathione concentration is associated with the absence of a specific transport system for cysteine, resulting in a shortened RBC life span (Young et al. 1975). Additionally, amino acid transporters expressed in mature RBCs may be the remnants of functional proteins required during the stages of erythropoiesis, or be involved in the inter organ transport of amino acids (Ellory 1987). Constantino (1913) was the first to report that amino acids, specifically glycine and asparagine, were transported by RBCs *in vitro*. In the 1960s, it was demonstrated that specific transport mechanisms are responsible for amino acid translocation in RBCs. A transport system for L-valine was suggested (Riessler et al. 1962), and this was followed by detailed studies of neutral amino acids transport in RBCs (Winter and Christhensen 1964). Since then, specific transport systems for amino acids have been identified and characterised in RBC membranes. In this chapter we will examine the role of RBC amino acid transport, mainly the role of alterations of cationic amino acid transport in uraemia, chronic heart failure, diabetes, sepsis and, in sickle cell disease, the importance of the transport of the amino acids precursors of glutathione synthesis.

27.2 Measurement of Transport Kinetics in Red Cells

The simpler morphological structure and metabolism of RBCs compared with most cells and the fact that they are easily obtained, have secured them as a model for the investigation of membrane transport. Kinetic experiments in RBCs provide an excellent tool to understand carrier transport characteristics of biological membranes in different tissues. The observed alterations in transport in RBCs are probably multifactorial and involve biochemical changes in the membrane (fluidity, membrane potential), a degree of *trans*-stimulation, changes in the expression of transporters at the erythroid cell level and disturbances of signalling pathways. The relevance of findings in RBCs to membrane transport in other tissues remains the subject of controversy, although frequently the results are supported by transport data from other cell types (Hendry 1992).

27.2.1 Transport Systems and Transporters

Lately, the use of new kinetic and molecular approaches has revealed a complex picture, involving distinct transporters possessing definite structure, and a site of expression and substrate specificity, with cDNA isolated for more than 20 mammalian amino acid transporters. Finding a correlation between tissue expression, transport characteristics and structure-function relationship is not an easy task. The association of transporter proteins with known amino acid transport systems is beginning to emerge, but a clear and detailed appropriate nomenclature is far from being achieved. Identifying the molecular structure of all amino acid transport systems and the key molecular mechanisms of amino acid transporters are challenges for the future in this field (Devés and Boyd 1998; Palacin et al. 1998). In this chapter, we will use the nomenclature of transport systems and, whenever necessary, include the associated transporter protein.

27.2.2 Role of L-Arginine Transport in Nitric Oxide Production

Recently, the interest in NO as a mediator involved in crucial physiological and pathophysiological mechanisms has highlighted the importance of the transport of its precursor L-arginine. Although normal intracellular levels of L-arginine are well above the K_m, and should be sufficient to saturate nitric oxide synthase in vascular endothelial cells, nitric oxide production can, nonetheless, be stimulated by exogenous L-arginine. In certain types of cells such as macrophages and platelets the correlation between L-arginine transport and NO synthesis is even more striking (Bogle et al. 1992; Mendes Ribeiro et al. 2001). This phenomenon, termed the "arginine paradox", has suggested the existence of a separate pool of L-arginine directed toward nitric oxide synthesis. In this chapter, we will emphasise alterations of L-arginine transport and their possible correlation with disturbances of NO synthesis. Although the synthesis of NO by RBCs it still a matter of controversy, RBCs participate in a major metabolic pathway for endogenously produced NO. In the plasma, NO forms nitrite and *S*-nitrosothiol. Nitrite is later oxidized by oxyhaemoglobin in RBCs and is eliminated in the urine as nitrate. NO may also react directly with oxyhaemoglobin forming nitrate and methehaemoglobin, and with haemoglobin, constituting nitrosylhaemoglobin (Wennmalm et al. 1993). Pawloski et al. (1998) have suggested that RBCs would release the NO linked to haemoglobin in the microcirculation under low oxygen tension. Because of this ability of haemoglobin and oxyhaemoglobin to bind to NO, the capacity of RBCs to produce NO is not universally accepted. However, it has been demonstrated that RBCs can synthesise NO from nitrovasodilators such as isosorbide dinitrate (Kosaka et al. 1994). Recently, several studies have suggested that RBCs possess both inducible NO-synthase (iNOS) and endothelial NO-synthase (eNOS) (Deliconstantinos et al. 1995; Jubelin and Gierman 1996; Chen and Mehta 1998). The release of NO by RBCs might be used to modulate their own physiology and platelet function (Chen and Mehta 1998). The metabolic utilisation of L-arginine requires the transport of this amino acid across the plasma membrane, and kinetic analysis has demonstrated that L-arginine enters human RBCs via two distinct transport systems: a high affinity, low capacity transport system which recognises L-leucine and L-lysine with comparable affinities, which was named y^+L, and a lower affinity, higher capacity transport system specific to cationic amino acids, y^+ (Devés and Boyd 1998). In human RBCs, y^+ activity is compatible with the transport properties translated by the gene CAT-1, which is ubiquitously expressed in all tissues investigated, except the liver. Unlike system y^+, system y^+L is peculiar in that it transports cationic amino acids independently of Na^+ and becomes an inorganic cation-dependent carrier for neutral amino acids (Devés et al. 1992). The gene coding for 4F2hc associated with y^+LAT may induce system y^+L in RBCs (Torrents et al. 1998; see Chap. 12).

27.3 Uraemia, Pathophysiology and Treatment

Chronic renal failure is characterised by a progressive and generally irreversible decline in the glomerular filtration rate, caused by a large number of diseases (Warnock 1996). Progressive impairment in renal function generates a clinical constellation of signs and symptoms known as uraemic syndrome (El Nahas and Winearls 1998). Uraemia is characterised by a state of systemic poisoning affecting the cardiovascular, neurologic, haematologic, endocrinologic and immunologic systems (Vanholder 1998). Malnutrition and acidosis may contribute to the pathophysiology of the uraemic syndrome (Vanholder 1998) that follows the retention of multiple compounds normally excreted by healthy kidneys, including metabolic products of protein and amino acids (Vanholder 1998). Several potential toxins that accumulate in uraemia have been identified, and in animal models they can reproduce some uraemic manifestations. However, uraemic syndrome is not only a result of decline in renal excretory function, since derangements of hormonal and enzymatic homeostasis also contribute to uraemic alterations (Vanholder 1998).

The severity of renal failure can be classified using an estimated glomerular filtration rate (GFR), which is satisfactory for most clinical indications and is calculated from serum creatinine together with the age, weight and sex of the patients (El Nahas and Winearls 1998). At end-stage renal disease (ESRD), when the GFR is less than 5 ml/min, renal replacement therapy is essential for the survival of the patients. Replacement therapy includes transplantation and two main types of dialysis: haemodialysis and peritoneal dialysis, mainly continuous ambulatory peritoneal dialysis (CAPD) (El Nahas and Winearls 1998). The role of peritoneal dialysis in renal replacement therapy is now well-established and represents about 15% of the world-wide dialysis population (Westman et al. 1996). Despite a lower solute clearance, CAPD is more efficient at removing fluid and controlling blood pressure than haemodialysis (Tattersall et al. 1998). Although comprehension of the complex multiorgan system derangements that compose uraemic syndrome has improved over the years, a reasonable understanding of the defects in specific cellular and sub-cellular functions has not yet been reached.

27.3.1 Transport Alterations in Uraemia

Alterations in cell membrane transport seem to contribute to the pathogenesis of the clinical uraemic syndrome (Swales 1988). The first observation that cellular Na^+, K^+-ATPase activity was reduced in ESRD patients on haemodialysis was made by Welt et al. (1968) in the 1960s, using RBCs as a model cell. Subsequently, this uraemic defect has been shown to occur in a wide variety of cells (Edmondson et al. 1975; Cotton et al. 1979; Fraser et al. 1985) and extended to other groups of uraemic patients, including undialysed chronic renal failure (CRF) patients and ESRD patients on peritoneal dialysis (Fervenza et al. 1989a). There is also evidence that membrane amino acid transport systems are altered in uraemia. The affinity of the transport system gly and the activity of the transport system ASC are reduced in human RBCs from ESRD patients on haemodialysis (Fervenza et al. 1990). In contrast, other transport systems such as L and T were unchanged in the same group of patients (Fervenza et al. 1990). An increase in lysine transport in uraemic patients was initially reported, with no discrimination between the y^+ and y^+L transport systems (Fervenza et al. 1989b).

27.3.2 L-Arginine Transport and Uraemia

Local conversion of L-arginine to NO in the kidney seems to be a prerequisite to normal kidney function. A wealth of studies shows that NO modulates the autoregulation of renal blood flow, glomerular surface area available for filtration, glomerulotubular feedback response and release of renin (Romero and Strick

1993). Inhibition of NO synthesis will affect the function of all of these mechanisms (Romero and Strick 1993). There is ample evidence suggesting that chronic renal failure is characterised by abnormal regulation of the synthesis of NO, both by the kidney and at the systemic level (Mendes Ribeiro et al. 2001).

Some important abnormalities in uraemia are associated with a disturbance in the L-arginine-NO pathway. NO is a potent inhibitor of platelet adhesion and aggregation. Increased synthesis of NO might explain prolonged bleeding time and bleeding tendency in uraemic patients (Remuzzi and Rossi 1996). A significant pathological feature in chronic renal failure patients is endothelial dysfunction. Uraemic patients exhibit endothelial cell damage and impairment in arterial endothelial-dependent relaxation that appears in the very early stages of the syndrome (Kari et al. 1997). It is suggested that endothelial dysfunction may involve abnormalities in both NO production and breakdown (Kari et al. 1997). This idea is supported by the fact that L-arginine infusions reverse endothelial dysfunction in ESRD patients on haemodialysis (Hand et al. 1998). Endogenous analogues of L-arginine including monomethyl-L-arginine (L-NMMA), asymmetric dimethyl-L-arginine (ADMA) and symmetric dimethyl-L-arginine (SDMA) are present in increased concentrations in plasma and blood cells from end-stage renal failure patients on regular haemodialysis (Vallance et al. 1992; Mendes Ribeiro et al. 1996, 1997a). Following a session of haemodialysis, methylarginine levels decrease significantly (MacAllister et al. 1996; Mendes Ribeiro et al. 1996, 1997a). L-arginine endogenous analogues compete with each other and with L-arginine for cationic amino acid transporters, but only L-NMMA and ADMA are effective inhibitors of NOS isoforms (MacAllister and Vallance 1998).

An important finding in uraemia is the altered metabolism of protein and amino acid, but a full understanding of the mechanism responsible has not yet been achieved. Several hypotheses have been proposed to explain this protein metabolism disturbance. These range from increased protein catabolism and low protein intake, to impairment of the renal metabolic and excretory functions together with uraemic toxin actions on the intermediate metabolism of amino acids, and the loss of protein and amino acids that occurs during dialysis (Bergstrom et al. 1990). However, the most likely explanation is that there is a link between hyperinsulinaemia seen in chronic renal failure and disturbance in protein metabolism (Castellino et al. 1992).

Apart from being a precursor for NO, the amino acid L-arginine is essential for the synthesis of proteins, urea, creatinine, other guanidino compounds, polyamines and ornithine, as well as affecting the release of hormones (Reyes et al. 1994). All of these compounds have been connected with the pathophysiology of uraemic syndrome (Vanholder 1998). Renal proximal tubule is the major site of L-arginine synthesis (Levillain et al. 1990) and most of synthesised L-arginine enters systemic circulation with only a small portion being utilised and metabolised to urea in the kidney. Thus, the kidney plays a important role in maintaining a constant supply of L-arginine for utilisation by other organs (Featherson et al. 1973).

Although studies have reported normal or increased levels of L-arginine in uraemic patients (De Deyn et al. 1986; Bergstrom et al. 1990), we have demonstrated that in patients with chronic renal failure including those not in treatment and those on CAPD or haemodialysis, plasma levels of L-arginine are significantly reduced compared with controls (Mendes Ribeiro et al. 1996, 1997, unpublished data; Brunini et al. 1999; Table 27.1). The transport system y^+ does not transport L-arginine exclusively, but also L-lysine, L-ornithine and other cationic amino acids (Devés and Boyd 1998). Plasma levels of L-lysine and L-ornithine are both reduced in undialysed and post-haemodialysis patients compared with controls, but only L-lysine is diminished in pre-haemodialysis and CAPD patients. Citrulline plasma concentration is approx. 3-, 2.5- and 2.6-fold higher in undialysed, pre-haemodialysis and CAPD patients, respectively, than in controls, but this normalises after a session of haemodialysis (Table 27.1). This is a common finding in chronic renal failure subjects (Reyes et al. 1994; Mendes Ribeiro et al. 1996), and traditionally it has been associated with the diminished L-citrulline uptake and low production of L-arginine by the failing kidney. An alternative mechanism to explain this enhanced L-citrulline plasma concentration would be that the increased production of nitric oxide yields an increased synthesis of citrulline as the by-product of the reaction.

Table 27.1. Amino acid concentration in plasma from controls and chronic renal failure patients not yet on dialysis (n=14), on CAPD (n=15) and on haemodialysis (n=20)

Amino acids	Controls [μM]	Undialysed [μM]	CAPD [μM]	Pre-HD [μM]	Post-HD [μM]
L-arginine	106 ± 8	80 ± 6*	54 ± 3*	74 ± 5*	61 ± 9*
L-citrulline	29 ± 7	90 ± 8*	76 ± 4*	74 ± 6*	39 ± 3
L-lysine	260 ± 42	34 ± 5*	81 ± 10*	94 ± 12*	62 ± 7*
L-ornithine	72 ± 8	29 ± 3*	65 ± 7	71 ± 11	36 ± 4*

Data denote the mean ± S.E.M., *P < 0.05, *CRF* chronic renal failure, *CAPD* peritoneal dialysis, *Pre-* and *Post-HD* before and after a session of haemodialysis.

In contrast with the reduced plasma availability, L-arginine transport capacity via y^+ system in RBCs taken from undialysed patients and uraemic patients on haemodialysis and CAPD is dramatically increased. The K_m values for L-arginine transport via y^+ system are also enhanced in RBCs from the same groups of patients. The results for V_{max} and K_m of the transport system y^+L in the RBCs of controls, undialysed and haemodialysis patients, did not differ significantly. In contrast, the K_m for L-arginine influx via y^+L system in RBCs from CAPD patients was dramatically reduced. The diffusional coefficient is not affected by uraemia (Mendes Ribeiro et al. 1997a, unpublished data; Brunini et al. 1999; Table 27.2).

The diminished L-arginine plasma concentration correlates negatively with plasma levels of creatinine ($r = 0.674$, $p < 0.05$) and urea ($r = 0.669$, $p < 0.05$) and positively with GFR ($r = 0.737$, $p < 0.05$) in undialysed CRF patients. The reduced L-arginine plasma levels in undialysed patients present a marked negative correlation with-arginine influx via system y^+ in RBCs ($r = 0.743$, $p < 0.05$). By contrast, there is no significant correlation between L-arginine plasma levels and L-arginine transport via system y^+L (Mendes Ribeiro, unpublished data).

The kinetic characteristics of the y^+L transport system would make it saturated at plasma L-arginine concentration, and this may be the reason why the y^+ system, modulates L-arginine uptake in uraemic RBCs. The changes in V_{max} and K_m observed in RBCs from CRF patients could be the consequence of changes in the plasma membrane characteristics of these patients. There are several reports of altered RBC lipid composition in uraemia (Panasenko et al. 1985; Girelli et al. 1992). It has been demonstrated that a particular lipid bilayer environment is required for the activity of membrane proteins (Deuticke and Haest 1987), and that perhaps the alterations present in uraemia affect y^+/CAT-1. Many alterations in several types of lipids have been associated with uraemic RBC membrane defects, including non-esterified fatty acids (Girelli et al. 1992) and lecithin (Bleiber 1980); their rapid adsorption to cell membrane and diffusion would allow them to interact with a wide range of membrane transporters (Hendry 1992). Interestingly, these alterations may affect the activity of some, but not all, membrane amino acid transport systems in RBCs (Fervenza et al. 1989b, 1990). A possible explanation for this finding is that the various transport systems either respond differently to changes in membrane lipid, or they are located in lipid domains that are modified to different extents by lipid changes (Yorek et al. 1984).

Table 27.2. Kinetic parameters for L-arginine influx into red blood cells from controls (n=10) and chronic renal failure patients not yet on dialysis (n=11), on CAPD (n=17) and haemodialysis (n=11)

	System y^+		System y^+L		Diffusion
	V_{max} [µmol/(l_{cells}·h)]	K_m [µM]	V_{max} [µmol/(l_{cells}·h)]	K_m [µM]	[1/h]
Controls	292 ± 19	54 ± 7	83 ± 15	35 ± 8	0.10 ± 0.02
Undialysed	459 ± 40*	89 ± 14*	143 ± 32	29 ± 6	0.11 ± 0.02
CAPD	539 ± 61*	130 ± 16*	90 ± 13	14 ± 2*	0.20 ± 0.03
Pre-HD	700 ± 92*	94 ± 15*	62 ± 6	18 ± 5	0.18 ± 0.07
Post-HD	627 ± 50*	81 ± 7*	55 ± 5	15 ± 3	0.09 ± 0.02

Data denote the mean ± S.E.M., *P < 0.05, *CAPD* peritoneal dialysis, *Pre-* and *Post-HD* before and after a session of haemodialysis.

Although there are a significant number of studies reporting widespread alterations of membrane transport in patients with uraemia, there are few studies on amino acid transport in animal models of uraemia. Unlike in humans, L-arginine influx via transport systems y^+ and y^+L in rat RBCs was not affected by chronic uraemia (Brunini et al. 1998). Kinetic analysis has shown that L-arginine is transported in rat RBCs by a variant of the y^+ and y^+L transport systems detected in humans. Moreover, rats with uraemia secondary to partial nephrectomy present with unaltered plasma L-arginine levels compared with controls, and under these conditions transport of L-arginine in RBCs is unaffected (Brunini et al. 1998).

27.3.3 Possible Mechanisms Explaining the Increased y^+ Transport Activity in Uraemia

Several studies have demonstrated that transport system y^+ and iNOS are co-induced by cytokines and lipopolysaccharide (LPS) in cells such as vascular smooth muscle cells and macrophages (Bogle et al. 1992; Durante et al. 1996). An increased concentration of circulating TNF-α is found in undialysed CRF, haemodialysis and CAPD patients (Herbelin et al. 1990; Pereira et al. 1994; Descamps-Latscha et al. 1995) that could induce expression of L-arginine transporter and iNOS in precursor cells of RBCs.

Much attention has been paid to endocrine dysfunction occurring in uraemia. Glucose intolerance is a common finding in uraemia and results primarily from peripheral tissue insensitivity to insulin, since plasma insulin levels are increased in the face of depressed glucose metabolism (Vanholder 1998). This endocrine disturbance is of potential relevance to amino acid metabolism since it will inhibit anabolic reactions including protein synthesis (El Nahas and Winearls 1998). Elevated glucose and insulin are known to be capable of increasing L-arginine uptake via transport system y^+(CAT-1) and nitric oxide production (Sobrevia et al. 1996). It is possible that activation of L-arginine transport via system y^+ seen in RBCs from uraemic patients is partially a consequence of the hyperglycaemia and hyperinsulinaemia present in chronic renal failure. This concept is supported by the fact that underlying metabolic abnormalities are not reversed completely by dialysis treatment, and no reversal of the up-regulation of L-arginine transport through the y^+ transport system in RBCs was observed in CRF patients on CAPD or on haemodialysis (Mendes Ribeiro et al. 1997; Brunini et al. 1999).

Another possible mechanism to explain the activation of system y^+ in CRF RBCs is the presence of a low extracellular concentration of L-arginine. There is evidence that low extracellular L-arginine up-regulates L-arginine transport in endothelial cells via system y^+ (Bogle et al. 1996) and amino acid starvation has been reported to increase CAT-1 mRNA threefold in Fao cells (Hyatt et al. 1997). L-arginine plasma levels are significantly reduced in undialysed CRF patients (25%), haemodialysis (40%) and CAPD patients (50%) compared with control values. Plasma L-arginine concentration is also strongly correlated to renal function indices in undialysed CRF patients. It is striking that in our data presented above an inverse correlation was found between plasma levels of L-arginine and L-arginine influx via transport system y^+ into RBCs in undialysed CRF patients. By contrast, a similar correlation was not found in CAPD or haemodialysis patients. Perhaps in this group of end-stage renal failure patients, plasma L-arginine depletion not only reflects the impairment of renal function but also losses of protein in the dialysate of CAPD and haemodialysis patients, thus disturbing any association between L-arginine uptake and plasma levels.

Although many factors are involved in the up-regulation of the transport system y^+ in uraemia, the results presented here strongly suggest that low concentration of plasma L-arginine is a key factor involved in the activation of L-arginine uptake in chronic renal failure. Assuming that they do produce NO, the finding that L-arginine transport is increased in RBCs, may be associated with elevated NO production in these cells. The increase in L-arginine uptake into RBCs may reflect a broad activation of L-arginine-NO pathway in CRF. It is possible that L-arginine transport activation in chronic renal failure serves to increase intracellular availability of L-arginine for the production of NO in conditions of reduced L-arginine, and increased L-arginine analogues plasma levels.

27.4 Chronic Heart Failure

Heart failure is a complex clinical syndrome broadly defined as a condition in which cardiac output is insufficient to maintain adequate perfusion of tissues (Packer 1992). Heart failure usually occurs following a loss of myocardial function, which in turn activates haemodynamic and neurohormonal mechanisms to preserve cardiac function. As a result of decreased myocardial contractility, a greater filling pressure is necessary to achieve the same stroke volume until a point is reached where this mechanism fails and further elevation of venous pressure reduces cardiac output (Packer 1992). Amongst the neurohormonal systems activated during chronic heart failure (CHF) are the renin-angiotensin-aldosterone system, sympathetic nervous system, endothelin and vasopressin leading to vasoconstriction and water and sodium retention (Coats and Poole-Wilson 1997).

The kidney plays a central role in the salt and water retention present in heart failure. A significant reduction in glomerular filtration rate accompanies impairment in cardiac output and is caused by reduced arterial pressure and by constriction of the preglomerular arterioles in response to increased sympathetic activity and angiotensin II production (Coats and Poole-Wilson 1997). This reduction in glomerular filtration rate reduces sodium delivery to the nephron and, associated with direct actions of angiotensin II on tubular sodium reabsorption, may cause sodium and water retention (Brandt and Burnett 1997). Whereas neurohormonal factors have been extensively studied, the impact of local factors regulating peripheral vasomotor tone and tissue perfusion in chronic heart failure is poorly defined.

27.4.1 L-Arginine-NO Pathway and Heart Failure

Clinical and experimental observations have demonstrated that chronic heart failure is associated with generalised impairment in endothelium-dependent vasodilation which may contribute to the progression of ventricular and vascular remodelling in this syndrome (Kaiser et al. 1989; Katz et al. 1997). There is also accumulating evidence that disturbances of NO availability contribute to abnormal endothelial cell function. NO is released both under basal conditions and upon stimulation by endothelial cells (Vanhoutte and Shimokawa 1989), which are able to express constitutive and inducible NOS (Fleming et al. 1997). NO is an endogenous local vasodilator and appears to be a major regulatory mechanism of peripheral vascular tone. Impaired endothelial function in heart failure may be due to decreased basal production or increased degradation of NO. On the one hand, decreased 24-hour urinary excretion of labelled nitrates in chronic heart failure patients and downregulation of eNOS gene expression in aortic endothelium in experimental models of chronic heart failure indicate reduction in basal release of NO (Comini et al. 1996; Katz et al. 1999), just as increased plasma levels of ADMA can also contribute to impaired NO-mediated vasodilation in chronic heart failure (Feng et al. 1998; Usui et al. 1998). On the other hand, the findings that chronic heart failure patients have elevated plasma levels of nitrate and exaggerated vasoconstrictor response to L-NMMA in systemic vascular resistance and in forearm blood flow suggest an enhanced vascular release of NO (Winlaw et al. 1994; Drexler et al. 1997). Improvements in endothelial function in chronic heart failure patients have been shown with some interventions that increase NO availability, such as radical scavenging by antioxidants (e.g. vitamin C), L-arginine infusions and physical training (Hornig et al. 1998; Hambrecht et al. 2000). Circulating cytokines, including TNF-α, IL-1 and IL-6, are increased in chronic heart failure (Katz et al. 1994; Lomni et al. 1997). The cause of cytokine activation in CHF remains unknown. It has been reported that TNF-α has a dual effect on the biosynthesis of NO, decreasing mRNA for the eNOS in endothelial cells (Yoshizumi et al. 1993), but inducing iNOS expression in vascular endothelial and smooth muscle cells (Knowles and Moncada 1994).

27.4.2 Amino Acid Plasma Profile in Chronic Heart Failure

In CHF patients, plasma L-arginine levels are halved compared with controls. Other cationic amino acids such as L-lysine and L-ornithine are also significantly reduced in chronic heart failure patients. Neutral amino acids, L-alanine and L-leucine, are likewise diminished in heart failure compared with controls. Considering the precursors of glutathione (L-cysteine, L-glycine and L-glutamate), L-cysteine was the only one for which the plasma concentration was higher in heart failure patients than controls (Hanssen et al. 1998; Table 27.3).

Table 27.3. Amino acid concentration in plasma from controls (n=14) and chronic heart failure patients (n=12)

Amino acids	Controls [μM]	CHF [μM]
L-alanine	516 ± 40	329 ± 32*
L-arginine	125 ± 5	59 ± 3*
L-citrulline	50 ± 8	52 ± 3
L-cysteine	84 ± 6	170 ± 17*
L-glutamate	167 ± 10	137 ± 21
L-glycine	250 ± 13	224 ± 26
L-lysine	196 ± 10	133 ± 13*
L-ornithine	111 ± 6	83 ± 9*

Data denote the mean ± S.E.M., *P < 0.05, *CHF* chronic heart failure.

The aetiology of the disturbance in the plasma amino acid profile in CHF, with most amino acids in lower concentration in this disease state compared with controls, is probably multifactorial and may involve increased activity of specific amino acid transport systems in different cell types, reduced renal function, inadequate dietary intake, increased catabolism and/or impaired absorption. The last three are risk factors for protein malnutrition, which is present in 30–50% of hospitalised heart failure patients (Sullivan and Hawthorne 1997). Our findings demonstrate a reduced supply of endogenous L-arginine and are consistent with reports of an improvement of blood flow and clinical conditions in chronic heart failure patients following L-arginine supplementation.

27.4.3 Alterations on L-Arginine Transport in Chronic Heart Failure

In the same CHF patients calculated L-arginine transport into RBCs was increased whilst total apparent K_m values were similar in controls and CHF patients. The V_{max} of L-arginine transport via y^+ system was markedly increased in RBCs taken from CHF when compared with controls. The K_m of transport system y^+ was not affected in patients with CHF. The V_{max} and K_m values of L-arginine transport via system y^+L were similar in patients with CHF and controls. The diffusional coefficient was unchanged in CHF (Hanssen et al. 1998; Table 27.4).

Table 27.4. Kinetic parameters for L-arginine influx into red blood cells from controls (n=10) and chronic heart failure patients (n=15)

	System y^+		System y^+L		Diffusion
	V_{max} [μmol/(l_{cells}·h)]	K_m [μM]	V_{max} [μmol/(l_{cells}·h)]	K_m [μM]	[1/h]
Controls	292 ± 19	54 ± 7	83 ± 15	35 ± 8	0.10 ± 0.02
CHF	490 ± 60*	65 ± 3	62 ± 8	18 ± 3*	0.09 ± 0.02

Data denote the mean ± S.E.M., *$P < 0.05$, *CHF* chronic heart failure.

The observed increase in V_{max} of L-arginine via system y^+ into RBCs with no apparent effect on K_m could be explained by an increase in the number of membrane transport sites per cell due to circulating factors present in the plasma of CHF patients. To investigate this, cross-incubation experiments were performed to test whether CHF plasma could induce the increased L-arginine transport via y^+ transport system in RBCs from CHF patients. Normal RBCs incubated in CHF plasma over a period of 3 hours showed no change in L-arginine transport via system y^+. Insufficient experiments were performed to be able to come to a final conclusion, but the present findings may suggest that alterations in y^+ transport system are not a result of circulating factors acting directly on mature RBCs, which do not have the machinery to produce membrane proteins. Perhaps the alterations arise from plasma factors affecting RBC development in the bone marrow.

Neuroendocrine activation described in patients with chronic heart failure may play a direct role in the up-regulation of L-arginine transport. Angiotensin II can stimulate the L-arginine uptake via system y^+ in rat cultured aortic smooth muscle cells and this stimulation was blocked by co-treatment with valsartan, an antagonist specific for angiotensin II receptor subtype-1. The up-regulation of system y^+ by angiotensin II is accompanied by increased expression of both murine cationic amino acid transporters (mCAT) 1 and 2B in these cells (Low and Grigor 1995). In contrast, chronically elevated noradrenaline (NA), in a concentration similar to that found in CHF patients (NA, 1 ng/ml), does not affect the L-arginine transport and NOS in human umbilical vein endothelial cells *in vitro* (Bachetti et al. 1998).

Up-regulation of transport system y^+ into RBCs could also be mediated by a process linked to the high circulating levels of cytokines observed in CHF. Simmons et al. (1996) have shown in rat cardiac myocytes that cytokines increase expression of mCAT-1, as well as mCAT-2A and mCAT-2B. They also demonstrated that the cytokine-induced production of NO was markedly dependent on L-arginine influx. Another potential mechanism involved in the up-regulation of system y^+ during chronic heart failure is hyperinsulinaemia. Patients with chronic heart failure have a reduced sensitivity to the actions of insulin on glucose metabolism and a compensatory increase in endogenous plasma insulin levels, probably reflecting neuroendocrine abnormalities (Houghton et al. 1998). A recent study by Sobrevia et al. (1996) show that insulin induces the L-arginine transport via system y^+ in human endothelial cells, and perhaps the high plasma levels of insulin during CHF, as in chronic renal failure, could contribute to activation of system y^+ in RBCs.

The increased L-arginine uptake seen in RBCs from CHF patients, despite low plasma concentration (transport rate of 204 $\mu mol/(l_{cells} \cdot h)$ in controls and 233 $\mu mol/(l_{cells} \cdot h)$ in CHF patients), may reflect a widespread compensatory response to the decreased plasma L-arginine concentration to sustain the increased production of NO (Mendes Ribeiro et al. 1998). Thus, increased release of NO in chronic heart failure may represent another failing counter-regulatory mechanism in the face of increased synthesis of vasoconstrictor mediators.

The up-regulation of NO synthesis in blood cells would further deplete the intracellular concentration of L-arginine, which, together with the presence of arginine metabolites within the cell, may trigger the activation of L-arginine transport. It is also possible that the same adaptive mechanism is present in endothelial cells yet, as in circulating blood cells, the compensatory increase in L-arginine transport via y^+/CAT-1 may not be sufficient to maintain an adequate synthesis of NO.

27.5 Possible Overlapping of Mechanisms Present in Both Chronic Heart Failure and Uraemia

Accumulation of L-arginine analogues in CHF may be due to impairment in renal function and might be involved in increased vascular tone and endothelial dysfunction present in this state. One exclusion criterion in the selection of patients with CHF was an abnormal plasma creatinine level. However, involvement of the kidney in heart failure is a slow and progressive process and the selected patients would certainly have some degree of deterioration in renal function, which would not be detected by creatinine measurement.

Both chronic heart and renal failure share some common aetiological mechanisms, particularly hypertension and atherosclerotic arterial disease, resulting in a certain degree of pathophysiological overlapping of these syndromes. Our findings demonstrate a reduced supply of endogenous L-arginine and are consistent with reports of an improvement of blood flow and clinical conditions in chronic heart failure patients following L-arginine supplementation (Koifman et al. 1995). Traditionally, the changes in membrane transport observed in uraemia have been associated with the presence of toxins that would alter membrane transport properties. The fact that a similar increase in transport capacity for L-arginine was observed in chronic renal and heart failure, argues against a specific effect of an uraemic toxin. These results are consistent with the view that similarity to uraemia increased L-arginine transport in chronic heart failure may be due to low plasma levels of L-arginine. It is possible that activation of L-arginine transport via system y^+ into RBCs, if this stimulation is an universal phenomenon, would represent a mechanism to increase L-arginine availability for putative increased basal nitric oxide production.

The failure of iNOS-mediated increases in NO synthesis to reverse the widespread vasoconstriction in chronic heart failure and hypertension in uraemia may be partially explained by an insufficient availability of L-arginine for eNOS in endothelial cells.

27.6 L-Arginine Transport Alterations in Diabetes

Diabetes mellitus is a metabolic disorder characterised by a lack of and/or a resistance to insulin leading to profound derangements of electrolyte, carbohydrate and lipid metabolism. The prevalence of early and accelerated development of microvascular complications is markedly increased among individuals with diabetes (Najemnil et al. 1999), but the mechanisms are not fully understood. Studies in animals (Kobayashi and Kamata 2001) and humans (Sobrevia and Mann 1997) have suggested that alterations in endothelial-NO-mediated relaxation may contribute to the pathophysiology of diabetic vascular complications (Stout 1987). The involvement of the L-arginine-nitric oxide pathway in this process is supported by the finding that administration of L-arginine ameliorates some of the changes observed in diabetes (Smulders et al. 1994). Gestational diabetes is associated with an activation of L-arginine transport (system y^+), which correlates with increased basal nitric oxide synthesis in human endothelial cells (Sobrevia et al. 1995). An up-regulation of L-arginine influx is also observed in RBCs from patients with diabetes mellitus. The enhanced rate of L-arginine transport into RBCs is mediated selectively by the transport system y^+ in diabetic patients of both types I and II (Mendes Ribeiro, unpublished data). Diabetic patients with the additional complication of chronic renal failure present much higher activation of system y^+. Red blood cells from uncomplicated and complicated diabetes mellitus patients present no change in the transport capacity of system y^+L and the diffusion component of the flux in RBCs (Mendes Ribeiro, unpublished data). Similarly to uraemia and heart failure, levels of L-arginine in the plasma of diabetic patients are dramatically reduced (Grill et al. 1992). Elevated glucose concentrations (5–25 mM) induce expression of both the cationic amino acid transport system y^+/CAT-2B and eNOS in human endothelial cells (Flores et al. 2001). It is possible that in diabetes mellitus, high levels of glucose associated with low L-arginine concentration in plasma, membrane changes and enhanced systemic cytokine levels (Lopez-Virella and Virella 1996) may trigger the increase in L-arginine transport in RBCs.

27.7 L-Arginine Transport and Sepsis

Sepsis can be defined as a systemic disease associated with the persistent presence of pathogenic microorganisms or their toxins in the blood. Septic shock develops in about 40% of septic patients and it is characterised as sepsis with significant hypotension. During the development of sepsis, the immune system is activated releasing a variety of mediators including TNF-α and IL-6 and IL-8 (Jaimes et al. 2001; Pinsky 2001). The inability of the body to down-regulate the immune response, and the failure of the endothelium to repair the damage itself leads to hypotension and causes the clinical manifestations of septic shock. Interleukin 6 plasma concentration correlates with a score of organic dysfunction and mortality in severe sepsis (Dougnac et al. 2001). The L-arginine-NO pathway has been associated with the pathogenesis of sepsis and its cardiovascular manifestations (Preiser et al. 2001). An accumulation of L-arginine analogues and a reduction of L-arginine concentration are present in the plasma of severe septic shock patients (Mendes Ribeiro et al. 1997b). The reduction in plasma levels of L-arginine in septic shock patients is more dramatic than that observed in other disease states (uraemia, heart failure and diabetes mellitus) and correlates with cyclic GMP levels. L-arginine transport via system y^+ is increased by 60% in RBCs taken from septic shock patients (Mendes Ribeiro et al. 1997b). The same activation of L-arginine transport by system y^+ is observed in human peripheral blood mononuclear cells concomitantly with higher levels of CAT-2 mRNA expression (Reade et al. 2001). The presence of high levels of circulating cytokines and endotoxin can stimulate the transport of L-arginine and NOS in septic patients. There is evidence that serum from patients with sepsis induces nitric oxide production (Fujita et al. 2000). It is likely that the increased uptake of L-arginine, if present in endothelial cells, enables the vasculature to produce more NO. This process depletes L-arginine plasma concentration, which may further activate L-arginine transport. In contrast to humans, RBCs taken from a sheep model of sepsis do not present an activation of system y^+ (Hanssen et al., unpublished data).

27.8 Amino Acid Transport Alterations in Sickle Cell Disease

Sickle cell disease is a genetic disorder caused by a single point mutation in the β-globin of the haemoglobin molecule, which polymerises when deoxygenated (Steinberg and Rogers 2001). If both parents carry the gene for abnormal haemoglobin (haemoglobin S) and transmit it to their child, the child develops clinical symptoms, which are categorised as sickle cell disease (Hb SS). Sickle RBCs are less deformable, have a higher membrane rigidity and show an increased adhesion to endothelial cells and increased RBC/RBC interaction (Serjeant 1992; Space et al. 2000). These cellular defects of sickle cells lead to vascular occlusion, which is the source of major morbidity and mortality of sickle cell patients by causing damage to organs and tissues (Serjeant 1992). Sickle cell disease also represents a state of acute anaemia caused by the increased fragility and enhanced interaction of sickle cells, abnormalities that are aggravated by increased oxidative damage in these cells. Sickle cell patients do not only have reduced plasma levels of vitamin E but also decreased total and reduced intracellular glutathione concentrations, the major antioxidant in RBCs (Chiu et al. 1982). Red cell glutathione is synthesised from its three constituent amino acids, L-cysteine, L-glycine and L-glutamate, as described previously. L-glutamine may also represent a potential substrate following its conversion to L-glutamate. Plasma levels of L-glycine and cysteine are increased in sickle cell patients, but L-glutamine and L-glutamate are unaltered (Kiessling et al. 2000). Transport capacity of L-glutamine is 8-fold higher and the affinity is 5-fold higher in sickle RBCs in comparison with normal RBCs (Niihara et al. 1997). Systems ASC and gly are activated in sickle RBCs and further stimulated by oxygenation (Kiessling et al. 2000). Therefore, low reduced gluthatione levels in sickle cells are not due to decreased substrate availability, and it has been suggested that the stimulation of these transport systems by oxygenation is secondary to protein phosphorylation (Kiessling et al. 2000). Recently, increasing interest has been focused on nitric oxide and its role in the pathogenesis and therapy of vascular complications in sickle cell disease (Gladwin and Schechter 2001). Delivery of NO bound to haemoglobin or in plasma may improve blood flow and haemoglobin saturation reducing ischemia-reperfusion injury (Gladwin and Schechter 2001). There is evidence that NO decreases the adherence of sickle RBCs to endothelium, the latter been implicated as an initiating event of vaso-occlusion (Space et al. 2000). It appears that children with sickle cell disease have lower concentration of L-arginine in plasma during vaso-occlusive crisis (Morris et al. 2000b). Adults with sickle cell disease in steady state present a reduced plasmatic concentration of L-arginine and an increased transport of L-arginine via system y^+ in RBCs compared with controls (Mendes Ribeiro, unpublished data). The increased L-arginine influx could be an up-regulation of transport to compensate for the lower plasma levels. However, this up-regulation is out of proportion since total L-arginine influx in sickle RBCs at the measured low plasma L-arginine level (49 μM) is approx. 130% of the total L-arginine influx at normal plasma L-arginine levels (78 μM) (Kiessling et al., unpublished data). These findings could reflect a state of increased nitric oxide production. Supplementation of L-arginine can induce an enhancement in NO production in sickle cell disease patients (Morris et al. 2000a). While direct evidence of NO therapy has not been reported, strategies that enhance NO activity might be therapeutic in sickle cell disease.

References

Arese M, Strasly M, Ruva C, Costamagna C, Ghigo D, MacAllister R, Verzetti G, Tetta C, Bosia A, Bussolino F (1995) Regulation of nitric oxide synthesis in uraemia. Nephrol Dial Transplant 10:1386-1397

Bachetti T, Comini L, Agnoletti L, Agnoletti L, Pedersini P, Gaia G, Cargnoni A, Bellet M, Curello S, Ferrari R (1998) Effects of chronic noradrenaline on the nitric oxide pathway in human endothelial cells. Basis Res Cardiol 93:250-256

Bergstrom J, Alvestrand A, Furst P (1990) Plasma and muscle free amino acids in maintenance hemodialysis patients without protein malnutrition. Kidney Int 38:108-114

Bleiber R, Eggert W, Reichmann G, Kunze D (1980) Erythrocyte and plasma lipids in terminal renal insufficiency. Acta Haematol 63:117-123

Bogle RG, Baydoun AR, Pearson JD, Moncada S, Mann GE (1992) L-arginine transport is increased in macrophages generating nitric oxide. Biochemical Journal 284:15-18

Bogle RG, Baydoun AR, Pearson JD, Mann GE (1996) Regulation of L-arginine transport and nitric oxide release in superfused porcine aortic endothelial cells. J Physiol 490:229-241

Brandt RR, Burnett JC (1997) Humoral control of the kidney during congestive heart failure: role of cardiac natriuretic peptides. In: Poole-Wilson PAP, Colluci WS, Massie BM, Chatterjee K, Coats AJS (eds) The heart failure. Church Livingstone, New York, pp 143-154

Braunwald E (1992) Pathophysiology of heart failure. In: Braunwald E (ed) Heart disease. A textbook of cardiovascular Medicine. WB Saunders, Philadelphia, pp 393-418

Brunini TMC, Roberts N, Yaqoob M, Ellory JC, Mann GE, Mendes Ribeiro AC (1998) L-arginine-nitric oxide pathway in a rat model of uraemia. J Physiol 506P:33P

Brunini TMC, Khoeler A, Yaqoob M, Ellory JC, Mann GE, Mendes Ribeiro AC (1999) Blood cells from patients on continuous ambulatorial peritoneal dialysis exhibit increased L-arginine-influx via the cationic amino acid transport system y^+. J Physiol 517P:52P

Cangiano C, Cardelli-Cangiano P, Cascino A, Ceci F, Fiori A, Mulieri M, Muscaritoli M, Barberini C, Strom R, Fanelli FR (1988) Uptake of amino acids by brain microvessels isolated from rats with experimental chronic renal failure. J Neurochem 51:1675-1681

Castellino P, Solini A, Luzi L, Barr JG, Smith DJ, Petrides A, Giordano M, Carroll C, DeFronzo RA (1992) Glucose and amino acid metabolism in chronic renal failure: effect of insulin and amino acids. Am J Physiol 262:F168-F176

Chen LY, Mehta JL (1998) Evidence for the presence of L-arginine-nitric oxide pathway in human red blood cells: relevance in the effects of red blood cells on platelet function. J Cardiovasc Pharmacol 32:57-61

Chiu D, Vichinsky E, Yee M, Kleman K, Lubin B (1982) Peroxydation, vitamin E and sickle cell anemia. Ann New York Acad Sci 393:323-335

Coats AJS, Poole-Wilson PA (1996) The syndrome of heart failure In: Weatherall DJ, Ledingham JGG, Warrel DA (eds) Oxford textbook of medicine, 3rd edn. Oxford University Press, Oxford, pp 2228-2237

Comini L, Bachetti T, Gaia G, Pasini E, Agnoletti L, Pepi P, Ceconi C, Curello S, Ferrari R (1996) Aorta and skeletal muscle NO synthase expression in experimental heart failure. J Mol Cell Cardiol 28:2241-2248

Constantino A (1913) Die Permeabilität der Blukörperchen für Aminosäuren. Biochem Z 55:411-418

Cotton JR, Woodward T, Carter NW, Knochel JP (1979) Resting skeletal muscle membrane potential as index of uremic toxicity. J Clin Invest 63:501-506

Davies CH, Harding SE, Poole-Wilson PA (1996) Cellular mechanisms of contractile dysfunction in human heart failure. Eur Heart J 17:189-198

Deliconstantinos G, Villiotou V, Stavrides JC, Salemes N, Gogas J (1995) Nitric oxide and peroxynitrite production by human erythrocytes. A causative factor of toxic anemia in breast cancer patients. Anticancer Res 15:1435-1446

Descamps-Latscha B, Herbelin A, Nguyen AT, Roux-Lombard P, Zinggraff J, Moynot A, Verger C, Dahmane D, De Groote D, Jungers P (1995) Balance between IL-1, TNF and their specific inhibitors in chronic renal failure and maintenance dialysis. J Immunol 154:882-892

Deuticke B, Haest CWM (1987) Lipid modulation of transport proteins in vertebrate cell membranes. Annu Rev Physiol 49:221-235

Devés R, Boyd CA (1998) Transporters for cationic amino acids in animal cells: discovery, structure, and function. Physiol Rev 78:487-545

Devés R, Angelo S, Chavez P (1993) *N*-ethylmaleimide discriminates between two lysine transport systems in human erythrocytes. J Physiol 468:753-66

De Deyn P, Marescau B, Lornoy W, Becaus I, Lowenthal A (1986) Guanidino compounds in uraemic dialysed patients. Clin Chim Acta 157:143-150

De Wardener HE, Macgregor GA (1980) Dahl's hypothesis that a saluretic substance may be responsible for a sustained rise in arterial pressure: its possible role in essential hypertension. Kidney Int 18:1-9

Dougnac A, Riquelme A, Calvo M, Andresen M, Magedzo A, Eugeni E, Marshall G, Gutierrez M (2001) Study of cytokines kinetics in severe sepsis and its relantionship with mortality and score of organic dysfunction. Rev Med Chil 129:347-358

Drexler H (1997) Endothelial factors. In: Poole-Wilson PAP, Colluci WS, Massie BM, Chatterjee K, Coats AJS (eds) The heart failure. Church Livingstone, New York, pp 173-186

Durante W, Liao L, Iftikhar I, Brien WEO, Schafer AI (1996) Differencial regulation of L-arginine transport and nitric oxide production by vascular smooth muscle and endothelium. Circulation Research 78:1075-1082

Edmondson RP, Hilton PJ, Jones NF, Patrick J, Thomas RD (1975) Leukocyte sodium transport in uraemia. Clin Sci Mol Med 49:213-216

Ellory JC (1987) Amino acid transport systems in mammalian cells. In: Yudilevich DL, Boyd CAR (eds) Amino acid transport in animal cells. Manchester University Press, Manchester, pp 106-119

El Nahas AM, Winearls CG (1996) Chronic renal failure and its treatment. In: Weatherall DJ, Ledingham JGG, Warrel DA (eds) Oxford textbook of medicine, 3rd edn. Oxford University Press, Oxford, pp 3294-3306

Featherson WR, Rogers QR, Freeland RA (1973) Relative importance of kidney and liver in synthesis of arginine by the rat. Am J Physiol 224:127-129

Feng Q, Lu X, Fortin AJ, Pettersson A, Hedner T, Kline RL, Arnold JM (1998) Elevation of an endogenous inhibitor of nitric oxide synthesis in experimental congestive heart failure. Cardiovasc Res 37:667-675

Fervenza FC, Harvey CM, Hendry BM, Ellory JC (1989) Increased lysine transport capacity in erythrocytes from patients with chronic renal failure. Clin Sci 76:419-422

Fervenza FC, Harvey CM, Hendry BM (1990) A study of erythrocyte amino acid transport in patients with chronic renal failure. Nephrol. Dial Transplant 5:594-599

Fleming I, Bauersachs J, Busse R (1997) Calcium-dependent and independent activation of the endothelial cells. J Vasc Res 34:165-174

Flores C, Aguayo C, Parodi J, Sugden D, Mann GE, Pearson JD, Sobrevia L (2001) Elevated D-glucose induces expression of the cationic amino acid transport system y^+/CAT-2B and endothelial nitric oxide synthase in human endothelial cells. J Physiol 533P:82P

Franscis GS (1998) Pathophysiology of the heart failure clinical syndrome. In: Topol EJ (ed) Textbook of cardiovascular medicine. Lipincott-Raven, Philadelphia, pp 2179-2202

Fraser CL, Sarnacki AP, Arieff AI (1985) Abnormal sodium transport in synaptosomes from brain of uraemic rats. J Clin Invest 75:2014-2023

Fujita H, Tomiyama J, Kudo H, Morita I, Murota S (2000) Sera from patients with sepsis induce nitric oxide production in vascular smooth muscle cells. Mediators Inflamm 9:285-287

Gladwin MT, Schechter AN (2001) Nitric oxide theraoy in sickle cell disease. Semin Hematol 38:333-342

Girelli D, Azzini M, Olivieri O, Guarini P, Trevisan MT, Lupo A, Bernich P, Panzetta G, Corrocher (1992) Red blood cells and platelet membrane fatty acids in non-dyalized and dyalised uremics. Clin Chim Acta 211:155-166

Grill V, Bjorkman O, Gutniak M, Lindquist M (1992) Brain uptake and release of amino acids in nondiabetic and insulin-dependent diabetic subjects: important role of glutamine release for nitrogen balance. Metabolism 4:28-32

Hagenfeldt L, Dahlquist G, Persson B (1989) Plasma amino acids in relation to metabolic control in insulin-dependent diabetic children. Acta Pediatr Scand 794:278-282

Hambrecht R, Hilbrich L, Erbs S, Gielen S, Fiehn E, Schoene N, Schuler G (2000) Correction of endothelial dysfunction in chronic heart failure: additional effects of exercise training and oral L-arginine. J Am Coll Cardiol 35:706-713

Hand MF, Haynes WG, Webb DJ (1998) Hemodialysis and L-arginine, but not D-arginine, correct renal failure-associated endothelial dysfunction. Kidney Int 53:1068-1077

Hanssen H, Brunini TMC, Conway M, Roberts NB, Mann GE, Mendes Ribeiro AC (1998) Activation of L-argnine transport in human red blood cells in chronic heart failure. Clinical Science 94:43-48

Hendry BM (1992) Membrane transport in uraemia. In: AEG Raine (ed) Advanced renal medicine. Oxford University Press, Oxford, pp 16-23

Herbelin A, Nguyen AT, Zingraff J, Urena P, Descamps-Latscha B (1990) Influence of uraemia and hemodialysis on circulating interleukin-1 and tumor necrosis factor. Kidney Int 37:116-125

Hillman RS, Finch CA (1996) General charactheristics of the erythron. In: Hillman RS, Finch CA (eds) Red cell manual. Davis Company, Philadelphia, pp 15-17

Hornig B, Arakawa N, Kohler C, Drexler H (1998) Vitamin C improves endothelial function of conduit arteries in patients with chronic heart failure. Circulation 97:363-368

Houghton AR, Harrison M, Perry AJ, Evans AJ, Cowley AJ (1998) Endogenous insulin and insulin sensitivity. An important determinant of skeletal muscle blood flow in chronic heart failure? Eur Heart J 19:476-480

Hyatt SL, Aulak KS, Malandro M, Kilberg MS, Hatzoglou M (1997) Adaptive regulation of the cationic amino acid transporter-1 (CAT-1) in Fao cells. J Biol Chem 272:19951-19957

Jaimes EA, Del Castillo D, Rutherford MS, Raij L (2001) Countervailing influence of tumor necrosis factor alpha and nitric oxide in endotoxemia. J Am Soc Nephrol 12:1204-1210

Jain SK (1989) Hyperglycaemia can cause membrane lipid peroxidation and osmotic fragility in human red blood cells. J Biol Chem 264:21340-21345

Javeshghani D, Magder S (2001) Regional changes in constitutive nitric oxide synthase and hemodynamic consequences of its inhibition in lipopolysacharide-treated pigs. Shock 16:232-238

Jubelin BC, Gierman JL (1996) Erythrocytes may synthesize their own nitric oxide. Am J Hypertens 12:1214-1219

Kari JA, Donald AE, Vallance DT, Bruckdorfer KR, Leone A, Mullen MJ, Bunce T, Dorado B, Deanfield JE, Rees L (1997) Physiology and biochemistry of endothelial function in children with chronic renal failure. Kidney Int 52:468-472

Kaiser L, Spickard RC, Olivier NB (1989) Heart failure depresses endothelium-dependent response in canine femoral artery. Am J Physiol 256:H962-H967

Katz SD (1997) Mechanisms and implications of endothelial dysfunction in congestive heart failure. Curr Opin Cardiol 12:259-264

Katz SD, Khan T, Zeballos GA, Mathew L, Potharlanka P, Knecht M, Whelan J (1999) Decreased activity of L-arginine-nitric oxide metabolic pathway in patients with congestive heart failure. Circulation 27:2113-2117

Kaye DM, Ahlers BA, Autelitano DJ, Chin-Dusting P (2000) *In vivo* and *in vitro* evidence for impaired arginine transport in human heart failure. Circulation 102:2707-2712

Kielstein JT, Boger RH, Bode-Boger SM, Schaffer J, Barbey M, Koch KM, Frolich JC (1999) Assymetric dimethylarginine concentrations differ in patients with end-stage renal disease: relantionship to treatment method and atherosclerotic disease. J Am Soc Nephrol 10:594-600

Knowles RG, Moncada S (1994) Nitric oxide synthases in mammals. Biochem J 298:249-258

Kobayashi T, Kamata K (2001) Effect of chronic insulin treatment on NO production and endothelium-dependent relaxation in aortae from established STZ-induced diabetic rats. Atherosclerosis 155:313-320

Koifman B, Wollman Y, Bogomolny N, Chernichowsky T, Finkelstein A, Peer G, Scherez J, Blum M, Laniado S, Iaina A (1995) Improvement of cardiac performance by intravenous infusion of L-arginine in patients with moderate congestive heart failure. J Am Coll Cardiol 26:1251-12566

Kosaka H, Tanaka S, Yoshii T, Kumura E, Seiyama A, Shiga T (1994) Direct proof of nitric oxide formation from a nitrovasodilator metabolised by erythrocytes. Biochem Biophys Res Commun 204:1055-1060

Levillain O, Hus-Citharel A, Morel F, Bankir L (1990) Localization of L-arginine synthesis along rat nephron. Am J Physiol 259:F916-F923

Lopez-Virella MF, Virella G (1996) Cytokines, modified lipoproteins and atherosclerosis in diabetes. Diabetes 45:40-44

Low BC, Grigor MR (1995) Angiotensin II stimulates system y^+ and cationic amino acid transporter gene expression in cultured vascular smooth muscle cells. J Biol Chem 270:27577-27583

Macallister RJ, Vallance P (1998) Endogenous inhibitors of nitric oxide synthesis: How important are they? Exp Nephrol 6:195-199

Maejima K, Nakano S, Himeno M, Tsuda S, Makiishi H, Ito T, Nakagawa A, Kigoshi T, Ishibashi T, Nishio M, Uchida K (2001) Increased basal levels of plasma nitric oxide in Type II diabetic subjects. Relashionship to microvascular complications. J Diabetes Complications 15:135-143

May RC, Clark AC, Goheer MA, Mitch WE (1985) Specific defects in insulin-mediated muscle metabolism in acute uremia. Kidney Int 28:490-497

Mendes Ribeiro AC, Roberts NB, Lane C, Yaqoob M, Ellory JC (1996) Accumulation of the endogenous analogue N^G-monomethyl-L-arginine in human end-stage renal failure patients on regular haemodialysis. Exp Physiol 81:475-481

Mendes Ribeiro AC, Hanssen H, Kiessling K, Roberts NB, Mann GE, Ellory JC (1997a) Transport of L-arginine and the nitric oxide inhibitor N^G-Monomethyl-L-arginine in human erythrocytes in chronic renal failure. Clin Sci 93:57-64

Mendes Ribeiro AC, Hanssen H, Young JD, Mann GE, Ellory JC (1997b) Transport of L-arginine in red blood cells from septic patients. Int J Microcirc Clin Exp 17:214P

Mendes Ribeiro AC, Brunini TMC, Yaqoob M, Ellory JC, Mann GE (1998) Peripheral mononuclear cells from uraemic patients exhibit an increased transport of L-arginine via system y^+ (CAT-1): implications for nitric oxide synthesis. J Physiol 506P:30P

Mendes Ribeiro AC, Brunini TMC, Yaqoob M, Aronson JK, Ellory JC, Mann GE (1999) Identification of system y^+L as the high affinity transporter for L-arginine in human platelets: up-regulation of L-arginine influx. Eur J Physiol 438:573-575

Mendes Ribeiro AC, Brunini TMC, Ellory JC, Mann GE (2001) Abnormalities in L-arginine transport and nitric oxide biossynthesis in chronic renal and heart failure. Cardiovasc Res 49:697-712

Morris CR, Kuypers FA, Larkin S, Sweeters N, Simon J, Vichinsky EP, Styles LA (2000a) Arginine therapy: a novel strategy to induce nitric oxide production in sickle cell disease. Br J Haematol 111: 498-500

Morris CR, Kuypers FA, Larkin S, Vichinsky EP, Styles LA (2000b) Patterns of arginine and nitric oxide in patients with sickle cell disease with vaso-occlusive crisis and acute chest syndrome. J Paediatr Hematol Oncol 22:515-520

Munoz M, Sweiry JH, Mann GE (1995) Insulin stimulates cationic amino acid transport activity in the isolated perfused rat pancreas. Exp Physiol 80:745-753

Najeminik C, Sinzinger H, Kritz H (1999) Endothelial dysfunction, atherosclerosis and diabetes. Acta Med Austriaca 26:148-153

Niihara Y, Zerez CR, Akiyama DS, Tanaka KR (1997) Increased red cell glutamine availability in sickle cell anemia: demonstration of increased active transport, affinity, and increased glutamate level in intact red cells. J Lab Clin Med 130:83-90

Packer M (1992) Pathophysiology of chronic heart failure. Lancet 340:88-92

Palacin M, Estevez R, Bertran J, Zorzano A (1998) Molecular biology of mammalian plasma membrane amino acid transporters. Physiol Rev 78:969-1054

Panasenko OM, Shalina RI, Azizova OA (1985) Structural changes in erythrocyte membranes in nephropathy. Biull Eksp Biol Med 99:434-437

Pawloski JR, Swaminathan RV, Stamler JS (1998) Cell-free and erythrocytic S-nitrosohemoglobin inhibits human platelet aggregation. Circulation 97:263-267

Pereira BJ, Shapiro L, King AJ, Falagas ME, Strom JA, Dinarello CA (1994) Plasma levels of IL-1 beta, TNF alpha and their specific inhibitors in undialysed chronic renal failure, CAPD and hemodialysis patients. Kiney Int 45:890-896

Pielichowski J, Kwiatkowska J, Zatonski W (1977) Erythrocyte lipids in chronic renal failure. Arch Immunol Ther Exp 25:213-217

Pinsky MR (2001) Sepsis: a pro- and anti-inflamatory disequilibrium syndrome. Contrib Nephrol 132: 354-366

Praiser JC, Zhang H, Vray B, Hrabak A, Vincent JL (2001) Time course of inducible nitric oxide synthase activity following endotoxin adiministraion in dogs. Nitric Oxide 5:208-211

Reade MC, Young JD, Boyd CAR (2001) Cationic amino acid transporter mRNA levels in peripheral blood mononuclear cells from patients with septic shock. J Physiol 535P:46P

Remuzzi G, Rossi EC (1996) Hematologic consequences of renal failure. In: Brenner BM (ed) The kidney. WB Saunders, Philadelphia, pp 2170-2186

Reyes AA, Karl I, Klahr S (1994) Role of L-arginine in health and in renal disease. Am J Physiol 267:F331-F346

Riesler P (1962) The kinetics of valine transport in the human erythrocyte. Expl Cell Res 27:577-580

Schmidt RJ, Domico J, Samsell LS, Yokota S, Tracy TS, Sorkin MI, Engels K, Baylis C (1999) Indices of activity of nitric oxide system in hemodialysis patients. Am J Kidney Dis 34:228-234

Shibazaki T, Fujiwara M, Sato H, Fujiwara K, Abe K, Bannai S (1996) Relevance of the L-arginine transport activity to the nitric oxide synthesis in mouse peritoneal macrophages stimulated with bacterial lipopolysaccharide. Biochim Biophys Acta 1311:150-154

Simmons WW, Closs EI, Cunningham JM, Smith TW, Kelly RA (1996) Cytokines and insulin induce cationic amino acid transporter (CAT) expression in cardiac myocytes. J Biol Chem 271:11694-11702

Smith CJ, Sun D, Hoegler C, Zhang X, Zhao G, Xu XB, Kobari Y, Pritchard K, Sessa WC, Hintze TH (1996) Reduced gene expression of vascular endothelial NO synthase and cyclooxygenase-1 in heart failure. Circ Res 78:58-64

Smulders RA, Stehouwer CDA, Olthof CG, Van Kamp GJ, Teerlink T, De Vries PM, Donker AJ (1994) Plasma endothelin levels and vascular efects of intravenous L-arginine infusion in subjects with uncomplicated insulin-dependent diabetes mellits. Clin Sci 87:37-43

Sobrevia L, Mann GE (1997) Dysfunction of the endothelial nitric oxide signalling pathway in diabetes and hyperglycaemia. Exp Physiol 82:423-452

Sobrevia L, Cesare P, Yudilevich DL, Mann GE (1995) Diabetes-induced activation of system y^+ and nitric oxide synthase in human endothelial cells: association with membrane hyperpolarization. J Physiol 489:183-192

Sobrevia L, Nadal A, Yudilevich DL, Mann GE (1996) Activation of L-arginine transport (system y^+) and nitric oxide synthase by elevated D-glucose and insulin in human endothelial cells. J Physiol 490:775-781

Space SL, Lane PA, Pickett CK, Weil JV (2000) Nitric oxide attenuates normal and sickle red blood cells adherence to pulmonary endothelium. Am J Hematol 63:200-204

Stathopulos PB, Lu X, Shen J, Scott JA, Hammond JR, McCormack DG, Arnold JM, Feng Q (2001) Increased L-arginine uptake and inducible nitric oxide synthase activity in aortas of rats with heart failure. Am J Physiol 280:H859-H867

Steinberg MH, Rodgers GP (2001) Pathophysiology of sickle cell disease: role of cellular and genetic modifers. Semin Hematol 38:299-306

Stout RW (1987) The endothelial cell in diabetes. Front Diabetes 8:116-124

Sullivan MJ, Hawthorne MH (1997) Nonpharmacologic intervention In: Poole-Wilson PAP, Colluci WS, Massie BM, Chatterjee K, Coats AJS (eds) The heart failure. Church Livingstone, New York, pp 617-633

Swales JD (1988) Blood pressure: from cells to populations. J Royal Coll Physicians 22:11-15

Swan JW, Walton C, Godsland IF, Clark AL, Coats AJ, Oliver MF (1994) Insulin resistance in chronic heart failure. Eur Heart J 15:1528-1532

Tattersall J, Farrington K, Greenwood R (1998) Adequacy of dialysis. In: Davison AM, Cameron JS, Grunfeld JP, Kerr DNS, Ritz E, Winearls CG (eds) Oxford textbook of clinical nephrology. Oxford University Press, Oxford, pp 2075-2088

Torrents D, Estevez R, Pineda M, Fernandez EL, Lloberas J, Shi YB, Zorzano A, Palancin M (1998) Identification and characterization of a membrane protein (y^+L amino acid transporter-1) that associates with 4F2hc to encode the amino acid transport activity y^+L. A candidate gene for lysinuric protein intolerance. J Biol Chem 273:32437-32445

Tunnicliff G (1994) Amino acid transport by human erythrocyte. Comp Biochem Physiol 108:471-478

Usui M, Matsuoka H, Miyazaki H, Ueda S, Okuda S, Imaizumi T (1998) Increased endogenous nitric oxide synthase inhibitor in patients with congestive heart failure. Life Sci 62:2425-2430

Vallance P, Leone A, Calver A, Collier J, Moncada S (1992) Accumulation of an endogenous inhibitor of nitric oxide synthesis in chronic renal failure. Lancet 339:572-575

Vanholder R (1998) The uraemic syndrome. In: Greenberg A (ed) Primer on kidney diseases. Academic Press, San Diego, pp 403-407

Vanhoutte PM, Shimokawa H (1989) Endothelium-derived relaxing factor and coronary vasoespasm. Circulation 80:87-98

Warnock DG (1996) Chronic renal failure. In: Benett CJ, Blum F (eds) Cecil textbook of medicine, 20th edn. WB Saunders, Baltimore, pp 556-576

Welt LG, Sachs JR, Mcmanus TJ (1964) An ion transport defect from uremic patients. Trans Assoc Am Physiol 77:169-181

Westman J, George S, Scheel PJ Jr, McMurray SD, Pulliam J (1996) Options for dialysis providers in a global capitated environment. Nephrol News Issues 10:26-31

Winter CG, Christensen H (1964) Migration of amino acids across the membrane of the human erythrocyte. J Biol Chem 239:872-878

Winlaw DS, Smythe GA, Keogh AM, Schyvens CG, Spratt PM, Macdonald PS (1994) Increased nitric oxide production in heart failure. Lancet 44:373-374

Yorek MA, Strom DK, Spector AA (1984) Effect of membrane polyunsaturation on carrier-mediated transport in cultured retinoblastoma cells: alterations in taurine uptake. J Neurochem 42:254-261

Yoshizumi M, Perrella MA, Burnett JC Jr, Lee ME (1993) Tumor necrosis factor down-regulates an endothelial nitric oxide synthase mRNA by shortening its half-life. Circulation Research 73:205-209

Young JD, Ellory JC, Tucker EM (1975) Amino acid transport defect in gluthatione-deficient sheep erythrocytes. Nature 254:156-157

28 Transgenic Models of Red Cell Disorders

George P. Patrinos and Frank G. Grosveld

MGC Department of Cell Biology and Genetics, Faculty of Medicine, Erasmus University Medical Center Rotterdam, PO Box 1738, 3000 DR, Rotterdam, The Netherlands

28.1 Introduction

Over the past two decades, the development of gene transfer technology in whole animals and the revolution of molecular genetics approaches have offered the investigators unprecedented opportunities. These include the study of mechanisms of gene expression *in vivo* and the development of animal models of human diseases. This has greatly advanced our knowledge of gene regulation and function, development and pathogenesis of disease.

This chapter describes the existing animal models to study the aetiology of red blood cell (RBC) diseases, such as RBC membrane and transport defects, haemoglobinopathies, haemophilia and various metabolic disorders. These animal models are obtained either by conventional transgenesis, or targeted disruption of the desired gene. This chapter also refers to the animal models, which helped elucidating the role of erythroid-specific transcription factors in haematopoiesis.

28.2 Production of Genetically Modified Animals

Although a number of different animal species have been employed in the transgenic technology, such as rats, pigs, cows, cattle and fish, the most commonly used mammal in the transgenic and gene targeting technology is the mouse (*Mus musculus*). The mouse has a rather short gestation time (19–20 days), and reaches the age of sexual maturity reasonably fast (6 weeks). Moreover the average litter size is 6–8 animals, which allows for quick generation and establishment of a transgenic line. In combination with the wealth of knowledge on the development of the mouse, which is closely related to human development, makes the mouse an attractive "substrate" to study gene regulation and resolve the aetiology of human diseases.

An overview of the technology available for the generation of transgenic animals and the merits of the different gene-transfer techniques, such as conventional transgenesis and conventional or tissue-specific gene targeting are discussed in this section.

28.2.1 Conventional Transgenesis

By definition, a transgenic animal contains a fragment of exogenous DNA (transgene) that has been physically inserted and integrated into the genome of all cells, including the germline, and thus can be successfully transmitted to the offspring.

The most common and well-characterized approach to generate transgenic animals is microinjection directly into the male pronucleus of a fertilized one-cell embryo. If appropriate equipment is available, microinjection can be achieved using an ultra thin glass capillary pipette (< 1 μm of tip diameter), filled with the transgene of interest-containing solution. Delivery of approx. 1–2 pl of the solution into the male pronucleus results in a visible increase of the pronuclear circumference. The surviving microinjected eggs are then incubated until they reach the two-cell stage and subsequently re-implantated into the oviducts of pseudopregnant foster mothers.

Eggs are obtained from hormonally super-ovulated female donors. The hormones used are pregnant mares' solution and human chorionic gonadotropin, both at a concentration of 10 units intra-peritoneally. The timing of the injection of the hormones is critical. The females are then bred to males overnight and the eggs are collected the next day from surgically removed oviducts. The pseudopregnant foster mothers are generated by breeding of female animals with vasectomized or genetically sterile males. Details on these procedures, the equipment and protocols can be found elsewhere (Hogan et al. 1986).

In practice, transgenes are integrated into the genome of 5% to 30% of the offspring. The identification of the transgenic animals can be done using restriction endonuclease digestion of genomic DNA, following by Southern blot analysis, dot blotting of undigested genomic DNA or Polymerase Chain Reaction (PCR). Genomic DNA is usually isolated from a small tail or ear biopsy sample.

Microinjected DNA fragments have integrate randomly in the host genome which can sometimes be a disadvantage. Random integration could for example lead to the inactivation of an essential gene or the activation of a gene that is normally silent, which both could lead to a phenotype not related to the properties of the transgene. More likely to be a problem are the fact that not all the regulatory sequences are known for the vast majority of genes which often results in poor or variegated or no expression of the transgene. A complicating factor in this is the fact that a transgene often integrates into a transcriptionally silent (heterochromatic) region, which can alter or severely depress the normal expression profile of the transgene. This difficulty can be circumvented by including in the construct for injection, sequences which are capable to drive position of integration-independent expression of the transgene. Such sequences are the "Locus Control Regions" (LCRs; Grosveld et al. 1987; Kollias and Grosveld 1992). An even safer way of mimicking the expression profile of a known gene is to integrate the transgene by homologous recombination into an endogenous mouse gene with the desired expression profile (see Sect. 28.2.2).

28.2.2 Gene Targeting

Gene targeting is referred to as the predetermined genetic modification in the germline. This resulted from the combination of two technologies, namely murine embryonic stem (ES) cell system and homologous recombination.

ES cells are derived from the inner cell mass of a blastocyst and remain undifferentiated under suitable tissue culture conditions. When these *in vitro* cultured ES cells are introduced into a blastocyst, they can continue their development and contribute by generating chimaeric animals that are capable of transmitting the ES cell genome to their offspring. Thus, genetic changes introduced into ES cells can be transferred into live animals by breeding the chimaeras and screening the offspring for the ES genotype.

Homologous recombination provides the means to alter the sequence of a chosen gene in a predetermined way and provides the capability of selectively altering genes in the mouse.

The most frequent application of gene targeting is the gene knockout, resulting in the disruption of the desired gene. Gene targeting requires a targeting construct. This is usually a plasmid containing:

(1) an upstream region of homology,
(2) a selectable marker gene for positive selection, which will disrupt the gene of interest by substituting part of its coding sequence. The most commonly used gene for positive selection is the neomycin gene, which renders the host resistant to the drug G418 (an aminoglycoside related to neomycin). This way ES cells in which the targeting construct will be integrated will survive in the presence of G418;
(3) a downstream homology region,
(4) a selectable marker gene for negative selection outside the area of homology. This marker gene will end up in the genome when the targeting construct will have been integrated randomly in the genome by non-homologous recombination, but is lost in case of homologous recombination in the target locus. The most frequently used marker gene is the herpes simplex thymidine kinase (TK) gene, which is used in conjunction with the drug ganciclovir.

The ES cells that integrate the targeting construct randomly into their genome retain and express the TK gene and consequently do not survive in the presence of ganciclovir. However, cells in which the targeting construct have been integrated homologously into the target sequence, the TK gene will be lost in the homologous crossover event and thus this ES cells will survive in the presence of ganciclovir. The correctly targeted ES cells can then be identified by Southern blot analysis or PCR. Examples of targeting constructs are given in Fig. 28.1. To ensure high homologous recombination frequencies, the targeting constructs should be isogenic, i.e. derived from the same mouse strain from which the ES cells are also derived (usually 129 Ola, derived from agouti-coloured animals). In addition, the length of the upstream and downstream homology region within the targeting construct is also important. In general, the longer the homology the higher the recombination frequency (Doi et al. 1992).

After integration the targeted ES cells are karyotyped to ensure that no aneuploidy has arisen as a result of the culturing procedures and subsequently intro-

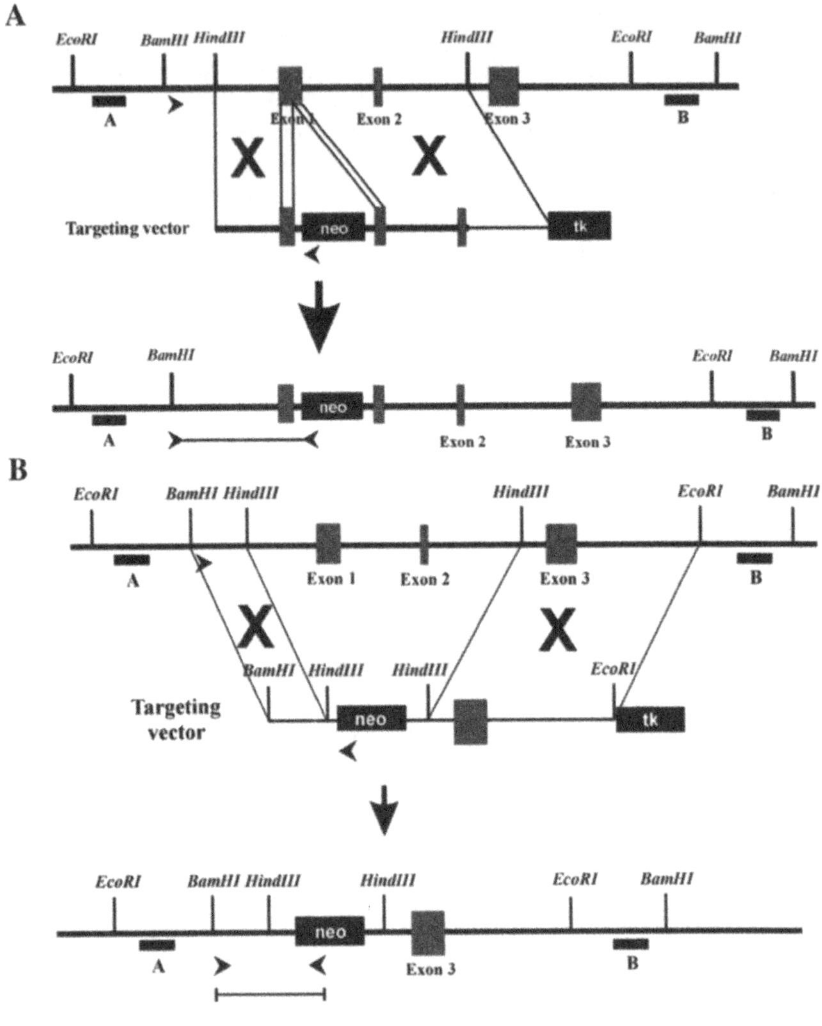

Fig. 28.1. Targeting strategies. Disruption of the gene of interest can result either by inserting a neomycin resistant gene into an exon sequence (insertional inactivation, (**A**)) or by substituting part of the coding sequences by a neomycin resistant gene (deletional inactivation, (**B**)). Both ways, the synthesis of a functional gene product is ablated. The existence of primer sets (black arrows) or probes (black boxes termed *A* and *B*), allows the identification of the correctly targeted clones either by PCR or by Southern blot analysis respectively

duced into blastocysts, isolated from donor female animals. This is usually done by microinjection, using a thin (4–5 μm tip diameter) glass pipette. Following the microinjection session, blastocysts are transferred surgically into the uterus of pseudopregnant foster mothers (Fig. 28.2). Blastocysts are derived from C57BL6 (black coloured) animals. This way, the resulting offspring will contain black animals, which will not be transgenic and chimaeric animals (black with white spots), which will be composed by a mosaic of black (C57BL6) and agouti (129) somatic and germline cells. An alternative to injecting targeted ES cells into blastocyst is the generation of aggregation chimaeras. In this case, 5–8-cell embryos (morulas) are isolated during the 2.5 day of gestation and treated with hyaluronic acid, which disturbs the *zona pelucida*. After the acidic treatment, the morulas are placed together with ES cells in a Petri dish as ordered arrays (Fig. 28.2). During the incubation period, the ES cells will penetrate and aggregate into the morula. The next day, the aggregated morulas can be surgically transferred into foster mothers. This approach is particularly convenient, since it circumvents the microinjection step, but does not work efficiently with a number of the available ES cell lines.

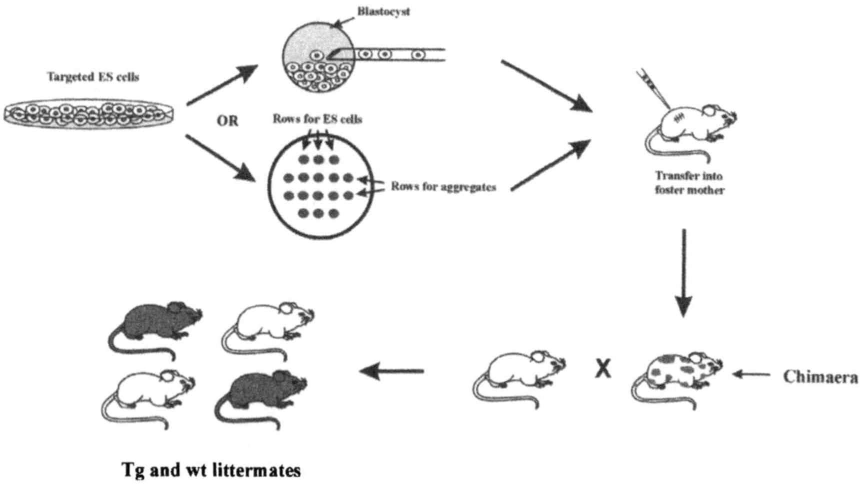

Fig. 28.2. Generation of knockout mice. Correctly modified (targeted) ES cells are cultured to establish a line and then transferred via microinjection into blastocysts or by aggregation into morulas. The blastocysts or morulas are then transferred into foster mothers which will give birth to chimaeric animals, which when bred to wild-type animals will give germline transmission to both wild-type (white-coloured) and transgenic (grey-coloured) animals

Recently, site-specific recombination as a means to inactivate genes has been increasingly employed. This type of recombination requires the presence of an exogenous protein, called a recombinase, which will mediate recombination at unique sequence motifs in the absence of any cofactors or additional proteins (reviewed in Kilby et al. (1993)). The most commonly used recombinase for this purpose is Cre-recombinase, a 38 kDa protein derived from P1 bacteriophage, which carries out site-specific recombination at two 34-bp DNA sequences, termed loxP sites. Similarly, Flp-recombinase, a 43 kDa protein derived from *Saccharomyces cerevisiae*, is also capable to perform site-specific recombination at two 34-bp Flp-specific sequences, termed as FRT sites.

Like the classical knockout, the targeting of a murine gene with loxP sites requires homologous recombination in ES cells and a targeting construct consisting of:

(1) an upstream homology region,
(2) a loxP site,
(3) the portion of the gene to be deleted,
(4) a loxP site,
(5) marker genes for positive and negative selection flanked by loxP or FRT sites,
(6) a downstream homology region (Fig. 28.3).

After transfection of ES cells and G418 selection, homologous recombinants are distinguished from random ones by Southern blot analysis. Once homologous recombinant ES cells have been identified, Cre recombinase is expressed transiently in order to eliminate the selectable marker genes. Alternatively, if the selectable marker genes are flanked by FRT sites, after the G418 selection, Flp recombinase can be expressed transiently in the ES cells to recombine out the selectable marker genes. The elimination of the selectable marker genes can be verified afterwards using southern blots or PCR in combination with gancyclovir selection. Whatever method is employed, once the final ES cell clones are identified, they can be used to generate chimaeric mice.

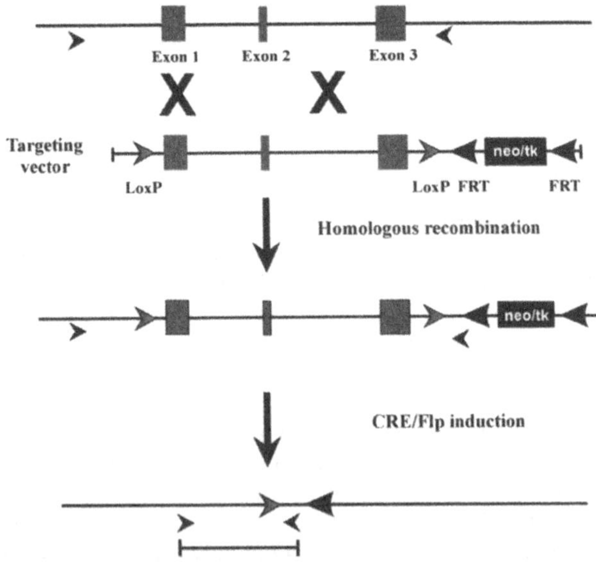

Fig. 28.3. Site-specific mediated recombination for gene knockout. In the targeting vector, the gene to be disrupted is flanked by *LoxP* sites and the positive and negative selectable markers by different target sequences, such as flp-specific *FRT* sites. By induction of Flp recombinase, the selectable marker sequences are lost and subsequently, by induction of Cre-recombinase expression, the gene itself is lost, leaving behind a singly *LoxP* and *FRT* site. The recombination event, which is indicative of a correctly targeted ES cell line, can be identified, either by PCR or Southern blot analysis

28.2.3 Tissue-Specific Gene Targeting

With the advent of gene targeting in murine ES cells, the use of knockout mice to study the effects of loss of gene function has become increasingly prevalent. However, there are a number of disadvantages with conventional gene targeting approaches, which can make the study of the resulting phenotype difficult or even impossible. Conventional gene targeting results in the inactivation of the gene product in all tissues and cells, which can be problematic for genes required developmentally, which are part of complex paracrine systems or exhibit a wide tissue-specific expression pattern. These limitations can be potentially overcome by using a system in which the loss of gene function is placed under spatial (tissue- and cell-specific) and/or temporal (stage-specific) control, thus restricting the loss of function to a specific tissue or developmental stage, respectively. The use of site-specific recombination to this respect is of particular importance, as it allows to the specific modification of target genes in a tissue- and/or developmentally-regulated manner. Since loxP or FRT targeted genes will be present in every cell, the ability to perform tissue- and developmentally specific recombination relies on the choice of regulatory sequences that are used to drive the expression of the recombinase (Cre or Flp, respectively). Alternatively the recombinase itself can be introduced into an endogenous (no essential) gene with the desired expression pattern. The latter strategy is particularly useful when several conditional recombination are planned in the same cell type or at the same time of development. Once the choice is made transgenic mice are generated with the appropriate Cre or Flp expression pattern and these are bred with mice with loxP sites in the gene of interest. The resulting offspring will include animals where the gene of interest will have been disrupted in a cell-specific manner, depending on the promoter, which drives the expression of the recombinase. Another important advance has been the development of inducible expression systems that rely on exogenous compounds that normally do not occur in mice, such as tetracycline, tamoxifen or ecdysone (Feil et al. 1997; Indra et al. 1999). In these system the transgene of interest or the recombinase (in case of a conditional recombination/deletion) are placed under the control of transcriptional regulators that are activated (or suppressed) by the addition of the exogenous compound, thus allowing temporal control over transgene expression or the genetic change.

28.3 Transgenic Models of Red Cell Membrane Disorders

The RBC membrane consists of a lipid bilayer and a multi-protein scaffolding, responsible for a number of physiological functions and mechanical properties of the cell.

These protein components of the RBC skeleton are interacting vertically with the lipid bilayer and transmembrane proteins to give the membrane its strength and integrity. They also interact horizontally with each other in the inner surface of the lipid bilayer. The main protein involved in the vertical interactions is band 3, the erythrocyte's anion exchanger 1 (AE1), which interacts with ankyrin, glycophorin A and proteins 4.1 and 4.2. α- and β-spectrins interact side-to-side to form heterodimers and subsequently with each other to form tetramers. The association of these tetramers to band 3 is mediated by ankyrin. Protein 4.2 binds also to band 3 and seems to facilitate the ankyrin-band 3 interaction. Multiple spectrin tetramers interact with actin, tropomyosin, tropomodulin and adducin to form complex junctions. Protein 4.1 binds to the protein glycophorin C and p55 and seems to increase the affinity of the spectrin-actin interaction (reviewed in Tse and Lux (1999)).

When interaction between these components of the RBC skeleton is disrupted at any contact point, this result in a loss of structural and functional integrity of the membrane, leading to a variety of clinical disorders. Defects in the vertical interactions may result in hereditary spherocytosis, while defects in the horizontal interactions may result in hereditary elliptocytosis or pyropoikilocytosis.

The genes of the major RBC skeleton components have been identified as well as a number of mutations leading to the disorders mentioned above. In the following sections the existing mouse models for the RBC membrane disorders will be discussed and the different defects due to the targeted mutations are summarized in Table 28.1.

28.3.1 Hereditary Spherocytosis

Hereditary spherocytosis (HS) is the most common cause of non-immune haemolytic anaemia. HS is clinically characterized by the presence of spherocytes in peripheral blood smears and with varying haemolysis and splenomegaly. Several surveys have established that HS is caused by defects in spectrins, ankyrin and band 3 and less frequently in protein 4.2. Chemically-induced spontaneous mutagenesis coupled to gene targeting technology has been employed to elucidate the role of these proteins in maintaining the erythrocyte's membrane scaffolding.

The murine band 3 gene had been the first to be disrupted, by replacing part of exon 9, the entire exon 10 and part of exon 11, including their intervening sequences by a neomycin resistance gene (Peters et al. 1996). Alternatively, a second group has inactivated the murine AE1 gene, by inserting a neomycin gene inside exon 3 to ensure erythroid- but not kidney-specific AE1 gene targeting (Southgate et al. 1996). Both targeting approaches resulted in a complete disruption of the murine AE1 gene. $AE1^{-/-}$ phenotype has a high mortality rate at the neonatal period and can be distinguished easily by the extreme pallor. This high mortality rate may be related to a high hypercoagulable state, which appears to originate from changes in the phospholipid composition of the RBC membrane as a result of the complete loss of band 3 (Hassoun et al. 1998b). Adult $AE1^{-/-}$ mice are smaller than their normal littermates and have severe haemolytic anaemia. Their haematological indices are dramatically decreased and the spleen is approx. twice as normal as their littermates. RBC morphology is markedly abnormal in $AE1^{-/-}$ mice with microcytosis, spherocytosis and frequent rod-like extensions in the membrane. In addition, $AE1^{-/-}$ RBCs are extremely fragile. These data indicate that band 3 deficiency results in spherocytosis and a severe haemolytic anaemia, which is evident at birth and persists throughout life.

Table 28.1. Summary of the existing mouse models for the RBC membrane disorders

Gene	Mutation	Phenotype	Effect on other membrane proteins									References
			Bd3	α-Sp	β-Sp	Ank	Pr 4.2	Pr 4.1	Gp A	Gp C	p55	
Bd 3	Sp (wan)	HS	-	ND	ND	ND	ND	ND	ND	ND	ND	Peters et al. 1998
Bd 3	GTg	HS	X	Nl	Nl	↓	-	Nl	-	Nl	Nl	Peters et al. 1996, Southgate et al. 1996
Pr 4.2	GTg	HS	↓↓	Nl	Nl	Nl	X	Nl	ND	Nl	Nl	Peters et al. 1999
Pr 4.1	GTg	HS	ND	↓	↓	↓	Nl	X	ND	↓↓	-	Shi et al. 1999
β-Sp	Sp (ja)	HS	Nl	-	-	Nl	ND	Nl	ND	ND	ND	Bodine et al. 1984
α-Sp	Sp (ha)	HS	Nl	↓	Nl	Nl	ND	Nl	ND	ND	ND	Bodine et al. 1984
α-Sp	Sp (sph)	HS	Nl	-	↓	Nl	ND	Nl	ND	ND	ND	Bodine et al. 1984
α-Sp	Sp (sph^{2BC})	HS	Nl	-	↓	Nl	ND	Nl	ND	ND	ND	Bodine et al. 1984
α-Sp	Sp (sph^{2J})	HS	Nl	-	↓	Nl	ND	Nl	ND	ND	ND	Bodine et al. 1984
Ank	Sp (nb)	HS	Nl	↓	↓	-	ND	Nl	ND	ND	ND	Bodine et al. 1984
α-Sp	Sp (sph^{Dem})	HE/HPP	Nl	↓↓	↓↓	↓↓	ND	ND	ND	ND	ND	Wandersee et al. 2001
β-add	GTg	HE/HPP	Nl	Nl	Nl	Nl	Nl	Nl	ND	ND	ND	Muro et al. 2000
E-Tmod	GTg	HE/HPP	ND	ND	ND	ND	ND	ND	ND	ND	ND	Sung et al. 2000

Bd 3: Band 3, *α-Sp*: α-spectrin, *β-Sp*: β-spectrin, *Ank*: Ankyrin, *Pr 4.1*: Protein 4.1, *Pr 4.2*: Protein 4.2, *Gp A*: Glycophorin A, *Gp C*: Glycophorin C, *β-add*: β-adducin, *E-Tmod*: Tropomodulin, *ND*: Not determined, *Nl*: Normal levels, *X*: Targeted inactivation, ↓: Reduced levels, -: Absent, *Sp*: Spontaneous, *GTg*: Gene targeting, *HS*: hereditary spherocytosis, *HE*: hereditary elliptocytosis, *HPP*: hereditary pyropoikilocytosis.

As for RBC membrane, $AE1^{-/-}$ RBCs have normal spectrin, ankyrin and protein 4.1 content, although protein 4.2 and glycophorin A are undetectable (Hassoun et al. 1998a). These data suggest that, in the absence of band 3, a normal RBC membrane skeleton is assembled with essentially normal architecture, as indicated by electron microscopy studies (Peters et al. 1996). In addition, the absence of protein 4.2 suggests that band 3 contains the only high affinity binding site for protein 4.2. The same scenario seems plausible for the interaction between band 3 and glycophorin A, which is also absent in $AE1^{-/-}$ RBCs (Hassoun et al. 1998a).

Apart from the targeted $AE1^{-/-}$ mice, there is also a spontaneous band 3 deficient mouse strain, known as the *wan* strain (Peters et al. 1998). This strain shares most of the features of the targeted $AE1^{-/-}$ mice, such as pallor, increased neonatal mortality rate, altered haematological indices, complete absence of protein 4.2 and reduced levels of ankyrin and most interestingly failure of effective reticulocytosis. The fact that more than 70% of targeted $AE1^{-/-}$ mice also present lack of reticulocytes suggests that is should be considered as the major mortality factor in band 3 deficient mice.

In order to address the role of protein 4.2 in RBC membrane skeleton, the murine protein 4.2 gene has been targeted by replacing a segment of the gene from intron 3 to part of exon 8 by a neomycin resistance gene (Peters et al. 1999). $4.2^{-/-}$ mice present mild spherocytosis with significantly reduced haematocrit and RBC count and slightly reduced Hb content.

Although $4.2^{-/-}$ RBCs are band 3-deficient, a normal membrane skeleton is assembled and normal amounts of spectrins, ankyrin, protein 4.1, p55 and glycophorin C are synthesized. The latter is a reminder of band $3^{-/-}$ RBCs. Both band 3 and protein 4.2 deficient mice provide evidence for lipid anchoring as a major determinant of membrane stability. According to that hypothesis, the lipid bilayer is stabilized by horizontal interactions between lipids and integral membrane proteins (IMPs), such as band 3. In the absence of IMPs (caused either directly (band $3^{-/-}$) or indirectly (protein $4.2^{-/-}$)), lipid anchoring is compromised, resulting in a spherocytic shape of the erythrocyte membrane.

The role of protein 4.1 in RBC membrane has been also assessed by means of targeted inactivation of the murine protein 4.1 gene, by substituting exons 2 to 4 by a combined lacZ/neomycin (β-geo) gene. $4.1^{-/-}$ mice are viable and fertile, but they present moderate haemolytic anaemia with decreased haematocrit, mean cell volume and mean cell haemoglobin (Hb) concentration, a situation indicative of cell fragmentation due to a mechanically compromised membrane skeleton. They also present enlarged spleens and increased reticulocyte counts. Interestingly, absence of protein 4.1 is accompanied by a complete loss of p55 and a dramatic reduction of glycophorin C (Shi et al. 1999). This is a reminiscent of the protein 4.1 deficiency in human RBCs. However, the amounts of both α- and β-spectrins are also reduced, a situation never observed in human 4.1-deficient RBCs. In addition, $4.1^{-/-}$ RBCs exhibit a spherocytic appearance, in contrast to the elliptical morphology of the human 4.1-deficient cells, probably as a result of the loss of spectrins and ankyrin. These data suggest that loss of protein 4.1 compromises membrane skeleton assembly.

There are a number of spontaneous mutations, resulting in haemolytic anaemia, due to deficient spectrin and ankyrin synthesis in the mouse. Bodine et al. (1984)

described six different spontaneous mutants, occurring at distinct loci. These mutant mice strains are: (1) normoblastosis (*nb*), due to a mutation which disable ankyrin synthesis, (2) jaundiced (*ja*) with defective β-spectrin synthesis, (3) haemolytic anaemia (sph^{ha}), (4) spherocytosis (*sph*), (5) spherocytosis 2BC (sph^{2BC}) and (6) spherocytosis 2J (sph^{2J}). All four *sph* mutations have been shown to affect the synthesis of α-spectrin.

In the homozygous state, the *nb* mutation result in the accumulation of almost half of the normal amount of spectrin in the membrane skeleton, although the levels of spectrin synthesis and mRNA is similar in *nb/nb* and normal reticulocytes. Apart from that, a relatively uniform membrane skeleton can be synthesized, which lacks band 3 tetramers (Yi et al. 1997). The latter implies that the absence of normal ankyrin is the main cause of spectrin deficiency from *nb/nb* RBC membranes. This mutation is a reminiscent of the protein 4.1 deficient mouse model. In addition, it seems plausible that ankyrin is required for the attachment of the skeleton to the lipid bilayer but is dispensable for the formation of a stable spectrin-based skeleton. The existence of a truncated form of ankyrin (150 kDa) in the *nb/nb* erythrocytes does not seem to compensate for the lack of the wild-type ankyrin (210 kDa), leading to a spherocytosis phenotype (White et al. 1990).

Homozygous *ja/ja* mice show no detectable β-spectrin synthesis, while the amount of α-spectrin is also reduced. Ankyrin, protein 4.1 and actin levels remain relatively unaltered. These data suggest that, even in the context of reduced spectrin levels, the assembly of the membrane skeleton can occur partially and that the reduced levels of α-spectrin can be due to the degradation of the unbound protein.

As for the mutations in the *sph* locus, it seems to be a different effect in the sph^{ha} and the other *sph* mutants. In particular, homozygous sph^{ha}/sph^{ha} animals synthesize normal amounts of α-spectrin, but very little is accumulated in the membrane, which probably suggests a rapid turnover of the mutant α-spectrin. On the contrary, the reticulocytes of all three different *sph* mutants do not contain functional α-spectrin mRNA and they neither synthesize α-spectrin in detectable levels nor they accumulate the latter in their membranes, despite the otherwise normal levels of the rest of the RBC membrane proteins. This fact implies that the *sph* mutants involve a molecular defect of α-spectrin at the transcriptional or post-transcriptional level.

28.3.2 Hereditary Elliptocytosis/Pyropoikilocytosis

Hereditary elliptocytosis and pyropoikilocytosis (HE/HPP) are a heterogeneous group of disorders with a wide spectrum of symptoms in humans, ranging from asymptomatic to severe haemolytic anaemia with poikilocytosis. The basic defect underlying most of the HE/HPP cases is the failure of spectrin heterodimers to form tetramers, which in turn are the necessary subunits to build the membrane network. The mutations responsible for the vast majority of HE/HPP involve proteins responsible for interaction in the horizontal dimension, namely the spectrins and protein 4.1. In addition, one would predict that adducin deficiency would also lead to HE/HPP, as adducin is localised at the spectrin/actin junction.

In order to shed more light into the role of these proteins and the consequences due to their deficiency, the effect of spectrin loss have been assessed in a mouse carrying a spontaneous mutation in the α-spectrin gene, known as *sph*^{Dem} (Wandersee et al. 2001). This mutation involves an in-frame deletion of exon 11, located at spectrin's self-association site, which in turn affect tetramerization. This is of particular importance, as all spectrin deficient mice known until now, have mutations that cause HS (see previous paragraph). The homozygous *sph*^{Dem}/*sph*^{Dem} phenotype has a high mortality rate, due to thrombosis in the heart or liver. Homozygous *sph*^{Dem} mice are jaundiced and exhibit hyperplastic erythropoiesis in spleen and bone marrow, splenomegaly and liver and kidney degeneration. They also exhibit reduced RBC counts, haematocrit and Hb concentration and increased mean cell volume. Blood smears from these mice are similar to those of humans with severe HE. Homozygous *sph*^{Dem} mice show extreme deficiency of both α- and β-spectrin, but comparable to the rest of the spontaneous spectrin mutants. However, distinct spectrin heterodimers and tetramers are not observed in *sph*^{Dem}/*sph*^{Dem} mice, suggesting that the spontaneous mutation ablates both heterodimer and tetramer structures.

Another mouse model for mild HE has been generated, by targeting the β-adducin gene. In particular, the gene has been disrupted by deleting exons 9 to 13 and replacing them by a neomycin resistance gene. β-adducin deficient mice are viable and fertile and exhibit no obvious phenotypical abnormality, except mild anaemia and splenomegaly. Blood smears from β-adducin deficient mice shows heterogeneity of erythrocyte shape, mainly elliptocytes and occasionally spherocytes. All major protein components of the membrane skeleton are present in β-adducin deficient RBC, but α-adducin levels are sharply decreased and γ-adducin levels are upregulated, probably in an attempt to rescue, at least partially, the absence of β-adducin (Muro et al. 2000). Thus, the altered levels of the adducin subunits affect the normal structure of the membrane skeleton, leading to symptoms of HE.

Targeted inactivation of protein 4.1 does not result in HE but rather HS, as described in detail in the previous paragraph.

Finally, a targeted deletion in the murine tropomodulin locus (E-Tmod) is lethal in the homozygous state, due to an arrest in definitive erythropoiesis and cardiac development. Homozygous animals die around 9.5–10.5 d.p.c., while the RBC membranes display reduced mechanical strength when compared to the wild type (Sung et al. 2000). These data are in agreement with partial haemolysis observed in erythroid cells from the yolk sac of E-Tmod$^{-/-}$ embryos.

28.4 Transgenic Models of Red Cell Transport Disorders

The haemoglobin concentration and volume of the erythrocyte is determined by its content of Hb, ions and water. Apart from the Hb content, which remains unaltered throughout the lifespan of the erythrocyte, the water and ion contents are under continuous regulation by the activity and interactions of several ion transport systems. These include the energy-driven Na^+/K^+ and Ca^{2+} pumps, the gradient-driven Na^+-K^+-$2Cl^-$ and K^+-Cl^- cotransport systems and Na^+/H^+ and anion exchangers and the K^+ (Gardos), voltage-gated and water channels (reviewed in Brugnara (1997); see also Chap. 4).

In this section, the existing animal models to study the erythrocyte's transport defects are analysed.

28.4.1 Gradient-Driven Systems

Members of the Na^+/H^+ exchanger (NHE) family have been shown to mediate the electroneutral exchange of Na^+ and H^+ across the plasma membrane. Until now, there have been six members of the NHE gene family. As the NHE isoforms are encoded by different genes, they are distinguished not only on the basis of amino acid sequence but also by their expression patterns and functional characteristics. NHE1 is more or less expressed ubiquitously, while isoforms NHE2–NHE6 exhibit a more restricted tissue distribution (Orlowski and Grinstein 1997; Numata et al. 1998). Due to its ubiquitous expression pattern, NHE1 is thought to be involved in the maintenance of pH homeostasis and cell volume regulation. The murine NHE1 gene has been inactivated by replacing part of the second exon and the adjacent intron sequences by the neomycin resistance gene. $NHE1^{-/-}$ animals are viable, but exhibit a significant reduction in weight, compared to heterozygous or wild-type littermates (Bell et al. 1999). This difference is persistent through adulthood. Although the majority of the organs of $NHE1^{-/-}$ mice are affected, as indicated by histological survey, there is no significant perturbation of systematic acid-base homeostasis as a result of the lack of NHE1 activity. The pH, P_{CO_2}, P_{O_2} and HCO_3^- values were not different between the $NHE^{-/-}$ animals and their $NHE^{+/-}$ or wild-type littermates.

In addition to ion exchangers, the basolateral Na^+-K^+-$2Cl^-$ cotransporter (NKCC1) is considered to play a major role in trans-epithelial chloride and fluid transport. NKCC1 gene has been inactivated by inserting the neomycin resistance gene inside exon 6. $NKCC1^{-/-}$ animals exhibit a reduction in body weight, while no significant alteration in the organs is observed by histological examination (Flagella et al. 1999). However, the mean arterial blood pressure of NKCC1 deficient mice is reduced and consistent intestinal haemorrhage, suggesting that NKCC1 deficiency may cause a vascular defect. The latter (coupled with the fact that during agonist stimulation of endothelial cells, NKCC1 activity is linked to that of Ca^{2+}-dependent K^+ channels) suggests that NKCC1 plays an important role in blood pressure regulation.

A defect in Na^+-K^+-$2Cl^-$ and K^+-Cl^- cotransporters, the NHE, and the Gardos channel is also seen in protein $4.2^{-/-}$ mice (Peters et al. 1999). A slightly increased activity of the K^+-Cl^- cotransporter is observed, but more interestingly a significantly increased activity of the Na^+-K^+-$2Cl^-$ cotransporter, the NHE and of the Gardos channel. These changes are most probably due to increased sensitivity to cell shrinkage and/or increased RBC membrane protein phosphorylation.

The latter notion has been strengthened by studying *fgr*$^{-/-}$*hck*$^{-/-}$ double mutant mice, which carry a targeted deletion of both *fgr* and *hck* kinases (Lowell et al. 1994). *Fgr* and *hck* consist of *Src*-kinase family members, which play a central role in regulation of haematopoietic cell functions (reviewed in Lowell and Soriano (1996)). RBCs, isolated from *fgr*$^{-/-}$*hck*$^{-/-}$ double mutant mice show no alteration in Na^+/K^+ pump or Na^+-K^+-$2Cl^-$ cotransport, but K^+-Cl^- activity is significantly higher compared to wild-type erythrocytes (De Franceschi et al. 1997), providing the first evidence that *Src*-family kinases are implicated in the regulation of RBC membrane transport.

Finally, the SAD mouse model for sickle cell anaemia (see Sect. 28.5.1) has been employed to study RBC membrane transport. SAD mouse RBCs have increased K^+-Cl^- cotransport (De Franceschi et al. 1995), while when treated with dietary Mg^{2+} supplements, the erythrocyte's Mg^{2+} and K^+ contents have been increased, accompanied by reduced K^+-Cl^- cotransport and dehydration (De Franceschi et al. 1996).

28.4.2 Stomatocytosis

Hereditary stomatocytosis is a rare inherited disorder of RBCs, that increases permeability to monovalent cations, leading to an increase of the intracellular cations accompanied by water movement into the cell and subsequent cell swelling. This leads to moderate to severe haemolytic anaemia as a result of the shortened RBC life span (see Chap. 21).

Deficiency of protein 7.2b or "stomatin" has been perceived as the main cause of hereditary stomatocytosis, as most but not all of the hereditary stomatocytosis individuals are protein 7.2b-deficient (Kanzaki and Yawata 1992). To investigate the potential role of protein 7.2b in hereditary stomatocytosis, the murine 7.2b gene has been inactivated by substituting the promoter and exon 1 of the murine 7.2b gene by a neomycin resistance gene (Zhu et al. 1999).

Protein 7.2b-deficient mice do not exhibit the phenotype seen in human RBCs with that defect. There is no haemolytic anaemia observed and the RBCs were normal in morphology, cell indices, hydration status, monovalent cation content and ability to translocate lipids. These findings suggest that protein 7.2b is not likely to play an important or direct role in cation transport in the RBC. However, it cannot be excluded that an abnormality of a putative partner binding protein rather than the lack of protein 7.2b itself is responsible for the phenotype observed in hereditary stomatocytosis. This is a reminder of the interaction between proteins 4.1, glycophorin and p55 (see also Sect. 28.3.1), where in the absence of either protein 4.1 or glycophorin C, p55 cannot be associated with the RBC membrane (Alloisio et al. 1993). In RBCs, band 7.2b is shown to co-immunoprecipitate with glucose transporter 1 and that overexpression of protein 7.2b results in a 30% reduction in the basal level of glucose transport (Zhang et al. 1999).

28.5 Transgenic Models of Other Red Cell Disorders

A variety of animal models have been generated for thalassemias and sickle cell disease, haemophilia A and B and various metabolic disorders, such as G6PD and

PKU/HPA. This chapter focuses on these animal models, which were among the first to be generated, and they are sill of great importance.

28.5.1 Sickle Cell Disease

A single base substitution of the human β-globin gene, leading to a amino acid change, is the cause for sickle cell disease, the first molecular defect identified in humans (see also Chap. 23). A large number of transgenic mice expressing Hb S has been described in the literature and has been employed to study the pathophysiology of the disease, and to optimize therapeutic agents or gene therapy protocols (reviewed in Nagel and Fabry (2001), and summarized in Table 28.2).

Table 28.2. Summary of the existing mouse models for sickle cell anaemia

Transgene	Mouse globin genes		Hb S [%]	References
	α	β		
LCR-α1-α1-β^S	+	+	10–30	Greaves et al. 1990
LCR-β^S + LCR-α1	+	+	50	Ryan et al. 1990
LCR-β^{SAD} + LCR-α2	+	+	19–26$^{(1)}$	Trudel et al. 1991
LCR-β^S + LCR-α2	+	–	75	Fabry et al. 1992a,b
LCR-α1 + LCR-Aγ-β^S	–	–	100	Pastzy et al. 1997
LCR + α1 + Gγ-Aγ-δ-β^S	–	–	100	Ryan et al. 1997
240 kb β^S-globin locus YAC	–	–	100	Chang et al. 1998
LCR-β^S/LCR-α2/HS4-Gγ-Aγ LCR-β^S/LCR-α2/HS3-HS4-Gγ-Aγ LCR-β^S/LCR-α2/LCR-Gγ-Aγ-3D$^{(2)}$	–	–	100	Fabry et al. 2001

$^{(1)}$ Hb SAD, $^{(2)}$ Sequences from the deletional HPFH-2 breakpoint (Arcasoy et al. 1997).

The first sickle cell mouse model generated (Greaves et al. 1990) contained a transgene with two human α-globin genes, flanked by an LCR and a human β^S gene. This animal has 83% Hb S, but surprisingly no obvious manifestations of disease and evidence of haemolytic anaemia like in humans. Ryan et al. (1990) generated a similar mouse model, expressing only 50% β^S-globin chains, while Fabry et al. (1992a,b) generated a β^S expressing mouse, but in the context of a murine βmaj gene deletion (Skow et al. 1983). The latter model had a moderate phenotype. In the following years, the SAD mouse has been generated. This mouse carries the SAD transgene, which contains three mutations in the same β-globin chain (β^{SAD}), namely the β^S, the $\beta^{S\text{-Antilles}}$ and $\beta^{D\text{-Los Angeles}}$ (Trudel et al. 1991, 1994). The last two mutations enhance polymer formation. This animal model has been used in RBC transport studies (see previous sections) and also to study the protective effect of γ-globin chains against polymer formation (Blouin et al. 2000).

The main disadvantage of these models is the existence of the mouse α-globin chains, which have been shown to partially inhibit polymerisation of Hb S (Roy et al. 1993). For this reason, a transgenic mouse model, expressing exclusively human β^S- and α-globin chains was the active objective of sickle cell anaemia research in the last years.

Such transgenic mouse models have been successfully generated. These lines contained a targeted disruption of the murine α- and β-globin genes and transgenes containing either an LCR linked to the human globin genes (Pastzy et al. 1997; Ryan et al. 1997), or a 240 kb yeast artificial chromosome (YAC), which contained the full human β-globin locus with β^S- rather than a wild-type β-globin gene (Chang et al. 1998). The latter model has the advantage that the structure of the human β-globin locus is preserved. In these models, the main feature is anaemia (although not as profound as in the average sickle cell anaemia patient) and the abundance of irreversibly sickled erythrocytes. However, these mice have a low MCHC, which protects these animals against polymer formation and allows them to survive until adulthood. MCHC values are normal to high in sickle cell anaemia patients.

In order to generate a mouse model, which mimics almost ideally the sickle cell anaemia phenotype in humans, the NYC1 mouse has been bred to a full murine globin gene knockout context. The latter mouse was subsequently bred to three different γ-globin gene-expressing transgenic mouse lines, depending on the level of expression (Fabry et al. 2001). It has been shown that the low γ-expressing mouse line had balanced globin chain synthesis and more interestingly higher MCHC values. The latter transgenic mouse model seems to faithfully reproduce the sickle cell anaemia phenotype in humans.

28.5.2 Haemoglobinopathies

Thalassemia in humans is a heterogeneous group of disorders involving haemoglobin synthesis, characterized by reduced or absent α- or β-like chains. This leads to a chain imbalance, depending on which type of chains is reduced in levels. A natural occurring mutant mouse strain was the first to be described, carrying a 3.7 kb deletion in the βmaj globin gene, which includes all exons and the 5' regulatory sequences (Skow et al. 1983). This mutation, in the homozygous state, results in anaemia and high reticulocytosis, but most of the homozygous animals are viable (Johnson and Lewis 1981). This is because the βmin gene transcription is elevated, leading to an increased β-globin translation. This naturally occurring mouse model for thalassemia provides insights into the effect of the globin chain imbalance on the membrane protein levels. Indeed, the electrophoretic profile of the membrane proteins shows a 25% decrease in spectrin and an overall decrease in the reactive thiol groups of membrane proteins, such as of ankyrin (49%) and spectrin (35%) (Rouyer-Fessard et al. 1990). In the same wavelength, Shehee et al. (1993) disrupted the mouse βmaj globin gene by homologous recombination. In particular, they inserted a neomycin resistance gene inside exon 2, which abolishes the production of functional β-globin chains from the βmaj gene. Homozygous animals are very pale and die perinatally. However, heterozygous mice are indistinguishable from normal littermates in both appearance and haematological indices.

The phenotypical differences in the latter and the previous mouse models are mainly due to the type of the βmaj genetic defect. In the latter case, the 5' regulatory sequences are intact and thus able to compete with those of the βmin gene for transcriptional activation from the LCR (Grosveld et al. 1987). However, in the mouse model of Johnson and Lewis (1981) the 5' regulatory sequences of the βmaj gene are missing and the βmin gene has no competition from the βmaj globin promoter.

Finally, Yang et al. (1995) generated a mouse line lacking both βmaj and βmin genes (see also Detloff et al. (1994)). Blood smears from heterozygous mice shows a clear anisocytosis and poikilocytosis with fragmented cells. In addition, haematological indices are severely anaemic with markedly elevated levels of reticulocytes, indicating a state of active erythropoiesis. Homozygous animals are extremely pale and die almost perinatally, within 12 hours after birth.

Mutations, mostly deletions, in the human α-globin locus result in perinatal lethal condition, known as Hb Bart's hydrops fetalis syndrome. Using gene targeting, both murine α-globin genes have been replaced by a neomycin gene ($Hba^{1\text{-}2\,del}$). Homozygous $Hba^{1\text{-}2\,del}$ animals die *in utero* at 14.5 d.p.c. and appear hydropic with large inclusion bodies and reduced RBC counts (Pastzy et al. 1995). Interestingly, when a 20% human α-globin gene-expressing mouse line, made using a LCR-α-globin gene construct, was bred into a $Hba^{1\text{-}2\,del}$ homozygous background, the lethal phenotype is rescued, although the survived mice were 15–35% smaller in weight and with globin chain imbalance. This experiment showed that human α-globin chains can compensate for the absence of the murine α-globin chains in $Hba^{1\text{-}2\,del}$ homozygous animals.

Hereditary persistence of fetal Hb (HPFH) is a heterogeneous group of disorders characterized by continuous expression of the fetal globin genes in adulthood. The cause of the, otherwise silenced, γ-globin genes expression in the adult stage are either point mutations in the γ-globin gene promoters or large deletions, removing the adult globin genes. In order to confirm this hypothesis, a transgenic mouse line has been generated, containing a human γ-globin gene, carrying an HPFH causing mutation in the promoter (Aγ-117 G→A) and a β-globin gene linked to the LCR (Berry et al. 1992). These transgenic mice show persistent γ-globin gene expression in the adult developmental stage, compared to transgenic mice carrying the same construct but with a wild-type γ-globin promoter, in which the γ-globin genes are silenced (see also Dillon and Grosveld (1991)). These authors have also shown that the HPFH causing mutation abolishes the binding of the GATA-1 transcription factor. The same results have been obtained using a transgenic mouse line carrying the same mutation but in the context of a 248 kb β-globin locus-containing YAC (Peterson et al. 1995).

28.5.3 The Haemophilias

Sex-linked haemophilia is a bleeding disorder characterized by a deficiency in factor VIII (haemophilia A) or factor IX (haemophilia B), which are components of the blood coagulation pathway (reviewed in Kazazian et al. (2001) and Pollak and High (2001)).

Haemophilia A is a severe X-linked clotting disorder, which occurs in 1 out of 5,000–10,000 males. Haemophilia A is due to mutations in the factor VIII gene, which encodes for factor VIII protein, an essential cofactor in the normal blood coagulation pathway. Using gene targeting approaches, the factor VIII gene has been inactivated. Two different gene targeted mice strains have been obtained, by inserting the neomycin resistance gene in either exon 16 or 17 of the factor VIII gene. The resulting phenotype in factor VIII-deficient male mice is much milder than the one observed in humans (Bi et al. 1995). The only disease phenotype observed is excessive bleeding and subsequent death in 2 hours, when a tail sample biopsy has obtained. In addition, affected female mice were able to proceed normally through pregnancy and delivery, suggesting either that factor VIII is dispensable for preventing bleeding after pregnancy or that an alternative factor VIII-independent pathway exists which promotes clotting in the absence of factor VIII (Bi et al. 1996).

Haemophilia B is a bleeding disorder, which afflicts 1 in 25,000–30,000 males and is responsible for almost 25% of all haemophilic patients. Haemophilia B results from mutations in the factor IX gene, responsible for the production of the factor IX protein. In order to obtain a mouse line for the evaluation of gene therapy protocols and for assessing the role of the normal factor IX protein, the murine factor IX gene has been inactivated. This has been done by substituting the promoter and the first three exons of the gene by the neomycin resistance gene. Factor IX deficient mice present similar phenotype to haemophilia B in humans. In other words, these mice survive well if no injury occurs, although sudden death cannot be ruled out during growth as a result of internal bleeding after normal fighting with their cage mates (Lin et al. 1997). In addition, when a tail sample biopsy is obtained from factor IX deficient mice, the wound should be cauterised to prevent massive bleeding and subsequent death. Relative clotting activity of plasma from factor IX deficient mice is reduced compared to normal mice. This is a reminder of factor VIII deficient mice, although the phenotype of the latter correlates with mild rather than a severe haemophilia A phenotype in humans.

28.5.4 Metabolic Disorders

Glucose 6-phosphate dehydrogenase (G6PD) is an important enzyme which catalyses the first step in the pentose phosphate pathway and provides reductive potential in the form of NADPH. G6PD is encoded by an X-linked house-keeping gene, which when mutated in human populations cause partial loss of G6PD activity and subsequently a variety of haemolytic anaemias, ranging from mild to severe (reviewed in Luzzatto et al. (2001), see also Chap. 22). To investigate the essential role of G6PD, the murine gene has been disrupted by inserting a neomycin resistance gene inside exon 3. Targeted ES cells are viable, but growth becomes impaired as cell density increases, suggesting that some catabolite(s) may limit or even inhibit growth. More interestingly, G6PD-deficient ES cells are very sensitive to diamide, a powerful sulfhydryl group oxidizing agent, which specifically causes the depletion of glutathione in RBCs. The same result was depicted in the presence of hydrogen peroxide. Based on these findings, it seems plausible that the primary physiological role of G6PD in mammalian cells is defence against oxidative stress. In addition, the supply of pentose is not an irreplaceable function of G6PD, as pentose is probably provided by the transketolase-transaldolase pathway, entirely bypassing G6PD. The latter suggests that G6PD function is dispensable for pentose synthesis (Pandolfi et al. 1995a).

Disorders of phenylalanine catabolism, resulting in phenylketonurea (PKU) and hyperphenylalaninemia (HPA) were among the first inherited metabolic disorders described in humans, due to defects in the phenylalanine hydroxylase (PAH) gene (Scriver and Kaufman 2001). For this reason, animal models with defined PKU and HPA mutations would be of extreme help in evaluating the features of these diseases. Using germline chemical mutagenesis with *N*-ethyl-*N*-nitrosourea, the murine PAH locus has been mutated, providing three different strains with defective phenylalanine hydroxylation (McDonald et al. 1990; Shedlovsky et al. 1993). These strains are orthologues of human PKU and HPA, presenting markedly reduced PAH activity. Injection of phenylalanine causes severe HPA and urinary secrition of phenylketones, while a low phenylalanine diet can partially reverse these abnormalities and ameliorate the phenotype.

Using also chemical mutagenesis, glucose phosphate isomerase (GPI) locus has been mutated (West et al. 1990), resulting in a lethal phenotype in the homozygous state at the embryonic developmental stage. Mutant GPI cells that are deficient in glycolysis grow slowly, using either the pentose phosphate shunt or transport, from the maternal blood, of energy sources other than glucose (such as glutamine) to bypass the block in glycolysis, caused by GPI deficiency, but this process does not seem adequate to support normal embryogenesis.

Another key enzyme in the glucose metabolic pathway is hexokinase II (HKII), a member of the hexokinase family of isoenzymes, which catalyses the phosphorylation of glucose (Glu) to Glu-6-P. Mutations in HKII gene have been observed in type 2 diabetic patients, resulting in reduced HKII mRNA levels and enzyme production with low Glu-6-P levels (Laakso et al. 1995).

To investigate the role of HKII, the murine gene has been disrupted by placing a neomycin resistant gene inside exon 4. $HKII^{+/-}$ mice were viable and fertile, but exhibit approx. half of the normal HKII mRNA levels and activity. However, the $HKII^{-/-}$ phenotype is lethal. $HKII^{-/-}$ embryos die around 7.5 d.p.c., presenting visible growth retardation (Heikkinen et al. 1999). This situation is a reminder of the GPI-deficient mice. HKII acts immediately upstream GPI in the glycolytic pathway, thus a defect in HKII activity would result in a similar phenotype, since the pentose shunt, which is functional even in the absence of GPI, would be completely blocked in HKII-deficient cells.

28.6 Growth and Differentiation Defects

This final section deals with the animal models to study the role of erythroid-specific transcription factors in haematopoiesis and RBC growth. The various developmental defects are summarized in Table 28.3.

Table 28.3. Summary of the existing mouse models for the RBC growth and differentiation defects

Gene	Viability [d.p.c.]	Haematopoietic defects		Other features	References
		Primitive	Definitive		
Epo	13.5–15.5	Impaired	Impaired	-	Wu et al. 1995
EpoR	13.5–15.5	Impaired	Impaired	-	Wu et al. 1995
GATA-1	10.5–11.5	Normal	Impaired	megakaryocytic development affected	Pevny et al. 1991
GATA-2	10.5–11.5	Impaired	Impaired	loss of erythroid progenitors	Tsai et al. 1994
GATA-3	11.5–12.5	Normal	Impaired	-	Pandolfi et al. 1995b
FOG-1	10.5–12.5	Impaired	Impaired	megakaryocytic arrest	Tsang et al. 1998
EKLF	15.5–16.5	Normal	Impaired	no β-globin expression	Nuez et al. 1995; Perkins et al. 1995

28.6.1 The Erythropoietin and the Erythropoietin Receptor

Erythropoietin (Epo) is the principal growth factor, which regulates the production of erythrocytes from committed erythroid progenitors (colony-forming unit-erythroid (CFU-E) and burst-forming unit-erythroid (BFU-E)), derived from a small group of pluripotent stem cells. Epo is remarkably specific for erythroid progenitors, owing to the tissue-specific expression of its receptor, namely the Epo receptor (EpoR).

The physiological importance or both Epo and EpoR has been established by eliminating gene function. Inactivation of the Epo gene has been achieved by substitution of exons 2–5 by a neomycin resistance gene. The same approach has been used for the targeted inactivation of the murine EpoR gene, by substituting exons 1–6 and the 5' flanking sequences.

Heterozygous $Epo^{+/-}$ or $EpoR^{+/-}$ mice appear normal and are viable and fertile with levels of haemoglobin, haematocrit, erythrocytes and leukocytes within the normal range (Wu et al. 1995).

However, the homozygous $Epo^{-/-}$ or $EpoR^{-/-}$ phenotype is embryonic lethal. Examination of the $Epo^{-/-}$ or $EpoR^{-/-}$ embryos at successive developmental stages showed that, although developed normally until 13.5 d.p.c., they appear very pale with no red colouring characteristic of the fetal liver, and they die between 13.5 and 15.5 d.p.c. Histological examination of sections from fetal livers from $Epo^{-/-}$ or $EpoR^{-/-}$ embryos showed that both BFU-E and CFU-E progenitors were significantly increased compared to $Epo^{+/-}$, $EpoR^{+/-}$ or wild-type embryos. Interestingly, there is also an impairment of the primitive erythropoiesis in the yolk sac, as a result of the targeted disruption of Epo or EpoR gene. $Epo^{-/-}$ or $EpoR^{-/-}$ embryos at 10.5 d.p.c. are very pale, allowing identification of the homozygous mutants, while their blood vessels have fewer erythrocytes than normal embryos.

In essence, these data suggest that Epo and EpoR have a pivotal role in both primitive and definitive erythropoiesis and further that the major function of Epo is to trigger proliferation and terminal differentiation of the late CFU-E population. On the other hand, Epo and EpoR are not required for erythroid lineage commitment or for the proliferation and differentiation of BFU-E to CFU-E progenitors.

28.6.2 The Erythroid-Specific GATA Transcription Factors

The GATA protein family consists of six related zinc finger transcription factors that recognize the consensus sequence $^{A}/_{T}$GATA$^{A}/_{G}$, a DNA motif first identified in the regulatory region of erythroid-expressed genes. The founding member of the GATA family of transcription factors, GATA-1, was initially identified as an erythroid nuclear protein that binds to GATA consensus sequences, found in globin gene promoters, enhancers and LCRs, through a conserved multifunctional domain consisting of two C4-type zinc fingers (reviewed in Orkin (2001)). From the six GATA proteins found so far, GATA-1, GATA-2 and GATA-3 are involved in haematopoiesis.

A mutation in the zinc finger region has been shown recently to cause dyserythropoietic anaemia and thrombocytopenia (Nichols et al. 2000), due to the abrogation of interaction between GATA-1 and its essential cofactor FOG-1 ("Friend Of GATA-1"; Tsang et al. 1997).

To further address the role of GATA-1 in haematopoiesis, the murine gene has been inactivated by insertion of a neomycin resistance gene immediately in front of exon 2 in a male (XY) murine ES cell line. GATA-1$^{-/-}$ embryos die around 10.5 and 11.5 d.p.c. due to severe anaemia (Penvy et al. 1991). In addition, lack of GATA-1 is accompanied by the production of erythroid precursors, both primitive and definitive (Weiss et al. 1994). GATA-1^{-} ES cells fail to contribute to the mature erythroid compartment in chimaeric mice, although they do so to all other tissues tested, including the white cell blood fraction (Pevny et al. 1991). However, when a normal GATA-1 transgene is re-introduced into a mutant ES cell, erythroid development is rescued (Simon et al. 1992). The latter suggest that the GATA-1^{-} phenotype is solely attributable to the loss of GATA-1 function. Interestingly, a requirement for GATA-1 also in megakaryocytic development has been established by deleting an upstream region in the GATA-1 locus, using site-specific recombination. Mice lacking megakaryocytic GATA-1 expression show an accumulation of megakaryocytes in spleen and bone marrow (Shivdasani et al. 1997). However, this targeted mutation preserved sufficient expression of GATA-1 in erythroid precursors to complete RBC maturation. These data reveal a critical role of GATA-1 in megakaryocytic development, while provide a nice example of generating tissue-specific knockout by inactivating regulatory elements.

Recently, the X-inactivation process has been exploited to study the effect of overexpression of GATA-1 in murine erythroid cells. Provided that GATA-1 is an X-linked transgene, it would be transcriptionally active in 50% of female erythroid precursors, while in males, where no X-inactivation occurs, all erythroid cells should overexpress GATA-1. Transgenic females for the X-linked GATA-1 transgene were indistinguishable from wild-type littermates, while male transgenics were anaemic and dead by 14.5 d.p.c. The number of erythroid precursors relative to nucleated erythrocytes was reduced more than 3-fold in transgenic males compared with controls. This shows that overexpression of GATA-1 in erythroid cells inhibits their differentiation, leading to lethal anaemia (Whyatt et al. 2000). This effect is intrinsic but cell non-autonomous, in other words, the cell exhibits a phenotype, which does not correspond to its genotype.

A unique feature among the GATA factors is that they seem to be crossregulated in erythropoiesis. This hypothesis was launched when all gene targets for GATA-1 in erythroid cells, such as embryonic and adult globins, erythroid Kruppel-like factor (EKLF, see Sect. 28.6.3), EpoR were expressed at near normal levels in GATA-1$^-$ proerythroblasts. These results suggested that other GATA family member(s) may be compensating in the absence of GATA-1. This was indeed the case for GATA-2, which was found to be upregulated about 50-fold in GATA-1$^-$ cells (Weiss et al. 1994). On the contrary, GATA-3 levels remained unaltered. The latter suggests that GATA-1 negatively regulates GATA-2 during normal erythroid maturation.

Targeted disruption of the GATA-2 gene revealed its fundamental role in erythropoiesis. GATA-2$^{-/-}$ embryos die around 10.5 and 11.5 d.p.c. by severe anaemia, which suggest a deficit in primitive erythropoiesis (Tsai et al. 1994). To examine the role of GATA-2 in definitive erythropoiesis, chimaeric mice were generated by injecting GATA-2$^{-/-}$ ES cells into wild-type blastocysts. Definitive erythropoiesis seems to be severely affected with a profound deficit of definitive haematopoietic stem or progenitor cells. These data strongly suggest that GATA-2 serves as a regulator of genes controlling growth factor responsiveness or proliferative capacity of early haematopoietic cells.

Transcription factor GATA-3, firstly identified by its sequence homology to GATA-1, seems to play also a role in haematopoiesis. The GATA-3 gene has been disrupted by substituting the second exon of the murine gene by a neomycin resistance gene. GATA-3$^{+/-}$ mice are viable and fertile. However, the homozygous GATA-3$^{-/-}$ phenotype seems to be lethal. GATA-3$^{-/-}$ embryos die around 11.5 and 12.5 d.p.c. and present, among other symptoms ranging from growth retardation to massive internal bleeding and severe deformities of the brain and spinal cord, gross aberrations in definitive haematopoiesis (Pandolfi et al. 1995b). An impaired fetal liver erythropoiesis is also seen in homozygous GATA-1$^{-/-}$ animals (Pevny et al. 1991). On the contrary, primitive haematopoiesis appears to be almost unaffected in homozygous GATA-3$^{-/-}$ animals.

FOG-1 is a zinc-finger GATA-1 interacting protein during erythroid and megakaryocytic differentiation, and whose expression profile parallels that of GATA-1. In order to assess its role as a GATA-1 cofactor, the FOG-1 gene has been disrupted by removing all the zinc-finger encoding exons. Heterozygous FOG-1$^{+/-}$ mice appeared healthy and fertile, whilst complete FOG-1 inactivation leads to an early- to mid-gestational lethal phenotype. Homozygous FOG-1$^{-/-}$ mice die around 10.5 and 12.5 d.p.c. of severe anaemia. Fetal livers of homozygous FOG-1$^{-/-}$ embryos are smaller than those of a wild type embryo and primitive and definitive erythroid cells of FOG-1$^{-/-}$ origin appear to be partially arrested, resembling proerythroblasts (Tsang et al. 1998). These data suggest that in the absence of FOG-1, both primitive and definitive erythropoiesis is severely impaired. The consequences of FOG-1 deficiency are similar to those observed in GATA-1 deficient mice, an observation, which strengthens the notion that GATA-1 and FOG-1 act in concert. However, there is no evidence for interaction between FOG-1 and GATA-2 or GATA-3 in erythroid cells. In addition, FOG-1 deficiency leads to a GATA-1 independent ablation of the megakaryocytic lineage at a very early stage. This unexpected phenotype is even more severe from the one observed when the GATA-1 upstream region is deleted (Shivdasani et al. 1997), probably suggesting that FOG-1 is critical at multiple points in the development and maturation of the megakaryocytic lineage.

28.6.3 Erythroid Kruppel-Like Factor

The EKLF is an essential component of the transcriptional activation machinery. EKLF shares the characteristics of the other members of the Kruppel family of transcription factors (Philipsen and Suske 1999). The most important one is the three zinc fingers, which bind to the sequence CCACACCCT and thus makes EKLF important for β-globin gene activation, as a result of its ability to bind to the CACC box in the β-globin gene promoter. In humans, mutations of the CACC box leads to reduced β-globin gene expression (Orkin et al. 1984).

The effect of EKLF on erythropoiesis was assessed by targeted inactivation of the EKLF gene. This has been done by two groups independently, either by insertion of a *lacZ* reporter and a neomycin resistance gene in the EKLF gene (Nuez et al. 1995), or by substituting part of exon and intron 2 by a neomycin resistance gene (Perkins et al. 1995).

EKLF heterozygous mice ($EKLF^{+/-}$) appeared to be healthy, fertile with normal haematocrit, mean cell volume and absolute and differential RBC counts. However, $EKLF^{-/-}$ phenotype is lethal. At the embryonic stage, $EKLF^{-/-}$ embryos appeared normal, indistinguishable from their $EKLF^{+/-}$ littermates, containing apparently normal nucleated embryonic RBCs in the yolk sac. However, upon switching from the embryonic to fetal erythropoiesis in the fetal liver, the absence of EKLF gives rise to a severe phenotypic effect. The $EKLF^{-/-}$ embryos become progressively more anaemic and die around 15.5 d.p.c.

Analysis of β-globin gene expression showed that the embryonic εy and βh1 RNAs are present in the circulating embryonic blood in the fetus, but only small amounts of mature βmaj RNA. Despite the presence of a consensus EKLF-type sites in their promoters, expression of the βh1, as well as of other erythroid genes, such as GATA-1, EpoR and porphobilinogen deaminase were normal (Perkins et al. 1995). In addition, the β-globin protein levels are dramatically reduced, which lead to a globin chain imbalance typical of thalassemia, since the α-globin chains levels remains unaltered. The latter suggest that the lack of EKLF leads to a highly selective deficiency of β-globin gene expression.

The stage-specific requirement for EKLF does not imply that EKLF is inactive or not expressed during yolk sac erythropoiesis, since EKLF transcripts are abundant in purified yolk sac and fetal liver erythroid cells and CACC-binding activity was similar in primitive and definitive erythroid cells but selectively missing from $EKLF^{-/-}$ cells.

EKLF seems also to be indirectly involved in the γ- to β-globin gene switching. Study of the expression pattern of the human globin genes transgenes showed that in a $EKLF^{-/-}$ environment, γ-globin gene expression is upregulated, probably due to the destabilization of the interaction between the LCR and the β-globin gene promoter (Wijgerde et al. 1996), but it is silenced normally. Fetal expression of human Aγ-globin gene corrects globin chain imbalance in $EKLF^{-/-}$ embryos but does neither prevent haemolysis nor prolong survival (Perkins et al. 2000). The $EKLF^{-/-}$ lethal phenotype is rescued only when bred into an EKLF overexpressing environment (van der Wijngaard and Grosveld, unpublished data).

Acknowledgement: Most of our own work has been supported by grants from the European Union and the Royal Dutch Organization (NWO). GPP is supported by the European Union (Contract # HPMFCT199900175).

References

Alloisio N, Dalla Venezia N, Rana A, Andrabi K, Texier P, Gilsanz F, Cartron JP, Delaunay J, Chishti AH (1993) Evidence that red blood cell protein p55 may participate in the skeleton-membrane linkage that involves protein 4.1 and glycophorin C. Blood 82:1323-1327

Arcasoy MO, Romana M, Fabry ME, Skarpidi E, Nagel RL, Forget BG (1997) High levels of human gamma-globin gene expression in adult mice carrying a transgene of deletion-type hereditary persistence of fetal hemoglobin. Mol Cell Biol 17:2076-2089

Bell SM, Schreiner CM, Schultheis PJ, Miller ML, Evans RL, Vorhees CV, Shull GE, Scott WJ (1999) Targeted disruption of the murine Nhe1 locus induces ataxia, growth retardation, and seizures. Am J Physiol 276:C788-C795

Berry M, Grosveld F, Dillon N (1992) A single point mutation is the cause of the Greek form of hereditary persistence of fetal haemoglobin. Nature 358:499-502

Bi L, Lawler AM, Antonarakis SE, High KA, Gearhart JD, Kazazian HH (1995) Targeted disruption of the mouse factor VIII gene produces a model of haemophilia A. Nat Genet 10:119-121

Bi L, Sarkar R, Naas T, Lawler AM, Pain J, Shumaker SL, Bedian V, Kazazian HH (1996) Further characterization of factor VIII-deficient mice created by gene targeting: RNA and protein studies. Blood 88:3446-3450

Blouin MJ, Beauchemin H, Wright A, De Paepe M, Sorette M, Bleau AM, Nakamoto B, Ou CN, Stamatoyannopoulos G, Trudel M (2000) Genetic correction of sickle cell disease: insights using transgenic mouse models. Nat Med 6:177-182

Bodine DM, Birkenmeier CS, Barker JE (1984) Spectrin deficient inherited hemolytic anemias in the mouse: characterization by spectrin synthesis and mRNA activity in reticulocytes. Cell 37:721-729

Brugnara C (1997) Erythrocyte membrane transport physiology. Curr Opinion Hematol 4:122-127

Chang JC, Lu R, Lin C, Xu SM, Kan YW, Porcu S, Carlson E, Kitamura M, Yang S, Flebbe-Rehwaldt L, Gaensler KM (1998) Transgenic knockout mice exclusively expressing human hemoglobin S after transfer of a 240-kb betas-globin yeast artificial chromosome: A mouse model of sickle cell anemia. Proc Natl Acad Sci USA 95:14886-14890

De Franceschi L, Beuzard Y, Brugnara C (1995) Sulfhydryl oxidation and activation of red cell K^+-Cl^- cotransport in the transgenic SAD mouse. Am J Physiol 269:C899-C906

De Franceschi L, Beuzard Y, Jouault H, Brugnara C (1996) Modulation of erythrocyte potassium chloride cotransport, potassium content, and density by dietary magnesium intake in transgenic SAD mouse. Blood 88:2738-2744

De Franceschi L, Fumagalli L, Olivieri O, Corrocher R, Lowell CA, Berton G (1997) Deficiency of *Src* family kinases *Fgr* and *Hck* results in activation of erythrocyte K/Cl cotransport. J Clin Invest 99:220-227

Detloff PJ, Lewis J, John SW, Shehee WR, Langenbach R, Maeda N, Smithies O (1994) Deletion and replacement of the mouse adult beta-globin genes by a "plug and socket" repeated targeting strategy. Mol Cell Biol 14:6936-6943

Dillon N, Grosveld F (1991) Human gamma-globin genes silenced independently of other genes in the beta-globin locus. Nature 350:252-254

Doi S, Campbell C, Kucherlapati R (1992) Directed modification of genes by homologous recombination in mammalian cells. In: Grosveld F, Kollias G (eds) Transgenic animals. Academic Press, London, pp 27-46

Fabry ME, Nagel RL, Pachnis A, Suzuka SM, Costantini F (1992a) High expression of human beta S- and alpha-globins in transgenic mice: hemoglobin composition and hematological consequences. Proc Natl Acad Sci USA 89:12150-12154

Fabry ME, Costantini F, Pachnis A, Suzuka SM, Bank N, Aynedjian HS, Factor SM, Nagel RL (1992b) High expression of human beta S- and alpha-globins in transgenic mice: erythrocyte abnormalities, organ damage, and the effect of hypoxia. Proc Natl Acad Sci USA 89:12155-12159

Fabry ME, Suzuka SM, Weinberg RS, Lawrence C, Factor SM, Gilman JG, Costantini F, Nagel RL (2001) Second generation knockout sickle mice: the effect of HbF. Blood 97:410-418

Feil R, Wagner J, Metzger D, Chambon P (1997) Regulation of Cre recombinase activity by mutated estrogen receptor ligand-binding domains. Biochem Biophys Res Commun 237:752-757

Flagella M, Clarke LL, Miller ML, Erway LC, Giannella RA, Andringa A, Gawenis LR, Kramer J, Duffy JJ, Doetschman T, Lorenz JN, Yamoah EN, Cardell EL, Shull GE (1999) Mice lacking the basolateral Na-K-2Cl cotransporter have impaired epithelial chloride secretion and are profoundly deaf. J Biol Chem 274:26946-26955

Greaves DR, Fraser P, Vidal MA, Hedges MJ, Ropers D, Luzzatto L, Grosveld F (1990) A transgenic mouse model of sickle cell disorder. Nature 343:183-185

Grosveld F, Van Assendelft GB, Greaves DR, Kollias G (1987) Position-independent, high-level expression of the human beta-globin gene in transgenic mice. Cell 51:975-985

Hassoun H, Hanada T, Lutchman M, Sahr KE, Palek J, Hanspal M, Chishti AH (1998a) Complete deficiency of glycophorin A in red blood cells from mice with targeted inactivation of the band 3 (AE1) gene. Blood 91:2146-2151

Hassoun H, Wang Y, Vassiliadis J, Lutchman M, Palek J, Aish L, Aish IS, Liu SC, Chishti AH (1998b) Targeted inactivation of murine band 3 (AE1) gene produces a hypercoagulable state causing widespread thrombosis *in vivo*. Blood 92:1785-1792

Heikkinen S, Pietila M, Halmekyto M, Suppola S, Pirinen E, Deeb SS, Janne J, Laakso M (1999) Hexokinase II-deficient mice. Prenatal death of homozygotes without disturbances in glucose tolerance in heterozygotes. J Biol Chem 274:22517-22523
Hogan B, Costantini F, Lacy E (1986) Manipulating the mouse embryo: A laboratory manual. Cold Spring Harbor Laboratory Press, New York
Indra AK, Warot X, Brocard J, Bornert JM, Xiao JH, Chambon P, Metzger D (1999) Temporally-controlled site-specific mutagenesis in the basal layer of the epidermis: comparison of the recombinase activity of the tamoxifen-inducible Cre-ER(T) and Cre-ER(T2) recombinases. Nucleic Acids Res 27:4324-4327
Johnson FM, Lewis SE (1981) Electrophoretically detected germinal mutations induced in the mouse by ethylnitrosourea. Proc Natl Acad Sci USA 78:3138-3141
Kanzaki A, Yawata Y (1992) Hereditary stomatocytosis: phenotypical expression of sodium transport and band 7 peptides in 44 cases. Br J Haematol 82:133-141
Kazazian HH, Tuddenham EGD, Antonarakis SE (2001) Hemophilia A: Deficiency of coagulation factor VIII. In: Scriver CR, Beaudet AL, Sly WS, Valle D (eds) The metabolic and molecular bases of inherited disease. McGraw-Hill, New York, pp 4367-4392
Kilby NJ, Snaith MR, Murray JA (1993) Site-specific recombinases: tools for genome engineering. Trends Genet 9:413-421
Kollias G, Grosveld F (1992) The study of gene regulation in transgenic mice. In: Grosveld F, Kollias G (eds) Transgenic animals. Academic Press, London, pp 79-98
Laakso M, Malkki M, Deeb SS (1995) Amino acid substitutions in hexokinase II among patients with NIDDM. Diabetes 44:330-334
Lin HF, Maeda N, Smithies O, Straight DL, Stafford DW (1997) A coagulation factor IX-deficient mouse model for human hemophilia B. Blood 90:3962-3966
Lowell CA, Soriano P (1996) Knockouts of *Src*-family kinases: stiff bones, wimpy T cells, and bad memories. Genes Dev 10:1845-1857
Lowell CA, Soriano P, Varmus HE (1994) Functional overlap in the *src* gene family: inactivation of *hck* and *fgr* impairs natural immunity. Genes Dev 8:387-398
Luzzatto L, Mehta A, Vulliamy T (2001) Glucose 6-Phosphate dehydrogenase deficiency. In: Scriver CR, Beaudet AL, Sly WS, Valle D (eds) The metabolic and molecular bases of inherited disease. McGraw-Hill, New York, pp 4517-4553
McDonald JD, Bode V, Dove WF, Shedlovsky A (1990) $Pah^{hph\text{-}5}$: A mouse mutant deficient in phenylalanine hydroxylase. Proc Natl Acad Sci USA 87:1965-1967
Muro AF, Marro ML, Gajovic S, Porro F, Luzzatto L, Baralle FE (2000) Mild spherocytic hereditary elliptocytosis and altered levels of alpha- and gamma-adducins in beta-adducin-deficient mice. Blood 95:3978-3985
Nagel RL, Fabry ME (2001) The panoply of animal models for sickle cell anemia. Br J Haematol 112:19-25
Nichols KE, Crispino JD, Poncz M, White JG, Orkin SH, Maris JM, Weiss MJ (2000) Familiar dyserythropoietic anaemia and thrombocytopenia due to an inherited mutation in GATA1. Nat Genet 24:266-270
Nuez B, Michalovich D, Bygrave A, Ploemacher R, Grosveld F (1995) Defective haematopoiesis in fetal liver resulting from inactivation of the EKLF gene. Nature 375:316-318
Numata M, Petrecca K, Lake N, Orlowski J (1998) Identification of a mitochondrial Na^+/H^+ exchanger. J Biol Chem 273:6951-6959

Orkin SH (2001) Transcription factors that regulate lineage desicions. In: Stamatoyannopoulos G, Majerus PW, Perlmutter RM, Varmus H (eds) The molecular basis of blood diseases. WB Saunders, Philadelphia, pp 80-102

Orkin SH, Antonarakis SE, Kazazian HH Jr (1984) Base substitution at position -88 in a beta-thalassemic globin gene. Further evidence for the role of distal promoter element ACACCC. J Biol Chem 259:8679-8681

Orlowski J, Grinstein S (1997) Na^+/H^+ exchangers of mammalian cells. J Biol Chem 272:22373-22376

Pandolfi PP, Sonati F, Rivi R, Mason P, Grosveld F, Luzzatto L (1995a) Targeted disruption of the housekeeping gene encoding glucose 6-phosphate dehydrogenase (G6PD): G6PD is dispensable for pentose synthesis but essential for defense against oxidative stress. EMBO J 14:5209-5215

Pandolfi PP, Roth ME, Karis A, Leonard MW, Dzierzak E, Grosveld FG, Engel JD, Lindenbaum MH (1995b) Targeted disruption of the GATA3 gene causes severe abnormalities in the nervous system and in fetal liver haematopoiesis. Nat Genet 11:40-44

Paszty C, Mohandas N, Stevens ME, Loring JF, Liebhaber SA, Brion CM, Rubin EM (1995) Lethal α-thalassaemia created by gene targeting in mice and its genetic rescue. Nat Genet 11:33-39

Paszty C, Brion CM, Manci E, Witkowska HE, Stevens ME, Mohandas N, Rubin EM (1997) Transgenic knockout mice with exclusively human sickle hemoglobin and sickle cell disease. Science 278:876-878

Perkins AC, Sharpe AH, Orkin SH (1995) Lethal β-thalassaemia in mice lacking the erythroid CACCC-transcription factor EKLF. Nature 375:318-322

Perkins AC, Peterson KR, Stamatoyannopoulos G, Witkowska HE, Orkin SH (2000) Fetal expression of a human Aγ globin transgene rescues globin chain imbalance but not hemolysis in EKLF null mouse embryos. Blood 95:1827-1833

Peters LL, Shivdasani RA, Liu SC, Hanspal M, John KM, Gonzalez JM, Brugnara C, Gwynn B, Mohandas N, Alper SL, Orkin SH, Lux SE (1996) Anion exchanger 1 (band 3) is required to prevent erythrocyte membrane surface loss but not to form the membrane skeleton. Cell 86:917-927

Peters LL, Swearingen RA, Gwynn B, Lux SE, Eicher EM, Washburn LL, Ciciotte SL, Goff-Andersen S (1998) Failure of effective reticulocytosis in a new spontaneous band 3 deficient mouse strain (C3H/HeJ-wan) and in a subset of targeted band 3 null mice (B6, 129 $AE1^{-/-}$) suggests the presence of genetic modifiers of band 3 function in red blood cells. Blood 92, suppl 1:301a

Peters LL, Jindel HK, Gwynn B, Korsgren C, John KM, Lux SE, Mohandas N, Cohen CM, Cho MR, Golan DE, Brugnara C (1999) Mild spherocytosis and altered red cell ion transport in protein 4.2-null mice. J Clin Invest 103:1527-1537

Peterson KR, Li QL, Clegg CH, Furukawa T, Navas PA, Norton EJ, Kimbrough TG, Stamatoyannopoulos G (1995) Use of yeast artificial chromosomes (YACs) in studies of mammalian development: production of beta-globin locus YAC mice carrying human globin developmental mutants. Proc Natl Acad Sci USA 92:5655-5659

Pevny L, Simon MC, Robertson E, Klein WH, Tsai S-F, D'Agati V, Orkin SH, Constantini F (1991) Erythroid differentiation in chimaeric mice blocked by a targeted mutation in the gene for transcription factor GATA-1. Nature 349:257-260

Philipsen S, Suske G (1999) A tale of three fingers: the family of mammalian Sp/XKLF transcription factors. Nucleic Acids Res 27:2991-3000

Pollak ES, High KA (2001) Hemophilia B: Factor IX deficiency. In: Scriver CR, Beaudet AL, Sly WS, Valle D (eds) The metabolic and molecular bases of inherited disease. McGraw-Hill, New York, pp 4393-4413

Rouyer-Fessard P, Leroy-Viard K, Domenget C, Mrad A, Beuzard Y (1990) Mouse beta thalassemia, a model for the membrane defects of erythrocytes in the human disease. J Biol Chem 265:20247-20251

Roy RP, Nagel RL, Acharya AS (1993) Molecular basis of the inhibition of beta s-chain-dependent polymerization by mouse alpha-chain. Semisynthesis of chimeras of human and mouse alpha-chains. J Biol Chem 268:16406-16412

Ryan TM, Townes TM, Reilly MP, Asakura T, Palmiter RD, Brinster RL, Behringer RR (1990) Human sickle hemoglobin in transgenic mice. Science 247:566-568

Ryan TM, Ciavatta DJ, Townes TM (1997) Knockout-transgenic mouse model of sickle cell disease. Science 278:873-876

Scriver CR, Kaufman S (2001) Hyperphenylalaninemia: Phenylalanine hydroxylase deficiency. In: Scriver CR, Beaudet AL, Sly WS, Valle D (eds) The metabolic and molecular bases of inherited disease. McGraw-Hill, New York, pp 1667-1724

Shedlovsky A, McDonald JD, Symula D, Dove WF (1993) Mouse models for human phenylketonurea. Genetics 134:1205-1210

Shehee WR, Oliver P, Smithies O (1993) Lethal thalassemia after insertional disruption of the mouse major adult beta-globin gene. Proc Natl Acad Sci USA 90:3177-3181

Shi ZT, Afzal V, Coller B, Patel D, Chasis JA, Parra M, Lee G, Paszty C, Stevens M, Walensky L, Peters LL, Mohandas N, Rubin E, Conboy JG (1999) Protein 4.1R-deficient mice are viable but have erythroid membrane skeleton abnormalities. J Clin Invest 103:331-340

Shivdasani RA, Fujiwara Y, McDevitt MA, Orkin SH (1997) A lineage-selective knockout establishes the critical role of transcription factor GATA-1 in megakaryocyte growth and platelet development. EMBO J 16:3965-3973

Simon MC, Pevny L, Wiles MV, Keller G, Costantini F, Orkin SH (1992) Rescue of erythroid development in gene targeted GATA-1-mouse embryonic stem cells. Nat Genet 1:92-98

Skow LC, Burkhart BA, Johnson FM, Popp RA, Popp DM, Goldberg SZ, Anderson WF, Barnett LB, Lewis SE (1983) A mouse model for beta-thalassemia. Cell 34:1043-1052

Southgate CD, Chishti AH, Mitchell B, Yi SJ, Palek J (1996) Targeted disruption of the murine erythroid band 3 gene results in spherocytosis and severe haemolytic anaemia despite a normal membrane skeleton. Nat Genet 14:227-230

Sung LA, Chu X, Vera C, Kim J, Sung K-LP (2000) Altered mechanical properties of erythroid cell membranes from E-tropomodulin -/- yolk sacs. Blood 96, suppl 1:593a

Trudel M, Saadane N, Garel MC, Bardakdjian-Michau J, Blouquit Y, Guerquin-Kern JL, Rouyer-Fessard P, Vidaud D, Pachnis A, Romeo PH, Beuzard Y, Costantini F (1991) Towards a transgenic mouse model of sickle cell disease: hemoglobin SAD. EMBO J 10:3157-3165

Trudel M, De Paepe ME, Chretien N, Saadane N, Jacmain J, Sorette M, Hoang T, Beuzard Y (1994) Sickle cell disease of transgenic SAD mice. Blood 84:3189-3197

Tsai F-Y, Keller G, Kuo FC, Weiss MJ, Chen J, Rosenblatt M, Alt FW, Orkin SH (1994) An early haematopoietic defect in mice lacking the transcription factor GATA-2. Nature 371:221-226

Tsang AP, Visvader JE, Turner CA, Fujiwara Y, Yu C, Weiss MJ, Crossley M, Orkin SH (1997) FOG, a multitype zinc finger protein, acts as cofactor for transcription factor GATA-1 in erythroid and megakaryocytic differentiation. Cell 65:109-119

Tsang AP, Fujiwara Y, Hom DB, Orkin SH (1998) Failure of megakaryopoiesis and arrested erythropoiesis in mice lacking the GATA-1 transcriptional cofactor FOG. Genes Dev 12:1176-1188

Tse WT, Lux SE (1999) Red blood cell membrane disorders. Br J Haematol 104:2-13

Wandersee NJ, Roesch AN, Hamblen NR, De Moes J, Van der Valk MA, Bronson RT, Gimm JA, Mohandas N, Demant P, Barker JE (2001) Defective spectrin integrity and neonatal thrombosis in the first mouse model for severe hereditary elliptocytosis. Blood 97:543-550

Weiss MJ, Keller G, Orkin SH (1994) Novel insights into erythroid development revealed through *in vitro* differentiation of GATA-1 embryonic stem cells. Genes Dev 8:1184-1197

West JD, Flockhart JH, Peters J, Ball ST (1990) Death of mouse embryos that lack a functional gene for glucose phosphate isomerase. Genet Res 56:223-236

White RA, Birkenmeier CS, Lux SE, Barker JE (1990) Ankyrin and the hemolytic anemia mutation, nb, map to mouse chromosome 8: presence of the nb allele is associated with a truncated erythrocyte ankyrin. Proc Natl Acad Sci USA 87:3117-3121

Whyatt D, Lindeboom F, Karis A, Ferreira R, Milot E, Hendriks R, De Bruijn M, Langeveld A, Gribnau J, Grosveld F, Philipsen S (2000) An intrinsic but cell-nonautonomous defect in GATA-1-overexpressing mouse erythroid cells. Nature 406:519-524

Wijgerde M, Gribnau J, Trimborn T, Nuez B, Philipsen S, Grosveld F, Fraser P (1996) The role of EKLF in human beta-globin gene competition. Genes Dev 10:2894-2902

Wu H, Liu X, Jaenisch R, Lodish HF (1995) Generation of committed erythroid BFU-E and CFU-E progenitors does not require erythropoietin or the erythropoietin receptor. Cell 83:59-67

Yang B, Kirby S, Lewis J, Detloff PJ, Maeda N, Smithies O (1995) A mouse model for beta 0-thalassemia. Proc Natl Acad Sci USA 92:11608-11612

Yi SJ, Liu SC, Derick LH, Murray J, Barker JE, Cho MR, Palek J, Golan DE (1997) Red cell membranes of ankyrin-deficient nb/nb mice lack band 3 tetramers but contain normal membrane skeletons. Biochemistry 36:9596-9604

Zhang JZ, Hayashi H, Ebina Y, Prohaska R, Ismail-Beigi F (1999) Association of stomatin (band 7.2b) with Glut1 glucose transporter. Arch Biochem Biophys 372:173-178

Zhu Y, Paszty C, Turetsky T, Tsai S, Kuypers FA, Lee G, Cooper P, Gallagher PG, Stevens ME, Rubin E, Mohandas N, Mentzer WC (1999) Stomatocytosis is absent in "stomatin"-deficient murine red blood cells. Blood 93:2404-2410

29 Red Cell Ageing

Augusta Brovelli and Giampaolo Minetti

Dipartimento di Biochimica "A. Castellani", Università di Pavia, via Bassi 21, 27100 Pavia, Italy

29.1 Introduction: the Properties of Mammalian Red Cells

The mammalian erythrocyte is unique among other cell types of the organism, being devoid of nucleus and other intracellular organelles. This cell lacks the transcriptional and translational apparatus for protein synthesis, and it therefore constitutes a good model for studying cell ageing. The mechanisms for the replacement or repair of damaged macromolecules are in fact absent, or minimal, thus permitting the analysis the age-related modifications of the cell and its constituents without major interference.

To perform its function of oxygen and carbon dioxide transport, the mammalian erythrocyte relies on a simple cellular architecture and molecular machinery. In its mature form this cell is constituted only of a cytosolic compartment and of the plasma membrane. All other intracellular compartments are lost during the maturation of the reticulocyte to the erythrocyte in the circulation (Rapoport 1986). Of the cytosolic proteins, 85% is represented by haemoglobin (Hb), while the remaining are enzymes of the metabolic apparatus. This consists of two major pathways: glycolysis and the pentose phosphate pathway; some minor, but essential pathways and biosynthetic activities are also present (Surgenor 1974).

As the reticulocyte matures to erythrocyte, the consumption of oxygen decreases from 50 to 1, and that of glucose from 7.5 to 1 μmoles/(ml_{cells}·h). The energy metabolism of circulating erythrocytes depends on the degradation of glucose through the glycolytic pathway, which provides ATP, 2,3-BPG (a negative allosteric modulator of the oxygen affinity of Hb) and NADH. ATP is important for the maintenance of the biconcave shape of the cell, it fuels the Ca^{2+}-ATPase and the Na^{+}, K^{+}-ATPase, it modulates the activity of several ion transporters, and it also plays a role in the cell defence against oxidants. NADH is used to reduce pyruvate to lactate and for the reduction of Fe^{3+} in methaemoglobin, to Fe^{2+} of fully functional Hb: cytochrome b_5 is an intermediate in this reaction.

The red cell antioxidant systems rely essentially on the pentose phosphate pathway, which provides reducing power as NADPH. Erythrocytes contain high concentrations (~2 mM) of glutathione (GSH), which functions as an antioxidant. The enzyme glutathione reductase uses NADPH to maintain a high ratio (~500) of glutathione to glutathione-disulfide (GSSG) inside the cell. One important function of GSH is the detoxification of hydrogen peroxide, which is reduced to water in a reaction catalysed by the enzyme glutathione peroxidase. The GSH/GSSG system also acts as a buffer to maintain in the reduced form the SH group of cysteines in cytosolic and membrane proteins. Whenever oxidative processes lead to a decrease in the GSH/GSSG ratio, GSSG is transported out of the cell by an ATP-dependent system. ATP also fuels the synthesis *de novo* of GSH. The enzymes superoxide dismutase and catalase detoxify the cell from superoxide anions and hydrogen peroxide, thus providing the erythrocyte with a defence against reactive oxygen species (ROS).

Minor pathways perform the synthesis and the metabolism of the pyridine nucleotides NADH and NADPH, the salvage of purine ribonucleotides, and the repair of membrane phospholipids. The erythrocyte also contains: carbonic anhydrase, protein-serine/threonine and -tyrosine kinases, and the corresponding phosphoprotein-phosphatases, monomeric and trimeric GTPases, calmodulin. Some of these species are cytosolic, some are bound to the plasma membrane and many can be found in both compartments in variable amounts, depending on the functional state of the cell. The plasma membrane is composed of approx. equal amounts (by weight) of lipids and proteins (see Fig. 29.1).

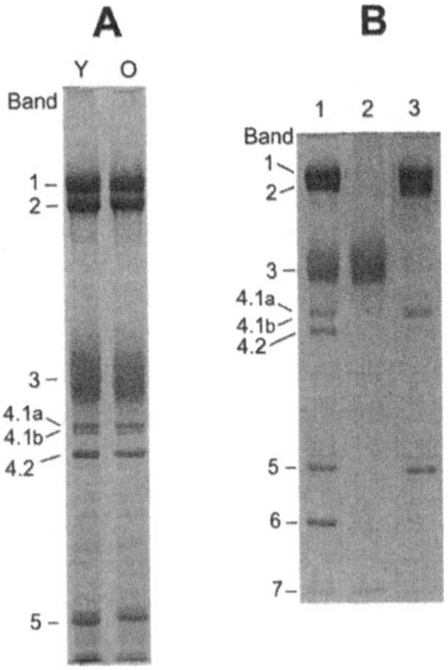

C

Lipid	*weight % of total lipid*
Cholesterol	23
Phosphatidylethanolamine	18
Phosphatidylcholine	17
Sphingomyelin	18
Phosphatidylserine	7
Glycolipid	3
Others	13

Fig. 29.1. Protein and lipid composition of the human erythrocyte membrane. SDS-PAGE and Coomassie blue staining of total proteins of ghost membranes prepared from young and old cells (**A**, lanes *Y* and *O*) and from total erythrocyte population (**B**, lane *1*). Integral proteins (**B**, lane *2*) and membrane skeleton proteins (**B**, lane *3*), prepared from ghost membranes as NaOH-insoluble material and Triton X 100-insoluble material, respectively. Bands 1, 2, 3, 5 and 6 correspond to α-spectrin, β-spectrin, anion exchanger, actin and glyceraldehyde-3-phosphate dehydrogenase, respectively. (**C**) Adapted from Cullis and Hope (1985)

The decline of several membrane constituents, including receptors, antigens, enzymes, transporters and other proteins, is observed during the maturation of the reticulocyte, and leads to a profound rearrangement of the membrane in the mature erythrocyte.

The high deformability of the erythrocyte is related to the cell's high surface area-to-volume ratio. Essential to the maintenance of normal deformability, which is critical for the oxygen transport function, is the precise regulation of cell water content. A cell with too little water will have a reduced deformability because of the high viscosity of the concentrated haemoglobin solution inside, while a cell with too much water will have a reduced deformability because of decreased area-to-volume ratio. It is clear that any process modifying the properties of the plasma membrane components or of cell water content can alter the deformability of the cell and, consequently, its life span in the circulation.

The majority of studies on red cell life span have been conducted on human erythrocytes; therefore we will deal in this chapter with results concerning human red cells, unless otherwise stated. Several authoritative reviews are available about mammalian erythrocyte ageing (Clark 1988; Bartosz 1990; Lutz 1990). These works describe and critically discuss the experimental methods adopted and the results obtained until the late eighties. In this chapter, after a history of the significant stages in the study of erythrocyte ageing up to the eighties, the focus will be on the methods of investigation and the results obtained in the last 15 years. Finally, a section will be dedicated to the topic of red cell storage in blood banks, where the ageing lesion and the storage lesion will be compared and discussed.

29.2 Red Cell Ageing in Mammals

In principle, the removal of red cells from the circulation can be an age-dependent (i.e. time-dependent) or a random process (independent on the time spent by the cell in the circulation). In humans, and in a number of other mammals, red cells are removed in an age-dependent manner. The first evidence against a random clearance of human erythrocytes became available after studies were performed in which human red cells of a donor were transfused in a recipient, whose erythrocytes differed antigenically, and then the survival of the donor's cells was followed. It was observed that the transfused cells were selectively cleared after a definite and characteristic time span (Ashby 1919).

Afterwards, cohort-labelling studies allowed comparing the mean life span of red cells in different mammals. For these experiments, a cohort of reticulocytes or bone marrow cells is labelled with isotopic methods, and its survival evaluated by measuring the persistence of labelled cells in the circulation. By administering to the animal a labelled compound (^{59}Fe, ^{14}C- or ^{15}N-glycine), which is incorporated only into the immature cells in the bone marrow, the compound is rapidly inserted into the haem molecule of reticulocytes. Since Hb constitutes most of the cell's proteins, nearly all the label remains inside the cell as part of the Hb molecule for the entire life of the cell in the circulation. In species whose erythrocytes display an age-dependent life span, the number of labelled cells in the circulation remains constant for a defined and characteristic period. The label then disappears rapidly, indicating that the cells of the tagged cohort have acquired characteristics that promote their removal for destruction. Data obtained from Shemin and Rittenberg (1946) using ^{15}N-glycine labelling, reported in Fig. 29.2A, demonstrated that human red cells have a relative uniform life span of about 120 days. In rabbits, a combination of time-dependent and random removal of erythrocytes was observed (Fig. 29.2B). In several other mammals, a similar behaviour was found, with a different contribution of the random and time-dependent cell death (Eadie and Brown 1953). Despite the presence of a different contribution of the random death component in different species, a characteristic red cell life span in various mammals was found, ranging from 40 to 160 days (see also Sect. 29.4, Fig. 29.4).

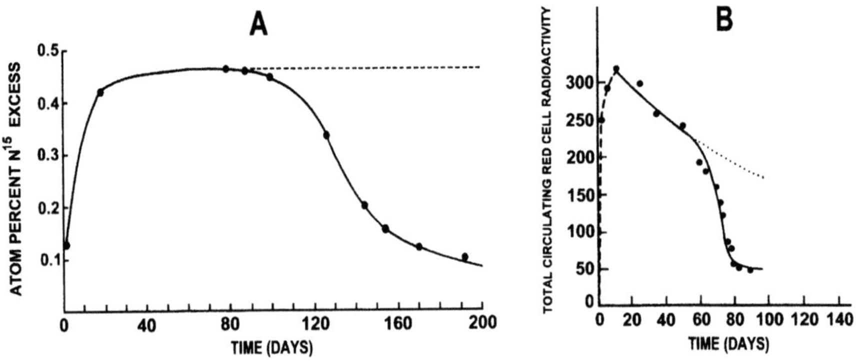

Fig. 29.2. *In vivo* survival of labelled cohorts of human (**A**) and rabbit (**B**) erythrocytes. Solid lines describe the amount of circulating labelled cells at different time points from the onset of the experiment. The S-shaped portion represents the component of time-dependent removal, and is present in both **A** and **B** curves. If random destruction alone were present, the curve would follow the dotted line. The slope of dotted line is zero in **A**, indicating the absence of random death in humans (adapted from Shemin and Rittenberg (1946)). In **B** the dotted line has a negative slope, indicating the presence of a random death component in rabbits. Adapted from Eadie and Brown (1953)

Two main questions then arose: where in the organism the old red cells are removed and what molecular processes lead to red cell death. Results from several cohort-labelling studies performed in animals suggested that all the reticulo-endothelial system, where phagocytes recognise and engulf the erythrocytes destined for destruction, might participate in the clearance, the major site for this process varying from species to species.

In humans, due to the slow rate of removal of physiologically aged red cells, several investigations focused on haematological disorders, where erythrocytes are destroyed at a faster rate. It was found that the spleen plays an important role in the removal of poorly deformable or metabolically impaired cells, since splenectomy ameliorates the survival capacity of erythrocytes in various haemolytic disorders. However, spleen removal has only a modest effect on the survival of normal red cells (Miescher 1956). Moreover, the only site where erythrophagocytosis of normal erythrocytes has been demonstrated in humans is the bone marrow (Marton 1970).

A novel insight into red cell ageing came from studies in which the cohort-labelling method was combined with the separation of the blood specimen into subpopulations of cells of different density. From such studies, conducted in normal human subjects and in animals (Borun et al. 1957; Piomelli et al. 1967), it appeared that the labelled erythrocytes became denser as they aged in the circulation.

Erythrocyte density is chiefly determined by intracellular water and Hb content. The latter appears to be constant throughout the red cell life (Lew et al. 1995). Water content is regulated by the amount of intracellular osmolytes, which, for the normal red cell, are represented mainly by Na^+, K^+, and Cl^- (Lee et al. 1984). An increase in cell density would result either from a decrease in total intracellular osmolytes leading to cell dehydration at constant membrane surface area, or from a balanced reduction in surface area and cytosol at constant Hb content, or from a combination of the two processes.

The relationship between cell age and cell density established by the above described experiments provided the rationale for a conspicuous number of studies to follow, in which the structural and functional properties of red cell subpopulations of different density were compared, on the assumption that density separation provided an effective means of isolating erythrocytes of different age.

29.3 Properties of Human Red Cells of Different Density

Many of the studies that describe modifications in the cellular properties of human erythrocytes during ageing were carried out with red cells purified from anticoagulated blood by removal of the buffy coat (the fluffy layer of white cells and platelets at the top of the column of erythrocytes sedimented by centrifugation), and by washing the pelleted cells. Several investigators have observed that this procedure does not completely eliminate white cell contamination and is the primary cause of artifacts. Therefore, filtration of blood through mixtures of α- and micro-crystalline cellulose was proposed as a method to eliminate white cells and platelets (Beutler et al. 1976). Omission of this step is the most likely source of errors and inconsistent results, in particular when functional studies of cell membrane proteins are carried out using ghost preparations or intact cells.

Studies on erythrocyte subpopulations were commonly performed by comparing the 10% least dense with the 10% densest cells obtained by centrifugation of the total population of erythrocytes, freed of white cells and platelets and resuspended in autologous plasma. Better results were obtained by separating the cells into several (4–6) subfractions on pre-formed density gradients of dextran, serum albumin, stractan, or on self-forming Percoll gradients (for review, see Lutz (1990)). With the centrifugation on density gradients, the lowest cross-contamination among cells of different densities was obtained.

From these studies (reviewed in Clark (1988)) it became clear that the progressive increase in cell density with cell age was due to the increase in the MCHC, which in turn depended on the loss of K^+ and of cell water. Reduction in cell volume and in the surface area of the membrane, evaluated as the decreased amount of phospholipids and cholesterol per cell, was also observed. All these modifications contribute to decrease cell deformability. Alteration of the asymmetric distribution of phospholipids in the membrane, with exposure of phosphatidylserine (PS) on the outer leaflet of the lipid bilayer, was also reported (see Chaps. 1, 2, and 22).

A number of other qualitative and quantitative changes in cellular constituents have been described: increased Na^+ content, decreased activity of superoxide-dismutase, glutathione-reductase and -peroxidase, decrease in the amount of membrane proteins like acetylcholinesterase, insulin receptor and Na^+, K^+-ATPase. Some of the observed alterations are sign of oxidative damage: increased Hb-spectrin crosslinking, methaemoglobin, membrane-bound hemichromes (products of Hb denaturation), lipid peroxidation, increased GSSG content and decreased GSH, cytochrome b_5-reductase and cytochtome b_5. Structural damage of cell proteins were also observed, like the racemization of Asp residues leading to an increased number of carboxylmethylation sites, and glycation of Lys residues, leading to the appearance of glycated Hb. Last, but not least, the decrease in oxygen delivery, in the concentration of 2,3-BPG and in intracellular pH was documented. Some of the modifications of the dense/aged red cells were observed in association with increased binding to the cell surface of autologous antibodies, which were able to stimulate the *in vitro* phagocytosis of modified cells (see below, this section). The debate on the role of the various cellular and membranous modifications in destining the aged red cell for destruction was reviewed in detail by Clark (1988). We will only present and discuss here the hypothesis proposed on the basis of the experimental evidence available.

In most of the studies where the metabolic efficiency of erythrocytes of different density was analysed, the ATP and GSH content per cell at stationary state showed a decrease in dense cells. Given the important role of the two compounds in the cell's energy supply and antioxidant defence, it was hypothesised that their decrease below a certain threshold would determine metabolic impairment and reduce the survival potential of the erythrocyte. This *metabolic hypothesis* was corroborated by evidence showing profound modifications in cell shape and deformability and the loss of membrane fragments as microvesicles occurring in ATP-depleted erythrocytes. Moreover, the decrease in GSH content correlated with that of cytochrome b_5 and with the increase in methaemoglobin content. It must also be remembered that the importance of a gradual ATP decline as a causative factor in red cell senescence has been questioned since a relatively constant ATP concentration was measured in red cells of different density (Kirkpatrick et al. 1979). Similar results were found concerning GSH concentration (Piccinini et al. 1995).

In 1975 M. Kay first reported on the presence of autologous antibodies bound to dense human erythrocytes, as revealed by electron microscopy. Having observed that the bound IgGs were effective in promoting red cell phagocytosis in an *in vitro* assay, Kay proposed a role for the auto-antibodies in the recognition and removal of senescent erythrocytes (Kay 1975). The *auto-immune hypothesis* gained in strength when IgGs were isolated from dense red cells (but not from light ones), that were able to recognise by immuno-blotting, a 62 kDa peptide derived from band 3. Several other groups confirmed the presence of anti-band 3 antibodies on dense erythrocytes.

Concerning the process(es) responsible for the exposure of the band 3 epitope(s) recognised by the auto-antibodies, various mechanisms were proposed: oxidation and proteolysis of band 3 (Kay et al. 1986), clustering of band 3 on the membrane and binding of the C3b component of the complement system to the auto-antibodies (Lutz et al. 1988), clustering of band 3 induced by the binding of hemichromes to the cytosolic domain of band 3 (Low et al. 1985). Data from another group revealed the existence in the normal serum of auto-antibodies directed against α-galactosyl moieties in membrane glycolipids of dense erythrocytes (Galili et al. 1984). Although much evidence exists on the increased binding of autologous antibodies to dense human red cells, their role in promoting the removal of senescent cells from the circulation has not been completely elucidated.

The appearance in the dense erythrocyte of properties concurring to the decrease in cell deformability led to the hypothesis that stiffening of the cells might be responsible for their clearance from the circulation. This was proposed in analogy to what was observed in a number of haematological disorders, in which premature cell death was ascribed to a loss of cell deformability leading to red cell entrapment in the spleen. Although there was little doubt about old cells being denser and less deformable than young ones, concerns were expressed about the correlation between cell age and density throughout the entire life span of erythrocytes. In fact, cells that have become dense independently of a regular ageing process can be considered functionally senescent since the higher the cell density the higher the loss of cell water and of cell deformability. In other words, whether red cells from a high-density subpopulation were also of uniformly old age was not known, due to the lack of a reliable cell-age parameter. Such lack represented a major difficulty for the evaluation of results of density-fractionation studies.

To estimate the age of density-separated subpopulations of erythrocytes, cell density itself was adopted as a criterion, as such or in combination with other markers, like the level of some enzymes (hexokinase, pyruvate kinase, glutamate-oxalacetate-transaminase), the amount of insulin receptors or the creatine content. Because the highest rate of change in the value of these parameters was observed during the maturation of the cells from the reticulocyte stage to young erythrocyte, they were good indicators of contaminating reticulocytes in the cell subpopulations, rather than reliable erythrocyte age parameters (Clark 1988). This was particularly evident when only the less dense and the densest subpopulations, each accounting approx. for 10% of the total population of erythrocytes, were compared.

When the total population of cells was separated in several subfractions of different densities and the subfractions compared to each other, a better evaluation was possible of the cell properties that changed abruptly at the end of the red cell life and that could be important for the removal of senescent cells. It is of interest to refer here to a single study in humans, in which red cell subpopulations obtained by density-separation on stractan gradients and constituting a progressively smaller percent of the total erythrocyte population, were labelled with ^{51}Cr and reinfused to measure their survival capacity (Ten Brinke and De Regt 1970). Results are reported in Table 29.1 and show that the higher the average cell density, the shorter the cell survival after reinfusion. Cells of extremely high density, representing less than 4% of total, had a residual survival of less than 24 h.

Table 29.1. *In vivo* survival of ^{51}Cr-labelled human red cells of the total erythrocyte population and of some selected subpopulations of relatively high and low cell density

	Relative fraction size [%]	$T_{1/2}$ (100% value at 20 min) [days]	$T_{1/2}$ (100% value at 24 h) [days]	Decrease radioactivity in blood between 20 min and 24 h [%]
Unfractionated erythrocyte population (n = 12)	100	17–29	20–30.5	7.60 ± 6.03
Relatively light erythrocyte fractions (n = 3)	6	28	29	4
	1.2	28	29	2.5
	3.0	25	26	2.7
Relatively dense erythrocyte fractions	20	7	24	40
	15	14–15	20	18
	8	< 1	13–14	90
	5	8	11	15
	4	6–10	9–12	10–26
	3.5	< 1	-	~0 after 20 min
	1.5	< 1	-	90
	1.5	< 1	-	80

Two values for survival half time ($T_{1/2}$: time needed to observe the reduction to one half of the originally infused population) were calculated, one by taking the circulating radioactivity 20 min after transfusion as the initial level, and the other by taking the radioactivity measured 24 h after transfusion. Adapted from Ten Brinke and De Regt (1970).

From these data it appeared therefore of paramount importance, in studies on red cell ageing, that high-efficiency separation methods were adopted, so as to isolate the few percent of cells having the lowest residual survival capacity. A reliable parameter for measuring cell age was also needed, to verify that dense subpopulations of erythrocytes were actually enriched in senescent cells.

29.4 Preparation of Red Cells of Well-Defined Cell Age

During the eighties, two innovative procedures were adopted to prepare *in vivo* aged erythrocytes using animal models: hypertransfusion in mouse (Mueller et al. 1987) and biotinylation in rabbit (Suzuki and Dale 1988). Aged red cells from mice were obtained using a modification of the Ganzoni hypertransfusion model (Ganzoni et al. 1971) in which erythropoiesis was suppressed for up to 56 days, i.e. nearly one life span of mouse erythrocytes. Starting with a large number of mice, half of the animals were sacrificed every 14 days and their blood transfused in the remaining mice, thus maintaining a high haematocrit and suppressing erythropoiesis in the recipients, and allowing the pre-existing cells to age *in vivo*. The operation was repeated several times until a relatively homogeneous population of circulating old erythrocytes was obtained, with a residual survival $T_{1/2}$ of less than 1 day (Fig. 29.3A). When the properties of these cells were examined, the most evident alteration was the increase in the ratio of membrane skeletal protein 4.1a to 4.1b (Fig. 29.3B). The measurement of 4.1a/4.1b ratios was thus proposed as a useful index of red cell age. The density of mouse erythrocytes displayed a small shift towards higher values during ageing *in vivo*, although a broad distribution of density values was measured (Mueller et al. 1987).

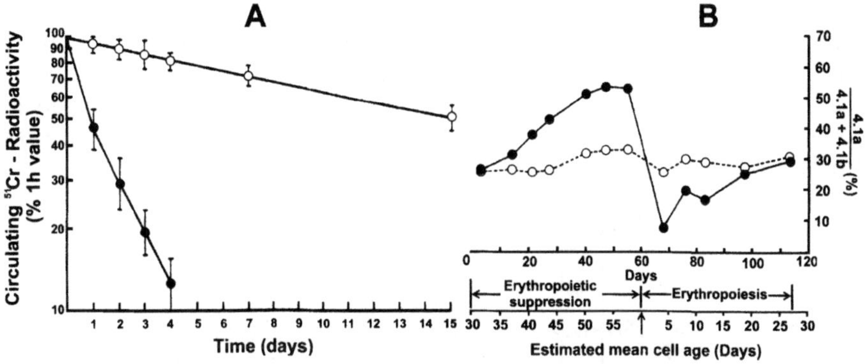

Fig. 29.3. Characteristics of *in vivo* aged mouse erythrocytes. (**A**) Residual survival of ^{51}Cr-labelled red cells from mice hypertransfused for 56 days (filled circles), compared to the survival of a normal red cell population (empty circles) The circulating radioactivity at each time point is given as the percent of the radioactivity circulating 1 hour after reinfusion of the labelled cells. (**B**) The increase in 4.1a/(4.1a+4.1b) values during erythropoietic suppression (filled circles), is followed by an abrupt decrease in the values after resumption of erythropoiesis and production of young cells. (empty circles) 4.1a/(4.1a+4.1b) values in control mice. Adapted from Mueller et al. (1987)

Suzuki and Dale introduced a new method, in which rabbit erythrocytes were biotinylated by reaction with *N*-hydroxysuccinimide biotin, forming a stable amide bond between biotin and amino groups on the surface of the cell. Biotinylated cells were preliminarily shown to have a normal *in vivo* survival of 60 days when reinfused into the rabbit. Recovery of biotinylated cells was achieved by allowing the erythrocytes taken from the animal at definite time points, to interact with avidin-coated polystyrene beads. Results obtained by this technique in the rabbit (Suzuki and Dale 1988) confirmed, as observed with the hypertransfusion model in mice, the lack of a strong correlation between cell age and cell density. It was also revealed that the activity of several glycolytic enzymes, previously used as a cell age index, did not change with ageing.

The most important result obtained from the characterisation of *in vivo* aged erythrocytes prepared with the above-described techniques was that measurements of 4.1a/4.1b ratios might serve as a useful index of erythrocyte age, independent on cell density. Protein 4.1 is a component of the erythrocyte membrane skeleton accounting for approx. 5% of total membrane proteins (see Fig. 29.1). In human erythrocytes it exists as two major electrophoretic forms: 4.1a (80 kDa) and 4.1b (78 kDa). The two polypeptides are closely related in sequence and function. As it was observed that the total amount of protein 4.1 remains constant throughout cell life, the increase in 4.1a/4.1b ratio during red cell ageing was supposed to be due to the conversion of protein 4.1b to 4.1a. It became soon clear that such conversion occurs through a deamidation reaction of two Asn residues located in the carboxyl terminus of the protein (Asn478 and Asn502). Protein 4.1 is synthesised as 4.1b, which is in fact the predominant form in immature erythroid cells such as reticulocytes and K562 cells. While Asn478 deamidates soon after the protein is synthesised, Asn502 deamidates at a slower rate. The difference in these two residues is sufficient to confer to the polypeptides a different electrophoretic mobility (Inaba et al. 1992) (see Fig. 29.1A). This post-translational modification occurs in many mammalian species. It was in fact observed that the 4.1a/4.1b ratio in erythrocytes from various mammals correlates closely to the life span of their red cells (Inaba and Maede 1988) (Fig. 29.4).

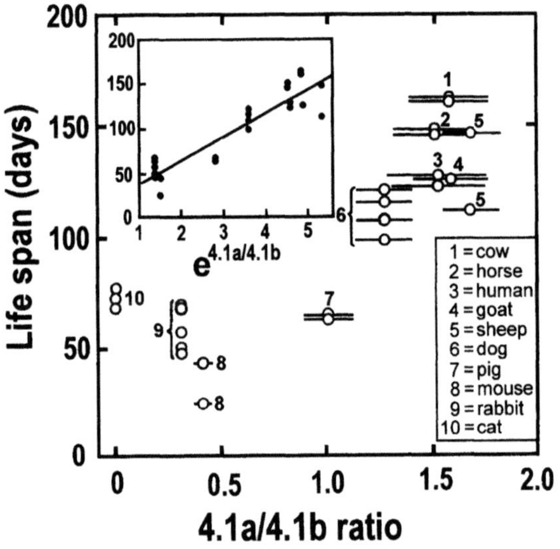

Fig. 29.4. Correlation between *4.1a/4.1b ratios* and erythrocyte life span in different mammals. In the inset the values of exp (4.1a/4.1b) are plotted against the mean life span values. The following regression formula was calculated:
Life span (days) = 26.3 exp (4.1a/4.1b) + 10.4 (r = 0.91). Adapted from Inaba and Maede (1988)

Studies on protein deamidation have shown that the instability of Asn residues correlates with polypeptide chain flexibility and that the rate of the deamidation reaction is sequence specific (Hejnaes et al. 1998). In the pH interval from 5–12 the non-enzymatic deamidation of Asn residues proceeds via a succinimide intermediate which spontaneously hydrolyses to produce L- and D-aspartyl and -isoaspartyl residues. At more acidic pH (1–2), direct hydrolysis of the side-chain amide generates Asp as the sole product. Studies on short peptides have shown that the deamidation of Asn residues through a succinimide intermediate increases with temperature, at elevated pH and ionic strength. Deamidation in proteins is influenced in the same ways and therefore the rate of deamidation can be minimised by working at neutral or slightly acidic pH, in buffers of low ionic strength and at low temperature. Since the deamidation of proteins is a non-enzymatic reaction and since the intracellular erythrocyte medium maintains a relatively constant composition, conversion of protein 4.1b to 4.1a occurs at a relatively constant rate under physiological conditions.

Deamidation of protein 4.1b to 4.1a is therefore an excellent marker that can be usefully adopted in studies on red cell ageing to characterise the absolute age of erythrocyte subpopulations obtained by density-separation.

29.5 The Ageing of Red Cells Reinvestigated

A milestone in studies about human red cell ageing was a paper published in 1992 (Waugh et al. 1992). In this work, the rheologic properties of rabbit biotinylated erythrocytes and of erythrocytes of hypertransfused mice, aged *in vivo* as described in Sect. 29.4, were examined. In both cases red cells were subjected to ektacytometry, a measure of the maximum cell elongation at different osmolalities, performed with a viscosimeter.

Aged red cells from both animal species displayed a decreased deformability due to cell dehydration and reduced cell surface area. Aged rabbit erythrocytes were also examined by micropipette methods, which revealed, with respect to the total population of cells, an average decrease in surface area of 10.5%, a decrease in cell volume of 8.4% and a rise in MCHC to 37 g/dl, while the membrane elasticity was found within normal values.

When human erythrocytes were examined with the same methods by these authors, the most dense cells separated on stractan density gradients corresponding to 1% or less of the total population of cells, and having an MCHC $\geq$ 37 g/dl, displayed rheologic properties similar to those observed in the *in vivo* aged red cells from rabbits and mice. Importantly, a considerably greater decrease in surface area (–20%) and volume (–17%) than those observed in rabbit were measured, probably because of the absence of random cell death in humans. The authors' conclusions were that by adopting stringent protocols for density gradient fractionation, that allow the isolation of the most dense cells, it is possible to obtain a population of human red cells having rheologic properties similar to those of senescent cells of other mammalian species.

Although the study carried out by Waugh and co-workers did not directly address the immediate cause of senescent cell removal, it validated the fractionation through high-efficiency density gradients as a procedure to isolate human senescent erythrocytes (the methods used to prepare *in vivo* aged red cells in mouse and rabbit cannot be adopted in humans).

On these bases, several groups have reinvestigated the characteristics of the densest erythrocytes, separated by means of more stringent density fractionation protocols. In most cases the absolute age of the various red cell subpopulations of different density was estimated with the new cell age parameter found in studies of the animal model: the 4.1a/4.1b ratio. The typical result obtained by fractionating human erythrocytes on discontinuous density gradients of stractan is shown in

Fig. 29.5. By adopting procedures similar to that reported in Fig. 29.5, data in favour of the autoimmune hypothesis were found by two groups. The first group showed an increased amount of IgGs bound to the densest red cells, recognising band 3 protein, α-galactosyl residues, and other, as yet unidentified epitopes (Sorette et al. 1991). The second group found that the 0.5% densest cells contained 6 times more membrane-bound Hb and 10 times more autologous IgGs than cells of lower density. A mixture of aggregated proteins, crosslinked via disulfide bridges, was also found, containing hemichromes, spectrin and band 3 (Kannan et al. 1991).

Density (g/mL)	Cell recovery (%) (n=6)	MCV (fL) (n=6)	MCHC (g/dL) (n=6)	4.1a / 4.1 b (n=5)
1.085	f1: ≃45	96.7 ± 7.1	29.9 ± 3.2	1.584 ± 0.172
1.095	f2: ≃45	92.8 ± 6.7	32.3 ± 3.2	2.343 ± 0.153
1.103	f3: 3.7 ± 2.4	89.3 ± 4.7	34.5 ± 2.5	3.113 ± 0.287
1.110				
1.115	f4: 1.0 ± 0.7	83.8 ± 5.0	37.2 ± 2.9	3.565 ± 0.938
	f5: 0.4 ± 0.3	80.5 ± 4.5	39.0 ± 2.3	3.667 (n=2)
1.137				

Fig. 29.5. Density-separation of human erythrocytes on a discontinuous stractan gradient. Schematic depiction of the distribution of cell fractions on a discontinuous density gradient: the age of red cells isolated in each fraction was characterised on the basis of the ratio between the amount of membrane proteins 4.1a and 4.1b. Adapted from Piccinini et al. (1995)

Another group described a disproportional loss, with respect to the loss of surface area, of the membrane proteins DAF (–40%) and CR1 (–60%) in the densest cells, and focused the attention on a possible interaction that might occur *in vivo* between red cells and the complement system (Lutz et al. 1992). DAF and CR1 are in fact, together with CD59, C8bp and glycophorin, a group of integral membrane proteins involved in the cell defence when erythrocytes, as bystanders during autologous complement activation, become the target of the complement system. The defence consists in the interference with the production and assembly of peptides of the complement system (components C5b, C6, C7, C8 and C9), constituting the membrane attack complex (MAC), a complex that, when successfully assembled, perforates the erythrocyte membrane. Should the defence fail to inhibit this assembly, the erythrocytes can still escape lysis by releasing membrane vesicles containing the MAC, thus getting rid of the regions of the membrane containing the pore-forming, lytic complex. This process appears to occur only in the presence of extracellular Ca^{2+} (Iida et al. 1991).

To account for the early observation that membrane lipids and proteins are lost during red cell life in the circulation, a process like membrane vesiculation was invoked (Morrison et al. 1974). *In vitro*, microvesiculation can be induced by a variety of conditions and treatments. However, the only physiological stimulus found so far, that is able to induce the release of membrane in the form of vesicles, is the treatment of the cells with a mixture of purified C5b, C6, C7, C8 and C9 (the final component of the complement cascade), in the presence of millimolar concentrations of Ca^{2+} (Iida et al. 1991).

Finally, in our laboratory we have recently found (Ciana 2000) in the most dense cell fraction with the highest 4.1a/4.1b value, representing between 1% and 3.5% of the total erythrocytes, a conspicuous increase in the concentration of intracellular Na^+, to values that are expected to stimulate by 25% over basal activity the Na^+/K^+ pump (O'Neill and Mikkelsen 1987). This confirms previous results obtained with less stringent protocols for cell fractionation or where an unsuitable cell age parameter was adopted (Cheng et al. 1984; Hentschel et al. 1986).

Although the investigations carried out in the last 10 years have not answered the questions as to what the modifications directly involved in red cell clearance might be, they have contributed to a better understanding of the cellular and molecular properties of the senescent erythrocyte.

Several studies carried out with model systems have provided evidence of what could be the molecular processes involved in determining the properties of senescent cells. Oxidation, transit in the microvasculature (Larsen et al. 1981), binding of substances like PGE_2 released by platelets (Li et al. 1996) and complement activation could relate to the loss of K^+ and consequently to cell dehydration, through activation of the K^+-Cl^- cotransport system (Brugnara and Tosteson 1987) and/or the Ca^{2+}-dependent K^+ channel (Gárdos 1958). Increased intracellular Na^+ concentrations would stimulate the Na^+/K^+ pump and aggravate cell dehydration (Clark et al. 1981; Lew and Bookchin 1986). All these processes would synergistically contribute to the reduction in cell volume and to the increase in MCHC. On the other hand, the decrease in surface area could result from membrane vesiculation, as a defence mechanism that cells are likely engage to escape complement attack (Halperin et al. 1989; Iida et al. 1991). Decreased surface area and increased cytoplasm viscosity (due to increased MCHC) (Stuart and Ellory 1988) also lower cell deformability with the consequence that aged cells are delayed in their transit through the filter networks of the spleen and bone marrow. This would improve red cell clearance, as the recognition by macrophages of the autoantibodies bound to the surface of aged erythrocytes would be facilitated. In this context it is worth reminding that the bone marrow and the spleen are the regions where the longest transit times for blood have been calculated (Leggett and Williams 1995).

29.6 Red Cell Storage in Blood Banks

The storage *in vitro* of human red cells, besides its evident practical relevance in transfusion medicine, was also considered by some authors as a model system to understand erythrocyte ageing. The properties of human erythrocytes preserved in blood banks at 4 °C in solutions containing different additives have been widely characterised in the total red cell population (Rapoport 1947; Wolfe 1985; Beutler 1996).

During storage, a progressive deterioration in the metabolic properties and in the membrane of the red cells occur, ranging from cell swelling to haemolysis, shape changes and loss of deformability. Membrane protein and lipid changes and shedding of membrane fragment as vesicles also occur. At 4 °C, glucose utilisation decreases to 1/50 of that at 37 °C, and the concentrations of metabolites like glucose, 2,3-BPG and ATP decline with time. Lactic acid accumulates, causing a pH shift from 7.2 to approx. 6.5 after 6 weeks of storage. The low temperature of storage also determines a decrease, to about 1/1000 of its activity at 37 °C, of the Na^+, K^+-ATPase, which has a high thermal coefficient. This causes an increase in intracellular Na^+ (+200%) and a decrease in intracellular K^+ (–40% to –60%) after 6 weeks of storage (Wood and Beutler 1967).

While most of the available data on the properties of stored erythrocytes come from investigations conducted on the total population of red cells, only a few studies have dealt with the different susceptibility of red cells of different age to the storage lesion. In one study, by comparing density-separated "neocytes" (50% less dense cells) with "gerocytes" (50% densest cells), that had been separately stored, an improved *in vivo* recovery both at 24 h and longer times after reinfusion was observed for neocytes, as compared to gerocytes (Keegan et al. 1992). The longer survival of neocytes was probably unrelated to their metabolic status, as no differences between them and gerocytes were found after 42 days of storage in any of the metabolic parameters evaluated: ATP, 2,3-BPG, glucose, lactate. Therefore, the precise reasons for the better performance of younger stored erythrocytes after reinfusion are still poorly understood.

After reinfusion, the metabolic alterations suffered by red cells during storage revert within 24 h and, although the inhibition of the Na^+/K^+ pump is rapidly relieved once the cells are returned to 37 °C, the recovery from the imbalance in monovalent cations occurs at a much slower rate; moreover, K^+ content is restored only partially. Therefore stored red cells will keep a reduced K^+ and water content, increased MCHC and impaired deformability even after reinfusion. This was documented by mimicking the reinfusion conditions by incubating the stored cells in autologous plasma for 24 h at 37 °C. It was also reported that rehydration of stored red cells before reinfusion, using the ionophore nystatin and buffers of appropriate ionic composition, could restore cell deformability of stored human erythrocytes and the *in vivo* survival times of stored rabbit red cells (Turner et al. 1987). Therefore it appears that after storage the survival of red cells is influenced by the same factors as in physiological ageing: whole cell deformability, cell hydration, cation content, and one could conclude that the most important lesion occurring during storage is the loss of K^+ consequent to the inhibition by cold of the Na^+, K^+-ATPase.

One of the issues that has not been sufficiently addressed so far is how red cells of different age respond to various situations or treatments, depending on the properties acquired during ageing. When even tentative approaches to these problems were made, a complex scenario was observed as, for example, human erythrocytes of different age do not react in the same way to storage conditions (Keegan et al. 1992; Minetti et al. 2001). Moreover, in a single study in dog it was found that cells of intermediate age maintain viability during storage in a better manner than do youngest and oldest cells (Gabrio and Finch 1956).

Only a few studies have analysed in detail the complement and functional state of the various known transport systems and the magnitude of the fluxes they regulate in human red cells of different age (reviewed in Bartosz 1990). Moreover the response of red cells of different age, *ex vivo* or after storage, to the hyperosmolality present in the vasa recta of the kidney has never been investigated.

A detailed analysis of the response of erythrocytes of different age to storage conditions and/or to the medium of the vasa recta would possibly give an insight into their capacities to (i) regulate their monovalent cation and water content, and (ii) to recover from the transient increase in ROS and H^+ concentrations, in each of the 170,000 times they cross the kidney during their circulatory life.

Acknowledgement: This work was supported by Ministero dell'Università e della Ricerca Scientifica e Tecnologica (MURST, PRIN), Italy.

References

Ashby W (1919) The determination of the length of life of transfused blood corpuscles in man. J Exp Med 29:267-281

Bartosz G (1990) Erythrocyte membrane changes during aging *in vivo*. In: Harris JR (ed) Blood cell biochemistry, vol 1, Erythroid cells. Plenum Press, New York, pp 45-79

Beutler E (1996) Principles of transfusion medicine. In: Rossi EC, Simon TL, Moss GS (eds) Principles of transfusion medicine. Williams and Wilkins, Baltimore, pp 51-60

Beutler E, West C, Blume KG (1976) The removal of leukocytes and platelets from whole blood. J Lab Clin Med 88:328-333

Borun ER, Figueroa WG, Perry SM (1957) The distribution of ^{59}Fe tagged human erythrocytes in centrifuged specimens as a function of cell age. J Clin Invest 36:676-680

Brugnara C, Tosteson D (1987) Cell volume, K transport and cell density in human erythrocytes. Am J Physiol 252:C269-C276

Cheng JT, Kahn T, Kaji DM (1984) Mechanism of alteration of sodium potassium pump of erythrocytes from patients with chronic renal failure. J Clin Invest 74:1811-1820

Ciana A (2000) Studio sul ruolo dell'età cellulare nelle modificazioni delle proprietà degli eritrociti conservati nelle banche del sangue. Ph.D. thesis, University of Pavia

Clark MR (1988) Senescence of red blood cells: progress and problems. Physiol Rev 68:503-554

Clark MR, Guatelli JC, White AT, Shohet SB (1981) Study on the dehydrating effect of the red cell Na^+/K^+ pump in nystatin-treated cells with varying Na^+ and water contents. Biochim Biophys Acta 646:422-432

Clark MR, Shohet SB (1985) Red cell senescence. In: Schrier SL (ed) The red blood cell membrane. WB Saunders, London, pp 223-257

Cullis PR, Hope MJ (1985) Physical properties and functional roles of lipids in membranes. In: Vance DE, Vance JE (eds) Biochemistry of lipids and membranes. The Benjamin Cummings Publishing, Menlo Park, pp 24-72

Eadie GS, Brown IW (1953) Red blood cell survival studies. Blood 8:1110-1136

Gabrio BW, Finch CA (1954) Erythrocyte preservation. I. The relation of the storage lesion to "in vivo" erythrocyte senescence. J Clin Invest 33:242-246

Galili U, Rachmilewitz EA, Peleg A, Flechner I (1984) A unique natural human IgG antibody with anti-alpha-galactosyl specificity. J Exp Med 160:1519-1531

Ganzoni AM, Oakes R, Hillman RS (1971) Red cell aging *in vivo*. J Clin Invest 50:1373-1378

Gárdos G (1958) The function of calcium in the potassium permeability of human erythrocytes. Biochim Biophys Acta 30:653-654

Halperin JA, Brugnara C, Nicholson-Weller A (1989) Ca^{2+}-activated K^+ efflux limits complemet-mediated lysis of human erythrocytes. J Clin Invest 83:1466-1471

Hejnaes K, Matthiesen F, Skriver L (1998) Protein stability in downstream processing. In: Subramanian G (ed) Bioseparation and bioprocessing, vol 2. Whiley-Vch, Weinheim, pp 31-65

Hentschel WM, Wu LL, Tobin GO, Anstall HB, Smith JB, Williams RR, Owen Ash K (1986) Erythrocyte cation transport activities as a function of cell age. Clin Chim Acta 157:33-44

Iida K, Whitlow MB, Nussenzweig V (1991) Membrane vesiculation protects erythrocytes from destruction by complement. J Immunol 147:2638-2642

Inaba M, Maede Y (1988) Correlation between protein 4.1a/4.1b ratio and erythrocyte life span. Biochim Biophys Acta 944:256-264

Inaba M, Gupta KC, Kuwabara M, Takahashi T, Benz EJ Jr, Maede Y (1992) Deamidation of human erythrocyte protein 4.1: possible role in aging. Blood 79:3355-3361

Kannan R, Yuan J, Low PS (1991) Isolation and partial characterization of antibody- and globin-enriched complexes from membranes of dense human erythrocytes. Biochem J 278:57-62

Kay MMB (1975) Mechanism of removal of senescent cells by human macrophages *in situ*. Proc Natl Acad Sci USA 72:3521-3525

Kay MMB, Bosman GJCG, Shapiro SS, Bendich A, Bassel PS (1986) Oxidation as a possible mechanism of cellular aging: vit E deficiency causes premature aging and IgG binding to erythrocytes. Proc Natl Acad Sci USA 83:2463-2467

Keegan TE, Heaton A, Holme S, Owens M, Nelson EJ (1992) Improved post-transfusion quality of density separated AS3 red cells after extended storage. Br J Haematol 82:114-121

Kirkpatrick FH, Muhs AG, Kostuk RK, Gabel CW (1979) Dense (aged) circulating red cells contain normal concentrations of adenosine triphosphate (ATP). Blood 54:946-950

Larsen FL, Katz S, Roufogalis BD, Brooks DE (1981) Physiological shear stress enhance the Ca^{++} permeability of human erythrocytes. Nature 294:667-668

Lee P, Kirk RG, Hoffman JF (1984) Interrelations among Na and K content, cell volume, and buoyant density in human red blood cell populations. J Membrane Biol 79:119-126

Leggett RW, Williams LR (1995) A proposed blood circulation model for reference man. Health Phys 69:187-201

Lew VL, Bookchin MR (1986) Volume, pH, and ion-content regulation in human red cells: analysis of transient behavior with an integrated model. J Membrane Biol 92:57-74

Lew VL, Raftos JE, Sorette M, Bookchin RM, Mohandas N (1995) Generation of normal human red cell volume, hemoglobin content, and membrane area distributions by "birth" or "regulation"? Blood 86:334-341

Li Q, Jungmann V, Kiyatkin A, Low PS (1996) Prostaglandin E_2 stimulates a Ca^{2+}-dependent K^+ channel in human erythrocytes and alters cell volume and filterability. J Biol Chem 271:18651-18656
Low PS, Waugh SM, Zinke K, Drenckhahn D (1985) The role of hemoglobin denaturation and band 3 clustering in red blood cell aging. Science 227:531-533
Lutz HU (1990) Erythrocyte clearance. In: Harris JR (ed) Blood cell biochemistry, vol 1, Erythroid cells. Plenum Press, New York, pp 81-120
Lutz HU, Fasler S, Stammler P, Bussolino F, Arese P (1988) Naturally occurring anti-band 3 antibodies and complement in phagocytosis of oxidatively stressed and in clearance of senescent red cells. Blood Cells 14:175-195
Lutz HU, Stammler P, Fasler S, Ingold M, Fehr J (1992) Density separation of human red blood cells on self forming Percoll gradients: correlation with cell age. Biochim Biophys Acta 1116:1-10
Marton F (1970) Erythrophagocytosis in the human bone marrow. Scand J Haematol 7:177-183
Miescher P (1956) Le mécanisme de l'érythroclasie a l'état normal. Rev Hémat 11:248-259
Minetti G, Ciana A, Profumo A, Zappa M, Vercellati C, Zanella A, Arduini A, Brovelli A (2001) Cell age-related monovalent cations content and density changes in stored human erythrocytes. Biochim Biophys Acta 1527:149-155
Morrison M, Michaels AW, Phillips DR, Choi SI (1974) Life span of erythrocyte membrane protein. Nature 248:763-764 p
Mueller TJ, Jackson CW, Dockter ME, Morrison M (1987) Membrane skeletal alterations during *in vivo* mouse red cell aging. Increase in the band 4.1a:4.1b ratio. J Clin Invest 79:492-499
O'Neill WC, Mikkelsen RB (1987) The role of pump number and intracellular sodium and potassium in determining Na,K pump activity in human erythrocytes. Metabolism 36:345-350
Piccinini G, Minetti G, Balduini C, Brovelli A (1995) Oxidation state of glutathione and membrane proteins in human red cells of different age. Mech Ageing Dev 78:15-26
Piomelli S, Lurinsky G, Wasserman LR (1967) The mechanism of red cell aging. I. Relationship between cell age and specific gravity evaluated by ultracentrifugation in a discontinuous density gradient. J Lab Clin Med 69:659-674
Rapoport S (1947) Dimensional, osmotic, and chemical changes of erythrocytes in stored blood. I. Blood preserved in sodium citrate, neutral, and acid citrate-glucose (ACD) mixtures. J Clin Invest 26:591-615
Rapoport SM (1986) The reticulocyte. CRC Press, Boca Raton
Shemin D, Rittenberg D (1946) The life span of the human red blood cell. J Biol Chem 166:627-636
Sorette MP, Galili U, Clark MR (1991) Comparison of serum anti-band 3 and anti-Gal antibody binding to density-separated human red blood cells. Blood 77:628-636
Stuart J, Ellory JC (1988) Rheological consequences of erythrocyte dehydration. Br J Haematol 69:1-4
Surgenor DMN (1974) The red blood cell, vol 1-2. Academic Press, New York
Suzuki T, Dale GL (1988) Senescent erythrocytes: isolation of *in vivo* aged cells and their biochemical characteristics. Proc Natl Acad Sci USA 85:1647-1651
Ten Brinke M, De Regt J (1970) ^{51}Cr-half time of heavy and light human erythrocytes. Scand J Haematol 7:336-341

Turner S, William AR, Rees JM (1987) The effects of reinflation on the survival of time-expired stored rabbit erythrocytes *in vivo*. Vox Sang 52:186-190

Waugh RE, Narla M, Jackson CW, Mueller TJ, Suzuki T, Dale GL (1992) Rheologic properties of senescent erythrocytes: loss of surface area and volume with red blood cell age. Blood 79:1351-1358

Wolfe LC (1985) The membrane and the lesions of storage in preserved red cells. Transfusion 25:185-203

Wood L, Beutler E (1967) Temperature dependence of sodium-potassium activated erythrocyte adenosine triphosphatase. J Lab Clin Med 70:287-294

30 Active and Passive Monovalent Ion Transport Association with Membrane Antigens in Sheep Red Cells: a Molecular Riddle

Peter K. Lauf

Department of Physiology and Biophysics, Wright State University School of Medicine, Dayton, OH 45435, USA

30.1 Introduction: a Well-Established Cellular Model

The occurrence in the animal kingdom of species (ruminants, canines) with red blood cells (RBCs) containing different concentrations of K^+ and Na^+ has fascinated a number of laboratories over the past century. One of the great discoveries in biology and physiology is the fact that active and passive cation transport necessary for the maintenance of the different cation steady states in ruminant RBCs is associated with membrane surface antigens through which the respective antibodies, by classic antigen/antibody reactions, modify transport. In particular, the sheep RBC system has served as a useful model to seek similar associations in other species, and in human pathophysiology, as reviewed earlier (Lauf 1975, 1978; Ellory and Tucker 1983; Dunham 1990).

This chapter reviews the present state and updates our understanding of the correlation between cation polymorphism and membrane antigens that exist in ruminant RBCs, especially in sheep. John V. Evans and Ben H. Rasmusen pioneered much of the genetic basis of the high potassium/low potassium (HK/LK) dimorphism (Evans 1954, 1956; Evans and King 1955) in sheep RBCs, noticed first by Abderhalden (1898) and subsequently classified further by Kerr (1937), and the association with blood group antigens (Rasmusen and Hall 1966; Ellory and Tucker 1969a), respectively. Evans and King (1955) concluded “the balance of sodium and potassium can be substantially modified by a single gene difference." During the early decades thereafter, investigators concentrated on defining the active and passive cation transport mechanisms responsible for cell volume homeostasis, and several reviews have dealt with this aspect (Tosteson 1966; Ellory and Tucker 1983; Dunham 1992; Lauf et al. 1992).

During the past two decades, the focus of investigation shifted toward analysis of the cation “leak” pathways, now defined as ouabain-resistant, furosemide-sensitive Cl^--dependent K^+ transport or K^+-Cl^- cotransport (COT) (recently reviewed by Lauf and Adragna 2000), and as channels and low ionic strength-induced $K^+(Na^+)/H^+$ exchange systems (Bernhardt 1994; Richter et al. 1997;

Kummerow et al. 2000), and the correlation with surface markers, in particular the L antigens, in LK sheep RBCs. These aspects as well as the most recent data on the nature of the antigens will be dealt with in this chapter. Table 30.1 contains an overview of the principal properties of transport and antigens in sheep RBCs and will be used throughout this text as reference.

Table 30.1. Genetic correlation between cation transport and membrane antigens in sheep red blood cells

Parameter	Property/unit	HK phenotype	LK phenotype	
genotype	cations/antigens	HK/MM homozygous	LK/LM heterozygous	LK/LL homozygous
antigens	M sites/cell	500–3000	500–1600	absent
	L sites/cell	absent	≤ 1200	≤ 1500
K^+	$mmol/l_{cells}$	93.3	20.1	16.6
Na^+	$mmol/l_{cells}$	15.9	82.9	84.0
Na^+/K^+ pump	$mmol/(l_{cells} \cdot h)$	0.63	0.16	0.16
K^+ "leak"	$mmol/(l_{cells} \cdot h)$	0.09	0.12	0.16
K^+-Cl^- COT	$mmol/(l_{cells} \cdot h)$	latent	n.a.	0.10
pump sites	^{3}H-Ouabain/cell	120	50	50–70
Na^+, K^+-ATPase	nmol $P_i/(mg_{protein} \cdot min)$	41	n.a.	11

Data modified from Tables 2 and 3 of Lauf (1978); *n.a.* not available.

30.2 Cation Polymorphism through Pumps and "Leaks"

In their classic paper on "regulation of cell volume by active cation transport in high and low potassium sheep RBCs", Tosteson and Hoffman (1960) established the pump-leak model, by which both HK and LK sheep RBCs maintain their volume by significant differences in both cation pumps and leaks commensurate with the different cation steady state concentrations. They suggested "that the molecular mechanisms of pump and leak may be closely related" (Tosteson and Hoffman 1960). Since then, much work has been done to clarify the Na^+/K^+ pump mechanism in many cellular systems from numerous species. Overwhelming evidence, however, has been accumulated that the molecular biology of both pumps and leaks are not closely related. Within the last twenty years, the nature of the leaks in these cells has become more transparent by the distinction of electroneutral co-transporters, exchangers and channels.

Figure 30.1 serves as a guide for important physiological properties and compares the major transport pathways of HK (A) and LK (B) sheep RBCs discussed in this chapter, in particular the Na^+/K^+ pumps and the various monovalent ion leak pathways dealt with in the coming paragraphs. Furthermore, Fig. 30.1 provides the membrane antigens functionally associated with monovalent cation transport, discussed in Sect. 30.3.

30.2.1 Na^+/K^+ Pumps

Commensurate with the high K^+ steady state, HK sheep RBCs possess Na^+/K^+ pumps and their corresponding Na^+, K^+-ATPases that, by all definitions such as internal Na^+ and external K^+ activation kinetics and ATP dependence, are normal (Tosteson 1966; Hoffman and Tosteson 1971). There are about 90 functionally active pumps/HK cell and the apparent affinity for ouabain is in the nanomolar range (Dunham and Hoffman 1971b; Joiner and Lauf 1978a). The molecular mechanism of these pumps has not been studied in detail but it is suspected that they are composed of alpha-1 type subunits and beta-1 and beta-3 type glycoproteins (Blostein and Grafova 1990; Hoffman, personal communication).

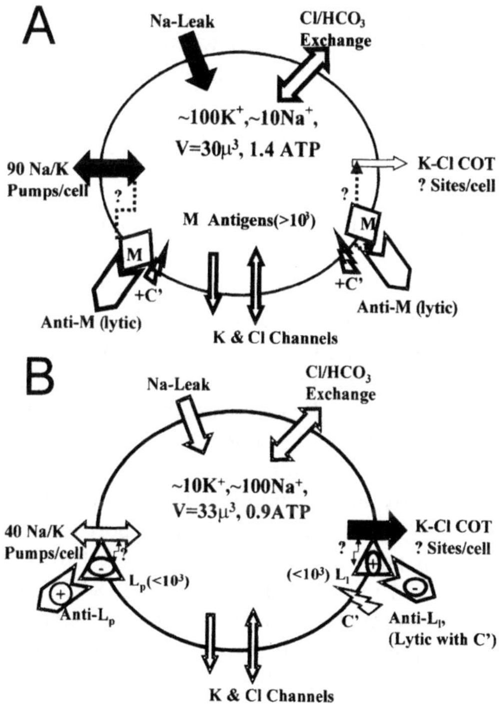

Fig. 30.1. (**A**) HK, MM, (**B**) LK, LL red blood cell. Numbers in front of K^+, Na^+ and *ATP* refer to mmol/l_{cells}. *C'* defines serum complement. See text for details. Modified from Lauf et al. (2001a) with publisher's permission

The Na^+/K^+ pump activity of LK sheep RBCs (see Table 30.1) is about or less than 1/4 of that in HK cells (Tosteson and Hoffman 1960), yet the number of ^{3}H-ouabain binding sites is only 1/2 of that in HK cells (Joiner and Lauf 1978a) commensurate with additional kinetic differences uncovered by Hoffman and Tosteson (1971). The membrane correlate of the Na^+/K^+ pump, the Na^+, K^+-ATPase in ghosts derived from LK RBCs, has also been found to be about 1/4 of the activity measured in ghosts from HK cells (Tosteson et al. 1965). The main kinetic difference is a pronounced cytoplasmic inhibition of the Na^+/K^+ pump by cell K^+ as shown by experiments using nystatin to replace K^+ with Na^+. Accordingly, lowering cellular K^+ and increasing Na^+ activates the Na^+/K^+ pump in LK cells and increases the rate of ouabain binding consistent with a Na^+-induced conformational change of the pump during its normal operational cycle (Joiner and Lauf 1978b). When all K^+ is replaced by Na^+, the pump activity is still only a fraction of that of Na^+-loaded HK cells (Lauf et al. 1970) consistent with the hypothesis that it is not solely the K^+ level that controls the rate of pumping in these cells (Glynn and Ellory 1972). Consistent with the flux studies, Whittington and Blostein (1971) showed that the Na^+, K^+-ATPase was more than an order of magnitude larger and less susceptible to K^+ inhibition in HK than in LK sheep RBC membranes. In addition, the response of HK and LK ATPases to oligomycin was different: K^+ decreased oligomycin inhibition in the former whereas oligomycin decreased K^+ inhibition in the latter (Blostein and Whittington 1973).

30.2.2 "Leaks"

Until the 1970s, the passive ouabain-resistant monovalent cation permeability of RBCs, in general, was lumped under the term "leak." Interestingly, LK sheep RBCs exhibit a much greater K^+ leak than HK cells (see Table 30.1) and this phenomenon was elegantly shown by Tosteson and Hoffman (1960) by placing both LK and HK cells into high K^+ media for a prolonged period: LK cells swelled by 15% and HK cells by only 3–4% over 24 hours. However, particularly from the work of Funder and Wieth (1967), Beauge and Adragna (1971), Wiley and Cooper (1976), and Kregenow (1980), it became clear that ouabain-resistant passive cation fluxes were coupled and did obey the Ussing flux ratio equation *only* in media with Cl^- replaced by other anions of the chaotropic series (Funder and Wieth 1967) suggesting Cl^--dependent cation transport such as the Na^+-K^+-$2Cl^-$ and K^+-Cl^- COT that depended on the gradients established by the Na^+/K^+ pump. The sheep RBC is a special model as it lacks the Na^+-K^+-$2Cl^-$ COT but possesses K^+-Cl^- COT as well as several channels defining the remaining fluxes as "ground permeability" (Ortiz-Carranza et al. 1997b). Because there have been two major reviews within which K^+-Cl^- COT has been discussed in terms of the overall molecular and metabolic regulation (Lauf et al. 1992; Lauf and Adragna 2000; Delpire and Mount 2002), this chapter will deal exclusively with aspects of the sheep erythrocyte, especially with the LK cells that, in contrast to HK cells (Fujise and Lauf 1987), exhibit K^+-Cl^- COT under isosmotic conditions.

K^+-Cl^- Cotransport

Ion Specificity, Driving Forces and Kinetics

K^+-Cl^- COT is a small but distinct ouabain-resistant, Cl^--dependent K^+ flux (bumetanide is not required as another specifier since there is no Na^+-K^+-$2Cl^-$ COT) that can be measured under isosmotic conditions in LK but not in HK sheep RBCs (Ellory and Dunham 1980; Dunham and Ellory 1981). The Cl^--dependence is characterized by data in Table 30.2. In LK sheep RBCs, only Br^- can replace Cl^- in the transport mechanism whereas all the other Hofmeister-series anions are either not transported and/or exert some inhibition, i.e., are not necessarily innocent by-stander ions (Ellory and Dunham 1980; Lauf 1984b, 1988c, 1991). It remains to be seen whether this anion sequence signals functional prevalence of a KCC3 iso-form-mediated K^+-Cl^- COT (see review by Delpire and Mount (2002)). Table 30.2. also lists the effect of cations studied thus far. Whereas Rb^+ is an accepted congener of K^+ (Lauf 1983), there is no direct evidence for transport of NH^+_4, Li^+ or H^+ (for protons see also in this section on chemical *Activators/Inhibitors*). However, all three cations exert inhibitory effects (Lauf et al. 1994, 2000; Ferrell et al. 2000; Lauf and Adragna 2000) that may involve sites other than the K^+/Rb^+ loading sites.

Table 30.2. Ion dependence of K^+-Cl^- cotransport in sheep erythrocytes

Ion	Transport	Inhibition	References
Cl^-	+	–	Dunham and Ellory 1981; Lauf 1984b
Br^-	+	–	Lauf 1988c; Lauf 1991
HCO_3^-	–	+	Dunham and Ellory 1981; Lauf 1984b
F^-, I^-	–	+	Lauf 1984b
NO_3^-, SCN^-	–	+	Dunham and Ellory 1981; Lauf 1984b; Lauf 1991
$CH_3SO_4^{2-}$	–	+	Lauf et al. 1994
K^+	+	–	Lauf 1983, 1988b
Rb^+	+	–	Lauf 1983
NH_4^+	–	+	Lauf et al. 2000
Li^+	–	+	Ferrell et al. 2000
H^+	–	–	Lauf et al. 1994; Lauf and Adragna 1998

Thermodynamically, intracellular K^+ cotransports Cl^- outwardly against the latter's gradient and extracellular Cl^- moves K^+ against its chemical gradient. Thus, the flux reversal point of the transporter, i.e., when the net driving forces are zero, follows the concentration of cellular Cl^-, the most permeable of the two ions whose shifts are brought about by changes in pH (Lauf and Adragna 1996). The lack of membrane potential effects in human RBCs (Brugnara et al. 1989; Kaji 1993) on K^+-Cl^- COT are consistent with these conclusions in sheep RBCs.

Kinetically, K^+-Cl^- COT exhibits low apparent affinities for K^+ and Cl^- (Delpire and Lauf 1991a,b). In general, apparent K^+ affinities are higher than those for Cl^-, and tend to increase somewhat when cells are treated with thiol reagents (see Table 30.3). In swollen LK RBCs, the binding order for both ions is random at the cytoplasmic aspect but ordered outside with Cl^- binding prior to K^+ (Delpire and Lauf 1991a). In this respect, the kinetic behaviour of K^+-Cl^- COT in LK sheep RBCs differs much from that reported for the Na^+-K^+-$2Cl^-$ COT in RBCs of other species where the cation (Na^+) binds/debinds first, followed by the anion (Miyamoto et al. 1986; Lytle et al. 1998). The overall advantage of such a kinetic behaviour is that the system is more sensitive to changes in cellular Cl^-. On the other hand, increases in the concentration of external K^+ may reverse a reduction of K^+-Cl^- COT due to a fall of external Cl^- concentrations (Delpire and Lauf 1991a).

Table 30.3. Activators of K^+-Cl^- cotransport in sheep erythrocytes

Compound	AC_{50} [mM]	Site of action	References
NEM	0.2	kinase thiols	Lauf and Adragna 1995; Lauf et al. 1995a; Logue et al. 1983
diamide	0.2	GSH, other thiols	Lauf 1988a; Adragna and Lauf 1997
MMTS	0.1	GSH, thiols	Lauf and Adragna 1995
CDNB	0.15	GSH	Lauf et al. 1995
H_2O_2	2.1	thiols, others	Bize and Dunham 1995
NH_2OH	0.1	haemoglobin	Lauf 1990
NO_2	< 1	GSH, thiols, Hb	Adragna and Lauf 1998
O_2	~30 mmHg	Hb? haem protein?	Campbell et al. 1998
hydralazine	< 1	?	Adragna et al. 2000
SNP, NO		cGMP pathway?	Adragna et al. 2000
straurosporine	< 0.001	Ser/Thr, Tyr-kinases	Flatman et al. 1996
Mg? (free)	< 0.001	kinases	Delpire and Lauf 1991c; Ortiz-Carranza et al. 1996; Dunham et al. 1993
Π_{osm}	n.a.	carrier, kinases	Dunham et al. 1993; Bergh et al. 1990
urea	~100	macromolecules	Dunham 1995

Modified from Lauf and Adragna (2000); *n.a.* not available.

Activators/Inhibitors

Cell swelling: K^+-Cl^- COT is stimulated by hypotonic cell swelling (Ellory and Dunham 1980; Dunham and Ellory 1981; Bergh et al. 1990) and by hyperbaric pressure (Hall and Ellory 1986) suggesting a complex cell volume sensing mechanism (see Table 30.3). This raises the possibility that LK RBCs may have a different volume than HK cells that would reflect the increased basal K^+-Cl^- COT activity. In a recent study (See et al. 1999), we analysed the osmotic properties of RBCs from three homozygous (LL), 3 heterozygous (LM) LK and 3 HK(MM) sheep. With the exception of one LM animal, LK cells exhibited a cell volume some 10% larger than that seen in HK cells, a finding confirming a report by Evans (1963) showing that LK cells are indeed larger than HK cells. These observations are at variance with Tosteson and Hoffman's (1960) report that there was no volume difference between the two genotypes. A volume difference between LK and HK cells is intriguing since, given an outwardly directed combined chemical driving force, K^+-Cl^- COT activation by the L_l antigen (see Sect. 30.3) should result in the opposite, i.e., at least the same or smaller cell volumes. The larger cell volume may be explained in terms of maturation, i.e., omission of a penultimate division (Lee et al. 1966), or may be due to a pump-leak imbalance caused by the L_p-inhibited Na^+/K^+ pump.

Chemical activators: A variety of chemical and ionic interventions known to activate Cl^--dependent K^+ fluxes in sheep RBCs are listed in Table 30.3. A group of thiol group modifying reagents, foremost *N*-ethylmaleimide (NEM, Lauf and Theg 1980a), causes activation either by lowering the protective redox potential and/or by targeting kinases (Lauf et al. 1995) that inhibit K^+-Cl^- COT through phosphorylation. This conclusion is corroborated by the fact that general kinase inhibitors such as staurosporine (Bize and Dunham 1994; Flatman et al. 1996) or removal of cellular Mg^{2+} by the ionophore A23187 and external chelators also activate K^+-Cl^- COT (Lauf 1985a; Lauf et al. 1992). That phosphorylation/dephosphorylation reactions regulate K^+-Cl^- COT activity was explained by Jennings and Al-Rohil (1990) in their elegant two state reaction model for K^+-Cl^- COT in rabbit RBCs and applied as a three state model to LK sheep RBCs by Dunham et al. (1993). Although the kinase has not yet been identified, there is sufficient evidence for the role of protein phosphatase 1, PP1A, in the activation of K^+-Cl^- COT (Bize et al. 1999). The stimulatory effect of low concentrations of urea has been interpreted in terms of a change in macromolecular crowding inhibiting the catalytic activity of the putative kinase or its regulator (Dunham 1995) thus favoring dephosphorylation. More recently, activation of K^+-Cl^- COT by oxygen (Gibson et al. 1998) as well as NO-derived/related compounds have suggested involvement of haemoglobin or haem proteins, respectively, in the regulation at the membrane level (Adragna and Lauf 1998; Adragna et al. 2000).

Inhibitors: There are three general classes of inhibitors of K^+-Cl^- COT in LK sheep RBCs (Table 30.4) and their action on the transporter or its regulatory machinery displays the complex properties of the transport system.

Table 30.4. Inhibitors of K^+-Cl^- cotransport in sheep erythrocytes

Compounds	IC_{50} [μM]	Site of action	References
furosemide	50 (with K^+_o)	external carrier aspect	Lauf 1984a
DIDS	3	external carrier aspect	Delpire and Lauf 1992
NEM	2,000	carrier	Lauf and Adragna 1995
DEPC	2,000	inside, His residues	Lauf and Adragna 1998
ocadaic acid	0.28, 0.040	PP1, PP2A	Flatman et al. 1996
calyculin	0.005	PP1A	Flatman et al. 1996
genistein	15	carrier? tyrosine kinase?	Flatman et al. 1996
tyrphostin B46	28	carrier? tyrosine kinase?	Flatman et al. 1996
KT5823	<< 1	protein kinase G	Adragna et al. 2000
Mn^{2+}>Ca^{2+}>Mg^{2+}	1–100	internal mechanism	Lauf 1985a,b; Dunham et al. 1993
protons	0.5	internal mechanism	Lauf and Adragna 1998
lithium	7	internal mechanism	Ferrell et al .2000
quinine/q-dine	230/2,240	internal carrier aspect	Adragna and Lauf 1994

Modified from Lauf and Adragna (2000).

The first group is comprised of chemicals like furosemide and bumetanide, also known to inhibit NKCC (absent in sheep RBCs), and DIDS, interfering with other membrane transporters such as Cl^- channels and anion exchangers. The inhibition of K^+-Cl^- COT by furosemide and DIDS is augmented by external K^+ ions probably through allosteric sites different from the ion transport sites (Lauf 1984a; Delpire and Lauf 1992). At high concentrations and at 37 °C, NEM inhibits through thiol groups that become deoccluded and available once K^+-Cl^- COT is activated by any means listed in Table 30.4 (Lauf and Adragna 1995), and diethylpyrocarbonate (DEPC) has been proposed to inactive K^+-Cl^- COT through His residues at the cytoplasmic aspect of the carrier (Lauf and Adragna 1998).

The second group lists inhibitors for two protein phosphatases, two tyrosine kinase and protein kinase G, providing the strongest evidence for the presence of a metabolon that closely controls the overall activity of the K^+-Cl^- COT system (Bize and Dunham 1994; Flatman et al. 1996; Ortiz-Carranza et al. 1997a).

The third group contains ions and compounds such as quinine/quinidine that inhibit at the cytoplasmic aspect probably because of their positive charge at the level of the transporter or at its regulatory moieties. Soluble and amphiphilic cations are known to inhibit K^+-Cl^- COT in human RBCs at the cytoplasmic membrane side (Sachs 1994; Sachs and Martin 1999). There must be more than one site through which divalent ions inhibit: When LK sheep RBCs are treated with NEM, both Mg^{2+} and Mn^{2+} fail to inhibit but not Ca^{2+}, Sr^{2+}, Zn^{2+}, and Cd^{2+} (Lauf 1985a,b). The molecular basis of this interesting difference is unknown. K^+-Cl^- COT in LK sheep RBCs responds to changes in pH between 6 and 8 in a biphasic manner with a maximum activity around pH 7 and minima at pH 6 and 8. The structural basis of the bell-shaped K^+-Cl^- COT activation is not yet understood (Lauf et al. 1996). Removal of cellular Mg^{2+} by A23187 as well as NEM treatment abolish the alkaline inactivation of K^+-Cl^- COT by creating a typical alkaline activation of the system (Lauf et al. 1996; Zade-Oppen and Lauf 1990). The remaining inhibition of K^+-Cl^- COT by protons occurs primarily at the cytoplasmic aspect and our hypothesis is that His residues at the cytoplasmic C-terminal could constitute some sort of proton sensor through which the transporter is turned off (Lauf and Adragna 1998; Lauf et al. 2001b).

Overall Regulation

As in other RBC models (Jennings and Al-Rohil 1990; Jennings 1999, for rabbit; Kaji 1986, for man; Parker et al. 1994, for dog), K^+-Cl^- COT of LK sheep RBCs is regulated by a metabolon of kinases and phosphatases (Bize and Dunham 1994; Lauf et al. 1995; Flatman et al. 1996) whose identities await further studies. According to the original two-state model by Jennings and Al-Rohil (1990), K^+-Cl^- COT activation from the resting (R) to the activated transporter (A) occurs through dephosphorylation by a serine/threonine (S/T) protein phosphatase (PP-1) and inactivation by a S/T-kinase-mediated phosphorylation (see Fig. 30.2). Both phosphatase and kinase are in equilibrium between inactive and active states, whereby inactivation may mean another phosphorylation reaction.

Calyculin and okadaic acid are pharmacological inhibitors of PP-1. There is sufficient evidence that PP-1 is under control of other kinases, possibly a src tyrosine kinase and other subsequent enzymes (Bize and Dunham 1994; Flatman et al. 1996). In Hck and Fgr src-kinase knock-out mice, RBC K^+-Cl^- COT is activated and resilient to NEM stimulation (De Franceschi et al. 1998).

Since genistein and tyrphostin B46 prevent the NEM and staurosporine-induced activation of K^+-Cl^- COT, we and others have proposed that these compounds and others such as diamide (Lauf 1988a), hydroxylamine (Lauf 1990), and NO (Adragna and Lauf 1999; Adragna et al. 2000) and low cellular Mg^{2+} (Delpire and

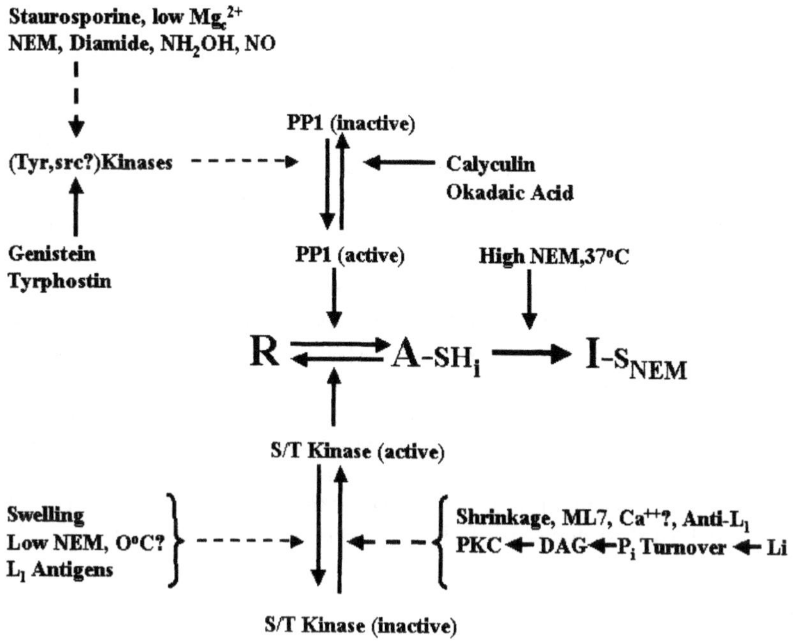

Fig. 30.2. Tentative K^+-Cl^- cotransport regulation in LK sheep erythrocytes

Lauf 1992; Dunham et al. 1993; Lauf et al. 1996) act through this "superkinase" system (Bize and Dunham 1994; Lauf et al. 1998; Kelly et al. 2001).

In contrast to PP-1, no specific inhibitor of the S/T-kinase is known. However, as explained in Jenning's two state model, swelling and NEM inactivate and shrinkage may activate this kinase. What does not seamlessly fit into this model is our finding that NEM exerts its purest activating effect at 0 °C (Lauf and Adragna 1998). The myosin light chain kinase (MLCK) inhibitor ML7 has been recently proposed to act on the S/T-kinase and not on MLCK since Ca^{2+} was not found to play a role (Kelly et al. 2001). However, Ca^{2+} introduced by A23187 is a powerful inhibitor of K^+-Cl^- COT (Lauf 1985a) and NEM does not abolish this effect (Lauf 1985a) suggesting a site, different from that controlled by Mg^{2+}, inactivated by NEM, perhaps calmodulin (Lauf 1985a), which in turn might activate a protein phosphokinase (PKA?) or the S/T-kinase directly. We have recently shown that Li^+ inhibits K^+-Cl^- COT, either due to a direct allosteric inhibition of K^+-Cl^- COT or to activation of PKC by diacylglycerate (DAG) that accumulates after Li^+ inhibition of the inositol monophosphorylase, a rate limiting enzyme in the sheep RBC's phosphoinositol metabolism (Ferrell et al. 2000).

Once K^+-Cl^- COT is activated to reach the A state by any stimulus (swelling, low NEM concentrations, diamide, hydroxylamine, staurosporine, Mg^{2+}-removal), a thiol group(s) is deoccluded (SH_i). High concentrations of NEM then irreversibly form adducts with this thiol(s) and K^+-Cl^- COT is likewise irreversibly inactivated (I-S_{NEM}) (cf. the RAI model of Lauf and Adragna (1998)).

Molecular Identity
At least four DNA isoforms (KCC1–4) encode for proteins carrying out K^+-Cl^- COT (Gillen et al. 1996; Holtzman et al. 1998; Hiki et al. 1999; Race et al. 1999; Su et al. 1999; Mount et al. 2000). An obvious question is whether the differences in K^+-Cl^- COT activity existing between LK and HK sheep RBCs can be explained by differences in isoforms and their quantitative presence in the plasma membranes, or whether different regulation and hence metabolons are the cause of the K^+-Cl^- COT dimorphism. Using rabbit antibodies against the C-terminal domain of rat KCC1 and the N-terminal domain of human KCC3 isoforms, we recently established immuno-histochemically the presence of both KCC1 and 3 isoforms in ghosts derived from all three sheep RBC genotypes (Lauf et al. 2001a). The KCC1 isoform, present more abundantly in the Western blots, exhibited a molecular weight of ~180 kDa as opposed to the KCC3 protein with ~145 kDa. These findings suggest that the differences in K^+-Cl^- COT activity reside in the regulation, i.e., the LK gene responsible for the low K^+ polymorphism expresses these properties somewhere at the regulatory level. Commensurate with this conclusion is the finding that ^{3}H-NEM labelling of the diamide-oxidized thiols, presumably including the K^+-Cl^- COT systems, is of similar magnitude in LK and HK sheep RBC ghosts (Lauf 1992).

Channels and Exchangers

Sheep RBCs are also unique in terms of ion channels. Whereas the fetal RBC has been reported to possess Gardos-type K^+ channels (Brown et al. 1978), these are absent in mature sheep RBCs of all genotypes. However, within the Cl^--independent, furosemide-insensitive "ground permeability" or "leak", there are cation and anion channels that, low in activity, are poorly studied in comparison to the K^+-Cl^- COT system.

A voltage-activated low selective cation channel is seen in HK but not in LK sheep RBCs exposed to low external Cl^- media, i.e., upon membrane depolarisation, and has been interpreted in terms of passive ion flow through the Na^+/K^+ pump (Halperin et al. 1990). We recently showed the presence in LK but not in HK sheep RBCs of at least two Cl^--independent pathways for K^+ and Na^+ transport activated by alkaline pH (Ortiz-Carranza et al. 1997b). The K^+ channel, like the Gardos K^+ channel (Gardos 1958), was activated by Ca_i^{2+} but, unlike the Gardos channel, also transported Na^+ and was not inhibited by quinine or charybdotoxin. The other alkaline pH-activated K^+ channel was inhibited by external divalent cations by quinine and by ATP depletion, clearly distinguishing it from a Gardos-type K^+_{Ca} channel (Ortiz-Carranza et al. 1997b).

Bernhardt's group showed the existence of low ionic strength-evoked cation "leaks" in human RBCs and explained them in terms of electroneutral $K^+(Na^+)/H^+$ exchangers responsible for the very low ground permeability left after inhibition of pump and COT fluxes (Richter et al. 1997; Kummerow et al. 2000). A similar "leak" has been found in LK sheep RBCs that is non-responsive to Cl^- replacement by NO_3^- and anti-L_1 inhibition (Erdmann et al. 1991). The absence of this low ionic strength-induced leak in HK sheep RBCs is intriguing and should be considered in terms of the presence of the M antigen in these cells (see Sect. 30.3.2 (*The M Antigen*)).

30.2.3 An Integrated View

In the foregoing Sects. 30.2.1 and 30.2.2, the picture emerges (see Fig. 30.1) that, ignoring here anion and cation exchangers such as the Na^+/Na^+ exchange prominent in these cells (Ryu et al. 1989), HK cells have 'normal' pumps and no K^+-Cl^- COT under physiological conditions, while LK cells have less pumps, that are in addition inhibited by K^+_i, and a measurable K^+-Cl^- COT. Given this situation, one would expect that HK cells have a smaller volume than LK cells. Based on our (See et al. 1999) and Evans' (1963) finding that HK sheep RBCs, however, are smaller than LK cells, neither pump nor outwardly directed K^+-Cl^- COT are able to stem the swelling tendency of LK cells, whereas the dehydration tendency of the Na^+/K^+ pump is evident in HK cells. Another paradox is that HK cells always have about 50% higher ATP and total Mg^{2+} levels than LK cells while the free Mg^{2+} concentrations are about the same (Fujise et al. 1988). Thus, with the Na^+/K^+ pump largely inhibited in LK cells, what is contributing to the lower steady state levels of ATP? Nevertheless, thermodynamically, there is more free energy of ATP in HK than in LK cells, which makes sense in light of the fundamental catalytic activity differences of the Na^+, K^+-ATPase between the two cell types.

What is the significance of the differences in ion channels and low ionic strength-induced "leaks" (K^+ or Na^+/H^+ exchangers) between HK and LK sheep RBCs? The fact that ion fluxes through the Na^+/K^+ pump and the K^+-Cl^- COT system are certainly qualitatively different, suggests that the LK gene plays out its effect at the metabolic and membrane levels affecting not only these two systems but also, directly or indirectly, various ion channels as well as the "ground permeability". Together, this means that a single allelic change must have vast consequences indeed for volume homeostasis in these RBCs, a perspective much in line with the general aspect of the pump-leak concept of Tosteson and Hoffman (1960) but not with the concept of Na^+/K^+ pump-leak interconvertibility of Dunham and Hoffman (1971).

30.3 M and L Membrane Antigen/Antibody Polymorphism

30.3.1 Genetic Basis

Rasmusen and Hall (1966) established the genetic basis of the M/L antigen polymorphism. In their original work, HK sheep RBCs were M-positive and LK cells either M positive or negative (m-cells) meaning that the heterozygous LK RBC is Mm (1966 nomenclature) or ML (more recent nomenclature). The association between these antigens and the cation types is shown in Table 30.1. The fact that the antigens segregate co-dominantly and the HK type is recessive has not yet been analysed in terms of chromosome/gene co-localization. An in depth discussion of the M and L blood group antigens has been published earlier (Lauf 1978) and Dunham and Blostein (1997) have surveyed the L antigens in relation to ion transport modulation. The M antigens occur in cattle and in goats also; however, as pointed out by Ellory and Tucker (1983) and others, the association with HK cattle or goat RBCs is by far less stringent. The L antigen, most likely only the L_p subtype, appears to be present on LK cattle and goat RBCs (Ellory and Maher 1977; Dunham and Ellory 1980; Ellory and Tucker 1983).

30.3.2 Molecular Properties of the M/L Antigens

The principal techniques for detecting the M and L antigens are the complement-dependent immune haemolysis assay (Rasmusen and Hall 1966; Lauf and Tosteson 1969), the activation of Na^+/K^+ pump fluxes (Ellory and Tucker 1969; Lauf et al. 1970) and inhibition of K^+-Cl^- COT-mediated ouabain-resistant K^+ or Rb^+ fluxes (Dunham and Ellory 1980; Lauf and Theg 1980b). Despite early and more recent attempts, the biochemical nature of the M and L antigens is hardly established due to at least five major features of the system: (i) low surface membrane density, (ii) epitope heterogeneity, (iii) lipid dependence, (iv) comparatively low affinities of the M and L antibodies for their respective membrane epitopes, and (v) cell maturation dependence.

The M Antigen

Based on ^{125}I-radiolabelled anti-M binding studies, there are 3,000–6,000 and 1,500–4,000 M antigenic sites in HK and heterozygous, M-positive LK sheep RBCs (Lauf and Sun 1976). A quantitatively similar number (2000/cell) was found by using haemocyanine-conjugated pig anti-sheep immunoglobulin antibodies (Tucker et al. 1976; Smalley et al. 1984), the number somewhat lower in LM cells. Thus, the density of the M sites is within the order reported for the human Rh blood group system. The antibody binding affinity K_d for anti-M is near 10^{-9} M and Scatchard plots revealed site heterogeneities with a range of K_d values (Lauf and Sun 1976).

Table 30.5 lists the molecular behaviour of sheep RBC surface membrane M antigen based on more recent studies done in this laboratory (Tang 1997). Although the M antigen survives ghosting, it is inactivated by freeze drying and boiling. Its protein nature is likely in light of the fact that extreme pH values, causing either complete uncoiling (> 10), or hydrolysis (< 3) and pronase treatment destroy the capability of M-positive ghosts to bind anti-M as tested by immune-haemolysis (Tang 1997). However, other proteolytic enzymes such as trypsin and chymotrypsin failed to affect the M antigen (Tang 1997). Another disappointment is the fact that widely used crosslinkers such as DIDS, DSP and glutaraldehyde were ineffective (Tang 1997). Furthermore, the lipid-dependence of the M antigen is born out by the finding that deoxycholate, butanol/water and chloroform/methanol completely destroy its anti-M binding capacity (Lauf and Tosteson 1969; Shrager et al. 1972; Tang 1997). However, Triton X 100 was less destructive and was used to solubilize residual M antigen activity (Wiedmer and Lauf

1981). Recovery rates from 25% (Smalley et al. 1983) to 46% (Wiedmer and Lauf 1981) were reported. Molecular weight estimates of Triton X 100-solubilized M antigen moieties range from > 150 to < 50 kDa indicating possible aggregation of smaller peptides into larger molecular weight complexes with perhaps better preserved epitopes (Wiedmer and Lauf 1981). The lipid dependence is certainly of more subtle nature since none of the phospholipases tested inactivated the M antigen (Tang 1997). Perhaps the M substance is a "lipoprotein" that is acylated with fatty acid residues not requiring intactness of the major membrane phospholipids. Although the role of carbohydrates has not been tested yet, neuraminic acid appears not to be involved in the formation of the antigenic site (Lauf et al. 1971).

The functional role of the M antigens is not established. In ML positive LK sheep RBCs, the M antigen appears to reduce somewhat the trypsin-induced inactivation of the L_p antigen (Dunham et al. 1980). Whether this observation is explainable in terms of an occlusion of the specific amide bonds (lysine or arginine), cleaved by trypsin, is unknown.

Table 30.5. Molecular behaviour of sheep erythrocyte surface M antigen

No or little effect	Inactivation
ghosting	lyophilization
freezing	heating 100 °C, 5 min
pH 3–10	pH > 10
chymotrypsin	pH < 3
trypsin	pronase at 37 °C, 90 min
phospholipase A_2	Triton X 100
phospholipase C	deoxycholate
sphingomyelinase	butanol/water
neuraminidase	chloroform/methanol
crosslinkers and others:	
diisocyanato-stilbene disulfonate (DIDS)	
dithio-*bis*-succinimidyl proprionate (DSP)	
glutaraldehyde	
N-ethylmaleimide (NEM)	
N-phenylmaleimide (NPM)	

The L Antigens

By both ^{125}I-radio-labelled anti-L IgG_1 and haemocyanine-coupled anti-sheep IgG, the total number of L antigens has been estimated to be in the order of 1000 or less/homozygous LK(LL) sheep RBC and even smaller in LK(LM) sheep RBCs, suggesting a gene dosage effect (Lauf and Sun 1976; Smalley et al. 1984). There are at least two if not three L-antigenic epitopes whose identities have been inferred from haemolysis and from pump and leak flux experiments.

L_{ly}, the "Haemolytic" Site

Like anti-M, anti-L haemolyses LK sheep RBCs (Rasmusen 1969; Lauf and Dessent 1973), but its complement-dependency is different, requiring rabbit rather than guinea pig serum (Lauf and Dessent 1973). The complement-dependent haemolysis is significantly enhanced when LK sheep RBCs are metabolically depleted. This phenomenon, reversible upon repletion of ATP, has been explained in terms of decreasing the inter-site distance that increases the statistical likelihood of adjacent IgG antibody molecules binding complement and thus haemolysing the cell (Lauf and Dessen 1973). If that is so, at least the lysis-producing L_{ly} sites tend to aggregate/disperse as a function of the metabolic state. It appears that LK goat and lamb RBCs lack these sites since anti-L_{ly} cannot be absorbed by these cells (Ellory and Tucker 1971; Lauf et al. 1977). The biochemical nature of the L_{ly} antigen is unknown. It is trypsin-resistant as anti-L_{ly} is absorbed by trypsinized mature LK sheep RBCs (Lauf et al. 1977).

L_p, the Na^+/K^+ Pump-Associated Site

Anti-L_p stimulates Na^+/K^+ pump flux in LL and LM LK sheep RBCs as well as LK goat and cattle RBCs (Ellory and Tucker 1969a; Lauf et al. 1970; Ellory and Carleton 1974). This unexpected and unusual effect involves the L_p epitope that was early recognized to be different from the L_{ly} site since trypsin inactivated the antibody effect (Lauf et al. 1971, 1977). There are about 600 L_p sites/mature LK cell (Smalley et al. 1984), thus exceeding the number of Na^+/K^+ pump sites by nearly an order of magnitude. The L_p antigen appears to be a moiety independent from the pump for several reasons: (i) trypsin inactivates the L_p antigen (Lauf et al. 1971, 1977; Dunham 1976a; Dunham et al. 1980); (ii) L_p not only acts as an inhibitor of endogenous Na^+/K^+ pumps but also reduces Na^+, K^+-ATPase activity after fusion of sheep kidney Na^+, K^+-ATPase with reticulocytes (Xu et al. 1992); and (iii) anti-L_p relieves the inhibition of exogenous rat kidney pumps by the L_p antigen (Xu et al. 1992). Anti-L_p stimulation of Na^+/K^+ pump flux is augmented when antibody binding is performed in cells that were K^+ loaded prior to K^+ depletion for the flux experiment. This effect has been attributed to a greater anti-L_p binding as consequence of stabilization of the pump-L_p antigen complex by cellular K^+ (Farquharson and Dunham 1986). This is a plausible explanation in light of the low binding affinity of anti-L_p, i.e., cellular K^+ would decrease K_d, a testable hypothesis.

The biochemical nature of the L_p antigen is similar to that of the M-antigen in terms of lipid dependency (Snyder et al. 1971) and low yield of solubilization (Smalley et al. 1983). This fact and the even lower K_d values for anti-L_p binding to its target epitopes have made it almost impossible to reproducibly identify its molecular nature. A 25 kDa polypeptide has been detected in ^{125}I labelled LK sheep RBC membranes immunoreacted with anti-L_p and immunoprecipitated with anti-sheep IgG (Pittman et al. 1990), a finding not yet repeated by others nor followed up further. Whether radiation inactivation detecting a 41 kDa moiety targeted a dimer of the immunoprecipitated version is unknown (Smalley et al. 1984).

L_l, the K^+-Cl^- Cotransport-Associated Site

Following the discovery in 1969 that whole anti-L serum stimulates Na^+/K^+ pump flux in LK sheep RBCs (Ellory and Tucker 1969a; Lauf et al. 1970), it took 8 years to really appreciate the fact that anti-L also lowered ouabain-resistant K^+ fluxes (Dunham 1976a,b; Lauf et al. 1977) which, in the absence of knowledge of the existence of transmembrane proteins (the fluid mosaic model of Singer and Nicolson of course changed that in 1971) were considered to occur through the lipid bilayer. The L_l antigen is an activator of K^+-Cl^- COT carried out by transmembrane proteins encoded by one or two of the KCC genes (see Lauf and Adragna 2000) because anti-L_l inhibits K^+-Cl^- COT. The number of L_l sites/cell has been given as about 850/cell but a quantitative correlation to the number of K^+-Cl^- cotransporters is missing since there is no specific label to assess the numbers of the latter. There is some likelihood that the L_{ly} and L_l properties are identical, but ultimate proof is missing since both properties, haemolysis of LK cells and reduction of K^+-Cl^- COT, are surviving trypsin treatment (Lauf et al. 1977). The biochemical nature of this interesting antigen is unknown.

30.3.3 The M/L Antibodies

By definition, anti-M and anti-L antisera, obtained by cross-immunization of LK sheep with HK sheep RBCs and HK sheep with LK sheep RBCs, contain polyclonal alloantibodies against the respective M and L antigens, as well as antibodies against unrelated surface antigens. Characterization of both M and L antibodies therefore includes contaminating antibodies unless they have been removed by specific absorption with a panel of different M or L negative sheep RBCs or the immunized animal possessed similar antigenic phenotypes. In contrast to the strongly haemolytic anti-M antibodies, the L antibodies are generally weakly haemolytic and require rabbit complement (Lauf and Dessent 1972), but exhibit functional diversity by affecting both Na^+/K^+ pump (anti-L_p) and K^+-Cl^- COT (anti-L_l) mediated K^+ fluxes in LK sheep RBCs.

Anti-M

The early work from this laboratory used an antibody that was primarily in the macroglobulin fraction and thus a powerful, high titer haemolytic reagent in the presence of guinea pig serum complement (Lauf and Tosteson 1969). Subsequent work used primarily IgG proteins separated by affinity chromatography to determine binding characteristics and nature of the M antigen (Lauf and Sun 1976; Wiedmer and Lauf 1981). Based on the elution profile, anti-M antibody activity was found in the IgG_l fraction adsorbing to DEAE sephadex, i.e., indicative of a complementary electro-positive M epitope charge (Tang 1997). Anti-M can be desorbed from HK sheep RBC membranes by either acid treatment or by butanol extraction (Lauf and Sun 1976).

Anti-L

Anti-L_p

The Na^+/K^+ pump stimulating anti-L_p antibody is an IgG_l type immunoglobulin signifying, like as in the case of anti-M IgG_l, the presence of positive charges on the L_p antigen (Snyder et al. 1971). Reduction and alkylation as well as papain treatment abolishes the physiological effect of anti-L_p IgG_l; however, pepsin digestion producing Fab_2 was without effect consistent with the hypothesis that binding to two L_p epitopes is required to produce the Na^+/K^+ pump stimulatory effect (Snyder et al. 1971). This finding was contradicted by others showing that monovalent Fab fragments were also stimulating Na^+/K^+ pump flux (Ellory et al. 1973; Bratcher et al. 1983) and cannot be explained by differences in total recovery (Dunham and Blostein 1997). In light of the fact that the L_p receptor may be dimeric to exert functional alignment and stimulation of the Na^+/K^+ pump, it is imperative this issue be resolved in the near future.

Anti-L_l

The exact immunoglobulin nature of the K^+-Cl^- COT inhibiting anti-L_l (most likely IgG as it is active after ammonium sulphate precipitation) has not been established despite the fact that early work on anti-L_p not only tested Na^+/K^+ pump fluxes but also ouabain-resistant K^+ "leak" transport which was not affected by the isolated immunoglobulins nor their monovalent and divalent derivatives (Snyder et al. 1971).

30.4 Functional Association of Antigens with "Pumps and Leaks"

Although there are two allelic products, the L and the M antigens, only the L substance is functionally associated with transport activities. This is pointed out in Fig. 30.1B, by the solid line connecting L antigens and pumps or K^+-Cl^- cotransporters in LK cells, and for the M antigens by the interrupted line in HK cells.

30.4.1 Na^+/K^+ Pump Activation by the L_p Antibody

Addition of anti-L serum stimulates the Na^+/K^+ pump in LK cells between 4–8-fold dependent on the source of the antiserum (Ellory and Tucker 1969; Lauf et al. 1970). This remarkable effect, due to the L_p-antibody (see Sect. 30.3.3), lifted the veil around the mechanism by which the Na^+/K^+ pump was inhibited in LK sheep RBCs. Contrary to early expectation, the L-antigen/L-antibody reaction did not confer HK properties on LK cells. The antibody did not modify the external K^+ affinity of the pump, but it modified the activation by internal Na^+ by reducing an inhibition by internal K^+ (Lauf et al. 1970). The site and the mechanism through which K^+ modifies the internal aspect of the Na^+/K^+ pump in LK cells was at the center of intense debate in the seventies and eighties. Based on kinetic interpretations of pump flux measurements (Lauf et al. 1970) and Na^+, K^+-ATPase determinations in RBC membranes of LK goats, Glynn and Ellory (1972) concluded that K^+ competes with Na^+ for the cytoplasmic Na^+ loading site of the pump. Joiner and Lauf (1978a) corroborated this interpretation by showing that anti-L_p increases the rate of ^{3}H-ouabain binding, without changing the number of Na^+/K^+ pump sites (Joiner and Lauf 1976). On the other hand, kinetic K^+ pump flux studies by Dunham's group showed that the effect of K^+_c is of non-competitive nature (Dunham 1976b; Dunham and Anderson 1987) explaining why in their hands the number of ^{3}H-ouabain binding sites increased somewhat following anti-L_p treatment. Corollary, Ellory et al. (1972) reported that anti-L_p reduced the K^+_c inhibition and increased the number of Na^+/K^+ pump sites in LK goat RBCs. Whatever the exact mechanism may be, a good working model is that by Farquharson and Dunham (1986) in which interaction of the pump with the antigen causes inhibition by K^+_c through an allosteric site. Anti-L_p binding to the L_p antigen dissociates it from the pump and thus removes allosterically K^+ from its site and relieves Na^+/K^+ pump inhibition:

$$L_p\text{-}Na^+/K^+\text{ pump-}K^+_c + \text{anti-}L_p \underset{H^+}{\overset{H^+}{\rightleftharpoons}} L_p\text{-anti-}L_p + Na^+/K^+\text{ pump} + K^+_c$$

There is no reason to assume *a priori* that the reaction of anti-L_p with the Na^+/K^+ pump is irreversible (Farquharson and Dunham 1986). Anti-L_p can be eluted from membranes by various techniques; although, experiments to show a return to increased Na^+/K^+ pump inhibition by K^+_c are missing. Thus, we have introduced into the above model the reversibility term. This becomes useful, because Ellory and Maher (1977) showed that increased pH reduces the K^+_c inhibition of the Na^+, K^+-ATPase in membranes of LK goat RBCs and consequently the activation by anti-L_p. This "cooperative" effect is represented in above equation by addition/removal of protons to the overall reaction rates.

In part, this model rests on the initial observation of Lauf et al. (1971, 1977) that trypsin destroys the L_p antigen without modifying the properties of the basic Na^+/K^+ pump flux in LK(LL) sheep RBCs, and was further supported by the approach of Xu et al. (1992) that the L_p antigen also inhibited extraneous rat kidney Na^+/K^+ pumps in sheep RBCs, and anti-L_p relieved that effect. The action of trypsin seems to be directly on the L_p antigen or its functional link to the Na^+/K^+ pump necessary for inhibition by K^+_c: in LK but not HK goat RBCs, trypsin *per se* stimulated Na^+/K^+ pump flux without changing the number of ^{3}H-ouabain binding sites (Dunham and Ellory 1980). Interestingly, there seems to be an effect of the M antigen on the hydrolysis of the L_p substance by trypsin because the latter does not fully abrogate the anti-L_p effect in LK(LM) sheep RBCs (Dunham et al. 1980), suggesting perhaps steric hindrance by the M antigen which in itself is trypsin-resistant (Tang 1997; see Sect. 30.3.2 (*The M Antigen*)).

30.4.2 K^+-Cl^- Cotransport Inhibition by the L_l Antibody

Overshadowed by the multi-fold stimulation of the Na^+/K^+ pump by anti-L_p, the presence of an inhibitory antibody invariably present in the same anti-L sera was first overlooked, or not understood, as in the early days of these discoveries, ouabain-resistant cation fluxes were explained as electrodiffusional in nature. With the separation of anti-L_l from anti-L_p by absorption to LK goat RBCs (Lauf et al. 1977) or to trypsinized LK sheep RBCs (Smalley et al. 1984), the effect of anti-L_l could be studied separately. When bound to LK sheep RBCs, anti-L_l reduces K^+-Cl^- COT by 30–60% dependent on the serum source and dilution used (Dunham and Ellory 1980; Lauf 1984; Lauf et al. 2001a) meaning the L_l antigen acts as an inhibitor of the system, the inverse situation to the Na^+/K^+ pump-L_p antigen interaction, in analogy to Sect. 30.4.1:

$$[L_l\text{-antigen}] + [K^+\text{-}Cl^- \text{ COT}] + [\text{anti-}L_l] \underset{k_r}{\overset{k_f}{\rightleftharpoons}} [L_l\text{-antigen-}K^+\text{-}Cl^- \text{ COT-anti-}L_l \text{ complex}]$$
(dephosphorylation) (phosphorylation)

In line with the overall regulation of K^+-Cl^- COT presented in Sect. 30.2.2 (*Overall regulation*) and in Fig. 30.2, the proposed forward rate process, k_f, causes activation of the volume-sensitive S/T kinase which in turn leads to inactivation (inhibition) of K^+-Cl^- COT through a phosphorylation reaction. The reverse rate process, k_r, the dissociation of the ternary complex upon removal of the L_l antibody or lowering its free concentration means dephosphorylation and activation of K^+-Cl^- COT. Accordingly, the L_l antigen may be part of a phosphatase that inactivates the S/T kinase or its regulator. We have recently shown that anti-L_l inhibits only a fraction of the staurosporine-induced K^+-Cl^- COT activity in LK sheep RBCs, which is in line with our general experience of incomplete inhibition of K^+-Cl^- COT after NEM stimulation (Lauf 1984b). These data actually may support the view of regulation of K^+-Cl^- COT by different kinases, staurosporine acting, according to Fig. 30.2, through a kinase that regulates a phosphatase not associated with the enzyme involved with the activity interconversion of the volume S/T kinase that is regulated by the L_l antigen.

30.4.3 Complement-Mediated Haemolysis by Anti-M and Anti-L

Addition of serum complement (C' in Fig. 30.1) to sheep RBCs with anti-M and anti-L antibodies adsorbed to their surface membranes (sensitization) causes membrane lesions through which small ions/solutes and water equilibrate resulting in immune haemolysis, an end stage at which the affected cells burst and release large macromolecules, such as haemoglobin, due to ever increasing colloid-osmotic pressure and limited membrane compliance (Sims and Lauf 1978). Obviously, the Na^+/K^+ pump is incapable of coping with the mainly Na^+ influx through such lesions, and K^+-Cl^- COT is too small to compensate for the rates of swelling, and finally, haemolysis commences.

Functionally, the M and L systems behave quite differently. Haemolysis by anti-M and usually guinea pig serum as complement source is rapid whether by IgM or IgG-type antibodies (Lauf and Tosteson 1969). In contrast, haemolysis by anti-L serum is sluggish and attempts to replace guinea pig by rabbit complement has only slightly improved the process (Lauf and Dessent 1972). This phenomenon is not related to the fact that the driving forces for Na^+ to enter the cell under colloidosmotic lysis are smaller for LK than for HK cells, since entirely different immunological reactions (Forssman antibodies plus guinea pig complement) yield very different rates in the two types of RBCs with LK(LL) cells being more susceptible than LK(LM) and HK(MM) cells (Bell et al. 1972).

The ATP depletion-augmented immune-haemolysis of LK sheep RBCs by anti-L and rabbit serum complement has been interpreted in terms of site-coalescence giving rise to a greater probability for complement activation through adjacent anti-L IgG_1 molecules (Lauf and Dessent 1973). It is not solidly established which of the two or three antibodies are participating in this process: anti-L_p, anti-L_l or anti-L_{ly}. Most likely, the latter two are involved since the haemolytic scores of anti-L_p are very low (Ellory and Tucker 1983).

30.5 Clonal *vs.* Maturational Changes of Transporters and Antigens

Soon after the discovery of the M and L polymorphism, it was shown that the cellular expression of the antigens changes in RBCs of lambs and anaemic sheep in conjunction with the membrane monovalent cation transport properties (Ellory and Tucker 1969a,b). It is important to distinguish clonal *vs.* maturational changes. Clonal changes are defined by the appearance of new cell populations, whereas maturational changes occur within the same cell population.

30.5.1 Clonal Cellular Changes in Lambs

All lambs are born with a fetal RBC population containing macrocytic HK-type RBCs with a volume of 36 μm^3 and fetal haemoglobin (Valet et al. 1978). However, within 40 days, the fetal haemoglobin is replaced by adult haemoglobin (Tucker et al. 1982), the cellular K^+ and Na^+ concentrations for genotypically LK animals approach those found in adult LK RBCs, and pump fluxes fall 3–4-fold accompanied by a 4-fold reduction in Na^+/K^+ pump sites as estimated by ^3H-ouabain binding (Lauf et al. 1978; Tucker et al. 1982). In addition, experiments replacing intracellular Na^+ with K^+ by the nystatin technique reveal the expected K^+ inhibition of the Na^+/K^+ pump (Tucker et al. 1982). This change in transport properties has been attributed to the fact that the HK type population is gradually replaced by an intermediate cell population with individual cell volumes of 28 μm^3 and haemoglobin C, and followed by the final haemoglobin A or B containing LK RBCs of 30 μm^3 (Valet et al. 1978). The progressive rise of both L and M antigens levels during the first 3–5 weeks after birth, as indicated by the rise in L and M antigen haemolysis titers (Ellory and Tucker 1969b; Valet et al. 1978), is related to the consecutive entrance of the three cell populations into the blood stream and accompanies the fall of cellular K^+ levels and Na^+/K^+ pump activity (Valet et al. 1978). Accordingly, anti-L_p fails to stimulate the HK-type Na^+/K^+ pump flux in the macrocytic cell population but does so in the final LK RBC population (Valet et al. 1978) commensurate with an increase of the number of L_p sites/cell with the appearance of cells containing adult haemoglobin (Tucker et al. 1982).

In contrast to the rise in the L_p antigenic sites, the L_l sites have been found to be constant in terms of cell surface area over the entire first period of development. Since the L_l antigen is an activator of the K^+-Cl^- COT, it would be of interest to know whether anti-L_l inhibits Cl^--dependent K^+ fluxes to the same degree in early *vs.* late lamb LK RBCs, i.e., to establish whether the operational link between flux and surface receptor is a function of cell populations.

30.5.2 Reticulocyte Maturation in Anaemic Sheep

Upon removal of circa 30% of the blood volume by severe bleeding within two days, sheep compensate for the blood loss within 5–7 days by producing large emergency reticulocytes, 2.5 times the normal mature RBCs (Kim et al. 1980) and containing haemoglobin C (Lauf and Valet 1983). This phenomenon offers a temporary window to study mainly maturational changes in antigen levels and cation transport activities.

Based on measurements of Na^+/K^+ pump fluxes and ^{3}H-ouabain binding, these emergency cells are by all means HK cells in both HK and LK animals (Lauf and Valet 1983). The Na^+/K^+ pump fluxes of these stress reticulocytes are 30 times higher than in mature RBCs (Kim et al. 1980) and relatively insensitive to changes in internal K^+ concentrations (Dunham and Blostein 1976), consistent with the observation that in LK sheep the kinetic properties of the Na^+, K^+-ATPase, in particular the inhibition by oligomycin, of early stress reticulocytes resembles more that of an HK enzyme than that of mature LK cells (Blostein et al. 1974). The number of the Na^+/K^+ pumps, increased within an order of magnitude as measured by ^{3}H-ouabain binding not quite to levels seen in HK sheep derived reticulocytes (Lauf and Valet 1983; Xu et al. 1994), surely includes LK elements (either some LK pumps on reticulocytes or all pumps under modification to LK pumps, or contamination with mature LK RBCs) since anti-L_p stimulated ouabain-sensitive K^+ influx (Kim et al. 1980).

In reticulocytes kept *in vitro*, the net content of Na^+, K^+-ATPase measured by antibodies against the alpha-subunits and by ^{3}H-ouabain binding (Blostein and Grafova 1990) decreases and the large K^+ pump fluxes decline within hours as the anti-L_p stimulation becomes more evident (Kim et al. 1980; Valet and Lauf 1983). As the numbers of pump sites/cell decrease (Lauf and Valet 1983; Blostein and Grafova 1990), the L_p antigen confers upon these pumps the inhibition by cellular K^+ and may even participate in the actual reduction of pump sites (Xu et al. 1994). Indeed, anti-L_p added to such cells in culture delays the Na^+/K^+ pump involution measured by ^{3}H-ouabain binding as well as by Western blots with anti-alpha subunit antibodies and reduces the sensitivity to cellular K^+ (Xu et al. 1994), and trypsin known to inactivate the association of the L_p antigen with the Na^+/K^+ pumps (Lauf et al. 1970) exhibits similar effects (Xu et al. 1993). This process is by no means stoichiometric since the number of L_p antigens/cell (from 600 to 700) does not change as much as that of the Na^+/K^+ pumps (from ~500 to < 100) commensurate with a much higher fractional stimulation of Na^+/K^+ pump flux by anti-L_p in mature cells as compared to reticulocytes (Xu et al. 1994). The maturation-induced increase in the L_p antigen/pump ratio has not yet been explained mechanistically in terms of the loss of pumps selective over that of antigens.

30.6 Overall Physiological Consequences and the HK/LK Dimorphism

Is the association of ion transport with membrane antigens an evolutionary advantage? The answer is probably no, as LK animals, with both RBC transporters and membrane antigens tightly associated, have not been found to display a selective survival disadvantage, in particular, as there is no evidence for a shortened RBC life span in these animals. This is in contrast to dogs with HK RBCs which, despite possessing active Na^+/K^+ pumps and K^+-Cl^- COT with volume regulatory capability (Fujise et al. 1997), are swollen and exhibit a tendency to intravascular haemolysis and haematuria and, therefore, may have been selected against animals with LK RBCs by man if not by nature (Fujise et al. 1997). The occurrence of a RBC cation polymorphism, rather than a clear-cut dimorphism, in cattle and goat (Ellory and Tucker 1970a,b), apparently less stringently associated with the M and L antigens, without evidence for a gross physiological disadvantage, is another point in case.

30.7 Membrane Antigens and Transport in Species Other than Ruminants

Although there are several membrane transporters that express antigenic epitopes in RBCs and other tissues, the M/L antigen association with Na^+/K^+ pump and K^+-Cl^- COT is the only system known in physiology where antigen/antibody reactions cause striking changes in transport activities. We suspected that the Rh antigen complex in human RBCs would be a good candidate; however, higher Na^+/K^+ pump and K^+ "leak" fluxes found in Rh_{null} cells (Lauf and Joiner 1976) rather signify other/additional underlying cytoskeletal and other defects. A protein related to the RhD antigen occurs in the kidney and has been speculated to be involved in ammonium transport (Marini et al. 2000). There is the Colton blood group antigen on aquaporin 1 (Agre et al. 1993) and the Kidd blood group on the RBC urea transporter (UT-B; Timmer et al. 2001), but there is no report that antibodies against these antigens do modify the underlying transport activities. The same conclusion applies to other transporters that host numerous epitopes such as band 3 protein (Tanner 1993).

Acknowledgement: I would like to thank many colleagues and students who have contributed to the data and concepts published in the papers referenced from this laboratory. Especially, I am indebted to my long-year collaborator, Professor Norma C. Adragna, who worked with me on many aspects of the K^+-Cl^- COT system (see review Lauf and Adragna (2000)). The National Institute of Health, for many years at Duke and Wright State Universities, the American Heart Association and the Ohio Research Challenge at Wright State University funded the research reviewed in this chapter. I also thank Diane Ebert, Glena Buchholtz, and especially, Donna Maas for their assistance in the preparation and template setting of this manuscript.

References

Abderhalden E (1889) Über den Blut Kalium Gehalt verschiedener Säugethiere. Hoppe-Seyler's Z Physiologische Chemie 25:65-115

Adragna NC, Lauf PK (1994) Quinine and quinidine inhibit and reveal heterogenity of K-Cl cotransport in low K sheep erythrocytes. J Membrane Biol 142:95-207

Adragna NC, Lauf PK (1997) Oxidative activation of K-Cl cotransport by diamide in erythrocytes from humans with red cell disorders, and from several mammalian species. J Membrane Biol 155:207-221

Adragna NC, Lauf PK (1998) Role of nitrite, a nitric oxide derivative, in K-Cl cotransport activation of low potassium sheep red blood cells. J Membrane Biol 166:157-167

Adragna NC, White RE, Orlov SN, Lauf PK (2000) K-C1 cotransport in vascular smooth muscle and erythrocytes: possible implication in vasodilation. Am J Physiol 278:C381-C390

Agre P, Preston GM, Smith BL, Jung JS, Raina S, Moon C, Guggino WB, Nielsen S (1993) Aquaporin CHIP: the archetypal molecular water channel. Am J Physiol 265:F463-F476

Beauge LA, Adragna N (1971) The kinetics of ouabain inhibition and the partition of rubidium influx in human red blood cells. J Gen Physiol 57:576-592

Bell DA, Leblond PF, Leddy JP, Lauf PK, LaCelle PL, Weed RI (1972) Immune hemolysis. J Immunol 108:467-474

Bergh CS, Kelley J, Dunham PB (1990) K-Cl cotransport in LK sheep erythrocytes: kinetics of stimulation by cell swelling. J Membrane Biol 117:177-188

Bernhardt I (1994) Alteration of cellular features after exposure to low ionic strength medium. In: Bauer J (ed) Cell electrophoresis. CRC Press, Boca Raton, pp 163-179

Bize I, Dunham PB (1994) Staurosporine, a protein kinase inhibitor, activates K-C1 cotransport in LK sheep erythrocytes. Am J Physiol 266:C759-C770

Bize I, Dunham PB (1995) H_2O_2 activates red blood cell K-C1 cotransport via stimulation of a phosphatase. Am J Physiol 269:C849-C855

Bize I, Muñoz P, Canessa M, Dunham PB (1998) Stimulation of membrane serine-threonine phosphatase in erythrocytes by hydrogen peroxide and staurosporine. Am J Physiol 247:C440-C446

Bize I, Güvenc B, Robb A, Buchbinder G, Brugnara C (1999) Serine/threonine protein phosphatases and regulation of K-Cl cotransport in human erythrocytes. Am J Physiol 277:C926-C936

Blostein R, Grafova E (1990) Decrease in Na^+-K^+-ATPase associated with maturation of sheep reticulocytes. Am J Physiol 259:C241-C250

Blostein R, Whittington ES (1973) Studies of high potassium and low potassium sheep erythrocyte membrane sodium-adenosine triphophatase. J Biol Chem 248:1772-1777

Blostein R, Whittington ES, Kuebler ES (1974) Na^+-ATPase of mammalian erythrocyte membranes: kinetic changes associated with postnatal development and following active erythropoiesis. Ann New York Acad Sci 242:305-316

Bratcher RL, Kanik-Ennulat CL, Logue PJ, Dunham PB (1983) Stimulation of the Na/K-pump in LK sheep erythrocytes by immunoglobulin fragments. Immunol Commun 12:565-571

Brown AM, Ellory JC, Young JD, Lew VL (1978) A calcium-activated potassium channel present in foetal red cells of the sheep but absent from reticulocytes and mature red cells. Biochim Biophys Acta 511:163-175

Brugnara C, Van Ha T, Tosteson DC (1989) Role of chloride in potassium transport through a K-Cl cotransport system in human red blood cells. Am J Physiol 256:C994-C1003

Campbell EH, Gibson JS, Ellory JC (1998) Oxygen-dependent K fluxes in sheep red cells. J Physiol 506:679-688

De Franceschi L, Fumagalli L, Olivieri O, Corrocher R, Lowell CA, Berton G (1997) Deficiency of Src family kinases Fgr and Kck results in activation of erythrocyte K/Cl cotransport. J Clin Invest 99:220-227

Delpire E, Lauf PK (1991a) Kinetics of Cl-dependent K fluxes in hyposmotically swollen low K sheep erythrocytes. J Gen Physiol 97:173-193

Delpire E, Lauf PK (1991b) Trans-effects of cellular K and Cl on ouabain-resistant Rb(K) influx in low K sheep red blood cells. Further evidence of asymmetry of K-Cl cotransport. Pflügers Arch 419:540-542

Delpire E, Lauf PK (1991c) Magnesium and ATP dependence of K:Cl cotransport in low K sheep red cells. J Physiol 441:219-231

Delpire E, Lauf PK (1992) Kinetics of DIDS inhibition of swelling-activated K-Cl cotransport in low K sheep erythrocytes. J Membrane Biol 126:89-96

Delpire E, Mount DB (2002) Human and murine phenotypes associated with defects in cation-chloride transport. Annu Rev Physiol 64:803-843

Dunham PB (1976a) Passive potassium transport in LK sheep red cells. Effects of anti-L antibody and intracellular potassium. J Gen Physiol 68:567-581

Dunham PB (1976b) Anti-L serum – two populations of antibodies affecting cation transport in LK erythrocytes of sheep and goats. Biochim Biophys Acta 443:219-226

Dunham PB (1990) K,Cl cotransport in mammalian erythrocytes. In: Reuss L, Russell JM, Szabo G (eds) Regulation of potassium transport across biological membranes. Univ Texas Press, Austin, pp 331-360

Dunham PB (1992) Ion transport in sheep red blood cells. Comp Biochem Physiol 102A:625-630

Dunham PB (1995) Effects of urea on K-Cl cotransport in sheep red blood cells: evidence for two signals of swelling. Am J Physiol 269:C1026-C1032

Dunham PB, Anderson C (1987) On the mechanism of stimulation of the Na/K-pump of LK sheep erythrocytes by anti-L antibody. J Gen Physiol 90:3-25

Dunham PB, Blostein R (1976) Active potassium transport in reticulocytes of high-K^+ and low-K^+ sheep. Biochim Biophys Acta 455:749-758

Dunham PB, Blostein R (1997) L antigens of sheep red blood cell membranes and modulation of ion transport. Am J Physiol 272:C357-C368

Dunham PB, Ellory JC (1980) Stimulation of the sodium-potassium pump by trypsin in low potassium type erythrocytes of goats. J Physiol 301:25-37

Dunham PB, Ellory JC (1981) Passive potassium transport in low potassium sheep red cells: dependence upon cell volume and chloride. J Physiol 318:511-530

Dunham PB, Hoffman JF (1971a) The number of Na^+:K^+-pump sites on red blood cells from HK and LK lambs. Biochim Biophys Acta 241:399-402

Dunham PB, Hoffman JF (1971b) Active cation transport and ouabain binding in high potassium and low potassium red blood cells of sheep. J Gen Physiol 58:94-116

Dunham PB, Klimczak J, Logue PJ (1993) Swelling activation of K-Cl cotransport in LK sheep erythrocytes: a three state process. J Gen Physiol 101:733-766

Dunham PB, Tucker EM, Simonsen E, Ellory JC (1980) Stimulation of active K^+ transport by anti-L antibodies in trypsin-treated low potassium sheep erythrocytes. J Gen Physiol 75:345-350

Ellory JC, Carleton S (1974) (Na^+-K^+)-activated ATPase in cattle erythrocytes. Biochim Biophys Acta 363:397-403

Ellory JC, Dunham PB (1980) Volume-dependent passive potassium transport in LK sheep red cells. In: Lassen UV, Ussing HH, Wieth JO (eds) Membrane transport in erythrocytes (Alfred Benzon symposium 14). Munksgaard, Copenhagen, pp 409-427

Ellory JC, Maher P (1977) A change in the internal affinity of LK goat red-cell sodium pumps induced by high pH. Biochim Biophys Acta 471:111-117

Ellory JC, Tucker EM (1969a) Stimulation of the potassium transport system in low potassium type sheep red cells by a specific antigen antibody reaction. Nature 222:477-478

Ellory JC, Tucker EM (1969b) Active potassium transport in the immature red cells of anaemic sheep. J Physiol 208:18-19

Ellory JC, Tucker EM (1970a) High potassium type red cells in cattle. J Agric Sci Cambridge 74:595-596

Ellory JC, Tucker EM (1970b) Active potassium transport and the L and M antigens of sheep and goat red cells. Biochim Biophys Acta 219:160-168

Ellory JC, Tucker EM (1971) The M-L blood group system and active postassium transport in sheep reticulocytes. Anim Blood Grps biochem Genet 2:77-87

Ellory JC, Tucker EM (1983) Cation transport in red blood cells. In: Agar NS, Board PG (eds) Red blood cells of domestic mammals. Elsevier, Amsterdam, pp 291-314

Ellory JC, Feinstein A, Herbert J (1973) Stimulation of active potassium transport in LK sheep red cells by monovalent fragments of anti-L antibody. Immunochemistry 10:785-787

Erdmann A, Bernhardt I, Pittman SJ, Ellory JC (1991) Low potassium-type but not high potassium-type sheep red blood cells show passive K^+ transport induced by low ionic strength. Biochim Biophys Acta 1061:85-88

Evans JV (1954) Electrolyte concentrations in red blood cells of British breeds of sheep. Nature 174:931

Evans JV (1956) The stability of the potassium concentration in the erythrocytes of individual sheep compared with the variability between different sheep. J Physiol 136:41-59

Evans JV, King JWB (1955) Genetic control of sodium and potassium concentrations in the red blood cells of sheep. Nature 176:171

Farquharson BE, Dunham PB (1986) Intracellular potassium promotes antibody binding to an antigen associated with the Na/K-pump of sheep erythrocytes. Biochim Biophys Res Commun 134:982-988

Ferrell CM, Lauf PK, Wilson BA, Adragna NC (2000) Lithium and protein kinase C. Modulators regulate swelling-activated K-Cl cotransport and reveal a complete phosphatidylinositol cycle in low K sheep erythrocytes. J Membrane Biol 177:81-93

Flatman PW, Adragna NC, Lauf PK (1996) Role of protein kinases in regulating sheep erythrocyte K-Cl cotransport. Am J Physiol 271:C255-C263

Fujise H, Lauf PK (1987) Swelling, NEM, and A23187 activate Cl^--dependent K^+ transport in high K^+ sheep red cells. Am J Physiol 252:C197-C204

Fujise H, Cruz P, Reo NV, Lauf PK (1991) Relationship between total magnesium concentration and free intracellular magnesium in sheep red blood cells. Biochim Biophys Acta 1094:51-54

Fujise H, Higa K, Nakayama T, Wada K, Ochiai H, Tanabe Y (1997) Incidence of dogs possessing red blood cells with high K in Japan and East Asia. J Vet Med Sci 59:495-497

Funder J, Wieth JO (1967) Effects of some monovalent anions on fluxes of Na and K, and on glucose metabolism of ouabain treated human red cells. Acta Physiol Scand 71:168-185

Gardos G (1958) Effect of ethylenediamine-tetraacetate on the permeability of human erythrocytes. Acta Physiol Acad Sci Hung 14:1-5

Gibson JS, Speake PF, Ellory JC (1998) Differential oxygen sensitivity of the KCl cotransporter in normal and sickle red blood cells. J Physiol 511:225-234

Gillen CM, Brill PS, Payne JA, Forbush III B (1996) Molecular cloning and functional expression of the K-Cl cotransporter from rabbit, rat, and human. J Biol Chem 271:16237-16244

Glynn IM, Ellory JC (1972) Stimulation of a sodium pump by an antibody that increases the apparent affinity of the sodium ions of the sodium-loading sites. In: Bolis L, Keynes RD, Wilbrandt W (eds) Role of membrane in secretory processes. North Holland, Amsterdam; American Elsevier, New York, pp 224-237

Hall AC, Ellory JC (1986) Effect of high hydrostatic pressure on 'passive' monovalent cation transport in human red cells. J Membrane Biol 94:1-7

Halperin JA, Brugnara C, Van Ha T, Tosteson DC (1990) Voltage-activated cation permeability in high-potassium but not low-potassium red blood cells. Am J Physiol 258:C1169-C1172

Hiki K, D'Andrea RJ, Furze J, Crawford J, Woollatt E, Sutherland GR, Vadas MA, Gamble JR (1999) Cloning, characterization, and chromosomal location of a novel human K^+-Cl^- cotransporter. J Biol Chem 274:10661-10667

Hoffman PG, Tosteson DC (1971) Active sodium and potassium transport in high potassium and low potassium sheep red cells. J Gen Physiol 58:438-466

Holtzman EJ, Kumar S, Faaland CA, Warner F, Logue PJ, Erickson SJ, Ricken G, Waldman G, Dunham PB (1998) Cloning, characterization, and gene organization of K^+-Cl^- cotransporter from pig and human kidney and *C. elegans*. Am J Physiol 275:F550-F564

Jennings ML (1999) Volume-sensitive K^+/Cl^- cotransport in rabbit erythrocytes: analysis of the rate limiting activation and inactivation events. J Gen Physiol 114:743-757

Jennings ML, Al-Rohil N (1990) Kinetics of activation and inactivation of swelling-induced K^+/Cl^- cotransport. Volume sensitive parameter is the rate constant for inactivation. J Gen Physiol 95:1021-1040

Joiner CH, Lauf PK (1975) The effect of anti-L on ouabain binding to sheep erythrocytes. J Membrane Biol 21:99-112

Joiner CH, Lauf PK (1978a) The correlation between ouabain binding and potassium pump inhibition in human and sheep erythrocytes. J Physiol 283:155-175

Joiner CH, PK Lauf (1978b) Modulation of ouabain binding and potassium pump fluxes by cellular sodium and potassium in human and sheep erythrocytes. J Physiol 283:177-196

Kaji D (1986) Volume-sensitive K transport in human erythrocytes. J Gen Physiol 88:719-738

Kaji D (1993) Effect of varying membrane potential on K-Cl transport in human erythrocytes. Am J Physiol 264:C376-C382

Kelly SJ, Thomas R, Dunham PB (2000) Candidate inhibitor of the volume-sensitive kinase regulating K-Cl cotransport: the myosin light chain kinase inhibitor ML-7. J Membrane Biol 178:31-41

Kerr SE (1937) Studies on the inorganic composition of blood. The relationship of potassium to the acid soluble phosphorous franction. J Biol Chem 117:227-235

Kim HD, Theg BE, Lauf PK (1980) Effect of anti-L on K influx and *in vitro* maturation. J Gen Physiol 76:109-121

Kregenow FM (1981) Osmoregulatory salt transporting mechanisms: control of cell volume in anisotonic media. Annu Rev Physiol 43:493-505

Kummerow D, Hamann J, Browning JA, Wilkins R, Ellory JC, Bernhardt I (2000) Variations of intracellular pH in human erythrocytes via $K^+(Na^+)/H^+$ exchanger under low ionic strength conditions. J Membrane Biol 176:207-216

Lauf PK (1975) Antigen-antibody reactions and cation transport in biomembranes: immunophysiological aspects. Biochim Biophys Acta 415:173-229

Lauf PK (1978) Membrane immunological reactions and transport. In: Giebisch G, Tosteson DC, Ussing HH (eds) Membrane transport in biology, vol I. Springer, Berlin, pp 291-348

Lauf PK (1983) Thiol stimulated passive K/Cl transport in sheep red cells. I. Dependence on chloride and external $K^+(Rb^+)$ ions. J Membrane Biol 73:237-246

Lauf PK (1984a) Thiol-dependent, passive K/Cl transport in sheep erythrocytes. IV. Furosemide inhibition and the role of external Rb^+, Na^+, and Cl^-. J Membrane Biol 77:57-62

Lauf PK (1984b) Thiol-dependent passive K^+ Cl^- transport in sheep red blood cells. VI. Functional heterogeneity and immunological identity with volume-stimulated $K^+(Rb^+)$ fluxes. J Membrane Biol 82:167-178

Lauf PK (1985a) Passive K^+-Cl^- fluxes in low K^+ sheep erythrocytes. Modulation by A23187 and bivalent cations. Am J Physiol 249:C271-C278

Lauf PK (1985b) K^+-Cl^- cotransport: sulfhydryls, divalent cations, and the mechamism of volume activation in a red cell. J Membrane Biol 88:1-13

Lauf PK (1988a) Thiol-dependent K:Cl transport in sheep red blood cells. VIII. Activation through metabolically and chemically reversible oxidation by diamide. J Membrane Biol 101:179-188

Lauf PK (1988b) Kinetic comparison of ouabain-resistant K:Cl fluxes (K:Cl [Co]-transport) stimulated in sheep erythrocytes by membrane thiol oxidation and alkylation. Mol Cell Biochem 82:97-106

Lauf PK (1988c) Volume and anion dependence of ouabain-resistant K-Rb fluxes in sheep red blood cells. Am J Physiol 255:C331-C339

Lauf PK (1990) Thiol-dependent passive K:Cl transport in sheep red blood cells. X. A hydroxylamine oxidation induced K:Cl cotransport blocked by diethylpyrocarbonate. J Membrane Biol 118:153-159

Lauf PK (1991) Foreign anions modulate volume set point of sheep erythrocyte K-Cl cotransport. Am J Physiol 260:C503-C512

Lauf PK (1992) Incorporation of ^{3}H-*N*-ethmaleimide into sheep red cell membrane thiol groups following protection by diamide-induced oxidation. Mol Cell Biochem 114:13-20

Lauf PK, Adragna NC (1995) Temperature-induced functional deocclusion of thiols inhibitory for sheep erythrocyte K-Cl cotransport. Am J Physiol 269:C1167-C1175

Lauf PK, Adragna NC (1996) A thermodynamic study of electroneutral K-C1 cotransport in pH- and volume-clamped low K sheep erythrocytes with normal and low internal magnesium. J Gen Physiol 108:341-350

Lauf PK, Adragna NC (1998) Functional evidence for a pH sensor of erythrocyte K-Cl cotransport through inhibition by internal protons and diethylpyrocarbonate. Cell Physiol Biochem 8:46-60

Lauf PK, Adragna NC (2000) K-Cl cotransport: properties and molecular mechanism. Cell Physiol Biochem 10:341-354

Lauf PK, Dessent MP (1973) Effect of metabolic state on immune-hemolysis of positive low potassium (LK) sheep red blood cells by iso-immune anti-L serum and rabbit serum complement. Immunol Commun 2:193-212

Lauf PK, Joiner CH (1976) Increased potassium transport and ouabain binding in human Rh_{null} red blood cells. Blood 48:457-468

Lauf PK, Sun WW (1976) Binding characteristics of M and L isoantibodies to high and low potassium sheep red cells. J Membrane Biol 28:351-372

Lauf PK, Theg BE (1980a) A chloride dependent K^+ flux induced by *N*-ethylmaleimide in genetically low K^+ sheep and goat erythrocytes. Biochim Biophys Res Commun 92:1422-1428

Lauf PK, Theg BE (1980b) *N*-ethylmaleimide enhances selectively passive K^+ permeability in low potassium sheep red cells. Adv Physiol Sci 6:285-291

Lauf PK, Valet G (1983) Na^+ K^+-pump and passive K^+ transport in large and small red cell populations of anaemic high and low K^+ sheep. J Cell Physiol 116:35-44

Lauf PK, Tosteson DC (1969) The M-antigen in HK and LK sheep red cell membranes. J Membrane Biol 1:177-193

Lauf PK, Rasmusen BA, Hoffman PG, Dunham PB, Cook P, Parmelee ML, Tosteson DC (1970) Stimulation of active potassium transport in LK sheep red cells by blood group-L-antiserum. J Membrane Biol 3:1-13

Lauf PK, Parmelee ML, Snyder JJ, Tosteson DC (1971) Enzymatic modification of the L and M antigens in LK and HK sheep erythrocytes and their membranes. J Membrane Biol 4:52-67

Lauf PK, Stiehl BJ, Joiner CH (1977) Active and passive cation transport and L antigen heterogeneity in low potassium sheep red cells. J Gen Physiol 70:221-242

Lauf PK, Shoemaker DG, Joiner CH (1978) Changes in K^+ pump transport and ouabain binding sites in erythrocytes of genetically low K^+ lambs. Biochim Biophys Acta 507:544-548

Lauf PK, Bauer J, Adragna NC, Fujise H, Zade Oppen AM, Ryu KH, Delpire E (1992) Erythrocyte K-Cl cotransport: properties and regulation. Am J Physiol 263:C917-C932

Lauf PK, Erdmann A, Adragna NC (1994) Response of K-Cl cotransport to pH, and the role of Mg in volume clamped low K sheep erythrocytes: three equilibrium states. Am J Physiol 266:C95-C103

Lauf PK, Adragna NC, Agar N (1995) Glutathione removal reveals kinases as common targets for K-Cl cotransport stimulation in sheep erythrocytes. Am J Physiol 269:C234-C241

Lauf PK, Ahmed S, Adragna NC (2000) NH^+_4 inhibits K-Cl cotransport on low K sheep red blood cells. J Gen Physiol 114:57a

Lauf PK, Zhang J, Delpire E, Fyffe REW, Mount DB, Adragna NC (2001a) Erythrocyte K-Cl cotransport: immunocytochemical and functional evidence for more than one KCC isoform in HK and LK sheep red blood cells. Comp Biochem Physiol 130A:499-509

Lauf PK, Zhang J, Gagnon KBE, Delpire E, Fyffe REW, Adragna NC (2001b) K-Cl cotransport: immunohistochemical and ion flux studies in human embryonic kidney (HEK293) cells transfected with full-length and C-terminal-comain-truncated KCC1 cDNAs. Cell Physiol Biochem 11:143-160

Lee P, Woo A, Tosteson CD (1966) Cytodifferentiation and membrane transport properties in LK sheep red Cells. J Gen Physiol 50:379-390

Logue P, Anderson C, Kanik C, Farquharson B, Dunham PB (1983) Passive potassium transport in LK sheep red cells. Modification by *N*-ethylmaleimide. J Gen Physiol 81:861-885

Lytle C, McManus TJ, Haas M (1998) A model of Na-K-2Cl cotransport based on ordered ion binding and glide symmetry. Am J Physiol 274:C299-C309

Marini AM, Matassi G, Raynal V, Andre B, Cartron JP, Chief-Zahar B (2000) The human Rhesus-associated RhAG protein and a kidney homologue promote ammonium transport in yeast. Nature Genetics 25:341-344

Miyamoto HT, Ikehara H, Yamaguchi K, Hosokawa TY, Masuya T (1986) Kinetic mechanism of Na^+, K^+, Cl^--cotransport as studied by Rb^+ influx into HeLa cells: effects of extracellular monovalent ions. J Membrane Biol 92:135-150

Mount DB, Mercado A, Song L, Xu J, George AL Jr, Delpire E, Gamba G (2000) Cloning and characterization of KCC3 and KCC4, new members of the cation chloride cotransporter gene family. J Biol Chem 274:16355-16362

Ortiz-Carranza O, Adragna NC, Lauf PK (1996) Modulation of K-Cl cotransport in volume-clamped low-K sheep erythrocytes by pH, magnesium and ATP. Am J Physiol 271:C1049-C1058

Ortiz-Carranza O, Adragna NC, Carnes L, Lauf PK (1997a) Two operational modes of K-Cl cotransport in low K^+ sheep red blood cells. Cell Physiol Biochem 7:251-263

Ortiz-Carranza O, Miller ME, Adragna NC, Lauf PK (1997b) Alkaline pH and internal calcium increase Na^+ and K^+ effluxes in LK sheep red blood cells in Cl^- free solutions. J Membrane Biol 156:287-295

Parker JC, Colclasure GC, McManus TJ (1991) Coordinated regulation of shrinkage-induced Na/H exchange and swelling-induced [K-Cl] cotransport in dog red cells. J Gen Physiol 98:869-880

Pittman SJ, Ellory JC, Tucker EM, Newbold CI (1990) Identification of a 25 kDa polypetide associated with the L antigen in low potassium-type sheep red cells. Biochim Biophys Acta 1022:408-410

Race JE, Makhlouf FN, Logue PJ, Wilson FH, Dunham PB, Holtzman EJ (1999) Molecular cloning and functional characterization of KCC3, a new K-Cl cotransporter. Am J Physiol 277:C1210-C1219

Rasmusen BA (1969) A blood group antibody which reacts exclusively with LK sheep red blood cells. Genetics 61:49

Rasmusen BA, Hall JG (1966) Association between potassium concentration and serological type of sheep red blood cells. Science 151:1551-1551

Richter S, Hamann J, Kummerow D, Bernhardt I (1997) The monovalent cation "leak" transport in human erythrocytes: an electroneutral exchange process. Biophys J 73:733-745

Ryu KH, Adragna NC, Lauf PK (1989) Kinetics of Na-Li exchange in high and low K sheep red blood cells. J Am Physiol 257:C58-C64

Sachs JR (1994) Soluble polycations and cationic amphiphiles inhibit volume-sensitive K-Cl cotransport in human red cell ghosts. Am J Physiol 266:C997-C1005

Sachs JR, Martin DW (1999) Role of polyamine structure in inhibition of K^+ Cl^- cotransport in human red cell ghosts. J Physiol 520:723-735

See JRH, Adragna NC, PK Lauf (1996) Osmotic resistance of low and high potassium sheep red blood cells: comparison with 'ROL' gene model. Blood 88:54a

Shrager P, Tosteson DC, Lauf PK (1972) Biochemical characterization of a lipid-dependent membrane protein antigen in HK sheep red cells. Biochim Biophys Acta 290:186-199

Sims PJ, Lauf PK (1978) Steady-state analysis of tracer exchange across the C5b-9 complement lesion in a biological membrane. Proc Natl Acad Sci USA 75:5669-5673

Smalley CE, Tucker EM, Ellory JC, Young JD (1983) The solubilization of the L and M antigens from sheep red cell membranes. Biochim Biophys Acta 733:283-285

Smalley CE, Tucker EM, Sepulveda I, Ellory JC (1984) Solubilization from sheep erythrocyte membranes of the potassium transport-associated antigens L_p, L_l and M, and the molecular size determination of the L_p antigen by radiation inactivation. Biochem Soc Trans 12:306-307

Snyder JJ, Rasmusen BA, Lauf PK (1971) The nature of the antibody in iso-immune anti-L sera affecting active potassium transport in LK sheep red cells. J Immunol 107:772-781

Su W, Shmukler BE, Chernova MN, Stuart-Tilley AK, De Francesci L, Brugnara C, Alper SL (1999) Mouse K-Cl cotransporter KCCl: cloning, mapping, pathological expression and functional regulation. Am J Physiol 277:C899-C912

Tang Y (1997) Biochemical characterization of the M antigen and antibody system in HK sheep erythrocytes. Ph.D. thesis, Wright State University

Tanner MJ (1993) The major integral proteins of the human red cell. Bailliere's Clin Haematol 6:333-356

Timmer RT, Klein JD, Bagnasco SM, Doran JJ, Gunn RB, Sands JM (2001) Localization of the urea transporter UT-B protein in human and rat erythrocytes and tissues. Am J Physiol 281:C1318-C1325

Tosteson DC (1966) Some properties of the plasma membranes of high potassium and low potassium sheep red cells. Ann New York Acad Sci 137:577-590

Tosteson DC, Hoffman JF (1960) Regulation of cell volume by active cation transport in high and low potassium sheep red cells. J Gen Physiol 44:169-194

Tosteson DC, Cook P, Blount R (1965) Separation of adenosine triphosphatase of HK and LK sheep red cell membranes by density gradient centrifugation. J Gen Physiol 48:1125-1143

Tucker EM, Ellory JC (1971) The M-L blood group system and active potassium transport in sheep reticulocytes. Anim Blood Grps Biochem Genet 2:77-87

Tucker EM, Ellory JC, Wooding FBP, Morgan G, Herbert J (1976) The number and specificity of L antigen sites on low potassium type sheep red cells. Proc R Soc London 194:271-277

Tucker EM, Smalley CE, Ellory JC, Dunham PB (1982) The transition from HK to LK phenotype in the red cells of newborn genetically LK lambs. J Gen Physiol 79:893-915

Valet G, Franz G, Lauf PK (1978) Different red cell populations in newborn, genetically low potassium sheep: relation to hematopoietic, immunnologic and physiologic differentiation. J Cell Physiol 94:215-228

Whittington ES, Blostein R (1971) Comparative properties of high potassium and low potassium sheep erythrocyte membrane sodium-activated adenosine triphosphatase. J Biol Chem 246:3518-3523

Wiedmer T, Lauf PK (1981) Properties of the M antigen solubilized from genetically high potassium sheep red cells. Membrane Biochem 4:31-47

Wiley JS, Cooper RA (1974) A furosemide-sensitive cotransport of sodium plus potassium in the human red cell. J Clin Invest 53:745-755

Xu Z-C, Dunham PB, Munzer JS, Silvius JR, Blostein R (1992) Rat kidney Na-K-pumps incorporated into low-K^+ sheep red blood cell membranes are stimulated by anti-L_p antibody. Am J Physiol 263:C1007-C1014

Xu Z-C, Dunham PB, Dyer B, Blostein R (1993) Differentiation of Na^+-K^+-pumps of low-K^+ sheep red blood cells is promoted by L_p membrane antigens. Am J Physiol 265:C99-C105

Xu Z-C, Dunham PB, Dyer B, Blostein R (1994) Decline in number of Na^+-K^+-pumps on low-K^+ sheep reticulocytes during maturation is modulated by L_p antigen. Am J Physiol 266:C1173-C1181

31 Comparative Physiology of Red Cell Membrane Transport

John S. Gibson

Department of Clinical Veterinary Medicine, University of Cambridge, Madingley Road, Cambridge CB3 0ES, UK

31.1 Introduction

The advantages of a comparative approach to red blood cell (RBC) membrane transport have been evident for many years. Indeed, Parker wrote that the "early history RBC transport physiology is written in the blood of dogs and cats" (Parker 1977). Many landmarks in research into RBC membrane transport involved animal tissue and the individuals who carried out the work include many of the major figures of the last fifty years or so. For example, appreciation of the sickling behaviour of RBCs from patients with sickle cell disease was anticipated by observations on deer RBCs (Gulliver 1840), some seventy years previously; the dimorphism of K^+ content of sheep RBCs was instrumental in development of the "pump-leak" hypothesis for cation balance and volume homeostasis (Tosteson and Hoffman 1960); much of our understanding on cellular volume regulation in response to anisotonic shock originated with work on RBCs from birds and fish (Kregenow 1981); and even the structure of haemoglobin was elucidated using horse RBCs (Perutz 1969). In many cases, however, RBCs from species other than human have been used as paradigms, rather than important to the comparative physiology of their donors. The comparative physiology of RBCs is an enormous subject, both very diverse and very rich. In addition, whilst this book concerns membrane transport, but other comparative aspects are equally interesting, for example metabolism, energy source or haemoglobin function (e.g. Nikinmaa 1990). A number of previous reviews have highlighted the value of a comparative approach (e.g., Davson and Daniell 1952; Dunker and Passow 1953; Whittam 1964; McManus 1967; Parker 1977; Ellory and Tucker 1983; Nikinmaa 1990). In the space available, of necessity, I only skim the surface, highlighting some areas of interest and ignorance, and concentrating mainly on cation balance and volume regulation.

31.2 Carnivores – Dog and Cat

Mature dog and cat RBCs, along with those of certain other mammalian groups such as bears (Willis et al. 1990), ferrets (Milanick 1989) and seals, are high Na^+-containing with low intracellular K^+ (Parker 1977). Typical intracellular cation contents are about 100 and 6 mM for Na^+ and K^+ respectively, compared to reciprocal levels in human RBCs.

K^+ content is therefore even lower than that found in low K^+-containing (LK) sheep RBCs (Dunham and Blostein 1997). Immature dog RBCs, however, resemble those of other species in having high K^+ content and Na^+/K^+ pumping which is lost as cells mature (Lee and Kirk 1982; Kirk et al. 1983; Maede and Inaba 1985). Not surprisingly, mature dog RBCs lack a functional Na^+/K^+ pump (Sha'afi and Lieb 1967; Parker 1973; Maede and Inaba 1985) and show no Na^+/K^+-stimulated ATPase activity (Chan et al. 1964). These transport characteristics appear similar to those in the cat (Parker and Hoffman 1965; Parker and Snow 1972; Romualdez et al. 1972; Parker 1973; Sha'afi and Naccache 1975). Physiologically, these cells regain normal volume following swelling by activating two systems, with much of the work on these species coming from the laboratory of John Parker (Fig. 31.1). First, they display a swelling-activated K^+-Cl^- cotransporter (KCC: Parker 1983a; Parker et al. 1990), but of course this cannot mediate net efflux of

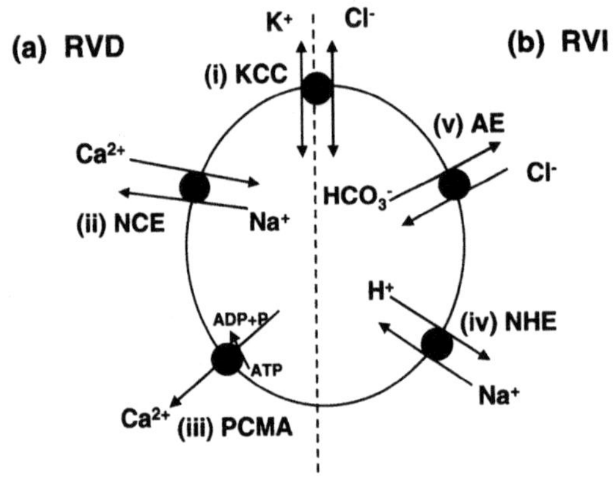

Fig. 31.1. Volume regulation in dog RBCs. Dog RBCs are usually low K^+-containing. (**a**) Regulatory volume decrease (*RVD*): The K^+-Cl^- cotransporter (*KCC*) is activated by swelling but does not take part in volume regulation since it is near electrochemical equilibrium – it may even have a small net inwardly directed gradient. Rather, loss of ions is achieved by activating the Na^+/Ca^{2+} exchanger (*NCE*), which transports Ca^{2+} ions into the cell in exchange for Na^+ (probably at a 1:3 stoichiometry), using the large inwardly directed Ca^{2+} gradient. The Ca^{2+} ions are then removed using metabolic energy in the form of ATP by the plasma membrane Ca^{2+} pump (*PMCA*). (**b**) Regulatory volume increase: On shrinkage, the Na^+/H^+ exchanger (*NHE*) mediates Na^+ entry and H^+ loss, driven by the outwardly directed H^+ gradient. This is accompanied by Cl^-/HCO_3^- exchange mediated by the anion exchanger (*AE*), which achieves net uptake of NaCl. There is little Na^+/K^+ pump activity in mature dog RBCs

KCl – it may even have an inwardly directed electrochemical gradient. In addition, at least in cats, the activity of this system is low. Second, they activate a Na^+/Ca^{2+} exchanger (Parker 1983b). This pathway mediates efflux of Na^+ coupled to influx of Ca^{2+}. The latter is then removed by the plasma membrane Ca^{2+} pump: the net effect being loss of Na^+. By contrast, shrinkage is accompanied by activation of a Na^+/H^+ exchanger (NHE: Parker 1983c; Parker and Castronova 1984), which may carry out Na^+ influx and H^+ efflux, as required for regulatory volume increase, because of the outward H^+ gradient (rather than an inward one for Na^+).

As for several RBC membrane transport systems, control in dog RBCs involves protein phosphorylation. In fact, there is co-ordinated regulation of swelling-induced K^+-Cl^- cotransport and shrinkage-induced Na^+/H^+ exchange. KCC is activated by dephosphorylation, NHE by phosphorylation (Parker 1992), as observed for KCC and NKCC in duck RBCs (Haas and McManus 1985; Starke and McManus 1990). There is evidence from dog RBC ghosts that protein phosphatase 1 (PP-1) is responsible for the activating dephosphorylation controlling KCC, or at least exogenous PP-1 can restore function of KCC in ghosts (Krarup and Dunham 1996). Finally, dog RBC ghosts also provided the evidence for macromolecular crowding as the sensor for volume regulation (see Chap. 9; Colclasure and Parker 1991, 1992).

Variant RBCs are observed in certain breeds of dogs, mainly those from Japan and East Asia, such as Shiba and Akita (e.g. Fujise et al. 1997). These dogs have mature RBCs with a high K^+ content and larger volume than the high Na^+ cells described above. K^+ leakage from these cells on storage, particularly at low temperature, which preferentially inhibits active K^+ sequestration, results in apparent high plasma K^+, or pseudohyperkalaemia, in stored clinical samples (Degen 1987). All dog reticulocytes have a high Na^+/K^+ pump capacity, with usual low K^+-phenotype losing almost all on maturity (Maede and Inaba 1985). High K^+ dog RBCs retain some Na^+/K^+ pump capacity and Na^+, K^+-ATPase activity (Maede et al. 1983; Maede and Inaba 1987; Fujise et al. 1991). In most cases, these animals also have high levels of reduced glutathione (GSH) in their RBCs and increased uptake of a number of amino acids including two GSH precursors, glutamine and glutamate (Maede et al. 1982, 1983). They are ascribed HK/HG, for high K^+, high glutathione. In addition, a further variant lacks high GSH but retains high Na^+/K^+ pump capacity and high K^+ content (Fujise et al. 1993, 1995), the HK/LG phenotype. Interestingly, these cells do not have a larger volume compared to the high Na^+ dog RBCs – elsewhere, we have summarised evidence linking depletion of GSH with cell shrinkage possibly via stimulation of KCC (see Chap. 9). Because of their high K^+ levels, and notwithstanding the high GSH levels, HK/HG dog RBCs are more susceptible to oxidative damage through ingestion of onions (Yamato and Maede 1992; Yamato et al. 1999), the active constituent of which appears to be sodium n-propylthiosulphate (Yamato et al. 1998). The molecular control of cation balance in dog RBCs, like that in sheep, awaits elucidation.

31.3 Herbivores – Horse and Deer

There are perhaps three main areas of interest in equine RBCs: cation balance, amino acid transport and O_2 sensitivity. Early work by Shaw showed that, despite maintaining high intracellular K^+ levels, the activity of the Na^+/K^+ pump was considerably lower than in humans (Shaw 1955). From the "pump-leak" model of cation balance, one would expect passive K^+ permeability to be low (Tosteson and Hoffman 1960). However, equine RBCs possess a K^+-Cl^- cotransporter (KCC) with a noticeably high capacity, many fold that of human RBCs (Gibson et al. 1993). The cotransporter participates in regulatory volume decrease (Gibson et al. 1993). There is minimal Na^+-K^+-Cl^- cotransport (Contreras et al. 1986). Maintenance of high intracellular K^+ requires down-regulation of KCC. We have shown that the transporter is markedly dependent on O_2 tension (Po_2's), stimulated by high Po_2's, inhibited by low ones (see Chap. 9) (cf. trout Borgese et al. 1991; Gibson et al. 1995; Speake et al. 1997). Furthermore, the transporter inactivates with time, maximal levels of activity require previous exposure to low Po_2's before O_2 activation. The precise dynamics of the O_2 response, the rate constants of activation and inactivation are not known. In addition, KCC in equine RBCs is inhibited by some constituent of plasma (Gibson et al. 1995). The response of the equine RBC to other blood parameters has been analysed in some detail (Speake et al. 1997). These features presumably keep the average activity of the transporter sufficiently low to balance active K^+ uptake via the Na^+/K^+ pump. Finally, although swelling-induced fluxes of amino acids are very small compared to KCC (Gibson et al. 1993), the amino acid transport phenotype of horses deserves comment. Fincham et al. (1987) elucidated a heterogeneity of amino acid transport, notably in thoroughbreds. Thus five distinct transport subgroups are present, apparently under the control of three co-dominant alleles (*s*, *h* and *l*). Allele *s* is silent; whilst *h* and *l* code for the presence of either high or low affinity, Na^+-independent transport of neutral amino acids, designated systems asc_1 and asc_2, respectively.

Of arteriodactyls, RBCs of sheep and cattle are reviewed elsewhere in this book (e.g., Chap. 30). Here, I shall discuss deer, which as well as being of interest *per se*, represent a species of steadily increasing importance economically. Deer RBCs are normally biconcave discs, typical of mammalian RBCs (for a review see Chapman (1975)). Under certain conditions, however, deer haemoglobin polymerises and the RBCs assume bizarre shapes including sickle, match stick and holly leaf forms (all covered under the term "sickle" in the following). Sickling occurs in most deer species, with the exception of Chinese water deer, reindeer and a few others (Chapman 1975; Butcher and Hawkey 1977; Taylor 1983). In fact, RBC sickling, which has now been intensively studied and well characterised in the human Hb S cells, was first described in cervidae, probably the result of a fixation artefact, some seventy years previously (e.g. Gulliver 1840 *vs.* Herrick 1910). In contrast to the situation with Hb S RBCs, sickling in deer cells is stimulated by high Po_2 and high pH, but exact conditions of sickling have not been thoroughly defined and it is not clear whether it can occur *in vivo*. Haemoglobin (Hb) is also important. In white-tailed deer (*Odocoileus virginianus*), the subject of most work on sickling, sickling is only seen in RBCs containing certain isoforms of Hb, e.g. Hbs II, III and IV, and not in others, e.g. V and VII (Kitchen et al. 1967). As in humans, sickling is not seen in fetal or Hb F-containing RBCs, which appear protected by the γ Hb. A similar pattern concerning Hb type and sickling propensity in other species of deer also seems to occur, although data is sparse.

In human sickle cells, transport is markedly abnormal, especially on deoxygenation when sickling occurs (see Chap. 23). The effect of sickling on deer RBCs is less clear. There have been few studies of the transport characteristics of deer RBCs apart from those of Lauf and colleagues (Adragna and Lauf 1988; Lauf et al. 1989), using a small number of immature American white-tailed deer (*Odocoileus virginianus*). The cells were of the normal high K^+-containing type, although with a low ATP content, and both K^+ "leak" and Na^+, K^+-ATPase activity were about a third those of human RBCs. In this limited study (of around six animals), no volume regulatory pathways were observed on changing extracellular osmolality and the cells appeared to lack a Ca^{2+}-activated K^+ channel (Gardos channel) in response to pharmacological loading with Ca^{2+} (via A23187). They showed no evidence of an *N*-ethylmaleimide (NEM)-stimulated KCC or Na^+/H^+ exchange activity. Passive and active transport in cells from these particular animals was thus very quiescent. A more detailed study, involving other species, would be of considerable interest.

31.4 Rodents – Mouse and Rat

The importance of understanding transport in these animals comes from their increasing use as the species of choice in transgenic work. For example, there are several models of transgenic mice expressing human sickle Hb, to investigate changes in membrane transport (see Chap. 28). To assess the appropriateness of these models, it is important to understand normal transport characteristics. Also, it should be appreciated that there may be significant differences between mouse strains. The most complete studies are those of Armsby et al. (1995) in mice, who investigated outbred CD1 and inbred C57/B6 mice, and Ihrig et al. (1992) in rat. Mice RBCs are the normal high K^+-containing type (Benos 1980), with $[K^+]$ and $[Na^+]$ of 135 and 5 mM, respectively. However, cell volume is relatively low (about 50 fl), particularly when compared to human RBCs (about 90 fl). This complicates comparisons of flux magnitudes, often expressed in terms of litre of cells, rather than cell number or surface area. The cation distribution is maintained by an active Na^+/K^+ pump, whose maximum activity is about 3-fold that of human (Benos and Tosteson 1980; Armsby et al. 1995; De Franceschi et al. 1997; Kazennov et al. 1998). The pump in mice, rats and some other rodents is less sensitive to ouabain than that in other mammals, and concentrations of about 1 mM are required for inhibition (Willis and Ellory 1983). Mouse RBCs also possess a Ca^{2+} pump, and again activity is several-fold higher than that in human RBCs (Romero et al. 1997a; Kazennov et al. 1998). Secondary active transporters include a modest Na^+/H^+ exchanger and Na^+-K^+-$2Cl^-$ cotransporter. A Ca^{2+}-activated K^+ transport (probably represented by the cloned mIK1: Vandorpe et al. 1998), or Gardos channel, is present, but of lower capacity compared to human (Armsby et al. 1995; Romero et al. 1997a). Like the channel in human RBCs, it is inhibited by clotrimazole (10 μM for 98% inhibition) and charybdotoxin (50 nM for 99% inhibition). K^+-Cl^- cotransport is also probably present. Thus there is a K^+ flux stimulated by swelling, staurosporine and NEM, and inhibited by calyculin and okadaic acid (Armsby et al. 1995; Su et al. 1999); a diamide-activated pathway has also been described and is inhibited partially by DIOA (Adragna and Lauf 1993). (Note that in rat RBCs, KCC is not stimulated by swelling: Ihrig et al. (1992)). Molecular studies have revealed a homologue of KCC1 in mouse erythroid tissue (Pellegrino et al. 1998; Su et al. 1999; Shmukler et al. 2000). As in other mammals, this pathway is O_2-sensitive (Gibson, unpublished observations), inhibited by Mg^{2+}-loading (De Franceschi et al. 1996) and regulated by serine-threonine and tyrosine phosphorylation (De Franceschi et al. 1997, 2001). It may also mediate regulatory volume decrease (Armsby et al. 1995), although this has been questioned (Romero et al. 1997b). However, the system is active at normal cell volumes and does not require additional stimulus cf. that in human RBCs (Armsby et al. 1995; Romero et al. 1997b). In addition, it is incompletely inhibited by substitution of Cl^- with NO_3^-, with the organic substitutes methylsulphate or methylsulphonate proving better inhibitors. In fact, the anion dependence differs markedly from those in other species such as human (Brugnara and Tosteson 1987) and sheep RBCs (Lauf 1983). This distinction is important when defining KCC activity in RBCs from this species. It might account for the apparent absence of KCC in some reports (Romero et al. 1993). In summary, these characteristics reproduce some of the features of human RBCs but there are subtle and potentially significant differences.

31.5 Lower Vertebrates

31.5.1 Birds

Much early work on the responses to anisotonicity was carried out on avian RBCs (Kregenow 1971, 1981). They have also been central to our understanding of chloride-coupled cation transporters (reviewed by Mount et al. (1999)). Along with nucleated RBCs from other lower vertebrates (from reptiles, amphibia and fish), they have been shown to respond to adrenergic stimulation (Palfrey and Greengard 1981), and, subsequently, they have been used to extend our understanding of regulation of membrane transporters by protein phosphorylation (Lytle 1997, 1998). The Na^+-K^+-Cl^- cotransporter (NKCC) of duck and turkey RBCs, in particular, responds to an impressive array of stimuli, including β-adrenergic agonists, shrinkage, protein kinase activators, protein phosphatase inhibitors and deoxygenation. The mechanism underlying the action of several of these stimuli has been explored in detail (e.g. Lytle 1997), but that by deoxygenation has received much less attention.

Deoxygenation-induced increase in avian RBC cation permeability was observed over fifty years ago (Orskov 1954; Tosteson and Robertson 1956; McManus 1967; Allen and McManus 1968; Palfrey and Greengard 1981). In many cases, this probably resulted from increased activity of NKCC, and this transporter is one of several which respond to physiological O_2 tensions (Gibson et al. 2000). We have recently investigated the O_2 response of NKCC in turkey RBCs (Muzyamba et al. 1999). Deoxygenation represented a very powerful stimulus, in fact more potent than maximal adrenergic stimulation and similar in effect to protein phosphatase inhibition with calyculin A. Although NKCC is regarded as a classical transport mediating regulatory volume increase, i.e. recovery in cell volume following shrinkage, in fact, after deoxygenation-induced activation, NKCC is almost insensitive to volume. In addition, response to adrenergic stimuli is small. The function of the O_2 response is unknown. Activation of NKCC could occur *in vivo*, e.g. in anaerobic white muscle, or following sympathetic activity. The transporter can operate at very high rates and may be therefore be involved in sequestration of plasma K^+ to prevent harmful elevation (Gibson et al. 2000).

31.5.2 Amphiuma

Red cells from this amphibian have proved a useful model for the study of volume regulation. Their large size, with a long diameter of > 60 μm and surface area of some 5000 μm^2, makes them amenable to puncture with microelectrodes for recording membrane voltage and conductance (Lassen 1974, 1977), although large shunt pathways have been reported between electrode and the cell membrane (Cala 1980a) which would short-circuit changes in potential difference (p.d.) and attenuate the effect of conductive ion movements. Cl^- permeability is several-fold that of K^+ (Hoffman and Laris 1974). It has been shown that regulatory volume decreases and increases occur via loss of K^+ and Cl^-, and gain of Na^+ and Cl^-, respectively (Cala 1980a,b). Ion movements appear to be electroneutral, but membrane p.d. and conductance are inadequate for conductive ion movement to account for the changes (Cala 1980a, 1983). Regulatory volume increase (RVI) is carried out by stimulation of an amiloride-sensitive Na^+/H^+ exchange (NHE), coupled with Cl^-/HCO_3^- exchange via band 3, with net gain of NaCl (Cala 1980a; Siebens and Kregenow 1985). In the presence of DIDS to inhibit band 3, Na^+ entry equates to H^+ efflux. *Amphiuma* red cell NHE also participates in acid-base balance, being stimulated by intracellular acidification. When stimulated by shrinkage, ion movement via NHE progresses until normal cell volume is restored, regardless of the concomitant acid-base changes; conversely, when stimulated by acidification, the exchanger mediates Na^+/H^+ exchange to restore normal pH, despite the ensuing cell swelling. Nevertheless, it appears that the same population of transporters carry out both functions, implying that they are coupled to two sensors and two sets of transduction apparatus. An isoform of NHE, homologous to NHE1, has been cloned from *Amphiuma* (McLean et al. 1999). Of more interest potentially is the mechanism of regulatory volume decrease (RVD; Fig. 31.2). Like RVI, this is thought to occur via electroneutral processes (Cala 1980a). This is purported to occur via a specific K^+/H^+ exchanger, with loss of K^+, and coupled again to band 3 mediating Cl^-/HCO_3^- exchange, to mediate KCl loss. The argument in support of this remains largely thermodynamic. Thus Cala has demonstrated that K^+ fluxes are proportional to $(\Delta\mu_K - \Delta\mu_H)$ – which implies K^+/H^+ exchange – rather than to $(\Delta\mu_K + zFE_m)$ or $(\Delta\mu_K + \Delta\mu_{Cl})$, implying movement via conductive K^+

channels or a K^+-Cl^- cotransporter, respectively. As expected for K^+/H^+ exchange, net K^+ transport falls to zero when $(\Delta\mu_K - \Delta\mu_H) = 0$. K^+ and conductance were stimulated by elevation of intracellular Ca^{2+}; swelling raised intracellular Ca^{2+} and RVD was inhibited by quinidine and phenothiazines (Cala 1983; Cala et al. 1986): all features of conductive K^+ channels. In addition, K^+ fluxes are necessarily measured using large numbers of cells, whilst data on electrochemical potential differences are derived from only a few (V.L. Lew, personal communication). In a related amphibium, *Necturus*, red cell RVD occurs through K^+ channels, activated by a Ca^{2+}-calmodulin-dependent mechanism (Bergeron et al. 1996). Whilst K^+/H^+ exchange is seen in mitochondria (Garlid 1978), and the gene for it has been cloned as NHE-7 (Numata and Orlowski 2001), it is a transport function, which appears to be confined to subcellular organelles. It has been described in the RBC plasma membrane but only under unusual conditions such as low ionic strength (Richter et al. 1997; Bernhardt et al. 1999). It has not been described for volume regulation in RBCs from any other species. Whilst a putative membrane protein has been described as around 85kDa, labelled with H_2DIDS, it is not clear whether this represents a KHE (Maldonado and Cala 1994). Specific inhibitors for a K^+/H^+ exchange, or definitive molecular identification, remain elusive.

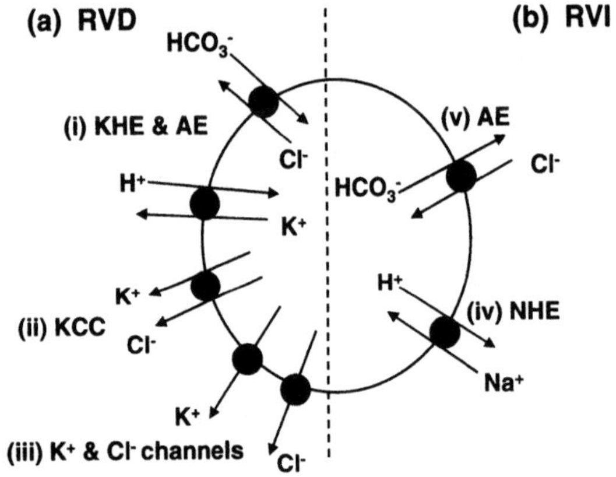

Fig. 31.2. Volume regulation in *Amphiuma* RBCs. (**a**) Regulatory volume decrease (*RVD*) may involve the unique K^+/H^+ exchanger (*KHE*) mediating K^+ loss, coupled to the familiar Cl^-/HCO_3^- exchanger (*AE*). Other possibilities would be the K^+-Cl^- cotransporter or separate conductive K^+ and Cl^- channels coupled through the membrane potential. (**b**) Regulatory volume increase (*RVI*) is more regular, involving coupled Na^+/H^+ exchanger (*NHE*) and Cl^- /HCO_3^- exchanger (*AE*) to mediate NaCl entry

31.6 Conclusion

Vertebrate RBCs show a rich diversity of transport phenotypes. Over the years, they have been extremely useful for investigating different membrane transport properties. Work has ranged from analysis of a particular transporter (e.g. the Na^+/K^+ pump), through cellular homeostasis (notably volume regulation) to integrated aspects with RBC transport participating in whole body physiology (notably O_2 carriage in teleosts – see Chap. 20). They continue to yield interesting information. Some gaps are manifest, particularly in our knowledge of reptilian and sub-human primate RBCs. The molecular control in transporter expression and activity, and their integrative physiology, represent future areas of study. It would be a pity, despite the existence of limited funding available, if comparative RBC research was neglected.

References

Adragna NC, Lauf PK (1988) Passive cation transport in reversibly sickling deer red blood cells. J Gen Physiol 92:44a

Adragna NC, Lauf PK (1993) Diamide activation as a tool to identify the most suitable animal model for KCl cotransport in sickle cell anaemia. Blood 82:648a

Allen DW, McManus TJ (1968) A non-metabolic role for oxygen in the control of passive cation permeability in the duck red cell. Biophys J 8:A125

Armsby CC, Brugnara C, Alper SL (1995) Cation transport in mouse erythrocytes: role of K^+-Cl^- cotransport in regulatory volume decrease. Am J Physiol 268:C894-C902

Benos DJ (1980) Intracellular analysis of sodium, potassium and chloride in mouse erythrocytes. J Cell Physiol 105:185-187

Benos DJ, Tosteson DC (1980) Cation movements across mouse red blood cells. Biochim Biophys Acta 601:167-179

Bergeron LJ, Stever AJ, Light DB (1996) Potassium conductance activated during regulatory volume decrease by mudpuppy red blood cells. Am J Physiol 270:R801-R810

Bernhardt I, Bogdanova AY, Kummerow D, Kiessling K, Hamann J, Ellory JC (1999) Characterization of the $K^+(Na^+)/H^+$ monovalent cation exchanger in the human red blood cell membrane: effects of transport inhibitors. Gen Physiol Biophys 18:119-137

Borgese F, Motais R, Garcia-Romeo F (1991) Regulation of Cl-dependent K transport by oxy-deoxyhemoglobin transitions in trout red cells. Biochim Biophys Acta 1066:252-256

Brugnara C, Tosteson DC (1987) Cell volume, K transport, and cell density in human erythrocytes. Am J Physiol 252:C269-C276

Butcher PD, Hawkey CM (1977) Haemoglobins and erythrocyte sickling in the artiodactyla: a survey. Comp Biochem Physiol A 57:391-398

Cala PM (1980a) Volume regulation by *Amphiuma* red blood cells: the membrane potential and its implications regarding the nature of the ion-flux pathways. J Gen Physiol 76:683-708

Cala PM (1980b) Volume regulation by *Amphiuma* red blood cells: the nature of the ion flux pathways. Fed Proc 39:379

Cala PM (1983) Cell volume regulation by *Amphiuma* red blood cells: the role of Ca^{2+} as a modulator of the alkali metal/proton exchanger. J Gen Physiol 82:761-784

Cala PM, Mandel LJ, Murphy E (1986) Volume regulation by *Amphiuma* red blood cells: cytosolic free Ca and alkali metal-H exchange. Am J Physiol 250:C423-C429

Chan PC, Calabrese V, Theil LS (1964) Species differences in the effect of sodium and potassium ions on the ATPase of erythrocyte membranes. Biochim Biophys Acta 79:424-426

Chapman DI (1977) Haematology of the deer. In: Archer RK, Jeffcott LB (eds) Clinical comparative haematology. Blackwell Sci Pub, Oxford, pp 345-364

Colclasure GC, Parker JC (1991) Cytosolic protein concentration is the primary volume signal in dog red cells. J Gen Physiol 98:881-892

Colclasure GC, Parker JC (1992) Cytosolic protein concentration is the primary volume signal for swelling-induced [K-Cl] cotransport in dog red cells. J Gen Physiol 100:1-10

Contreras A, Martinez R, Daves R, Marusic T (1986) An unusual pattern of Na^+ and K^+ movements across the horse erythrocyte membrane. Biochim Biophys Acta 856:388-391

Davson H, Danielli JF (1952) The permeability of natural membranes. Cambridge University Press, Cambridge

De Franceschi L, Beuzard Y, Jouault H, Brugnara C (1996) Modulation of erythrocyte potassium chloride cotransport, potassium content, and density by dietary magnesium intake in transgenic SAD mice. Blood 88:2738-2744

De Franceschi L, Fumagalli L, Olivieri O, Corrocher R, Lowell CA, Berton G (1997) Deficiency of Src family kinases Fgr and Hck results in activation of erythrocyte K/Cl cotransport. J Clin Invest 99:220-227

De Franceschi L, Villa-Moruzzi E, Fumagalli L, Brugnara C, Turrini F, Motta R, Veghini E, Corato C, Alper SL, Berton G (2001) K-Cl cotransport modulation by intracellular Mg in erythrocytes from mice bred for low and high Mg levels. Am J Physiol 281:C1385-C1395

Degen M (1987) Pseudohyperkalaemia in Akitas. J Am Vet Med Assoc 190:541-543

Dunham PB, Blostein R (1997) L antigens of sheep red blood cell membranes and modulation of ion transport. Am J Physiol 272:C357-C368

Dunker E, Passow H (1953) Zwei Arten des Anionenaustausches bei den roten Blutkörperchen verschiedener Säugetiere. Pflügers Arch 256:S446-S456

Ellory JC, Tucker EM (1983) Cation transport in red blood cells. In: Agar NS, Board PG (ed) Red blood cells of domestic animals. Elsevier, Amsterdam, pp 291-314

Fincham DA, Manson DK, Paterson JYF, Young JD (1987) Heterogeneity of amino acid transport in horse erythrocytes: a detailed kinetic analysis of inherited transport variation. J Physiol 389:835-409

Fujise H, Yamada I, Masuda M, Miyazawa Y, Ogawa E, Takahashi R (1991) Several cation transporters and volume regulation in high-K dog red blood cells. Am J Physiol 260:C589-C597

Fujise H, Mori M, Ogawa E, Maede Y (1993) Variant of canine erythrocytes with high potassium content and lack of glutathione accumulation. Am J Vet Res 54:602-606

Fujise H, Hamada Y, Mori M, Ochiai H (1995) Na-dependent glutamate transport in high K and high glutathione (HK/HG) and high K and low glutathione (HK/LG) dog red blood cells. Biochim Biophys Acta 1239:22-26

Fujise H, Higa K, Nakayama T, Wada K, Ochiai H, Tanabe Y (1997) Incidence of dogs possessing red blood cells with high K in Japan and East Asia. J Vet Med Sci 59:495-497

Garlid KD (1978) Unmasking the mitochondrial K/H exchanger: Swelling-induced K^+ loss. Biochem Biophys Res Commun 83:1450-1455

Gibson JS, Ellory JC, Culliford SJ, Fincham DA (1993) Volume-sensitive KCl co-transport and taurine fluxes in horse red blood cells. Exp Physiol 78:685-695

Gibson JS, Godart H, Ellory JC, Staines H, Honess NA, Cossins AR (1995) Modulation of K^+-Cl^- cotransport in equine red blood cells. Exp Physiol 79:997-1009

Gibson JS, Cossins AR, Ellory JC (2000) Oxygen-sensitive membrane transporters in vertebrate red cells. J Exp Biol 203:1395-1407

Gulliver G (1840) Observations on the blood corpuscles of certain species of the genus Cervus. Philosophical Magazine 17:327-330

Haas M, McManus TJ (1985) Effect of norepinephrine on swelling-induced potassium transport in duck red cells. J Gen Physiol 85:649-667

Herrick JB (1910) Peculiar elongated and sickle-shaped red blood corpuscles in a case of severe anaemia. Arch Int Med 6:517-521

Hoffman JF, Laris PC (1974) Determination of membrane potentials in human and *Amphiuma* red blood cells by means of a fluorescent probe. J Physiol 239:519-552

Ihrig I, Schönheit C, Häussner W, Bernhardt I (1992) Characterisation of the potassium influx in rat erythrocytes. Gen Physiol Biophys 11:377-388

Kazennov AM, Maslova MN, Matskevich YA, Rustamov FA, Shalabodov AD (1998) Species variability of erythrocyte transport ATPases in mammals. Comp Biochem Physiol B 119:169-175

Kirk RG, Andrews SB, Lee P (1983) The correlation of composition and morphology during the high to low potassium transition in single erythropoietic cells. J Membrane Biol 76:281-287

Kitchen H, Putnam FW, Taylor WJ (1967) Hemoglobin polymorphism in white-tailed deer: subunit basis. Blood 29:867-877

Krarup T, Dunham PB (1996) Reconstitution of calyculin-inhibited K-Cl cotransport in dog erythrocyte ghosts by exogenous PP-1. Am J Physiol 270:C898-C902

Kregenow FM (1971) The response of duck erythrocytes to nonhemolytic hypotonic media. Evidence for a volume-controlling mechanism. J Gen Physiol 58:372-395

Kregenow FM (1981) Osmoregulatory salt transporting mechanisms. Control of cell volume in anisotonic media. Annu Rev Physiol 43:493-505

Lassen UV (1974) Calcium-related hyperpolarisation of the *Amphiuma* red cell membrane following micropuncture. J Membrane Biol 108:125-144

Lassen UV (1977) Electrical potential and conductance of the red cell membrane. In: Ellory JC, Lew VL (ed) Membrane transport in red cells. Academic Press, London, pp 137-172

Lauf PK (1983) Thiol-dependent passive K/Cl transport in sheep red cells. I. Dependence on chloride and external K^+ (Rb^+) ions. J Membrane Biol 73:237-246

Lauf PK, Wilkie K, Stefko J, Adragna NC (1989) Steady state ion and flux parameters of sickling deer red blood cells. Ann New York Acad Sci 565:428-429

Lee P, Kirk RG (1982) Study of maturation of membrane transport function in red blood cells by X-ray microanalysis. J Membrane Biol 67:103-111

Lytle C (1997) Activation of the avian erythrocyte Na-K-Cl cotransport protein by cell shrinkage, cAMP, fluoride, and calyculin-A involves phosphorylation at common sites. J Biol Chem 272:15069-15077

Lytle C (1998) A volume-sensitive protein kinase regulates the Na-K-2Cl cotransporter in duck red blood cells. Am J Physiol 274:C1002-C1010

Maede Y, Inaba M (1985) (Na,K)-ATPase and ouabain binding in reticulocytes from dogs with high K and low Na erythrocytes and their changes during maturation. J Biol Chem 260:3337-3343

Maede Y, Inaba M (1987) Energy metabolism in canine erythrocytes associated with inherited high Na and K-stimulated adenosine triphosphatase activity. Am J Vet Res 48:114-118

Maede Y, Kasai N, Taniguchi N (1982) Hereditary high concentration of glutathione in canine erythrocyte associated with high accumulation of glutamate, glutamine, and aspartate. Blood 59:883-889

Maede Y, Inaba M, Taniguchi N (1983) Increased Na,K-ATPase activity, glutamate, and aspartate uptake in dog erythrocytes associated with hereditary high accumulation of GSH, glutamate, glutamine, and aspartate. Blood 61:493-499

Maldonado HM, Cala PM (1994) Labeling of the *Amphiuma* erythrocyte K^+/H^+ exchanger with H_2DIDS. Am J Physiol 267:C1002-C1012

McLean LA, Zia S, Gorin FA, Cala PM (1999) Cloning and expression of the Na^+/H^+ exchanger from *Amphiuma* RBCs: resemblance to mammalian NHE1. Am J Physiol 276:C1025-C1037

McManus TJ (1967) Comparative biology of red cells. Fed Proc 26:1821-1826

Milanick MA (1989) Na/Ca exchange in ferret red blood cells. Am J Physiol 256:C390-C398

Mount DB, Mercado A, Song L, Xu J, George AL, Delpire E, Gamba G (1999) Cloning and characterization of KCC3 and KCC4, new members of the cation-chloride cotransporter gene family. J Biol Chem 274:16355-16362

Muzyamba MC, Cossins AR, Gibson JS (1999) Regulation of Na^+-K^+-$2Cl^-$ cotransport in turkey red cells: the role of oxygen tension and protein phosphorylation. J Physiol 517:421-429

Nikinmaa M (1990) Vertebrate red blood cells. In: Bradshaw SD, Burggren W, Heller HC, Ishii S, Langer H, Neuweiler G, Randall DJ (eds) Zoophysiology, vol. 28. Springer, Berlin

Numata M, Orlowski J (2001) Molecular cloning and characterization of a novel (Na^+, K^+)/H^+ exchanger localized to the trans-Golgi network. J Biol Chem 276:17387-17394

Orskov SL (1954) The potassium absorption by pigeon blood cells. Acta Physiol Scand 31:221-229

Palfrey HC, Greengard P (1981) Hormone-sensitive ion transport systems in erythrocytes as models for epithelial ion pathways. Ann New York Acad Sci 372:291-309

Parker JC (1973) Dog red blood cells, adjustment of salt and water content *in vitro*. J Gen Physiol 62:147-156

Parker JC (1977) Solute and water movement in dog and cat red blood cells. In: Ellory JC, Lew VL (eds) Membrane transport in red cells. Academic Press, London, pp 427-465

Parker JC (1983a) Hemolytic action of potassium salts on dog red blood cells. Am J Physiol 244:C313-C317

Parker JC (1983b) Passive calcium movements in dog red blood cells: anion effects. Am J Physiol 244:C318-C323

Parker JC (1983c) Volume-responsive sodium movements in dog red blood cells. Am J Physiol 244:C324-C330

Parker JC (1992) Volume-activated cation transport in dog red cells: detection and transduction of the volume stimulus. Comp Biochem Physiol A 102:615-618

Parker JC, Castronova V (1984) Volume-responsive sodium and proton movements in dog red blood cells. J Gen Physiol 84:379-401

Parker JC, Hoffman JF (1965) Interdependence of cation permeability, cell volume and metabolism in dog red blood cells. Fed Proc 24:589

Parker JC, Snow RL (1972) Influence of external ATP on permeability and metabolism of dog red blood cells. Am J Physiol 223:888-893

Parker JC, McManus TJ, Starke LC, Gitelman HJ (1990) Coordinated regulation of Na/H exchange and K-Cl cotransport in dog red cells. J Gen Physiol 96:1141-1152

Pellegrino CM, Rybicki AC, Musto S, Nagel RL, Schwartz RS (1998) Molecular identification of erythroid K:Cl cotransporter in human and mouse erythroleukemic cells. Blood Cells, Molecules, and Diseases 24:31-40

Perutz MF (1969) The haemoglobin molecule. Proc Roy Soc (B) 173:113-140

Richter S, Hamann J, Kummerow D, Bernhardt I (1997) The monovalent cation "leak" transport in human erythrocytes: an electroneutral exchange process. Biophys J 73:733-745

Romero J, Schwartz RS, Fabry ME, Nagel RL, Canessa M (1993) Deoxygenation of red cells from transgenic mice with high HbS stimulates a charybdotoxin-sensitive K^+ efflux and forms dense cells. Clin Res 41:307a

Romero JR, Fabry ME, Suzuka S, Nagel RL, Canessa M (1997a) Red blood cells of a transgenic mouse expressing high levels of human hemoglobin S exhibit deoxy-stimulated cation flux. J Membrane Biol 159:187-196

Romero JR, Fabry ME, Suzuka SM, Costantini F, Nagel RL, Canessa M (1997b) K:Cl cotransport in red cells of transgenic mice expressing high levels of human hemoglobin S. Am J Hematol 55:112-114

Romualdez A, Sha'afi RI, Lange Y, Solomon AK (1972) Cation transport in dog red cells. J Gen Physiol 60:46-57

Sha'afi RI, Lieb WR (1967) Cation movements in the high sodium erythrocyte of the cat. J Gen Physiol 50:1751-1764

Sha'afi RI, Naccache P (1975) Sodium and calcium transport in cat red cells. J Cell Physiol 85: 655-664

Shaw TL (1955) Potassium movements in washed erythrocytes. J Physiol 129:464-475

Shmukler BE, Brugnara C, Alper SL (2000) Structure and genetic polymorphism of the mouse KCC1 gene. Biochim Biophys Acta 1492:353-361

Siebens AW, Kregenow FM (1985) Volume-regulatory responses of *Amphiuma* red cells in anisotonic media. The effect of amiloride. J Gen Physiol 86:527-564

Speake PF, Roberts CA, Gibson JS (1997) Effect of changes in respiratory blood parameters on equine red blood cell K-Cl cotransporter. Am J Physiol 273:C1811-C1818

Starke LC, McManus TJ (1990) Intracellular free magnesium determines the volume regulatory set point in duck red cells. FASEB J 4:A818

Su W, Shmukler BE, Chernova MN, Stuart-Tilley AK, De Franceschi L, Brugnara C, Alper SL (1999) Mouse K-Cl cotransporter KCC1: cloning, mapping, pathological expression, and functional regulation. Am J Physiol 277:C899-C912

Taylor WJ (1983) Sickled red cells in the Cervidae. Adv Vet Sci Comp Med 27:77-98

Tosteson DC, Hoffman JF (1960) Regulation of cell volume by active cation transport in high and low potassium sheep red cells. J Gen Physiol 44:169-194

Tosteson DC, Robertson JS (1956) Potassium transport in duck red cells. J Cell Comp Physiol 47:147-166

Vandorpe DH, Shmukler BE, Jiang L, Lim B, Maylie J, Adelman JP, De Franceschi L, Cappellini MD, Brugnara C, Alper SL (1998) cDNA cloning and functional characterization of the mouse Ca^{2+}-gated K^+ channel, mIK1. J Biol Chem 273:21542-21553

Whittam R (1964) Transport and diffusion in red blood cells. Edward Arnold, London

Willis JS, Ellory JC (1983) Ouabain sensitivity: diversity and disparities. Current Topics Membrane Transport 19:277-280

Willis JS, Nelson RA, Vilaro P, Zhao Z (1990) Membrane transport of sodium ions in erythrocytes of the American black bear, *Ursus americanus*. Comp Biochem Physiol A 96:91-96

Yamato O, Maede Y (1992) Susceptibility to onion-induced hemolysis in dogs with hereditary high erythrocyte reduced glutathione and potassium concentrations. Am J Vet Res 53:134-137

Yamato O, Hayashi M, Yamasaki M, Maede Y (1998) Induction of onion-induced haemolytic anaemia in dogs with sodium n-propylthiosulphate. Vet Rec 142:216-219

Yamato O, Hayashi M, Kasai E, Tajima M, Yamasaki M, Maede Y (1999) Reduced glutathione accelerates the oxidative damage produced by sodium n-propylthiosulfate, one of the causative agents of onion-induced hemolytic anemia in dogs. Biochim Biophys Acta 1427:175-182

Index

Printing: Mercedes-Druck, Berlin
Binding: Stein+Lehmann, Berlin

GPSR Compliance

The European Union's (EU) General Product Safety Regulation (GPSR) is a set of rules that requires consumer products to be safe and our obligations to ensure this.

If you have any concerns about our products, you can contact us on ProductSafety@springernature.com

In case Publisher is established outside the EU, the EU authorized representative is:

Springer Nature Customer Service Center GmbH
Europaplatz 3
69115 Heidelberg, Germany

Batch number: 10755320

Printed by Printforce, the Netherlands